CONTRACTILE MECHANISMS IN MUSCLE

ADVANCES IN EXPERIMENTAL MEDICINE AND BIOLOGY

Recent Volumes in this Series

Volume 164
THROMBOSIS AND CARDIOVASCULAR DISEASES
Edited by Antonio Strano

Volume 165
PURINE METABOLISM IN MAN – IV
Edited by Chris H. M. M. De Bruyn, H. Anne Simmonds, and Mathias M. Müller

Volume 166
BIOLOGICAL RESPONSE MODIFIERS IN HUMAN ONCOLOGY AND IMMUNOLOGY
Edited by Thomas Klein, Steven Specter, Herman Friedman, and Andor Szentivanyi

Volume 167
PROTEASES: Potential Role in Health and Disease
Edited by Walter H. Hörl and August Heidland

Volume 168
THE HEALING AND SCARRING OF ATHEROMA
Edited by Moshe Wolman

Volume 169
OXYGEN TRANSPORT TO TISSUE – V
Edited by D. W. Lübbers, H. Acker, E. Leniger-Follert, and T. K. Goldstick

Volume 170
CONTRACTILE MECHANISMS IN MUSCLE
Edited by Gerald H. Pollack and Haruo Sugi

Volume 171
GLUCOCORTICOID EFFECTS AND THEIR BIOLOGICAL CONSEQUENCES
Edited by Louis V. Avioli, Carlo Gennari, and Bruno Imbimbo

Volume 172
EUKARYOTIC CELL CULTURES: Basics and Applications
Edited by Ronald T. Acton and J. Daniel Lynn

A Continuation Order Plan is available for this series. A continuation order will bring delivery of each new volume immediately upon publication. Volumes are billed only upon actual shipment. For further information please contact the publisher.

CONTRACTILE MECHANISMS IN MUSCLE

Edited by

Gerald H. Pollack

University of Washington School of Medicine
Seattle, Washington

and

Haruo Sugi

Teikyo University School of Medicine
Tokyo, Japan

PLENUM PRESS • NEW YORK AND LONDON

Library of Congress Cataloging in Publication Data

Main entry under title:

Contractile mechanisms in muscle.

(Advances in experimental medicine and biology; v. 170)
"Proceedings of the second international symposium on Cross-bridge mechanisms in muscle contraction, held in June of 1982, in Seattle, Washington"—Verso t.p.
Includes bibliographical references and index.
1. Muscle contraction—Congresses. I. Pollack, Gerald H. II. Sugi, Haruo, 1933– III. Series. [DNLM: 1. Muscle contraction—Congresses. 2. Muscles—Physiology—Congresses. W1 AD559 v.170 / WE 500 I605 1982c]
QP321.C73 1984 591.1′852 83-26950
ISBN 978-1-4684-4705-7 ISBN 978-1-4684-4703-3 (eBook)
DOI 10.1007/978-1-4684-4703-3

Proceedings of the second international symposium on Cross-Bridge Mechanisms in Muscle Contraction, held in June of 1982, in Seattle, Washington

Softcover reprint of the hardcover 1st edition 1984
A Division of Plenum Publishing Corporation
233 Spring Street, New York, N.Y. 10013

PREFACE

Prior to the emergence of the sliding filament model, contraction theories had been in abundance. In the absence of the kinds of structural and biochemical information available today, it has been a simple matter to speculate about the possible ways in which tension generation and shortening might occur. The advent of the sliding filament model had an immediate impact on these theories; within several years they fell by the wayside, and attention was redirected towards mechanisms by which the filaments might be driven to slide by one another.

In terms of identifying the driving mechanism, the pivotal observation was the electron micrographic indentification of cross-bridges extending from the thick filaments. It was quite naturally assumed that such bridges, which had the ability to split ATP, were the molecular motors, i.e., that they were the sites of mechanochemical transduction. Out of this presumption grew the cross-bridge model, in which filament sliding is presumed to be driven by the cyclic interaction of cross-bridges with complementary actin sites located along the thin filaments.

Although some thirty years have elapsed since the sliding filament theory was first proposed, the cross-bridge mechanism of contraction remains a matter for debate and speculation. A variety of cross-bridge models has been proposed to account for contraction and there is a significant minority view that none of these models is satisfactory. Thus, the present status of the field is somewhat reminiscent of that just prior to the emergence of the sliding filament theory, where competing models formed the basis of considerable debate; except that at the present time the debate is more or less restricted to which of the various cross-bridge models is most consistent with the data.

Since most experiments on muscular contraction are currently designed, carried out and interpreted within the framework of Huxley and Huxley-Simmons type cross-bridge models, we felt it would be useful to consider the validity of the basic assumptions underlying these models. In 1978, one of us (H.S.) organized a symposium in Tokyo, at which time various results were rather thoroughly discussed on the above basis, and we published the proceedings under the title of "Cross-bridge Mechanisms in Muscle Contraction" (ed. H.E. Sugi and G.H. Pollack, University of Tokyo Press/University Park Press, 1979). The unusual interest in this volume encouraged us to organize the present symposium in Seattle in 1982, as a sequel to the Tokyo symposium.

As in the earlier symposium, most papers describe experiments on intact, skinned or glycerinated muscles, where the three-dimensional filament-lattice structures have been preserved. Biochemical experiments on actomyosin solution systems have been rather intentionally excluded to maintain this focus. This omission is not meant to imply that the organizers regard such experiments as unuseful; quite the contrary. We did feel, however that such work is well covered at various other symposia, whereas much of the material contained herein is not. Thus we have attempted to emphasize the relationship between structure and function.

The preceedings of the symposium have been divided into two parts. The first, entitled *Structural Dynamics*, focuses on structural aspects of the contractile process, particularly on dynamic structural changes in thick filaments and cross-bridges. Controversial areas are not only incorporated but emphasized. These include, for example, questions of whether thick filaments shorten during contraction and questions of whether there is indeed any significant rotation of cross-bridges during tension development. New methods are brought to bear on these and other questions. The second part, entitled *Mechanics, Energetics, and Molecular Models*, builds on the structural information in the first part in attempting to relate findings of mechanical and energetic experiments to molecular models. Again, the emphasis is on new and controversial topics.

Both parts contain three types of presentation that have been woven into a cohesive framework. These are: full oral presentations, posters, and short, extemporaneous presentations that were later formalized into short communications. The organizational format is by topic. Thus, each section of the book may contain a number of topically related contributions of variable length.

More than half the time of the symposium was devoted to discussion. Many participants came away from the meeting expressing the view that the discussions were among the most stimulating and informative of any symposium they had attended. Thus, we have endeavored to preserve the pristine nature of the interchanges as much as possible. Only repetitious or vague comments were deleted, and only minor reorganizations were required to preserve continuity. The short poster presentations also contain discussion sections; these are summaries, prepared by the presentor, of the most interesting and provocative comments surrounding their poster.

This seminar was originally planned as one of the United States - Japan Cooperative Seminars, under the auspices of the National Science Foundation and the Japan Society for the Promotion of Science, for a limited number of U.S. and Japanese participants. Owing to generous additional support from Teikyo University, Rigaku Denki, Inc., the Center for Advanced Studies of the University of Washington, and the American Cyanamid Company, we were able to invite many additional participants, particularly Europeans. Although there were still others we would like to have been able to invite, the complement of scientists included many distinguished scientists from around the world as well as younger investigators who we felt had something significant to contribute.

-Gerald H. Pollack
-Haruo Sugi

ACKNOWLEDGEMENTS

The editors express their sincere thanks to the National Science Foundation and the Japan Society for the Promotion of Science for the basic financial support of this symposium.

Our thanks are also due to Dr. Shoichi Okinaga, President of Teikyo University, to Mr Hikaru Shimura, President of Rigaku Denki, Inc. and to Professor Robert F. Rushmer, Director of the Center for Advanced Studies at the University of Washington as well as to the American Cyanamid Company, for their very generous additional financial support which enabled us to invite so many distinguished investigators from various countries.

We owe a debt of gratitude to Toni Ameslav, Joanne Brubaker, Joy Crist and Eileen Swanberg who spent countless hours transcribing the tapes of the discussions and painstakingly retyping all the manuscripts, and especially to Kathy Altringham, our Editorial Assistant, who assumed major responsibility for preparing the text for camera-ready duplication.

G.P.
H.S.

CONTENTS

PART I: STRUCTURAL DYNAMICS

STRUCTURE OF THE MYOFILAMENTS

Introduction 3

Symmetry and Self-Assembly in Vertebrate A-Filaments 5
A.J. Rowe and M.C. Maw

Image Analysis of the Complex of Actin-Tropomyosin and Myosin Subfragment 1 21
T. Wakabayashi, C. Toyoshima and E. Katayama

A-Band Mass Exceeds Mass of its Filament Components by 30-45% 29
M.K. Reedy and C. Lucaveche

Myofilament Diameters: An Ultrastructural Re-Evaluation 47
T.F. Robinson and L. Cohen-Gould

DO THICK FILAMENTS SHORTEN?

Introduction 65

Limulus Striated Muscle Provides an Unusual Model for Muscle Contraction 67
M.M. Dewey, P. Brink, D.E. Colflesh, B. Gaylinn, S-F. Fan and F. Anapol

Dynamic Laser Light Scattering of Papain-Treated Thick Filaments from *Limulus* Striated Muscle in Suspension 89
S-F. Fan, M.M. Dewey, D. Colflesh, P. Brink and B. Chu

Structure of *Limulus* and Other Invertebrate Thick Filaments 93
R.J.C. Levine, R.W. Kensler, M. Reedy, W. Hoffman, S. Davidheiser and R.E. Davies

Cinematographic Studies on the A-Band Length Changes During Ca-Activated Contraction in Horseshoe Crab Muscle Myofibrils 107
H. Sugi and S. Gomi

Contraction Bands: Differences Between Physiologically vs. Maximally Activated Single Heart Muscle Cells 119
J. Krueger and B. London

Structural Studies of Glycerinated Skeletal Muscle. A-Band Length and Cross-Bridge Period in ATP-Contracted Fibers 135
P. Dreizen, L. Herman and J.E. Berger

X-RAY DIFFRACTION APPROACHES TO STRUCTURAL DYNAMICS

Introduction 159

Time Resolved X-Ray Diffraction Studies of Cross-Bridge Movement, and Their Interpretation 161
H.E. Huxley

On the Possibility of Interaction Between Neighbouring Crossbridges 177
R.T. Tregear and M. L. Clarke

Cross-Bridge States in Invertebrate Muscles 185
J.S. Wray

Factors Affecting the Equatorial X-Ray Diffraction Pattern from Contracting Frog Skeletal Muscle 193
H. Tanaka, H. Hashizume and H. Sugi

Effect of Stretch on the Equatorial X-Ray Diffraction Pattern from Frog Skeletal Muscle in Rigor 203
H. Tanaka, H. Sugi and H. Hashizume

Structural Studies of Muscle During Force Development in Various States 207
L.C. Yu, T. Arata, A.C. Steven, G.R.S. Naylor, R.C. Gamble and R.J. Podolsky

Muscle Crossbridge Positions from Equatorial Diffraction Data: An Approach Towards Solving the Phase Problem 221
J. Squire and J. Harford

Configurations of Myosin Heads in the Crab Striated Muscle as Studied by X-Ray Diffraction 237
K. Wakabayashi, K. Namba and T. Mitsui

On the Intensity Reversal of the "Tropomyosin Reflexions" in X-Ray Diffraction Patterns From Crab Striated Muscle 251
Y. Maéda

STRUCTURAL BASIS OF FORCES IN RESTING MUSCLE

Introduction 267

Cross-Bridge Attachment in Relaxed Muscle 269
M. Schoenberg, B. Brenner, J.M. Chalovich, L.E. Greene and E. Eisenberg

Cytoskeletal Matrix in Striated Muscle: The Role of Titin, Nebulin and Intermediate Filaments 285
K. Wang

Connecting Filaments, Core Filaments and Side Struts: A Proposal to Add Three New Load-Bearing Structures to the Sliding Filament Model 307
A. Magid, H.P. Ting-Beall, M. Carvell, T. Kontis and C. Lucaveche

THE COMPOSITION OF THE INTRACELLULAR MILIEU

Introduction 331

^{31}P NMR Studies of Resting Muscle in Normal Human Subjects 333
D.R. Wilkie, M.J. Dawson, R.H.T. Edwards, R.E. Gordon and D. Shaw

Intracellular pH and Energy Changes in Muscle 349
N.A. Curtin

Change in Fixed Charge in the Thick Filament Lattice of *Limulus* Striated Muscle with Sarcomere Shortening 353
P. Brink and M.M. Dewey

A New Method to Measure Intracellular Diffusible Elemental Concentration 359
D.W. Maughan

Ion Concentrations Surrounding the Myofilaments 365
T. Iwazumi

DO CROSS-BRIDGES ROTATE DURING CONTRACTION?

Introduction 371

The Nature of the Actin - Cross-Bridge Interaction 373
K.C. Holmes and R.S. Goody

Cross Linking Studies Related to the Location of the Rigor Compliance in Glycerinated Rabbit Psoas Fibers: Is the S-2 Portion of the Cross-Bridge Compliant? 385
K. Tawada and M. Kimura

Angles of Fluorescently Labelled Myosin Heads and Actin Monomers in Contracting and Rigor Strained Muscle Fiber 397
T. Yanagida

Muscle Cross-Bridges: Do They Rotate? 413
R. Cooke, M.S. Crowder, C.H. Wendt, V.A. Barnett and D.D. Thomas

An Actomyosin Motor 429
H. Shimizu

CONCLUDING DISCUSSION

Structural Dynamics: Implications for Contractile Mechanisms 439

PART II: MECHANICS, ENERGETICS AND MOLECULAR MODELS

LENGTH-TENSION RELATIONS

Introduction 453

Changes in Intracellular Ca^{2+} Induced by Shortening Imposed During Tetanic Contractions 455
G. Cecchi, P.J. Griffiths and S. Taylor

Sarcomere Length Changes in Single Frog Muscle Fibers During Tetani at Long Sarcomere Lengths 473
J.D. Altringham and G.H. Pollack

Length-Tension-Velocity Relationships Studied in Short Consecutive Segments of Intact Muscle Fibres of the Frog 495
K.A.P. Edman and C. Reggiani

Force - Sarcomere-Length Relation and Filament Length in Rat Extensor Digitorum Muscle 511
H.E.D.J. ter Keurs, A.R. Luff and S.E. Luff

Some Specific Predictions and Experiments on Single Myofibrillar Mechanics 527
T. Iwazumi

1kgf/cm^2 - The Isometric Tension of Muscle Contraction: Implications to Cross-Bridge and Hydraulic Mechanisms 531
R. Tirosh

General Discussion: Length-Tension Relations 541

ACTIVATION OF THE MYOFILAMENTS

Introduction 549

Calcium Sensitivity is Modified by Contraction 553
A.M. Gordon, E.B. Ridgway and D.A. Martyn

Changes in $[Ca^{2+}]_i$ Induced by Rapid Cooling of Single Skeletal Muscle Fibres Treated with Low Concentration of Caffeine 565
S. Kurihara, M. Konishi and T. Sakai

Contractile Responses to MgATP and pH in a Thick Filament Regulated Muscle: Studies with Skinned Scallop Fibers 569
R.E. Godt and J.L. Morgan

Formation of Calcium-Parvalbumin Complex During Contraction. A Source of "Unexplained Heat"? 573
J.M. Gillis, D. Thomason, J. LeFevre and R.H. Kretsinger

TENSION TRANSIENTS AND STIFFNESS

Introduction 583

Symmetric and Asymmetric Processes in the Mechano-Chemical Conversion in the Cross-Bridge Mechanism Studied by Isometric Tension Transients 585
H. Shimizu and H. Tanaka

A Comparison of Muscle Stiffness Measurements Obtained with Rapid Releases or Stretches of Frog Semitendinosus Fibers 601
B.H. Bressler and L.A. Dusik

Tension Transients in Skinned Muscle Fibres of Insect Flight Muscle and Mammalian Cardiac Muscle: Effect of Substrate Concentration and Treatment with Myosin Light Chain Kinase 605
J.C. Rüegg, H.J. Kuhn, K. Güth, G. Pfitzer and F. Hofmann

Tension Transients in Single Isolated Smooth Muscle Cells 617
D.M. Warshaw and F.S. Fay

Sarcomere Length and Force Changes in Single Tetanized Frog Muscle Fibers Following Quick Changes in Fiber Length 623
H. Sugi and T. Kobayashi

Analysis of Mechanical Behavior of Muscle by a Multi-Sarcomere Model 637
K. Nishiyama

The Kinetics of Cross-Bridge Attachment and Detachment Studied by High Frequency Stiffness Measurements 641
G. Cecchi, P.J. Griffiths and S. Taylor

The Role of Ca^{2+} in Cross-Bridge Kinetics in Chemically Skinned Rabbit Psoas Fibers 657
M. Kawai, P.W. Brandt, and R.N. Cox

Muscle Stiffness Changes During Isometric Contraction in Frog Skeletal Muscle as Studied by the Use of Ultrasonic Waves 673
I. Hatta, Y. Tamura, T. Matsuda, H. Sugi and T. Tsuchiya

General Discussion: Interpretation of Stiffness Measurements 687

INFLUENCE OF MYOFILAMENT LATTICE DIMENSIONS ON CONTRACTILE FUNCTION

Introduction 693

Force Response to Width and Length Perturbations in Compressed Skinned Skeletal Muscle Fibers 697
D.W. Maughan and M.R. Berman

Lateral Shrinkage of the Myofilament Lattice in Chemically Skinned Muscles During Contraction 711
I. Matsubara, Y. Umazume and N. Yagi

Effect of Lattice Spacing on Cross-Bridge Orientations in Relaxed Crab Muscle 721
Y. Maéda

Isotonic Contraction of Temp-Step Activated Muscle Fibers with Varied Tonicity: Effects of Cell Volume and the Degree of Activation 725
J. Gulati and A. Babu

Changes in Mechanical Properties in Osmotically Compressed Skinned Muscle Fibers of Frog 731
T. Tsuchiya

CONTRACTION DYNAMICS

Introduction 737

Stretch of Contracting Muscle Fibres: Evidence for Regularly Spaced Active Sites Along the Filaments and Enhanced Mechanical Performance 739
K.A.P. Edman, G. Elzinga and M.I.M. Noble

Velocity Sensitivity of Yielding During Stretch in the Cat Soleus Muscle 753
T.R. Nichols

Force-Velocity Relation and Stiffness in Frog Single Muscle Fibres During the Rise of Tension in an Isometric Tetanus 757
C.A. Lorenzini, F. Colomo and V. Lombardi

Stepwise Shortening: Evidence and Implications 765
G.H. Pollack, R. Tirosh, F.V. Brozovich, J.W. Lacktis, R.C. Jacobson and T. Tameyasu

A Proposed Mechanism of Contraction in Which Stepwise Shortening is a Basic Feature 787
G.H. Pollack

Stepwise Changes in Crossbridge State and Sarcomere Length: Do Lattice Constraints Play a Critical Role? 793
C.J. Ritz-Gold and C.M. Gold

Distilled Water-Induced Contractions in Dehydrated and Skinned Muscle Fibers 797
R. Natori

CARDIAC MUSCLE MECHANICS

Introduction 805

Modeling of Cardiac Muscle Contraction Based on the Cross-Bridge Mechanism 807
H. Mashima and K. Kabasawa

The Dependence of Force and Velocity on Calcium and Length in Cardiac Muscle Segments 821
D.A. Martyn, J.F. Rondinone and L.L. Huntsman

Non-Uniformity of Contraction and Relaxation of Mammalian Cardiac Muscle 837
P.R. Housmans, L.H.S. Chuck, V.A. Claes and D.L. Brutsaert

Transient Length Responses of Heart Muscle in Ba^{2+}-Contracture to Step Tension Reductions 841
Y. Saeki, T. Shibata, C. Sato and K. Yanagisawa

ENERGETICS

Introduction 847

Dependence of the Shortening Heat on Sarcomere Length in Fibre Bundles from Frog Semitendinosus Muscle 853
K. Yamada and K. Kometani

The Effect of Shortening on Energy Liberation and High Energy Phosphate Hydrolysis in Frog Skeletal Muscle 865
E. Homsher, M. Irving and T. Yamada

The Dependence on the Distance of Shortening of the Energy Output from Frog Skeletal Muscle Shortening at Velocities of V_{max}, ½ V_{max} and ¼ V_{max} 883
T. Yamada and E. Homsher

Simultaneous Heat and Tension Measurements from Single Muscle Cells 887
N.A. Curtin, J.V. Howarth, J.A. Rall, M.G.A. Wilson, and R.C. Woledge

CONCLUDING REMARKS

Concluding Remarks 903
H.E. Huxley

PARTICIPANTS 911

INDEX 915

PART I: STRUCTURAL DYNAMICS

STRUCTURE OF THE MYOFILAMENTS

INTRODUCTION

The subject of myofilament structure, introduced in the following four contributions, appears as a recurring theme throughout this volume. This is not surprising, since this topic is intimately related to our understanding of the crossbridge mechanism in muscle contraction. What may be surprising, however, is the current reinvestigation, with improved methodology, of such aspects of myofilament structure as filament dimensions and the three-dimensional geometry of the S-1-thin filament complex, which, only recently were considered to be well established. Here, these issues share the podium with such acknowledged unsolved problems as the packing of component myosin molecules within the thick filaments and the nature of the constraints on freedom of crossbridge movement within the overlap zone of the filament lattice.

It now appears fairly certain that, in relaxed muscle, the crossbridge mass occupies a considerable volume of space between the shafts of the thick filaments and the surrounding thin filaments, and contributes to a significant increase in the apparent diameter of the thick filaments themselves. This has been suggested by the osmotic and X-ray diffraction studies of Millman, Racey & Matsubara (Biophys. J. 33: 189-202, 1981) and is visually evident in the rotary-shadowed images of freeze-cleaved fish muscle obtained by Franzini-Armstrong (personal communication). In this series of presentations, Robinson & Cohen-Gould's transmission electron microscope study of de-embedded sections of frog skeletal and rat cardiac muscle demonstrates just this point.

The implications of such an invasion of interfilament space by "relaxed" crossbridges (and their proximity to the thin filaments) for the freedom of movement of myosin heads during a contraction cycle are obvious: large-scale movements, involving the swinging of longer regions (S-2) of myosin molecules away from the thick filament backbone, seem, very unlikely. Rather, the evidence suggests that crossbridge movement occurs entirely by positional shifts among domains located either within or very close to the myosin heads. T. Wakabayashi's elegant three-dimensional reconstructions of negatively stained thin filament-S-1 complexes, imaged by low-dose electron microscopy, provide visualization of several distinct regions of S-1, which vary in their proximity to actin monomers. Later in this volume, the relative movements that occur between different S-1 regions during a contractile cycle and the relationship of these regions to the actin-binding site and other biochemically identified fragments of S-1 will be discussed.

Additional constraints on crossbridge movement are predicted by Reedy & Lucaveche's finding of 30-45% more mass in the A-bands of both insect flight muscle and vertebrate skeletal muscles than can be accounted for by the expected mass of the myofilaments, as determined by STEM electron scattering measurements (Reedy, Leonard, Freeman & Arad, J. Mus. Res. Cell Motil. 2: 45-64, 1981). This remains an unexplained result, although it may be related to increasingly frequent reports of sarcomeric cytoskeletal elements (See Magid, Ting-Beall, Carvell, Contis & Lucaveche, and Wang, 1982, this volume)

In the face of this new concept of a "crowded" filament lattice, it is amazing that Rowe and Maw have been able to disassemble and reassemble thick filaments within intact A-bands of skinned skeletal muscle fibers by varying the ionic strength and preventing escape of molecular constituents. Their studies show distinct regional differences along the thick filament, with respect to aggregation properties. Whether these results reflect differences among the myosin molecules, themselves, or the effect of regionally-present, thick filament-associated proteins (e.g. C-protein) or the directional influence of cytoskeletal structures, has yet to be determined.

The way in which myosin molecules assemble to form thick filaments is still a lively research problem. Although evidence is rapidly accruing in favor of a three-fold rotational symmetry for the crossbridge arrangement on the surface of vertebrate thick filaments, as predicted by Squire (J. Mol. Biol. 77: 291-323, 1973), the packing of myosin rods within the filament backbone remains a subject for speculation. The frayed filaments produced by Rowe and Maw support a model for the filament shaft in which the subunits run parallel to the filament's long axis, as earlier suggested by Pepe (Prog. Biophys. Mol. Biol. 22: 77-96, 1971) and Wray (Nature 277: 37-40, 1979), rather than one in which they are twisted. Nevertheless, more work must be done in order to relate subfilament structure to the arrangement of myosin rods.

Thus, it is indeed an exciting time for students of myofilament structure. Our various attempts to unravel the organization and interacton of the molecular assemblies that generate and control the crossbridges are now beginning to come together to reveal a physiologically valid picture of the contractile apparatus. The enthusiasm of these investigators is amply evident throughout this volume. The first four papers, while differing enormously in their approaches to the analysis of myofilament structure, serve both to exemplify some current insights into this subject and as an overture, providing tantalizing hints of the works that follow.

-Rhea J.C. Levine

SYMMETRY AND SELF-ASSEMBLY IN VERTEBRATE A-FILAMENTS

Arthur J. Rowe and Maria C. Maw

Department of Biochemistry, University of Leicester, Leicester LE1 7RH, U.K.

ABSTRACT

The myosin-containing A-filaments of vertebrate skeletal muscle contain 294 myosin molecules packed to give overall 3-fold rotational symmetry, as illustrated by the fraying of the filament into 3 sub-filaments (Maw and Rowe, 1980). Further studies on slightly frayed filaments are consistent with a highly linear arrangement of these sub-filaments, at least in the major part of the cross-bridge region where sub-filaments can be observed. Isolated filaments have an unusual hydrodynamic property in the form of an anomalous frictional increment. This property is as yet unexplained; it may possibly be related to flow-induced cyclic movements in the myosin heads.

Self-assembly of A-filaments *in vitro* to correct length and width has yet to be achieved. We have found however that under certain exactly defined conditions a very accurate reconstruction of both filaments and A-band can be accomplished *in situ*. Solubilisation of the myosin in chloride-free medium and maintenance of a high local myosin concentration are absolute requirements. Reconstruction is either abolished or modified by preglycerolation or at longer sarcomere length. It is argued that these results suggest a role for the actin filament lattice in myosin assembly, and imply that myosin assembly in the M-line region may be separable from myosin assembly in the cross-bridge region.

The concept of the cross-bridge remains a continuing enigma. Few can doubt that in vertebrate skeletal muscle there are structures connecting the thick, myosin-containing (A-) filaments to the thin, actin-containing (I-) filaments, or that the conformation of these structures is modulated in some fashion during contraction; yet beyond their biochemical identification as myosin 'heads' - in a loosely defined sense - our hard knowledge concerning these structures has remained surprisingly sparse and subject to dispute. Their axial translation along the filament is established as 14.3 nm by X-ray diffraction (Huxley and Brown, 1967) but this may be only an average probably over 3 × 143 nm, of a more complex local pattern (Squire and Harford, 1982). The functional cross-bridge may comprise one, two or even four myosin heads (below): the geometrical description of the individual cross-bridge in any defined physiological state remains highly obscure, although *changes* can be followed with considerable time-resolution (Huxley et al., 1981) and the highly ordered flight muscles of insects provide valuable clues (Reedy, Holmes and Tregear, 1965): the order of the lattice on which the cross-bridges lie continues to be disputed: and finally, the possible routes for assembly of the filaments have been obscure, since attempts to re-form filaments from stable myosin either *in vitro*, or *in situ* (to give an A-band) have met with only very partial success.

It will doubtless be some time before all or even most of the above issues are fully elucidated. Recent findings in our laboratory do however clarify certain important aspects. In particular our own and other recently published work shows that the order of the cross-bridge helix and the related number of myosin heads per axial repeat can now be regarded as settled beyond reasonable doubt. We now also show that under defined conditions A-filaments and indeed complete A-bands in muscle fibres can be re-assembled with correct length and packing. This finding demonstrates that in principle the myosin filament system possesses the property of self-assembly.

Structural studies can never by themselves define the mechanism of muscular contraction. Physiological mechanisms may well however be suggested by a more detailed knowledge of structure, and certainly the structural properties of the cross-bridge and the cross-bridge lattice are a powerful constraint on the types of contractile mechanisms which can be hypothesised. In addition to our results concerning filament symmetry and assembly, we discuss very briefly below a well-defined but little remarked property of A-filaments which may have considerable significance in relation to cross-bridge function.

Symmetry of the Filament and Cross-bridge Lattice

It had early been considered that the cross-bridges lay on a 2-start helical lattice, with a true repeat of 6 axial periods (Huxley and Brown, 1967). The exact myosin content of muscle has proved difficult to determine, but on the suppositions that (a) the number of myosins per axial period is integral, and (b) all cross-bridges are equivalent (e.g. suppositions such as 50% '2 myosin cross-bridges', 50% '1 myosin cross-bridges' can be discounted) then only the figure of 4 myosins per axial repeat is at all plausibly compatible with both the evidence and the model. A detailed possible model for myosin packing in the filament shaft to give a cross-bridge lattice of this type has been described (Pepe, 1967; Pepe, Ashton and Dowben, 1981). The association of two myosins with each cross-bridge required by this model was apparently supported by the evidence adduced in favour of a high degree of reversible dimerisation in solutions of myosin at high ionic strength (Godfrey and Harrington, 1970; Herbert and Carlson, 1971). Whilst these solutions are obviously non-physiological, such a finding implies that $K_D \gg K_i \cdots K_j$ in the polymerisation of myosin (M) by addition of monomers

$$M + M \underset{K_D}{\rightleftharpoons} (M)_2 + M \underset{K_i}{\rightleftharpoons} (M)_3 \,.....{}_{K_j} (M)_n$$

and hence that at physiological ionic strength the concentration of free monomer would be so extremely low that the kinetic unit for further assembly would of necessity be the dimer.

However, as pointed out by Squire (1973), the X-ray evidence was always compatible with a 3- or even a 4-start cross-bridge helix, depending upon assumptions concerning the radial distribution of filament mass. The overwhelming balance of recent evidence indeed now supports models based upon a 3-start cross-bridge helix. Direct measurements of mass/unit length in isolated filaments yields a value of (n=) 3 myosins/ 14.3 nm by hydrodynamics (Emes and Rowe, 1978a) and by STEM (scanning transmission electron microscopic) 'weighing' (Lamvik, 1978; Reedy and Leonard, 1981). This value is also the approximate average of the various disparate estimates yielded by quantitative SDS-PAGE of myofibrils (reviewed in Emes and Rowe, 1978a). Only particle counting in electron micrographs, a technically difficult procedure to validate, has yielded a single, modern estimate of n=4 (Morimoto and Harrington, 1974).

The value n=3 is of course compatible only with a 3-start cross-bridge helix. It implies the existence of three-fold rotational symmetry in any filament cross-section; and this has been confirmed by the demonstration that the filaments can fray into 3 sub-filaments in the cross-bridge region (Maw and Rowe, 1980), and by the observation of triangular profiles in transversely sectional tissue and of the attachment of 3 M-line bridges at a single level (Luther and Squire, 1978). Squire (1973, 1981) has considered a family of models in which the cross-bridges arise from myosin tails packed (unlike Pepe's model) into a coiled array. The type of 3-fold symmetry suggested by an extended study of filaments frayed into sub-filaments is not however compatible with such models. These sub-filaments (Figure 1) are produced from filaments pre-adsorbed onto a carbon film. Especially where the sub-filaments are held together at their tips by 'end-filaments' (Trinick, 1981), the rotation necessary to 'uncoil' an initially coiled structure to give co-planar sub-filaments would be unlikely to be complete, to say the least. Yet in a careful statistical survey we have found that in at least 98% of cases frayed filaments are co-planar, showing no evidence of any previous coiling, even when they have frayed but a small distance apart (Figure 1). In the small number of cases where the sub-filaments cross each other, this invariably involves sub-filaments with free ends, i.e., without end-filaments, and gives no indication of being other than a random drying-down effect. Our conclusion is thus that the sub-filaments must be packed into a strictly linear array in the cross-bridge region.

An apparent additional conclusion to be drawn from the demonstration of sub-filaments is that the myosin molecules cannot be identical in their environments. A minimal hypothesis would be that two types of myosin-myosin tail packing exist. Such conclusions however are not necessarily required to be drawn. The sub-filaments could be *generated* by fraying, if this were to start from the filament tip, where only three myosins are present in cross-section. A direct and inevitable conclusion

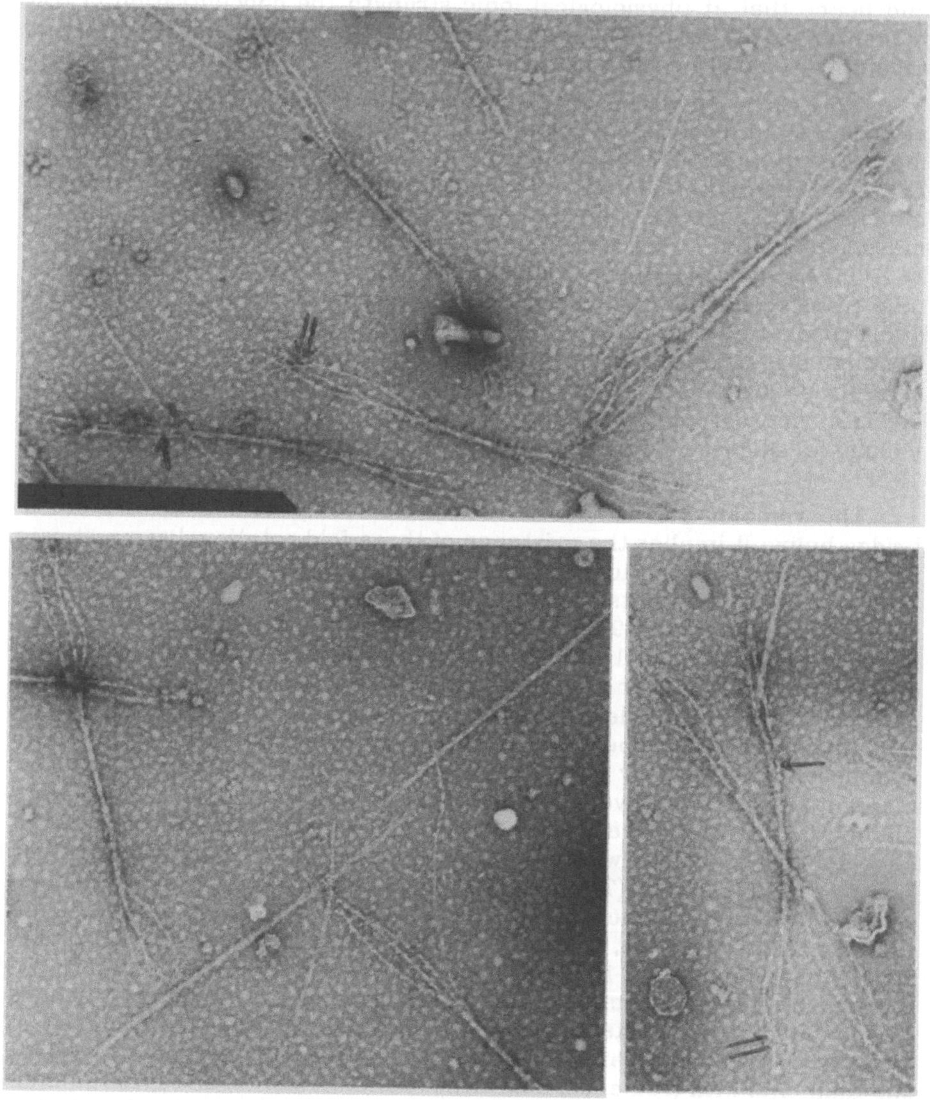

Figure 1: Electron micrographs of frayed filaments (Maw and Rowe, 1980) showing coplanarity of the sub-filaments even when only slightly separated (arrows). Very occasional 'cross-overs' in the substrate plane are seen (double arrows).

is that all models in which the myosin tails are 'coiled' throughout the filament shaft must be wrong; though a small degree of axial tilt or slew within each sub-filament is allowable, and indeed necessary on the assumption of an axial repeat of 43 nm for the lattice points. In revised models the concept of the dimer as a building unit can also now be discarded. There are insuperable objections to accepting the postulate that myosin in true solution is largely dimerised (Emes and Rowe, 1978b), and the recent demonstration by Davis et al. (1982) that myosin dimers, formed under conditions (lower ionic strength) when they indubitably will exist, are readily visualised in the electron microscope, must surely finally dispose of the 'myosin dimer' hypothesis. The axial stagger of 14.3 nm seen in these dimers is a definitive result which must be incorporated into any new model of filament shaft structure.

Flexibility of the Cross-Bridge Lattice

Whether viewed as a cause or consequence of contraction, conformational changes in the cross-bridge lattice are universally accepted as occurring during the contractile cycle. *In vitro* the cross-bridges appear to be at least sensitive to environmental conditions (Trinick and Elliott, 1982), though a genuine correlation with physiological conditions has not proved easy to demonstrate. We wish to draw attention to a clearly defined property of isolated filaments which may be surprisingly informative in this context.

The sedimentation coefficient of highly purified A-filaments is 132 ± 2 S (Emes and Rowe, 1978a; Figure 2), in agreement with an earlier less precise estimation using zone velocity sedimentation (Trinick, 1973). This value in common with the values determined for synthetic myosin filaments (Emes and Rowe, 1978a), implies a uniquely and inexplicably high frictional ratio - about 70% higher than can be computed for any static model, rigid or flexible, of a particle of this general mass, size and shape (J.M. Rallison, and A.J. Rowe; unpublished). Actin filaments, by

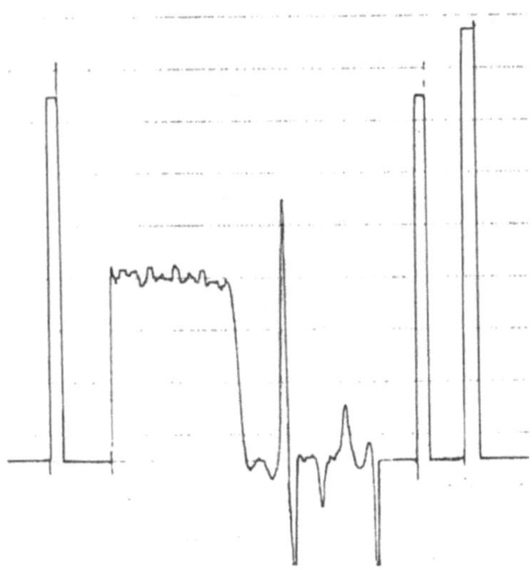

Figure 2: A trace recorded by the u-v scanner showing the NADH boundary formed during active enzyme centrifugation. The trace records absorbance at 340 nm (vertical direction) against radial position (horizontal direction). The sample was purified A-filaments layered onto a reaction mixture. Speed: 12330 rev/min; temperature: 22.0°C. The direction of sedimentation is from right to left.

contrast, show a completely normal behaviour (Johnson, Napper and Rowe, 1961). No certain explanation is yet apparent, but it may be that shear in sedimentation - of the same order of magnitude, incidentally, to the fluid shear over the filament surface in contraction - induces cyclical changes in the cross-bridge lattice, leading to additional energy dissipation (cf. the recent demonstration of shear-induced cyclical motion in bacterial flagella (Hotani, 1982)). The implication of such a hypothesis for models of the contractile cycle would be considerable.

Self-Assembly of the Filament and Cross-Bridge Lattice

Is the A-filament, and indeed the whole A-band, of vertebrate skeletal muscle a system capable of self-assembly? Or are we to interpret the marked lack of success over the years in reconstructing these structures to *in vivo* dimensions as indirect evidence for some type of post-assembly modification either of myosin or of some other essentially involved protein? The question is of importance not only for our understanding of filamentogenesis and turnover, but also for our apprehension of contraction as a broadly based biological event, including systems such as smooth muscle and cell movement where myosin filaments may have a less than permanent existence.

Moreover, the definition of a system in which dimensionally correct filaments and A-bands could be produced from soluble components should provide valuable approaches to understanding the nature of the

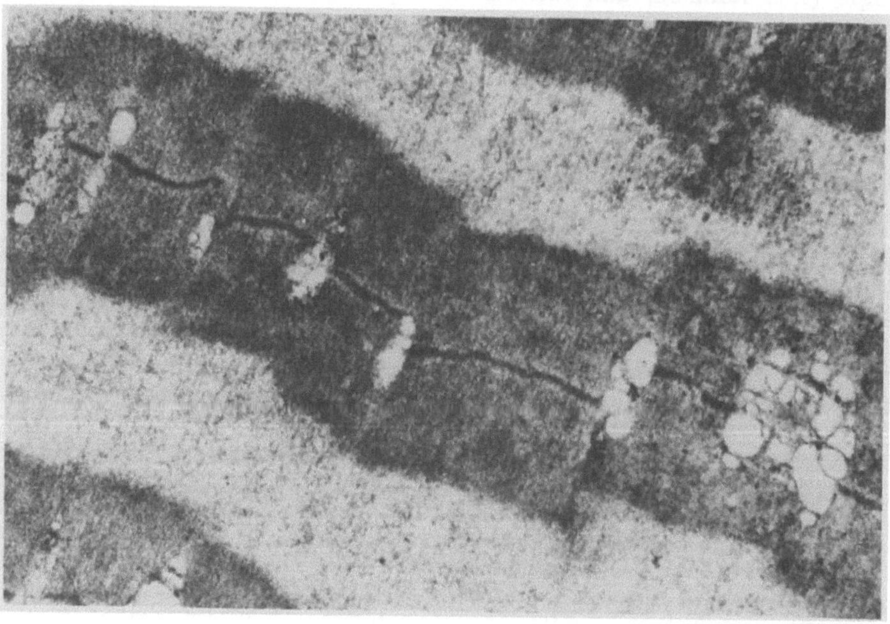

Figure 3: A longitudinal section of a rabbit psoas muscle fibre, pre-incubated in 0.266 M potassium phosphate. Only Z-disks and I-segments remain as clearly defined structures.

Figure 4: A longitudinal section of a rabbit psoas muscle fibre, reconstructed by dialysis to physiological ionic strength after incubation in potassium phosphate (as in Figure 3). A closely apposed artificial membrane was used throughout to maintain the myosin concentration within the fibre.

protein packing in the final structure. We have approached this problem (Maw and Rowe, 1979; Maw, 1982), by attempting *in situ* reconstruction of the filaments after total solubilisation of myosin filaments in 0.266 M potassium phosphate solution, a solvent in which myosin is very highly soluble, yet which unlike chloride- containing media causes remarkably little perturbation of the Z-discs or I-band structure (Mihalyi and Rowe, 1965; Maw, 1982). The complete loss of all A-filament structure has been monitored by electron and light microscopy and by X-ray and optical diffraction techniques. 'Ghost' fibres are produced, in which the solubilised myosin is now largely present in the very dense I-band region (Figure 3). Under certain conditions it is possible to return the fibres to

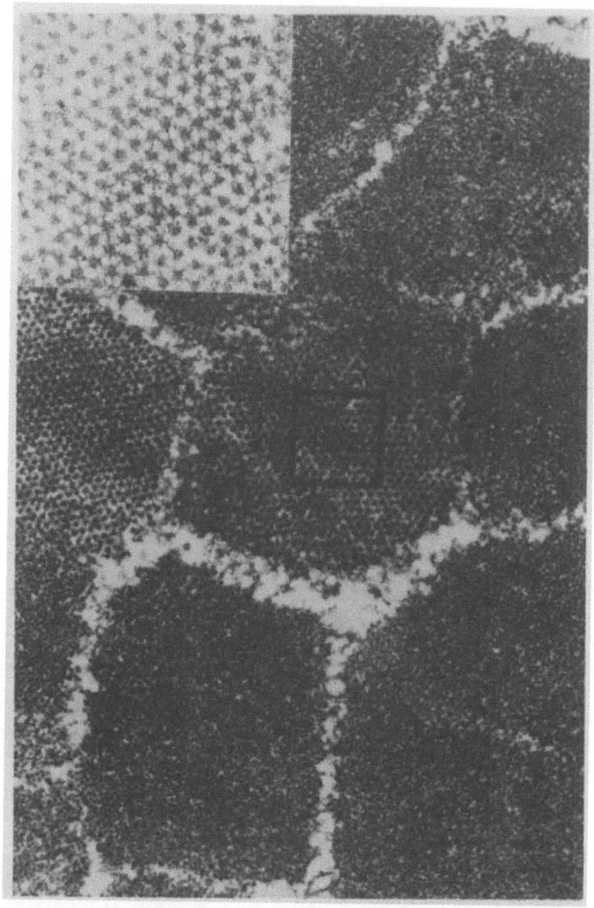

Figure 5: A transverse section (cf. Figure 4), showing in particular the ordered packing in the M-line region (insert).

physiological ionic strength and achieve a reconstruction of A-filaments organised into A-bands. The principal necessary conditions are:

1. Leakage of soluble myosin from the fibre must be avoided. In the case of rabbit psoas fibres this has been achieved by a technique in which an artificial semipermeable membrane (visking tubing) is tightly apposed around the fibre prior to solubilisation. If this precaution is not taken, reconstruction can still occur, but the A-bands formed will be 'mini A-bands', in which the A-band width will be limited by the quantity of residual myosin available. This finding incidentally supports the normal supposition that the A-filament grows from its centre outwards. The results of a successful reconstruction are shown in Figure 4. The M-line lattice appears to return, both as seen in longitudinal (Figure 4) and transverse (Figure 5)

Figure 6: Longitudinal section - rabbit psoas muscle fibre reconstructed at sarcomere length of about 2.6 μm.

section. A similar reconstruction can be achieved in frog muscle fibres (Byrne and Rowe, unpublished) without the apposition of an artificial membrane.

2. The sarcomere length must be ≦2.6 μm. At longer sarcomere lengths reconstruction of a modified type occurs, in which separate half-A-bands are seen, located at the periphery of the I-bands (Figure 6). Filamentous material can be seen projecting from the I-bands (Figure 7). An M-line pattern - poorly marked in rabbit fibres but better defined in rat fibres - can often be seen.

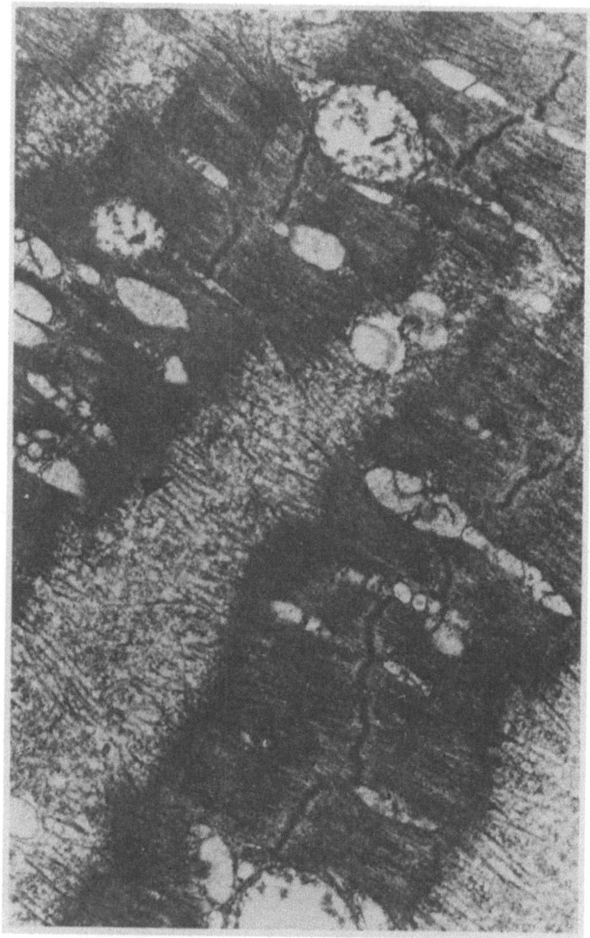

Figure 7: Longitudinal section - rabbit psoas muscle fibre reconstructed at sarcomere length of about 3.0 μm. Note projecting filaments (arrowed).

3. Reconstruction is totally abolished if the fibres are glycerolated or calcium depleted prior to solubilisation. The latter effect can be reversed by re-addition of calcium. On the other hand, reconstruction appears to work equally well from fresh samples (ATP present) or from fibres which have been permitted to go into rigor prior to solubilisation.

We do not yet have a comprehensive theory to explain all these and other related effects which we have studied. We can however very definitely conclude that the A-filament and the A-band lattice are capable of self-assembly - hypotheses such as assembly of a myosin precursor molecule prior to post-translational (and post assembly) modification can

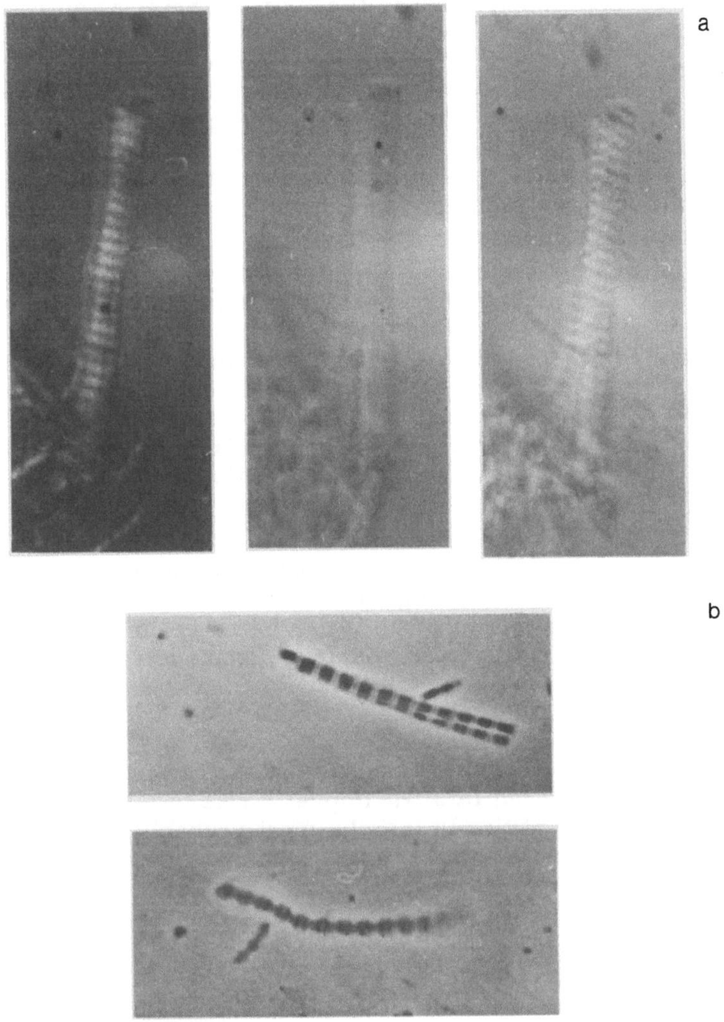

Figure 8: Optical micrographs of myofibrils and small myofibrillar bundles, showing (a) Solubilisation in 0.266 M potassium phosphate followed by reconstruction. Polarisation optics used to demonstrate loss and regain of A-band anisotropy. (b) 'Split' A-bands on reconstructed at sarcomere length 2.67 μm (cf. normal reconstruction at 2.47 μm, above).

clearly be rejected. Detailed evidence can be adduced (Maw, 1982; Maw and Rowe, to be published) in favour of the reconstructed A-filaments being of correct length, width and molecular packing. Why can this be achieved in our system, when it so conspicuously cannot be effected *in vitro* (Emes and Rowe, 1978a)? Clearly the very high local myosin concentration - not readily reproduced *in vitro* may be significant. Alternatively, the results at longer sarcomere lengths suggest an intimate involvement of the I-band lattice, with assembly commencing from the I-band periphery and continuing towards the Z-line. These latter results also imply a lack of necessary coupling between assembly in the M-line/ bare zone region, and in the cross-bridge region, since the latter can apparently proceed in a normal fashion independently of the former. It may even be that at shorter sarcomere lengths (<2.6 μm) filament assembly proceeds via initial assembly of two half- filaments which then fuse into a re-assembled M-line, composed of the characteristic M-line protein and a minimal number - presumably for symmetry reasons an integral multiple of six - myosin molecules.

As an alternative system for the study of re-assembly, we have also employed light microscopy of myofibrils, or small bundles of myofibrils. Here it is not possible of course to maintain maximal myosin concentration, and thus the reconstructed A-bands tend to be short. However, very rapid dynamic experiments can be performed, and by careful flow of solvent media under a cover-slip it is possible to achieve both solubilisation and reconstruction within seconds, thus minimising the loss of myosin by diffusion. Polarisation microscopy is particularly useful here, in confirming that A-band anisotropy is entirely lost on the solubilising medium, but is largely regained on reconstruction (Figure 6).

In conclusion, whilst it is obvious that the *in vivo* assembly of myosin filaments does not occur from high ionic strength, the definition of a system in which A-filaments and A-bands can be re-assembled from solubilised components, to give a system which in ultrastructural terms and also in its contractile properties (Byrne and Rowe, to be published) closely resembles normal muscle, provides us with both a probe for filament structure and potential insights into *in vivo* assembly mechanisms. The strikingly successful reconstruction attained, as compared with the 'synthetic filaments' produced *in vitro*, implies a role in A-band assembly for the other structural entities in muscle, most probably the I-filament lattice but not excluding other as yet less well defined cytoskeletal elements.

ACKNOWLEDGEMENTS

This work was supported by grants from the Medical Research Council.

REFERENCES

Davis, J.S., Buck, J. and Greene, E.P. (1982). The myosin dimer: an intermediate in the self-assembly of the thick filament of vertebrate skeletal muscle. FEBS Letters 140: 293-297.

Emes, C.H. and Rowe, A.J. (1978a). Frictional properties and molecular weight of native and synthetic myosin filaments from vertebrate skeletal muscle. Biochim. Biophys. Acta 537: 125-144.

Emes, C.H. and Rowe, A.J. (1978b). Hydrodynamic studies on the self-association of vertebrate skeletal muscle myosin. Biochim. Biophys. Acta 537: 110-124.

Godfrey, J.E. and Harrington, W.F. (1970). Self-association in the myosin system at high ionic strength. I. Sensitivity of the interaction to pH and ionic environment. Biochemistry 9: 886-893.

Herbert, T.J. and Carlson, F.D. (1971). Spectroscopic study of the self-association of myosin. Biopolymers 10: 2231-2252.
Hotani, H. (1982). Micro-video study of moving bacterial flagellar filaments. III. Cyclic transformation induced by mechanical force. J. Molec. Biol. 156: 791-806.
Huxley, H.E. and Brown, W. (1967). The low-angle X-ray diagram of vertebrate striated muscle and its behaviour during contraction and rigor. J. Molec. Biol. 30: 383-434.
Huxley, H.E., Simmons, R.M., Tarvgi, A.R., Kress, M., Bordas, J. and Koch, M.H.J. Millisecond time-resolved changes in X-ray reflections from contracting muscle during rapid mechanical transients, recorded using synchrotron radiation. Proc. Natl. Acad. Sci. (U.S.A.) 78: 2297-2301.
Lamvik, M.K. (1978). Muscle thick filament mass measured by electron scattering. J. Molec. Biol. 122: 55-68.
Luther, P.K. and Squire, J.M. (1978). Three-dimensional structure of the vertebrate muscle M-region. J. Molec. Biol. 125: 313-324.
Maw, M.C. (1982). A-filaments: Structure and reconstruction. Ph.D. Thesis (Leicester).
Maw, M.C. and Rowe, A.J. (1979). Reconstruction of the A-band and A-filaments of rabbit psoas muscle after dissolution in high ionic strength solution. J. Ultrastruct. Res. 69: 142-143.
Maw, M.C. and Rowe, A.J. (1980). Fraying of A-filaments into three sub-filaments. Nature 286: 412-414.
Mihalyi, E. and Rowe, A.J. (1965). Studies on the extraction of actomyosin from rabbit muscle. Biochem. Z. 345: 267-285.
Morimoto, K. and Harrington, W.F. (1974). Substructure of the thick filament of vertebrate striated muscle. J. Molec. Biol. 83: 83-97.
Pepe, F.A. (1967). The myosin filament I. Structural organisation from antibody staining observed in electron microscopy. J. Molec. Biol. 27: 203-225.
Pepe, F.A., Ashton, F.T. and Dowben, P. (1981). The myosin filament VII. Changes in internal structure along the length of the filament. J. Molec. Biol. 145: 421-440.
Reedy, M.K., Holmes, K.C. and Tregear, R.T. (1965). Induced changes in orientation of the cross-bridges of glycerinated insect flight muscle. Nature 207: 1276-1280.
Reedy, M.K., Leonard, K.R., Freeman, R. and Arod, T. (1981). Thick myofilament mass determination by electron scattering measurements with the scanning transmission electron microscope. J. Musc. Res. Cell. Motil. 2: 45-64.
Squire, J.M. (1973). General model of myosin filament structure III. Molecular packing arrangements in myosin filaments. J. Molec. Biol. 77: 291-323.
Squire, J.M. (1981). *The Structural Basis of Muscular Contraction* Plenum, New York and London.
Squire, J.M. and Harford, J.J. (1982). Fine structure of the A-band in cryosections III. Cross-bridge distribution and the axial structure of the human C-zone. J. Molec. Biol. 155: 467-494.
Trinick, J.A. (1973). A-filaments from rabbit skeletal muscle. Ph.D. Thesis (Leicester).
Trinick, J.A. (1981). End-filaments: A new structural element of vertebrate skeletal muscle thick filaments. J. Molec. Biol. 151: 151-156.
Trinick, J.A. and Elliott, A. (1982). Electron microscopy of myosin filaments. J. Microscopy 126: 151-156.

DISCUSSION

REEDY: Does TMV sediment at a very much different rate than the filaments that you and Emes were exploring?

ROWE: No. So far as we've been able to review the sedimentation of expanded linear structures ranging from TMV through DNA and other things, classical hydrodynamics explains the mass per unit length within the sort of error that you'd expect. Myosin filaments seem to be totally unique. I'd also say that the method used, which is active enzyme sedimentation, picks out the heaviest species that you've got, thereby excluding the possibility of it being a slow-moving artifact that we're looking at.

HUXLEY: On the same subject, could I ask, then, how far the cross-bridges would have to extend out sideways to trap solvents in order to slow down sedimentation? Is that a possible mechanism without them actually moving?

ROWE: That would have almost no effect, there is an effect of only a few percent if the cross-bridges extend themselves. I should say there's a popular view among biochemists, that hydrodynamics is something that is (a) done by biochemists; and (b) stopped being done about 15 years ago; and both attitudes are quite incorrect. We've seen major advances the last few years. There are now methods derived by John Rallison of Cambridge and Howard Brenner of New York which can unequivocally yield the frictional properties for something like a myosin filament and, according to John Rallison, the answer is certainly that there is no static structure you can produce *however* you've designed the cross-bridge extension which will account for the observed properties.

COOKE: I find your explanation of the slow sedimentation a little bewildering. If you allow the cross-bridges, including the S-2 region, to splay out from the thick filament, you're sedimenting really quite a bushy piece of broccoli. I don't see why allowing the cross-bridges to cycle would slow it down any further. Do you allow the entire S-2 region to reach away from your filament?

ROWE: Well, the answer is that it doesn't matter from the point of view of calculating the frictional ratio; it doesn't matter within a few percent whether you assume the cross-bridges are extended or not. De la Torre & Bloomfield's paper (Garcia de la Torre and Bloomfield, Biochemistry 19: 5118-5123, 1980) shows that in the case of a myosin molecule, the configuration of the cross-bridge and its flexibility affect the frictional drag *much* less than the layman might feel. What I *am* suggesting is that some form of cross-bridge cycling could break down the laminar flow regime on which all these calculations are based, and thereby, if you have turbulent flow, you can get that sort of frictional increment

VERDUGO: Is this coefficient dependent on the ionic strength?

ROWE: Well, not within the fairly narrow limits with which we can vary it. If you raise it much the myosin dissolves. If you lower it very much, filaments tend to fray apart.

VERDUGO: If you clip the heads, is there a big change in the sedimentation coefficient?

ROWE: That's a nice experiment, which has some technical difficulties. To clean-clip to get S-1 using chymotrypsin, you have to do a rather minimal clip. But then only a fairly small number of the heads come off. So I'm not sure if biochemically it is do-able.

POLLACK: You've shown elegantly that the filament frays into three linear strands. How do you account for the helical arrangement of the cross-bridges -- by just assuming that there's a slight axial displacement of one strand relative to the other?

ROWE: No, we assume that within each sub-filament -- I'm reluctant to go into describing the model we have because it would take a long time -- we suppose each sub-filament to be made of three protofilaments, each protofilament being staggered by 14.3 nm with respect to the one adjacent to it. The sub-filaments are, therefore, fully in register with each other.

TREGEAR: You said as an off-the-cuff remark that you didn't think there was much twisting within the sub-filaments. Could you expand on that, please?.

ROWE: Well, it's a little difficult to produce something that will plausibly fray apart to give an *apparently* smooth structure (though I stress the word apparently), if the myosin tails are very strongly tilted. There is, in any case, a limitation I think that is *defined*, on the amount of tilt or slew you put on the myosin tail as determined by the wide angle X-ray pattern.

T. WAKABAYASHI: Does the myosin filament fray into three in the presence of the actin filament, I mean in the rigor complex of thick and thin filaments?

ROWE: Yes, it does.

HUXLEY: There is the very nice technique of Dr. Rowe's, showing the existence of subfilaments. I wondered if anyone has had any experience in applying it to other muscles than vertebrate striated muscle and have seen anything as a result.

ROWE: Many other filaments seem to show remarkable reluctance to fray apart. We're making some efforts to try to get the Lethocerus filaments and scallop filaments to fray apart. Lethocerus, insofar as they do, and after fairly exhaustive calcium pre-depletion of the muscle, to a certain extent look like four strands, but it's difficult to be certain, because of it's tiresome habit -- if it does come apart it comes apart completely all the way down. The great thing about vertebrate muscle is that it sticks together in the middle, so that you can see where all the strands come from. If you've just got a mass of individual frayed subfilaments, you can't do anything.

LEVINE: I think that it would be very hard to do with most invertebrate filaments because they have a core. You are likely to strip the myosin off when you want to really divide the core into segments so that you have true subfilaments.

HUXLEY: Yes, but that's assuming that there is a core and a peripheral ring. Doing the experiment may provide a good demonstration that that structure is likely to be right.

TIROSH: Did you check tension generation in the reconstituted myofibrils?

ROWE: The tension experiments we've done have all been with frog sartorius muscle. The pictures I showed were rabbit psoas. In rabbit psoas, reconstruction gives something extremely fragile -- it looks nice structurally -- but it's pretty delicate. In the frog we get 90% of tension recovery, and indeed it compares quite well with a control muscle which is stored for the same period. In other words, the time constants of the twitch do change, but they don't look significantly worse in the reconstructed muscle than in the muscle stored for the equivalent period.

NOBLE: Do you get the tension back if you've reconstituted it at about 2.6 μm sarcomere length?

ROWE: Good question. We haven't done that experiment yet.

REEDY: Does the reconstruction require that the extraction not only occur at this critical sarcomere length but that there be no perturbation? Can you stretch after extraction, restore the sarcomere length, and then reconstruct, or is there a suggestion there's some fragile element that's disturbed by the longer length?

ROWE: 2.6 μm is a *critical* length. Reconstruction occurs quite well below that length. The experiment you suggest would be a little difficult one to do. By the time we've got these things into artificial membranes and solubilised them in phosphates, I'm not quite sure what happens if you pull on them, but I think it will be fairly nasty.

WANG: Continuing on the subject of reconstitution, I seem to notice on your micrographs at the low sarcomere length, below 2.6 μm, you do have residual filamental structure left in the gap, whereas in the highly stretched one, there seems to be much less of this material. Is that your impression also?

ROWE: Yes. It's very difficult to be at all certain about the residual filamental structure. We see, for example, sometimes the slightly ghostly remains of an M line, but by no means always, and its presence doesn't seem to be associated with the reconstruction. Samples where we don't see one reconstruct apparently well. Longitudinal filaments certainly can often be seen bridging the gap where the A-band was, but we've no evidence as to what these residual filaments are composed of. And I know there will be presentations later (Ed: cf. Magid, Wang) which will suggest, of course, that there are important structural elements that are not myosin and not actin.

HOMSHER: Is ATP present at all times during the reconstruction?

ROWE: No, it's not necessary. We found you *can* do a reconstruction with ATP present, and indeed what we're doing now is experiments on myofibrils in which we get the entire solubilisation and reconstruction done within about 30 or 40 seconds. Indeed, we're currently making a movie of it happening, and there is certainly ATP present all the time there. But we have let the thing go into rigor even for a day or two and succeeded in effecting a reconstruction. But there *are* quite a few ways of abolishing the reconstruction, not all of which we understand. Predepleting the muscle calcium will inhibit reconstruction, though reversibly, if you re-add calcium. Glycerolation of the muscle *totally* and apparently irreversibly stops reconstruction. For that we have no very clear explanation.

IMAGE ANALYSIS OF THE COMPLEX OF ACTIN-TROPOMYOSIN AND MYOSIN SUBFRAGMENT 1

Takeyuki Wakabayashi, Chikashi Toyoshima and Eisaku Katayama*

Department of Physics, Faculty of Science,
**Department of Pharmacology, Faculty of Medicine, University of Tokyo, Hongo 7-3-1, Bunyo-ku, Tokyo 113, Japan*

ABSTRACT

1. A three-dimensional image of the "rigor" complex of actin-tropomyosin-S1 was reconstituted from both low dose (10 electrons/Å²) and high dose (>500 electrons/Å²) electron microscopic images of specimens embedded in unbroken and unbacked stain sheets of uranyl acetate over the holes of perforated carbon films.

2. Myosin S1 shows multi-domain submolecular structure as has been earlier observed in actin-S1 (Wakabayshi & Toyoshima, 1981) and actin-heavy meromyosin (Katayama & Wakabayashi, 1981). The morphological unit of the actin-tropomyosin-S1 was found to be composed of at least three domains (domains A, B and D) and three regions (C, E and H).

3. A myosin S1 molecule has a complex shape, which cannot be represented by a simple rod with one major axis. The shape of S1 should be approximated by at least two rods.

4. The domain D is identified as the main part of S1. The angle between the major axis of this domain and the axis of actin helix was about 72°, which is almost right angle.

5. The angle between the axis of actin helix and major axis of the region E, which is less bulky than the domain D and makes no contact with actin, is much smaller than the value for the domain D.

6. The resolution of reconstituted images from both high and low dose micrographs was improved so that the radial resolution became about 15 Å and the axial one became about 25 Å. Due to the improvement of resolution in both the radial and axial direction, all major domains A, B and D split into two domains, i.e. into A1 and A2, B1 and B2, and D1 and D2 respectively.

7. Though unambiguous assignment of actin is not yet achieved by us, it can be confirmed that a S1 molecule interacts morphologically with actin at two sites (Wakabayashi & Toyoshima, 1981).

INTRODUCTION

To understand the molecular mechanism of muscle contraction on the structural basis, the three-dimensional image of actin-containing filaments and those decorated with S1 have been reconstituted by many workers (Moore, Huxley & DeRosier, 1970; Spudich, Huxley & Finch, 1972; Wakabayashi, Huxley, Amos & Klug, 1975; Toyoshima & Wakabayashi, 1979; Seymour & O'Brien, 1980; Wakabayashi & Toyoshima, 1981, Katayama and Wakabayashi, 1981; Taylor & Amos, 1981; Vibert & Craig, 1982).

Moore et al. (1970) showed that the angle between helix axis of thin filament and long axis of S1 is about 45°. However, it was found that angle between the actin helix axis and the main axis of the major part of S1 (domain D) is not so small as described by Moore et al. (1970) but is almost right angle (Toyoshima & Wakabayashi, 1979; Wakabayashi & Toyoshima, 1981). Also it was reported for the first time that the S1 shows the multi-domain structure in actin-S1 (Wakabayashi & Toyoshima, 1981) and in actin-heavy meromyosin complex (Katayama & Wakabayashi, 1981). The morphological interaction between myosin and actin is shown to be at least two-sited (Wakabayashi & Toyoshima, 1981).

Recently, Taylor and Amos (1981) pointed out that the assignment of actin by Moore et al. is wrong and they proposed a new assignment. However, Wakabayashi and Toyoshima (1981) reserved the conclusion because the reconstituted images were complicated and the resolution of the images was worse than 20 Å. Also the presence of the forbidden reflection on the first layer-line indicates that the preservation of the helical symmetry is not perfect and that the images reconstituted by assuming this symmetry should be interpreted cautiously.

The resolution was improved by reconstructing the three-dimensional image of actin-tropomyosin-S1 complex from both low dose and high dose images (Toyoshima & Wakabayashi, unpublished results). The results of the image analysis of actin-tropomyosin-S1 supported the three points described above, i.e., non-tilted configuration of rigor complex, multi-domain structure of decorated thin filaments and two-site or multi-site morphological interaction between S1 and actin.

However, the reconstituted image became more difficult to interpret since it became more complicated because of the improvement of the resolution. Also, we could not suppress the forbidden reflection on the first layer-line. These factors prevented us from unambiguous identification of actin, tropomyosin and S1 in the complex structure.

Other approaches which provide more direct evidence are necessary for the unambiguous identification.

MATERIALS AND METHODS

To reconstitute the actin-tropomyosin, 100 μg of G-actin/ml was polymerized over night in the presence of 25-30 μg of tropomyosin/ml by the addition of 50 mM NaCl, 10 mM sodium phosphate (pH 7.6), 5 mM $MgCl_2$ and 0.2 mM dithiothreitol.

Rigor complex of actin-tropomyosin-S1 was formed by mixing 160-200 μg of S1/ml and the actin-tropomyosin (50 μg actin/ml) and the mixture was left to stand for sufficient time for the consumption of ATP originally contained in G-actin preparation.

All of the specimens analysed digitally were embedded in unbroken stain sheets stretched over the holes. Throughout the work no backing of carbon by vacuum deposition was used. Specimens were examined on JEM-100B, -100C, -100CX according to the various stage of the study. Low dose work (about 10 e/Å ^{2}) was carried out at 100 kV on a JEM-100CX with anti-contamination device and a minimum dose system using side-entry goniometer and Kodak Electron Image Film 4463. Calibration of electron doses were estimated by Kodak 4463 films and Faraday cage kindly lent by JEOL. For high dose work, Fuji FG film and Agfa-Gebart 23D-56 film were used.

The image analysis was done as previously described (Wakabayashi & Toyoshima, 1981) using HITAC M200-H at the Computer Centre of the University of Tokyo. To accomodate longer filaments, bigger Fourier transform program (256 X 1024), which Dr. Linda Amos kindly sent to us, was used. Layer-line data of the individual filaments were extracted from the computed transform and its two sides were averaged as done by Taylor & Amos (1981). After this step, the averages of low dose images of six filaments and high dose images of five filaments are calculated.

RESULTS AND DISCUSSION

The image analysis and three-dimensional image reconstruction of high dose micrographs were done as described previously (Wakabayashi & Toyoshima, 1981). In the case of low dose images, the step of examination of the micrographs by eye was difficult and was sometimes skipped when they were too fuzzy. The meridional spot with 1/27.5 Å spacing was clearly observed and was incorporated into the three-dimensional reconstruction. The incorporation of this meridional spot facilitated the separation of neighboring S1 molecules in the reconstituted images.

A myosin S1 molecule has a complex shape and cannot be represented by the main axis only. The axis through the domain E is more tilted than the main axis. However, it should be noted that the domain E is very small in volume and low in density and does not bind to actin. Also the domain E locates at the outermost part of the whole structure. This part is affected by the forbidden meridional spot and may be less reliable than the inner parts. The angle between the axis of helical symmetry and the main axis of the main part of S1 was about 72°, which is almost perpendicular as pointed out by Toyoshima & Wakabayashi, (1979).

End on view of untwisted models of the averaged reconstituted image from electron micrographs of actin-tropomyosin-S1 is presented in Fig. 1(A). The lowest contour level of these models was chosen so that the model shows the most reliable parts of the reconstituted image. The total volume of the model is about 71% of the actin-tropomyosin-S1 complex

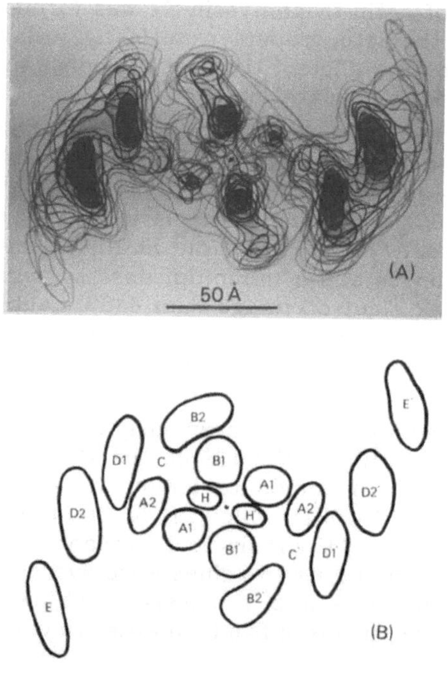

Figure 1: An end-on view showing the multi-domain structure of the *untwisted* transparent model of a portion of the three-dimensional image reconstituted from six low dose electron micrographs of actin-tropomyosin-S1(A) and its schematic diagram (B). The depth of the horizontal slices shown is about 60 Å in the axial direction.

assuming that the partial specific volume is 0.76 and that the molecular weight of S1 and actin are 110,000 and 42,000 repectively and that the weight ratio of tropomyosin to actin is 1:4.5.

Multi-domain structure of the complex can also be seen in the actin-tropomyosin-S1 complex (Fig. 1(A)). Each region could be labelled in the similar fashion as shown schematically in Fig. 1(B). The resolution of the present models shown in Fig. 1 is about 15 Å radially and 25 Å axially and is better than previous works (Moore et al., 1970; Toyoshima & Wakabayashi, 1979; Wakabayashi & Toyoshima, 1981; Taylor & Amos, 1981) and one can now recognize sub-domain structure in the major domains A, B, and D, i.e. these major domains split into subdomains A1/A2, B1/B2 and D1/D2 respectively. Also new domain H emerged. This domain locates between the subdomains A1 and B1. The domain H, which exists only when tropomyosin was added to actin-S1 and is elongated along the helix axis, can be tropomyosin. But the addition of tropomyosin made the thin filament more straight and the resolution of the image became better at the same time. Therefore, the domain H may be the very intricate feature of actin-S1 which can be reconstituted only when good resolution is achieved.

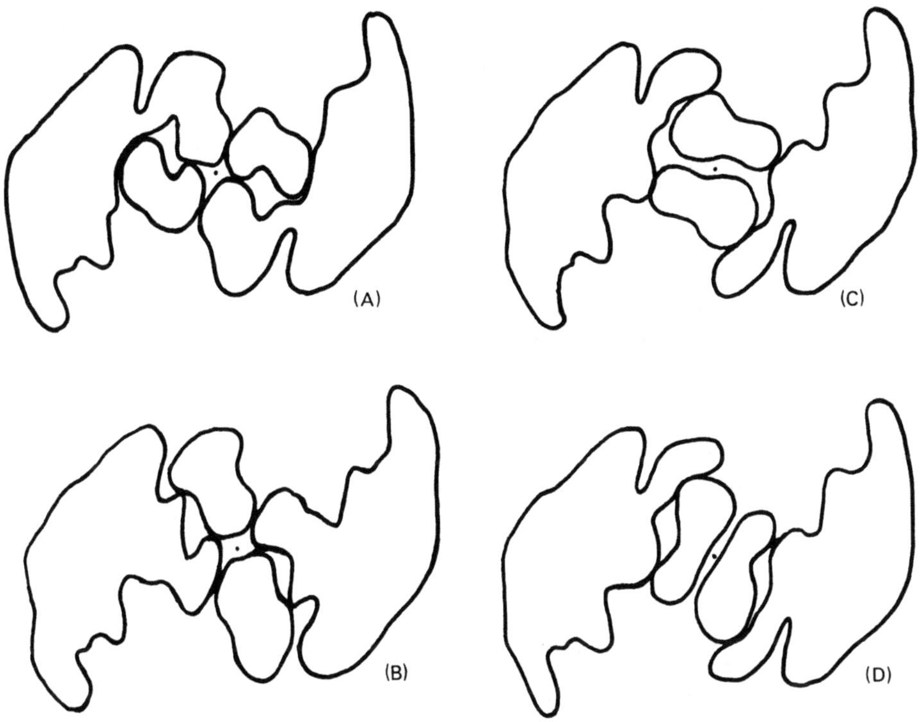

Figure 2: Schematic diagram of Fig. 1 showing the possible spatial relation of actin and myosin S1 in actin-tropomyosin-S1. In (a), (b), (c) and (d), domain A(A1+A2), domain B(B1+B2), subdomains B1+A1' and subdomains B1+A1 are assumed to be actin, respectively. Note that in each of four arrangements, S1 and actin make morphological contact at two or more sites and in some cases S1 binds two actin molecules.

The domain A is lower in density in actin-tropomyosin-S1 than in actin-S1. The domain B is larger in actin-tropomyosin-S1 than in actin-S1. This difference is more obvious in a low dose image. Also the radius from the helix axis to the centre of B1 subdomain is smaller in comparison with that in actin-S1. The radii to the centre of domain A1 and domain B1 are almost the same and are about 20 Å. The volume of A1 and B1 is about one half of that of actin monomer. Therefore A1+B1, A1+B1' (for definition of B1' see Fig. 1(B)), A1+A2, or B1+B2 is more satisfactory as the candidate of actin in view of the volume.

There is a hole or a low density region surrounded by domains A, B, C and D. Because of the presence of this hole, it is independent of the choice of actin candidates that there are two or more morphological actin-binding sites per S1 molecule as proposed previously (Wakabayashi & Toyoshima, 1981). Fig. 2 shows schematically this situation.

Fig. 2(A) and Fig. 2(B) show the situation according to the type A and type B assignment by Wakabayashi & Toyoshima (1981), respectively. In these types of assignment the domain A(A1+A2) and the domain B(B1+B2) are assigned as actin and there are three morphological contact regions per S1 molecule. Fig. 2(C) and Fig. 2(D) show the two kinds of the modified B type assignment, in which B1+A1' and B1+A1 is assigned as actin and S1 poses two morphological contact regions. In these modified B type assignments, the subdomain B2 may be added to a part of actin.

Except for the case shown in Fig. 2(D), S1 binds two actin molecules, both in homo- and hetero-strand. This feature is proposed by Mornet, Bertrand, Pantel, Audemard and Kassab (1981). But we do not exclude the possibility of B1+A1 type assignment until the molecular nature of the crosslinked material is clarified. The other factor which makes the interpretation of the reconstituted image more difficult is the presence of strong forbidden meridional reflection on the first layerline presumably due to non-symmetrical distribution of staining reagent. At the present stage of structural analysis, therefore, we cannot identify actin or tropomyosin. To overcome the difficulties described above, we are collecting the data of tilt series hoping to cope with the forbidden reflection and are doing experiments using monoclonal anti-S1 antibody to show the S1 part more directly.

REFERENCES

DeRosier, D.J. & Moore, P.B. (1970). Reconstruction of three-dimensional images from electron micrographs of structures with helical symmetry. J. Mol. Biol. 52: 355-369.

Katayama, E. & Wakabayshi, T. (1981). Three-dimensional image analysis of the complex of thin filaments and myosin molecules from skeletal muscle. III. The multi-domain structure of actin-heavy meromyosin complex. J. Biochem. 90: 703-714.

Moore, P.B., Huxley, H.E. & DeRosier, D.J. (1970). Three-dimensional reconstruction of F-actin, thin filaments and decorated thin filaments. J. Mol. Biol. 50: 279-295.

Mornet, M., Bertrand, R., Pantel, P., Audemard, E. & Kassab, R. (1981). Structure of the actin-myosin interface. Nature 292: 301-306.

Seymour, J. & O'Brien, E.J. (1980). The position of tropomyosin in muscle thin filaments. Nature 283: 680-682.

Spudich, J.S., Huxley, H.E. & Finch, J.T. (1972). Regulation of skeletal muscle contraction. II. Structural studies of the interaction of the tropomyosin-troponin complex with actin. J. Mol. Biol. 72: 619-632.

Szent-Gyorgyi, A. (1951). *Chemistry of Muscular Contraction*, 2nd ed., Academic Press.

Taylor, K.A. & Amos, L.A. (1981). A new model for the geometry of the binding of myosin crossbridges to muscle thin filaments. J. Mol. Biol. 147: 297-324.

Toyoshima, C. & Wakabayashi, T. (1979). Three-dimensional image analysis of the complex of thin filaments and myosin molecules from skeletal muscle. I. Tilt angle of myosin subfragment-1 in the rigor complex. J. Biochem. 86: 1887-1890.

Vibert & Craig, R. (1982). Three-dimensional reconstruction of thin filaments decorated with a Ca^{2+}-regulated myosin. J. Mol. Biol. 157: 299-319.

Wakabayashi, T., Huxley, H.E., Amos, L.A. & Klug, A. (1975). Three-dimensional image reconstruction of actin-tropomyosin complex and actin-tropomyosin-troponin T-troponin I complex. J. Mol. Biol. 93: 477-497.

Wakabayashi, T. and Toyoshima, C. (1981). Three-dimensional image analysis of the complex of thin filaments and myosin molecules from skeletal muscle. II. The multi-domain structure of actin-myosin S1 complex. J. Biochem. 90: 683-701.

DISCUSSION

KAWAI: Does your domain analysis change if you add small amounts of S-1 instead of full amounts of S-1?

WAKABAYASHI: We're trying, but it is very difficult to get good symmetry when we partially decorate the thin filament.

KAWAI: What happens if you add small quantities of troponin?

WAKABAYASHI: We did not do that experiment yet.

SUGI: Did you prepare the actin-S-1 complex with calcium or without calcium?

WAKABAYASHI: The calcium concentration was not controlled with EGTA, so it's nominally free.

SUGI: As you know, the properties of the rigor linkage depend on the absence or presence of calcium.

HUXLEY: You didn't say anything about where you thought tropomyosin might be located.

WAKABAYASHI: At first we thought this new H domain (Fig. 1) must be the tropomyosin because we didn't see it in the actin-S-1 complex, but when we put in tropomyosin, it emerged. But the difference between the resolution of the actin-S-1 and the actin-tropomyosin-S-1 is rather large. With the actin-S-1 our radial resolution is just about 20 Å, and with actin-tropomyosin-S-1 it is 15 Å. So we checked by restricting our resolution by limiting the area in reciprocal space just within the 20 Å resolution. In that case we couldn't see the H domain anymore. So, this H domain might also be there in the system of actin-S-1 if we could attain as good resolution as with actin-tropomyosin-S-1, in which case its appearance would be simply a result of improved resolution. Therefore, we are, at the moment, uncertain which is tropomyosin.

TREGEAR: Does the Kassab result (that you get one S-1 binding two actin monomers) help you at all in assigning where the actin is and where the subfragment-1 is?

WAKABAYASHI: Yes, in Kassab's complex, the weight ratio of S-1 to actin is smaller in comparison with that in the actin-S-1 complex. So if we could reconstitute that cross-linked complex, the density of S-1 should be smaller and this could help the assignment.

HUXLEY: Linda Amos, Ken Taylor, Ken Holmes and I have done further reconstruction in a sort of cookbook method of selecting amongst the various reconstructions which differ somewhat from one particle to another, selecting those which seem to agree most closely with the X-ray diffraction, and then using those to give, I hope, at least a different average structure. That does seem to show the two-point contact of the S-1 with the actin filament. But could I ask in your sort of marvelous compromise where both are in the actin, how does that agree with other information you may have on the shape of the actin monomers?

WAKABAYASHI: In the model of actin-tropomyosin, actin is elongated in the azimuthal direction. If B1+A1 or B1+A1' is assigned as actin, both types agree with this feature of actin.

HUXLEY: What about the average radius of the sub domains on that? What figure do you get for that?

WAKABAYASHI: The radii to A1 and B1 are the same and are around 20 Å.

HUXLEY: What about the low-dose monomers? What's the best value now for the total overall length of the S-1? It looks slightly longer than earlier results.

WAKABAYASHI: Yes, almost 200 Å, and we confirmed the total length by directly shadowing the monomer of S-1 and myosin.

GULATI: In your bent S-1 model, how far would the S-1 stick out from the actin, in the longest dimension?

WAKABAYASHI: A radius of 100 Å.

SUGI: Could you make some speculation relevant to the cross-bridge mechanism of contraction?.

WAKABAYASHI: In 1979, we reconstructed the actin-S1 complex. Even though this complex is believed to be in the rigor state, the angle between the helix axis and major axis of S-1 was about 75° and was not 45°. So I would like to point out the possibility that the S-1 head might move in the horizontal plane which is normal to helix axis rather than in the vertical plane as usually assumed. In the former case, if S-1 head rotates in the horizontal plane by 100°, then it can pull the actin filament by about 70 Å. Another possibility is a change in the shape of S-1 itself, i.e., bending or flipping of the tail part (e.g., the domain E in Fig. 1).

A-BAND MASS EXCEEDS MASS OF ITS FILAMENT COMPONENTS BY 30-45%

M.K. Reedy and C. Lucaveche

Department of Anatomy, Duke University Medical Center, Durham, NC 27710

ABSTRACT

We have measured filament lattice spacing in fibrils using X-ray diffraction, and find that STEM-determined mass/length values reported for myofilaments should give 13% w/v as the filament protein concentration in the lattice of the AO-band (filament overlap zone) of both insect flight muscle (IFM) and vertebrate skeletal muscle (VSkM). This is well below the actual mass concentration of AO-bands as measured by immersion refractometry of detergent-washed rigor myofibrils under the phase microscope. This technique identifies an immersion fluid of suitable refractive index (RI) for matching out all image contrast between background and the selected cross-band. We used RI fluids in which the effective RI matching component is a large-particle solute (Percoll or hemocyanin) excluded by the filament lattice. The measured RI indicates that protein concentration in AO-bands is 16-17% in Lethocerus and other IFM fibrils including CAF-digested fibrils, and is 18-19% in rabbit VSkM fibrils. On IFM fibrils we also measured absolute buoyant density in Percoll as ≥1.042; this supports the value for mass concentration as determined by RI. The mass discrepancy between fibrils and filaments does not seem to arise from faults in the methods used. We therefore accept the STEM-determined mass/length values for thick filaments which indicate 4 myosins/crown in IFM and 3 in VSkM, and we believe there is considerable extra nonfilament material (concentration: 30-50 mg/ml) between the filaments in fibrils. In stretched VSkM, it is the H-bands, not the I-bands, which have an excess over filament mass content. The extra mass has not been identified.

INTRODUCTION

Rigor crossbridges of insect flight muscle (IFM) are so regularly arranged that counting them in electron micrographs is easy. In transverse sections 10-20 nm thick, the grouping in a single 14.5 repeat along each thick filament is a set of four attached bridges making a "flared X" (Reedy, 1968), requiring at least four myosin heads, and thus at least two myosin molecules. We have used quantitative microscopy to determine the mass and possible myosin content of thick filaments. Now this has confronted us with a discrepancy which appears soluble only if we can identify non-filament components which account for 1/4 or more of A-band mass. This must be material which does not remain associated with dispersed and isolated myofilaments, and which indeed is not readily identified with structures known from EM, X-ray diffraction, or with proteins familiar from SDS-PAGE gels.

Bullard and Reedy (1973) estimated myosin as 57.5% w/w of IFM fibrillar protein, while Reedy, Bahr and Fischman (1973) measured total protein concentration in the filament overlap zone (AO-band[1]) of whole fibrils as 17.1% w/v, inferring that the 57.5% myosin must be confined there at a concentration of 62% w/w or 235 μM, sufficient for 6 two-headed myosins per crown (per 14.5 nm repeat). However, the mass per unit length of individual IFM thick filaments, as measured with the scanning transmission electron microscope (STEM), can accommodate only 4 myosins/crown (Reedy et al., 1981). We calculated (Table 1) that 4 myosins/crown should give an AO-band protein concentration of only 0.13 g/ml or 13%.

We tried to guess how the measurements from whole fibrils could come to be in such striking conflict with those from isolated filaments. Reedy et al. (1973) had used interference microscopy and quantitative electron microscopy, carefully measuring fibrils one by one to determine diameter and total mass. Both methods indicated an identical (average) total mass when applied together to the same set of 29 fibrils. But mass concentration estimates depended on volume, and so might have been too high if measurements of the 2.5-3.5 μm diameters of cylindrical fibrils were somehow biased 0.3-0.4 μm low.

We finally decided to remeasure AO-band dry mass concentration to see if it might be 13% after all instead of 17%, by using a different method which did not require diameter measurements, which allowed rapid comparison and evaluation of many fibrils in the microscope field of view, and which can establish protein concentration to 0.5% or better. We used Barer's (1966) adaptation of immersion refractometery to the phase contrast microscope, determining RI of fibrillar cross-bands by trying immersion fluids of different RI to identify the RI which matched out all image contrast between cross-band and background. The specific refraction increment of proteins is generally very near 0.18 ml/g, meaning that RI of a protein solution or structure increases by 0.0018 for each 1% w/v increase in concentration. We also measured filament lattice spacing directly by X-ray diffraction of the fibrils. Our new measurements with these methods, taken here from papers submitted and in preparation (Reedy and Lucaveche, 1983) convinced us that the discrepancy between filaments and fibrils is a definite finding in IFM, and can be demonstrated in material from vertebrate skeletal muscle (VSkM) as well.

---

[1]We propose "AO-band"as an easily recognized abbreviation for "filament overlap zone", to replace the less obvious "A*-zone" (Huxley and Hanson, 1957).

METHODS

Washed myofibrils in suspension, including some digested selectively with CAF to remove Z-lines, were prepared as in Reedy et al. (1981). They were mounted under a coverslip in a channel defined by parallel grease stripes, and observed by green light (546nm) using a Zeiss Neofluar 100× phase contrast objective. Refractive index matching fluids were drawn through this channel by filter paper wicks, taking precautions to ensure that fluids were not diluted by preceding buffer or concentrated by evaporation before they reached and immersed the fibrils being observed and photographed.

In order to measure the averaged RI of filaments plus any other material within the lattice at a known filament spacing, we chose immersion fluids whose effective RI comes from solute particles too large to penetrate the 12-14 nm interstices between filaments of the AO-band. Percoll (17 nm, but shows hydrodynamic and exclusion volume behavior of a 30-35 nm sphere; Laurent et al., 1980) and purified Limulus hemocyanin (abbreviated Hc; a gift from M. Brenowitz and J. Bonaventura; measured diameter 22-26 nm) were concentrated under pressure or by centrifugation and made up in standard carrier buffer (100 mM KCl, 5 mM NaN_3, 20 mM MOPS buffer, pH 6.8; 10 mM $CaCl_2$ was included in Hc solutions to inhibit dissociation into subunits). Concentration was determined and adjusted according to RI measurements to 0.0002 on a B & L Abbe refractometer, and was expressed in percent PE (protein equivalent), as that of a protein solution having the same RI. Hc solutions up to 27% PE and Percoll up to 21% PE were used. Effective RI of an RI matching fluid was determined by Airfuging an aliquot and subtracting supernatant RI from total RI before calculating the effective percent PE of the fluid. RI of such supernatants was usually just that of carrier buffer, 1.3345.

Colloid osmotic pressure of RI fluids in the AO-band matching concentration range was measured by a Wescor 4100B membrane osmometer.

For X-ray diffraction, fibrils were lightly centrifuged to eliminate larger segments and bundles, then packed in quartz capillaries at 3000 g to form a pellet. These were diffracted 3-10 hr using a 50 cm camera on a rotating anode source as described (Magid & Reedy, 1980).

RESULTS

Matching experiments are shown in Figures 1-3, control experiments in Figures 4-5, and major findings in Table 1.

AO-bands from IFM fibrils of waterbug, flies and bumblebees all showed RI matching in Percoll or Hc solutions of 16-18% PE. The best established value, from repeated experiments with waterbug (Lethocerus) fibrils, was 16.5-17% (Fig 1). 13% PE solutions, corresponding to the calculated concentration of filament proteins, always fell far short of matching AO-bands. CAF-digested waterbug fibrils lost Z-band material, but RI of the AO-band did not change (Fig 2). The CAFed fibrils were the same batch used to make filaments cited by Reedy et al. (1981) in a "Note

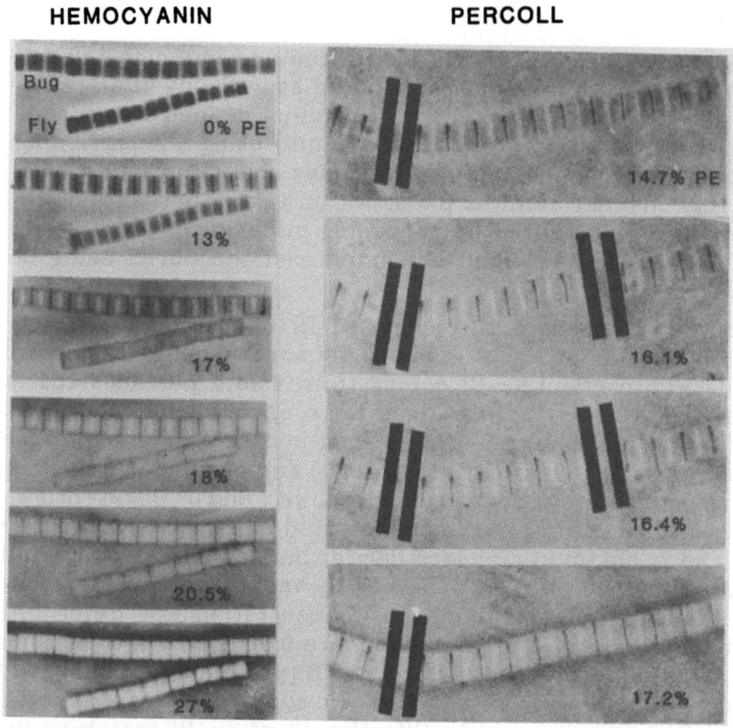

Figure 1: Phase contrast image of IFM fibrils from Lethocerus and Sarcophaga goes through reversal of contrast when immersed in higher concentrations of Limulus hemocyanin, but fully recovers original appearance when restored to buffer (as shown in Fig. 3). Refractive index of AO-band in indicated by contrast matching in Hc of 17%-18% PE (protein equivalent: see text). In micrographs of Percoll series, black strips isolate individual Lethocerus AO-bands to help demonstrate matching in 16.4% PE, show undermatching (positive or dark contrast) in lower RI fluids and show overmatching (negative contrast) in higher RI fluids. Dark outline often bounds lateral edge of matched AO-bands in Hc, but not in Percoll.

added in proof". These filaments gave mass/length values in the STEM that showed less variation and were actually closer to 4 myosins/crown than most other batches measured for that report.

Only Percoll could be used with VSkM (rabbit) fibrils (Fig 3), because in Hc, VSkM cross-band appearance was unstable during immersion and not fully reversible upon return to buffer. VSkM AO-bands were matched in 18-19% PE Percoll at sarcomere length (SL) 2.3 μm. H- and I-bands at SL 2.9 μm matched respectively in 13-15% and 4-5% PE. "A-band shortening" (up to 0.3 μm) and other contrast-dependent band length changes were noted at constant sarcomere length in the higher RI fluids.

X-ray "powder patterns" from fibrils showed sharp dense rings for the strongest equatorial reflections (Fig. 4), allowing thick filament spacing (d_m) and unit cell cross section to be calculated (Table 1). Since IFM d_m is 57.5 nm in fibrils, these measurements were necessary to prevent

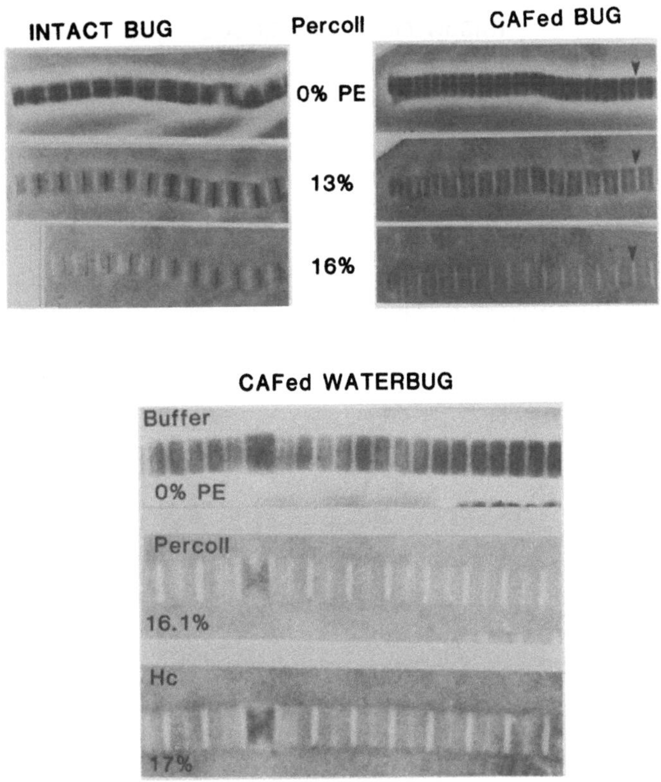

Figure 2: CAF digestion of Lethocerus IFM fibrils to remove dense Z-band material (arrowheads) aided filament preparation for STEM mass/length determinations. However, side-by-side comparisons in mixed suspension of CAFed and undigested fibrils (upper series) showed same AO-band RI match point for both, thus indicated no loss of A-band material in CAF. Lower series shows perfect match near 16% PE in Percoll, slight overmatch at 17% PE in Hc. Hc generally showed equal or slightly (@0.5%) higher match point than Percoll on same fibrils, regardless of sequence.

the error of estimating unit cell volume 12-15% too low (and estimating filament protein concentration too high by 13-18%) which would have resulted from using the 53 nm value we usually found in oriented Lethocerus whole fiber bundles. VSkM did not show such a discrepancy between fibrils and fibers. Volume changed with SL; VSkM rigor fibrils showed only 7% reduction in unit cell cross section accompanying a 26% lengthening of SL from 2.3 to 2.9 μm.

Colloid osmotic pressure was 4-9 mm Hg for 16.5-19.5% PE solutions of Percoll and Hc which gave an RI match for AO-bands. No osmotic squeezing of even the largest IFM fibrils was measurable on photographs, even in 27% PE Hc or 21% Percoll. In contrast, trials of dextrans and PVP at 15% PE (100-200 mm Hg, regardless of molecular weight) cause visually obvious squeezing of both IFM and VSkM fibrils.

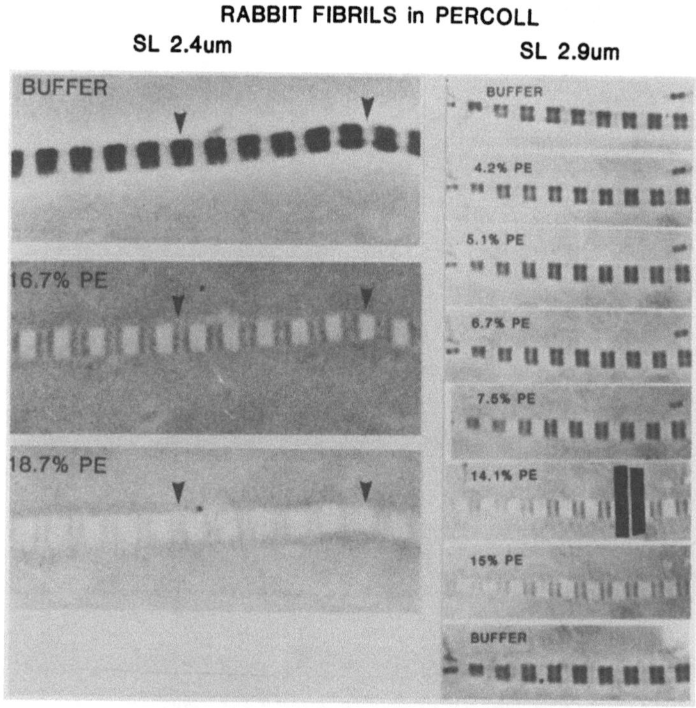

Figure 3: AO-bands in rabbit psoas VSkM fibrils at sarcomere length (SL) 2.4 μm remain obviously undermatched in 16.7% PE Percoll, appear matched or slightly overmatched at 18.7% PE. Note appearance of A-band shortening (left arrowhead) and I-band lengthening (right arrowhead) near contrast match point. Fibrils at SL 2.9 μm show I-band matching near 4.2% PE, H-band matching at 14% PE or higher. Exclusion of Percoll by H- and I-bands was not expected, but stability of match points during prolonged immersion and their level relative to AO-band indicates good values.

X-ray measurements showed that 4 days soaking with daily stirring in 17% PE Percoll caused no change in lattice spacing of fiber bundles of IFM (Fig.4). Hc caused 2% shrinkage in 1 day, 5% in 5 days, corresponding to lattice volume reductions by 4% and 10% respectively. Similar experiments with packed fibrils and with oriented VSkM fibers have failed, because the muscle reflections from these were too weak to show up over the intense central scattering from the colloidal silica Percoll particles.

Thin sections of IFM fibrils fixed after immersion in 17% PE matching fluids showed in the EM that particles were completely excluded by the AO-band lattice, although Hc did penetrate the H-band (Fig. 5). VSkM fibers soaked 4 days in Percoll, then fixed and cross-sectioned, did not show generalized permeation, but did show up to 20% of some fibril cross sections apparently invaded by localized clusters of low-contrast Percoll particles, especially in H and I bands.

The findings that long exposure allowed Hc to squeeze the IFM lattice and allowed Percoll to invade the VSkM lattice outside the AO-band

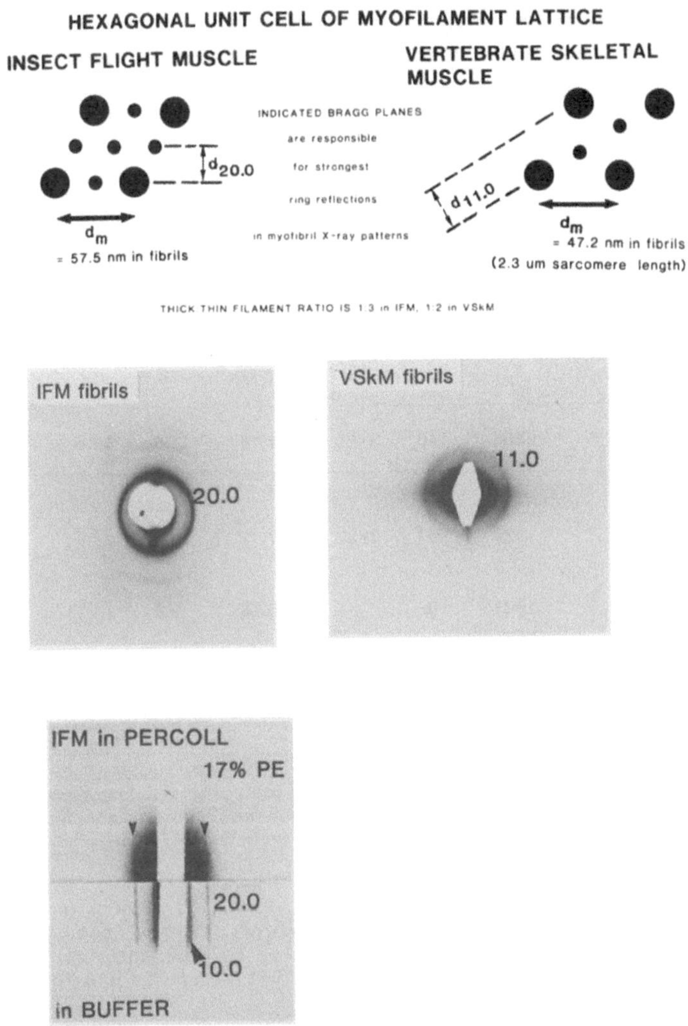

Figure 4: X-ray diffraction of myofibrils indicates filament spacing and unit cell cross-section area as indicated. Oriented fiber bundle shows no change in lattice spacing after 4 days in stirred 17% PE Percoll. (Fixed and cross-sectioned, same bundle showed penetration of Percoll throughout inter-fibrillar space but not into fibrils).

require caution and more control experiments, but do not yet establish that these problems affected fibrils during the much briefer procedure of immersion refractometry.

Table 1: (from Reedy & Lucaveche, in preparation)

	Thick Filament Spacing in Fibrils (X-ray)	Unit Cell Cross-sect. Area	Thick Fil. Mass/ Length in Unit Cell (STEM)	Thin Fil. Mass/ Length in Unit Cell (Calculated)	Total Fil. Protein Concn	Total Fil. Protein Concn	Total Measured Protein Concn (RI)	Excess over Fil. Mass Concn	Missing or Non-Fil. Mass Fraction of Lattice
Footnotes	(1)	(2)	(3)	(4)	(5)	(6)	(7)	(8)	(9)
Units	nm	nm^2	kd/nm	kd/nm	daltons per nm^3	g/100 ml	g/100 ml	percent	percent
IFM AO-band	57.5	2863	146	3×25.6	77.8	12.9	16.5	28	22
VSkM AO-band (SL 2.3 μm)	47.2	1929	104	2×22.5	77.2	12.8	18.5	45	31
VSkM H-band (SL 2.9 μm)	45.6	1801	104	——	57.8	9.6	14	46	31
VSkM I-band (SL 2.9 μm)	45.6	1801	–	2×22.5	25	4.2	~4.6	~10	~9

Footnotes:

(1) Rigor fibrils in same carrier buffer used for RI matching. Spacing calculated from 500 mm camera length and also calibrated from 14.5 nm IFM meridional from oriented IFM fibers. Thick filament separation d_m was derived from IFM $d_{20.0}$ lattice spacing by $d_m = 2(d_{20.0})/\sin 60°$. For VSkM, $d_m = 2(d_{11.0})$.

(2) Area = $(\sin 60°)(d_m)^2$.

(3) Mass/length used corresponds to the nearest integral number of 470 kd myosins/crown (plus accesory proteins) favored by STEM measurements, 4 in IFM (complemented by 11% paramyosin) and 3 in VSkM (with 5.5% C{etc.}-proteins)(Lamvik 1978: Reedy et al., 1981).

(4) Calculated by taking actin as 42 kd per 2.73 nm in VSkM, actin-arthrin average as 45.3 kd per 2.76 nm in IFM, and taking troponin-tropomyosin as 275 kd per 38.5 nm in VSkM (Lamvik, 1978) but 354 kd per 38.7 nm in IFM (Bullard et al., 1973; Bullard, personal communication).

(5) Filament mass, in kd per 1 nm length of unit cell, divided by volume of 1 nm thickness of unit cell. Thus, $[146 + (3 \times 25.6)]$ kd in 2863 nm^3 gives 77.8 daltons/nm^3.

(6) Derived by this conversion factor: 1 g/100 ml = 6.023 daltons/nm^3.

(7) Results of immersion refractometry in present study, assuming a specific refraction increment of 0.18 ml/g for deriving protein concentration from RI.

(8) Discrepancy expressed by taking filament mass concentration as 100% of known material, and designating non-filament fraction as percentage in excess over 100%.

(9) Smaller percentage will express same discrepancy, if we take measured total cross-band mass as 100%, and designate non-filament fraction as "missing" mass.

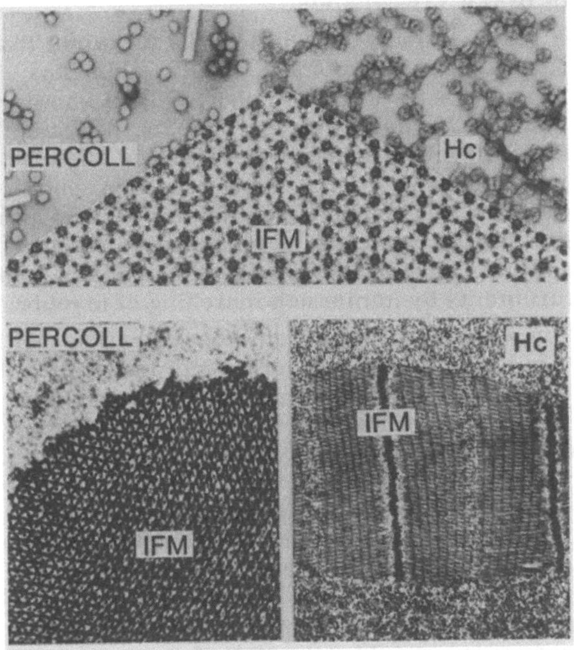

Figure 5: Above, composite of three electron micrographs shows why relative size of Percoll and hemocyanin particles should exclude them from Lethocerus IFM lattice. Lattice is enlarged to show fresh muscle spacing (53 nm) rather than reduced spacing (45 nm) typical of fixed, embedded muscle. Lower EMs demonstrate expected exclusion of effective RI solute particles in 17% PE fluids from AO-band lattice of single IFM fibrils, fixed in glutaraldehyde-tannic acid after 30 min. immersion in RI fluids. Hc (but not Percoll) penetrates H-band and possibly I-band regions.

DISCUSSION

A-Band Length Changes

This subject may be important in other sessions of this symposium, so we will touch upon it before proceeding with our major Discussion. Huxley and Niedergerke (1958) found that A-band length (by densitometry of photographs) varied by less than 0.05 μm at fixed sarcomere length when contrast in the interference microscope was reversed by changing compensator setting. We have now observed length changes up to 6 times greater than this in the phase contrast image of the A-band and I-band during immersion of fibrils in higher RI fluids. Sarcomere length was constant. The changes were completely reversible on washing out the immersion fluid (Fig 3). We conclude that contrast-dependent optical effects can produce illusory changes in A-band length, and could be a major reason for some reported examples of A-band shortening.

Protein Content of IFM Cross-Bands

Immersion refractometry with IFM fibrils of known filament spacing from Lethocerus, Musca, Sarcophaga and Bombus has confirmed the finding with Sarcophaga by Reedy et al. (1973) in showing that protein concentration in the filament overlap zone or AO-band, estimated to within ±0.5% from refractive index, is 16-17%. The effective RI of the preferred matching fluids was due to large particles, shown by EM to be excluded from the filament lattice, and shown by X-ray diffraction not to reduce filament lattice spacing significantly. Non-penetrating, non-squeezing immersion fluids are necessary and sufficient to obtain reliable native RI measurements by immersion matching of myofibril cross-bands.

The question whether total protein concentration can be accurately derived by assuming a specific refraction increment of 0.18 ml/g for all fibrillar proteins seems well-answered by the agreement Reedy et al. (1973) found between interference microscopy (where 0.18 ml/g was taken) and EM mass determinations (which depend on electron scattering, not on refraction increment). Nevertheless, we recently devised another independent test. We used a Mettler-Paar density meter (accurate to 5-6 decimal places) to measure the density of the well-defined isopycnic band which IFM fibrils produce on self-forming gradients when centrifuged in Percoll of about 3-4% PE. Assuming a partial specific volume for all proteins as 0.75, the density of AO-bands (which dominate average IFM sarcomere density) should be 1.0325 if they were 13% protein, but 1.0425 if they were 17% protein. We found density values (corrected for buffer density of 1.004) indicating an average sarcomere protein concentration of 17%-19% from Sarcophaga (1.0422), Musca (1.0432), and CAFed Lethocerus (1.048).

Thus IFM AO-band protein concentration is clearly around 16-17% (entered in Table 1 as 16.5%), rather than 13%, and concentration of myofibrillar proteins can be reliably found from RI, using a refraction increment of 0.18 ml/g.

Protein Content of Vertebrate Muscle Cross-Bands

Refractometry of the rabbit VSkM AO-band indicates a protein concentration of 18-19%, even higher than IFM. This is the expected sum from the indicated I-band (4-5%) and H-band (13-15%) values. The indicated H:I mass ratio is near 3.1, which conflicts with 2.17 as found by Huxley and Hanson (1957), but needs a few more supporting experiments before we are prepared to insist on it. Among those must be satisfactory control experiments that rule out both squeezing and penetration by matching fluids, like our controls with IFM. Nonetheless, our preliminary mass thickness profiles, from EMs of unstained rabbit sarcomeres like that published from an insect sarcomere (Reedy et al., 1973), do already support an H:I ratio near 3 rather than 2.20. We should note that 2.31 is the H:I ratio expected for 3 myosin/crown on the basis of filament masses alone. If we did not trust the STEM measurements which indicate 3 myosins/crown in rabbit thick filaments, we could argue that an observed ratio near 3.05 "proves" a VSkM thick filament model with 4 myosins/crown. Instead, we believe it indicates extra non-filament matter in the A band.

Table 1 shows that the H-band, rather than the I-band, exhibits the greatest excess over expected filament protein concentration, suggesting that most of the excess material remains associated with thick filaments when thin filaments are withdrawn. The difficulty with stretching IFM muscle leaves us free only to speculate that such an association might also occur in insect. But most of all, it leaves us wondering why such material does not remain associated with thick filaments when they are weighed in the STEM.

Our VSkM values are from washed fibrils, thus not comparable with the values from intact whole fibers measured by Huxley and Niedergerke (1958). As these authors noted, soluble sarcoplasmic proteins help to account for the mean protein concentration of 24% w/v they found in frog fibers. Such solutes were expected to be more excluded from A-band than I-band, thus giving 7.1% w/v as the observed difference between these, rather than 10-11% as found in our washed fibrils.

Mass/Length of Thick Filaments by STEM

If the measurements of AO-band protein concentration are really without fault, then they indicate enough mass to allow thick filaments at the observed lattice spacings to contain 6 myosins/crown in IFM and over 5 in VSkM. But the STEM measurements of mass/length for thick filaments are not easy to fault either. Thick filaments from waterbug averaged 138-160 kilodaltons/nm, while those from rabbit averaged 94-99 kd/nm, in the several experiments published (Lamvik, 1978; Reedy et al., 1981). These are near ideal values of 146 kd/nm for 4 myosins/crown in IFM and of 104 kd/nm for 3/crown in VSkM. There are several reasons for confidence in such STEM measured values:

1) The CAF digestion treatment used to prepare IFM fibrils for dispersion into filaments used for STEM measurements did not remove protein from thc AO-band, according to our immersion refractometry here.

2) Filaments cross-linked before de-ionized water washing were compared with unfixed filaments subject to prolonged water desalting. Both sets averaged within 5% of the mass/length value for 4 myosins/crown. (Note added in proof to Reedy et al., 1981).

3) Other STEM measurements (unpublished, with J. Wall) of waterbug filaments which were desalted quite differently, by freeze-drying from 10mM ammonium acetate, gave an average mass/length about 138 kd/nm, ±10%, supporting 3.8 myosins/crown.

4) Recalculation by Bullard (in preparation; personal communication) of the myosin:actin ratio in equivalent filament lengths as estimated from SDS-PAGE gel scans has now taken account of arthrin, a 55 kd isomorph of actin present 1:3 with 42 kd actin in Lethocerus thin filaments. The previous calculations based on the gel stain ratio (Bullard and Reedy, 1973) defined a crown as equivalent to 3 × 14.5 nm thin filaments, taking 15.8 monomers of "45" kd actin as the measure of a crown of myosin. By

this measure, the 2.83 ratio of myosin heavy-chain (201 kd) to actin indicated 5 myosins/crown. If Bullard now takes 11.8 actins of 42 kd (+ 4 arthrins) instead, this indicates 3.5 myosins per crown, rather than 5.

5) STEM measurements on Limulus and tarantula thick filaments, which clearly have fourfold rotational symmetry and a 4-stranded helical surface array of crossbridges, indicate mass/length values near 4 myosins/crown (personal communications, M. Dewey & J. Wall, and R. Levine, R. Freeman and W. Hofmann).

Mass/Length of Thin Filaments by STEM

Mass/length of native thin filaments containing just actin and regulatory proteins should lie near 22.5 (VSkM) or 25.6 kd/nm (IFM), as calculated in Table 1. For our purpose, the important result from the few STEM measurements available is that thin filaments are less heavy, not more heavy than expected. Native thin filaments from VSkM give about 0.24 the mass/length of thick filaments, judging from their use in calibrating thick filament mass (Lamvik, 1978). Those from IFM have averaged about 1/8-1/6 the value for IFM thick filaments (or 18-24 kd/nm), while insect F-actin and F-arthrin calibrated by fd bacteriophage gave about 12.5 and 17 kd/nm respectively, *versus* 15.4 and 20.2 kd/nm expected (Freeman and Leonard, 1981; and personal communication).

Filament Component Concentration and Excess Mass in the Unit Cell

As Table 1 shows, once we have measured the spacing (columns 1 & 2), mass (columns 3 & 4) and number of thick and thin filaments in the unit cell (Fig. 4), we can easily calculate the concentration of filament components (columns 5 & 6). Since we can find no reason to doubt the values used in these calculations, we must argue that the filament protein concentration in the AO-band is at or just less than 13% in both IFM and VSkM, while total protein concentration is unmistakably a good deal higher (columns 7 & 8). Just how much higher depends on how we express the discrepancy (column 7 *versus* column 8).

The extra material is lost from view at the moment when we disperse CAFed IFM fibrils by shearing, converting them to a mixed filament suspension. Weighed by STEM, these filaments are found wanting in the extra mass. Pelletted in the Airfuge, they leave nothing in the supernatant which can be concentrated by TCA precipitation and run on SDS-PAGE gels except for a little thin filament protein. Nevertheless, we continue to suspect that part of the extra material might be high molecular weight proteins (titins, nebulins) as identified with up to 10-20% w/w of VSkM fibrils by Wang et al. (1979, and this symposium). However, on gels made from Lethocerus fibrils up to now by K. Wang (personal communication), only about 5% of staining appears as 300-600 kd bands, and no other unfamiliar bands appear. The extra material may not all be protein. Polysaccharide is not indicated, because no PAS staining is seen in whole IFM fibrils. Nucleic acid is not indicated, for its total mass in VSkM (Winick and Noble, 1966) could add only 1.5% to fibrillar mass. We were therefore puzzled at first by a high A_{260}/A_{280} ratio from NaOH-solubilized fibrils of IFM and VSkM, until the full UV absorption spectrum showed only a smooth slope rising from 270 nm to peak near 225 nm, with no peak at 260 nm.

Clearly, the whole issue must remain in some doubt until the extrafilamentous substances are identified biochemically. Part of this doubt must be to wonder how any such amount of material could be structured so it remains undetected while out in the open between the filaments. It has no recognized effect on either equatorial or axial features of X-ray diffraction. It answers no mysteries raised by 20 years of examining negatively stained filament suspensions. In the IFM unit cell, it represents half again the mass of the one thick filament, or twice the mass of the three thin filaments, yet heavily stained sections show the interstices of the lattice devoid of anything but crossbridges. How could crossbridges swing and filaments slide during thirty years of scientific scrutiny without betraying the steric hindrances and major functions that ought to arise from such a quantity of interfilament material? Despite such doubts, our physical findings make its presence seem highly probable. It if is truly present, such features as the interfilament struts proposed by Magid (this symposium) may suggest structure and function for some of this material. While we work to resolve these doubts, we can take some comfort that missing matter is not a problem restricted to muscle. Cosmologists are currently troubled on a far grander scale, finding that the gravity necessary to hold spinning galaxies to their observed shape requires at least ten times more matter than astronomers have detected with their telescopes (J. Silk, Nature 297: 102).

ACKNOWLEDGEMENTS

Supported by NIH Grant AM 14317. We thank Leo Cordova for skilled assistance with X-ray diffraction, Dr. R.W. McIntyre for use of his colloid osmometer, and Dr. J. Reynolds for use of her density meter.

REFERENCES

Barer, R. (1966) Phase Contrast and Interference Microscopy in Cytology. In Physical Techniques in Biological Research. Vol IIIA (edited by Pollister, A.W.) pp. 1056. New York: Academic Press.

Bullard, B. and Reedy, M. K. (1973) How Many Myosins per Cross-Bridge? II. Flight Muscle Myosin from the Blowfly, *Sarcophaga bullata*. In Cold Spring Harbor Symp. Quant. Biol. 37: 423-428.

Bullard, B., Dabrowska, R. and Winkelman, L. (1973) The Contractile and Regulatory Proteins of Insect Flight Muscle. Biochem. J. 135: 277-286.

Davies, H.G. and Thornburg, W. (1960) The Specific Refraction Increment of Chrystalline Protein. II. Alpha-Chymotrypsinogen. Biochim. Biophys. Acta 37: 25-33.

Freeman, R. and Leonard, K. (1981) Comparative mass measurement of biological macromolecules by scanning transmission electron microscopy. J. Microscopy, 122: 275-286.

Huxley, A.F. and Niedergerke, R. (1958) Measurement of the Striations of Isolated Muscle Fibres with the Interference Microscope. J. Physiol. 144: 403-425.

Huxley, H. E. and Hanson, J. (1957) Quantitative Studies on the Structure of Cross-Striated Myofibrils. I. Investigations by Interference Microscopy. Biochim. Biophys. Acta 23: 229-249.

Lamvik, M. K. (1978) Muscle Thick Filament Mass Measured by Electron Scattering. J. Mol. Biol. 122: 55-68.

Laurent, T. C., Ogston, A. G., Pertoft, H. and Carlsson, B. (1980) Physical Chemical Characterization of Percoll. II. Size and Interaction of Colloidal Particles. J. Colloid Interface Sci. 76: 133-141.

Magid, A. and Reedy, M. K. (1980) X-ray Diffraction Observations of Chemically Skinned Frog Skeletal Muscle Processed by an Improved Method. Biophys. J. 30: 27-40.

Reedy, M. K. (1968) Ultrastructure of Insect Flight Muscle. I. Screw Sense and Structural Grouping in the Rigor Cross-Bridge Lattice. J. Mol. Biol. 31: 155-176.

Reedy, M. K., Bahr, G. F. and Fischman, D. A. (1973) How Many Myosins per Cross-Bridge? I. Flight Muscle Myofibrils from the Blowfly, *Sarcophage bullata.* Cold Spring Harbor Symp. Quant. Biol. 37: 397-421.

Reedy, M. K., Leonard, K. R., Freeman, R. and Arad, T. (1981) Thick Filament Mass Determination by Electron Scattering Measurements with the Scanning Transmission Electron Microscope. J. Musc. Res. and Cell Motil. 2: 45-64.

Wang, K., McClure, J. and Tu, A. (1979) Titin: Major myofibrillar components of striated muscle. Proc. Natl. Acad. Sci. USA 76: 3698-3702.

Winick, M. Noble, A. (1966) Cellular Response in Rats during Malnutrition at Various Ages. J. Nutrition 89: 300-306.

DISCUSSION

WILKIE: Can I ask a very naive question and that is, where you say A-band mass, you don't actually mean "mass" do you?

REEDY: What would you have me say?

WILKIE: Whatever you mean -- what *do* you mean?

REEDY: I mean total material that's responsible for the refractive index which is higher than the proteins in the filaments could give. In one case (Reedy et al., 1973) I mean total material which I've measured in whole fibrils by electron scattering, which again is greater than could be accounted for by the filaments, themselves.

WILKIE: By what criteria do you call it mass? It's confusing for people like me who are peripheral to the field to see the word "mass" used really under circumstances where it's inappropriate.

REEDY: I appreciate the criticism. I'll excuse myself, but not very much, by saying that I borrowed it from the interference microscopists of the 50's, who were prone to talk about dry mass concentration, so it comes naturally to the tongue in this sort of work. I felt that there were several converging kinds of evidence: specific gravity evidence, electron scattering evidence, and refractive index evidence, which make it difficult to find any common factor apart from excess material or mass that was simple to talk about.

WAKABAYASHI: When we see single myofibrils of glycerinated rabbit skeletal muscles, the phase contrast image observed by an Olympus microscope using a 40× objective looks very natural, but when I use 100× objectives the contrast becomes very sensitive to the focus and is sometime reversed. So I wonder what type of objective are you using to do the experiment?

REEDY: Always oil, either 63× or 100× Zeiss Neofluoar positive contrast oil, and always searching carefully to judge the best focus. We also see contrast reversal upon slight defocus, but fine detail seems never to be as clear as in true focus. The dense Z-band in insect flight muscle is also a helpful guide to the correct contrast. But of course there is little or no contrast to undergo reversal when refractive index-matching fluids are near the match point. When A-overlap-band contrast is very low, I think we can only see the in-focus image. So I see no chance for contrast reversal by defocus to be misleading us.

KREUGER: Is there a way of ruling out the possibility that there might be some slight shift in the length of the A-band? You're really talking about mass from the A-band volume.

REEDY: Yes. The fibrils are in rigor, and the change in appearance as I immerse them in the refractive index matching medium is totally reversible. As you wash away the immersion medium, the fibrils go immediately back to their normal appearance. There's no apparent penetration of this material from ends *or* sides. I would say one thing though, as a side comment, because I think there are people here interested in A-band shortening -- that the contrast changes here bring about an apparent A-band shortening of 20% or so in vertebrate muscle. It's interesting to see how solid-looking A-bands get apparently shorter under conditions where the change must be contrast-dependent entirely. There's no change in overlap, and no change in sarcomere length.

HUXLEY: In the case of the refractive index measurements, I presume you must have looked into whether you get or don't get any "funny effects" on the observed refractive index, depending on the packing density of the particles. Are there any anomalous effects of that kind?

REEDY: The only ones I know were published by Davies and Thornburg (1960) who did a couple of interference microscope experiments on crystals. They determined that, opposite to salts and amino acids, the refractive index of alpha chymotrypsinogen and beta lactoglobulin *increased* when they were densely packed in crystals. But the refractive index increase was only on the order of 3 to 7%, even when they tried to account for all possible effects of birefringence by measuring along different crystal axes, etc. So I considered for a while whether .0018 might be the wrong refractive increment to use, but I now believe that the refractive index results are supported by the electron scattering results we got before (Reedy et al., 1973). There we got agreement between these two *very* different kinds of measurements -- interference measurements that require *assuming* a refractive increment to estimate the protein concentration, versus electron density that was calibrated against polystyrene spheres used as mass standards. The second measurement which agrees with our refractometry is this isopycnic banding. Those are two measurements where the refractive index has nothing to do with the estimation of dry mass.

HUXLEY: Yes, OK. In the case of the refractive index of a birefringent myofibril, how much do the two refractive indices differ? If you did your measurements with polarized light, with different orientations, would you get the same result?

REEDY: I think they would differ a little bit. I did that once (Reedy et al.,1973) with an interference microscope. I would expect a fibril that gave me 17% with the electric vector of polarized light parallel to the fibril, to give me 16% with the electric vector perpendicular.

HUXLEY: So your excess mass is considerably more than the refractive index difference between fast and slow directions in the A-band. I wasn't sure.

TER KEURS: The excess mass in the overlap zone -- does it depend on ionic strength of the solutions?

REEDY: I think not. I restricted fibril work to the near-physiologic ionic strength. When you play with ionic strength you invite variations in filament lattice spacing. This would change the refractive index, and each ionic strength would require a separate X-ray measurement of the filament spacing. I would expect the total mass to remain constant, but refractive index and mass concentration to be lower, if you used higher ionic strength without dissolving anything.

HOMSHER: Have you made the same kinds of experiments after you've dissolved out the thick filaments?

REEDY: I could but I haven't. The only material that's readily adaptable to that sort of thing is the flesh fly. It allows the myosin to be dissolved quite easily, but for one reason or another I haven't done it. For one thing, I felt that the thin filaments alone were likely not to be so able to exclude these refractive index matching materials. We saw in some long experiments that they partly penetrated the I-band in vertebrate muscle during four-day soaks. We embedded and sectioned such muscles and saw some little islands of particles invading the I-band. And so, not wishing to deal with confusion, I just haven't gotten around to any myosin extracting experiments on the fly.

ROWE: The quantity of soluble proteins in muscle -- we used to call it the myogen fraction -- must be known. If it were selectively localized to a considerable extent in the overlap region, would that acount for the results?

REEDY: Let me answer that by saying that A. F. Huxley and Niedergerke published on the A-band refractive index in 1958. They found that whole frog fibers had a total solids concentration of something like 24%, substantially higher than I'm giving here. They thought that an excess of soluble material was probably in the I-band, because the difference between the A- and I-band (between the overlap zone and the I-band) was only 7 grams per hundred ml. I'm getting something like 11, which is what they estimated the *filament* protein differences ought to be, based on the work of Hugh Huxley and Jean Hanson. So yes, there is a myogen fraction of sarcoplasmic proteins, but we've done our best to wash it off here. Why should it cling? Yet something "clings". This is the mystery.

HUXLEY: I think you said that when you've looked for the possible presence of some soluble proteins and you spin down the dispersed filaments, that there's nothing left in the supernatant. Have you taken the pellet and run a gel on it?

REEDY: No, we ran gels on the entire filament suspension.

HUXLEY: And there was nothing present other than myosin and the usual proteins - paramyosin, etc?

REEDY: Yes - actin, myosin and paramyosin dominated the bands -- there wasn't much else.

HUXLEY: Can I ask a second question? All your problems about this seem to be based to some extent on the assumption that the STEM results are correct. Could you speculate a bit on possible problems there?

REEDY: I've listed about five reasons in my manuscript for being confident of the STEM results. One of them is that the actin-myosin ratio, the kind of thing that Frank Pepe does with vertebrate muscle, has been refigured by Bullard to take into account this heavy actin isomorph named arthrin. Our 1972 Cold Spring Harbor results (Bullard and Reedy, 1973) had suggested there were five myosins per crown, which we felt supported other calculations giving six. Now, the readjusted ratio, taking into account the heavy actin, gives four, so there's gel-staining evidence supporting the STEM result. As far as the STEM techniques, the note added in proof to the paper we did in Heidelberg (Reedy et al., 1981) represents our best experiments. Here we tried washing, desalting filaments for an hour or more in deionized water. And while it would do terrible things to Arthur Rowe's rabbit psoas filaments, it does nothing at all to the insect. They retain the same mass per unit length that they retain if you cross-link them in relaxing buffer with glutaraldehyde before water desalting. So, given those two experiments, both matching and giving mass/length data with a scatter of only about five percent, I don't expect future STEM results to wander very far from 4 myosins/crown on insect.

Finally, I find support from the work by Rhea Levine, Bob Freeman and Waltraud Hofmann (in preparation) at the STEM in Heidelberg, and similar work by Maynard Dewey and Joe Wall at the STEM in Brookhaven, who have measured the *Limulus* thick filament and the tarantula filament, where there is genuine morphological evidence for a four-stranded helix with four-fold rotational symmetry at each level. They find the mass per unit length, corrected for 20% paramyosin, agrees with four myosins per crown. So in a case where there's morphological support, the mass results agree. And all I can say is that it is unfortunate that so far we don't have good enough morphology from isolated filaments of insect flight muscle to verify the four-fold structure which the STEM seems to indicate. We're well-convinced that STEM results leave room for no more and no less than four myosins per crown. That's as far as I can answer that.

MYOFILAMENT DIAMETERS: AN ULTRASTRUCTURAL RE-EVALUATION

Thomas F. Robinson and Leona Cohen-Gould

Cardiovascular Research Laboratories, Department of Medicine, and Department of Physiology and Biophysics, Albert Einstein College of Medicine, Bronx, NY 10461

ABSTRACT

In situ, ultrastructural measurements of diameters of contractile filaments in skeletal and heart muscle differ considerably from those previously reported. Past measurements have been made in thin, transverse epoxy sections that were non-specifically stained with heavy metal salts to overcome background scattering of epoxy polymer. In our images from transverse de-embedded sections, the hexagonal lattice has some considerable differences from that seen in epoxy sections. Muscle samples from rat atrium and frog sartorius were fixed, dehydrated, embedded in polyethylene glycol, and sectioned. Sections were de-embedded in graded polyethylene glycol/ethanol, mounted on coated grids, critical point dried, and viewed in the electron microscope without staining. The backbone diameters of thick filaments were measured in the M band region and have an average value, after correction for shrinkage, of 25 nm. Thin filament diameters range from 6.5-9.5 nm. In regions of overlap of thin and thick filaments, the thick filament profiles varied from circular to asymmetric; diameters range up to 36 nm and yield eccentricity ratios varying from 1.5 to 1.0 (circular profiles). Portions of thick filaments touch or partially envelope neighboring thin filaments. The relative contributions of cytoskeletal components to these images of overlap regions remains to be determined, but the backbone diameters in glycerinated frog sartorius are not significantly different from control samples. The present results are consistent with those reported for rotary shadowed thick filaments; from recent experiments in muscles whose myofilament lattice is osmotically compressed; and with estimates of A band mass. This lattice geometry yields relatively low surface-to-surface distances between filaments. Steric considerations and their implications for cross bridge theory are discussed.

INTRODUCTION

In vertebrate striated muscle the configurations of myofilaments of the sarcomere have been extensively studied by a variety of methods. Electron microscopy has been an especially important technique in measuring the size and disposition of lattice structures. H.E. Huxley (1972) described ultrastructural studies made of thin sections of epoxy resin-embedded tissue, stained with heavy metal salts, that yielded the following measurements: thick filaments, 10-12 nm in diameter and 1.55 μm length; and thin filaments, 5-7 nm in diameter and 0.975 μm in length. The use of heavy metal staining of epoxy sections is widespread and has been necessary in order to enhance the electron density of biological material relative to the random electron scattering from the epoxy resin itself; however, the specificity of the heavy metal staining has not been well characterized.

In the present study, we have obtained images of the myofilament lattice from sections of tissue that were de-embedded according to a method recently developed by Wolosewick and Porter (1979) and hence required no staining. We have found that these images give filament diameters roughly twice those seen in epoxy-embedded sections. The present results are consistent with those reported from several studies performed with considerably different techniques. We discuss the implications of the resultant lattice geometry to the sliding filament cross-bridge mechanism.

MATERIALS AND METHODS

Rat atrial trabeculae and frog sartorius muscles were fixed in a solution of 6% glutaraldehyde in 0.1 M cacodylate buffer (pH 7.3) for 3 hours at room temperature. The sartorius muscles were then further dissected into small fiber bundles and placed in fresh fixative for an additional hour. Half of the samples were fixed in 0.1 M cacodylate buffer containing 6% glutaraldehyde and tannic acid (0.5% for the rat atrial muscles and 4.0% for the frog sartorius). The tissues were rinsed in plain buffer, post-fixed in 1.0% OsO_4 in buffer and rinsed again. They were then dehydrated in a graded alcohol series, infiltrated and embedded in Spurr's low viscosity resin (Spurr, 1969). Some samples, after dehydration, were infiltrated and embedded in polyethylene glycol (*PEG*), a 4:1 mixture of MW 4000:6000 (Wolosewick and Porter, 1979; Wolosewick, 1980).

Additionally, frog sartorius muscle was tied at rest length and placed in a solution of glycerine and buffer (1:1, v:v) for 1 hour at 4°C. The glycerine-buffer mixture was replaced with similar fresh solution, and the tissue was returned to 4°C overnight. The muscle was then transferred to a solution of 90% glycerine in buffer and stored at -20°C for at least one month. It was then slowly brought to room temperature, fixed, and embedded as described above.

Epoxy blocks were sectioned at 0.1-0.15 μm on a Reichert Ultracut, mounted on uncoated grids, and stained with uranyl acetate (Stempak and Ward, 1964) and lead citrate (Venable and Coggeshall, 1965). PEG blocks were cut at 0.1-0.15 μm on the Ultracut using dry glass knives. The sections were then de-embedded by transferring them, using a single hair brush, to poly-l-lysine-coated grids immersed in a dilute solution of PEG in alcohol (less than 5% w/v). The solution was gradually changed to

100% alcohol and the grids were then critical point dried (Wolosewick, 1980; Robinson, 1980). All grids were examined in a JEOL 100CX electron microscope at an accelerating voltage of 100 Kv.

The magnifications of the negatives were verified using a calibration grid having a grating replica of 54,864 lines/inch (Ladd Research Industries, Burlington, VT.). Print magnifications were calculated by point-to-point comparisons with features in the negative. All measurements were made using either a Peak 4× Anastigmatic Loupe with a calibrated reticule attachment, or a Keuffel and Esser precision ruler. Filament diameter measurements were made from both calibrated prints and negatives of epoxy embedded and de-embedded samples. Thick filament diameters were measured in M band and overlap regions, and thin filaments were measured in overlap and I band. Only filaments with circular profiles were included in the data. In addition to filament diameters, measurements were also made of the center-to-center distances of thick filaments (nearest neighbor) and thick-to-thin filaments. From these data, surface-to-surface distances could be ascertained, using simple geometry (diagram in Figure 6).

As a control study for the PEG samples, a frog sartorius muscle fiber was embedded in PEG, sectioned, de-embedded, and photographed in the electron microscope. The remaining block was then de-embedded and re-embedded in Spurr's epoxy resin. This was accomplished by gently dissecting the fiber out of the PEG block and immersing it in double distilled water overnight at room temperature. It was then placed in fresh water, and warmed to 40°C for 2 hours, after which it was dehydrated in alcohols and embedded in Spurr's resin in the usual manner.

RESULTS

Electron micrographs of transverse sections of epoxy-embedded rat atrial muscle are shown in Figures 1-3. Representative regions of the sarcomere, Z band, I band, overlap regions, and M bands, are included Figure 1. The section was post-stained with uranium and lead salts before it was photographed. The filaments are well aligned. A similar region from an unstained section of the same epoxy is shown in Figure 2. Contrast is low, indicating that the background scattering of the matrix epoxy resin is significant compared to that of the fixed muscle. The use of tannic acid in the primary fixative enhances preservation and post-staining contrast (Figure 3). The diameters of the filaments appear larger in this preparation than in standard fixation and epoxy embedment, but they are not as high as those measured in the de-embedded samples (Figure 4).

Measurements of filament diameters in de-embedded samples were made only where the overall preservation of the lattice is comparable to that seen in epoxy sections. Where sections of tissue were strained during dissolution of the polyethylene glycol embedding matrix, gaps appear mainly between myofibrils, occasionally between filaments within a single myofibril, but never within an individual filament. The absence of strain was used as a criterion for the suitability of any region for measurements.

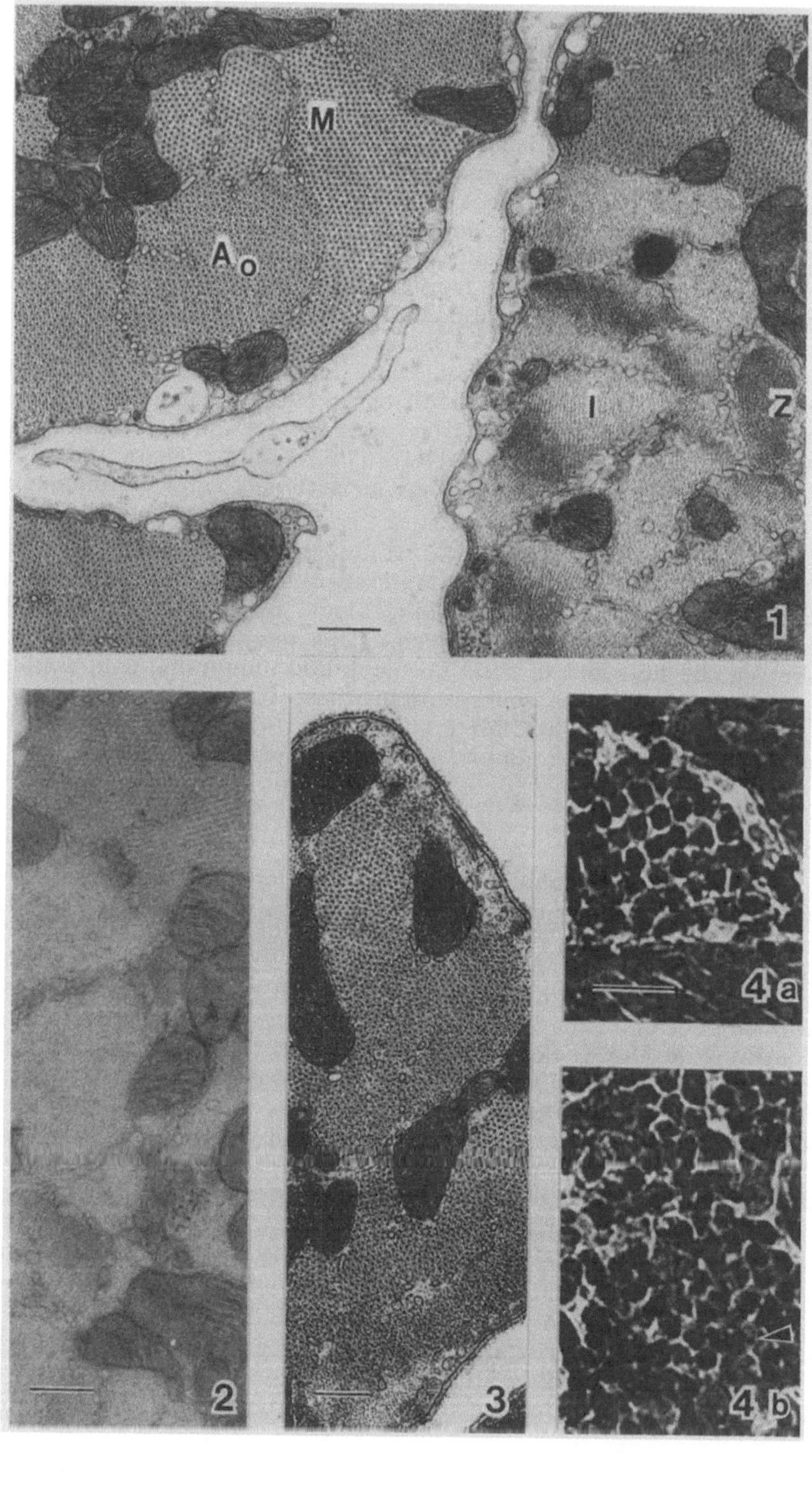
M
Ao
I
Z
1
2
3
4a
4b

Figure 1: Electron micrograph of transverse section of epoxy embedded rat atrial trabecula. Regions of A, M, I, and Z band are readily apparent. Post stained with uranium and lead salts. Bar = 0.3 μm.

Figure 2: Electron micrograph of an unstained section from the same epoxy block used for Figure 1. The low contrast demonstrates that background scattering from the epoxy embedding matrix is significant compared to scattering from the fixed muscle. Bar = 0.3 μm.

Figure 3: Electron micrograph of a transverse epoxy section of a rat atrial trabecula that was fixed in a solution containing tannic acid. Material in the lattice has enhanced preservation and contrast, and filament diameters appear somewhat larger than in standard epoxy-enbedded muscle. Bar = 0.3 μm.

Figure 4: High magnification electron micrographs of a de-embedded transverse section of a rat atrial trabecula fixed in a solution containing tannic acid. Bar = 0.1 μm.
(a) Fullness of thick filament backbone diameters and their hexagonal arrangement are most clearly evident in the M band region.
(b) In the overlap region thin filaments and thick filament backbones can be seen as regions of high electron density within a nearly continuous array of material in the lattice.

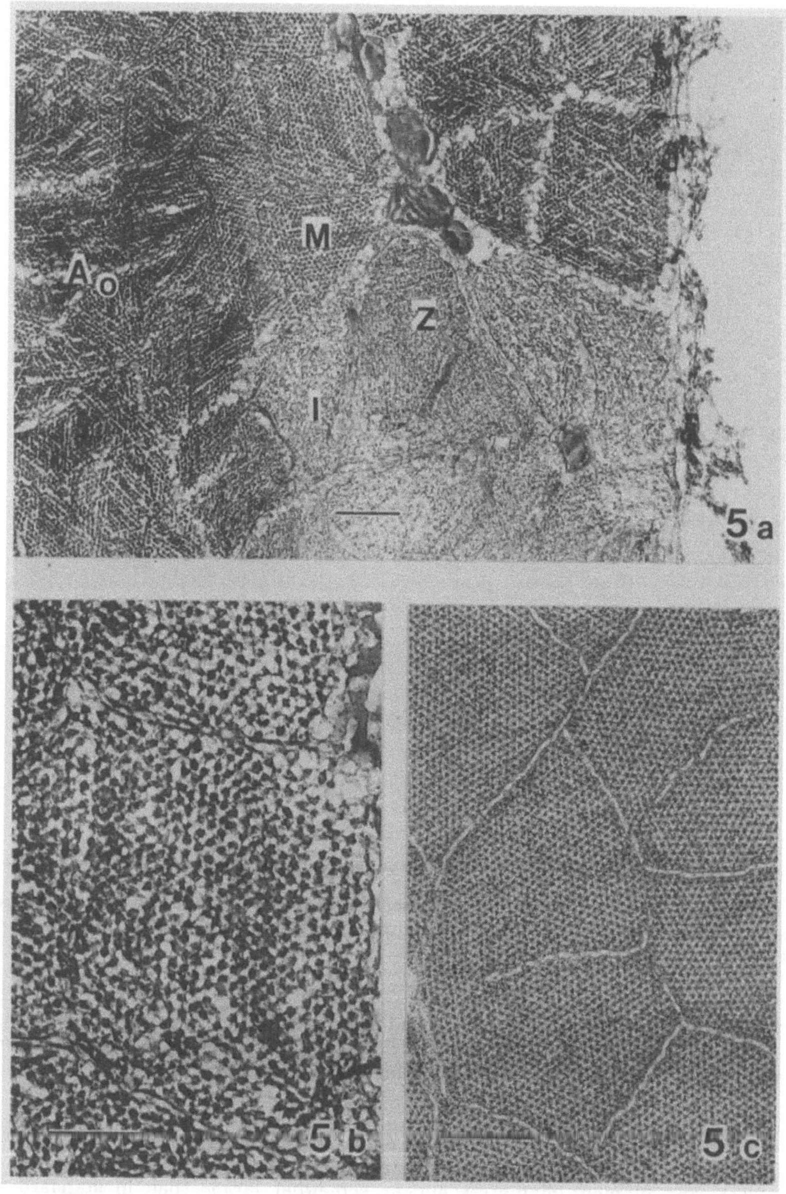

Figure 5: Electron micrographs of transverse sections of frog sartorius muscle.
(a) De-embedded section showing the good overall preservation suitable for measurements of lattice parameters. Appearance of Z, I, M and A_o (A overlap region) bands is easily correlated with that in epoxy sections. Note extreme density of overlap region. The disposition of sarcoplasmic reticulum surrounding myofibrils as well as mitochondria and extracellular collagen is normal. Bar = 0.4 μm.
(b) M band region from Figure 5a. Bar=0.2 μm.

Figure 5a is from a de-embedded section of frog sartorius muscle that was not *en bloc* or post-stained. Regions of Z band, I band, overlap, and M band are seen, as well as mitochondria, sarcolemma, and extracellular collagen. The disposition of filaments, myofibrils, and organelles is similar to that seen in paired epoxy samples. Figure 5b is an enlargement of the M band region, where measurements of thick filament backbone diameters were made. Overlap regions are filled with material (lower part of Figure 4), and measurements of thick filament profiles were made with the aid of transparent overlays of the hexagonal array of filaments. Material seems to envelope the thin filament (Figure 4; bottom part of diagram, Figure 6). In many cases the shafts of the filaments are more dense than the intervening material. This is true also in glycerinated fibers, in which the membranes have been disrupted and soluble proteins lost.

Table 1: Myofilament backbone diameters (nm)*.

	Thick Filament	Thin Filament
de-embedded normal:	25	8.5
de-embedded glycerinated:	22	7

* includes addition of 5% shrinkage correction factor.

Measurements of filament diameters are listed in the Table. These values include an incremental 5% correction factor for shrinkage during preparation (Page and Huxley, 1963; Page, 1974; Robinson, et al., 1981). The center-to-center distances of thick filaments in the M band and of thin filaments near the Z band are comparable to those in the paired epoxy samples. The question of swelling of filaments during the de-embedding process has also been directly addressed. The relevant observations are that the samples underwent both primary and secondary fixation and were never observed in a dis-assembled state. We also performed the control experiment of dissolving the polyethylene glycol embedding matrix out of some blocks from which sections and micrographs had been taken and re-embedding them in epoxy. Electron micrographs of resultant epoxy sections, stained by standard heavy metal techniques, gave images virtually identical to those of samples processed only by the epoxy resin technique (Figure 5c); that is, no evidence of swelling could be detected.

(c) Stained section from the same muscle used in Figures 5a & b, re-embedded in epoxy following de-embedment. Lattice structure is the same as in paired samples embedded in epoxy only. Center-to-center distances of thick filaments are the same in epoxy embedded, de-embedded and epoxy re-embedded control cases. Bar = 0.4 μm.

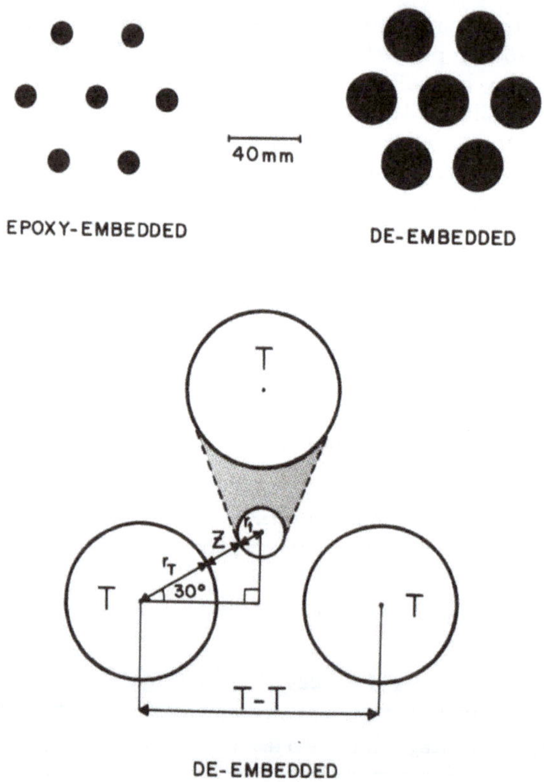

Figure 6: Schematic representation of the myofilament lattice in transverse section. Top panels show thick filament profiles in the M band region. The scale drawing is based on a center-to-center distance of 40 nm; thick filament diameters are taken as 12 nm in the epoxy-embedded case and 25 nm for de-embedded. The lower panel represents images from de-embedded lattice in the overlap region within the thin filament in trigonal position. Thin filament diameter taken as 8.5 nm; r_T = radius, thick filament; T-T = center-to-center distance between nearest neighbor thick filaments. Shaded region represents a very commonly seen feature, presumably cross bridge material that is less electron dense than the neighboring filament backbones.

DISCUSSION

The sizes of the myofilaments relative to their lattice spacings at a selected sarcomere length is represented diagrammatically in Figure 6. Using the values from the diagram as an example, Z, the surface-to-surface distance between thin filament and thick filament backbone can be calculated:

$$\cos 30° = [(T-T)/2] / (r_T + Z + r_t)$$

$$Z = [(T-T)/2] / \cos 30° - r_T - r_t$$

For T-T, center-to-center spacing of thick filaments = 40 nm; r_T, radius of thick filament = 12.5 nm; and r_t, radius of thin filament = 4.2 nm;

Z=6.4nm.

The small surface-to-surface distance is further diminished as a function of increasing sarcomere length. Given the recent estimates of the size of the globular S1 head portion of myosin, 6×19 nm (Elliott and Offer, 1978), an implication of these structural parameters is a severe restriction of cross bridge rotation if there is no change in filament dimensions with contractile state or, possibly, another type of motion of the cross bridge.

Because this re-estimation of filament lattice parameters has such important ramifications for the cross bridge-sliding filament model of muscle contraction, it is especially important to discuss the control experiments for these structural studies and to consider the results of several independent types of experiments, performed by other investigators, that are consistent with the present results. The occurrence of shrinkage of muscles during preparation for electron microscopy has been long recognized and well studied for longitudinal features, such as sarcomere length and I and A band widths (Page and Huxley, 1963; Page, 1974; Robinson, et al., 1981). Shrinkage is minimized in muscles that are restrained from shortening during fixation. Single cells from frog skeletal muscle (Eisenberg and Mobley, 1975) and rat heart (Bishop, et al., 1982) undergo net changes of volume of 5% or less. In the absence of explicit documentation of correlated x-ray and electron microscopic data regarding the degree of shrinkage of center-to-center distances of filaments at stages from living state to embedment, we have added a correction factor of 5%. This factor far outweighs inaccuracies in microscope and micrograph calibrations (Brown, 1978). A larger shrinkage factor yields even higher values for diameters of thin filaments and thick filament backbones.

A second consideration of vital importance in the assessment of the present measurements is the effect of dissolution of the polyethylene glycol from sections during the de-embedding process. Does swelling occur? We have minimized the effects of forces of dissolution during the procedure by employing both primary and secondary fixaton and by placing sections in a graded series of polyethylene glycol and solvent, finishing, for example, in 100% ethanol after the sections have adhered to the grid. Measurements on critical point-dried sections were performed only in regions where overall structural order was well preserved and inter-filament spacing was comparable to that in the paired epoxy-embedded samples. We have also dissolved the polyethylene glycol out of blocks from which de-embedded sections had been used for micrographs and re-embedded the tissue in epoxy. Micrographs of the resultant sections show the same inter-filament spacings as seen in the de-embedded images and paired epoxy samples. Furthermore, the images of filaments yield the same dimensions as those in the paired epoxy controls.

The present results are consistent with measurements made with other techniques. For thin filament diameters Franzini-Armstrong (1973) reported measurements after various fixation regimes for vertebrate striated muscle of 8.5 nm in studies made of the Z band. This value falls within our range of measurements for de-embedded samples and is larger than the 5-7 nm measured with previous techniques (Huxley, 1972). She and Varriano-Marston have also used a modified freeze fracture method to obtain striking images of the myofilament lattice that are in overall agreement with our findings (Personal communications, 1981, 1982). Trinick and Elliott (1979) obtained remarkable images of rotary shadowed, isolated thick filaments of vertebrate striated muscle (rabbit psoas) and obtained a value on the same order as ours: 29 nm for the width of the backbone. No mention was made in that paper of a correction factor for the thickness of the platinum coating. If the stated value is uncorrected, and we subtract from each side of the filament the thickness of 2 nm quoted by Heuser and Salpeter (1979) for similar rotary shadowing, a resultant backbone width of 25 nm is in even closer agreement with the values from de-embedded samples.

X-ray diffraction measurements of lattice spacings as a function of osmotic pressure have yielded charge diameters for thick filaments of glycerol-extracted muscles of 32 nm for rabbit psoas and 31 nm for frog leg muscle (Millman and Nickel, 1980). Millman, et al. (1981), found diameters in living frog sartorius or semitendinosus that varied from 28-31 nm, depending on sarcomere length. If we make a 5 nm allowance to represent an average contribution of compressed crossbridges, we obtain 25 + 5 = 30 nm-diameters. A direct estimate of crossbridge size from de-embedded samples, however, is unavailable at this point because we have not yet identified all components of the material abundant in pictures of transverse sections of the lattice.

The large amount of material seen in our images of the lattice is also consistent with measurements of the mass of the A band made by Reedy and Lucaveche (1982). The contribution of fixed soluble proteins to the images of filaments appears to be small. Filament diameters are only slightly reduced in glycerinated sartorius muscle, and part of that difference is probably due to partial extraction of the structural proteins themselves. Future experiments will hopefully permit correlation of some of the structures in our images with specific cytoskeletal structures, such as those described by Wang (1982) and Magid (1982), and with crossbridges themselves.

ACKNOWLEDGEMENT

We thank Dr's E.H. Sonnenblick, R. Kinne, C. Franzini-Armstrong, K.R. Porter, J. Wolosewick, G.H. Pollack, H. Gale and J. Gulati, and Mr. B. Gruenwald for helpful discussion. We are grateful to Ms. Renee Remily for the illustration, Mr. David Hays for photographic assistance and Ms. Kathleen Daly for word processing assistance. This work was supported by NIH Research Grant #HL-24336, a New York Heart Association Grant-in-Aid, and an NIH Research Career Development Award #HL-00568 (to TFR).

REFERENCES

Brown, L.M. (1978). Calibration of a commercial electron microscope with a grating replica to an acuracy of better than 1%. J. Microsc. 113 (pt2): 149-160.

Eisenberg, B.R. and Mobley, B.A. (1975). Size changes in single muscle fibers during fixation and embedding. Tissue and Cell 7(2): 383-387.

Elliot A. and Offer G. (1978). Shape and flexibility of the myosin molecule. J. Molec. Biol. 123: 505-519.

Franzini-Armstrong, C. (1973). The structure of a simple Z line. J. Cell Biol. 58: 630-642.

Gerdes, A.M., Kriseman, J. and Bishop, S.P. (1982). Morphometric Study of Cardiac Muscle: The problem of tissue shrinkage. Lab. Invest. 46(3): 271-274.

Heuser, J.E. and Salpeter, S.R. (1979). Organization of acetylcholine receptors in quick-frozen, deep-etched, and rotary-replicated *Torpedo* postsynaptic membrane. J. Cell Biol. 82: 150-173.

Huxley, H.E. (1972). Molecular basis of contraction in cross-striated muscles. In: *The Structure and Function of Muscle,* 2nd ed., vol. 1, pt. 1, 301-387, Bourne, G.H. (ed.) Academic Press. New York and London.

Magid, A., Ting-Beall, H.P., Carvell, M., Kontis, T. and Lucaveche, C. (1982). Connecting filaments, core filaments, and side struts: A proposal to add three new load-bearing structures to the sliding filament model. (This volume).

Millman, B.M. and Nickel, B.G. (1980). Electrostatic forces in muscle and cylindrical gel systems. Biophys. J. 32: 49-63.

Millman, B.M., Racey, T.J. and Matsubara, I. (1981). Effects of hyperosmotic solutions on the filament lattice of intact frog skeletal muscle. Biophys. J. 33: 189-202.

Page, S.G. and Huxley, H.E. (1963). Filament lengths in striated muscle. J. Cell Biol. 19: 369-390.

Page, S.G. (1974). Measurements of structural parameters in cardiac muscle. In: *The Physiological Basis of Starlings Law of the Heart.* Ciba Foundation Symposium 24 (new series) 13-30 Elsevier Exerpta Medica, North Holland.

Reedy, M.D. and Lucaveche, C. (1982). A-Band mass exceeds mass of its filament components by 30-45%. (This volume).

Robinson, T.F. (1980). Lateral connections between heart muscle cells as revealed by conventional and high voltage transmission electron microscopy. Cell and Tiss. Res. 211(3): 353-359.

Robinson, T.F., Hayward, B.S., Krueger, J.K., Sonnenblick, E.H., and Wittenberg, B.A. (1981). Isolated heart myocytes: ultrastructural case study technique. J. Microsc. 124 Pt. 2: 135-142.

Spurr, A.R. (1969). A low viscosity epoxy resin embedding medium for electron microscopy. J. Ultrastruc. Res. 26: 31-43.

Stempak, J.G. and Ward, R.T. (1964). An improved staining method for E.M. J. Cell Biol. 22: 697-701.

Trinick, J. and Elliott, A. (1979). Electron microscope studies of thick filaments from vertebrate skeletal muscle. J. Molec. Biol. 131: 133-136.

Venable, J.H. and Coggeshall, R. (1965). A simplified lead citrate stain for use in electron microscopy. J. Cell Biol. 25(2) Pt. 1: 407.

Wang, K. (1982). Are there new types of longitudinal filaments in the sarcomeres of striated muscles? (This volume).

Wolosewick, J.J. and Porter, K.R. (1979). Polyethylene glycol (PEG) and its application in electron microscopy. J. Cell Biol. 83: 303a.

Wolosewick, J.J. (1980). The application of polyethylene glycol (PEG) to electron microscopy. J. Cell Biol. 86: 675-681.

DISCUSSION

POLLACK: A number of people who have observed that thick filaments can shorten have also observed that they get fatter as they shorten. Is it possible that your procedure has somehow resulted in a shortening of a segment of the thick filament which has then gotten fatter?

ROBINSON: We want to monitor diameters but it's been impossible to take serial cross-sections in the de-embedded samples. With the state of the technique so far, we're lucky to catch 10% of the sections. So I can't rule out local shortening during fixation or something.

MAGID: There might be shortening after its been cut.

ROBINSON: Well, that's one reason we did the re-embedded samples. In the longitudinal sections I can say that the thick filament length is in the range of 1.5 to 1.6 μm.

MAGID: No, I meant shortening in the depth of the section, as if the section were a slab that got thinner and blown out sideways.

ROBINSON: But we haven't seen any increase in the spacings between the filaments. You'd expect to see that too, right?

MAGID: Yes.

ROBINSON: The filaments' center to center spacings obtained with embedded, re-embedded and de-embedded samples have all been comparable.

TREGEAR: Why do you think you're seeing just the diameter of the filaments and not also the cross-bridges around the outside?

ROBINSON: In the overlap region there is a great mass of material. That's why we confine measurements to the M-band.

REEDY: Have you done any optical diffraction of these cross- sections?

ROBINSON: Not yet.

REEDY: Probably that's one of the things that will make or break the issue of how lifelike things are, because there is X-ray diffraction available from *living* muscle. Based on some experience in my laboratory with cross-sections of insect muscle, diffraction of them shows the proper distribution of diffracted intensity for the different states -- rigor, relaxed in ATP; rigor, relaxed in AMP-PNP. That's not with your procedure but the standard conventional procedures. That was one of our reasons for congratulating ourselves that we had the real thing. It's going to be a test that has to be applied in some way to yours.

The other thing I'd say is that it's terribly difficult to look at these micrographs and then look at some of the photographs I've seen of beautiful freeze fractured muscles, prepared by people like John Heuser and Clara Franzini-Armstrong, and say "Where is the extra width?" It looks so empty between the filaments in their freeze-fractured, freeze-etched material that it's terribly difficult to take care of either my problem (extra mass) or yours when you look at those pictures.

COOKE: Just to amplify on Mike Reedy's comments, Heuser and I looked at both insect flight muscle and rabbit psoas. I forget the exact diameter that we measured, but very definitely it was not as wide as you've seen. I believe it agreed with the conventional embedded material though ours were neither embedded nor fixed.

ROBINSON: It brings to mind observations on isolated filaments. Trinick and Elliott (J. Mol. Biol. 131: 133-136, 1979) have beautiful rotary shadowed pictures of the isolated thick filaments. They mentioned just in passing that the diameter that they had at the backbone of the filament was 29 nm. They didn't say whether that corrected for the platinum shadow thickness or not, but it was an interesting finding. Also, if you look at some of the estimates that Millman and co-workers put out recently, the numbers are in fair agreement.

WANG: How efficient or how complete is the de-embedding? You say the PEG is water soluble, or ethanol-soluble. Could it be that some residue is left?

ROBINSON: That's a good question. We're doing a couple of things to test this. For one, we're taking stereo pairs, and find that we can see PEG when it's left. The second thing is, we started to shadow some of these. If we have a true surface, rotary shadowing should enhance the appearance of the surface, and it did so. I think if it were not a true surface, that sectioned PEG would have a smooth surface, and give just a light background.

HUXLEY: I'm still not completely clear about where you're measuring the filament diameter. You quoted a diameter for the *actin* filament, so obviously you're measuring that in the overlap zone.

ROBINSON: No, we're measuring that near the Z-band because the overlap zone is filled with material, and you can't be sure what is or is not thin filament.

HUXLEY: If you try and measure thick filament diameter in the overlap zone, how does that compare with thick filament diameter in the M region?

ROBINSON: In the overlap regions there are structures that look like components of Wankel engines, where there seems to be material from the thick circle to the thin circle. The diameters are about 24 nm with additional contributions from the cross-bridges thrown in, so that adds up to the upper thirties. It's really a question of where to cut off the measurement. The material looks very elongated. That's why we made a point of measuring in the M-band, at least to have some starting point for what is the thick filament backbone diameter.

HUXLEY: If I could just make one further comment on this -- when Ken Taylor was working in Cambridge for a while, he was looking at cross-sections of frozen muscle, and in that case there was a considerable variability in the apparent diameter of the filaments. Sometimes they looked normal, and sometimes they looked much wider, although the spacing between the filaments didn't change very much. He got the distinct impression that there was a variable amount of collapse of filaments without any sideways expansion of the lattice as a whole. I just wonder to what extent this may be a problem.

ROBINSON: That's interesting because that's not what we see. The filament diameters are quite uniform in our samples. But the filaments in longitudinal sections should have less of this possible collapsing effect.

HUXLEY: Yes, of course. What did you see there?

ROBINSON: The diameter measurements we've got so far are in the twenties, but they weren't sections that I'm happy with. We are striving to get better samples.

PODOLSKY: You are de-embedding with a solution with a very low ionic strength. Is there any way of putting a little salt in it to see whether the filaments go back to more conventional sizes in that case?

ROBINSON: That's an interesting idea. But I doubt that there is swelling during the de-embedding because re-embedded samples look like the controls. If there is a change, it would have to be a reversible one.

PODOLSKY: Do you add any salt when you re-embed it?

ROBINSON: No. That's just going from ethanol back to the epoxy resin, so it's not seeing the buffer at all.

LEVINE: When you say that you re-embed samples and they look like controls, do you mean that they look like epoxy-embedded controls?

ROBINSON: Yes.

TER KEURS: Suppose the properties of the filaments you have described are the true natural properties. What would the consequences of such a structure be for diffusion of ions in between the filaments? There's hardly any space any more.

ROBINSON: I don't have any data on that. You can see when the filaments are aligned perfectly that there's some space between filaments -- it's reduced, but there is space in between them.

REEDY: There are some estimates of bound water in myofibrils based on work on the osmotically active fraction, that sort of thing. All estimates of bound water are in the neighborhood of 30% -- that is, of the total volume in the myofibril something like 50 to 70% is freely exchangeable and is pretty available to exchange between water and exchanging solutes. I've had some experience with the interference microscope measuring and estimating the same thing. It's something like 65%. Albumin will penetrate about 65% of the volume of the fibril; something like 35% of it is unavailable to serum albumin. It's my impression from insect fibrils. I don't know about vertebrate muscle.

TREGEAR: I'd like to ask Dr. Robinson or Dr. Huxley whether the X-ray diffraction of frog muscle is compatible with these observations, or can you rule out the tremendous filling of the space from the equatorial X-ray diffraction?

HUXLEY: As you know, the reconstruction of equatorial patterns would usually come with two reflections. It gives you a very low resolution view. I don't want to pre-empt what people might be saying later on, but according to his abstract, Dr. Squire has improved cross-sectional reconstruction with perhaps more reasonable phases for the higher order reflections. I don't know whether he'd like to comment at this point on what he thinks the diameter of the filaments is.

SQUIRE: I think the only comment is that I find it very hard, I think, to model the X-ray diffraction measurements from such large filaments. That's all I would say.

HUXLEY: John Haselgrove and I, from time to time tried modeling the X-ray data and usually we finish up with a filament of 100 Å or so in diameter, with cross-bridges going out to 130 Å, and if in these de-embedded sections one is looking at the overall envelope of the cross-bridges, it wouldn't surprise me in the slightest if one were seeing something that was 250 Å in diameter. The question, what's the diameter in the M-line region, where there are no bridges, is what the issue is about, and clearly the X-ray data won't be able to tell you very much about that.

ROBINSON: Most of the literature I've seen describes shrinkage in longitudinal features. Can you say from the X-ray and the EM what it would be laterally, because I only applied five percent correction for shrinkage during the process? Maybe I should have applied a bigger factor.

HUXLEY: The lateral shrinkage between filaments is much higher than five percent.

ROBINSON: Do you have an estimate? That would change the spacings and pictures relative to the living state.

HUXLEY: As I recall, in the X-ray the spacing in frog sartorius is about 400 Å filament to filament. It's my recollection in the earlier embedments that it was more like 300 to 350 Å by the time you looked at sections. With better fixation you might get that up a bit, but there is a substantial amount of shrinkage sideways.

ROBINSON: That would not only increase the apparent size of the filaments, but the spaces between them.

DO THICK FILAMENTS SHORTEN?

INTRODUCTION

Prior to the early fifties, contraction was widely believed to be mediated by shortening of the A-band substance. This view was upset by the publication, in the same issue of the journal, *Nature*, of two papers disputing that view (A.F. Huxley and Niedergerke, 173: 971, 1954; H.E. Huxley and Hanson, 173: 973, 1954). Out of these seminal observations grew the sliding filament theory, which is now widely accepted. The underlying essence of this theory is that filaments of essentially constant length slide by one another, and that the driving force arises out of the action of the cross-bridges distributed along the length of the thick filament.

Lately, a germ of doubt has arisen about the firmness with which the concept of constant thick filament length has been established. The reasons for this doubt fall into two main categories. First, a series of observations carried out over the past decade on the horseshoe crab, *Limulus*, principally in the laboratories of Maynard Dewey and Rhea Levine, have shown that these thick filaments may shorten under certain conditions; in particular, it appears that shortening of the sarcomere below its natural resting length is brought about by, or accompanied by, shortening of the thick filaments. A variety of light microscopic, fluorescence, and electron micrographic methods have supported this phenomenon.

The second germ of doubt has arisen in studies of vertebrate muscle. A recent review (Pollack, Physiol. Rev. 1983) has turned up the interesting observation that an unexpectedly large number of papers published since the early fifties have apparently confirmed the old view that A-bands, or thick filaments, tend to shorten during contraction; in fact, the preponderance of the literature appears to point in this direction. The existence of this largely unknown body of literature has raised the question of whether in vertebrate muscle, as in *Limulus*, there may be conditions which permit thick filament shortening. This question has triggered considerable debate (cf. Sugi and Pollack, Cross-Bridge Mechanism in Muscle Contraction, U. Tokyo Press / Univ. Park Press, 1979; Ingels, Molecular Basis of Force Development in Muscle, Palo Alto Med. Res. Fndtn., 1979). Are these observations genuine, or do they reflect a consistent artifact creeping into the experimental observations? Are the classical findings as secure as is generally thought? Does thick filament shortening occur, but only under non-physiological conditions?

The following group of presentations is devoted to this issue. The first, presented by Maynard Dewey, reviews recent observations on

Limulus. Particular attention is given to the requirements for, and characteristics of, thick filament shortening. A model is developed in which both filament shortening and sliding are taken into account. A particularly interesting aspect of this work is the observation that thick filament shortening is associated with a profound increase of fixed charge on the thick filaments.

A second, brief communication from the same laboratory describes experiments using dynamic laser light scattering, again on *Limulus* muscles. This technique is useful in studies of cross-bridge motion. It appears that the high-frequency intensity fluctuations attributed to cross-bridge motion that occur when calcium is infused fail to occur if the S-1 portion of the myosin molecule has been enzymatically cleaved.

In an attempt to understand the structural basis underlying contraction, Rhea Levine has produced a series of micrographs of isolated thick filaments of *Limulus* and other chelicerate arthropods. The quality of the micro-graphs is outstanding, and one can easily visualize the helical pattern of cross-bridges along the surface of the filaments. It appears that thick filament shortening in *Limulus* does not disrupt the 14.3 axial repeat of the bridges, raising the interesting question of what kinds of molecular changes might occur to give rise to thick filament shortening.

The paper presented by Sugi describes experiments on the Japanese horseshoe crab, *Tachypleus*. Although changes in A-band width were found during contraction, these were apparently associated with misalignment of thick filaments rather than to genuine changes in the lengths of the filaments. Sugi concludes that in the kinds of physiological conditions under which the present experiments are carried out, thick filaments in *Tachypleus* do not undergo length changes; that such length changes occur only under non-physiological activating conditions. He raises the question of whether thick filament shortening in *Limulus* also occurs only under non-physiological activating conditions.

Krueger's contribution makes use of high resolution interference microscopy to study A-band widths in isolated cardiac muscle cells. Krueger finds some activating conditions in which sarcomere shortening is associated with a closing up of the I-band and formation of the expected contraction bands. On the other hand, this does not always occur; in intact, electrically stimulated specimens, sarcomere shortening is associated with A-band shortening. Since A-band shortening is seen only under some conditions, it is argued that this cannot be an artifact arising out of limited resolution of the microscope.

Finally, in another study of vertebrate muscle fibers, Dreizen finds modest shortening of thick filaments as sarcomeres shorten to lengths below 2.0 μm. This shortening appears to be at least in part the result of N lines encroaching on the ends of the thick filament and causing a disarray of cross-bridges at either end of the filament.

LIMULUS STRIATED MUSCLE PROVIDES AN UNUSUAL MODEL FOR MUSCLE CONTRACTION

M.M. Dewey, P. Brink, D.E. Colflesh, B. Gaylinn, S-F. Fan* and F. Anapol

Department of Anatomical Sciences, School of Medicine, Health Sciences Center
State University of New York at Stony Brook, Stony Brook, New York, 11794
** Shanghai Institute of Physiology, Academia Sinica, Shanghai, China*

ABSTRACT

Although nearly two decades have passed since de Villafranca (1961) described A-band shortening, controversy persists. Here we will review the data which has been amassed since de Villafranca's description. We will conclude that A-bands and thick filaments shorten during sarcomere shortening in *Limulus* striated muscle. Further we will suggest that two machines operate in this muscle: a tension generating sliding filament system and a tension generating thick filament shortening system. Also we will suggest a mechanism of force generation of the filament shortening system and provide evidence for a cycling bridge mechanism for this muscle.

Fiber Type and Thick Filament Length

Apparently a major point of question has arisen over whether the telson levators of *Limulus* are a homogeneous population of fibers or whether there are mixed fibers such as described by Franzini-Armstrong (1970) in the crab. In personal discussions with Professors Rhea Levine and Hugh Huxley both have described different muscle fiber types in the telson muscles and suggest that they represent a relatively small percentage of the total population. We have extensively reanalyzed our published work and laboratory notes and can find no evidence for such heterogeneity. We note that lengths of thick filaments measured in longitudinally sectioned fixed muscle by electron microscopy were 4.9 ± 0.45 μm (Dewey, Levine and Colflesh, 1973). The distribution was decidedly not bimodal. Also while the distribution is remarkedly narrow it is broad enough to encompass a small population of fibers with thick filaments ≃ 4.2 μm in length but not significantly shorter. Similarly, thick filaments isolated from glycerinated muscle bundles with rest length A-bands ≃ 4.5

± 1.0 μm were 4.4 ± 0.5 μm (Dewey, Walcott, Colflesh, Terry and Levine, 1977). Again the distribution was not bimodal. Clearly, if thick filaments shorten upon activation, the handling of the muscle preparation prior to glycerination or fixation is critical. Walcott & Dewey (1980) described the effects of various treatments on tension development of small muscle bundles. In the case of K^+ stimulation, maximum tension was produced with 300 mM K^+ saline. Application of solutions with lower K^+ concentrations produced contractions which developed less tension and smaller amplitude responses were obtained with 25 and 50 mM K^+. Application of EGTA containing (5 mM) relaxing solution at rest induced a significant contracture similar in magnitude to that produced by a 10 mM K^+ saline. Addition of a 50% glycerol solution (v/v, in relaxing solution) or a 5% glutaraldehyde fixative produced 6 to 15% of the tension produced by a 300 mM K^+ saline. In all cases tension development was induced by the bathing medium. Thus any of these treatments might induce some decrease in thick filament length. In fact thick filaments isolated from relaxed living muscle are uniformly 4.0 to 4.2 μm (see Dewey et al., 1977, and Levine, Kensler, Reedy and Hofman, this volume). It is largely for this reason that we have chosen to perform our studies on filaments isolated by means of glycerol gradients from extensively relaxed muscle (tied at maximum fiber length).

In these preparations we can obtain thick filaments 4.9 to 5.2 μm in length. These latter lengths are probably more reflective of the thick filament length in the resting muscle. In general, however, thick filaments isolated by this means are 4.2 μm in length and thus are probably partially shortened. We would suggest that the value of A-band and thick filament lengths be taken from measurements made on electron micrographs of longitudinally sectioned material, i.e. 4.9-5.2 μm (sarcomere lengths greater than 7.2 μm).

Additionally, we have attempted to determine by cytochemical techniques (Anapol & Dewey, unpublished results) whether different fiber types occur in this muscle. We have demonstrated myosin ATPase activity and have found no reason to suspect heterogeneity among fibers. Further we have stained for NADH tetrazolium reductase and succinic dehydrogenase activity. In transverse sections we have observed fibers with apparent increased enzymatic activity. However, from studying serially sectioned material it is clear that there is increased enzymatic activity at the tapered terminal ends of fibers. Whether this represents increased specific activity in these regions or simply reduced volume of contractile machinery is uncertain. No distinctly different fiber types were observed with the techniques for oxidative enzymes.

Fourier Analysis of Refractive Index Along Sarcomeres at Various Sarcomere Lengths

Fourier reconstructions of the refractive index along the sarcomere were calculated using the relative intensities of layer lines from optical diffraction patterns of small bundles of telson levator muscles in various conditions (Dewey, Blasie, Levine, and Colflesh, 1972; Dewey, Colflesh and

Blasie, 1973; Dewey, Levine, Colflesh, Walcott, Brann, Baldwin and Brink, 1978; Dewey, Colflesh, Brink, Fan, Gaylinn and Gural, 1982). Living muscle bundles were stretched to various sarcomere lengths above rest length or were electrically stimulated and allowed to shorten below rest length. At each sarcomere length a laser diffraction pattern was obtained. Single glycerinated fibers were activated by addition of ATP, Mg^{2+} and Ca^{2+} in a bathing medium and allowed to shorten incrementally; a laser diffraction pattern was photographed at each attained sarcomere length. Additionally, muscle fibers were glycerinated and single fibers dissected and shortened incrementally; other fibers were fixed and embedded at different sarcomere lengths and diffraction patterns obtained. A helium/neon laser was used. Diffraction patterns with 10 layer lines or more could be obtained from fibers with sarcomere lengths 10 μm or longer. At short sarcomere lengths an ultraviolet laser was employed to obtain at least three layer lines from fibers with sarcomeres below 4.0 μm. The latter cases were done on glycerinated and fixed fibers only. All possible phase combinations were used to calculate Fourier transforms of sarcomeric structure. In each case the numbers of orders used to make the calculations were equal to the sarcomere length in microns minus one. Thus, for example, seven orders were used to calculate the transforms of sarcomeres 8 μm long. Phases were selected by requiring that the chosen transform exhibit Z, A, and I bands and an H-band when appropriate. It was of interest to note that for sarcomeres below $\simeq$ 7.0 μm only a unique transform met these requirements. Below $\simeq$ 7.0 μm A-band shortening is due to thick filament shortening. Figure 1 illustrates transforms for sarcomeres 9.85 μm and below. Note the decrease in A-band width with

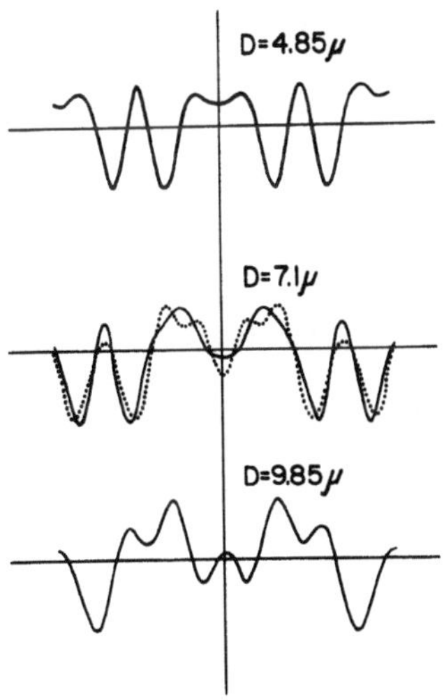

Figure 1: Fourier transforms of sarcomeres with lengths of 9.85 μm, 7.1 μm and 4.85 μm Note the shoulders on the A-band of the long sarcomere. We interpret this as reflecting the misalignment of thick filaments in the A-band. Note that the thick filaments are aligned at $\simeq l_0$ (no shoulders present). Also note that at the short sarcomere length an I-band is present. There is a linear decrease in A-band length with sarcomere length.

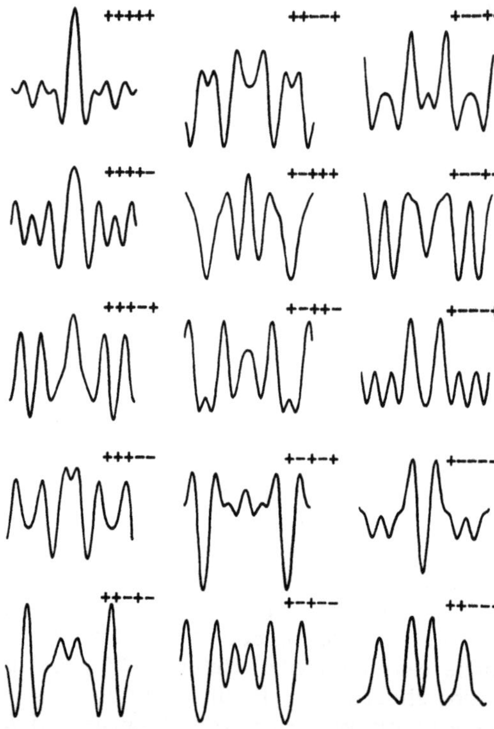

Figure 2: All of the unique transforms calculated (out of 2^5) for a sarcomere of 6.7 μm. Note only one transform meets the requirements for selection, A-, I- and Z-bands. The phases are +−−+−. Thus the phases are unique for all sarcomere lengths below l_0

decreasing sarcomere length. We interpret the shoulders in the A-band of the long sarcomere as reflecting the misalignment of thick filaments (see Figure 6). Figure 2 demonstrates all the transforms and phase combinations for a sarcomere of 6.7 μm. Notice that only the transform with phases +−−+− meets the criteria for selection. It is also interesting and expected that if only the phases are used, not including the relative intensities, similar transforms are obtained with identical A-band widths. The A-band widths were determined by measuring half-heights of the A-bands in the transforms. Analysis of the A-band widths showed that A-band width varied linearly from 7.8 μm to 3.2 μm as sarcomeres shorten from 12.0 μm to 4.5 μm. These observations correlated remarkably well with measurements of A-band and sarcomere lengths from light, phase and interference microscopy (Dewey et al., 1978).

Video Observations on Shortening of Single Glycerinated Fibers

Single glycerinated telson levator fibers were dissected (fibers were glycerinated in 100 mM KCl, 5 mM EGTA, 5 mM Tris, 2 mM $MgCl_2$, 1 mM DTT and deionized glycerol 50% v/v for a minimum of 10 days) and placed in relaxing solution (100 mM KCl, 50 μM EGTA, 2 mM $MgCl_2$, 1 mM DTT, and 5

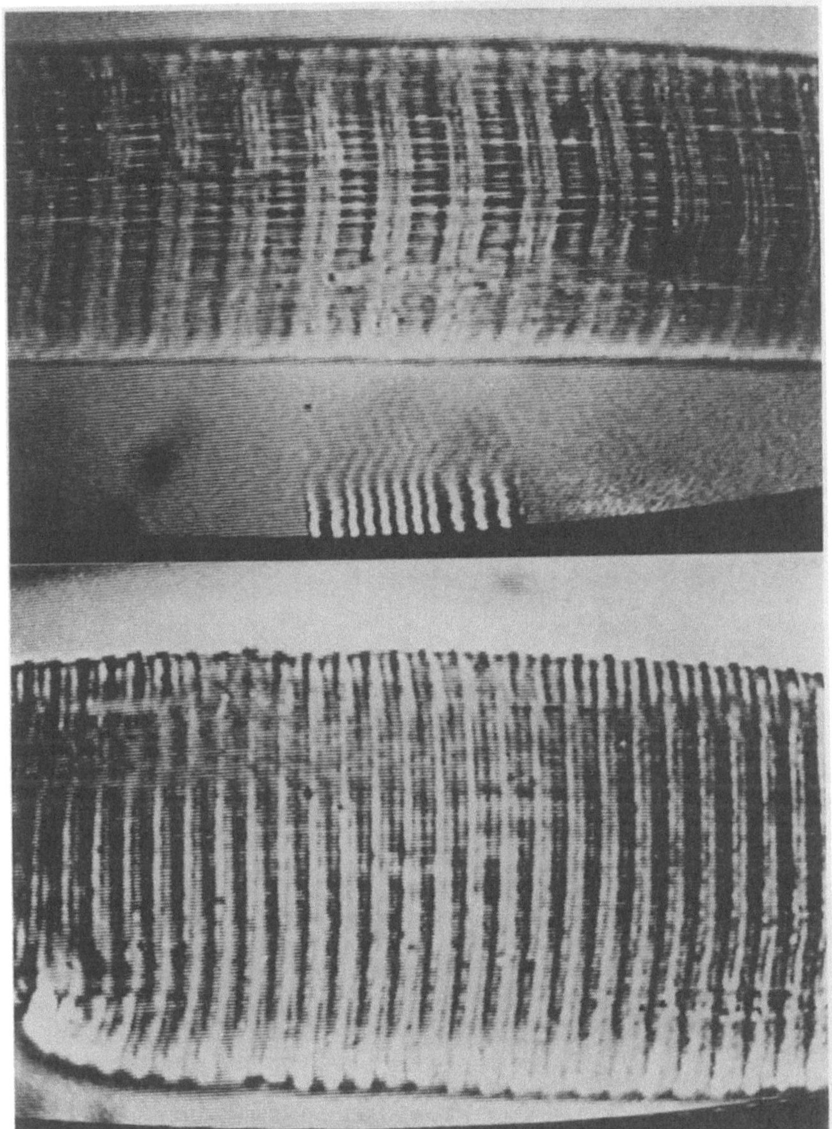

Figure 3: Photograph of stopped-tape on video-screen. A single glycerinated fiber in 50% relaxing solution/50% deionized glycerol with sarcomere lengths of $\simeq$ 13.0 μm is shown in panel a. Following addition of buffered Ca^{2+} activating solution under coverslip, the fiber shortened. The final sarcomere length attained is $\simeq$ 6.5 μm.

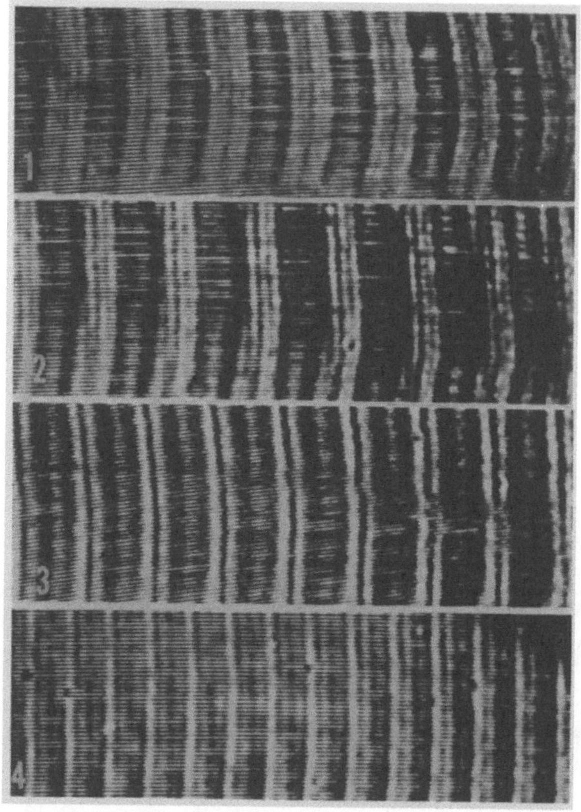

Figure 4: Photographs of stopped-tape on video-screen showing enlarged regions of the fiber in Figure 3. See text for description.

mM ATP and 50% deionized glycerol, v/v) on a slide and cover-slipped. A SONY video camera was mounted on a Zeiss Universal microscope equipped with phase and Normarski optics. An activating solution (identical to the relaxing solution but containing 5 mM $CaCl_2$) was added and the solution drawn under the cover-slip by filter paper absorption on the opposite side. Figures 3 and 4 illustrate views from the videotape of a single fiber before and after activation. The sequence of shortening was followed in real time with the fiber first being activated on the right hand side. Figure 4 shows a series of views photographed from the video screen with the videotape stopped. At the top, (panel 1) the sarcomeres are stretched beyond overlap; margins of the I-band can be identified. Panel 2 illustrates the first stages of sarcomere shortening. Notice, on the right hand side, the A-band has widened to engage the one-half of the I-band. This is presumably a function of misalignment of thick filaments. In panel 3, on the right hand side, the sarcomere has shortened to occlude the H-band. In panel 4 the fiber has shortened maximally. Note the presence of distinct Z, I and A-bands. The fiber began with sarcomeres of ≃ 15.0 μm and shortened to have sarcomeres of ≃ 6.5 μm. The A-band shortens from ≃ 6.8 μm in panel 2 (thick filaments misaligned) to ≃ 3.0 μm in panel 4. We believe that shortening of this single fiber clearly demonstrates A-band and thick filament shortening.

Changes in Sarcomere and Filaments as a Function of the Length Tension Curve

The relationship between A-band length, thick filament length and diameter, and I-band length to the length tension curve of *Limulus* striated muscle is shown in Figure 5 (Walcott and Dewey, 1980). Figure 6, in stick drawing form, illustrates the gross structural features of the sarcomere as it shortens from 10.3 μm to 4.8 μm. Proportions are to scale; note how the changes here match those seen in Figure 4. The changes shown in this diagram were derived from several studies (Dewey et al., 1972; Dewey, Colflesh and Blasie, 1973; Dewey, Levine and Colflesh, 1973; Dewey et al., 1977; Dewey et al., 1978; Dewey et al., 1982; Levine, Dewey and de Villafranca, 1972; Walcott and Dewey, 1980). Changes that occur in the sarcomere as it shortens from ≃ 10.0 μm to l_0 (≃ 7 μm) include increasing numbers of thick and thin filaments overlapping and increasing numbers of bridges along any one thick filament that can interact with thin filaments. We have assumed that the thin filaments are of uniform length (Dewey et al., 1977) but some thin filaments may be longer. This might in part explain the tension development at lengths where overlap should not occur. In addition, fragments of actin scattered within the thick filaments would allow tension transmission from Z line to Z line. We have seen evidence for the latter but not the former (Walcott and Dewey, 1980). Thus during this phase of sarcomere shortening, sliding of thick and thin filaments past each other occurs as in the model for vertebrate

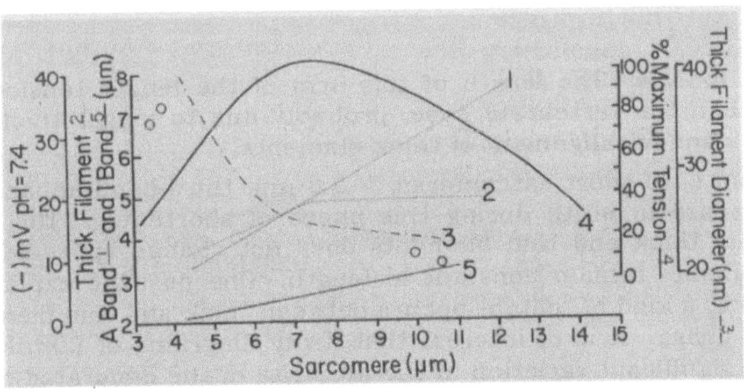

Figure 5: A number of relevant structural parameters, as well as, tension and Donnan potentials are plotted as a function of sarcomere length. 1. A-band length as determined by measurements from light, phase and interference microscopy, electron microscopy and Fourier transforms of optical diffraction patterns (Dewey et al., 1972; Dewey, Colflesh and Blasie, 1973; Dewey, Levine and Colflesh, 1973; Dewey et al., 1977; Dewey et al., 1978; Dewey et al., in press). 2. Thick filament length (Dewey, Levine and Colflesh, 1973). 3. Thick filament diameter (Dewey et al., 1978). 4. Active tension (Walcott and Dewey, 1980). 5. I-band length (Dewey et al., 1972; Dewey, Colflesh and Blasie, 1973). Open circles are Donnan potential measurements (Dewey et al., in press).

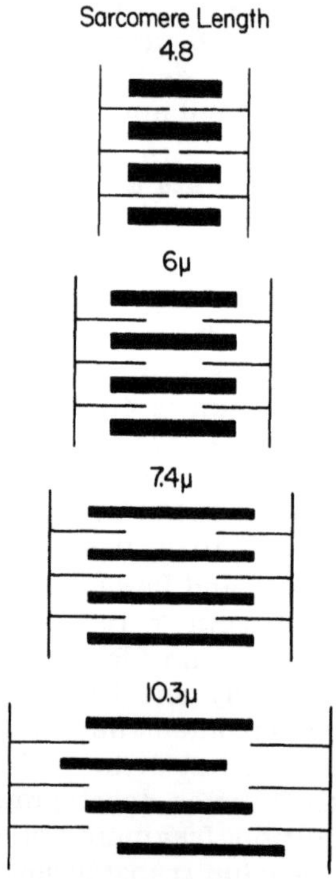

Figure 6: Stick diagram illustrating the relationships between thick and thin filaments in *Limulus* striated muscle at *in vivo* range sarcomere lengths. Thick filament shortening and thickening are shown. Note the presence and constancy of the I-band at all sarcomere lengths.

striated muscle. The length of this arm of the length tension curve is longer than the vertebrate case, probably due to longer thick and thin filaments and misalignment of thick filaments.

From l_0 to short sarcomeres, $\simeq$ 3.0 μm, the I-band apparently does not decrease in width during this phase of shortening. The degree of overlap of thick and thin filaments does not change appreciably, if the thin filaments remain constant in length. One possible explanation for this is that a kind of "catch" occurs between thick and thin filaments during this phase. It is of interest that X-ray diagrams of *Limulus* muscle show no significant variation in the intensity of the decorated actin layer lines in fibers with sarcomeres from $\simeq$ 4 to 12 μm (Wray, Vibert and Cohen, 1974). The only additional evidence we have which would suggest this possibility is from cytochemical studies (Anapol and Dewey, unpublished results). Cytochemical localization of myosin ATPase is greatly reduced in sections of muscle which have been maximally shortened prior to freezing. Its localization is strong in the A-bands of fibers with rest or longer length sarcomeres prior to freezing. In shortened muscle, reaction product of myosin ATPase was localized to regions in the center of the A-band while in muscle with long sarcomeres the reaction product was distributed throughout the A-band. This might suggest that myosin ATPase sites are unavailable in the lateral regions of the A-band at short sarcomere lengths.

Attention should be drawn to the fact that ≃ 30% of maximum tension is obtained in fibers with sarcomere lengths of ≃ 14 μm or 3.0 μm. This is a surprisingly broad range and could not happen if thick filaments remained at a constant length of 4.9 μm. Further, this argues that the thick filament shortening must be tension generating. To develop tension over this broad a range, tension generated by the thick filaments must be equal to or greater than the force generated by actin and myosin. Mechanical studies are sorely needed to confirm that this muscle works by two machines: sliding filaments and shortening thick filaments.

Requirements for In Vitro Shortening of Isolated Thick Filaments

We have determined the ionic and energy requirements for shortening isolated thick filaments *in vitro* (Brann, Dewey, Baldwin, Brink and Walcott, 1979). Thick filaments were isolated from briefly glycerinated or fresh, thoroughly relaxed, muscle bundles. Such filament preparations show an 80% reduction of actin as compared to the original homogenate (determined by SDS gel electrophoresis and microdensitometry). A calcium activation curve was determined by varying the ration of Ca^{2+} to EGTA in the presence of a constant concentration of ATP and Mg^{2+}. Maximal slope of the line (which implies maximum rate of change of length) of filament shortening occured at an estimated pCa of 5.5. It is important to note that lower calcium levels gave filaments of intermediate length which suggests that Ca^{2+} acts as a trigger. Since the shortening curve is sigmoidal, it suggests that there are multiple binding sites for Ca^{2+} (Dewey et al., 1978; Dewey et al., 1982).

An activation curve for ATP was also determined. The Ca^{2+}/EGTA ratio was kept constant and the concentration of ATP varied. Concentrations of 1 mM ATP and higher gave maximal shortening. ATP hydrolysis is essential for shortening since filaments did not shorten in the presence of adenylyl imidodiphosphate with sufficient Ca^{2+} present to activate the process.

The requirements for thick filament shortening then include Ca^{2+} (pCa ≃ 5 to 6), Mg^{2+}, and ATP. These requirements are consistent with those reported to be "physiological" levels for activation and tension development in other muscles (Fabiato and Fabiato, 1975; Kushmerick and Davies, 1969; Stephenson, Wendt and Forest, 1981). It is also of interest that purified thick filaments shorten so that actin is not essential for shortening.

Phosphorylation of Thick Filament Proteins During Filament Shortening and Sarcomere Shortening

We have examined the pattern of protein phosphorylation in glycerinated and skinned *Limulus* muscle. This analysis was prompted by earlier work (Brann et al., 1979) which reported that isolated thick filaments that are shortened can be relengthened by alkaline phosphatase treatment. We anticipated that paramyosin phosphorylation might be involved in filament shortening. This seemed reasonable since paramyosin has been demonstrated to be phosphorylated in molluscan muscle (Achazi, 1979; Radlick and Johnson, 1982). Our results in *Limulus* striated muscle reject this idea. Paramyosin is not phosphorylated under conditions where thick filaments shorten. Instead an unidentified thick filament associated 35 k dalton peptide is dephosphorylated when filaments shorten.

Radioactive γ-^{32}P–ATP was used to identify proteins phosphorylated in glycerinated or skinned muscle under various conditions. Glycerinated muscle bundles were soaked in a relaxing solution to remove the glycerol. These bundles were then either allowed to shorten in ^{32}P labeled activating solution (relaxing solution containing 4 mM EGTA and 8 mM $CaCl_2$) or soaked in ^{32}P labeled relaxing solution. The muscles were then quickly frozen, denatured and homogenized, electrophoresed on SDS polyacrylamide gels and autoradiographed.

Sellers (1981) has reported that isolated *Limulus* myosin can be phosphorylated on two of its three light chains by a calcium-dependent myosin kinase. Figure 7 demonstrates that this phosphorylation of myosin light chains also occurs when glycerinated muscle bundles are calcium activated.

When muscle bundles were incubated in labeled relaxing solution in the absence of calcium the myosin light chains were not phosphorylated but other proteins were (see Figure 8, column c). The proteins that incorporate phosphate preferentially in the absence of calcium include one of approximately 100 k dalton chain weight. The 100 k dalton band seemed likely to be paramyosin. Closer examination (Figure 9) shows

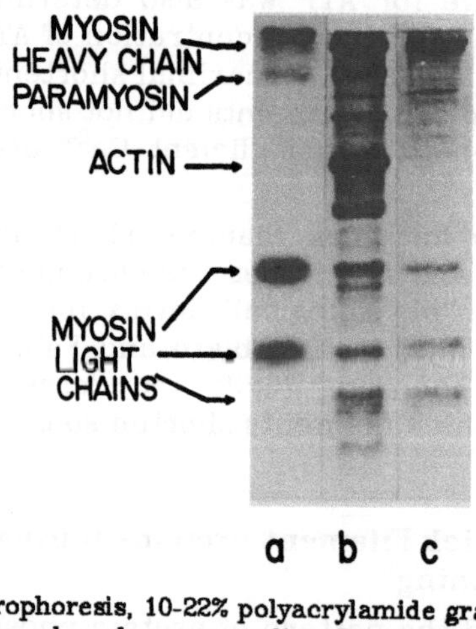

Figure 7: SDS slab gel electrophoresis, 10-22% polyacrylamide gradient. Columns (a) and (b) show respectively the radiograph and corresponding Coomassie blue staining from Ca^{2+} activated ^{32}P labeled, glycerinated *Limulus* muscle. Column (c) shows the stained gel of column purified *Limulus* myosin.

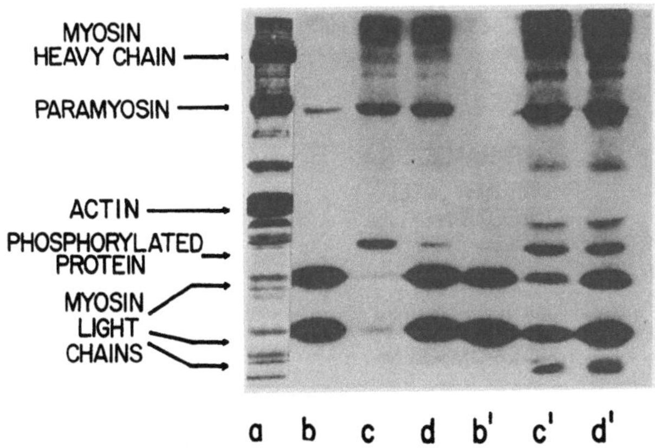

Figure 8: ^{32}P labeled, glycerinated *Limulus* muscle, SDS slab gel electrophoresis and autoradiography on a 10-18% polyacrylamide gradient. Column (a) shows the Coomassie blue staining pattern, for example, of radiograph column (b). Columns (b)-(d) are radiographs of long sarcomere, long A-band muscle. Columns (b')-(d') are radiographs of K^+-contracted, long sarcomere, short A-band muscle samples (see text). Columns (b) and (b') were labeled in activating solution, (c) and (c') were labeled in relaxing solution and (d) and (d') were labeled in activating solution after previous labeling in relaxing solution (see text for composition of solutions).

that on lower percentage gels the 100 k dalton band does not co-electrophorese with purified paramyosin. The mobility of this band phosphorylated or dephosphorylated remains different from paramyosin and it does not co-purify with paramyosin. Under other conditions paramyosin may be phosphorylated.

When muscle that has been ^{32}P labeled in the absence of calcium is then allowed to shorten in a solution containing calcium and γ-^{32}P-ATP myosin light chains are phosphorylated, the 100 k dalton band remains phosphorylated and the 35 k dalton band is specifically dephosphorylated (Figure 8, column d).

To determine how these phosphorylations and dephosphorylations might relate to thick filament shortening, we examined samples where the A-bands had already been shortened before the muscle was allowed to shorten in length. These samples were prepared either by potassium contracting living muscle tied isometrically or ATP, Mg^{2+} and Ca^{2+} activating glycerinated muscle also held isometrically. This produced muscle with long sarcomeres and short A-bands (which remained short during glycerination in the case of live muscle) (Dewey et al., in press). The ^{32}P labeling of this long sarcomere short A-band muscle is shown in Figure 8, columns b', c', and d' (as compared to b, c and d for muscle not A-band shortened before labeling). In short A-band muscle the 35 k dalton band does not dephosphorylate as it does in the long A-band and thus this dephosphorylation may correlate with thick filament shortening.

In addition, isolated thick filaments were analysed for phosphorylation. Thick filaments were isolated by centrifugation on a step gradient in

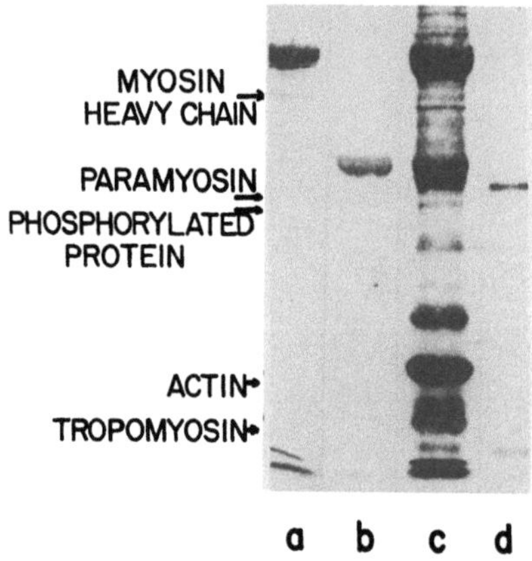

Figure 9: SDS slab gel electrophoresis, 7.5% polyacrylamide. The various columns contain (a) column purified *Limulus* myosin; (b) purified *Limulus* paramyosin; and (c) glycerinated *Limulus* muscle ^{32}P labeled in relaxing solution (see text); (d) radiograph of column (c).

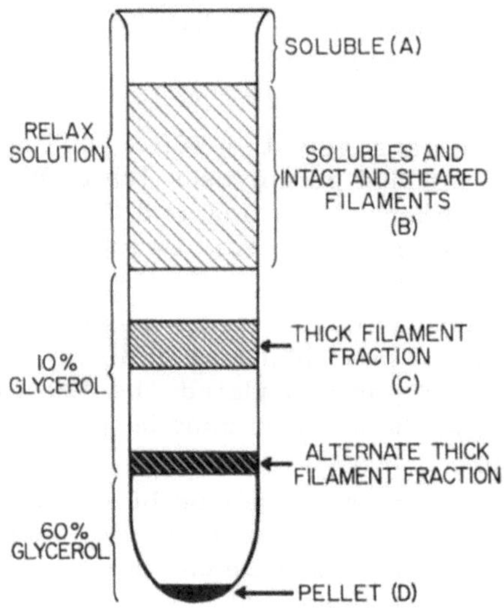

Figure 10: Diagram showing the isolation of thick filaments on a glycerol gradient. Fractions correspond to those in Figure 11.

relaxing solution containing 0, 10, and 60% glycerol (Figure 10). Each fraction from the gradient was tested for incorporation of ^{32}P in both the presence and absence of calcium. Figure 11 shows that myosin light chain kinase activity is found in the upper gradient fractions but is lost in the isolated thick filaments probably due to the extraction of calmodulin which is soluble (Cheung, 1980). Kinase activity for the 35 k dalton band is retained in all gradiont fractions, but calcium specificity is lost in some

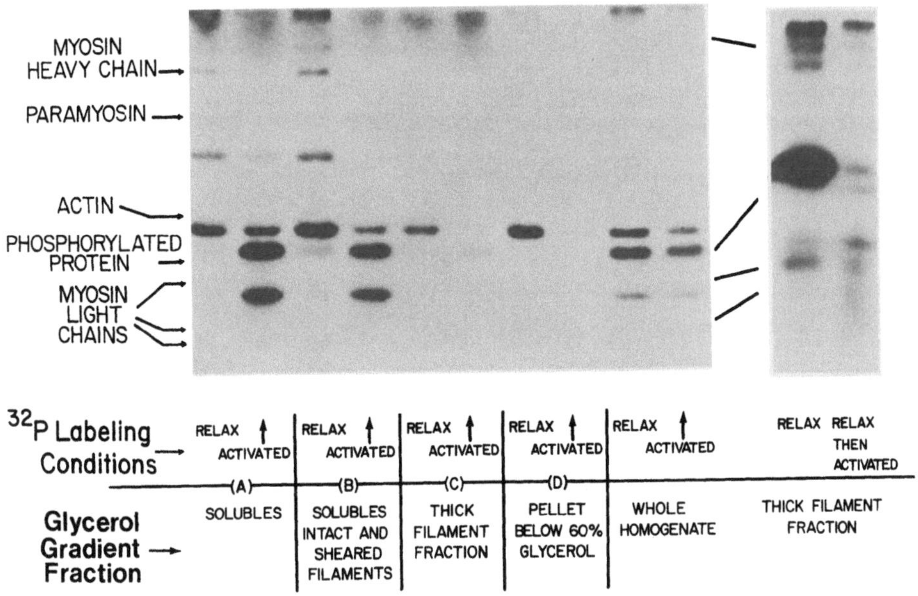

Figure 11: Radiograph showing ^{32}P labeling of various fractions during the purification of *Limulus* thick filaments.

of the upper fractions. The thick filament preparation maintains a calcium inhibited phosphorylation of this band and a calcium activated dephosphorylation (Figure 11, far right column). This data suggests that the dephosphorylation of a thick filament associated protein is related to filament shortening.

An interesting observation occured when glycerinated muscle bundles were incubated with only micromolar amounts at ATP. As can be seen in the left half of Figure 12, labeling in rigor solution (containing less than 2 μM ^{32}P-ATP) resulted in very little phosphorylation except for a band just below paramyosin (as seen in Figure 9). Unexpectedly, when labeled rigor solutions contained calcium a new phosphorylation occured not seen under any other conditions. This new band on the radiographs appeared to coincide with paramyosin and its phosphorylation in rigor-Ca^{2+} solution also occurred in skinned muscle and in isolated thick filaments.

Information Pertinent to Understanding the Mechanism of Thick Filament Shortening

Thick filament mass

We have measured the mass of isolated thick and thin filaments using the dedicated STEM Biotechnology Resource at Brookhaven National Laboratory in collaboration with Drs. J. Wall and B. Panessa-Warren (Dewey et al., in press). The mass is measured by comparing the electron scattering signal per unit length of unstained filaments with that from

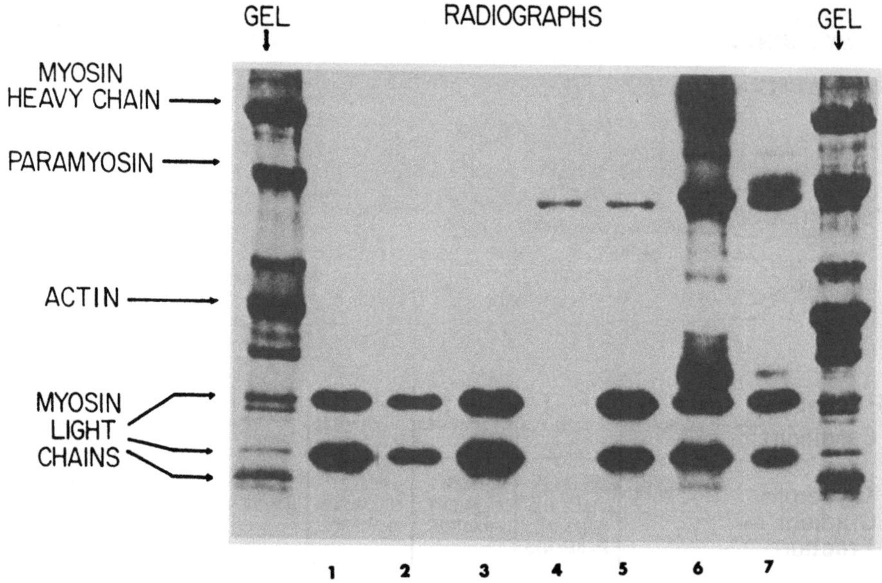

Figure 12: Radiographs from muscle incubated under the following conditions: 1. labeled activating solution; 2. unlabeled activating solution; then labeled activating solution; 3. labeled activating solution, then labeled relaxing solution; 4. labeled rigor solution; 5. labeled activating solution, then labeled rigor solution; 6. labeled relaxing solution, then labeled rigor solution; 7. labeled rigor-calcium solution.

tobacco mosaic virus particles in the same image. The results are preliminary. The mass of actin filaments taken in the same field as the thick filaments was 2040 daltons/Å (SD = 340 daltons/Å, n = 73). We calculate the weight of actin filaments assuming the published weights for *Limulus* troponins (Lehman, 1982) to be 1996 daltons/Å. Thick filaments were isolated from muscle bundles with long sarcomeres and were long (3.9 to 4.3 μm). Their mass was determined to be 16,154 dalton/Å with a standard deviation of 974 daltons/Å, n=18. Filaments isolated long were treated with calcium and ATP to shorten them (2.9 to 3.2 μm). Mass of these shortened thick filaments was determined to be 20,374 dalton/Å (SD = 6571, n = 75). These data suggest that the total mass of a long thick filament is in the range of 640 mega-daltons and does not change as the filament shortens. Thus, it is likely that shortening is not a process of dissolving. The high standard deviation of the shortened filaments may be due to non-uniform shortening along the filament. This is currently being analysed. It is of interest that using this data and assuming the paramyosin/myosin heavy chain ratio, 0.48, is correct (Levine, Elfvin, Dewey and Walcott, 1976) and allowing for myosin light chain weight and assuming that no other major proteins occur in the filament, the number of myosin molecules can be calculated. Upon calculation the number of myosin molecules per filament is ≃ 1000. Assuming a thick filament of 4.2 μm, with a 0.3 μm bare zone, a three or four stranded model is consistent. If the paramyosin/myosin heavy chain ratio were 0.32, then a four stranded model would be suggested (Kensler and Levine, 1982; Stewart, Kensler and Levine, 1981). Here we have assumed that the myosin subunit repeat does not change as the filament changes length which has been established by X-ray diffraction (Millman, Warden, Colflesh and Dewey, 1974; Wray et al., 1974) and optical diffraction (Dewey et al., 1978).

Specific ATPase activity of long and short filaments

We believe that the following experiments also argue against filament dissolution during shortening. Filaments were isolated long. Samples were shortened by addition of Ca^{2+} and ATP. Both long and short samples were extensively dialysed against relaxing solution. The calcium level was then raised to pCa of 5.5 (EGTA/Ca = 1.0) and myosin ATPase activity determined by pH stat following centrifugation and resuspension of the filaments. This pCa level is not high enough to maximumly shorten the filaments. Both samples gave specific activities of 4 to 5 nM/min/mg protein. Activities are low as these filaments have very little actin present. This argues that the filaments are not dissolving nor are heads being "buried" during shortening.

Donnan potentials

Recently Elliott and Bartels (1982) and Naylor (1982) have treated theoretically the calculation of charge-concentration in extended polyelectrolyte gels such as muscle and have concluded that the reported measurements are "legitimate".

We have measured Donnan potentials in glycerinated *Limulus* fibers with both long and then shortened sarcomeres (Dewey, et al., in press). Potentials were measured by placement of 3 M KCl microelectrodes (5 to 20 MΩ) into the muscle matrix and measuring the potential differences between it and a silver-silver chloride reference electrode in the bathing medium surrounding the muscle bundle. In long sarcomeres it was easy to differentiate between Donnan potentials of A and I bands. In short sarcomeres A-I overlap regions were measured. In K+ acetate buffer, A-bands in long sarcomeres had Donnan potentials ranging from -5 mV to -12 mV. Following shortening of the fiber with Ca^{2+}, ATP and Mg^{2+}, A-band potentials ranged from -20 to -25 mV. By varying the pH (see Brink and Dewey Abstract, this volume), the isoelectric point of the A-band was measured. Following shortening there was not only an increase in the negative potential of the A-band but an alkaline shift in its isoelectric point. The calculated increase in the negatively charged protein concentration was dramatic, from 36.4 mM to 114.6 mM. As others have reported (Bartels and Elliott, 1981), we observed no significant difference in fixed charge in frog sartorius muscle between long and short sarcomeres. Thus during sarcomere shortening in *Limulus* striated muscle a highly significant change occurs that does not occur in the vertebrate muscles so far studied. This charge change could serve as a force generating system by bringing about protein rearrangement which causes thick filament shortening.

Quasi-Elastic Light Scattering of Suspensions of Isolated Limulus Thick Filaments

Quasi-elastic light scattering has been used to investigate the dynamics of macromolecules in solution (Chen, Chu and Nossol eds., 1981; Chu, 1979). Further Fujime (Fujime et al., 1981) and Carlson (Newman and Carlson, 1980) with coworkers have made extensive studies of thin filaments and muscle fibers (Haskell and Carlson, 1981). In collaboration with Professor B. Chu of the Department of Chemistry, The State University of New York at Stony Brook, Stony Brook, New York, we have begun an analysis of dilute suspensions of isolated thick filaments from *Limulus* striated muscle by photoelectron count autocorrelation function of light scattered as a function of scattering angle (Fan, Dewey, Colflesh and Chu, in press; Kubota, Chu, Fan, Dewey, Brink and Colflesh, submitted). By using the cumulants method of data analysis, the average linewidth over large ranges of KL (up to 120, where K and L are the magnitude of the momentum transfer vector and the length of the thick filament, respectively), we were able to visualize the flexibility of isolated thick filaments in the long, short, and rerelaxed (isolated filaments shortened and then dialysed against relaxing solution) states. In doing this we noted a dramatic increase in the average linewidth denoting the present of additional high frequency components from suspensions in which the filaments were activated i.e., Ca^{2+}, Mg^{2+} and ATP were present. These high frequency components must arise from a new internal motion within the filaments upon activation. If activated (thus shortened), filaments are then dialysed against relaxing solution these high frequencies disappear and the average linewidth decreases to close to that of the relaxed filaments. Since *Limulus* striated muscle is both thick and thin filament regulated (Lehman and Szent-Gyorgyi, 1975), we attributed this highly increased internal motion to cross bridge cycling in the presence of Ca^{2+} and Mg^{2+} and ATP. The ratio of the average linewidth of relaxed to that of the activated filaments, as K approaches zero, is consistent with that calculated with the values of filament lengths and diameters from electron micrographs of negatively stained long and shortened filaments. We have subsequently looked at a number of controls which further support this contention. Vanadium ions alone or with Ca^{2+} in the presence of Mg^{2+} and ATP do not result in these increased internal frequencies (Fan et al., in press). Heat denaturation in the presence of Ca^{2+}, Mg^{2+} and ATP leads to linewidths consistent with relaxed filaments (Fan et al., in press). Here we report that brief treatment of the isolated filaments with papain (see Fan et al., abstract this volume) which leads to $\simeq$ 90% removal of heads, results again in linewidths consistent with relaxed filaments. From these observations we are increasingly confident that we will be able to study, in detail, various parameters of regulation of cross-bridge cycling.

SUMMARY

In *Limulus* striated muscle in addition to a sliding filament mechanism similar to that proposed for vertebrate striated muscle during contraction, a second mechanism exists. In this striated muscle there is a set of thick and thin filaments which interdigitate and slide past each other over nearly one-half of the excursion of the sarcomere. Over the remaining one-half the filaments do not slide past each other but the thick filaments themselves shorten. As the thick filaments shorten there is no change in the subunit repeat of myosin along the thick filament but a profound increase in the fixed charge of the thick filaments. We propose that the filament sliding and filament shortening both do work. Any model for *Limulus* muscle must include crossbridge cycling to account for shortening and tension development.

REFERENCES

Achazi, R.K. (1979). Phosphorylation of molluscan paramyosin. Pflugers Arch. 379: 197-201.

Bartels, E.M. and Elliott, G.F. (1981). Donnan potentials from the A-band and I-band of skeletal muscle, relaxed and in rigor. J. Physiol. 317: 85-86.

Brann, L., Dewey, M., Baldwin, A., Brink, P. and Walcott, B. (1979). Requirements for *in vitro* shortening and lengthening of isolated thick filaments of *Limulus* striated muscle. Nature 279: 256-257.

Cheung, W. (1980). Calmodulin plays a pivotal role in cellular regulation. Science 207: 19-27.

Chen, S.H., Chu, B. and Nossol, R., eds. (1981). *Scattering Techniques Applied to Supramolecular and Non-Equilibrium Systems.* p.928, Plenum Press, New York.

Chu, B. (1979). Dynamics of macromolecular solutions. Phys. Scripta 19: 458-470.

de Villafranca, G.W. (1961). The A- and I-band lengths in stretched or contracted horseshoe crab muscle. J. Ultrastruct. Res. 5: 109-115.

Dewey, M.M., Blasie, K., Levine, R.J.C. and Colflesh, D. (1972). Changes in A-band structure during shortening of a paramyosin-containing striated muscle. Biophys. J. 612: 82a.

Dewey, M.M., Colflesh, D. and Blasie, J.K. (1973). Changes in A-band structure during shortening of a paramyosin-containing striated muscle. Part II. Biophys. J. 13: 184a.

Dewey, M.M., Levine, R.J.C. and Colflesh, D.E. (1973). Structure of *Limulus* striated muscle. The contractile apparatus at various sarcomere lengths. J. Cell Biol. 58: 574-593.

Dewey, M.M., Walcott, B., Colflesh, D.E., Terry, H. and Levine, R.J.C. (1977). Changes in thick filament length in *Limulus* striated muscle. J. Cell. Biol. 75: 366-380.

Dewey, M.M., Levine, R.J.C., Colflesh, D., Walcott, B., Brann, L., Baldwin, A. and Brink, P. (1978). Structural changes in thick filaments during sarcomere shortening in *Limulus* striated muscle. In: *Cross-bridge Mechanism in Muscle Contraction* Edited by H. Sugi and G.H. Pollack, University of Tokyo Press, Tokyo.

Dewey, M.M., Colflesh, D., Brink, P., Fan, S-F., Gaylinn, B. and Gural, N. (1982). Structural, functional and chemical changes in the contractile apparatus of *Limulus* striated muscle as a function of sarcomere shortening and tension development. In: *Basic Biology of Muscles: A Comparative Approach,* Edited by B. Twarog, R.J.C. Levine and M.M. Dewey, Raven Press, New York. Vol. 37, pp. 53-72.

Elliott, G.F. and Bartels, E.M. (1982). Donnan potential measurements in extended hexagonal poly-electrolyte gels such as muscle. Biophys. J. 38: 195-200.

Fabiato, A. and Fabiato, F. (1975). Effects of magnesium on contraction activation of skinned cardiac cells. J. Physiol. (London), 249: 497-517.

Fan. S-F., Dewey, M.M., Colflesh, D.E. and Chu, B. (June-July 1982, in press). The application of laser light scattering to the study of biological motions. NATO Advanced Study Institute, Maratear, Italy.

Franzini-Armstrong, C. (1970). Natural variability in the length of thin and thick filaments in single fibers from a crab, *Portunus depurator.* J. Cell Sci. 6: 559.

Fujime, S., Maeda, T. and Ishiwata, I. (1981, Sept 8-10). F-actin and thin filament of muscle studied by laser light scattering. *Proc. of the Cambridge Conference on "Biomedical applications of laser light scattering".*

Haskell, R.C. and Carlson, F.D. (1981). Quasi-elastic light-scattering studies of single skeletal muscle fibers. Biophys. J. 33: 39-62.

Kensler, R.W. and Levine, R.J.C. (1982). An electron microscopic and optical diffraction analysis of the structure of *Limulus* telson muscle thick filaments. J. Cell Biol. 92: 443-451.

Kubota, K., Chu, B., Fan, S-F., Dewey, M.M., Brink, P. and Colflesh, D.E. (submitted for publication J. Mol. Biol.). Quasi-elastic light scattering of *Limulus* thick myofilament suspensions in relaxed (long), activated and rerelaxed (short) states.

Kushmerick, M.J. and Davies, R.C. (1969). The chemical energetics of muscle contraction. II. The chemistry, efficiency and power of maximally working sartorius muscles. Proc. R. Soc. Lond. (Biol.), 174: 315-353.

Lehman, W. (1982). The location and periodicity of a troponin-T-like protein in the myofibril of the horseshoe crab *Limulus polyphemus*. J. Mol. Biol. 154: 385-391.

Lehman, W. and Szent-Gyorgyi, (1975) Regulation of muscular contraction. Distribution of actin control and myosin control in the animal kingdom. J. Gen. Physiol. 66: 1-30.

Levine, R.J.C., Dewey, M.M. and de Villafranca, G.W.(1972). Immunohistochemical localization of contractile proteins in *Limulus* striated muscle. J. Cell Biol. 55: 221-235.

Levine, R.J.C., Elfvin M., Dewey, M.M. and Walcott, B. (1976). Paramyosin in invertebrate muscles. II. Content in relation to structure and function. J. Cell Biol. 71: 273-279.

Millman, B.M., Warden, W.J., Colflesh, D.E. and Dewey, M.M. (1974). X-ray diffraction from glycerol-extracted *Limulus* muscle. Fed. Proceedings 33: 5:1333, Abs. 622.

Naylor, G.R.S. (1982). Average electrostatic potential between the filaments in striated muscle and its relation to a simple Donnan potential. Biophys. J. 38: 201-204.

Newman, J. Carlson, F.D. (1980). Dynamic light scattering evidence for the flexibility of native muscle thin filaments. Biophys. J. 29: 37-47.

Radlick, L. and Johnson, W.H.J. (1982). Characterization of paramyosin phosphokinase. Biophys. J. 37: 36a.

Sellers, R. (1981). Phosphorylation-dependent regulation of *Limulus* muscle myosin. J. Biol. Chem. 256: 9271-9278.

Stephenson, D.G., Wendt, I.R. and Forest, Q.G. (1981). Non-uniform ion distributions and electrical potentials in sarcoplasmic regions of skeletal muscle fibers. Nature 289: 690-692.

Stewart, M., Kensler, R.W. and Levine, R.J.C. (1981). Structure of *Limulus* telson muscle thick filaments. J. Mol. Biol. 155: 781-790.

Walcott, B. and Dewey, M.M. (1980). Length-tension relation in *Limulus* striated muscle. J. Cell Biol. 87: 204-208.

Wray, J.S., Vibert, P.J. and Cohen, C. (1974). Cross-bridge arrangements in *Limulus* muscle. J. Mol. Biol. 88: 823-830.

DISCUSSION

HUXLEY: What I wanted to say was more in the nature of a short comment rather than a specific question. In the case of frog sartorius muscle, one has a particularly favorable system for looking at the relationship between A-band length and sarcomere length for the reason that at a given muscle length there's a rather sharp uniformity of sarcomere lengths; and, also, the ends of the A-band are delineated extremely sharply because the A-filaments within an A-band are of extremely uniform length. So it's relatively easy to get accurate measurements of the A-band length. When one does that, one generally finds -- and I think maybe this morning we may hear exceptions to this at short sarcomere lengths -- one generally finds a very constant A-band length. About ten years ago, Clara Franzini-Armstrong published an interesting paper (J. Cell Sci. 6: 559-592, 1970), which I think may not be as generally known as it deserves to be, on some studies of arthropod muscles in which she observed that this constancy of sarcomere length and A-band length at a given muscle length, which has almost been taken for granted in vertebrate striated muscle, didn't seem to apply to these muscles. In muscles which were fixed at a constant length, and both within a single fiber and between different fibers, it was possible to find both sarcomeres and A-bands in which considerable, apparently natural, differences in lengths occurred.

For that reason, over the last year or two, a co-worker of mine, Elizabeth Rutherford, and I did spend some time having a look at *Limulus* muscles to see if there was any evidence of the same sort of natural variability of A-band and sarcomere lengths. What we found was that in muscles which were fixed in the relaxed state at a given length, there was indeed a surprisingly large variation in both the sarcomere length and the A-band length within given fibers and between one fiber and another fiber. In particular, there was an obvious division into two quite distinctive fiber types, one with a much longer sarcomere length and A-band length than the others, which could be distinguished from each other not merely by these length differences but also by characteristic differences in the thickness of the Z-line. In addition, even within the same fiber it was often possible to find, in localized regions quite close together, that there were substantial variations in both sarcomere length and in A-band length. Thirdly, instead of the very sharply defined ends of the A-bands, which are charactersitic of vertebrate striated muscle, it was obvious that the A-bands were much fuzzier at the ends in these muscles. This fuzziness was not due to a misalignemnt of the A-filaments because it was possible in well-fixed specimens to see extremely sharply delineated M-lines. So the variation in the fuzziness of the A-band was due to local variation of the individual thick filament length within a given A-band, Elizabeth Rutherford was able to check this by extremely laborious serial sectioning through complete A-bands. Very substantial variations of A-filament length, on the order of 10% or 20%, were found within a single A-band. Moreover, we found that in muscles that were fixed at different lengths, we could still see approximately the same range of A-band lengths, but at the same time there were variations in I-band lengths, and at shorter lengths the I-bands on the whole were shorter.

Now, I'm not saying that this, in itself, is evidence that A-band shortening doesn't take place. I think the only way to find whether A-band shortening takes place is to look at a given fibril and follow whatever happens in a very closely observed series of stages, from a long sarcomere length to a shorter sarcomere length, so that you can see exactly what happens to each stage. But I do think that there is this difficulty of possible natural variability, and I think one should be on one's guard when one looks at different samples in the electron microscope.

The illustration (Fig. D-1) shows two adjacent fibers in a muscle which may have been slightly shortened as you can see from the slight waviness of the membranes. If you observe the sarcomeres on the right, they have a very distinctly different appearance from those on the left. They have much thicker Z-lines and the A-bands are about 50% longer than those on the left. That's an example of one of the problems with this type of muscle. I think it is clear that there are somewhat different fiber types present in this muscle, the longer sarcomere one being in the minority; but as well as that, as I say, within a given fiber, we did find what looks like evidence for the same sort of *local* variability as Franzini-Armstrong found in other arthropod muscles.

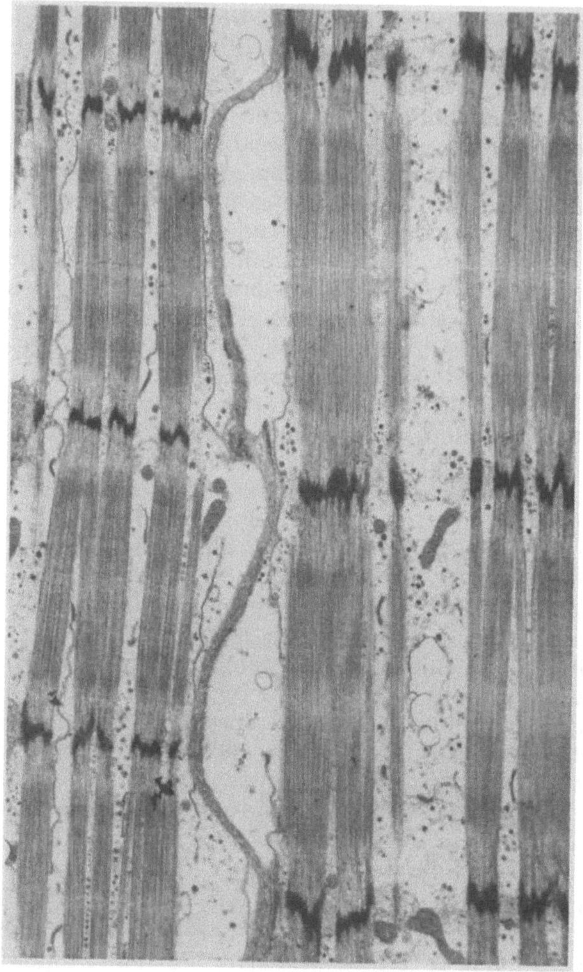

Figure D-1: Electron-micrograph of *Limulus* telson muscle, showing two different fibre types (recognizable by their characteristically different Z-line structures) having very different A-band lengths.

DEWEY: I don't see that the evidence we presented is in direct conflict with that. Let me say this: if you go back and look at the isolated thick filaments in our previous papers, you'll see that their lengths are normally distributed in each case (long and shortened). There is no bimodal distribution. One would expect, if there were several fiber types, say being nearly equal or three-quarters of the total fiber number, that there would be overlap in the standard deviation between the measurements of the two types of filament, but this was not the case.

Further, we can isolate the thick filaments, and they are long, about 4.2 μm (measurements give a bell-shaped distribution). These filaments incrementally shorten *in vitro* by changing the calcium concentration. There are at least two different kinds of observations on this -- light scattering and the measurements of filament length by negative staining. Filament shortening seems to be a real phenomenon, and that doesn't deny the fact that there *might* be another fiber type there. We simply have *not* had the evidence for it. If you go back and look at our Fourier analysis study of refractive index, all of our calculations came from the higher orders -- 11th and 12th -- which means there must be really very high ordering in this muscle. And so the population of fibers that would be uniquely different, shorter, say, must be small.

HOMSHER: Has anyone made an observation on what the A-bands do in *living* muscle fibers? It seems you're always working with fixed muscles or glycerinated muscles.

DEWEY: No. The Fourier analysis was done on single, living fibers, or bundles of two or three fibers -- stimulating them and allowing them to shorten, then doing the laser work. The length-tension curve was done with the laser on living specimens. The video taped demonstrations of A-band shortening were done on single isolated glycerinated fibers.

HOMSHER: Using direct optical measurements?

DEWEY: Yes. We've done it with interference and phase microscopy, using microdensitometry to measure the A-bands. Currently John Haselgrove and I are looking at optical diffraction of isolated filaments that have been shortened *in vitro*. It is premature, but I think we do see some changes. Dr. Levine is also looking at this, and she may say something about it a little later.

KRUEGER: Would you care to elaborate on the mechanism by which you think vanadium might prevent shortening of the thick filament?

DEWEY: The experiment done with light scattering of preshortened filaments with vanadium shows no increase in Γ. Further, the requirements for shortening that we have determined so far do include ATP (above 1 millimolar) -- presumably hydrolysis of ATP -- magnesium (above 1 millimolar), and calcium at pCa 5.5. If one then clips the heads with papain, the thick filaments don't shorten. So the shortening is somehow related to the myosin ATPase.

DYNAMIC LASER LIGHT SCATTERING OF PAPAIN-TREATED THICK FILAMENTS FROM LIMULUS STRIATED MUSCLE IN SUSPENSION

S-F. Fan, M.M. Dewey, D. Colflesh, P. Brink and B. Chu*

Department of Anatomical Sciences, Health Sciences Center,
**Department of Chemistry, State University of New York at Stony Brook,*
Stony Brook, New York 11794

ABSTRACT

Using quasielastic light scattering we have previously shown an increase in high frequency internal motion of isolated thick filament upon activation. This we have attributed to cross-bridge motion. Here we show that after cleavage of the S_1 moiety of myosin from isolated filaments with papain, calcium ions no longer activate the isolated filaments to produce high-frequency motions.

Cyclic cross-bridge motions may be one of the key points in understanding the mechanism of muscular contraction. In earlier work we have measured the photoelectron count autocorrelation function of light scattered by *Limulus* thick filament suspensions as a function of scattering angle in relaxed (long), activated and re-relaxed (short) states (Kubota et al., in press). Upon activation by calcium ions, a dramatic increase in the average linewidth Γ at high KL values was observed. For monodisperse semi-flexible filaments of length L and KL>3, high-frequency internal bending and twisting motions of the filament contribute towards an increase in Γ values such that in a plot of Γ versus K^2 the curve bends upward (Maeda and Fujimi, 1981) where K is the magnitude of the momentum transfer vector. For suspensions of Limulus thick filaments, the extra increase in Γ had been attributed to cross-bridge motions. Later we obtained further evidence to support this assertion (Fan et al., unpublished): 1. a similar increase in Γ by Congo Red (Fan, 1964) rather than by calcium ions, 2. suppression of the increase of Γ by calcium ions either by heat denaturation of the cross-bridges (Huxley, 1972; Fan and Wen, 1979) or by treating the filaments with vanadate (Goodno, 1979; Goodno and Taylor, 1982), which acts as a myosin ATPase inhibitor.

Here we will present another independent observation which supports our contention. After cleavage of the S_1 moiety of myosin from isolated filaments with papain, calcium ions no longer activate the filaments to produce high-frequency motions.

The method of isolation of thick filaments and the light scattering spectrometer used were the same as that described previously (Kubota et al., in press). Cross-bridges were cleaved by using papain (Koninz, et al., 1965; Yamanaotoik and Schiao, 1980). Papain (0.046 mg/ml, final concentration) was added directly to the filament suspensions. After seven minutes the digestion was stopped by adding iodoacetate (final concentration 1.5 mM). As judged from electron micrographs of negatively stained and angle shadowed filaments and SDS gel electrophoretic studies, filaments were nearly, 95%, free of S_1.

We recapitulate some results of our earlier findings (Kubota et al., in press) as shown in Figure 1. Curve a represents the results obtained from a suspension of the relaxed (long) filaments in relaxing solution (100 mM KCl, 1 mM $MgCl_2$, 1 mM ATP, 5 mM EGTA, 5 mM Tris buffer, pH 7.4). Curve b represents the results obtained from a suspension of activated filaments after the relaxed (long) filament suspension had been dialyzed against an activating solution (100 mM KCl, 1 mM $MgCl_2$, 5 mM $CaCl_2$, 1 mM ATP, 5 mM Tris buffer, pH 7.4). The inset is a plot of Γ_a/Γ_r with subscripts a and r denoting filaments in activated and relaxed (long) states, respectively. When $K^2=10\times10^{10}cm^{-2}$, the value of $\Gamma_a/\Gamma_r \simeq 2.3$. Figure 2 shows a plot of $\Gamma_{r,p}/\Gamma_r$ where the subscript p denotes a suspension of filaments which has been treated with papain. Both suspensions were in relaxing solution as denoted by the subscript r. The diffusion constant D of a rigid rod in solution can be calculated according to (Newman et al., 1977):

$$D=k_BT\ 1n(2L/d)-1/2(1.46-7.4(1/1n(2L/d)-0.34)^2-4.2(1/1n(2L/d)-0.39)^2)\ /4\pi\eta L \quad (1)$$

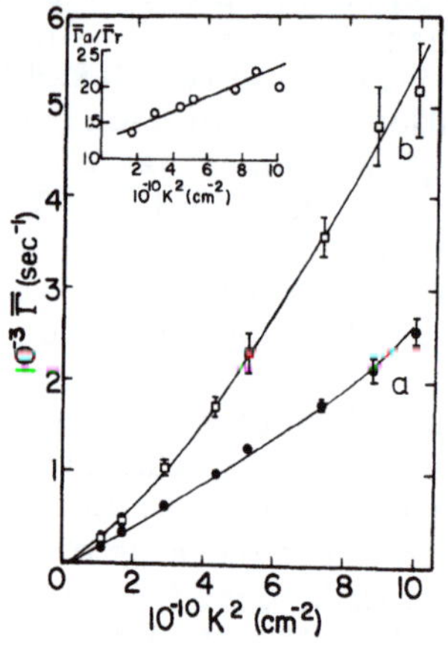

Figure 1: Plot of Γ versus K^2 for Limulus thick filaments suspensions. The fitting was a 2nd order cumulants fit. (a) Filaments dialyzed against relaxing solution. (b) Filaments dialyzed against activating solution. Average values and the standard deviation of four experiments. Inset: plot of Γ_a/Γ_r versus K^2.

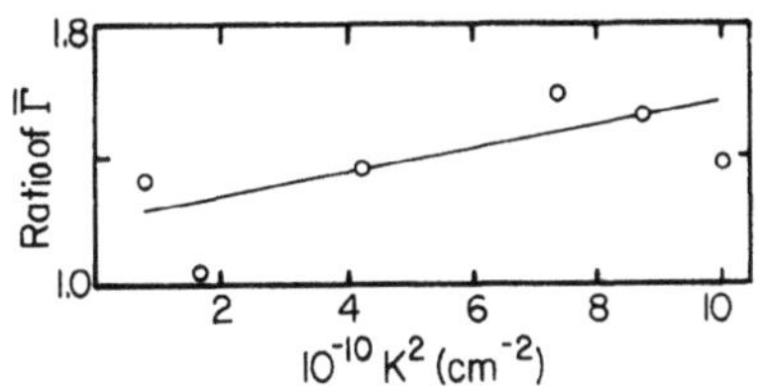

Figure 2: Plot of the ratio of Γ of the sample treated with papain to that untreated versus K^2. Both suspensions were in relaxing solution.

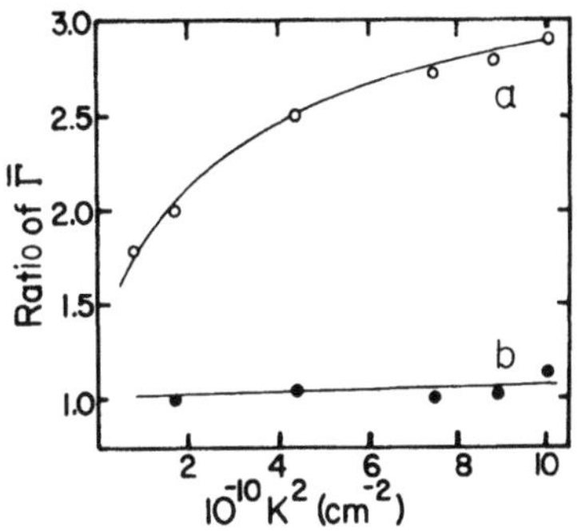

Figure 3: (a) Plot of the ratio of Γ of the sample untreated with papain to that treated versus K^2. Both suspensions are in activating solution. (b) Plot of the ratio of Γ of the sample in activating solution to that in relaxing solution. Both suspensions had been treated with papain.

where k_B, T and η have the usual meaning and d is the diameter of the filament. As $K \to 0$, $\Gamma = DK^2$. The limiting value of $\Gamma_{r,p}/\Gamma_r$ as $K \to 0$ shows that the effective diameter of the treated filaments had been decreased by about 40%, provided that papain treatment had not changed the filament length. Electron microscopy of negatively-stained and angle-shadowed filaments established that the filaments did not change length. The positive slope in Figure 2 shows that after papain treatment the filament flexibility had increased. In Figure 3, curve a shows a plot of $\Gamma_a/\Gamma_{a,p}$ versus K^2 where $\Gamma_{a,p}$ denotes those Γ values from a suspension of papain treated filaments activated with calcium ions. The values of $\Gamma_{a,p}$ are much smaller than those of Γ_a at higher KL values indicating clearly that the excess high-frequency motions had practically disappeared. Although we are not sure what value $(\Gamma_a/\Gamma_{a,p})$ has in the limit as K approaches zero, it seems highly probable that the value is greater than one, implying the papain treatment of the activated filaments had caused an increase in the effective diameters. This conclusion was strengthened by the fact that $\Gamma_{a,p}/\Gamma_{r,p} \simeq 1$ even if we take into account the difference in the lengths of the filaments in relaxed (long) and activated states, according to Equation (1). From Equation (1) $\lim_{K \to 0}(\Gamma_a/\Gamma_{a,p}) \simeq 1.1$, is in reasonable agreement with the observed tendency of $\Gamma_a/\Gamma_{a,p}$ as $K \to 0$. The increase in the effective diameter of the shortened filaments after papain treatment could mean that papain has not only cleaved the cross-bridges but also caused a loosening of the shortened filaments.

REFERENCES

Fan, S-F, (1964). Shortening in Congo Red solution of myofibrils isolated from glycerinated muscle fibers. Scientia Sinica. 13: 692-693.

Fan, S-F. and Wen, Y-S. (1979). Concerning the binding sites of myofibril with Congo Red and dichroism change with myofibril length of Congo Red stained glycerinated sartorius muscle fibers. Acta Physiol. Sinica. 31: 227-238.

Goodno, C.C. (1979). Inhibition of myosin ATPase by vanadate ion. Proc. Natl. Acad. Sci. U.S.A. 76: 2629-2634.

Goodno, C.C. and Taylor, E.W. (1982). Inhibition of actomyosin ATPase by vandate. ibid., 79: 21-25.

Huxley, H.E. (1972). Structural changes in the actin and myosin containing filaments during contraction. Cold Spring Harb. Symp. Quant. Biol. 37: 361-376.

Koninz, D.P., Mitchell, F.R., Niekei, T. and King, C.M. (1965). The papain digestion of skeletal myosin A. Biochemistry. 4: 2373-2381.

Kubota, K., Chu, B., Fan, S-F., Dewey, M.M., Brink, P. and Colflesh, D. Quasi-elastic light scattering of suspensions of *Limulus* thick myofilaments in relaxed (long), activated and re-relaxed (short) states. Submitted to J. Mol. Biol.

Maeda, T. and Fujimi, S. (1981). Effect of filament flexibility on the dynamic light scattering spectrum with special reference to fd virus and muscle thin filaments. Macromolecules. 14: 809-818.

Newman, J., Swinney, H.L. and Day, L.A. (1977). Hydrodynamic properties and structure of fd virus. J. Mol. Biol. 116: 593-606.

Yamanaotoik, K. and Schiao, T. (1980). Substructure of myosin subfragment (as revealed by digestion with proteolytic enzymes). J. Biochem. (Tokyo). 87: 219-226.

STRUCTURE OF *LIMULUS* AND OTHER INVERTEBRATE THICK FILAMENTS

Rhea J.C. Levine, Robert W.Kensler, Mary Reedy * Waltraud Hoffman +, Sandra Davidheiser ++ and Robert E. Davies ++

The Medical College of Pennsylvania Philadelphia, PA 19129
**Duke University School of Medicine Durham, NC 27710*
+Max-Planck Institute für Medizinische Forschung Heidelberg, West Germany
++University of Pennsylvania School of Veterinary Medicine Philadelphia, PA 19104

ABSTRACT

We have demonstrated remarkable similarity among the skeletal muscles of chelicerate arthropods with respect to the cross-bridge arrangement on the surface of their thick filaments. The latter, gently isolated from the muscles of three representative species (*Limulus* telson, tarantula leg and scorpion leg and tail) have been examined by electron microscopy and optical diffraction using both negatively stained and unidirectionally metal shadowed preparations. The filaments are highly periodic and produce clear and detailed diffraction patterns. The cross-bridge projections form integral surface helices, with an axial spacing of 14.5 nm between adjacent crowns and a major axial repeat every 43.5 nm. We have demonstrated previously that *Limulus* filaments are four-stranded and analysis of both electron micrographs and their transforms, as well as optical reconstructions of the arachnid filaments is consistent with their also having a four-start surface helix, which is right-handed in all cases. Of all those examined, thus far, only *Limulus* thick filaments have been demonstrated to change length under various conditions. Shortened *Limulus* filaments isolated from K^+-stimulated fibers retain the 43.5 nm axial repeat periodicity and 14.5 nm axial spacing between crowns. In preliminary analysis of negatively stained and metal shadowed preparations, we see no systematic change with respect to screw or rotational symmetry in short as compared with long filaments. A few of the former have a very slightly increased diameter (3-4 nm) in the middle of each filament arm. This region often shows disorder on optical transforms. From our results we cannot rule out the possibility that disaggregation and reaggregation of thick filament proteins accompany the changes in length of *Limulus* thick filaments.

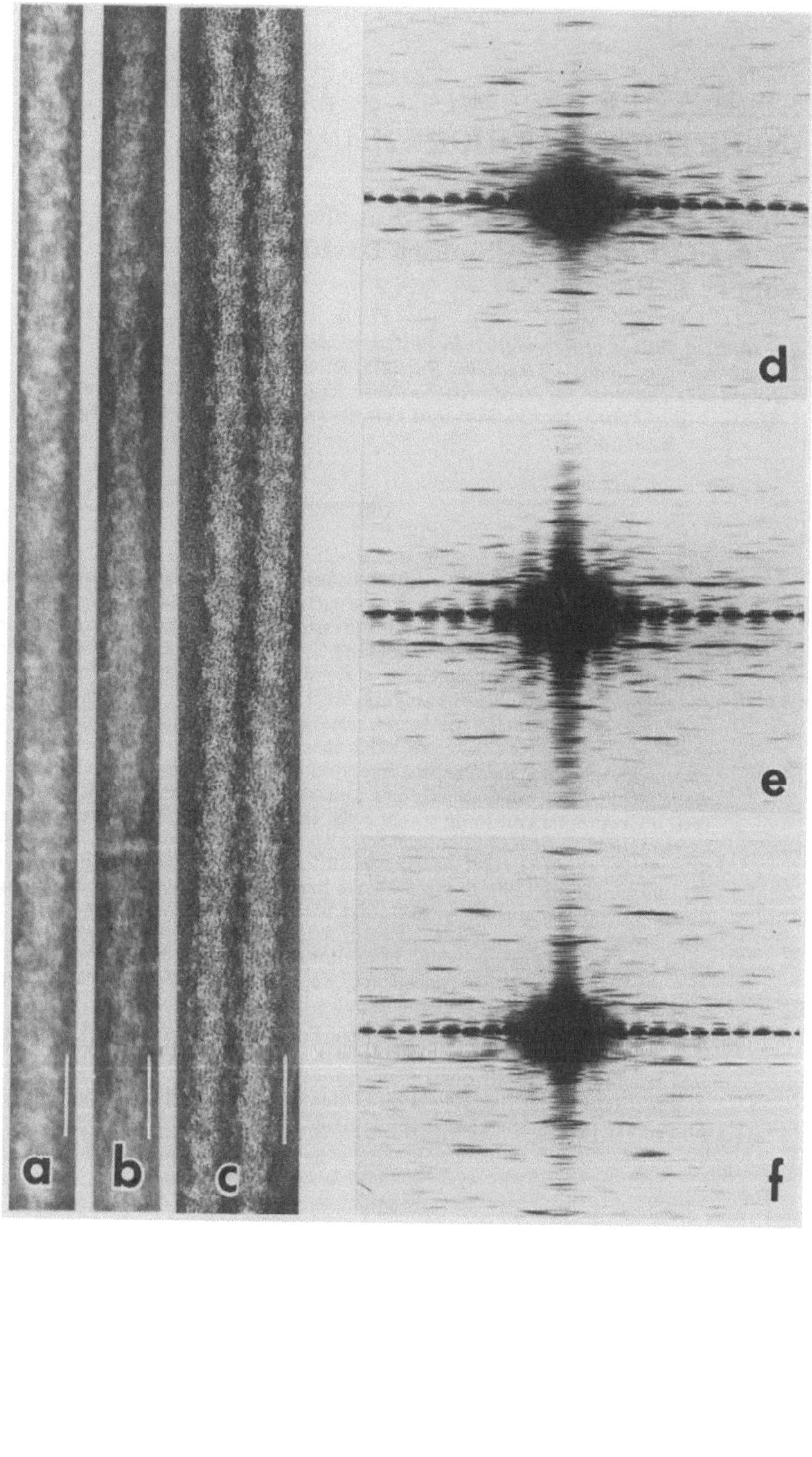
a
b
c
d
e
f

INTRODUCTION

For any given muscle, the structural features of the contractile apparatus that affect the physiology of actin-myosin interaction include: the latice array of thick and thin myofilaments, the number of thin filaments that surround each thick filament and the number and arrangement of crossbridges (myosin heads) on the surface of the thick filaments. Most of these features have been determined for a variety of muscle types by the techniques of X-ray diffraction of living and chemically skinned fibers and electron microscopy of sectioned tissue. Information regarding the number of crossbridges per "crown" on thick filaments, however, has remained inaccessible. Recently, using computed and optical transforms obtained from electron micrographs of negatively stained and metal shadowed filament preparations, we have established this parameter for *Limulus* thick filaments, isolated in their "long" conformation from unstimulated telson muscle (Stewart, Kensler & Levine, 1981; Kensler & Levine, 1982a, b; Levine, Kensler, Stewart and Haselgrove, 1982). Vibert & Craig (1982) have since employed these techniques to determine the crossbridge array on scallop striated muscle thick filaments.

The success of these studies on invertebrate thick filaments may be related to their size (length and diameter) which is greater than that of vertebrate thick filaments. The larger dimensions of the invertebrate filaments may act to stabilize their structure, thus permitting examination of relatively long straight segments having many structural repeats. Our results were also dependent on the highly ordered crossbridge array present on *Limulus* thick filaments (Wray, Vibert & Cohen, 1974), which is retained during the preparative procedure (Stewart et al., 1981; Kensler & Levine, 1982 a, b).

Chelicerate Arthropods Have Similar Thick Filaments

Limulus is an ancient arthropod genus belonging to the Subphylum Chelicerata, which also includes the arachnids (scorpions, spiders and mites) and the pycnogonids (sea spiders; Barnes, 1969). Following Wray's (1982) observation that X-ray diagrams of tarantula muscle are very similar to those of *Limulus*, we undertook an investigation of the structure of the thick filaments isolated from the muscles of two other chelicerate arthropods: tarantula (leg muscle) and scorpion (leg and tail muscle). The chelicerate muscles, like those of many arthropods (Elfvin, Levine & Dewey, 1976; Levine, Elfvin, Dewey & Walcott, 1976) have long sarcomeres

Figure 1: High power electron micrographs of negatively stained chelicerate arthropod thick filaments and optical diffraction patterns obtained from them. The marked resemblance among all of these filaments is apparent with respect to both their appearance and the optical transforms, which display similar layer line spacings. Bars = 50 nm on all electron micrographs. a) Central region (including the bare zone) of long *Limulus* thick filament. Magnification: ×225,000. b) Portion of tarantula thick filament. Magnification: ×240,000. c) Portions of two scorpion thick filaments. Magnification: ×250,000. d), e) and f) are optical transforms of a), b) and c), respectively.

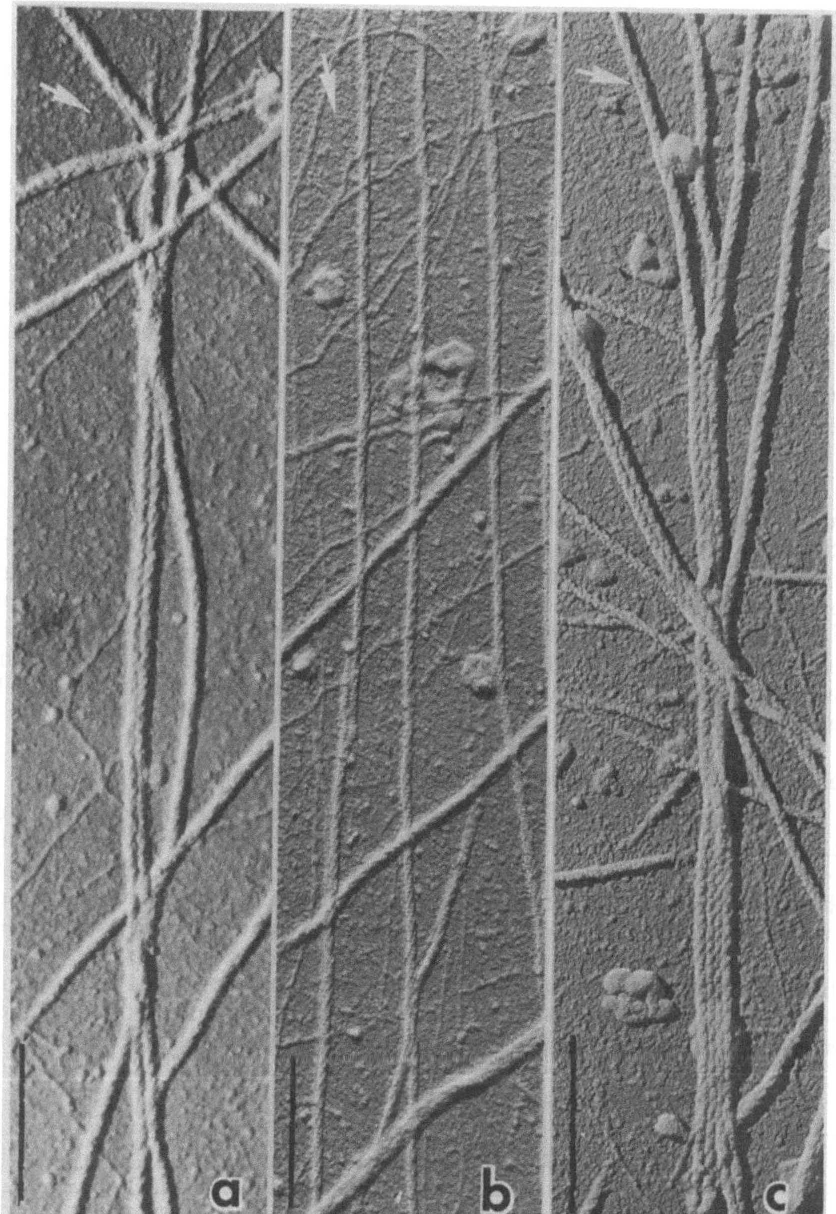
a
b
c

and A-bands and a ring of nine to twelve thin filaments surrounding each thick filament (Sherman & Luff, 1971). Negatively stained or metal shadowed thick filaments isolated from both tarantula and scorpion muscles are highly ordered and display crossbridge arrangements that closely resemble that we reported for long *Limulus* thick filaments (Kensler, Levine, Reedy & Hofman, 1982; Figure 1). Interestingly, thick filaments isolated from the long sarcomere muscles of such non-chelicerate arthropods as crustaceans and insects lack a highly periodic surface structure (Kensler et al., 1982) and exhibit distinctly different crossbridge arrays (Wray, 1982; Wray, Vibert & Cohen, 1975).

The chelicerate thick filaments that we have examined display surface periodicity that repeats at every 43nm: at every third level of crossbridges. Images of these filaments produce sharp optical transforms with layer lines indexing on orders of a 1/43.5 nm^{-1} helical repeat. Meridional reflections index on orders of 1/14.5 nm^{-1}, which is the axial rise between successive crowns of bridges (Figure 1). Estimation of the rotational symmetry of these filaments was based on calculations of the radius of the greatest crossbridge mass from the subsidiary maxima of the 1/14.5 nm^{-1} meridional reflections and measurements of the radial positions of the primary maxima on the first and fourth layer lines on optical transforms. Calculated values (5.5 = *Limulus*, 5.36 = tarantula) fall very close to the expected maximum (5.32) for a J_4 Bessel function describing a four-stranded myosin helix on the filaments' surfaces. This estimate has been confirmed for *Limulus* from phase information retrieved from the computed transforms and computer reconstructions of the negatively stained filaments (Stewart et al., 1981), and is supported for the other chelicerates by the symmetrical positions of stain-excluding structures on images of negatively stained filaments, as well as their striking resemblance to *Limulus* filaments (Kensler et al., 1982). Although individual myosin heads are not resolved in negatively stained preparations, mass measurements made at the EMBL facility in Heidelberg are consistent with four myosin molecules per crown, or one per crossbridge on long *Limulus* filaments (Levine, Hofmann & Freeman, unpublished observation).

All of the chelicerate filaments display a prominent right-handed surface helix after unidirectional shadowing with either platinum or platinum-carbon (Kensler & Levine, 1982b; Kensler et al., 1982; Levine and Kensler, 1982; Figure 2). The arrangement of surface subunits on shadowed filaments is most apparent on those filaments which are oriented nearly parallel to the shadowing source. Each subunit presumably

---

Figure 2: Medium-power electron micrographs of chelicerate thick filaments unidirectionally shadowed at an ~30° angle with platinum-carbon. Arrows indicate direction of shadowing source. Note the striking right-handed helical surface structure of all of these filaments. The subunit array is most clearly delineated on filaments oriented nearly parallel to the shadowing source. Bars =0.5 μm on all electron micrographs. a) *Limulus* long filaments. Magnification: ×45,000; b) tarantula filaments. Magnification: ×41,400; c) scorpion filaments. Magnification: ×46,000.

consists of two S-1s, but these cannot be resolved on shadowed filaments, either. In fact, in addition to the presence of information from only one filament surface, which makes shadowed filaments unsuitable for analysis of helical parameters by either computer reconstruction or optical diffraction, resolution of surface structure is worse on shadowed than on negatively stained filaments. The subunit organization on shadowed filaments from all species closely resembles the crossbridge arrangement detailed on the one-sided computer reconstruction of negatively stained long *Limulus* thick filaments, with respect to both number and orientation (Stewart et al., 1981). This similarity is also apparent when optically filtered images of the shadowed filaments are compared with this computer reconstruction.

Although tarantula and scorpion thick filaments are approximately as long as those isolated from unstimulated *Limulus* muscle (~4.2 μm average length), the spider filaments have a smaller shaft diameter (19-20 nm tarantula vs. 23.5 *Limulus*) and a lower paramyosin:myosin molar ratio (0.31 tarantula vs. 0.49 *Limulus*) than *Limulus* filaments (Levine, unpublished observation). Nevertheless, the centers of mass of the crossbridges lie at the same distance (~4 nm) from the surfaces of both *Limulus* and tarantula filaments (Stewart et al., 1981; Levine, Kensler,

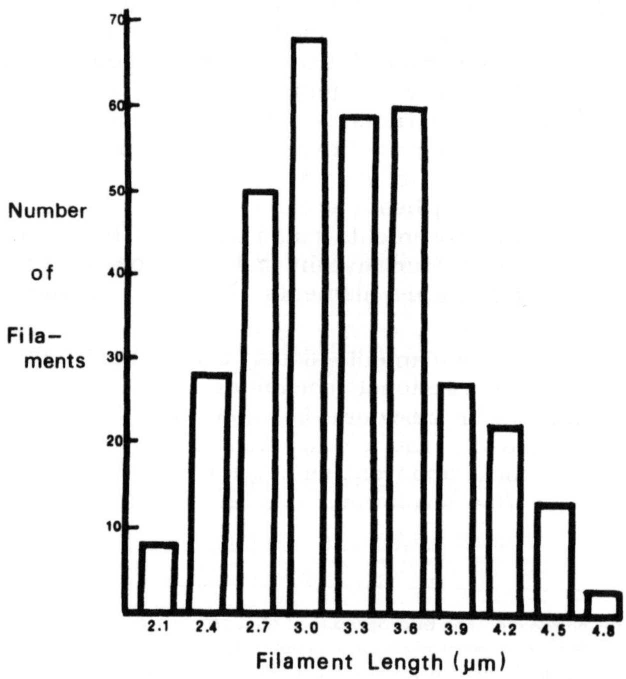

Figure 3: Histogram showing length distribution of thick filaments isolated from a K^+-stimulated intact *Limulus* muscle fiber bundle. Mean filament length is 3.1 μm. Some long filaments are always present.

Reedy & Hofmann, unpublished observation). These parameters have not yet been determined for scorpion thick filaments but, based on their images in both negatively stained and metal shadowed preparations and the diffraction patterns obtained from these images, we are confident that they are structurally very similar to the other chelicerate filaments.

Short Limulus Thick Filaments

Of the thick filaments examined from chelicerate arthropod muscles, thus far only those of *Limulus* have been demonstrated to undergo a conformational change during muscle activity (Dewey, Walcott, Colflesh, Terry & Levine, 1977; Dewey, Levine, Colflesh, Walcott, Brann, Baldwin & Brink, 1979). In our hands, short filaments are isolated preferentially from intact *Limulus* muscle fiber bundles following stimulation (by K^+ or electrically) either under unloaded conditions or when the fibers are stretched beyond overlap. Filaments isolated from such preparations exhibit either a bimodal length distribution, with one population averaging ~3.0 μm and a second averaging ~4.0 μm, or a normal distribution with a mean length of ~3.0-3.2 μm (Figure 3). The point is, that some long filaments are *always* present in preparations from stimulated muscle, while filaments under 3.6 μm are *extremely rare* in preparations from unstimulated muscle (mean length 4.2 μm, range 3.6 to ≧ 5.0 μm). It is not yet clear whether the longer filaments in stimulated preparations are isolated from fibers that have not been successfully activated or whether they represent a "shortened" filament population isolated from an additional fiber type that we recently discovered in *Limulus* telson muscle (Davidheiser, Levine & Davies, 1982). The newly observed fibers are much smaller in diameter, richer in mitochondria and have longer A and I bands than the majority of *Limulus* telson fibers. While filament length has not been measured *in situ* specifically in these fibers it is likely that they posses longer thick filaments than the rest of the *Limulus* fibers and when isolated, these filaments may contribute to the population at the longer end of the length spectrum. Preliminary estimates from cross-sections suggest that this additional fiber type comprises ≦ 18% of the muscle cross-sectional area (Davidheiser et al., 1982).

We have studied only those short filaments which display a distinct surface periodicity, a central bare zone and two tapered ends. In negatively stained preparations, the bare zones of short filaments are less well-defined and appear shorter and "rougher" than those we commonly observe on long filaments. Preliminary measurements (Figure 4) suggest that a decrease in the length of the bare zone on the order of 50 nm (from ~200 nm to ~150 nm) may occur during filament shortening. Frequently, regions that appear wider and less obviously periodic than the rest of the filament are seen in the middle of each arm of the short filaments. Otherwise, these resemble long filaments with respect to diameter and surface periodicity (Figure 5).

It is obviously of great interest to examine the apparently "swollen" regions of the short filaments in order to determine whether a significant change in the helical array of crossbridges occurs at these sites. There

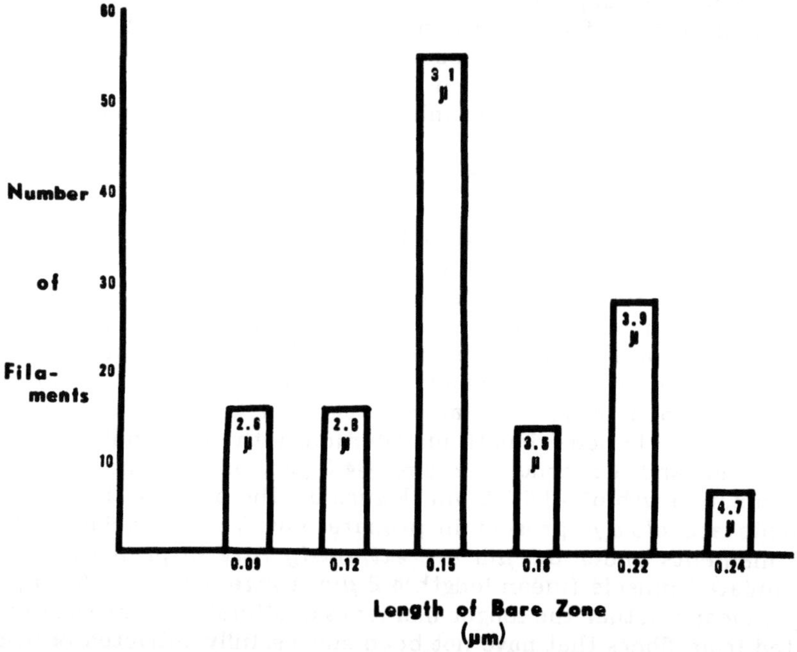

Figure 4: Histogram showing length distribution of bare zones on thick filaments isolated from a K$^+$-stimulated intact *Limulus* muscle fiber bundle. Numbers in each bar represent the mean filament length of the population having the bare zone length indicated beneath the bar on the abscissa.

are, however, equally obvious difficulties in achieving this goal. First, the swollen segment may encompass relatively few helical repeats. Thus, even if it were possible to isolate just this region for optical diffraction, the resolution of the transform obtained from such a short length would be greatly decreased over that obtained from a longer segment with many repeats. Second, the enlarged diameter of this region produces a curvature to the filament edges, which also lessens the quality of the

---

Figure 5: Short *Limulus* thick filaments, isolated from K$^+$ stimulated intact muscle fiber bundles. a) Medium-power electron micrograph of negatively stained filament. Note the "swollen" segments in the middle of the filament arms, and the "rough" appearance of the bare zone. Region from which the optical transform (b) was taken is indicated by the double-arrowed bar. Simple bars = 0.5 μm on all electron micrographs. Magnification: ×57,500. b) Optical transform of region indicated by the double-arrowed bar in (a). The layer lines index on orders of ~1/43.5 nm^{-1} and the pattern is qualitatively similar to that in Figure 1d. c) Medium-power electron micrograph of platinum-carbon shadowed short filaments. Note the similarity in helical structure to the filaments in Figure 2a. Arrow indicates direction of shadowing. Magnification: ×46,000. d) Higher power electron micrograph of one arm of a shadowed, short filament, used for the optical reconstruction in (e). Arrow indicates direction of shadowing. Magnification: ×76,600. e) Optically filtered image of filament arm shown in (d). Note helical array of subunits. At least 4 to 5 are visible on this photographed reconstruction. 6 to 7 are visible on the original, but are obscured by contrast here.

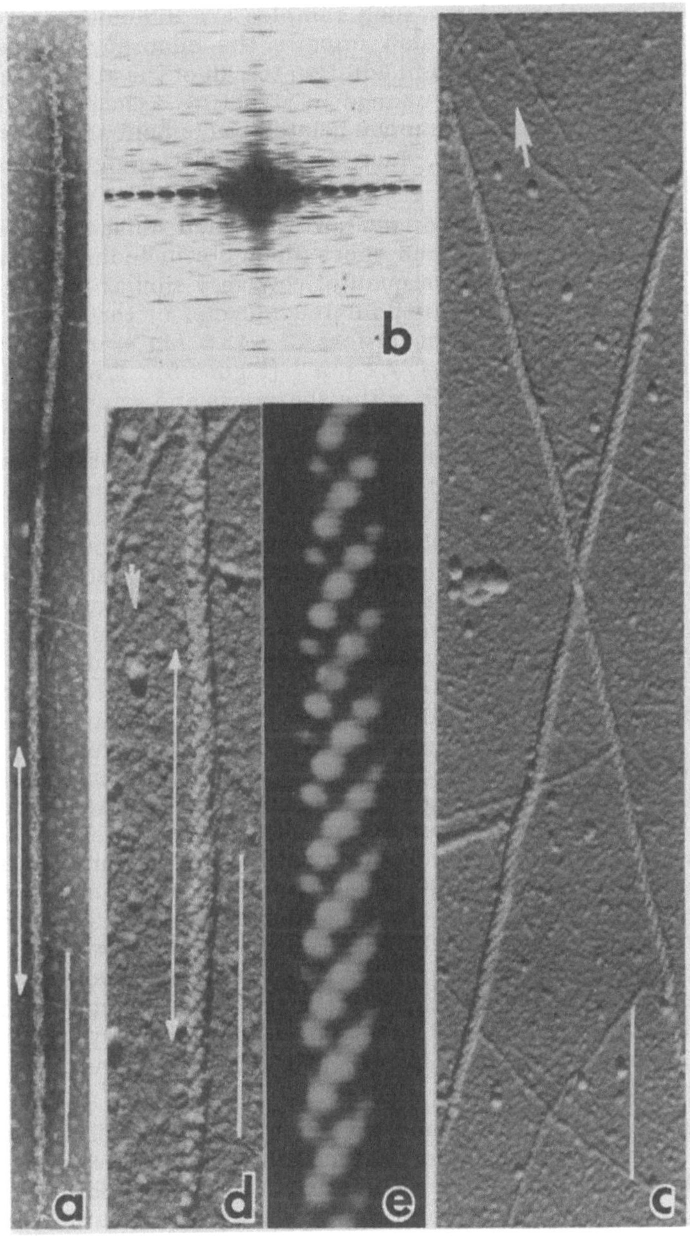
b
a
d
e
c

transform. Third, filament segments selected for diffraction on the basis of relative straightness and clear periodicity usually consist of both the swollen segments and the adjacent region having "long" filament structure. Diffraction patterns from such samples are difficult to interpret, having extended reflections which impairs the accuracy of measurements. Finally, because of - and in addition to - all of the above, it may be impossible to ascertain if any change in structure associated with the swollen segment is a real phenomenon related to filament shortening or if it is an artifact of preparation, such as severe flattening of the upper filament surface by the stain.

With these caveats in mind, we have obtained diffraction patterns from images of negatively stained short filaments including the swollen segments. These transforms are qualitatively very similar to those from long filaments isolated from unstimulated muscle, in that they have a series of layer lines indexing on orders of ~43.5 nm^{-1} and meridional reflections at 1/14.5 nm^{-1} and 1/7.2 nm^{-1} (Figure 5). Transforms were selected for analysis on the basis of the sharpness and symmetry of their reflections and a resolution out to at least 1/7.2 nm^{-1} on the meridian. Preliminary calculation of the radius of the greatest crossbridge mass from the subsidiary maxima on the third layer line indicates that it is the same as for the long filaments, averaging 15.5 ± 1.3 nm S.D. (range 13.6 to 17.0 nm) on transforms of ten short filaments. The rotational symmetry of the short filaments was estimated from the calculated crossbridge radius and measurement of radial positions of the primary maxima on the first and fourth layer lines for each of these filaments. Individual values ranging from 4.6 to 6.5 were calculated. The average value, 5.4 ± 0.59 S.D., is very close to the predicted maximum of a J_4 Bessel function, and to the values calculated for both long *Limulus* and tarantula filaments. Admittedly, relatively few short filaments have been analyzed in this way. Nevertheless, except for some indication of a very slight increase in diameter (~3-4 nm) in the swollen segments of some short filaments (Figure 5 - radius 17.0 nm), we have been unable to discern a systematic structural change with respect to either the rotational or screw symmetry accompanying filament shortening. It should be mentioned, however, that approximately one half of the short filaments selected for diffraction yielded transforms showing sufficient disorder in the swollen regions (absence of subsidiary maxima on the third layer line, non-systematic discrepancies in the radial positions of the primary maxima on the first and fourth layer lines) to make them unsuitable for analysis. Even on these transforms, however, the helical repeat period remains the same as it is on the long filaments.

Unidirectionally metal shadowed short *Limulus* filaments also display the right handed surface helix of subunits we have observed on the other thick filaments we have studied (Figure 5). Both by direct observation and optical filtration of appropriately oriented filaments, we can resolve six to seven subunits per helical strand, even on swollen segments, which again is consistent with a four-stranded myosin helix.

Our preliminary observations of the organization of myosin

crossbridges on the surface of *Limulus* thick filaments shortened *in situ* do not support the hypothesis that a significant increase in diameter accompanies a decrease in filament length. At the very most, only a 3-4 nm increase in diameter is seen in and only in some short filaments. Neither have we found evidence of an increase in the number of crossbridges per crown in the short filaments. Clearly, we have analyzed the surface structure of relatively few filaments in a systematic way and much additional work, including computer image reconstruction, must be done in order to verify these observations. Nevertheless, it is puzzling and somewhat disappointing not to be able to discern obvious surface-structural differences between the long and short filaments that would easily account for the decrease in length that we consistently observe. It may be that the evidence of disorder in the myosin helix that we see on transforms including the swollen segments of many of the short filaments represents the structural change that we seek. If filament shortening is accomplished with a shift from an ordered to a disordered structure we may not be able to determine the precise change in myosin packing that occurs. If there is really no change in the myosin helix with filament shortening, however, we are left with the unsatisfying possibility that disaggregation and reaggregation of myosin and paramyosin may produce the observed conformational change in *Limulus* thick filaments.

ACKNOWLEDGEMENTS

The authors wish to thank H. King and K. McGlynn for technical and K. Golden for secretarial assistance. We are indebted to Drs. R. Craig, R. Freeman, J. Haselgrove, P. Vibert and J. Wray for helpful discussion and expert advice. This work was supported by USPHS grants GM 07475, AM 30442, AM 14317 and HL 15835 to the Pennsylvania Muscle Institute and Projekt GO 28414-7 from the Federal Republic of Germany.

REFERENCES

Barnes, R.D. (1969). The Chelicerates. In: *Invertebrate Zoology*. 2nd Ed. Philadelphia: W.B. Saunders Co., Chapt. 13.

Davidheiser, S., Levine, R.J.C. and Davies, R.E. (1982). Two different fiber types in *Limulus* telson muscle. Fed. Proc. 41: 1522 (abstract).

Dewey, M.M., Levine, R.J.C., Colflesh, D.E., Walcott, B., Brann, L., Baldwin, A. and Brink, P. (1979) Structural changes in thick filaments during sarcomere shortening in *Limulus* striated muscle. In: *Crossbridge Mechanism in Muscle Contraction* pp. 3-22, ed. Sugi, H. and Pollack, G.H. Tokyo: University of Tokyo Press.

Dewey, M.M., Walcott, B., Colflesh, D.E., Terry, H. and Levine, R.J.C. (1977). Changes in thick filament length in *Limulus* striated muscle. J. Cell Biol. 75: 366-380.

Elfvin, M., Levine, R.J.C. and Dewey, M.M. (1976). Paramyosin in invertebrate muscles. I. Identification and localization. J. Cell Biol. 71: 261-272.

Kensler, R.W. and Levine, R.J.C. (1982a). An electron microscopic and optical diffraction analysis of the structure of *Limulus* telson muscle thick filaments. J. Cell Biol. 92: 443-451.

Kensler, R.W. and Levine, R.J.C. (1982b). Determination of the hand of the crossbridge helix on *Limulus* thick filaments. J. Mus. Res. Cell. Motil. 3: 349-362.

Kensler, R.W., Levine, R.J.C., Reedy, M. and Hofman, W. (1982). Arthropod thick filament structure. Biophys. J. 37: 34a (abstract).

Levine, R.J.C., Elfvin, M., Dewey, M.M. and Walcott, B. (1976). Paramyosin in invertebrate muscles. II. Content in relation to structure and function. J. Cell Biol. 71: 273-279.

Levine, R.J.C. and Kensler, R.W. (1982) Platinum shadowing of *Limulus* thick filaments. Biophys. J. 37: 50a (abstract).

Levine, R.J.C., Kensler, R.W., Stewart, M. and Haselgrove, J. (1982). Molecular organizaton of *Limulus* thick filaments. In: *Basic Biology of Muscles: A Comparative Approach* pp. 37-51, ed. Twarog, B.M., Levine, R.J.C. and Dewey, M.M. New York: Raven Press.

Sherman, R.G. and Luff, A.R. (1971). Structural features of the tarsal claw muscles of the spider *Eurypelma marxi* Simon. Can. J. Zool. 49: 1549-1556.

Stewart, M., Kensler, R.W. and Levine, R.J.C. (1981). Structure of *Limulus* telson muscle thick filaments. J. Mol. Biol. 153: 781-790.

Vibert, P.J. and Craig, R. (1982). Three-dimensional reconstruction of scallop thick filaments. Biophys. J. 37: 266a (abstract).

Wray, J.S. (1982). Organization of myosin in invertebrate thick filaments. In: *Basic Biology of Muscles: A Comparative Approach.* Ed. Twarog, B.M., Levine, R.J.C. and Dewey, M.M. New York: Raven Press.

Wray, J.S., Vibert, P.J. and Cohen, C. (1974). Cross-bridge arrangements in *Limulus* muscle. J. Mol. Biol. 88: 343-348.

Wray, J.S., Vibert, P.J. and Cohen, C. (1975). Diversity of cross-bridge arrangements in invertebrate muscles. Nature (Lond.) 257: 561-564.

DISCUSSION

POLLACK: You mentioned that you get a specific region along the thick filament that seems to be swollen. Is it possible to infer that the shortening of the thick filament then occurs at a specific *region* along the thick filament, possibly where it gets thicker?

LEVINE: Getting thicker is sort of subjective. I can't guarantee that what looks thicker really is, because when you go ahead and measure on those diffraction patterns where you can, the radial location of the mass of the cross-bridge is not, on average, different from what it is on the long thick filaments. In some cases the disorder is very great in that area and it may be that there is a loosening of the lattice.

HUXLEY: Are all the examples you showed actually from muscles which were contracted in high potassium?

LEVINE: One of those was electrically stimulated and the other was potassium stimulated.

HUXLEY: Were the helical parameters that you measured, i.e., the 143 Å and the 429 Å spacings, the same or different in the two types of filament?

LEVINE: There is no difference in spacing between short filaments and long filaments, as long as they had measurable diffraction patterns. Nor was there a difference between filaments isolated from muscles that had been stimulated with high potassium and those stimulated electrically. The only thing we were *possibly* thinking of is that there may have been a *small* diameter change on the order of 4 nm, which is possible from some of the measurements. Also, there may have been an increase in strandedness -- the number we actually came out with was 5.4, which is very, very close to 5.32, which is the Bessel function expected for a 4-stranded structure.

HUXLEY: The thick filaments are shorter, but have they shortened?

LEVINE: Well, I don't know. There seems to be a change in the filament surface appearance that is *really noticeable*.

HUXLEY: Yes, but one is talking about filaments that are supposed to have shortened from five microns down to three microns.

LEVINE: Most of the short filaments we looked at were about 3.2 μm and the mean longs run around 4.1 μm (see- histogram figure in paper).

HARRINGTON: You've demonstrated shortening *in vitro*. Is all the mass associated with the filaments? In other words, are you throwing mass off at the edge of the filaments? Is there myosin in the supernatant, for example?

LEVINE: We had less success in working with filaments shortened *in vitro* because we evidently had some contaminating actin left. Unless you catch it really fast after adding Ca^{2+}, it's become a mass of actomyosin. We do not believe we could catch *in vitro* shortening fast enough because the thick and thin filaments were found meshed together very frequently. On those we *have* been able to look at, we have no disagreement with the distribution of filament lengths: it's very similar to what I've just shown on the short filaments isolated from stimulated muscle.

We also attempted mass measurements of thick filaments with the STEM. We made STEM measurements of mass/crown on the shortened filaments from stimulated fibers and also on long filaments. The short filaments came out having less mass per crown than the long ones did. And on our negatively stained preparations (preparations for STEM mesurements have to be floated on distilled water so you won't get ions contaminating the filaments) there seems to be a loosening of structure in the shortened filaments because we have seen very scattered cross-bridge structure on those that we examined. This explains why the shortened filaments would be *decreased* in mass per unit length. So I think there is a loosening of filament structure that has gone on. Now if thick filament shortening can occur by molecules falling off a specific area - myosin and paramyosin coming off the ends, that's possible and I can't rule it out -- I don't like it very much, but I can't rule it out.

DEWEY: You can. We've repeatedly looked at the supernatants in our isolated filaments, and the myosin is not there. So you run a TCA precipitate and you get very low molecular weight things. The myosin comes down with both the long and the short filaments.

HARRINGTON: But the myosin could have been taken off the ends and reassembled into additional similar filaments very rapidly. In other words, if the length of the filament is established under the given ionic conditions, what you may be doing is taking mass off the long filaments and making shorter filaments from that.

DEWEY: That's possible, but it doesn't seem very likely.

ROWE: Could I just make one point on that. It should, I would have thought, have been fairly simple to test the hypothesis that extra filaments are formed simply by using some type of EM counting device and a constant concentration of long and short filaments, checking the number in a given area. You should neither have created nor destroyed the number of filaments per unit volume.

WINEGRAD: Are those *in vitro* shortenings with calcium reversible? That is, can you reverse the shortening *in vitro* by removing calcium?

DEWEY: The only way we've been able to relengthen the filament *in vitro* is by the treatment with alkaline phosphatase. That's published some years ago. They relengthen -- they go back to about 4.5 to 4.9 μm something like that. You can repeatedly do it -- shorten them, lengthen them, take the same sample, shorten it again and it will relengthen.

CINEMATOGRAPHIC STUDIES ON THE A-BAND LENGTH CHANGES DURING Ca-ACTIVATED CONTRACTION IN HORSESHOE CRAB MUSCLE MYOFIBRILS

Haruo Sugi and Setsuko Gomi

Department of Physiology, School of Medicine, Teikyo University, Itabashi-ku, Tokyo 173, Japan

ABSTRACT

Cinematographic recordings of sarcomere shortening were performed on glycerinated horseshoe crab muscle myofibrils during Ca-activated contraction. When the preparations at slack length (sarcomere length, 7-9 μm) were locally activated with iontophoretically applied Ca ions, the A-band length did not change appreciably while the activated sarcomeres shortened linearly with a velocity similar to the maximum shortening velocity measured on intact muscle fibers. If, on the other hand, previously stretched preparations (sarcomere length, 11-14 μm) were locally activated, the A-band length first increased by 40-50% and then shortened to the initial length, while the activated sarcomeres continued to shorten. These results indicate that the thick filament shortening may not be associated with the physiological sarcomere shortening; the transient A-band lengthening with long initial sarcomere lengths may result from the transient misalignment of the thick filaments followed by their realignment, implying that the force exerted by the cross-bridge is not constant but may vary according to its past history.

INTRODUCTION

Horseshoe crab striated muscle is known to exhibit shortening of the A-band associated with shortening of the thick filament under some conditions, and this phenomenon has been studied extensively by many investigators (de Villafranca and Marschhaus, 1963; Dewey, Levine and Colflesh, 1973; Dewey, Baldwin, Colflesh, Walcott and Brink, 1977; Dewey, Levine, Colflesh, Walcott, Brann, Baldwin and Brink, 1979). The thick filament shortening suggests the possibility that, contrary to the general view that contraction in striated muscle results only from a relative sliding between the thick and thin filaments (Huxley & Niedergerke, 1954; Huxley and Hanson, 1954), it serves as an additional mechanism involved in producing muscle force and motion. In the above literature, however, the A-band and the thick filament shortening seems not to have been studied under the physiological condition; in de Villafranca and Marshhaus's experiments, the ATP-induced sarcomere shortening was extremely slow, taking many seconds, while the A-band and the thick filament length measurements have been made by Dewey and his co-workers after fixing muscles with glutaraldehyde.

The present experiments were undertaken to examine whether the thick filament shortening actually takes place during the physiological contraction, by cinematographically recording sarcomere shortening in glycerinated horseshoe crab muscle myofibrils under various conditions.

METHODS

All experiments were performed with telson depressor muscles isolated from Japanese horseshoe crabs (*Tachypleus tridentatus*). The muscles were glycerinated by the method of Tanaka, Tanaka & Sugi (1979), and bundles of myofibrils (20-50 μm in diameter and about 1 cm in length) were dissected from the glycerinated muscle fibers in a relaxing solution containing 100 mM KCl, 3mM EGTA, 5 mM $MgCl_2$, 4 mM ATP and 10 mM histidine (pH 6.8), and mounted in a thin layer of the relaxing solution between a slide and a coverslip, one end being glued to the slide with Aron alpha adhesive (Sankyo) while the other end being glued to a glass rod carried by a micromanipulator. The preparation was observed under a Nikon phase-contrast microscope (Nikon DM 40 objective, 40×, n.a. 0.64), and locally activated to contract with iontophoretically applied Ca ions through a glass pipette (diameter, 5-15 μm) filled with 1 M $CaCl_2$. The strength and the duration of current pulses applied to the pipette was 1 μA and 1-3 sec respectively. The initial sarcomere length of the preparation, which could be varied with the micromanipulator, was fairly uniform along its length. The local contractions were recorded on Kodak 7247 or Fuji Minicopy films with a 16 mm cine-camera (Blex or Locam) at 64-100 frames/sec (Figs. 1, 2 and 4). The length changes of a single sarcomere and its A-band were measured during the course of contraction either directly on the cinefilm with a film motion analyser (Vanguard), or indirectly after microdensitometer (Narumi, type NLM-D3) tracings of the cinefilm with similar results. All experiments were performed at room temperature (18-22°C).

RESULTS

Absence of the A-band Length Changes During Local Contraction at Short Initial Sarcomere Length

Fig. 1 shows typical results of the experiments in which the myofibril bundle preparations were locally activated with iontophoretically applied

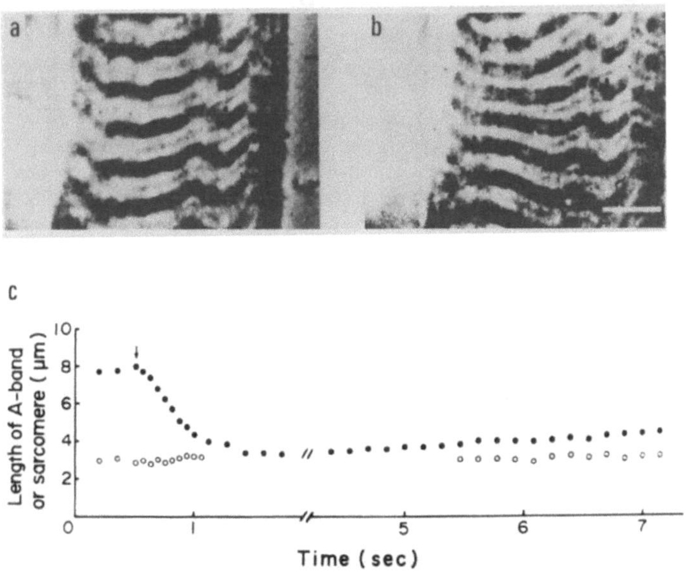

Figure 1: Sarcomere shortening in horseshoe crab muscle myofibril bundle preparation at a short initial sarcomere length (about 8 μm). a and b: Selected frames from a cinefilm of the preparation at rest (a) and during local contraction produced by iontophoretic application of Ca ions (b). Scale bar, 10 μm. c: Time course of length changes of a single sarcomere (filled circles) and its A-band (open circles) during local contraction. Arrow indicates the beginning of Ca application. Note the constancy of the A-band length during marked sarcomere shortening and subsequent relaxation. Modified from Sugi and Gomi (1981).

Ca ions at their slack length (Fig. 1a, b), the initial sarcomere length ranging from 7 to 9 μm (Sugi and Gomi, 1981). As shown in Fig. 1c, no appreciable change in the A-band length was observed during the course of resulting local contraction and subsequent relaxation. The early sarcomere shortening from 7-10 to 3.5-4 μm was taken up only by the I-band shortening, while the A-band length (3-4 μm) remained virtually unchanged. With further sarcomere shortening to less than 3.5 μm, the A-band was no longer visible because of the contraction band formation (Hodge, 1956). On relaxation, however, the A-band became again visible with sarcomere lengths above 3.5 μm, and its length remained unchanged until the completion of relaxation.

The above local contractions involved at most ten adjacent sarcomeres, while the rest of the preparation (about 1 cm in length) was at the slack length. Under such conditions, the activated sarcomeres are considered to shorten against a very small load (Podolsky, 1964; Sugi, 1974). As a matter of fact, the velocity of linear sarcomere shortening at the early phase of local contraction (about 8 μm/sec) was similar to the maximum shortening velocity (about 1 length/sec) determined on intact muscle fibers at their slack length (Gomi and Sugi, 1979), indicating that the preparation was in a good physiological condition.

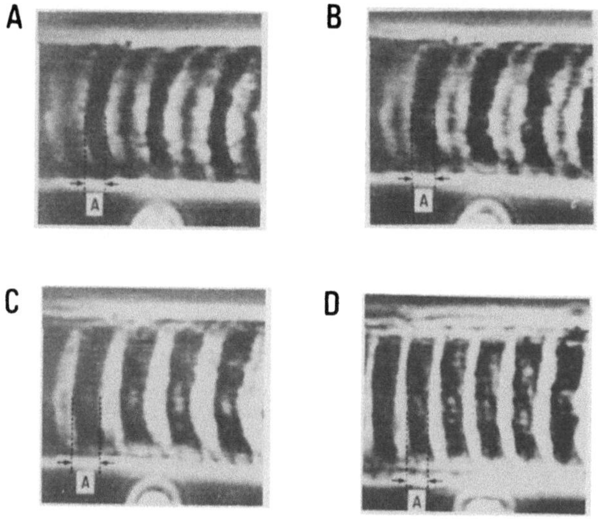

Figure 2: Selected frames from a cinefilm showing the transient A-band lengthening during local contraction at a long initial sarcomere length (about 13 μm). A: preparation at rest. B, C and D: frames taken at 0.25 sec, 0.9 sec and 2.8 sec after the beginning of local contraction respectively. Scale bar, 15 μm. The A-band is indicated in each frame. Note the lengthening of the A-band followed by shortening.

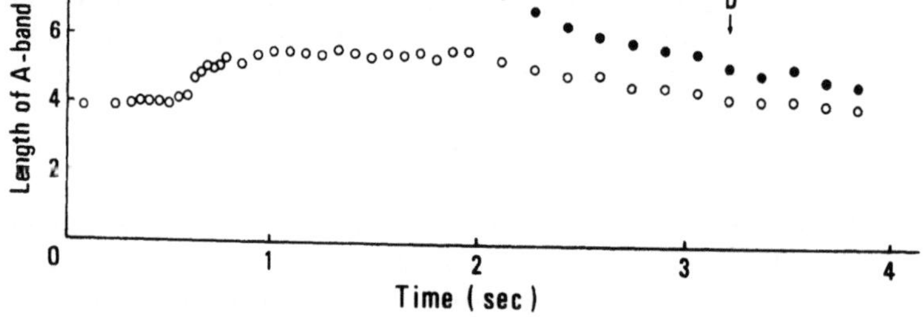

Figure 3: Time course of length changes of a single sarcomere (filled circles) and its A-band (open circles) during local contraction at a long initial sarcomere length (about 13 μm). Data points were obtained from the cinefilm shown in Fig. 2. Arrows indicate the times when the frames A, B, C and D in Fig. 2 were taken. Note the initial transient A-band lengthening by about 40%, while the sarcomere shortens linearly.

Transient A-Band Lengthening During Local Contraction at Long Initial Sarcomere Lengths

The experiments were further performed in which the preparations were previously stretched beyond their slack length, so that Ca-activated local contractions were produced at the initial sarcomere length of 11-14 μm. The A-band length in the stretched preparation was 3.5-4.5 μm, being only slightly larger than that in the preparation at its slack length (see Fig. 5). A typical result is shown in Figs. 2 and 3. In response to iontophoretically applied Ca ions, the A-band exhibited initial lengthening (by 40-50%) which was followed by subsequent shortening to its initial length, while the activated sarcomeres only showed continuous shortening (Fig. 2). As shown in Fig. 3, the A-band length first increased from 4 to 5.5 μm, at the beginning of the local contraction, and the increased A-band length was maintained until the sarcomere shortened from 13 to 7 μm; with further sarcomere shortening, the A-band started to shorten to its initial length. In spite of the above transient A-band lengthening, the sarcomere shortening at a long initial length also took place with a nearly constant velocity (Fig. 3) as was the case in the sarcomere shortening at a short initial length (Fig. 1C). During relaxation, the A-band length did not change appreciably until the sarcomere restored its initial length.

Increase in the A-Band Length During Isometric Force Generation

Fig. 4 shows typical results of the experiments in which the distance between the fixed ends of the preparation was made very short (100 μm or less), so that all the sarcomeres within the short segment could be activated fairly uniformly by iontophoretic application of Ca ions (Fig. 4a, b). The initial sarcomere length ranged from 7 to 11 μm. Since there were no non-activated sarcomeres which could be readily stretched by the activated ones, the speed and extent of the resulting sarcomere shortening were very limited compared to the experiments described previously. As shown in Fig. 4c, the A-band length increased gradually by about 10% during the course of slow, small sarcomere shortening, reaching a steady value when the sarcomere shortening stopped due to the balance in force between the activated sarcomeres.

DISCUSSION

Lack of Evidence for the Thick Filament Shortening During the Physiological Contraction in Horseshoe Crab Striated Muscle

The present experiments have shown that (1) the A-band does not shorten significantly during sarcomere shortening from 7-9 μm (Fig. 1); (2) the A-band does not shorten but exhibits transient lengthening (by 40-50%) during sarcomere shortening from 11-14 μm (Figs. 2 and 3); and (3) the A-band length increases (by about 10%) when sarcomeres are activated in a nearly isometric condition (Fig. 4). Thus, the A-band shortening was never observed in the present study. Since the myofibril bundle preparations used are believed to be in a good physiological condition,

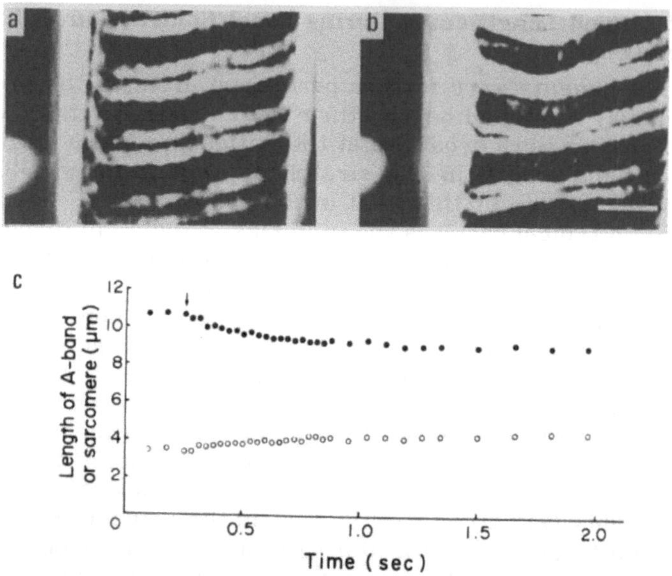

Figure 4: Increase in the A-band length when a very short segment of myofibrils with fixed ends was activated fairly uniformly with Ca ions. a and b: Selected frames from a cinefilm of resting preparation (a) and of contraction in response to Ca ions (b). Scale bar, 10 μm. c: Time course of length changes of a single sarcomere (filled circles) and its A-band (open circles) during the slow and small sarcomere shortening at the middle of the segment. Note that the A-band length increases gradually while the sarcomere shortens slowly. Modified from Sugi and Gomi (1981).

the present results indicate that the thick filament shortening may not be associated with the physiological sarcomere shortening.

According to Dewey and his coworkers (Dewey et al., 1973; 1979), the A-band length in relaxed horseshoe crab muscle increases nearly twofold with increasing sarcomere length from 6 to 12 μm, partly by the thick filament lengthening and partly by increasing degree of misalignent of the thick filaments within the A-band. They measured the A-band and the thick filament lengths after fixing the muscle with glutaraldehyde at various clamped muscle lengths. In the present study, we measured the A-band length under the phase-contrast microscope when the myofibril bundle preparation was stretched to a variable extent, and found a much smaller sarcomere length dependence of the A-band length (Fig. 5). Since glutaraldehyde causes activation of the contractile mechanism in horseshoe crab muscles (Gomi and Sugi, unpublished), it seems possible that the marked A-band lengthening such as shown in Figs. 2 and 3 takes place when the muscle with long sarcomere lengths is fixed with glutaraldehyde to result in an apparent marked increase in the A-band length.

Dewey et al. (1973, 1979) have shown that the increased A-band length at long sarcomeres is largely due to the misalignment of the thick filaments within the A-band. Thus, the transient A-band lengthening

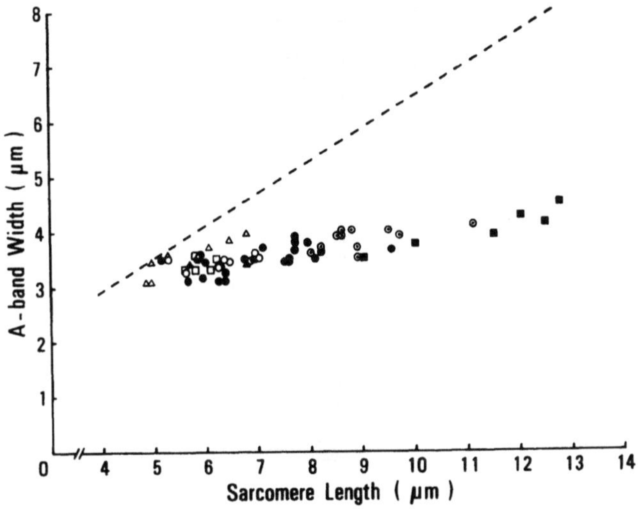

Figure 5: Relation between the A-band and the sarcomere lengths in resting myofibril bundle preparations. Different symbols indicate different preparations. Broken line shows the A-band length versus sarcomere length relation reported by Dewey et al., (1973). The preparations were stretched and released slowly in a random fashion.

observed at long sarcomere lengths (Figs. 2 and 3) may be accounted for in terms of the thick filament misalignment followed by realignment. This will be discussed later in connection with the mode of operation of the cross-bridges responsible for muscle contraction.

The increase in the A-band length in nearly isometrically activated sarcomeres (Fig. 4) seems to be explained, on the other hand, as being due to the elastic extension of the thick filaments (possibly at the H-zone, Sugi and Suzuki, 1980) by the force generated in the activated sarcomeres, since the degree of the A-band lengthening (about 10%) can well account for the extension of the series elastic component in intact muscle fibers at P_0 (Sugi and Gomi, 1981).

Mode of Operation of the Cross-Bridges as Suggested by the Transient A-Band Lengthening

The transient A-band lengthening observed at long sarcomere lengths may be qualitatively explained in the following way (Fig. 6):

(1) At long sarcomeres, the overlap between the thick and thin filaments is very small, with some degree of misalignment of the thick filaments within the A-band (Fig. 6A);

(2) On activation, the thick filament misalignment increases as the two filaments slide past each other, to results in the A-band lengthening, while the activated sarcomere shortens continuously (Fig. 6B, C);

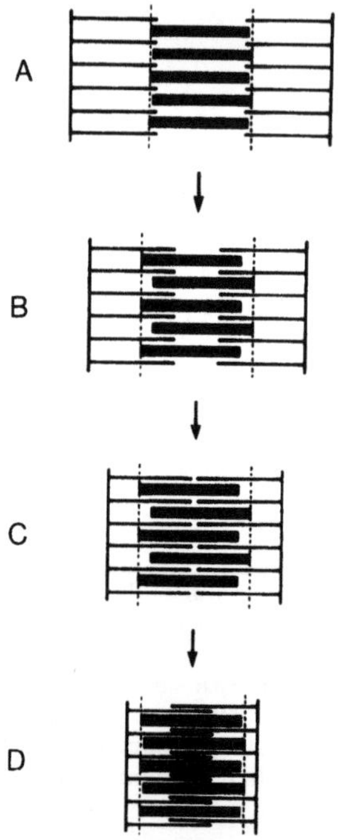

Figure 6: Diagram illustrating the transient A-band lengthening in terms of the thick filament misalignment and realignment. For further explanations see text.

(3) The realignment of the thick filament takes place with further sarcomere shortening to result in the A-band shortening to its initial length (Fig. 6D).

It is generally assumed that each cross-bridge exerts the same force during contraction, so that the isometric force in a muscle fiber is proportional to the amount of overlap between the filaments (e.g., Gordon, Huxley and Julian, 1966). At long sarcomere lengths, a slight misalignment of the thick filaments causes a relatively large difference in the amount of overlap between the two halves of each thick filament, and the thick filament misalignment increases on activation due to the difference in the number of operating cross-bridges; one half of the thick filament with more overlap than the other exerts more force, so that the thick filament tends to move towards the stronger side (Fig. 6A, B). Marked thick filament misalignment may readily take place in arthropod muscles which show no distinct M-line structures connecting the adjacent thick filaments at the center of the A-band. When a certain degree of the thick filament misalignment is reached, further development of the filament misalignment stops (Fig. 6C) probably due to a mechanism of longitudical stability, with which one half of the thick filament with small overlap can resist the force exerted by the other half with large overlap (e.g., Edman, Elzinga and Noble, 1978).

In this connection, it is of interest that, after the thick filament misalignment reaches its maximum, the thick filament realignment takes place while the sarcomere shortening is still in progress (Fig. 6D). This indicates that one side of the thick filament with small overlap can exert more force than does the other side with large overlap, so that the thick filament moves to return to its initial position at the center of the sarcomere. Thus, the thick filament realignment implies that the mode of operation of the cross-bridges is not constant, but changes markedly depending on their past history in such a way that a small number of the cross-bridges not only can resist the force exerted by a large number of the cross-bridges, but also can generate a larger force to pull the thick filament back to its initial central position.

ACKNOWLEDGEMENTS

We wish to thank T. Taniguchi, M. Kusama and M. Goto for their technical assistance in taking the cinematographic records.

REFERENCES

de Villafranca, G.W. and Marschhaus, C.E. (1963). Contraction in the A-band. J. Ultrastruct. Res. 9: 156-165.

Dewey, M.M., Levine, R.J.C. and Colflesh, D.E. (1973). Structure of *Limulus* striated muscle. The contractile apparatus at various sarcomere length. J. Cell Biol. 58: 574-593.

Dewey, M.M., Levine, R.J.C., Colflesh, D.E., Walcott, B., Brann, L., Baldwin, A. and Brink, P. (1979). Structural changes in thick filaments during sarcomere shortening in *Limulus* striated muscle. In: *Cross-bridge Mechanism in Muscle Contraction*, pp. 3-22, ed. Sugi, H. and Pollack, G.H. Tokyo: University of Tokyo Press and Baltimore: University Park Press.

Dewy, M.M., Walcott, B., Colflesh, D.E., Terry, H. and Levine, R.J.C. (1977). Changes in thick filament length in *Limulus* striated muscle. J. Cell Biol. 75: 366-380.

Edman, K.A.P., Elzinga, G. and Noble, M.I.M. (1978). Enhancement of mechanical performance by stretch during tetanic contractions of vertebrate skeletal muscle fibres. J. Physiol. 281: 139-155.

Gomi, S. and Sugi, H. (1979). Mechanical properties of horseshoe crab skeletal muscle. Zool. Mag. 88: 509.

Gordon, A.M., Huxley, A.F. and Julian, F.J. (1966). The variation in isometric tension with sarcomere length in vertebrate muscle fibres. J. Physiol. 184: 170-192.

Hodge, A.J. (1956). The fine structure of striated muscle. J. Biophys. Biochem. Cytol. 2: Suppl. 131-142.

Huxley, A.F. and Niedergerke, R. (1954). Interference microscopy of living muscle fibres. Nature, Lond. 173: 971-973.

Huxley, H.E. and Hanson, J. (1954). Changes in the cross-striations of muscle during and stretch and their structural interpretation. Nature, Lond. 172: 973-976.

Podolsky, R.J. (1964). The maximim sarcomere length for contraction of isolated myofibrils. J. Physiol. 170: 110-123.

Sugi, H. (1974). Inward spread of activation in frog muscle fibres investigated by means of high-speed microcinematography. J. Physiol. 242: 219-235.

Sugi, H. and Gomi, S. (1981). Changes in the A-band width during contraction in horseshoe crab striated muscle. Experientia 37: 65-67.

Sugi, H. and Suzuki, S. (1980). Extensibility of the myofilaments in vertebrate skeletal muscle as studied by stretching rigor muscle fibres. Proc. Japan Acad. 56B: 290-293.

Tanaka, H., Tanaka, M. and Sugi, H. (1979). The effect of sarcomere length and stretching on the rate of ATP splitting in glycerinated rabbit psoas muscle fibers. J. Biochem. Tokyo 86: 1587-1593.

DISCUSSION

LEVINE: I wonder, can you give me, Dr. Sugi, the standard A-band width? I was confused by the fact that, when shortening started from long sarcomere lengths, the final A-band width finally attained was around 4 μm (Fig. 3), while the A-band remained around 3 μm during the course of shortening from the resting sarcomere length (Fig. 1).

SUGI: In preparations at slack length, the sarcomere length was mostly 7-8 μm and the A-band width was about 3 μm. If you stretch the sarcomeres beyond 12-13 μm, the A-band width increased from 3 to about 4 μm (Fig. 5). I would like to emphasize that the above sarcomere length dependence of A-band width is very small compared to Dewey's published literature.

LEVINE: I thought that you were going from long sarcomeres, that is, from 13 or 14 μm down to less than 7 μm (Fig. 3). Then you had another series of experiments where you started from 7 or 8 μm and went down further (Fig. 1). It seemed to me that, in the two different instances, the A-band width at 7 μm differed.

SUGI: The A-band width does show some variation at the same sarcomere length. It is 3 μm in some fibers, and 3.5 or 4 μm in others. This may probably be due to some variation in the degree of thick filament misalignment within the A-band.

CURTIN: In your experiments, is there evidence that there is actual ATP turnover during shortening? Do you have a system for reconstituting ATP in your solutions?

SUGI: An ATP regenerating system was not included in our solutions. Since the diameter of the myofibrillar bundle was very thin, 20 μm or less, I think that the regenerating system is not essential in our experiments.

HUXLEY: Responding to the earlier point, the other possibility that might explain why you see different A-band widths at the same sarcomere length is that the A-filaments really are genuinely different in length in different fibers that you happen to look at.

SUGI: Yes, that is possible.

POLLACK: I am wondering whether some of the differences that you observe relative to Dewey's observations might have to do with differences of load. For example, are the conditions in your local activation experiments such that the activated region of the fiber is shortening basically without load?

SUGI: No. At long sarcomere lengths, there is considerable resting tension in the preparation against which local shortening takes place.

TREGEAR: Did you do experiments in which you started from very long sarcomere length and went down to contraction bands? If so, does the A-band length appear to stay constant throughout the sarcomere shortening?

SUGI: The duration of the current pulse for Ca application was 1-3 seconds, so that relaxation always started several seconds after the onset of Ca application. For this reason, together with the presence of considerable resting tension providing resistance against rapid and extreme sarcomere shortening, the Ca-activated long sarcomeres did not shorten enough that the contraction band formation could take place.

TIROSH: Could you comment on the possibility that myosin filaments are rather flexible?

SUGI: In our opinion, the thick filaments may be flexible in a sense that they can be a source of the series elasticity in muscle.

CECCHI: You said that in the preparation there is a lot of compliance, much more than in frog muscle. Does this mean that the compliance is not all in the cross-bridges?

SUGI: It is a very important question. I think that the compliance in horseshoe crab muscle is too large to be explained by cross-bridge compliance. Much of it may originate from the H-zone compliance or from thick filament misalignment. I have evidence for the H-zone compliance in vertebrate skeletal muscle, too.

POLLACK: If the H-zone or some other region of the thick filament is compliant, it is difficult for me to understand how it could sustain a load during normal contraction without changing its length.

SUGI: I think that some region of the thick filament in horseshoe crab muscle may be extended by about 10% when the force rises from zero to the maximum value. This doesn't necessarily contradict the idea of force generation by cross-bridges.

LEVINE: I just want to make a general comment. I know you are working with a different animal, and even a different muscle, but when we take filaments out of unstimulated muscle, the length range we get is from 3.5 μm up to over 5 μm. We have never seen unbroken filaments shorter than 3.6 μm.

HUXLEY: What is the shortest A-band length that you are measuring?

SUGI: The shortest was about 3 μm.

DEWEY: That's about right.

HUXLEY: Those are in fully-shortened sarcomeres, where the I-band disappears and where you are beginning to get a sign of contraction bands. Right?

SUGI: Yes. At slack length of the preparation, the A-band width is about 3 μm, and it remains the same until immediately before contraction band formation.

HUXLEY: So, the A-bands are probably squashed into contraction bands.

MAGID: When you stretched myofibrils in rigor, they showed an increase in the A-band width. Did you ever look at the effect of releasing the stretched rigor preparations to see whether the A-band width recovered?

SUGI: When we stretched rigor preparations, sometimes rupture took place out of the microscopic field, and we observed rapid elastic restoration of A-band width which had been increased by stretch.

MAGID: Do you think that the elastic restoring force is in the non-overlap zone of the A-filaments?

SUGI: Yes, I tend to think so, based on the electron microscopic evidence that the thick filament is extensible only at the H-zone (see Fig. D-2 in General Discussion on Structural Dynamics).

KRUEGER: I just wanted to return to something Jerry Pollack asked. Is the myofibrillar bundle producing a significant tension at the time when the contraction bands are forming?

SUGI: In these experiments, I started from slack length and only a very small part of the preparation was locally activated with Ca. Therefore, the activated sarcomeres shortened against a very small resistance to stretch in the non-activated part. As a matter of fact, the sarcomeres shortened at a velocity equal to the maximum shortening velocity at zero load observed in intact fibers (Fig. 1).

BRESSLER: I'd like to ask Professor Sugi how fast was your release?

SUGI: The force step required about two to five msec for completion, so it's not so rapid. In the case of a force-step it cannot be so rapid as in a length step. Anyway, in intact fibers the muscle shortening velocity is relatively slow, one length per second at 20°C, and so the imposed step is still much faster than the maximum shortening velocity. So, I think it's all right.

EDMAN: In a contractile system where we have repeated cycles of activity of cross-bridges having a relatively short working distance one would expect to have the same tension at a given sarcomere length, wherever one starts the contraction; this we found holds true of the vertebrate skeletal muscle. I wonder, have you checked that point in this *Limulus* muscle where you seem to have some thick filament shortening?

SUGI: I measured the length-tension relation in *Limulus* as Dewey did, and my result is very similar to his result already published (Walcott & Dewey, Cell Biol. 87: 204-208, 1980). A rather flat length-tension curve is obtained. The plateau of maximum tension occurs between seven and twelve or so microns.

EDMAN: Yes, but is the tension the same at a given sarcomere length on the ascending limb, independent of where you initiate the contraction? It should forget the starting length.

SUGI: In several experiments we looked into this, but the results were not so clear.

CONTRACTION BANDS: DIFFERENCES BETWEEN PHYSIOLOGICALLY VS. MAXIMALLY ACTIVATED SINGLE HEART MUSCLE CELLS

John W. Krueger and Barry London

Depts. of Medicine and Physiology/Biophysics, The Albert Einstein College of Medicine, 1300 Morris Park Ave., Bronx, NY 10461

ABSTRACT

High resolution interference and phase microscopy were used to inspect the striations' appearance in shortening rat heart cells. Isolated cells were treated with detergent so that shortening could be graded by addition of calcium. Upon activation sarcomeres shortened to form (a) contraction densities in the middle of the A band at 1.7 um, (b) disappearance of the I bands and (c) phase brightening of the A bands at 1.6 um, and (d) dense C_z contraction bands at shorter lengths. These changes are totally consistent with the uniform sliding of myofilaments of previously accepted fixed dimensions. However, the striated patterns differed significantly in *intact* cells which were electrically stimulated to shorten. *Here* individual A bands remained distinct, without phase brightening or contraction band formation despite sarcomere shortening to less than the length of the A band as measured in the unstimulated cell. Maximal activation of intact cells by barium contracture elicited the full sequence of striation changes (a-d) seen in the chemically skinned cells. Light diffraction analysis gave comparable interpretation, i.e., the protein within the shortened sarcomere in the physiologically activated cardiac cell is more narrowly distributed than expected for thick filaments of fixed dimensions. These optical differences may reflect the restricted presence of the globular myosin heads at the *ends* of the cardiac sarcomere. This situation would explain the narrow range of the cardiac length-tension relation.

INTRODUCTION

Sliding of myofilaments (Huxley & Hanson, 1954; Huxley & Niedergerke, 1954) is now widely accepted as the mechanism of shortening in vertebrate striated muscle fibers. *Uniform*, aligned sliding of myofilaments of constant length then well accounts for the observed changes in a striation's appearance at longer sarcomere lengths. At short sarcomere lengths uniform sliding should result in the formation of contraction bands (C_z), i.e., localized increments in refractive index signaling the abutment of the ends of the thick filament at the z lines. However, contraction bands are not *always* seen at the short length predicted by the model (Huxley & Niedergerke, 1958), and the reason is not clear. The absence of contraction bands can be explained by buckling of myofibrils (Huxley & Gordon, 1962), but it may also be that the striated morphology of the shortened myofibril depends on activation.

Internal buckling is not seen in the actively shortened cardiac cell (Krueger, Forletti & Wittenberg, 1980), and so contraction must be graded within the cardiac myofibril. 50% of the cardiac sarcomere's functional range (Pollack & Krueger, 1976) lies below those lengths (i.e.,<2.0 μm) where *inactive* myofilaments are thought to resist sarcomere shortening (Brown et al., 1970). Thus, the striated appearance of the cardiac sarcomere holds information about (a) how activation affects the contractile lattice as well as (b) the constraints which might be placed upon the sliding filament model under these conditions.

In our initial studies, individual striations in isolated cardiac cells were observed by cinemicroscopy (40×, na 0.70 obj.) (Krueger et al., 1980). Surprisingly, the appearance of individual striations did not change despite uniform shortening to sarcomere lengths as short as 1.4 μm, i.e., lengths at which some evidence of contraction band should have been detected. This observation differs from expectations based on the accepted value of 1.55 to 1.65 μm for the length of the cardiac thick filament (Page, 1974). Therefore, we have extended our studies to clarify the microscopy of the shortened cardiac sarcomere.

METHODS AND RESULTS

Single cardiac muscle cells were enzymatically isolated from ventricles of adult (200-350 g) rats using procedures described elsewhere (Haworth, Hunter, & Berkoff, 1980). These cells were bathed in a physiological salt solution (composition [in mM]: NaCl 110, $NaHCO_3$ 25, KCl 4.8, $MgSO_4.7H_2O$ 1.2, KH_2PO_4 1.2, Glucose 11.0, Na Hepes Buffer, pH 7.3) which contained 0.5 to 2.0 mM Ca^{2+}.

The appearance of individual striations was inspected by phase contrast (100×, na 1.25 obj., na 0.70 cond.) and Jamin-Lebedeff Interference (100×, na 1.0 obj.) microscopy. . .i.e., forms of microscopy equivalent to those used in the original description of contraction band formation in skeletal muscle sarcomeres (Huxley & Hanson, Huxley & Niedergerke, 1954). The striations' image was recorded on 16 mm cine film or by videomicroscopy. Image blurring was eliminated by stroboscopic illumination synchronized with the sample rates of the television or cine cameras (60 to 200 Hz, respectively). An important feature of videomicroscopy is that image contrast could be maximized. This enabled phase contrast visualization of z bands at sarcomere lengths as short as 1.9 μm, H zones in detergent-treated cells stretched to sarcomere lengths >2.3 μm, and it also permitted use of full condensor aperture to maximize resolution with the interference system.

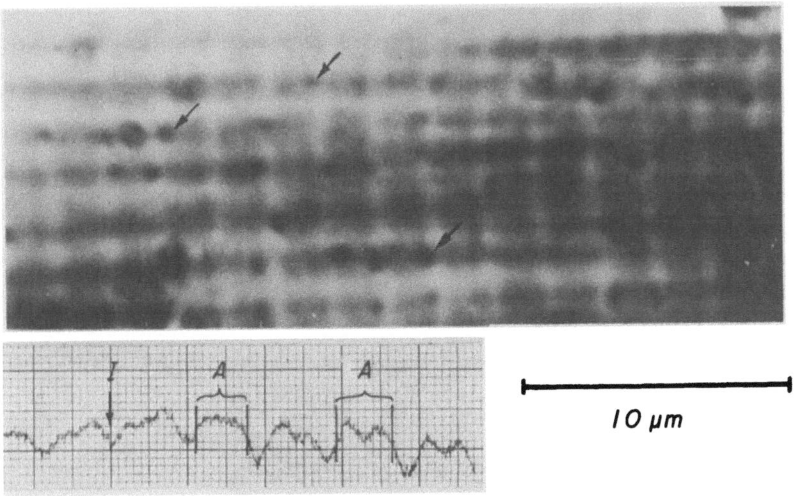

Figure 1: Interference micrograph of striations near edge of an intact heart cell. A-bands are dark, I-bands light. S.L.=1.95 µm. The faint spots (arrows) confined to the I-bands may be openings of transverse tubules, made visible on end by presence of refractive index solution. (Insert; uncalibrated). Intensity profile of sarcomeres in another cell viewed with the T.V. system. Dimensions of A-band, here bright, were taken as distance between the estimated mid-points of the peak A/I intensity changes. Areas of lowered protein concentration occur in middle of A-band which are also visible in micrograph. (S.L.=1.97 µm).

Fig. 1 illustrates an interference micrograph of the striations in an intact cell bathed in a refractive index matching solution (Hill, 1977). The intensity at any point in the image with this form of microscopy can be related to the relative disposition of protein. Measurements of A band length gave values of 1.43 ± 0.13 µm (N=6) and 1.55 ± 0.13 µm (N=9) with the compensator set to darken or brighten the A bands, respectively, at a sarcomere length of 1.97 µm. Phase contrast microscopy gave comparable dimensions in other cells (1.42 ± 0.07 µm, N=31). These values are consistent with previous high resolution measurements in rat isolated cardiac myofibrils (Fabiato & Fabiato, 1976), and indicate that full activation and sarcomere shortening should lead to C_z contraction band formation at sarcomere lengths of 1.6 to 1.5 µm.

Appearance of Striations in Chemically Skinned Cells.

Figure 2 illustrates the sequence of changes in the striated appearance of enzymatically isolated cells where shortening was graded by iontophoretic deposition of Ca^{2+}. Here isolated cells were bathed for >1 hr in 0.5% (w/v) sol'n of a nonionic detergent (Brij 58) also containing (in mM): KPr 140, Imidazole 5, MgAcetate 2, Na_2ATP 5, K_2EGTA 5, pH 7.2. Activation elicited sarcomere shortening from rest lengths of 1.90 ± 0.05 µm (N=37) to form invariably (a) contraction densities in the middle of the A bands at $1.7 < sl < 1.8$ µm, (b) disappearance of I bands and (c) brightening of A bands at sl=1.6 µm, and (d) dense C_z contraction bands at lengths shorter

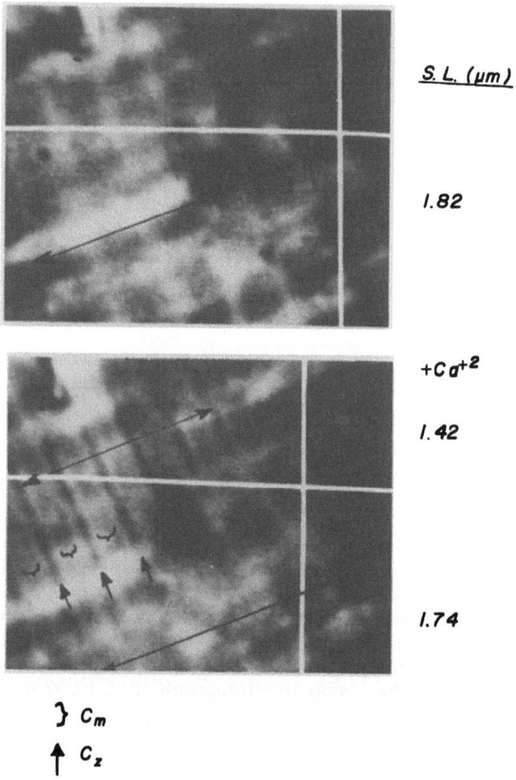

Figure 2: Phase contrast appearance of sarcomeres in detergent-treated cells. (Top) Micropipet for iontophoresis lies over an I-band in resting cell. (Bottom) Regional activation results in clear formation of contraction densities in center of A-band (C_m bands) and the more dense C_z contraction bands at ends of A-band. S.L. between the arrows is shown at the right. The existence of C_m bands suggests that thin filament lengths are quite uniform.

than 1.6 μm. At least two of these changes do not appear to be limited by microscopic resolution, since a comparable sarcomere length dependence of (c) & (d) were found with a 40×, na 0.60 obj lens. Brightening of only one half an A band could be seen at the edge of the activated group of sarcomeres, so that the occurrence of (c) cannot be due to longitudinal compression of the thick filament lattice.

Sarcomeres in the directly activated cells relengthened rapidly at the cessation of the 10-50 nA iontophoretic current, thereby allowing multiple activation and relaxation cycles. The lengths at which C_z contraction bands formed did not appear to depend upon prior activity providing that the sarcomeres did not shorten to <1.4 μm. Inclusion of a regenerating system (5 mM creatine phosphate & 70 U/ml CPK) appeared to speed relengthening but did not affect the results. Nor was the sarcomere length at which C_z contraction bands formed altered by variation of ionic strength (20<KPr<200 mM). (The rationale for these considerations become apparent in the next section.)

These results show that the appearance of the striations in the directly activated, freely permeable cell - which others have in part described in isolated cardiac myofibrils (Fabiato & Fabiato, 1976) - is entirely consistent with shortening by the sliding filament mechanism in its simplest form.

Striation Patterns in Cells with Intact Membranes.

Fig. 3 illustrates the appearance of the striations in cells bathed in physiologically salt solutions which were stimulated to shorten with a 1-2 ms electrical pulse. The striation patterns in these physiologically activated cells differ in several important respects from that sequence seen in the directly activated sarcomeres. Individual A bands remained distinct and could be followed without phase brightening during shortening. With interference microscopy, the retardation of the A bands (but *not* the I bands) changed during shortening suggesting protein shifts in A band alone. Finally, C_z contraction bands *did not form* despite sarcomere shortening equivalent to that seen in the directly activated sarcomere.

These "negative" findings, coupled with a 45% decrease in the protein content of the detergent-treated cell (refer to discussion), led us to ask if contraction bands could be seen in *intact* rat heart cells were they to exist. Intact cells were bathed in the physiological salt solution without added Ca^{2+}, and activated by iontophoretic deposition of Ba^{2+}. Low barium currents caused first a localized, gradable, and reversible shortening of sarcomeres within the intact cell. Higher iontophoretic currents caused rapid, synchronized and slowly reversible shortening throughout the whole cell superimposed on a localized contracture. This was followed by progressive and irreversible shortening. The shortening was characterized by loss of I bands, and phase brightening of A bands, and complete contraction band formation (Fig 3b) which characterized shortening in the detergent-treated cells. Importantly, interferometry showed that the mass increased at the Z-line region during contracture, and that the mean protein content of the cells did not fall. The most attractive explanation for the *absence* of contraction bands in the electrically stimulated cell is that activation affects the disposition of protein within the contractile lattice at short lengths.

Short A Bands in Detergent-Treated Cells

The converse situation was also explored: Detergent treated cells were preshortened in controlled calcium solutions (Fabiato & Fabiato, 1979). There appeared to be a great variation in the striations' appearance and length. Distinct A bands remained in some cells shortened to sarcomere lengths of 1.6 to 1.4 μm (Fig. 4). (Interferometry ruled out the possibility of a spurious phase reversal of Z band contrast.) Calcium iontophoresis then caused sarcomere shortening but contraction bands formed at respectively shorter sarcomere lengths than those seen before, thereby indicating that the protein disposition within the A band was initially narrowed.

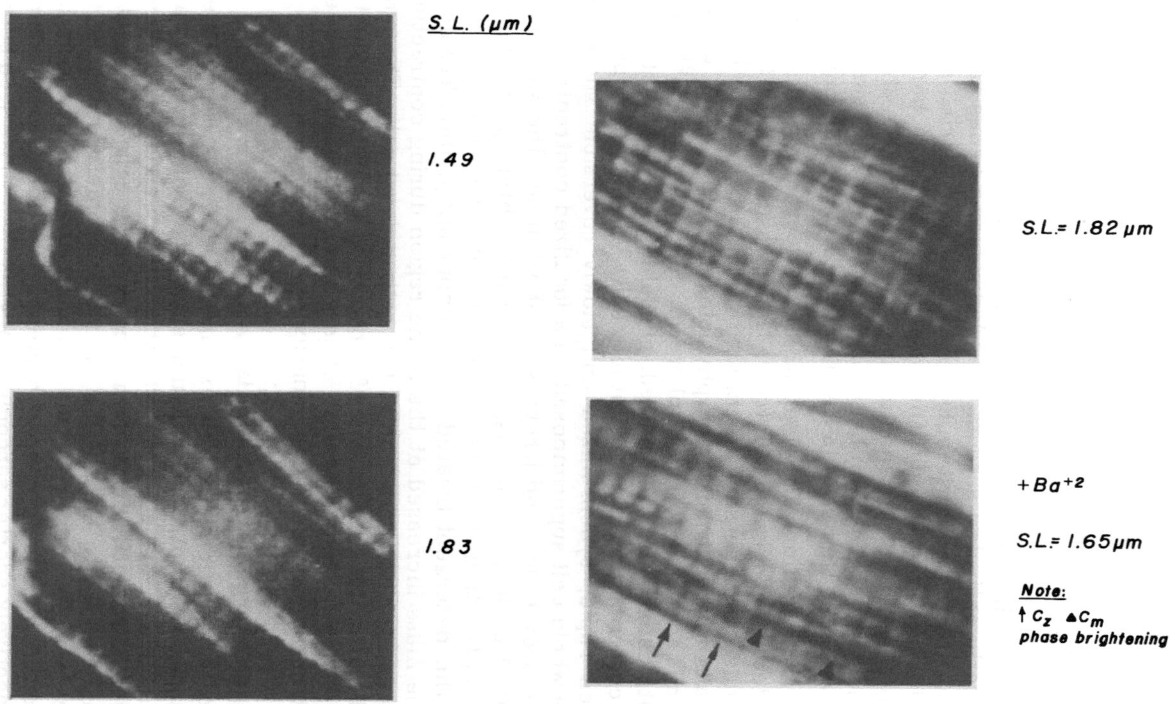

Figure 3: Striation patterns in intact cells. *Left*: Interference microscopy of striations at S.L.=1.83 and 1.49 μm during electrically stimulated contraction in the intact cell. Individual A-bands are bright and remained so throughout shortening and relengthening cycle (Negative compensation). *Right*: Field of sarcomeres in intact cell before and after shortening induced with Ba^{2+} iontophoresis in low Ca^{2+} solutions. Change in A-band intensity, C_m and C_z contraction bands are present (Phase contrast).

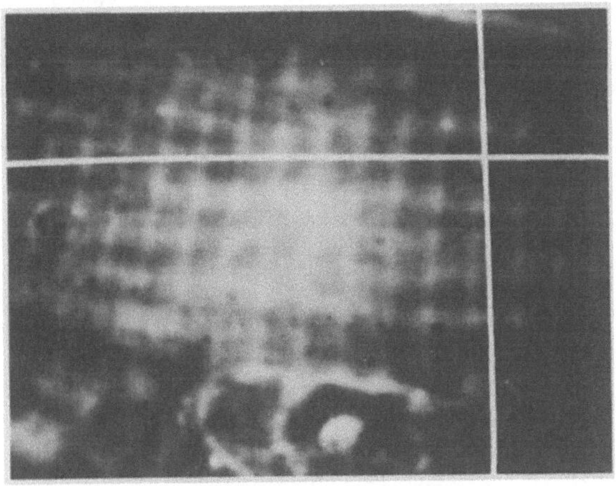

Figure 4: Phase contrast appearance of striations in skinned cell shortened in controlled calcium solution (pCa 6.75) which show no contraction bands and clear differences in the A/I widths (1.48<S.L.<1.62 μm).

Light Diffraction Analysis

Diffractometry provides a means of inferring the protein disposition in the sarcomere which is not confined to the specific striations examined within the highly magnified image section - estimated here to be <1 μm thick. An optical model (Fujime, 1975) of sliding filaments of fixed length predicts that the intensity of the first order spectra should diminish to near zero at sarcomere lengths of 1.6-1.5 μm provided that the contribution from thin filament scattering is less than 0.35 that of the thick filaments Moreover, double overlap of thin filament should not prevent the ultimate loss of the 1st order intensities even at higher scattering ratios.

The behavior of intensity spectra from directly activated, detergent treated cells fully supported this prediction (Fig. 5). Chemically skinned cells were cooled to 2-4°C with a thermoelectric stage to insure slow and fully uniform activation upon addition of calcium. While some variation was observed between individual first orders, each diminished to zero as the respective sarcomere length approached 1.6 μm and then, as predicted, rose to more intense values with further shortening.

This length-dependence of the diffraction intensities of skinned cells differs strikingly from the diffraction behavior observed in intact cells. The first order diffraction intensities from intact cardiac muscle and single cardiac muscle cells do not approach zero at short sarcomere lengths (Krueger et al., 1980). Volume diffraction effects (Rudel & Zite-Ferenczi, 1980) might explain the cell to cell variability of first order spectral intensity, but do not appear to explain the failure of the first order spectra from intact cells and muscle to disappear at the preaccepted length of the A band. Consequently, both direct microscopy in cells and diffraction in cells and muscle indicate that the respective distribution of protein or

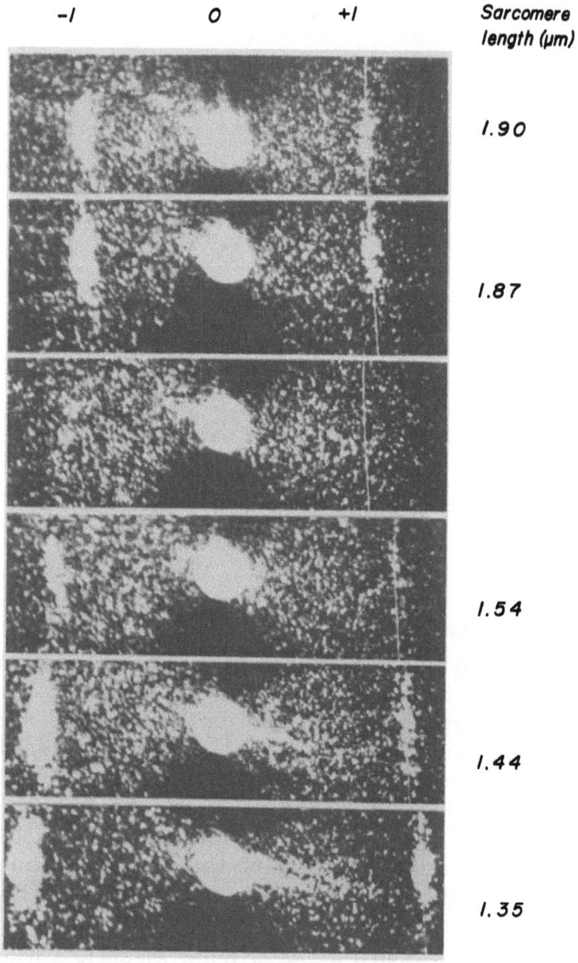

Figure 5: Appearance of diffraction spectra during shortening in a slowly activated detergent-treated cell. Individual first orders (±1) sometimes vary in intensity but always disappeared as their respective S.L. approached 1.6 μm and reappeared with intensity progressively increasing with further shortening (Numbers refer to average S.L.). First order diffraction intensities in physiologically activated cells do not disappear.

scattering elements within the physiologically shortened sarcomere is different than one would predict from a homogeneously activated and uniformly sliding population of myofilaments of previously accepted fixed dimensions.

DISCUSSION

Like others (Huxley & Niedergerke, 1958), we have seen that living vertebrate muscle can shorten to sarcomere lengths less than the measured length of the A band without visible contraction band formation. However, our observations differ in two important respects from their findings. First, differences in A/I band widths are still resolveable at short lengths (Fig. 4) and second, we saw no indication of internal buckling. Since buckling occurs at these sarcomere lengths in the compressed inactive heart cell (Krueger et al., 1980), we think that the striations' appearance directly characterizes the activated contractile lattice.

The protein concentrations determined by interferometry were 28.4 ± 4.5 (N=26) and 13.1 ± 3.5 (N=24) g/100 ml, respectively, in the intact and detergent-treated cells. Correcting the former for an increase in the mean cell thickness observed in the detergent-treated cells gave 23.6 g/100 ml. Thus 13.1/23.6 = 0.55 of the cardiac cell protein appears to be insoluble, a value in agreement with the volume occupied by the cardiac myofibrils (54%; Anversa, Loud, Giacomelli, & Wiener, 1981). Aside from the fibrils, mitochondria are the major refractile component in the cardiac cell. It is unlikely that the mitochondria have the laterally aligned arrangement which characterizes the striations we observe. Shortened striations were observed in the detergent-treated cells where the mitochondria probably do not exist, and this explanation would not account for the Ba^{2+} contracture's affect on the striation patterns in the intact cells.

Ba^{2+} does not appear to activate the myofilaments directly in the experimental solutions we employed. Ba^{2+} precipitates phosphate, and so the maximal activation we observed may be related to ATP depletion. Individual detergent-treated cells shortened with full contraction band formation when washed with solutions containing 10 mM EGTA, 3 mM EDTA, without added ATP, Mg^{2+} or Ca^{2+}. (Addition of ATP and Mg^{2+} to these preshortened cells relengthened sarcomeres.) We could not detect any difference by elevating free Mg^{2+} to 3 mM (8 mM total). Thus, complete lack of substrate does not account for the absence of contraction bands in the intact cells.

A band 'shortening' has been described in invertebrate muscle (Dewey, Levine, Colflesh, Walcott, Brann, Baldwin & Brink, 1979), in isolated myofibrils (Hasselbach, Somer & v. Graf, 1975), and in glycerinated skeletal and cardiac fibers of vertebrates (Herman & Dreizen, 1971). Our results indicate that it also occurs within the physiological setting of intact muscle cells, but the precise mechanisms may not be at all similar. We cannot at present determine whether the absence of contraction bands reflects changes which are confined to the terminal projections of the thick filament (Pepe, 1967) or represents intrinsic shortening mechanisms.

Significance for Cardiac Muscle Function

Only 20% of the complete fall in cardiac muscle tension (Pollack & Krueger, 1976) can be attributed to steric hindrance arising from double overlap of thin filaments in the center of the sarcomere (Gordon et al., 1966). However, our results are consistent with the notion that the ultrastructural basis for extreme shortening in the partially activated sarcomere is not as simple as that predicted by the sliding filament mechanism in its simplest form.

A feature that we commonly noticed with the interference microscope was that the center of the cardiac A band appeared optically less dense than its ends (Fig. 1). *There* this may represent the absence of myosin projections. Approximately 65% of the mass of the individual myosin molecule is concentrated on the moveable S-1 globular head of the molecule which projects away from the thick filament backbone. Due to reduced molecular packing, individual myosin projections are thought to be free to 80 nm at each tapered end of the thick filament (Pepe, 1967). This would account for a 'free' space of up to 0.16 μm at the ends of the partially activated, shortened sarcomere. The absence of contraction bands in physiologically activated cardiac muscle may then reflect the lack of the S-1 myosin heads in the regions of the z bands (Fig. 6). If so, then the actual region for unhindered cross bridge interaction would be appreciably restricted over that formerly assumed from the sliding filament mechanism. In this way, the striated appearance of the shortened sarcomere provides an explanation for the steep shape and narrow range of the cardiac length-tension relation in cardiac muscle.

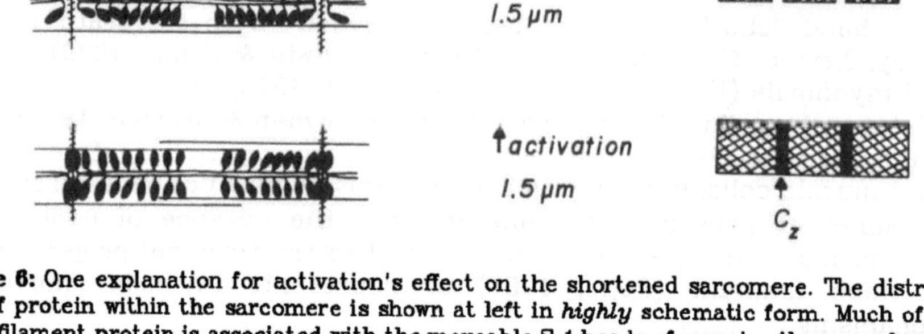

Figure 6: One explanation for activation's effect on the shortened sarcomere. The distribution of protein within the sarcomere is shown at left in *highly* schematic form. Much of the thick filament protein is associated with the moveable S-1 heads of myosin. Absence of active sites and/or stickiness of actomyosin complex would then reduce the number of myosin heads at the ends of the shortened sarcomere. The corresponding optical appearance of the striations is shown at right.

ACKNOWLEDGEMENTS

We thank Drs. A.L. Sorenson & E.H. Sonnenblick for advice and C. Cuadrado, R. Glassman, and R. Smith for assistance. This work was supported, in part, by HL 21325 and HL 18824 (NIH), an Established Fellowship from the New York Heart Association (JWK), and a medical student research program (BL).

REFERENCES

Anversa, P., Loud, A.V., Giacomelli, F. and Wiener, J. (1978). Absolute morphometric study of myocardial hypertrophy in experimental hypertension. II. Ultrastructure of myocytes and interstitium. Lab. Invest. 38: 597-609.

Brown, L., Gonzalez-Serratos, H. and Huxley, A.F. (1970). Electron microscopy of frog muscle fibre in extreme passive shortening. J. Physiol. 208: 86-88P.

Fabiato, A. and Fabiato, F. (1976). Dependence of calcium release, tension generation and resting forces on sarcomere length in skinned cardiac cells. Eur. J. Cardiol. 4/suppl. 13-27.

Fabiato, A. and Fabiato, F. (1979). Calculator programs for computing the composition of the solutions containing multiple metals and liquids used for experiments in skinned muscle cells. J. Physiol. Paris 75: 463-505.

Fujime, S. (1975). Optical diffraction study of muscle fibres. Biochim. Biophys. Acta 3799: 227-238.

Dewey, M.M., Levine, R.J.C., Colflesh, D., Walcott, B., Brann, L., Baldwin, A. and Brink, P. (1979). Structural changes in thicck filaments during sarcomere shortening in *Limulus* striated muscle. In: *Cross-Bridge Mechanism in Muscle Contraction*, 3-19, eds. Sugi, H. and Pollack, G.H., Baltimore: Univ. Park Press.

Gordon, A.M., Huxley, A.F. and Julian, F. (1966). The variation in isometric tension with sarcomere length in vertebrate muscle fibres. J. Physiol. 184: 170-192.

Hasselbach, W., Somer, J.R. and v.Graff, H. (1975). A-band shortening in contracted skeletal muscle fibrils. Fed. Proc. 34: 474.

Haworth, R., Hunter, D.R. and Berkoff, H.A. (1980). The isolation of Ca^{2+} resistant myocytes from the adult rat. J. Molec. Cell Cardiol. 12: 715-724.

Herman, L. and Dreizen, P. (1971). Electron microscopic studies of skeletal and cardiac muscle of a benthic fish. I. Myofibrillar structure in resting and contracted muscle. Am. Zool. 11: 543-557.

Hill, L. (1977). A-band length, striation spacing and tension change on stretch of active muscle. J. Physiol. 226: 677-685.

Huxley, A.F. and Gordon, A.M. (1962). Striation patterns in active and passive shortening of muscle. Nature 193: 280-281.

Huxley, A.F. and Niedergerke, R. (1954). Structural changes in muscle during contraction: Interference microscopy of living muscle fibres. Nature 173: 971-973.

Huxley, A.F. and Niedergerke, R. (1958). Measurement of the striations of isolated muscle fibres with the interference microscope. J. Physiol. 144: 403-441.

Huxley, H.E. and Hanson, J. (1954). Changes in the cross striations of muscle during contraction and stretch and their structural interpretation. Nature 173: 973-976.

Krueger, J.W., Forletti, D. and Wittenberg, B.A. (1980). Uniform sarcomere shortening behavior in isolated cardiac muscle cells. J. Gen. Physiol. 76: 587-607.

Page, S.G. (1974). Measurement of structural parameter in cardiac muscle. CIBA Foundation Symposium 24: 11-26, Elsevier, Amsterdam.

Pepe, F.A. (1967). The myosin filament II. Interaction between myosin and active filament observed using antibody staining in fluorescent and electron microscopy. J. Molec. Biol. 27: 227-236.

Pollack, G.H. and Krueger, J.W. (1976). Sarcomere dynamics in intact cardiac muscle. Eur. J. Cardiol. 4/suppl., 53-65.

Rüdel, R. and Zite-Ferenczi, F. (1980). Efficiency of light diffraction by cross striated muscle fibers under stretch and during isometric contraction. Biophys. J. 30: 507-516.

DISCUSSION

WINEGRAD: In the intact cells that you've stimulated, you point out areas where no contraction bands appear. Is this true across the entire transverse section of the fiber?

KRUEGER: Yes. The pattern of striations appeared absolutely uniform. In some cases where intact cells shortened to appreciably less than sarcomere lengths of 1.5 μm we did find the beginnings of changes in the intensity of the A-band. But we have to look hard to find the contraction bands, rather than to find the cases where they don't occur.

POLLACK: In the case of the intact fibers, where you don't generally see contraction bands, are you implying that sarcomere shortening is concomitant with A-band shortening and, if so, about how much?

KRUEGER: I'm hesitant to use the word "shortening." It might be more precise if we say there are shifts in what we detect as the width of the A-band, either as a result of misregistration of individual filaments, or by slight shifts in protein components projecting out from the thick filament. If there is true shortening, I would estimate it to be on the order of 0.2 μm. But I'm hesitant to say how much "shortening" there is since this requires precise measurement of I-band dimensions.

POLLACK: But you do retain an I-band despite contraction down to very short sarcomere lengths.

KRUEGER: We retain bright regions at the ends of the sarcomeres which were I-bands at one time.

HUXLEY: You said you were dealing primarily with shortenings over a range which was a bit less than perhaps one would normally be concerned with. Have you looked at the effect of slightly larger sarcomere lengths, and at what sarcomere lengths do these apparent A-band shortenings set in in the intact fiber preparation?

KRUEGER: These are unattached heart cells where the normal rest length for the sarcomere is about 2 μm. At only slightly longer sarcomere lengths, rat cardiac cells strongly resist stretch. The range here encompasses the lower 50% of the cardiac length-tension range.

HUXLEY: Am I right in thinking that in cardiac muscle, when it is activated physiologically, there is a very steep ascending limb of the length-tension curve which is not present if you activate them directly with calcium?

KRUEGER: That's true.

HUXLEY: So what you're looking at is partly active, for some reason generating much less tension than you'd normally get with calcium.

KRUEGER: That's our basic premise -- that partial activation does, in fact, govern the disposition of protein within the shortened sarcomere.

HUXLEY: Well, Huxley and Niedergerke (1958) reported that when they looked at frog muscles at sarcomere lengths less than 2 μm, they also saw some apparent shortening of A-bands, and they could still see A- and I-bands separately at much shorter sarcomere lengths than they would have expected on a simple basis. Then there was a paper in the early 60's by Al Gordon (Huxley and Gordon, *Nature* 193: 280-281. 1962) showing that in that case there appeared to be some shearing effect, so that what you really had were obliquely sheared A-bands that were passively shortened.

KRUEGER: I'd invite Al Gordon to comment specifically: as I recall their results, they had myofibrillar buckling.

GORDON: What we found was the distinction between active and passive shortening. When the fiber was activated along the edge, we saw formation of contraction bands, and at the center of the fiber we saw apparent shortening of the A-bands. The center also was wavy. The waviness implied that it was not actively shortening, it was being pushed on. So I think the distinction is active versus some sort of passive shortening. The question is whether that is characteristic of what you're dealing with in the partially-activated fiber, where the A-band is being pushed.

KRUEGER: But there's another question, Al. Was there an intervening "gray" zone where the I-bands persisted without internal buckling?

GORDON: Remembering 20 years back, the gray zone has become grayer.

POLLACK: I'd like to return to Hugh's comment with regard to the relevance of the Huxley and Neidergerke observations. They observed an apparent shortening of A-bands with preservation of I-bands, which they ascribed to limited aperture of their microscope. I was just wondering whether that sort of possibility is ruled out in your particular set of observations, since you did find, in some instances that the I-bands *did* disappear.

KRUEGER: This was exactly the point of the experimental design -- to show that we could observe both disappearance and non-disappearance of the I-band, with the same microscopic system, where no buckling is seen.

REEDY: I haven't got clear in my mind all the conditions under which one fails to see the anomalous A-band shortening, but it's clear that the condition under which one *does* see the apparent A-band shortening and the preservation of I-bands as sarcomeres shorten is in the intact fiber, electrically stimulated. I have two comments on that. One is that your observation of what apparent change in A-band length, of how it depends on contrast as you change the compensator setting in the interference microscope, seems to match *exactly* the amount of shortening found similarly by Huxley and Niedergerke, 0.05 or 0.06 μm, as you switch from positive contrast to negative contrast. I want to point out that this contrast modification I've been performing on fibrils produces an apparent A-band shortening that's five or six time greater than that. It's a reversible one. It's strictly contrast dependent. Secondly, because of the situation that the intact fiber has all this soluble protein in it, I can't help wondering what might happen in permeabilized fibers that are re-exposed to soluble protein, like a high concentration of serum albumin or something like that. I'm very much fishing, because I don't understand what is going on. I'm simply wondering if an attempt to recover that condition of high soluble protein might in any way alter the apparent contrast behavior. I'm thinking of proteins flushing in and out of the interfilament space as charge conditions alter and cross-bridge behavior alters. But I'm very vague on particulars.

KRUEGER: We were very concerned about the soluble protein question until we were able to maximally activate with barium, at which time we measured the mean refractive index of the cells and found, in fact, that there was no protein lost. So, I'm less concerned about it than I was initially, although that's certainly a possibility.

REEDY: In other words, in the presence of barium, there was not apparent A-band shortening, if I understand you?

KRUEGER: That's right, in the intact cells. And I want to emphasize that we have reason to think the barium does not activate the filaments directly. As to the other part of your question, again, I'd emphasize that we see this A-band shortening with several different kinds of microscopy including differential interference, phase contrast, and straight interference microscopy.

TAYLOR: Am I correct in understanding that you're suggesting that the individual myofibrils forcibly re-extend themselves to the resting length at the end of stimulation? If this is correct, I wonder if you could speculate on why this is different from what happens in highly permeable skeletal muscle fibers, frog fibers in particular.

KRUEGER: Permeablized cardiac muscle cells, uniformly shortened with full contraction bands, relengthen (Biophys. J. 365a, 1982). I don't really know what the precise difference is. There may be cytoskeletal differences, although I think we might be removing those with the detergent. Cardiac cell diameter is considerably smaller than the single skeletal muscle fiber preparation (enhancing calcium efflux), and the cardiac myofibrils interconnect. Size itself and the myofibrillar uniformity of activation and relaxation may play a role.

WINEGRAD: There's possibly an important difference between skeletal muscle and cardiac muscle in this passive shortening. When Al Gordon and Andrew Huxley described this passive shortening, they stimulated so that only the superficial myofibrils were activated and there was passive shortening in the core. This meant that there was a difference in location between the region of active and the region of passive shortening. Now one of the characteristics of cardiac muscle is that you cannot achieve more than about 50 or 60% of the maximum calcium activated force by the most optimal conditions of stimulation of the intact cell. If that means that there is now a mixture of active and passive shortening *throughout* the intact cell, then you might see a shortening of the A-band or a configuration of sarcomeres similar to what Gordon and Huxley saw *without* the buckling, because it's now uniformly distributed and volume constraints prevent the waviness from occurring. So I think one has to bear this in mind when looking at the differences between calcium activation, where you're fairly certain that you're getting activation of all force generators, versus stimulation of the intact cell, where you're almost certainly getting no more than 50% or so of force generators.

KREUGER: If what you speculate about the mixture of active and passive shortening is true, then what we've *shown*, which is an important point, is that the distribution of this activity must exist *within* the myofibrils of cardiac muscle rather than between fibrils as occurs in skeletal muscle.

TAYLOR: I'd like to follow up on my question about the mechanism for restoring forces in his preparation. You suggested that the explanation might be due to the difference in relative size, but there was a report from Parsons and Porter several years ago (Science 153: 426-427, 1967) that this was observed even in isolated myofibrils. The point I was driving at is -- is it conceivable that this could, in fact, represent something like a reversible cross-bridge cycle?

KRUEGER: When we looked at relaxation we suggested that the force which relengthens the single heart cells may *not* be present during shortening, and it is therefore not equivalent to a compressively stored elastic energy. That force *may* also represent a repulsive action between the thin filaments, which are inactivated as a result of the shortening. Paul Edman has made a relevant observation and might like to comment.

EDMAN: I'm not quite sure what you mean.

KRUEGER: Well, you showed a constant unloaded velocity of shortening down to 1.6 μm sarcomere length, which implies no internal loading.

EDMAN: Well, it's certainly so in skeletal muscle, that the maximum velocity of shortening is independent of sarcomere length down to that region, which would be a rather strong evidence, I think, in favor of the idea that you have no force working against sliding down to that sarcomere length.

MAGID: John, isn't it true that when you apply a little pulse of calcium to get local activation, you get a small spot of shortening, and that sarcomeres in series with this region are passively lengthened?

KRUEGER: Yes, when *small* regions are activated.

MAGID: I should like to point out to you that the passive tension of rat cardiac myofibrils was measured carefully by Alex and Francois Fabiato (J. Gen. Physiol. 72, 667-699, 1978). It's *very* much higher than it is in fibrils from skeletal muscle. So that a little bit of local shortening would be stretching sarcomeres in series. The passive tension would build up along the myofibril much greater in the case of cardiac muscle than it would in the case of skeletal muscle. And so you might expect to see, as indeed has been observed, differences in the behavior on the relaxing side, when the effect of the local pulse of activation wanes away, and passive tension restores resting sarcomere length.

KRUEGER: Well, Alex found a parallel elasticity which rose sharply at about 2.3 or 2.4 μm. This is a longer sarcomere length than we're dealing with here.

MAGID: 2.4 μm was as far as they could stretch them. At that point the passive tension equalled the maximum active tension. My point is that any stretch above slack length will store strain energy.

KRUEGER: I think you're alluding to something we see -- rapid relengthening. That may be assisted by a pushing force exerted by sarcomeres in parallel.

MAGID: No, I meant a pulling force by sarcomeres in series.

KRUEGER: But the cells are not attached. They're free. I don't think in fact you're stretching any sarcomeres in uniformly shortened cells.

REEDY: There is another piece of behavior that's reminiscent of this business of active lengthening. Jean Hanson (J. Bioph. Cytol., 2, 691-710, 1956) observed some years ago that when you teased fly myofibrils from the flight muscle in a rigor buffer and then exposed them to a relaxing solution, they often showed a little lengthening, four or five percent, and I guess that was reversible when they returned to rigor. Aronson (J. Cell. Biol. 13: 33, 1962) did more on this. I've seen that sort of thing in fly fibrils and occasionally in others, and my impression is that the rigor induction sometimes produces shortening that loads something compressively in a sarcomere which then restores the sarcomere length when it's again relaxed.

POLLACK: Returning to the A-band shortening, I wonder if you have any electron micrographs that you've looked at to see whether any structural changes are concomitant with the lack of contraction band formation?

KRUEGER: I'd like to turn this over to Tom Robinson, who has looked at the morphology of the shortened heart cells. He published a micrograph showing short A-bands (Fig. 4, Cell Tissue Res. 216: 231-251, 1981).

POLLACK: Were there any changes in thick filament length?

ROBINSON: With nominal calibration of the scope, thick filaments were in the 1.5 μm range in the relaxed cells. They may be a little shorter in some of these contracted cells, but I haven't pursued this issue using proper calibration.

STRUCTURAL STUDIES OF GLYCERINATED SKELETAL MUSCLE. I. A-BAND LENGTH AND CROSS-BRIDGE PERIOD IN ATP-CONTRACTED FIBERS

Paul Dreizen, Lawrence Herman* and Jacob E. Berger[+]

Biophysics Program and Department of Medicine, State University of New York, Downstate Medical Center, Brooklyn, New York
**Department of Anatomy, New York Medical College, Valhalla, New York*
[+]Center for Crystallographic Research, Roswell Park Memorial Institute, Buffalo, New York

ABSTRACT

An electron microscope study is reported of structural changes during ATP-induced contraction of glycerinated rabbit psoas. In the absence of ATP, A-band length is constant at sarcomere lengths above 1.9 μm, with average length of 1.54 μ. In ATP-treated fibers, A-band length is also constant at sarcomere lengths above 2.0 μm, but the apparent length of A-band decreases to approximately 1.3 μm, as sarcomere length decreases from 1.9 μm to 1.5 μ. The occurrence of short A-bands cannot be attributed to crumpling of thick filaments against Z-lines, since I-bands remain patent; nor to the presence of heterogeneous filaments, since resting muscle does not show comparable heterogeneity, nor to compressive artifacts, which are minor when knife edge is oriented parallel with fiber axis during microtomy. The decrease of A-band length appears related, at least in part, to disarray of terminal cross-bridges as the thick filaments encroach upon the N-line, a structure which becomes evident within the I-band during contraction of glycerinated fibers. In preliminary studies, optical transforms of A-bands from individual sarcomeres reveal a characteristic myosin layer-line pattern as low as 1.5 μm sarcomere length. A cross-bridge repeat of 143 Å is obtained for sarcomeres above 1.6 μm length; however, an appreciable proportion of sarcomeres in the range from 1.5 μm to 1.9 μ length generate meridional reflections less than 143 Å, and as low as 130 Å.

INTRODUCTION

It is widely accepted that contraction of vertebrate striated muscle involves interdigitation of thick and thin filaments, without change in filament length or cross-bridge interval (Huxley and Hansen, 1954; Huxley and Niedergerke, 1954; Hansen and Huxley, 1955; Huxley and Brown, 1967; Hansen, 1968). The earliest studies involved phase-contrast data (Huxley and Hansen, 1954) and interference microscopic data (Huxley and Niedergerke, 1954), which show constant A-band length for glycerinated muscle under conditions of stretch, rest, and contraction. Early electron microscopic studies did not show gross change of filament length during extension and contraction of glycerinated muscle (Hansen and Huxley, 1955; Huxley, 1957). Later, in a careful electron microscopic study of glycerinated fibers and fresh fibers of rabbit psoas and other vertebrate muscles, Page and Huxley (1963) showed that A-band length is constant at sarcomere lengths from 2.0 μm to 3.5 μ, and minor differences of filament length were attributed to shrinkage during tissue preparation.

All of these studies demonstrate convincingly that filament length remains constant over a wide range of sarcomere length, but there is some ambiguity concerning filament length of glycerinated muscle at sarcomere lengths below 2.0 μm, that is, approximately 90% rest length. For example, Huxley (1965) has described occasional instances of short thick filaments in muscle allowed to contract after previous extension. There have been reports of electron micrographs showing thick filament shortening during contraction of vertebrate muscle (Sjostrand and Jagendorf-Elfvin, 1967; Samasudova and Frank, 1971; Samasudova, Lyudkovskaya, and Frank, 1972), but most investigators have not regarded these reports seriously.

The classical X-ray diffraction studies of living muscle relate mostly to relaxed, isometric contracted, and rigor states, with limited studies on muscle shortened by 5 or 10% (Huxley and Brown, 1967; Huxley, 1967; Haselgrove and Huxley, 1973; Haselgrove, 1975). There have been no other significant X-ray studies on vertebrate skeletal muscle during more extensive shortening, perhaps related to the finding that layer-line patterns are blurred and cross-bridge periodicities are lost during usual X-ray diffraction studies of markedly shortened muscle.

The structural arrangement of cross-bridges has also been examined by means of optical diffraction of electron micrographs in preliminary studies on vertebrate striated muscle (O'Brien, Bennett, and Hansen, 1971), and more recently on isolated A-segments (Craig, 1977) and *Limulus* muscle (Kensler and Levine, 1982). However, this method has not yet been used to examine in a comprehensive way ATP-contracted fibers of vertebrate striated muscle.

Our interest in this matter has a somewhat exotic origin, going back to previously reported studies (Herman and Dreizen, 1971) on skeletal and cardiac muscle of *Coryphaenoides*, a benthic fish captured at 2,200 m depth in the off-shore waters of the Galapagos archipelago. In electron microscopic studies of glycerinated fibers of *Coryphaenoides* muscle, we found that ATP-induced contraction may be accompanied by shortening of the A-band, by as much as 0.2 μm below resting length. These findings were of course obtained under unique conditions, on muscle fibers from an animal living at 200 atmospheres pressure and about 5°C, and one may question whether the rapid ascent from the ocean depths to sea level has severe consequences on myofilament organization and interactions. However, the earlier work led us to reinvestigate a more conventional model, namely, ATP-induced contraction of glycerinated rabbit psoas. The present paper presents electron microscopic and optical diffraction studies which demonstrate an apparent shortening of A-bands during unloaded contracton of glycerinated fibers, and relate these findings to an end-effect involving cross-bridges at the A-band edges, with a preliminary suggestion of changes in cross-bridge repeat within the A-band interior.

METHODS

Elongated strips of rabbit psoas were excised, tied to a polyethylene rod, and extracted in 50% glycerol - 50% phosphate buffer (0.1M KCl, 6.7 mM potassium phosphate, 1 mM $MgCl_2$, 1 μM $CaCl_2$, pH 7.0) at 4°C, with four changes over 48 hours (Huxley, 1963). The glycerinated fibers were stored in 50% glycerol - 50% phosphate buffer for periods from 10 days to 4 months. Just prior to use, a portion was transferred to 15% glycerol - 85% phosphate buffer at 4°C for 1 hour. The fibers were shredded and homogenized for 2 minutes at 4°C, using a Sorvall Omnimixer, and incubated for 1 hour in 15% glycerol - 85% phosphate buffer, in the presence or absence at ATP, as noted.

For polarizing microscopic observations, the fibers were brought to room temperature and examined using an inverted Reichert microscope with Xenon light source and filter with maximal intensity at 5400 Å. Length measurements are based on a Kellner micrometer eyepiece, calibrated against a 1 mm-stage micrometer. Values of sarcomere length were obtained as the average for 10 to 20 sarcomeres in series.

For electron microscopic analysis, the fibers were fixed with 8 volumes of glutaraldehyde (Polysciences Inc., Rydal, Pa.), buffered with 0.1 M sodium cacodylate, pH 7.4, for 1 hour (Sabatini, Bensch, and Barnett, 1963). The homogenate was centrifuged at 15,000 rpm for 15 minutes. The pellet was washed 3 times in 0.2 M sucrose-0.1 M sodium cacodylate, pH 7.4, for ten minutes. The samples were dehydrated in a graded series of alcohols, and embedded in Epon (Luft, 1961). All procedures were done at 4°C.

Ultra-thin sections were cut on an LKB Ultratome, with the blocks oriented so that knife edge was parallel with fiber axis. Sections were obtained at thicknesses from approximately 400 Å (silver color) to approximately 1,200 Å (gold color). The sections were mounted on uncoated Athene grids, stained with methanolic uranyl acetate (Stempak and Ward, 1964) or aqueous uranyl acetate and lead citrate (Reynolds, 1963), and examined in a Philips 300 electron microscope. The microscope was routinely calibrated against a diffraction grating replica with 54,864 lines per inch (E.F Fullam and Company, Schenectady, New York). Calibration photographs were obtained at different positions within the grid and at different times during the course of an experiment.

Optical diffraction studies were done on selected micrograph negatives, at magnifications ranging from 14,000 to 22,000, with masking of an entire sarcomere, or parts of a sarcomere. The micrographs were analyzed with an oil-cell diffractometer with helium-neon laser (Spectra-Physics), as previously described (Klug and Berger, 1964; Berger, Zobel, and Engler, 1966; Berger and Harker, 1967).

RESULTS

Polarizing Microscopic Observations.

In the absence of ATP, glycerinated muscle strips contain sarcomeres of length from 1.7 to 2.6 mm. Approximately 90% of the sarcomeres have lengths between 2.0 and 2.5 μm (Fig. 1). The average length is 2.23 μ (±0.19 μm SD), as determined for 5,380 sarcomeres in 250 myofibrils. Fibers tend to have sheets of myofibrils with sarcomeres of nearly equivalent length in register, so that extensive sampling of fibers is required.

On addition of ATP, the sarcomere distribution is shifted to lower lengths (Fig. 1), for example, 2.04 μm (±0.27 μ SD) in 0.25 mM ATP, 1.67 μ (±0.49 μm SD) in 1.0 mM ATP, and 1.39 μ (±0.46 μm SD) in 5.0 mM ATP. The wide distribution of sarcomere lengths at each ATP concentration reflects an increased proportion of short sarcomeres, as well as the persistence of some sarcomeres at near rest length.

The ATP-treated fibers show a gradient of sarcomere lengths from the fiber surface, where sarcomere length may approach 0.9 μm, to myofibrils within the interior of the fiber, where sarcomere length may remain at 2.2 μm, or greater. The gradient of sarcomere lengths within a single fiber is presumably due to incomplete diffusion of ATP during the period between addition of ATP and microscopic observations. At ATP concentrations of 1.0 mM, or greater, many intact fibers, and almost all disrupted myofibrillar bundles, form compact globular structures, with loss of striation pattern and birefringence.

Electron Microscopic Observations.

Longitudinal sections of glycerinated fibers in the presence and absence of ATP were selected for length measurements on the basis of strict criteria, including the occurrence of sheets of intact sarcomeres with characteristic A-band and M-line; the appearance of knife marks showing that the section was cut parallel with filament axis; symmetric I-bands and half A-bands; the absence of buckling or bending along the myofibrillar axis; and continuity of thick filaments along the A-band in high-power micrographs. Most measurements were done on micrographs at magnifications ranging from 16,000 to 20,000, permitting observations on multiple sarcomeres within each micrograph.

Fig. 2 summarizes data from 3 preparations of glycerinated rabbit psoas, each of which was treated in 15% glycerol - 85% phosphate buffer containing no ATP, 0.02 mM ATP, 0.1 mM ATP, and 0.2 mM ATP. In the

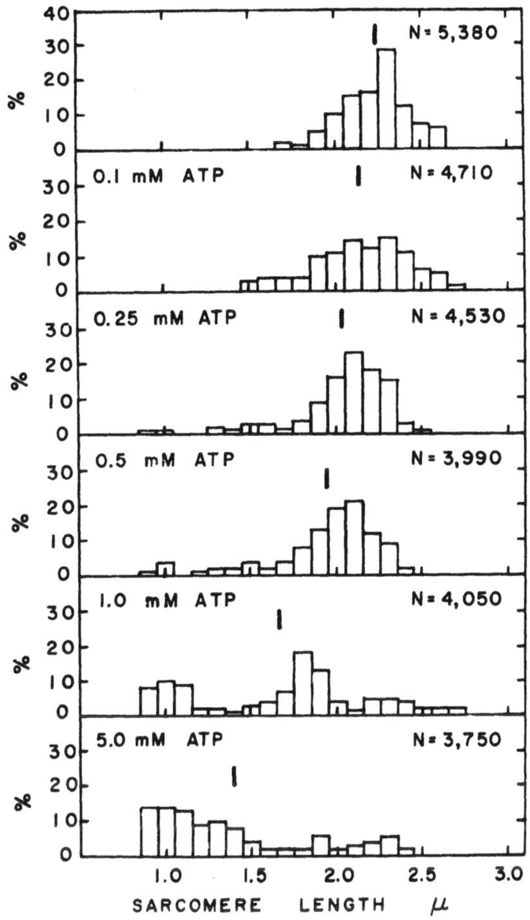

Figure 1: Polarizing microscopic observations of sarcomere lengths of glycerinated rabbit psoas muscle fibers treated at different ATP concentrations for 1 hour at 5°C. At each ATP concentration, ordinate shows distribution of sarcomere lengths, as percentage of total sarcomeres (N); vertical bar shows average length.

absence of ATP, approximately 90% of the sarcomeres have lengths of 1.9 μm to 2.3 μ, and the distribution of sarcomere lengths is comparable with that observed in polarizing microscopic measurements of identically treated fibers. A-band length is constant over the range from 1.9 μm to 2.3 μ sarcomere length, and the average length of A-band is approximately 1.54 μm. The length of I-band varies directly with sarcomere length. In fibers treated with 0.02 mM ATP, average sarcomere length is slightly less, but A-band length remains constant at sarcomere lengths from 2.0 to 2.3 μm.

Fibers treated with 0.1 mM ATP and 0.2 mM ATP show considerable heterogeneity, with a marked increase in the proportion of sarcomeres

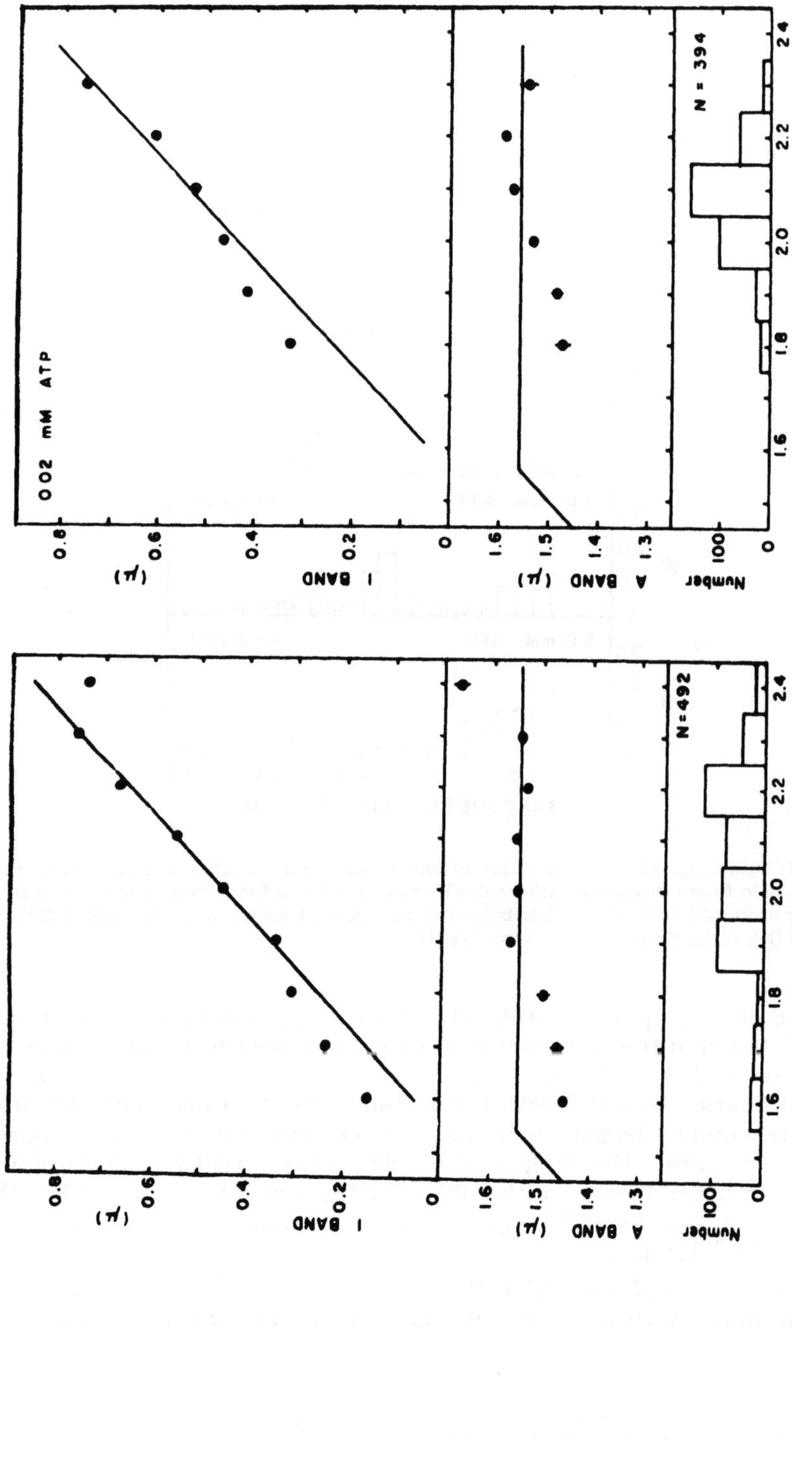
0.02 mM ATP
N = 394
N=492
I BAND (μ)
A BAND (μ)
Number
SARCOMERE LENGTH (μ)

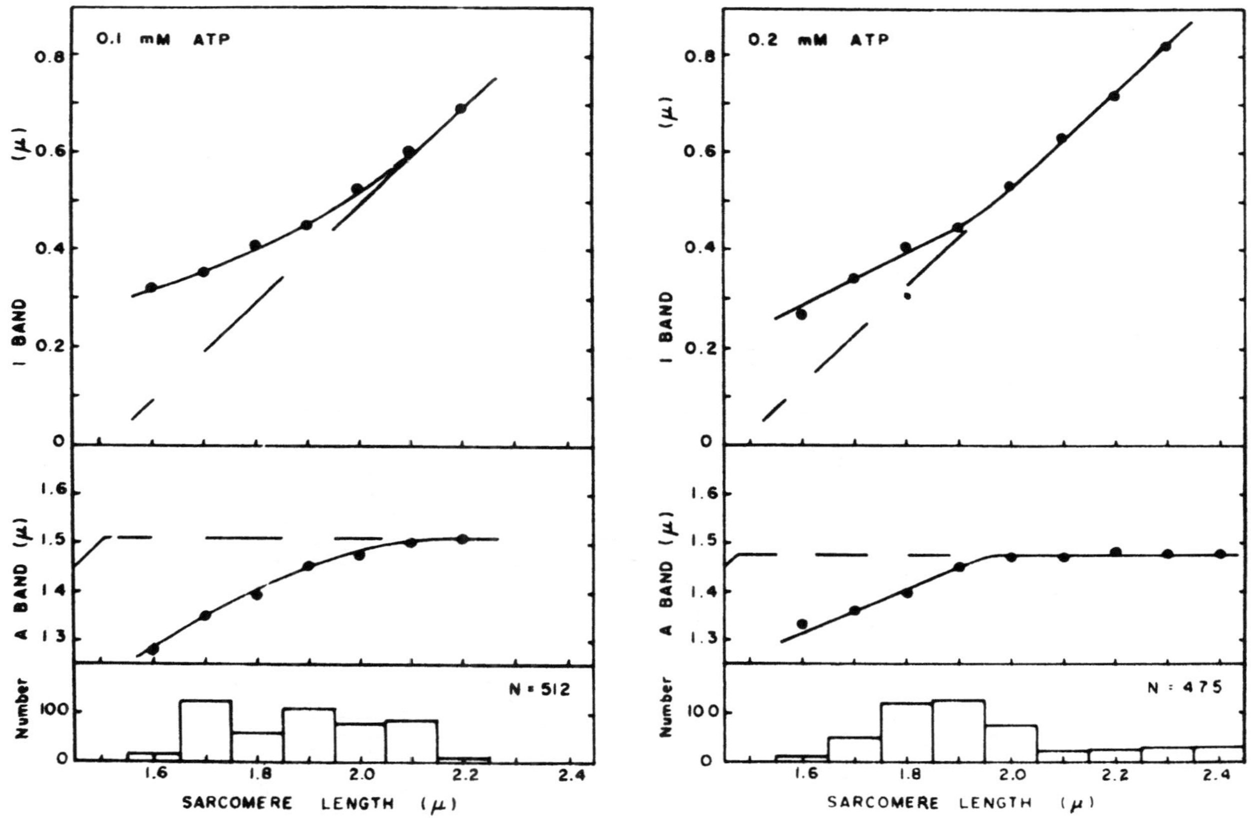

Figure 2: Average lengths of I-band and A-band, and frequency distribution of measured sarcomeres plotted against sarcomere length, for glycerinated rabbit psoas fibers treated at different ATP concentrations for 1 hour at 5°C. Number (N) of measured sarcomeres is indicated. Measurements are based on electron micrographs in which knife marks confirm orientation of long axis of myofibrils parallel with knife edge; electron micrographs are excluded in which knife marks are absent or at angles between 15° and 75° with respect to filament axis. Extrapolated lines show expected values, assuming thick and thin filaments interdigitate without change in length.

having length less than 2.0 μm. In sarcomeres above 2.0 μ length, A-band length is constant, as in the absence of ATP. However, as sarcomere length decreases below 1.9 μm, the apparent A-band also undergoes progressive decrease to approximately 1.3 μm at a sarcomere length of 1.6 μ. Significantly, there may be confusion between the A-band edge and the N-line, a structure which lies between the Z-line and the A-band (Page, 1968; Franzini-Armstrong, 1970; Yarom and Meiri, 1971). The N-line becomes especially prominent in shortened sarcomeres, and the A-band appears to merge with the N-line at sarcomere lengths below 1.6 μm. I-band length decreases with sarcomere length throughout, but the curve of I-band vs. sarcomere length shows a break and levels off at sarcomere lengths below 1.9 μm, so that some I-band appears to remain patent at 1.6 μm sarcomere length. According to these findings, the apparent length of A-bands may shorten by as much as 15% during ATP-induced contraction of glycerinated muscle fibers.

One immediately questions whether this apparent shortening represents an artifact, and several plausible effects were considered. First, errors in calibration are unlikely, since calibration is based on measurements of diffraction grating replicas throughout each series of microscopic observations, nor were there significant time-dependent changes in voltage during instrumental use. Second, shrinkage, at least of this magnitude, appears unlikely. Although common during the early days of electron microscopy, shrinkage appears to be minimal during appropriate fixation of tissues with glutaraldehyde. Moreover, there was no significant variation of A-band length in longitudinal sections of resting muscle, as might be expected if shrinkage during fixation were to account for the apparent shortening of A-bands.

Finally, shortening of A-bands to the extent observed cannot be attributed to compressive artifacts during sectioning. Tissue was routinely sectioned with knife edge parallel with the fiber axis, in order to preserve filament length and avoid compression of myofilaments along their longitudinal axis. Proper sectioning technique may be confirmed by examination of low-power electron micrographs, which show residual knife marks *perpendicular* to the knife edge. In routine sections, most myofibrils are oriented parallel with knife edge, but occasional myofibrils are oriented perpendicular, or nearly perpendicular, to the knife edge. Differences in myofibrillar orientation with respect to knife edge are especially severe in ATP-treated fibers, and polarizing microscopic observations on freely contracting fibers show that myofibril orientation varies as a result of local contraction of some regions along axes different from the major fiber axis. Although micrographs in which knife marks are parallel, or nearly parallel, with myofibrillar axis may obviously be excluded from consideration, many micrographs do not contain visible knife marks. Consequently, attempts were made to estimate the error in A-band length which might result from maximal compression of filaments during sectioning.

In a set of experiments on fibers at different ATP concentrations, the same block was successively cut with knife edge parallel with fiber axis, and knife edge perpendicular to fiber axis. Low-power surveys show

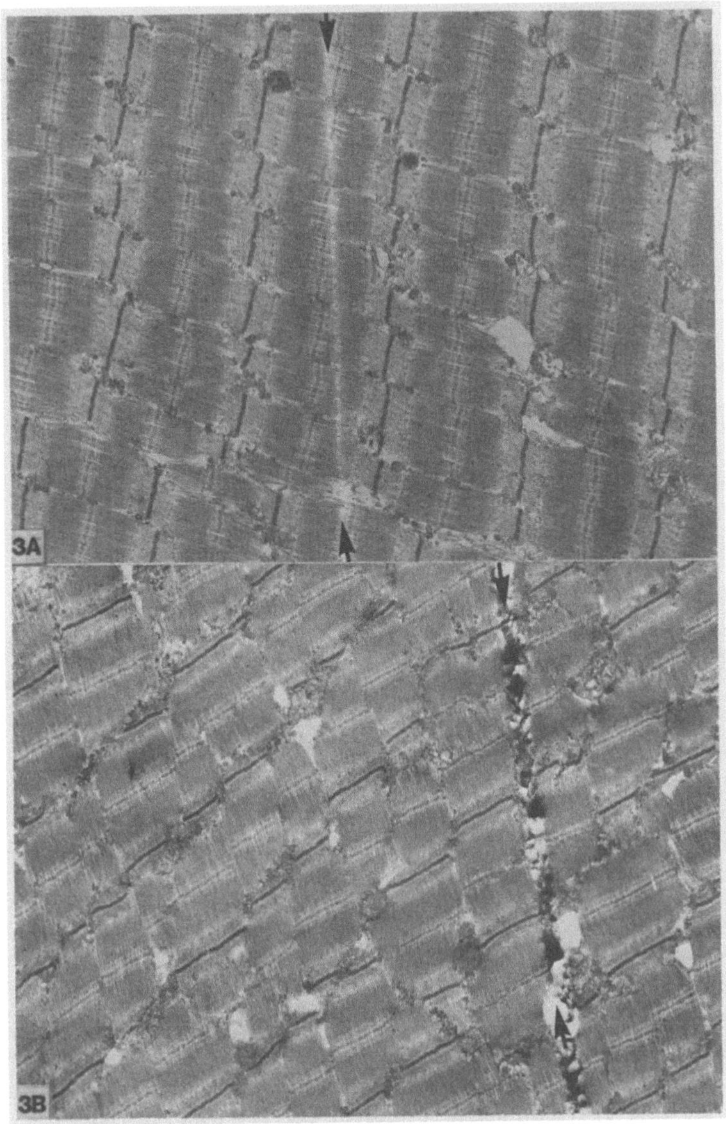

Figure 3: Longitudinal sections of glycerinated rabbit psoas cut with knife edge parallel with fiber axis (*a*) and perpendicular to fiber axis (*b*). Opposing arrows indicate knife marks, which are perpendicular to myofibrillar axis in (*a*), and parallel with myofibrillar axis in (*b*).

myofibrils oriented parallel with the knife edge (Fig. 3, upper) and myofibrils oriented perpendicular to the knife edge (Fig 3, lower). On analysis of sarcomeres from correctly-sectioned tissue and incorrectly-sectioned tissue, A-bands show progressive decrease in length as sarcomere length is diminished below 2.0 μm, but the sarcomeres sectioned

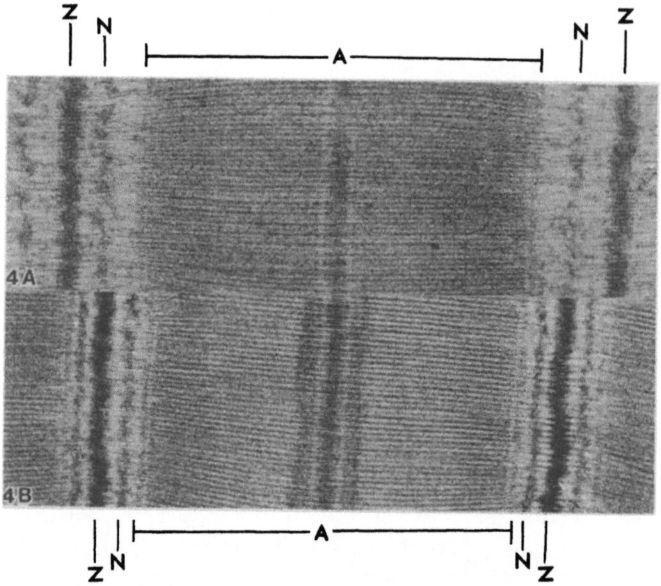

Figure 4: Electron micrographs of longitudinal sections (approximately 1,200 Å thickness) of glycerinated rabbit psoas fibers treated with 0.1 mM ATP for 1 hour at 5°. (*a*), sarcomere length 2.0 μ, A-band length 1.6 μ; (*b*), sarcomere length 1.8 μ, A-band length 1.4 μ.

with knife edge perpendicular to myofibrillar axis have A-bands of length approximately 0.5 μm less than A-bands of correctly sectioned tissue, irrespective of sarcomere length. Thus, maximal compression of myofilaments during sectioning, as would occur during perpendicular orientation of knife edge with respect to myofibrillar axis, is manifest by 3 to 4% shortening of A-bands. Minor errors in orientation of knife edge might be expected to cause correspondingly less shortening of A-bands.

These considerations would suggest that the occurrence of A-band shortening during ATP-induced contraction of glycerinated muscle represents a real phenomenon. Some plausible explanations follow:

(1) The terminal ends of thick filaments may undergo disarray during shortening, with apparent shortening of the A-band. In fact, there is evidence for this explanation. Fig. 4 shows 2 representative sarcomeres from a fiber sample treated with 0.1 mM ATP. Both sarcomeres show N-lines between the Z-line and the edge of the A-band. In the shorter sarcomere (1.8 μm), the N-line lies at approximately the same position as the edge of A-band in the longer sarcomere (2.0 μm), and there is a gap between the N-line and the edge of the shorter A-band. Although the A-N gap region shows decreased intensity as compared with the intensity of the AI overlap zone, some filaments in the A-N gap region appear contiguous with filaments from the A-band. This would suggest that at least part of the apparent shortening of A-band results from one or more sets of terminal cross-bridges falling out of register, as the ends of thick filaments

encroach upon the N-line region during progressive shortening of a sarcomere.

This phenomenon might be related to the spreading apart of the thin filament lattice during sarcomere shortening, so that cross-bridges bend more sharply in reaching out for actin sites. Moreover, as the terminal ends of thick filament approach the N-line, this structure may interfere with the usual interaction between cross-bridges and actin sites in the AI overlap zone. Finally, the thin filament lattice is hexagonal in the AI overlap zone (Huxley, 1957, 1960); however, the Z-band forms a tetragonal network (Knappeis and Carlsen, 1962; Huxley, 1963; Franzini-Armstrong and Porter, 1964), and in the adjacent region of the I-band thin filaments form a tetragonal lattice (Squire, 1981). Consequently, there is a zone in the general neighborhood of the N-line where the thin filament lattice is transformed from tetragonal to hexagonal, and it is plausible that this transformation might interfere with the ordinary rigor pattern of cross-bridge attachments to actin, so as to cause apparent disarray of cross-bridge binding sites along the thin filaments.

(2) Cross-bridges must extend outwards as the thin filament lattice widens during shortening of sarcomeres. This might result in apparent shortening between cross-bridges in opposite A-band halves. As an upper estimate to this effect, shortening of sarcomeres from 2.2 μm to 1.6 μm would increase interfilament distance by about 17%, assuming constant volume behavior. Taking a length of 600 Å for myosin S2, the A-band would show an apparent shortening of about 150 Å, but this is far less than the observed shortening of up to 0.25 μm.

(3) A change of orientation between cross-bridge attachments to actin sites in relaxed (or rigor) and ATP-induced contracted states would not account for the extent or the length-dependency of A-band shortening, as observed.

(4) ATP might possibly lead to shortening of the thick filament core, with secondary change in the helical organization of cross-bridges. This explanation seems unlikely because, among other reasons, fibers which were allowed to contract spontaneously *prior* to glycerination also exhibit an apparent shortening of A-bands in short sarcomeres.

(5) There might be rearrangement of cross-bridges during ATP-induced contraction of glycerinated fibers, with shortening of the cross-bridge repeat.

Optical Diffraction Studies.

In order to determine whether the presumptive changes involving the terminal cross-bridges of thick filaments are accompanied by any alterations of cross-bridge organization within the interior of thick filaments during ATP-induced contraction, transforms were obtained on micrographs of longitudinal sections of individual sarcomeres, or their A-bands, from glycerinated muscle fibers as described above. In general, we found that myosin layer-line patterns and meridional reflections are obtained for A-bands from sarcomeres which are shortened to as low as 1.5 μm, or less. This is in contrast with X-ray diffraction studies of living

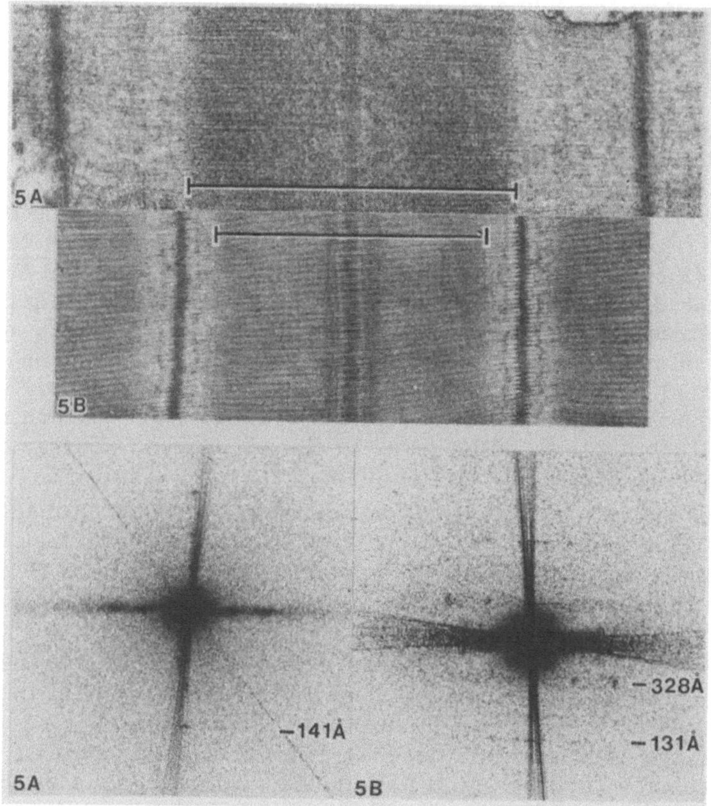

Figure 5: Electron micrographs and corresponding optical transforms of longitudinal sections of glycerinated rabbit psoas. *a*, fiber stretched prior to homogenization; sarcomere length, 2.7 μm; A-band length, 1.55 μm. *b*, fiber treated in 0.01 mM ATP for 1 hour at 5°C; Sarcomere length, 1.64 μ; A-band length, ca. 1.36 μ.

muscle, where the myosin layer-line patterns are blurred, or obliterated, when sarcomeres shorten by 5 to 10% of rest length (Huxley and Brown, 1967). The difference in results between the two procedures may have several possible causes. Shortening may result in variation of individual sarcomere axis about the predominant fiber axis. The optical transform procedure does minimize this effect by allowing diffraction of individual sarcomeres, or their A-bands, on an appropriate axis. In addition, examination of a large number of fibers from a particular preparation indicates considerable heterogeneity with respect to sarcomere length, and also with respect to the attachment between cross-bridges and actin filaments, as indicated by the transform pattern. This is not surprising, since the method here employed for ATP-induced shortening does generate time and concentration gradients such that large fibers may possibly contain some sarcomeres in rigor state at rest length, others which are actively contracting in the presence of ATP, and still others in rigor state *after* shortening and perhaps local depletion of ATP. While this feature might be undesirable for bulk physiological studies, where a single response is desirable, the heterogeneity so obtained surveys a variety of myofibrillar responses, with straightforward correlation between the morphological features on micrographs and their periodicities on optical transforms.

Fig. 5A shows a sarcomere from a glycerinated fiber subjected to stretch prior to homogenization, in the absence of ATP. Sarcomere length is 2.7 μm, and A-band length is 1.55 μm. The optical transform (A) of this A-band shows a meridional reflection at 141 Å, and other weak meridional reflections. These are raw data, and the difference between 141 Å and 143 Å is within our experimental error. The [10] reflection is slightly more intense than the [11] reflection. In other transforms of unstretched "resting" muscle, in the absence of ATP, there is usually a strong [11] reflection and 370 Å layer-line, consistent with rigor attachment of cross-bridges. Fig. 5B shows a sarcomere from a glycerinated fiber in the presence of 0.01 mM ATP. Sarcomere length is 1.65 μm, and A-band length is approximately 1.36 μm. There is overlap of I-filaments, which are approximately 1.0 μm in length, and the N-line is visible between Z-line and A-band. The A-band contains a central region of near uniform density for filaments and cross-bridges, extending to *ca* 1.30 μm length. Just beyond this central region lies another less dense region, extending to *ca* 1.36 μm, which appears to contain cross-bridges roughly but not precisely in register in a direction perpendicular to the filament axis. The optical transform (B) of the A-band of this sarcomere shows a characteristic rigor pattern; however, the first layer-line is at 335 Å, and the meridional reflection is at 131 Å. The ratio between these periods is approximately 2.6, as obtained on X-ray diffraction of rigor muscle (Huxley and Brown, 1967).

We are presently examining optical transforms of sarcomeres from glycerinated muscle, in the presence and absence of ATP, and some preliminary observations should be noted. In general, the transforms show well-defined major axial and equatorial reflections, but nowhere near the detail obtained on X-ray diffraction of living vertebrate muscle (Haselgrove and Huxley, 1973) or optical diffraction of *Limulus* micrographs (Kensler and Levine, 1982). Equivalent myosin layer-line patterns are obtained for the entire A-band, and right and left hand halves of an A-band. We have not yet obtained significant layer-line patterns from the A-N gap region, possibly due to the relatively little scattering material in this narrow region. The overall findings with respect to sarcomere length are as follows:

(1) In sarcomeres of length above 2.0 μm, the cross-bridge repeat is 143 Å. Rigor-like patterns are usually obtained with a layer-line *ca* 370 Å, except in relaxing solution, where some transforms show a weak first layer-line and reversal of the [11]/[10] ratio.

(2) In sarcomeres of length from 1.4 μm to 1.9 μm, a variety of patterns are obtained. Some transforms, especially at the upper end of this range, above 1.6 μm sarcomere length, show a layer-line *ca* 370

Å and a meridional reflection at 143 Å. Other transforms show meridional reflections to as low as 130 Å, and a similar continuum for the first layer-line, but the ratio between these two periods is usually close to 2.6. Often, there are weak reflections, and on occasion no reflections at all, although with improved selection of sarcomeres for diffraction, this last group has been reduced.

(3) In sarcomeres of length from 0.8 μm to 1.2 μm, there are no significant axial reflections in transforms of contraction bands or the residual A-bands.

The most significant of these observations, other than the actual demonstration of axial repeats during or, more properly, following shortening of glycerinated muscle fibers to as low as 1.6 μm sarcomere length, is of course the occurrence of a short meridional repeat during ATP-induced contraction. At present, we are uncertain whether this observation represents a real change in cross-bridge organization during ATP-induced shortening of glycerinated fibers, or whether the short periods result from some as yet unrecognized artifact. In this respect, our greatest concern involves a possible divergence between the plane of section and the myofibrillar axis. The transform would then sample a projection of the A-band, and thereby foreshorten the axial period between cross-bridges. However, a section which is off-axis to the extent required for such shortening to occur would presumably cut through multiple filaments along the entire length of an A-band, and this has not yet been observed, nor are short meridional spacings found in myofibrils close to resting length.

ACKNOWLEDGEMENTS

This work was supported by research grants from the National Institutes of Health, the New York Heart Association, and the American Heart Association.

REFERENCES

Berger, J.E., Zobel, C.R. and Engler, P.E. (1966). Laser as light source for optical diffractometers; Fourier analysis of electron micrographs. Science 153: 168-170.

Berger, J.E. and Harker, D. (1967). Optical diffractometer for production of Fourier transforms of electron micrographs. Rev. Sci. Inst. 38: 292-293.

Craig, R. (1977). Structure of A-segments from frog and rabbit skeletal muscle. J. Mol. Biol. 109: 69-81.

Franzini-Armstrong, C. and Porter, K.R. (1964). The Z-disc of skeletal muscle fibers. Z. Zellforsch. 61: 661-672.

Franzini-Armstrong, C. (1970). Details of the I-band structure as revealed by the localization of ferritin. Tissue and Cell. 2: 327-338.

Hanson, J. and Huxley, H.E. (1955). The structural basis of contraction in striated muscle. Symp. Soc. Expt. Biology. 9: 228-264.

Hanson, J. (1968). X-ray diffraction of muscle. Quart. Rev. Biophysics. 1: 177-216.

Haselgrove, J.C. and Huxley, H.E. (1973). X-ray evidence for radial cross-bridge movement and the sliding filament model in actively contracting skeletal muscle. J. Mol. Biol. 77: 549-568.

Haselgrove, J.C. (1975). X-ray evidence for conformational changes in the myosin filaments of vertebrate striated muscle. J. Mol. Biol. 92: 113-143.
Herman, L. and Dreizen, P. (1971). Electron microscopic studies of skeletal and cardiac muscle of a benthic fish. I. Myofibrillar structure in resting and contracted muscle. Amer. Zoologist. 11: 543-557.
Huxley, A.F. and Niedergerke, R. (1954). Structural changes in muscle during contraction. Interference microscopy of living muscle fibers. Nature. 173: 971-973.
Huxley, H.E. and Hanson, J. (1954). Changes in the cross-striations of muscle during contraction and stretch and their structural interpretation. Nature. 173: 973-976.
Huxley, H.E. (1957). The double array of filaments in cross-striated muscle. J. Biophys. Biochem. Cytology. 3: 631-647.
Huxley, H.E. (1960). Muscle Cells. In: *The Cell*, Vol. 4, pp. 365-481, ed. Brachet, J. and Mirsky, A.R. New York, Academic Press.
Huxley, H.E. (1963). Electron microscopic studies on the structure of natural and synthetic protein filaments from striated muscle. J. Mol. Biol. 7: 281-308.
Huxley, H.E. (1965). Structural evidence concerning the mechanism of contraction in striated muscle. In: *Muscle*. pp. 3-28. Paul, W.M., Daniel, E.E., Kay, C.M., and Monkton, G. Oxford, Pergamon Press.
Huxley, H.E. and Brown, W. (1967). The low-angle X-ray diagram of vertebrate striated muscle and its behavior during contraction and rigor. J. Molec. Biol. 30: 383-434.
Huxley, H.E. (1968). Structural difference between resting and rigor muscle; evidence from intensity changes in the low-angle equatorial X-ray diagram. J. Molec. Biol. 37: 507-520.
Kensler, R.W. and Levine, R.J.C. (1982). An electron microscopic and optical diffraction analysis of the structure of *Limulus* telson muscle thick filaments. J. Cell Biol. 92: 443-451.
Klug, A. and Berger, J.E. (1964). An optical method for the analysis of periodicities in electron micrographs, and some observations on the mechanism of negative staining. J. Mol. Biol. 10: 565-569.
Knappeis, G.G. and Carlsen, F. (1962). The ultrastructure of the Z-disc in skeletal muscle. J. Cell Biol. 13: 323-335.
Luft, J.H. (1961). Improvements in epoxy resin embedding methods. J. Biophys. Biochem. Cytology. 9: 409-414.
O'Brien, E.J., Bennett, P.M. and Hanson, J. (1971). Optical diffraction studies of myofibrillar structure. Phil. Trans. Roy. Soc. Lond. B. 261: 201-208.
Page, S.G. and Huxley, H.E. (1963). Filament lengths in striated muscle. J. Cell Biology. 19: 369-390.
Page, S.G. (1968). Fine structure of tortoise skeletal muscle. J. Physiol. 197: 709-715.
Reynolds, E.S. (1963). The use of lead citrate at high pH as an electron-opaque stain in electron microscopy. J. Cell Biology. 17: 208-212.
Sabatini, D.D., Bensch, K.G. and Barnett, R.J. (1963). Cytochemistry and electron microscopy. The preservation of cellular structures and enzymatic activity by aldehyde fixation. J. Cell Biology. 17: 19-58.
Samosudova, N.V. and Frank, G.M. (1971). Change in the ultrastructure of contractile apparatus of striated muscle under toxic contraction. Biophysika. 16: 244.
Samosudova, N.V., Lyudkovskaya, R.G. and Frank, G.M. (1972). Ultrastructural studies of slow and intermediate isolated frog muscle fibers under toxic contraction. Biophysika. 17: 1055.
Sjostrand, F.S. and Jagendorf-Elfvin, M. (1967). Ultrastructural studies of the contraction-relaxation cycle of glycerinated rabbit psoas muscle. I. The ultrastructure of glycerinated fibers contracted by treatment with ATP. J. Ultrastruct. Research. 17: 348-378.
Squire, J. (1981). *The Structural Basis of Muscular Contraction*. New York, Plenum Press.
Stempak, J.G. and Ward, R.T. (1964). An improved staining method for electron microscopy. J. Cell Biology. 22: 697-701.
Yarom, R. and Meiri, U. (1971). N-lines in striated muscle: a site of intracellular Ca^{2+}. Nature. 234: 254-256.

DISCUSSION

PODOLSKY: One of the explanations that is often offered for an extra band formation at about 1.6 μm is that the Z-line is a square lattice and that the thick filaments form a hexagonal lattice, and those two lattices are incompatible. So at sarcomere lengths of about 1.6 or a little bit more, you'd expect the two lattices to collide with each other and form some sort of interaction. Is that a *possible* explanation of your results?

DREIZEN: I would include that explanation as one of the possible causes of terminal changes in the thick filament. It seems clear that the cross-bridge arrangement at the terminal ends of each thick filament is altered as the thick filaments approach the N-line region. Now whether or not the thick filament is trying to penetrate a new lattice, or the filaments are moving apart, or there is a bending as John Krueger described, or there is some other explanation -- I think is uncertain. In any case, these terminal changes do not account for the optical diffraction observations, where the changes occur throughout the interior of the A-band.

POLLACK: In the electron micrograph that you showed to demonstrate that some changes were occurring at the ends of the thick filament, in the regions just flanking the M region, the cross-bridges appeared to be somewhat darker than the remainder of the cross-bridges along the rest of the filament. Was this a consistent finding?

DREIZEN: The darkening is probably due to overlapping I-bands. This region varies in appearance, depending on the extent of shortening.

ROWE: Can I make a point? When you start building models, the *length* you assign to the thick filament depends very much on the cross-bridge configuration you assign, especially near the filament tips. The myosin molecule is enormous, really, at light microscope dimensions. From the end of the head to the tip of the tail, you've got almost 0.2 μm per filament. That is one-eighth in round figures of the length of the entire filament, isn't it?

DREIZEN: You're absolutely right. In order to interpret A-band length from EM measurements you have to consider these changes in the terminal end of the thick filament, including the elbows between S-1 and S-2 and between LMM and HMM. What happens here during shortening? The end myosins may unravel, the cross-bridges may reach out, and there may be lattice changes. One cannot simply look at electron microscope measurements of apparent A-band length, but must also look at optical transforms which provide information on cross-bridge arrangement away from the terminal ends of the thick filament.

HUXLEY: How did you eliminate the possibility that the fiber you were looking at in the EM wasn't somewhat tilted on the plane of view, so it wasn't parallel, as it were, to the grid but tilted at an angle to the grid?

DREIZEN: We tried very carefully to avoid any measurements on oblique sections. Keeping to strict criteria, we did not find it difficult to recognize oblique sections, and we did not take any measurements on them.

HUXLEY: The thing that surprised me most about that was taking the 143 period down to 131. I don't find it surprising that you see A-band shortening under some circumstances. Dr. Edman and I did some experiments a very long time ago looking at single glycerinated fibers shortening under load in ATP, precisely for this reason, to see whether in loaded shortening there was any sign of something different happening to the A-bands. What we found was that in most cases the A-bands remained constant in length. So, in fact we never bothered publishing the work. But in one or two cases there were changes in A-band length, and these were associated with changes in apparent density of the A-bands, particularly at the ends, but not necessarily always at the ends. These were the exceptions, and we ascribed these to cases in which cross-bridge cycling was not taking place in a normal way. The cross-bridges were getting stuck onto thin filaments in these glycerinated muscles, and we were getting a certain amount of compression of parts of the A-band. Nevertheless, I would have thought that over most of the A-band we would still see a 143 period, and the fact that you pick up a 130 period -- I would be a little bit worried about whether you might be getting a slightly oblique section through the structure. I agree that obviously there could be other explanations.

DREIZEN: I would agree that slightly oblique sections would be hard to exclude, even with the greatest of care. But we did not see short periods in resting state sarcomeres as might be expected if oblique sections were to explain our observations. On addition of ATP we induced a high proportion of short sarcomeres, and these were the ones which showed the change. Also, I don't want to leave the impression that short sarcomeres always gave diffraction patterns with nice sharp layer lines, because we did see considerable heterogeneity, and often the layer line patterns were not good or might not be found.

HUXLEY: Where do you think the 326 layer line comes from?

DREIZEN: I'm not yet sure, but there are several possible explanations. One is that the whole cross-bridge assembly undergoes rearrangement, that we're picking up the attachments to the actin filament.

HUXLEY: So you think the *actin* filaments are also shortening, as well as the A filaments?

DREIZEN: No, so far we have no evidence for that. I'm suggesting that the arrangement of cross-bridges along the thin filaments might be different.

But I think the key question -- which I can't answer with any certainty -- is whether or not when the electron microscope sections yield significant findings as compared with live muscle. The sections have a built-in set of problems. There is no way around this question, since the ultimate reference for the transform is the section itself. The advantage of using EM sections for the optical transforms is also obvious in that you can look at individual sarcomeres, whereas in live muscle you are looking at a whole spectrum of responses with considerable heterogeneity, at least in fibers shortened below 2 μm, and it's not that easy to interpret everything on the basis of average values of distance and intensity for a large number of sarcomeres.

HUXLEY: Did I understand you to say that you saw these changes in A-band length when you did the experiments in 0.1 mM ATP, but that the changes were very much less marked when you did it in 0.5 mM ATP?

DREIZEN: No. The distribution of sarcomere lengths shifted progressively with increasing ATP, but the A-band length at a given sarcomere length did not vary greatly as ATP was increased.

HUXLEY: But didn't the break in your curve occur at a lower sarcomere length and a higher ATP concentration?

DREIZEN: I did show plots at 0.1 mM and 0.2 mM ATP. There is a slight, about 0.05 μm, difference but I don't think this is significant.

HUXLEY: Were these single glycerinated fibers or bundles?

DREIZEN: They were glycerinated strips that were shredded and homogenized in a Sorvall Omnimixer before adding ATP. We did not attempt to fix their length. This procedure yielded a spectrum varying from very thin microscopic fibrils to large fibers. The same fiber preparations were used for polarizing microscopic and electron microscopic observations.

HUXLEY: But the electron microscope pictures were taken on bigger fiber bundles --

DREIZEN: Yes. The sections were obtained from visual-sized fibers, because we wanted to align the knife edge with the fiber axis.

TREGEAR: Maybe I missed this, but didn't you have a back-up system to keep the ATP maintained?

DREIZEN: No, we did not. From just gross calculations of the ATP turnover there was no significant depletion.

TREGEAR: Well, gross calculations can be misleading, of course, if you get rapid hydrolysis at a particular point, as Dr. Huxley was perhaps wondering too.

HUXLEY: But the point was that you were not looking at intact fibers. If I understand correctly, you were looking at a mixed fibril preparation.

DREIZEN: Yes.

TREGEAR: But sometimes those mixed fibril preparations can be very thick. That is the point I was getting at. If one had some repeat observations with backup system, this would clear it.

DREIZEN: How would the presence or absence of a backup system alter the morphological observation?

TREGEAR: Well, you might not get the same at corrected 0.2 mM ATP.

DREIZEN: I did show essentially similar findings at 0.1 mM and 0.2 mM ATP, allowing for the differences in sarcomere length distribution. In general, we did see the same phenomena at different starting concentrations of ATP, so I'm not that much concerned as to the precise ATP concentration in the particular experiment.

HUXLEY: I think the point Richard is getting at is that with *normal* ATP concentration, if you're looking at single myofibrils they contract to very short sarcomere lengths. So clearly you're working with ATP concentrations which are different from normal, so that they produce relatively weak contraction in which presumably a fair number of rigor links are present to start the muscle shortening. That doesn't necessarily produce the effect you see, but it seems that you are looking at a slightly unusual system.

DREIZEN: I would argue what is "normal" ATP. Certainly, with respect to physiological state, 0.2 mM ATP is much less than, say, 4 mM ATP. However, from a biochemical point of view, which I find not unreasonable, there's no indication for different kinetic mechanisms at 0.5 mM ATP and 5 mM ATP. In fact, perfectly good kinetic studies are done at ATP concentrations much lower than 0.2 mM; albeit with a backup system.

HUXLEY: But it's certainly different in the amount of shortening of fibrils at different ATP level.

DREIZEN: Yes. Our intent was to take an intermediate range of ATP, in order to generate a spectrum of normal to very short sarcomeres. We especially wanted this, so that we would have an internal control over possible fixation artifacts. At higher ATP, as you noted, there are many very short sarcomeres, below 1.6 μm, and not many normal sarcomeres, so we did not like to do measurements under those conditions. In a way, we were seeking the very heterogeneity that from your viewpoint may be objectionable. But I would agree that your point warrants further examination.

EDMAN: I should like to make a further comment on the A-filament shortening at low ATP concentrations. When glycerinated muscle fibers are contracted and allowed to shorten at low ATP concentrations (about 0.4 mM) one obtains a length-tension relation which is quite different from that seen in intact living fibers. Shortening of the glycerinated fibers leads to a steady decrease in isometric force, even if the area of overlap between the A and I filaments is increased by the shortening (Edman, K.A.P., in *Biochemistry of Muscle Contraction*, Ed. J. Gergely, Little, Brown & Co., pp. 345-353. Boston 1964). It may be that the turnover rate of the cross-bridges is quite low under these conditions, so that a high proportion of the cross-bridges get into a "holding" position and create a braking force. Such bridges may affect the configuration of the thick filament. For instance, the ends of the thick filaments may be deformed (compressed or bent) if the end bridges do not dissociate properly. This might account for the observed decrease in A-band width after sarcomere shortening.

DREIZEN: I can't exclude that explanation, although I don't really like it. Suppose that the terminal cross-bridges become fixed, either in rigor or due to some geometric constraint, but the interior cross-bridges keep on working. Then, I think you'd expect typical 143/365 repeats for the interior cross-bridges, but an end-effect with crumpled filaments, maybe a contraction band, before the I-band gets obliterated. Also, in your 1964 paper, the anomalous length-tension curves were obtained in glycerinated muscle contracting in 0.8 mM and 6 mM ATP, so this effect seems more likely related to differences between living muscle and glycerinated muscle, and not to low ATP concentration. In any case, our own observations are also referable to glycerinated muscle, and not living muscle, and the concurrence of anomalous length-tension curve and A-band shortening is certainly interesting.

HARRINGTON: I would like to follow up Paul's comment with a question. Were you sure that in the experiments you did with increasing ATP concentration, that the divalent metal ion concentration was also sufficient to saturate the ATP? Because otherwise you'd get, of course, dissociation of the filaments due to ATP.

DREIZEN: The EM series at different ATP levels were all done in buffer containing 1 mM $MgCl_2$ and 1 μM $CaCl_2$, so that magnesium was in excess of ATP concentration in these experiments. We also looked at glycerinated fibers in relaxing solution, where cross-bridges do dissociate from thin filaments, and on occasion other cation conditions, but we have not yet explored cation effects in a systematic way.

KRUEGER: I want to return to the question of low ATP levels. We found that we could create very nice contraction bands and shortening in the skinned heart cell by washing away the ATP and the magnesium. It puzzled me because I expected them *not* to shorten in rigor, but I believe Bremel and Weber pointed out that you heighten activity under these circumstances. So I don't think that lowering the ATP would account for the absence of contraction band formation.

HUXLEY: But that's starting with *high* levels of ATP.

KRUEGER: That's correct.

HUXLEY: But if you *never* have high levels of ATP you may not ever get sufficient ATP in the middle.

POLLACK: Some comments have been made regarding the *small* amount of A-band shortening that has been observed. I want to show two slides that demonstrate that under appropriate circumstances, fairly large amounts of A-band shortening are seen. The first one was done by Dr. Tameyasu in my lab. We've been investigating both fish and frog myofibrils. Figure D-1 shows fish -- rigor and contraction. You can see that as the fiber contracts the A-band shortens very substantially. Figure D-2 is from Joe Sommer and Hasselbach, who have kindly given me permission to show it. The specimen is rabbit psoas, and ATP is being infused from the left side. You can see the gradient of A-band width from the left side to the right. The point of showing these slides is not necessarily to suggest that there's anything physiological about these kinds of contractions, but only to show that under certain conditions it is possible to see changes of A-band width that are substantially larger than the 10 or 15% that was being discussed earlier.

HUXLEY: Is that a picture of one of Hasselbach's experiments where the fibril is embedded in a gel? In those cases the interdigitating filaments are not free to slide; they are all glued together.

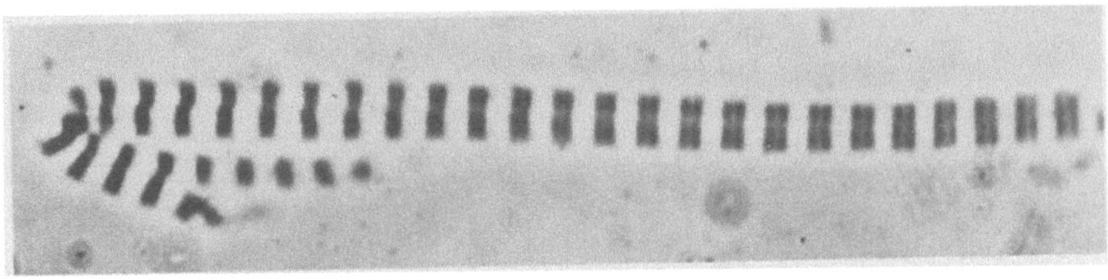

Figure 1: Isolated myofibrils taken from trout myotomal muscle. Top specimen is in rigor, while the bottom one has been activated by infusion of ATP. Note narrowing of A-bands in lower panel.

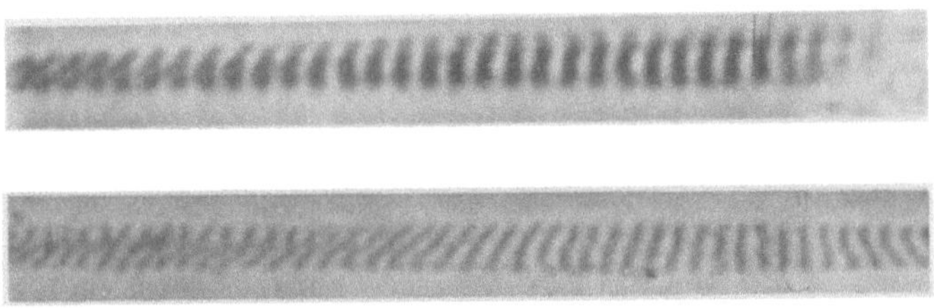

Figure 2: Isolated myofibrillar bundle taken from rabbit psoas muscle (courtesy of J. Sommer and W. Hasselbach). Photomicrograph was taken during the period of infusion of ATP. Note the gradient of A-band width as ATP diffuses along the length of the myofibril from left to right.

POLLACK: He does have experiments in which the fibril is enmeshed in a fibrin clot. This particular specimen I showed was freely suspended in solution.

MAGID: Gerry, I would like to comment on a source of trouble in the cover slip kind of perfusion experiments in the last slide you showed (Fig. D-2). I tell you this from my own experience. After one sets up a fibril under the phase contrast microscope, studies it for a while, defines the field, gets focus, and makes a few micrographs initially under control conditions -- by the time you irrigate with any solution at all, you always run the risk that the drop at the cover slip edge has dried out during these preliminary operations. What you pass under the cover slip is not a low ionic strength or physiological ionic strength solution, but one which has artifactually been raised to a very high ionic strength. I have oftentimes passed a normal relaxing solution under the cover slip and said, "Damn!" while I watched the A-bands wash away. I think what happened was the formation, under uncontrolled conditions, of a Hasselbach-Schneider type solution, that is, a high salt concentration with a plasticizing substance; this can just dissolve the myosin off the ends of the thick filaments. So that in interpreting that *last* kind of picture you have to be very careful that that kind of thing has not gone on.

X-RAY DIFFRACTION APPROACHES TO STRUCTURAL DYNAMICS

INTRODUCTION

The ultrastructure of striated muscle is an extremely elegant example of structure as the basis of function. The remarkably regular arrangement of the thick and thin myofilaments in each sarcomere provides material eminently suitable for X-ray diffraction studies.

The X-ray diffraction patterns from striated muscle result from various periodic structures in each sarcomere; the equatorial reflections are due to the filament-lattice structure, while the meridional reflections and layer lines are due to the periodic structures along the thick and thin filaments. X-ray diffraction methods have the great advantage that the filament structures can be studied at the molecular level in living muscle preparations. In order to investigate the molecular mechanism of contraction, it is essential to obtain information about the structural changes of the myofilaments during contraction. At an early stage of X-ray diffraction studies on striated muscle, it was found that the position of the meridional reflections did not change during contraction (Elliott et al., Nature, 206: 1357, 1965; H.E. Huxley et al., Nature 206: 1358, 1965.) This result, together with electron microscopy of sarcomeric structures (Hanson & Huxley, Symp. Soc. Exp. Biol. 9: 228, 1955), disproved the idea that muscle contraction was due to the folding up or coiling of linear macromolecules, and established the experimental basis of the current sliding filament mechanism.

Recently, the time resolution of X-ray diffraction studies has been markedly improved by the development of the position-sensitive proportional counter, which has supplanted film and is now widely used to record X-ray diffraction patterns. Thus, time-resolved X-ray diffraction studies are being performed intensively by many investigators to provide information about dynamic muscle structural changes during contraction. Through use of extremely intense X-ray sources derived from synchorotron radiation, changes in the X-ray diffraction pattern during rapid mechanical responses of contracting muscles can be studied with a time resolution of 1 msec or less. The papers of Huxley, Hashizume, Podolsky and Tanaka are concerned with the dynamic structural changes in contracting frog skeletal muscle.

It is also important to study the detailed static muscle structures in relaxed and rigor states, since this kind of work may also lead to the understanding of molecular mechanism of contraction. The papers of Tregear, Wray, Wakabayashi and Maeda are related to the static structures of various invertebrate skeletal muscles which, in some respects, are more suitable for this kind of study than vertebrate skeletal muscles. The great disadvantage of X-ray diffraction methods, on the other hand, is that the diffraction pattern does not give direct information about the structures studied; sophisticated mathematical treatments of the X-ray data are necessary to obtain definitive results, since the diffraction pattern does not contain phase information. This problem is discussed in detail in the paper of Squire.

TIME-RESOLVED X-RAY DIFFRACTION STUDIES OF CROSS-BRIDGE MOVEMENT AND THEIR INTERPRETATION

H.E. Huxley

MRC Laboratory of Molecular Biology, Hills Road, Cambridge, CB2 2QH, U.K.

ABSTRACT

The purpose of these studies has been to obtain information about the structural behaviour of the cross-bridges during contraction. Since there are so few reflections still present in the part of the X-ray diagram produced by cross-bridges in a contracting muscle they cannot on their own give a detailed picture. However, they can give information of a more general nature - much in the same way as measurements of tension may do, for example - and the patterns can also tell us what structural regularities are no longer present during contraction. The experiments which I will describe have been carried out in nearly all cases on frog sartorius muscles using synchrotron radiation as an intense X-ray source. The necessary facilities were provided by the European Molecular Biology Laboratory Outstation on the storage ring DORIS at DESY Hamburg. The results to which I will refer have in many cases already been described in papers published or in press (Huxley, 1979; Huxley, Faruqi, Bordas, Koch and Milch, 1980; Huxley, Simmons, Faruqi, Kress, Bordas and Koch, 1981; Huxley, Faruqi, Kress, Bordas and Koch, 1982), to which reference may also be made for experimental details.

Changes During Contraction

The studies fall into two broad classes - ones in which the changes in pattern as between resting and isometrically contracting muscles were observed, and ones in which rapid length changes were imposed on a contracting muscle, and the effect of the X-ray diagram recorded. The overall change during isometric contraction takes the form of changes in both the equatorial and layer-line regions of the pattern, but since equatorial changes have been discussed in many previous publications (Huxley, 1968; Haselgrove and Huxley, 1973; Huxley, 1975; Huxley and Haselgrove, 1976; Podolsky, St. Onge, Yu and Lymn, 1976; Sugi, Amemiya and Hashizume, 1977, 1978; Matsubara and Yagi, 1978) I will refer to them only very briefly.

i) Equatorial changes

The [10] reflection is reduced in intensity by a factor usually between 1.5 and 2 times during isometric contraction, whereas the [11] reflection increases in intensity by a factor of about two. The changes indicate that a substantial proportion of the cross-bridges have moved to the vicinity of the thin filaments. To a first approximation these changes follow the same time course as tension development and decay, which would be consistent with a contraction mechanism in which a close physical interaction between myosin heads and actin was needed. Since no changes are seen in the equatorial pattern from frog muscles stimulated at sarcomere lengths of which actin and myosin filaments no longer overlap, it is presumed that the changes in cross-bridge position in a normal contracting muscle depend on attachment to actin of bridges which are undergoing sufficient Brownian motion, (even in a resting muscle) to bring them transiently in contact. If the onset of the changes at the start of contraction are observed in more detail, it is apparent that the intensity changes take place faster than tension development, suggesting a possible two step mechanism in which a cross-bridge first attaches in a non-tension generating configuration, and has to undergo some further change before it begins to exert a sliding force between the filaments. Such a mechanism may also help to explain the rather unexpected finding that little change is seen in the equatorial pattern as between an isometrically contracting muscle and one shortening at half-maximum load. This finding could indicate that the same number of attached bridges, in the same configurations, were present in each case, but that only half of them were developing tension when shortening was taking place, i.e. there was a delay between attachment and pulling so that when filament sliding was producing continuous detachment and reattachment of bridges, only half of the attached ones had time to develop tension.

ii) Axial changes

The changes in the off-meridional myosin layer line pattern (i.e. the off-meridional reflections arising from the 429 Å approximately helical repeat of the cross-bridges) take the form of a large decrease in intensity during contraction and a recovery during relaxation. As in the case of the equatorials, the changes have approximately the same time course as tension development and its decay, but do occur slightly faster than tension during the onset of activity. The first layer line at 429 Å decreases to about 20% of its resting value. This residue is visible even with completely unfatigued muscles, so it probably does not arise from inactive regions of the muscle. Rather, it may reflect a fraction of cross-bridges which at any given moment are not attached or, more probably, a portion of even attached cross-bridges which still conforms approximately to the myosin symmetry. Since the drop in intensity is less on some of the higher layer lines, it would appear that the residual regular structure is different in nature from that in resting muscle. No new off-meridional layer-line reflections make their appearance during contraction, although strong reflections indexing on the actin helical systems are seen in rigor muscles.

Clearly, then, the regular helical arrangement of the cross-bridges around the backbone of the thick filaments is disrupted during contraction. This is what we should expect their behaviour to be if the cross-bridges have to depart from the average positions in order to find specific attachment sites on actin. The fact that no 'labelled actin' pattern is visible shows that the cross-bridges are not all attached in the same configuration to actin monomers. This would arise if the myosin heads attached in one configuration and then progressively changed to a different configuration as the actin and myosin filaments moved past each other. Since several hundreds of angstroms of internal shortening takes place during tension development even in an 'isometric' contraction, we would expect always to find a wide distribution of cross-bridge configurations to be present, especially if many bridges became attached even when tension was low, early in tension development. Alternatively, or additionally, one could conceive that there might be a substantial amount of azimuthal flexibility present either in the attachment of the cross-bridge (or of a large part of it) to actin, or in the positioning of the actin monomers themselves. It should be made clear, however, that the result does not exclude the possibility that a portion (say one sixth) of each cross bridge was firmly fixed to actin and that the "hinges" were located within the myosin heads themselves.

The meridional reflections exhibit a more complicated behaviour than the off-meridional regions of the myosin layer lines. The most prominent of these reflections in a resting muscle is of course the 143 Å meridional reflection which arises from the axial spacing of the sets of cross-bridges along the length of the myosin filaments. Since there is a 'forbidden' meridional reflection on the second layer line at 215 Å it is clear that there must be some perturbation of the regular cross-bridge repeat. During contraction, the 143 Å meridional reflection first decreases in intensity, then recovers to a measured value which is variable from one muscle to another, but may be even higher than at rest (and is almost always higher in fresh muscles when changes in width of the reflection are taken into account). During relaxation, a large decrease in measured intensity takes place, to a value always much lower than the resting intensity, and the intensity then finally returns to its resting value with a delay of several hundred milliseconds behind tension decay. A substantial part of this fall and delayed return arises because of delay in the width of the reflection across the meridian returning to its resting value, coupled with some other process which brings about an *increase* in the intensity of the 143 Å reflection while the muscle is fully active.

This may at first seem paradoxical in view of the large decrease that occurs in the off-meridional layer lines, presumably generated by the same structures. In principle, the explanation is straightforward and is contained in the evidence - that is, there is a large amount of azimuthal and/or radial disorder during contraction plus an improvement in the axial order. Clearly this could only come about if there were factors present in the resting muscle which weakened the 143 Å meridional reflection. We have already referred to the perturbation indicated by the presence of a 215 Å meridional reflection and we should now refer to the fact that this reflection almost completely disappears during contraction. As Yagi, O'Brien and Matsubara point out, this could supply more intensity for the 143 Å meridional reflection. However, we have already been obliged to postulate a large amount of disorder in the attachment of the cross-bridges to actin, and we have accounted for it in terms of the axial change in position of the attachment sites as the actin moves along carrying bridges at various stages of their working stroke with it. If this is the case, how can we still get a strong 143 Å reflection?

iii) Overall interpretation of isometric patterns

We believe the key to the apparent paradox lies in the fact that while the attached cross-bridges may be disordered with respect to the actin helix, they are necessarily "disordered" in such a way that the ends of the cross-bridges near the myosin still remain attached to the myosin backbone and developing tension. That is, the region of the cross-bridges near to the $S_1 - S_2$ junction must remain approximately fixed in an axial direction, although it may move radially and azimuthally in order to allow the cross-bridge to attach to an actin monomer at a nearby axial level. Since the maximum extensibility of S_2 is about 40 Å (Ford, Huxley and Simmons, 1977) (and may be less) this places a similar limit on the maximum axial excursion of the proximal end of S_1 in cross-bridges which are developing tension. Indeed, if the force-position behavior of a cross-bridge was rather constant over most of its working stroke, the average excursion in an isometrically contracting or slowly shortening muscle could be very small indeed. Hence this part of the cross-bridge could give rise to a strong 143 Å meridional reflection, providing that any shortening was sufficiently slow to keep the tension close to that present during an isometric contraction. In a resting muscle, on the other hand, in which the cross-bridges are free to move about under Brownian motion, a considerable part of their mass could be greatly disordered and even the proximal region could be more disordered axially than when all the cross-bridges are under tension, though now retaining the imprint of helical symmetry.

The 215 Å meridional reflection would on this basis arise from some weak perturbation of the regular axial repeat in a resting muscle, which was overcome when tension was applied to the bridges and their positions became determined by a regular 143 Å axial repeat of the myosin backbone.

Effect of Rapid Length Changes

The results described above seem to make sense in terms of a mechanism by which tension is developed by attached cross-bridges - myosin heads - on actin, and in which myosin heads may be found attached to actin in a range of configurations. However, while the direct, positive evidence that attachment is necessary for contraction seems very strong indeed, we were without direct experimental evidence about the crucial feature of such a mechanism - namely what longitudinal movement of cross-bridges, if any, occurs when the actin filaments move along? The difficulty about obtaining such evidence lay in the fact that in normal steady shortening one would expect to always be seeing a mixture of cross-bridge states. It was for this reason that my colleagues and I (Huxley et al., 1981) investigated the behaviour of the X-ray diagrams from contracting muscles which had been subjected to small but very rapid changes in length (i.e. taking about 1 millisecond).

We found that when either a quick release or a quick stretch was applied, a large decrease (down to 20% or less of the original intensity) took place in the intensity of the 143 Å meridional reflection. In the case of releases the drop in intensity was delayed behind tension by about 0.5 milliseconds. This was followed by a fairly rapid recovery to about two-thirds of the original intensity and lasting about six milliseconds, i.e. taking place much more rapidly than tension recovery. This was followed by a slower recovery towards the original contracting level. In the case of stretches, there was virtually no delay behind the tension change, but no rapid recovery took place, only a slow return to the previous level, following approximately the tension time course.

Interpretation of Transient Changes

We have described elsewhere a number of reasons why we do not believe that these effects arise from some trivial artefact but do indeed provide evidence for a change in the longitudinal configuration of the cross-bridges. Given that the effects are genuine, they show us that whatever structural feature of the cross-bridge arrangement it is that gives rise to the strong 143 Å meridional reflection in contracting muscle, that feature is greatly weakened when movement of actin filament past myosin filament takes place by an amount of the order of 100 Å. Thus, clearly, longitudinal motion of the cross-bridges must have taken place.

Once again though, there are what at first appear to be paradoxical aspects of the results. Though our observations are incomplete, we have so far seen no sign of the momentary appearance of a rigor-like pattern in the sense of one giving a fairly strong 385 Å layer-line. Thus we may have synchronized the cross-bridges to the extent that most of them have been brought to a state where they do not generate tension, but we have not forced them all into the configuration they adopt in rigor. There are several possible explanations for this. It could be that the bridge at the end of its working stroke is in a state different from that of rigor, possibly one with considerable azimuthal flexibility. It could be that since many bridges are moved well past the end of their strokes, the S_2 moieties of the bridges are under compression and are therefore distorting the attachment to actin. Or it could be that effects will begin to show up when we can measure the higher actin layer-lines.

Another 'paradoxical' finding is that there is very little sign of a change in the intensity of the equatorial reflections during quick releases or stretches. Our observations so far would not have detected changes of less than 10%, but it might be surprising at first that changes in cross-bridge configuration which produced so large a change in axial distribution of scattering matter would produce such a small one equatorially. However, the observation does in fact provide an important clue to the nature of the change, for it shows us that the movement of matter which decreases the 143 Å intensity is taking place predominantly in an axial direction.

Possible Models

We have already suggested that in a contracting muscle this reflection arises predominantly from the proximal ends of the cross-bridges, which are kept in close register as long as the bridges are under tension. However, immediately after a quick release which has brought a large number of bridges well past the end of their working stroke, a very different situation will obtain. The bridges may now experience more Brownian motion; alternatively, the S_2 moietys may become compressed, to variable extents depending on how far beyond the end of the stroke the different bridges have been moved. In either case, the close register would be lost, at a rate dependent on how fast the configurational change in the cross-bridge took place. This could account for the 0.5 msec delay (at 5°C) of the structural change behind the tension fall. A contributory factor to the decrease in intensity could be the change in orientation of the attached bridges which will occupy zones probably less than 143 Å in length and hence could make a contribution from their more distal parts to the 143 Å meridional reflection. However, we suspect this may be a smaller effect than the other we have suggested.

The early rapid recovery of intensity could arise if rapid detachment of cross-bridges took place, followed by reattachment again, further out along the actin, but initially in a non-tension-generating state. The effect of stretch on the pattern could be explained along similar lines, on the basis that the S_2 moietys were progressively stretched, by varying amounts, as more and more bridges were taken past the normal 'upper end' of their working stroke.

We think that contraction mechanisms in which large and active length changes occur in S_2 throughout the working stroke (Harrington, 1979) are not readily compatible with the observation of a very strong 143 Å meridional reflection in an isometrically contractingg muscle, nor with our observation that the change in head configuration is delayed behind the elastic phase of tension fall in a quick release. On the other hand, one can conceive of a 'hybrid' cross-bridge model which would be more difficult 'hybrid' cross-bridge model which would be more difficult to eliminate on these grounds. In such a model, initial cross-bridge attachment would take place in the *angled* configuration, and would be followed by a full shortening of the S_2 region, which would bend the cross-bridge to its 'perpendicular' configuration. The S_2 regions would remain at approximately constant length during the cross-bridge stroke, which would be produced by the relief of the elastic deformation in the bridges. One would have to assume that this elastic change was partly damped. The apparent structural behaviour of the mechanism during this part of the cycle would then be very similar to that which would result if active change occurred in the head of the cross-bridge during tension development, and so the presence of a strong meridional 143 Å reflection and its reduction in intensity during a quick release could be equally well explained. However, the subsequent behaviour, i.e. the partial rapid recovery of intensity when cross-beidges detached and reattached again in an initially non-tension-generating state would not be well explained by the 'hybrid' model.

CONCLUSION

These are all somewhat tentative arguments, for a great deal of experimentation remains to be done, as well as a great deal of computing of the effect of such changes on partially ordered arrays of cross-bridges of still somewhat uncertain shape. Nevertheless, we believe that the central conclusions - that cross-bridge attachment does take place, and that a change in the longitudinal configuration of the myosin heads does take place when filament sliding occurs - are now rather firmly based. However, our results so far provide little information about the actual nature of the structural change in the attached heads.

REFERENCES

Ford, L.E., Huxley, A.F., & Simmons, R.M. (1977). Tension responses to sudden length changes in stimulated frog muscle fibres near slack length. J. Physiol. 269: 441-515.

Harrington, W.F. (1979). On the origin of the contractile force in skeletal muscle. Proc. Natl. Acad. Sci. USA 76: 5066-5070.

Haselgrove, J.C. & Huxley, H.E. (1973). X-ray evidence for radial cross-bridge movement and for the sliding filament model in actively contracting skeletal muscle. J. Mol. Biol. 77: 549-568.

Huxley, H.E. (1968). Structural difference between resting and rigor muscle; evidence from intensity changes in the low-angle equatorial X-ray diagram. J. Mol. Biol. 37: 507-520.

Huxley, H.E. (1975). The structural basis of contraction and regulation in skeletal muscle. Acta. Anat. Nippon 50: 310-325.

Huxley, H.E. & Haselgrove, J.C. (1976). The structural basis of contraction in muscle and its study by rapid X-ray diffraction methods. In: *Myocardial Failure*, pp. 4-15. Springer, Berlin.

Huxley, H.E. (1979). Time resolved X-ray diffraction studies on muscle. In: *Cross-bridge Mechanism in Muscle Contraction*, (Sugi, H. & Pollack, G.H. eds., pp. 391-401. University of Tokyo Press, Tokyo.

Huxley, H.E., Faruqi, A.R., Bordas, J., Koch, M.H.J. & Milch, J.R. (1980). The use of synchrotron radiation in time-resolved X-ray diffraction studies of myosin layer-line reflections during muscle contraction. Nature 284: 140-143.

Huxley, H.E., Simmons, R.M., Faruqi, A.R., Kress, M., Bordas, J. & Koch, M.H.J. (1981). Millisecond time-resolved changes in X-ray reflections from contracting muscle during rapid mechanical transients, recorded using synchrotron radiation. Proc. Natl. Acad. Sci. USA 78: 2297-2301.

Huxley, H.E., Faruqi, A.R., Kress, M. Bordas, J. & Koch, M.H.J. (1982). Time-resolved X-ray diffraction studies of the myosin layer-line reflections during muscle contraction. J. Mol. Biol. 158: 637-684.

Matsubara, I. & Yagi, N. (1978). A time-resolved X-ray diffraction study of muscle during twitch. J. Physiol. 278: 297-307.

Podolsky, R.J., St. Onge, R., Yu, L. & Lymn, R.W. (1976). X-ray diffraction of actively shortening muscle. Proc. Natl. Acad. Sci. USA 73: 813-817.

Sugi, H., Amemiya, Y. & Hashizume, H. (1977). X-ray diffraction of active frog skeletal muscle before and after a slow stretch. Proc. Japan Acad. 53B: 178-182.

Sugi, H., Amemiya, Y. & Hashizume, H. (1978). Time-resolved X-ray diffraction from frog

skeletal muscle during an isotonic twitch under a small load. Proc. Japan Acad. 54B: 559-564.
Yagi, N., O'Brien, E.J. & Matsubara, I. (1981). Changes in thick filament structure during contraction of frog striated muscle. Biophys. J. 33: 121-138.

DISCUSSION

Editor's note: The reader may wish to refer to Huxley et al., J. Mol. Biol. 158: 637-684, 1982, where some data and figures relevant to this discussion have been published.

HOLMES: What do you think happens to the bridges on a quick stretch?

HUXLEY: Well, I think there's a range of movements over which the cross-bridges can change their configuration relatively freely, relatively "hygienically." When they're pulled to the upper end of that range it becomes much more difficult to alter their tilt. So that further stretch then produces a much greater extension of the S-2. Thus, in a similar way to quick release, you draw out the bunch of cross-bridge origins (i.e., S-1 - S-2 junctions) in the longitudinal direction.

KAWAI: In the twitch experiment, you showed that the 143 Å intensity increases while the 429 decreases. I still don't understand why one increases and the other decreases.

HUXLEY: The suggestion I was making was that the 429 decreases because that is generated by the helical arrangement of the cross-bridges around the backbone of the thick filaments. In a relaxed muscle, no doubt there's a lot of Brownian motion going on, but there's still sufficient structure present to give rise to a 143/429 helical arrangement. When the bridges are attached to actin in contraction, they can no longer conform to the myosin helical parameters, but the axial positions of the ends of the S-1's near the S-1 - S-2 junction are determined by the lengths of the S-2. No doubt the myosin heads will move azimuthally and radially in order to *attach* to actin, but if they're attached to actin in a number of different configurations they won't label the actin helix in a regular manner, so you won't see a labelled actin helix diffraction pattern. All you will see is whatever residual structural regularity there is, and the only residual structural regularity arises from the fact that the cross-bridges are fastened to the myosin backbone at 143 Å axial intervals. That repeat will remain.

KAWAI: But if that remains then the helical structure on myosin remains too.

HUXLEY: No, it doesn't, because if you believe the S-2 is flexibly attached so that it can swing around the backbone of the myosin filament, and if the S-2 is 400 or 500 Å long, it can move around by a considerable angle before it makes any perceptible difference whatsoever to the axial position of the end of it near S-1.

VERDUGO: I hope you'll comment on the increase in the intensity of 143 Å meridional reflection. If I understand you well, the interaction of the cross-bridges should be a random, one during contraction. Now, wouldn't you therefore expect a random Marcovian space distribution of the sequence? It seems to me that you are interpreting your data as though each cross-bridge is undergoing the same change at the same time.

HUXLEY: No. The cross-bridges are built into the backbone of the thick filament. I personally believe that in vertebrate striated muscle that's a very stable structure. They're built in with an axial period which is, on average 143 Å, and I don't think that's something that's changing. It's that periodicity which is maintained during contraction and which I think may be *emphasized* during contraction by the reduction of the Brownian motion of a portion of the cross-bridge structure when they become attached to actin. This would cause the increase of the 143 Å intensity.

GULATI: As I understand your explanation, Dr. Huxley, you make a fair assumption that during contraction the attached bridges have a steady state distribution of various angular configurations. These different populations of bridges are momentarily *aligned* into a 45° (rigor-like) configuration on the application of an appropriate amount of quick release. If so, you should have detected the characteristic rigor-pattern in these time resolved studies. But you don't. Since the amount of quick release in these experiments was similar to the presumed working stroke of a cross-bridge, your findings seem to raise the possibility that the bridge conformation reached at the end of the working stroke is different from that in rigor. Would you please comment.

HUXLEY: That's a very good question, and it was a great surprise to us -- I mean one of our hopes was that we *would* bring everything up to a rigor configuration, then we'd get a nice momentary flash of the rigor pattern, and everyone would be happy. But in point of fact we've looked for it as hard as we can -- so far unsuccessfully. We haven't exhausted the experimental possibilities, and on the weak layer line reflections we would do better with two-dimensional counters. Even so, 1 millisecond time resolution on diffuse layer lines is pushing the technique. But so far we certainly haven't seen any great sign of this type of pattern. From that I think there are three possible explanations. One is that the end of the working stroke is different from the rigor configuration. It may be different in the sense that it's totally different or it may be different in the sense that it has much more azimuthal flexibility in it. The second possibility is that when you come to the end of the working stroke, you then begin to compress the S-2 component of myosin, (and there's evidence from Huxley and Simmons that the S-2 does resist compression). So that the population of bridges which are taken past the end of their working stroke are not just floating freely but they're compressing S-2; so it's possible that they are distorted sufficiently from a rigor-like configuration so that their diffraction is weakened. But I agree, it's a very interesting and significant fact, or apparent fact!

RITZ-GOLD: Given the difference in actin and myosin periodicities, would this mean that in rigor there's some periodic pattern of S-2 distortion?

HUXLEY: Well, I suspect that during rigor there's plenty of time for the structure to anneal, so that bridges can adjust themselves to the most energetically favorable configurations so that they can settle down more or less all the same configuration on the S-2.

PODOLSKY: You've explained things in terms of bridge rotations. If so, wouldn't you expect the equatorials to change? I remember from your paper that they don't change.

HUXLEY: Yes, the equatorials either don't change, or they change very little. Again, I think that's a very significant observation, but I think that there are two things I'd like to say about it. One is that the explanation for the change in intensity of the 143 reflection is that this change is now arising, not from what's happening to the myosin near the actin, but what's happening to the myosin head near the myosin backbone. That is now a linear movement, so that is not going to give any equatorial changes whatsoever.

There's *still* the question of why you do not see very large equatorial changes if you are changing the tilt. If you have a wide distribution of cross-bridge configurations on the actin, and if a considerable part of the mass of the cross-bridge is close to the actin, then the change in the equatorial pattern is not going to be very large when you produce this net change in tilt. I'm as concerned about this as you (and I guess everyone else) -- why is it that one sees so little change on the equator? It doesn't seem to me inconceivable that most of the mass of the cross-bridge is near the actin, that the cross-bridges are distributed over a wide range of configurations and that there is a very substantial lateral temperature factor on the actin. Thus, it may be that small changes in radial density distribution near the actin will, in fact, not have all that much effect on the equator and that the only thing that affects the equator are some *gross* changes. Could I just make one final point? You said I'm talking about changes in tilt. I'm talking about change in the configuration of the attached myosin, and I think that there's obviously very good reason to believe, as Ken Holmes has suggested, that there's some sort of a nose cone within the myosin head near the actin, and that may move very little relative to actin. But all these changes, if they exist, are configurational changes in the rest of the myosin.

INGELS: With regard to that nose-cone, could you comment on an interpretation on that small intensity dip in the meridional 143 at the onset of contraction?

HUXLEY: I think that during the onset of contraction, when you've got filament sliding taking place (and also during the decay of contraction, when you've got all sorts of internal length readjustments taking place in the muscle) there are all sorts of reasons why the cross-bridges shouldn't be very cleanly lined up. So the intensity would be reduced during these transition stages. But I think the important thing is what are they doing during a nice, steady isometric contraction when things are not moving about; and what are they doing when you impose a very controlled movement on them. As to what happens at the beginning and end of contraction, I think these are much more complicated processes

CECCHI: You said, if I understood well, that if you apply the stretch at five or six milliseconds after the first quick release, you get a new drop of intensity.

HUXLEY: Right.

CECCHI: Does this moment correspond to phase three of recovery described by Ford, Huxley and Simmons (J. Physiol. 269: 441-515, 1977)? Is it the same time?

HUXLEY: Yes, it's about that time. Perhaps I should say that the phase two of that recovery is mostly over within the first millisecond. We're doing most of our experiments, for technical reasons, around 5°-6°C. So the second phase is taking place relatively rapidly. I think we generally don't resolve it. The only sense in which we may be resolving it is that the intensity change as we see it is delayed by about half a millisecond on the length change.

CECCHI: So you're about at a time where the tension recovery is very low, I suppose.

HUXLEY: Yes, the tension recovery is still very slow; I'm obliged to say the cross-bridges are detaching and re-attaching because I believe that we can now move them again to drop the intensity by restretch. But I'm also obliged to say that they're not yet in a tension generating state, and that some further rate process has to occur first.

MATSUBARA: Dr. Huxley, in one of your slides I saw a graph in which the intensity of the meridional reflection at 14.3 nm was plotted against tension changes caused by stretch and release. I wonder if you could comment on that because I have a comparable graph obtained under a slightly different condition.

HUXLEY: Yes, what the slide was concerned with was the initial drop in intensity of the 143 meridional reflection immediately following a quick release or a quick stretch, and it showed that the larger the extent of release the larger the drop in intensity. And to a very rough approximation it was linear. Similarly on the other side of the starting point, the larger the amount of stretch you apply to the muscle, the larger the drop in intensity was. In slightly more detail, we do have a little evidence that for very, very small releases there is, in the main, a very, very small increase in intensity, but in the first approximation it's a couple of straight lines roughly centered on no length change.

MATSUBARA: Our graph represents data obtained from muscles during *slow* release and stretch. The size of the length change was fairly large, 7% of muscle length. The length change was completed in a second. We plotted the intensity change of the meridional reflection at 14.3 nm against the tension change caused by stretch or release. In other words, the abscissa is the size of the intensity fall and the ordinate is the size of the tension change. We obtained linear relations during both stretch and release. This agrees with the linear relation which you have shown us. But, in our graph, the slope of the line obtained during stretch was different from that obtained during release. For a given amount of intensity fall, the size of tension change was greater during stretch than during release. Such an asymmetrical feature was different from what you have obtained. This might mean that the origin of the intensity fall during slow length changes is somewhat different from that during quick length changes. There are at least two possible mechanisms for the intensity fall of the meridional reflection, namely rotation and axial shift of myosin heads.

GILLIS: You are implying that the behavior of the cross-bridges is not just a mirror image when you stretch or when you release. I think there are some mechanical results which go in the same direction. When you plot the relationship between speed of release against tension, or speed of stretch against extra tension, the relationships between speed and tension are not the same if you stretch or if you release. So what we have here is also not a mirror image.

HUXLEY: I think, as I indicated, the explanation I've preferred and the one I think is likely to account for more of the change, is the explanation based not on cross-bridge rotation but on either extension or compression of the S-2, and there's no reason why that should be symmetrical about the midpoint.

POLLACK: Ichiro, I'm a little bit puzzled about the interpretation of the releases. Your releases are slow, while Hugh's are very rapid. Yet the intensity changes are roughly comparable. I'm not sure I understand your interpretation of the diminution of intensity, and how you would contrast that to what Hugh is saying.

MATSUBARA: When the thick and thin filaments slide relative to each other, the myosin heads attached to actin will be shifted along the filament axis, towards the M-line during release and towards the Z-line during stretch. This will disorder the axial alignment of myosin heads, leading to a fall in the meridional intensity. There are also other factors which can decrease the intensity. The simplest way to explain the asymmetrical feature we observed is to assume that the main cause of the intensity fall during a slow length change may be axial shift of heads, and that the tension may be asymmetrically related to the degree of axial shift (according to the direction of shift). Our observation does not agree with the symmetrical relation shown by Dr. Huxley. What I wanted to point out was that the discrepancy might be due to a difference in the main cause of the intensity fall.

POLLACK: But is it clear that the intensity should continue to decrease as the change of length progresses? Does that follow along your line of interpretation?

MATSUBARA: To answer that question we need calculations based on a more detailed model. We have not done such calculations.

COOKE: Question to Hugh Huxley. If the explanation for the drop in the 143 Å is the translation of bridges which are *not* rotating, if you want to translate by 14 nm/ half sarcomere, shouldn't you tend to realign the bridges, and so not get such a strong drop in the intensity of 143?

HUXLEY: No. It depends on what you think the working stroke is. But if you think it's, say, 100 to 140 Å, it means that if you give a release of 140 Å, that you will then spread out your bunch of bridges into a line 140 Å long, because the ones that were at the beginning of their stroke will then be at the end of their stroke, so that the end of that cross-bridge near S-2 won't have moved. The ones at the end of the stroke, couldn't twist or whatever they do anymore, will have moved 140 Å so you will then get the absolute maximum spreading out of those ends and the maximum drop in intensity.

COOKE: But the increase in 143 which occurs during the twitch is best explained by cross-bridges that bunch close to their original origins on the myosin.

HUXLEY: Right.

COOKE: It's now argued that those are now bunched at their original 143's which allow them to translate 143; then they're back where they started.

HUXLEY: No. No. If you think of a muscle during very slow isotonic shortening, when the 143 is still very strong -- if you look at the distribution of bridges on any cross-bridge model, you would expect to find them uniformly distributed through whatever their working range is.

COOKE: Certainly.

HUXLEY: But the ends near the S-2 will still be bunched, because all those bridges are generating tension. When you apply a very rapid release, then after a millisecond or so -- a la Huxley and Simmons -- the bridges near the beginning of that stroke, having had the tension removed from them, can rotate, and that tight bunch gets spread out into a long line.

COOKE: So this model requires cross-bridge rotation.

HUXLEY: Oh, absolutely.

COOKE: In the power stroke.

HUXLEY: Of course!

KAWAI: Can you explain that phenomenon in terms of cross-bridge detachment/reattachment, i.e., when you change the muscle length then cross-bridge heads may transiently come off to assume a relaxed configuration. Therefore, you don't see as intense a 143 Å reflection. To be consistent with your data, we would have to assume that the detachment can happen in either stretch or release. After a moment, the 143 Å reflection can recover because of reattachment.

HUXLEY: The problem is that the intensity drops extremely quickly, within half a millisecond of the release. So you're going to have extremely rapid detachment. Also, if you're going to explain the changes on the basis of detached cross-bridges being able to move a lot, you then get into difficulties explaining why the drop is reversible when you immediately apply a restretch.

WINEGRAD: In these quick releases and then quick stretches where the effect was reversible, did you vary the amount of release so that you had shorter releases, where it's unlikely that you necessarily had detachment, and longer releases where you were more *likely* to have detachment of cross-bridges, to see whether there was a certain amount of release beyond which the quick stretch did not produce an immediate reversal of the 143 reflection?

HUXLEY: No. All of our releases were of the order of 50 to 100 Å. Presumably there's going to be *some* detachment if you wait a little while. But they weren't very short releases like 10 or 20 Å per half sarcomere.

SQUIRE: One of the features of the relaxed x-ray diffraction pattern from frog is that the first layer line is very much stronger than the other layer lines. That is normally explained by the fact that the bridges are probably pointing along the 429 Å helix, if you like, on the myosin filament. If the cross-bridges are doing that, they're also going across the 143 Å space. So the same feature is making the 429 Å layer line strong and the 143 weak, relatively. Now, if as you suggest, the bridges become centered around the small spheres on the helix, that could account for the increase in the 143 and a reduction in the 429 without the bridges having to move off the helix. If they just become localized but still on the helix, both of those observations could be explained. Now, you'd also expect, in addition to the 143 getting stronger, that some of the other layer lines could get stronger, because originally they're also such that the bridges lie across the appropriate helices. So you might expect variable changes in intensity on the other layer lines apart from the 429. So my first question is, do you see any differences in changes in intensity on the different layer lines or are they all in the same direction?

HUXLEY: They're all in the same direction. They all get less, but some get less more than others.

SQUIRE: Yes. I think that's significant. If the other layer lines get weaker but by differing amounts I think that the cross-bridge origins have got to be moved around a bit on the helix, whereas the observation on the 143 and the 429 could be explained if they don't move off the helix. So I think the point is that if you look at those layer lines, they might tell you how much the bridges are moving off the helix. There's another rider, and that is, have you looked at the 72 Å reflection?

HUXLEY: Yes.

SQUIRE: And does it change like the 143 or --

HUXLEY: No. The 72 Å reflection decreases in intensity by a much smaller extent, and it does not respond in any very obvious way to quick releases. I suspect a good deal of the 72 comes not from cross-bridges but from the backbone.

TIROSH: I'd like to address the same issue, but with respect to the attachment-detachment of the cross-bridge from the actin. As far as I understand there are two questions addressed in these studies. One has to do with the reattachment of the cross-bridge to the actin, and the other one with mass transfer from the myosin core towards the actin. The question is, in rigor we have good correlation. We have at the same time an indication of mass transfer towards the actin from the equatorial reflections, and from the meridional reflections we get an indication there is real interaction with the actin.

Now it was hoped, I understand, to get the same results during contraction. Since it's already consistent for rigor, it's nice to keep on with this line of interpretation and say that as long as we have the 14.3 nm reflection, it indicates that the periodicity of the myosin is preserved rather than that of the actin. Therefore, I address again the question, does the persistence of the 14.3 nm spacing for the meridional reflection indicate a dissociation of the cross-bridge from the actin? If so, this correlates with the possibility that most of the time the cross-bridges in the active state are detached from the actin.

HUXLEY: I think since you don't see a labeled actin pattern in an actively contracting muscle, obviously it's not straightforward to determine what proportion, if any, of the cross-bridges are attached to actin. I think one of the strong arguments to show that cross-bridges really *are* attached to actin is that the change in the equatorial pattern that you see during contraction is dependent on actin and myosin overlapping, and that if you stretch muscle so there's no overlap, then you don't see this change in the equatorial reflections. So you do need the presence of actin alongside for this change in the radial arrangement or position of the cross-bridges to take place. As for the arguments about what proportion of cross-bridges are attached at any one moment, it is difficult, certainly as far as the equatorial x-ray results are concerned, to arrive at a reliable figure. One line of argument is that since you can produce such a large change in the 143 reflection by moving the actin filaments along 100 Å, you must be influencing a rather large number of cross-bridges. Maybe it's a common sense argument -- you can always construct other arguments about free cross-bridges which are being influenced by the attached ones. But I think the whole question of what proportion of bridges are attached is going to come up in discussion of later probes, and I think perhaps we might more properly pursue it then.

ON THE POSSIBILITY OF INTERACTION BETWEEN NEIGHBOURING CROSSBRIDGES

R.T. Tregear and M.L. Clarke*

ARC Institute of Animal Physiology, Babraham, Cambridge CB2 4AT
**MRC Cell Biophysics Unit, 26-29 Drury Lane, London WC2B 5RL*

ABSTRACT

Demembranated insect or rabbit striated muscle fibres at equilibrium (i.e. in the absence of ATP hydrolysis) were modified either by substituting ethylene glycol for water or by adding AMPPNP. The resultant states observed by electron microscopy and X-ray diffraction appeared to contain a mixture of at least two distinct types of crossbridge, which were not randomly mixed. The crossbridges held tension for a great deal longer than they remained attached to actin in solution. In the presence of AMPPNP the muscle fibres relaxed at a critical glycol concentration. These properties indicate that the crossbridges interacted with one another.

INTRODUCTION

We have attempted to elucidate the mechanism of action of crossbridges by studying the properties of demembranated muscle at equilibrium, i.e. in the absence of ATP hydrolysis. The equilibrium of such muscle can be altered either by addition of an unhydrolysed ATP analogue or by substitution of polyhydric alcohols for water in the bathing solution; a combination of the two factors causes relaxation of the muscle (Tregear, Clarke, Marston, Rodger, Bordas & Koch, 1982). Many of the properties seen can be interpreted in terms of independent crossbridge action (Marston, Rodger, & Tregear 1976) but certain of them apparently cannot. The purpose of the present paper is to expose this evidence in order to indicate the limits of validity of the elementary crossbridge hypothesis.

Much of the evidence cited has already been published or is in the course of full publication. It will therefore not be described in full here; only those facts which bear directly on crossbridge interaction are given.

METHODS

Glycerol-extracted rabbit psoas or *Lethocerus* dorsal longitudinal flight muscle fibres were prepared and retained at -20° in 50% glycerol for several weeks before use. Rabbit back muscle was used for the production of actin or subfragment-1.

All measurements were made at low ionic strength, 10°, pH 7.0 in the presence of Mg^{2+} and the absence of phosphate or Ca^{2+} (50 mM KCl, 5 mM PIPES, 5 mM EGTA, 6 mM $MgCl_2$, 2 mM NaN_3, up to 2 mM AMPPNP; 1 mM glucose, 10 μg/ml hexokinase and up to 100 μg/ml AP_5A were also added to the AMPPNP solution). Ethylene glycol, up to 50%, was substituted for solvent water, without alteration of solute concentrations.

X-ray diffraction patterns were obtained from large bundles of insect fibre on the high-intensity synchrotron X-ray sources at EMBL, in collaboration with Drs. C.D. Rodger, M. Koch, and J. Bordas (Tregear, et al., 1982). Electron micrographs of sets of insect filaments, isolated by the action of protease (Reedy, Leonard, Freeman & Arad, 1981), were obtained by a variant on the raft technique of Meisner and Beinbrech (1979); for details see Clarke (1982). Mechanical measurements were made on single rabbit and insect fibres, avoiding passage though solution surfaces by a flooding technique (Clarke, 1982). Nucleotide binding on rabbit and insect fibres was measured by the technique of Marston (1973). Acto-subfragment-1 binding was measured in collaboration with Dr. S.B. Marston, by the method of Marston & Weber (1975).

RESULTS & DISCUSSION

The mechanical behaviour of a muscle fibre in rigor, whether insect or rabbit, consists of a near-linear steady tension rise with extension, and thus a near-constant *stiffness* over a considerable range of tension (Marston, et al., 1976). The stiffness is only slightly dependent on the method of measurement; raising the frequency of an applied oscillation from 0.2 to 110 Hz raised the measured stiffness of insect fibres by less than 30%. Thus the mechanics are extremely simple, and approximate to the behaviour of a spring. The spring probably resides in the crossbridges rather than the filaments, since extension of the filaments has not been detected in insect muscle (Miller & Tregear, 1972), and the stiffness of active frog muscle is closely related to filament overlap (Ford, Huxley & Simmons, 1981). The crossbridges of insect muscle in rigor have been clearly visualised as angled bars in electron micrographs (Reedy, 1968) and shown by X-ray diffraction to exist in groups at each turn of the actin helix (Holmes, Tregear & Barrington-Leigh, 1980).

When the equilibrium condition of either a vertebrate or insect muscle in rigor is modified by addition of an unhydrolysed nucleotide, the tension drops (Kuhn, 1973). The fall is considerable, often amounting to some half of the original tension, or three quarters of the stored energy. The process parallels the binding of nucleotide at myosin's enzymatic site (Marston, et al., 1976), and its magnitude increases with the expected fit of the nucleotide into the protein (Marston, Tregear, Rodger & Clarke, 1979). Thus the stored energy in the springs appears to depend precisely on the nucleotide binding to the crossbridges. The stiffness of the muscle is not altered at all by this process, so that if the springs are in the crossbridges their number must be maintained (with the proviso that one in each pair of heads could detach if the spring were in the combined tail).

The crossbridge structure alters when insect muscle binds AMPPNP. Electron micrographs of longitudinal sections taken by the Reedys and their co-workers show crossbridges at 14 nm intervals, apparently at 90° to the filaments, as well as the 45°, 38 nm-spaced, chevrons which are so prominent in rigor (Reedy, 1968); the two sorts of the crossbridge are related to each other in a unique pattern characteristic of AMPPNP-treated muscle, and do not resemble a mixture of rigor and relaxation (Reedy, Reedy & Goody, 1981). Clarke (1982) obtained preparations of separated unfixed parallel thick and thin insect filaments. When AMPPNP was present such filament rafts showed crossbridges of both types, at 90° at 14 nm intervals or at 45° at 38 nm intervals. Sometimes the structure of one form was maintained over several sets of crossbridges along one filament, but both sorts of crossbridge array were also seen on the same filament.

The X-ray diffraction pattern of insect muscle is changed by AMPPNP. The lattice spacing does not alter, but the intensities of the inner equatorials change to values midway between rigor and relaxation. These changes are rapid, reversible and proportional to nucleotide binding (Goody, Barrington-Leigh, Mannherz, Rosenbaum & Tregear, 1976); ADP produces a slighter change (Marston et al., 1976). Thus the motion of the crossbridges across the filament lattice, like the stored energy, appears to be directly related to the quantity and quality of the nucleotide binding.

The layer lines of the X-ray diffraction pattern also change in a characteristic fashion when AMPPNP binds (Barrington-Leigh, Goody, Hofman, Holmes, Rosenbaum & Tregear 1977). The intensities of those lines based on the actin helix reduce, while that based on the myosin helix in the thick filament increases. In one experimental series the actin-based 38 nm layer line fell to some one-half of its intensity in rigor while the myosin-based 14 nm layer line rose to somewhat less than a half of its relaxed value (Table 1A), while the row-line sampling of both remained very sharp (Tregear, Milch, Goody, Holmes & Rodger, 1979). It follows that there probably were extended myosin- and actin-based groupings within the muscle, axially because the intensities of the two reflections were greater than that expected from a random existence of two types of crossbridge (when, for a 1:1 population, each amplitude would be one-half and thus each intensity one-quarter of its maximal value) and laterally because the lattice sampled the reflections. Thus the groupings seen in electron micrographs may well exist in the unfixed, intact lattice.

The crossbridge array in the presence of AMPPNP, although mixed, appears to be stable, for the structure will store energy for a long time. The tension in a stretched muscle, either rabbit or insect, did not appreciably fall over 15 hours (Clarke & Tregear, 1982). Myosin heads binding AMPPNP detach from actin in solution in less than a second

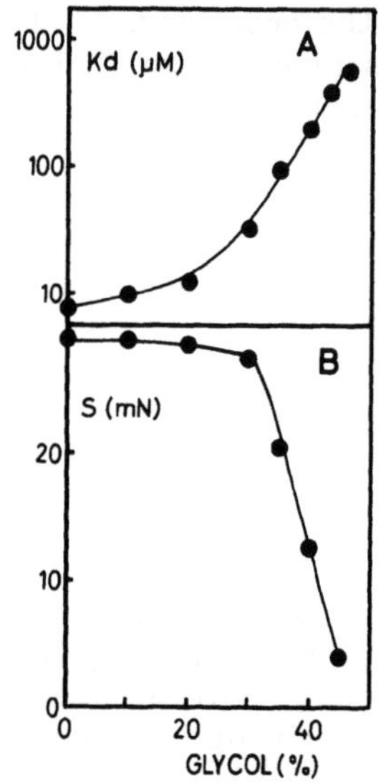

Figure 1: The effect of substituting ethylene glycol for water in the solution bathing rabbit muscle proteins and glycerol-extracted fibres. 1 mM Mg AMPPNP present throughout.
A. The dissociation constant of acto-subfragment-1 (Clarke, Marston & Tregear, 1981).
B. The stiffness of single psoas fibres held at constant tension (Clarke, 1982).

Table 1: The intensity of diffraction from the actin and myosin-based layer lines X-ray diffraction pattern of insect flight muscle in two experiments.

Condition	Intensity (fraction of 4$\bar{1}$$\bar{5}$0) of: 31 peak on 38 nm 1.1	00 peak on 14 nm 1.1
A: addition of AMPPNP in aqueous solution (Tregear et al., 1979)		
Rigor	0.52	0.32
+ 0.5 or 1 mM Mg AMPPNP	0.31	1.10
+ 5 mM Mg ATP	0.13	3.20
B: substitution of ethylene glycol for 50% of the solvent water, followed by addition of AMPPNP (Tregear et al., 1982)		
Rigor	0.44	1.05
50% glycol	0.26	2.06
+ 2 mM Mg AMPPNP	0.12	5.36
Aqueous + 5 mM Mg ATP	0.11	5.11

(Marston, 1982) so that either the axial or lateral array of the crossbridges appears to stabilise their existence.

Substitution of ethylene glycol for water in the bathing medium of insect or rabbit muscle fibres in rigor has a similar mechanical effect to that of AMPPNP; the tension falls but the stiffness remains unaltered

(Clarke, 1982). The lattice spacing of insect muscle falls by approximately 5% on addition of 50% glycol (Tregear, et al., 1982). This resembles the effects of dextran, pH and ionic strength on rabbit muscle seen by Arata and Podolsky (1982). The intensities of the various diffraction maxima from insect muscle change in the same way on glycol substitution as they do when AMPPNP is added; the relative intensities of the 14 nm and 38 nm intensities are again too high to fit to a mixture of independently assorted relaxed and rigor crossbridges, although the 14 nm intensity is less than half that in relaxation (Table 1B). Thus glycol substitution appears to induce a similar structural change in the muscle, without the specific effects of nucleotide at the enzymatic site.

Substitution of ethylene glycol for water in the bathing medium of a muscle fibre which is already binding AMPPNP causes a further reduction in the tension, without loss of stiffness, until at a critical level of glycol the stiffness falls abruptly to the relaxed level (Fig. 1.). Insect muscle requires only some 15% glycol to cause "equilibrium relaxation" while rabbit muscle requires 40% glycol (Tregear, et al., 1982). The effect of ethylene glycol on the dissociation constant of rabbit acto-subfragment 1 in the presence of AMPPNP is more graded than that on muscle fiber stiffness (Fig. 1A). The dissociation constant changes by a factor of ten over the range necessary to change from 10% to 90% relaxation (Fig. 1) which is one-tenth the amount expected on independent action of the crossbridges. Co-operative protein association has also been found for regulated actin and sub-fragment-1 binding AMPPNP (Williams & Greene, 1982).

Once fully relaxed in this manner, the enzymatic site of the myosin appears to bind AMPPNP tightly and to release it only very slowly, as if the conformation had become tightened up (Tregear, et al., 1982); active-site trapping can also be caused by cross linkage of sulfhydryls (Wells, Sheldon & Yount, 1980).

CONCLUSIONS

The structural changes when AMPPNP is added to insect flight muscle do not readily fit with the notion of independent action of crossbridges. A simple crossbridge rotation, which would account for the mechanical change, would not obviously produce either the arrays of 14 nm images seen in the electron micrographs, or the 14 nm intensification in the X-ray diffraction pattern. The simple explanation of these observations is that the crossbridges form two, or more, distinct types of array. The extension of the array across the myofibril indicates that thin filament structure is also involved. The retention of tension by the muscle could be accounted for by such crossbridge interaction, as could the criticality of ethylene glycol content in causing its relaxation.

These observations have been selected for specific conditions where crossbridges do not act independently in generating tension and storing energy. They should therefore be treated with caution. It may be both that insect flight muscle is peculiarly adapted to show crossbridge interaction and that equilibrium conditions tend to make all muscles do so. Thus the general hypothesis may still hold for most active muscles. The present observations do, however, indicate that crossbridge interaction does occur under some conditions.

REFERENCES

Arata, T. and Podolsky, R.J. (1982). Crossbridge flexibility derived from the influence of lattice spacing on mechanical properties of muscles fibres in rigor. Biophys. J. 37: 362a.

Barrington-Leigh, J., Goody, R.S., Hofman, W., Holmes, K.C., Rosenbaum, G. and Tregear, R.T. (1977). In: *Insect Flight Muscle* ed. R.T. Tregear. Amsterdam: North-Holland pp 137-146.

Clarke, M.L., Marston, S.B. and Tregear, R.T. (1980). An attempt to assess the biochemically effective actin concentration in rabbit skeletal muscle. J. Muscle Res. Cell. Motil. 1: 447-448.

Clarke, M.L. (1982). The attachment of myosin heads to actin in the presence and absence of an unhydrolysable analogue of ATP. D. Phil. Thesis, Oxford.

Clarke, M.L. and Tregear, R.T. (1982). Tension maintenance and crossbridge detachment. FEBS Letters (in press).

Ford, L.E., Huxley, A.F. and Simmons, R.M. (1981). The relation between stiffness and filament overlap in stimulated frog muscle fibres. J. Physiol. 311: 219-249.

Goody, R.S., Barrington-Leigh, J., Mannherz, H.G., Tregear, R.T. and Rosenbaum, G. (1976). X-ray titration of binding of β, γ, imido ATP to myosin in insect flight muscle. Nature, 262: 613-615.

Holmes, K.C., Tregear, R.T. and Barrington-Leigh, J. (1980). Interpretation of the low angle X-ray diffraction from insect flight muscle in rigor. Proc. Roy. Soc. B. 207: 13-33.

Kuhn, H.J. (1973). Transformation of chemical energy into mechanical energy by glycerol-extracted fibres of insect flight muscle in the absence of nucleosidetriphosphate hydrolysis. Experientia, 29: 1086-1088.

Marston, S.B. (1973). Kinetic studies of the contractile mechanism of muscle. D. Phil. Thesis; University of Oxford.

Marston, S.B. (1982). The rates of formation and dissociation of actin-myosin complexes. Biochem. J. 203: 453-460.

Marston, S.B. and Weber, A.M. (1975). The dissociation constant of the actin-heavy meromyosin subfragment-1 complex. Biochemistry 14: 3868-3873.

Marston, S.B., Rodger, C.D. and Tregear, R.T. (1976). Changes in muscle crossbridges when β, γ - imido-ATP binds to myosin. J. Mol. Biol. 104: 263-276.

Marston, S.B., Tregear, R.T., Rodger, C.D. and Clarke, M.L. (1979). Coupling between the enzymatic site of myosin and the mechanical output of muscle. J. Mol. Biol. 128: 111-126.

Meisner, D. and Beinbrech, G. (1979). Alterations of crossbridge angle induced by β, γ - imido-adenosine-triphosphate. Electron microscope and optical diffraction studies on myofibrillar fragments of abdominal muscles of the crayfish *Orconectes limosus*. Eur. J. Cell. Biol. 19: 189-195.

Miller A. and Tregear R.T. (1972). Structure of insect fibrillar flight muscle in the presence and absence of ATP. J. Mol. Biol. 70: 85-104.

Reedy, M.K. (1968). Ultrastructure of insect flight muscle. I. screw sense and structural grouping in the rigor cross-bridge lattice. J. Mol. Biol. 31: 155-176.

Reedy, M.K., Leonard, K.R., Freeman, R. and Arad, T. (1981). Thick myofilament mass determination by electron scattering measurements with the scanning transmission electron microscope. J. Musc. Res. Cell. Motil. 2: 45-64.

Reedy, M., Reedy, M.K. and Goody, R.S. (1981). Crossbridge structure in rigor and AMP.PNP states of insect flight muscle. Biophys. J. 33: 22a.

Tregear, R.T., Milch, J., Goody, R.S., Holmes, K.C., and Rodger, C.D. (1979). X-ray diffraction of insect flight muscle. In: *Crossbridge Mechanism in Muscle Contraction*, pp 407-423, ed. by H. Sugi and G. Pollack, University of Tokyo Press.

Tregear, R.T., Clarke, M.L., Marston, S.B., Rodger, C.D., Bordas, J., and Koch, M. (1982). A study of demembranated muscle fibres under equilibrium conditions. In: *Basic Biology of Muscles: a Comparative Approach*. Ed B. Twarog and R. Levine. Raven Press, NY..

Wells, J.A., Sheldon, M. and Yount, R.G. (1980). Magnesium nucleotide is stoicheometrically trapped at the active site of myosin and its active proteolytic fragments by thiol cross-linking reagents. J. Biol. Chem. 255: 1598-1602.

Williams, D.L. and Greene L.E. (1982). Comparison of the effect of tropomyosin and tropomin-tropomyosin on acto-Sl. Biophys. J. 37: 50a.

DISCUSSION

DREIZEN: As I recall, about 20 years ago Brahms and Kay reported experiments on the effects of ethylene glycol on myosin ATPase. I wonder whether that could acount for at least part of your results?

TREGEAR: Yes, there is something, not in Brahms and Kay, but in some more recent work by Travis and Hillaire which has relevance; the high concentrations of glycol have a considerably greater effect than the low concentrations. I suppose my defense to that is twofold. One is quantitative, namely that the change in dissociation constant is only a factor of 10 in the range over which relaxation occurs where you would expect it would have been much greater. The other is that the insect muscle relaxes at a much *lower* concentration of glycol where glycol has very little effect on the dissociation constant.

RITZ-GOLD: If you increase the concentration of AMP-PNP slowly, do you see any sudden changes as you might expect from a possible phase transition? Does magnesium pyrophosphate have any such effect?

TREGEAR: I haven't looked at pyrophosphate. As you may know, the AMP-PNP effect on its own appears to be totally linear, that is to say, the tension drop that you get is proportional to the amount found at the binding site, as far as one can say of such things. They both give proper binding constants of around 50 to 100 μM. So I'm afraid that that doesn't fit in with my hypothesis. I've concentrated all along anyway on the *discrepancies* from linearity. There are a lot of things that go linear.

REEDY: In the pictures that Mary Reedy has produced much more than I from the material that Roger Goody and I prepared, we could see, in our best examples of unprocessed images, a *reasonable* number of bridges that looked as if they were making an approximation of a 380 repeat of something chevron-like. Likewise, we could see some stripes. But it looked like there might be something else not completely associated with those. So we are still juggling the question that seems more strongly put by the filtered, averaged images we looked at, namely when we see these two different conformational populations of a 145 element and a sort of active coordinated element, we don't know for sure whether we're looking at two populations of cross-bridges or whether we're looking at two ends, i.e., whether each cross-bridge is divided into two domains in which one end expresses its loyalty to the myosin repeat and the other expresses its loyalty to the actin repeat. This domain idea versus the two population idea is one we deliberately tried to keep in the air. It sounds like you're coming down in favor of the two populations of cross-bridges idea, and I just want to identify that point.

TREGEAR: Yes, that was what the experiment was intended to test. The results, as far as they go, tend to give that conclusion. I'm not totally convinced of the conclusion, but that was exactly what we hoped to find out.

SCHOENBERG: In your insect muscle you showed that when you added about one millimole of AMP-PNP to the rigor muscle, you got a drop of tension and then tension seemed to be maintained for a rather long period of time. Two questions. One is: was there calcium present in those solutions, the second is - did you try the experiment with higher concentrations of AMP-PNP?

TREGEAR: No, there wasn't calcium present. I understand the significance of your question because of the work from your group. We do intend to go back and do a lot of work relative to calcium now. It wasn't until your work actually came out in the biophysical abstracts of this year that I got the point.

SCHOENBERG: What about the higher concentration of AMP-PNP?

TREGEAR: Well, we have used higher concentrations -- Maxine Clarke, actually, in her thesis work (D. Phil. Thesis, Univ. of Oxford, 1982). She found no change in response up to five millimolar AMP-PNP. There is a difficulty, of course, because you've got contaminating ADP$-NH_2$ which may produce contaminating ATP as a result.

WRAY: Richard, I've been very concerned with the effect of glycol on the lattice shrinkage and I've been wondering along the following lines. When Lethocerus muscle fibers relax, the filament lattice shrinks slightly compared to rigor. Conversely, one might expect that if you shrink the rigor lattice, you would influence the equilibrium between the attached and detached states so as to favor detachment. Could the "equilibrium relaxation" found with glycol reflect partly the effect of glycol in shrinking the lattice? The apparent cooperativity you observe is to be expected if at a certain critical glycol concentration the compressive force becomes greater than the array of attached bridges can resist.

TREGEAR: Yes, of course. How could I say otherwise to a precise idea of this sort? It opens what I hope will be debated later this week, the relationship between this work and that of others. We have a spectrum here between AMP-PNP which does not have an effect on the spacing, as far as we can determine, and glycol which does have an effect on the spacing and interacts with the AMP-PNP in a rather specific way, as I've just shown; later I think we have papers from Dr. Podolsky, for example, on the effect of dextran and pH, and ionic strength, I believe, as well, which *all* have a rather similar effect in the sense that they change the tension in a rigor muscle. They also change the spacing, and these two are very closely related, and clearly here's a puzzle. How do we connect all those data together, because at one end of the spectrum we've got AMP-PNP which is having a mechanical effect that does *not* have a lattice effect and is binding to an enzymatic site; this appears to be a specific effect. At the other end of the spectrum you've got things electrically active, and having their effect by changing the spacing. I don't know the simple answer to your question.

CROSS-BRIDGE STATES IN INVERTEBRATE MUSCLES

John S. Wray

Max Planck Institute for Medical Research, Heidelberg, West Germany

ABSTRACT

Arguments are presented for doubting whether the effect of AMPPNP on insect flight muscle in rigor signals a reversion of the power stroke of attached cross-bridges. Instead, the effect of this nucleotide on insect and other muscles may be better explained in terms of the behavior of detached bridges. Knowledge of events in the detached half of the contractile cycle may nevertheless be relevant to understanding the mechanism of energy transduction.

The classic study of insect flight muscle by Reedy, Holmes and Tregear (1965) has suggested a specific structural basis for force generation and filament sliding in muscle. According to this view, myosin cross-bridges lie approximately perpendicular to the filaments in relaxed muscle; following attachment to the thin filaments they change to a more angled conformation characteristic of rigor, and then detach and return to their resting state. H.E. Huxley (1983) has suggested that direct evidence for cross-bridge swinging in active muscle is provided by his recent synchrotron measurements of the 143 Å X-ray reflection from contracting frog muscle. Other authors have questioned whether the conformation of myosin heads can change while they are attached to actin, implying some quite different structural change such as shortening of the S2 part of the molecule (Harrington, 1979). An important approach to this problem is provided by the effects of ATP analogs on the structure of insect flight muscle (Goody, Holmes, Mannherz, Barrington-Leigh and Rosenbaum, 1975). In particular, the effects of the analog AMPPNP have been interpreted as indicating a reversion of the cross-bridge conformation to an earlier stage in the power stroke (Tregear, Milch, Goody, Holmes and Rodger, 1979). I discuss this conclusion here in the light of further experimental results. It is useful first to emphasize briefly certain features of the structure of muscle in relaxed and rigor states.

Structure of Relaxed Muscle

Myosin heads in the relaxed state have nucleotide bound to them in the form of the hydrolysis products of ATP, and are believed to be largely or entirely prevented from attaching to actin. The diversity of structure adopted by relaxed bridges in different muscles is well known (Wray, 1982). To differing extents in different muscles, relaxed bridges form an ordered array whose symmetry reflects the symmetry of the myosin filament backbone. It is less clear how this cross-bridge order is maintained: probably it is due partly to the connection through S2 to the backbone surface lattice, and partly to direct interactions between bridges. This order is independently modifiable by such factors as pH, ionic strength and interfilament distance. In the absence of ATP, cross-bridges bind to actin unless prevented from doing so by, for example, stretching the muscle to non-overlap. Induction of iodoacetate rigor in non-overlap frog muscles was reported by Haselgrove (1975) to cause loss of the characteristic resting X-ray pattern. Stretched relaxed glycerinated *Limulus* muscle shows a similar loss of the relaxed pattern on removing ATP, with the added advantage that the effect can be controlled and reversed. At least in these muscles, therefore, cross-bridges move substantially when the hydrolysis products of ATP are no longer bound. It would be interesting to know how biochemical and structural states are linked in the course of this transition.

Structure of Rigor Muscle

Rigor muscle poses more complex problems of structural description, requiring consideration of both actin and myosin filaments and their geometrical relationship. A natural starting point for such a discussion is the structure of *Lethocerus* flight muscle in rigor, for which a widely accepted description was given by Offer, Couch, O'Brien & Elliott (1981). This model developed the concept of target areas (Reedy, 1968) in the light of the idea that the two myosin heads of one molecule may attach to different thin filaments because actin monomers occur on neighboring thin filaments in this lattice with favorable separation and orientation. Such two-filament binding would predominate only if further requirements were met (Offer et al. 1981), namely the matching of helical tracks which endows this muscle with its specialized stretch-activation properties (Wray, 1979).

Application of these ideas allows one to specify the pattern of marking of the actin lattice with only minor ambiguities. Offer et al. have suggested a likely pattern of interconnection between the myosin surface lattice and the actin target areas (see their Fig. 3b). One important feature of this representation is the significant fraction of heads that remain unbound, for steric reasons. Another is the pattern of large axial distortions required for attachment, a consequence of the incommensurate relationship between the axial spacings of myosin and actin filaments. Attached cross-bridges must vary in their axial distortion by up to ±60 Å, almost regardless of the precise model presupposed for the structure. As noted by Reedy & Garrett (1977), this periodic distortion manifests itself in insect muscle as a longer repeat (equal to 1160 Å, or three 'double chevrons').

The structure of other invertebrate muscles in rigor cannot be described so simply as for insect flight muscle, but the same target area attachment probably persists and is manifest as broadly similar rigor X-ray patterns (Wray, Vibert & Cohen, 1978). Rigor in other muscles may differ from that in insect muscle in showing less precise chevron structure, less predominant two-filament binding, and perhaps fewer non-binding heads.

Effects of AMPPNP

AMPPNP has a dramatic mechanical effect on insect flight muscle in rigor, namely a reversible decrease of rigor tension by about 40%, with negligible loss of stiffness (Marston, Rodger and Tregear, 1976). The X-ray pattern from the muscle in the presence of AMPPNP differs from the rigor pattern. At 4°C these differences comprise changes in the 'outer layer lines' arising from the decoration of the actin helical structure by myosin heads, in the strong inner reflections whose intensity depends on the distribution of mass within the entire unit cell, and in the meridional reflections indexing on 145 Å corresponding to the distribution of density on the myosin filament periodicity. These changes have been interpreted intuitively as meaning that cross-bridge density in AMPPNP is no longer actin-based as in rigor but has moved so as to take up the myosin-based symmetry (Tregear et al., 1979). Many specific models of such a change can be envisaged; while these have not so far been investigated by model calculations based on the X-ray data, it is attractive to suppose that cross-bridges in AMPPNP have reverted to a less angled conformation, enhancing the 145 Å reflection and reducing muscle tension.

However, two types of observation suggest caution over such a conclusion. First, although the mechanical effect of AMPPNP is observed at room temperature (Kuhn, 1981) as clearly as at 4°C, the changes in the X-ray pattern are greatly reduced (Goody et al., 1975). Secondly, other muscles behave differently from *Lethocerus* (Wray, MS in preparation). Thus the pattern from *Limulus* muscle in rigor (Wray et al., 1974) is not observably changed by AMPPNP at either temperature, while that from crayfish muscle (Wray et al., 1978) shows merely some enhancement of the 145 Å series of reflections, to a smaller extent than for insect muscle. *Limulus* muscle fibers showed a smaller (approximately 10%) loss of tension in AMPPNP. The asynchronous flight muscle of bumble bees, on the other hand, responds to AMPPNP in the same way as *Lethocerus*. These results suggest that asynchronous insect flight muscles behave in a manner that is not typical of less specialized invertebrate muscles, and that the large effects observed with AMPPNP are not common to all muscle types. Instead, structural and mechanical changes may be independent of each other. The natural question arises whether these diverse effects can be subsumed under a common scheme for the mode of action and significance of AMPPNP.

AMPPNP-State as Modified Rigor

As noted in the last section, the modification of the rigor structure corresponding to the mechanical effect of AMPPNP in insect flight muscle is observed fully only at low temperature. Conversely, the large structural effect seen at low temperature may have no necessary mechanical counterpart. Evidently the sought-for structural correlate of the mechanical effect involves either only a small part of the cross-bridge, or a disordered part, or only a small population of bridges.

The structure of muscle in rigor is such that bridges must necessarily be structurally heterogeneous. They are therefore likely to be mechanically heterogeneous also, so that the tension in a muscle in rigor cannot be uniformly distributed but must be borne by a small fraction of bridges. Mechanical observations on muscles in rigor and rigor-like states should therefore reflect the behavior of some minority of the attached bridges that are bearing a disproportionate amount of tension. Knowledge of structure allows, and even demands, that structural and mechanical effects be assigned to particular sites in this inhomogeneous structure. In rigor insect muscle, sites of particularly large strain occur within each 1160 Å period of 'good' and 'bad' attachments. If, as is reasonable on kinetic grounds, the effect of AMPPNP were to modify strained bridges preferentially, a large drop in tension could result with only a small change in structure. The nature of the change is uncertain: it might be an increased flexibility, a conformational change, or merely detachment. The point is that most of the attached bridges would be little changed by the binding of AMPPNP. The immediate structural consequence of AMPPNP binding for cross-bridge structure would then be inaccessible to observation, since too small a fraction of cross-bridges is involved.

On this view, the striking structural changes caused by AMPPNP at low temperature would be largely independent of the changes underlying the loss of tension. A plausible explanation would be a change in the conformation of the unattached rigor bridges, returning to a form more like that in relaxed muscle (that is, more ordered and more perpendicular to the filament). It is not to be expected that the state of individual detached bridges is identical to that in relaxed muscle, where the order probably requires systematic interactions with a full complement of detached neighboring heads. It would be this recovery of the relaxed-like conformation that shows a marked dependence on temperature. The modification of the outer layer lines indicates some change in the majority of attached bridges as a result of the binding of AMPPNP; possibly increased flexibility permits slight readjustment to ease any intramolecular strain caused by two-filament binding. Finally, the dramatic changes in the inner layer lines could well result from the significant redistribution of mass within the unit cell caused by the recovery of relaxed-like structure, though the relevant transform calculations have not yet been done.

An important test of these proposals is how far they can explain the electron microscopic appearance of insect muscle in AMPPNP (Reedy,

Reedy and Goody, 1981). The persistence of modified flared X's, and of target areas in actin sections, fit directly with the model, while the loss of chevron detail could reflect the increase in flexibility envisaged as the main change from rigor in the majority of the cross-bridge population.

The results from *Limulus* and crayfish muscles fit well with these ideas. The response to AMPPNP in a particular muscle type will depend on the distribution of strain among cross-bridges following induction of rigor, and the visibility of the structural change will depend on the number of detached bridges, on the ability of the nucleotide to re-establish an ordered cross-bridge state, and on whether two-filament or one-filament binding predominates. The diversity of muscle structure is especially marked in regard to the organization of detached cross-bridges, and the diversity of response to AMPPNP is perhaps more naturally understandable in terms of the behavior of detached bridges than in terms of diversity in the nature of the power stroke itself.

DISCUSSION

The above arguments suggest a working hypothesis for the action of AMPPNP. According to this hypothesis, the nucleotide does not in general cause a reversion of the cross-bridges to another attached conformation, and its effects are not evidence for cross-bridge rotation. The effects of AMPPNP on attached bridges are obscure, but in any case not uniform. The observed structural effects relate instead to unattached heads. The variability of the response of different muscles, and of insect flight muscle at different temperatures, reflects the behavior of these free heads. Could such effects nevertheless have relevance to understanding contraction?

Clearly myosin heads must change structure following detachment from actin in such a way as to complete the contractile cycle. This detached transition might proceed by an entirely different pathway; but it is also possible that it involves a reversal of the switch mechanism that comprises the power stroke, and if so the question of the transduction mechanism could be stated alternatively in terms of events in the detached part of the cycle. The results above are relevant to describing this reverse transition. Removal of nucleotide from relaxed *Limulus* cross-bridges changes their state (in some way that cannot yet be defined), so that they no longer take up their regular position and conformation on the myosin filament surface. Experiments on stretched muscles of other types will show how far all cross-bridges behave similarly. The behavior of unattached cross-bridges may offer clues to the cross-bridge mechanism that are easier to interpret than results on necessarily heterogeneous attached states. Behind this and other recent approaches (e.g. Shriver and Sykes 1981) lies a conception of the transduction step as a property of the myosin molecule and filament, not just a change in the conformation of myosin heads relative to actin.

ACKNOWLEDGEMENTS

I am grateful to Drs. K.C. Holmes and R.S. Goody for their constant advice and criticism.

REFERENCES

Goody, R.S., Holmes, K.C., Mannherz, H.G., Barrington-Leigh, J. & Rosenbaum, G. (1975). X-ray studies of insect flight muscle with ATP analogues. Biophys. J. 15: 687-705.

Harrington, W.F. (1979). On the origin of the contractile force in skeletal muscle. Proc. Natl. Acad. Sci. USA. 76: 5066-5070.

Haselgrove, J.C. (1975). X-ray evidence for conformational changes in the myosin filaments of vertebrate striated muscle. J. Mol. Biol. 92: 113-143.

Huxley, H.E. (1983). This volume.

Kuhn, H.J. (1981). The mechanochemistry of force production in muscle. J. Muscle Res. Cell Motil. 2: 7-44.

Marston, S.B., Rodger, C.D. and Tregear, R.T. (1976). Changes in muscle cross-bridges when β,γ-imido ATP binds to myosin. J. Mol. Biol. 104: 263-276.

Offer, G., Couch, J., O'Brien, E. and Elliott, A. (1981). Arrangement of cross-bridges in insect flight muscle in rigor. J. Mol. Biol. 151: 663-702.

Reedy, M.K. (1968). Ultrastructure of insect flight muscle. J. Mol. Biol. 31: 155-176.

Reedy, M.K., Holmes, K.C. and Tregear, R.T. (1965). Induced changes in orientation in the cross-bridges of glycerinated insect flight muscle. Nature 207: 1276-1280.

Reedy, M.K. and Garrett, W.E. (1977). Electron microscope studies of insect flight muscle in rigor. In: *Insect Flight Muscle*, pp. 115-136, ed. R.T. Tregear. Amsterdam: Elsevier North Holland.

Reedy, M.C., Reedy, M.K. and Goody, R.S. (1981). Cross-bridge structure in rigor and AMPPNP states of insect flight muscle. Biophys. J. 33: 22a.

Shriver, J.W. and Sykes, B.D. (1981). Phosphorus-31 Nuclear Magnetic Resonance Evidence for two conformations of myosin subfragment-1 nucleotide complexes. Biochemistry 20: 2004-2012.

Tregear, R.T., Milch, J.R., Goody, R.S., Holmes, K.C. & Rodger, C.D. (1979). The use of some novel X-ray diffraction techniques to study the effect of nucleotides on cross-bridges in insect flight muscle. In: *Cross-bridge Mechanism in Muscle Contraction*, ed. Sugi, H. and Pollack, G.H., pp. 425-440. Tokyo: Univ. of Tokyo Press.

Wray, J.S. (1979) Filament geometry and the activation of insect flight muscles. Nature 280: 325-326.

Wray, J.S. (1982). Organization of myosin in invertebrate thick filaments. In: *Basic Biology of Muscles: A Comparative Approach*, ed. Twarog, B.M., Levine, R.J.C. and Dewey, M.M., New York: Raven Press.

Wray, J.S., Vibert, P.J. and Cohen, C. (1974). Cross-bridge arrangements in *Limulus* muscle. J. Mol. Biol. 88: 343-348.

Wray, J.S., Vibert, P.J. and Cohen, C. (1978). Actin filaments in muscle: pattern of tropomyosin/troponin attachments. J. Mol. Biol. 124: 501-521.

DISCUSSION

REEDY: I'd like to repeat for re-emphasis and confirmation what idea you're putting forth. It seems that you're adopting the view that Offer and Elliot are essentially close to correct in estimating that maybe half the bridges are loose in rigor in insect muscle and, given that, that the major structural changes in the rigor pattern that are induced by adding AMP-PNP at 4°C are a response of those unattached, loose heads.

Whereas the interesting bridges, the ones we'd like to know about, are the ones that remain attached in rigor and AMP-PNP, but their response is somewhat hidden, swamped out in the X-ray pattern by the fact that the X-ray pattern changes are dominated by the response of unattached heads.

WRAY: Yes, I'm suggesting that neither the mechanics nor the structural change necessarily relates directly to what we want to know, which is the effect of the analog on the bridges that remain attached. The mechanical change could be an effect on some of the attached bridges that are maximally strained. Either they're coming off or they're swinging back.

REEDY: By maximally strained do you mean those that are maximally force-bearing in the rigor tension state?

WRAY: Yes.

HUXLEY: Is it reasonable to think that the attached bridges would be sort of divided into two groups, one which were not much strained and which were maximally strained? Wouldn't you think there would be some sort of annealing process so that the strains tend to average out?

WRAY: The geometry of the final structure seems to require large inhomogeneities. You can argue it is possible to modify the pattern of attachment in order to make things a little more uniform. But in order to do that you end up with something that doesn't look consistent with available structural evidence, such as the "double chevron" appearance.

GULATI: In the muscles out of overlap, when you put them in rigor, do you get actin-based reflections or myosin-based reflections?

WRAY: Well, the cross-bridges have no access to actin, and very little order remains at all. There's a weakened 145 Å reflection, and no off meridionals. I'm thinking basically of *Limulus* here.

GILLIS: Is it true that the mismatching of the periodicities of the actin and the myosin could give regions where a certain number of cross-bridges can attach to the thin filament, and when the constraints get too demanding they cannot attach any longer, thereby giving the rationale for these sets of cross-bridges that Richard Tregear has spoken of?

HOLMES: I don't think that would be a natural explanation.

KAWAI: Coming back to this non-overlap experiment, do you see the same thing with vertebrate skeletal muscles, and what do you think is the mechanism?

WRAY: I haven't done that experiment, but a comparable one was reported by John Haselgrove. The disadvantage of his experiment was that he did it on intact muscle by putting it into iodoacetate rigor, and it's very hard to know what exactly has happened in those circumstances. It would be important to know what the result is in skinned muscle.

HUXLEY: I don't think Haselgrove's was a totally uncontrolled experiment. I mean, you get perfectly good labelled actin patterns when you look at iodoacetate rigor with normal overlap, but with no overlap you don't get myosin layer lines.

WRAY: What worried me was that unattached cross-bridges are very sensitive to changes in their environment, and it's important in that experiment to know that when you see nothing, it's not the pH or something that has changed.

KAWAI: How do you explain the disappearance of the 145 Å in rigor in *Limulus* at non-overlap?

WRAY: That's indeed a good question which I can't answer, but it remains a very dramatic fact that cross-bridges behave completely differently when they have lost their nucleotide.

TREGEAR: What happened when you added the AMP-PNP back in *Limulus* at non-overlap?

WRAY: Nothing.

TREGEAR: Then why do you assert that something would happen when you added it at overlap? I mean you've just done the control, got the wrong results, and then persist in fitting your hypothesis!

WRAY: No, the result on *Limulus* is not a control on my suggestion about *Lethocerus.* I presume that muscles do vary in the way their detached bridges depend on nucleotide, which in fact strikes me as more satisfying than that their attached bridges behave differently.

FACTORS AFFECTING THE EQUATORIAL X-RAY DIFFRACTION PATTERN FROM CONTRACTING FROG SKELETAL MUSCLE

H. Tanaka, H. Hashizume* and H. Sugi

Department of Physiology, School of Medicine, Teikyo University, Itabashi-ku, Tokyo 173, Japan
+Research Laboratory of Engineering Materials, Tokyo Institute of Technology, Nagatsuda, Midori-ku, Yokohama 227, Japan

ABSTRACT

Changes in the equatorial X-ray diffraction pattern from tetanized frog sartorius muscles (*Rana catesbiana*) were studied by use of time-resolved data collection technique (time resolution, 0.5 sec) to give information about the dynamic properties of the cross-bridges. No significant changes in the intensity ratio of two equatorial reflections ($I_{1,0}/I_{1,1}$) were observed when isometrically contracting muscles were slowly stretched by 5-6%, in spite of marked force changes. The intensity ratio also showed no significant changes when the load on isometrically contracting muscles was suddenly increased from P_0 to 1.2-1.5 P_0 to produce isotonic muscle lengthening. Closer examination of the data indicated that a small decrease in the value of $I_{1,1}$ was caused by both slow stretch and isotonic lengthening. Because of the scatter of experimental plots in $I_{1,0}$, the effect of small change in $I_{1,1}$ on the intensity ratio fell within the range of accuracy of measurement. It is suggested that no marked changes in myosin head orientation or in the number of the cross-bridges in the vicinity of the thin filaments take place in response to slow stretches or isotonic lengthening, and that the decreased regularity of the filament lattice may produce the change in $I_{1,1}$.

INTRODUCTION

Vertebrate skeletal muscles exhibit two prominent equatorial X-ray reflections, 1,0 and 1,1, which arise from the hexagonal arrays of the thick and thin filaments. When a muscle contracts isometrically, the intensity of 1,0 reflection ($I_{1,0}$) decreases while that of 1,1 reflection ($I_{1,1}$) increases, each by about a factor of two (Huxley, 1968; Haselgrove & Huxley, 1973). This has been interpreted by Huxley and his co-workers as being due to the radial movement of the cross-bridges from the vicinity of thick filaments towards that of the thin filaments (Huxley & Brown, 1967; Huxley, 1968; Haselgrove & Huxley, 1973).

However, there are many other possible causes leading to the changes in equatorial reflections. For example, a large change in the equatorial intensities may occur as a consequence of a more precise localization of the thin filaments at the trigonal points in the hexagonal thick filament lattice when they become cross-linked to the thick filaments (Huxley, 1953, 1957). Another possibility is that the equatorial reflections come entirely from the A band because only in that region both filaments are regularly arranged in the hexagonal lattice; the thin filaments out of the A band are irregularly arranged so that they do not contribute to the equatorial intensities (Elliott, Lowy & Worthington, 1963). On the other hand, computer modeling work of Lymn (1978) have shown that different cross-bridge orientation may give different numerical values for $I_{1,0}$ and $I_{1,1}$, while the number of attached cross-bridges remains the same.

When a muscle is slowly stretched during isometric tetanus, the force rises to a peak, and decays towards a new isometric level which is higher than the normal isometric force at the same length (Abbott & Aubert, 1952; Sugi, 1972). On the other hand, when a load greater than the isometric force P_0 is applied on a muscle during isometric tetanus, the muscle is lengthened by the applied load. These perturbations imposed on tetanized muscles are expected to cause some changes in the number of attached cross-bridges, in the orientation of attached cross-bridges or in the regularity of the thick and thin filament lattice structures, which may be detected as the changes in the equatorial intensities.

The present experiments were undertaken to give information about the dynamic behavior of the cross-bridges in response to slow stretch and isotonic lengthening by recording the equatorial X-ray diffraction patterns with the time-resolved data collection technique.

METHODS

The sartorius mucle dissected from the bullfrog *Rana catesbiana* (slack length, 4-5 cm) was mounted vertically in an experimental chamber with two Mylar windows. The pelvic end was clamped to a strain guage (UT-500, Shinko), while the tibial end was connected to a vibrator (G-21-010, Shinken) or to one arm of an isotonic lever. Constant-velocity slow stretches were applied with the vibrator controlled by a feedback circuit. Quick increases in load above P_0 were applied by removing the stop of the isotonic lever, so that the muscle was lengthened by the load above P_0 attached to the opposite arm of the lever. The initial sarcomere length was adjusted to 2.2-2.4 μm by the diffraction pattern of He-Ne laser light. The muscle was continuously perfused with precooled Ringer solution (115 mM NaC1 2.5 mM KC1, 1.8 mM $CaC1_2$, pH 7.2 by $NaHCO_3$) at 4-6°C, and tetanized with 3 sec train of supramaximal 2 msec pulses at 20 Hz given through a multi electrode-assembly. At 1 sec after the onset of stimulation, slow stretches (5-6% of L_0 at about 0.1 L_0 /sec) or isotonic loads of 1.2-1.5 P_0 were applied to the muscle. This procedure was repeated 5 times at 3 min intervals unless otherwise stated.

The X-ray source was a rotating-anode generator (RU-200, Rigaku) with a fine focus (1 × 0.1 mm) on a copper target. The generator was operated at 40 kV with a tube current of 30 mA, and a camera with a bent-crystal monochromater was used to produce a line focused beam. Distributions of X-ray scattered from muscle were detected by a linear position sensitive proportional counter (Rigaku) with an external delay line for position encoding. The specimen to counter distance was 45 cm. The counter operated with 90% Xe and 10% CO_2 mixture at 1 atm, providing a spatial resolution of about 0.15 mm.

Data were stored in computer memory to give 15 diffraction patterns of 0.5 sec time resolution. The data collection started at 1 sec before the onset of tetanic stimulation. In each diffraction pattern thus obtained, the background was drawn by eye and the area under each peak was

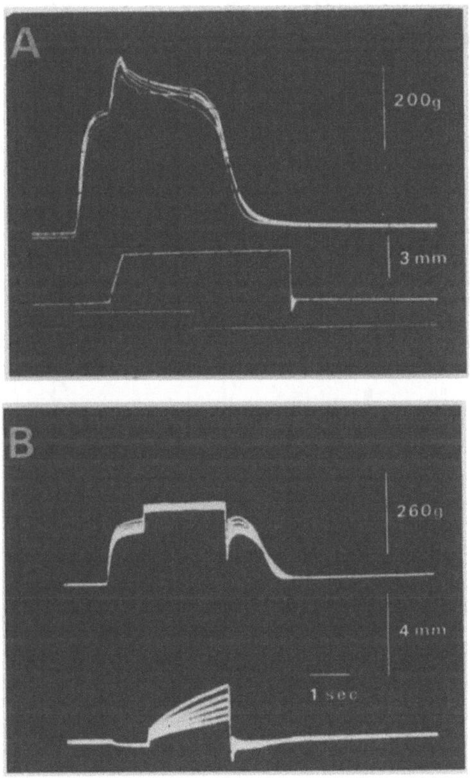

Figure 1: A: Superimposed records of force changes in response to a slow stretch applied to an isometrically tetanized muscle. The upper traces show the force changes, while the middle and lower traces show the length changes and the period of stimulation respectively. Stimulation was repeated 5 times. The initial sarcomere length was 2.3 μm. B: Superimposed records of force and length changes when the load on an isometrically contracting muscle was increased from P_0 to 1.2-1.5 P_0 to produce isotonic lengthening. The upper and lower traces show the force and the length changes respectively. The initial sarcomere length was 2.3 μm.

measured. Only the average values of $I_{1,0}$ or $I_{1,1}$ at both sides of the diffraction pattern were used in the present experiments.

RESULTS

Effect of Slow Stretch

Fig. 1A shows a typical time course of force and length changes when a muscle was subjected to a slow stretch during isometric tetanus. The force rose a peak which was reached at the end of stretch, and then decayed towards a new isometric level higher than the initial level. The peak isometric force attained by slow stretches was 1.4-1.6 P_0, while the excess tension above the initial isometric level at the end of stimulation was 0.20-0.25 P_0. Fig. 2 exhibits the changes in $I_{1,0}$, $I_{1,1}$ and the ratio $I_{1,0}/I_{1,1}$ during the experiments as shown in Fig. 1A. Before stimulation, the intensity ratio was about two. On stimulation, the intensity ratio

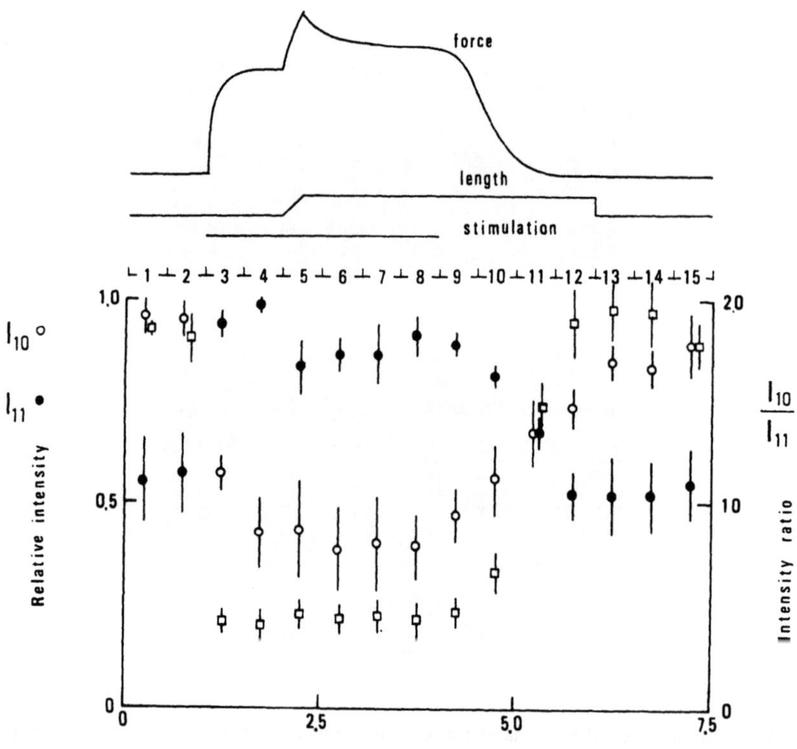

Figure 2: Changes in $I_{1,0}$ (open circles), $I_{1,1}$ (filled circles) and $I_{1,0}/I_{1,1}$ (squares) before, during and after a slow stretch applied to an isometrically tetanized muscle. Each data point represents mean ± S.D. from 5 muscles. Force and length records are also shown in the upper part. The data were collected during 15 different phases of 0.5 sec duration, which are also indicated and numbered in the upper part.

decreased rapidly to a low value which was maintained until the end of stimulation. The decrease in the intensity ratio was induced by a decrease in $I_{1,0}$ and an increase in $I_{1,1}$.

Table 1: Changes in the equatorial reflection intensities at various phases of mechanical response.

Phase number	4	5	6	7	8
Muscles subjected to slow stretches (N=5)					
$I_{1,0}/I_{1,1}$	0.36 ± 0.06	0.46 ± 0.11	0.37 ± 0.06	0.41 ± 0.09	0.38 ± 0.09
$I_{1,0}$	0.42 ± 0.09	0.43 ± 0.12	0.38 ± 0.11	0.40 ± 0.12	0.39 ± 0.08
$I_{1,1}$	0.99 ± 0.02	0.83 ± 0.07	0.86 ± 0.04	0.86 ± 0.08	0.91 ± 0.05
Muscles simply tetanized isometrically (N=5)					
$I_{1,0}/I_{1,1}$	0.39 ± 0.08	0.42 ± 0.07	0.43 ± 0.04	0.45 ± 0.07	0.42 ± 0.05
$I_{1,0}$	0.43 ± 0.09	0.46 ± 0.07	0.47 ± 0.06	0.49 ± 0.08	0.48 ± 0.05
$I_{1,1}$	0.96 ± 0.04	0.97 ± 0.04	0.94 ± 0.04	0.96 ± 0.04	0.98 ± 0.02

Values are mean ± SD. The values of $I_{1,0}$ and $I_{1,1}$ are expressed relative to the maximum values obtained in each experiment. The phases of data collection are illustrated in the upper part of Fig. 2 together with the force and length changes.

In the upper part of Table 1 are summarized the average values of $I_{1,0}/I_{1,1}$, $I_{1,0}$ and $I_{1,1}$ obtained from five muscles at various phases of the stretch experiment as illustrated in Fig. 2. The inital phases of contraction (phases 1-3 in Fig. 2) were excluded, since attention was focused on whether slow stretches caused some changes in the equatorial reflections. A small increase in $I_{1,0}/I_{1,1}$ at the phase of stretch (phase 5) was statistically insignificant ($P{\sim}0.2$), indicating that the intensity ratio was constant within the range of accuracy of measurement in spite of large force changes in response to slow stretches. The variation in the values of $I_{1,0}$ was also within the range of accuracy of measurement ($P>0.6$). On the other hand, the value of $I_{1,1}$ at the phase before stretch was significantly larger than those in other phases ($P{\sim}0.02$).

In the lower part of Table 1, the corresponding values obtained from five muscles which were simply tetanized isometrically for 3 sec without stretch are shown, the isometric force being well maintained. All the value showed no significant changes during the maintained isometric force generation ($P>0.3$).

The experiments were also made in which five muscles were isometrically tetanized at two different lengths corresponding to the length before stretch and that after stretch. The isometric force level decreased by 5-10% when the muscle was tetanized at the longer length, with a tendency of $I_{1,0}$ to increase and that of $I_{1,1}$ to decrease (Elliott et al., 1963). However, no significant differences in the values of $I_{1,0}/I_{1,1}$, $I_{1,0}$ and $I_{1,1}$ were observed between the muscles tetanized at the short length and those tetanized at the long length (P~0.2), indicating that the equatorial reflection intensities are insensitive to muscle length at least within the range studied (Podolsky, St. Onge, Yu & Lymn, 1976; Sugi, Amemiya & Hashizume, 1978).

Effect of Isotonic Lengthening

Fig. 1B shows typical force and length changes when a load of 1.2-1.5 P_0 was suddenly applied to an isometrically tetanized muscle to produce isotonic lengthening. The load was removed at the end of stimulation, so that the duration of isotonic lengthening was 2 sec. The distance of isotonic lenthening was 5-10% of L_0. The amount of load relative to P_0 increased as the isometric force decreased each time the muscle was

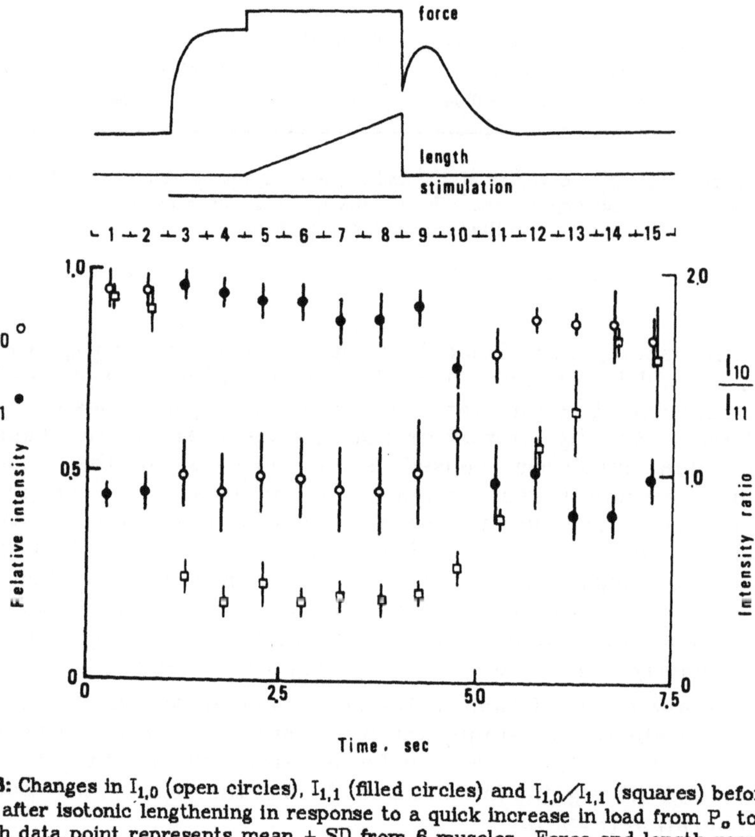

Figure 3: Changes in $I_{1,0}$ (open circles), $I_{1,1}$ (filled circles) and $I_{1,0}/I_{1,1}$ (squares) before, during and after isotonic lengthening in response to a quick increase in load from P_0 to 1.2-1.5 P_0. Each data point represents mean ± SD from 6 muscles. Force and length records are also shown in the upper part. The data were also collected during 15 different phases of 0.5 sec duration.

stimulated, and as a result both the velocity and amount of isotonic lengthening increased about twofold at the end of the experiment. Fig. 3 shows the changes in $I_{1,0}$, $I_{1,1}$, and the ratio $I_{1,0}/I_{1,1}$ during the experiments as shown in Fig. 1B. The decrease in the intensity ratio on stimulation was also by a decrease in $I_{1,0}$ and an increase in $I_{1,1}$. In the upper part of Table 2, the average values of $I_{1,0}/I_{1,1}$, $I_{1,0}$ and $I_{1,1}$ obtained from six muscles before and during the isotonic lengthening are shown. Both the intensity ratio $I_{1,0}/I_{1,1}$ and $I_{1,0}$ before lengthening were not significantly different from those during lengthening ($P>0.5$), while the value of $I_{1,1}$ during lengthening was significantly smaller than that before lengthening ($P\sim0.02$).

Table 2: Changes in equatorial reflection intensities under various conditions.

Muscles subjected to isotonic lengthening (N=6)		
	before lengthening (phase 3-4)	during lengthening (phase 5-8)
$I_{1,0}/I_{1,1}$	0.41 ± 0.06	0.44 ± 0.09
$I_{1,0}$	0.47 ± 0.10	0.47 ± 0.12
$I_{1,1}$	0.97 ± 0.02	0.90 ± 0.05
Muscles at rest (N=4)		
	control length	stretched length
$I_{1,0}/I_{1,1}$	1.98 ± 0.08	2.47 ± 0.09
$I_{1,0}$	0.94 ± 0.02	0.92 ± 0.03
$I_{1,1}$	0.93 ± 0.05	0.72 ± 0.03

The phases of data collection are indicated in the upper part of Fig. 3 together with the force and length changes.

The experiments were also done in which five muscles were simply tetanized isometrically for 3 sec without isotonic lengthening. The values of $I_{1,0}/I_{1,1}$, $I_{1,0}$ and $I_{1,1}$ did not change significantly during the maintained isometric force generation ($P>0.5$).

When the muscle at rest was stretched to the same extent as in the isotonic lengthening experiment in Fig. 1B, the value of $I_{1,0}/I_{1,1}$ at the stretched muscle length was definitely greater than that at the control length (Elliott et al., 1963; Haselgrove & Huxley, 1973) as shown in the lower part of Table 2. The value of $I_{1,0}$ did not change significantly by stretch, while that of $I_{1,1}$ decreased definitely as shown in Table 2.

DISCUSSION

The present experiments have shown that, when an isometrically contracting frog skeletal muscle was slowly stretched by 5-6% or isotonically lengthened under a load of 1.2-1.5 P_o, the intensity ratio $I_{1,0}/I_{1,1}$ did not change significantly in agreement with the previous reports (Sugi, Amemiya & Hashizume, 1977; Amemiya, Sugi & Hashizume, 1979; Yagi & Matsubara, 1977), while $I_{1,1}$ showed a small decrease. On the other hand, the values of $I_{1,0}$ exhibited considerable scatter due to the difficulty in measuring them accurately, so that the effect of the small decrease in $I_{1,1}$ on the intensity ratio was masked by the wide range of accuracy of measurement. These results may, however, be taken to indicate that no marked changes in the number of the cross-bridges in the vicinity of the thin filaments or in the orientation of attached cross-bridges, since these kinds of changes are expected to cause reciprocal changes between $I_{1,0}$ and $I_{1,1}$ to be readily detected by the changes in the intensity ratio. Naylor & Podolsky (1981) and Tanaka et al. (1982) observed no significant changes in $I_{1,0}/I_{1,1}$ ratio when glycerinated rabbit muscle fibers and frog sartorius muscles are stretched in rigor state.

It has been known that $I_{1,0}/I_{1,1}$ increases with increasing muscle length (Haselgrove & Huxley, 1973). The dependence of the intensity ratio on muscle or sarcomere length is much less marked in isometrically contracting muscles than in resting muscles (Podolsky et al., 1976; Sugi et al., 1978). In the present study, the increase in the intensity ratio by stretching a resting muscle was due to the decrease in $I_{1,1}$ while $I_{1,0}$ did not change significantly. Elliott et al., (1963) suggested that the thin filament lattice is only regularly arranged at the A-band in resting muscles. On this basis, the relative independence of $I_{1,0}/I_{1,1}$ and $I_{1,1}$ on muscle length in isometrically contracting muscles may result from the thin filament lattice becoming more regular everywhere due to the cross-links between the thick and thin filaments so that $I_{1,1}$ is no longer sensitive to the amount of overlap between the filaments. It is also possible that during isometric contraction, the sarcomere spacings become so irregular that they can no longer be controlled effectively by changing muscle length.

Concerning the small decrease of $I_{1,1}$ in response to slow stretches or isotonic lengthening, one possibility may be that the thin filament lattice is made less regular when a tetanized muscle is slowly stretched or isotonically lengthened.

REFERENCES

Abbott, B.C. and Aubert, X.M. (1952). The force exerted by active striated muscle during and after change of length. J. Physiol. 117: 77-86.

Amemiya, Y., Sugi, H. and Hashizume, H. (1979). X-ray diffraction studies on the dynamic properties of cross-bridges in skeletal muscle. In: *Cross-bridge Mechanism in Muscle Contraction*, ed. Sugi, H. and Pollack, G.H., pp. 425-443, Tokyo: University of Tokyo Press and Baltimore: University Park Press.

Elliott, G.F., Lowy, J. and Worthington, C.R. (1963). An X-ray and light diffraction study of the filament lattice of striated muscle in the living state and in rigor. J. Molec. Biol. 6: 295-305.

Haselgrove, J.C. and Huxley, H.E. (1973). X-ray evidence for radial cross-bridge movement and for the sliding filament model in actively contracting skeletal muscle. J. Molec. Biol. 77: 549-568.
Huxley, H.E. (1953). X-ray analysis and the problem of muscle. Proc. Roy. Soc. B141: 59-62.
Huxley, H.E. (1957). The double array of filaments in cross-striated muscle. J. Biophys. biochem. Cytol. 3: 631-648.
Huxley, H.E. and Brown, W. (1967). The low angle X-ray diagram of vertebrate striated muscle and its behavior during contraction and rigor. J. Molec. Biol. 30: 383-434.
Huxley, H.E. (1968). Structural difference between resting and rigor muscle; evidence from intensity changes in the low-angle X-ray diagram. J. Molec. Biol. 37: 507-520.
Naylor, G.R.S. and Podolsky, R.J. (1981). X-ray diffraction of strained muscle fibers in rigor. Proc. Natl. Acad. Sci. U.S.A. 78: 5559-5563.
Podolsky, R.J., Onge, R.St., Yu, L. and Lymn, R.W. (1976). X-ray diffraction of actively shortening muscle. Proc. Natl. Acad. Sci. U.S.A. 73: 813-817.
Sugi, H. (1972). Tension changes during and after stretch in frog muscle fibers. J. Physiol. 225: 237-253.
Sugi, H., Amemiya, Y. and Hashizume, H. (1977). X-ray diffraction of active frog skeletal muscle before and after a slow stretch. Proc. Japan Acad. 53B: 178-182.
Sugi, H., Amemiya, Y. and Hashizume, H. (1978). Time-resolved X-ray diffraction from frog skeletal muscle during an isotonic twitch under a small load. Proc. Japan Acad. 54B: 559-564.
Tanaka, H., Hashizume, H. and Sugi, H. (1982). Effect of stretch on the equatorial diffraction pattern from frog skeletal muscle in rigor. (This volume).
Yagi, N. and Matsubara, I. (1977). Equatorial X-ray reflections from contracting muscle after an applied stretch. Pflüger Arch. 372: 113-114.

DISCUSSION

MATSUBARA: Your observation that the equatorial intensities in tetanized muscle are relatively insensitive to the sarcomere length are in contradiction with those by Haselgrove and Huxley (J. Mol. Biol. 77: 549-568, 1973) and by us (Yagi and Matsubara, Pflügers Archiv. 372: 113-114, 1977). I wonder how you explain the difference?

HASHIZUME: First of all, I would like to point out that our result is in agreement with that of Podolsky's group at NIH.

HUXLEY: Wait a minute. I think there may be two separate issues getting confused here. The question is whether we're talking about large static changes observed at different sarcomere lengths, or the effect of the actual changing length, itself, during contraction. What is the extent of the length change that you are referring to?

HASHIZUME: In our experiment it is about 5 or 6% of the muscle length.

HUXLEY: Yes. Well, in the Huxley and Haselgrove experiments, these were over ranges of sarcomere lengths between two microns and three microns, and they were static experiments. We weren't stretching the active muscle. We were simply looking at the isometric pattern at different sarcomere lengths. So I don't think there's any contradiction about *those* experiments anyway. Your experiment, Ichiro, is a different one, I suspect.

MATSUBARA: Well, in our case, we did much the same experiments as Hashizume, giving slow stretch and slow release to the active muscle.

By changing the sarcomere length by about 10% we found a significant change in the intensity ratio.

HASHIZUME: I'm talking about the insensitivity in muscle isometrically tetanized at different sarcomere lengths. This is a static experiment. If we plot the intensity on the ordinate and sarcomere length on the abscissa, then the dependence of the intensity ratio to the muscle length or sarcomere length is weak. This is what I mean by insensitivity.

EDMAN: Does this method have potential to decide about the number of cross-bridges which are attached during contraction? Can you distinguish whether bridges are just very close to the I filament or actually attached to it?

HASHIZUME: This is a difficult question. We have no measure of whether the cross-bridges are attached to, or just in the vicinity of, the thin filaments. As far as we take the rather naive assumption that the intensity ratio represents, in one way, the number of cross-bridges in the vicinity of the thin filament, then the slow twitch and the lengthening doesn't change appreciably the number of cross-bridges in the vicinity of the thin filaments. That is the only conclusion which I can draw from our results.

HUXLEY: Well, that would agree with Dr. Podolsky's results and with mine, that relatively slow shortening also doesn't have an effect.

GILLIS: So the mystery of the excess tension remains complete! There's no reason why the tension should increase by 20% when you stretch a muscle. You don't find any explanation in the cross-bridges.

HASHIZUME: Yes, there *is* an enhanced tension, but the intensity ratio is unchanged. We are then obliged to say that the individual cross-bridges generate greater tensions in one way or another.

EFFECT OF STRETCH ON THE EQUATORIAL X-RAY DIFFRACTION PATTERN FROM FROG SKELETAL MUSCLE IN RIGOR

H. Tanaka, H. Sugi and H. Hashizume*

Department of Physiology, School of Medicine, Teikyo University, Itabashi-ku, Tokyo 173, Japan
**Research Laboratory of Engineering Materials, Tokyo Institute of Technology, Nagatsuda, Midori-ku, Yokohama 227, Japan*

Changes in the equatorial X-ray diffraction pattern during stretch of frog sartorius muscle in rigor were studied. No significant changes in $I_{1,0}/I_{1,1}$ were observed by stretching the rigor muscles. The value of $I_{1,1}$ was, however, found to decrease reversibly by stretch, while the value of $I_{1,0}$ was so small that its changes by stretch fell within the range of accuracy of measurement.

Lymn (1978) suggested that different orientation of cross-bridges gave different numerical values for $I_{1,0}$ and $I_{1,1}$, assuming the same number of attachments. Experiments were undertaken to investigate the effect of stretch on the rigor muscles in which some changes in cross-bridge orientation might be expected, and then a considerable change in $I_{1,0}/I_{1,1}$ might be expected. The muscles were repeatedly tetanized at L_0 until no appreciable force development was seen on stimulation, and then treated with 1 mM iodoacetic acid for more than 12 hours. A time resolved data collection technique was used, in which 16 diffraction patterns of 0.5 sec resolution were obtained during 8 sec period of data collection (Tanaka, Hashizume & Sugi, 1982). After 2 sec on the trigger pulse of data collection, the muscles were stretched by 1-3% L_0 for 3 sec, and then released. This cycle was repeated 10 times at 10 sec intervals to obtain reasonable photon statistics for each diffraction pattern. Fig. 1 shows $I_{1,0}/I_{1,1}$, $I_{1,1}$ and $I_{1,0}$ as a function of time, and the average values of $I_{1,0}/I_{1,1}$, $I_{1,1}$ and $I_{1,0}$ at various phases are shown in Table 1. As shown in Table 1, the value of $I_{1,1}$ was found to decrease significantly during stretch ($P < 0.001$), and returned reversibly to the initial value when the muscle was released. On the other hand, the value of $I_{1,0}$ was so small that its changes by stretch fell within the range of accuracy of measurement, thus resulting in no significant change in $I_{1,0}/I_{1,1}$ (Naylor & Podolsky,

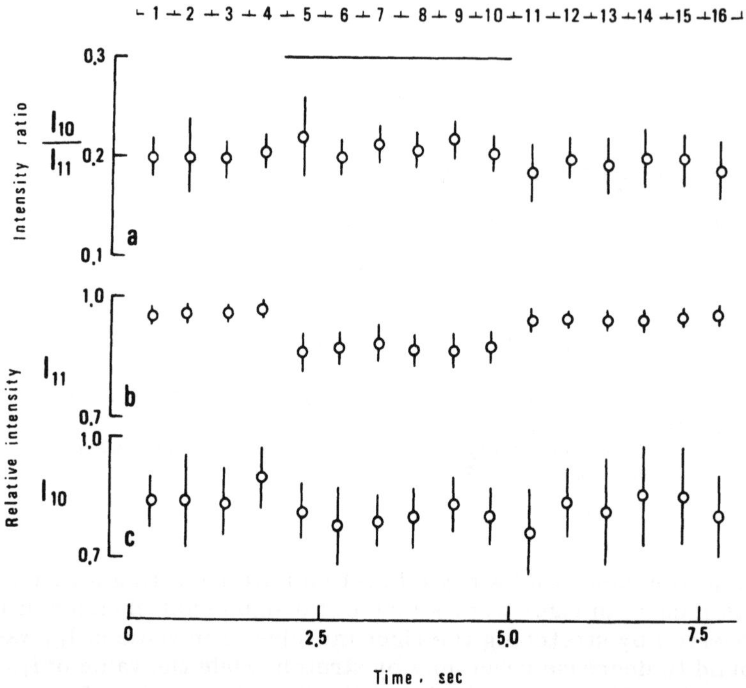

Figure 1: Time course of change in the equatorial diffractions from muscles in rigor. a: $I_{1,0}/I_{1,1}$, b: $I_{1,1}$, c: $I_{1,0}$. Each point represents a mean ± SD from 10 muscles. $I_{1,1}$ and $I_{1,0}$ were expressed relative to the maximum values obtained from each muscle. The horizontal bar represents the period of stretch. The data were collected during 16 different phases of 0.5 sec duration, which are indicated and numbered in the upper part.

Table 1: Effect of stretch on the equatorial diffractions from muscles in rigor.

	Muscles relaxed (phase 1-4)	Muscles stretched (phase 5-10)	Muscles relaxed (phase 11-16)
$I_{1,0}/I_{1,1}$	0.20 ± 0.02	0.21 ± 0.02	0.20 ± 0.02
$I_{1,1}$	0.97 ± 0.01	0.87 ± 0.05	0.95 ± 0.01
$I_{1,0}$	0.86 ± 0.07	0.81 ± 0.06	0.83 ± 0.08

Values are mean ± SD (N = 10). $I_{1,1}$ and $I_{1,0}$ were expressed relative to the maximum values obtained from each muscle. The phases of data collection are indicated in the upper part of Fig. 1 together with the period of stretch.

1981). A computational approach was made in which we approximated both filaments as simple cylinders of uniform electron density. A decrease in the regularity of the filament lattice structures was modeled by smearing out the filament mass into cylinders of larger radii and correspondingly lower densities. Thus it was possible to obtain a satisfactory agreement between the computed and the experimental results.

REFERENCES

Hashizume, H., Tanaka, H. and Sugi, H. (1982). Factors affecting the equatorial X-ray diffraction pattern from contraction and rigor skeletal muscles. (This volume).

Lymn, R.W. (1978). Myosin subfragment-1 attachment to actin. Biophys. J. 21: 93-98.

Naylor, G.R.S. and Podolsky, R.J. (1981). X-ray diffraction of strained muscle fibers in rigor. Proc. Natl. Acad. Sci. U.S.A. 78: 5559-5563.

DISCUSSION

HUXLEY: Did you stretch muscles at rest?

TANAKA: Yes. In the case of relaxed muscles $I_{1,1}$ decreased by about 20% and a small change in $I_{1,0}$ was observed. This resulted in a 25% increase in $I_{1,0}/I_{1,1}$ by stretch of about 2% L_0. As these values show, the effect of stretch on the equatorial reflection intensities in rigor was less marked than in fibers at rest. This indicates that the rigor muscles have cross-linkages between the thick and thin filaments, and that the regularity of the filament lattice structures in rigor may be better than that in the resting fibers.

HUXLEY: I once undertook a similar experiment, but you showed more details. Did you find a dependence of the decrease in $I_{1,1}$ on the amount of stretch?

TANAKA: With an increase in the amount of stretch, a further decrease in $I_{1,1}$ was observed. But, we need further experiments to elucidate a relation between the amount of stretch and the change in $I_{1,1}$.

NOBLE: You could not obtain any evidence which suggested the rotation of crossbridges by an X-ray method. Do you have confidence that crossbridges rotate? If you use other techniques, for example, a fluorescent technique, do you think you can get an appreciable change which suggests the rotation of crossbridges?

TANAKA: I do not expect so.

STRUCTURAL STUDIES OF MUSCLE DURING FORCE DEVELOPMENT IN VARIOUS STATES

Leepo C. Yu, Toshiaki Arata, Alasdair C. Steven, Geoffrey R.S. Naylor*, Ronald C. Gamble+, and Richard J. Podolsky

Laboratory of Physical Biology, National Institute of Arthritis, Diabetes, and Digestive and Kidney Diseases, National Institutes of Health, Bethesda, Maryland 20205
**Physical Chemistry Laboratory, Oxford University, South Parks Road, Oxford, England OX1 3QZ*
+California Institute of Technology, Pasadena, California 91125

ABSTRACT

Structural studies concerned with force generating mechanisms in striated muscle fibers in different states are described.

The first study deals with fibers in the "rigor" state, where ATP is absent and all the myosin heads form cross-bridges with the actin-containing filaments. In this state large axial forces are developed when ionic strength is reduced to very low levels. At the same time, the fiber expands radially, as indicated by both X-ray diffraction and light microscopy. Comparison of the latter two measurements indicates that force is developed in part because of differences in the lateral expandability of different parts of the sarcomere. Thus, under these conditions, force appears to be modulated by factors that operate at the filament rather than the cross-bridge level.

The second study deals with the location of the myosin heads in the relaxed state, and the mass movement that takes place when the fiber is physiologically activated. By using the intense X-ray source at the Stanford Synchrotron Radiation Laboratory, five equatorial reflections were recorded for both the relaxed and the activated state, and a spatial resolution of 100 Å was obtained. Analysis of the data indicates that (1) in the resting state, myosin heads protrude out from the thick filaments and extend toward the thin filaments, and (2) upon activation, significant loss of mass occurs only in the region peripheral to the thick filament backbone through a movement that has a pronounced azimuthal component. The latter movement can be taken as the overall shift in myosin subfragment 1 during the cross-bridge cycle.

INTRODUCTION

We would like to describe several structural studies concerned with force generation in striated muscle fibers. The first of these has to do with one of the mechanisms by which the force developed by muscle fibers in the rigor state is modulated by ionic strength, pH, and osmotic pressure. This question is of interest because these parameters can produce large force changes in rigor fibers which are sometimes discussed in relation to physiological force generation. Another study deals with the distribution of mass within the myofilament lattice of resting frog muscle fibers, and the change in this distribution that takes place following physiological activation.

Rigor State

The simplest state of the actomyosin system is that in which ATP is absent. In this "rigor" condition, all the myosin heads form cross bridges with the actin-containing filaments (Cooke & Franks, 1980; Lovell & Harrington, 1981). The force developed by a fiber in this state depends on the strain imposed upon it as well as on the composition of the bathing solution. This behavior is shown in Figure 1. The upper trace shows the force record of a fiber bundle of an ionic strength of 50 mM; the lower trace shows the myofilament lattice spacing.

The bundle was first strained about 1%, which produced a steady

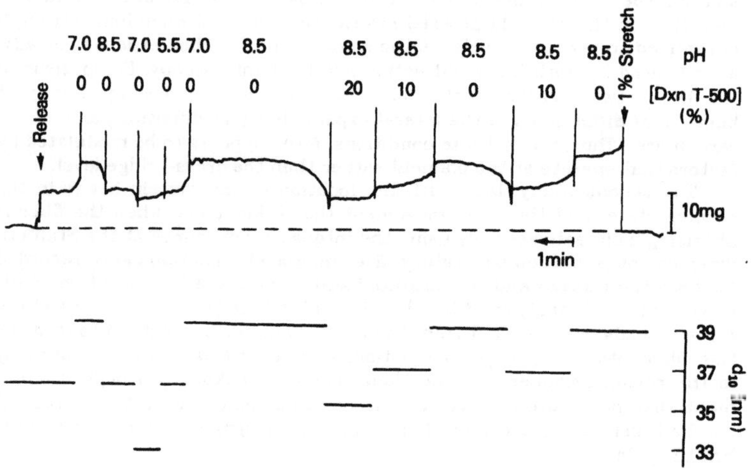

Figure 1: Effect of lattice spacing on isometric force. *Upper panel:* Isometric force record from a glycerinated psoas fiber bundle in solutions where lattice spacing was controlled with Dextran T-500 (right side) or change in pH (left side). Bundle diameter, 100 μ. *Lower panel:* Influence of solution composition on lattice spacing. X-ray diffraction measurements were made in a separate experiment with the solutions used in the upper panel but with a 200 μ diameter bundle. Solution composition for pH 8.5: 40 mM triethanolamine HC1, 10 mM NaC1, 0.1 mM $MgCl_2$; pH 7.0: 40 mM imidazole, 30 mM NaC1, 0.1mM $MgCl_2$; pH 5.5: 40 mM Na acetate, 30 mM NaC1, 0.1 mM $MgCl_2$. Temperature, 40°C.

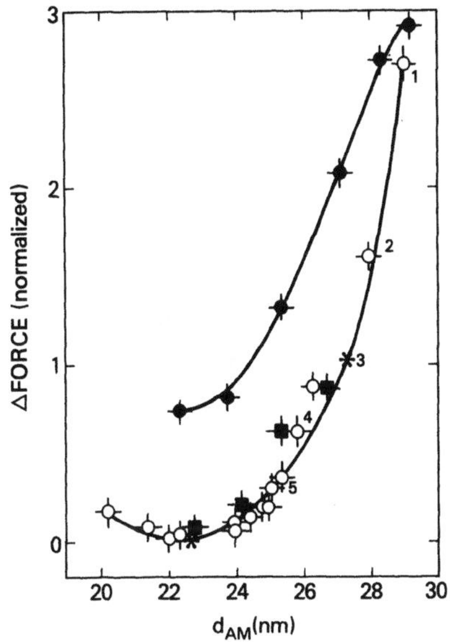

Figure 2: Effect of lattice spacing and ionic strength on isometric force at sarcomere length 2.2-2.3 μm. Lattice spacing was controlled by addition of 0-50% Dextran to solution at γ = 0.005 M, pH 8.5 (●) or γ = 0.025 M, pH 8.5 (■). pH was also varied between 5.5 and 8.5 at γ = 0.005–0.15 M (O); points 1,2,3,4 and 5 are for pH 8.5 at γ = 0.005, 0.015, 0.015, 0.05, and 0.075, respectively. Temperature = 40°C. For each preparation, force was measured relative to the value at γ = 0.15 M, pH 5.5 (star); the unit ΔF was taken as the force increment when the solution was changed to γ = 0.025 M, pH 8.5 (star), which came out to be 130 ± 10 g/cm^2 in magnitude, where the cross sectional area refers to that of the fiber in the relaxed state. Each point is the average value of several determinations under the same conditions; bars give the standard error of the mean.

force of about 20 mg. When 10% dextran T-500 was added to the bathing solution, which compressed the filament lattice, the force dropped markedly. The force recovered when the dextran was withdrawn. Similar decreases in force could be produced by lowering the pH of the bathing solution, which also caused the filament spacing to decrease (Figure 1, left), or by increasing the ionic strength at a given pH (data not shown).

Over a considerable range of solution composition, force is a single valued function of filament spacing. This is shown in Figure 2, where the open circles are data at different pH values and ionic strengths, and the closed points are solutions containing various concentration of dextran. For each preparation, force was measured relative to the value at γ = 150 mM, pH 5.5 (lower asterisk). The unit of force increment was taken as the force when the ionic strength was lowered to 25 mM and the pH raised to 8.5 (upper asterisk), which caused the lattice spacing to increase from about 23 nm to 27 nm.

The first thing we see is that the force is a very non-linear function of filament spacing. We also note that the electrostatic changes produced by changes in ionic strength and pH (open circles) can generally be

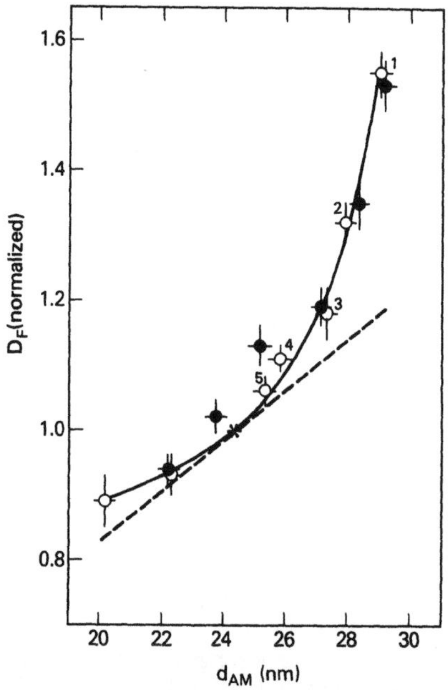

Figure 3: Relation between fiber diameter D_F, and lattice spacing, d_{AM}, in rabbit psoas fibers in rigor. Lattice spacing and fiber diameter were controlled by addition of 0-50% Dextran to solutions at $\gamma = 0.005$ M, pH 8.5 (●). Also, the pH was varied between 5.5 and 8.5 at $\gamma = 0.005-0.05$ M (○); composition of solutions for numbered points given in legend to Figure 2. Temperature = 40°C. The data are normalized relative to the values at $\gamma = 0.05$ M, pH 7.0 (star). Dotted line shows case for which the relative changes in lattice spacing and fiber diameter are equal. Sarcomere length = 2.2-2.3 μm.

counteracted by adding various amounts of dextran to the 25 mM, pH 8.5 solution (closed squares). However when the lattice was expanded a great deal, by lowering the ionic strength to 5 mM, the original force is no longer completely restored by addition of dextran, even though the lattice spacing is returned to the initial value. In this condition the force apparently depends on more than lattice spacing alone.

What causes these force changes? One factor is suggested by the data shown in Figure 3. This experiment was carried out to see whether the increase in fiber diameter in a low ionic strength, high pH solution (measured by light microscopy) is proportional to the increase in myofilament lattice spacing (measured by X-ray diffraction). We would expect this to be the case in general only if the fiber were isotropic. However we know that the sarcomere is highly differentiated axially, and it would be surprising if the region that gives rise to the strong equatorial reflections would expand radially in exactly the same way as other parts of the sarcomere.

In Figure 3, the abscissa is the filament spacing, obtained from the equatorial X-ray diffraction pattern, and the ordinate is the fiber diameter, measured by light microscopy. The broken line is the expected relation if the two parameters were proportional and the myofibrils always remained cylindrical in shape. The actual results, however, deviate markedly from this line when the filament spacing exceeds 26 nm, and they indicate that sarcomeres undergo significant distortion in this range of spacing.

For relatively large lattice spacings, the relation between fiber diameter and spacing (right side of Fig. 3) resembles that between force and spacing (right side of Fig. 2). This suggests that changes in the rigor force may be caused in part by the distortion of the sarcomere. In fact, if we define distortion as the difference beteween the fiber diameter and diameter of the region that gives rise to the diffraction pattern, and plot the rigor force at a given fiber length against this function, we find the relation shown in Figure 4. The correlation at the higher force levels (normalized force between 1 and 3) suggests that over a certain range of filament spacing force may be linearly related to sarcomere distortion. It is interesting that this part of the curve back extrapolates to a finite force at zero distortion. This suggests that distortion is not the only factor involved in force development in the rigor fibers, and that another effect, unrelated to distortion, may also affect force development in the rigor state. The latter effect has been associated with the geometry of the cross-bridge in the rigor state, and has been described elsewhere (Podolsky, Naylor & Arata, 1982).

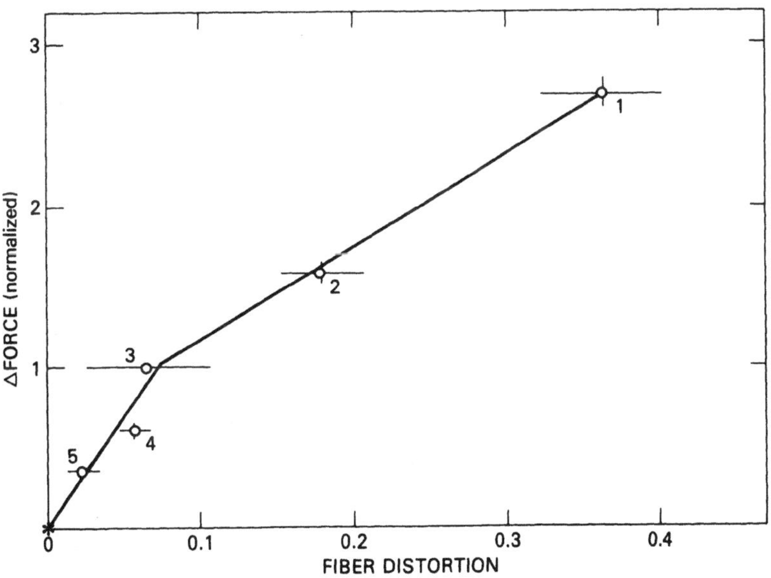

Figure 4: Relation between isometric force and fiber distortion. Distortion, defined as the difference between the fiber diameter and the diameter of the region that gives rise to the diffraction pattern, is taken from Figure 3. The corresponding force is taken from Figure 2. The numbered points are the same in the three figures; the solution compositions are given in the legend to Figure 2.

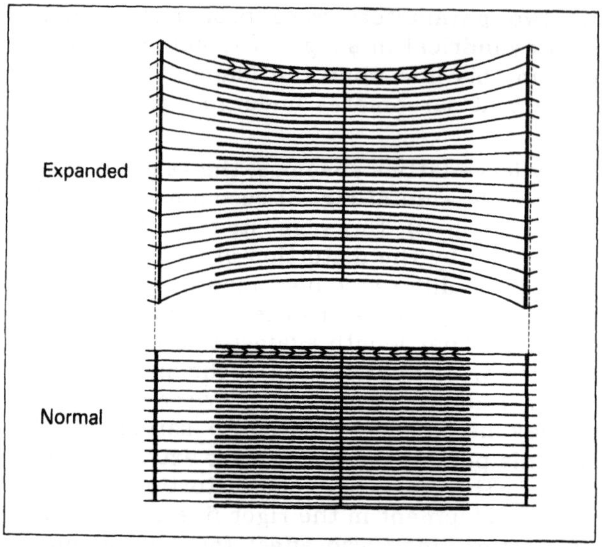

Figure 5: Possible shape of normal and radially expanded sarcomere. The filaments in the normal sarcomere (bottom) are assumed to be axially oriented. The expanded sarcomere (top) may expand more at the Z line than at the middle of the A band; the equatorial X-ray diffraction pattern presumably arises from the axially oriented filaments in the middle of the A band.

Figure 5 shows the kind of sarcomere distortion that could occur when the sarcomere expands laterally. The lower drawing shows the sarcomere at normal ionic strength, where the shape is presumably cylindrical. The upper drawing is the electrostatically stressed sarcomere. If the equatorial reflections were produced by the region near the middle of the A band, and the fiber diameter were set by the region near the Z line, the edge of the myofibril would be scalloped. (Another possibility is that the M regions sets the fiber diameter and that the equatorial reflections are given by the region near the edges of the A band.) This kind of scalloping would be difficult to detect in the light microscope, because for a 1 μ diameter fibril, the depth of the valley would be less than 0.2 μ, which is at the limit of resolution of the light microscope. However, by blending the fibers and isolating the myofibrils, we found that the swelling in a low salt, high pH solution does take place at the myofibril level, and that isolated myofibrils and glycerinated fibers swell to almost the same extent in low ionic strength, high pH solutions.

If we accept the model shown in the upper part of Figure 5, we see that force will be generated by the tendency of the sarcomere to shorten when the peripheral filaments are bent outward, since the axial length of these filaments would become shorter than that of the straight filaments in the middle of the sarcomere. The amount of shortening depends on the actual shape of the sarcomere, but it can be shown that it would be about 1% of the sarcomere length, which is also the magnitude found experimentally (Arata & Podolsky, in preparation).

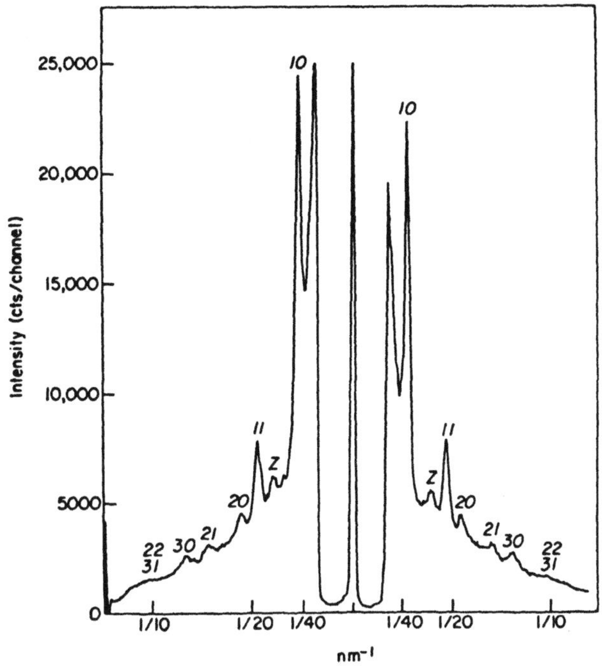

Figure 6: X-ray equatorial reflections from sartorius muscle at sarcomere length 2.2 μm. Various orders of reflection are shown as labeled (from Yu, Lymn, and Podolsky, 1977).

The experiments we have described provide evidence that rigor force can be modulated by factors that operate at the filament level and that originate from the non-isotropic properties of the sarcomere. Although these phenomena do not seem to be directly related to the physiological force generation mechanism, they are of interest as potential sources of information about the charge distribution along the myofilaments, and in relation to the mechanical properties of both the myofilaments and the cross-bridges in the rigor state[1].

Relaxed and Activated States

The spatial resolution that can be obtained by analyzing equatorial X-ray diffraction data for muscle fibers depends on the number of reflections included in the analysis. Figure 6 shows a diffraction pattern for resting frog muscle that we published several years ago (Yu, Lymn & Podolsky, 1977). This was obtained with an exposure of about an hour, using a rotating anode X-ray generator and an electronic position-

[1] It is interesting in this connection that similar force responses are seen when the sarcomere length is increased to 4.0 μ, where overlap of the myosin and actin filaments is presumably no longer present. The persistence of the force response may be due to the presence within the sarcomere of connecting filaments that are associated with the two major filaments and which provide axial continuity when sufficiently strained.

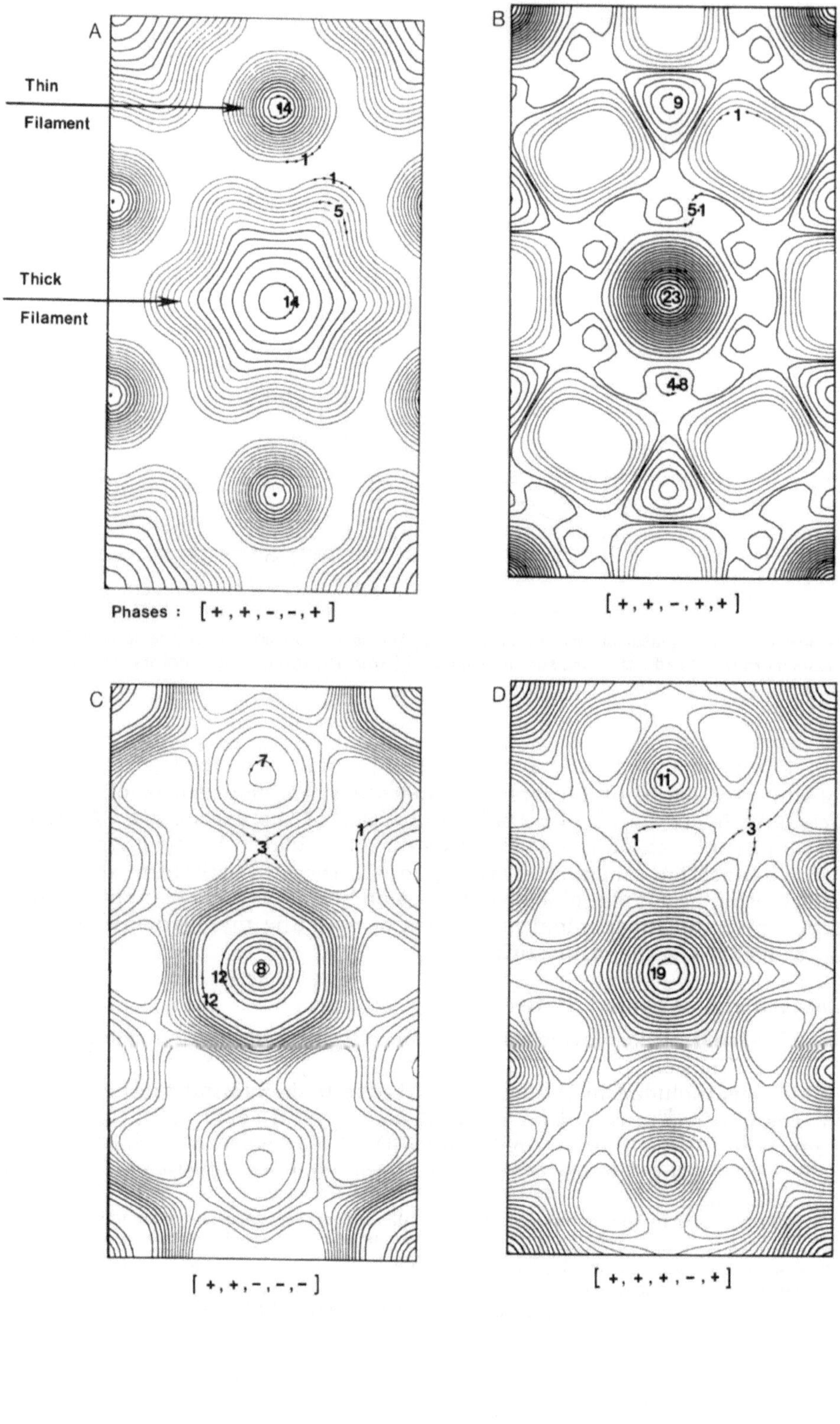
A
Thin
Filament
Thick
Filament
14
1
1
5
14
Phases : [+ , + , − , − , +]
B
9
1
5·1
23
4·8
[+, +, −, +, +]
C
7
1
3
8
12
12
[+, +, −, −, −]
D
11
1
3
19
[+, +, +, −, +]

sensitive detector. In addition to the strong *10* and *11* reflections, the intensity of three higher order reflections, *20, 21*, and *30*, could easily be measured. By including these high orders in the analysis of the diffraction patterns, the spatial resolution is improved about two-fold.

To obtain detailed information from these data, one has to deal with the phase question. For five orders there are 32 possible phase combinations. After careful selection based on known results for the relative masses of thick filaments and thin filaments, we came down to four most likely phase combinations for the resting state (Yu et al. in preparation). The corresponding mass distributions are shown in Figure 7.

Model A employs a combination of phases first considered by Haselgrove, Stewart, & Huxley (1976) and shows myosin heads close to the backbone of the thick filament, with a sharply defined 6-fold symmetry. However, we now view this model as unlikely on the following grounds: the bulk of evidence currently available (Maw & Rowe (1980), Squire (1981)) indicates that the myosin filament in axial projection should have 9- or 18-fold rotational symmetry. Consequently the projection of the distribution of myosin heads should, at the resolution shown here, appear as an annulus surrounding the core of the myosin filament. Model A and D are not compatible with this constraint, but Model B and C are generally consistent with this higher symmetry. However C has a low density myosin core which is difficult to reconcile with electron microscope observations Thus Model B, whose thick filament has a solid core surrounded by an almost azimuthally uniform distribution of density is our favored model for resting muscle structure.

To obtain similar data from activated muscle, we used the intense X-ray source available at the Standard Synchrotron Radiation Laboratory. To get sufficient exposure for the active state, we took time cuts from many activity cycles. The intensity data were stripped from the background by a non-linear least squares technique that took into account the systematic increase in width of the reflections according to the diffraction order.

Figure 7: Maps of average density distribution in resting frog sartorius muscle viewed in axial projection. These maps were obtained by Fourier synthesis of experimentally determined amplitudes together with several different sets of plausible phases. Analysis of the intensity profile of equatorial X-ray diffraction yielded amplitudes for the *10,11,20,21* and *30* reflections of the myofilament lattice. Assuming centrosymmetry, the phases must be real (+ or -), and 32 (=2^5) combinations are possible. Constraints imposed by electron microscope observations, biochemical quantitation, and the distribution of protein species among the thick and thin filaments respectively reduce the number of feasible sets of phases to four (A, B, C, D). The corresponding Fourier synthesis are shown here, with uniform normalization such that each map contains the same total amount of mass, and they are displayed according to a uniform set of equally spaced contours (except as indicated in B), covering a common dynamic range (1-23, in arbitrary units).

Table 1: Reflection intensities (I_{hk}) normalized with respect to transmitted beam intensity.

State	I_{10}	I_{11}	I_{20}	I_{21}	I_{30}
Rest	0.71	0.27	0.04	0.05	0.05
	σ = (0.10)	(0.06)	(0.01)	(0.02)	(0.01)
Active	0.42	0.60	0.05	0.03	0.03
	σ = (0.06)	(0.15)	(0.01)	(0.02)	(0.02)

Data are average values of four experiments with frog sartorius muscle at 20°C.

Averaged data from four experiments are given in Table 1. The first two columns show the well-known reciprocal changes in the intensities of the *10* and *11* reflections upon activation. The next column shows that the intensity of the *20* reflection remains almost the same. This differs from the behavior of *21* and *30* reflections, which appear to decrease in intensity upon activation.

In phasing the reflections from active muscle, the same lines of argument were applied as in the case of resting muscle. For the active state, however, the higher order reflections which dictate the compatibility of a given model with the high degree of symmetry attributed to the myosin filament are overall less intense, and therefore the resulting arguments are less conclusive. Moreover, in the active state, presumptive interactions with the six actin filaments which surround any given myosin filament might be expected to contribute some effect of symmetry breaking.

In order to visualize the redistribution of mass within the unit cell which takes place upon activation, we constructed difference maps between all reasonable models for both the active and the resting states, with appropriate relative normalization. In essence, we find that all the possible difference maps reduce to two qualitatively distinct patterns of mass transfer, according to whether or not the phases of any of the five reflections have been changed upon activation. Two of these are shown in Figure 8.

The contours on the left side of Figure 8 show the *change* in mass distribution in the transition to the active state if the phases are the same in the relaxed and active states (the phase set of model B in both cases). The mass shift is rather subtle in this case. The total mass within the thick filament backbone is effectively unchanged, the density of the cloud surrounding the thick filament decreases, and the thin filament appears to expand. The mass shift, which amounts to about 20% of the S1 mass has a strong azimuthal component. The constancy of the *20* intensity in this case does not indicate constancy of a particular region in the unit cell. Rather it says something about the nature of the mass shift; that is, it defines the way S1 moves toward actin. An increase with the *20*

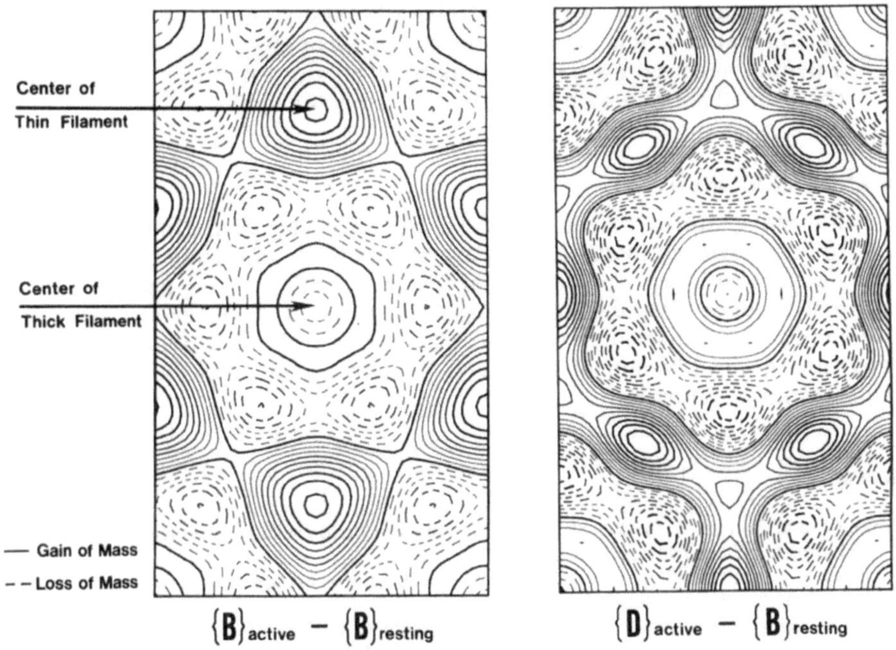

Figure 8: Redistribution of mass within the myofilament lattice upon activation of resting muscle. These are difference maps between the respective axial projections of density which correspond to the active and resting states. Dashed contour lines denote regions of the unit cell of the myofilament lattice which have incurred a net loss of mass upon activation and solid lines indicate regions of net gain. All contour lines represent equally spaced increments in density. The constraint of conservation of total mass upon activation has been imposed. A few different maps are candidates for both states (cf.Fig.7). These yield essentially only two distinct "difference maps" according to whether or not there is a change of phase in any of the five equatorial reflections under consideration. The difference picture "$B_{active}-B_{resting}$" typifies the case for isophasic activation, whereas "$D_{active}-B_{resting}$" is representative of the case in which phase changes are involved (see Fig. 7 for individual maps of resting states B & D). Note the apparent accretion of mass around the location of the thin filament in each case, which derives from the region peripheral to the thin filament core in

intensity would have made the shift more azimuthal and less radial, while a decrease would indicate more of a uniform radial transfer of mass.

The contours on the right show the mass difference map if the phases in the active state correspond to model D. The qualitative changes are the same as on the left, although in this case the mass shift is greater and amounts to about 40% of the S1 mass.

The present findings can be summarized as follows:

(1) In the resting state, myosin heads appear to protrude out from the thick filaments and extend towards the thin filaments.

(2) Upon activation, significant loss of mass occurs only in the region peripheral to thick filament backbone through a movement that has a pronounced azimuthal component. This movement can be taken as the overall shift in S1 during the cross-bridge cycle.

ACKNOWLEDGEMENT

We are grateful to the Stanford Synchrotron Radiation Laboratory (SSRL) for providing facilities for part of this work, and to Dr. B.L. Trus for help with computing and image processing. SSRL is supported by the National Science Foundation through the Division of Materials Research, and the National Institutes of Health through the Biotechnology Resource Program in the Division of Research Resources, in cooperation with the Department of Energy.

REFERENCES

Cooke, R. and Franks, K. (1980). All myosin heads form bonds with actin in rigor rabbit skeletal muscle. Biochemistry 19: 2265-2269.

Haselgrove, J.C., Stewart, M. and Huxley, H.E. (1976). Cross-bridge movement during muscle contraction. Nature 261: 606-608.

Lovell, S.J. and Harrington, W.F. (1981). Measurement of the fraction of myosin heads bound to actin in rabbit skeletal myofibrils in rigor. J. Mol. Biol. 149: 659-674.

Maw, M.C. and Rowe, A.J. (1980). Fraying of A-filaments into three subfilaments. Nature 286: 412-414.

Podolsky, R.J., Naylor, G.R.S. and Arata, T. (1982). Cross-bridge properties in the rigor state. In: *Basic Biology of Muscles: A Comparative Approach*, pp. 79-89, ed. Twarog, B.M., Levine, R.J.C. and Dewey, M.M. New York: Raven Press.

Squire, J. (1981). *The Structural Basis of Muscular Contraction.* New York: Plenum Press.

Yu, L.C., Lymn R.W. and Podolsky, R.J. (1977). Characterization of a non-indexible equatorial X-ray reflection from frog sartorius muscle. J. Mol. Biol. 115: 455-464.

DISCUSSION

HOLMES: Even with the 1,0 peak, if you assume you have a hexagonal lattice, there are two superimposed crystallographically distinct sorts of reflections, the 1,0 and the 0,1, in that peak. The situation becomes progressively worse the further out you go. How does one deal with that situation in this kind of calculation?

PODOLSKY: We assumed that the measured intensity contains equal contributions from the different reflection planes. Various arguments can be put forward for doing this. For example, in the case of the 2,1 and the 1,2 reflections, the 2,1 reflection makes the projection of the myosin heads tilt azimuthally in one direction, while the 1,2 produces a tilt in the opposite sense. If these two contributions are equal, the tilts balance out and you have a symmetrical structure in projection. If the 2,1 and 1,2 contributions were unequal, the projection of the thick filament would be asymmetric, which is unreasonable. Does that make sense?

HOLMES: Yes, it does.

KAWAI: I think that mass is lost from the core of the thick filament. How do you account for that?

PODOLSKY: It turns out that the mass loss is only a few percent of the total thick filament mass. That is not a very significant change.

TREGEAR: You used an argument for applying a high order of symmetry to the thick filament. Could you use the same argument for the thin filament? If so, can you explain the three-fold symmetry that you have at the trigonal positions in your model B for example.

PODOLSKY: The symmetry arguments clearly apply to both filaments. However, the three-fold symmetry of the mass distribution at the thin filament locations in model B is probably due to spillover from the mass cloud associated with each of the three adjacent thick filaments.

HUXLEY: Going back to the first part of your talk, I recall that you had an earlier argument which was that the flexible part of S-2 must be rather short (Podolsky, R.D. in *Basic Biology of Muscles: A Comparative Approach*, ed. by B.M. Twarog, R.J.C. Levine, and M.M. Dewey. Raven Press, New York, pp. 79-89, 1982). Do these new considerations about the asymmetrical expansion of the lattice mean that the flexible part of S-2, i.e., the part that can bend away from the backbone, can now be longer?

PODOLSKY: No, I don't think so. According to the data in Figure 3, expansion of the lattice seems to take place in several phases. In the first phase where d_{AM} is between 22 and 26 nm, the fiber diameter increases in proportion to the lattice spacing, which implies that sarcomere swelling is isotropic. The relation between fiber length (or isometric force) and lattice spacing in this phase of the swelling could be explained if the flexible part of S-2 were relatively short (Podolsky et al., 1982). Where d_{AM} swells beyond 26 nm, fiber diameter increases more than lattice spacing, which suggests that sarcomere distortion occurs. In this range of lattice spacing the flexible part of S-2 may lengthen, or part of the non-flexible S-2 detaches from the backbone of the thick filament.

HUXLEY: Can you remind me -- how much of the S-2 did you include in the flexible part, that is, how much was free to lean out in from the thick filament?

PODOLSKY: It was about 3 nm in projection in the plane containing the actin and the myosin filaments. If you take into consideration that the structures are three-dimensional, the flexible part of S-2 could be as long as 10 nm.

HUXLEY: So, therefore, the remaining 30 or 40 nm of the S-2 you would say was permanently stuck to the backbone.

PODOLSKY: Yes. Is that all right?

HUXLEY: It's all right by me!

HARRINGTON: I don't like it. I would like to point out that we have experiments that are not consistent with that. These are the cross-linking experiments which demonstrate, we think, that the arm is moving out when the pH is raised from 7 to 8.5 (Ueno & Harrington, J. Mol. Biol. 149: 619-640, 1981). And additionally, enzyme hydrolysis experiments suggest that the hinge is opening, or that the region between LMM and HMM is opening to the enzyme when the pH is raised (Ueno & Harrington, Proc. Nat. Acad. Sci. 78: 6101-6105, 1981). So S-2 is not lying in the same orientation at the high pH as it is at pH 7. Those experiments are still under discussion between Dick (Podolsky) and me. We have something of a contradiction there in our interpretation.

HUXLEY: Could I ask another question? On the question of having a ring of cross-bridges around the backbone which didn't show six-fold symmetry -- I mean that's fine, and I'd be very happy with nine-fold or eighteen-fold symmetry. But it seems to me that when you were looking at the transition to the attached state, the plot of what density had moved went back to a six-fold symmetry once again. Was that just because of the presence of six actins?

PODLOSKY: I think that a great deal of it comes from that. When you look at the active state, there is an imposition of six-fold symmetry on it because of the actin interaction.

MATSUBARA: May I ask one question about this six-fold symmetry? Is it possible that as you cut off the data, that you could get six-fold symmetry simply because of a cutoff effect?

PODOLSKY: We looked at some of the cutoff effects. They seemed to change the dimensions that you get for the filaments, but they do not change the symmetry a great deal.

MUSCLE CROSSBRIDGE POSITIONS FROM EQUATORIAL DIFFRACTION DATA: AN APPROACH TOWARDS SOLVING THE PHASE PROBLEM

John Squire and Jeffrey Harford

Biopolymer Group, Imperial College, London SW7, England

ABSTRACT

Following a discussion of the problems involved in the analysis of X-ray diffraction data from muscle, a description is given of a possible procedure for solving the phase problem in the case of equatorial diffraction data. The approach involves the use of the Patterson Function which can be determined unambiguously from the observed diffracted intensities. The method is tested using five different muscle-like model density distributions for which the correct phases can be calculated directly. It is then applied to the equatorial X-ray diffraction data from relaxed frog sartorius muscle where it selects a phase set which is also the most likely to be correct on the basis of other available data on frog muscle. This phase set gives rise to a Fourier synthesis map in which the crossbridges form a uniform shelf of density around the myosin filament backbones. Possible lateral movements of the crossbridges from this relaxed configuration in active and rigor muscle are discussed. The approach to solving the phase problem is now being applied to data from fish muscle, insect flight muscle and crab muscle. It should also have its application to other fibrous materials apart from muscle.

INTRODUCTION

X-ray diffraction studies of muscle clearly have the potential to reveal the nature of the myosin crossbridge movements which are involved in force generation. Changes have been observed in the diffracted intensity from muscles in different static states (e.g. relaxed and rigor) or while active (e.g. Huxley and Brown, 1967; Yu, Hartt and Podolsky, 1979; Huxley et al., 1980; 1981) and these changes are thought to be largely due to crossbridge movements. However, in the past, the problems involved in interpreting the observed diffraction patterns, or the changes between them, have been formidable. In the case of vertebrate skeletal muscle this has been for two main reasons. First, there is the problem inherent to all X-ray diffraction studies that when the diffraction pattern is recorded only one-half of the information necessary to reconstruct an image of the diffracting object is available. As described below, both the diffracted amplitudes and the relative phases of the diffracted beams are needed to do this, but only the amplitudes can be determined experimentally. This is the well-known phase problem in X-ray diffraction. Second, there has been the problem in vertebrate muscle that the 'unit cell' (including the symmetry and three-dimensional arrangement of the myosin filaments) in the overlap region of the A-band where force is actually generated has until recently been ill-defined. Attempting to understand the diffraction patterns from vertebrate muscles in detail has therefore been akin to trying to solve the structure of a globular protein using X-ray crystallography but without knowing the basic symmetry of the unit cell of the protein crystal.

Recently we have been able to define the three-dimensional geometry of two types of vertebrate skeletal muscle; frog muscle with a myosin filament superlattice (Luther and Squire, 1980) and fish muscle with a simple myosin filament lattice (Luther, Munro and Squire, 1981). We have also provided strong evidence that vertebrate myosin filaments have 3-fold rotational symmetry (Luther et al., 1981); a result consistent with data from other sources (e.g. Maw and Rowe, 1980; Stewart et al., 1981). More recently, Drs. P.K. Luther (this laboratory) and A.R. Crowther (MRC Laboratory of Molecular Biology, Cambridge) have confirmed directly, by three-dimensional reconstruction of tilted A-band cross-sections of fish muscle, that these myosin filaments have the symmetry of the Dihedral Point Group 32 (i.e. a 3-fold rotation axis along the filament axis, perpendicular to three 2-fold axes in the plane of the middle of the M-band (M1; Luther, Crowther and Squire, 1982). Thus the symmetry of the vertebrate A-band unit cell is now becoming much clearer. For this reason it is timely to turn to the other basic problem in X-ray diffraction studies; that of solving the phase problem. This paper discusses one possible approach to the solution of the phase problem in the case of centrosymmetric structures. It has particular application to the equatorial X-ray diffraction data from muscle.

Equatorial Diffraction Studies

If a diffraction pattern is recorded from a structure with order in three dimensions, then each observed diffraction peak hkl will have an intensity I(hkl) which, after suitable correction, can be used to yield a structure amplitude $|F(hkl)|$ (= $(I(hkl))^{1/2}$). The structure factor F(hkl) is in general a complex quantity given by $F(hkl) = |F(hkl)| \exp(i\alpha)$ where α is the relative phase of that particular diffracted beam. Since the intensity of the peak (I(hkl)) is the only thing recorded experimentally and this is equal to $|F(hkl)|^2$, no information is available directly from the diffraction pattern about the value of the phase angle α. This is the phase problem. It is a problem because, in order to reconstruct the electron density $\rho(xyz)$ in the diffracting object, it is necessary to know F(hkl) in both amplitude and phase since:

$$\rho(xyz) \propto \sum_{hkl} F(hkl) \exp(-2\pi i (hx + ky + lz)) \quad (1)$$

where x, y and z are fractional coordinates in the unit cell of the diffracting object.

Fortunately in the case of centrosymmetric structures in which identical densities occur at (xyz) and at (-x,-y,-z), equation 1 simplifies to:

$$\rho(xyz) \propto \sum_{hkl} (\mp) \,|F(hkl)|\, \cos 2\pi (hx + ky + lz) \quad (2)$$

where the only ambiguity is the sign of each term of the series (i.e. α is either 0° or 180°). Although the structure in the vertebrate A-band as a whole may not be centrosymmetric, it is very likely that in projection down the fibre axis it will appear so, at least at low resolution. This projected density $\rho(xy)$ is given by:

$$\rho(xy) \propto \sum_{hk} (\mp) \,|F(hk0)|\, \cos 2\pi (hx + ky) \quad (3)$$

where the hk0 reflections are those on the equator of the diffraction pattern. Thus analysis of the equatorial diffraction data from muscle gives data about the A-band density projected down the fibre axis and it invariably assumes (reasonably) that this projected density is centrosymmetric. In addition, since several different diffracting planes (parallel to the fibre axis) contribute to each observed reflection, an assumption is usually made that each plane is contributing equally to the observed peak. In the case of vertebrate muscle this requires two additional asumptions to be made about its structure, both of which are reasonable. One is that the structure has 6-fold symmetry around the fibre axis and the other is that there are 6 mirror planes along planes of the type $(11\bar{2}0)$ in the hexagonal unit cell (i.e. the two-dimensional unit cell has space group p6m). At the limited resolution in most muscle equatorial studies (often about 100 Å) these assumptions are readily justifiable (see Squire, 1981).

Thus Fourier synthesis (Equation 3) of the electron density of the structure projected down the axis, can be calculated from the observed $|F(hkl)|$ values if the sign (phase) of each term in the series is known. Since this is not normally so, the usual practice is to carry out 'trial and error' synthesis using all of the possible combinations of sign for the N reflections involved. Since the phases are relative, the first term in the series (usually h=1, k=0) can be taken to be positive. There are therefore 2^{N-1} possible combinations of sign for the N reflections, each of which is combined with the observed structure amplitudes $|F(hkl)|$ to give a possible electron density distribution ρ_n. In the past the 'correct' map has invariably been chosen somewhat subjectively on the basis of knowing from other sources of data what the basic features of the map ought to look like. Usually a large proportion of the possible maps can be rejected by this means, but often much uncertainty remains and several quite distinct maps cannot be ruled out. It is therefore of great importance to find objective selection procedures to give the correct map. We have attempted this as described below.

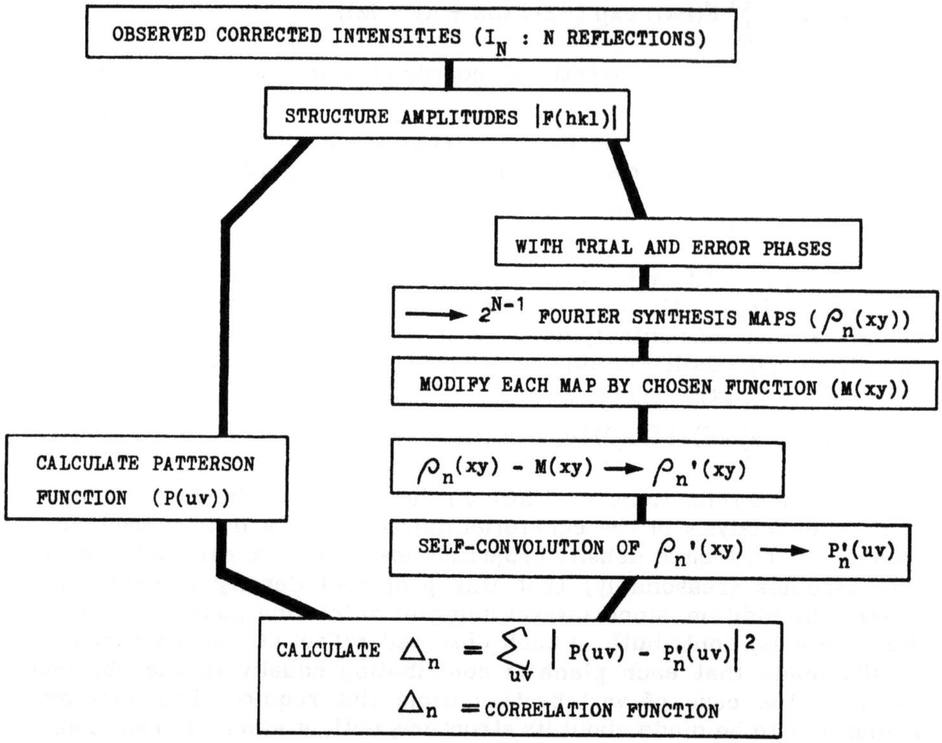

Figure 1: Schematic flow diagram to indicate the derivation of the correlation function Δ_n as described in the text.

A Possible Phase-Selection Procedure

Figure 1 indicates schematically the logic of a possible phase-selection procedure for equatorial diffraction data. The observed corrected intensities I(hk0) are used to give structure amplitudes |F(hk0)| which in turn are used in two ways. One way (right hand path in Figure 1) is to generate possible Fourier maps $\rho_n(xy)$, as described above, using the 2^{N-1} different phase combinations for the N observed reflections. The other way (left hand path) is to calculate the Patterson function

$$(P(uv) \propto \sum_{hk} |F(hk0)|^2 \cos 2\pi (hu + kv))$$

which can be calculated directly from |F(hk0)| without any knowledge of the phases. The Patterson function P(uv) contains information about the electron density distribution in the unit cell, since it is defined alternatively as

$$P(uv) \propto \int_0^1\int_0^1 \rho(xy)\rho(x + u, y + v)dxdy$$

This is the self-convolution or auto-correlation of the unit cell density distribution.

A similar self-convolution P'(uv) can be calculated from each of the 2^{N-1} maps produced by trial and error Fourier synthesis and if these maps are unmodified they will each yield *identical* self-convolutions $P_n'(uv)$ which are also identical to P(uv). However if these Fourier maps are modified in some way (by a function M(xy): Figure 1) then $P_n'(uv)$ will in general be different from P(uv) and a *correlation function* $\Delta_n = |P(uv)-P_n'(uv)|^2$ can be defined. The problem then is to find a suitable modifying function (M(xy)) which will yield a correlation function Δ_n that can be used to aid selection of the correct Fourier synthesis map.

Fortunately there are some grounds on which to define the function M(xy). Previously in muscle studies the 'correct' Fourier maps have been selected on the basis of the known positions and relative sizes and masses of the muscle filaments. This approach is very useful, but there is also other information about the structure which is not normally used. We also know that the electron density everywhere in the unit cell is positive and in general this will be true in the correct Fourier synthesis map even though the limited resolution may give rise to some small negative excursions in the electron density. However this is only true if the important origin peak (giving F(000)) is included in the Fourier synthesis and since this peak cannot normally be measured it is inevitably excluded. If it was included it would add a uniform density background to each Fourier synthesis map and this would raise to positive values most, if not all, of the density in the correct map. In the absence of F(000), the calculated electron density maps $\rho_n(xy)$ all have both positive and negative density regions. Bearing these various points in mind, we have worked on the

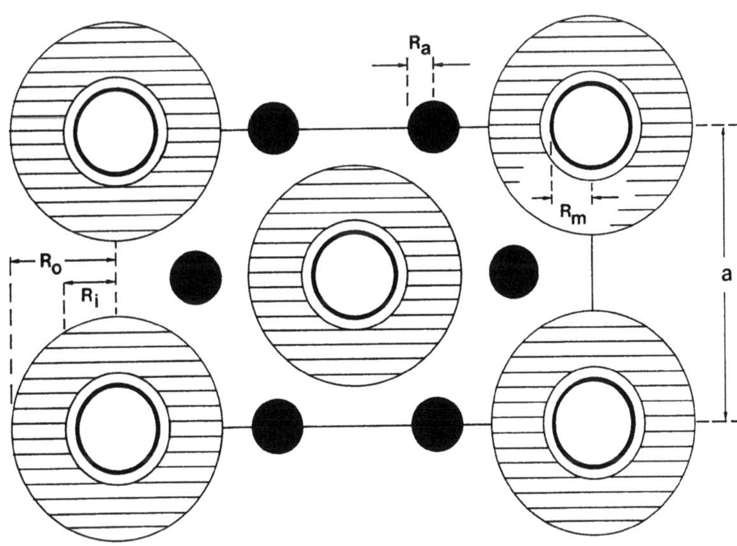

Figure 2: Unit cell used for the model calculations discussed in the text. Values of the various parameters used for models 1 to 5 are given in Table 1.

assumption that if a uniform density cut-off is used as M(xy), only low density features will be cut from the correct density map whereas important negative peaks could well be cut from the incorrect maps. The latter event would in turn have a very serious effect on the self-convolution P'(uv). As described in the next section, we have applied this approach to several muscle-like model structures and have been encouraged by the results so far.

Table 1: Muscle-like Model Structures (a is 400 Å). Values for the filament parameters defined in Figure 2 for the muscle-like models 1 to 5.

MODEL	R_m	D_m	R_a	D_a	R_i	R_o	D_c
MODEL 1	70	9	45	7	-	-	-
MODEL 2	80	9	40	3	-	-	-
MODEL 3	50	5	45	1	-	-	-
MODEL 4	70	8.2	50	8	100	140	5.4
MODEL 5	70	8.2	50	8	70	110	7.2

(R_m , R_a , R_i and R_o in Å ; D_m , D_a and D_c in arbitrary units)

Table 2: Values of the phase factors for the 16 sets used in the Fourier synthesis and Patterson synthesis calculations discussed in the text.

SET NUMBER	PHASES				
1	1	1	1	1	1
2	1	1	1	1	-1
3	1	1	1	-1	1
4	1	1	1	-1	-1
5	1	1	-1	1	1
6	1	1	-1	1	-1
7	1	1	-1	-1	1
8	1	1	-1	-1	-1
9	1	-1	1	1	1
10	1	-1	1	1	-1
11	1	-1	1	-1	1
12	1	-1	1	-1	-1
13	1	-1	-1	1	1
14	1	-1	-1	1	-1
15	1	-1	-1	-1	1
16	1	-1	-1	-1	-1

The sequence of reflections is 10,11,20,21 and 30.

Phase-Selection with Model Structures

We have tested this approach to muscle equatorial diffraction data using models of the form indicated in Figure 2. These models contain a distribution of myosin-like filaments (hollow rings) and actin-like filaments (solid circles) which is similar to that in vertebrate skeletal muscle. The shaded regions represent crossbridge-like density. Table 1 lists the parameters used in five models for the thick filaments (radius R_m, density D_m), thin filaments (radius R_a, density D_a) and crossbridges (inner radius R_i, outer radius R_o, uniform density D_c). Table 2 shows the phase combinations used in Fourier synthesis involving five reflections (10, 11, 20, 21, 30). Figures 3 and 4 show the distributions of correlation functions Δ_n for these five models and 16 phase sets in the case where negative density in $\rho_n(xy)$ is set equal to zero (i.e. M(xy) removes all negative density). Since we are using model structures we can also calculate

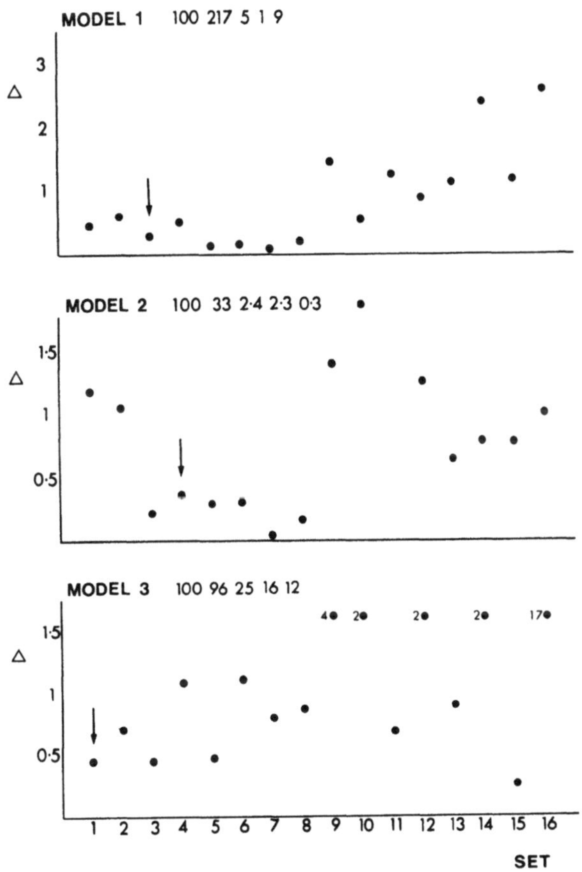

Figure 3: Values of Δ_n (scaled) plotted as a function of phase set number (n) for models 1,2 and 3. The numbers above each plot and after the model number are the approximate intensities of the five reflections (10, 11, 20, 21 and 30) scaled to that of the 10 which was set at 100 in arbitrary units. Arrowed positions correspond to the correct phase sets.

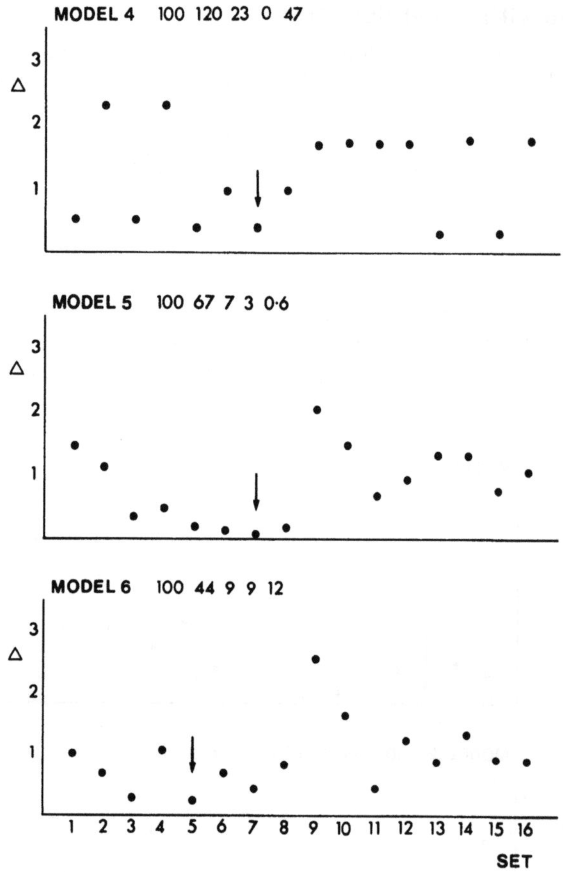

Figure 4: Similar diagrams to those in Figure 3 but for the models 4,5 and 6.

the structure factor completely. We therefore know which is the correct phase set; in each case it is arrowed in Figures 3 and 4. From these results, together with a study of the Fourier maps themselves (not shown), various conclusions can be reached:

(i) Since the basic filament arrangement, as in muscle, can be assumed to be known, several maps can immediately be excluded as being unreasonable. In all cases here this applies to maps 9 to 16 where the 11 phase is -1.

(ii) Of the maps that remain, the correct map always has a relatively low correlation function. In several cases (Models 3, 4, and 5) the correlation function for the correct map is the lowest of those not excluded by (i).

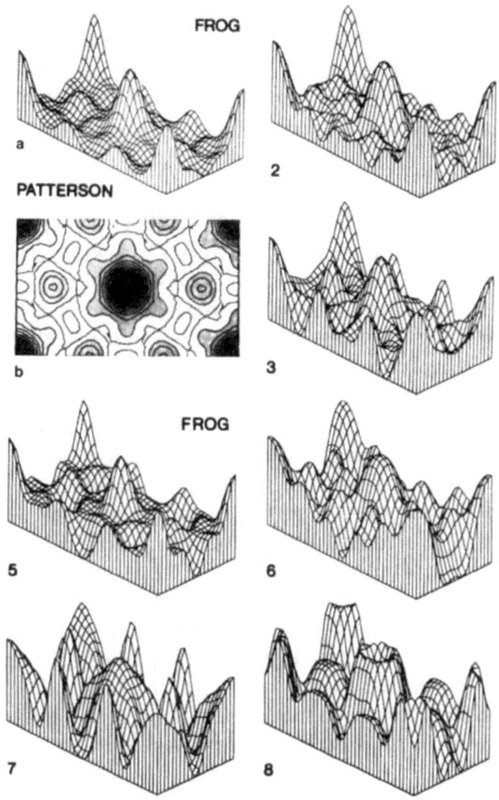

Figure 5: Various density maps calculated from the data in Model 6 from frog sartorius muscle. Maps (a) and (b) are respectively a three-dimensional view and a shaded contour map of the Patterson function (P(uv)). Maps 2, 3, 5, 6, 7 and 8 are three- dimensional views of the Fourier synthesis maps with the corresponding n number. The distribution of filaments is as in Figure 2, with the myosin filaments at the corners and the middle.

(iii) In the cases where the Δ_n value is not the lowest for the correct map (i.e. models 1 and 2), all of the maps with low values of Δ_n are rather similar. This is because the intensities of some of the reflections are very low and altering their phase has only a minor effect on the calculated density map. Note that in model 3 all reflections are appreciable and the method goes straight for the correct map (except for set 15 which can be excluded on other grounds) whereas in model 4 the 21 reflection is weak and the Δ_n values for maps 5 and 7 are virtually identical. Similarly in model 5 the 30 reflection is weak and Δ_n for set 7 is close to that of set 8. In this latter case the 21 peak is also fairly weak, so that sets 5, 6, 7 and 8 all give low values for Δ_n and their Fourier maps are very similar.

In view of these promising conclusions we believe that this method might be of value at least in selecting the phases of reflections which are relatively strong. In any case it is such reflections which dominate the Fourier synthesis maps. We have only just started exploring alternative forms of M(xy) and hope that the method can be improved. However even as it stands it should be able to help in muscle studies. For this reason we have applied it to data from vertebrate skeletal muscle as described below.

Equatorial Diffraction Data from Frog Muscle

In order to produce results comparable with previous studies of the equatorial diffraction pattern from frog sartorius muscle (e.g. Lymn and Cohen, 1975; Haselgrove, Stewart and Huxley, 1976) we have used the intensities from frog muscle reported by Lymn and Cohen. However our own equatorial data from both frog sartorius muscle and fish muscle are very similar to these published values. Our analysis of the Lymn and Cohen data is actually represented as Model 6 in Figure 4. All of the first five reflections are appreciable in intensity and the method would immediately suggest that model 5, with the lowest Δ_n value is the correct density map. However map 3 has a Δ_n value which is also relatively small. These two maps are shown in three-dimensional representation in Figure 5, as are maps 2, 6, 7 and 8 and the Patterson function P(uv) from these same reflections. It is interesting to study these various maps subjectively to see if maps 3 and 5 are reasonable choices. It is easy to see that they are. Maps 2 and 6 have hollow actin filaments of very large diameter. Map 7 (the one selected by Haselgrove et al., 1976) has actin peaks of larger density than the myosin peaks (not smaller as expected). Map 8 has very hollow looking myosin filaments, but it is easy to show that if

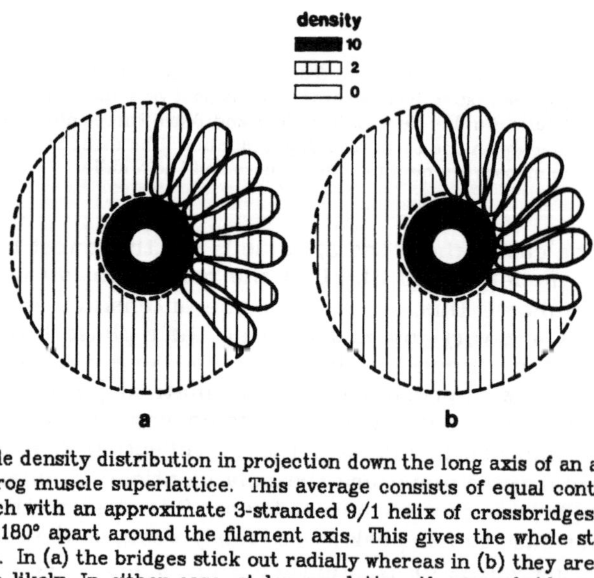

Figure 6: Probable density distribution in projection down the long axis of an average myosin filament in the frog muscle superlattice. This average consists of equal contributions from two filaments each with an approximate 3-stranded 9/1 helix of crossbridges on it but with two orientations 180° apart around the filament axis. This gives the whole structure an 18-fold rotation axis. In (a) the bridges stick out radially whereas in (b) they are slewed azimuthally as is more likely. In either case, at low resolution, the crossbridges will produce a smooth shelf of electron density around the myosin filament backbone (black ring). Approximate densities are shown. From Squire, 1981.

there was say a 40 or 50 Å diameter hole down the middle of the filament as can sometimes be seen in electron micrographs of frog muscle (e.g. Luther and Squire, 1978), this would not show up at all at the resolution being used here (about 130 Å). The hole in Map 8 must be an artifact. Thus maps 5 and 3 are left as being possible from this subjective point of view. Finally, since the crossbridges on vertebrate myosin filaments are probably arranged on an approximate 3-stranded 9/1 helix (Squire, 1972), there would be 9 crossbridge directions on one half of one myosin filament or 18 directions on the average myosin filament in the frog muscle superlattice (which is made up from filaments with one or two orientations 180° apart; Luther and Squire, 1980). One would therefore expect the crossbridges to form a fairly uniform shelf of density around the myosin filament backbone as indicated in Figure 6. This is just what is observed in Figure 5, map 5, which also gave the lowest Δ_n value. There are therefore good reasons for believing both that model 5 represents the correct electron density distribution for relaxed frog sartorius muscle and also that our phase-selection method has been useful in this case.

As discussed in Squire (1981), this conclusion implies that when a vertebrate muscle is activated, the crossbridges must swing by variable amounts azimuthally and by only a small amount radially in order to interact with actin. They then become partly actin centred in active muscle and largely actin centred in rigor muscle (Haselgrove and Huxley, 1973) during which further radial movement is involved. This is indicated schematically in Figure 7. Further analysis of high resolution equatorial diffraction data from relaxed, active and rigor muscles using our phase-selection procedure should help to define the nature of these movements in more detail.

CONCLUSION

We have chosen to start our investigation of phase-solving techniques for muscle equatorial data using relatively simple models, and muscle

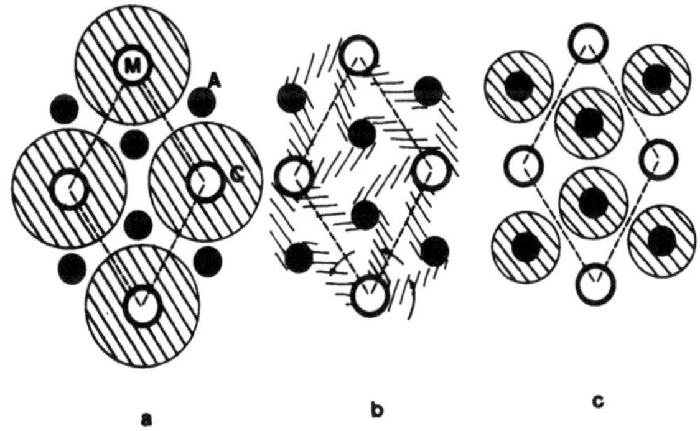

Figure 7: Possible density changes in the vertebrate muscle unit cell when projected down is between (a) relaxed muscle, (b) active muscle and (c) rigor muscle. For details

data which can also be analyzed subjectively, as described in the last section, with a fair degree of confidence. The true test of the approach will come in studies of muscle at higher resolution or in preparations such as rigor insect flight muscle where there are alternative published crossbridge models for the same muscle state (Squire, 1972; 1981; Freundlich, Luther and Squire, 1980; Offer, Couch, O'Brien and Elliott, 1981). We are now in the process both of testing our method further using different model structures and different forms of M(xy) and also of processing the equatorial diffraction data from fish muscle, insect muscle and crab muscle. We hope by this means to be able to provide relatively unambiguous conclusions about the crossbridge positions in different muscles and states. However, it is appropriate to note finally that there is no reason why our phase-selection method should be used only in muscle studies. The method is completely general and could find useful applications in studies of other fibrous structures.

ACKNOWLEDGEMENTS

We are indebted to the Medical Research Council and the Muscular Dystrophy Association of America for supporting this work and to Drs. A. Freundlich and P.K. Luther for helpful discussions.

REFERENCES

Freundlich, A., Luther, P.K. and Squire, J.M. (1980). High-voltage electron microscopy of crossbridge interactions in striated muscle. J. Muscle Res. 1: 321-343.

Haselgrove, J.C., and Huxley, H.E. (1973). X-ray evidence for radial crossbridge movement and for the sliding filament model in actively contracting skeletal muscle. J. Mol. Biol. 77: 549-568.

Haselgrove, J.C., Stewart, M. and Huxley, H.E. (1976). Cross-bridge movement during muscular contraction. Nature 261: 606-608.

Huxley, H.E. and Brown, W. (1967) The low-angle X-ray diagram of vertebrate striated muscle and its behavior during contraction and rigor. J. Mol. Biol. 30: 383-434.

Huxley, H.E., Faruqi, A.R., Bordas, J., Koch, M.H.J. and Milch, J.R. (1980). The use of synchrotron radiation in time-resolved X-ray diffraction studies of myosin layer-line reflections during muscle contraction. Nature 284: 140-143.

Huxley, H.E., Simmons, R.M., Faruqi, A.R,. Kress, M., Bordas, J. and Koch, M.H.J. (1981) Millisecond time-resolved changes in X-ray reflections from contracting muscle during rapid mechanical transients, recorded using synchrotron radiation. Proc. Nat. Acad. Sci. 78: 2297-2301.

Luther, P.K. and Squire, J.M. (1978). Three-dimensional structure of the vertebrate muscle M-region. J. Mol. Biol. 125: 313-324.

Luther, P.K. and Squire, J.M. (1980). Three-dimensional structure of the vertebrate muscle A-band. II. The myosin filament superlattice. J. Mol. Biol. 141: 409-439.

Luther, P.K., Crowther, A.R. and Squire, J.M. (1982). Three-dimensional structure of the M-band in fish muscle. Biophys. J. 37: 51a.

Luther, P.K., Munro, P.M.G. and Squire, J.M. (1981). Three-dimensional structure of the vertebrate muscle A-band. J. Mol. Biol. 151: 703-730.

Lymn, R.W. and Cohen, G.H. (1975). Equatorial X-ray reflections and cross arm movement in skeletal muscle. Nature 258: 770-772.

Maw, M.C. and Rowe, A.J. (1980). Fraying of A-filaments into three subfilaments. Nature 286: 412-414.

Offer, G., Couch, J., O'Brien, E.J. and Elliott, A. (1981). Arrangement of cross-bridges in insect flight muscle in rigor. J. Mol. Biol. 151: 663-702.

Stewart, M., Ashton, F.T., Lieberson, R. and Pepe, F.A. (1981). The myosin filament. IX. Determinations of subfilament positions by computer processing of electron micrographs. J. Mol. Biol. 153: 381-392.

Squire, J.M. (1972). General model of myosin filament structure. II. Myosin filaments and cross-bridge interactions in vertebrate striated and insect flight muscles. J. Mol. Biol. 72: 125-138.

Squire, J.M. (1981). The structural basis of muscular contraction. Plenum Publishing Corp., New York.

Yu, L.C., Hartt, J.E. and Podolsky, R.J., (1979). Equatorial X-ray intensities and isometric force levels in frog sartorius muscle. J. Mol. Biol. 132: 53-68.

DISCUSSION

HOLMES: John, on the back of an envelope I've worked out that what you're actually doing is maximizing all the cross-correlation functions between the function you choose and whichever set of phases you have to choose. That's what it amounts to. So clearly the thing is going maximalize for just that reason. So, when you say a particular published model doesn't agree with your selected Fourier map, what you're saying is "it doesn't agree with the model you put in."

SQUIRE: No. You don't put in a model.

HOLMES: Yes, you do. You put in this function which you use; that function is already a model.

SQUIRE: What I haven't said is that there are reasons for choosing the function.

HOLMES: Well, of course, but my point is that that isn't very much better than just taking a model, is it?

SQUIRE: Except that it works.

HOLMES: Well, it works in the sense that you say, if I choose a function and then maximize something with it, then I get the answer out which is the same as the function, which is not very surprising because that's what you've just done. The point I'm trying to make is that your procedure actually maximalizes the cross-correlation function between the function with those various signs and the function you put in. So it's not surprising that it comes out the same.

SQUIRE: No. What I'm saying is that what the function you're putting in is doing is removing a uniform density from the Fourier synthesis. That's all. There's no variation of density within the modifying function.

HOLMES: But you said you took your density and took away something.

SQUIRE: Yes.

HOLMES: What did you take away?

SQUIRE: We took away all the density in each Fourier synthesis that was negative. In other words in calculating $P'_n(uv)$, we didn't include anything in $\rho_n(xy)$ that was below zero. So, if you're in the right phase set, what you are doing is knocking out low density features, whereas if you're in the wrong phase set you're knocking out negative peaks that are rather important.

HOLMES: Then I misunderstood you. Can I get this straight, because I think it is a bit important to know what you're doing. Didn't you say that you maximalize some index, which was the density with lots of different signs minus some model density?

SQUIRE: Right. We use the Patterson function as a reference (Fig. 1). You calculate it directly.

HOLMES: What is M(x,y)?

SQUIRE: M(x,y) is a modifying function that can be chosen in order to be sensitive to the correct phase set. But all this amounted to in all the cases I've shown is just a cut off level in the Fourier synthesis. Everything that was negative in the Fourier synthesis was made equal to zero. So you don't put in anything about the density distribution. You just accept that in the right map the electron densities will be largely positive. If you do that you only remove low density information whereas in the wrong maps, where negative peaks are important, their removal is rather disastrous.

HOLMES: Yes, that's right.

SQUIRE: So you're not putting in a model at any point.

TREGEAR: John, you mentioned the phase is changing on going to activation. Would it be possible to get a test of that by going to partial activation and, presumably, the reflections must go through zero at some point. They must become less before they become more if they're changing phase. In other words, would partial activation show the blanking out of some of the reflections? Is that a reasonable experiment to attempt?

SQUIRE: It probably would be sensible to try and get the right phase for active or rigor muscle and see if the phases actually change.

HUXLEY: Are you suggesting, Richard, that if you look at a muscle during the onset of activation, that some of the reflections become zero?

TREGEAR: Well, I was hoping that if they're changing sign that they might get less -- they might not ever go through zero, I'm sure. We've got little chance of getting them absolutely down to zero.

SQUIRE: They might get less before they get more.

TREGEAR: Do they?

HUXLEY: Well, the 1,1 certainly doesn't. I think it would be very strange if the phases of the 1,0 and the 1,1 changed.

TREGEAR: But what about the outer reflections?

SQUIRE: They may well change.

COOKE: Can you get some information on the phases at least in the rigor case by adding subfragment-1 and then requiring that added intensity end up on the actin?

SQUIRE: Yes, I guess you could do that.

COOKE: That would then give you some information on the phases in the active muscle, since contraction is maybe some intermediate between rigor and relaxation.

SQUIRE: Yes, it's probably worth a try.

ROWE: John, you observed, am I correct, a rotation in the three-fold structure of the myosin backbone as you go through the M-line.

SQUIRE: Yes.

ROWE: Is that of the same sort of order as suggested by Trinick and Cooper's electron micrography (J. Mol. Biol. 141: 315-321, 1980)?

SQUIRE: It doesn't look like a rotation in the sense that you're talking about. But what you see on one side is that there's a triangular structure (see Figure D-1), and what you see on the other side is another triangular structure. But the triangle on one side tends to disappear into a more circular profile at M-1 and re-emerges as a triangular structure on the other side. You don't see something twisted in the sense that you talked about.

PODOLSKY: It wasn't clear to me, John, what your argument was for the reasonableness of the phase changes during activation. Could you go through that again a little more slowly? Why do you think the phases should change?

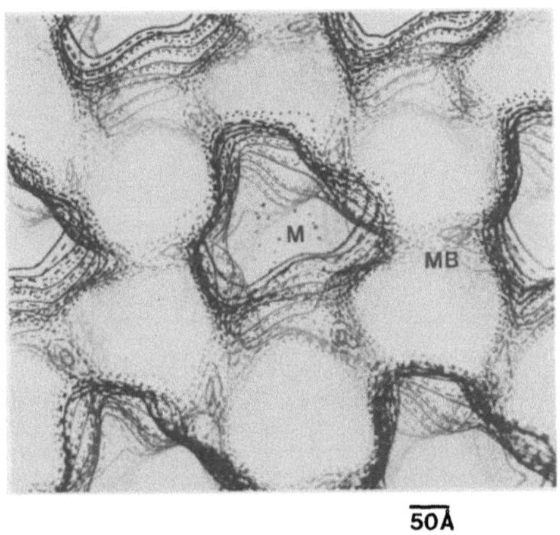

Figure D-1: Three-dimensional reconstruction of the M-band in fish muscle obtained by Fourier inversion of the Fourier transform data computed from about 50 different tilted views of a single transverse section about 700 Å thick. The reconstruction is viewed in a direction that is nearly parallel to the myosin filament axis. M: Myosin filament backbones, MB: M-bridges. It can be seen that the structure not only has very clear 3-fold rotation axes coincident with the myosin filament long axes but also that there are clear 2-fold rotation axes in the plane of the page (perpendicular to the 3-fold axes) as required for the dihedral point group 32. Also at the top of the reconstruction the myosin filament backbone appears rather triangular in shape and that the bottom part of the reconstruction (fainter lines) is similar but rotated by 40°C from the top part. (Reconstruction by Drs. P.K. Luther and R.A. Crowther).

SQUIRE: Well, I think if you start playing around with Fourier transforms and Fourier synthesis, you'll find that you can take a large -- well, a lot of very different intensities for the various reflections, but if you keep the phases the same, the density will look very much the same. This is something we discovered from using a lot of different models, because if you look at all the sets that have the same phase, but different amplitudes, they look rather similar.

PODOLSKY: I think that's only partly true. In the calculations that we tried, I think I mentioned to you, if you take the 2,0 reflection and if you change its amplitude but keep its phase the same, you get rather big changes in the redistribution of mass. If the amplitude had gone down, then redistribution of mass would have been radial, whereas if amplitude had gone up, then the redistribution of mass would have been more azimuthal. This is always with the same phase.

SQUIRE: I didn't mean there would be no change, I'm just saying they're more sensitive to phase than to amplitude. You might phase differently and find a very big change in density distribution.

PODOLSKY: I think it's sensitive to both of them.

SQUIRE: Of course, but, for example, in the trial and error method of Fourier synthesis only the phases are altered and very different density maps are produced.

CONFIGURATIONS OF MYOSIN HEADS IN THE CRAB STRIATED MUSCLE AS STUDIED BY X-RAY DIFFRACTION

Katsuzo Wakabayashi, Keiichi Namba and Toshio Mitsui

Department of Biophysical Engineering, Faculty of Engineering Science, Osaka University, Toyonaka, Osaka, Japan 560

ABSTRACT

The configurations of myosin projections in striated muscles from the marine crab, *Portunus trituberculatus* were described in the relaxed and rigor states at the full overlap length of the thin and thick filaments. The crystallographic period of the thick filament is 101.5 nm (14.5 nm × 7) and the thick filament has four-fold rotational symmetry. In the relaxed state, the myosin projections sit about 19 nm from the thick filament axis, lying just between the surface of the thick filament backbone and that of the thin filament. They have an elongated structure with the length of 10 nm ~ 12 nm and a maximum axial thickness of about 4 nm. They are tilted axially by 20° ~ 30° to the thick filament axis. The configuration of the resting projections sensitively depends on the ionic strength and pH of the solution.

In the rigor state, myosin heads are bound periodically to the thin filaments (Namba, Wakabayashi & Mitsui, 1980); four myosin heads attach in groups every 38.3 nm to successive actin molecules of each strand of F-actin. Most of the bound myosin head is incorporated in the thin filament with the centre of gravity 2.8 nm from the thin filament axis. They are inclined at about 30° to and slewed round the thin filament axis.

INTRODUCTION

Recently, several authors have reported that myosin projections attach periodically to the thin filaments in the rigor state of insect flight muscle (Holmes, Tregear & Barrington Leigh, 1980; Offer, Couch, O'Brien & Elliott, 1981) and crustacean muscles (Wray, Vibert & Cohen, 1978; Maeda, Matsubara & Yagi, 1979; Namba et al., 1979, 1980), and the configuration of attached myosin heads to the thin filament has been described (Holmes et al., 1980; Namba et al., 1980; Offer et al., 1981). Schemes for the structure of the thick filaments in various muscles have also been proposed (Huxley & Brown, 1967; Squire, 1972, 1975; Wray,

Vibert & Cohen, 1975; Wray, 1979; Haselgrove, 1980) but there is a considerable variety in the structure of the thick filaments between different muscles.

Our present concern is how the myosin projections are arranged in the relaxed state. It has been suggested that the high intensity of the 14.5 nm meridional reflection relative to the off-meridional reflection in the relaxed state of insect flight muscle would be due to myosin projections lying almost perpendicularly to the thick filament axis (Squire, 1972). On the other hand, Wray et al. (1975) suggested that smaller ratio of the meridional to off- meridional intensities of the 14.5 nm layer line in the patterns from relaxed muscles of limulus, lobster and scallop indicates that the resting projections are tilted to thick filament axis. Wray et al. (1975) and Squire (1975) and Haselgrove (1980) have done model calculations to find a configuration of myosin projections which generates the observed layer-line profiles. In the relaxed state, however, the meridional reflections due to the thick filaments do not arise solely from the myosin projections but also include the contribution from the thick filament backbone, since the meridional reflections are still observed in the rigor state. Thus, the interpretation of the layer-line intensity data cannot be done straightforwardly.

In this article, we present a way to examine the configuration of resting myosin projections, and report our results on the symmetry and configuration of the myosin projections of crab muscle in the relaxed state. In conjunction with this, we discuss briefly the structure of myosin heads attached to the thin filaments in the rigor state.

MATERIALS AND METHODS

Striated muscle from the hind leg of the marine crab *Portunus trituberculatus*, was used for this study. Muscle fibers were glycerinated at the sarcomere length of 5.5 μm. They were examined at 4°C in the relaxed state in a solution containing 100 mM KCl, 10 mM $MgCl_2$, 2 mM EGTA, 5 mM Na_2ATP and 10 mM histidine- HCl (pH 7.0), and in the rigor state in the same solution but without Na_2ATP. X-ray patterns were taken using a mirror-monochromator camera (Huxley & Brown, 1967) with a glass mirror and a germanium monochromator (5°-cut to (111)). The wavelength of X-rays was 1.5405 Å. The microfocus X-ray generator (Rigaku Denki, FR-B) was run at 50 kV and 60 mA. Sakura cosmic-ray films and CEAVERKEN Reflex 25 X-ray films were used to record the diffraction patterns. The specimen-to-film distance was 45 cm. The X-ray intensity data were obtained by the same procedures described by Namba et al. (1980).

RESULTS & DISCUSSION

Layer-line Diffraction Patterns

Figure 1 shows diffraction patterns from the relaxed state (a) and the rigor state (b) of crab striated muscle at the full overlap length of the thin and thick filaments. In both states, diffraction patterns revealed

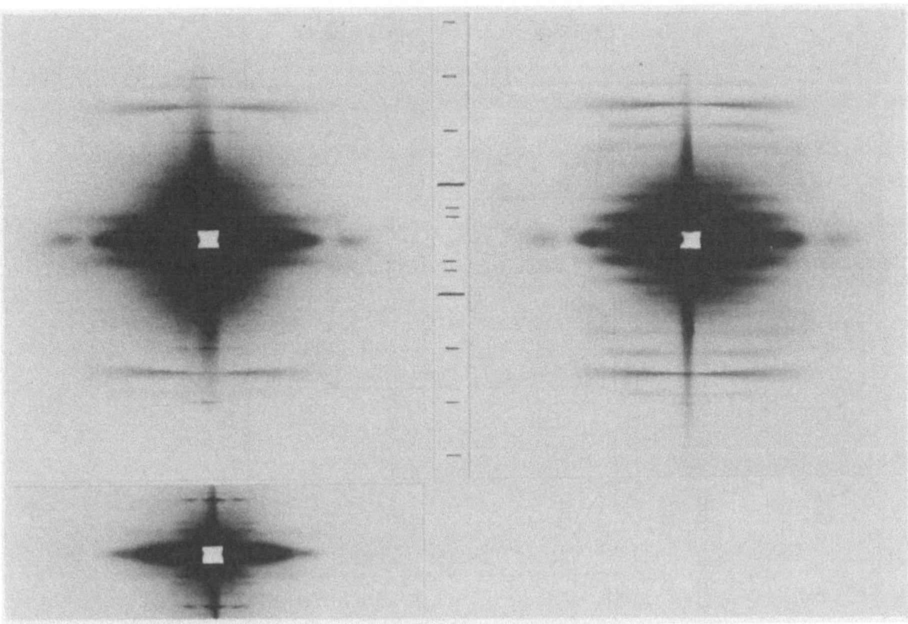

a b

Figure 1: X-ray diffraction patterns from crab (*Portunus trituberculatus*) striated muscles. (a) X-ray pattern in the relaxed state; (b) in the rigor state. The layer lines due to the thick filaments are marked by bars, in which the long bars denote the 14.5 nm meridional layer lines. In (a), the small photograph shows only the centre of the pattern after a short exposure.

distinct layer lines due to the thick and thin filaments. The layer lines with periods of 101.5 nm and 76.5 nm correspond to the thick and thin filaments, respectively. The two sets of layer lines are well-resolved without overlap and are not affected by the sampling effect due to the hexagonal filament lattice except on the equator as reported by Namba et al. (1979, 1980). In the relaxed and rigor states, layer lines from both filaments show specific changes in their intensities but the periods are respectively common to the relaxed and rigor states.

Diffraction pattern from the thick filaments in the relaxed state in Fig. 1(a) consists of several strong meridional layer lines at orders of 14.5 nm, with axial spacings of 14.5 nm, 7.24 nm, 4.83 nm and 3.62 nm, and two non-meridional layer lines with axial spacings of 33.7 nm and 25.4 nm. The meridional reflections are sharply localized along the filament axis, showing little indication of broadening in this direction, and can be observed up to the tenth order of 14.5 nm. The spacing of all these layer lines shows that the crystallographic period of the thick filament is 101.5 nm (14.5 nm × 7). Figure 2 shows the intensity distributions of the layer lines due to the thick filaments in the relaxed and rigor states from Fig. 1. Hereinafter the intensity distributions of the meridional layer lines will

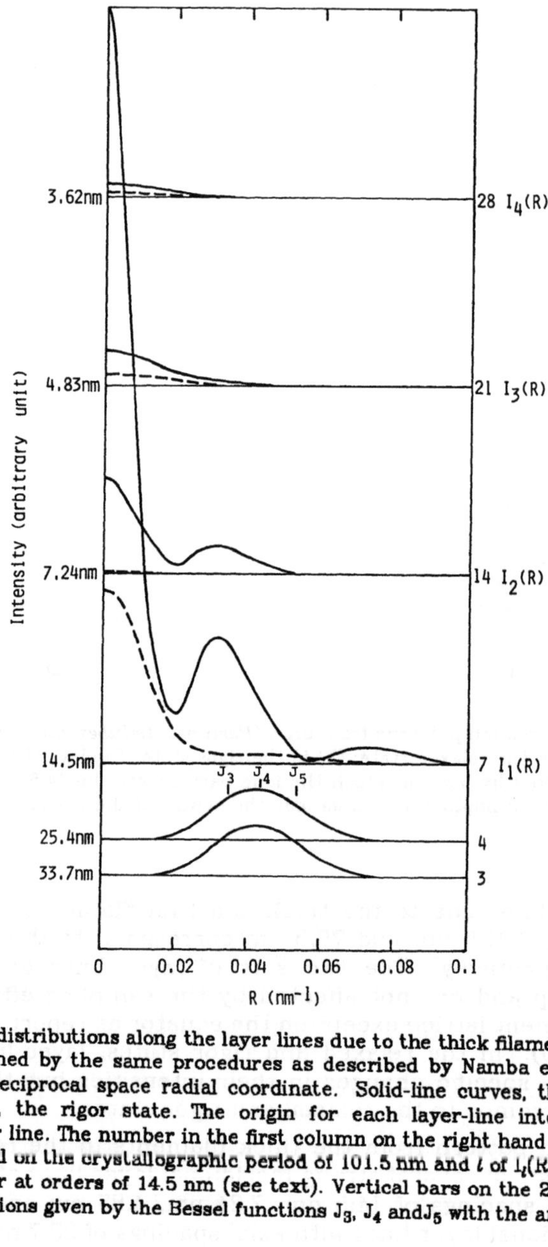

Figure 2: Intensity distributions along the layer lines due to the thick filaments. The intensity data were obtained by the same procedures as described by Namba et al. (1980). The abscissa R is the reciprocal space radial coordinate. Solid-line curves, the relaxed state; dashed-line curves, the rigor state. The origin for each layer-line intensity is on the corresponding layer line. The number in the first column on the right hand side is the layer-line number indexed on the crystallographic period of 101.5 nm and l of $I_l(R)$ is the meridional layer-line number at orders of 14.5 nm (see text). Vertical bars on the 25.4 nm layer line show the peak positions given by the Bessel functions J_3, J_4 and J_5 with the argument of $2\pi r_g R$ where r_g=19.0 nm.

be referred to as $I_l(R)$ where l is the meridional layer line number at orders of 14.5 nm and R the reciprocal space radial coordinate. In the relaxed state (solid-line curves in Fig. 2), $I_1(R)$ (14.5 nm) has a meridional maximum and two off-meridional subsidiary maxima associated with the meridional reflection, and $I_2(R)$ (7.24 nm) has a meridional maximum and one off-meridional subsidiary maximum. $I_3(R)$ (4.83 nm) and $I_4(R)$ (3.62 nm) have no apparent subsidiary maxima and only diffuse shoulders. In $I_1(R)$ and $I_2(R)$ the maxima and minima are at the same position, and the intensities between peaks do not fall to zero. The two non-meridional layer lines at 33.7 nm and 25.4 nm have similar intensity profiles to each other. In the rigor state (dashed-line curves in Fig. 2), all these layer lines become weak and the subsidiary maxima of the meridional reflections disappear, as do the non-meridional layer lines but the meridional reflections remain.

Symmetry of the Thick Filament

The reduction (and in many places virtual extinction) in the intensity of the thick filament layer lines by the transition to the rigor state is caused by the destruction of the helical arrangement of resting myosin projections around the thick filament backbone as a result of their attachment to the thin filaments. The diffraction remaining in rigor comes from the thick filament backbone, revealing its structure (Huxley & Brown, 1967; Squire, 1972; Haselgrove, 1980). The reflections which disappear in the transition to rigor arise solely from the myosin projections arranged helically around the thick filament backbone. So the symmetry of the helical arrangement of resting projections and also of the thick filament backbone can be elucidated from the peak positions of the off- and the non-meridional reflections, since the diffraction from the backbone is not large enough to cause the peak positions to deviate from those in the diffraction from the projections alone. The radial position of the resting projections was approximately derived from the positions of the first minima and the first subsidiary maxima in $I_1(R)$ and $I_2(R)$, which come from the Bessel function J_0. In solid-line curves of Fig. 2, the minima occur at $R=0.020\ nm^{-1}$ and the maxima at $0.032\ nm^{-1}$. From these values, the radius of centre of gravity of the resting projections, r_g was determined to be about 19 nm from the thick filament axis, this being in good agreement with that found in the Fourier map of the axial projection (see below). The intensity peaks of the two non-meridional layer lines occur at almost the same position of $R=0.043\ nm^{-1}$. Using the value of the radial position r_g = 19 nm, the peak positions obtained with the Bessel functions $J_n\ (2\pi r_g R)$ for n=3,4 and 5 are shown on the profile of the 25.4 nm layer line in Fig. 2. The peak value from J_4 (or J_{-4}) agrees most closely with the observed peak. Thus it is likely that the thick filament of crab muscle has four-fold rotational symmetry, indicating that four myosin molecules form projections on a level, with each level separated by 14.5 nm along the thick filaments. Figure 3 represents a surface lattice of the thick filament derived from the observed layer-line spacings and the Bessel functions contributing to the layer lines. In summary, the thick filament of crab muscle may have myosin projections arranged in a four-stranded helix with 28 residues in 3 turns, pitch 134.8 nm and axial repeat 14.5 nm.

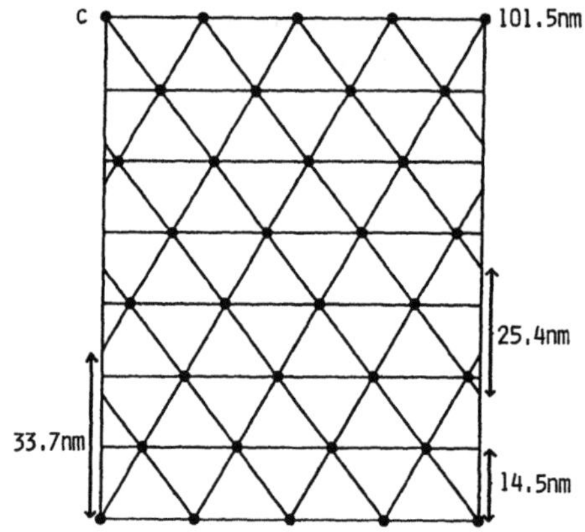

Figure 3: Surface lattice of the thick filament. c denotes the crystallographic period.

Configuration of Myosin Projections in the Relaxed State

In an attempt to elucidate the configuration of resting projections, the diffraction from the backbone cannot be disregarded (see Fig. 5 and Haselgrove, 1980), though the major part of the diffraction may come from the resting projections. In practice, it is impossible to extract the intensity profiles due only to the projections from the intensities observed in the relaxed state. The quantities which can be obtained experimentally are $|F_{relax}|^2$ and $|F_{rigor}|^2$ but not $\mathbf{F}_{relax}$ and $\mathbf{F}_{rigor}$ (we denote respectively the structure factors of the thick filament in the relaxed and rigor states as $\mathbf{F}_{relax}$ and $\mathbf{F}_{rigor}$). $\mathbf{F}_{relax}$ corresponds to the sum of the structure factors of the backbone $\mathbf{F}_b$ and that of the resting projection $\mathbf{F}_x$. $\mathbf{F}_{rigor}$ corresponds to $\mathbf{F}_b$ if all myosin projections are bound to the thin filaments in the rigor state as is the case of crab muscle (see below). Therefore, $\mathbf{F}_x$ ($= \mathbf{F}_{relax} - \mathbf{F}_{rigor}$) or $|F_x|$ cannot be obtained even if the intensity data in both states are available. We can only determine the range of $|F_x|$, of which the upper limit is $|F_{relax}| + |F_{rigor}|$ and the lower limit $|F_{relax}| - |F_{rigor}|$ (see Fig. 5). Thus a quantitative analysis of the amplitude data cannot be done since they are not determined definitely. However, it is possible to extract some information on the structure by characterizing the features of the diffraction patterns.

To elucidate the configuration of resting myosin projections, we concentrate here on the meridional layer lines. In the diffraction pattern of relaxed crab muscle (Fig. 2), the number of subsidiary maxima associated with the meridional reflections on $I_l(R)$ decreases with increasing the layer-line number, l. The intensities between peaks of $I_1(R)$ and $I_2(R)$ do not fall to zero. The meridional maxima of $I_3(R)$ and $I_4(R)$ decay smoothly and disappear along the layer line, although extended tails were observed. To characterize these features, which are predominantly governed by the resting projections, we calculated $I_l(R)$ for the appropriate structure of the thick filament. In the calculation the backbone was described by a uniform cylinder and so its transform makes no

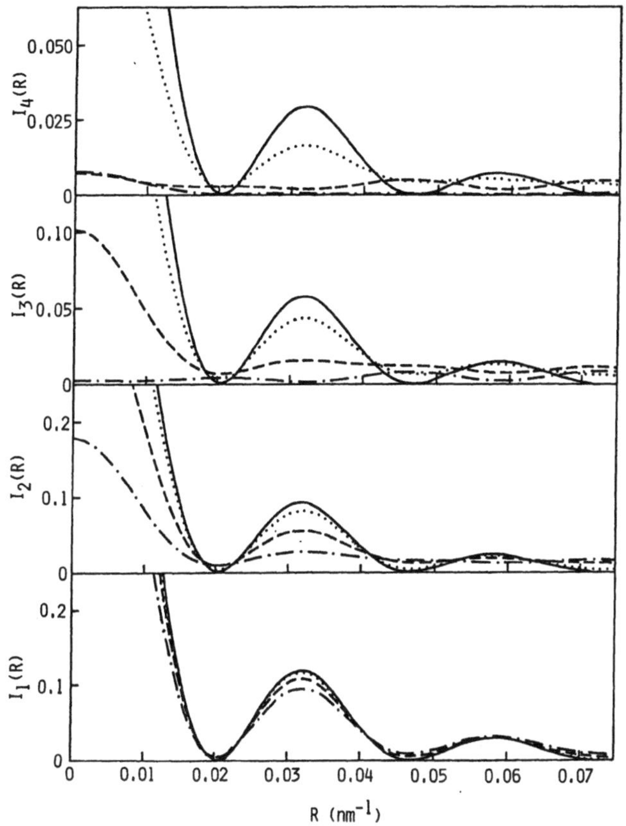

Figure 4: Changes of intensity distributions on the meridional layer lines calculated by varying the axial tilt of myosin projections. Myosin projections were described as an elongated rod 11 nm long and 3 nm thick. The angle of axial tilt is defined as the angle between the long axis of myosin projection and the plane normal to the fibre axis (and as zero for a projection normal to the filament axis)(Wray et al, 1975). l of $I_l(R)$ is the same as in Fig. 2. Intensities are normalized so that $I_1(0)$ is unity. Axial tilt: ————, 0°; ··········, 10°; ------------, 20°; ——·——, 30°.

contribution to $I_{l\neq0}(R)$. The myosin projection was modelled as a single rod, since in Fig. 2 there is no indication of any axial separation of two myosin heads as suggested by Wray et al. (1975). Variables used for the projections were the angle of axial tilt (see legend of Fig. 4), the thickness along the fibre axis and the length along its long axis. The radial position of the projections from the thick filament axis was fixed as 19 nm. We need not consider the azimuthal twist and the width of the projection in the plane normal to the filament axis, since we are considering the intensities of meridional layer lines which include only a contribution from J_0. So the calculated intensity corresponds to the transform of part of a disc or cone with appropriate thickness generated by cylindrically averaging a projection from the backbone. Figure 4 shows changes in $I_1(R)$ to $I_4(R)$

calculated by varying the axial tilt of projections 3 nm thick and 11 nm long. According to Fig. 4, increase in the angle of tilt decreases the intensities of the subsidiary peaks on each layer line as well as the meridional ones and simultaneously makes intervening minima between peaks shallow, resulting in the disappearance of the subsidiary maxima. This feature reflects the asymmetry of a polar structure induced by tilting a myosin projection having an elongated structure. In Fig. 5, the rate of decrease in intensity at the maxima and that of increase at the minima by tilting become larger with increasing l, so that the angle of tilt at which the subsidiary peaks disappear depends on l. On the other hand, to see the effect due to disorder in the arrangement of resting projections, we included in the calculation a disorder term through the Debye-Waller factor similarly to Haselgrove (1980). The effect of disorder decreases the intensity almost uniformly along each layer line, since the change of scattering angle along the meridional layer line is small. The subsidiary maxima did not completely disappear with increasing l. The absence of clear zero intensity minima between peaks on $I_l(R)$ could not be explained by such disorder. Thus the difference in intensity distribution of the different layer lines indicates that the myosin projection has an elongated structure and is tilted axially along the filament backbone.

Our calculations showed that the variation of off-meridional intensities on $I_l(R)$ due to tilting is very insensitive to the change in the thickness but depends on the length of the projection. In the calculation of Fig. 4, a myosin projection was constructed by nine overlapping spheres of radius 1.5 nm spaced at intervals of 1.0 nm. In Fig. 4, two subsidiary peaks are distinct in $I_1(R)$ even when the angle of tilt is varied up to 30°. In $I_2(R)$ the second subsidiary peak disappears at 20° and the first one has almost disappeared at 30°. In $I_3(R)$ and $I_4(R)$ the first subsidiary peak is distinct up to 10° but it disappears at 20°. In the observed $I_l(R)$ of Fig. 2, two subsidiary peaks in $I_1(R)$ and one subsidiary peak in $I_2(R)$ are present, but no subsidiary peak is present in $I_3(R)$ and $I_4(R)$. The comparison with Fig. 5 suggests that a tilt 20° of the myosin projection explains these observed features well. When the projection was constructed by eight spheres with the same interval, a tilt 30° explained the observed features in a similar way. But for any arrays except for eight and nine, they could not be reproduced. Thus our calculations reveal that the resting projections are axially tilted 20°~30°. Using these values of tilt, we estimated the axial thickness of the projection. It was determined by varying the radius of overlapping spheres so that $|F_x|$ on the meridian ($|F_l(0)|$) lies in the range of $|F_x|$ shown with a vertical bar for each layer line in Fig. 5, where the ranges of $|F_x|$ were derived using the values of $|F_{relax}|$ and $|F_{rigor}|$ from $I_1(0)$ to $I_4(0)$ in Fig. 2. The calculated $|F_l(0)|$ for the projection with 3 nm and 4 nm thickness at tilts of 20° for the nine array and 30° for the eight array are shown in Fig. 5. For the nine sphere array, the maximum thickness was about 4 nm and the full length was 12 nm. For the eight one, it was about 3.5 nm and the full length 10.5 nm. This small value of the thickness was close to that obtained for the head of the myosin molecule isolated from rabbit skeletal muscle by Elliott & Offer (1978).

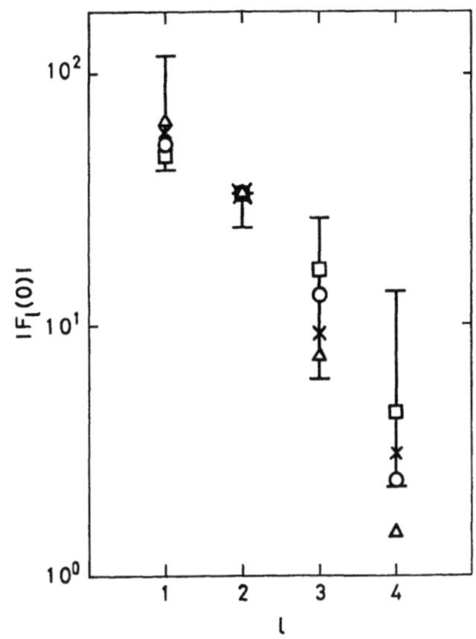

Figure 5: Ranges of the absolute values of the structure factors of resting projections on the meridian derived from $I_l(R)$ in Fig. 2. The abscissa l is the same as in Fig. 2. For each layer line, the upper limit is $\{I_{relax}(0)\}^{0.5} + \{I_{rigor}(0)\}^{0.5}$ and the lower limit $\{I_{relax}(0)\}^{0.5} - \{I_{rigor}(0)\}^{0.5}$. $|F_l(0)|$ calculated for the thickness 3 nm (□) and 4 nm (○) of the projection with nine overlapping spheres at the axial tilt 20° and for the thickness 3 nm (×) and 4 nm (Δ) of the projection with eight spheres at the tilt 30° are scaled so as to coincide at the upper limit of $|F_2(0)|$

The resting myosin projections of crab muscle at full overlap sit about 19 nm from the thick filament axis and 8.5 nm away from the backbone surface, and the thick filament had four-fold rotational symmetry. According to Wray et al. (1975), in lobster muscle they sit about 14 nm from the axis and 3.5 nm away from the surface, and the thick filament has six-fold rotational symmetry; in limulus muscle, they are at about 16.5 nm and at 6.7 nm, and the thick filament has three or four-fold symmetry; in scallop muscle, they are at 17 nm and at 5.7 nm, and the thick filament has six or seven-fold symmetry. Thus the symmetry of the thick filament of crab muscle is different from that of these muscles, and the myosin projections lie at a fairly large radius.

Recently we have found that the stable position of the resting projections sensitively depends on the ionic strength and pH of the solution. Figure 6 shows the changes in the radial position (r_g) of the resting projections induced by varying the concentration of KCl and pH of the solution. With decreasing KCl concentration at low temperature, the projections shift away from the thick filament surface and at 0 mM concentration, they move very close to the thin filament surface. On the other hand, they shift back to the thick filament backbone when the pH value is decreased. In addition, these changes in ionic strength and pH are accompanied by specific changes in the non-equatorial layer-line intensity as well as in the equatorial intensity, suggesting that certain changes

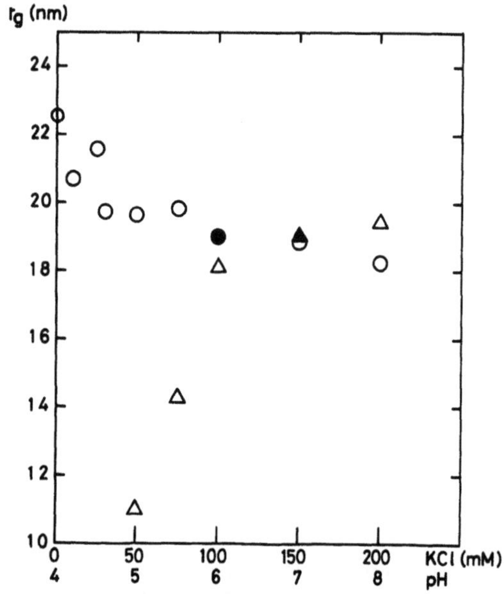

Figure 6: Changes in the radial position (r_g) of the resting projections induced by changing the concentration of KCl (O) (at pH=7.0) and pH (Δ) (at 100 mM KCl) of the relaxing solution at temperature 4°C. Filled symbols: pH=7.0, 100 mM KCl.

of the configuration of the projection take place in response to ionic changes in the solution. Detailed interpretations will be reported elsewhere.

The angle of axial tilt of the resting projection of crab muscle has been shown here to be 20°~30°. As mentioned in Introduction, Wray et al. (1975), Squire (1975) and Haselgrove (1980) have analysed the patterns from relaxed crustacean muscles and frog sartorius muscle. One similarity in all these muscles is that the resting projections are tilted to the filament axis; 30° for limulus, 20° for scallop and small tilt for lobster, and ca. 30° for frog sartorius. In the insect flight muscle, a perpendicular orientation of the projection has been shown by electron microscopy of sectioned muscles (e.g. Reedy et al., 1965).

Configuration of Myosin Heads in the Rigor State

In the rigor state, a series of layer lines indexing on a period of 76.5 nm (38.25 nm x 2) due to the thin filaments become markedly strong (Fig. 1(b)). This intensity enhancement arises from myosin heads attached to the thin filaments in the rigor state. Namba et al. (1979,1980) derived the structure of the thin filament. (The structure of the thin filament in the rigor state is briefly outlined in conjunction with the configuration of myosin heads bound to the thin filaments.) The myosin heads are attached periodically to the thin filaments with the period of the thin filament: four myosin heads are tightly bound in groups every 38.25 nm to successive actin molecules of each strand of F-actin. The bound myosin heads have very compact conformation and most of the head is incorporated in the thin filament with the centre of gravity 2.8 nm from the thin filament axis. They are inclined at about 30° to and slewed round the axis. The inclination angle (defined here in the same way as in Fig. 4) of bound myosin heads was close to that of the rabbit skeletal muscle in rigor derived by the measurement of electron paramagnetic resonance (EPR) (Thomas & Cooke, 1980). Recent EPR measurement has shown that virtually all of the myosin heads in a rigor crab muscle are attached to actin at full overlap at almost the same orientation to the filament axis as that in rabbit skeletal muscle (Thomas & Wakabayashi, unpublished data).

Electron Density Distributions in the Axial Projection of Muscle

Wakabayashi & Namba (1981) derived the electron density distributions of muscle in the relaxed and rigor states projected along the fibre axis. They assigned the relatively high electron density region centered at about 19.6 nm from the thick filament axis to the resting projections. The resting projections lie in between the surface of the thick filament backbone (diameter, ca. 21 nm) and that of the thin filament (diameter, ca. 8.4 nm), having a radial width of about 11 nm. Their estimation of the mass associated with myosin projections suggested that there would not be more than four projections on every 14.5 nm level along the thick filament in the relaxed state. This is consistent with the rotational symmetry of the thick filament derived from the present layer-line analysis. Further, the number of myosin heads around the thick filament backbone in the relaxed state is in fairly good agreement with the number of myosin heads bound to the thin filaments in the rigor state, suggesting that all myosin projections bind to the thin filaments in going from the relaxed to the rigor states as in the case of rabbit skeletal muscle (Thomas & Cooke, 1980; Cooke and Franks, 1980). This implies that the diffraction remaining in rigor may be equated with the diffraction from the thick filament backbone, although Haselgrove & Reedy (1979) discussed the possibility that myosin heads attached to the thin filaments of insect flight muscle could contribute to the meridional reflections of 14.5 nm order observed in the rigor state.

ACKNOWLEDGMENT

We thank Mr. T. Hayashi for his assistance. Thanks are also due to Dr. D.D. Thomas for the EPR measurements and to Dr. G. Stubb for his critical reading of the manuscript.

REFERENCES

Cooke, R. and Franks, K. (1980). All myosin heads form bonds with actin in rigor rabbit skeletal muscle. Biochemistry 19: 2265-2269.

Elliott, A. and Offer, G. (1978). Shape and flexibility of the myosin molecule. J. Mol. Biol. 123: 505-519.

Haselgrove, J.C. and Reedy, M.K. (1979). Modeling rigor cross-bridge pattern in muscle I.

Initial studies on the rigor lattice of insect flight muscle. Biophys. J. 24: 713-728.
Haselgrove, J.C. (1980). A model of myosin crossbridge structure consistent with the low-angle X-ray diffraction pattern of vertebrate muscle. J. Mus. Res. Cell Motil. 1: 177-191.
Holmes, K.C., Tregear, R.T. and Barrington Leigh, J. (1980). Interpretation of the low-angle X-ray diffraction from insect flight muscle in rigor. Proc. R. Soc. Lond. B207: 13-33.
Huxley, H.E. and Brown, W. (1967). The low-angle X-ray diagram of vertebrate striated muscle and its behaviour during contraction and rigor. J. Mol. Biol. 30: 383-434.
Maeda, Y., Matsubara, I. and Yagi, N. (1979). Structural changes in thin filaments of crab striated muscle. J. Mol. Biol. 127: 191-201.
Namba, K., Wakabayashi, K. and Mitsui, T. (1979). The structure of thin filaments of crab striated muscle in the rigor state. In: *Cross-bridge Mechanism in Muscle Contraction*, pp. 445-470, ed. Sugi, H. and Pollack, G.H. University of Tokyo Press.
Namba, K., Wakabayashi, K. and Mitsui, T. (1980). X-ray structure analysis of the thin filament of crab striated muscle in the rigor state. J. Mol. Biol. 138: 1-26.
Offer, G., Couch, J., O'Brien, E. and Elliott, A. (1981). Arrangement of cross-bridges in insect flight muscle in rigor. J. Mol. Biol. 151: 663-702.
Reedy, M.K., Holmes, K.C. and Tregear, R.T. (1965). Induced changes in orientation of the crossbridges of glycerinated insect flight muscle. Nature 207: 1276-1280.
Squire, J.M. (1972). General model of myosin filament structure II. Myosin filaments and cross-bridge interactions in vertebrate striated and insect flight muscles. J. Mol. Biol. 72: 125-138.
Squire, J.M. (1975). Muscle filament structure and muscle contraction. Ann. Rev. Biophys. Bioeng. 4: 137-163.
Thomas, D.D. and Cooke, R. (1980). Orientation of spin-labelled myosin heads in glycerinated muscle fibres. Biophys. J. 32: 891-906.
Wakabayashi, K. and Namba, K. (1981). X-ray equatorial analysis of crab striated muscle in the relaxed and rigor states. Biophys. Chem. 14: 111-122.
Wray, J.S., Vibert, P.J. and Cohen, C. (1975). Diversity of cross-bridge configurations in invertebrate muscles. Nature 257: 561-564.
Wray, J.S., Vibert, P.J. and Cohen, C. (1978). Actin filaments in muscle: pattern of myosin and tropomyosin/troponin attachments. J. Mol. Biol. 124: 501-521.
Wray, J.S. (1979). Structure of the backbone in myosin filaments of muscle. Nature 277: 37-40.

DISCUSSION

SCHOENBERG: I think it is very interesting that you showed that at low KCl concentration the myosin projection seems to be out more near the actin filament, because I will be talking about mechanical measurements that we made which suggest that at low ionic strength the cross-bridges actually are attached and interacting with the actin filaments quite strongly. Those are measurements made on stiffness. I should also mention that Bernhard Brenner and Leepo Yu, who are working with Richard Podolsky, also have similar X-ray evidence using skinned rabbit psoas muscles, that at low ionic strength the cross-bridges seem to be closest to the actin filaments. So that would agree with your findings.

K. WAKABAYASHI: Our data at low KCl concentration were obtained at low temperature, 4°C. At room temperature the situation is different from that at 4°C.

SCHOENBERG: The experiments that I am talking about, on rabbit psoas muscle, were also done at 5°C. What is the difference?

K. WAKABAYASHI: Studies on crab muscle done with Dr. Yanagida showed that at low temperature no tension was developed, but the value of stiffness was not zero (consistent with your result). But at room temperature the tension did develop and the stiffness increased similarly to that in rabbit muscle (Yanagida et al., J. Biochem. 92: 407-412, 1982).

ROWE: Could I just recall in connection with the remarks you made about the effect of temperature, it is one of the older observations that the myosin-actin interaction, or at least the number of myosins combined with actin, which I think is a more correct way to put it, is strongly influenced by temperature. When you go down in temperature, you do get far fewer myosins bound to actin.

K. WAKABAYASHI: Thank You. Our mechanical measurements do suggest that.

MAEDA: I would like to ask you about configurational changes at different pH. You mentioned that the configuration of myosin heads may be changed. Did you measure the lattice spacing at lower pH?

K. WAKABAYASHI: We measured the lattice spacing. It gradually decreased with lowering pH and decreasing KCl concentration.

MAEDA: So, by configurational change, do you mean that you observed the change of just the tilting of myosin heads by pushing back the thin filament, or a real change of shape of the myosin head?

K. WAKABAYASHI: The myosin heads actively move back to the thick filament backbone when lowering pH and actively move near the thin filament when lowering KCl concentration, though the lattice shrinkage occurs in any case. Thus, the lattice shrinkage has no direct effect on the configurational change if present on lowering pH. When decreasing KCl concentration, some myosin heads form crossbridges. So the attached myosin heads may be affected by the lattice shrinkage. Now we are analyzing the X-ray data. At present we have no definitive conclusion about that.

TREGEAR: Have you related the strength of the equatorial pattern to your layer-line pattern? Can you say how well-ordered the crossbridges are relative to the backbone in the relaxed muscle?

K. WAKABAYASHI: We have not related them quantitatively. But analysis of the equatorial data (Wakabayashi & Namba, Biophys. Chem. 14: 111-122, 1981) suggested that the radial extent of the electron density corresponding to the myosin projections in the axial projection is consistent with the configuration of myosin heads derived by the layer-line analysis. We have judged the orderliness of myosin projections from the general appearance of the layer-line reflections. As for the general appearance of the diffraction pattern with decreasing KCl concentration and pH, the layer line reflections, particularly the meridional layer lines, become strong. In the case of the decrease of pH, the two non-meridional layer lines become weak and almost disappear at a pH of 5. So these would accompany the changes of orderliness of myosin projections around the thick filament backbone as well as their configurational changes. Thus, in the normal, relaxed condition, the resting projections are arranged regularly relative to the backbone to a degree which the average orientation is maintained.

HOLMES: May I go on with that question? If you compare the zero layer line or some other part of the diffraction patterns with the higher layer lines, does it come out right in your model calculations? Or is it weaker than it should be?

K. WAKABAYASHI: I have not done the detailed model calculations to interpret the intensities of all the observed layer lines as well as the equatorial reflections. I have explained how the features of the intensity distributions on the meridional layer lines which could not be eliminated by disorder are related to the configuration of myosin projections. Maybe the configuration of myosin projections derived in the present work in combination with a certain degree of disorder would explain the weakness and/or the disappearance of the other parts of layer-line diffraction. In particular, I think that the relative intensities of the inner two or three equatorial reflections should be interpreted by such configuration of myosin projections in the lattice formed by the thick and thin filaments.

HOLMES: That would be very interesting.

ON THE INTENSITY REVERSAL OF THE "TROPOMYOSIN REFLEXIONS" IN X-RAY DIFFRACTION PATTERNS FROM CRAB STRIATED MUSCLE

Yuichiro Maéda

Max-Planck-Institute for Medical Research, Department of Biophysics, Jahnstrasse 29, D-6900 Heidelberg, FRG.

ABSTRACT

X-ray diffraction patterns have been recorded from skinned single fibres obtained from crab leg muscle, and the outer parts of layer-lines indexed as orders of 38.5 nm, which have been assigned to tropomyosin and actin, were examined. Fibres at normal length in the presence of ATP (γ-S), and over-stretched fibres in rigor solutions, show no intensity reversal of the 2nd and the 3rd layer-lines when the Ca^{2+} concentration is raised. The results are discussed with reference to the mechanism of Ca^{2+} regulation.

Fibres in the presence of Mg-ADP and vanadate ion (V_i) and fibres pre-treated at low pH, though generating no substantial tension at high Ca^{2+} concentration, give rise to rigor-like patterns which are dependent on Ca^{2+} concentration.

INTRODUCTION

The mechanism of calcium regulation of muscle contraction mediated via the troponin (TN) - tropomyosin (TM) complex has been explained by the steric blocking model (H. Huxley, 1972; Haselgrove, 1972; Parry & Squire, 1973) which suggests that the position of TM molecule in the groove of the two-stranded actin helix controls actin-myosin interaction; in the absence of Ca^{2+}, TM takes up a position where it blocks the binding of myosin to the thin filament whereas when Ca^{2+} binds to TN, the TM moves towards the center of the groove, enabling myosin to interact with actin.

The model was proposed on the basis of an intensity reversal of the outer parts ("tropomyosin reflexions") of layer-lines assigned to the thin filaments; in relaxed state, the intensity of the outer part of the second layer-line (at an axial spacing of 19 nm) is much weaker than the 3rd (at 12.8 nm), which was interpreted as indicating that TM is located at the edge of the groove ("off position"), whereas in the activated state, the second becomes much stronger than the 3rd, which was explained by a shift of TM towards the center of the groove ("on position").

It is generally thought that binding of Ca^{2+} to a TN molecule alone causes the shift of a TM molecule to which the TN binds, even when there is no actomyosin interaction. If this is the case, in the thin filaments which are free from interaction with the myosin heads, the translocation of TM should take place in the same range of Ca^{2+} concentration in which tension generation is regulated. Up to the present time, however, there has been no definitive evidence on this point. Experiments reported here were originally designed to determine a range of Ca^{2+} concentration controlling the TM position. For that purpose, X-ray diffraction patterns were recorded from skinned single fibres of crab leg muscle, which were soaked in various solutions in which one might expect to prevent the myosin heads from interacting with the thin filaments even at high Ca^{2+} concentration > 10^{-5} M. In this communication, it is suggested that there is no intensity reversal of the "TM reflexions", if the actomyosin interaction is successfully inhibited. Moreover, under some conditions under which tension generation were almost inhibited even at high Ca^{2+} concentration, cross-bridges were formed and the formation was Ca^{2+} sensitive, as judged from the intensified outer actin layer-lines (rigor pattern). Apparently the intensity reversal of the "TM reflexions" accompanies the appearance of the rigor pattern.

Crab muscle was used for the following reasons: (1) In the X-ray diffraction pattern, the "TM reflexions" are relatively strong and these are spatially isolated from thick filament reflexions. (2) It is easy to obtain chemically skinned single fibres which are large enough for recording detailed X-ray diagrams, and which can be stretched up to no overlap between the thin and thick filaments.

MATERIALS AND METHODS

Muscle Fibres

Leg muscles were dissected from crab *Portunus puber*, and treated in skinning solutions containing 0.5% Triton X-100 for 1-3 hours at 4°C. Fibres were used within 3 days after dissection.

X-Ray Cameras

X-ray diffraction patterns were obtained using conventional mirror-monochrometer cameras (constructed by Dr. J. Wray) with specimen-film distances of either 12, 30 or 45 cm on an Elliott GX 13 rotating anode X-ray generator.

Solutions

In all solutions, pH was adjusted to 6.8-7.0 at 4°C. The ATP relaxing solution contained (in mM): potassium methansulphonate (KMS), 80; $MgSO_4$, 10; tris-(hydroxymethyl)-aminomethane (Tris), 20; maleic acid, 20; ATP-Na_2, 4; NaN_3, 2. For the skinning solutions detergent was added to the relaxing solution. The rigor-G solution contained: KCl, 50; $MgCl_2$, 5; EGTA, 4; PIPES, 10; NaN_3, 2. For the rigor-Ca solution 5 mM $CaCl_2$ was added to the normal rigor solution. The ATP (γ-S)-G solution (Goody, Mannherz, Holmes, Barrington Leigh & Rosenbaum, 1975) contained; adenosine 5'-0-(3-thiotriphosphate) [ATP(γ-S)] (Goody & Eckstein, 1971; supplied by R. S. Goody and M. Isakov), 13; $MgSO_4$, 14; KMS, 13; EGTA, 2; NaN_3, 1.3; Tris, 13; maleic acid, 13; 1,4 dithioerythrit (DTE), 4; AP_5A (Lienhard & Secemski, 1973; supplied by R. S. Goody and M. Isakov), 1 mM as a specific adenylate kinase inhibitor; acid phosphatase 0.1 mg/ml to degrade ATP and ADP produced. The ATP (γ-S)-Ca solution contained 3.7 mM $CaCl_2$ in addition. The ADP-V_i-G solution was the same as the normal relaxing solution except that it contained: ADP, 4 instead of ATP and in addition, vanadate (V_i), 4 ; hexokinase, 0.1 mg/ml together with glucose, 1 mM to degrade ATP, and AP_5A, 1. The ADP-V_i-Ca solution contained 5 mM $CaCl_2$ in addition. Stock solutions of V_i were prepared from Na_3VO_4 as described by Goodno (1982).

RESULTS

1. Fibres in Relaxed and Rigor States

X-ray diffraction patterns (Fig. 1a) obtained from chemically skinned single fibres in the normal relaxing solution show that outer part of the 2nd order layer-line at an axial spacing of about 19 nm is much weaker than that of the 3rd order (at 12.8 nm). Fibres in the rigor solutions at normal length (with sarcomere lengths of 6-7 μm) give rise to patterns (Fig. 1b) with 2nd layer-lines which are now stronger than the 3rd, irrespective of Ca^{2+} concentration. Radial spacing of these intensity

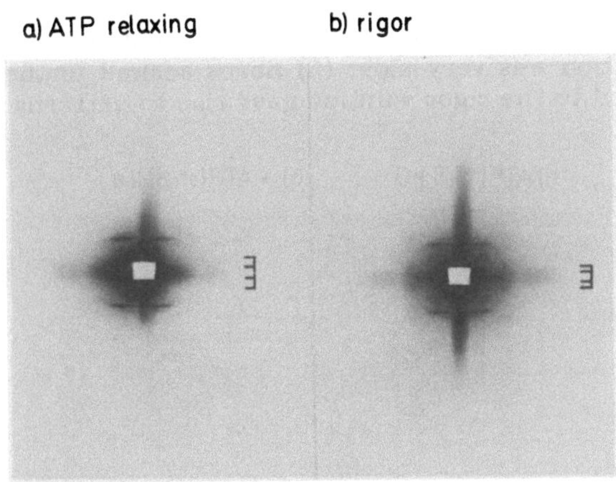

Figure 1: X-ray diffraction patterns obtained from single fibres of skinned crab leg muscle (*Portunus puber*), *a*, in ATP relaxing solution, *b*, in rigor solution. Specimen-film distance 12 cm. The outer part of the 2nd order layer-line is weaker in *a* and stronger in *b* than that of the 3rd. Bars indicate axial positions of the 1st and the 3rd layer-lines in *a*, and the 1st and the 2nd layer-lines in *b*.

peaks are measured to be 1/6.7 nm (on the 1st layer-line at 38 nm), 1/4 nm (on the 2nd) and 1/3.3 nm (on the 3rd). These results are highly consistent with those obtained from frog and other muscles (Vibert et al., 1972; Haselgrove, 1972; Huxley, 1972).

These reflexions have been assigned to TM and the reversal of intensities on the 2nd and 3rd has been interpreted in terms of a shift of TM from the edge to the centre of the actin groove.

2. Fibres in ATP(γ-S) Solution

In order to examine the effect of Ca^{2+} concentration independently of the effect of cross-bridge formation on the TM position, the thin filaments should be kept not only intact but also free from interaction with the myosin heads, since the actomyosin interaction alone could shift TM to the centre in the groove irrespective of Ca^{2+} concentration. Among several conditions which had been investigated, two were found to be satisfactory, namely fibres at normal sarcomere length in the presence of Mg-ATP(γ-S) and fibres at non-overlap length (longer than 10.5 μm) in rigor state.

Fibres in the presence of 13-15 mM of Mg-ATP(γ-S) at low Ca^{2+} concentration give rise to X-ray diffraction patterns indistinguishable to the normal ATP relaxed patterns including the myosin reflexions indexed as orders of 14.5 nm. The "TM reflexions" (Fig. 2a) were also similar to those from ATP relaxed fibres, namely the 2nd layer-line was weaker than the 3rd. At higher concentrations of Ca^{2+}, even at 1 mM, the 2nd layer-line was not intensified (Fig. 2b), but remained weaker than the 3rd.

It is possible that no intensity reversal of the TM reflexion was observed because the thin filaments became insensitive to Ca^{2+} in the ATP(γ-S) solutions. However, this is unlikely, because (1) contraction of fibres which are soaked first in the ATP(γ-S) solution then returned to the normal relaxing solution was Ca^{2+} sensitive, though the time course of tension generation was very slow; (2) fibres soaked in the ATP(γ-S) and then transferred to the rigor solution gave rise to patterns where the 2nd

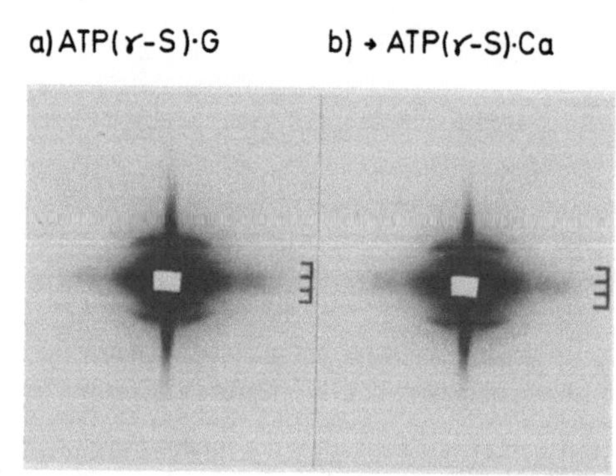

Figure 2: Fibres in the presence of ATP(γ-S), *a*, at low < 10^{-8} M, *b*, at high (1 mM) Ca^{2+} concentration. The outer part of the 2nd layer-line is weaker than the 3rd in both cases. Exposures were made on the same fibre in succession.

layer-lines were stronger than the 3rd, indicating that TM was not removed from the thin filament and that TM did not lose the ability of altering its position; (3) the meridional TN reflexions (Maeda, Matsubara & Yagi, 1979; Maeda, 1979a) were not decreased in intensity but were observed up to an axial spacing of 3 nm, indicating that at least most of the TN was not removed in the presence of ATP(γ-S).

It is reported that, at 20 mM ATP(γ-S), stiffness of insect flight muscle is reduced to the value for relaxed state (Barrington Leigh, Holmes, Mannherz, Rosenbaum, Eckstein & Goody, 1972). In the case of the crab fibres, however, at 20 mM X-ray diagrams were poor, showing that the thick filaments were dissolved and the thin filaments were disoriented. This might be ascribable to the effects of polyphosphates, since the same concentration of ATP or pyrophosphate brought about similar results. At 10 mM ATP(γ-S) (at an ionic strength of 80-150 mM), on the other hand, the rigor patterns (Maeda et al., 1979, Maeda, 1979a) were observed irrespective of Ca^{2+} concentration, namely layer-lines which are indexed as orders of 77 nm were intensified and extended to the meridian. This is explained by cross-bridge formation, which put additional material along the actin based helices. The results are consistent with relatively high stiffness of the insect flight muscle at this concentration.

3. Stretched Fibres in Rigor Solutions.

Skinned crab muscle fibres can be stretched to a sarcomere length of 11 μm where no overlap would be expected between the thin and the thick filaments. Patterns from stretched fibres soaked in the rigor solutions were very sensitive to sarcomere length in the range of 9.5-10.5 μm. At sarcomere lengths shorter than 10.3 μm, relatively strong rigor patterns were always accompanied by 2nd TM layer-lines which were much stronger than the 3rd. The TM patterns remained the same when the Ca^{2+} concentration was increased to 1 mM. Sarcomere lengths longer than 10.5 μm resulted in no rigor patterns, and in addition the TM reflexions became poor, namely not only the 2nd but also the 3rd layer-lines were not clearly observed, which may be due to disordering of the thin filaments on removal from the thick filament lattice. In this range of sarcomere lengths, high Ca^{2+} concentration did not cause significant enhancement of the 2nd layer-lines, which could be taken as additional evidence for no intensity reversal. However, the possibility exists that an enhancement would be compensated by disorientation of the thin filaments, leading to no observed change. In the range of 10.3-10.5 μm, despite remnants of the rigor patterns, both layer-lines were observed and the 2nd layer-lines were about as weak as the 3rd, showing that the numbers of cross-bridges formed were large enough to keep the thin filaments in good orientation, and small enough to keep the 2nd layer-line relatively weak (Fig. 3a). High Ca^{2+} concentration did not cause significant intensification of the 2nd layer-lines, and the 3rd remained visible (Fig. 3b).

If ATP is present at high Ca^{2+} concentration (the normal activating solution), even the over-stretched fibres broke at either end, probably

a) stretched·rigor·G b) → stretched·rigor·Ca

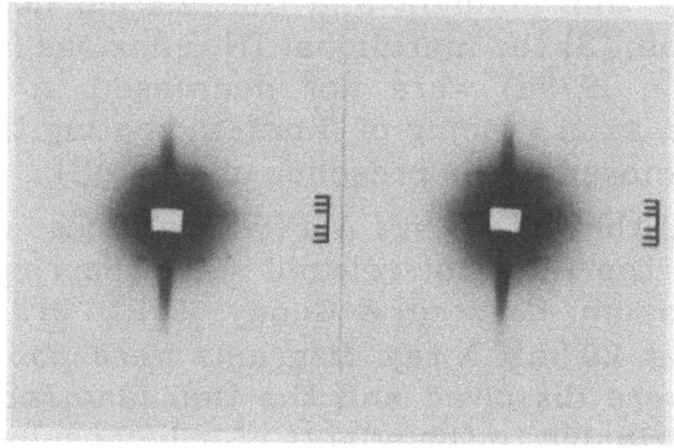

Figure 3: Stretched fibres (with sarcomere length of 10.3 μm) in rigor solutions, *a*, at low and, *b*, at high Ca^{2+} concentration. Exposures were made on the same fibre in succession. In *a* and *b*, the 2nd layer-lines are visible, but not much stronger than the 3rd which are also visible in the original. It was not possible to reproduce these patterns to show the 3rd layer-lines.

due to tension generated at short sarcomeres under the loops of thread fixing the ends of the fibres.

The muscle contains thin filaments about 2 μm long from the Z-line to the tip and thick filaments 4.4 μm long (Maeda, 1978; Hofmann & Maeda, unpublished results). Therefore no overlap would be expected with the sarcomere length longer than 8.6 μm. Nevertheless considerable cross-bridge formation was observed even at 10.3 μm, which might be explained by inhomogeneity in sarcomere length and/or in filament length.

4. Fibres in ADP-V_i Solution

In the course of the present work, the effect of vanadate ions (V_i) were investigated to see if it would completely prevent actomyosin interaction at high Ca^{2+} concentration. It transpired that, although tension generation is prevented, cross-bridges are formed. Therefore, the system is not suitable for studying the effect of Ca^{2+} on TM position. However, this cross-bridge formation is Ca^{2+} sensitive.

V_i is known as an inhibitor of myosin- and actomyosin-ATPase (Goodno, 1979; Goody, Hofmann, Reedy, Magid & Goodno, 1980; Goodno & Taylor, 1982). It is thought that V_i works as an analogue of phosphate (P_i), and forms a stable myosin-ADP-V_i complex.

In the presence of Mg-ATP (4 mM) and V_i (4 mM), X-ray patterns at low Ca^{2+} were indistinguishable from the normal ATP relaxed pattern. At high Ca^{2+}, even if fibres were pretreated in the ATP-V_i solution at low Ca^{2+} concentration for one day, some contraction appeared to be induced and the fibres were degraded.

In the presence of Mg-ADP (4 mM) and V_i (4 mM), patterns at low Ca^{2+} (Fi. 4a,c) are again very similar to the normal relaxed patterns. At high Ca^{2+}, however, degradation of the fibres did not happen, instead,

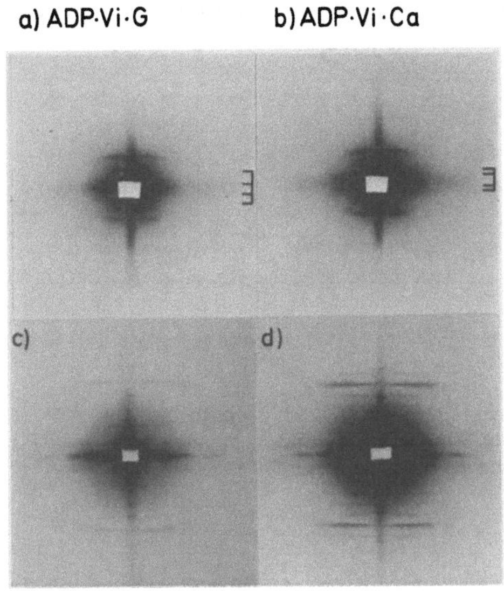

Figure 4: Fibres in the presence of ADP and V_i, *a* and *c*, at low and *b*, and *d*, at high Ca^{2+} concentration. Specimen-film distances, 12 cm in *a* and *b*, 30 cm in *c* and *d*. Independent experiments.

patterns (Fig. 4b, d) were rigor like, indicating cross-bridge formation. The rigor pattern was accompanied by an intensification of the 2nd tropomyosin layer-line. The changes of the patterns were reversible, namely when Ca^{2+} was decreased, the rigor patterns disappeared and the relaxed patterns reappeared, if the fibres had been exposed to high Ca^{2+} not longer than half a day.

5. Fibres Treated at Low pH

Cross-bridge formation which is dependent on Ca^{2+} concentration were also observed with fibres which was pretreated at low pH. Fibres were first treated at pH 5 for 30 min. then returned to pH 7, in the presence of ATP and EGTA throughout the procedures. The fibres were not degraded when Ca^{2+} was raised, showing that no substantial tension was generated. X-ray patterns showed that, at low Ca^{2+} the structure resembles the normal relaxed state, and at high Ca^{2+} cross-bridges were formed, accompanied by the intensified 2nd order tropomyosin layer-line. This change was again reversible (Fig. 5).

When fibres were pretreated at a pH higher than 5.3, the fibres still kept contractility. On the other hand, fibres pretreated at pH lower than 4.5 did not show any response to Ca^{2+}, that is, no tension nor rigor patterns were generated.

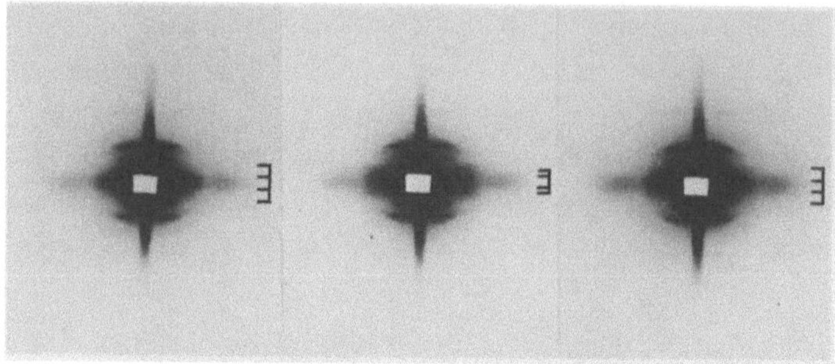

Figure 5: Fibers pretreated at ph 5 and *a*, put into the ATP relaxing solution and *b*, transferred into the activating solution, and, *c*, returned to the ATP relaxing solution. Exposures were made in succession on the same fibre.

DISCUSSION

The results obtained from fibres at normal length in the ATP(γ-S) solution, and from overstretched fibres in the rigor solutions showed that the intensity reversal of the "TM reflexions" was not caused by raising Ca^{2+} concentration alone. Three questions arise: (1) does the TN-TM complex actually exist in the crab muscle, (2) are the "TM reflexions" ascribable to TM molecules, and (3) does Ca^{2+} binding to TN alone cause the TM shift?

(1) The muscle used here could be regulated by a myosin-linked regulatory system, in which case, Ca^{2+} concentration alone would not affect the structure of the thin filaments. However, this is unlikely. Leg muscle of *Carcinus maenus*, which belongs to the same family as the species used here, contains an actin-linked regulatory system but not a myosin-linked system (Lehman & Szent-Gyorgyi, 1975). The muscle used in the present experiments contains TN and TM (Maeda, 1978 and unpublished results).

(2) A possible and interesting interpretation of the results is that binding of Ca^{2+} to TN alone does not cause the TM shift. This interpretation is consistent with recent analyses of the cooperative properties of thin filaments (Greene & Eisenberg, 1981; Nagashima & Asakura, 1982). These analyses showed that actin can exist in two forms "off" and "on", and Ca^{2+} and S1 act as allosteric effectors, shifting the equilibrium between the two forms. The equilibrium is in favour of "off" state, even at high Ca^{2+}, if no S1 is bound, and transition from "off" to "on" occurs only if some actins interact with S1. At a low Ca^{2+} concentration the equilibrium is much favored towards "off" form, and the transition is much more cooperative. The results presented here, therefore, would imply that the TM shift is cooperative not only at low Ca^{2+} but also at high Ca^{2+} concentration, or in other words that the "allosteric constant" (equilibrium constant between the weak-binding and strong-binding forms of the

thin filaments), although modified by Ca^{2+} , is still in favour of the weak-binding form even at saturating Ca^{2+} concentrations. It should be pointed out that this does not necessarily contradict the essential elements of the steric blocking model, but only a particular detail. This will be discussed in more detail in a forthcoming publication.

On the other hand, the results presented here are not consistent with published X-ray diffraction studies of frog muscle (Haselgrove, 1972) and of reconstituted thin filaments oriented in capillary (Gillis & O'Brien, 1975).

(3) The discrepancy described above raises the question whether so called "TM reflexions" from crab muscle are attributable solely to components of the thin filaments, that is, actin and TM molecules.

In crustacean muscles in rigor, as well as in insect flight muscle in rigor, the myosin heads bind to the thin filaments with the characteristic repeat distance of 38.5 nm, which is an intrinsic structural periodicity of the thin filaments (Wray, Vibert & Cohen, 1978; Maeda, 1979a; Maeda, 1979b; Holmes, Tregear & Barrington Leigh, 1980). Accordingly, cross-bridges formed could contribute to the intensities along these reflexions, which are at orders of 38.5 nm. Therefore, the intensities and the reversal would be attributable not only to TM (and actin) and the TM shift, but also to the cross-bridge formation. At the present time, no experimental evidence is available concerning this point, therefore the question remains open. I should emphasize that, if this applies to crab muscle, the same possibility must also be considered for other muscles such as those from frog, because the random attachment of the myosin heads known in the frog muscle also contains a helical symmetry with 38.5 nm axial periodicity, contributing to the reflexions. At the present time, a definitive distinction between these possibilities can not be made. The simplest explanation appears to be that Ca^{2+} alone does not cause TM movement. The observation that conditions which lead to cross-bridge formation, even if only to a limited extent, cause intensity reversal of the "TM reflexions" supports the idea that TM does not move until a number of cross-bridges are formed. However, the results require caution in interpreting X-ray patterns obtained in the presence of Ca^{2+}. In particular, not only tension generation but also rigor patterns are dependent on Ca^{2+} concentration, and these two can be separated from each other under some conditions, for example by the use of vanadate or by treatment at a low pH.

ACKNOWLEDGEMENTS

I am grateful to Dr. John Wray for the use of his cameras and suggesting the low pH treatment, to Dr. Roger Goody for many helpful discussions and suggesting the use of ATP(γ-S), to Prof. Kenneth Holmes for discussions and encouragements, and to Mrs. Marija Isakov for technical assistance. I also thank the Max-Planck-Gesellschaft and the Yamada Science Foundation for support.

REFERENCES

Barrington Leigh, J., Holmes, K.C., Mannherz, H.G., Rosenbaum, G., Eckstein, F. & Goody, R. (1972). Effects of ATP analogs on the low-angle X-ray diffraction pattern of insect flight muscle. Cold Spring Habor Symp. Quant. Biol. 37: 443-447.

Gillis, J.M. & O'Brien, E.J. (1975). The effect of calcium ions on the structure of reconstituted muscle thin filaments. J. Mol. Biol. 99: 445-459.

Goodno, C.C. (1979). Inhibition of myosin ATPase by vanadate ion. Proc. Natl. Acad. Sci. USA. 76: 2620-2624.

Goodno, C.C. (1982). Myosin active-site trapping with vanadate ion. In: *Methods of Enzymology*, ed. Frederikson, D.W. & Cunningham, L.W., pp. 116-123. New York: Academic Press.

Goodno, C.C. & Taylor, E.W. (1982). Inhibition of actomyosin ATPase by vanadate. Proc. Natl. Acad. Sci. USA. 79: 21-25.

Goody, R.S. & Eckstein, F. (1971). Thiophosphate analogues of nucleotide di- and triphosphates. J. Amer. Chem. Soc. 93: 6252.

Goody, R.S., Hofmann, W. Reedy, M.K., Magid, A. & Goodno, C.C. (1980) Relaxation of glycerinated insect flight muscle by vanadate. J. Muscle Res. Cell Motility. 1: 198-199 (a conference abstract).

Goody, R.S., Holmes, K.C., Mannherz, H.G., Barrington Leigh, J. & Rosenbaum, G. (1975). Cross-bridge conformation as revealed by X-ray diffraction studies of insect flight muscles with ATP analogues. Biophys. J. 15: 687-705.

Greene, L.E. & Eisenberg, E. (1980). Cooperative binding of myosin subfragment-1 to the actin-troponin-tropomyosin complex. Proc. Natl. Acad. Sci. USA. 77: 2621-2620.

Haselgrove, J.C. (1972). X-ray evidence for a conformational change in the actin-containing filaments of vertebrate striated muscle. Cold Spring Harbor Symp. Quant. Biol. 37: 341-352.

Holmes, K.C., Tregear, R.T. & Barrington Leigh, J. (1980). Interpretation of the low angle X-ray diffraction from insect flight muscle in rigor. Proc. R. Soc. Lond. B207: 13-33.

Huxley, H.E. (1972). Structural changes in the actin- and myosin-containing filaments during contraction. Cold Spring Harbor Symp. Quant. Biol. 37: 361-376.

Lehman, W. & Szent-Gyorgyi, A.G. (1975). Regulation of muscular contraction. J. Gen. Physiol. 66: 1-30.

Lienhard, G.E. & Secemski, I.I. (1973). P^1, P^5–Di (adenosine-5') pentaphosphate, a potent multisubstrate inhibitor of adenylate kinase. J. Biol. Chem. 248: 1121-1123.

Maeda, Y. (1978). Optical and X-ray diffraction studies on crab leg striated muscle. Ph.D. Thesis, Nagoya University.

Maeda, Y. (1979a). X-ray diffraction patterns from molecular arrangements with 38 nm periodicities around muscle thin filaments. Nature (London). 277: 670-672.

Maeda, Y. (1979b). Arrangement of troponin and cross-bridges around the thin filaments in crab leg striated muscle. In: *Cross-bridge Mechanism in Muscle Contraction*, ed Sugi, H. & Pollack. G.H., pp.457-468. Tokyo: University Tokyo Press.

Maeda, Y., Matsubara, I. & Yagi, N. (1979). Structural changes in thin filaments of crab striated muscle. J. Mol. Biol. 127: 191--201.

Nagashima, H. & Asakura, S. (1982). Studies on cooperative properties of tropomyosin-actin and tropomyosin-troponin-actin complexes by the use of *N* ethylmaleimide-treated and untreated species of myosin subfragment-1. J. Molec. Biol.

Parry, D.A.D. & Squire, J.M. (1973). Structural role of tropomyosin in muscle regulation: analysis of the X-ray diffraction patterns from relaxed and contracting muscles. J. Molec. Biol. 75: 33-55.

Vibert, P.J., Haselgrove, J.C., Lowy, J. & Poulsen, F.R. (1972). Structural changes in actin-containing filaments of muscle. J. Mol. Biol. 71: 757-767.

Wray, J., Vibert, P. & Cohen, C. (1978) Actin filaments in muscle: pattern of myosin and tropomyosin/troponin attachments. J. Mol. Biol. 124: 501-521.

DISCUSSION

GILLIS: The evidence for the tropomyosin shift due to Ca^{2+} alone comes from systems in which there are actin, tropomyosin, and the full troponin system (Gillis & O'Brien, 1975). In that case I think there is no doubt that the tropomyosin shift can be produced by adding Ca^{2+} and reversed by removing Ca^{2+}. The work of Wakayabashi, Huxley, Amos & Klug (J. Mol. Biol. 93: 477-797, 1975) is on the same line. But in your fibers, what is the situation of the troponin? Is the troponin as effective as in vertebrate skeletal muscle, or has the muscle another type of control system?

MAEDA: It is very likely that this muscle is regulated solely by the tropomyosin-troponin system. A troponin-like protein has been isolated from the muscle. The protein consists of three components, and the total molecular weight is larger than troponin from vertebrate skeletal muscles. Tropomyosin is also larger (Maeda, 1978). According to Lehman & Szent-Györgi (1975), crab leg muscle has only the actin-linked regulatory system. However, it does not necessarily mean that the particular muscle which is used in the present work contains no myosin-linked regulatory system, because, according to their report, some other types of crustacean muscle are regulated by both systems and crustacean muscles are highly inhomogeneous as far as regulatory systems are concerned.

TIROSH: I want to make a point about the effect of calcium concentration on properties of thin filaments. Ishiwata & Fujime (J. Mol. Biol. 8: 511-522, 1972) measured flexibility changes of thin filaments under various conditions using quasielastic light scattering methods. I have also undertaken similar experiments using thin filaments reconstituted from actin and natural tropomyosin (extracted as a complex of tropomyosin plus troponin) in the presence of HMM and Mg-ATP, especially at very low protein concentrations (to be prepared for publication). We found that there are really two states of the actin-natural tropomyosin complex, which seem to be stable only in relatively narrow ranges of Ca^{2+} concentration; one is at about 10^{-8} M and the other is at 10^{-6} M. Outside these ranges (either higher, lower, or in between) the system is unstationary. Therefore, I think it is important in this kind of experiment to be more specific about the buffer conditions of the Ca-EGTA that you used. Otherwise, you would get a wide spectrum of experimental results.

MAEDA: In the experiments presented here, low Ca^{2+} means in the presence of 4 mM EGTA, while high Ca^{2+} patterns were taken in the presence of 1 mM EGTA and 5 mM $CaSO_4$. For the stretched fibers, I also recorded patterns in the presence of equimolar EGTA and $CaSO_4$ and I got the same results. I have not obtained ATP(γ-S) patterns in the presence of equimolar EGTA and $CaSO_4$.

TIROSH: All my measurements were made in the presence of Mg-ATP, under physiological conditions, at least from this point of view.

MAEDA: In my case too.

T. WAKABAYASHI: I would like to make two comments. The first thing is that, when I did 3-D reconstructions from electron micrographs of the the actin-tropomyosin complex compared with the actin-tropomyosin-troponin complex, the intensity of the second and the third layer-lines were similar to each other; the difference was just about 10% (Wakabayashi et al., 1975). I suppose it would be very difficult to detect that kind of difference in X-ray patterns. Therefore, I think it is very important to *measure* the intensities in your pattern, not by eye, to make sure that you do not detect any difference. In addition, in the reconstructed three-dimensional models, the part of the tropomyosin molecule that binds to actin does not move in either system. The change of the position of tropomyosin takes place in the part which does not find actin. Therefore, to show up that sort of difference, it is important to see the whole pattern, not only the second and the third layer-lines but other layer-lines, as well. In my case, in transforms of electron micrographs, the amplitude of the eighth layer-line relative to the seventh layer-line changes in the actin-tropomyosin complex compared with the actin-tropomyosin-troponin complex.

The second point is about cooperativity. According to Nagashima & Asakura (1982), almost all actin sites for interaction with S-1 are allosterically inhibited and are in the "off" state. This is not only in regulated thin filaments in the absence of Ca^{2+}, but also in the presence of Ca^{2+}. But the important point to which you did not refer is the difference in the allosteric constant L (the equilibrium constant for "off"/"on" transitions). L is very different in the actin-tropomyosin complex (so-called classical "on" system) compared with the actin-tropomyosin-troponin complex in the absence of Ca^{2+} (so-called classical "off" system). In the latter, L is more than a hundred times larger than in the former. Therefore, even though the actin sites are almost in "off" state, yet this difference in L should be in the structure itself.

MAEDA: I agree. According to their results, L is 37 for the actin-tropomyosin complex, 9 for the regulated thin filament in the presence of Ca^{2+} and much larger (10,000) for the regulated thin filament in the absence of Ca^{2+}. Even in the activated state the equilibrium is favored toward the "off" state. However, it does not necessarily mean that the structure of the thin filament must be the same in the activated state and in the relaxed state, because L is different in the two states of the thin filament.

HUXLEY: Perhaps I could just make a comment on that. I think that there are two experimental disagreements in the sense that two systems seem to be behaving in different ways. The one you mentioned, the experiment that Haselgrove did (Haselgrove, 1972), with a very stretched semi-tendinosus muscle (frog), in which he reported that he could see a difference with and without Ca^{2+}. There was an experiment which I did (Huxley, 1972) in which the myosin was denatured by heat, by heating the muscle to about 45°C. I am saying that under those conditions, even though myosin is denatured and all the myosin reflections are swept out, there did seem to be such a shift. It seems to me there is something a bit strange about the positions that tropomyosin takes up in the helix, because there has always been that strange business about the *Mytilus* muscle, which I think is supposed not to have a troponin system, and yet in which the resting pattern seems to show tropomyosin in the outer region of the groove. So, perhaps conceivably, tropomyosin is slightly different between different species.

STRUCTURAL BASIS OF FORCES IN RESTING MUSCLE

INTRODUCTION

Each of the following contributions, in some sense, concern the nature of unstimulated muscle, that is, the properties of relaxed myofibrils. The first report describes efforts to probe for attached cross-bridges in relaxed skinned fibers using very high speed stretches -- complete in about 150 μsec. The second paper describes an important new class of fibrous myofibrillar proteins. These very high molecular weight proteins (VHMWP) still remain in myofibrils extracted of most myosin and actin. This makes them good candidates as components of the additional sarcomere structures proposed in the third article. In that paper, mechanical and structural responses to stretching and swelling forces are described which can not be understood in terms of the present-day sliding filament model for the sarcomere. The authors offer a new model for striated muscle which incorporates three additional structures.

In the first paper, Schoenberg, Brenner, Chalovich, Greene, and Eisenberg describe experiments design to detect and characterize myosin cross-bridges which might be attached to actin under resting conditions, that is, in the absence of ionized calcium and in the presence of MgATP. They were stimulated to try these experiments with skinned fibers by the finding (Chalovich et al., J. Biol. Chem. 256: 575-578, 1981; Chalovich and Eisenberg, J. Biol. Chem. 257: 2432-2437, 1982) that synthetic actin filaments containing the regulatory proteins, tropomyosin and troponin, still bound soluble myosin "heads" (subfragment-1) even though calcium was absent. The ATPase activity of these attached S-1's was inhibited however. These biochemical results present a strong challenge to a simple steric blocking model for muscle regulation; Schoenberg et al.'s study extends this challenge to a structured system of contractile proteins. Not only are cross-bridges found attached under calcium-free conditions (at low ionic strength at least), but also by analogy to the inhibition of the ATPase, these bridges produce no force.

Wang's work, reported in the second article, continues along what has been, until recently, an obscure path in muscle biochemistry: studies on the nature of the residual "stroma" or "ghost" which remains after most of the contractile proteins have been extracted. This problem was probed earliest by Guba and his school who isolated a 110 Kdl protein they call "fibrillin" from rabbit muscle (Guba et al., Acta Biochim. et Biophys. Acad. Sci. Hung. 3: 353-363, 1968). They think it constitutes the "basic filamentary structure" of the fibril. Based on work with harshly-extracted fibrils and skinned fibers, Maruyama and his coworkers reopened the question of what protein(s) constitute the myofibril stroma in 1976. They consider it to exist as a random three-dimensional network of an elastic protein they call "connectin" (Maruyama et al., Nature 262: 58-60, 1976; Maruyama et al., J. Biochem. 82: 317-337, 1977) Wang and his colleagues have shown that these new structural proteins include at least five VHWMP's. Extreme insolubility, remarkable size, and marked proteolytic susceptibility have complicated study of these proteins. Three of these have purified and partially characterized, a doublet of about 10^6 dl named "titin(s)" and a 600 Kdl peptide called "nebulin". Maruyama et al., have since shown that titin(s) is (are) a major constitutent of "connectin" (J. Biochem. 89: 701-709, 1981). Efforts to localize them within the myofibril by immunologic methods led Wang to conclude that titin and nebulin are the major components of a "continuous longitudinal filament" which connects Z-lines.

In the third contribution, Magid, Ting-Beall, Carvell, Kontis, and Lucaveche pursue the problem of a muscle scaffolding from a new direction. They were drawn to the problem because the rubberband-like elasticity of relaxed skinned fibers seemed inexplicable in terms of the orthodox two-filament model for the sarcomere. Indeed, Natori (Jikei. Med. J. 1: 119-126, 1954) in his pioneering description of skinned fibers showed that they were as stiff as the single fibers from which they had been dissected. These authors have studied the sarcomere length dependence of resting tension in doubly-skinned fibers (no surface coat, negligible internal membranes). Since they found substantial forces at filament non-overlap, leading to thick filament strain with large stretches, they conclude that vertebrate muscle shares the organization of insect flight muscle, that is, it too has elastic connecting filaments which attach the thick filaments to the Z-line. Because myosin solvent dramatically reduced resting tension and increased passive extensibility without affecting muscle continuity, they also concluded that an extensible core, mechanically continuous with the connecting filaments, supports the myosin of the thick filament. This connecting filament-core filament proposal is reminiscent of the "basic filamentary structure" advanced by Guba and the "gap filament" notion of Locker and Leet (J. Ultrast. Res. 52: 157-172, 1976). Those approaches have not gained currency, perhaps because no mechanical evidence was given in their support. Beyond the problem of longitudinal elasticity, Magid et al. deal also with radial elasticity in the A-band. They consider that it resides in "side-struts", extensible links between the thick filaments. With the Z-line, their model proposes the minimum number of elements needed to form a compact three-dimensional stroma upon which the array of contractile proteins is ordered. How the proteins discovered by Wang fit into this scheme can not be answered at present. In this regard, it is interesting that a connecting filament protein purified from honeybee flight muscle by Saide (J. Mol. Biol., 153: 661-679, 1981) is, like titin, rich in proline.

-Alan Magid

CROSS-BRIDGE ATTACHMENT IN RELAXED MUSCLE

Mark Schoenberg, B. Brenner, J. M. Chalovich*, L.E. Greene* and E. Eisenberg*

Laboratory of Physical Biology, National Institute of Arthritis, Diabetes, and Digestive and Kidney Diseases
**Laboratory of Cell Biology, National Heart, Lung and Blood Institute*

ABSTRACT

We have measured the stiffness of relaxed, skinned rabbit psoas fibers at 5°C in low ionic strength relaxing solution (μ = 0.02 M) by stretching the fibers and measuring the resulting force and sarcomere length changes. This stiffness is very dependent upon the velocity of stretch. With very slow stretches (0.5% of fiber length in > 30 ms), it is almost negligible but with stretches as fast as 0.5% of fiber length in 150 μs, the stiffness approaches 1/3 that of the rigor fiber. This stiffness is also very sensitive to ionic strength, being reduced more than 20-fold at an ionic strength of 0.17 M. This ionic strength sensitive stiffness scales with the amount of overlap between the actin and myosin filaments which strongly suggests that it is due to attached cross-bridges. The speed dependence suggests that the attached cross-bridges are not statically attached but in rapid equilibrium between attached and detached states. Experiments with adenylyl-imidodiphosphate suggest that the rates of attachment and detachment depend upon nucleotide.

INTRODUCTION

The activity of vertebrate skeletal muscle is regulated by the binding of Ca^{2+} to troponin (see Ebashi and Endo, 1968) which results in a shift of the position of the tropomyosin molecule on the actin filament. According to the steric blocking model (Huxley, 1972; Haselgrove, 1972; Parry and Squire, 1973), this shift in tropomyosin regulates the activity of the cross-bridge cycle by controlling the binding of the cross-bridge to actin. However, it has been shown (Chalovich et al., 1981; Chalovich and Eisenberg, 1982; Wagner and Giniger, 1981), that *in vitro*, at low ionic strength, troponin-tropomyosin regulates the actin-activated ATPase rate but has little effect on the binding of myosin subfragment-1 (S1) to actin in the presence of adenosine triphosphate (ATP). Binding to regulated actin occurs to the same degree both in the presence and absence of Ca^{2+}.

This suggests that troponin-tropomyosin can regulate the cross-bridge cycle by affecting a kinetic step other than the attachment step, which in turn suggests that at least under certain conditions cross-bridges might be attached to actin in a relaxed muscle.

In order to determine whether at low ionic strength a relaxed muscle fiber has attached cross-bridges, as suggested by the biochemical findings, we measured muscle stiffness, currently the best measure of cross-bridge attachment. When the fiber was stretched with a relatively slow step taking 1-2 ms to complete, only a small and transitory force was measured, indicating little stiffness. However, the stiffness was found to be time dependent and when the muscle was stretched more rapidly, a stiffness approaching 1/3 that of a rigor fiber was measured. The relationship of this stiffness with the amount of overlap between the actin and myosin filaments suggests it is due to attached cross-bridges.

METHODS

Single skinned rabbit psoas fibers were prepared by a method similar to that of Eastwood et al., 1979 (see Brenner, 1983). The fibers, 3-6 mm in length, were mounted between a displacement generator and a force transducer using cyanoacrylate glue. The displacement generator was a modified galvanometer movement and was capable of stretching the fiber segments 0.5% in 100-200 μs. The force transducer was a modified AE 801 (AKERS, Horten, Norway). It was 0.15 mm thick and 3 mm in length. The

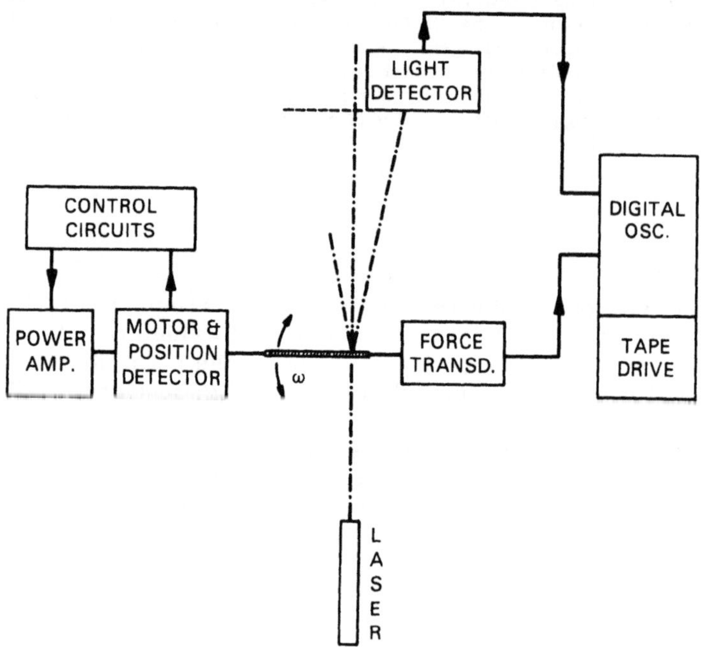

Figure 1: Block diagram of experimental setup.

beam was extended by a small carbon fiber to which the muscle fibers were glued. The active area of the beam was coated with RTV 60 to keep the elements from shorting. The transducer had a resolution of 0.5 dyne and a natural frequency of 12-15 kHz with the fibers mounted and in solution.

Sarcomere displacement was monitored using a Schottky barrier photodiode as the position sensitive element (see Brenner et al., 1983). A laser beam, about 800 μm in length was projected upon the fiber about 1 mm from the force transducer. This reduced artifacts due to the finite transmission time for mechanical disturbances (Schoenberg et al., 1974) and yet avoided errors due to disturbance of the striation pattern very near the attachment site. The bandwidth of the sarcomere length detector was > 30 kHz.

Stiffness was measured by injecting a pulse of current into the displacement generator to rapidly stretch the fiber segment while recording sarcomere length and force in a digital oscilloscope (Nicolet 1090, Nicolet, Madison, Wisconsin). The slope of the force vs sarcomere length data yielded the muscle stiffness. Often, 9 records were computer averaged to increase the signal to noise ratio. A block diagram of the setup is shown in Figure 1.

RESULTS

Behavior of Rabbit Skinned Psoas Fibers at Low Ionic Strength.

Skinned rabbit psoas fibers, at room temperature tend not to relax at low ionic strength. Furthermore, unlike frog fibers (Gulati, 1981), relaxation is not aided by increasing free Mg^{2+}. However, at 5°C, skinned rabbit psoas fibers do relax completely in a solution with 1 mM ATP, 3 mM $MgCl_2$, pCa > 8, ionic strength, μ = 0.02 M. These fibers do not actively shorten and they have very little resting tension. Upon addition of Ca^{2+}, pCa = 4.8, and a phosphocreatine kinase backup system (see Brenner, 1983), skinned rabbit fibers exert a force of about 1.3 $Mdyn/cm^2$ and have unloaded shortening velocities of 0.7 muscle lengths/s. Upon removal of Ca^{2+}, the fibers completely relax. Therefore at 5°C, low ionic strength, skinned rabbit psoas fibers behave much as they do at normal (μ = 0.17 M) ionic strength.

Muscle Stiffness

In order to look for evidence of cross-bridge attachment, we measured fiber stiffness. When we stretched a relaxed muscle fiber in low ionic strength solution (μ = 0.02 M), using a step taking 1-2 ms to complete, the fiber did not appear stiff (Fig. 2). However we found the stiffness in low ionic strength relaxing solution was time dependent. Fig. 3, which shows several force-displacement traces for different velocities of stretch, illustrates that as the muscle fiber is stretched more rapidly, it appears stiffer. With the fastest velocity stretch, 1/2% of fiber length in 150 μs, (the displacement now being proportional to t^3 where t is time),

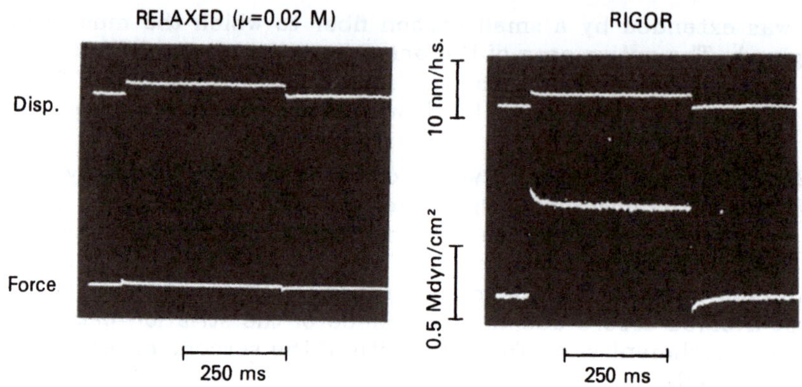

Figure 2: Step stretch of a relaxed and rigor fiber in low ionic strength solution (μ = 0.02 M). Step required 1-2 ms to complete. In the rigor fiber, the relaxation in the force trace was accompanied by a proportional relaxation in the displacement trace. Note apparent absence of significant stiffness in the relaxed fiber. [From Brenner et al., 1982.]

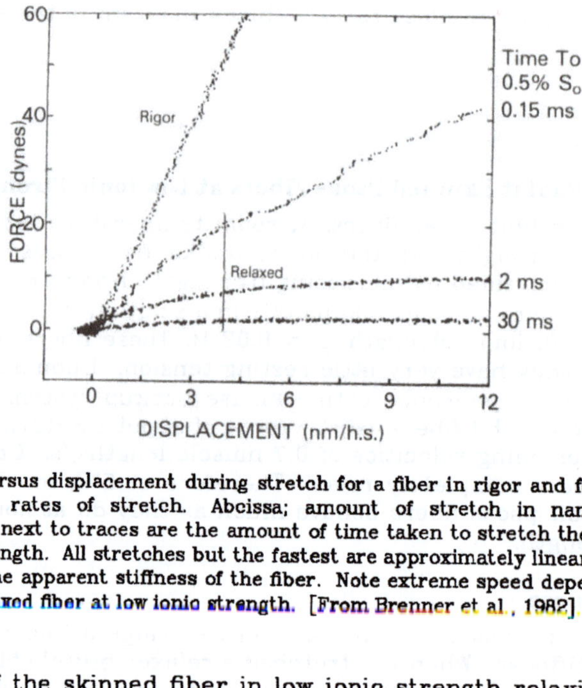

Figure 3: Force versus displacement during stretch for a fiber in rigor and for the relaxed fiber at different rates of stretch. Abcissa; amount of stretch in nanometers/half-sarcomere. Times next to traces are the amount of time taken to stretch the fiber by 0.5% above its resting length. All stretches but the fastest are approximately linear. The slope of each trace gives the apparent stiffness of the fiber. Note extreme speed dependence of the stiffness of the relaxed fiber at low ionic strength. [From Brenner et al., 1982]

the stiffness of the skinned fiber in low ionic strength relaxing solution was as large as 1/3 that of a rigor fiber.

The stiffness of the resting muscle measured with rapid stretches ("rapid stiffness") was very sensitive to ionic strength. Fig. 4 shows that the stiffness in normal ionic strength relaxing solution (μ = 0.17 M), measured with the fastest stretch, is less than 1/20 of that measured in low ionic strength solution. The stiffness of the rigor fiber was nearly independent of speed of stretch or ionic strength.

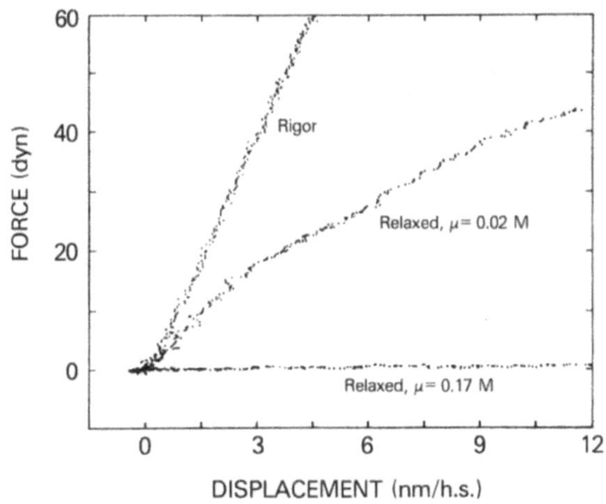

Figure 4: Effect of ionic strength upon the stiffness of a relaxed muscle. Stretches done at a speed of approximately 0.5% in 0.15 ms. Rigor trace included as a reference. The slope of each trace gives the fiber stiffness.

Relationship Between Muscle Stiffness and Myofilament Overlap

To see whether the rapid stiffness we measured is due to the presence of attached cross-bridges, we measured this stiffness as a function of sarcomere length. Fig. 5 shows that at low ionic strength, the rapid stiffness (solid circles) approximately scales with the amount of overlap between the actin and myosin filaments. At zero overlap, S.L. = 3.8 μm, the stiffness is less than 20% of the stiffness measured at full overlap (S.L. = 2.4 μm). The stiffness remaining at zero overlap, as Fig. 5 shows, is largely insensitive to ionic strength (compare solid circles and triangles). The resting tension in skinned rabbit fibers is also not very sensitive to ionic strength (unpublished observations) which suggests that the remaining stiffness at zero overlap may be related to the passive structures responsible for the resting tension. The ionic strength sensitive stiffness (open circles, Fig. 5) gives an almost perfect correspondence with overlap, the best straight line fit through these points extrapolating to zero at 3.75 μm, in excellent agreement with the known lengths of the myofilaments in rabbit skeletal fibers. This provides very strong evidence that the large rapid stiffness measured in a relaxed muscle at low ionic strength is related to the presence of attached cross-bridges. It should be noted that at low ionic strength, these fibers, like normal relaxed fibers, are quite extensible when stretch slowly.

Stiffness as a Function of Nucleotide.

Our results thus far have suggested that cross-bridges bound to actin behave very differently in ATP solution and in rigor. One major difference between these conditions is the very different affinity of S1 for actin in the presence and absence of ATP. In order to see whether this difference in affinity could account for the difference in behavior we observe, we

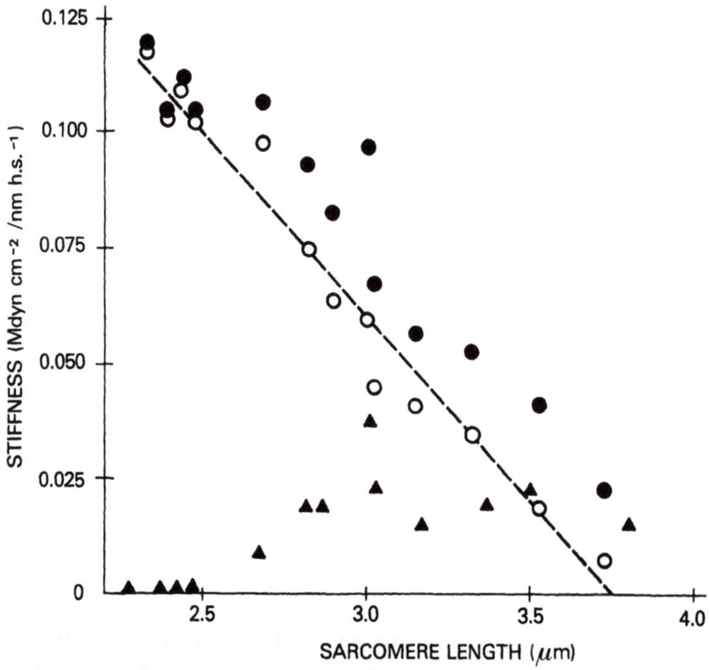

Figure 5: Rapid stiffness of a relaxed fiber as a function of sarcomere length. Closed circles, stiffness in low ionic strength solution (μ = 0.02 M); closed triangles, stiffness in normal ionic strength solution (μ = 0.17 M); open circles, difference between closed circles and closed triangles. Dashed line is best straight-line fit through open circles. Note that the ionic strength sensitive component of the stiffness, open circles, scales almost perfectly with the amount of overlap between actin and myosin filaments. [From Brenner et al., 1982.]

wanted to study an additional condition where the binding constant of S1 to actin is the same as it is in our low ionic strength ATP solution. Since this was impossible in rigor, we studied the behavior of the muscle in the presence of the ATP analogue adenyl-5'-yl-imido-diphosphate (AMP-PNP). Since ionic strength has been shown to weaken the binding of S1 to actin, we used KCl to adjust the *in vitro* binding constant of S1 to actin in the AMP-PNP solution to be equal to that found in the low salt ATP solution. Figure 6 shows the response of a single fiber segment to stretch in rigor, in low salt relaxing solution, and in 4 mM AMP-PNP solution. The mechanical response to stretch is clearly different. The fiber is stiffer in AMP-PNP solution and the stiffness for a stretch of 0.5% in 10 ms is already maximal (it does not increase with increasing speeds of stretch). With ATP, as both Fig. 6 and Fig. 3 show, the stiffness with a stretch of 0.5% in 10 ms is considerably less than the stiffness obtained with faster stretches. As a control, we varied the KCl concentration in the AMP-PNP solution and found that the difference between the behavior of the muscle in AMP-PNP and low salt ATP solutions was not due to the difference in salt concentration.

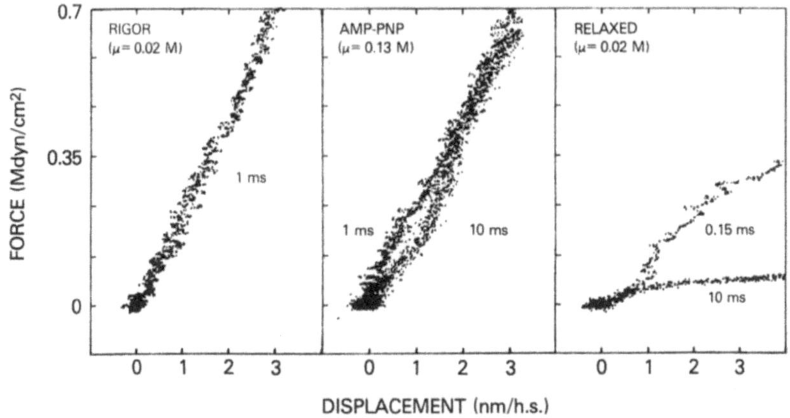

Figure 6: Stiffness of a muscle in rigor and in the presence of different nucleotides. Ordinate, force; abscissa amount of stretch (nm/half-sarcomere). Slope of force-displacement trace gives fiber stiffness. The times next to each trace, a measure of the speed of stretch, give the amount of time necessary to stretch the muscle 0.5%. Rigor stiffness was independent of speed of stretch. The magnitude of the stiffness in AMP-PNP solution measured with a 10 ms stretch was similar to that of rigor fibers and did not increase with faster stretches. Stiffness of the relaxed fiber in low ionic strength solution (0.02 M) measured with a 10 ms stretch was much lower than that of the rigor fiber and increased to about 1/3 that of the rigor fiber with a 0.15 ms stretch. The ionic strength of the AMP-PNP solution (4 mM AMP-PNP, 5 mM $MgCl_2$, 1 mM CaEGTA, 10 mM imidazole, 100 mM KCl) was chosen as to make the *in vitro* binding constant of Sl to actin similar to that in the low ionic strength relaxing solution.

We also found, as Lymn (1975) has remarked, that in 4 mM AMP-PNP solution, just as in low salt ATP solution, skinned rabbit psoas fibers are quite extensible when stretched slowly, although the rate of stretch with AMP-PNP needs to be much slower than in ATP solution. All these results suggest that under conditions where the *in vitro* binding constant for S1 to actin is about the same, cross-bridges attach and detach more rapidly when complexed with ATP than when complexed with AMP-PNP.

A Simple Model.

Fig. 3 shows that at slow velocities of stretch the relaxed muscle exhibits a "viscous" sort of behavior (magnitude of force response dependent upon velocity and independent of amount of stretch) while at high velocities of stretch the muscle exhibits more "spring-like" behavior (increased force with increased *amount* of stretch). It is of interest to see whether in theory, attached cross-bridges (which are in rapid equilibrium between attached and detached states), might qualitatively give this type of behavior. To do this, we employed a very simple model, using the fewest parameters necessary to describe the system. It was assumed that there was one attached state for the cross-bridge, AMN, and one detached state, MN. The attached cross-bridge was assumed to have a linear stiffness and it was assumed that a cross-bridge could attach to only a single actin site in seven.

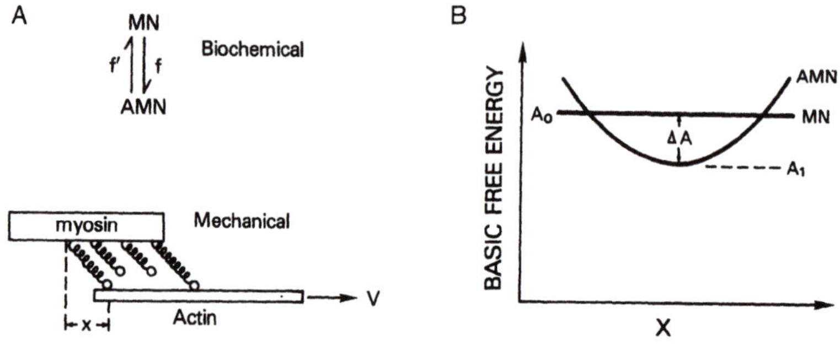

Figure 7: A simple cross-bridge model of the relaxed muscle. A. Schematic of model showing the two states, MN and AMN with attachment rate f and reverse rate f', and cross-bridges in rapid equilibrium between these two states. B. Free energy diagrams of both states, showing definition of free energy terms.

The model is shown schematically in Fig. 7A. Myosin cross-bridges are continually attaching and detaching from actin, with rate constants f and f' respectively, as the filaments are stretched past each other. The parameter f may be chosen and the parameter f' obtained from the basic free energies (Fig. 7B) of the states MN and AMN (Hill, 1974), according to the equation

$$f(x)/f'(x) = \exp\left[(\Delta A - \tfrac{1}{2}Kx^2)/kT\right] \quad (1)$$

where k is Boltzmann's constant, T, temperature, K the spring constant, and where the difference in free energy between the two states is assumed to be $\Delta A - \frac{1}{2}Kx^2$ where ΔA is the difference in minimum free energy.

The model illustrated in Fig. 7, which is easily computed using a PDP-10 computer (Digital Equipment Corp., Maynard, MA), successfully duplicates the behavior of the muscle shown in Fig. 3; it gives "viscous" behavior for slow velocities of stretch and elastic behavior for high velocities. We found, however, that, because of the large number of parameters involved, the data in Fig. 3 can be approximated with a wide range of parameter values. Therefore it will be necessary to model a wider range of data to uniquely determine the cross-bridge model parameters. However, preliminary modeling work based solely on the data of Fig. 3 suggests that the model cross-bridge stiffness or the binding constant or possibly both may be lower than the values usually assumed in models explaining the behavior of active frog muscle (Huxley and Simmons, 1971; Eisenberg, et al., 1980). It is not clear if this choice of parameters is necessary to explain the behavior of the resting muscle or whether the unusual parameters are a consequence of the simplicity of the model. A more realistic model would include the possibility of more than one actin site per seven being available for attachment and might also require more states.

DISCUSSION

The finding that the stiffness of a relaxed muscle fiber in low ionic strength ATP solution roughly scales with the amount of overlap between actin and myosin filaments strongly suggests that this stiffness is due to the presence of attached cross-bridges. The fact that this stiffness differs with speed of stretch suggests that the cross-bridges are not statically attached but in rapid equilibrium between attached and detached states. Both of these findings are in agreement with the biochemical findings of Chalovich et al., 1981, and Chalovich and Eisenberg, 1982, that under conditions similar to those studied here, myosin S1 binds to regulated actin equally well in the presence and absence of Ca^{2+} and that the binding is rapidly reversible.

Since the apparent stiffness of the relaxed muscle varies with speed of stretch, the behavior of the muscle is, in essence, "viscous"-like and we may ask whether the results are compatible with a simple Stokes type of myoplasmic viscosity? There are three reasons for ruling out a simple Stokes type of viscous drag between the filaments. First, from the known viscosity of the myoplasm, the calculated viscous drag between the filaments is several orders of magnitude lower than the "apparent" viscosity in our experiment. The filaments would need a surface separation of less than 1 Å to account for the measured forces. Second, with moderate constant velocity stretches, a constant force is attained only after several milliseconds. For a Stokes viscous interaction, the force should become steady within the response time of our force transducer, about 0.1 ms. Third, preliminary experiments suggest that stiffness is less sensitive to velocity at higher velocities than at lower ones. If a Stokes viscous interaction were important, the apparent stiffness would always be proportional to the velocity of stretch. The ability of the muscle to exhibit viscous-like behavior at low rates of stretch and more spring-like behavior at rapid rates is therefore more likely to be related to attached cross-bridges, the viscous behavior at lower rates of stretch being due to the fact that the cross-bridges attach and detach during the stretch at lower velocities as suggested by the model of Fig. 7.

The model of Fig. 7 also illustrates that one can explain the low stiffness of the relaxed skinned fiber at normal ionic strength (Fig. 4) as well as the apparent resting viscosity of relaxed intact frog fibers (see Ford, Huxley and Simmons, 1971; Schoenberg, 1980) in two alternative ways. The data may be explained either by a decrease in the strengh of S1-actin binding, ΔA, relative to the value in low ionic strength solution, or by a change in the attachment and detachment rates (increase in f and f'). A consequence of the former explanation is that much fewer cross-bridges are attached at normal ionic strength whereas the latter explanation allows the number of attached bridges to remain similar in low and in normal ionic strength solutions. One way of distinguishing between these two possibilities is to estimate the attachment/detachment rates by determining at which velocity of stretch the apparent stiffness no longer increases with stretch. Unfortunately this data is difficult to obtain at normal ionic strength and not yet available; it may be necessary to turn to techniques such as X-ray diffraction to determine if there are attached cross-bridges present at normal ionic strength.

Already published experiments using X-ray diffraction and e.p.r. spectroscopy have been interpreted to imply that cross-bridges are not attached in the resting muscle under normal ionic strength conditions (Thomas and Cooke, 1980; Thomas et al., 1980; Huxley, 1968). It would be of interest to repeat these measurements at low ionic strength, where we have detected stiffness, to see whether these techniques can detect the weakly attached cross-bridges described here or whether possibly they detect only the more strongly bound bridges found in rigor and activated muscles. On the basis of theory one would certainly expect that X-ray diffraction is capable of detecting both weakly and strongly attached bridges.

That at low ionic strength cross-bridges are attached in the relaxed muscle without producing active force or shortening suggests that the steric blocking model (Huxley, 1972; Haselgrove, 1972; Parry and Squire, 1973) is not the mechanism keeping the muscle relaxed in this instance. Rather, relaxation seems to be brought about by blocking a kinetic step which is subsequent to attachment but prior to the force generating step. This agrees with the biochemical findings of Chalovich et al. (1981) and Chalovich and Eisenberg (1982) who further suggested the possibility that the phosphate release step was the one inhibited.

The simplest explanation for the similarities and differences between the behavior of the fibers in ATP and in AMP-PNP solution is that both behaviors are due to attached cross-bridges, with rapid equilibrium between attached and detached states. Under conditions where the *in vitro* binding constant of S1 to actin is similar, the attachment and detachment rates, f and f', appear to be much slower in AMP-PNP solution than in low ionic strength ATP solution.

In summary, cross-bridges appear to be attached in a relaxed muscle at low ionic strength. Although the mechanical measurements are compatible with large numbers of cross-bridges being attached in a relaxed muscle at normal ionic strength if they are in extremely rapid equilibrium with their detached state, most existing evidence seems to be against this possibility. Regardless of whether or not cross-bridges are attached at normal ionic strength, there clearly seems to be a mechanism besides the classic steric blocking mechanism for keeping a muscle relaxed.

REFERENCES

Brenner, B. 1983. A technique for stabilizing the striation pattern in fully activated skinned rabbit psoas fibers. Biophys. J. 41: 99-102.

Brenner, B., Schoenberg, M., Chalovich, H.M., Greene, L.E. & Eisenberg, E. 1982. Evidence for cross-bridge attachment in relaxed muscle at low ionic strength. Proc. Natl. Acad. Sci., U.S.A. 79: 7288-7291.

Ebashi, S. & Endo, M. 1968. Calcium ion and muscle contraction. Prog. in Biophys. 18: 123-183.

Chalovich, J.M. & Eisenberg, E. 1982. Inhibition of actomyosin ATPase activity by troponin-

tropomyosin without blocking the binding of myosin to actin. J. Biol. Chem. 252: 2432-2437.
Chalovich, J.M., Chock, P.B. & Eisenberg, E. 1981. Mechanisms of action of troponin·tropomyosin. J. Biol. Chem. 256: 575-578.
Eastwood, A.B., Wood, D.S., Bock, K.L. & Sorenson, M.M. 1979. Chemically skinned mammalian skeletal muscle 1. The structure of skinned rabbit psoas. Tissue Cell 11: 553-556.
Eisenberg, E., Hill, T.L. & Chen, Y. 1980. Cross-bridge model of muscle contraction. Quantitative analysis. Biophys. J. 29: 195-227.
Gulati, J. 1981. Cross-bridge turnover during Ca-free, non-rigor, contraction in skinned muscle fibers. Biophys. J. 33: 839.
Haselgrove, J.C. 1972. X-ray evidence for a conformational change in the actin-containing filaments of vertebrate striated muscle. Cold Spring Harbor Symp. Quant. Biol. 37: 341-352.
Hill, T.L. 1974. Theoretical formalism for the sliding filament model of contraction of striated muscle, Part I. Prog. Biophys. Molec. Biol. 28: 267-340.
Huxley, A.F. & Simmons, R.M. 1971. Proposed mechanism of force generation in striated muscle. Nature, London, 233: 533-538.
Huxley, H.E. 1968. Structural difference between resting and rigor muscle; evidence from intensity changes in the low-angle equatorial X-ray diagram. J. Mol. Biol. 37: 507-520.
Huxley, H.E. 1972. Structural changes in the actin- and myosin-containing filaments during contraction. Cold Spring Harbor Symp. Quant. Biol. 37: 361-376.
Lymn, R.W. 1975. Low-angle X-ray diagrams from skeletal muscle: The effect of AMP-PNP, a non-hydrolyzed analogue of ATP. J. Mol. Biol. 99: 567-582.
Parry, D.A.P. & Squire, J.M. 1973. The structural role of tropomyosin in muscle regulation: Analysis of X-ray diffraction patterns from relaxed and contracting muscles. J. Mol. Biol. 75: 33-55.
Thomas, D.D. & Cooke, R. 1980. Orientation of spin-labeled myosin heads in glycerinated muscle fibers. Biophys. J. 32: 891-906.
Thomas, D.D., Ishiwata, S., Seidel, J.C. & Gergely, J. 1980. Submillisecond rotational dynamics of spin-labeled myosin heads in myofibrils. Biophys. J. 32: 873-890.
Schoenberg, M. 1980. The stretch response of resting and activated skeletal muscle. Biophys. J. 39: 1729.
Schoenberg, M., Wells, J.B. & Podolsky R.J. 1974. Muscle compliance and the longitudinal transmission of mechanical impulses. J. Gen. Physiol. 64: 623-642.
Wagner, P.D. & Giniger, E., 1981. Calcium-sensitive binding of heavy meromyosin to regulated actin in the presence of ATP. J. Biol. Chem. 256: 12647-12650.

DISCUSSION

HUXLEY: Not so much a question as a comment. We, too, were naturally very concerned with Chalovich and Eisenberg's apparent demonstration a year or two ago that cross-bridges were likely to be connected all the time in relaxed muscle, although his experiments were done at lower ionic strengths. Dr. Padron in our lab looked at this using the equatorial X-ray diffraction generated by frog muscles skinned by the Magid-Reedy procedure and found that, although in an ionic strength around 0.15 M one obtains a perfectly good relaxed pattern in a relaxing solution, when the ionic strength is dropped to 50 mM, one obtains a rigor type equatorial pattern. This seems to me to suggest that Eisenberg's conclusion may only apply at lower ionic strength, and not under physiological conditions.

SCHOENBERG: Yes, I think that's true. In fact, for the same reason, Bernhard Brenner and Leepo Yu in collaboration with Richard Podolsky, have made those same measurements that you describe, using single

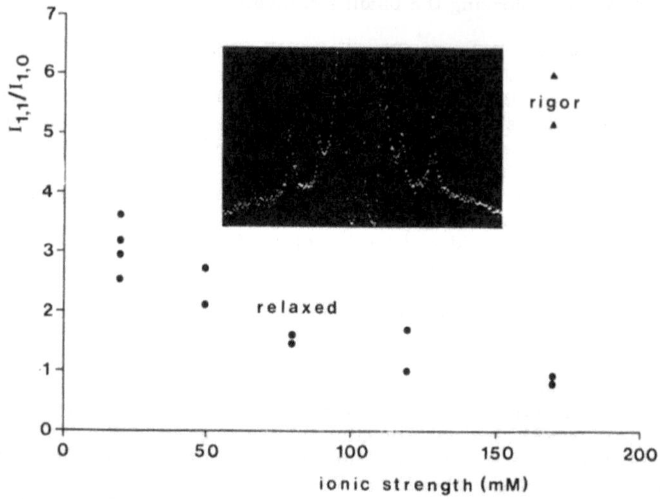

Figure D-1: Equatorial X-ray intensity ratio, $I_{1,1}/I_{1,0}$, as a function of ionic strength, obtained by B. Brenner, L.C. Yu, and R.J. Podolsky from a single skinned rabbit psoas fiber. (●) relaxed, (▲) rigor. Sarcomere length, 2.3 μm, fiber diameter at 170 mM, 170 μm; solution composition as in preceding paper of Schoenberg, et al.

rabbit psoas fibers, and get similar results (see Fig. D-1).

TANAKA (Hiroaki): Is there any resting tension exerted in the relaxed condition at normal ionic strength?

SCHOENBERG: There's very little resting tension in these fibers. There's almost none that you can see at normal sarcomere lengths, about 2.5 μm. If you stretch the fibers to 3.8 μm -- if you do the stretch very slowly and wait a while, there is, in fact, *still* very little resting tension. If you just take the muscle and very rapidly stretch it from 2.5 μm to 3.5 μm, you do get lots of resting tension which then decays with time.

POLLACK: There is an observation which we've made, which I want to mention because I think it supports your conclusion that there may be some actomyosin interaction in the resting state, in this case at normal ionic strength. We've been doing releases of unstimulated fibers and it turns out, at least to *my* surprise (not to the surprise of some others), that the unloaded velocity of shortening is on the order of two or two and a half muscle lengths per second, which is rather close to the V_{max} of the stimulated fiber. So from this it seems reasonable to conclude that the velocities of unloaded shortening are fairly close in unstimulated and stimulated fibers – suggesting that the same sort of interactions that go on in the stimulated fibers might also be going on in the unstimulated fibers.

SCHOENBERG: That's pretty interesting. I think that one of the problems which remains is to quantify these effects. It seems that most techniques are beginning to illustrate now that at low ionic strength the cross-bridges indeed seem to be attached in a relaxed muscle, and at higher ionic strengths fewer cross-bridges seem to be attached. An interesting question that remains is to quantitate exactly how many cross-bridges might be attached at normal ionic strength. I think Dave Thomas has something to say about that. He's got his hand up.

THOMAS: Roger Cooke and I will describe later a somewhat more molecular method of measuring the fraction of cross-bridges attached. We find that with ionic strength on the order of 0.15 to 0.2 M, the fraction is negligible, maybe it's 5 to 10% at the most. But as we drop the ionic strength, the fraction goes up. We also can put limits on the rate, at low ionic strength, at which those two populations, attached and unattached, can be exchanging. I don't want to go into detail on that now, but at least it appears that they must be exchanging slower than on a microsecond time scale. So we may be able to quantitate some of these effects.

SCHOENBERG: The kind of rates that we get, on the basis of some *extremely* preliminary modeling is something like 200 per second, which is fairly slow. But that number may be very model dependent and we're just beginning to look into that sort of question.

EDMAN: I think this problem was taken up by D.K. Hill (J. Physiol. 199: 637-684, 1968). He used slow stretches and measured the force response as a function of velocity of stretch. He concluded that a small fraction of bridges might be attached. So there seems to be the possibility of cross-bridges being formed in relaxed muscle at the ionic strength that exists inside living fiber. Do your data agree?

SCHOENBERG: Yes. I've thought a lot about the relationship between D.K. Hill's results and ours because I have some experiments that I reported on about a year or two ago that were similar to the kinds of experiments that Ford, Huxley, and Simmons did in rapidly stretching intact fibers and looking at the kinds of responses one gets. The question is how those responses that you get when you stretch very rapidly relate to the kinds of things that D.K. Hill measured with slow stretches. I have a very tentative hypothesis: I think both kinds of experiments give good evidence for attached cross-bridges. D.K. Hill found that the stiffness measured with slow stretches seemed to correlate with the resting tension at sarcomere length about 2.1 μm. The stiffness we measure with rapid stretches does not correlate with resting tension. What I imagine could be happening in the resting muscle is that, as in an active muscle, there is possibly more than one attached state. Using jargon similar to that used in discussing active muscle; there could be a rapidly attaching and detaching "90°" state and a longer lasting, "45°" force-producing state. The very few cross-bridges in the latter state could be responsible for the response to slow stretch (which correlates with the resting tension) whereas the behavior in response to very rapid stretches would be determined mainly by cross-bridges in the "90°" state.

HOMSHER: Your length changes didn't go past 12 nm. If you do, do you see breaks in the force-stretch curve?

SCHOENBERG: Yes. In fact, we've also been able to model this with our simple cross-bridge model. If you stretch the muscle, the increased strain in the cross-bridge, according to the simplest model, will cause the cross-bridges to detach more rapidly, and then instead of getting a nice linear increase of force with stretch, the response starts to level off and you can reach a more or less steady-state force level. We see this in our experiments and, as I've said, one can imagine a simple cross-bridge model which gives this kind of behavior.

HOMSHER: How far do you have to stretch or release to see this?

SCHOENBERG: It depends upon the velocity of the stretch. It's in the neighborhood of 12 nm per half sarcomere.

KAWAI: As you increase the speed of stretch, does stiffness plateau, or does it still go up?

SCHOENBERG: At all speeds we've been able to measure, the stiffness increases with increasing speed of stretch, although the increase is proportionately much less at higher speeds (Biophys. J. 41: 32a, 1983).

GORDON: I wanted to comment on the attached cross-bridges at low ionic strength. Bob Godt, others, and I (J. Gen. Physiol. 62: 550-574, 1973) showed that in low ionic strength solutions, skinned frog fibers at room temperature could generate substantial forces in the absence of Ca^{2+}, as much as the Ca^{2+}-activated force in 0.17 M ionic strength solution. So substantial cross-bridge attachment and active force is produced in a low ionic strength solution at room temperature. In your case, are the muscles kept relaxed by going to lower temperature?

SCHOENBERG: That's right.

GORDON: So the question is, what is the temperature doing? In the absence of Ca^{2+}, low ionic strength and low temperature lead to attached cross-bridges in a relaxed muscle; low ionic strength at room temperature leads to attached cross-bridges in an active muscle.

SCHOENBERG: Right. Jag Gulati has done some work on this in frog fibers, and we've done some on rabbit fibers. You can activate these fibers in the absence of calcium simply by raising the temperature. What we think is going on --- although we haven't spent a lot of time exploring it in detail --- but our working hypothesis is that when you raise the temperature what you're doing is increasing the rate at which cross-bridges go from what we loosely called the 90° state into the 45° state, into the force-producing state. I think what you begin to get is the kind of cooperative effect that Annemarie Weber spoke about. Once you start to get a few of these 45° cross-bridges, you should be able to turn the filament on in the absence of calcium. You find that in the absence of calcium, as I'm sure you know, the muscle behaves beautifully. You can do quick releases and you can see the same kinds of transients that Ford, Huxley, and Simmons reported. It's very clear that these muscles get activated without calcium being present.

YANAGIDA: Did you check the dependency of the concentration of free magnesium ion, because at low ionic strength the interaction of the regulatory protein with actin is greatly affected by free magnesium.

SCHOENBERG: Yes, we did look at that. In general, what we try to do is to keep the free magnesium at about 2 mM, which I think is generally believed to keep the regulatory proteins attached and in working order. I know from Jag Gulati's results which he reported at the Biophysical Society (Biophys. J. 33: 83a, 1981) that in the frog you can see very big effects of change in the magnesium ion concentration. The rabbit doesn't seem to be as sensitive to the magnesium ion. For example, he found that at room temperature you could get the muscles to be relaxed at low ionic strength by increasing the free magnesium to 1 mM. That's not true for the rabbit.

YANAGIDA: I measured the ATPase activity of the relaxed fiber and found that it increased by 50% when the ionic strength was decreased.

SCHOENBERG: Was this also done at low temperature?

YANAGIDA: Yes.

SCHOENBERG: We haven't done that on a muscle fiber. We made certain, of course, to vary the ATP concentration and the amount of backup system to make sure we weren't having any ATP depletion problems.

YANAGIDA: Compared with the ATPase activity in the presence of calcium, the ATPase value in the absence of calcium is always very low.

T. WAKABAYASHI: My question concerns the dependence of your stiffness in low ionic strength solution upon calcium ion concentration. How Ca^{2+} sensitive is it? The reason I ask is that calcium regulation of the tropomyosin system (in actomyosin in suspension) begins to fail at low ionic strength or low ATP concentration.

SCHOENBERG: Most of the experiments that I showed were done at 20 mM ionic strength. If you do the experiment at 50 mM ionic strength, then you can do the experiments with a backup system, so you can add calcium and not worry about ATP depletion. Under those conditions, if I remember correctly, when you add calcium you get about a -- I think a two or three-fold increase in stiffness, upon increasing the calcium from pCa 8 to pCa 4.8.

GULATI: In the frog fibers, you mentioned the calcium-free tension goes away if you raise the magnesium concentration.

SCHOENBERG: We don't have any data -- it's all yours.

GULATI: Yes, sure. But you also said you do see the calcium free tension in rabbit psoas at 2 mM free magnesium under your conditions. I've done some preliminary experiments to see why you don't see the magnesium effect. One possibility that I considered was that you're using 20 mM ionic strength, whereas my experiments with frog were done at 50 mM ionic strength. So in the experiment I did, I used exactly your solutions and raised the free magnesium from 2 mM to 4 mM, and kept the ionic strength constant. I found under those conditions you could also reverse the tension and ---

SCHOENBERG: So if you go up to 4 mM free magnesium you eliminate the calcium free tension at room temperature?

GULATI: It does seem to come down.

SCHOENBERG: In your earlier experiments you only used one mM. We went to your ionic strength and tried 2 mM and it didn't work. I think it's very interesting that at 4 mM free magnesium it does seem to work.

GULATI: On another issue, you mentioned in passing that you were able to "lock" these bridges, even though they're attaching and detaching very rapidly. You were able to lock them in the stressed state. Did I understand that correctly?

SCHOENBERG: No. What I'm saying is that if you try to measure the stiffness of the cross-bridges, and if they're falling off while you're stretching the muscle, you're not going to be able to measure their stiffness. You can only measure the stiffness if they remain attached while you're pulling on them so you can stretch them out. So the idea of being "locked" was just sort of a figurative way of speaking -- that if you stretch rapidly enough, none of the cross-bridges will have time to detach during your stretch, so you will measure the true cross-bridge stiffness.

COLOMO: Did you try to change the relation between fiber stiffness and velocity of stretch by changing the temperature or by adding some drug?

SCHOENBERG: We haven't looked very much with temperature because of this problem of the fiber becoming active at higher temperature. We have, though, changed the nucleotide and looked at this question. We've studied this with AMP-PNP as well as with pyrophosphate. In both of these cases, you get responses similar to those that you see with ATP, but on a much, much slower time scale. For example, with either ATP, AMP-PNP or pyrophosphate present, when you stretch very slowly the muscle doesn't appear very stiff at all. But if you stretch rapidly, it does appear stiff. With AMP-PNP or pyrophosphate, you get a stiffness that is about as stiff as that of the rigor fiber.

CYTOSKELETAL MATRIX IN STRIATED MUSCLE: THE ROLE OF TITIN, NEBULIN AND INTERMEDIATE FILAMENTS

Kuan Wang

Clayton Foundation Biochemical Institute, Department of Chemistry and Cell Research Institute, University of Texas at Austin, Austin, Texas 78712

ABSTRACT

In this chapter, first I will briefly describe the molecular properties of titin and nebulin – two extremely large, myofibrillar proteins -- and discuss their distribution and organization in the sarcomere. Although these novel proteins are major myofibrillar components of a wide range of striated muscles, they have escaped the attention of muscle biochemists until very recently. As I shall point out below, biochemical studies of these proteins have been unexpectedly challenging; many standard techniques had to be modified before they became capable of handling such giant proteins. In addition, our structural studies of these proteins have encountered a challange of a different nature: how to explain their distribution in the sarcomere according to the currently accepted two filament sarcomere model, because these proteins do not appear to be thick or thin filament-associated regulatory or anchoring proteins. These studies have led us to reexamine the question of whether continuous, longitudinal filaments exist within the sarcomere of striated muscle. I will attempt to integrate our results, as well as available literature data, within the framework of a hypothetical sarcomere model which incorporates an elastic filamentous matrix consisting of titin and nebulin as additional sarcomere constituents. Finally, I will very briefly mention our recent findings that an extensive three dimensional network of intermediate (10 nm) filaments, distinct from titin and nebulin, is intimately associated with the sarcomere of adult striated muscle.

I believe that the recognition of the existence of two sets of sarcomere-associated cytoskeletal filaments within adult striated muscle fibers may be a significant step toward resolving some of the unsettled questions in muscle mechanics such as those that have been discussed in this meeting

A Third Longitudinal Filament of the Sarcomere: Early Observations

The possible existence of filamentous constituents within the sarcomere in addition to thick and thin filaments has been recognized since the 1960's. Electron micrographs frequently showed slender (2-6 nm), longitudinal filaments crossing the gap regions of those sarcomeres in which A band has been extracted away or of overly-stretched sarcomeres in which thick and thin filaments no longer overlap (Huxley and Peachy, 1961; Carlesen et al., 1961; Page and Huxley, 1963). The detailed organization of the third filament in the sarcomere could only be speculated and was assigned somewhat arbitrarily (since these "gap filaments" were never unambiguously demonstrated in unextracted or in unstretched sarcomeres). A number of proposed localizations have described the third filament as: connecting the ends of adjacent thin filaments (S-filament: Huxley and Hanson, 1954); connecting the ends of thick and thin filaments (Sjostrand, 1962); connecting the ends of thick filaments to Z lines (C-filament: Auber and Couteaux, 1963; Garamvologyi and Kerner, 1966; Guba et al., 1968); connecting adjacent Z lines in between thick and thin filaments (T-filament: McNeil and Hoyle, 1967; Walcott and Ridgeway, 1967; Hoyle, 1968); or as the core of thick filaments (core filament: Guba et al., 1968). The only common theme among these widely divergent early three-filament models is that the third filament provides continuity within the sarcomere.

The proposal of thin filament continuity (S-filament) was subsequently ruled out by the observation of the double-overlap of thin filaments in highly shortened sarcomeres (see Huxley, 1972). The concept of thick filament continuity (C-filament, core filament) in vertebrate skeletal muscle was challenged by filament counting experiments which demonstrated the absence of additional independent filaments in the I band (see Huxley, 1972). It is significant however, that a continuity between the ends of thick filaments and adjacent Z lines was clearly demonstrated in insect flight muscle (see Pringle, 1978). In the 1968 Symposium on Muscle (Ernst and Straub, 1968) where various three filament models were discussed, H.E. Huxley commented, "I think it seems very likely that some type of filament is visible here, but whether this represents a genuine third type of filament in the muscle, or whether it represents thin tapered extensions of the thick filaments or whether it represents some structure formed between the ends of thick and thin filaments during fixation, still seems undecided." (Huxley, 1968).

The question of the existence of the third filament lay dormant until in the early 1970's when a series of highly interesting and revealing papers were published by Locker and his colleagues detailing the behavior of gap filaments in highly stretched beef muscle (see Locker and Daines, 1980). A novel model was proposed in which a gap filament serves as the core of two thick filaments in adjacent sarcomeres, linking them through the Z line and passing between thin filaments (Locker and Leet, 1976). Although the validity of this model remains to be tested, their micrographs of gap filaments and the elongation and dislocation of thick filaments under strain are highly suggestive of the existence of an extensible longitudinal connection between thick filaments in vertebrate muscle.

Besides these ultrastructural clues of sarcomere continuity, some aspects of the mechanical and structural behaviors of resting muscle also suggest the presence of structural continuity within the muscle cell. According to the two-filament sarcomere model, structural and mechanical continuity of the contractile elements is brought about only by the dynamic and transient interactions of two sets of discontinuous filaments in active muscle. This model, therefore, encounters difficulty explaining the considerable contractile elements-related resistance to passive stretch (i.e. resting tension) of relaxed muscle where thick and thin filament interactions are expected to be absent (see Magid, this volume). The two-filament model also encounters difficulty explaining why A bands remain centrally located between regularly spaced Z lines when resting muscle is stretched even to a length where thick and thin filaments no longer overlap. On the other hand, a three-filament model in which a set of parallel elastic filaments is present to maintain sarcomere continuity could easily explain these behaviors.

Titin and Nebulin: Novel Myofibrillar Proteins

Despite the ultrastructural evidence and biophysical considerations which favor the existence of a third filament in the sarcomere, the lack of a firm *molecular* basis -- the identification of unique proteins associated specifically with such filaments -- probably has discouraged serious pursuits. Earlier, Guba and colleagues (1968) isolated a protein ("fibrillin") from ATP-ascorbic acid-extracted muscle residues and claimed it as the component of the ~4 nm residual filaments. However, no further characterizations or confirmation of fibrillin have been reported since. Recently, Maruyama and colleagues (1977), while pursuing the question of muscle elasticity, obtained an insoluble elastic protein gel after extracting vertebrate and invertebrate myofibrils with a series of media including KI, phenol, SDS, NaOH or boiling urea. This harshly extracted muscle residue was considered as representing an elastic protein and was designated as "connectin." (It is now known that the "connectin" fraction is a heterogeneous mixture of predominantly denatured, degraded and aggregated titin and nebulin, plus variable amounts of myosin, actin, connective tissues and intermediate filament protein (cf. Fig. 2d) (K. Wang, in preparation). Based on the studies of this mixture of residual proteins, Maruyama proposed that "connectin" forms a network of criss-crossing elastic filaments within and surrounding each sarcomere (Maruyama, 1980; see also Locker and Dianes, 1980, King and Kurth, 1980).

Independently, we began our study of titin and nebulin in 1976, as a result of our search for smooth muscle filamin-like protein in striated muscles (Wang et al., 1975). We noticed that although striated muscles contained little filamin, they contained major amounts of very high molecular weight proteins which have not previously been identified. As can be seen in Fig. 1 where an SDS gel pattern representative of a wide range of vertebrate striated myofibrils is shown, nearly all of the major protein bands below myosin heavy chain have been identified as either thick or thin filament proteins or as anchoring proteins at Z lines and M lines. In contrast, the three major protein bands above myosin heavy chain, which together constitute 12 to 15% of the total protein, have received little attention (Etlinger et al., 1976; Porzio and Pearson, 1977;

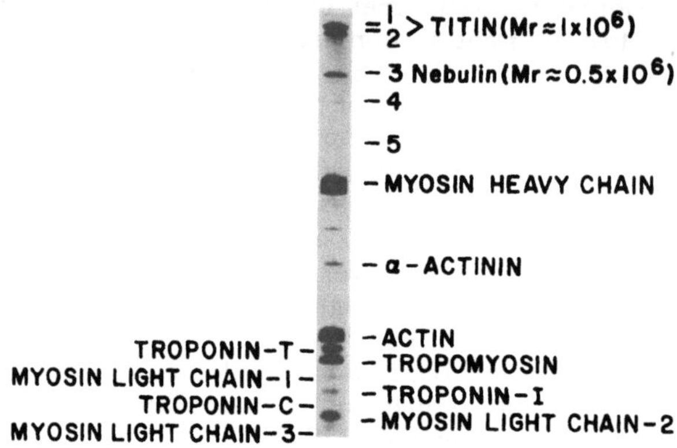

Figure 1: Major myofibrillar proteins of rabbit skeletal muscle. Exhaustively-washed rabbit myofibrils were denatured, reduced and then resolved on a high porosity SDS-polyacrylamide gel. Note that titin and nebulin are giant proteins with very low electrophoretic mobilities.

Wang and McClure, 1978). We therefore set out to characterize these proteins. We designated the top doublet collectively as titin (derived from "titan": anything of great size) because of their titanic size and because these two bands (designated T_1 and T_2) are immunologically, and possibly chemically, related (Wang et al., 1979). The third band, distinct from titin, is referred to as nebulin because it is found associated with the N_2-line of the sarcomere -- a nebulous striation within the I-band (Wang and Williamson, 1980; Wang, 1981). Their sizes are approximately: 1 x $10^6 M_r$ (titin) and 0.5 x $10^6 M_r$ (nebulin), determined by comigration with the hexa-heptamer and the trimer of crosslinked myosin heavy chains, respectively, under dissociating conditions on high porosity SDS gels. Proteins of this size range do not enter SDS gels commonly used for normal-sized proteins. We have found, however, that even when a suitable high porosity gel system was used, reproducible gel patterns were obtained only when strict precautions were taken to avoid the extremely rapid proteolytical degradation during muscle preparation, as well as the irreversible aggregation of these proteins during SDS gel sample preparation (Wang, 1982; Wang and Palter, in preparation).

Another nagging technical difficulty resulted from the insolubility (or insolubilizability) and aggregation of titin and nebulin. Intact titin and nebulin are not extractable by a wide range of solutions commonly used to dissolve thick and thin filaments and therefore represent the major residual proteins of extracted myofibrils (Fig. 2). To solubilize these proteins, it was necessary to use denaturants such as SDS, guanidinium chloride or urea. However, we were surprised to find out that SDS did not consistently and quantitatively dissolve titin, unless the ionic strength of the myofibril suspension was appreciably lowered. When SDS was added to the myofibril at higher (e.g., near physiological) ionic strength, the majority of titin frequently remained as an elastic gel while other myofibrillar proteins, including nebulin, were readily dissolved. As a

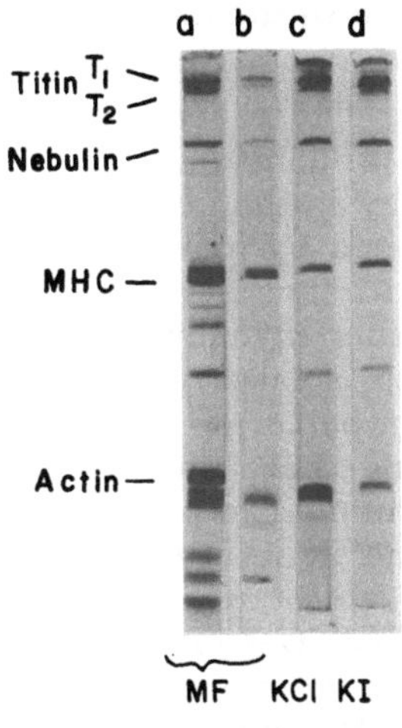

Figure 2: Resistance of titin and nebulin to extraction. Rabbit skeletal myofibrils (a and b, two loadings) were sequentially extracted by 0.6 M KCl and then 0.6 M KI to remove the bulk of thick and thin filaments. The residues at each step were analyzed by SDS gel electrophoresis. Note that titin and nebulin are the major residual proteins in the KCl (c) and KI (d) residues.

result, titin bands were either absent or greatly diminished in gel patterns of such SDS samples (Wang, in preparation). The SDS-insoluble titin gel undoubtedly must have been seen by many muscle biochemists but was probably discarded as uninteresting "connective tissues."

To summarize, we have encountered unexpected technical difficulties in nearly every step of the seemingly straightforward task of obtaining reproducible SDS gel patterns: extremely rapid degradation of titin and nebulin during sample preparation and storage, inconsistent and erratic solubilization of titin by SDS, and the extremely low electrophoretic mobility of titin and nebulin in commonly used SDS gels. These technical difficulties are perhaps largely responsible for the evasiveness of these proteins.

After having overcome most technical difficulties, we then proceeded to purify and characterize titin and nebulin from SDS solubilized myofibrils with gel filtration chromatography, taking advantage of their large sizes, as well as a novel differential solubility in high ionic strength SDS solution (Wang, 1982). T_1, T_2 and nebulin were successfully purified. The amino acid compositions of titin and nebulin are clearly distinct from other myofibrillar proteins, while T_1 and T_2 have very similar, if not identical, composition. It is worth noting that titin is enriched in proline (8-9 mole percent) -- an α-helix-breaking residue, suggesting that native titin may be more flexible than myosin heavy chain (~2% proline). Although we consider it possible that polypeptides of such giant sizes may consist of multiple subunits covalently linked together by non-disulfides, we have so far found no side-chain cross-linkers.

Titin and Nebulin: Distribution in the Sarcomere

The question of how titin and nebulin are arranged in the sarcomere has been approached using fluorescent antibody staining techniques (Wang et al., 1979; Wang and Williamson, 1980). Since these proteins are fairly antigenic, monospecific antibodies were successfully prepared in animals and then characterized by immunoblots and solid phase immunoassays. It was found that titin and nebulin are immunologically distinct from each other and from other myofibrillar proteins, whereas T_1 and T_2 cross-react. When applied to isolated myofibrils, these antibodies gave highly reproducible staining patterns with standard fluorescent techniques. However, interpreting these patterns was far from straightforward. A bewilderingly large number of patterns was obtained which could not easily be correlated to the structural features of the sarcomere. For example, monospecific titin antibody consistently labeled a wide zone centered on the edge of A bands (A-I junction). Yet in many, but not all, myofibrils it also stained one or more of the following loci: Mid-A (M line?), Z lines and throughout the entire A band. Furthermore, even with myofibrils with similar overall staining loci, the relative intensity, shape and dimension of these fluorescent bands varied significantly from one myofibril to the next. For this reason, we were able to reach only some general conclusions in our early studies. We concluded that the majority of titin is preferentially distributed within the A band. However, it was noted that the staining at the A-I junction, in fact extended considerably into the I band, in sharp contrast to the labeling patterns of myosin which are always confined within the boundaries of the phase dense A band. This observation, as well as the resistance of titin to extraction by A band solvents, suggested to us that titin, despite its A band distribution, may not be an A-band protein in the classical sense.

An important clue for interpretation of titin patterns came from our immunofluorescence studies of nebulin: anti-nebulin stained subregions of the I band corresponding in location and in behavior to the N_2-line -- the nebulous narrow striation which flanks either side of the Z line (Wang and Williamson, 1980). The N_2-line is an obscure, yet highly interesting, structure because it exhibits a novel sarcomere length-dependent translocation: when the sarcomere is lengthened, the N_2-line translocates in such a way that it maintains the same proportional distance to the adjacent M line and Z line, as if it is associated with or is part of an *elastic* longitudinal structure anchored at both M lines and Z lines. Futhermore, the N_2-line is located in the I band region where thin filaments alter packing geometry from a square lattice at the Z line to a hexagonal array near the A-I junction (Franzini-Armstrong, 1970). Despite these puzzling and novel features, little work has been done specifically addressing N_2-lines, perhaps because they are not always detected by electron microscopy. Our work has identified, for the first time, a major myofibrillar protein, nebulin, as an N_2-line-associated protein. Perhaps more significant for us was that this finding encouraged us to look seriously into the possibility of interpreting fluorescent labeling patterns of both titin and nebulin in terms of more unorthodox sarcomere models which differ from the current two-filament model by the inclusion of an additional set of elastic longitudinal filaments (see above).

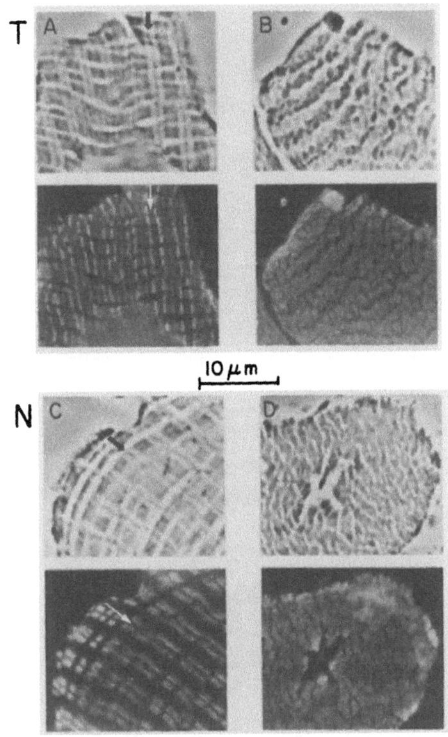

Figure 3: Immunofluorescent localization of titin and nebulin in superthin frozen sections of rabbit skeletal muscle. Fresh, unfixed rabbit skeletal muscle was frozen in liquid N_2, sectioned at -80°C to 0.1-0.3 μm thick sections and stained with anti-titin (T) and anti-nebulin (N). A pair of phase contrast (top) and fluorescent (bottom) micrographs are presented for each longitudinal view (A and C) and cross-sectional view (B and D). Arrows indicate the position of Z lines. Note the uniform labeling of cross-sections by both titin and nebulin in B and D.

Our task of interpreting labeling patterns has also been greatly facilitated by using more advanced techniques which yielded better structural correlations. This was accomplished by (1) localizing titin and nebulin in frozen sections of fresh muscle; (2) applying double fluorescence techniques to localize simultaneously two different proteins in the same sarcomere (Wang et al., 1982); and (3) correlating fluorescence patterns with other optical images such as that of polarized optics. Presently, a detailed investigation of the relative distribution of titin, nebulin, myosin and C-protein in both isolated myofibrils and in superthin (0.1-0.3 μm) frozen sections of fresh muscle tissue has been completed (Wang and Williamson, in preparation). Some representative results are presented in Figs. 3 and 4. A composite summary diagram containing all observed banding patterns and their longitudinal (or axial) positions is presented in Fig. 4.

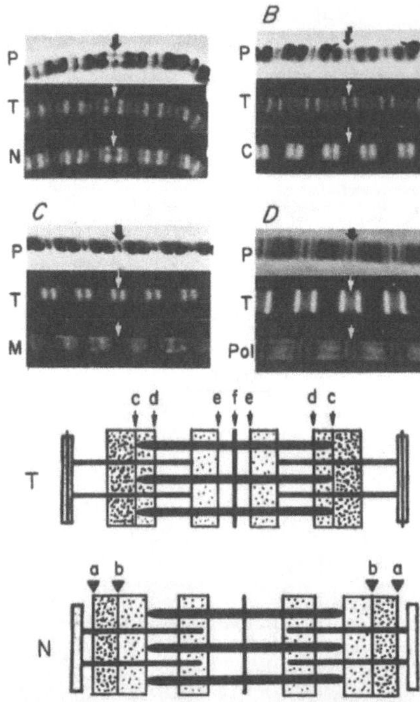

Figure 4: Dual-immunofluorescent localization and optical image correlation of titin and nebulin in isolated rabbit myofibrils. (A, B, and C): Isolated rabbit skeletal myofibrils were stained simultaneously with a pair of selected antibodies to reveal the distribution of two proteins in the same sarcomere. Arrows indicate the position of Z lines. (*A*) A titin (T) and nebulin (N) pair. (*B*) A titin (T) and C-protein (C) pair. (*C*) A titin (T) and myosin (M) pair. (*D*) A comparison of phase contrast (P), titin fluorescence (T) and polarized light (Pol) images of the same myofibril. Note that the titin is localized on a pair of phase dense, negatively-birefringent lines in the I band. See text for details. (*Bottom*) Two composite diagrams incorporating all observed staining zones of titin (T) and nebulin (N) are superimposed on the two-filament sarcomere model. The density of dotted zones approximates relative staining intensity. Note that not all staining zones appear on all myofibrils at all times. The boundary of each zone is labeled from a to f. These axial loci are also depicted in the sarcomere model in Fig. 6.

Labeling studies of frozen sections indicated that many of the staining patterns of isolated myofibrils are indeed observed in longitudinal frozen sections (Fig. 3A and 3C). Both anti-titin and anti-nebulin uniformly labeled the cross-sectionnal areas of myofibrils (Fig. 3B and 3D), indicating that both are distributed *within* the sarcomere.

The advantage of the double labeling technique can be appreciated by the examples given in Fig. 4. Although the pair of labeling patterns of titin and nebulin are very similar (Fig. 4A), closer examination reveals small but significant differences in detailed axial locations. Such subtle differences would have been impossible to pin-point with confidence by comparing two singly-labeled samples. In Fig. 4B, a titin-C-protein pair shows that the pair of weaker titin zones within the A band (e-zones of Fig.

4B) may in fact overlap somewhat with the C-protein domains. Similarly, in Fig. 4C, a titin-myosin pair demonstrates that titin does indeed label structure beyond the boundary of myosin domain, ruling out the possibility that the wider titin staining is simply an optical artifact of comparing images of two different optics.

Our attempts to correlate images of fluorescence optics and polarized optics have brought some surprises. We observed that the locations of titin and nebulin seem to correspond with several *negatively* birefringent lines (when A band is positively birefringent) in these myofibrils. An example is shown in Fig. 4D, where titin labels two phase dense lines on either side of the A band. These positions coincide precisely with the pair of negatively birefringent lines in the I band region of the polarized image (Fig. 4D, bottom panel). Because these and other negatively birefringent lines in the A band were also visible in unlabeled sarcomeres, they were not created by the extra mass of attached antibody molecules. Similar observations of correspondence have also been made using anti-nebulin antibodies. These results suggest that titin and nebulin-containing structures may contribute significantly to the optical properties of the sarcomere. This, in turn, raises the hope that such non-invasive optical methods may be useful to study the dynamic organization of titin and nebulin in living muscle.

Titin and Nebulin: Proposed Components of the Third Filament

For reasons discussed in previous sections, we have interpreted antibody labeling patterns of titin and nebulin on the basis of a speculative three-filament sarcomere model containing a longitudinal elastic filament (Wang and Ramirez-Mitchell, 1979; Wang and Williamson, 1980; Wang and Ramirez-Mitchell, 1981a, 1981b; Wang, 1982). We proposed that (Fig. 6) titin and nebulin are the major, if not the exclusive, components of a third set of longitudinal filaments which connect Z lines from within the sarcomere. The third filament is intrinsically elastic and/or extensible except when and where it interacts at certain loci with other inextensible sarcomeric structures such as thick or thin filaments, M or Z lines. Furthermore, this filament is fragile and is exceedingly prone to mechanical or proteolytical degradation, resulting in discontinuous segments.

Based on these hypothetical features of a third filament, our fluorescent staining data were interpreted to mean that the third filament system consists of two distinct, and perhaps non-overlapping, molecular domains: One domain contains all the titin molecules. This titin domain overlaps with the A band, but extends beyond the A-I junction and terminates at locus b (presumably corresponding to the N_2 line). A second domain, containing nebulin, extends from locus b to locus a (presumably corresponding to the N_1 line near the Z line).

Most of our staining patterns can be understood based on this two-domain model and on the fragility of the third filament. For example, anti-titin frequently stained the entire A-band weakly, but stained A-I junctions brightly. This difference may be due to the limited accessibility of antibody in the crowded thick filament lattice except near the tapered

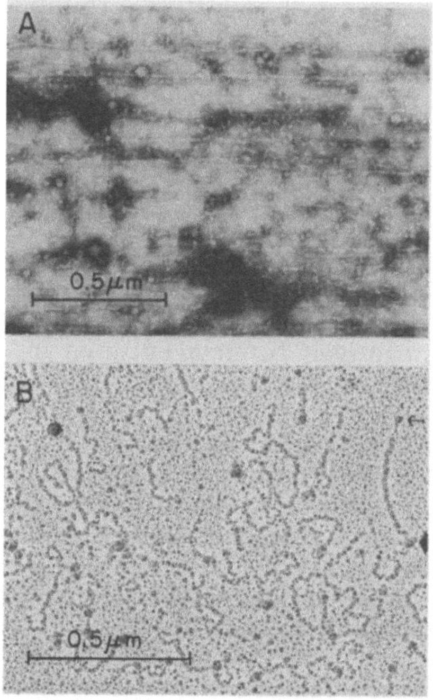

Figure 5: Molecular morphology of titin. (*A*) Negatively stained filamentous gel. Rabbit titin (T_1 and T_2) was purified by SDS gel filtration, aggregated into a filamentous gel by removing SDS and negatively stained with uranyl acetate. Note the abundance of slender filaments (2-10 nm in diameter). (*B*) Rotary-shadowed native titin. Native rabbit T_2 was purified in the absence of denaturants, sprayed onto mica and rotary shadowed by platinum at low angle. Note the length and the flexibility of the fibrous molecule. Arrows point to the globules frequently seen near the ends.

thick filament ends. The occasionally observed staining of titin and nebulin at locations other than the specified domains (e.g. titin on N_2-line in Fig. 4D) may be explained if mechanical or proteolytic damage of the third filament has allowed the elastic material to retract and to accumulate near certain loci where further translocation is inhibited. To illustrate this idea further, if the third filament is severed close to locus a in Fig. 6, then nebulin would appear concentrated on locus b, giving rise to narrow N_2-line labelings observed previously (Wang and Williamson, 1980). If, for some reason, the third filament at locus b is not anchored, then nebulin would appear on the edge of A band (i.e., locus c), as is occasionally observed (Wang and Williamson, in preparation). Similarly, if the filament is severed between b and c, then titin would also appear on either the N_2-line (locus b), or even on the N_1-line (locus a, which is indistinguishable from Z line by light microscopy) if locus b fails to anchor the filament. Again, such patterns have occasionally been observed. It follows that, depending on the extent and location of filament damage and the amount of filament fragment that is dissociated and released from the sarcomere, the relative intensity and dimension of each staining band would vary accordingly.

It should be noted that the self-consistent interpretation described above is only one of many schemes which one can devise, some of which could exclude the presence of a third filament. However, we found it difficult to interpret these results in terms of the two-filament model without introducing many *ad hoc* structural assumptions which clearly contradict the well-established structural characteristics of thick and thin filaments. A major support for the third filament interpretation came from our recent morphological studies of purified titin (Fig. 5). We have shown previously that titin purified in the presence of denaturants, tends to form a filamentous gel (Fig. 5A) when denaturants are removed or when protein concentration is increased (Wang and Ramirez-Mitchell, 1979). We have recently purified native T_2 without the use of denaturants and have studied its morphology using low-angle rotary shadowing techniques (Wang and Ramirez-Mitchell, in preparation). As shown in Fig. 5B, fibrous structures with convoluted contours measuring between 0.4 to 0.9 μ are frequently seen. Although detailed analysis is still in progress, the tendency for titin polypeptides to form extraordinarily long and flexible filaments is clearly established. It is significant that, if associated end to end, such extended molecules could easily form a filament long enough to span the entire titin domain (of more than 1.6 μm) in our sarcomere model.

A Speculative Three-Filament Sarcomere Model: Assumptions and Implications

Although our antibody localization data have revealed the fragility of the filaments and the axial distribution of their component proteins, little concrete structural evidence is currently available on the detailed arrangement and molecular interactions of the third filament within the sarcomere. However, it is perhaps instructive to speculate and to discuss how the third filament might be arranged in the sarcomere, thereby enabling us to focus on relevant issues to be resolved in the future. In Fig. 6 a speculative sarcomere model incorporating our current working hypotheses is presented. Several salient features are worth noting.

First, we have assumed that the third filament, composed mainly of titin and nebulin, is *external* to thick and thin filaments. Furthermore, the third filament may *interact*, under certain (unspecified) conditions, with thick and thin filament components. Thus, the present model differs from several other three-filament models in that no "core filament" of thick filament is assumed to exist. The structural continuity of the third filament with thick filaments is assumed in this model to result from the interaction of the third filaments with external components of the thick filaments. To emphasize this association, the third filament is depicted as helically winding around the thick filaments. Other arrangements, such as straight filament, is of course possible but less appealing, given the helical nature of the thick filament structure. This novel idea may also be applicable to the third filament-thin filament interaction. However, since

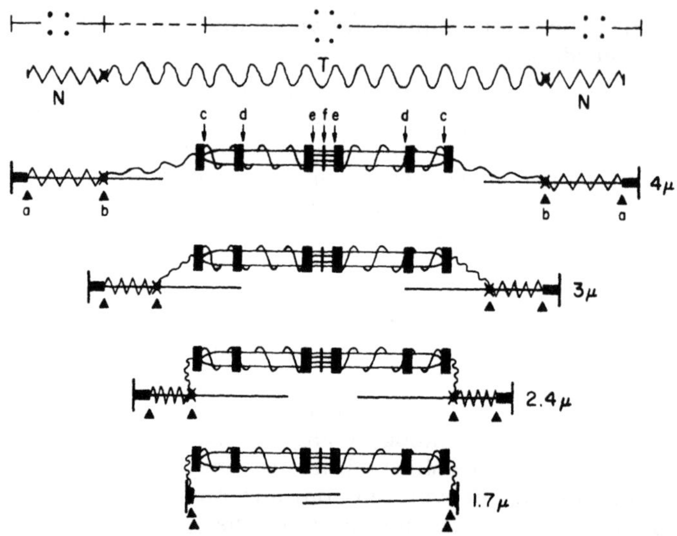

Figure 6: Schematic representation of a hypothetical three-filament sarcomere at various lengths. In this model, we postulate that (1) the third filament constitutes a three-dimensional lattice containing both longitudinal and transverse elements; (2) the lattice structure conforms to the packing geometry of thick and thin filaments with which it interacts, i.e., a hexagonal lattice in the A band (·:·:·), a tetragonal lattice between N_2-line (b) and Z line (a) (::) and an intermediate one in the transition zone between A-I junction (c) and N_2-line (b) (---); (3) the longitudinal third filament is continuous from Z line to Z line and consists of two alternating filament domains, each containing either titin (T) or nebulin (N); (4) the longitudinal third filament associates dynamically with the parallel thick and thin filaments and is anchored at the Z lines (a) and at the M line (f); (5) the longitudinal third filament is intrinsically elastic along the entire length. However, in the presence of thick and thin filaments, only the I band domain (a to c) changes length when the sarcomere shortens or elongates; (6) the longitudinal third filaments are linked to each other laterally by transverse elements at a few specific loci, e.g., at the N_2-line (b) and at the A-I junction (c). Note that the axial loci a to f are assigned according to antibody localization data as summarized in Fig 4. A more detailed discussion is presented in the text.

thin filaments are rarely, if ever, dislocated or elongated in overly stretched sarcomeres, a weaker interaction or a different mode of interaction must exist between the third filament and thin filament. Such an association would explain why the third filament has been difficult to detect as a distinct ultrastructural entity in normal length muscle, or why filament counting experiments have provided no support for the presence of additional filaments. According to this model, both the thick filament and a portion of the thin filaments in the I-band region ultrastructurally defined in the micrographs are *composite filaments* of the third filament with the myosin and actin filaments, respectively.

Such an association would also explain why the third filament in the gap regions of overly-stretched sarcomeres (i.e., gap filaments) appear to connect the ends of "thick" filaments to the ends of "thin" filaments.

Second, the third filament is assumed to be intrinsically elastic unless it is anchored or is interacting dynamically with less extensible structures in the sarcomere. Z lines and M lines are assumed to be strong anchor points for the third filament. Several loci along the length, inferred from our antibody data discussed above, may also be loci of anchorage or strong interaction. It should be emphasized that a dynamic or even cyclic interaction with thick and thin filaments along the entire length of the third filament is also assumed to occur. This concept of *dynamic association* is useful in explaining how the third filaments shorten or elongate without changing the length of thick and thin filaments when sarcomeres change length. As illustrated in Fig. 6, we have assumed that a preferential coiling or uncoiling of the third filament domains outside of the A band (from a to c) occurs under normal ranges of stress. The third filament in the A band domain may not shorten appreciably, if at all, due to a tighter association with thick filaments. In this sense, this model is formally analogous to the C-filament of insect flight muscle.

Third, and perhaps more importantly, the "longitudinal" third filament discussed so far is envisaged as a component of a more elaborate three-dimensional, elastic lattice existing within the sarcomere. Although we have emphasized the longitudinal orientation and the continuity of the third filament by depicting it as a single unbranched filament in Fig. 6, it should be pointed out that transverse filaments or branching structures may also exist. Such transverse linkages may exist at one or more of the axial loci that we have discussed above. In particular, we consider it likely that at the A-I junction (locus c) and at the N_2-line (locus b), transverse elements are arranged in a manner that demarcates a hexagonal lattice in the A band region and a tetragonal one in the N_2 to Z line region. In other words, we propose that the third filament is organized into a three dimensional elastic lattice that can be considered as serially-connected hexagonal and tetragonal cages alternating with a transition zone in between. It seems reasonable to assume that transverse linkages are present near the boundaries to help maintain the lattice geometry. In particular, the need for transverse structure at the N_2-line becomes clear when the packing geometry of thin filaments is considered: thin filaments are tetragonally anchored at the Z line and undergo a gradual transition at the N_2-line before assuming the hexagonal array in the overlap region. Given the flexibility of the thin filament, this orderly transition in packing is most likely accomplished by some structural guidance provided by the transverse structure at the N_2- line.

It should be clear from the above discussion that this three filament model is speculative in nature. However, this model is a useful one because many structural features and predictions are experimentally testable. Furthermore, this model in its present form, can provide a conceptual framework with which a large number of seemingly unrelated and puzzling structural data in the literature may be interpreted and integrated. While a detailed discussion is clearly beyond the scope of this chapter, perhaps it is justified to mention a few examples below to illustrate how the model can be used.

One puzzling structural feature of the sarcomere, frequently

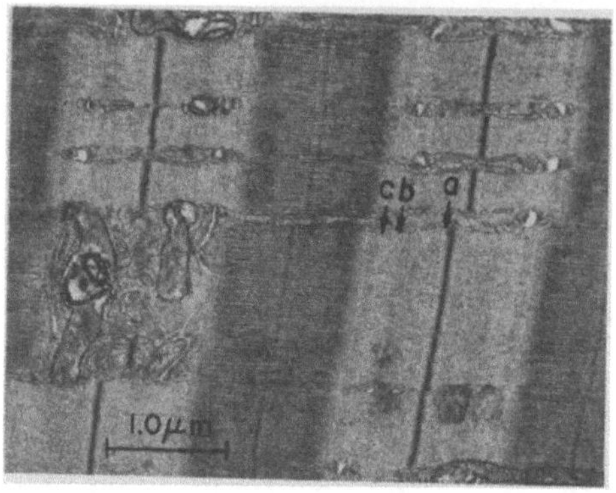

Figure 7: I-band substructure. The variation of electron density within the I band of a rat LOMO skeletal muscle is illustrated in this micrograph (courtesy of Dr. Roger McCarter). Note the higher density in regions between a and b, and near c. Such substructures in the I band may reflect the organization of the third filament lattice in this region. Loci a, b, and c appear to coincide approximately with those deduced in Fig. 4 (also, c.f. Fig. 6).

observed and ignored is the presence of electron dense zones within the I band of relatively long sarcomeres (see Fig. 7 for a typical example; note the dense zones between a and b and at c). This nonuniformity of I band density cannot be explained by the known distribution of thin filament components. It is therefore reasonable to assume that it resulted from the contribution of the third filament lattice. If this is true, then the data suggest that the third filament has a nonuniform or heterogeneous mode of interaction with thin filaments in the I band region. This conclusion agrees well with the predictions of the present model since both molecular composition and packing geometry of the third filament are assumed to vary in the I band region.

Another example concerns the interpretation of cytological localization studies of calcium binding sites in striated muscle when pyroantimonate was used as a calcium precipitating agent (e.g., see Yarom and Meiri, 1971). It has been repeatedly observed that the calcium precipitates are concentrated on the N_2-line and on the A-I junction in relatively long sarcomeres. Again, this distribution was puzzling because it cannot be interpreted based on the known distribution of calcium binding proteins of thick and thin filaments, we suggest that the distribution of the precipitates mainly reflects the organization of the third filament lattice. This conclusion raises the exciting possibility that calcium ions may bind to and modulate the structure and function of the third filament lattice.

In a similar fashion, the model can be used to predict that the third filament may contribute to the mass of the A band (Reedy and Lucaveche, this volume); may add to the diameter of the thick filament in the A band (Robinson and Cohen-Gould, this volume); may serve as a scaffold in the reconstitution of the A filaments *in situ* (Rowe and Maw, this volume); may yield, upon fragmentation, the "end filament" of isolated thick filaments (Trinick, 1981); may contribute, upon coiling, to the formation of contraction bands in short sarcomeres (Krueger and London, this volume); may provide the binding sites to proteins which have been localized to A-I junctions, such as AMP deaminase (Ashby et al., 1979); and may contribute to X-ray diffraction patterns of striated muscles (see Haselgrove and Rodger, 1980).

Sarcomere-Associated Intermediate Filaments: A Fourth Filament

As described previously, the early observations that certain longitudinal filaments appear to maintain the continuity and elasticity of residual structure in KI-extracted myofibrils provided one of the first clues that additional filamentous constituents may be present in the sarcomere. These extracted myofibrils, because of the ease of preparation, have since been a popular subject in which to search for additional filaments. Indeed, several recent reports claimed that the longitudinal residual filaments which span the gap between adjacent Z discs represent the third (gap) fialment (see dos Remedios and Gilmour, 1978). Prompted by these reports, we decided to apply more refined structural techniques to study the fine structure and spatial organization of these residual filaments in rabbit myofibrils (Wang and Ramirez-Mitchell, 1983). Our results have allowed us to conclude, rather unexpectedly, that the longitudinal residual filaments have the characteristic morphology of the intermediate (10 nm) filament -- a cytoskeletal filament distinct from thick, thin, and third filaments. Moreover, using critical point drying technique to preserve the spatial relationship of residual filaments and structures, we found that these parallel, longitudinal filaments appear to *envelope* the sarcomere by connecting the peripheries of successive Z discs (Fig. 8). In other words, these longitudinal filaments are too thick and are in the wrong place to be the sought-after third filaments which are 2-6 nm thick and exist *within* the sarcomere. We have been able to explain this discrepancy in interpretation by our findings that KI treatment somehow causes titin and nebulin to be dislodged from the original loci and then translocate and accumulate near the remains of Z discs, leaving behind the more resistent intermediate filament in the gap region (Wang, in preparation).

Equally significant was our observation that these longitudinal intermediate filaments appear to be an integral component of an extensive, three-dimensional network of intermediate filaments which connect all Z lines and M lines, both longitudinally as well as transversely in striated muscle, as shown in a schematic diagram in Fig. 8B. These observations are rather unexpected, because intermediate filaments, despite their abundance and three-dimensional distribution in smooth muscle and in developing striated muscle, have generally been assumed to form only a transverse filamentous system in adult skeletal muscle, maintaining the lateral adhesion of registered myofibrils (see Lazarides, 1980). The existence of the longitudinal components and the intimate association of the network with sarcomeres are, therefore, significant new findings which have many interesting implications in muscle mechanics. We have proposed that this cytoskeletal lattice, by connecting all force-bearing sarcomeric structures such as Z lines and M lines in directions both

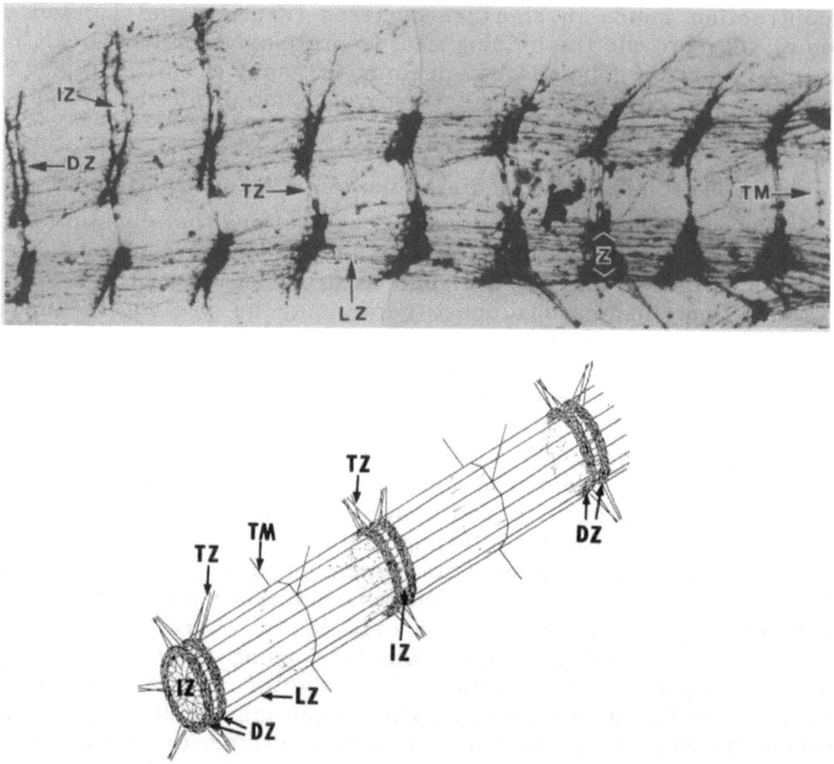

Figure 8: Sarcomere-associated intermediate filament lattice. (*Top*): The organization of the intermediate-filament network is revealed in the KI-extracted rabbit myofibrillar residue. Note the presence of longitudinal and transverse filaments. *TZ*: transverse filaments connecting Z to Z. *TM*: transverse filaments connecting M to M. *LZ*: longitudinal filaments connecting Z to Z. *Z*: Z structures. *DZ*: doublet structure of highly disintegrated Z structures. *IZ*: internal filaments in the gaps of doublet Z structures. (*Bottom*): A highly schematic diagram of sarcomere-associated intermediate filament lattice. In this model, intermediate filaments associated with two adjacent sarcomeres are depicted. This diagram emphasizes only the gross organizational principles of these filaments and is not drawn to scale. The disposition of these filaments relative to other sarcomere elements is discussed elsewhere (Wang and Ramirez-Mitchell, 1983).

parallel and perpendicular to muscle fibers axes, may play a structural role in tension transmission. Although it is generally assumed that muscle tension is longitudinally transmitted through the unbroken chains of serially-connected sarcomeres of each individual myofibril to the myotendon junction, the organization of this network suggests that sarcomeric tension may be transmitted both parallel and perpendicular to the fiber axis, even when certain individual sarcomeres fail to generate and/or transmit tension. In other words, this highly integrated network may provide an effective mechanism to allow muscle to by-pass defective sarcomeres (Wang and Ramirez-Mitchell, 1983). The contribution of this network to muscle continuity and to muscle elasticity and compliance is clearly worth exploring.

Given the intimate sarcomere-association, it is proposed that intermediate filaments should perhaps be considered, for many purposes, as an integral sarcomeric constituent and that the longitudinal filaments be considered as the *fourth filament*.

Cytoskeletal Matrix: Structural Continuity in Striated Muscle

To summarize, the structural studies and considerations, as briefly described in this chapter, appear to favor the notion that the cytoplasm of striated muscle cell contains a highly elaborate cytoskeletal matrix within which thick-thin filaments and other cellular organelles are integrated structurally and functionally. The matrix consists of at least two independent but interconnected lattices operating at two different levels: at the cellular and organelle level. The intermediate filament lattice appears to envelope all sarcomeres and to link together sarcomeres and membraneous organelles such as plasma membrane, nucleus, mitochondria, sarcoplasmic reticulum, etc.. At the sarcomere-myofilament level, the third filament (titin-nebulin) lattice may provide continuity and support for the interdigitating arrays of discontinuous myofilaments.

While the concept of interconnected cytoskeletal matrix is well established in current understandings of the structure of nonmuscle cytoplasm (see Cold Spring Harbor Symposium, 1981), so far only a limited effort has been attempted to apply this concept to address the question of structural continuity within striated muscle. We believe that although the details of our proposed model are likely to be refined with further investigation, the basic proposal of the existence of structural continuity in striated muscle, beyond the level of crossbridge interactions, is sound and may be a significant first step toward providing a more complete structural basis for understanding muscle behavior. To emphasize the possible contribution of these lattices in muscle mechanics, we propose that the sarcomere should be considered as consisting of two sets of continuous filaments linking adjacent Z lines, in addition to the discontinuous thick and thin filaments.

The functional roles of these two lattices are only speculative now: they may have a scaffolding role in stabilizing and anchoring the myofibril, thick and thin filament lattice, and associated membraneous organelles; they may have mechanical roles of bearing at least part of the resting tension of the myofibrils; and they may have regulatory roles in the assembly and turnover of the myofibril. The third filament lattice may modulate the interaction of thick and thin filaments by facilitating geometrical transition of thin filaments, or may even participate directly in the regulation of cross-bridges movement.

Clearly our studies raise more questions than they answer, but perhaps they point to some interesting new directions that future work on muscle structure and function might take.

ACKNOWLEDGEMENTS

I wish to thank present and former members of my laboratory: P. Louro, J. McClure, D. Palter, L. Somerville, A. Tu and C.L. Williamson for joining me in the pursuit of these elusive proteins. In particular, I thank R. Ramirez-Mitchell of the Cell Research Institute for his important contributions to our ultrastructural studies. I thank Dr. R. McCarter of the University of Texas Medical Branch at San Antonio for the use of his micrograph and for many helpful discussions.

This work is supported in part by grants from USPHS AM 20270, CA 09182 and a grant from the American Heart Association, Texas Affiliate, Inc.

REFERENCES

Ashby, B., Frieden, C., and Bischoff, R. (1979). Immunofluorescent and histochemical localization of AMP deaminase in skeletal muscle. J. Cell Biol. 81: 361-373.

Auber, J., and Couteaux, R. (1963). Ultrastructure de la strie Z dans des muscules de dipteres. J. Microscopy 2: 309.

Carlesen, F., Knappeis, G.G., and Buchtal, F. (1961). Ultrastructure of the resting and contracted striated muscle fiber at different degrees of stretch. J. Biophys. Biochem. Cyto. 11: 91-117.

dos Remedios, C., and Gilmour, D. (1978). Is there a third type of filament in striated muscle? J. Biochem. 84: 235-238.

Ernst, E. and Straub, F.B. (1968). Symposium on Muscle. Budapest, Akademiai Kiado.

Etlinger, J.D., Zak, R., and Fischmann, D.A. (1976). Compositional studies of myofibrils from rabbit striated muscle. J. Cell Biol. 68: 123-141.

Franzini-Armstrong, C. (1970). Details of the I-band structure as revealed by the localization of ferritin. Tissue and Cell 2: 327-338.

Garamvolgyi, N., and Kerner, J. (1966). The ultrastructure of the insect flight muscle fibril ghost. Acta Biochim. et Biophys. Acad. Sci. Hung. 1: 81-88.

Guba, F., Harsanyi, V., and Vajda, E. (1968). The muscle protein fibrillin. Acta Biochim. et Biophys. Acad. Sci. Hung. 3: 433-440.

Haselgrove, J.C., and Rodger, C.D. (1981). The interpretation of X-ray diffraction patterns from vertebrate striated muscle. J. Musc. Res. Cell Motil. 1: 371-390.

Hoyle, G. (1968). In: *Symposium on Muscle* ed. Ernst, E. and Straub, F.B. pp.34 Budapest; Akademiai Kiado.

Huxley, A.F., and Peachey, L.D. (1961). The maximum length for contraction in vertebrate striated muscle. J. Physiol. 156: 150-165.

Huxley, H.E., and Hanson, J. (1954). Changes in the cross striations of muscle during contraction and stretch and their structural interpretation. Nature 173: 973-976.

Huxley, H.E. (1968). In: *Symposium on Muscle* ed. Ernst, E. and Straub, F.B. p. 247. Budapest; Akademiai Kiado.

Huxley, H.E. (1972). Molecular basis of contraction in cross-striated muscles. In: *The Structure and Function of Muscle*, Bourne, G.H. Ed. 2nd edition, Volume 1 pp. 301-387, Academic Press, London.

King, N.L. and Kurth, L. (1980). SDS gel electrophoresis studies of connectin. In: *Fibrous Proteins: Scientific, Industrial and Medical aspects.* Eds. Parry, D.A.D., and Creamer, L.K. Vol. 2, pp. 57-66, Academic Press, London.

Lazarides, E. (1980). Intermediate filaments as mechanical integrators of cellular space. Nature 283: 249-256.

Locker, R.H., and Leet, N.G. (1976). Histology of highly stretched beef muscle. II. Further evidence on the location and nature of gap filaments. J. Ultrastruc. Res. 55: 157-172.

Locker, R.H., and Dianes, G.J. (1980). Gap filaments - the third set in the myofibril. In: *Fibrous Proteins: Scientific, Industrial and Medical Aspects.* eds. Parry, D.A.D., and Creamer, L.K. Vol. 2, pp 43-55, Academic Press, London.

Maruyama, K., Matsubara, S., Natori, S.R., Nonomura, Y., Kimura, S., Ohashi, K., Murakami, F., Handa, S. and Eguchi, G. (1977). Connectin, and elastic protein in muscle. Characterization and function. J. Biochem. 82: 317-337.
Maruyama, K., Kimura, S., Ohashi, K., and Juwano, Y. (1980). Connectin, an elastic protein of muscle. In: *Fibrous Proteins: Scientific, Industrial and Medical Aspects.* eds. Parry, D.A.D., and Creamer, L.K., Volume 2, pp. 30-36. Academic Press, London.
McNeil, P.A., and Hoyle, G. (1967). Evidence for superthin filaments. Am. Zoologist 7: 483-498.
Organization of the Cytoplasm, Cold Spring Harbor Symp. Quant. Biol. vol. 46. Cold Spring Harbor Laboratory, New York.
Porzio, M.A., and Pearson, A.M. (1977). Improved resolution of myofibrillar proteins with sodium dodecyl sulfate-polyacrylamide gel electrophoresis. Biochim. Biophys. Acta 490: 21-34.
Pringle, J.W.S. (1978). Stretch activation of muscle: function and mechanism. Proc. Roy. Soc. London B. 201: 107-130.
Sjostrand, F.S. (1962). The connection between A- and I-band filaments in striated frog muscle. J. Ultrastruct. Res. 7: 225-246.
Trinick, J. (1981). End filaments, a new structural element of vertebrate skeletal muscle thick filaments. J. Mol. Biol. 151: 309-314.
Walcott, B., and Ridgeway, E.B. (1967). The ultrastructure of myosin-extracted striated muscle fibers. Am. Zoologist 7: 499-503.
Wang, K., Ash, J.G., and Singer, S.J. (1975). Filamin, a new high molecular weight protein of smooth muscle and non-muscle cells. Proc. Natl. Acad. Sci. USA 72: 4483-4486.
Wang, K. and McClure, J. (1978). Extremely large proteins of vertebrate striated muscle myofibrils. J. Cell Biol. 79: 334a.
Wang, K., McClure, J., and Tu, A. (1979). Titin: major myofibrillar components of striated muscle. Proc. Natl. Acad. Sci. USA 76: 3698-3702.
Wang, K., and Ramirez-Mitchell, R. (1979). Titin: possible candidate as components of putative longitudinal filaments in striated muscle. J. Cell Biol. 83: 389a.
Wang, K. and Williamson, C.L. (1980). Identification of an N_2-line protein of striated muscle. Proc. Natl. Acad. Sci. USA 77: 3254-3258.
Wang, K., and Ramirez-Mitchell, R. (1981a). Myofibrillar connections. The role of titin, N_2-line protein and intermediate filaments. Biophys. J. 33: 21a.
Wang, K., and Ramirez-Mitchell, R. (1981). Myofibrillar connections: The role of titin, nebulin, and intermediate filaments. J. Cell Biol. 83: 389a.
Wang, K. (1981). Nebulin, a giant protein component of N_2-line striated muscle. J. Cell Biol. 91: 355a.
Wang, K., Feramisco, J.R., and Ash, J.F. (1982). Fluorescent localization of contractile proteins. Methods in Enzymol. 85: 514-562.
Wang, K., and Ramirez-Mitchell, R. (1983). A network of transverse and longitudinal intermediate filaments is associated with the sarcomeres of adult vertebrate skeletal muscle. J. Cell Biol. 96: 562-570.
Yarom, R. and Meiri, U. (1971). N lines in striated muscle: a site of intracellular Ca^{2+}. Nature New Biology 234: 254-255.

DISCUSSION

CURTIN: You started out by saying that about 10% of the total protein is not accounted for by the usual proteins in the muscle fiber. So do your new proteins quantitatively come up to 10%?

WANG: Right. They come up to about 15% of the total myofibrillar proteins.

ROWE: Do you think titin forms the end filaments that are seen in isolated thick filaments by Trinick?

WANG: Yes. This is certainly a real possibility. Both titin and nebulin may be components. In fact, my model would predict that unless there is breakage of these third filaments, we can't isolate thick filaments. This has been our experience. I think it has been yours too.

ROWE: Indeed, yes. Another point is that titin is very susceptible to postmortem nicking and proteolysis. The reason why *we*, in isolated filament preparations, have never seen anything heavier than the myosin heavy chain is that they seem to get nicked.

WANG: Yes. It could also be that titin is lost or dissociated during filament isolation. Perhaps I can use this chance to comment on the difficulties in detecting titin on gels. First of all, titin is not always solubilized by SDS unless the condition is right. If SDS is added to myofibrils in physiological salt solution, titin frequently aggregates into a gel and will not dissolve in SDS. This titin gel must have been observed by many, but was discarded as "connective tissue." The trick we use is to wash myofibrils in either deionized water or a low salt, high pH buffer to swell the myofibril before adding SDS. Only then does everything dissolve. Therefore, if one doesn't see a titin band on SDS gels, it doesn't necessarily follow that titin is absent or nicked. It could simply be that titin is not dissolved. Perhaps it is right at the bottom of the test tube! Whenever we saw gel-like material in SDS samples we invariably lost some titin.

KRUEGER: I caught an implication that you had found these proteins in some kind of cardiac muscle.

WANG: Yes, we have seen them in chicken, rabbit and rat hearts. But, we have been working mainly on rabbit back muscle and chicken breast muscle. Some general surveys were also done on rabbit psoas, frog and insect skeletal muscles, and we have looked at water bug flight muscle supplied by Dr. Reedy.

KRUEGER: Was there any functional relationship of the two proteins to the fast muscles or slow muscles?

WANG: Oh well, we don't really know. Our work has not reached that stage of sophistication yet. We chose our muscle based mainly on biochemical convenience. Functional relationships are certainly worth looking at.

HUXLEY: Mike, do these proteins help your A-band measurements?

WANG: Dr. Reedy has a problem of extra mass, and I have a problem of missing structure!

REEDY: I think it would be very surprising indeed if these two problems wouldn't partially solve one another. But the problem is, firstly, that the higher molecular weight things that Dr. Wang saw in gels of insect fibrils and filaments don't *yet* include the very highest ones or the same amounts of material that have been demonstrated repeatedly in his gels of vertebrate muscle. Secondly, it is hard to imagine why these things don't stay entangled with the filaments if they stay entangled with the fibrils. That's just where the problem stands at the moment.

WINEGRAD: We've been looking at gels of rat heart and finding material right in the range that you've described for titin. When the rat heart is treated with cyclic AMP, a great deal of phosphate is incorporated.

WANG: That's very interesting. We have been looking at titin and nebulin phosphorylation in frog skeletal muscle. They appear to be major phosphoproteins.

HOMSHER: Given that these proteins account for, say 10% of the fiber weight, and given the protein weight and the dimensions of the protein, would it fit your model -- would there be enough protein to fit in?

WANG: Oh, yes indeed. The model I presented took into consideration these molecular properties. However, the detailed organization of these filaments is based mainly on antibody localization data. The very audacious proposal that these filaments may wrap around thick or thin filaments, is of course sheer speculation. But I think it is a possibility worth thinking about.

MAGID: In your paper (Proc. Natl. Acac. Sci. USA 76: 3678-3702, 1979) on titin, you showed some work trying to characterize the proteins using antibodies, which is very important for localization work. I was concerned that the Ouchterlony gel of titin did not show a band against dissolved myofibrils. I was wondering what methods you've used *since* then to characterize these antibodies, and particularly, I am concerned about whether or not titin may be a doublet or two distinct elements of sarcomere structure.

WANG: Let me answer your last question first. In our studies of titin degradation, we noticed a very characteristic conversion of T_1 to T_2 upon proteolysis. We have recently isolated T_1 and T_2 individually and found them to have very similar amino acid composition. We have prepared antibodies to each band and found them cross-reactive. Based on these results, I think we can say that T_1 and T_2 are related. Most of our immunoassays are now performed with immunoblots and solid phase assays. These binding assays are more reliable and more informative than Ouchterlony tests, which rely on diffusion and precipitation, so negative results could also mean that antigens either do not diffuse or that the antibody-antigen does not precipitate.

GILLIS: Some years ago, Sally Page and I did some experiments mixing, in a glycerinated fiber, phosphate first, and then lead salts, to see where the phosphate was going. The fact was it was going to the N line and also on the A-I junction, exactly where your proteins are found.

WANG: Exactly!

GILLIS: As if there was transverse structure near these locations which bound or trapped the lead phosphate.

WANG: Yes, in fact this is the reason why I have some reservations about the calcium-antimonate localization pattern, because the distribution of precipitates may be caused simply by trapping of mobile precipitants. In other words, the pattern may indicate either calcium binding sites or non-specific diffusion barriers. I think the N_2–line region almost has to contain some types of transverse linkage, because the N_2 line is located on a transition zone where thin filaments change their packing from tetragonal to hexagonal. A structure at this point is certainly one way to accomplish this.

[illegible]

[illegible] of the [illegible]

Dr. G.: [illegible] nurse [illegible] question [illegible] degradation, we noticed a very different ratio [illegible] conversion of [illegible] to [illegible] upon [illegible] proteolysis. We have recently isolated [illegible] individually [illegible] them to have very similar amino acid compositions. We have generated antibodies to each band and found them cross-reactive. [illegible] there [illegible] we [illegible] say [illegible] and [illegible] are related. [illegible] are now performing [illegible] composition of [illegible] these [illegible]

[illegible] results could also mean that antigens simply do not diffuse, or that the antibody antigen does not precipitate.

Dr. [illegible]: Some years ago, [illegible] and I did some experiments [illegible] using glycerinated [illegible] first, and then [illegible] to see while the [illegible] was going. The [illegible] it was going [illegible] also on the [illegible] proteins are [illegible].

[illegible]

[illegible] bound or [illegible] the [illegible]

[illegible]

CONNECTING FILAMENTS, CORE FILAMENTS, AND SIDE-STRUTS: A PROPOSAL TO ADD THREE NEW LOAD-BEARING STRUCTURES TO THE SLIDING FILAMENT MODEL

Alan Magid, H.P. Ting-Beall, Melanie Carvell, Theda Kontis, and Carmen Lucaveche

Department of Anatomy, Duke University Medical Center, Durham, North Carolina 27710

ABSTRACT

This report concerns structural forces in resting muscle and proposes three additions to the sliding filament model to account for these mechanical properties. The proposal includes: *connecting filaments* (C-filaments) which connect the ends of each thick filament to the neighboring Z-lines, *core filaments* which support the myosin of the thick filament and which attach to the C-filaments, and *side-struts* which bind the thick filaments together along their length and restrict their radial movement. C-filaments would act as the parallel elastic element and transmit the passive tension to the thick filaments.

Isolated myofibrils (mechanically-skinned and detergent-treated frog semitendinosus fibers) when stretched progressively showed exponentially-increasing passive tension which did not disappear when filament overlap was exceeded, but continued to rise. SL was monitored with a HeNe laser. Passive tension phasically exceeded 3×10^5 N/M^2.

Electron microscopy (thin-sectioned and freeze-fracture/deep-etch specimens) of non-overlap fibers showed orderly fibril structure with clear separation of A- and I-bands. In the gap between them could be seen filaments, 40-50 Å in diameter, connected to the thick filament ends. Unlike actin, these filaments did not become decorated by myosin S-1.

Equatorial X-ray measurements showed that stretching relaxed skinned muscles squeezed the thick filaments closer; this radial compression continued beyond filament overlap.

Extreme stretch of fibers caused the thick filaments to strain several-fold.

Treatment of non-overlap fibers with a high ionic strength pyrophosphate myosin solvent caused a large drop in passive tension and stiffness, but no change in SL was detected nor was myofibril continuity detectably affected.

Non-overlap fibrils, when treated with elastase, released A-segments which retain three-dimensional coherency. Deep-etch EM's of non-overlap fibers disclosed abundant structures (about 75 Å) wide attaching adjacent thick filaments.

INTRODUCTION

Understandably, most of this symposium concerns active muscle force and its underlying basis. Our work looks instead at the mechanical and structural properties of resting myofibrils. This work was initially undertaken because informal observations of skinned fibers indicated that they possess a robust rubber band-like long-range elasticity difficult to reconcile with the customary sliding filament model for the sarcomere. The present-day two-filament model includes no specific structure with which this parallel elastic component can be easily associated. This problem with basic muscle theory did not exist in the earliest sliding filament models (Hanson & Huxley, 1953, 1955; Huxley and Hanson, 1954). Those authors were concerned, in part, with the structural basis of resting elasticity and proposed it resided in "S filaments". These were imagined to elastically connect the thin filaments associated with one Z-line to those of the next, and, further, to attach to the thick filaments at some point in the H-band. This proposal was later withdrawn (Huxley, 1966).

In our work aimed at finding where the passive elasticity resides we have used skinned fibers and muscles and have combined the techniques of force recording, laser diffraction, light and electron microscopy, and X-ray diffraction. Because skinned fibers swell we were also led to inquire on what basis a stable limit to swelling is produced. Our results have led to proposing that the sliding filament model be extended by incorporating two additional axial load-bearing structures and an additional radial load-bearing element.

MATERIALS AND METHODS

Muscles. In the course of this work, fibers from frog, rabbit, human, waterbug, bee, fly, and cricket muscle have been studied. However, all the results reported here were obtained with *Rana pipiens* semitendinosus dorsal head (DHST).

Solutions. Ringer solution contained (mM): 115 NaCl, 2.5 KCl, 2 $MgCl_2$, 2.0 $CaCl_2$, 10 NaPi (pH 7.2). Standard relaxing solution comprised (mM): 75 KOAc, 15 KPi (pH 7.0), 5 MgOAc, 5 ATP, 5 EGTA. It included 0.5% Triton X-100 when demembranation was desired. Rigor buffer omitted ATP and increased KOAc to 90 mM. Other solutions will be described where appropriate.

Fiber preparation, force recording, and sarcomere length measurement. Fibers were separated in mineral oil and skinned with forceps under relaxing solution as previously described (Magid, 1974). Improved morphology was found when muscles were first depolarized with rigor solution to prevent the contracture when bundles were cut. Fibers were attached to the force transducer (Pixie silicon strain gage) and length vibrator (Ling 101) with miniature stainless steel clips or with cellulose nitrate/acetone cement. The force transducer was carried on a micrometer screw and sarcomere length (SL) changes were produced by turning the screw manually. SL was determined by passing a HeNe laser beam horizontally through the fiber and read from a calibrated screen 37 mm (second order) or 380 mm (first order) distant. In some experiments, diffraction patterns were recorded with a motorized 35 mm camera.

Electron microscopy. Fibers were stretched on special "holddown" slides, or on the transducer and then fixed for electron microscopy (EM) in 2.5 or 5% glutaraldehyde in relaxing or rigor solution, post-fixed in carbodiimide/ethylene diamine (Reedy, 1976) or osmium, embedded in Araldite, and cut with diamond knives. Sections (ranging downward to 150 Å) were stained with permanganate and lead. For freeze fracture/deep-etch, fibers (fixed only in glutaraldehyde) were split longitudinally into thin bundles (~10 μm), rinsed in deionized water, mounted between thin copper foils, ultra-rapidly frozen in liquid propane (-185°C) and stored under LN_2 (Costello, 1980). Specimens were fractured, etched (1 hr. at -95°C, 5×10^{-7} torr) and rotary shadowed with platinum-carbon in a Balzers 360 freeze/etch unit.

Light microscopy. Fibers and fibrils were studied by phase contrast or darkfield microscopy using Zeiss 10× or 100× (oil) achromats on a WL stand. Micrographs were recorded on 35 mm film.

X-ray diffraction. Triton-skinned DHST muscles were prepared as previously described for sartorius muscles (Magid and Reedy, 1980). Thick filament spacing and mean SL were measured by methods given there.

A segments. A segments were liberated from non-overlap fibrils by digestion with elastase (Sigma, type III) at 1 μg/ml with 2-4 μg/ml soybean trypsin inhibitor added to block trypsin-like activity (Brokaw, 1980) in 4 mM MOPS, pH 7.0. Digestion times from 1 to 20 min. at 20.0°C were used. Digestion was stopped with 1 mM PMSF. Sonication (in an ice bath) after digestion markedly improved yields. Non-overlap fibrils were prepared by blending (in a Sorvall Omnimixer) muscle which had been stretched while in Ringer's to 4.0-5.0 μm SL and then extracted overnight in 100 KOAc, 20 KP_i, 5 EDTA, 0.5% Triton X-100. Homogenization buffer comprised (mM): 100 NaOAc, 15 MOPS buffer (pH 7.0), 5 EGTA, 5 azide, and 0.5 DTT.

RESULTS

Passive Tension in Relaxed Myofibrils

A typical set of tension traces resulting from the step-wise stretch of a skinned fiber is shown in Fig. 1a. The responses were almost purely elastic for small stretches, but stress relaxation became more prominent with increasing strain. Laser patterns obtained during stress relaxation showed that SL remained virtually constant and thus the drop in tension must reflect events occurring within the sarcomere (data not shown).

The steady values of tension reached at each SL are plotted in Fig. 1b. Passive tension increased monotonically with SL, but no discontinuity was seen at SL 3.65 μm about where thick-thin filament overlap is lost. Since neither the sarcolemma nor internal membranes nor filament interaction can be bearing this substantial passive tension, some other feature of myofibrillar structure must be.

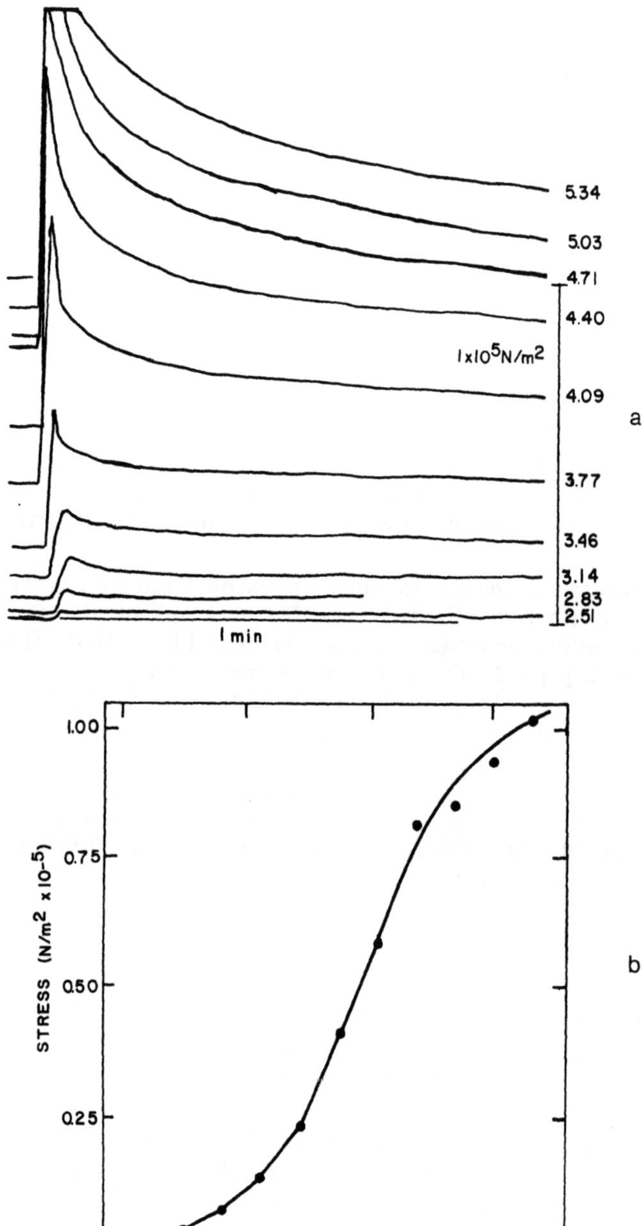
5.34
5.03
4.71
4.40
1x10^{5}N/m^2
4.09
3.77
3.46
3.14
2.83
2.51
1 min
a
1.00
0.75
0.50
0.25
0.00
STRESS (N/m^2 x10^{-5})
2.0
3.0
4.0
5.0
SARCOMERE LENGTH (μm)
b

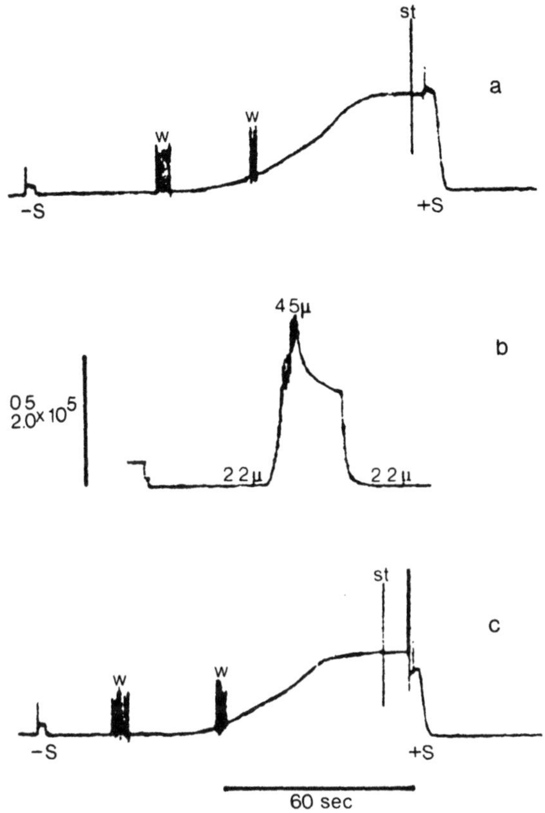

Figure 2: Recovery of contractility after non-overlap. (a) Control rigor and stiffness (± 0.5%). W indicates wash artifact; -S, the depletion of ATP. (b) Force trace resulting from stretch to 4.5 μm and release to SL_0. The vertical bar calibrates 0.5 and 2.0×10^5 N/m^2 for top and bottom, and middle traces respectively. (c) Retesting of rigor tension and stiffness after non-overlap. Almost complete recovery was seen. (From Magid, unpublished).

Recovery of Contractility After Non-Overlap

Figure 2 shows an experiment designed to discover the effect of stretch to non-overlap and release to slack length upon a fiber's ability to develop force. In (a) contractility was tested with a substrate-withdrawal contracture ("rigor" tension) and the associated stiffness estimated with a ±0.5% sinusoidal length change. In (b) the relaxed fiber was stretched

---

Figure 1: Long-range elasticity of myofibrils in relaxing solution. (a) Tracings from tension records produced by step length changes. The final SL of each step can be read to right of each trace. The final tension after stress relaxation is indicated by the starting level of the next record. The peak tension for the final record extrapolates to about 3×10^5 N/m^2. Temp. 22°C. Time bar indicates zero tension. (b) Steady tensions vs. SL. The curve is fit by eye. (From Magid, in preparation).

briefly to SL 4.5 μm and released to slack length. The passive elasticity mechanism restored SL to its initial equilibrium value (2.2 μm). When retested (c), rigor tension and stiffness were restored to 85-90% of their initial value, indicating that mechanically-effective interdigitation had been largely restored.

Electron Microscopy of Non-Overlap Fibers

When fibers stretched to non-overlap like that in Fig. 1 were examined under the electron microscope, they presented an orderly appearance of repeating A- and I-bands separated by light "gap" bands (Fig. 3a). A regular Z-line-to-Z-line spacing and a central position of the A-bands between them are seen; note that these are cardinal features of the model in Figs. 9 and 10. Although the thick filaments are slightly staggered axially, A-bands remain compact enough to be confident that

Figure 3: Longitudinal sections from non-overlap skinned fibers. (a) Survey view from a fiber soaked briefly in myosin S-1 to label actin filaments. This has resulted in a radial gradient of I-band density, but the "gap" bands are not labeled. (b) View from same fiber where several clear examples of A-band-Z-line parallelism are evident. (c) Very thin section from another fiber showing fine filaments extending from the thick filaments. Examples which can be followed from one gap band across the A-band to the other are indicated with pairs of arrowheads. Actin filaments show S-1 decoration; connecting filaments do not. Scale: (a) 10 μm; (b) 2.5 μm; (c) 0.5 μm. (From Magid, in preparation).

actin-myosin interaction has been abolished. This fiber was soaked briefly in myosin sub-fragment-1 to label actin filaments. It is clear that this produced a radial gradient of I-band density due to S-1 labeling, but no additional density appears in the gap band.

Detailed comparison of the shape of Z-lines and the edges of adjacent A-bands often reveals marked parallels in shape (Fig. 3b). This is most easily understood if there are direct links from the A-band to the Z-line.

In high magnification views (e.g., Fig. 3c) fine filaments (40 to 50 Å in diameter) can be seen extending from the tapering ends of the thick filaments, crossing the gap zone, and disappearing in the tangle of filaments in the I-band. That these are not actin filaments somehow pulled out of the I-band is established by their failure to form "arrowheads", conspicuous on the actin filaments. It is apparent that the only structures which could be bearing the passive tension are these *connecting filaments*, exposed in the gap between A- and I-bands of overstretched myofibrils.

X-Ray Diffraction of Stretched Skinned Muscles

Comparing the lattice spacing data from Triton-skinned muscles differing greatly in passive stiffness gives two clear indications of the operation of connecting filaments in these specimens (Magid, 1981). (1) The stiffer sartorius (note that no data could be obtained from this

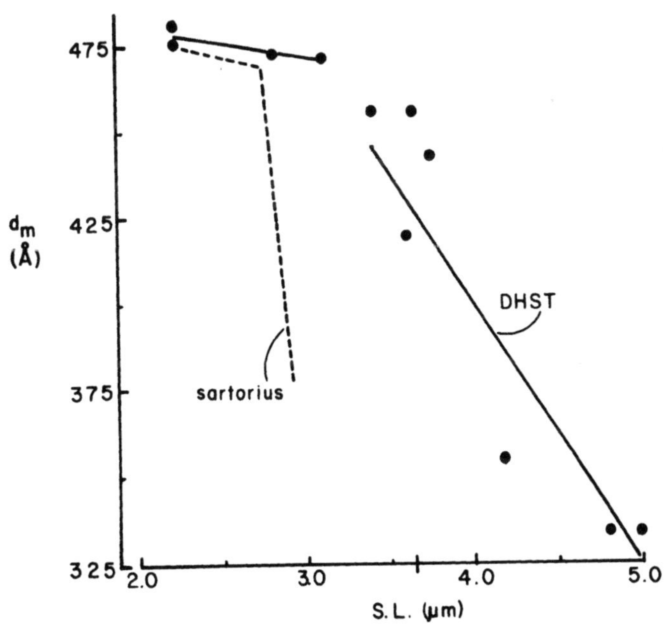

Figure 4: A-band compression produced by stretch. Center-to-center thick filament spacing (d_m) falls as SL is increased. Dashed line for sartorius is from Magid and Reedy (1980). (From Magid, 1981).

muscle type beyond SL 3.0 μm because passive tension became sufficient to tear the tendinous ends) showed a far steeper compression relation which appeared at a shorter SL than the more compliant DHST (Fig. 4). More crucial, however, is (2) the finding that lattice compression continued well *beyond filament overlap*.

Long-Range Elasticity After Myosin Extraction

To investigate whether the connecting filaments stop at the ends of the thick filaments or continue through it in some fashion as *core filaments*, fibers were stretched to non-overlap and treated with a Hasselbach-Schneider (H-S) solution to dissolve myosin. If the thick filament backbone comprises only the LMM tails as has been proposed (Huxley, 1963), then myosin removal ought to bring about a profound loss of sarcomere order and myofibril continuity. This was not seen. Rather, passive tension dropped markedly and quickly (Fig. 5) to about 10-20% of that before extraction; but, neither diffraction fringe spacing nor fringe width changed appreciably. Fibers were much more compliant after H-S extraction, but SL continued to increase in parallel with fiber length (data not shown). These findings are expected if a non-myosin elastic core underlies the myosin of the thick filament.

Microscopic study confirmed the mechanical behavior. Phase contrast images of non-overlap fibers treated to dissolve myosin show retention of sarcomere order (Fig. 6a,c). Optical diffraction of the light micrographs (b,d) reveals no change in spacing.

EM of the same fiber shows, in a survey view (6e), that thick filaments are no longer seen, but a filamentous residuum occupies their normal place. Further, the I-segments are especially dense, probably owing to the binding of dissolved myosin there. At higher power (f) patches of material (most likely myosin) can be seen clinging in places to a much thinner filament. This microscopic evidence that much myosin remains behind after H-S treatment is consistent with a similar finding using SDS

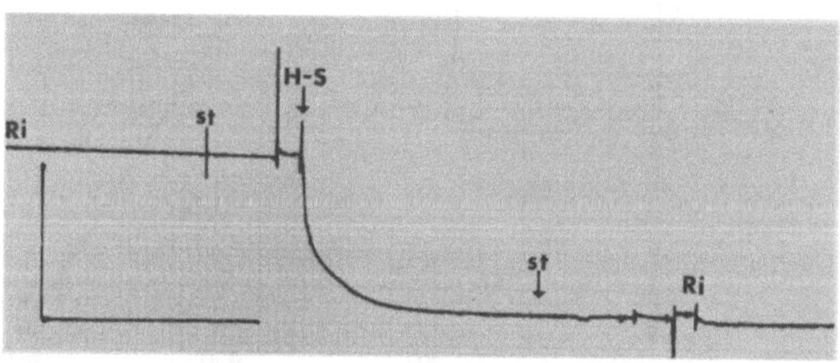

Figure 5: Effect of myosin solvent on passive tension. Fiber was stretched to SL 4.2 μm in relaxing solution and then transferred to rigor buffer (Ri). st is ± 0.7% stiffness test. At "H-S" fiber was exposed to myosin solvent. Both tension and stiffness fell markedly. Return to Ri does not restore tension and therefore response is not due to solution effects on connecting filaments. Horizontal: 60 sec. and zero tension; vertical: 0.5 N/m^2. Myosin solvent comprised (mM): 600 KCl, 5 NaPPi, 5 $MgCl_2$, 1 EGTA, 20 KPi, pH 7.0. (From Magid, unpublished).

Figure 6: Light and electron micrographs showing that myosin depletion does not destroy long-range myofibril structure. (a) Fiber stretched to SL 6.8 μm, phase contrast, in Triton-relaxing solution. (b) Laser diffraction of *micrograph* in (a) (obtained on 1.5 m optical bench). (c) Same after myosin extraction. (d) Optical diffraction of (c). Layer line intensity has changed, but neither spacing nor line width has. (e) Survey EM of the same fiber. I-bands are very dense and A-bands no longer have intact filaments. (f) High power detail. Residual filaments are evident. (g) Unextracted control fiber at SL 6.8 μm. Thick filaments have strained due to high stress. Core filaments seem visible. Whether this process is reversible is under investigation. Scale: (a,c) 100 μm; (e) 10 μm; (f,g) 1.0 μm. (From Magid, in preparation).

gel electrophoresis (Ullrick, Toselli, Chase, and Dasse, 1977). Study of an unextracted fiber (6g) stretched to the same SL as that in Fig. 6a-f reveals that if sufficient passive stress is applied to the thick filaments, they can *strain* appreciably, in this case, to almost three times normal size. These strained filaments appear as an interrupted layer of myosin upon an underlying core. The taut appearance of these filaments contrasts with the slack appearance of the extracted filaments.

Radial Elasticity of Relaxed A-Band

The results discussed so far have concerned structures which resist longitudinal stress and produce lengthwise continuity. Next, observations relating to side-forces between relaxed thick filaments are presented. Non-overlap fibrils provide a useful material for studying the properties of isolated A-segments by light microscopy. When stuck to a coverslip, their responses to various solvents flowing past can be followed and recorded. Fig. 7 shows a non-overlap fibril representative of the type usually abundant in our preparations (see MATERIALS AND METHODS). When the ionic strength of the bathing medium was lowered about 75-fold (increasing the electrostatic swelling pressure), the isolated A-segments swelled about 2-fold to a new stable limit and then returned to almost the original width in control saline (Fig. 7b,c).

The A-segments remained roughly rectangular indicating that the thick filaments remained approximately parallel to each during these radial movements. Figure 7d shows a typical field of the A-segments released when non-overlap fibrils were treated with elastase. Since they

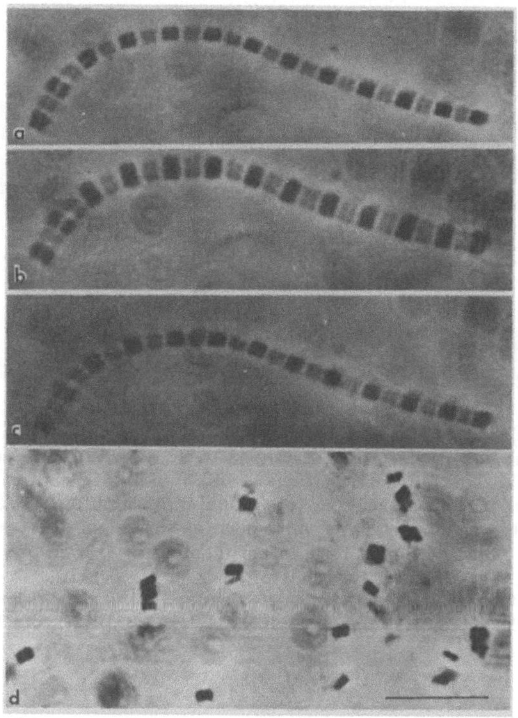

Figure 7: Non-overlap fibril and isolated A-segments. (a) Isolated fibril in homogenization buffer; SL 4.6 μm. Endogenous proteolysis weakened longitudinal elasticity and removed most Z-line density. (b) Swollen in 3.76 mM MOPS, pH 7.0. (c) In control saline; some irreversible lateral strain of A-segments occurred. (d) Typical field of A-segments isolated by elastase digestion of non-overlap fibrils. Note that they differ little from A-bands seen *in situ*. Scale 10 μm. (From Kontis and Magid, unpublished).

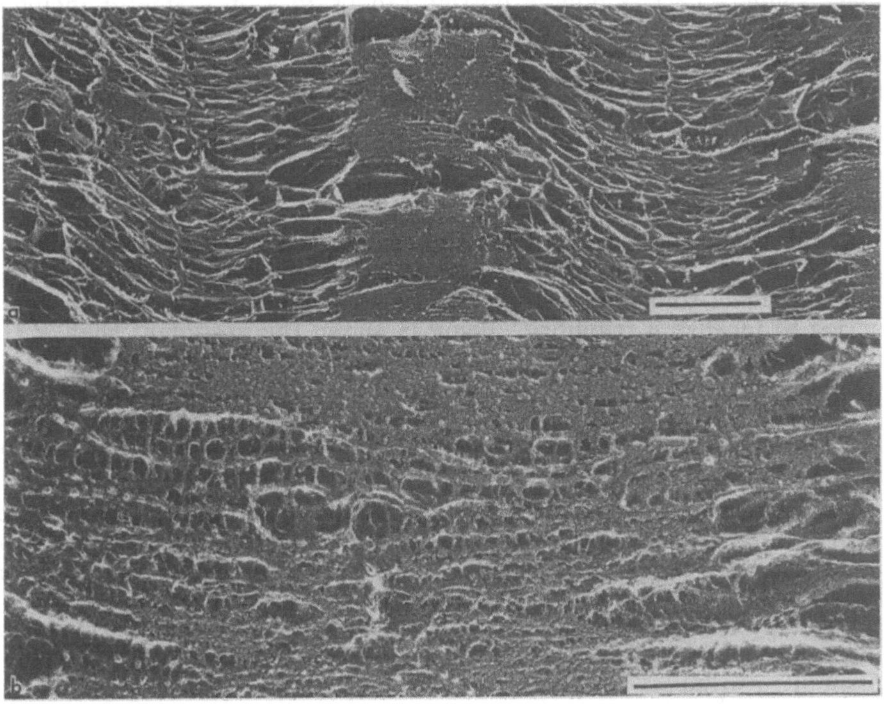

Figure 8: Freeze-fracture/deep-etch EM's of non-overlap fiber. (a) Survey view. C-filaments are clearly seen attached to A-bands. (b) High power view of A-band of fiber in (a). Many structures which laterally connect neighboring thick filaments are apparent. It is expected that the radial stiffness of the relaxed A-band resides in some or all of these. Negative prints. Scale: (a) 1.0 μm; (b) 0.5 μm. (From Magid, Ting-Beall, and Lucaveche, unpublished).

retain 3-dimensional coherency after being freed from the fibril, the connecting filaments cannot be bundling the thick filaments together in a fibril.

Freeze-Fracture/Deep-Etch Images of Isolated A-Segments

Heuser and co-workers (e.g., Heuser and Kirschner, 1980) have recently pioneered an EM technique which provides an alternative to thin-sectioning of embedded material. This method permits seeing the specimen in depth and thus avoids the ambiguities which superimposition of specimen details in sectioned materials most always produces. Fig. 8a shows a survey view of a non-overlap fiber prepared by the deep-etch technique. The wavy appearance is an expression of the passive tension tending to shorten the fiber. The sinuous I-segments and connecting filaments contrast with the straight appearance of the isolated A-segments. Fig. 8b is a higher magnification view showing that abundant structures link the thick filaments together sideways along their length. Although these features present a range of appearances, an abundant class shows a narrow middle portion (about 75 Å in diameter including the platinum layer) and wider portions at either end where they contact the filaments.

DISCUSSION

Sarcomere Models and Their Behavior

The conclusions drawn from the results given here are summarized in Figs. 9 and 10. The upper panel of Fig. 9 illustrates the present-day thick filament-thin filament sarcomere model using two sarcomeres in series to idealize a myofibril. Below is shown the argument that, if this model were stretched, the A-bands would not move because they are not longitudinally connected and thus no passive force could act on them. The lower pair of diagrams demonstrates how an elastic connecting filament (C-filament) would change the model's behavior (the side-struts shown will be discussed later).

It should be evident that such a structure would imbue the myofibril with a number of properties: 1. Rest sarcomere length (SL_0) would be established by the zero tension length of the C-filaments (assuming no cross-bridge activity). 2. A restoring force will arise when a fiber is stretched from SL_0 to which it will return when released (provided no internal yielding has occurred). 3. Passive tension will not require filament overlap. 4. During passive length changes, SL (the Z-line-to-Z-line distance) will remain uniform. 5. The A-bands will remain centered between the Z-lines because here, they experience an equal and opposite force. 6. A third filament type attached to the ends of the thick filaments will be seen in the gap between non-overlapping A- and I-segments. All these expected properties are manifested in the results given in Figs. 1-4.

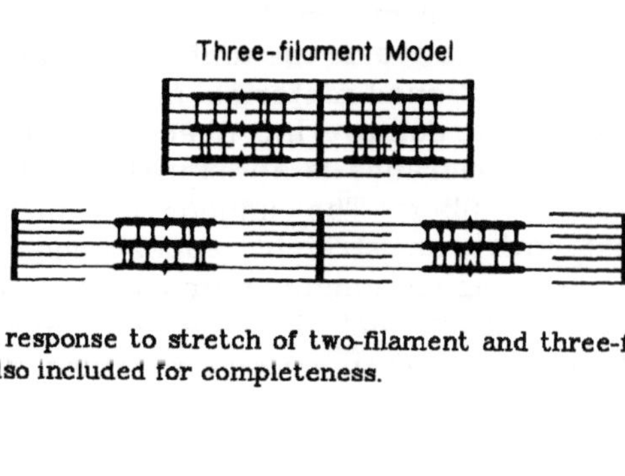

Figure 9: Comparison of response to stretch of two-filament and three-filament sarcomere models. Side-struts are also included for completeness.

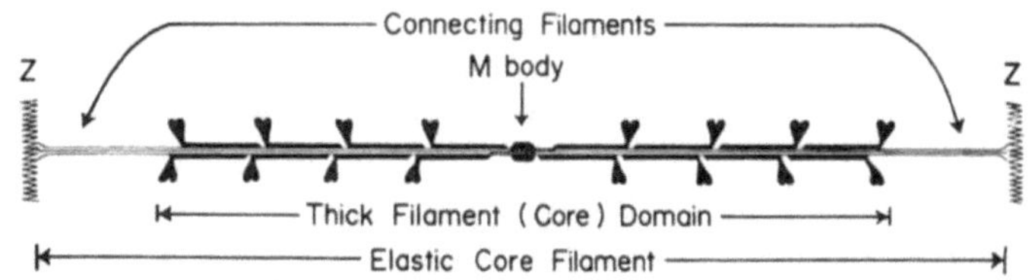

Figure 10: Proposed scheme for organization of connecting filament and thick filament. The M-body is hypothesized to be a bipolar organizing center from which the core filaments originate. Unspecified is the strandedness of the core and connecting domains, details of their attachment to each other, and C-filament attachment to the Z-line. The myosin molecules are drawn to approximate scale for a 2 μm sarcomere.

Passive Tension in Intact Fibers

The initial portion of the sarcomere stress-strain relation shown in Fig. 1b is well fit by an expression derived by Sten-Knudsen (1953) from intact single fiber data:

$$\sigma = \frac{E_0}{\alpha}(e^{\alpha\varepsilon} - 1)$$

where σ is stress, E_0, the initial modulus, α, an empirical constant, and ε, the fiber strain. Sten-Knudsen also observed decreased curvature at long lengths in frog fibers, as seen in Fig. 1b. If one takes ε as sarcomere strain $\frac{(SL-SL_0)}{SL_0}$ then α was 4.06 and E_0 was 9×10^3 N/m^2. Applying the same analysis to Rapoport's (1973) data for intact single DHST fibers yields α equals 3.87 and E_0 equals 8.7×10^3 N/m^2, essentially identical to the skinned fiber results. This means most or all of the passive tension of single DHST fibers resides in the myofibrils themselves. To answer whether this result extends to other fiber types and to whole muscles requires further study.

Passive Extensibility Differences Among Fiber Types

Asynchronous insect flight muscle is notable for high resting stiffness and stretch activation. White and co-workers (1967; White, Wilson and Thorson, 1978) have suggested that C-filaments (named such by him) are responsible for these properties. The general belief (Pringle, 1978) is that C-filaments are a specialized evolutionary invention peculiar to asynchronous muscles. However, vertebrate muscles (in particular heart muscle) also manifest isometric tension oscillation after a quick stretch and can do oscillatory work (Steiger, 1977). Perhaps stretch activation by means of the connecting filament underlies increased force sometimes seen with decreasing overlap (Endo, 1973; Edman, Elzinga and Noble, 1978). If operative, this effect may contribute to stabilizing the force-length relation.

Beyond changes in E_0, the C-filament model suggests how differences in passive extensibility among different fiber types might arise. If C-filament length were a smaller fraction of rest sarcomere length in two different muscles, then an equal amount of sarcomere strain would yield unequal amounts of C-filament strain. For instance, insect flight muscles have quite short I-bands at rest (and thus equivalently short C-filaments) compared to many vertebrate skeletal muscles and are correspondingly stiffer. Rat heart myofibrils exhibit high resting stiffness (Fabiato and Fabiato, 1978) and correspondingly short I-bands [SL_0 about 1.83 μm in isolated cells (Roos, Brady and Tan, 1982)].

"Gap Filaments"

Since their initial description by A.F. Huxley and Peachey (1961), "gap filaments" have been described often in the literature (e.g., Sjostrand, 1962; Page and Huxley, 1963; McNeill and Hoyle, 1967). Though these "third filaments" were given various interpretations, they were never discussed in terms of a C-filament model. Recently, Locker and Leet (1975, 1976) published micrographs of overstretched beef fibers which they interpreted as supporting a model for gap filaments which envision a direct link from the thick filament to the Z-line but *at one end only*. Since this model can not account for symmetrical thick filament strain (Fig. 6g), it would seem untenable. Ullrick et al. (1977) were stimulated by this reappearance of the "third filament" controversy to repeat H. Huxley's 1960 rabbit psoas filament-counting experiments and to compare vertebrate to insect muscle. They confirmed the presence of C-filaments in insect muscle, but could not demonstrate them in frog or lizard muscle and thought them absent. Two factors probably contribute to the evident unreliability of filament counts of vertebrate muscle cross-sections: 1) The two kinds of filaments in the rather long I-band of vertebrates become clumped together during tissue processing, obscuring their profiles, and 2) the C-filament substance accepts heavy-metal stain only poorly. This latter problem is circumvented by the deep-etch technique where structures are revealed by platinum sputtering.

Elastic Core to the Thick Filament

Figure 10 summarizes the conclusions about the longitudinal organization of the sarcomere, based in part on results from myosin extraction experiments. That kind of experiment has been repeated by many investigators since Huxley and Hanson (1954) used it on myofibrils under the light microscope to localize myosin to the A-band. The microscopic result is always of the same general form: isolated I-Z-I "brushes" linked in strings. Through the light microscope little is seen that might hold the fibril together, but EM has shown fine filaments in the space between I-segments. Reported diameters range from 2 to 13 nm, 4-5 being most commonly described. They have variously been called: T-filaments (Walcott and Ridgeway, 1967), fibrillin filaments (Guba, Harsanyi, and Vajda, 1968), connectin filaments (dos Remedios and Gilmour, 1978), and intermediate filaments (Price and Sanger, 1979; Wang, this volume). Both Guba et al. and Locker and Leet (1976) think them core filaments.

The drop in passive tension when myosin is extracted (Fig. 5) is interpreted to indicate that the myosin coat normally stiffens the extensible core so that thick filament length remains about constant during movement. However, Haselgrove (1975) reported that passive stretch of DHST to non-overlap increased meridional thick filament spacing slightly. Further, it is possible that thick filament shortening in *Limulus* telson muscle (Dewey, Walcott, Colflesh, Terry and Levine, 1977) may reflect an elastic core.

In general, efforts to form thick filaments from myosin solutions lead to long disordered aggregates which usually lack a distinct bare zone (Koretz, 1979). Evidence that a core structure seems necessary for thick filament reconstitution comes from several recent successes using the stroma of myosin-extracted fibers or isolated thick filaments as the site of assembly (Tawada, Yoshida and Morita, 1976; Niederman and Peters, 1980; Rowe and Maw, this volume).

Radial Elasticity of A-Band

It is well known that fibers swell when they are skinned. Maughan and Godt (1979, 1980) have analyzed this phenomenon and have concluded that elastic (structural) forces are important for setting a limit to swelling. Although van der Waals forces have been thought important in balancing electrostatic repulsion between filaments, computations show that they are about 500 times too small to oppose the swelling pressure manifested when fibers are skinned (Maughan and Godt, 1980; Magid and Reedy, 1980). Torsional elasticity measurements of resting fibers (Sten-Knudsen, 1953) also indicate the presence of permanent radial cross-links in muscle. In Fig. 8b, abundant "side-struts" can be seen linking thick filaments side-by-side. Since these images exhibit a range of appearances, there may be several classes of structures visible there. Of particular concern is the possibility that some of them might represent glutaraldehyde-induced adhesion of myosin heads to neighboring filaments. Additional study using unfixed fibers and myosin-extracted fibers are necessary to be more certain about the morphology of the lateral elements of the A-band.

Composition of the Elastic Stroma

Maruyama and co-workers (Maruyama, Matsubara, Natori, Nonomura, Kimura, Ohashi, Muraakami, Handa, and Eguchi, 1977) have described an elastic fibril stroma material which they designate connectin. They regard it as the basis of passive elasticity and believe it is organized as a random net which diffusely pervades the sarcomere. Wang and his colleagues (Wang, McClure and Tu, 1979; Wang and Williamson, 1980; Wang, this volume) have described a family of high molecular weight proteins which they believe form the "N_2-line" and other structures. Further work will be necessary before an association between the structures proposed in this paper and particular proteins can be made with any confidence.

Generality of the Elastic Stroma Model

Important questions raised by the work discussed here, based on frog muscle, concern how far the model can be generalized. The connecting filament seems to be a general property of all striated muscle; they all manifest long-range elasticity. Preliminary study of isolated insect thick filaments and skinned fibers show a residual core after myosin extraction treatment (Magid, unpublished). It seems important to find out if the connecting-core filament longitudinal organization has a functional analog in smooth muscles and in non-muscle actomyosin motility systems. Further, the possible presence of side-struts in the non-vertebrate A-bands needs study.

ACKNOWLEDGEMENTS

Dr. M.K. Reedy generously lent laboratory facilities (supported by NIH grant AM 14317) and offered useful criticism and advice throughout these studies. The work was funded additionally by an MDA postdoctoral fellowship and research grants and NIH grants AM27763 and HL20749.

REFERENCES

Brokaw, C.J. (1980). Elastase digestion of demembranated sperm flagella. Science 207: 1365-1367.

Costello, M.J. (1980). Ultra-rapid freezing of thin biological samples. Scan. Elec. Microsc. 2: 361-370.

Dewey, M.M., Walcott, B., Colflesh, D., Terry, H. and Levine, R.J.C. (1977). Changes in thick filament length in *Limulus* striated muscle. J. Cell Biol. 75: 366-380.

Dos Remedios, C.G. and Gilmour, D. (1978). Is there a third type of filament in striated muscles? J. Biochem. 84: 235-238.

Edman, K.A.P., Elzinga, G. and Noble, M.I.M. (1978). The effect of stretch of contracting muscle. In: *Cross-bridge Mechanism in Muscle Contraction*, eds. Sugi, H. and Pollack, G.D. University of Tokyo Press, Japan.

Endo, M. (1972). Length dependence of activation of skinned muscle fibers by calcium. Cold Spring Harbor Symp. Quant. Biol. 37: 505-510.

Fabiato, A. and Fabiato, F. (1978). Myofilament-generated tension oscillations during partial calcium activation and activation dependence of the sarcomere length-tension relation of skinned cardiac cells. J. Gen. Physiol. 72: 667-699.

Guba, F., Harsanyi, V. and Vajda, E. (1968). Ultrastructure of myofibrils after selective protein extraction. Acta Biochim. et Biophys. Acad. Sci. Hung. 3: 433-440.

Hanson, J. and Huxley, H.E. (1953). Structural basis of the cross-striations in muscle. Nature 172: 530-532.

Hanson, J., and Huxley, H.E. (1955). The structural basis of contraction in striated muscle. Symp. Soc. Exptl. Biol. 9: 228-264.

Haselgrove, J.C. (1975). X-ray evidence for conformational changes in the myosin filaments of vertebrate striated muscle. J. Mol. Biol. 92: 113-143.

Heuser, J.E. and Kirschner, M.W. (1980). Filament organization resolved in platinum replicas of freeze-dried cytoskeletons. J. Cell Biol. 86: 212-234.

Huxley, A.F. and Peachey, L.D. (1961). The maximum length for contraction in vertebrate striated muscle. J. Physiol. 156: 150-165.

Huxley, H.E. and Hanson, J. (1954). Changes in the cross-striations of muscle during contraction and stretch and their structural interpretation. Nature 173: 973-976.

Huxley, H.E. (1963). Electron microscope studies on the structure of natural and synthetic protein filaments from striated muscle. J. Mol. Biol. 7: 281-308.

Huxley, H.E. (1966) Remarks during a general discussion. *Symposium on Muscle*. Ernst, E. and Straub, F.B., eds. Akademiai Kiado, Budapest.

Koretz, J.F. (1979). Structural studies of synthetic filaments prepared from column-purified myosin. Biophys. J. 27: 423-432.

Locker, R.H. and Leet, N.G. (1975). Histology of highly-stretched beef muscle. I. The fine structure of grossly stretched single fibers. J. Ultrast. Res. 52: 64-75.

Locker, R.H. and Leet, N.G. (1976). Histology of highly stretched beef muscle. II. Further evidence on the location and nature of gap filaments. J. Ultrast. Res. 55: 157-172.

Magid, A. (1974). The relationship of MgATP to force generation in striated muscle. Ph.D.

thesis. Univ. Washington, Seattle.
Magid, A. and Reedy, M.K. (1980). X-ray diffraction observations of chemically skinned frog skeletal muscle processed by an improved method. Biophys. J. 30: 27-40.
Magid, A. (1981). Interfilament forces in relaxed detergent-skinned frog skeletal muscle. Biophys. J. 33: 226a.
Maruyama, K., Matsubara, S., Natori, R., Nonomura, Y., Kimura, S., Ohashi, K., Murakami, F., Handa, S. and Eguchi, G. (1977). Connectin, an elastic protein of muscle. J. Biochem. 82: 317-337.
Maughan, D.W. and Godt, R.E. (1980). A quantitative analysis of elastic, entropic, electrostatic, and osmotic forces within relaxed skinned muscle fibers. Biophys. Struct. Mech. 7: 17-40.
McNeill, P.A. and Hoyle, G. (1967). Evidence for super-thin filaments. Amer. Zool. 7: 483-498.
Niederman, R. and Peters, L. (1980). Bare zone myosin molecules determine myosin filament length. J. Cell Biol. 87m: 267a.
Page, S.G. and Huxley, H.E. (1963). Filament lengths in striated muscle. J. Cell Biol. 19: 369-390.
Price, M. and Sanger, J.W. (1979). Intermediate filaments connect Z-discs in adult chicken muscle. J. Exptl. Zool. 208 (2): 263-269.
Pringle, J.W.S. (1978). The Croonian lecture, 1977. Stretch activation of muscle: function and mechanism. Proc. Roy. Soc. Lond. B. 201: 107-130.
Rapoport, S.I. (1973). The anisotropic elastic properties of the sarcolemma of the frog semitendinosus muscle fiber. Biophys. J. 13: 14-36.
Reedy, M.K. (1976). Preservation of X-ray patterns from frog sartorius muscle prepared for electron microscopy. Biophys. J. 16: 126a.
Roos, K.P., Brady, A.J. and Tan, S.T. (1982). Direct measurement of sarcomere length from isolated cardiac cells. Am. J. Physiol. 242, H68-H78.
Sjostrand, F.S. (1962). The connections between A- and I-band filaments in striated frog muscle. J. Ultrast. Res. 7: 225-246.
Steiger, G.J. (1977). Stretch activation and tension transients in cardiac, skeletal, and insect flight muscle. In: *Insect Flight Muscle*, ed. Tregear, R.T. North Holland, Amsterdam. pp. 221-268.
Sten-Knudsen, O. (1953). Torsional elasticity of the isolated cross-striated muscle fiber. Acta Physiol. Scand. 28: supp. 104.
Tawada, K., Yoshida, A. and Morita, K. (1976). Myosin-free ghosts of single fibers and an attempt to re-form myosin filaments in the ghost fibers. J. Biochem. 80: 121-127.
Ullrick, W.C., Toselli, P.A., Chase, D. and Dasse, K. (1977). Are there extensions of thick filaments to the Z-line in vertebrate and invertebrate striated muscle? J. Ultrast. Res. 60: 263-271.
Walcott, B. and Ridgway, E.B. (1967). The ultrastructure of myosin-extracted striated muscle fibers. Amer. Zool. 7: 499-504.
Wang, K., McClure, J. and Tu, A. (1979). Titin: major myofibrillar components of striated muscle. Proc. Nat. Acad. Sci. USA 76 (8): 3698-3702.
Wang, K. and Ramirez-Mitchell, R. (1979). Titin: possible candidate as components of putative longitudinal filaments in striated muscle. J. Cell Biol. 83: 2128a.
Wang, K. and Williamson, C.L. (1980). Identification of an N_2 line protein of striated muscle. Proc. Nat. Acad. Sci. USA 77: 3254-3258.
White, D.C.S. (1967). D. Phil. Thesis. Oxford University.
White, D.C.S., Wilson, M.G.A. and Thorson, J. (1978). What does relaxed insect flight muscle tell us about the mechanism of active contraction? In: *Cross-Bridge Mechanism in Muscle Contraction*. Eds., Sugi, H. and Pollack, G.H. University of Tokyo Press, Japan.

DISCUSSION

GODT: Do you think the sort of icicle-like strut things could be the titin that Dr. Wang sees?

MAGID: As I've discussed with Dr. Wang this morning at breakfast, the question of which proteins are where in this model is an entirely open one. As far as what proteins might make up the struts, I really have no idea whatsoever. They could be very large proteins, considering their size. They're about 75 Å in diameter and about, let us say, 400 Å long. If that's a single protein that would be on the order of 500,000. But I should say that I have no idea what protein might make up this structure.

GODT: My suggestion was that if you, in fact, could demonstrate that they were titin you could call them "stalactitin."

(Laughter)

MAGID: In deciding which direction to put the slides in the gate of the projector, I had to decide between stalactites and stalagmites. I'm not enough of a geologist to know which is which.

WANG: At the risk of getting in trouble -- I think the question here is really whether the titin is inside the thick filaments or outside. Now suppose it's on the outside. Any structural features based on titin could then be what you've seen there, struts for example. But another possible interpretation of this is that A segment isolation involves breakage of some of these connecting filaments. Some of the debris may then accumulate along the length of the thick filaments, either on the zones which we've seen or somewhere else, and now will give you a much bigger structure there.

MAGID: Like those putative struts? Well, let me make plain that those images come *not* from enzymatically isolated A segments but from A segments isolated just by simply stretching a living fiber or a freshly skinned fiber to non-overlap and then fixing it at once. So I think the chances for degradation are very small.

WANG: But the sample has been fractured. I think the point I'm trying to make is any shear forces are generating a lot of trouble. This has shown up in our frozen section studies. If you freeze the tissue and section it, you can see changes occurring.

MAGID: That's true in any new technique, just as the knife passing through an embedded specimen distorts it. One has to be concerned, and indeed one *sees*, that the fracture (this is a fracture at about -150°C) is definitely going to fracture the ice and fracture the specimen and, for the few microseconds that the fracture surfaces are pulling apart, distort the specimen in some way. I think some of the distortion that I see results from the fracture process. I think the fracture itself represents a very crude mechanical experiment, not easy to interpret.

REEDY: I'd like to address an important question, one that seemed important to me for many months as I confronted this story of yours, and ask you to respond to something. Many of us are familiar with preparing stretched fibers and myofibrils from psoas muscle for class work and various demonstrations, and we get a rigor myofibril that's stretched well beyond the point at which you show elasticity beginning in stretched skinned fibers. Yet those free isolated fibrils retain the same sarcomere length that we first imposed on them. In the light of such retracting forces, how do you explain that?

MAGID: I don't have direct data on the question. Perhaps I could see the fourth to the last slide (Fig. 7a). This will show a non-overlap fibril.

What Dr. Reedy is getting at -- one would expect that if you stretched an elastic structure, you couldn't obtain an image like this; you couldn't have a free myofibril that was in non-overlap without something holding it stretched. Indeed, when I first set out to get this I naturally knew it was impossible to do it, but I went after it nonetheless. What must be happening in order to have something like this (this non-overlap fibril was suspended in buffer before becoming stuck to the coverslip) is that the connecting filaments must have lost a lot of their passive tension. These are made from frog semitendinosus muscles that have been incubated overnight in a Triton-rigor buffer after setting the sarcomere length to non-overlap. When one goes ahead and tries to make myofibrils from these, in a substantial number of experiments -- I would say more than 80% -- one gets results something like those in Fig. 7a. On the other hand, in some experiments, especially in preparations made after only a few hours in Triton, most of the fibrils *don't* look like this but look like fibrils that have been slammed back together by the passive tension, where the I-bands are caught up against the A-bands or the thin filaments splay out to the side. They do not re-enter the A-bands far because of rigor. But they do express some passive tension. So I think that in the stretched, glycerinated psoas, during glycerination, during storage, this stuff being very prone to proteolysis, become broken. It's possible that they get nicked on a molecular level and lose their passive tension, whereas enough structure survives to maintain some continuity.

REEDY: If you were to stretch a whole fiber on a transducer to non-overlap and then put it into rigor and release it, what happens to the sarcomere length?

MAGID: The fiber reshortens. With a laser you can see the sarcomeres shorten, which is a surprising thing in the absence of ATP; of course, we know it is forbidden. However, it *does* occur. EMs of insect show that what is happening is what might be expected. The passive tension is causing the sarcomeres to shorten all right, but the thin filaments do not re-enter the A-band; the filaments do not slide. They pile up at the edge of the A-band and splay out to the side of the fibril. So the passive tension *can cause* sarcomere shortening in the absence of filament sliding. That phenomenon may be something that some of you may have found in your work that you ought to be aware of. It was brought to my attention by, I should say, Roger Goody and Waltraud Hoffman, and the Reedy's and we studied the phenomenon in some detail last summer.

REEDY: It was first noted by someone in Hungary ---

MAGID: -- oh, that's right, Trombitas and Tigyi-Sebes (In: *Insect Flight Muscle* ed. R. Tregear, North-Holland, Amsterdam, 1977)

REEDY: Yes, in the red book on insect muscle. They stretched insect fibers way out, put them into rigor and the I-bands collapsed upon release, shortening the sarcomeres without filament sliding.

MAGID: We became aware of that result after we'd done all the experiments.

MAUGHAN: One of the most solid-seeming pieces of evidence you present for the connecting third filaments is that in over-stretched fibers the pronounced irregularity of the A-band matches almost perfectly that of the I-band. I was wondering if you could comment on this and also if you could comment on whether you found conditions where the irregularities *don't* match up, and if you have independent evidence under these conditions that the connecting filament is missing.

MAGID: I don't. The slide which I did not show to which Dr. Maughan is referring is a detail (Fig. 3b) from the picture I showed you of the non-overlap frog fiber where you can very often see places where the Z line has a characteristic wiggle in it, and the edge of the A-band next to it has a characteristic wiggle in it, and the two wiggles are almost superimposeable. The most direct interpretation of that, of course, is that the Z line is attached to the edge of the A band. Why these misalignments between thick filaments *should* occur -- why *does* it occur I can't say for sure, but one possibility is that in the course of isolation of the fiber or in the course of life of the animal from which the fiber was taken some of the connecting filaments were proteolysed. If you recall my model for the A-band, thick filaments are lashed together sideways, so there's no need for each and every one of them to have connecting filaments attached to the Z line. If some of them are broken, when you pull, then the thick filaments will naturally experience an unequal force. They will *pretend* to pull out of the A band towards the Z line that's pulling on them, yet they'll be *held* back in the A-band by the radial structures that bundle the thick filament together. I think that kind of phenomenon is a possible basis of this staggering of filaments in the A-band and why the Z line and A-band match up very well.

TREGEAR: As you've already mentioned, Trombitas has looked at similar sorts of observations in insect muscle. He found filaments very nicely. It was pointed out that the nice images were only gotten when the muscle was pulled on, when the whole thing was strained. The question which Pringle raised, and which was never satisfactorily settled, was whether these structures -- these connecting filaments -- were always there in an organized form or whether they self-organized when you pulled the muscle. I think that's a question I hope you will address, because though it may sound absurd, it's not.

MAGID: My belief is that, in order for strain to produce stress, there has to be something there.

TREGEAR: I think you misunderstood the nature of the comment that Pringle made, which was *not* that there wasn't any material there; indeed, there must be a gel there. The issue is whether the gel becomes orientated as a result of the strain. If that were true, then the observations would have to slightly differently interpreted. You wouldn't be looking for the same thing.

MAGID: I think the answer comes from an experiment that Jean Hanson published in 1956 using fly myofibrils. She found that, under her conditions of isolation, the thin filaments often would break off from the Z lines, and in the gap between the broken I band and the Z line was found some birefringent material. I imagine that what she was seeing there were the connecting filaments that would normally go into the A band.

But the birefringence was present in this unloaded case, where fibrils settle from suspension onto a coverslip. So I would say that directly shows that in unfixed material, there is some oriented filamentous substance in that position.

HUXLEY: There is a rather strange fact that it's very easy to stretch semitendinosis fibers so that there's no overlap, whereas it's difficult to stretch sartorii beyond about three microns. I wonder if either you or Professor Wang have any data on the relative amounts of these additional proteins that might be connected with that difference.

WANG: We've looked at gels in frog sartorius muscle in conjunction with phosphorylation studies. The banding pattern as well as the amount is the same as the rabbit and chicken. But we never looked at semitendinosus.

MAGID: Dr. Huxley, I'd like to remark on that. I'm naturally concerned about the differences in passive stiffness between different fiber types. Of course, there are *striking* differences, say, in insect flight muscle, comparing that to semitendinosus muscle. I think the connecting filament model gives an easy approach to thinking about this. The differences in stiffness could arise for two reasons. One is that the connecting filament substance has a different modulus in different cells, -- I think that's important; I also think that something else is important, the idea of sarcomere proportion.. In sarcomeres where the A-band is long relative to rest sarcomere length, we should expect short connecting filaments. This is like the case of insect flight muscle. For a given degree of sarcomere strain, connecting filament strain (and therefore passive tension) will be greater than in muscles which have relatively long I-bands at slack length, that is, long connecting filaments. Comparing frog semitendinosus and sartorius, slack sarcomere length is about 2.2 in the former whereas it is more like 2.0 in the latter. On this basis, the connecting-filament model would predict that sartorius should be stiffer, which it is.

POLLACK: Al, the basic difference between your approach and Dr. Wang's is that you propose that the filament is in the core and Dr. Wang proposes the filament runs basically along the outside of the thick filament. Are these the essential differences?

MAGID: I would say that Dr. Wang's model is -- I haven't had a chance to study it in printed form -- *very* much more complicated than mine in many respects, and I don't think that it really warrants making detailed comparison. I think he is saying that something wraps around the outside of the thick filament. Although I did not show the results I have done experiments using myosin solvents of graded ionic strength on non-overlap skinned fibers. These produce a graded drop in pasive tension, a graded increase in extensibility, coupled with graded removal of myosin from the edge of the A-band. EM images of these partially extracted thick filaments show the "core" filaments emerging, in every case, from the center of the filament stub, never from one edge. Also, I find it easier to imagine that thick filament reconstitution, of the kind Arthur Rowe showed, might occur by recoating a naked core filament than by myosin somehow reinsinuating itself within a wraparound structure. However, Dr.

Wang and I agree that a definitive experiment to distinguish a core model from a wraparound model would be to see if these treatments produce a similarly-graded appearance of antibody labeling using antisera to putative core filament subunits.

THE COMPOSITION OF THE INTRACELLULAR MILIEU

INTRODUCTION

The function of the contractile or regulatory proteins depends critically on their chemical environment. Consequently, there are a variety of experiments under way that are designed to measure the exact composition of muscle cytosol. Ionic H, Ca, Mg and MgATP, for example, are particularly important quantities inasmuch as these ions, as mediators and substrate, profoundly affect the rate and magnitude of force developed by muscle fibers. *In vivo* measurements are important, especially in trying to distinguish pathological and normal states and in seeking the causes for particular myopathies.

A knowledge of concentrations and distributions of cytoplasmic elements and their ionic forms is also important to those who design experiments using skinned muscle fiber preparations with removed or disrupted sarcolemma. An important practical question is: What constitutes an *in vivo*-like chemical environment and what concentrations of salts should be used in mixing the solutions which bathe the skinned fiber? Natori faced this problem when he invented the skinned fiber preparation thirty years ago. Since then there have been many, often elegant, refinements of the bathing solutions (some of the participants in this symposium have contributed to these refinements). There are still, however, a number of uncertainties, some of which are addressed by the following papers.

In the first of these presentations, Professor Wilkie and his collaborators assess the concentrations of phosphorus-containing compounds using ^{31}P-nuclear magnetic resonance. The application of NMR to muscle physiology is relatively recent, but its spectacular non-invasive results makes it a widely appreciated research tool. Wilkie describes an extension of the application of their NMR techniques from normal to diseased intact muscle from humans.

Intracellular pH is relevant to questions about energy available to the contractile system because intracellular pH affects the amount of energy produced by the chemical reactions. Dr. Curtin, in her talk, discusses the measurement of pH by means of the ion-selective electrode. A quarter of the energy in a contraction comes from as yet unidentified sources. Estimation of intracellular pH under a variety of conditions that affect mechanical response and heat output of the muscle may therefore shed additional light on the chemical reactions that are involved in muscle contraction.

Dr. Dewey and his collaborators are one of several groups attempting to study, by means of microelectrodes, the fixed charges and the distribution of ions associated with fixed charges within muscle. Dewey describes some experiments designed to probe the A- and I-bands of glycerinated (functionally skinned) muscle, both relaxed and in rigor. The changes in Donnan potential observed in *Limulus* muscle as sarcomeres shorten are a clear indication of a decrease in charge with shortening sarcomeres; these changes are not, however, seen in frog muscle. They postulate that the change in potential could represent either exposure of fixed charge groups with shortening or adsorption of charge.

I introduce a new method of estimating the concentration of diffusible elements in muscle cytoplasm. Tiny liquid droplets are equilibrated against fibers freshly skinned under oil, and the concentrations of the elements diffusing into the droplet from the fiber fluid are determined by electron probe microanalysis of the liquid samples. Magnesium, calcium and phosphorus are three examples of many elements that can be analyzed using this sampling and X-ray spectroscopic method.

Dr. Iwazumi's talk reminds us once again that the ionic environment surrounding the contractile and regulatory proteins is not homogeneous, but a dynamic, heterogeneous soup whose various constituents, fixed and mobile, interact with one another. Indeed, the definition of such terms as "concentration" and "mobility" become blurred in the microenvironment. Iwazumi attempts to come to grips with the collective evidence bearing on the spatial charge distribution in muscle, presenting his view of how mobile ions are partitioned in the intracellular fluid medium.

Aside from the ponderous theoretical questions, those who work with the skinned fiber preparation are faced with an immediate, practical problem. Assuming some success in measuring *in vivo*, average values of diffusible elements and their moieties (by, e.g., the NMR or electron probe techniques), an important question is: What are the concentrations to be used for different types of experiments? Take magnesium as an example. In a recent paper (Biophys. J. 35: 385-392, 1981), Dr. Robert Godt proposed that, because of the influence of the electrostatic field associated with the net negative charge on the myofilaments (at neutral intracellular pH), the concentration of divalent cations like Mg^{2+} in the immediate vicinity of the filaments is approximately 25 times greater than the concentration in the bulk fluid phase between the filaments. Thus, if one wants to study the force-pCa curve in skinned fibers, what is the concentration of Mg^{2+} to be used in the bathing media? If, in intact fibers, the free Mg^{2+} concentration in the vicinity of the myofilaments is around 25×0.2 mM = 5 mM as the Godt calculation and my results (presented elsewhere at this symposium) suggest, then is 5 mM the relevant free Mg^{2+} concentration, and 0.2 mM the one to be used in solution? The same sort of questions exist for other ions, including Ca^{2+}. Issues of this type are of profound importance to the many practitioners of skinned fiber research.

-David Maughan

^{31}P NMR STUDIES OF RESTING MUSCLE IN NORMAL HUMAN SUBJECTS

D.R. Wilkie, M.J. Dawson, R.H.T. Edwards, R.E. Gordon*, and D. Shaw*

University College London, School of Medicine, London, England
**Oxford Research Systems, Oxford, England*

ABSTRACT

Study of human tissues using ^{31}P Topical Magnetic Resonance is completely atraumatic; it allows simultaneous measurement of the concentrations of many important metabolites and of intracellular pH. In some critical situations, TMR yields more accurate results than those obtained by chemical analysis of tissue biopsies. We have shown that TMR can be calibrated to obtain *quantitative* measurements in human subjects. We have also shown that theories of control of glycolysis based on regulation by key metabolites of rate-limiting enzymes are inconsistent with the observed changes in intact muscle.

Recent technical advances employing wide-bore superconducting magnets with a localized sensitive volume have made it possible to obtain high quality nuclear magnetic resonance (NMR) spectra from selected areas in intact human subjects; this variation of NMR has been given the name "Topical Magnetic Resonance" (TMR), after the Greek word *τοποσ* meaning "a place" (Gordon, Hanley, Shaw, Gadian, Radda, Styles, Bore & Chan, 1980). The technique has already been employed in studies of the limb muscles of normal subjects and patients with muscle disorders (reviewed by Edwards, Dawson, Wilkie, Gordon & Shaw, 1982a). However, most published studies so far have been qualitative in nature and the false impression has been given that TMR is unsuitable for *quantitative* analysis of metabolite concentrations.

By an extension of methods we devised some years ago for work on isolated muscles (Dawson, Gadian & Wilkie, 1977) we have validated the use of TMR for noninvasive measurement of *absolute* concentrations of energetically important phosphorus metabolites (P-metabolites) and pH in muscles of the forearm and lower leg of normal subjects. These studies

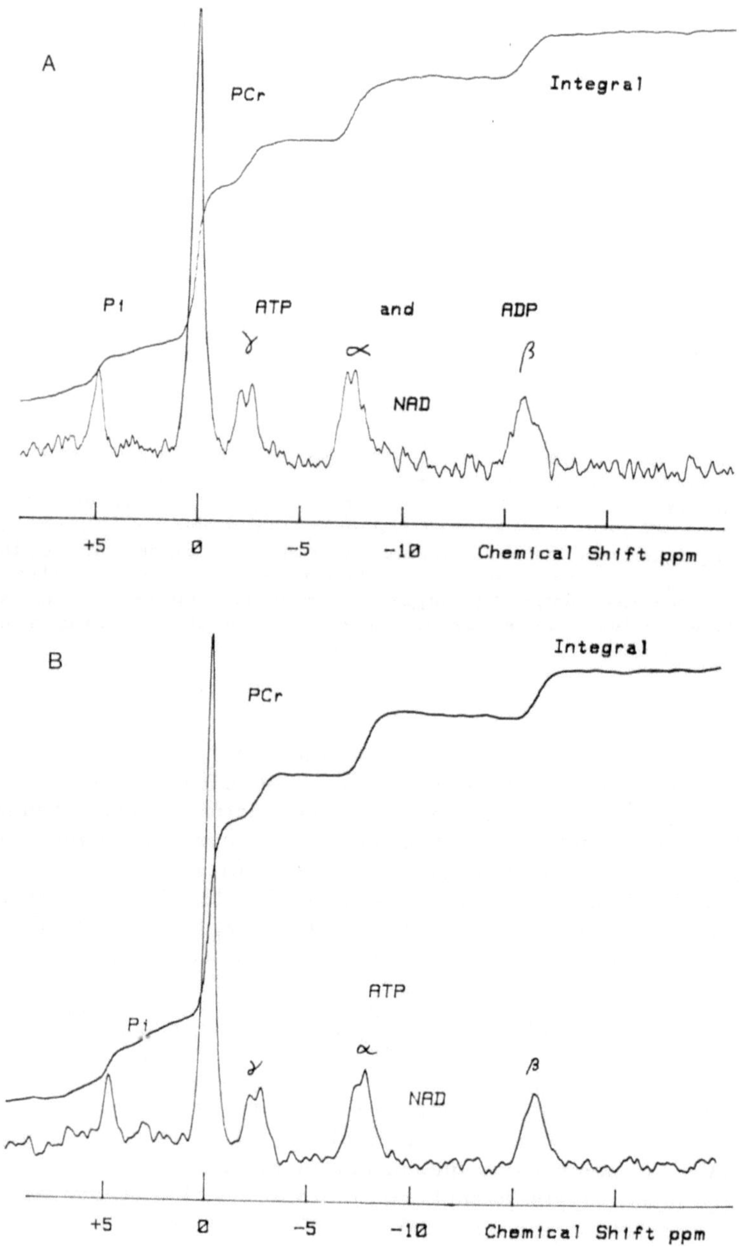
A
PCr
Integral
Pi
ATP
and
ADP
γ
α
β
NAD
+5
0
-5
-10
Chemical Shift ppm
B
Integral
PCr
ATP
Pi
γ
α
β
NAD
+5
0
-5
-10
Chemical Shift ppm

show that in some critical instances TMR yields more accurate results than those obtained by even the best means of sampling muscle metabolite concentrations now available, the needle biopsy method, and have pointed out the need for re-examination of many of the biochemical results obtained on human tissues by conventional methods. We have found that present theories of the control of glycolysis based on the regulation of rate-limiting enzymes by key P-metabolites (Newsholme & Start, 1973; Atkinson, 1977) do not account for the biochemical behaviour of intact muscle. This conclusion also calls into question current concepts of the control of glycolysis in other tissues.

The Technique

The NMR technique requires that the object of interest be placed in a powerful and very uniform magnetic field (B_0). At suitable intervals a brief pulse of radio frequency magnetic field (20-80 μsec at 32.5 MHz in the present experiments), is applied at right angles to B_0. This field (B_1) excites resonance in the atomic nuclei of ^{31}P, which subsequently emit a weak radiofrequency signal for a few msec. The precise frequency of the signal depends on the chemical environment of the various nuclei, leading to the "chemical shift" shown along the x-axis of Fig. 1; this makes it possible to identify the various compounds present. TMR differs from conventional NMR in that the magnetic field is shaped so that it is uniform within a localized "sensitive volume" outside which it changes very rapidly. The TMR spectrometer employed in this work was designed and built by Oxford Research Systems.

Under suitable conditions the areas of the resonances obtained in NMR spectra (*not* the peak heights, if the shapes of the peaks differ) are directly proportional to the concentration of compounds from which they are derived. However, this proportionality cannot be taken for granted in biological experiments. Two new problems arise in TMR studies that have not been encountered before: 1) In order to produce spectra that lend themselves to quantitative interpretation it is essential to apply spectral

Figure 1: Spectra obtained from a 20 cm-bore magnet with a B_0 field of 1.89 tesla, and containing profiling coils which, together with additional field shaping by a surface B_1 coil (4 cm in diameter) defines a sensitive volume of approximately 22 cm^3. The x-axis is frequency in p.p.m. (parts per million deviation from reference frequency); the y-axis is signal intensity. The areas under the peaks are related to the concentrations of metabolites present, as described in the text. (A) The subject's arm was located by a limb-holder and grip force transducer (two duralumin cylindrical pillars, 25 mm diameter, mounted with their axes 44 mm apart) in such a way that the sensitive volume of the spectrometer was 6 cm distal to the ulnar tuberosity and sampled the muscular parts of the flexor carpi radialis and palmaris longus. Special care must be taken in constructing the limb support because of the physiological necessity not to apply pressure to the nerves, arteries or veins that service the limb. The spectrum was obtained in approximately 27 min (800 pulses of 80 μs duration at 2 s intervals), and was enhanced by 10 Hz line broadening and convolution differencing. October 8, 1980. Subject M.J.D. (B) Lower leg of the same subject. 22 cm^3 volume at fleshiest part of gastrocnemius muscle. Spectrum was accumulated in approximately 34 min (1024 pulses of 45 μs duration at 2 s intervals), and was enhanced by 10 Hz line broadening. Subject M.J.D.

enhancement techniques. We have used convolution differencing to minimize the peak-broadening effects of the inhomogeneities of the B_0 and B_1 fields and of P in bone, which because it is bound, gives a wide peak (Campbell, Dobson, Williams & Xavier, 1973). 2) Because we are dealing with an intact human limb, and not merely an isolated muscle, we could not assume that muscle is the only tissue that contributes significantly to the spectra.

Some caution is therefore required in approaching the problem of absolute calibration.

Calibration of the Resting Spectrum

Fig. 1 shows ^{31}P TMR spectra obtained from a resting forearm (A) and gastrocnemius muscle (B) of the same subject. Clearly-resolved peaks can be observed, indicating the presence of Pi, PCr and ATP. The β and γ peaks do not differ significantly (P=0.85) confirming that the concentration of *free* ADP, which lacks the β resonance, is small so we have taken their mean as the best estimate of [ATP]. NAD appears as a highly significant shoulder (t=10, P <<< 0.0005) to the right-hand side of the α-adenosine peak. The characteristic fine structure of the ATP peaks, doubling of the α- and γ-phosphorous peaks and tripling of the β-peak is evident. This fine structure is always observed in solution but is rarely discernable in an intact biological preparation. The hexose and triose phosphates, which resonate together at + 7.5 ppm at physiological pH, are too low in concentration to be observed. Although it is not certain from spectrum 1A alone, the summed results on seven spectra show a significant (P < .0005) peak at + 3.5 ppm, the phosphodiester region (labelled PdiE in Table 1), which has been shown to have special interest for the study of muscle disease (Chalovich, Burt, Danon, Glonek & Bárány, 1979). Spectrum 1B of the gastrocnemius muscle is very similar to the forearm spectrum (1A), but shows a more pronounced peak in the phosphodiester region.

The 2 s pulse interval used in these experiments does not allow complete spin lattice (T_1) relaxation between successive pulses and we have corrected for the differential loss of signal that results. This effect was evaluated in separate experiments in which the peak areas for PCr and Pi were compared to that of ATP in spectra obtained with 2 or 20 s pulse intervals. Such comparison yields an estimate of the "saturation factors" (1.43 for PCr and 1.30 for Pi) by which the respective peak areas must be multiplied in order that the result be proportional to metabolite concentrations. A check on the accuracy of the saturation factors determined for PCr and Pi is shown in Fig. 2, which illustrates the effect of a 59 min period of ischaemia, followed by 30 min of recovery. After correction for saturation, the PCr and Pi peak areas were converted to concentrations by the method detailed below. At no time does the estimated change in [PCr] differ significantly from the opposite change in [Pi]. Since the changes are large and since the only source of Pi in this experiment is PCr hydrolysis, this result illustrates that our calculations of the *relative* concentrations of these two compounds must be approximately correct.

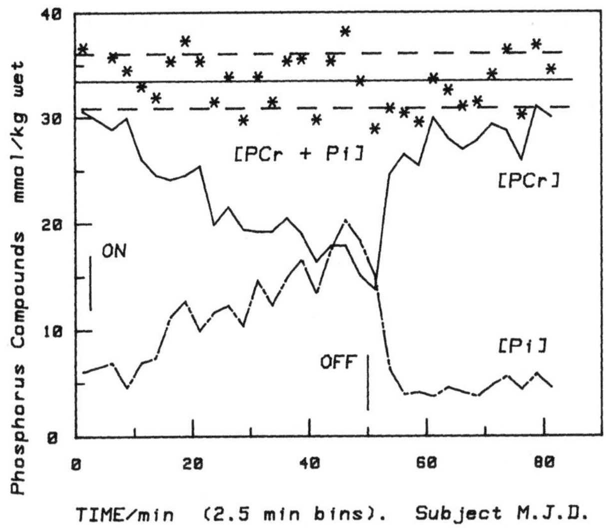

Figure 2: Changes in PCr (solid line) and Pi (interrupted line) concentrations during and following 59 min of ischaemia in the right arm. A sphygmomanometer cuff was blown up very rapidly to 300 mm Hg (subject's systolic pressure is 120 mm Hg) using a pressurized nitrogen cylinder. This procedure is important to avoid venous distension and petechial haemmorhages which occur otherwise. The pressure was then lowered to 180 mm Hg so as to avoid direct pressure damage to the nerves.

Spectra were averaged over 2.5 min periods and concentrations were calculated by the "ATP standard" method described in the text. The sum ([PCr] + [Pi]) is denoted by *; the horizontal solid and dashed lines are the mean and SD. Correlation between ([PCr] + [Pi]) and time during the 59 min of occlusion was not significant ($P > 0.05$) and the Run Test (Bendat & Piersol, 1971) was used to determine that there is indeed no trend indicating either an increase or a decrease in ([PCr] + [Pi]) at any time over the 90 min of the experiment. Subject M.J.D.

We conclude that neither saturation nor spectral enhancement introduces detectable errors into our estimates of the relative concentrations of these two metabolites.

We must now ask what contribution could other tissues (skin, fat, blood, bone) or variations in blood flow, and especially blood volume, make to the results that we describe? The contribution from skin is negligible, since it represents a maximum of about 8% of the total sample volume and contains only 1-4 mmol kg^{-1} of P in all forms (Spector, 1956). The ^{31}P TMR spectrum of bone contains only a single very broad peak derived from bound P; this is eliminated by convolution differencing. No correlation was found in our experiments between skinfold thickness (a rough measure of surface fat) and the observed variations in phosphorus signal. The presence of 2, 3 diphosphoglycerate resonating at + 5.95 and + 6.65 ppm serves as a marker for red blood cells (Henderson, Costello & Omachi, 1974); no trace of this marker is observable in our spectra, indicating that the contribution of blood to any of our spectral peaks is below the limit of resolution (<0.5 mmol l^{-1}). We have also undertaken separate plethysmographic experiments which showed that the changes in blood volume during the procedure shown in Fig. 2 were extremely small and that even when large changes in blood volume were induced by first emptying and then engorging the arm, no differences could be detected in the ^{31}P spectrum.

Thus in spite of our best efforts we are unable to detect a contribution to our spectra which can be attributed to other tissues or to alterations outside the muscle tissue itself. For the purpose of further analysis we must therefore regard the relative peak areas, after correction for saturation, as a measure of the relative concentrations of the P-metabolites in muscle. The accuracy of this assumption is tested further in the next section.

Phosphorus Metabolites in Resting Muscle

Absolution calibration must rest ultimately upon comparison of peak areas with the results of chemical analysis of muscle extracts, exploiting the fact that both techniques have their unavoidable but complementary uncertainties. There are at least two ways of making this comparison. They are independent of one another; they yield essentially the same results.

The most accurate chemical measurements now available for human muscle metabolites have been obtained on needle biopsy samples (Edwards, Harris & Jones, 1981); the major source of error associated with this method arises from slow freezing which causes artifactual PCr breakdown. In studies of metabolic changes in frog muscles it has been found that accurate estimate of [PCr] and [Pi] requires a freezing time of 100 msec or less (Kretzschmar & Wilkie, 1969; Kretzschmar, 1970), whereas the time required to freeze the needle biopsy samples is at least 6 seconds, nearly one hundred times too long. For this reason the [PCr], [Cr] and [Pi] determined for human muscle biopsy samples have long been suspect. There is no reason, however, to suspect measurements of the *sum* of the P-containing metabolites, or [ATP] determined by biopsy in non-fatigued muscle; ATP is in equilibrium with PCr via the creatine kinase reaction (Gadian, Radda, Brown, Chance, Dawson and Wilkie, 1981), the equilibrium constant being such that it ensures great stability of [ATP] except in extreme fatigue.

Column 1 of Table 1 shows the best chemical estimates available of the concentrations of P-metabolites contained in extracts from biopsy material. In column 2 of Table 1 we have set the ATP-peak area in the TMR spectrum equal to the [ATP] as determined by independent needle biopsies and have calculated the concentrations represented by the other peak areas accordingly. The basis of the second method of calibration is to set the total area of the TMR spectrum (after appropriate correction for saturation effects) equal to the mobile P determined by chemical analysis (i.e. 3 × [ATP] + [PCr + Pi] + 2 × [NAD = NADH] = 49.5 mmol/kg wet). These results are shown in column 3 of Table 1.

Table 1: Composition of resting human muscle

[P-metabolite]	Biopsy mean±SE(n) mmol kg^{-1}wet	^{31}P TMR [ATP] assumed constant at 5.5 mmol kg^{-1} mean±SE(n=7) mmol kg^{-1}wet	^{31}P TMR. Total integral assumed constant at 49.5 mmol kg^{-1} mean±SE(n=7) mmol kg^{-1}wet
ATP	5.5 ± 0.07(81)	≡ 5.5	5.1 ± 0.10
PCr	17.4 ± 0.19(81)	29.0 ± 0.69	27.4 ± 0.23
Pi	10.0 ± ? (3)	4.4 ± 0.33	4.3 ± 0.27
PCr + Cr	28.6 ± 0.28(81)	-	-
PCr + Pi	31.6 ± 3.27(11)	33.4 ± 0.77	31.7 ± 0.29
P di E	-	0.7 ± 0.09	0.8 ± 0.09
NAD + NADH	0.7	1.0 ± 0.09	0.9 ± 0.07

Needle biopsy estimates of [ATP], [PCr], [PCr + Cr] are from (Harris, Hultman & Nordesjo, 1974); [Pi] from (Sahlin, Palmskog & Hultman, 1978). Open biopsy determinations of [PCr + Pi] are from (Chalovich et al., 1979). [NAD + NADH] seems not to have been reported in human biopsies; this value is from fast glycolyzing porcine muscle (Kastenschmidt, 1970).

Table 1 shows that for our seven forearm spectra there is close agreement on [PCr + Pi] as determined by chemical analysis or by either of our TMR calibration methods. The reliable biopsy result [PCr + Cr] has been included to show that it is not significantly exceeded by the TMR estimates of [PCr], which would have shed doubt on the latter. The important differences to be noted in Table 1 are that, compared with the TMR results, chemical estimates of [PCr] are too low and those of [Pi] and [Cr] too high. There is also a significant difference ($0.01 > P > 0.005$) between [ATP] estimated chemically and by the "total phosphorus" TMR technique, when tested by a "strict" t-test (Diem, 1962) that takes appropriate account of the differences in sample size and standard deviation. Presumably the difference is genuine, but it is small and does not alter our conclusions. The good agreement between the two TMR calibration method tends to verify the underlying assumptions of both.

Muscle pH

One of the great advantages of the nuclear magnetic resonance technique is that the Pi peak position is dependent upon pH, and after appropriate calibration allows determination of intracellular pH. The chemical shift of Pi in relation to PCr was + 4.84 ± 0.017 ppm (SE) in the seven forearm spectra, leading to an estimate of the pH of resting forearm muscle of 7.08 ± 0.036 (SE, n=7). A similar value, 7.15, was obtained in one gastrocnemius spectrum. This agrees well with the value of 7.08 + 0.03 (SD, n = 12) obtained on human muscle homogenates (Edwards et al., 1981).

We conclude that our estimates of P-metabolites and pH by TMR agree well with those by needle biopsy where such agreement can be expected, and that the disagreement between the two techniques results from artifacts associated with the needle biopsy method. Thus TMR offers powerful advantages for many types of study though we are fully aware that the methods of absolute calibration described above cannot be used uncritically in studies of fatigued or diseased muscle (Edwards, Wilkie, Dawson, Gordon & Shaw, 1982b; Edwards et al., 1982a). We are at present devising methods of using ^{1}H and ^{13}C in conjunction with ^{31}P for absolute calibration in cases of diseases, such as Duchenne dystrophy, in which there is replacement of muscle by fatty-fibrous tissue.

Importance of Correct [PCr], [Cr] and [Pi] Measurements

The errors in measurement by needle biopsy method are not merely of importance in themselves, but also lead to gross errors in estimation of free [ADP] (proportional to [Cr]/[PCr]); [AMP] (proportional to $([Cr]/[PCr])^2$) and the free-energy change for ATP hydrolysis. At present these quantities can only be determined by calculation from the ratios PCr/Cr or PCr/(Pi × Cr). The unfortunate mistake is often made by biochemists that [ADP] or [AMP] determined in tissue extracts are reliable guides to the availability of these substances to take part in metabolic reactions; thus incorrect assumptions are made concerning the control of key enzymes *in vivo* (Wilkie, 1981a). A failure to realize that almost all of the ADP measured chemically after perchloric acid extraction has been stripped from known binding sites, e.g. actin and myosin, has lead to spurious arguments in favour of metabolic compartmentation in skeletal muscle (Wilkie, 1981b) and in the heart (Dawson, 1983).

We have published estimates of free [ADP] and free-energy change for ATP hydrolysis in frog muscle (Dawson, Gadian & Wilkie, 1978; Dawson, Gadian & Wilkie, 1980). However, we have not attempted to determine these quantities in the present study because it would require biopsy estimates of total [PCr + Cr] from the same muscles. Biopsy studies on the forearms of normal subjects are widely held to be unethical.

Glycolysis in Resting Ischaemic Muscles

Although the experiment shown in Fig. 2 was done primarily to check the calibration procedure, it also yields useful biological results. The changes in pH during the ischaemic period were small but highly significant ($r = 0.613$, $n = 23$, $0.01 > P > 0.001$), and were in the *alkaline* direction; this is shown in Fig. 3A. Since this anaerobic muscle shows no significant change in the concentrations of sugar P involved in glycolysis, it is reasonable to believe that only two *net* changes of quantitative significance can be occurring: hydrolysis of PCr (proton-absorbing) and build-up of lactic acid (proton-producing) at the expense of glycogen

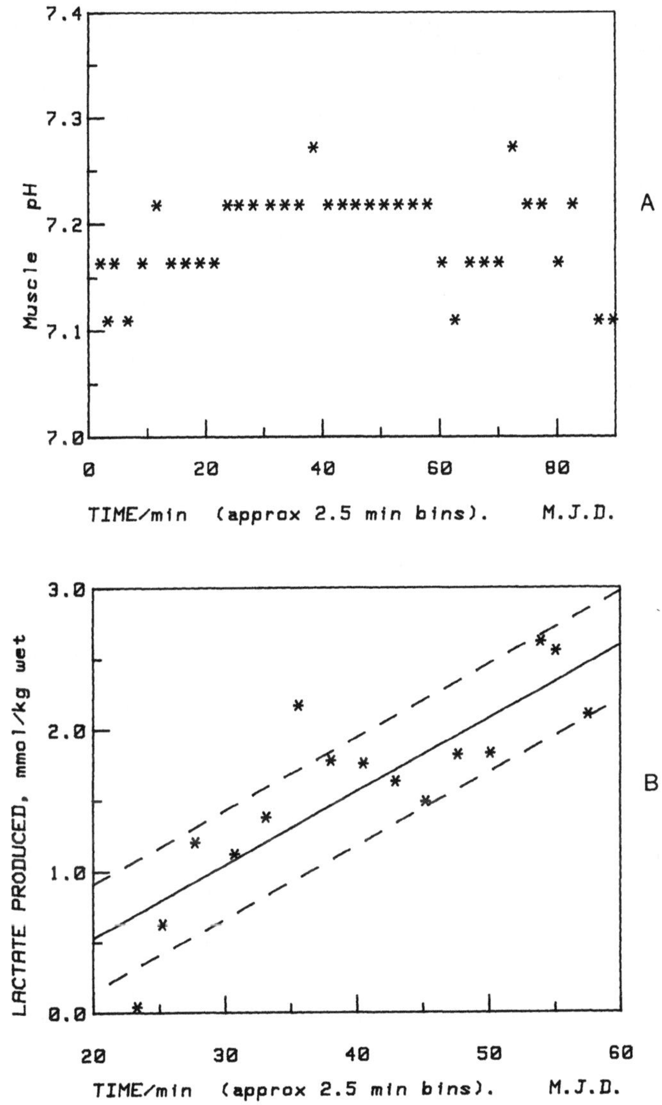

Figure 3: Changes in pH (A) and [Lactic Acid] (B) during the same experiment as shown in Fig. 2. (A): pH=pK+log $\frac{\sigma-3.31}{5.65-\sigma}$ where σ is the chemical shift (in p.p.m.) of Pi with respect to PCr; the chemical shift of $H_2PO_4^-$=+3.31 p.p.m. and of HPO_4^{2-}=+5.65 p.p.m.. Our best estimate of the pK for $H_2PO_4^{1-} \rightleftharpoons HPO_4^{2-}+H^+$ is 6.8. The "stepped" changes in pH arise from the digitization of resonance frequencies. (B) Note that this refers only to the period 20-60 min following application of the sphygmomanometer cuff. The solid line is the regression of [Lactic Acid] upon time [LA^-] = 0.530 + 0.0516 × T; n = 15; r = 0.8243; P << 0.001). The dashed lines are 95% confidence limits.

breakdown. We have previously shown (Dawson et al., 1978) that the amount of lactic acid (LA^-) produced can be calculated from information about changes in $[H^+]$ and [Pi]. During the last 40 min of ischaemia there is no change in $[H^+]$, so the capacity of the fixed buffers can be disregarded, and LA^- build-up can be calculated from the following equation:

$$\Delta[LA^-] = \Delta \left\{ \frac{[Pi][H^+]}{K_{Pi} + [H^+]} \right\} \tag{1}$$

Fig. 3B shows that the rate of LA^- formation is approximately constant over the 40 min interval at about 3 mmol kg^{-1} hr^{-1}, a value which agrees well with chemical analysis during similar periods of ischaemia (Haljamae & Enger, 1975).

Contraction and Recovery

In this experiment the subject maintained a maximal voluntary contraction for 3 min, during which the force declined from 146 N to 55 N. Fig. 4 shows that PCr dropped to less than ½ of its initial level as a result of the contraction and pH fell dramatically, by a full pH unit. Neither pH nor PCr recovered measurably during the post-contractile ischaemic period, a result which is consistent with earlier reports that metabolic recovery does not occur in ischaemic muscle (Harris, Edwards, Hultman, Nordesjo, Nylind & Sahlin, 1976). [Pi] decreased slightly, a change which was mirrored by an approximately equal increase in [sugar P]. When occlusion was terminated, recovery of force development was rapid and approximately exponential, being more than 90% complete within 4 min.

In order to convert ΔpH as a result of contraction into ΔLA^- it is necessary to know what buffers are present. Histidine residues on the proteins in human muscle provide 36 mmol kg^{-1} (Block & Weiss, 1956) with pK=6.5. Since the [carnosine] is low and variable we have neglected it and calculated ΔLA^- from the equation.

$$\Delta LA^{-1} = \Delta \left\{ \frac{36 \times 10^{-3}[H^+]}{3.162 \times 10^{-7} + [H^+]} + \frac{[Pi][H^+]}{1.585 \times 10^{-7} + [H^+]} \right\} \tag{2}$$

The result is that during the 3 min isometric contraction $\Delta LA^- \approx 34$ mmol kg^{-1}. The estimate used in equation 2 that there are approximately 36 mmol kg^{-1} of fixed buffer present with pK ≈ 6.5, accords well with needle biopsy studies in which [Pi], $[LA^-]$ and pH were estimated in human skeletal muscle (ischaemic and exercised)(Sahlin, Harris & Hultman, 1975) and ischaemic dog ventricular muscle (Opie, 1976). Equation (2) may thus be regarded as generally applicable to mammalian striated muscle.

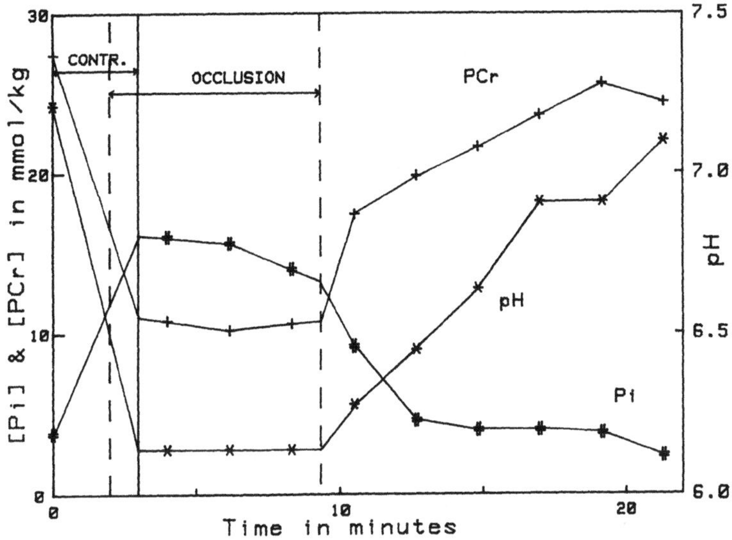

Figure 4: Changes in PCr (+), Pi, (#) and pH (*) as a result of maximum voluntary isometric contraction. After 2 min of contraction a sphygmomanometer cuff was inflated as described in the legend to Fig. 2; contraction was continued for one additional minute and ischaemic conditions were maintained for a further 6 min. This procedure was adopted after preliminary investigations showed that inflating the cuff earlier in contraction yields a lower tension-time integral, presumably as a result of the effect of the pressure on nerves.

Changes in P-metabolites were calculated by assuming the total metabolite phosphorus remains unchanged during the course of the experiment (see text). Spectra were averaged in 2 min bins, beginning when the contraction was terminated. There was no apparent change in [ATP] during the course of this experiment, but there was an approximately 3-fold increase in [sugar P] by the end of the ischaemic period, which did not return appreciably toward normal by the end of the experiment.

When occlusion was terminated recovery of force development was studied by measuring peak force roughly every 20 s in a brief (1-2 s) test contraction. 50% of the initial force was restored within 2 min and more than 90% within 4 min after the restoration of circulation.

Control of Glycolysis

It is widely believed that glycolysis is regulated by [ADP], [AMP] or calculated quantities dependent upon them, such as 'phosphate potential' or 'adenylate charge'; these theories are presented as facts in well-known textbooks (Lehninger, 1975; Stryer, 1981). ADP and AMP are activators of the rate-limiting enzymes, phosphorylase and phosphofructokinase. However, the required concentrations of these substances are dependent upon a number of different factors, and no attempt seems to have been made to reproduce *in vivo* conditions in the *in vitro* experiments on which these biochemical theories were based.

The present results on ischaemic and contracting muscle show that such a regulatory mechanism does not normally operate in human skeletal muscle: in spite of similar changes in [Cr]/[PCr] and thus similar (large) changes in [ADP] and [AMP], the rate of glycolysis initiated by ischaemia is 200 to 300-fold less than that accompanying a maximum voluntary contraction. We therefore conclude that whatever activates glycolysis in contracting muscle is closely associated with the events of the contraction itself, e.g. Ca^{2+} release from the sarcoplasmic reticulum. We have reached a similar conclusion as a result of studies of isolated frog muscle (Dawson, Gadian & Wilkie, 1980).

CONCLUSIONS

We have used ^{31}P TMR as a completely non-invasive and atraumatic method of making accurate measurements of important P-metabolites and intracellular pH in superficial muscles of the limbs of normal human subjects. We are satisfied that at the signal-to-noise ratio achieved, other tissues make a negligible contribution to the ^{31}P TMR spectra obtained by the methods described in this paper and that no serious errors are introduced by our signal-processing methods. We conclude that ^{31}P TMR of intact muscle is a much more reliable method of estimating [PCr] and [Pi] than is needle biopsy followed by extraction and chemical analysis. As a result of these findings, we have stopped using needle biopsy as a means of measuring [P-metabolites] in the muscles of patients or normal subjects, although we continue to use the biopsy method for enzyme analysis and histochemical studies.

Our study of contraction and ischaemia shows that present theories based upon control of glycolysis by changes in [ADP], [AMP] or quantities dependent upon them do not explain the biochemical behaviour of intact muscle. This is the first test in human subjects of these control theories, which are based solely on *in vitro* evidence, and the result has altered our assumptions concerning both normal muscle and the nature of some metabolic disease processes. It remains to be seen whether [AMP] or [ADP] have a regulatory role on glycolytic rate in tissues other than muscle.

Our own studies, and those of others, show that TMR can be used to obtain a characteristic "fingerprint" of muscle metabolic diseases. It may well be that enzyme defects in muscle are more prevalent than has been suspected; complaints of muscle pain are common, but mass screening has not, until now, been possible because of the trauma and expense of conventional biopsy methods. T.H. Huxley (Huxley, 1885) wrote "What an enormous revolution would be made in biology if physics or chemistry could supply the physiologist with a means of making out the molecular structure of living tissues...... At the present moment the constituents of our own bodies are more remote from our ken than those of Sirius, in this respect". That revolution is now upon us.

REFERENCES

Atkinson, D.E. (1977). In: *Cellular Energy Metabolism and its Regulation.* Academic Press. New York.

Bendat, J.S. and Piersol, A.G. (1971). In: *Random Data: Analysis and Measurement Procedures.* Wiley-Inter Science, New York. pp. 122-125.

Block, R.J. and Weiss, K.W. (1956). In: *Amino Acid Handbook: Methods and Results of Protein Analysis.* Charles Thomas. p. 343. Table 5.
Campbell, I.D., Dobson, C.M., Williams, R.J.P. and Xavier, A.V. (1973). Resolution enhancement of protein PMR spectra using the difference between a broadened and a normal spectrum. J. Mag. Res. 11: 172-181.
Chalovich, J.M., Burt, C.T., Danon, M.J., Glonek, T. and Bárány, M. (1979). Phosphodiesters in muscular dystrophies. Ann. N.Y. Acad. Sci. 317: 649-668.
Dawson, M.J., Gadian, D.G. and Wilkie, D.R. (1977). Contraction and recovery of living muscle studied by ^{31}P nuclear magnetic resonance. J. Physiol. 267: 703-735.
Dawson, M.J., Gadian, D.G. and Wilkie, D.R. (1978). Muscular fatigue investigated by phosphorus nuclear magnetic resonance. Nature 274: 861-866.
Dawson, M.J., Gadian, D.G. and Wilkie, D.R. (1980). Studies of the biochemistry of contracting and relaxing muscle by the use of ^{31}P n.m.r. in conjunction with other techniques. Phil. Trans. R. Soc. Lond. B289: 445-455.
Dawson, M.J. (1983). Nuclear magnetic resonance. In: *Cardiac Metabolism.* Ed. A.J. Drake-Holland and M.I.M. Noble. Wiley and Sons Ltd. pp. 309-337.
Diem, K. (1962). *Documenta Geigy Scientific Tables.* 6th edition. Geigy Pharmaceutical Co. Ltd. Manchester. p. 172.
Edwards, R.H.T., Harris, R.C. and Jones, D.A. (1981). The biochemistry of muscle biopsy in man: clinical applications. Recent Adv. in Clin. Biochem. 2: 243-269.
Edwards, R.H.T., Dawson, M.J., Wilkie, D.R., Gordon, R.E. and Shaw, D. (1982a). Clinical use of nuclear magnetic resonance in the investigation of myopathy. Lancet, March 27, 1982, pp. 725-731.
Edwards, R.H.T., Wilkie, D.R., Dawson, M.J., Gordon, R.E. and Shaw, D. (1982b). Measurement of muscle pH and intermediary metabolism by ^{31}P topical magnetic resonance (TMR) in normal subjects and patients with myopathy. *European Society for Clinical Investigation.* Annual Meeting Luxembourg, 15-17 April, 1982.
Gadian, D.G., Radda, G.K., Brown, T.R., Chance, E.M., Dawson, M.J. and Wilkie, D.R. (1981). The activity of creatine kinase in frog skeletal muscle studied by saturation-transfer nuclear magnetic resonance. Biochem J. 194: 215-228.
Gordon, R.E., Hanley, P.E., Shaw, D., Gadian, D.G., Radda, G.K., Styles, P., Bore, P.J. and Chan, L. (1980). Localization of metabolites in animals using ^{31}P topical magnetic resonance. Nature 287: 736-738.
Haljamae, H. and Enger, E. (1975). Human skeletal muscle energy metabolism during and after complete tourniquet ischemia. Ann. Surg. 182: 9-14.
Harris, R.C., Hultman, E., and Nordesjö, L.-O. (1974). Glycogen, glycolytic intermediates and high-energy phosphates determined in biopsy samples of musculus quadriceps femoris of man at rest. Methods and variance of values. Scand. J. Clin. Lab. Invest. 33: 109-120.
Harris, R.C., Edwards, R.H.T., Hultman, E., Nordesjö, L-O., Nylind, B. and Sahlin, K. (1976). The time course of phosphorylcreatine resynthesis during recovery of the quadriceps muscle in man. Pflugers Archiv. 367: 137-142.
Henderson, T.O., Costello, A.J.R. and Omachi, A. (1974). Phosphate metabolism in intact human erythrocytes: determination by phosphorus-31 nuclear magnetic resonance spectroscopy. Proc. Nat. Acad. Sci. 71: 2487-2490.
Huxley, T.H. (1885). Presidential address. Proc. Roy. Soc. 39: 294.
Kastenschmidt, L.L. (1970). The metabolism of muscle as a food. *Physiology and Biochemistry of Muscle as a Food.* E. Briskey, Cassens, Marsh. Univ. of Wisconsin Press. Vol. 2, pp. 735-753.
Kretzschmar, K.M. and Wilkie, D.R. (1969). A new approach to freezing tissues rapidly. J. Physiol. 202: 66-67P.
Kretzschmar, K.M. (1970). Energy production and chemical change during muscular contraction. Ph.D. Thesis. Univ. of London. p. 126 and Table 3.
Lehninger, A.L. (1975). *Biochemistry.* 2nd Ed. Worth Publishers Inc., New York, N.Y. p. 425.
Newsholme, E.A. and Start, C. (1973). *Regulation in Metabolism.* John Wiley and Sons, London.
Opie, L.H. (1976). II Metabolic regulation in ischemia and hypoxia. Supp. 1. Circ. Res. 38: 1-52 - 1-174.
Sahlin, K., Harris, R.C. and Hultman, E. (1975). Creatine kinase equilibrium and lactate content compared with muscle pH in tissue samples obtained after isometric exercise.

Biochem. J. 152: 173-180.
Sahlin, K., Palmskog, G. and Hultman, E. (1978). Adenine nucleotide and IMP contents of the quadriceps muscle in man after exercise. Pflugers Arch. 374: 193-198.
Spector, W.S. Editor (1956). *Handbook of Biological Data.* Wright Air Development Center Technical Report 56-273. United States Air Force, Wright-Patterson Air Force Base, Ohio.
Stryer, L. (1981). *Biochemistry.* 2nd ed. W.H. Freeman and Co., San Francisco. p. 299.
Wilkie, D.R. (1981a). Comment At Royal Society Meeting "The Enzymes of Glycolysis: Structure, Activity and Evolution". Oct. 16, 1980. Phil. Trans. R. Soc. Lond. B293: 40-41.
Wilkie, D.R. (1981b). Shortage of chemical fuel as a cause of fatigue: studies by nuclear magnetic resonance and bicycle ergometry. *CIBA Foundation Symposium No. 82, Human Muscle Fatigue: Physiological Mechanisms.* Pitman Medical, London.

DISCUSSION

MAUGHAN: Professor Wilkie, you show a phosphocreatine concentration around 17.4 mmol/kg wet for the given biopsy. That's quite a bit lower than you've found previously in frog muscle, I believe. Is that correct?

WILKIE: That's right. It has long been suspect because unless you go to extreme lengths to freeze frog muscle or other tissue very rapidly, artifactual breakdown of phosphocreatine occurs. The best conditions are given by a thin muscle such as the frog sartorius, starting with it at 0° and then freezing it extremely rapidly. We used -- I think that everyone now uses -- the so-called hammer apparatus that Merlin Kretzschmar and I designed so that freezing takes place in 100 msec or perhaps less. The case of resting human muscle sampled by needle biopsy is very much worse. The muscle must be stimulated by the biopsy, which cuts fibers, and the freezing takes six seconds. The low estimate of phosphocreatine in human muscle has long been suspect to all of us "frog people" and I think suspect with good reason, as we now see from our non-invasive NMR studies.

HOUSMANS: Professor Wilkie, I would like to know if you think that in the foreseeable future it will be possible with your technique to look non-invasively at the way in which heart functions as a muscle and how it fails as a muscle in patients.

WILKIE: Well, we hope so. The magnet you saw was a simple 20 cm diameter cylinder. It will accommodate an adult human limb and the heads of premature newborn babies. We now have such a machine just starting work at the University College School of Medicine. In addition, a joint application has been approved for a 60 cm superconducting system to be situated at the John Radcliffe Hospital, Oxford. With this 60 cm system we hope, with good reason, to be able to focus on the heart and thus to examine the changes that happen during normal and abnormal cardiac action.

SHIMIZU: You have not mentioned yet about the relaxation time of the NMR spectrum. Are you now interested in the difference of the relaxation times between normal and dystrophic persons, in particular in the relaxation time of the proton?

WILKIE: We do not have such a plan at present because the problem would be better approached by NMR imaging techniques. To avoid confusion I should explain that the term "relaxation time" that he is using has a NMR meaning which bears no relation to the mechanical relaxation time of the muscle that I was talking about earlier. In the NMR context there are two relaxation times involved. One of these, T_2, is the time it takes for the signal that has been provoked by the radio frequency pulse to disappear. Strictly speaking, it becomes incoherent in the optical sense, so you can't detect it any longer. I imagine that relaxation time of which you are thinking is T_1. This is concerned with the rate at which energy is lost from the excited nucleii to the surroundings; for this reason T_1 is called the spin-lattice relaxation time.

EDMAN: I was pleased to see the conclusions you drew from these experiments. I think, you are the first to conclude that fatigue is due to accumulation of products.

WILKIE: Yes.

EDMAN: It could be tested in a simple preparation, still intact and living. The pattern of the changes you get in velocity of shortening relative to tension decrease can be completely simulated or imitated by raising one of the products, the hydrogen ions.

WILKIE: Yes. There are two aspects to my answer. One is that I've repeatedly asked the people who have actomyosin systems set up *in vitro* in their labs to imitate the concentrations of the products in a fresh muscle and a fatigued muscle, and to find out the effect on the rate of ATP hydrolysis. Unfortunately, so far I haven't persuaded anyone to do it. The second experimental approach is to reduce the internal pH using carbon dioxide.

EDMAN: As I said, it can be simulated in the intact fiber, as we have shown.

WILKIE: Yes, I have seen your interesting results on that subject. During the past couple of months we have carried out the complementary NMR experiments. Exposure to CO_2 adequate to lower the internal pH to 6.56 can be seen by NMR to have no effect on any metabolite other than H^+. The effect on the isometric contraction is very striking. Development of force is only slightly affected whereas mechanical relaxation is delayed and its time course is very much slowed down. The effects are completely reversed when the CO_2 is replaced by O_2.

PODOLSKY: You pointed out that the ^{31}P signal was much smaller than the ^{1}H signal in the dystrophic boy. Is that because there's less muscle in the boy, or because the state of the water is different?

WILKIE: Essentially, it's because there's less muscle and more fat. In a way that was a bad slide to show to demonstrate the advantages of NMR because clinically, Duchenne's dystrophy makes itself tragically obvious. What you see pathologically is a replacement of muscle tissue by fatty fibrous tissue. This change can now be monitored using NMR, which should help in patient care. It's the enzyme defects that are most interesting to look at by NMR because they are so infernally difficult to diagnose by conventional methods.

INTRACELLULAR pH AND ENERGY CHANGES IN MUSCLE

Nancy A. Curtin

Department of Physiology, Charing Cross Hospital Medical School, London W6 8RF, England.

ABSTRACT

Intracellular pH (pH_i) and buffering power of unstimulated sartorius muscle (frog) were measured with pH-sensitive microelectrodes. The mean pH_i for muscle was about 0.3 units more acid than the extracellular pH in the range 7.0 to 7.2. The buffer power was 44.0 ± SE 7.1 mM/pH unit for 19 fibres with pH_i 6.83 ± SE 0.05.

Force is produced during crossbridge cycles in which the energy from chemical reactions is converted into mechanical work and heat. What chemical reactions are involved? The cleavage of ATP and the creatine kinase reaction, which resynthesizes ATP, are accepted as the immediate source of part of this energy, but the rest, which can amount to as much as a quarter of the energy in a contraction, comes from another, as yet unidentified, source (reviewed by Curtin & Woledge, 1978).

Intracellular pH (pH_i) is relevant to questions about energy because pH_i affects the amount of energy produced by known reactions, and hydrogen ions may participate in unidentified reactions.

For these reasons pH_i is being studied using pH-sensitive microelectrodes (the recessed-tip design of R.C. Thomas, 1978). Their sensitivity and selectivity properties are similar to those of the 'macro' electrodes used with pH meters because the sensitive element, pH glass 0150, is the same. Experiments were done on unstimulated sartorius muscle (*R. temporaria*) in Ringer's solution containing 95 mM NaCl, 3.7 mM KCl, 2.0 mM $CaCl_2$ and 10 mM buffer (HEPES or $NaHCO_3$ + nominal 5% CO_2) at room temperature. In a few cases the buffer was 50 mM and the NaCl, 55 mM. A pH microelectrode and a conventional KCl-filled microelectrode were inserted in the same fibre on the surface of the muscle. The output of each electrode was recorded along with the difference between them; this difference is proportional to the pH.

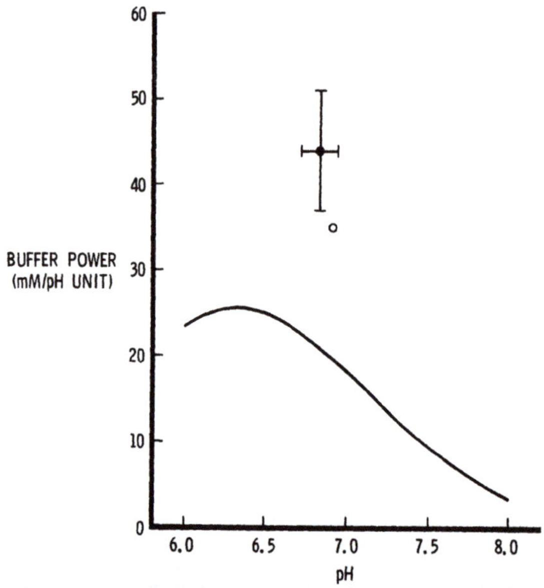

Figure 1: The relation between buffer power and pH. The line is that expected for the following mixture of buffers at 20°C: histidine, 36 mM, pK 6.15; carnosine, 14 mM, pK 6.91; phosphate, 2.0 mM pK 7.03. Chemical analysis has shown that these buffers exist in muscle. Buffer power depends on pH because a buffer is more effective the nearer the pH is to its pK. The open point is from Bolton & Vaughan-Jones (1977) and the filled point is the mean value ± 1 SE of the results reported here.

The values of pH_i for fibres in an individual muscle are quite consistent; a typical mean for 9 different fibres in a muscle is pH_i 6.74 ± SD 0.069. Somewhat more variation is found between the values of fibres from different muscles. Nevertheless, for 7 muscles that have been examined in detail, the mean pH_i of the muscle was always at least 0.3 pH units lower than the extracellular pH in the range 7.03 to 7.14. The enthalpy produced by ATP cleavage and the creatine kinase reactions at this pH_i, is smaller than the value of -34 kJ/mol usually used in energy balance studies. So the energy from these reactions may explain an even smaller part of the heat and work produced during contraction than has been supposed. However, if pH_i changes during contraction this should be taken into account too.

The intracellular buffering power, which indicates the effectiveness of the buffers, has been measured by adding a known amount of H^+ to the cells (by equilibrating them with a known P_{CO_2}) and observing the change in pH_i. For 19 fibres with a mean pH_i of 6.83 ± SE 0.05, the buffering power was 44.0 ± SE 7.1 mM/pH unit. See Fig. 1. This agrees with the results of Bolton & Vaughan-Jones (1977), but is higher than that calculated from literature values for the chemical content of muscle (see Curtin & Woledge, 1978). One possible explanation for this discrepancy is that additional buffers exist in the muscle. Energy from their reactions may contribute to the energy production by muscle.

REFERENCES

Bolton, T.B. and Vaughan-Jones, R.D. (1977). Continuous and direct measurement of intracellular chloride and pH in frog skeletal muscle. J. Physiol. 270: 801-833.
Curtin, N.A. and Woledge, R.C. (1978). Energy changes and muscular contraction. Physiol. Rev. 58: 690-761.
Thomas, R.C. (1978). Ion-sensitive Intracellular microelectrodes. London: Academic Press.

DISCUSSION

A common question about intracellular pH and energy balance was how big a pH change would be required to explain the 40 mJ/g wet weight energy gap in isometric contraction. If the hypothetical pH change were to have its effect solely by changing the molar enthalpy for phosphocreatine, the intracellular pH would have to change on the order of 0.5 pH units in the alkaline direction. PCr splitting does absorb hydrogen ions but in the light of the buffering power of muscle, it seems unlikely that such a large pH change would occur.

A more likely way that the intracellular pH could be involved with the energy gap is to do with buffer reactions. Measurements of buffer power with pH microelectrodes in related experiments suggest that the buffer power is greater than the values used in past energy balance calculation. This suggests than an additional buffer exists in muscle. If an additional buffer does exist, the heat of its reaction has not been included in energy balance calculations and may contribute to the energy balance.

CHANGE IN FIXED-CHARGE IN THE THICK FILAMENT LATTICE OF LIMULUS STRIATED MUSCLE WITH SARCOMERE SHORTENING

P. Brink and M.M. Dewey

Department of Anatomical Sciences, State University of New York at Stony Brook, Stony Brook, New York 11794

ABSTRACT

A highly significant increase in fixed-charge occurs in the A-bands of *Limulus* striated muscle following activation and sarcomere shortening. This is in striking contrast to vertebrate striated muscle. Treatment with either papain or alkaline phosphatase reduces this fixed-charge. The increase in charge may serve as a motive force in thick filament shortening.

The measurement of Donnan potentials within A- and I-bands of *Limulus* muscle has shown it to behave in striking contrast to frog sartorius or rabbit psoas (Bartels and Elliott, 1981) with respect to myofibril shortening. In vertebrate muscles as sarcomeres shorten the A-band and I-band potentials remain relatively constant showing only a small increase, if any. Isotonically shortened sarcomeres of *Limulus* (< 3.8 μm) on the other hand show striking increases in the magnitude of the Donnan potential (Bartels and Elliott, 1981; Dewey, et al., 1982; Elliott et al., 1978). Here we describe the measurement of Donnan potentials in K^+ acetate buffer over a range of 4.0-7.4. Acetate has been used because of its buffering capacity over the pH range used. All measurements were made in a rigor solution containing 50 mM K^+ acetate and 50 μM EGTA. The pH of solutions was adjusted by addition of acetic acid or KOH. Sarcomere shortening was effected by placing glycerinated muscle bundles in an activating solution which contained 100 mM KCl, 1 mM $CaCl_2$, 1 mM $MgCl_2$, 5 mM ATP and 10 mM buffer. To insure equilibration of the ions following activation, repeated rinsings in rigor solution followed. Additionally the effects of two enzymes were studied. The enzymes were papain and alkaline phosphatase. The measurements were carried out as previously described (Bartels and Elliott, 1981 and Dewey et al., 1982). Papain and alkaline phosphatase were introduced into the bathing media in which the muscle fibers were suspended. Ten minutes were allowed for digestion in the case of papain (pH= 7.4) before rinsing the muscle in rigor solution which contained 1 mM iodoacetic acid. The muscle was then repeatedly rinsed in normal rigor solution. All muscles were glycerinated as previously decribed (Dewey et al., 1982). Alkaline phosphatase was prepared as described by Brann et al., 1979 and measurements were made while the enzyme was present in the rigor solution. The pH of the bathing rigor solution was tested before and after measurements.

The Donnan potentials in K^+ acetate for long sarcomeres (10.0 μm) at pH 7.4 were negative in sign and ranged in magnitude from 5 mV to 12 mV. These values are in the same range as those seen for Tris buffer at a similar pH (Dewey et al., 1982). No attempt was made to distinguish A-band from I-band but from a previous report (Dewey et al., 1982) the smaller of the potentials was undoubtedly generated by the I-band while presumably the A-band gave the larger. The potentials were then measured at pH 6.0, 5.0, and 4.0 as Figure 1 shows. The isoelectric point was 5.1, very similar to that in Tris at 5.0 (Dewey et al., 1982). Isotonic shortening of *Limulus* glycerinated muscle bundles results in sarcomere lengths less than 3.8 μm. Measurements were made on the shortened bundles at the various pHs. The data are shown in Figure 1 also. Once again shortening caused the potential to increase in magnitude as has been observed in Tris buffer (Dewey et al., 1982). The isoelectric point was shifted in the alkaline direction such that the isoelectric point was ≃5.7. The triangles at pH 6.0 and 5.0 are means (n=11) for Donnan potentials measured on short and long sarcomeres of frog sartorius muscle. At pH =6.0 no distinction could be made for long versus short sarcomeres. At pH=5.0 the long sarcomere potential was slightly larger in magnitude

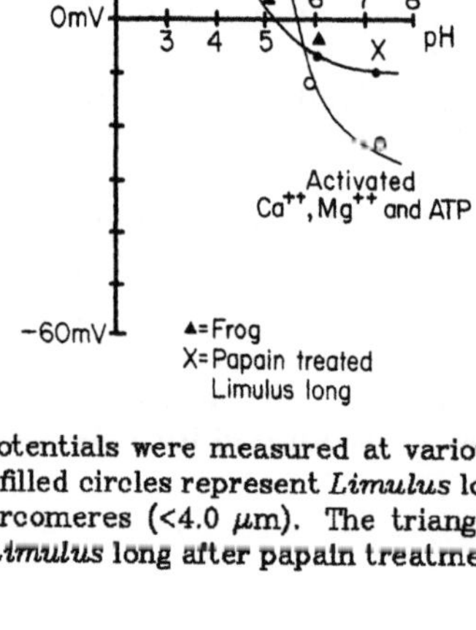

Figure 1: Donnan or matrix potentials were measured at various pHs (4.0-7.4) in 50 mM K^+ acetate and 50 μm EGTA. The filled circles represent *Limulus* long sarcomeres (10 μm) while the open circles are short sarcomeres (<4.0 μm). The triangles represent frog sartorius, both long and short. The x is *Limulus* long after papain treatment.

(+8 mV versus +5 mV). There was no indication of any significant shift in isoelectric point. Each point on the graph represents an n of at least 8 to a maximum of 15.

The potentials are a reflection of the fixed charge density within the muscle matrix. It is possible to calculate the fixed charge concentration (in mM) from the Donnan potential data (Bartels and Elliott, 1981; Elliott et al., 1978 and Elliott and Bartels, 1982) the equation:

$$\text{protein}^{(-)}1 = 2(A^{+}B^{-}) \sinh(EF/RT)$$

can be used over the range of potentials seen here to calculate the charge concentration (Pr^- mM). Table 1 gives the calculated charged protein concentration for *Limulus* at pH=7.4 in rigor solution.

Table 1

Rigor Solution		Donnan Potential	IEP	$(Pr)^-$
50 mM K acetate				
50 μM EGTA	long	-9.1 mV	5.1	36.4 mM
	short	-25.0 mV	5.7	114.6 mM
papain	long	-5.2 mV		20.8 mM
50 mM KCl				
50 μM EGTA				
10 mM Tris	long			
	A-band	-12.0 mV	5.0	51.4 mM
	I-band	-4.8 mV	5.1	18.9 mM
	short sarcomere			
	A+I-band	-32.0 mV	4.2	180.1 mM

IEP = isoelectric point
$(Pr)^-$ = concentration of negative charges on protein

The difference in A-band and I-band charge concentration is also seen in rabbit psoas muscle (Bartels and Elliott, 1981) in rigor solution but is absent in relaxing solution. Bartels and Elliott (1981) speculated that the effect may be due to anion binding to the thick filament. The alteration of pK with buffer in *Limulus* from an acid shift in Tris with shortening to an alkaline shift with shortening in acetate also supports the idea of absorption or binding of anions onto the filaments. In both cases (acetate and Tris) the charge concentration is significantly increased for short sarcomeres. The increase can not be explained by addition of A- and I-bands alone (Tris). The lattice spacing of *Limulus* has been shown to remain constant over a large sarcomere range (Millman et al., 1974).

Therefore, one might conclude that the potential increase observed in shortened sarcomeres is generated by exposure or accumulation of fixed charge groups on the surface of the thick filaments or some other sarcomeric structure. It should be noted that at physiologically relevant ion concentrations the Donnan potential is sufficiently masked such that no potentials at any significant magnitude can be measured. The role of this charge however is not necessarily the establishment of a potential between matrix and cytoplasm but, if effector at all, to alter protein structure, for example, thick filament structure (i.e. length). This assumes of course that the charge is causal which has yet to be demonstrated.

Finally both papain and alkaline phosphatase had the effect of reducing the observed Donnan potential. Papain digestion was only performed on muscle bundles with long sarcomeres at pH=7.4. A ten minute digestion caused significant reduction in the observed potential (50% reduction). Electron micrographs of isolated filaments, platinum shadowed, similarly treated reveal a loss of what is thought to be the myosin heads. Experiments on short sarcomeres are presently being done. In general, alkaline phosphatase reduced the Donnan potentials. But short sarcomeres were more resistant to the enzyme treatment than were the long sarcomeres. This could be an access related problem or it may point to a real biochemical difference in the number of phosphate groups accessible to alkaline phosphatase in the shortened case (Dewey et al., 1982). Since alkaline phosphatase cleaves phosphate groups, it will be of some interest to see if pre-treated (papain) filaments lacking myosin heads are affected by the phosphatase or not.

REFERENCES

Bartels, E.M. and Elliott, G.F. (1981). Donnan potentials from the A-band and I-band of skeletal muscle, relaxed and in rigor. J. Physiol. 317: 85-86.

Brann, L., Dewey, M., Baldwin, A., Brink, P. and Walcott, B. (1979). Requirements for *in vitro* shortening and lengthening of isolated thick filaments of *Limulus* striated muscle. Nature 279: 256-257.

Dewey, M., Colflesh, D., Brink, P., Fan, S-F., Gaylinn, B. and Gural, N. (1982). Structural, functional, chemical changes in the contractile apparatus of *Limulus* striated muscle as a function of sarcomere shortening and tension development. In: *Basic Biology of Muscles: A Comparative Approach.* Ed. by Twarog, B., Levine, R. and Dewey, M. Raven Press, New York, vol. 37, pp. 53-72.

Elliott, G.F., Naylor, G.R.S. and Woolgar, A.E. (1978). In: *Ions in Macromolecular and Biological Systems.* Ed. by Everett, D.H. and Vincent, B. pp. 329-339, Bristol Scientechnica Press.

Elliott, G.F. and Bartels, E.M. (1982). Donnan potential measurements in extended hexagonal poly-electrolyte gels such as muscle. Biophys. J. 38: 195-200.

Millman, B.M., Warden, W.J., Colflesh, D.E. and Dewey, M.M. (May 1974). X-ray diffraction from glycerol-extracted *Limulus* muscle. Biophysical Soc. Meetings. Fed. Proceedings 33/5, 1333, Abst 622.

DISCUSSION

GODT: You are talking about potentials that were measured from the myofibrils, and you call them "Donnan potentials." A number of individuals have measured these potentials. If they are, in fact, Donnan potentials then one can calculate myofilament charge densities. Some have gone so far as to infer from these calculated charge densities something about the state or conformation of the contractile proteins, or something about cross-bridges. Clive Baumgarten and I have looked into this a little bit more carefully, and using potassium selective microelectrodes inside of skinned fibers. If the potential really is a Donnan potential, then it's an equilibrium potential and the ions, under all conditions, should be in equilibrium. In fact, at low pH (pH6 and pH5) in ATP containing solutions, potassium is *out* of equilibrium. So I would suggest that it's a potential all right, but it's probably *not* purely a Donnan potential.

By the way, we find that when you take away the ATP, potassium goes into equilibrium. So our working hypothesis is that the potential is *not* a Donnan *equilibrium* potential. Rather, in the presence of ATP, it's a *steady-state* potential, possibly the superposition of a Donnan equilibrium sort of potential and a diffusion potential. If so, going from the potential that you measure with a conventional micro electrode to ideas about what's happening on the molecular scale is very perilous indeed.

DEWEY: I will have to disagree with that. They have been standardly called Donnan potentials. I think the significant thing is that there are changes which are dependent upon shortening, pH, and ion concentration in the extracellular fluid. The potentials are stable in rigor solution for days, which clearly demonstrates the system is in equilibrium. We accept a definition of equilibrium as a condition in which all constituent particles are at rest or in an unaccelerated motion.

GODT: Well, if under *any* circumstance an ion is not in equilibrium, it cannot be a pure Donnan potential. Because a Donnan potential is an equilibrium potential, and if a potential is not an equilibrium potential it cannot be Donnan. It is a potential, of course, which can be measured, but what the basis of it is I'm not really sure myself.

A NEW METHOD TO MEASURE INTRACELLULAR DIFFUSIBLE ELEMENTAL CONCENTRATION

David Maughan

Department of Physiology, University of Vermont, Burlington, VT 05401

I was asked to comment about a new technique that I've been developing that allows analysis of very small samples of fluid from muscle cells. I'm trying to estimate the elemental composition of muscle fluid, as close to the *in vivo* condition as possible.

The method is take out a whole muscle from a frog, blot and place it under mineral oil, remove a muscle fiber, and mechanically skin it while it's still under the oil. The fiber is relaxed, as close to an *in vivo* condition as possible. I place on the fiber (the bundle of myofibrils bathed in the oil envelope) a drop of water that's small compared to the fiber itself, so that the dilution effects are very small (Fig. 1a and 1b). A pipette is used to apply the droplet of water (about 0.2 nanoliters, about 100 to 1000 times smaller than the skinned fiber segments). This procedure resembles, in reverse, the sort of procedure that Richard Podolsky used a number of years ago, in which he activated a skinned fiber with a micropipette filled with calcium. This is simply going the other way, i.e. letting the intracellular milieu equilibrate with the small sample. The sample water taken from an adjacent drop (Fig. 1a) contains 0.25 M sucrose so that the sample water is isosmotic with the fiber fluid and 50 micromolar EGTA, so that there's no contaminant calcium to activate the system. At this point I allow the droplet to equilibriate with the skinned fiber for various periods of time, storing sequential droplets in the shank of the pipette (Fig. 1c and c').

After equilibration, I remove the pipette, freeze it, and ship it to Boston where, at the Biotechnology Resource Facility under the direction of Claude Lechene, X-ray spectroscopy is performed on the sample using the wavelength dispersive system (Fig. 1d and e). Professor Lechene has been very generous in allowing me to use the facility (I also acknowledge the help of Christopher Recchia, who did the experiments for me in Vermont). From the X-ray spectroscopy, we can expect a resolution of about 0.1 millimolar concentration for the following elements: potassium, chlorine, sodium, magnesium, calcium, sulphur, and phosphorus. Those

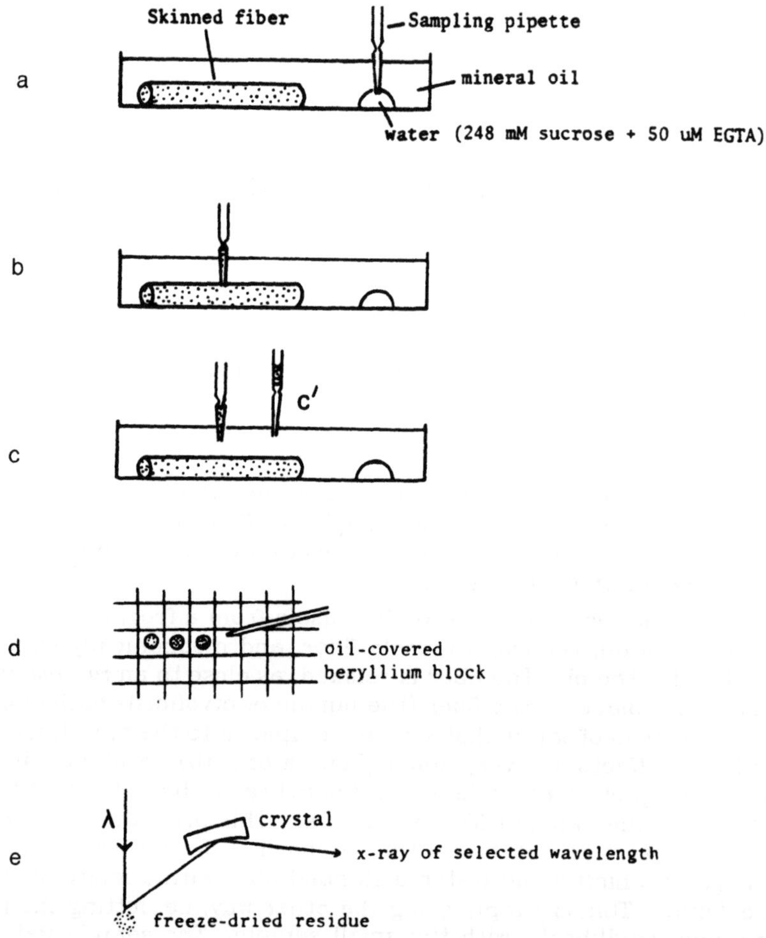

Figure 1: Electron probe microanalysis of intracellular fluid from a single muscle fiber. The fluid is sampled using a method of equilibrating a tiny droplet of isosmotic fluid with the fluid of a fiber freshly skinned under oil (a-c). The sample is analyzed using a method of X-ray spectroscopy of the freeze-dried residue (d-e). The droplets are expelled onto an oil-covered beryllium block; the oil is removed by washing with m-xylene. The samples (along with droplets of standard solutions of known concentration) are frozen, freeze-dried, and analyzed with a Cameca microprobe.

are the primary elements, but I'll just discuss magnesium, calcium, and phosphorus. I'm picking these out because these have been of interest today in this session.

Taking magnesium as the first example, we allow the fiber fluid to equilibrate with the small droplets for 0, 0.5, 1, 2, 5, and 10 minutes. The equilibration uptake curves (plotting sampling time against the concentration of magnesium) indicates that within 5-10 minutes we have full equilibration (Fig. 2). The value comes out to be 5.2 mM (24 experiments with frog muscle fibers); this is the concentration of diffusible magnesium

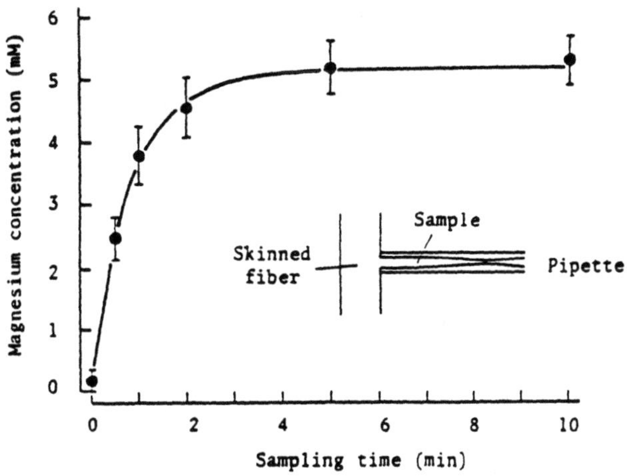

Figure 2: Magnesium analysis of liquid samples from frog muscle fibers. Magnesium concentration estimated from electron probe microanalysis of pipette fluid residues calibrated against residues from standard solutions using the same volumetric pipette (as in Fig 1d & e). The zero minute sample is that of the water/sucrose/EGTA solution alone. Mean values (±1 SEM) pooled from 24 experiments. Sample chamber volume is approximately 0.2 nl. Note that the uptake of Mg is complete within 5-10 min. The final value, about 5.2 mM, is probably close to that of the *in vivo* total diffusible Mg concentration in the frog muscle cytosol.

inside a muscle cell. Free magnesium or magnesium bound to ATP or creatine phosphate have to add up to 5.2 mM.

For calcium we get concentrations that are around 0.8 mM. From the value of the diffusion constant which we're able to extract from these curves, it appears that this calcium comes out very slowly. It comes out as if it were bound to a protein of molecular weight of 15-25 kD.

The other element of interest is phosphorus and that works out to be approximately 45 mM which, if you add up all the creatine phosphate, ATP hexomonophosphates and other moieties inside the muscle, comes out quite close to the value expected (Conway, Physiol. Rev. 37: 84-132, 1957).

DISCUSSION

NOBLE: What is the total ionic strength?

MAUGHAN: I haven't done that calculation, but it's certainly in excess of 0.1 M. There is some unexplained variability in the sodium, potassium and chloride concentrations (ions which contribute appreciably to the ionic strength). The numbers we're getting out are somewhat different than the current values.

MAGID: First of all I want to congratulate you, David, on a very elegant application of a favorite preparation of mine, the skinned fiber. Two things I want to ask. What precautions do you take that the Ringer's solution bathing the fibers (before you dissect the single fibers) does not come along with the fiber? How do you get rid of that little thin layer of saline that clings to the fiber? And the second question is - do you think that if you describe the 5.2 mM of magnesium as being the diffusable ion, some of it is diffusing while bound to, for instance, parvalbumin? Or do you think that this is the free level of ionized magnesium in the cell?

MAUGHAN: Well, to answer the second question first, certainly this concentration of 5.2 mM includes *all* diffusible magnesium, including free magnesium ion. Taking values of total ATP from the literature, the value of Professor Wilkie and colleagues, for example, and taking the value of parvalbumin from Gillis' work, and taking creatine phosphate concentration which is also given by Professor Wilkie, plus the estimated affinity constants of magnesium bound to these different species, I can break that 5.2 mM down to about 4.2 mM magnesium bound to ATP, 0.8 mM magnesium bound to creatine phosphate and parvalbumin, and 0.2 mM free magnesium.

MAGID: Very low.

MAUGHAN: Yes. And to answer the first question - what you do is strip the membrane from the muscle fiber, and as you strip it off you also carry off the extracellular water, i.e., any residual extracellular fluid that's contained on the outside. As a test for possible contamination, I included cobalt in the Ringer's solution. There was no detectable cobalt in the samples. If there had been any appreciable contamination from the extracellular fluid, cobalt would have been detected.

LEVINE: How do your values compare with the results of X-ray microanalysis?

MAUGHAN: It's hard to compare because the X-ray microanalysis, if you're referring to Somlyos' work, focuses on spots that are 50 nanometers or greater in diameter I believe, so more than just intracellular fluid is analyzed.

LEVINE: 87 Å is what he uses ... that's his lowest.

MAUGHAN: If that's their minimum, their absolutely smallest spot, that would still include the material that's bound to the filaments as well. It's impossible on the basis of their experiments to distinguish between diffusable magnesium for example, and magnesium bound to the filaments. I want to emphasize that these are complementary techniques; from both sets of data one should be able to extract out information as to how things are compartmentalized inside the muscle.

WILKIE: First off, it's a marvellous technique, and we all want to know the diffusable concentrations of these things. What do you make of the difficulties that you yourself mentioned -- the ones with potassium and sodium? What do you expect from the fact that there really should be a membrane potential of say 90 millivolts, and if you've got so much potassium outside there ought to be a calculable concentration inside? So, with regard to these discrepancies, I'd like to know how big they are and how much to worry about them.

MAUGHAN: I'm reluctant to comment on these because of the wide variation in values we're getting in potassium from fiber to fiber in comparison to the variation that we get with magnesium and the other elements. For potassium, I get a value of around 80 mM which is low, very low in comparison to the values we expect from electrophysiology. One possibility is that there may be some movement of ions across the membrane when I dissect the muscle out, so that there is an inward flux that would tend to load the cell with sodium and chloride and extract potassium as a result of the depolarization. I'm hoping that by doing some additional experiments to reduce the extracellular space, we'll remove the discrepancies. On the other hand, we may merely be observing natural variation.

WILKIE: It is true, isn't it, that these are skinned fibers?

MAUGHAN: Yes. It's about 10 minutes between the time we killed the frog and the time we skinned the fiber.

WILKIE: Are they mechanically skinned?

MAUGHAN: They're mechanically skinned.

WILKIE: Well of course in that case the conditions are different -- there is in fact no longer an 80 mV potential difference across where the membrane would be.

MAUGHAN: That's true.

WILKIE: Therefore there is no discrepancy.

MAUGHAN: Yes, but the value of this technique lies in measuring the concentration which the fiber had just prior to skinning. We assume that the extracellular fluid surrounding that fiber is small, so that ionic movement across the membrane wouldn't significantly affect the elemental concentration. Even though the membrane potential is zero, it wouldn't matter. Nevertheless, we're trying to test and correct for this possibility. If little or no correction is needed, then there's some real discrepancies between our values and literature values for sodium, potassium, and chloride.

NOBLE: If this discrepancy is real it raises some profound implications.

ION CONCENTRATIONS SURROUNDING THE MYOFILAMENTS

Tatsuo Iwazumi

Department of Physiology, University of Texas Medical Branch, Galveston, TX 77550

My point of view is a radical departure from everybody elses. Cellular water has been looked at rather extensively. The more we look at it, the more we are confused. The kind of explanation I give here might have been overlooked in the past, but I think it provides clear and consistent physical interpretations. Let's assume we have a dipole on the protein molecule in ionic water. The positive and negative charges are separated over a distance of about 100 Å. This is shown in Figure 1. The energy required to deposit these charges would come from ATP. As you can see right away the small region between charges experiences a very high electric field strength. If we have a significant potential gradient, then the electric field energy rapidly increases (Field Energy = $1/2\ \varepsilon E^2$). Note that it is proportional to the dielectric constant. Now, the dielectric

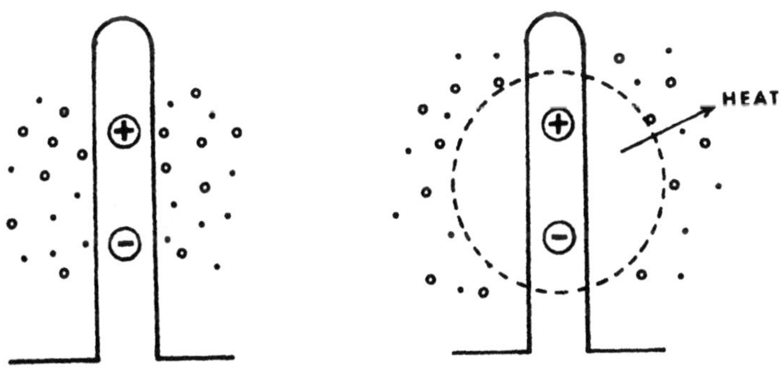

Figure 1: *Left*: At the moment a dipole appeared on myosin head when ATP bound to it. *Right*: Some moment later (a second or two), water molecules in the vicinity of the dipole assume ice-like structure which is energetically advantageous because of increased dielectric constant. The consequential entropy reduction releases heat and drives solute out of the region, which in turn increases Debye length, helping polarization of water molecules further.

constant of liquid water is about 80. When water freezes and becomes solid, the dielectric constant increases to about 90. This might sound odd because intuitively the dielectric constant of ice should be less than 80 due to more restricted rotational motion of water molecules. But actually, the opposite is true.

So the point is, if you apply a high electric field gradient to liquid water, the water molecules see it energetically advantageous to assume an ice-like structure. In other words, the water molecules in the vicinity of the dipole tend to assume the ice-like structure, and as a result their entropy diminishes. This entropy reduction brings about three consequences. First of all, solute tries to escape from the ice-like region. Second, an entropic heat is liberated. And third, because of ionic redistribution, mass transfer occurs from one region to another. When this process continues, what we end up with is a small region around the dipole in which the ionic concentration is much reduced, which means the Debye length in this region is much, much longer. In my estimate, the Debye length in the vicinity of the dipole is about 100 Å when the dipole acquires the highest possible energy, and this is approximately the same distance between the positive and negative charges.

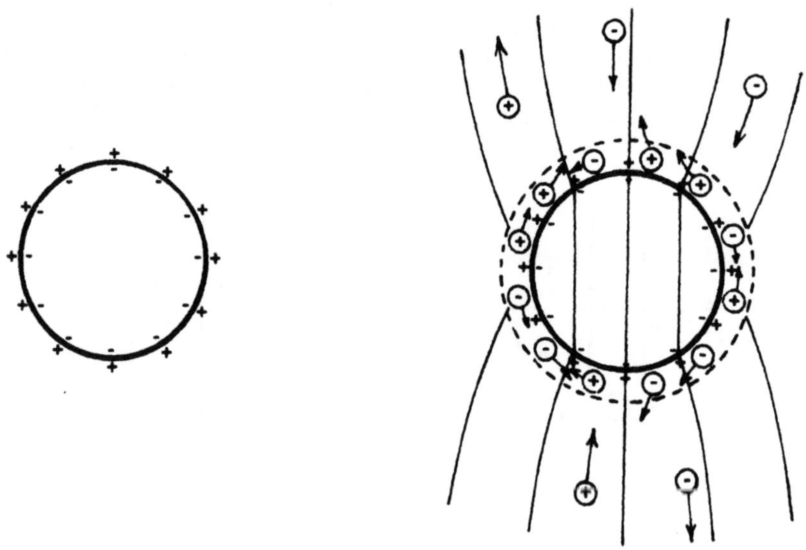

Figure 2: *Left*: Cross sectional view of a cylindrical macromolecule (actin filament) with negative surface charges and their counter ions (mostly Mg^{2+}) in aqueous medium. Such double charge layer gives rise to a very high "effective" dielectric constant even though the dielectric constant of the molecule in dry state is very low. *Right*: Boundary condition violation leading to formation of an anisotropic high conductive layer on the surface. Ions following the field lines strike the surface while the field lines penetrate the solid body. As a result, ions accumulate on the surface until they form an electrically neutral high conductive layer on the surface thus establishing new boundary conditions consistent with the high dielectric constant of the body.

OK, if that happens, the question is where those excluded ions go? They don't just go out to infinity. At this point, let's digress a little before going into more detail. Suppose we have in an aqueous medium a molecule which has an ionic double layer on its surface (Fig. 2 left). Let me emphasize that this molecule is different and separate from the one with a dipole. When a macromolecule has negative surface charges, a layer of positive ions forms the double layer. Such molecules have a very high so-called effective dielectric constant in the aqueous medium. The value of the effective dielectric constant is a function of the surface charge density and the size and shape of the macromolecule, but it is in the range of hundreds to millions. Now, when a high dielectric body is placed in an electric field, obviously we know that field lines converge to such a high dielectric body.

Now, let's think about ion's motion in the surrounding medium. The ion sees a field gradient, so it tries to follow the field line (Fig. 2 right), but it soon hits a solid body. The field lines, of course, penetrate through the solid body. Then what should this ion do? This is a violation of the boundary condition. When this happens, the only way the ion could save the situation is to form an electrically neutral anisotropic ion layer of high conductance, so that the ionic currents can flow on the surface of the solid body. The field exerts a pressure on the ions, so that the ions tend to concentrate on the dielectric bodies. This phenomenon has long been known as surface conductance.

So, I argue that sodium, potassium, and chloride ions tend to accumulate on the protein surface when an electric field is applied. Now this phenomenon should not be confused with chemical binding. It has nothing to do with chemistry. It's just originated from violated boundary conditions; the ions disperse as soon as the electric field is removed. These ions must be potassium, sodium and chloride, because they provide greater electric conductance per unit concentration than other ionic species in the muscle cell; therefore, they are preferentially accumulated on the surface of protein (and between K and Na, K is preferential). In other words, these ions appear to be excluded from the intracellular water.

Therefore, what I expect to happen is something like this: when you look at the intracellular ionic content of living muscle, or maybe some other cells, I expect at the sarcomere length of something like two microns, where the overlap is the greatest, that means the largest proportion of protein surface is under the influence of electric fields, the sodium, potassium and chloride ion concentrations appear to be much less than their textbook values. If you stretch fiber all the way to non-overlap length, I expect the measured concentrations of these ions to approach textbook values. Then if you let the cell die, by depleting ATP, you get exactly the same values.

DISCUSSION

MAUGHAN: I just should like to note that my experiments were done at around 2.2 μm, slack length, so I really don't have any information that would be pertinent to this test that Dr. Iwazumi is proposing.

NOBLE: Are you stimulated to do the no-overlap experiment or the rigor experiment?

MAUGHAN: The rigor test, I guess, would be a little bit difficult. You have to wait quite a while before the muscle would go into rigor. You might get irreversible structural changes.

INGELS: I haven't talked with Dr. Iwazumi about this previously but I would certainly wholeheartedly support this concept that the very high charges in muscle are certainly going to redistribute the ions (that we know in double layer theory) and order the water about the high charge regions. These charges may be quite high, so it was interesting that was one of the assumptions I had to make in 1965, was that the ionic strength was less than isotonic. In fact I went ahead and picked a value about one order of magnitude less, 0.01 rather than 0.1. And things lined up, so this certainly is a feature I'm sure you'll have to deal with at some time.

DO CROSS-BRIDGES ROTATE DURING CONTRACTION?

INTRODUCTION

The following papers are all concerned with the conformational changes which occur in myosin cross-bridges during contraction. A number of studies have established that the force of muscle contraction is generated by a cyclic interaction in which the heads of the myosin molecule attach to sites on the thin filament and subsequently exert a force on the thin filament which translates it by approximately 10 nm towards the center of the sarcomere. However, the nature of the changes in the conformation of the actomyosin complex, which are responsible for force generation, remain unknown. The initial indication that these conformational changes may involve an alteration in the angle of the myosin cross-bridge came from electron micrographs of thin sections of insect flight muscle. Both electron micrographs and small angle X-ray diffraction patterns suggested that in rigor muscle the cross-bridges were attached to the actin filaments in an angled configuration while in relaxed muscle they assumed a different configuration and did not interact strongly with the actin. In contracting muscle the X-ray diffraction patterns indicated that the cross-bridges were considerably disordered. These early studies, which helped establish the moving cross-bridge hypothesis, have now been extended considerably by new X-ray techniques and these recent developments are discussed by Dr. H.E. Huxley in an earlier session, and by Dr. K. Holmes in this session.

Our understanding of crossbridge conformation has been greatly extended by the use of extrinsic probes, both fluorescent and paramagnetic, that can be attached to specific sites of the myosin head. The spectra of these probes can be used to determine the angle between the probe and the fiber axis, and thus one can obtain information on the angular configurations of the portion of the myosin head to which the probe is attached. Two of the following papers describe studies in which probes are used to determine whether the angle of the myosin head changes during the generation of force. Dr. Yanagida has used a fluorescent nucleotide bound to a specific site on the myosin head and Drs. Thomas and Cooke have used paramagnetic probes bound to a reactive sulfhydryl on the myosin head.

Although the most popular models of crossbridge function have envisioned an alteration of the orientation of the myosin head while attached to actin, other possibilities have also been explored. Force could be generated by the shortening of some segment of that portion of myosin which attaches the myosin head to the core of the thick filament. This hypothesis has been explored by using bifunctional reagents which can generate crosslinks within the thick filament, as discussed by Dr. Tawada.

One traditional method of understanding complex biological phenomena involves the reconstitution of a functional system from selected purified components. Early reconstitution experiments showed that a mixture of actin and myosin filaments could generate force, and subsequent work has extended these experiments to actin and myosin subfragments. The ability of these reconstituted systems to generate force provides information on both the location of the force generating element and on its mode of action. Dr. Shimizu outlines the latest results obtained with these reconstituted systems.

The nature of the events which lead to force generation have been difficult to unravel. The difficulties arise because the events occur transiently in complex filament structures, and because the asynchronous nature of cross-bridge action only allows the observation of an ensemble of various different cross-bridge states. In spite of these difficulties, steady progress has been made in refining our understanding of cross-bridge action. In this session new data provides further insight into the mechanism of force generation.

-Roger Cooke

THE NATURE OF THE ACTIN CROSS-BRIDGE INTERACTION

K.C. Holmes and R.S. Goody

Max-Planck Institute, Department of Biophysics, Jahn Str. 29, Heidelberg, West Germany D-6900

ABSTRACT

Evidence from sequence studies and from proteolysis suggests that S1 consists of three domains. Cross-linking studies show that one S1 can bind to two actin monomers which may lie in different strands of the actin long helix. The S1-actin interaction comprises two states "weak" and "strong". We suggest there are distinct hinged binding sites, "weak" and "rigor", of which only the rigor site is sensitive to tropomyosin control. If one takes the weak binding domain to be a "nose-cone" which is attached to the rest of the S1 by a flexible covalent hinge allowing the rigor link to be formed independently a number of structural phenomena observed in fibres may be explained.

INTRODUCTION

In 1969 H.E. Huxley, reviewing X-ray diffraction and electron microscopic investigations of frog and insect flight muscle, suggested that the cross-bridges, which are flexibly attached to the myosin filament at their proximal ends, rotate as a rigid body about their point of attachment to the actin filament (the "rolling bridge" hypothesis) thereby moving the actin filament past the myosin filament (the "power stroke").

A.F. Huxley (1974) categorised the kinds of acto-myosin interactions which would be consistent with the rolling bridge hypothesis (Fig. 1). He further argued that in order to explain the rapid interconversion between states which is deduced from the response to rapid length changes at least three bound states should exist. Below we assemble reasons for investigating the second of A.F. Huxley's categories in which the conformational changes are internal to the cross-bridge. We set out arguments which suggest that one should consider modulated interactions between hinged domains as the basis of the functionally inferred polymorphism of the myosin cross-bridge.

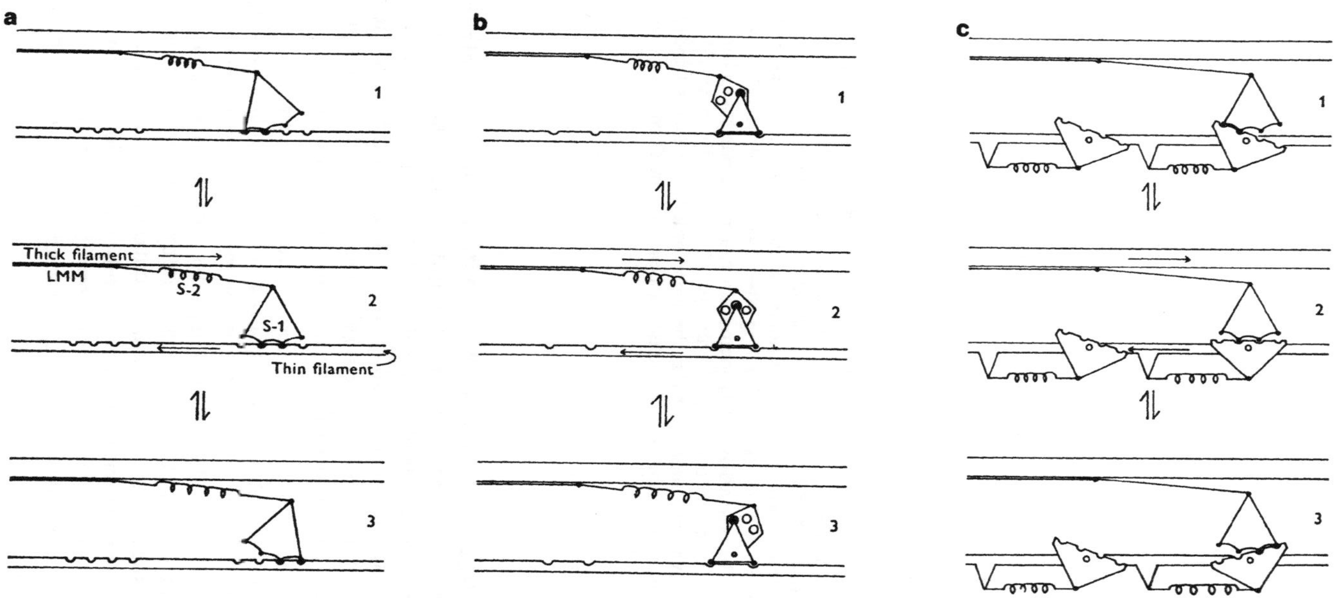

Figure 1: Three possible realisations of the rolling bridge (A.F. Huxley, 1974): a. The myosin head rolls on the surface of the actin filament using in turn a succession of binding sites; b. The myosin head binds to the actin with a fixed configuration but undergoes internal rearrangements during the power stroke; c. An alteration within the thin filament produces the power stroke.

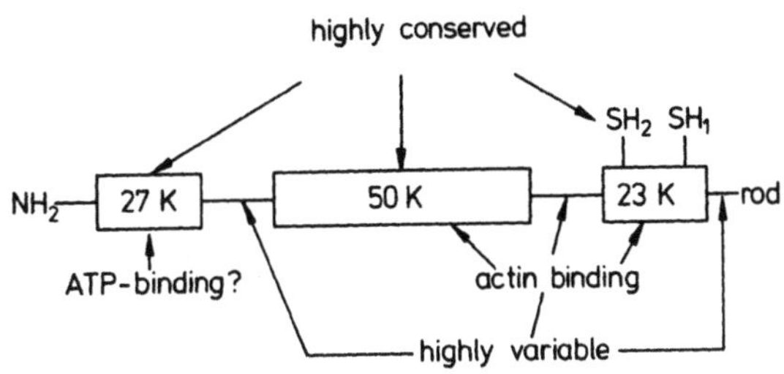

Figure 2: The domain structure of S1 as revealed by tryptic digestion and sequence studies (see text). Three domains may be distinguished having molecular weights 27K, 50K, and 23K Daltons in order from the n-terminus. The c-terminal domain contains the reactive SH groups SH1 and SH2. The n-terminal domain very likely contains the nucleotide binding site.

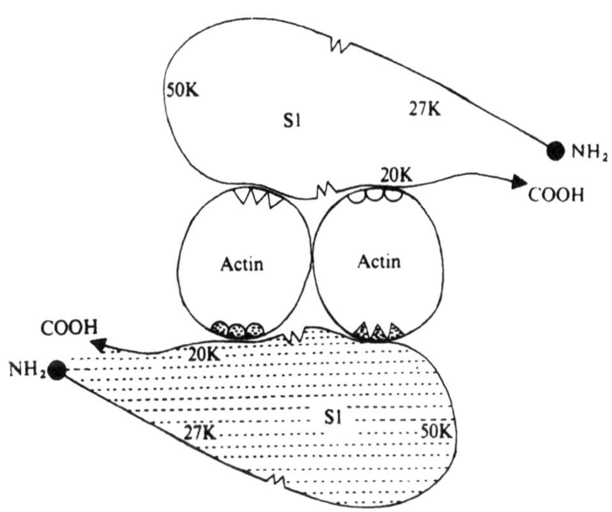

Figure 3: Two domains of S1 cross-link to two distinct actin monomers. The diagram shows how this might come about (Mornet et al., 1981b).

Domains in S1

Besides the two obligatory light chains, myosin S1 very probably contains a number of domains in the heavy chain. Evidence from tryptic digestion (e.g Mornet et al., 1981a; Yamamoto and Sekine, 1979) points to the existence of three domains of molecular weight 27K, 50K and 23K in order from the N-terminus. Between the domains are linker regions of molecular weight 1-2K Daltons which are labile to trypsin. Sequence studies on nematode body wall myosin (Karn et al., 1982) and on two other myosins from nematode (Karn, personal communication) show regions which are highly conserved separated by short regions which are highly variable. The conserved regions correspond rather well with the domains revealed by triptic cleavage (Fig. 2). Moreover, cross-linking studies (Mornet et al., 1981b) show that the 50K and 23K domains but not the 27K domains cross-link to actin. Furthermore these authors demonstrate that the two domains cross-link two distinct actin monomers, (Fig. 3).

Studies by Green, Chalovich, and Eisenberg (1982) show that the S1-actin binding exists in two states depending upon whether ATP or the direct products of hydrolysis (ADP+Pi) are bound to the enzyme or not. With ATP or products the binding is typically weak. Furthermore, the two states may be distinguished by whether or not they are sensitive to control by Ca-troponin-tropomyosin. The weak state is not Ca-sensitive whereas the strong (rigor) binding is Ca-sensitive. This observation would be readily explained if there were two distinct binding sites, one sensitive to tropomyosin control and one not.

On the basis of their cross-linking studies Yamamoto and Sekine (1979) conclude that the domains can also move relative to each other. In the absence of actin the 50K domain cross-links strongly to the 23K (21K) domain and weakly to the 27K (26K) domain. On binding to actin the cross linking to the 23K domain is further strengthened and the cross-linking to the 27K domain further attenuated. The binding to actin is accompanied by reciprocal changes in the susceptibility of the linker regions to tryptic digestion (Fig. 4).

Walker et al. (1982) have discovered a notable sequence homology extending over some 20 residues between the flexible loop encompassing the AMP binding site of adenyate kinase (Sachsenheimer and Schulz, 1977; Pai et al., 1977) and a part of the 27K domain of nematode S1. Just this region is highly homologous with the rabbit sequence (Elzinga, personal communication). This is probably the best evidence yet available for believing the 27K fragment to contain the nucleotide binding site of myosin.

Proton NMR shows internal flexibility in the S1 (Highsmith et al., 1979). Moreover this flexibility vanishes on binding to actin. Prince et al.

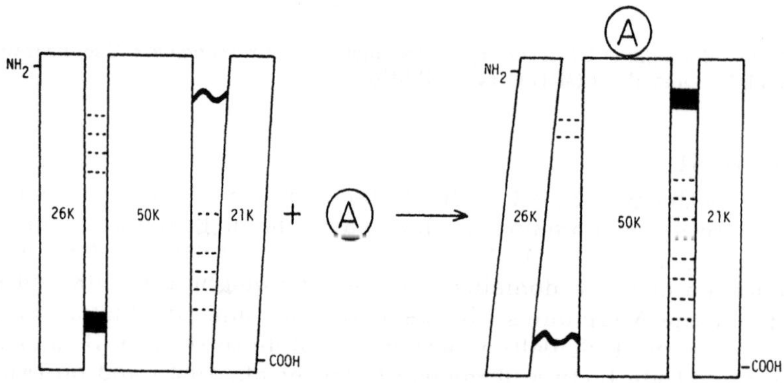

Figure 4: Schematic representation of the conformational change in S1 produced by the binding to actin (Yamamoto and Sekine, 1979). Actin produces changes in the cross-linkability of the 27(26)K and 50K domains and the 50K and 23(21)K domains (shown by the number of joining lines) and in the susceptibility of the 27-50 and 50-23 linkers to proteolytic digestion (solid line is trypsin resistant).

(1982) conclude that at least part of the flexibility arises in the A1 light chain.

Structural Evidence for S1 Binding to Two Actins in Rigor

Decorated Actin

The structure of decorated actin as revealed by negative staining was the subject of the now classical reconstruction of a particle with helical symmetry by Moore, Huxley, and DeRosier (1970). Taylor and Amos (1981) have recently re-examined this structure and suggest that the relationship between the myosin and actin molecules may be rather different from that proposed by Moore et al.. In particular, a modified Taylor-Amos geometry (Amos et al., 1982 - see below) is compatible with the binding of one S1 to two actins on neighbouring strands of the actin double helix as has been suggested by Mornet et al. (1981b).

Rigor Insect Flight Muscle Bathed in Exogenous S1

Studies on insect flight muscle in rigor saturated with exogenous S1 have been made (Barrington Leigh et al., 1977; Holmes et al., 1982). Whereas rigor insect diffraction patterns of insect muscle give rise to outer (actin-based) layer-lines having a ladder-like appearance (Holmes et al., 1980) arising from the periodic marking of the actin helix, the diffraction pattern from insect muscle incubated with rabbit-S1 shows the outer layer-lines to have the form of a helix cross, which is to be expected from decorated actin. The diffraction patterns from such specimens are easier to interpret than from native muscle because the symmetry of the actin helix is more fully expressed. The measured intensities agree well with those calculated by Taylor and Amos from negatively stained decorated actin. Therefore it should be possible to compare and even to combine X-ray and e.m. data. The electron micrograph has the advantage that it yields phases in the Fourier transform. The X-ray diffraction data are more accurate since they average over a very large sample which remains hydrated and is not subject to radiation damage. Thus from the electron micrograph we can obtain the phases and from X-ray fibre diagrams we can obtain the amplitudes. This combination of data should be more accurate than that derived solely from electron micrographs.

Amos et al. (1982) have used the X-ray data as a criterion for the selection of relatively undistorted electron microscope images. Further they have corrected the first layer-line data in the diffraction pattern calculated from the electron micrographs to make them more like the X-ray data in intensity. The three dimensional reconstruction computed from this data shows density connecting part of the image interpreted as the S1 region to the second actin helix, much as can be inferred from the work of Mornet et al. (1981b).

Holmes et al. (1982) have fitted a simple model consisting of a sphere for actin and a roughly cylindrical figure of revolution for S1 (Elliott and Offer, 1978) to the composite data by means of a least squares procedure. Fig. 5 shows two views of the filament generated from the refined

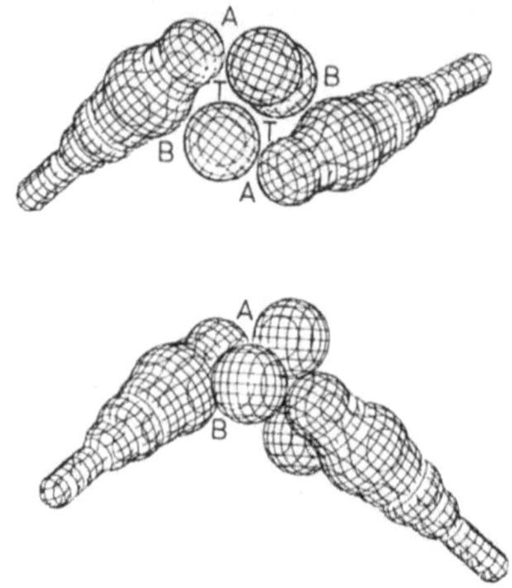

Figure 5: The result of fitting a simple model of actin and S1 to a combined data set derived from X-ray diffraction and electron microscopy (Holmes et al., 1982). Views of three actin monomers and two S1's are shown viewed along and at right angles to the actin filament axis. Note the elongated S1 appears to span the two strands of the actin helix. This geometry generates two distinct actin-S1 binding sites (A and B) on separate actin monomers (cf. Fig. 3). The tropomyosin (T) lies between the two binding sites. The S1, which is markedly slewed with respect to the actin filament axis, seems to slope down at an angle of about 35° and follows the trace of the 5.9 nm genetic helix.

coordinates. Note that the cross bridge slews across the actin helix so that it is in contact with two neighbouring actin monomers on different actin long-helices. It takes on approximately the angle of the actin monomers (ca 35°) and in fact lies more or less along the genetic helix of the actin filament.

This calculation appears to support the modified Taylor-Amos geometry. Moreover it appears to provide a structural basis for the observations of Mornet et al. (1981b).

The Active State

Changes on Activation

In frog muscle the change from resting to active state is characterised by some movement of mass towards the actin leading to the changes on the equator of the diffraction pattern. This effect is accompanied by the weakening of the myosin layer-line (Huxley and Brown, 1967) whereas the 14.32 nm meridional reflection becomes even stronger (Huxley et al., 1982) than in resting muscle. Noteworthy is the fact that the weakening of the myosin layer-lines is not accompanied by an increase in the strength of the rigor (actin based) layer-lines. Huxley et al. (1982) estimate that in the active state the 36 nm layer-line is at least 30 times weaker than the layer-line in rigor.

The fact that the 14.32 nm reflexion becomes stronger on going from the resting to the active state argues for this reflexion arising at least in part from actin-bound cross-bridges. In point of fact a 14.32 nm repeat does not exist in the actin structure. However, it could be achieved in an average sense by binding to actin monomers irregularly every second (11 nm) or third (16.5 nm) site but only if a considerable degree of lateral (azimuthal) freedom in the angle of binding of the cross-bridge to actin can be tolerated.

If it is true that the 14.32 nm reflexion *does* arise from actin-bound cross bridges, we have to explain why at the same time the most typical actin layer-line at 36 nm (corresponding to the actin long helix) remains invisible. The most readily available explanation is azimuthal disorder. Apparently, although in active muscle as judged from the equatorial intensities a goodly proportion of the cross-bridges are attached, the mass of cross bridge *stereospecifically* related to actin (i.e. with the actin helix geometry) in the active state is rather small.

X-ray Transient Response

Huxley et al. (1981) (and these preceedings) report that large changes in the 14.32 nm reflexion are observed on quick release or quick stretch of active frog muscle. The changes in the 14.32 nm reflexion may be categorized under three headings: They are fast, reversible and symmetric. One deduction which may apparently be drawn from these observations is that the actin-bound cross-bridges *can* display a 14.32 nm period, as was suggested above. The myosin geometry overrides the actin geometry and dictates the z-heights of the bridges (cf. rigor).

Of the classes of hypothesis which may be invoked to explain the extinction of the 14.32 nm on quick release or quick stretch the most readily understandable would be a change in angle of the cross-bridge, as has been suggested by Huxley et al. (1981). Simple trial calculations with an elongated cross-bridge show that this is possible if the magnitude of the movement is large enough (larger than the 35° inclination characteristic of rigor).

Cross-Bridge Flexibility

The geometry of the filaments and their arrangement in a lattice lead to considerable distortions in the cross-bridges. These distortions may be roughly classified as longitudinal or azimuthal. Typically in rigor the distortions are mostly longitudinal whereas in active muscle the distortions are inferred to be largely azimuthal. These changes apparently reflect changes in cross-bridge flexibility in the two states. In rigor a rigid stereospecificity is imposed on the actin-cross-bridge union leading to the majority of cross-bridges taking up the actin symmetry and pulling away from the myosin filament enough cross-bridge tail to make this possible. In active muscle the 14.5 nm registration of the cross-bridges is preserved through a high degree of flexibility in the actin-cross-bridge union. This is one of the most noteworthy differences between the active state and the rigor state.

Results with Probes

We mention paranthetically the results obtained with probes (Cooke and Thomas - this symposium; Yanagida - this symposium) attached respectively to the rective SH1 group (23K domain) and to the ATP binding site (27K domain). Both show that in the active state some part of the S1 moity remains bound to actin with a rigor-like geometry and, moreover, that no alternative kind of bound state can be distinguished.

A Model for the Power Stroke

The fact that the components display high binding constants puts a limit on the minimum surface area for an acto-myosin contact or for postulated intra-myosin domain contacts. The concept of a hinge actually *at* the acto-myosin contact seems, therefore, rather unlikely. On the other hand, well documented examples of protein hinges involving designed-in flexibility of the protein main-chain (e.g. tobacco bushy stunt virus) are known. Moreover, we know already of at least two hinges in the myosin backbone.

The fact that the rigor geometry is not detected in active muscle (Huxley et al., 1982) puts considerable restraints on the degree to which the rigor configuration can be represented in the cross-bridge cycle. In view of the results of Huxley et al. it is necessary to assume that the greater part of the power stroke can be accomplished *without* the cross bridge having to take up the strict geometry of the rigor state. This in turn implies that some internal rearrangement of the S1 moity is capable of generating force without recourse to the rigor interaction. However, in contraction, those bridges which become mechanically discharged use the steric freedom provided by the unloading of the connections to the myosin filaments to bind to actin with the full rigor geometry for a short time. In this period the products of hydrolysis are released and a new ATP molecule can be bound leading rapidly to the dissociation of the cross-bridge from actin.

One is led rather naturally to the following kind of hypothesis. In the active state a small domain (the "nose cone") first binds weakly to the actin filament (Fig. 6). This is joined to the body of the S1 by a hinge flexible in the azimuthal direction (with the hinge axis roughly parallel to the actin filament). The binding triggers a mechanism allowing a rearrangement of the domains of S1 producing an elastically strained configuration which does work on the acto-myosin lattice (the "power stroke") leading possibly to the formation of the second actin binding *if the geometry is favourable.* Then the myosin cross bridge transitorily reaches the rigor state (Fig. 5). However, when the second binding site is reached, the products are rapidly released, making the nucleotide binding site free for a fresh ATP molecule which in turn causes the cross-bridge to let go of the actin filament and return to the starting position.

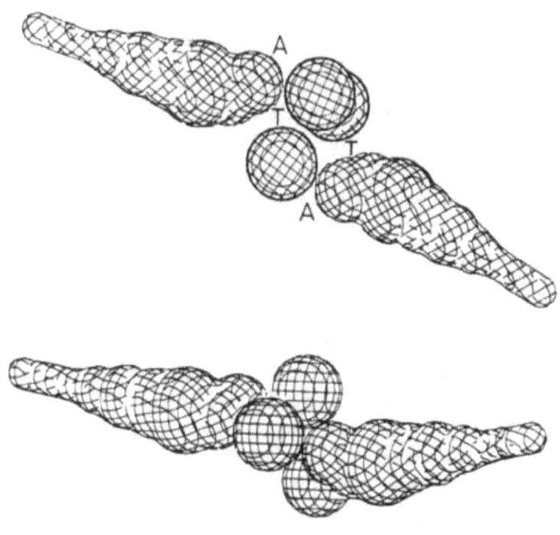

Figure 6: A possible initial binding mode of the S1 to actin in which only the A site is formed. A bend in the S1 in the neighbourhood of the A site is depicted. It is envisaged that this part of the S1 is flexible so as to allow the nose cone (A-site) to bind in the variety of geometrical situations arising between actin and myosin filaments. The power stroke is envisaged as a rearrangement of the S1 so as to produce the necessary displacement terminated by two filaments (A and B site) binding as in rigor (Fig. 5).

SUMMARY

We suggest that the following list of cross-bridge properties can account for much of the currently available evidence:

1. In the active state for the duration of the power stroke only one actin binding site is present (the nose cone). Thus the rest of the S1 does not have to conform to the decorated actin geometry and no rigor diffraction pattern is produced. Rearrangement of the three domains of the S1 and the two light chains takes place during the power stroke.

2. At the end of the power stroke the bridge reaches a state which is energetically similar to the rigor state. A transition to the rigor state enables the cross-bridge to discharge its nucleotide and start a new cycle.

3. In rigor practically all cross bridges bind to the actin via the nose cone but for geometrical reasons not all can achieve the rigor link (in insect only 50%). One tends to see preferentially those bridges which have taken up the rigor geometry.

4. AMPPNP shifts the equilibrium towards the less stereospecific nose cone binding, the extent depending on prevailing conditions.

5. The nose cone contains (or is intimately associated with) the SH1 group which therefore has a fixed geometrical relationship to actin.

Necessarily this list is incomplete and conjectural. We hope, however, that it will provide a basis for discussion and may serve as a stimulus for experiment.

REFERENCES

Amos, L.A., Huxley, H.E., Holmes, K.C., Goody, R.S. and Taylor, K.A. (1982). Structural evidence that myosin heads may interact with two sites on f-actin. Nature. 299: 467-469

Barrington Leigh, J., Goody, R.S., Hofmann, W., Holmes, K., Mannherz, H.G., Rosenbaum, G., and Tregear, R. (1977). The interpretation of X-ray diffraction from glycerinated insect flight muscle fibre bundles: new theoretical and experimental approaches. In: *Insect Flight Muscle*, Proceedings of the Oxford Symposium 3rd-5th April, 1977. (ed. R.T. Tregear), North-Holland Amsterdam. pp. 137-143.

Elliott, A. and Offer, G. (1978). The shape and flexibility of the myosin molecule. J. Mol. Biol. 123: 505-519.

Greene, L.E., Chalovich, J., and Eisenberg, E. (1982). SH1-SH2 cross-linked myosin subfragment-one (S-1): an "analogue" of S-1. ATP. Biophys. J. 37: 265a.

Highsmith, S., Akasaka, K., Konrad, M., Goody, R., Holmes, K.C., Wade-Jardetzky, N., and Jardetzky, O. (1979). Internal motions in myosin. Biochemistry 18: 4238-4244.

Holmes, K.C., Goody, R.S., and Amos, L.A. (1982). The structure of S1-decorated actin filaments calculated from X-ray diffraction data with phases derived from electron-micrographs. Ultramicroscopy 9: 37-44

Holmes, K.C., Tregear, R.T., Barrington Leigh, J. (1980). Interpretation of the low angle X-ray diffraction from insect muscle in rigor. Proc. R. Soc. Lond. B. 207: 13-33.

Huxley, A.F. (1974). Muscular contraction. J. Physiol. 243: 1-43.

Huxley, H.E. (1969). The mechanism of muscle contraction. Science 164: 1356-1366.

Huxley, H.E. and Brown, W. (1967). The low-angle X-ray diagram of vertebrate striated muscle and its behavior during contraction and rigor. J. Mol. Biol. 30: 383-434.

Huxley, H.E., Faruqi, A.R., Kress, M., Bordas, J., and Koch, M.H.J. (1982). Time-resolved X-ray diffraction studies of the myosin layer-line reflexions during muscle contraction. J. Mol. Biol. 158: 637-684

Huxley, H.E., Simmons, R.M., Faruqi, A.R., Kress, M., Bordas, J., and Koch, M.H.J. (1981). Millisecond time-resolved changes in X-ray reflections from contracting muscle during rapid mechanical transients, recorded using synchrotron radiation. Proc. Natl. Acad. Sci. U.S.A. 78: 2297-2301.

Karn, J., McLachlan, A.D., and Barnett, L. (1982). The unc-54 myosin heavy chain gene of *Caenorhabditis elegans:* genetics, sequence, structure. In: *Control of Muscle Development.* Cold Spring Harbor Laboratory of Quantitative Biology, Pearson, M., Epstein, H. (eds). (in Press).

Moore, P.B., Huxley, H.E. and DeRosier, D.J. (1970). Three-dimensional reconstruction of f-actin, thin fragments and decorated thin filaments. J. Mol. Biol. 50: 279-295.

Mornet, D., Betrand, R., Pantel, P., Audemard, E., and Kassab, R. (1981a). Proteolytic approach to structure and function of actin recognition site in myosin heads. Biochemistry 20: 2110-2120.

Mornet, D., Betrand, R., Pantel, P., Audemard, E., and Kassab, R. (1981b). Structure of the actin-myosin interface. Nature 292: 301-306.

Pai, E.F., Sachsenheimer, W., Schirmer, R.H. and Schulz, G.E. (1977). Substrate positions and induced-fit in crystalline adenylate kinase. J. Mol. Biol. 114: 37-45.

Prince, H.P., Trayer, H.R., Henry, G.D., Trayer, I.P., Dalgarno, D.C., Levine, B.A., Cary, P.D., and Turner, C. (1981). Proton nuclear-magnetic-resonance spectroscopy of myosin subfragment 1 isoenzymes. Eur. J. Biochem. 121: 213-219.

Sachsenheimer, W. and Schulz, G.E. (1977). Two conformations of crystalline adenylate kinase. J. Mol. Biol. 114: 23-36.

Taylor, K.A. and Amos, L.A. (1981). A new model for the geometry of binding of myosin crossbridges to muscle thin filaments. J. Mol. Biol. 147: 297-324.

Walker, J.E., Saraste, M., Runswick, M.J. and Gay, N.J. (1982). Distantly related sequences in the α- and β- subunits of ATP synthase, myosin, kinases and other ATP-requiring enzymes and a common nucleotide binding fold. EMBO Journal 1: 945-951

Yamamoto, K. and Sekine, T. (1979). Interaction of myosin subfragment-1 with actin. J. Biochem. 86: 1855-1881.

DISCUSSION

BRESSLER: I'm just curious about the possibility that one head could bind to one filament and the other head to the other filament. I'm sure you've considered that.

HOLMES: Well, this is the basis of Offer and Elliott's suggestion (Nature, 271: 325-329, 1978), isn't it? In insect rigor it probably does happen. I think there are a number of people in this room who would carry on a debate about how *much* it happens, but I don't think anyone would actually doubt that *sometimes* it does happen. On the other hand, I don't think it has to happen. I'm sure there are plenty of situations where it doesn't happen and the muscle contracts perfectly well.

TREGEAR: You asigned the strong binding site to the 50 K and the weak actin binding site to the 21 K. Why was that?

HOLMES: Roger Cooke or Dave Thomas will tell you why.

THOMAS: Both in Dr. Yanagida's talk and in the material Roger and I will present, we'll report negative evidence for rotational motion for two out of three of those domains, namely, the one you indicated as the ATP binding domain and the SH1 containing domain.

HUXLEY: In this model you've described, do you think it would be possible that in a contracting muscle you had both the weak and the strong binding sites made, but with still sufficient flexibility in the rest of the head to give you the relative motion? In other words the bridge could have gone to its end point, still not having moved very much, and then moving as filament sliding takes place.

HOLMES: I think that's rather a nice thought. Perhaps the cross-bridge gets into an unstrained rigor configuration before the products come off. Then we need a signal for release of products.

TIROSH: I want to mention that experiments of crosslinking of S-1 with two neighboring monomers on actin as reported by Kassab's group (Mornet et al., Nature, 292: 301-306. 1981) have been repeated in Dr. Oplatka's laboratory and confirmed. Moreover, a few days ago I heard Dr. Oplatka reporting that S-1 cross-linked to F-actin can give rise to super-precipitation (Oplatka, Levi and Friedman in: "Biological Structures and Coupled Flows" Proc. Katchalsky Mem. Symp., Rehovoth, 1982, in press). Secondly, I want to mention also my experiments with Drs. Low and Oplatka (paper in preparation) using laser light in a very dilute solution of actin in the presence of heavy meromyosin and MgATP. Although you would expect dissociation between actin and HMM, the information suggests that there is a binding of the filaments.

COOKE: Before we start accepting the evidence for two sites, I would like to inject a note of caution. Because S-1 crosslinks with two actins, that does not mean that there is necessarily a *functional* interaction with two actins. Many years ago I checked the binding of subfragment-1 to actin. At low concentrations we found that it bound one actin with a fairly high affinity, and *only* one actin.

HOLMES: Could I answer that before Hugh Huxley? The results of Mornet et al. don't seem to be absolutely and completely consistent, but I hope this is just a trivial problem. The results for the two actins joined with one S-1 seem fine. It comes out with the correct molecular weight. But you would expect higher polymers of some sort. The next species one would predict from just looking at the geometry of the actin helix would be *three* actins and *two* S-1's. The molecular weight of the putative band Mornet et al. get is said to correspond with *two* actins and *two* S-1's. From the geometry I don't see how it could possibly form. It's possible that the gel calibration is a little bit out. Since the molecular weights of the two species are rather similar, one could mistake one for the other. Nevertheless, the note of caution is timely. We have to be careful. However, the "two actin binding sites" is a very attractive concept, and I think we all ought to do a few experiments to try and check it out.

K. WAKABAYASHI: I'd like to ask you a question concerning the mode of binding of cross-bridges in the rigor state of insect flight muscle. Do you now accept the model proposed by Offer's group?

HOLMES: I think we generally would accept the marking scheme of Offer and Elliott (loc. cit.) as a *basis* for thinking about insect flight muscle in rigor. However, I think this is less than the whole truth. From the point of view of comparing calculated and observed X-ray intensities, the Offer-Elliott scheme gives a perfectly satisfactory fit (Offer et al., J. Mol. Biol., 151: 663-702. 1981).

CROSS-LINKING STUDIES RELATED TO THE LOCATION OF THE RIGOR COMPLIANCE IN GLYCERINATED RABBIT PSOAS FIBERS: IS THE SII PORTION OF THE CROSS-BRIDGE COMPLIANT?

Katsuhisa Tawada and Michio Kimura

Department of Biology, Faculty of Science Kyushu University, Fukuoka 812, Japan

ABSTRACT

The muscle tension generation model of Huxley and Simmons (1971) postulates an independent elastic element in the cross-bridge. This elastic structure was tentatively placed in the SII portion of the cross-bridge in the model. To check this assumption, we fixed the SII portion onto the surface of the thick filament in glycerinated rabbit psoas fibers in rigor by chemically cross-linking with dimethyl suberimidate, and compared the stiffness of the cross-linked fibers with that of the fibers before cross-linking. The stiffness was determined by measuring the tension increment upon stretching a fiber segment in rigor. The contribution of the end compliance was found to be small.

Cross-linking increased the rigor stiffness by 20 to 30%. Almost the same amount of the stiffness increase was also observed at a sarcomere length where there was no overlap between the thin and thick filaments, and in a fiber segment cross-linked in relaxing solution. Therefore, the 20 to 30% increase of the stiffness is not caused by the fixation of the SII portion onto the thick filament but caused by the cross-linking of some parallel elastic components. Since the rigor stiffness before cross-linking is almost proportional to the overlap between thick and thin filaments, we conclude that the muscle stiffness in rigor does not originate in the SII portion but reflects some compliance of the head portion of the cross-bridge.

INTRODUCTION

The myosin cross-bridge is compliant (Civan and Podolsky, 1966; Huxley and Simmons, 1971). Ford, Huxley and Simmons (1981) showed that most of the compliance of active intact muscle fibers is in cross-bridges.

The tension generation model of muscle proposed by Huxley & Simmons (1971) contains an assumption of a compliant element in the cross-bridge. In the model, they tentatively placed the compliant element in the SII portion of the cross-bridge. As Huxley (1974) pointed out, the structure and the nature of the cross-bridge compliance are not yet known. To elucidate these points, we have started to study the stiffness of glycerinated rabbit psoas fibers in rigor.

In the paper, we will first show that the compliance of muscle fibers in rigor mostly originates in cross-bridges as well. The myosin cross-bridge consists of the SI and SII portions. To see whether the SII portion in the cross-bridge is compliant, we fixed the entire SII portion onto the surface of thick filaments in rigor fibers with a chemical cross-linking technique (c.f., Ueno and Harrington, 1981), and compared the values of the cross-bridge compliance before and after the SII fixation. We will show that the SII fixation did not make the cross-bridge stiffer and therefore the SII portion is not compliant in rigor fibers.

MATERIALS AND METHODS

Fiber Preparation

Glycerinated rabbit psoas fibers were prepared by a conventional method or by that of Eastwood et al. (1979). Sometimes, fibers "chemically skinned" by a relaxing solution (Eastwood et al., 1979) were used for experiments before glycerination. Glycerinated rabbit psoas fibers were used within 3 months after preparation.

Apparatus and Procedure

The apparatus consisted of a servo motor with a displacement sensor (a home-made servo system made of a speaker voice coil; or an optical scanner G-100PD, General Scanning Inc.), a force transducer (AE 801, Mikro-electronikk), and a position sensor of the diffraction light (United Detector Technology, SC-10) from a helium-neon gas laser (N.E.C., Model GLG-5220, λ = 632.8 nm). The compliance of the servo systems and that of the force transducer were small enough.

One end of a segment of a single fiber or a small bundle of fibers was tied to the force transducer, and the other end, to the servo motor, with surgical silk under a microscope at low temperature below 10°C. Command signals from a pulse generator were fed into a servo controller for quickly changing the length of muscle fiber segments. The step time for the length changes was set to 3 ms in the present experiments to circumvent difficulties due to mechanical oscillation of the motor arm and the transducer beam.

Laser diffractometry to detect small changes in sarcomere length was carried out as described by Naylor and Podolsky (1981).

Cross-linking of Glycerinated Rabbit Psoas Fibers

Glycerinated rabbit psoas fibers were cross-linked with dimethyl suberimidate (DMS) at 4°C, according to the method of Sutoh and Harrington (1977). Rigor fibers were cross-linked in a rigor solution containing 5.5 mM DMS. Relaxed fibers were cross-linked in a relaxing solution containing 5.5 mM DMS. The pH of these solutions was 7.35 at 4°C.

The time course of the cross-linking of the SI and the rod of myosin in muscle fibers was followed by SDS-polyacrylamide gel electrophoresis (SDS-PAGE), according to the method of Chiao and Harrington (1979).

When the time course of the stiffness change during the cross-linking reaction was followed, the stiffness was measured in the cross-linking solution. The pH of the cross-linking solution is different from that of rigor or relaxing solutions without DMS. However, the pH difference (0.35 pH unit) is not crucial for the stiffness measurements because the rigor stiffness is not sensitive to pH in this range.

Solutions

Rigor solution: 80 mM KCl, 40 mM imidazole (pH 7.0 at 4°C), 5 mM EDTA.

Relaxing solution: 80 mM KCl, 40 mM imidazole (pH 7.0 at 4°C), 5 mM EGTA, 5 mM Na_2ATP, 2 mM $MgCl_2$.

RESULTS

We determined the stiffness of glycerinated rabbit psoas fibers in rigor (or in relaxed state) by measuring tension increment at one end of a fiber segment upon giving a quick stretch to the other end of the segment. Lower trace in each panel in Fig. 1 shows tension response of a fiber segment in rigor to stretches before cross-linking.

Fig. 2 shows the relation of the tension increment to the amplitude of stretch. The relation is almost linear. The stress-strain relations are linear if the slack of fiber segments is taken out by application of steady stretching force to the segments. The average Young's elastic modulus of the rigor stiffness, which is given by the slope of the relation, was $2.5\times10^2\pm0.1\times10^2$ (S.E.M., N=14) kg/cm^2. Almost the same value for the stiffness was obtained with three different preparations, i.e., fibers glycerinated in rigor, fibers glycerinated in relaxed state and fresh "chemically skinned" fibers. A similar value for the rigor stiffness has been reported by Güth and Kuhn (1978).

The rigor stiffness did not change when pH was varied from 7.0 to 8.5 in the presence of 80 mM KCl + 40 mM buffer at 15°C.

Small Contribution of End Compliance

The elastic property of muscle near the tying knots in a fiber segment (end compliance) may be different from the true elastic property of the muscle fiber. We checked the contribution of end compliance in the rigor-stiffness measurement by studying the dependence of the Young's

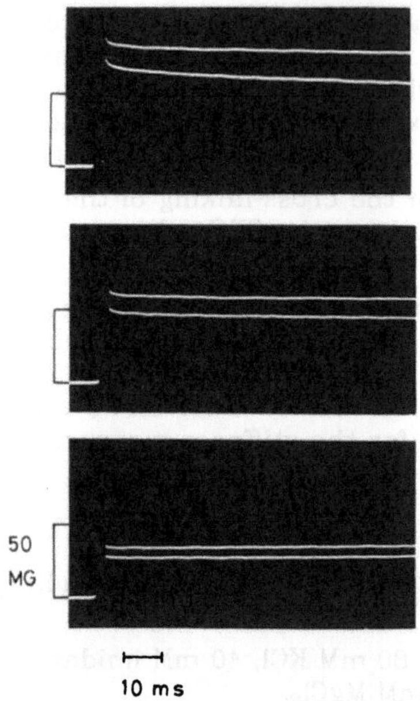

Figure 1: Tension changes due to stretch of a single fiber segment in rigor before (lower trace) and after cross-linking (upper trace). The length changes given at one end of the fiber segment were 0.66% muscle length (top panel), 0.46% muscle length (middle panel), and 0.26% muscle length (bottom panel). Muscle segment length = 0.32 cm, sarcomere length = 2.3 μm, 4°C.

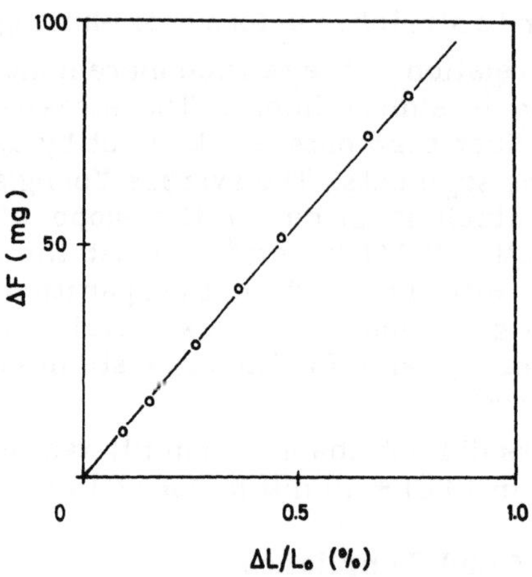

Figure 2: The stress-strain relation of a single fiber segment in rigor before cross-linking. The data were taken from the records (lower trace in each panel) in Fig. 1 and other records of the same fiber segment.

elastic modulus on the fiber segment length at a sarcomere length with full filament-overlap (Details of the method will be published elsewhere.) and also by the laser diffractometry of the sarcomere length change.

Both the two methods gave quantitatively the same results. With a fiber segment of 1 cm length, 93% of a small length change given at one end of the segment went to the sarcomere at its middle portion. With a 1/2 cm fiber segment, 86% of a given length change went to the sarcomere and with a 1/3 cm segment, 80% of a given length change went to the sarcomere. Since the length of fiber segments used in the study was about 0.3 to 0.5 cm or longer, the contribution of end compliance was small.

Dependence of Rigor Stiffness on the Sarcomere Length

To determine the distribution of sarcomere compliance of rigor fibers between cross-bridges and other structures, we studied the dependence of the rigor stiffness on the sarcomere length.

Resting tension of glycerinated rabbit psoas fibers becomes prominent at sarcomere lengths greater than 2.7 μm. Correspondingly, resting stiffness of the fibers appeared above this sarcomere length. The resting stiffness was almost constant at the sarcomere lengths greater than 3.0 μm and its value was about 10% of the rigor stiffness at a sarcomere length with full filament-overlap. The resting stiffness took almost the same value even at sarcomere lengths far beyond the end of the filament-overlap, suggesting that the resting stiffness originates from some parallel elastic components rather than the interaction between cross-bridges and thin filaments in relaxed fibers under our solution condition.

The stiffness of a fiber segment in rigor with no filament-overlap was the same as the resting stiffness of the fiber segment at the same sarcomere length, suggesting that the elastic property of the parallel elastic components in rigor fibers is the same as in relaxed fibers. Thus, the resting stiffness was subtracted from the stiffness value of a fiber segment in rigor at each sarcomere length for measurement and the corrected value of rigor stiffness is shown as a function of sarcomere length in Fig. 3. No correction was necessary below the sarcomere length of 2.7 μm because of no resting stiffness.

The corrected rigor stiffness was constant at the sarcomere lengths between 2.1 and 2.4 μm. Above the sarcomere length of 2.4 μm, the rigor stiffness began to decrease with the decrease of the overlap between thick and thin filaments. The slope is fairly linear. This fact indicates that the sarcomere compliance of rigor fibers is mostly the cross-bridge compliance. (A more detailed analysis for the determination of the sarcomere compliance distribution will be published elsewhere.)

Effect of Cross-linking on the Rigor Stiffness

Fig. 4 schematically illustrates how the SI and rod portions of myosin are cross-linked onto the surface of thick filaments.

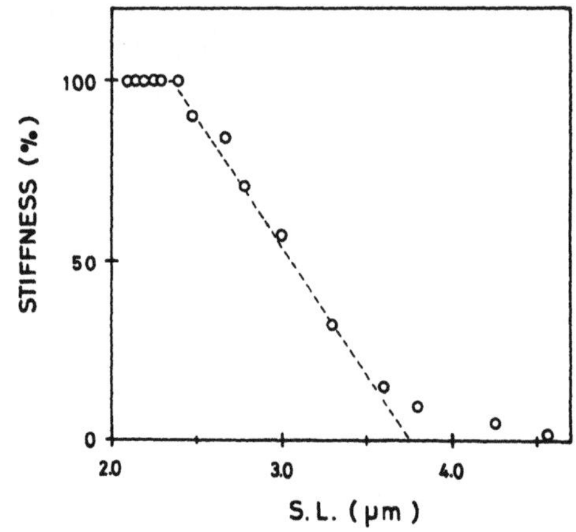

Figure 3: Dependence of rigor stiffness of a single fiber segment on the sarcomere length. The initial length of the fiber segment at the shortest sarcomere = 0.30 cm. 13°C

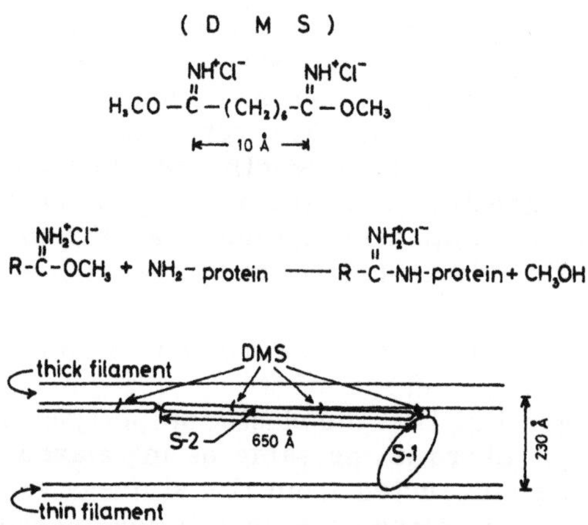

Figure 4: Schematic illustration of cross-linking of SI and rod portions of myosin in muscle. The length of the cross-linker (DMS) is about 10 Å and short enough for the purpose of our experiments, because muscle fibers are stretched by more than 100 Å per half sarcomere for the stiffness measurement.

The SI portion is not cross-linked to and is not induced to detach from, thin filaments by DMS cross-linking of rigor fibers (Ueno and Harrington, 1981; Kimura, unpublished data).

The broken lines in Fig. 5 show time courses of cross-linking SI and rod portions in glycerinated fibers in rigor and in relaxed state. Within 5 h after the start of the cross-linking reaction, the entire SII portion of more than 80% of the cross-bridges was cross-linked onto the surface of thick filaments.

The upper trace in each panel in Fig. 1 shows tension response to quick stretches of a single fiber segment in rigor after cross-linking for 8 h with DMS. The lower traces show tension response to quick stetches of the same fiber segment before the cross-linking. The stress-strain relation after the cross-linking was linear as that before the cross-linking was. The rigor stiffness increased only by 25% after the cross-linking. This fact immediately concludes that the SII portion is, if it is compliant, not the only compliant element of the cross-bridge in rigor fibers.

If the small stiffness increase observed after the extensive DMS cross-linking (Fig. 1) were due to the fixation of the SII portion onto the surface of the thick filaments, the stiffness increment would be smaller in proportion to the decrease of the filament overlap and there would be no increase in the stiffness at sarcomere lengths with no filament-overlap. However, this was not the case as shown in Fig. 5.

Fig. 5 shows time courses of the rigor stiffness increase during the cross-linking of two single fiber segments of different sarcomere lengths: one with full filament overlap and the other with no filament overlap. The time courses of the stiffness increment were very similar to each other. A similar increment of the rigor stiffness was observed with another fiber segment with half filament overlap. The stiffness of a single fiber segment in relaxing solution also increased by a similar amount after DMS cross-linking, although the fiber segment remained to be relaxed after the cross-linking.

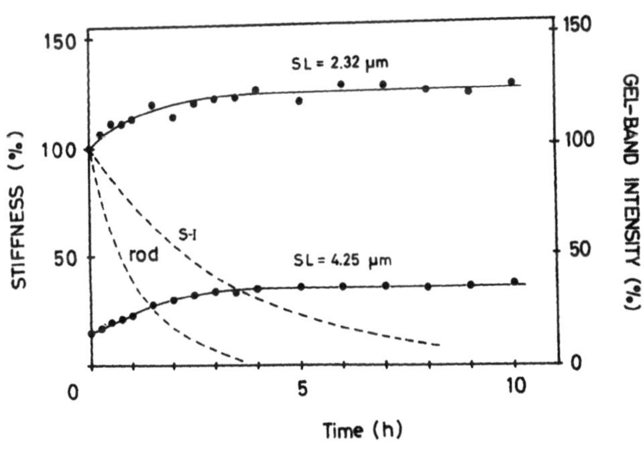

Figure 5: Time courses of cross-linking SI and rod segments of myosin in rigor and relaxed fibers (broken lines), and of stiffness increase of two single-fiber segments in rigor at different sarcomere lengths during cross-linking at 4°C. Top curve: sarcomere length = 2.32 μm (0.31 cm segment length); bottom curve: sarcomere length = 4.25 μm(0.465 cm segment length). The time courses of cross-linking SI and rod segments were followed by measuring the amount of uncross-linked SI and rod segments with SDS-PAGE.

CONCLUSION

From these results, we conclude that the small stiffness increase after DMS cross-linking of rigor ribers was produced by cross-linking of some structures other than cross-bridges (probably parallel elastic components) in muscle. In other words, the fixation of the SII portion onto the surface of the thick filament did not make the cross-bridge stiffer: the SII portion is not compliant in rigor fibers.

DISCUSSION

This study showed (a) that the compliance observed with rigor fibers is mostly the cross-bridge compliance, and (b) that the SII portion of the cross-bridge is not compliant.

The SII portion of the cross-bridge in rigor fibers is in close contact with the thick filament surface at neutral pH but moves away from the surface when the pH is raised to alkaline pH (Ueno and Harrington, 1981). Nonetheless, the stiffness of rigor fibers did not change when pH was raised from 7.0 to 8.5 as described in the paper. Thus, the reason why the SII fixation did not make the cross-bridge stiffer is not that the SII portion has been tightly bound to the thick filament surface in rigor fibers already before the cross-linking of the SII portion with DMS.

Since the cross-bridge consists of the SI portion and the SII portion, an inference from our two results is that the SI portion is compliant or the angle of the SI binding to thin filaments is not rigid. However, paramagnetic resonance (Cooke, 1981) and X-ray (Naylor and Podolsky, 1981) studies of rigor fibers showed that the SI head bound to thin filaments does not rock in rigor fibers when the fibers are stretched. Therefore, the most likely explanation for the cross-bridge compliance is to assume that the proximal domain in the SI head behaves as a stretchable spring or a bendable spring (Fig. 6). Since the SI head has a "pear-like" shape with a thinner proximal end (Elliott and Offer, 1978), the X-ray study by Naylor and Podolsky (1981) may not have detected such structural changes as those occurring in the proximal domain of the SI head in rigor fibers.

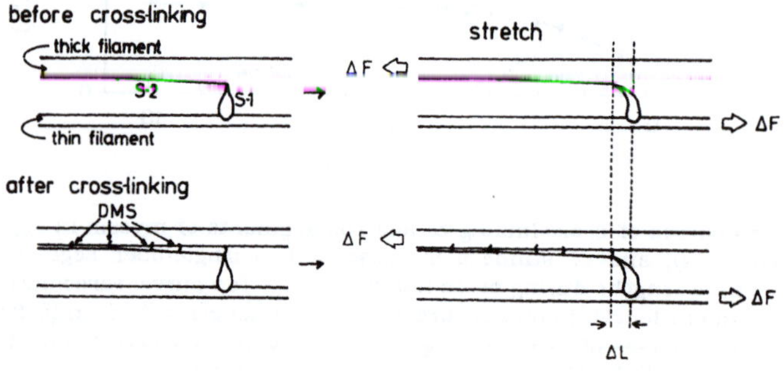

Figure 6: The model of the cross-bridge compliance in rigor fibers.

Analyzing the physicochemical aspects of the muscle force generation model proposed by Huxley and Simmons (1971), Eisenberg & Hill (1978) suggested that the assumption of an independent compliant element in series with the cross-bridge head is unlikely and the cross-bridge head itself is more likely compliant. We believe that the cross-bridge compliance resides in the proximal domain in the cross-bridge head. Since the active site for the ATPase activity of myosin also resides in the head, the cross-bridge ATPase activity may be more directly coupled with structural changes in the compliant proximal domain of the head.

REFERENCES

Chiao, Y.-C.C. and Harrington, W.F. (1979). Cross-bridge movement in glycerinated rabbit psoas muscle fiber. Biochemistry 18: 959-963.

Civan, M.M. and Podolsky, R.J. (1964). Contraction kinetics of striated muscle fibers following quick changes in load. J. Physiol. 184: 511-534.

Cooke, R. (1981). Stress does not alter the conformation of a domain of the myosin cross-bridge in rigor muscle fibers. Nature 249: 570-571.

Eastwood, A.B., Wood, D.S., Bock, K.L. and Sorenson, M.M. (1979). Chemically skinned mammalian skeletal muscle. 1. The structure of skinned rabbit psoas. Tissue. Cell. 11: 553-556.

Eisenberg, E. and Hill, T.L. (1978). A cross-bridge model of muscle contraction. Prog. Biophys. Mol. Biol. 33: 55-80.

Elliott, A. and Offer, G. (1978). Shape and flexibility of myosin molecule. J. Mol. Biol. 123: 505-519.

Güth, K. and Kuhn, H.J. (1978). Stiffness and tension during and after sudden length changes of glycerinated rabbit psoas fibers. Biophys. Struct. Mech. 4: 223-236.

Huxley, A.F. and Simmons, R.M. (1971). Proposed mechanism of force generation in striated muscle. Nature 233: 533-538.

Huxley, A.F. (1974). Review lecture muscular contraction. J. Physiol. 243: 1-43.

Naylor, G.R.S. and Podolsky, R.J. (1981). X-ray diffraction of strained muscle fibers in rigor. Proc. Natl. Acad. Sci. U.S.A. 78: 5559-5563.

Sutoh, K. and Harrington, W.F. (1977). Cross-linking of myosin thick filaments under activating and rigor conditions. A study of the radial disposition of cross-bridges. Biochemistry 16: 2441-2449.

Ueno, H. and Harrington, W.F. (1981). Cross-bridge movement and the myosin hinge in skeletal muscle. J. Mol. Biol. 149: 619-640.

DISCUSSION

EDMAN: Suppose you've tied the S-2 very closely to the myosin filament, and then widened the lattice by raising the pH.

What happens? Do you reach a degree of swelling where you can't produce rigor bridges because the bridges can't swing out far enough?

TAWADA: I do not know.

HOMSHER: Following the same line of questioning, if the filament lattice swells as the fibers shorten, if the S-2's are really tied down, your treated fibers should cease shortening much longer sarcomere lengths than non-crosslinked fibers, since the bridges would not be able to reach the thin filaments.

TAWADA: But the filament lattice does not expand with shortening in glycerinated fibers.

MARTYN: Once the cross-bridges are crosslinked, can you then impose a swelling condition while the lattice is swollen, or is the lattice restrained by the crosslinking?

PODOLSKY: We tried one experiment like that, and if I remember the answer, the radial expansion was much less in the crosslinked fiber. Is that right, Bill?

HARRINGTON: Yes, that's correct.

MAGID: I have a similar result with glutaraldehyde.

GORDON: I have a question about the conditions used for the crosslinkage. Did you try crosslinking, say, either in the presence or absence of ATP, so that, for example, the bridges could be either free to move during the crosslinking process or attached onto the thin filament as in rigor.

TAWADA: Yes. We crosslinked muscle fibers in rigor as well as in the relaxed state. The time constants for the cross-link were the same in both cases (Fig. 5).

GORDON: Were the stiffness changes the same in the two cases?

TAWADA: Yes, if you cross-link in relaxed fibers, stiffness increases about 40% in relaxing solution. When you transfer cross-linked relaxed fiber to a rigor solution, you get almost the same rigor stiffness with the fiber as with fibers crosslinked in a rigor solution.

MAGID: I'm concerned about where the stiffness change that you see with crosslinking is occurring. I think that your evidence is fairly strong that it's not occurring in the S-2 or in the rigor bridge. My question is, do you mechanically remove the sarcolemma by dissection, or is it present in these preparations? The reason I ask is that I did a few experiments, not very many, and found that when the sarcolemma was still present (having just triton-skinned) I saw a much bigger increase in stiffness fixing with glutaraldehyde than one sees when the sarcolemma has been mechanically removed. Do you take it off?

TAWADA: Yes. I compared the stiffness of rigor fibers with and without the sarcolemma. The stiffness was the same.

SUGI: You showed your sarcomere length vs. stiffness relation and concluded that the compliance originates from the cross-bridge. But Ford, Huxley and Simmons (1981) showed that half the compliance was from the filaments, while the other half was from the bridges.

TAWADA: I'm worried about this myself. I borrowed the equations from the Appendix of Dr. Ford, Huxley and Simmons and applied them to my results. The analysis shows that more than 90% of the rigor stiffness is in the cross-bridges, and less than 10% is in the filaments.

SUGI: As far as I understand, this relation in your paper stands on the assumption that the isometric force should be proportional to the amount of overlap, and this is also a matter of argument.

TAWADA: No, that is not necessary. But we do have to assume some relation between the filament overlap and the sarcomere length. That's a fundamental assumption.

HOLMES: Can you relax the cross-linked fibers?

TAWADA: If you cross-link rigor fibers for a couple of hours you can, but not if you cross-link them very extensively. But if you cross-link the relaxed fiber, the fiber remains relaxed.

HUXLEY: But if you cross-link a relaxed fiber, will it then contract?

TAWADA: Initially, for the initial two or three or four hours they can contract. Eventually, they lose contractility.

HUXLEY: But are you saying that you can have a fiber in which the S-2 is firmly cross-linked to the backbone, and that will contract?

TAWADA: No, they don't. That's probably because the muscle is too stiff.

HUXLEY: Why?

TAWADA: Because I've showed that the cross-linked, relaxed fiber has a significant stiffness -- the Young's elastic modulus is about 40% of the value in rigor fibers. So if you observe shortening of the fibers after cross-linking, the shortening velocity becomes very slow, but finally, if you wait for a long time, you see the complete contraction of the muscle. But the shortening velocity becomes slower with fibers cross-linked for longer time, and eventually the fiber loses contractile activity. This is if you cross-link fiber too long, eight hours for example.

EDMAN: Isn't that another way of saying that you can't loosen up the bridges again, or that you probably can't dissociate the bridges again by putting in ATP; not, at least, to the same extent as would occur normally?

TAWADA: I don't know. But I have a comment. Fibers cross-linked in rigor solution lose contractile ability as do fibers cross-linked in relaxing solution, although no rigor cross-bridge can be formed in the latter fibers. This is because we can stretch the fiber by hand from a sarcomere length of 2.4 μm to 4.2 μm in relaxing solution. On the other hand, you can't stretch fibers cross-linked in rigor, even in a relaxing solution.

KAWAI: I think what we're asking is how selective this cross-linking procedure is. In other words, regarding selectivity -- is it limited to the S-2 portion of the cross-bridge?

TAWADA: No. The rod and S-1 of myosin are both cross-linked. Initially the rod portion and finally proximal end of S-1 are cross-linked.

PODOLSKY: I think that many people are trying to find out if the S-2 is cross-linked to the myosin filament, whether under those conditions the ability of the cross-bridge to turn over in a normal way can be demonstrated.

TAWADA: I don't know, because muscle loses contractility eventually.

PODOLSKY: But I think if you cross-link for only about four hours, you have substantial cross-linking, and then under those conditions you retain contractility.

TAWADA: Yes, but the muscle shortens with very, very slow velocity.

PODOLSKY: And does it develop full active force?

TAWADA: I didn't measure force. I simply observed shortening of the muscle under the microscope.

PODOLSKY: I think it might be an interesting experiment if you kept the length constant and waited for a long time to see whether force develops.

ANGLES OF FLUORESCENTLY LABELLED MYOSIN HEADS AND ACTIN MONOMERS IN CONTRACTING AND RIGOR STAINED MUSCLE FIBER

Toshio Yanagida

Department of Biophysical Engineering, Faculty of Engineering Science, Osaka University, Osaka, Japan

ABSTRACT

The measurement of polarized fluorescence from a fluorescent ATP or ADP analog, ε-ATP or ε-ADP (1, N^6 -Etheno-ATP or ADP) bound to myosin heads in a glycerinated muscle fiber revealed that during isometric contraction at low concentrations of Mg-ε-ATP the bound nucleotides are highly oriented with respect to the fiber axis, and their mean angle is almost the same as that in rigor. Furthermore, the polarization of tryptophan fluorescence did not change when the fiber was transferred from a rigor state to an active state at low ATP concentrations. Thus, if a rotation of myosin heads occurs during contraction, it seems to be very limited. The angle of bound ε-ADP is not changed by passive stretching of the fiber in the presence of ε-ADP.

By using the same technique, it was found that orientation of a phalloidin-FITC complex (Ph-FITC) specifically bound to F-actin in a glycerinated muscle fiber is significantly changed when the fiber is activated from relaxation or rigor to contraction. No change in orientation of Ph-FITC bound to F-actin is induced by simple addition of Mg-ADP or by passive stretching of the fiber in the absence of ATP. These results suggest that a rotation or a distortion of actin monomers occurs during contraction, which is involved in the process of active tension development.

INTRODUCTION

Muscle contraction is due to interaction between actin and myosin molecules coupled with hydrolysis of ATP. It has been long expected that a rotational movement of myosin heads on an actin filament causes a translation between the two filaments (Huxley, 1969; Huxley and Simmons, 1971). However, till now no direct evidence for such a movement has been obtained. In the sliding theory, F-actin has been often treated as a simple rigid rod. Many studies, however, have shown that the actin filament is not rigid but flexible, and the flexibility changes with changes in the physiological states of muscle (see Oosawa, 1980). The flexibility of F-actin or thin filament is almost enough to explain the muscle compliance (Oosawa, 1974). As pointed out by A.F. Huxley (1974), a rotation of myosin heads in Huxley and Simmon's model may be replaced by that of actin monomers. Thus, it is still unknown whether movement of myosin heads or that of actin monomers (or both) is involved in the process of active tension generation.

To investigate movements of both actin monomers and myosin heads, we have measured the polarized fluorescence from fluorophores attached to actin monomers and myosin heads in rigor, relaxation, and activation states of glycerinated muscle fibers, and also during passive stretching of the fiber. The method of polarized fluorescence, which was first used by Morales and co-workers to study the conformation of myosin heads in the fibers, is a powerful means of detecting changes in orientation of fluorescently labeled actin and myosin heads.

In the case of myosin, fluorescent analogs of ATP and ADP; i.e., ε-ATP and ε-ADP were used. ε-ATP can replace ATP in the myosin ATPase reaction (Onishi, et al., 1973) and induce 70 to 80% of tension in comparison with ATP (Yanagida, 1981a). On the other hand, F-actin in the fiber was labeled with a phalloidin-FITC complex. Phalloidin is a bicyclic peptide (MW. 788 dalton) which binds specifically to actin monomers (Lengsfeld, et al., 1974), without any change in such properties of F-actin as activation of myosin ATPase or interaction with regulatory proteins (Danker, et al., 1975).

We have found that during isometric contraction in the presence of low ε-ATP concentration, nucleotides bound to myosin heads are highly oriented and their angle with respect to the fiber axis is almost the same as that of bound ε-ADP in rigor state. Furthermore, no change in the polarized fluorescence from tryptophan residues in the fiber was observed when the fiber was transferred from rigor to activation at low ATP concentrations. On the other hand, the polarized fluorescence from F-actin bound Ph-FITC showed a significant change during isometric contraction. When the fiber was stretched in the rigor state, neither polarized fluorescence from myosin ε-ADP nor that from F-actin·Ph-FITC were changed. Based on these results, the mechanism of tension generation is discussed below.

Part of the results concerning the angle of ε-ATP and ε-ADP bound to myosin heads in a fiber has been published elsewhere (Yanagida, 1980, 1981).

MATERIALS

Fibers of striated muscle were obtained from rabbit psoas muscle. Glycerinated muscle fibers with sarcomere length (SL) from 2.0 μm to 3.8 μm were prepared according to the method previously described (Yanagida, 1981a). A single fiber with a thickness of 40 μm to 80 μm was dissected from the glycerinated fibers and fixed with a tape and collodion in a quartz cell with a dimension of 10 mm × 10 mm.

The composition of a rigor solution was 100 mM KCl, 10 mM phosphate buffer, pH 7.0, 5 mM $MgCl_2$. ε-ATP and ε-ADP were synthesized and purified by the method previously described (Yanagida, 1981a). An ε-ATP regenerating system of PC and PCK was used to keep [ε-ATP] constant. When ε-ADP and/or ADP were used, 100 μM P_1, P_5 deadenosine pentaphosphate was added. Phalloidin-FITC, synthesized according to the method described by Wulf et al., (1979) was a gift from Dr. Wieland.

METHODS

Polarized fluorescence from muscle fibers was measured with a microspectrophotometer as described by Yanagida and Oosawa (1978). The fluorescence due to ε-ATP or ε-ADP bound to myosin (in the presence of ε-ATP or ε-ADP in solution) was obtained from the difference of fluorescence of a fiber in a solution containing ε-ATP or ε-ADP and in another solution containing a more than 10-fold concentration of ATP or ADP over that of ε-ADP or ε-ATP. ε-ATP or ε-ADP were excited at 325 or 335 nm (±0.2 nm) and the emitted light was collected above 360 nm. Tryptophan and Ph-FITC were excited at 295 and 470 nm (+0.2 nm), respectively, and the emitted light was detected with optical filters. Four components of polarized fluorescence (${}_{\parallel}I_{\parallel}$, ${}_{\parallel}I_{\perp}$, ${}_{\perp}I_{\parallel}$, ${}_{\perp}I_{\perp}$) were measured by this apparatus: The subscripts $\parallel$ and $\perp$ denote the direction of polarization parallel and perpendicular to the fiber axis, respectively. The former subscripts are concerned with the incident light and the latter with the emitted light. The degree of fluorescence polarization, $p_{\parallel}$ and $p_{\perp}$ were defined as: $p_{\parallel} = ({}_{\parallel}I_{\parallel} - {}_{\perp}I_{\parallel})/({}_{\parallel}I_{\parallel} + {}_{\parallel}I_{\perp})$, and $p_{\perp} = ({}_{\perp}I_{\perp} - {}_{\perp}I_{\parallel})/({}_{\perp}I_{\perp} + {}_{\perp}I_{\parallel})$. Depolarization due to scattering and birefringence of a rigor muscle fiber (SL = 2.2 μm, d = 50 μm) was about 3% at 325 to 400 nm and about 1.5% at 470 to 550 nm. When a fiber was transferred from relaxation to rigor and relaxation to activation solutions, changes in depolarization at 470 to 550 nm were about 0.7% and 0.2% respectively.

The angles of absorption and emission dipoles of bound fluorophores relative to F-actin axis, φ_A, φ_E, (Fig. 1) were determined from the components of polarized fluorescence as described by Yanagida and Oosawa (1978). The half width of the angular distribution, $\Delta\vartheta_{1/2}$ was also calculated.

When the effect of length change of the fiber on polarized fluorescence was examined, a single glycerinated fiber was mounted to a stainless steel needle, connected to the electromagnetic coil of a loud speaker on one end, and the other was fixed with a tape and collodion. A stepwise change of the length was applied to the fiber by the electromagnetic coil and the polarized fluorescence was monitored. Under the same conditions the tension of the fiber was measured. The changes in polarized fluorescence and tension were recorded with a signal processor (7T07A, Sanei Sokki Co., Japan).

Tension of fibers was measured with a tension detector made with a semi-conductor transducer element (AE 801 Aksjeseiskopet Micro Elktronikk Co., Norway, resonance frequency 7 kHz). Sarcomere length was determined by a laser light (Ne-He) diffraction method.

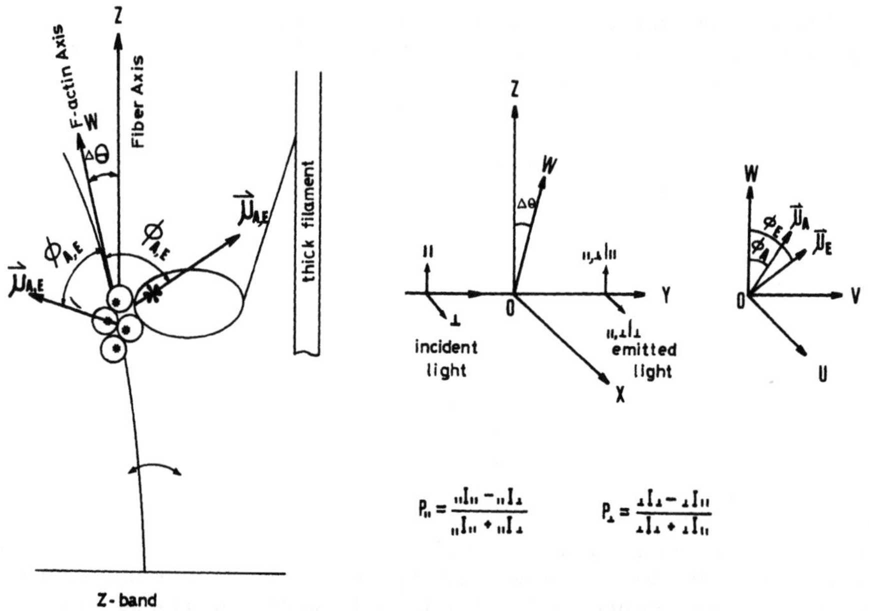

Figure 1: Diagram explaining the polarized components of fluorescence, the angles of absorption and emission, φ_A and φ_E, and $\Delta\vartheta$. φ_A and φ_E are angles between directions of absorption and emission dipoles ($\vec{\mu}_A$, $\vec{\mu}_E$) and F-actin axis, respectively, and $\Delta\vartheta$ is the angle between F-actin and the fiber axis. Oxyz and Ouvw are the laboratory and the molecular frames, respectively.

RESULTS

Angles of ε-ATP and ε-ADP Bound to Myosin Heads in Muscle Fiber.

(i) in the presence of ε-ADP.

The maximum amount of ε-ADP bound to myosin heads in the fiber (SL = 2.1 μm) and the dissociation constant (Kd) of the nucleotide, obtained at 25°C, were 265 μM and 335 μM respectively. At low concentrations of added ε-ADP ($<K_d$), the values of $p_{\parallel}$ and $p_{\perp}$ from the bound nucleotide, were -0.27 and +0.48, respectively, which were close to the values characteristic for an ordered array of fluorophores. ε-ADP bound to myosin heads is highly oriented relatively to the fiber axis (Fig. 2). Calculated values of φ_A, φ_E and $\Delta\vartheta_{1/2}$ were 69°, 66° and 20°, respectively. When the fiber was stretched until the overlap was lost between thick and thin filaments, the values of $p_{\parallel}$ and $p_{\perp}$ ($p_{\parallel}$ = +0.18, $p_{\perp}$ = +0.38) were close to those predicted for completely disordered array of fluorophores (p = +0.32).

As the concentration of free ε-ADP in solution exceeded K_d, $p_{\perp}$ slightly decreased but $p_{\parallel}$ considerably increased. The orientation of bound ε-ADP was made heterogeneous (see Yanagida, 1981a for detail).

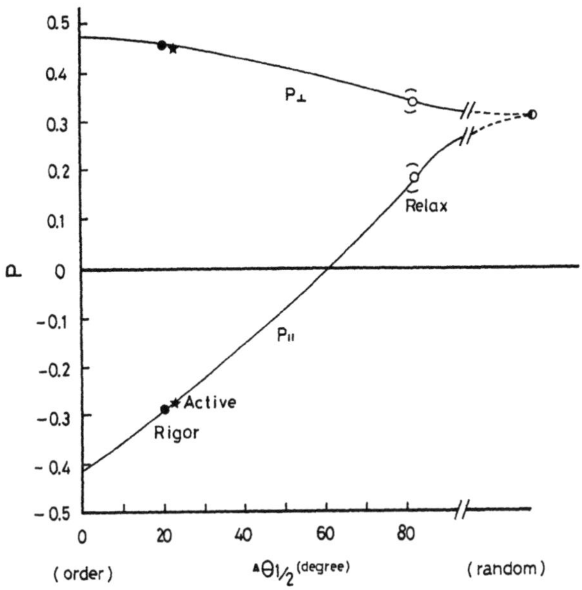

Figure 2: Theoretically calculated $P_{\parallel}$ and $P_{\perp}$ at $\varphi_A = 69°$ and $\varphi_E = 66°$. $P_{\parallel}$ and $P_{\perp}$ are plotted as a function of the half-maximum width of the angular distribution of $\Delta\vartheta$, $\Delta\vartheta_{0.5}$. (●), (O) and (∗) are experimental values from ε-ADP bound to myosin in rigor state and ε-ATP (or ε-ADP) in relaxed and active states, respectively. Calculations were performed according to the method previously described. (Yanagida and Oosawa, 1978).

(ii) during contraction and in relaxed state

The measurement of binding of ε-ATP to the fiber (SL = 2.1 μM) showed that the maximum binding and the dissociation constant were 214 μM and 224 μM at 25°C, respectively. As the concentration of ε-ATP in a rigor solution (containing sufficient amount of ε-ATP regenerating system - 8 mM PC and 700 units PCK/ml) increases, the isometric tension of a single glycerinated fiber increased and reached a peak in the presence of about 200 μM ATP almost independently of Ca^{2+} concentration. With further increase of the ATP concentration, the tension decreased to 80% of the peak tension at pCa = 4 and to a level of relaxation at pCa > 9. The values of $p_{\parallel}$ and $p_{\perp}$ from the bound nucleotides during contraction at less than 100 μM ε-ATP were -0.23 and +0.48, respectively, which is almost the same as the values obtained in the presence of ε-ADP (at concentrations lower than K_d). These results indicated that the nucleotides bound to myosin heads are highly ordered even during tension development and their angles are almost the same as those in the presence of ε-ADP. As the concentration of free ε-ATP exceeded 100 μM, the values of p deviated from the values predicted for the ordered array of the fluorophores. This is probably because the number of myosin heads dissociated from thin filaments, which underwent thermal motion (Thomas et al., 1980), increased with increase in the ε-ATP concentration. In the relaxed state

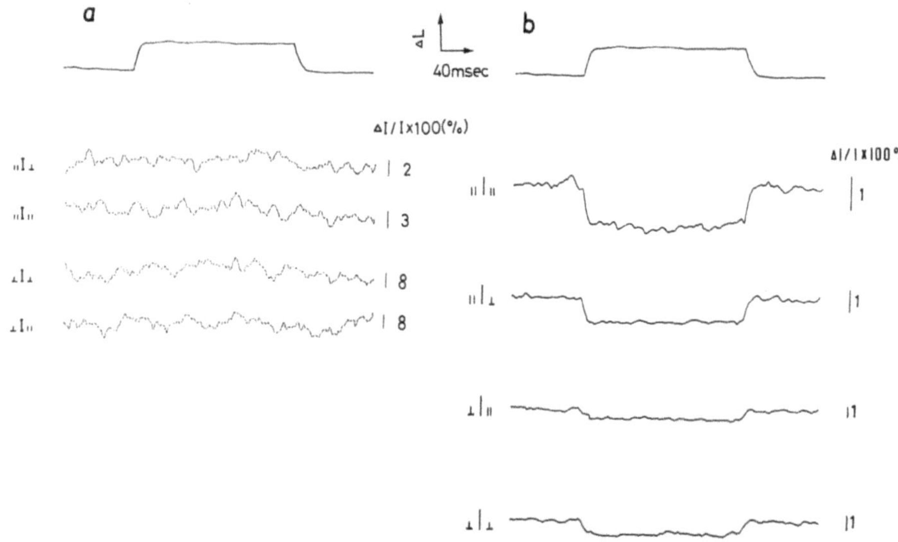

Figure 3: a) Changes in polarized fluorescence from ε-ADP bound to myosin heads in the fiber when suddenly stretched by 1%. Changes in four components of polarized fluorescence were represented as percentages of fluorescence intensities before stretch. Traces show averages of 20 repeated measurements with a time resolution of 10 ms. b) Changes in polarized fluorescence from Ph-FITC bound to F-actin when suddenly stretched or released by 1% in rigor state. Changes in I were represented as in (a). Traces show averages of 20 repeated measurements with a time resolution of 10 ms. The ratios $_{\parallel}I_{\perp}/_{\parallel}I_{\parallel}$ and $_{\perp}I_{\parallel}/_{\perp}I_{\perp}$ did not change.

(1 mM ε-ATP, pCa > 9), $p_{\parallel}$ and $p_{\perp}$ were +0.20 and +0.38, respectively, indicating almost completely random orientation of nucleotides bound to myosin heads (Fig. 2). When the sarcomere length was increased to 3.7 μm, the values of p were close to those in the relaxed state, independent of the concentration of ε-ATP and Ca^{2+}.

(iii)during passive stretching of the fiber in the presence of ε-ADP

The polarized fluorescence from ε-ADP bound to myosin heads in a glycerinated single fiber (SL = 2.1 μm) was measured when the fiber was quickly stretched by 1% of the total length. As shown in Figure 3a, no change in the polarized fluorescence was induced by passive stretching.

Polarized Fluorescence from Trytophan in a Fiber During Isometric Contraction.

In a rigor state, the values of $p_{\parallel}$ and $p_{\perp}$ were 0.310 ± 0.004 and 0.130 ± 0.003, respectively. No appreciable changes in $P_{\parallel}$ and $P_{\perp}$ were observed when a fiber was transferred from a rigor to a contracting state at low ATP concentrations (<10 μM, at PC = 8 mM and PCK = 300 units/ml) at 20°C.

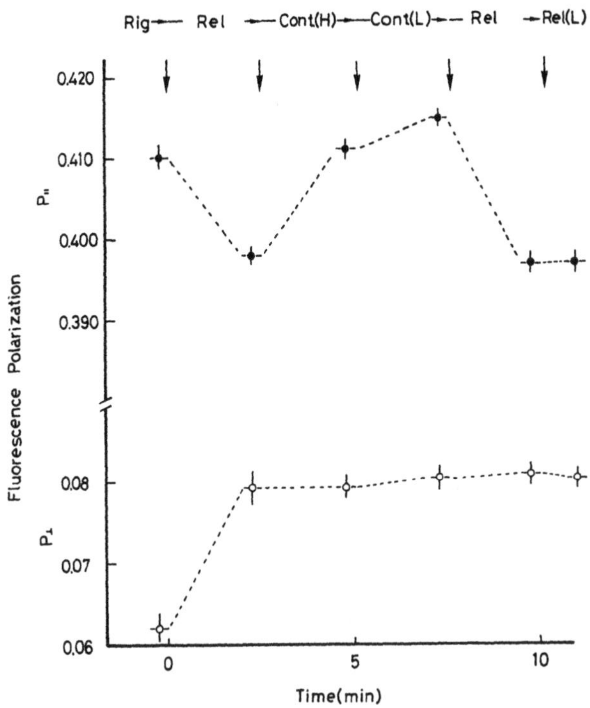

Figure 4: Fluorescence polarization from Ph-FITC bound to F-actin in a single glycerinated muscle fiber. Rig, Rel, Cont (H), Cont (L)-bathing solutions, in which the fiber was subsequently immersed. Rig-rigor solution; Rel-rigor solution + 5 mM ATP + 1 mM EGTA; Cont (H)-rigor solution (100 mM KCl) + 5 mM ATP + 0.1 mM $CaCl_2$, Cont (L)-rigor solution (50 mM KCl) + 5 mM ATP + 0.1 mM $CaCl_2$.

Angles of Ph-FITC Bound to F-actin in a Muscle Fiber

(i) In rigor, relaxed and active state

The binding of Ph-FITC to a single glycerinated muscle fiber was performed by immersing the fiber in a rigor solution containing 5 μM Ph-FITC. The binding was saturated after 20 min incubation at 6°C. Maximum amount of Ph-FITC specifically bound to actin was approximately 1 mole of the toxin per one mole of the protein.

The binding of Ph-FITC to actin interfered neither tension nor its regulation by Ca^{2+}

Fluorescence intensities from Ph-FITC incorporated into a single fiber were very strong and stable; S/N >300 with time constant of 2s. Therefore reliable values of the four components of polarized fluorescence could be obtained within 7 s and, because of such a short time of measurement, the photobleaching effect was negligibly small.

Fluorescence polarization $p_{\parallel}$ and $p_{\perp}$ from Ph-FITC bound to a single glycerinated muscle fiber at 6°C in rigor, relaxed and active states are shown in Figure 4. With SL = 2.2 μm. The values of $p_{\parallel}$ and $p_{\perp}$ in rigor were

0.410 (±0.0015) and 0.062 (±0.002), respectively. When the fiber was immersed in the relaxation solution (i.e. rigor solution + 5 mM ATP + 1 mM EGTA), $p_{\parallel}$ decreased to 0.398 (±0.001) and $p_{\perp}$ increased to 0.079 (±0.003). The extent of this change was independent of the KCl concentration from 50 to 100 mM. When the fiber was subsequently immersed in the activating solution (i.e. rigor solution + 5 mM ATP + 0.1 mM $CaCl_2$), $p_{\parallel}$ increased to 0.411 (±0.001), while $p_{\perp}$ hardly changed. In this case the extent of the change in p was dependent on the ionic strength: when the concentration of KCl in activating solution was decreased from 100 to 50 mM, $p_{\parallel}$ increased further. It is noteworthy that such a decrease in KCl concentration simultaneously results in an increase in isometric tension of the fiber.

Addition of 1 mM ADP instead of ATP to a rigor solution affected neither $p_{\parallel}$ nor $p_{\perp}$.

Polarized fluorescence from the fiber immersed in activating solution was dependent on the sarcomere length: with increase in sarcomere length, the extent of the change in p decreased and only a little change was observed when SL was 3.8 μM.

$p_{\parallel}$ and $p_{\perp}$ measured during isometric contraction were very stable: when the fiber was kept in activating solution for 10 min, no change in $p_{\parallel}$ and $p_{\perp}$ was noticed. Furthermore, when the cycle of contraction-relaxation was repeated four times in the same fiber, very reproducible values of $p_{\parallel}$ and $p_{\perp}$ were obtained. Thus, during the experiments, the overall structure of the fiber was well preserved.

After removal of myosin with a Hasselbach-Schneider solution, $p_{\parallel}$ and $p_{\perp}$ did not change during transition of the fiber from a rigor solution to relaxing or activating solution.

Total fluorescence intensity from the bound Ph-FITC was hardly affected by interaction with myosin in the presence and in the absence of ATP.

(ii) During the passive stretching of the fiber

The changes in the four components of polarized fluorescence from F-actin bound Ph-FITC were measured when a single fiber was stretched by 1% of the total length in rigor at 6°C. The stepwise length change of 1% resulted in tension of about 2 kg/cm^2. The effect of stretching on the intensities of the four components is shown in Figure 3b. An observed small (about 1%) decrease in the absolute values of the intensities of all the components was due to the decrease in the number of fluorophores illuminated by the incident light beam in the stretched fiber. However, the ratio of $_{\parallel}I_{\perp}/_{\parallel}I_{\parallel}$ and $_{\perp}I_{\parallel}/_{\perp}I_{\perp}$ and hence the polarized fluorescence, remained essentially constant. In the presence of ADP, the same result was obtained.

DISCUSSION

Molecules of ε-ADP bound to myosin heads in a glycerinated rabbit psoas muscle fiber are highly oriented with respect to the fiber axis in a rigor solution containing ε-ADP (Fig.2). The small angular distribution observed here ($\Delta\vartheta_{1/2} = 20°$) is almost the same as that of spin labels attached rigidly to a thiol group (SH_1) on a myosin head in a glycerinated muscle fiber in a rigor state and in the presence of ADP ($\Delta\vartheta_{1/2} = 15°$) (Thomas and Cooke, 1980). Since the extent of angular distribution is similar to that of F-actin, or thin filaments in a fiber ($\Delta\vartheta_{1/2} = 16°$) (Yanagida and Oosawa, 1978, 1980), the disorder of orientation of ε-ADP or spin labels bound to myosin heads relative to the fiber axis in a rigor state may be caused by the flexibility of F-actin, i.e. the angle of myosin heads relative to the F-actin axis may be rigidly fixed (Fig. 1).

Nucleotides bound to myosin heads are also highly oriented during isometric contraction at low concentrations of ε-ATP and their angle relative to the thin filament axis is almost the same as that of bound ε-ADP in a rigor state (Fig. 2). At low concentrations of ε-ATP, the dissociated state of myosin heads has only a very small contribution. In the present measurements all states myosin with bound nucleotide contribute to the polarized fluorescence, except for myosin forming rigor complex with actin without nucleotide. One of the possible interpretations of the obtained results is that the angle of myosin heads does not change during contraction. The second one is that if rotational movement of myosin heads occurs, the life time of the actomyosin-(ε)-ADP state, i.e., the end state of rotation, is much longer than others. This possibility is not likely because the life time of actomyosin-ADP is not longer than other states (Tonomura, 1972). The third possibility is that the angle change occurs within the myosin head but the nucleotide binding site is separated from the domain of the head where the angle changes. Recent EPR measurement of spin labels attached to a thiol group (SH_1) on myosin heads show that the probes on myosin heads attached to thin filaments during isometric contraction have the same orientation as that in a rigor state (Cooke et al., 1982).We also have found that the orientation of 1,5-IAEDANS attached mainly to the thiol groups on myosin heads does not change during tension development at low concentrations of ATP (<10 μM) (Kuranaga and Yanagida, unpublished). Furthermore, we found that the polarized fluorescence from tryptophan does not change during isometric contraction at low ATP. (This does not necessarily exclude small angle change because rigor actomyosin complexes also contribute to tryptophan fluorescence and tryptophan is not specific for myosin head). Thus, if a rotation of myosin heads occurs during contraction, it seems to be very limited.

On the other hand, polarized fluorescence from Ph-FITC bound to F-actin significantly changed during isometric contraction. It is interesting that the values of fluorescence polarization during isometric contraction are different from those in a rigor and a relaxed state and not intermediate between the two latter values. Ph-FITC binding interfered with neither the interaction of actin with myosin nor regulatory proteins. Furthermore, total fluorescence intensity from bound Ph-FITC was hardly affected by interaction with myosin. Therefore, it is very likely that the change in polarized fluorescence during contraction is not caused by a direct interaction of myosin with Ph-FITC bound to F-actin, but is due to a

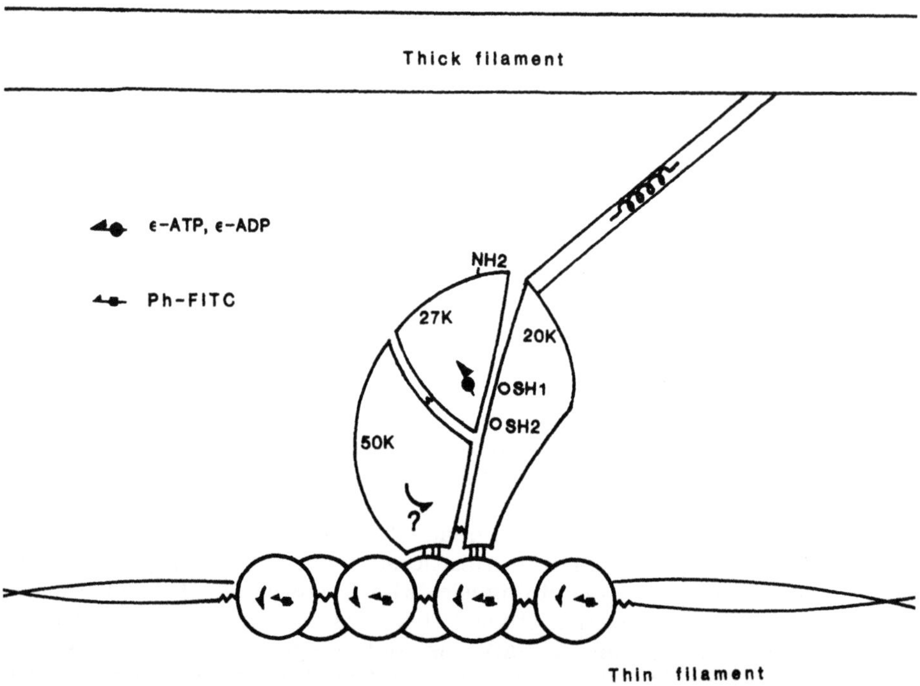

Figure 5: Schematic diagram for explanation of orientation of fluorescently labeled myosin heads and actin monomers.
(a) Interaction between myosin heads and actin monomers in a rigor state (±ADP). Heavy chain of myosin head (S-1) consists of three domains (NH_2-27K--50K--20K------rod), which are joined by flexible junctions, and binds to two actin monomers which are on two strands of the actin helix with 20K and 50K domains attached (Mornet et al., 1981). Nucleotide binding site locates at the 27K domain (Szilagyi et al., 1979) and thiol groups (SH_1 – SH_2 at the 20K domain (Balint et al., 1978). The nucleotide binding site and thiol groups are expected to be close to each other (Sekine and Yamaguchi, 1984). The positions of nucleotide binding site and thiol groups on S-1 are not known and therefore are drawn for convenience here. Ph-FITC is bound to a domain of the actin monomer far from both myosin and regulatory protein binding sites.
(b) Interaction of myosin heads and actin monomers during tension development. The angle of nucleotide (probably ε-ADP-P_i) on myosin heads is almost the same as that in a rigor state (+ ε-ADP)(Yanagida, 1980,1981a). Cooke et al. (1982) have found that spin labels attached to SH_1 of myosin heads on thin filaments have the same angle as in the rigor state. Thus, at least the orientation of 20K and 27K domains does not change during tension development. The orientation of 50K domain may change but its angle is not thought to be large because the polarized fluorescence from tryptophan which distribute to S-1 did not change.
(c) The state of myosin heads and actin monomers in passively stretched fibers. Neither ε-ADP attached to myosin heads nor Ph-FITC attached to actin monomers change their orientation. It has been reported that spin labels on SH_1 of myosin heads also do not change their angle (Thomas and Cooke, 1980; Cooke, 1981; Cooke et al., 1982).

conformational change in actin. Recently we have shown that polarized fluorescence from ε-ADP incorporated into F-actin in a ghost fiber irrigated with myosin and native tropomyosin changes during isometric contraction (Yanagida, 1981). On the other hand, no change in polarized fluorescence was induced by passive stretching of the fiber in a rigor solution or by simple addition of ADP. These results suggest that a rotation and/or distortion of actin monomers takes place during contraction. Conformational changes in actin (local distortion and/or rotation of the monomers) may easily affect the state of the intermonomer bond and/or the helical structure, and induce tension in the F-actin filament. Thus, it is likely that the movement of actin is involved in the process of tension generation.

ACKNOWLEDGEMENTS

I thank Prof. F. Oosawa, Dr. E. Prôniewicz-Nakayama and Dr. D.D. Thomas (University of Minnesota, USA) for stimulating discussion and critical reading of the manuscript. I also thank Prof. Wieland (Max Planck Institute, West Germany) for a gift of phalloidin-FITC.

REFERENCES

Balint, M., Wolf, I., Tarcsafavi, A., Gergely, J.P., Sreter, A. (1978). Location of SH-1 and SH-2 in the heavy chain segment of meromyosin. Arch. Biochem. Biophys. 190: 793-799.

Cooke, R. (1981). Stress does not alter the conformation of a domain of the myosin cross-bridge in rigor muscle fibers. Nature 294: 570-571.

Cooke, R., Crowder, M., and Thomas, D.D. (1982). Measuring cross-bridge angles with paramagnetic probes in rigor, relaxed and contracting muscle fibers. In: *Cross-bridge Mechanisms in Muscle Contraction.* eds. G.H. Pollack and H. Sugi.

Danker, P., Low, J., Hassellach, W. and Wieland, T. (1975). Interaction of actin with phalloidin: Polymerization and staloilization of F-actin. Biochim. Biophys. Acta 400: 407-414.

Huxley, A.F. (1974). Review lecture Muscle Contraction. J. Physiol. 243: 1-43.

Huxley, A.F. and Simmons, R.M. (1971). Proposed mechanism of force generation in striated muscle. Nature 233: 533-538.

Huxley, H.E. (1969). The mechanism of muscular contraction: Recent stuctural studies suggest a revealing model for cross-bridge action at variable filament spacing. Science 164: 1356-1366.

Lengsfeld, A.M., Low, J., Wieland, T., Dancker, P. and Hassellbach, W. (1974). Interaction of phalloidin with actin. Proc. Natl. Acad. Sci. USA 71: 2803-2807.

Mornet, D., Bertrand, R., Pantel, P., Audemard, E. and Kassab, R. (1981). Structure of the actin-myosin interface. Nature 292: 301-306.

Onishi, H., Ohtsuka, E., Ikehara, M. and Tonomura, Y. (1973). Energy transfer from tryptophan residues to a fluorescent ATP analog, 1, N^6-Ethenoadenosine triphosphate, bound to H-meromyosin. J. Bioch. 74: 435-450.

Oosawa, F. (1977). Actin-actin bond strength and the conformational change of F-actin. Biorheology 14: 11-19.

Oosawa, F. (1980). Dynamics of Actin Filament. In: *Muscle Contraction: Its Regulatory Mechanisms.* Ebashi, S., Maruyama, K. and Endo, M., eds. Japan Science Societies Press/Springer-Verlag. pp. 165-172.

Sekine, T. and Yamaguchi, M. (1963). Effect of ATP on the binding of N-ethylmaleimide to SH groups in the active site of myosin ATPase. J. Biochem. 54: 196-198.

Szilagyi, L., Balint, M., Sneter, F.A. and Gergely, J. (1979). Photo affinity labelling with an ATP analog of the N-terminal peptide of myosin. Biochem. Biophys. Res. Comm. 87: 936-945.

Thomas, D.D. and Cooke, R. (1980). Orientation of spin-labeled myosin heads in glycerinated muscle fibers. Biophys. J. 32: 891-906.

Thomas, D.D., Ishiwata, S.I., Seidel, J.C. and Gergely, J. (1980) Rotational dynamics of spin-labeled myosin heads in myofibrils. Biophys. J. 32: 873-890.

Tonomura, Y. (1972). *Muscle Proteins, Muscle Contraction and Cation Transport.* Tokyo:

University of Tokyo Press.
Wulf, E., Deboben, A., Bautz, F.A., Faulstich, H. and Wieland, T. (1979). Fluorescent phallotoxin, a tool for the visualization of cellular actin. Proc. Natl. Acad. Sci. USA 76: 4498-4502.
Yanagida, T. (1980). The angles of cross-bridges bound to nucleotide at various concentrations of Mg-(ε)-ATP In: *Muscle Contraction: Its Regulatory Mechanisms* Ebashi et al., eds. pp. 165-172.
Yanagida, T. (1981a). Angles of nucleotides bound to cross-bridges in glycerinated muscle fiber at various concentrations of ε-ATP, ε-ADP and ε-AMPPNP detected by polarized fluorescence. J. Mol. Biol. 146: 539-560.
Yanagida, T. (1981b). Rotational motion of actin monomers in a muscle fiber during contraction. VII International Biophys. Cong. Abs. p. 274.
Yanagida, T. and Oosawa, F. (1978). Polarized fluorescence from ε-ADP incorporated into F-actin in a myosin-free single fiber: Conformation of F-actin and changes induced in it by heavy meromyosin. J. Mol. Biol. 126: 507-524.
Yanagida, T. and Oosawa, F. (1980). Conformtional changes in their filaments in muscle fibers induced by Ca^{2+} ions. J. Mol. Biol. 140: 313-320.

DISCUSSION

TREGEAR: You found that about 200 μM εATP was needed to get complete binding. That is quite a high concentration compared to what is required with ordinary ATP to get binding.

YANAGIDA: That's right. According to almost the same measurements as you did, the dissociation constant of εATP binding to actomyosin in muscle fiber was 200 μM.

TREGEAR: Oh, good. That's fine. That disposes of that point. The second point was that your main conclusion, I think, was that the εATP in the active muscle is quite rigid -- the fluorophore does not turn around. But you made one remark while you were going on which I would move that you could amplify. If I understand you -- you said that if you used a *high* εATP concentration, then a lot of the εATP was *not* so rigid. Was that a correct statement?

YANAGIDA: Yes. At high εATP concentrations the polarized fluorescence showed partial disorientation of bound εATP. That's because the number of myosin heads in the dissociated state from actin increases at high εATP.

TREGEAR: So your first experiment was done at what concentration of εATP?

YANAGIDA: Less than the dissociation constant.

TREGEAR: I see. So when you draw the conclusion that the cross-bridge does not turn around during contraction, I'm not quite clear why you make that conclusion from the low concentration work.

YANAGIDA: At low εATP the dissociated state of myosin has only a very small contribution. However, at low ATP, the εATP binding to actomyosin is the rate limiting step. So at low ATP, a number of myosin heads form a rigor complex. In this experiment, the fluorescence comes from only myosin bound with nucleotide. Therefore, we can selectively obtain it, or we can eliminate the contribution from the actomyosin rigor complex. That's a most important point in this experiment. That's why I used the fluorescent ATP. I think that the present measurements at low εATP detect attached states of myosin in the actomyosin ATPase reaction which are the same as those at high ATP except for the actomyosin rigor complex.

TREGEAR: Yes. I know we will hear shortly from Dr. Cooke, and I expect the nature of the discrepancy between the conclusions will become obvious -- I mean, to avoid unnecessary discrepancies between your work and theirs.

COOKE: Richard, I don't think there is much discrepancy between his work and ours.

RUEGG: You mentioned that there was no change in the polarization of the tryptophan fluorescence when you went from the relaxed state into a contracted state. Was there a change when you went into rigor?

YANAGIDA: When the fiber went from a relaxed state into a contracted state at high ATP, the polarized fluorescence was changed a little, but significantly. It's consistent with the work of Morales' lab (Dos Remedios et al., J. Gen. Physiol. 59: 103-120, 1972). The small change was observed when the fiber went from a relaxed state at high ATP into rigor, but was not observed when a fiber went from a contracted state at low ATP into rigor.

HUXLEY: I'm not quite sure that I understood one point in your argument. You show that there isn't a change in the orientation of the fluorescentally-labelled nucleotide. But it seems to me that doesn't exclude the possibility that some change in conformation takes place after the nucleotide is dissociated away from the cross-bridge during a stroke, because then there's nothing to indicate it.

YANAGIDA: If a new state of actomyosin without nucleotide, different from a rigor state, exists *in vivo*, that possibility can't be excluded. But this idea seems to be inconsistent with kinetic studies of actomyosin ATPase in solution.

KRUEGER: Why does actin twinkle? Why was there this fluctuation of intensity down the thin filament? When I looked at your movie of spontaneously bending thin filaments, I was intrigued by the fact that it appeared to blink in various regions.

YANAGIDA: I've not paid attention to that point. I'm now not sure whether the intensity fluctuation has some significance or is just artificial.

INGELS: What were the diameters of those filaments?

YANAGIDA: In this case, we see a fluorescent image of F-actin, so we cannot estimate the actual diameter from it. However, since this filament is a single F-actin-fluorophore complex, its diameter should be similar to that of unlabelled F-actin.

HARRINGTON: Do you get the same result with the actin in the presence of S-1?

YANAGIDA: The bending motion does seem to be affected by the interaction with S-1. It becomes flexible.

HUXLEY: What happens when you add ATP?

YANAGIDA: I haven't done enough experiments to answer your question. The bending motion seems to be modulated. However, the active "swimming" is probably not induced.

RITZ-GOLD: I was wondering about the modulatory effect of tropomyosin and troponin in the presence or absence of calcium, on the actin mobility.

YANAGIDA: Its flexibility in the presence of calcium increases by 30%.

RITZ-GOLD: But it's more rigid without calcium?

YANAGIDA: That's right.

PODOLSKY: Dr. Yanagida, you said that you could evaluate the elasticity of actin filaments from the thin filament bending fluctuations on the film you shot. I wondered what the axial compliance was of those filaments and how they compared with actin in muscle?

YANAGIDA: The estimated elasticity was, of course, for bending motion. F-actin has a double-stranded structure. However, if we assume F-actin is isotropic, we can estimate Young's modulus from the elasticity for bending motion. If we stretch F-actin having length 1 μm by the same force as that during active tension development, it would be stretched by about 100 Å.

PODOLSKY: If that were the case, then all of the elasticity produced in a quick release or stretch would be in the actin. Is that right?

YANAGIDA: No, the situation is not so simple. If we take into account the elasticity of F-actin, the total elasticity of the sarcomere will be very complicated. I have not simulated it yet.

PODOLOSKY: But what if the thin filament were a simple spring?

YANAGIDA: The total sarcomeric elasticity would then largely arise from F-actin. However, I can't answer clearly what extent of the total elaticity can be explained by that of F-actin estimated here, because the sarcomere structure is not simple.

FAY: Is the bending motion that you see at all a function of the fluorochrome concentration, and can you see similar bending with darkfield illumination rather than fluorescence methods? I was also wondering if, to some extent, the bending you saw was a function of the label that you were using, that it had somehow perturbed the structure. What type of controls were there?

YANAGIDA: Previously, I have measured the flexibility of F-actin in ghost fibers by using another method. The estimated value of flexibility was almost the same as that estimated here. Furthermore, many people have also shown that F-actin is flexible. The flexibility is also almost the same as the present value.

MAEDA: But is it possible to see single actin filaments under the darkfield microscope in order to check your fluorscent-labelled result?

YANAGIDA: Probably impossible now, because a light scattering is greatly dependent on the mass, and F-actin is too thin to see it under present darkfield microscopy. In this sense, I can't compare the flexibility of labeled F-actin directly with that of unlabeled one.

BRESSLER: I was just going to develop your theme about elasticity in other places a little. I think we have to be very mindful of elastic structures showing up in other areas, like for example, in connecting filaments Alan Magid has been discussing. We need to know the stiffness of that filament in relation to the stiffness of the bridges.

SCHOENBERG: We've done an experiment similar to Magid's. He's done it better, but I can at least perhaps answer your question. That is, that you can affect the resting tension very easily by extracting with high ionic strength, 300 to 400 mM, Hasselbach-Schneider's solution. I believe this is supposed to affect connectin, and it has rather dramatic effects on resting tension. I think this is in good agreement with Alan Magid's results as well as his explanation of the resting tension, if that answers your question.

HOLMES: Can I come back to the other thing which Dr. Yanagida has raised, that using actin as a flexible element in this whole system, I think there is a constraint from all this, which arises from all the X-ray folks, including perhaps even Richard Tregear, who have looked at active muscle. Huxley and Brown did a lot of work on this. I don't believe that the 59 Å reflection which is from the ordinary background of the actin, has ever moved by more than a quarter to a half of a percent; nor has it got particularly disoriented. Is that right, Hugh?

HUXLEY: Yes.

HOLMES: So you really don't have much freedom of design, do you?

YANAGIDA: A 100 Å stretch of a thin filament represents a change of 1%, only a little more than you mentioned. I'm not sure such a change would have been resolved.

MUSCLE CROSS-BRIDGES: DO THEY ROTATE?

Roger Cooke, Mark S. Crowder, Christine H. Wendt*, Vincent A. Barnett* and David D. Thomas*

Department of Biochemistry and Biophysics and the Cardiovascular Research Institute, University of California, San Francisco, California 94143
**Department of Biochemistry, University of Minnesota Medical School, Minneapolis, Minnesota 55455*

ABSTRACT

We have used electron paramagnetic resonance (EPR) spectroscopy to monitor the orientation of spin labels attached specifically to a reactive sulfhydryl on the myosin heads in glycerinated rabbit psoas skeletal muscle. Previous work has shown that the paramagnetic probes are highly ordered in rigor muscle and display a random angular distribution in relaxed muscle (Thomas and Cooke, 1980). Addition of ADP to rigor fibers caused no spectral changes, while addition of AMPPNP or PPi increased the fraction of disordered probes. We show here that the application of stress to fibers in the presence of ADP, AMPPNP or PPi causes no change in their spectra. During the generation of isometric tension approximately 80% of the probes display a random angular distribution as in relaxed muscle while the remaining 20% are highly oriented at the same angle as found in rigor muscle. In each of the above cases the spectrum consists of two components, one highly ordered as in rigor and one highly disordered. Saturation transfer EPR has shown that the ordered component is rigid while the disordered component is mobile on the microsecond time scale (Thomas, Ishiwata, Seidel and Gergely, 1980). These data lead to the conclusion that the disordered spectral component arises from myosin heads that are detached from actin while the ordered component comes from heads that are attached to actin. The observation that the ordered component displays an identical angular distribution under all conditions indicates that its orientation is not linked to force generation.

INTRODUCTION

Studies of muscle structure using electron microscopy and X-ray diffraction led to the suggestion that force generation was the result of a change in the orientation of myosin heads attached to actin filaments (Reedy, Holmes and Tregear, 1965; Huxley, 1969). This hypothesis has been supported by the finding that the myosin molecule has the flexibility required for rotation of the myosin heads (Mendelson, Morales and Botts, 1973; Thomas, Seidel, Gergely and Hyde, 1975). In addition there are a number of studies which indicate that the orientation of myosin heads changes when the state of the muscle is altered (Aronson, 1969; dos Remedios, Millikan and Morales, 1972; Borejdo and Putnam, 1977; Yanagida, 1981). Recently, two results have suggested that changes in crossbridge orientation occur during contraction: 1) The polarization of the fluorescence of a probe attached to the myosin head was found to fluctuate during contraction but not during relaxation or rigor (Borejdo, Putnam and Morales, 1979), 2) time resolved X-ray diffraction patterns suggest that the attached cross-bridges of a contracting muscle rotate following step changes in muscle length (Huxley, Simmons, Faruqi, Kress, Bordas, and Koch, 1981). Taken together this evidence all supports the hypothesis that myosin heads rotate during contraction. However, in spite of an intense effort, definitive evidence for changes in the orientation of myosin heads while attached to actin has not been found. In the present paper we describe studies in which spin labels, attached to a reactive sulfhydryl (SH_1) on the myosin heads, have been used to monitor crossbridge orientation in a variety of conditions.

The angular distribution of an ensemble of spin labels can be determined from their EPR spectrum. The spectrum of a nitroxide radical is split into three lines by the hyperfine interaction between the unpaired electron and the nitrogen nucleus, and the ability to sense orientation is due to the dependence of both the hyperfine interaction and the g tensor on the orientation of the radical relative to the external field of the spectrometer. The spin Hamiltonian that describes these interactions is well understood and the EPR spectrum expected for a given angular distribution of spin labels can be easily calculated with good accuracy. The comparison between calculated and observed spectra can then be used to determine the angular distribution of the spin labels with little ambiguity to within a resolution of a few degrees, limited by the Lorentzian line width of the three spectral lines. In addition to the information provided on probe orientation by conventional EPR, saturation transfer EPR (STEPR) spectra allow one to measure probe motions over a wide range of correlation times (10^{-6} to 10^{-3} seconds).

Both the conventional EPR spectra and STEPR spectra have been used to study the orientation and mobility of nitroxide spin labels attached to myosin heads in fibers, in myofibrils and in solution (Thomas and Cooke, 1980; Thomas et al., 1975; Thomas et al., 1980). In the present paper, we first summarize the results obtained in these investigations of rigor and relaxed states then describe recent results obtained from fibers during the generation of isometric tension.

METHODS

Glycerinated rabbit psoas fibers were prepared by tying thin strips of muscle, 2-3 mm diameter, to supports and bathing them in a solution of 50% v/v of glycerol and rigor buffer (0.12 M KCl, 5 mM $MgCl_2$, 2 mM EGTA,

20 mM TES, pH 7.0). After 24 hours incubation at 0° the solution was exchanged and the fibers stored at 0° for up to several months. In some preparations the EGTA concentration was raised to 10 mM, but these fibers had properties similar to the regular preparations. Fibers were labeled with a spin label analog of maleimide (MSL) by procedures similar to those of Thomas and Cooke (1980). After washing with rigor buffer they were reacted with 0.5 mM MSL for 10 minutes at 0° in rigor buffer adjusted to pH 6.5. Approximately 50-70% of the heads were reacted at the SH_1 as judged by extraction of myosin and measurement of ATPase activity in the presence of EDTA. Spectra were obtained at 25°C with a Varian E109E spectrometer interfaced to a Northstar computer. The E238 cavity was modified to accept a capillary (1.2 mm inside diameter) parallel to the applied magnetic field. Approximately 100 fibers, in bundles of about 10, were secured on either end with surgical thread and pulled into the capillary, filling it almost completely. Solution was flowed past the fibers at a rate of 0.3 ml/min. This forces fluid past the fibers at a velocity of 2-5 cm/sec, aiding perfusion of the bundles. No difference in spectra were seen when the bundle size or flow rate was increased or decreased by a factor of 2. Fiber tensions were measured using an Akers solid state force transducer. Single fibers were dissected and glued to the force transducer at one end and to a motor capable of fast length changes (General Scanning) at the other end. The stiffness of the fibers was measured by oscillating one end of the fibers by a small amplitude 0.1-0.3% of the fiber length and measuring the oscillation in tension generated at the other end. The tension and stiffness of rigor, relaxed, and active fibers were measured under conditions similar to those used to obtain spectra.

RESULTS

Rigor Fibers

Our previous investigations have shown that the spectra of spin labels attached to the myosin heads of rigor fibers consist of essentially three sharp lines, when the magnetic field is aligned along the fiber axis (Thomas and Cooke, 1980; and see Fig. 1). These spectra depend on the orientation of the fiber relative to the magnetic field, and spectra obtained with the fiber perpendicular to the magnetic field have broad peaks which resemble those of a more random distribution of probes. These spectra indicate that the probes are highly oriented with respect to the fiber axis with an angular distribution that can be approximated with a spherically weighted gaussian distribution of 15-17° full width at half maximum, centered at 82° relative to the fiber axis. This angular distribution is identical to that obtained with labeled myosin subfragment-1 diffused into unlabeled fibers, showing that the probe orientation is determined by the bond between actin and myosin (Thomas and Cooke, 1980). STEPR spectra show that the probes are rigid on the millisecond time scale (Thomas et al., 1980). The spectrum of rigor fibers stretched to a sarcomere length of 3.6 microns resembles that of an isotropic

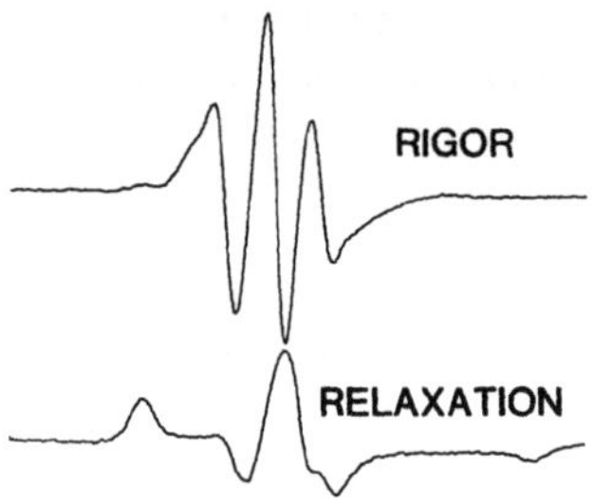

Figure 1: The EPR spectra of spin labels (MSL) attached to the myosin heads of glycerinated rabbit psoas fibers in states of rigor and relaxation. Fibers in rigor (top spectrum) were bathed in "rigor buffer": 0.12 M KCl, 5 mM $MgCl_1$, 2 mM EGTA and 20 mM MOPS, pH 7.0, at 25°C. Relaxation (bottom spectrum) was achieved by addition of 5 mM ATP, 20 mM CP and 1.0 mg/ml creatine kinase to the rigor buffer. In this and following figures, the derivative of absorption is plotted against the magnetic field, and the spectral baseline is 100 gauss wide.

Figure 2: EPR spectra of spin labels attached to the myosin heads of fibers in the states of relaxation (dashed line), and contraction (solid line). Contraction was induced by the addition of 2 mM $CaCl_2$ to the relaxed fibers as described in the text.

distribution of probes and STEPR spectra show that probes now have mobility on a 10 microsecond time scale (Barnett and Thomas, 1982). We conclude that at full overlap of filaments, all myosin heads are rigidly attached to actin with similar orientations, at least in the vicinity of the SH_1. The spectra of rigor insect flight muscle lead to nearly identical conclusions (Thomas, Cooke, and Barnett, 1982).

Relaxed Fibers

Addition of ATP in the absence of Ca induces relaxation which produces a random distribution of probe orientations (Fig.1) and rotational motion in the 10 microsecond time range, (Thomas, et al., 1980). Similar rotational mobility is observed for isolated myosin filaments (Thomas et al., 1980) and for non-overlap regions of stretched fibers (Barnett and Thomas, 1982), indicating that the myosin heads in relaxed fibers are most probably detached from the actin filaments. We conclude that the myosin heads in relaxed fibers are executing large angle Brownian rotations in microseconds. The exact amplitude of the rotations is difficult to assess because the angle of the probe relative to the myosin head is unknown. It is possible that the long axis of the myosin head is somewhat constrained and that rotations about this axis account for some of the randomness of the probes. A quantitative consideration of this problem has led Wilson and Mendelson (1982) to conclude that the most constrained axis of the myosin head must at least have freedom of mobility within a cone whose full angle is approximately 40 degrees. This maximum degree of constraint can only be obtained when the axis of the probe is aligned along a second axis of the myosin which has no constraints on its motion. A small fraction of ordered probes is seen at 25°C (Fig. 1 and 2), but not at 0°C (Thomas and Cooke, 1980), and this fraction may correspond to a small fraction (5-10%) of attached heads, as has been suggested by studies of S-1 and actin in solution (Chalovich and Eisenberg, 1982).

Rigor Fibers Plus ADP, AMPPNP and PPi

Previous work has shown that the application of a static stress does not change the angle of spin labels attached to rigor fibers (Cooke, 1981). This study has now been extended to fibers in the presence of ADP, AMPPNP, and PPi. The addition of ADP to the fibers in rigor buffer (plus the myokinase inhibitor AP_5A) causes no change in the EPR spectrum. The additional application of a static stress of 0.1 N/mm^2, approximately one half the active isometric tension, caused no change in the spectra. In particular, the measured splitting between the three hyperfine lines showed no change within an experimental accuracy of 0.5 Gauss, corresponding to an upper limit on the possible changes in the mean angle of the gaussian distribution of less than 1 degree. The addition of 1 mM AMPPNP or PPi produces a two component spectrum in which approximately half of the probes are oriented in the rigor-like angular distribution while the other half are disoriented as in relaxation (Thomas and Cooke, 1980). It has been shown that the disordered component of this spectrum is associated with mobility (Barnett and Thomas, private communication). In addition, the fraction of mobile heads has been shown to correspond to the fraction of dissociated heads as determined by sedimentation (Ishiwata, Seidel, and Gergely, 1979). The application of stress, 0.1 N/mm^2, to these fibers, caused no change in the angular distribution of the rigor-like component in the presence of either ligand. Nor did stress alter the amplitudes of either spectral component within experimental accuracy, $\pm 10\%$. The rationale for this experiment is straightforward, stress on the fiber exerts a torque on the cross-bridge which could be expected to alter its orientation. The above experiments show that this torque does not change the orientation of the spin labels in any of the states that can be achieved with the above ligands. Although two of the ligands induce disorder and mobility in a fraction of the probes, the equilibrium between this disordered, and presumably detached, state and the ordered state is also not shifted by stress.

Isometric Tension

Addition of both ATP and calcium to fibers produces isometric tension and the EPR spectrum shown in Fig. 2. This spectrum is a superposition of the spectra obtained in rigor and relaxation. When 81% of the spectrum in relaxation is subtracted from that in tension (Fig. 2), the difference spectrum is identical to that of rigor fibers (Fig. 3). Thus, under isometric tension, 19% $\pm$ 3% of the myosin heads have probes oriented at the same angle as in rigor muscle, while the probes on the

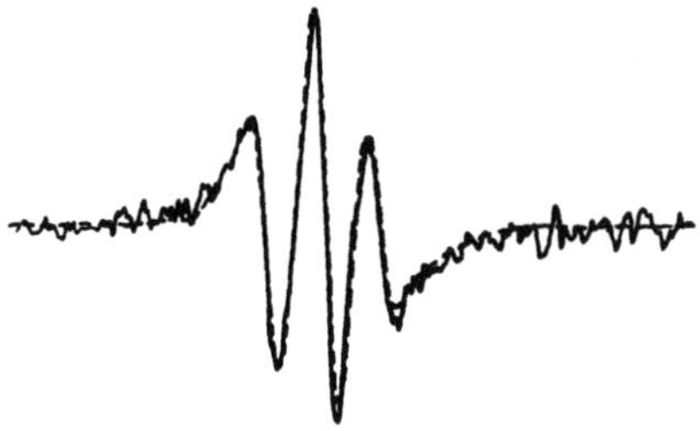

Figure 3: A comparison between the EPR spectrum of rigor fibers (dashed line), and the difference spectrum obtained by subtracting 81% of the spectrum of relaxed fibers from that of contracting fibers (solid line).

remaining heads are highly disoriented, with a spectrum that is identical to that of relaxed fibers. The most dramatic aspect of Fig. 1c is that no new probe angles, different from those seen in rigor or relaxation, are observed. The resolution of the spectrum is high; if more than 10% of the probes had assumed a new well-defined angle differing from the rigor angle by more than 10°, this would have been detected. Figure 4 shows the effect of adding the spectra from two populations of probes which have different angular distributions.

One possible interpretation of the spectra of Fig. 2 is that the rigor component represents probes on myosin heads that are not fully perfused with ATP. We have addressed this question by varying the concentration of ATP or creatine phosphate (CP). The fraction of heads that do not form rigor bonds was measured from the EPR spectra as a function of the ATP concentration. The fraction of disoriented probes increases as the concentration of ATP increases; but this fraction approaches only about 75% as the concentration of the substrate approaches infinity. In similar experiments, in which the concentration of CP was varied, increasing CP caused a decrease in the fraction of rigor bonds, but this effect reached a plateau between 5 and 10 mM CP. The fact that the fraction of rigor bonds can be described by a hyperbolic function which approaches a non zero limit as the substrate concentration is raised is a strong indication that this fraction is not a result of inadequate substrate.

The labeling procedure resulted in a 10-20% decrease in the active tension generated by the fibers. This decrease is probably due to reactions of MSL at sites other than SH_1, because labeling of the fibers with the spin label analog of iodoacetamide (IASL), which is more specific for the SH_1, resulted in no loss of tension. Unfortunately, the mobility of IASL is affected by nucleotides, so that the present studies must be performed with MSL. To determine whether the reaction with MSL was inducing rigor bonds during muscle contraction, we measured the stiffness of control and labeled fibers. There was no detectable change in the stiffness of active, relaxed or rigor fibers induced by the reaction with MSL.

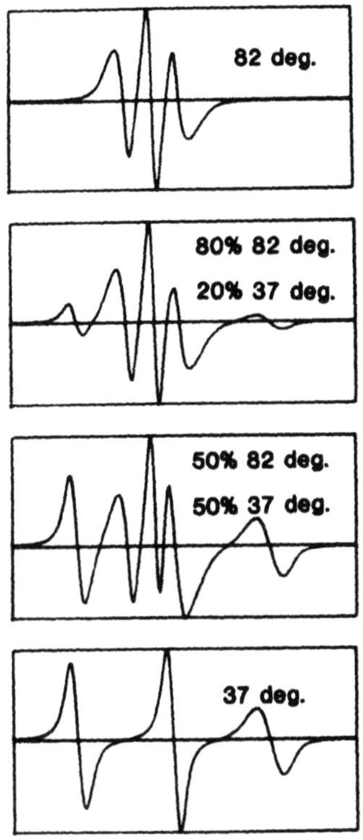

Figure 4: Simulated EPR spectra that would result from linear combinations of two populations of probes that have different angular distributions. One population, shown in the top spectrum, has a gaussian distribution of angles centered at 82° (the angle seen for MSL attached to fibers in rigor); and the other population, shown in the bottom spectrum, has a gaussian distribution centered at 37° (a 45° change from 82°). The full width at half maximum of both gaussian distributions is 15°.

CONCLUSIONS

The present paper focuses on the use of conventional EPR to measure the orientation distribution of probes. EPR is not only sensitive to probe orientation, but also to the rate of change of orientation (rotational motion) within the observed distribution. However, the conventional EPR spectrum is sensitive to this motion only if it occurs within a time less than 1 microsecond, and the spectra of MSL-labeled fibers clearly show that no sub-microsecond motions occur, whether in rigor, relaxation, or isometric tension. Thus, the probes (and, presumably, the heads) are orientationally static on the nsec time scale. In other work, we have used saturation transfer EPR (ST-EPR) a technique that is sensitive to rotational motion in the *microsecond* time range (correlation times as long as 1 msec can be measured; see Thomas et al., 1976). We have found that the probes on myosin heads do indeed undergo significant microsecond rotational motion under some conditions in myofibrils (Thomas et al., 1980). When combined with the conventional EPR data, a remarkably simple picture emerges, strongly suggesting that the probed region of the myosin head exists in two states with respect to orientation and motion, (a) a state in which all of the probes are immobile (on the microsecond scale) and highly oriented (disorder ~15°) with respect to the fiber axis, and (b) a state in which probes are mobile (in the range of 1-10 microseconds) and highly disoriented (disorder >90°). In rigor at full overlap, all probes are in state (a), immobile and oriented, and in relaxation or at zero overlap, all probes are in state (b), mobile and disoriented. Thus it appears that state (a) corresponds to attached heads and state (b) corresponds to detached heads. During contraction, or in the presence of AMPPNP or PPi, each head spends a significant portion of its time in each state, apparently corresponding to the portion of time it spends attached or detached. The spectra appear to be a linear combination of the two extreme states, with no evidence for either an intermediate *orientation* or rate of *motion*.

Several arguments lead to the conclusion that the rigor-like component of the spectra obtained during force generation represents a state in the normal cycle of the cross-bridge. First, the data argues that the rigor-like component does not arise because of an insufficient supply of substrate. Second, the only other spectral component, state b, is characterized by a high degree of motion and disorder and is most probably due to probes on heads that are detached from actin. Thus by elimination the states in which myosin forms a specific and rigid bond with actin during its contractile cycle must give rise to the rigor-like spectral component. This leads to the the conclusion that 20% of the heads are attached during contraction and that they are attached with probes in the rigor configuration.

The conclusion that 20% of the myosin heads are attached during contraction has been compared to values obtained by other methods. An estimate can be made of the fraction of heads attached in ATP splitting power-strokes during isotonic contraction:

$$N(v) = R(v)x/v$$

where N(v) is the fraction attached, x is the length of the power-stroke, R(v) is the ATPase rate and v is the velocity (muscle length/sec) (Curtin, Gilbert, Kretzschmar and Wilkie, 1974). We here define the power-stroke as that portion of the cycle in which myosin exerts a force on actin as the actin moves through a distance, x. Using data for frog muscle (Curtin et al., 1974), $R(0.3)=3.7\ sec^{-1}$, estimating x=1.2% of the half-sarcomere length and assuming that the heads of myosin operate independently in the generation of tension (Cooke and Franks, 1978), we obtain a value for N(0.3) of 14%. Extrapolation to zero velocity suggests N(0) in frog muscle is approximately 25%, which is close to the value obtained here by EPR for rabbit muscle. The fraction of attached heads in active muscle can also be estimated by relationships connecting the force exerted per cross bridge with the work performed by the muscle. The efficiency of frog muscle has been found to be in excess of 50% so that approximately 30 kJ/mole of useful work must be derived from the splitting of each ATP. Assuming that in each cross-bridge power-stroke an average force F is exerted through a distance of 10 nm the magnitude of F can be calculated to be 5 pN. From the known concentration of myosin heads, the known force generated by a fiber, and the assumption that the heads operate independently in the generation of force (Cooke and Franks, 1978), one calculates again that 25% of the heads are generating tension during isometric tension generation or during low velocity contraction. A study of the EPR spectra of IASL, a probe whose motion relative to the myosin head is increased by bound nucleotides, also reached the conclusion that about 20% of the myosin heads were attached during contraction (Arata and Shimizu, 1981). It thus appears that 25% is a reasonable estimate for the fraction of heads attached during isometric contraction. However higher estimates have been inferred from measurements of X-ray diffraction and stiffness (Haselgrove and Huxley, 1973; Goldman and Simmons, 1977).

Assuming that the disordered probes are on heads detached from actin, we conclude that all heads that are attached to actin during force generation have probes that are essentially at the same orientation, an orientation identical to that for myosin subfragment-1 bound to actin. This state is thought to represent the actomyosin state of lowest free energy, which occurs at the end of the power-stroke in the cyclic interaction which produces force. This conclusion is not consistent with a model in which a rigid myosin head generates force by a change in orientation, unless it is assumed that the predominant attached state is at the end of the power-stroke. It appears unlikely that such an assumption is reasonable. In the model presented by Huxley and Simmons (1971), the rapid recovery of tension following a step decrease in length is explained by the rotation of attached crossbridges through a portion of their power-stroke. This model is supported by time-resolved X-ray diffraction changes observed during the rapid recovery (Huxley, et al., 1981), and by the observation of a rapid 12 nm shortening following a step decrease in length (Barden and Mason, 1978). Although there are many possible molecular explanations of these and other transient phenomena, they strongly suggest that the predominant attached states during isometric tension are *early* in the power-stroke, not at the end. Thus the oriented probes in our labeled fibers are apparently bound to a portion of the myosin head that does not rotate during the power-stroke. This conclusion is supported by our EPR spectra obtained when stress was applied to a fiber. If the production of force is coupled to cross-bridge orientation in some state of the cyclic actomyosin interaction the application of force should produce changes in orientation when the cross-bridges are in that state. Three states studied here, rigor, rigor plus ADP, and rigor plus AMPPNP, are all thought to mimic states found in this cycle and in one case, rigor + AMPPNP, work can be obtained from the system by alterations in the ligand concentration. Yet the application of force causes no change in the probe angles in any of these states. Taken together with the data from contracting muscle these observations suggest that the orientation of the probed region of myosin is not coupled to force production. In principle, it is possible that the probe is oriented relative to the myosin head at just such an angle so that myosin rotation does not affect the

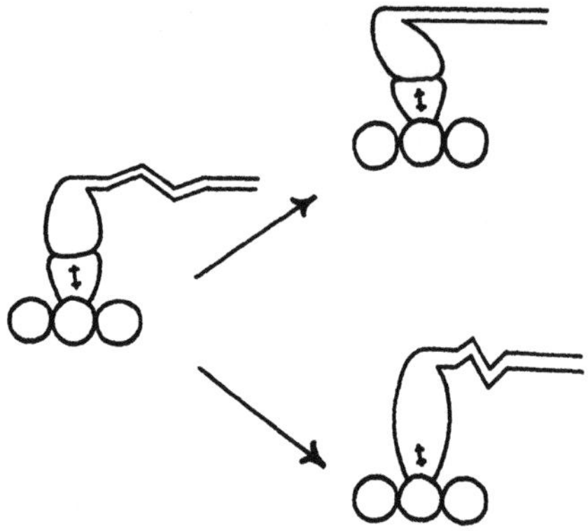

Figure 5: Two simple models of the cross-bridge power-stroke that are compatible with the present results. The cross-bridge on the left is attached at the beginning of its power-stroke. On the right, above, the cross-bridge has translated the actin by 12 nm via the rotation of a domain distal from actin. On the right, below, the actin has been translated via a shortening of a region of the myosin. In both models, the spin label, indicated by the double arrow, has not altered its orientation.

probe orientation. However, this is extremely unlikely in light of the high resolution of the spectra: in all the cases studied, either during the contractile cycle or in the presence of nucleotide analogs, neither the center of the rigor-like distribution nor its width differs from that of rigor by more than 1 degree.

The hypothesis that force is generated by the change in orientation of a rigid myosin is not compatible with our conclusion that the orientation of the spin label does not change during the power-stroke. However several other simple models can explain this data. Force could be generated by the rotation of one domain of myosin while a second domain, containing the SH_1, remains rigidly attached to actin. Alternatively, some region of the myosin could shorten during the power-stroke. These two possibilities are illustrated in Fig. 5.

ACKNOWLEDGMENT

We thank Vivian Siegel and Toshio Yanagida for assistance with preparations and EPR experiments, and Marilyn Hersh for help with the manuscript. This work was supported in part by grants to R.C. from the USPHS (HL16683, AM30668 and a Research Career Development Award AM00479); and to D.D.T. from the USPHS (GM27906, and a Research Career Development Award AM00851), NSF (PCM-8004612), Muscular Dystrophy Association, and the American Heart Association.

REFERENCES

Arata, T. and Shimizu, H. (1981). Spin label studies of actin-myosin-nucleotide interaction in contracting glycerinated muscle fibers. J. Mol. Biol. 151: 411-437

Aronson, J.F., and Morales, M.F. (1969). Polarization of tryptophan fluorescence in muscle. Biochemistry 8: 4517-4522.
Barden, J., and Mason, P. (1978). Muscle cross-bridge stroke and activity revealed by optical diffraction. Science (Wash. D.C.). 199: 1212-1213.
Barnett, V.A., and Thomas, D.D. (1982). STEPR of spin labeled fibers; dependence on sarcomere length. J. Mol. Biol., submitted.
Borejdo, J. and Putnam, S. (1977). Polarization of fluorescence from single skinned glycerinated rabbit psoas fibers in rigor and relaxation. Biochem. Biophys. Acta 459: 578.
Borejdo, J., Putnam, S., Morales, M.F. (1979). Fluctuations in polarized fluorescence: Evidence that muscle cross-bridges rotate repetitively during contraction. Proc. Natl. Acad. Sci. USA 76: 6346.
Cooke, R. (1981). Stress does not alter the conformation of a domain of the myosin cross-bridge in rigor muscle fibers. Nature 294: 570-571.
Cooke, R. and Franks, K. (1978). Generation of force by single-headed myosin. J. Mol. Biol. 120: 361.
Curtin, N.A., Gilbert, C., Kretzschmar, K.M. and Wilkie, D.R. (1974). The effect of the performance of work on total energy output and metabolism during muscular contraction. J. Physiol. 238: 455.
Chalovich, J.M. and Eisenberg, E. (1982). Inhibition of actomyosin ATPase activity by troponin without blocking the binding of myosin to actin. J. Biol. Chem. 257: 2432.
Dos Remedios, C.G., Millikan, R.G.C., Morales, M.F., (1972). Polarization of tryptophan fluorescence from single striated muscle fibers. J. Gen. Physiol. 59: 103-120.
Eisenberg, E., Hill, T.L. and Chen, Y. (1980). Cross-bridge model of contraction: quantitative analysis. Biophys. J. 29: 195-227.
Goldman, Y.E. and Simmons, R.M. (1977). Active and rigor muscle stiffness. J. Physiol. 269: 55P-57P.
Haselgrove, J. and Huxley, H.E. (1973). X-ray evidence for radial cross-bridge movement and for the sliding filament model in actively contracting skeletal muscle. J. Mol. Biol. 77: 549.
Huxley, H.E. (1969). The mechanism of muscular contraction. Science (Wash. D.C.) 114: 1356-1366.
Huxley, A.F. and Simmons, R.M. (1971). Proposed mechanism of force generation in striated muscle. Nature 233: 533-538.
Huxley, H.E., Simmons, R.M., Faruqi, A.R., Kress, M., Bordas, J. and Koch, M.H. (1981). Millisecond time-resolved changes in X-ray reflections from contracting muscle during rapid mechanical transients, recorded using synchrotron radiation. Proc. Natl. Acad. Sci. USA 78: 2297.
Ishiwata, S., Seidel, J. and Gergely, J. (1979). Regulation by Ca ions of cross-bridge attachment in myofibrils studied by STEPR. Biophys. J. 25: 19a
Mendelson, R.A., Morales, M.F. and Botts, J. (1973). Segmental flexibility of the S-1 moiety of myosin. Biochemistry 12: 2250-2255.
Mendelson, R.A. and Wilson, M.G.A. (1982). Three dimensional disorder of dipolar probes in a helical array: application to muscle crossbridges. Biophys J. (in press).
Podolsky, R.J. and Nolan, A.C. (1972). Muscle contraction transients, cross-bridge kinetics, and the Fenn Effect. Cold Spring Harbor Symp. Quant. Biol. 37: 661.
Reedy, M.K., Holmes, K.C. Tregear, R.T. (1965). Induced changes in orientation of the crossbridges of glycerinated insect flight muscle. Nature (Lond.) 207: 1276-1280.
Thomas, D.D., Seidel, J.C., Gergely, J. and Hyde, J.S. (1975). Motion of subfragment-1 in myosin and its supramolecular complexes: saturation transfer electron paramagnetic resonance. Proc. Natl. Acad. Sci. USA 72: 1729-1733.
Thomas, D.D. and Cooke, R. (1980). Orientation of spin-labeled myosin heads in glycerinated muscle fibers. Biophys. J. 32: 891-906.
Thomas, D.D., Ishiwata, S., Seidel, J.C. and Gergely, J. (1980). Submillisecond rotational dynamics of spin-labeled crossbridges in myofibrils. Biophys. J. 32: 873-890.
Thomas, D.D., Barnett, B. an Cooke, R. (1982). Spectra of spin labels attached to insect flight muscle. J. Muscle Research and Cell Motility, (in press).
Yanagida, T. (1981). Angles of nucleotides bound to cross-bridges in glycerinated muscle fibers at various concentrations of ε-ATP, ε-ADP and ε-AMPPNP detected by polarized fluorescence. J. Mol. Biol. 146: 539.

DISCUSSION

PODOLSKY: How long does it take to make a measurement in the contracting state? Does the force stay up throughout the entire period?

COOKE: The fiber exerts a tension of about two kilograms per square centimeter. The measurement takes about two to four minutes and the fibers maintain tension for that long. I might say that the tension is measured in single fibers on a tensiometer, while the fibers from which the spectra are taken are in a small capillary with a fairly large bundle of fibers and they're perfused; the fluid is forced past them from one end of the capillary, so we can't measure tension *while* we're doing spectra.

PODOLSKY: I guess you know it's something to worry about, because in making similar experiments on an X-ray bench, you have a bundle of fibers, then the laser pattern of the bundle tends to deteriorate with time, and after about a minute is quite deteriorated. This effect is stronger, the bigger the bundle.

COOKE: We've looked at the effect of bundle size, and we've pared them down to, say, five fibers per bundle in many instances, or 20 fibers per bundle in several instances, and we don't see any difference in the spectrum.

THOMAS: We can scan the first two peaks in the spectrum which tell us the ratio of the two components in five or ten seconds. Recently, we've been doing repeat scans to make sure there's no deterioration. It's constant for five or ten minutes, and reversible too. We can go back to rigor.

RALL: If you assume cross-bridges are working asynchronously, it's not clear to me why there would be an EPR signal in the first place.

COOKE: I don't know why that precludes a signal. What we see, essentially, is a "snapshot" of the different states in the contractile cycle. So, for instance, if there's a transient intermediate which takes a few milliseconds, you will not see a signal from that intermediate.

RALL: How long is the snapshot exposure?

THOMAS: It's a tenth of a microsecond.

COOKE: So, if it moves faster than a tenth of a microsecond, we would get an average.

RALL: But what if, during the snapshot, the cross-bridges were in a variety of different orientations?

COOKE: Well, they *are* in a variety of *two* orientation ensembles, one of which is highly ordered, and the other of which is wagging around. In the latter, for instance, the bridges are in a variety of orientations and we see a smear.

RALL: But they could very well be attached states.

COOKE: Well, I was arguing that they're not attached for two reasons. They're so disordered and they appear to be moving.

THOMAS: But that's a good point. The only thing we're measuring directly here is the orientation of the bridges in the snapshot, and it's the conclusion that one fraction is attached and the other is detached that's less direct.

RALL: Yes.

THOMAS: But this conclusion is based on control experiments and what seems to be the most straightforward interpretation.

RALL: I guess I'm really asking how good that conclusion is.

COOKE: No, we cannot eliminate the possibility that there are other interpretations of the data. Let me tell you what you have to assume if you wish to have heads attached at orientations different from the rigor orientation. You have to assume that the domain of myosin which contains our probe is moving through random angles in terms of microseconds. That's *exactly* the same movement that one sees, for instance, in isolated myosin filaments. So you have to say that the attachment is essentially invisible, both for the measurement of orientation and the measurement of motion.

GULATI: I've been measuring isometric forces in intact fibers while varying their volumes by varying the tonicity of the solution. We are able to shrink these fibers from a surface to surface separation of thick filaments of 13.2 nm down to 6.1 nm, and we were able to conclude from the data that the force made by each cross-bridge should be constant. Since we have come down to such narrow spaces between filaments, I'd like to say that the simplest conclusion was, that if the cross-bridge had to rotate during isometric contraction as one unit, it should experience steric hindrances, but it doesn't seem to. So the nose cone type of model that Ken Holmes proposes, and that you put up, I think that kind of data would go along with such an hypothesis.

SHIMIZU: A few years ago Dr. Arata and I performed a similar experiment with EPR by using a glycerinated muscle bundle under various substrate concentrations. I think this was the first spin-labelled EPR study of glycerinated muscle fibers in a contracting state. From the analysis of the EPR spectra, we found that under normal conditions -- I think millimolar magnesium ATP -- about 30 to 35% of the cross-bridges were attached to thin filaments. The number of these cross-bridges was proportional to the magnitude of tension, and these force-generating cross-bridges gave an EPR spectrum of strongly immobilized spin labels. In addition to these cross-bridges, there was another kind of attached cross-bridges of which the EPR spectra were of a weakly immobilized one, and the amount of which was independent of tension. Therefore, our observation supports the presence of two kinds of attached cross-bridges, as proposed by three-state models of muscle contraction based on physiological studies of muscle in tension transients. Under physiological conditions, almost equal numbers of cross-bridges are distributed in these two attached states. So that about 60 to 70% of cross-bridges are in the attached states in isometric contraction. This estimation is about the same as that from X-ray measurements.

PODOLSKY: I think the best estimate of the number of attached heads is stiffness measurements, and those are, I guess, 60%.

THOMAS: The stiffness measurements generally estimate that a much higher fraction of bridges are attached. I think that this rests on the assumption that the stiffness of an active bridge is the same as the stiffness of a rigor bridge. That might not be a good assumption.

PODOLSKY: I think that assumption, though, is the basis of *every* estimate.

THOMAS: No, not for the EPR's. One important point for the EPR is that every spin is counted, i.e., whether a spin is rotating at this angle or that angle it gives the same intensity. So if you count in the EPR spectrum that a given component is 20%, the uncertainty is not in the magnitude of the component but in whether that component really corresponds to attached heads or not. But we don't have any uncertainty about the relative contributions of the different spins in different states. I think that's a problem with the stiffness, that there may be different contributions in different kinds of bridge states.

PODOLSKY: I think I meant different *other* estimates.

VERDUGO: You mentioned random fluctuations of angle that you observed in windows of time in the range of one tenth of a microsecond. Is that compatible with the expected random fluctuations of a head that size?

COOKE: There's some misunderstanding. The window is due to the technique. That is, if one is looking at two populations of probes, and they stay either in state A or state B longer than several microseconds, then we see two populations. If they pop back and forth faster than that time, you can't see the populations; they merge into one. So we're not saying anything fundamental about the fluctuations of the heads when we say the time window is one tenth of a microsecond.

THOMAS: Well, actually, there are two issues. One is how rapidly are these two distributions exchangeable with each other, going from the oriented, presumably attached, to the disoriented, presumably detached state. That's what he's speaking about now. But that jumping back and forth has to be slower than on the order of a microsecond. The second issue is the rate of rotational motion of the disordered bridges. We have measured that directly, and it's on the order of one to ten microseconds. They're rotating through 90° on that time scale, and that's within the range that theoreticians get when they try to simulate such motion.

VERDUGO: I was wondering whether the probe could be attached to a portion of the head that is indeed fluctuating at the observed rate, while the whole big piece is not doing so.

THOMAS: That's certainly possible. For it to be that slow, it would have to be a fairly large domain. But the only direct evidence we have is that when we crosslink isolated subfragment-1 that has been spin-labelled to beads that provide a solid support with only an average of about three to five crosslinks per S-1, that completely immobilizes the probe. That's why I say if it is internal motion within S-1 it's probably a large domain.

YAMADA: You mentioned that the ESR signal that you get during the isometric contraction can be explained by the superposition of the rigor and the relaxed signals. Is it true that there are no additional components?

COOKE: That's right. There's only two components. There was smear but it was in a different context. It was describing the disordered component which is an isotropic distribution of angles.

HOLMES: Three or four years ago we made a proton NMR measurement on the complete S-1 and, somewhat surprisingly, it does give a fairly sharp spectrum. This was repeated and checked out and made into a rather thorough study by Stefan Highsmith and his collaborators (Biochemistry 18: 4238-4244, 1979). There is apparently some part of the S-1 which is relatively flexible. Now, it depends what you think it is. There's work being done in the Birmingham group (Prince et al., Eur. J. Biochem. 121: 213-219, 1981) who think that this flexibility resides partly in the A light chain. We didn't have such precise ideas about were it was. Our guess was that it corresponded to a piece of molecular weight around 20,000, perhaps a hinged domain. Now the interesting thing is, that when you bind S-1 to actin fibers, this mobility vanishes. So there's another indication that the flexibility *within* S-1 can exist and moreover, that this flexibility is modulatable by binding to actin.

COOKE: That's a good point.

T. WAKABAYASHI: I have a question to David Thomas. I am not familiar with EPR theory. Can you detect or exclude the possibility that myosin can rotate in the horizontal plane when you look at different angles?

THOMAS: No, we can't exclude that. There are some motions like that which wouldn't be detected, because all we can detect is the probe's projection on the fiber axis. So a purely azimuthal rotation would not be detected.

REEDY: I'd just like a rough estimate as to how long it may be before the spin label technique can do quick release experiments, like the quick release X-ray experiments that Hugh Huxley's group at the synchrotron source has done.

COOKE: If NIH would give us more money. We have a ways to go before we have millisecond resolution.

THOMAS: We're just now getting to where we can do a single fiber in a time-averaged experiment over several minutes. We may be able to obtain sufficient signal-to-noise by averaging several hundred transients.

AN ACTOMYOSIN MOTOR

H. Shimizu

Faculty of Pharmaceutical Sciences, The University of Tokyo, Hongo, Bunkyo-ku, Tokyo 113, Japan

ABSTRACT

I would like to report some results obtained by Yano, Yamamoto and myself on a novel system (Yano et al., 1982) we have named the actomyosin motor in which a rotor with attached F-actin rotates in a specific direction, driven by the ATP-splitting interaction with active fragments of myosin, heavy meromyosin or subfragment-1, in a solution containing MgATP. The actomyosin motor is not only interesting as a new kind of motor made of biological material but also, as a stream cell (Yano, 1978; Yano et al., 1978; Yano & Shimizu, 1978; Shimizu & Yano, 1978; Shimizu, 1979), is suitable for the study of chemo-mechanical coupling by actin and active fragments of myosin. Active motion of the motor was observed in almost 100% of the experiments, when carefully performed.

The structure of the rotor is schematically shown in Fig. 1: it has an axis of rotation on which a float, and three transverse bars with attached blades, are fixed with wax. The axis of rotation and the transverse bars are made of glass capillaries. On each terminal of the transverse bars, a blade of thin mica sheet is fixed, and only a specific side of the blade is coated with teflon. The layer of teflon is further coated with poly-L-lysine. The total weight of the rotor is about 300 mg. The size of the float is adjusted to suspend the rotor in the solution.

Brown and Spudich (1979) found that G-actin purified from Dictyostelium discoideum polymerizes on poly-L-lysine with a specific polarity. We found that under suitable conditions G-actin of rabbit skeletal muscle polymerizes in a similar manner on the surface of poly-L-lysine coated teflon. This was verified with an electron microscope, by observing the arrow head structure made by HMM and F-actin. This technique was applied to fix polarized F-actin on the blades of the rotor. As schematically shown in Fig. 1, the arrow head points toward the surface of the blade. Troponin and tropomyosin were attached to the F-actin to examine the calcium-regulation of the rotation. The movement of the rotor was observed from the top of the actomyosin motor with a stereo-

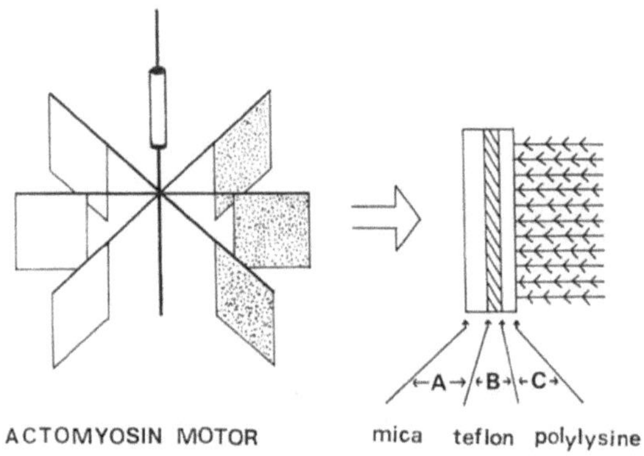

Figure 1: A schematic representation of the rotor. The axis of rotation and the three transverse bars of the blades are made of glass capillaries of about 0.3 mm in diameter and 4.0 cm in length. They are attached to each other with wax. On each terminal of the transverse bars, a blade of mica (1 × 1 cm, 50 μm in thickness), indicated by A, is fixed. The layer B in the figure denotes a thin sheet of teflon with a thickness of about 10 μm. The surface of the teflon sheet is further coated with poly-L-lysine, denoted by C. Partial coating of the other side of the blade strongly interferes with the rotation. The float is prepared by cutting a straw made of polyethylene pipe, (2.7 and 3 mm inner and outer diameters, respectively), to about 1.5 cm in length and fixing it to the axis of rotation with wax. F-actin is fixed as follows, pointing the arrow head toward the surface of the blade. The blades were incubated for 3 hours at 0°C in a poly-L-lysine solution (5 mg/ml, 0.4 N KOH) and were washed well with a solution composed of 0.1 M KCl, 0.5 mM $MgCl_2$ and 2 mM imidazole buffer (pH 7.6). They were then immersed in a G-actin solution (0.25 mg/ml G-actin, 1 mM $MgCl_2$, 2 mM imidazole buffer (pH 7.6), 0.2 mM ATP), and the polymerization of G-actin was initiated by adding a KCl solution (0.5 M KCl, 1 mM $MgCl_2$, 2 mM imidazole buffer) of 20% in v/v. After incubation for 1 hour at 20°C the surface of the blades were coated with F-actin (about 6 μg/cm^2).

microscope, Olympus type X, with x80 at 20°C and recorded with a video camera (Ikegami CTC-2100) connected to a video recorder.

In the presence of a sufficient amount of heavy meromyosin (HMM) and substrate, the rotor spontaneously turned in a direction uniquely related to the side of the blades coated with F-actin, as shown in Fig.2. That is, in the direction of the arrow head. Thus, HMM pushes F-actin in the direction of the blade; probably by moving the fluid in the counter direction. The direction of the motion of HMM is identical to that of cross-bridge movement on the thin filaments in muscle fibers. In order to obtain information on the mechanical property of the rotor by reproducing the same initial condition as often as possible, slight rotation was manually given to the rotor in the counter direction at the start of the experiment. The influence of weak mechanical perturbation given to the rotor at that time soon vanished and the rotor came to a complete stop within several minutes. Spontaneous rotation then started in the opposite direction and continued for 30-40 minutes as denoted by the open circles in Fig. 3. However, such spontaneous rotation was not observed when calcium was absent from the solution, as shown by the solid circles in Fig. 3. The initial counter rotation was not needed to start the spontaneous rotation.

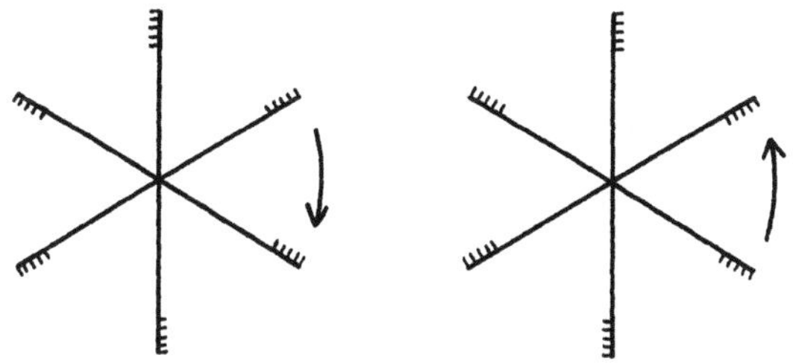

Figure 2: The direction of the spontaneous rotation of the rotor. The direction was uniquely related to the direction of the F-actin fixed on the surface of the blades.

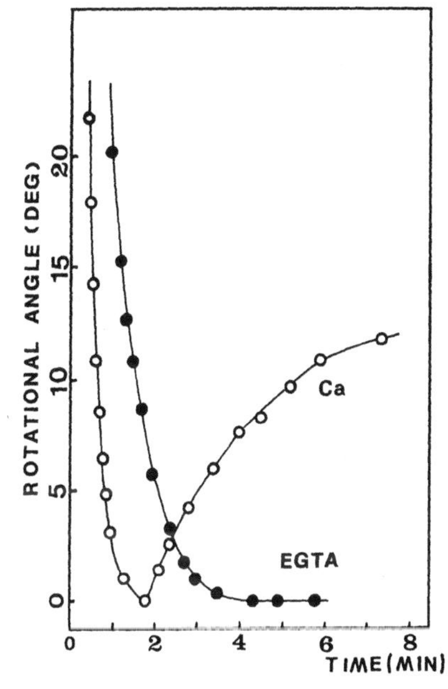

Figure 3: Rotation of a rotor under the presence (open symbols) and absence (solid symbols) of calcium when an initial rotation is given in the counter specific direction. The ordinate denotes the angle between one of the blades and the axis of a space-fixed coordinate system whose origin is placed on the axis of rotation. The abscissa represents the time in minutes. The solution contained 2.0 mg/ml HMM, 90 mM KCl, 2 mM ATP, 4 mM $MgCl_2$, and 10 mM Mops (pH 7.0). Either 30 μM calcium or 1 mM EGTA is added as occasion demands.

The system must be carefully arranged to protect it from convection, since this may move the rotor. Continuous illumination from the top of the system with a tungsten light was effective in suppressing convection caused by evaporation from the surface of the liquid. In our motor the coated side of the blades tends to be hotter than the other side due to heat liberated by the actomyosin ATPase activity. Brownian movement of molecules will be greater on the hotter side than on the other side. Therefore, the hotter side might push the colder side, giving rise to

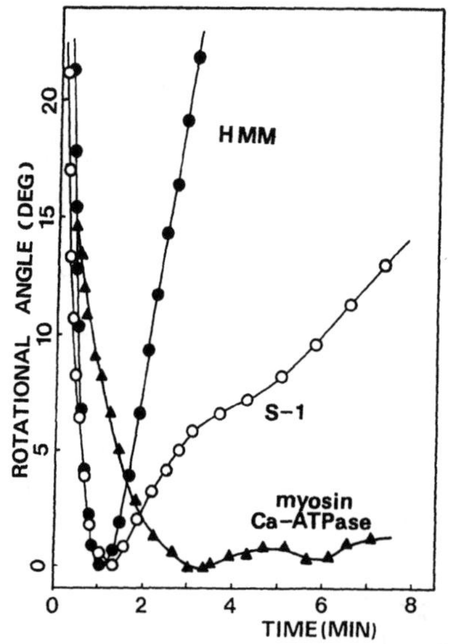

Figure 4: The effect of heat produced by ATP hydrolysis on the rotation of the rotor. See the text for detailed explanation. Myosin was fixed to the blades (in the experiment denoted by solid triangles) as follows. Micro-crystals of cellulose were uniformly fixed to the blades with adhesive (Araldite, CIBA-GEIGY). Fixed cellulose was activated by cyanogen bromide as reported previously (Yano et al., 1978). At first, light meromyosin (LMM) was fixed on activated cellulose powders to avoid the denaturation of myosin owing to direct interaction with active sites. The activated blades were immersed in a LMM solution containing 5 mg/ml LMM, 0.4 M KCl, and 10 mM Tris-Cl (pH 7.4) for half an hour at 0°C. After washing several times with a buffer solution (50 mM KCl, 10 mM Mops (pH 7.0)), they were immersed in a solution of soluble myosin (5 mg/ml myosin, 0.4 M KCl, and 50 mM Mops (pH 7.0)). The ionic concentration of the solution was gradually reduced by adding distilled water. In low ionic concentrations, myosin took a filamentous form incorporating with the LMM attached to the blades. The total amount of protein (LMM + myosin) fixed per blade was about 20 μg. The calcium ATPase activity of the fixed myosin was measured in a solution containing 50 mM KCl, 1 mM $CaCl_2$, 10 mM Mops (pH 7.0), and 2 mM ATP. In the measurement of rotor movement, 0.9% sucrose was added to the solution to adjust the viscosity to the level of the other case. Solid circles denote the rotation produced by HMM, which is essentially the same as that indicated by open circles in Fig. 3 (a) except the HMM concentration is 1.0 mg/ml. Open circles denote the rotation driven by S-1 in place of HMM. The solution consisted of 2.5 mg/ml S-1, 2 mM ATP, 4 mM $MgCl_2$, 10 mM Mops (pH 7.0), 30 μM $CaCl_2$, and 45 mM KCl.

rotation of the motor in the spontaneous direction. However, the result shown in Fig. 4 indicates that this is not the case for the present system. Solid circles in Fig. 4 represent the spontaneous rotation of the rotor. Solid triangles denote the rotation when myosin was fixed in place of F-actin on the specific side of the blades. The calcium ATPase activity of the fixed myosin was adjusted to be more than ten times larger than the acto-HMM ATPase activity in the other case. Therefore, the temperature difference between the two sides of the blades should be considerably larger than in the case of acto-HMM ATPase. As shown in the figure, only small and irregular movements of the rotor were observed after the decay of the initial rotation. Such movements were clearly distinguishable from the large and definite rotation denoted by solid circles for the case of acto-HMM ATPase. Thus, the spontaneous rotation of the actomyosin motor may be concluded to be active and driven by an ATPase. The present results are consistent with those on active streaming observed in our stream cell where HMM produces streaming in the direction of the tail of the arrow head. By using the actomyosin motor, we have studied the case where active movement is generated in the cross-bridge. Open circles in Fig. 4 show the rotation of the rotor when HMM in the MgATP solution was replaced by subfragment-1, (S-1), prepared from myosin by chymotryptic digestion. The purity of the preparation was checked with SDS PAGE. As is shown in this figure, S-1 with F-actin also has the ability to rotate the rotor. Thus, we are able to conclude that the active motion occurs in the S-1 part of cross-bridge.

REFERENCES

Brown, S.S. and Spudich, J.A (1979). Nucleation of polar actin filament assembly by a positively charged surface. J. Cell. Biol. 80: 499-504.

Shimizu, H. (1979). Dynamic cooperativity of molecular processes in active streaming, muscle contraction and subcellular dynamics. Adv. Biophys. 13: 195-278.

Shimizu, H. and Yano, M. (1978). Studies of the chemo-mechanical conversion in artificially produced streamings. III. Dynamic cooperativity. A new cooperativity in actomyosin system with a polarized arrangement of F-actin. J. Biochem. 84: 1093-1102.

Yano, M. (1978). Observation of steady streamings in a solution of Mg-ATP and acto-heavy meromyosin from rabbit skeletal muscle. J. Biochem. 83: 1203-1204.

Yano, M. and Shimizu, H. (1978). Studies of the chemo-mechanical conversion in artificially produced streamings. II. An order-disorder phase transition in the chemo-mechanical conversion. J. Biochem. 84: 1087-1092.

Yano, M., Yamada, T. and Shimizu, H. (1978). Studies of the chemo-mechanical conversion in artificially produced streamings. I. Reconstruction of a chemo-mechanical system from acto-HMM of rabbit skeletal muscle. J. Biochem. 84: 277-283.

Yano, M., Yamamoto, Y. and Shimizu, H. (1982). Actomyosin motor. Nature. In press.

DISCUSSION

TIROSH: I want to mention two things. First, I am very happy to see this kind of experiment. And second, I want to refer to similar experiments that I did ten years ago, and additional ones recently published, where we have shown active streaming of soluble actomyosin solution (Tirosh and Oplatka, J. Biochem. (Japan) 91: 1435-1440, 1982).

WILKIE: Yes, but you must admit he's added a thing or two. for example, by adding the regulating system and showing that if you add the regulating system you get control.

TIROSH: I did it too, and it has been reported.

WILKIE: Good for you!

EDMAN: Jerry Pollack is probably a bit shy about this, so I'll ask instead. Do you see a stepwise rotation?

SHIMIZU: No, I have not. I have some opinion about Dr. Pollack's experiment but this is only in my interpretation. That phenomenon would be related to the shortening of filaments connecting Z-bands to each other as mentioned by Dr. Wang at this symposium. I should tell you that Professor Maruyama in our country found the same kind of filament. He called it connectin. I think that stepwise shortening results from the competition between contractile force of actomyosin and the passive resistance of connectin which has a nonlinear elasticity.

KRUEGER: Could a "ciliary" model at least qualitatively account for the rotation if you put in the sort of dimensions that you might expect?

SHIMIZU: Yes, we can calibrate the motive force for our cross-bridge, but I didn't present any material on that, because I didn't think that this is the correct place to report these things.

TER KEURS: Did you actually measure the driving force?

SHIMIZU: No, I haven't because we've been busy establishing a good system.

TER KEURS: But it should, in principle, be possible.

SHIMIZU: Yes, I agree.

TER KEURS: Any guess about the force level?

SHIMIZU: No, but I have no doubt that the motive force is in a measurable range. But I think it is rather difficult to obtain the motive force per heavy meromyosin because we are not sure whether or not all the F-actins on the blade contribute equally to the motive-force generation.

POLLACK: Could you provide some sort of explanation of how you would relate your findings to the present view on how cross-bridges work?

SHIMIZU: I have several speculations about that, but the idea I report is the rotation of myosin heads on the actin, which was reported by Professor Hugh Huxley, and if we assume that kind of motion, we can explain why this rotor can move.

WILKIE: It seems to me that if it works in the way that Dr. Shimizu suggests, which is the most plausible, that is to say that the S-1's attach onto the thin filament and then create an active movement, an active torque, and then drop off again, they have to be attached at two points, because if you imagine them attached at just one point there is no way that you can exert a torque. And so I was asking, does he think it's a single S-1 that's got two attachment sites, or could it be that there are two S-1's on a cross-bridge. No one at this meeting so far seems to have discussed why there are two sites, what the two are doing, whether they are the same, whether they are attached to different places on the same thin filament, or anything about them. Is it a single S-1 that's forming, or do you have two of them associated together in the way that the two S-1's are on a cross-bridge?

SHIMIZU: I am sorry I cannot quite understand what you mean. We used S-1 prepared from myosin by chymotryptic digestion and purified by means of column chromatography. The S-1 purity was checked by SDS-polyacrylamide gel electrophoresis.

TREGEAR: You were wondering whether there would be two of them together.

WILKIE: I *was* wondering about that. Certainly you need two points, otherwise it's like somebody pulling on a rope while standing on one foot.

TREGEAR: Holmes has just suggested that there are two attachments on different actins with one S-1, hasn't he?

WILKIE: Yes, but is it true?

INGELS: Could I suggest that you might not need two points. Say you had an S-1 with a charge on the end of it, and a charge halfway that matched another charge. The S-1 would float in, with a one-point attachment. Now the *fields*, interact, not *things*, and the S-1 rows.

WILKIE: That's not a bad idea.

TIROSH: I would like to raise a consequence of this model, namely that with rowing you'd expect that the fluid in the vicinity of the actin filaments will move in the opposite direction. But in fact there is a beautiful demonstration by Dr. Fujime who showed motion of actin filaments in the presence of HMM. Well, they moved in a well defined direction, and the water in the vicinity of these filaments goes the *same* direction. So this is a problem.

WILKIE: Yes, but you're taking the analogy of rowing a boat. Because when you sit in the boat, your frame of reference is the boat but you think it's the water that's moving. But if you were rowing the boat with perfect efficiency, the water wouldn't be moving at all; only the boat. It's a question of your frame of reference.

TIROSH: But the water in the above experiments is moving ... it's moving along with the filaments. That's the problematic point for a rowing mechanism. But this phenomenon was predicted by the hydrodynamic approach to the mechanochemical function of the actomyosin system, where soluble actin filaments are expected to be carried by active streaming that is presumably generated by vectorial flux of energetic protons due to ATP hydrolysis, rather than by a conformation rowing mechanism.

NOBLE: If Dr. Ingel's suggestion is right, there should be a great dependence of the kinetics of this motor on ionic strength of the fluid. Have you investigated this?

SHIMIZU: Yes, we did. Only under physiological conditions did our motor move.

HOMSHER: If you raise the temperature, does the motor spin more rapidly?

SHIMIZU: Yes, it does.

REEDY: It seems to me that before you can say very much about what's happening here you really need to know the polarity, whether the actin is being pushed or pulled, and which way the arrowheads go. The arrowheads were drawn pointing away from the actin as if microplate were the Z-line. Do you know that with confidence?

SHIMIZU: Yes, yes, yes! We observed the polarity by observing the arrowhead structure with an electron microscope.

REEDY: All right. So it's like pushing on the A-band from the Z-band.

SHIMIZU: Yes, that's right.

INGELS: Could I just make one comment ... you were talking about the calculations. I have done some calculations about the length of the S-1 and 2 charges that were 9 Å apart. This gives you a 20-1 lever arm disadvantage. When you do that you get precisely the maximum force in muscle.

WILKIE: Really? I'm impressed.

MAGID: Dr. Shimizu, have you identified conditions that will let you bind actin of opposite polarity?

SHIMIZU: Yes. We are now pursuing this.

CONCLUDING DISCUSSION

CONCLUDING DISCUSSION

Structural Dynamics: Implications for Contractile Mechanisms

NOBLE: We would now like to have a general discussion about structure and I would suggest that we should be thinking not just in anatomical terms. We should be thinking in terms of the relevance of this to the mechanisms of muscle physiology. So perhaps I could set the ball rolling by reminding you of the study that Dr. Rowe presented in which, following dissolution of the myosin and reconstitution, he could regrow, as it were, the filament lattice, and if this was done at a very long sarcomere length, he obtained a sarcomere, which if I remember correctly, had a gap between the two half sarcomeres.

There are certain structural considerations about this, which I think Dr. Rowe will want to talk about, e.g., what might happen to the core filaments, if they exist. But I would like to suggest to you that this is a very nice model for physiological experiments.

MAGID: Hugh Huxley, in an appendix to a paper with Jean Hanson (Symp. Soc. Exp. Biol. 9: 228-264, 1955), reported results from experiments where he also tried to reconstitute thick filaments after extraction. And though there were no micrographs given, either light or electron micrographs, the verbal description seems very much like that which we have just seen. What was seen was density forming near the Z-line, and upon activation with ATP a movement of mass up toward the Z-line. That's the direction of the movement of myosin that one would expect from an ordinary sliding filament model, except where, rather than having an A-band that was intact, which would normally draw the thin filaments into the A-band, there are very compliant links between the edges of the half thick filaments and so the half A-band is drawn up toward the Z-line, stretching the bare, core filaments. Now those experiments were never, so far as I know, reported in detail anywhere, only as an appendix to the paper I spoke of. Dr. Huxley is here, and he might recall those experiments.

HUXLEY: Yes, I can't remember which of us did those experiments and described them, but they're certainly ones that I also have done from time to time. They're not done in a very different way from Dr. Rowe's experiments; they simply consist of extracting myosin from glycerinated fibrils, and then if you add back S-1, so that you get a large increase in density, and then add ATP, then you simply lose the S-1 again and it just goes up into solution. But if you add back whole myosin at high ionic strength so it's in solution, then you again get a large increase in density of these I segments, and then if you lower the ionic strength and add ATP you get a retraction of parts of that density back towards the Z-line. And then, in fact, you can add more myosin and go through the process several times, forcing more and more myosin back up against the Z-line, so the system behaves in the way you might expect it to, but I doubt very much if one is going to learn a great deal by studying the system in that way. Quite likely there will be a lot of compression and disorientation of the myosin filaments which have been formed between the actins up against the Z-line, so it may be useful for some things, but its not a completely clean system.

MAGID: I would like to ask Professor Rowe about the experiments when you say that you see a reconstitution of twitch tension. To me that is a very surprising result, that the entire physiological mechanism that produces action potentials, releases calcium and all of that, is intact after such a severe treatment. Can you describe these experiments?

ROWE: Yes. As far as we can see the membrane becomes severely damaged and leaky to the extent where even the myoglobin can leak out into the outer solution. SDS gels of that outer solution (this is the potassium phosphate solution and it's concentrated down), reveal that very little mysoin gets out. The reconstruction as we know from the light microscope, occurs fairly rapidly: It can be seen in the sartorius muscle within minutes and in the pectoralis muscle it can be within 20 or 30 seconds. If, however, you then incubate for a matter of several hours, it seems that the membrane reanneals at least to the extent necessary for excitability. We can then do the normal sort of physiological experiments of which we've done a limited number. The results seem to be comparable with the behavior of a muscle that has been stored in Ringer's solution for the same amount of time. I should say it fatigues very rapidly. You have to leave it quite a while before you get another stimulation, as you might expect.

NOBLE: What about the role of the connecting filaments in reconstruction?

ROWE: I'm reasonably confident there's some form of cytoskeletal network within a muscle fiber, and whether you regard it as a core down the middle or a sort of network, a multi-string shopping bag all around the outside, I think something that is there, and it might well be enough to hold those two half filaments together to produce contraction. But I'm a little bit nervous about this experiment. I'm scared the thing might actually contract.

NOBLE: Don't be scared! Don't be scared! It's a crucial experiment.

TIROSH: I'm sorry to say but it's clear to me that I'm perhaps not making myself understood. We have already done these experiments using HMM rather than filamentous myosin in order to check the cross-bridge hypothesis, namely if there really is a need for a continuous structure for force generation (Oplatka et al., J. Mech. Cell Mot. 2: 295, 1974). We have gotten contraction with heavy meromyosin reincorporated to glycerinated fibers from which myosin had been removed. The level of force is comparable to that obtained with reincorporated myosin instead of HMM, but in both cases the reconstituted force is much smaller than that obtained with intact myosin. However, the reincorporated fibers shrink during contraction. Thus, the level of force (as carefully distinguished from tension) is a great deal smaller. However, if we take into account tension it seems to be quite comparable to $1Kg/cm^2$ (Borejdo and Oplatka, Biochem. Biophys. Acta. 440: 241-258, 1976) of the actual tension rather than force is the main issue of my presentation at this symposium.

MAGID: I find that it's very difficult to remove all the myosin from the thick filaments. I think we've seen a gel at this meeting showing that myosin is a very stubborn item. It may not necessarily be located in its normal location; it may not be confined to the thick filament. Much of it seems to relocate and rebind to the actin filaments.

On a related issue, I'd like to make a point about what provides continuity between Z lines in striated muscle. Kuan Wang has said that keratin-like filaments going from Z line to Z line might be doing this. Keratin filaments, in fact, are extremely stiff whereas, in fact, at slack length one finds that myofibrils isolated from frog's semitendinosus fibers are quite extensible. On the other hand, myofibrils from vertebrate heart or insect flight muscle are very stiff. I'd like to point out that the model I'm proposing, the connecting filament model, makes it very easy to understand, or at least to think about, these very large differences in passive stiffness without invoking anything very special.

Figure D-1 shows two rather extreme versions of a simple three-element model for passive tension (and muscle continuity). The case for a half-sarcomere is illustrated. Two inextensible elements, a half-thick-filament of length of A/2 and a half-Z-body of length Z/2 are in series with an elastic element of slack length, C_0. Thin filaments, playing no role in passive tension, have been omitted for clarity. The top model, which might represent resting insect flight muscle, has a short C-filament

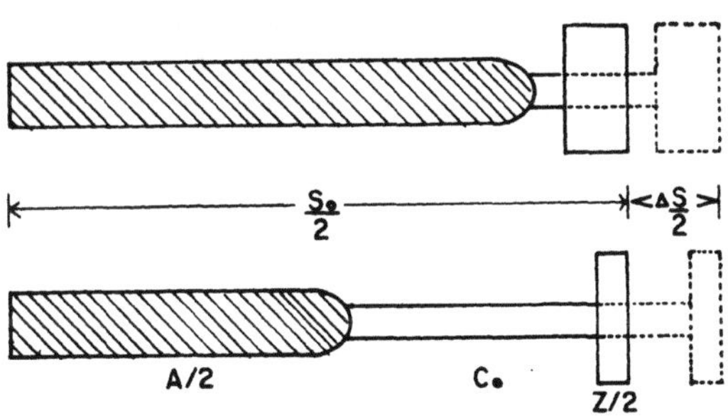

Figure D-1: Diagram comparing passive C-filament strain in two half-sarcomeres of very different proportions.

connecting a long A-filament and a long Z-disc. Below is a model proportioned more like vertebrate leg muscle (heart muscle would fall somewhere between these extremes). Shown with dashed lines is a 15% stretch applied to both sarcomeres. This representation should make it clear that in this model sarcomere strain is applied entirely to the C-filament. For the insect-like sarcomere this will result in a 300% C-filament extension but only 37.5% for the other. That is, even if the moduli of both C-filaments were the same, the insect-type would appear stiffer. It is convenient to define the ratio $\alpha = S_o/2C_o$. α is the factor by which sarcomere strain should be multiplied to yield C-filament strain. For the examples shown $\alpha = 20$ and 2.5 for the top and bottom respectively.

Using this approach for comparing two different frog muscles, one finds that sartorius muscle, where the sarcomere length at rest is 1.95 μm, is stiffer than semitendinosus muscle where the sarcomere length is on the order of 2.2 μm. The sartorius' presumably shorter connecting filaments would be very much stiffer than semitendinosus. The muscle EDL from the rat (cf. ter Keurs, this symposium) has a very long rest sarcomere length, 2.65 μm. I believe it is an extremely compliant preparation. I think this is also the case for *Limulus* telson muscle. Insect flight muscle, which has an almost invisibly short I band, and appropriately short connecting filaments, is notorious for its stiffness. So, I think that on the basis of a very simple kind of analysis one has an approach to thinking about the very broad range of resting elasticity one sees in various muscle fibers.

SUGI: I think there may be an additional element that needs to be considered in interpreting what happens when a sarcomere is stretched. I would like to present our experimental results on the extensibility of the myofilaments, which are normally regarded to be completely rigid in the sliding filament contraction models. We studied the myofilament extensibility by stretching glycerinated rabbit psoas muscle fibers in the rigor state and examining the resulting extension of sarcomere structures under an electron microscope.

Though stretches of rigor fibers produced successive yielding of weakest sarcomeres, the length of remaining intact sarcomeres was fairly uniform and definitely longer than that of control non-stretched sarcomeres. As shown in Figure D-2, the stretch-induced increase in sarcomere length was taken up by the extension of the H-zone and the I-band, while the amount of overlap between the thick and thin filaments did not change appreciably with stretches of less than 20%. The thick filament extension at the H-zone was localized in the bare region, whereas the thin filament extension at the I-band appeared to take place uniformly along the filament length, indicating that the thick filament is very rigid except for the bare region. Though the myofilament extension with large stretches are irreversible and plastic in nature, we have evidence that the myofilament extension with small stretches is reversible and elastic. Our tentative estimate of the degree of contribution of the myofilament elasticity to the instantaneous elasticity of muscle fibers suggests that a large part of the instantaneous elasticity may arise from the myofilament elasticity.

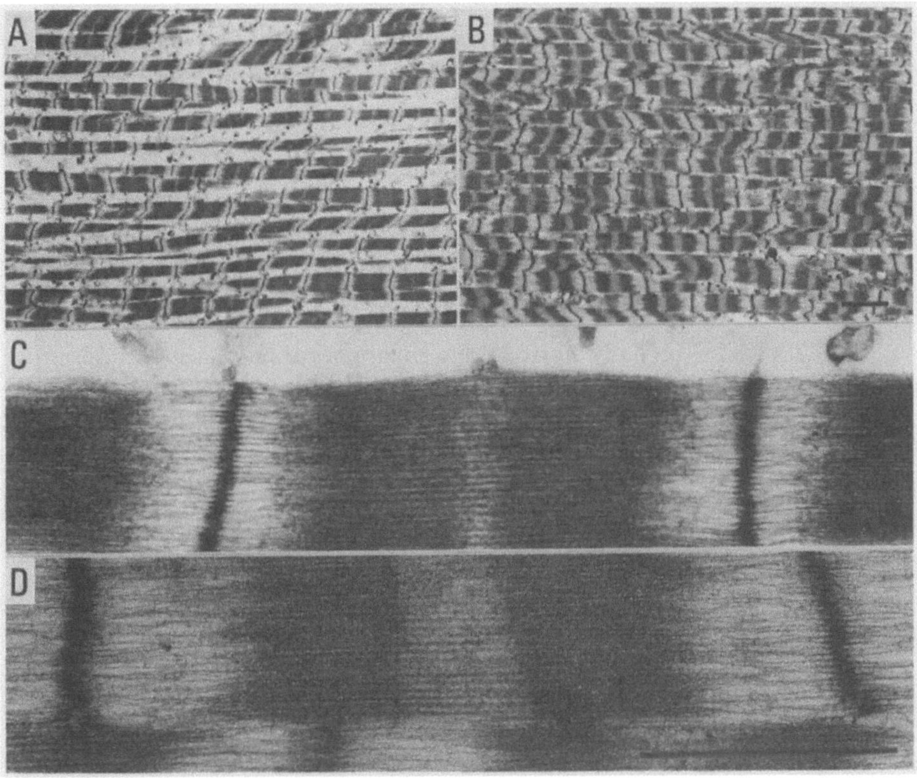

Figure D-2: Electron micrographs of rabbit psoas muscle myofibrils showing the extensibility of the myofilaments. A and B: Longitudinal sections of myofibrils in the control nonstretched part (A) and the stretched part (B) of glycerinated fibers in rigor state. Bar, 2 μm. C and D: High-magnification view of nonstretched (C) and stretched (D) sarcomeres showing stretch-induced extension of the thick filaments at the H-zone and of the thin filaments at the I-band. Bar, 1 μm.

MAGID: In interpreting mechanical/structural experiments of the kind Professor Sugi has shown, I think it is probably useful to keep in mind the evidence and ideas that I have put forward. In the first case there are two kinds of filaments, both about the same dimension, in the I-band of muscle -- thin filaments and connecting filaments. This means that when you see filamentous material spanning the I-band, you cannot, without independent chemical evidence, assume that what you are seeing are actin filaments. You see an elongation of the I-band in such pictures. It's possible to imagine that the thin filaments have broken but that sarcomere continuity persists because of the presence of the connecting filaments.

And the second, I was also gratified to see that Professor Sugi could show elongation of the A filament. This is consistent, and in fact expected from the idea that I have that the thick filament possesses an elastic core, a non-myosin core. I have shown in my poster that you can easily stretch resting frog fibers to S.L. 7 μm, the connecting filaments evidently producing enough stress to cause strain of thick filaments to a length about three times what one would normally see *in vivo*.

ROWE: One possibility though, for those stretched H zones could be in terms of the suggestions I made earlier. I showed some micrographs which implied that a possible mode of assembly of the thick filaments is that they comprised two half filaments, which fused together, with a resulting M-line. Now I did notice that the M-line of Dr. Sugi's stretch pictures appears to have gone, so there may be a little slippage of the myosin filaments. That, I think, could be a possible explanation of his result, without having to invoke core filaments.

SUGI: Yes. I already noticed the disappearance of the M-line, and I already submitted this work to some journal. The objection was that the M-line was missing. I repeated the experiment under many conditions, but stretch tended to eliminate the M-line structure.

KRUEGER: Commenting to both Alan Magid and Professor Sugi, are you sure that at least some of that motion or the strain that you might attribute to a change in core, might not be due to a shift of mass around the core? Maybe not all of it, but at least part of it.

SUGI: I expect that the marked elongation of the H-zone may be due to stabilizing forces in the rod portions of the myosin molecules.

MAGID: I noticed in the heavily stretched A-bands that I've observed, very often it's difficult to recognize the M-line, although in Fig. 6g (Magid et al., this volume) there are, indeed, places where an M-line can be recognized, and thick filament strain is occurring symmetrically at both ends.

KAWAI: Have you checked the spacings of myosin and actin by X-ray?

SUGI: No, I have not.

MAGID: On that point, I'd like to remind the group of a fact which appeared in a paper by John Haselgrove in 1975 (Haselgrove, J.C., J. Molec. Biol. 92: 113-143, 1975) concerning the behavior of stretched frog muscles. He reported that when living frog muscles are stretched to lengths at which one would expect non-overlap, the meridional periods, one at 143 Å and 72 Å, increase spacing by a small fraction of a percent. He made quite sure that these differences were real. He also reported in that paper that in the rigor transition, the spacing increases even a bit more. But it's possible to produce a small change in thick filament spacing, as Haselgrove reported, by passive stretch alone. Whether or not this supports the notion of a core to the thick filament or not I don't know, but I consider it easy to interpret in those terms.

NOBLE: I think we should move on to another topic which is raised by Dr. Robinson's paper, where, in cross-sections, he shows almost complete filling up of the interfilamentous spaces with electron dense material. Obviously this has tremendous implications for the way the muscle would work. If he's right, one can imagine there would be

Figure D-3: Electron micrograph of deep-etched rigor insect flight muscle which was quick frozen after glycerination and aldehyde fixation. A portion of one half of a sarcomere is shown. The Z line is out of the picture to the right and the M line is out of the picture to the left. The fracture plane has passed just above the "myac" layer defined by Reedy (J. Mol. Biol. 31: 155, 1968). (Bar = 100 nm).

considerable diffusion barriers through the tiny spaces he showed. I think Dr. Cooke has some slides which he thinks disagree with Dr. Robinson.

COOKE: John Heuser developed some rather elegant techniques for electron microscopy and we collaborated on a project which examined rigor and relaxed insect flight muscle. For those of you who aren't familiar with the technique, it involves fast-freezing the sample, fracturing the sample, freeze etching, and then plating -- this is a platinum replica that's used in the transmission electron microscope. I want to show these because (1) they address several points that were raised at this conference, and (2) I think they're very pretty pictures.

Figure D-3 shows an insect flight muscle in rigor. You can see the thick filaments here having a diameter considerably less than Dr. Robinson suggested. This sample has not gone through either, embedding or staining. And yet the thick filament diameters are similar to what is seen

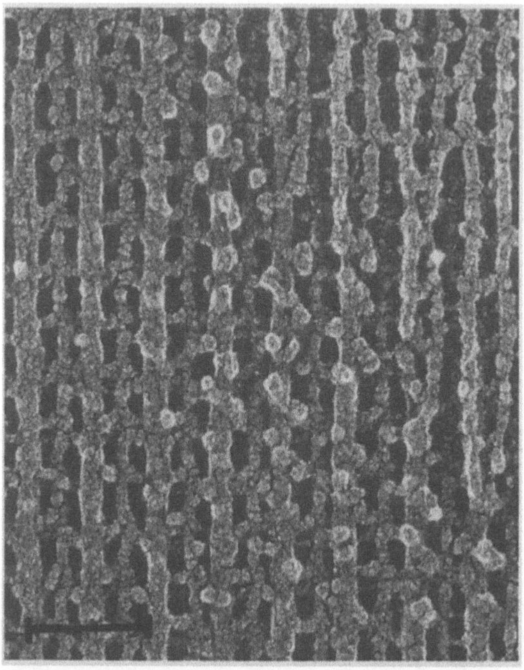

Figure D-4: Electron micrograph of glycerinated rigor insect flight muscle. The fracture plane passed from one "myac" layer to another, revealing a left handed two start helical array of projections on the surface of the thick filaments. (Bar = 100 nm).

by other techniques. You can also see the cross-bridges running from thick filaments to thin filaments, very similar to electron micrographs taken much earlier using negative staining by Mike Reedy. I think one difference between these electron micrographs and those taken with negative stain is that the cross-bridges don't make such a prominent angle with the fiber axis. Nonetheless, there is a polarity to these structures; there is a quantitative difference, not a qualitative difference between the angles found here and those seen in negative stain. The angles of the cross-bridges vary from about 90° down to about 50°. Also, we rarely see connections going from one thick filament to another thick filament in insect flight muscle.

Figure D-4 shows a longitudinal section from a rigor insect flight muscle. You can see the myosin, and the actin filaments and the cross-bridges between them. Near the middle of the micrograph you can see that the plane of the fracture has lifted above the myosin filaments and you can see the helical arrays of what was probably myosin heads. These heads form a left-handed helical array that Mike Reedy has found in cross sections of rigor insect flight muscle.

Finally, Figure D-5 shows relaxed insect flight muscle. Here you can see there are few connections between the actin and the myosin

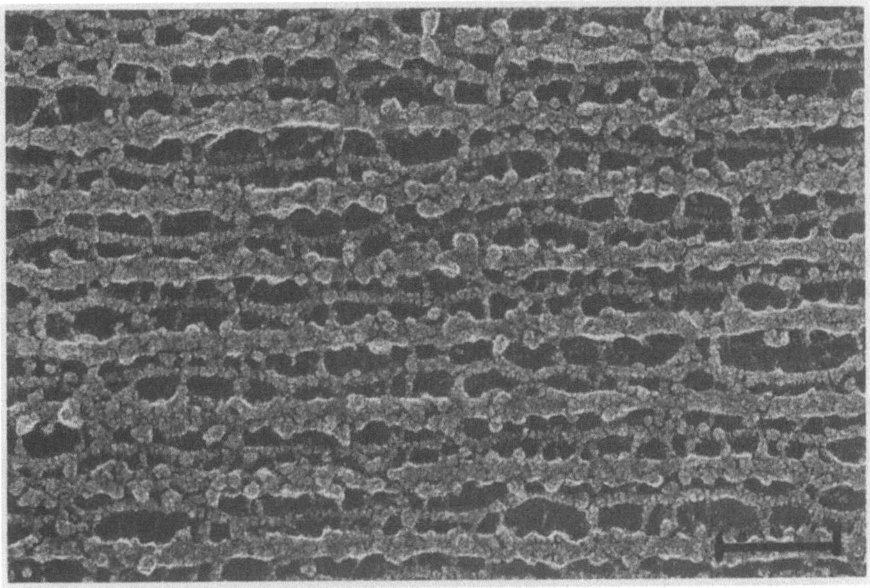

Figure D-5: Relaxed insect flight muscle which was bathed in a buffered salt solution containing 2 mM EGTA and 5 mM Mg-ATP immediately prior to quick freezing. (Bar = 100 nm).

filaments, far fewer than seen in the rigor muscle. The thick filaments appear quite lumpy. We think the lumps again represent cross bridges which now have been caught unattached to the actin and forced back down on the thick filaments during sample preparation. I don't know whether the few cross-bridges that are seen represent functional attachments between myosin and actin. It could be that cross-bridges flop around and when the muscle is frozen, some cross-bridges are found apparently attached to the actin.

KRUEGER: In the rigor insect flight muscle preparation, is there any soluble protein loss when you put the muscle into rigor? I wonder, in Dr. Robinson's case, if he is seeing only the diameter of myosin filaments or filaments plus a contribution from other proteins?

ROBINSON: From the images all I can say is that there's an electron density present that gives the pattern. I had a question for Dr. Cooke. The images were beautiful. I was wondering if you had a method for assessing possible shrinkage in your preparation?

COOKE: Well, we can measure, for instance, the myosin-myosin distance and we can measure crossbridge-to-crossbridge distance, and there doesn't appear to be much shrinkage.

REEDY: Based on X-ray spacing?

COOKE: Yes, compared to X-ray spacing.

CANTINO: I just wanted to comment that since these are glycerinated it might be good to consider the possibility of proteolysis due to that procedure, because Dr. Wang has told me that he gets some proteolysis after glycerination.

COOKE: That's certainly a possibility.

MAGID: I, myself, have started to do freeze-fracture, deep-etch in muscle, and showed some pictures in my poster. The filament dimensions that I've measured from skinned frog fibers, are about 180 Å for the thick filament, and about 85 Å for the actin filament. If one allows, say 15 Å for the platinum coat, then that brings us down to almost precisely what's been seen in published thin sections. But I should also say that in my work, the fibers were not in a saline; they were frozen in deionized water to avoid salt artefacts. Therefore they had to be fixed lightly in gluteraldehyde before being ultrarapidly frozen.

COOKE: We haven't seen much difference between samples that are glutaraldehyde fixed before freezing and those that aren't. There's somewhat better order with the gluteraldehyde fixed samples, but we also freeze them in the saline (50-100 mM KCl). If you don't etch very deeply, the salt doesn't adversely affect the quality of the micrographs.

REEDY: Al, do you think there are "struts" in insect flight muscle? Heuser has provided me with freeze-etch images from relaxed insect fibers and they do not seem to show anything which could reliably be identified as struts. Can you comment?

MAGID: I've said nothing about this because I don't know. I have the impression from some experiments that there are such structures. But I agree, I've looked at Heuser's pictures and there seems to be very little sign of such a thing. I wanted to raise the caution, repeat the caution, that Marie Cantino has made, which is that the specimens were glycerinated. They've been shipped around the country several times. I was one of the shipping legs of the journey of these beasts. The high molecular weight materials that I believe to be important elements in the elastic stroma of muscle are indeed very prone to proteolysis. Hours can make a difference at room temperature. This proteolysis is difficult to inhibit. So I think that the failure to find something may be nothing more than that. I think the question has to be left open.

KRUEGER: Have you looked at this in intact muscle? It might be more relevant to what Dr. Robinson has shown.

COOKE: No we haven't. If you look at insect muscle the problem is that the soluble proteins gum up onto everything and you don't get a very clear picture.

LEVINE: Roger, I notice you said that you had a left-handed helix. Are you really sure about that, with all the transfer of film and replicas and everything? Because all the filaments that we've looked at so far, and there are really quite a variety of them, had right-handed helices.

COOKE: I feel relatively confident about that. We checked the actin helix of a phage head and that was found to be correct. In some of these cases you can see both the actin helix and the myosin helix in the same electron micrograph. In those cases it agrees with Mike Reedy.

REEDY: I wanted to place on record a few remarks that I'm afraid are going to make it discouraging to interpret mechanical experiments done on subcellular preparations. I've talked with Dr. Wang about his experience in preparing both myofibrils and myofilaments, and we decided it would be worthwhile placing on record a warning that the simplified preparations of myofibrils and myofilaments, from which we reconstruct our models of how muscle works, are slightly incomplete in that some proteolysis seems to be necessary before these can be liberated. My most widely shared experience is one that Dr. Wang and I have both had, that when we've learned to prepare myofibrils that are extremely native in their behavior, that respond to relaxing solution without any shortening and so on, that when you blend these in a relaxing solution you don't get filaments. If we let them get a little bit old, or we don't take the same precautions against proteolysis they still look good, they still may be able to relax reasonably well in relaxing solution, but they are incapable of liberating filaments in the way that's become famous since Hugh Huxley first did it in 1963. The second point has to do with difficulties in liberating myofibrils and perhaps Dr. Wang should make it himself.

WANG: Yes, thank you. I believe that the difficulties that we had experienced in preparing myofibrils may be attributed to the presence of an intermediate filament network. These filaments appear to connect all myofibrils together. Therefore, in order to prepare myofibrils, we need to break all of the transverse connections between adjacent myofibrils. When we used fresh muscle and added protease inhibitors to buffers to minimize proteolysis, we obtained after blending, small bundles of curved myofibrils still linked together. If blended further, we did obtain some single myofibrils, but never the kind of straight, nice-looking ones that we saw in the literature. Instead, most of the single myofibrils had abnormal phase patterns and frequently were curved or even zig-zag in shape.

However, when we used glycerinated muscle or left intact muscle to age in the cold for a while, then only very gentle blending was needed to release a lot of pretty myofibrils with sharp phase contrast patterns. My interpretation is that these interfibrillar connections must be very robust and unless they are weakened or damaged by, say, proteolysis, it is difficult to break them without at the same time also altering the internal organization of the sarcomere. I think this is an interesting conclusion because a robust cytoskeletal network, especially the longitudinal components, may well contribute to the mechanical properties of muscle, e.g., compliance.

I also wish to comment that intracellular proteolysis may vary widely in speed and in extent in various muscles and in various buffers. In fact, I am inclined to think that unintentional proteolysis may be an essential feature of several frequently used procedures. For example, most glycerination procedures include as the first step incubation of muscle at 4°C for up to several days before lowering the temperature to -20° C. We have found that titin, nebulin and intermediate filaments were degraded rapidly at this step. Similarly, I think proteolysis may also be a prerequisite for the release and purification of sarcomere components such as A-segments, thin filaments and thick filaments. Again, our experience is that blending fresh muscle in relaxing buffers does not work well, but aged or glycerinated muscles work beautifully. This behavior is difficult to explain in terms of the two filament sarcomere model, but can easily be explained if we assume that thick filaments are connected to Z-discs by the third filament. In this model, A-segments can be released only after the third filaments and the longitudinal intermediate filaments are broken.

ALTRINGHAM: I have done some work on fish skinned fibers and one of the big problems with fish proteins is that they are notoriously labile. An interesting feature is that it's also relatively easy to dissect out myofibrils, possibly due to rapid proteolysis of these Z-Z connections.

CECCHI: I want to say one thing about how strong the connection is between the Z lines because it seems to me that Dr. Wang suggests that perhaps the compliance is associated with this. I would say that if I understood well, we should see how strong they are by just stretching the passive fiber. Am I right?

WANG: These filaments are *surrounding* the myofibril.

REEDY: I would like just to summarize. Our warning is to the people who do mechanics that proteolytically labile elements not recognized in the biochemical preparations may be present in their preparations, and our warning is to the structuralist and biochemist that their preparations may be lacking things which are present in intact muscle.

PART II: MECHANICS, ENERGETICS AND MOLECULAR MODELS

INTRODUCTION

The shape of the length-tension relation has always been regarded as a critical issue *vis a vis* the validity of the cross-bridge theory. The classical work published by Gordon, Huxley and Julian (J. Physiol. 184: 170, 1966) demonstrated that the measured and predicted shapes were remarkably coincident; so much so, that this work has become one of the pivotal pieces of evidence supporting the cross-bridge theory.

A study by ter Keurs et al. (J. Gen. Physiol. 72: 565, 1978) challenged this result. In contrast to the "classical" curve, in which tension fell off linearly with stretch above 2.2 μm sarcomere length, the length-tension relation in this more recent study remained relatively flat over a considerable degree of stretch before finally falling off at highly extended lengths. The differences lay more in interpretation than in results; for when ter Keurs and colleagues plotted their data in the same way as Gordon et al., virtually identical results were obtained.

The differences revolved around the interpretation of "creep," the slow rise of tetanic tension that follows the initial rapid rise. Gordon and colleagues felt this was an artifact; i.e., that it was caused by a population of sarcomeres shortening and "climbing up" the length-tension relation, thereby giving the slow creep of tension. On this presumption they plotted the tension measured prior to the onset of the slow-rise, where measurements were assumed to have been free of such artifact. Ter Keurs et al., failing to find such shortening populations in their fibers, concluded that the slow rise of tension was not an artifact, but a natural part of the tension generation process, and thus, that the tensions attained subsequent to it were the appropriate ones to consider. A plot of such steady-state tensions gave the relatively flat length-tension curve.

These differences stirred vigorous debate (Sugi and Pollack, Cross-Bridge Mechanism in Muscle Contraction, Univ. Tokyo Press/ Univ. Park Press, 1979; Ingels, Molecular Basis of Force Development in Muscle, Palo Alto Med. Res. Fndtn., 1979), and, as is evident from the pages that follow, inspired considerable experimentation. All of the studies are concerned with some aspects of the descending limb of the length-tension relation. In Taylor's presentation, for example, the question of whether the contractile apparatus is saturated with calcium is considered. One of the assumptions in the analysis of length-tension relations at long sarcomere lengths is that activation is at saturation level over the entire descending limb.

The presentation of Altringham describes experiments using single fibers in which sarcomeres at the terminal regions shorten and stretch the remaining sarcomeres in the central region. The measured length-tension relation is extremely flat, falling off to only about 50% of maximum tension at the point of no-overlap. But the authors are divided on their interpretation, one claiming consistency with the "classical" curve, the other not.

The paper by Edman presents a new surface marker method by which it is possible to hold a segment of a single fiber at constant length during a tetanus. Except for a slightly sigmoidal tendency, the shape of the descending limb of the length tension relation derived from such segments agreed quite well with the results of Gordon et al. (1966). The paper also shows reasonable consistency of isometric tension levels among segments along the fiber, while the corresponding values of maximum velocity of shortening varied considerably.

Ter Keurs and colleagues examine the length-tension relation in bundles of mammalian fibers in which the sarcomere length could be clamped using the laser diffraction method. The plateau of the length-tension curve extended to 2.65 μm sarcomere length. Although filament lengths in this preparation were found to be somewhat longer than in frog muscle, giving a "theoretical" curve shifted rightward from the classical curve, the measured tensions were still somewhat to the right of those expected.

Iwazumi presents the results of length-tension experiments obtained using single myofibrils, a breakthrough. Every sarcomere can be visualized during all stages of tension development. Under the conditions measured, which appear to be less than fully saturated, tension *increases* with increasing sarcomere length. The "descending" limb ascends.

Tirosh discusses the implications of the variety of shapes of the length-tension relation with regard to his hydraulic mechanism of muscular contraction.

CHANGES IN INTRACELLULAR Ca^{2+} INDUCED BY SHORTENING IMPOSED DURING TETANIC CONTRACTIONS

Giovanni Cecchi, Peter J. Griffiths* and Stuart Taylor+

Istituto di Fisiologia Umana, Viale Morgagni 63, Florence, Italy
**University Laboratory of Physiology, Oxford, OXI 3PT England*
+The Department of Pharmacology, Mayo Foundation, Rochester, MN 55905 USA

ABSTRACT

Calcium transients, monitored by aequorin, and force were recorded simultaneously during tetanic contractions of isolated frog skeletal muscle fibers. Quick length changes were applied to the fibers during contractions at sarcomere lengths on the descending limb of the length-tension relationship. Previous experiments showed that regulatory Ca^{2+} binding sites are apparently saturated during a plateau of tetanic force development at these sarcomere lengths. However, quick releases of greater than 4 to 5% of fiber length produced a momentary fall in the calcium transient that followed a time course similar to the redevelopment of force. The fall in the Ca^{2+} transient after a release was maximum at striation spacings about half way along the descending limb (2.6-2.7 μm), which suggests it is not related to an increase in the number of Ca^{2+} binding sites distributed uniformly along the filaments. The effect was absent or barely detectable when highly stretched fibers were released during contraction. The fall in the Ca^{2+} transient was unrelated to the time during a tetanus that a release was made or to the velocity of the release. One explanation of these results is that complexes between actin and myosin are broken by a sudden reduction of length, and as they reform during the recovery of force the affinity of troponin for Ca^{2+} increases. Quick stretch had no effect on the rapid decay of Ca^{2+} transients, but stretch increased peak force and slowed relaxation for almost a second after the end of stimulation. Evidently the decrease in the rate of relaxation produced by stretch is unrelated to changes in the amount of Ca^{2+} released or the rate of Ca^{2+} removal, which supports suggestions that the kinetics of muscle relaxation are determined by more than one mechanism. The apparent increase in the overall duration of mechanical activity after stretch probably results from the longitudinal inhomogeneity in the duration of activity - known to occur during relaxation - coupled with the decreased compliance of stretched fibers.

INTRODUCTION

Changes in muscle fiber length *prior* to stimulation have been shown to influence the apparent amount of free calcium in the myoplasm during subsequent twitch and tetanic contractions (Blinks, Rüdel and Taylor, 1978). If a fiber is stretched beyond the length at which it is just taut (i.e., slack length), the size of a Ca^{2+} transient during contraction increases slightly, but with further stretch it decreases markedly (Blinks et al., 1978). When a fiber is stimulated and allowed to shorten below slack length the relative size of an aequorin signal is also diminished (Blinks et al., 1978). Force measured simultaneously with the aequorin signal changes size with changes in length in roughly the same fashion, although the maxima for light and force rarely occur at the same length. A further difference between the two is that mechanical responses are prolonged at stretched lengths, although the duration of a Ca^{2+} transient is not influenced appreciably by fiber length. These observations do not act as a guide to a simple explanation for the relation between muscle length, Ca^{2+} uptake by the sarcoplasmic reticulum, and relaxation (Blinks et al., 1978). Hence we have further explored the question of whether stretch increases activation in light of earlier suggested answers (Edman and Kiessling, 1971; Taylor, 1974; Edman, 1974). This was done for tetanic contractions only, by imposing quick length changes *during* and immediately *after* stimulation. We then interpreted our results within the framework of an idealized model of muscle contraction in which identical sites of interaction are uniformly distributed along the thick and thin myofilaments (A.F. Huxley, 1980; H.E. Huxley, 1982).

We also considered the possibility that shortening itself mobilized calcium from the contractile proteins when the thick and thin filaments slide along each other during activity (Edman and Kiessling, 1971). Our studies were limited to stretched fibers because previous experiments without quick length changes showed that calcium-induced activation is normally less than maximum below slack length along the ascending limb of the length-tension relationship, and approaches maximum only along the descending limb (Lopez, Wanek and Taylor, 1981). When the level of free calcium during contraction is enhanced by treating fibers with chemical potentiating agents, the qualitative similarities between the length dependence of calcium transients and force development are lessened. The size of a Ca^{2+} transient is increased by potentiators, and although the fall of force is slowed, the rate of fall of light is unaffected. In the chemically potentiated state, extra calcium has a great effect on maximum force at short lengths, a small effect on force at lengths near the optimum for filament overlap, and virtually no effect on force at striation spacings along the descending limb of the length-tension relationship. These observations suggest that regulatory Ca^{2+} binding sites are saturated during the plateau of tetani elicited along the descending limb, unlike the variable situation along the ascending limb where we would expect the effects of a quick length change to be capable of two or more interpretations. In this study we found that shortening along the descending limb produced a fall in the Ca^{2+} transient, a result that might be explained by an increased affinity of the regulatory proteins for Ca^{2+} during the formation of cross-bridges (Bremel and Weber, 1972).

The changes in the relationships among fiber length, intracellular Ca^{2+}, and force development after treatment with chemical potentiating agents are presumably due solely to changes in the ability of the sarcoplasmic reticulum (SR) to release Ca^{2+} as the rate of Ca^{2+} uptake into the SR is apparently unaffected under these circumstances (Lopez et al., 1981). In other circumstances, the association between light and force signals can be explained in another way. For example, an increase in the frequency or in the duration of repetitive stimulation, or a decrease in temperature slow the decay of the aequorin response and also slow the decay of mechanical responses (Blinks et al., 1978). In these cases, relaxation may be related to the rate of Ca^{2+} uptake into the SR, rather than the amount of calcium released. However, this is one among several possible explanations previously considered but not tested (Blinks et al., 1978). Here we describe some further observations and reaffirm the idea that mechanisms underlying changes in muscle relaxation are perceivably of more than one kind (Taylor, 1974; Blinks et al., 1978; Pagala, 1980).

METHODS

Skeletal muscle fibers were isolated intact from tibialis anterior muscles of the frog, *Rana temporaria*. They were mounted on the stage of an inverted microscope that held a temperature controlled chamber. All the experiments described here were performed at 15°C. In the cold, the presence of parvalbumins in these frog fibers might significantly buffer the influence of an experimental perturbation that would otherwise tend to alter the rate of fall of the calcium transient as well as the tetanus relaxation (Gillis, 1980). Parvalbumins evidently are freely soluble in the cytoplasm (Gillis, Piront and Gosselin-Rey, 1979) and at 4°C appear to serve as the initial "sink" for Ca^{2+} at the end of a tetanus (Somlyo, Gonzalez-Serratos, Shuman, McClellan and Somlyo, 1981). Hence we chose to study fibers at a moderately warm temperature (i.e.,15°C) near the range where SR activity should be relatively high, but not warm enough to eliminate the difference between the rate of decline of aequorin responses from intact amphibian fibers and the maximum rate with which aequorin responses can be quenched *in vitro* (Blinks et al., 1978; Cannell, 1982). Length-dependent changes in intracellular Ca^{2+} transients due to changes in SR reuptake should have been detected more readily at this temperature (15°C) than at a temperature where SR activity may be negligible (Gillis, 1980).

We examined the falling phase of Ca^{2+} transients in tetani with regard to possible distortion introduced by (i) the recording equipment and (ii) the kinetics of the aequorin reaction with Ca^{2+}, as already described (Taylor et al., 1982). These corrections did not appreciably alter the decay of the tetanic light response compared to the "true" intracellular Ca^{2+} transient. It should be noted that this procedure ignores the possible effects of $[Ca^{2+}]$ gradients during relaxation. We previously speculated about the possible relations among (i) the rate of decline of aequorin responses, (ii) the fall of the "true" intracellular $[Ca^{2+}]$. (iii) the redistribution of Ca^{2+} between release sites and reuptake sites, and (iv) the existence of gradients of [Ca^{2+}] in the myoplasm (Blinks et al., 1978). One interpretation of some of the evidence that was the basis for these speculations (Winegrad, 1968; 1970) is now in doubt (Somlyo et al., 1981). Ca^{2+} previously thought to be bound to sites of reuptake by SR, may be bound to low molecular weight soluble proteins in the cytoplasm (Somlyo et al., 1981). However, these experiments were performed in the cold, where SR activity may be very low (Gillis, 1980). Thus, we have not attempted to analyze the kinetics of the aequorin transients observed in our present experiments in a way other than already described (Taylor et al., 1982), until questions about possible [Ca^{2+}] gradients produced by SR activity at moderately warm temperatures (e.g., 15°C) are answered.

The small tabs of tendon at each end of a fiber from the tibialis anterior muscle were connected to a capacitance gauge transducer (with a natural frequency of 50 kHz and compliance of 0.05 μm/mN) and a length-step generator, respectively, as previously decribed (Cecchi, Griffiths and Taylor, 1982; 1983). The fibers were stimulated transversely with high frequency dc pulses of brief duration to develop fused tetani. They were also microinjected with the calcium-sensitive photoprotein aequorin which enabled us to simultaneously monitor a relative change in intracellular free calcium and contractile force by methods already described (Blinks et al., 1978). Fibers from the tibialis anterior muscle are relatively short and can be injected at several spots along their length to obtain an apparently uniform longitudinal and circumferential distribution of aequorin (Trube, Lopez and Taylor, 1981). Constant velocity releases and stretches were made at times from the onset of tetanic stimulation up to about a second after the last stimulus. The velocity of release ranged from 1.2 to 25.2 muscle lengths per second; the velocity of stretch was, without design, somewhat slower. The amplitude of the releases ranged from about 2.5% to 10% of a fiber's slack length; the amplitudes of the stretches were the same. Releases and stretches were made only over the range of striation spacing from the length where filament overlap was optimal (about 2.2 μm) to lengths along the descending limb of the relation between filament overlap and tetanic force (about 3.1 μm).

RESULTS

Figure 1 illustrates the influence of rapid shortening on intracellular Ca^{2+} during a tetanus. A constant velocity release caused a rapid drop in force followed by its redevelopment as stimulation was continued. Light emission from the aequorin injected fibers began to drop after the release was complete and force had begun to redevelop (area C in the expanded part of Figure 1). The drop in light emission reached its limit at the time that force had nearly completed its recovery. Subsequently, if stimulation was continued, light emission also began to recover and the aequorin signal continued with the same time course it followed when no release was imposed on a fiber. Possible causes of the momentary fall in Ca^{2+} that comes to mind are transitory effects of the quick length change on the amplitude or duration of an action potential, or on coupling

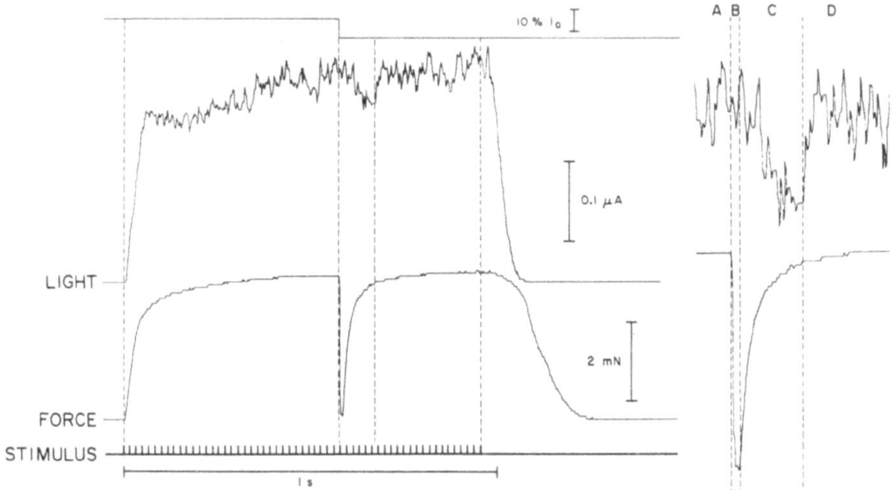

Figure 1: The change in light emission and force production in response to a rapid release imposed during the plateau of a tetanic contraction. Fiber (19.ii.81) isolated from the tibialis anterior muscle of a frog and studied at 15°C. The release began from a striation spacing of 2.68 μm. When held just taut the fiber length (l_0) was 5.8 mm at 2.2 μm striation spacing. The stimulus frequency was 60 Hz, and the velocity of shortening was a constant 10.2 fiber lengths per second. The entire records are shown on the left side of the figure and an expanded view of the changes during the release are shown on the right. The duration of period B in the inset is 6.41 milliseconds. There is no apparent deviation in the light response during the time it takes to complete the length step, which is one indication that the later slower change in light is not by itself the result of displacing parts of the fiber that are emitting light.

between action potentials and Ca^{2+} release from SR (e.g., Hennekes, Kaufman, Lab and Steiner, 1977). But the change shown in Figure 1 lasts less than a hundred milliseconds, whereas other changes that suggest shortening somehow interrupts excitation-contraction coupling persist for the duration of stimulation and do not depend upon time (Taylor and Rüdel, 1970).

The amplitude of the light drop depended on several factors. A release imposed on a fiber contracting at slack length (2.2 μm) produced a detectable light drop. But when a release was made from a striation spacing at which aequorin light emission during stimulation was greatest (2.6 μm - 2.7 μm) the amplitude of the light drop was also greatest. This is not explained by an effect related only to the interaction of identical sites distributed uniformly along the myofilaments (e.g., Fuchs, 1978). One would expect the relative size of the drop in light emission to be largest near slack length if this were the answer. But we have not determined the degree of uniformity among striation spacings in the injected regions during and immediately after a release from different lengths. Therefore, our interpretation of this result is uncertain. Contractions that began at stretched lengths might have been intrinsically less uniform than those near slack length (Julian and Morgan, 1979).

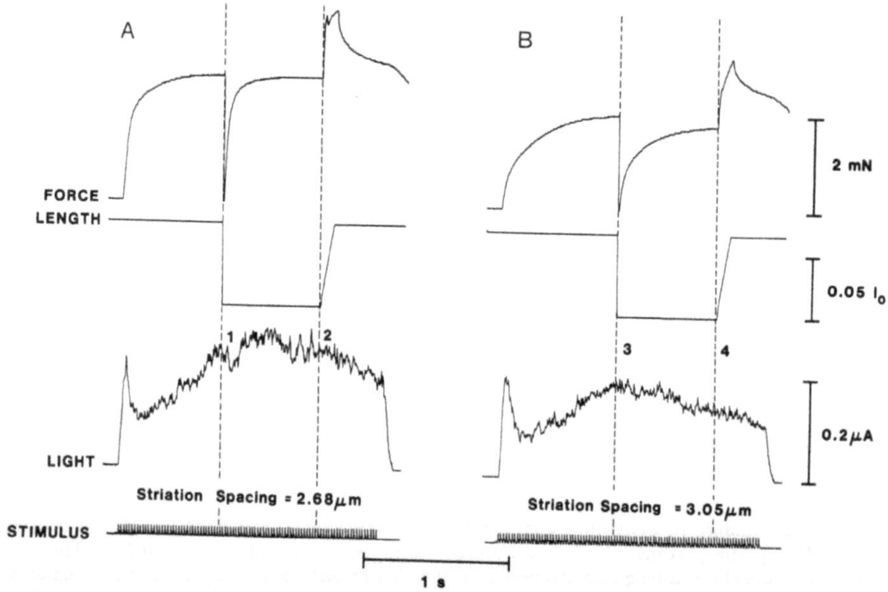

Figure 2: The influence of quick releases and stretches on tetanic responses at different striation spacings. This is the same fiber shown in Figure 3, but the velocity of release was higher (8.13 fiber lengths per second). Responses in column A began from a striation spacing of 2.68 μm. Those in column B began from a striation spacing of 3.05 μm. A release during the plateau of a tetanus at a moderately stretched length (1) produced a momentary drop in light. A stretch at this length (2) had no detectable effect. A release during a tetanus at a high degree of stretch (3) had no detectable effect, nor did a stretch (4).

The size of the largest light drop was about 10-20% of the maximum signal (Figure 2A). When a release was made from a length further along the descending limb where filament overlap was considerably less than optimal (e.g., 60 to 70% less in Figure 2B) the light drop was either not detectable or absent. The overall size of a Ca^{2+} transient was larger at shorter lengths on the descending limb than at stretched lengths (Figure 2B), as previously reported (Blinks et al., 1978). Furthermore, the overall depression of the aequorin response produced by extreme stretch (Figure 2B) usually persisted for minutes after a fiber was released (Blinks et al., 1978). Therefore, the length-dependent changes in Ca^{2+} transients attributed to effects on the ability of the action potential to uniformly depolarize membranes involved in excitation-contraction coupling or effects on the ability of the SR to release Ca^{2+} were in the direction opposite to the changes produced by a quick release during stimulation and were much slower to recover (Figure 2).

The amplitude of the release also influenced the size of a light drop (Figure 3). Releases that were roughly 3% or less of the fiber length produced no detectable change in the light signal. Releases larger than about 4 to 5% of fiber length in amplitude, sizes that would be expected to produce detachment of cross-bridges and significant displacement of the

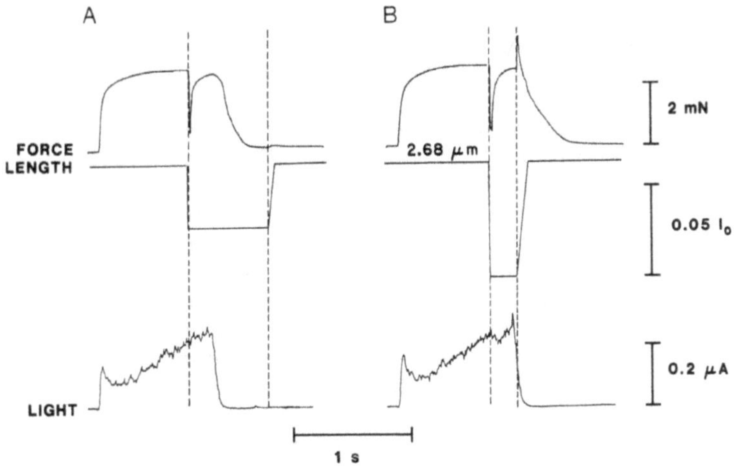

Figure 3: The influence of quick releases of different amplitudes on tetanic responses. Fiber (18.ii.81) from the tibialis anterior of a frog studied at 15°C. Both releases began from a striation spacing of 2.68 μm. When held just taut the standard fiber length (l_0) was 7.4 mm at 2.2 μm striation spacing, the stimulus frequency was 50 Hz. This figure is reproduced with permission from Taylor et al. 1982, where it is Figure 8. The velocity of shortening was a constant 2.93 fiber lengths per second rather than the value given in error in Taylor et al. (1982). A dashed line is added to show the time of the stretch, which was slower than the release in all our experiments. The smaller stretch (3% of l_0) occurred about 500 msec after the last stimulus and produced a small rise and fall in force (A). The size of the force increase produced by stretch was progressively larger at longer striation spacings, but was never accompanied by a detectable change in the light record. The larger stretch (6% of l_0) occurred about half-way through the final decay phase of the aequorin transient and is not associated with the spike of light at the end of the transient (B).

filaments relative to one another, were necessary to produce a detectable light drop regardless of the other factors. An additional feature to note is that a stretch imposed during stimulation (points 2 and 4 in Figure 2) or after stimulation on the falling phase of an aequorin response (Figure 3B) did not produce a detectable change in the light record, although the force response was markedly affected. We did not average aequorin responses to study the possibility that there might be a very small effect of stretch (Snowdowne and Lee, 1980).

The size of a light drop seemed unrelated to the time during the tetanus that a release was made. The size of the light drop was also unrelated to the velocity of the release, at least in the range of velocities from about 3 to 25 fiber lengths per second at 15°C. The independence of a light drop from the velocity of release and the lack of any detectable effect of a stretch on light emission also indicate that the light drop was not merely due to movement of spots where aequorin was injected along a fiber.

DISCUSSION

Among the factors that have been suggested to influence the time course of relaxation after a brief period of stimulation, other than those associated with SR function (Ca^{2+} release and reaccumulation), are the following: 1) speed of dissociation between cross-bridges and sites of interaction on thin filaments, 2) increased affinity of troponin for calcium resulting from the formation of actin-myosin complexes (Bremel and Weber, 1972), 3) rate of calcium binding by proteins (e.g., parvalbumins) outside the myofilaments (Gillis, 1980), and 4) changes in ATP concentration or affinity for ATP hydrolysis (Edwards, Hill and Jones, 1975; Dawson, Gadian and Wilkie, 1980). Our data are relevant to only the first two factors, and we have considered the first elsewhere in this volume (Cecchi, Griffiths and Taylor, 1983). With regard to the second factor, our observations allow some response to the questions of whether quick length changes create new Ca^{2+} binding sites on the contractile filaments or change the affinity of those sites available (Bremel and Weber, 1972; Weber and Murray, 1973). We infer that our findings are, for the most part, consistent with the idea that the affinity of troponin for Ca^{2+} may increase after a quick release as a result of the formation of actin-myosin complexes (Bremel and Weber, 1972). A change in affinity might momentarily alter the balance among calcium ions associated with aequorin and those associated with other Ca^{2+} binding proteins. Consistent with this idea is the fact that a drop in the Ca^{2+} transient is just detectable after quick releases from lengths with little overlap, and becomes bigger as releases are made nearer to maximum overlap.

Rapid shortening imposed at the onset of stimulation or at any other time during tetanic stimulation produced a temporary decline in the aequorin response. The decline was complete in about 70 to 100 milliseconds at 15°C, which is much slower than the rate of fall of the response at the cessation of stimulation. This change is not likely to be caused by a length-dependent alteration in the Ca^{2+} releasing ability of the sarcoplasmic reticulum, because effects seemingly associated with the SR do not occur in the same direction nor do they immediately reverse when a stretched fiber is released (Blinks et al., 1978). The light drop reaches its limit when the recovery of tension is nearly complete, which is consistent with the idea that cross-bridge attachment and the associated changes in calcium-binding affinity of the contractile proteins may cause this effect. One would expect an increase in the aequorin signal when cross-bridges are broken by the release if the opposite effect occurred and if, for example, the depressant effect of shortening on force production is due to an effect on the affinity of the binding sites for calcium in the regulatory protein complex (Edman, 1980). But we never observed a change that suggested a decrease in Ca^{2+} binding affinity.

No reduction in light emission was produced by relatively small releases. The releases associated with a light drop are clearly larger than those required to just discharge the force in an attached cross-bridge, and the subsequent changes in force must involve something more than an adjustment between attached states only (Huxley and Simmons, 1971).

A length change of at least 4% was required to produce an observable effect, and this is well beyond the range within which one would expect to impose a length change that did not cause cross-bridge detachment. Small length perturbations imposed on one end of a muscle fiber to measure properties such as stiffness (Cecchi et al., 1983) are, therefore, not associated with a change in myoplasmic free Ca^{2+}.

It has been suggested from experiments on giant barnacle muscle fibers, where a step decrease in length is followed by a rise in the aequorin signal, that the affinity of Ca^{2+} binding sites involved in regulating force production might be altered by changing the size of the overlap zone (Gordon and Ridgway, 1978). Our data are not the same as those obtained with barnacle fibers, but are quite similar to those previously reported for frog skeletal muscle (Allen, 1978). Frog skeletal muscle evidently has only actin-linked regulation, whereas Ca^{2+} binding to myosin may be important in the regulation of barnacle muscle contraction (Lehman and Szent-Györgyi, 1975). Hence calcium bound to myosin is one of the possible sources of the extra light produced by barnacle muscle (Gordon and Ridgway, 1978). We never observed an extra burst of light, but we did not examine the effects of releases on the rate of decay of an aequorin transient.

We also considered the possibility that stretch induces a change in Ca^{2+} affinity of the regulatory sites and this change is correlated in time with the slow relaxation of stretched muscle. Stretches on the decay phase of an aequorin response produced no detectable effect on light emission from our frog fibers. We conclude that the slowing of the rate of relaxation that is induced by stretch has no relation to myoplasmic Ca^{2+}, and has an origin different from the slowing of relaxation induced by other means (Edwards, Hill and Jones, 1975; Blinks et al., 1978; Sandow and Zeman, 1979; Dawson, Gadian and Wilkie, 1980; Pagala, 1980). For example, changes induced by repetitive stimulation, by prolonging the duration of repetitive stimulation, by changing the temperature, and in post-tetanic responses are all associated with changes in amplitude and rate of decay of Ca^{2+} transients (Blinks et al., 1978). On the other hand, the slower than normal relaxation of force in the presence of chemical potentiators is also associated with a change in size of a Ca^{2+} transient, but not with a change in its rate of decay (Lopez et al., 1981). Likewise, no such changes in Ca^{2+} transients are associated with quick stretches, as shown here, and fibers held at stretched lengths prior to stimulation produce reduced Ca^{2+} transients that decay at essentially the same rate also (Blinks et al., 1978).

An alternative explanation for the observations suggesting that the duration of mechanical activity is increased by stretch has already been proposed (Taylor, 1974). Longitudinal inhomogeneity in the duration of activity could account for the stretch effect. Such inhomogeneity has been demonstrated in several ways (e.g., Huxley and Simmons, 1970; Edman and Flitney, 1977; Cecchi et al., 1983, Figure 3). When a muscle is stretched and becomes stiffer, the individual segments of cells with intrinsically shorter cycles of activity would be expected to yield less

(Taylor, 1974). Accordingly, force developed by segments with longer cycles of activity will be transmitted more effectively to the tendon attachments, and force will only appear to last longer. The same may be true for stretches imposed before relaxation begins, as it seems the number of cross-bridges displaced from their rest positions is the same before and after stretch during a tetanus (Yagi and Matsubara, 1977). Calcium transients fall rapidly towards pre-stimulus levels during the isometric phase of relaxation (Blinks et al., 1978), much more rapidly than the apparent rate of cross-bridge detachment, and this relates to the possibility that cross-bridge dissociation occurs by some process other than removal of calcium bound to the contractile filaments (Gordon and Ridgway, 1978). The facts that the decay of force is markedly slowed by stretches imposed on the tetanus plateau as well as by stretches imposed during the isometric phase of relaxation with no associated effect on the decay of a Ca^{2+} transient (Figures 2 and 3), are consistent with the idea that Ca^{2+} sequestration usually occurs very rapidly and the initial phase of relaxation occurs by some process other than the removal of calcium from the myofilaments (Gordon and Ridgway, 1978). Stiffness after the cessation of excitation outlasts the production of force during the isometric phase of relaxation, and one possible explanation is that cross-bridges remain attached during this period but have a reduced ability to generate force (Cecchi et al., 1983). Moreover, the long-lived state of cross-bridge attachment during relaxation in the absence of a large intracellular $[Ca^{2+}]$, might be similar to the large resistance to stretch that is a prominent feature of molluscan muscle behavior (Twarog, 1976). The same phenomenon may occur to some lesser degree in vascular smooth muscle, skeletal muscle of invertebrates and perhaps in vertebrate skeletal muscle also (Twarog, 1976). The control of contraction by Ca^{2+} in molluscan muscle is exerted directly by its binding to myosin (Lehman and Szent-Györgyi, 1975; Twarog, 1976). Furthermore, the very slow return of cross-bridges to their resting position (Matsubara and Yagi, 1978), might be another indication of a process associated with some event other than removal of Ca^{2+} from troponin. On the other hand, the slow return of cross-bridges could be connected with Ca^{2+} removal from myosin in frog muscle too. Aequorin light emission at a very low level has been detected throughout the entire phase of relaxation in frog skeletal muscle (Snowdowne and Lee, 1980) as well as in barnacle muscle (Ashley and Lignon, 1982), which is consistent with the latter possibility.

ACKNOWLEDGEMENTS

This work was supported in part by fellowships from the Muscular Dystrophy Association of America Inc. and the Minnesota Heart Association (to G.C. and P.J.G.), and by grants-in-aid from the Minnesota Heart Association and U.S. Public Health Service (NS 14268). We are indebted to L.A. Wanek for assistance with the experiments, to L.A. Wanek, S. Winegrad and M.K.D. Pagala for helpful criticism of the manuscript, and to L.F. Wussow for preparation of the figures and the manuscript.

REFERENCES

Allen, D.G. (1978). Shortening of tetanized skeletal muscle causes a fall of intracellular calcium concentration. J. Physiol. 275: 63P.

Ashley, C.C. and Lignon, J. (1982). Aequorin responses during relaxation of tension of single muscle fibres stimulated by voltage clamp. J. Physiol. 318: 10-11P.

Blinks, J.R., Rüdel, R. and Taylor, S.R. (1978). Calcium transients in isolated amphibian skeletal muscle fibres: Detection with aequorin. J. Physiol. 277: 291-323.

Bremel, R.D. and Weber, A. (1972). Cooperation within actin filament in vertebrate skeletal muscle. Nature 238: 97-101.

Cannell, M.N. (1982). The effect of temperature on the rate of fall of intracellular calcium during relaxation of frog skeletal muscle. J. Physiol. 340: 27P.

Cecchi, G., Griffiths, P.J. and Taylor, S.R. (1981). Decrease in aequorin response of skeletal muscle produced by quick release during tetanic contraction. Proc. VII Internat. Biophys. Congr. pp. 177.

Cecchi, G., Griffiths, P.J. and Taylor, S.R. (1982). Muscular contraction: Kinetics of crossbridge attachment studied by high frequency stiffness measurements. Science 217: 70-72.

Cecchi, G., Griffiths, P.J. and Taylor, S.R. (1983). The kinetics of cross-bridge attachment studied by high frequency stiffness measurements. This volume.

Dawson, M.J., Gadian, D.G. and Wilkie, D.R. (1980). Mechanical relaxation rate and metabolism studied in fatiguing muscle by phosphorus nuclear magnetic resonance. J. Physiol. 299: 465-484.

Edman, K.A.P. (1974). In: *The Physiological Basis of Starling's Law of the Heart.* Ciba Foundation Symposium, vol. 24, American Elsevier, New York, pp. 112-115.

Edman, K.A.P. (1980). Depression of mechanical performance by active shortening during twitch and tetanus of vertebrate muscle fibres. Acta Physiol. Scand. 109: 15-26.

Edman, K.A.P. and Kiessling, A. (1971). The time course of the active state in relation to sarcomere length and movement studied in single skeletal muscle fibres of the frog. Acta Physiol. Scand. 81: 182-196.

Edman, K.A.P. and Flitney, F.W. (1977). Non-uniform behaviour of sarcomeres during isometric relaxation of skeletal muscle. J. Physiol. 276: 78-79.

Edwards, R.H.T., Hill, D.K. and Jones, D.A. (1975). Metabolic changes associated with the slowing of relaxation in fatigued mouse muscle. J. Physiol. 251: 287-301.

Fuchs, F. (1978). On the relation between filament overlap and the number of calcium-binding sites on glycerinated muscle fibers. Biophys. J. 21: 273-277.

Gillis, J.M. (1980). The biological significance of muscle parvalbumins. In: *Calcium-binding Proteins: Structure and Function.* edited by F.L. Siegel, E.Carafoli, R.H. Kretsinger, D.H. MacLennan and R.H. Wasserman. Elsevier North-Holland Inc., pp. 309-311.

Gillis J.M., Piront, A. and Gosselin-Rey, C. (1979). Parvalbumins. Distribution and physical state inside the muscle cell. Biochim. Biophys. Acta 585: 444-450.

Gordon, A.M. and Ridgway, E.B. (1978). Calcium transients and relaxation in single muscle fibers. Eur. J. Cardiol. 7/Suppl.: 27-34.

Hennekes, R., Kaufmann, R., Lab, M. and Steiner, R. (1977). Feedback loops involved in cardiac excitation-contraction coupling: Evidence for two different pathways. J. Mol. Cell. Cardiol. 9: 699-713.

Huxley, A.F. (1980). Reflections on muscle. In: *The Sherrington Lectures*, volume XIV, 111 pp., Princeton Univ. Press, Princeton, NJ.

Huxley, H.E. (1982). Guest Lecture. The Mechanism of force production in muscle. In: *Disorders of the Motor Unit.* edited by Schotland, D.L. John Wiley and Sons, New York, pp. 1-11.

Julian, F.J. and Morgan, D.L. (1979). Intersarcomere dynamics during fixed-end tetanic contractions of frog muscle fibres. J. Physiol. 293: 365-378.

Lehman, W. and Szent-Györgyi, A.G. (1976). Regulation of muscle contraction: Distribution of actin control and myosin control in the animal kingdom. J. Gen. Physiol. 66: 1-30.

Lopez, J.R., Wanek, L.A. and Taylor, S.R. (1981). Skeletal muscle: Length dependent effects of potentiating agents. Science 214: 79-82.

Pagala, M.K.D. (1980). Effect of length and caffeine on isometric tetanus relaxation of frog sartorius muscles. Biochim. Biophys. Acta 591: 177-186.

Sandow, A and Zeman, R.J. (1979). Tetanus relaxation. Temperature effects and arrhenius analysis. Biochim Biophys. Acta 547: 27-35.

Snowdowne, K.W. and Lee, N.K.M. (1980). Subcontracture concentrations of potassium and stretch cause an increase in the activity of intracellular calcium in frog skeletal muscle. Fed. Proc. 39: 1733.

Somlyo, A.V., Gonzalez-Serratos, H., Shuman, H., McClellan, G. and Somlyo, A.P. (1981). Calcium release and ionic changes in the sarcoplasmic reticulum of tetanized muscle: An electron-probe study. J. Cell Biol. 90: 577-594.

Taylor, S.R. (1974). Decreased activation in skeletal muscle fibres at short lengths. In: *The Physiological Basis of Starling's Law of the Heart.* Ciba Foundation Symposium, vol. 24, American Elsevier, New York, pp. 93-116.

Taylor S.R., Lopez, J.R., Griffiths, P.J., Trube, G. and Cecchi, G. (1982). Calcium in excitation-contraction coupling of frog skeletal muscle. Can. J. Physiol. Pharmacol. 60: 202-209.

Taylor, S.R. and Rüdel, R. (1970). Striated muscle fibers: Inactivation of contraction induced by shortening. Science 167: 882-884.

Trube, G., Lopez, J.R. and Taylor, S.R. (1981). Calcium transients in asymmetrically activated skeletal muscle fibers. Biophys. J. 36: 491-507.

Twarog, B.M. (1976). Aspects of smooth muscle function in molluscan catch muscle. Physiol. Rev. 56: 829-838.

Weber, A. and Murray, J.M. (1973). Molecular control mechanisms in muscle contraction. Physiol. Rev. 53: 612-673.

Winegrad, S. (1968). Intracellular calcium movements of frog skeletal muscle during recovery from tetanus. J. Gen. Physiol. 51: 65-83.

Winegrad, S. (1970). The intracellular site of calcium activation of contraction in frog skeletal muscle. J. Gen. Physiol. 55: 77-88.

Yagi, N. and Matsubara, I. (1977). Equatorial x-ray reflections from contracting muscle after an applied stretch. Pflugers Arch. 372:113-114.

DISCUSSION

KURIHARA: What happens if you just stretch during the tetanus?

TAYLOR: We were unable to detect any change in the aequorin response produced by a stretch at any phase of a tetanus. But we have not looked for an effect of stretch with signal averaging or with the increased gain that others have used (Snowdowne and Lee, 1980; Ashley and Lignon, 1982; Connell, 1982).

EDMAN: When you stimulated the fibers with an AC field on the ascending limb, were the differences in the aequorin signal you recorded due to a change in frequency or intensity of the stimulation?

TAYLOR: First of all, we have not yet looked at the effects of AC field stimulation on the ascending limb. When we did vary the frequency of D.C. stimulation there was a substantial increase in luminescence but little or no increase in force of an already fused tetanic contraction if a fiber was at least slightly stretched (Blinks et al., J. Physiol. 277: 291-323, 1978). On the other hand, the plateau value of fused tetanic force was markedly increased by increasing the stimulus frequency if a fiber was allowed to shorten (Taylor, et al., Can. J. Physiol. Pharmacol. 60:489-502, 1982).

EDMAN: I wonder if you can rule out the possibility that the changes were due to differences of the inward spread of activation?

TAYLOR: In fact, that's what I believe is the explanation. The fibers showed no change in uniformity of the striation spacing after only a few brief tetani at short lengths. The frequency-dependent increase in tetanic force on the ascending limb is not detected unless one exceeds the lowest frequencies of stimulation that produce a fused plateau of force at relatively warm temperatures (15°C).

SUGI: I would like to mention briefly some experiments I've done on myofibrillar bundles from glycerinated rabbit psoas muscle. The experiments are related to the ascending limb of the length-tension curve. The myofibril bundle preparation was about 20 μm in diameter and 100-200 μm in length, so that the striation spacings were more clearly visible from end to end. One end of the preparation was fixed in position, while the other end was connected to a force transducer having a relatively large compliance, so that the preparation shortened auxotonically by 10-20% until exerting a steady isometric force when activated with Ca.

A typical result is shown in Fig. D-1A. In the relaxed state, there was already some degree of sarcomere length non-uniformity, ranging from 2.2 to 2.9 μm in this particular case. When the preparation was first half-maximally activated at pCa 6.5, and then maximally activated at pCa 5.5, it shortened progressively against increasing auxotonic load. It was always observed that some segments shortened so much that the striation spacings could no longer be observed, giving the average sarcomere

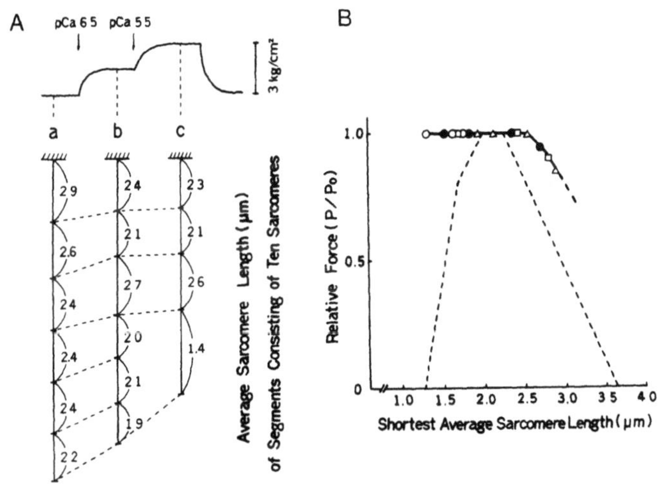

Figure D-1: A: Typical example of sarcomere length changes in each elementary segment along the entire length of the glycerinated rabbit psoas muscle myofibril bundle preparation during the course of Ca-activated auxotonic shortening. The average length of 5 sarcomeres is shown for each segment in relaxed state (a) and in contracted states at pCa 6.5 (b) and pCa 5.5 (c). In a,b, and c, the top and the bottom segments represent the segment nearest the fixed end and that nearest the other end connected to the force transducer respectively. The force development during auxotonic contraction is also shown at the top of the figure. B: Relation between the maximum isometric force at pCa 5.5 and the shortest average sarcomere length during the isometric force generation, when the preparation was previously stretched at various degrees beyond slack length. Each symbol represents results obtained from the same preparation. Broken line is the sarcomere length-force relation of intact muscle fibers by Gordon et al. (1966).

lengths of 1.4-1.2 μm, while the preparation was exerting a large steady isometric force of about 3 Kg/cm^2.

The sarcomere length changes during the auxotonic contraction were completely reverisble, and can be repeated many times. If you plot the steady isometric force relative to the maximum value against the shortest average sarcomere length, the ascending limb of the sarcomere length-force curve is virtually absent; there is a plateau from 1.2 to 2.0 μm (Fig. D-1B). It has been pointed out that the ascending limb observed on intact fibers mainly results from an incomplete activation at short sarcomere lengths (Taylor & Rüdel, Science, 1970) and the present results agree with the work of Schoenberg & Podolsky, Science, 1972) that Ca-activated skinned fibers can generate a large force at short sarcomere lengths. My conclusion is that the ability of Ca-activated myofibrils to generate the maximum force may not be impaired by double overlap of the thin filaments or collision of the thick filaments against the Z-band.

NOBLE: With regard to this flat part to the left, with no ascending limb. I'm not quite sure what the classical theory would predict, anyway. Do you have any comments about this?

EDMAN: I believe that the flat portion of the length-tension relation which you obtain below 2.0 μm sarcomere length could, in large measure, be due to sarcomere non-uniformity. What probably happens is that in the beginning of the contraction you get a wide dispersion of sarcomere lengths. That is, some sarcomeres may shorten to, say, 1.0 μm, whereas others shorten much less. Those sarcomeres which end up at a very short length are not able to actively produce the force which is finally produced by the preparation. These short sarcomeres are being stretched and there is force enhancement by stretch in these sarcomeres. The tension produced by the preparation will be set by the strongest sarcomeres, and the length of these sarcomeres may actually be quite close to 2.0 μm. In other words, the experiment you presented does not tell us that the short sarcomeres are able to *actively* produce the tension that is recorded in the preparation.

SUGI: But it's auxotonic shortening, so short sarcomeres *should* shorten actively against increasing load.

EDMAN: No, I would think that the short sarcomeres you observed have reached their short length very early on during contraction, that is, when the force is still low. These sarcomeres will be slowly stretched as tension rises in the preparation.

TER KEURS: But I think Fabiato and Fabiato (Nature, 1977; J. Gen. Physiol. 72: 667-699, 1978) looked at that problem, and activated even smaller fragments of muscle, both cardiac muscle and I think also skeletal muscle. The Fabiatos proposed both in their paper in *Nature* and in more recent work that the relationship between force and sarcomere length, where you would expect an ascending limb, is rather flat. This measurement suggested that there is less dispersion of sarcomere lengths than the dispersion we have seen here. This aspect thus seems of less importance than suggested.

NOBLE: One should certainly expect that an ascending limb would not be particularly unstable, while the descending limb might.

KRUEGER: I wanted to support Professor Sugi indirectly by challenging Paul Edman's interpretation of extreme shortening. In some conditions, unattached single heart cells can shorten to an average sarcomere length of one micron or so. The striation patterns appear uniform and neither the cell nor the fibrils appear buckled. Therefore I do not think that heterogeneity (where some sarcomeres are shortened by longer neighbors) can explain extreme shortening.

EDMAN: I'm simply suggesting that you look at the possible development of striation pattern irregularities before you state that fibers produce this much tension, because this appears to me to be a possibility.

SUGI: All sarcomeres along the length of the preparation were always clearly visible at half-maximum at pCa 6.5. In this condition, the isometric force attained was 1.5 to 2.0 Kg/cm^2 which is a fairly high tension. Then, some sarcomeres shortened further against increasing load starting from that high level at pCa 5.5. Therefore I think that short sarcomeres can develop the maximum force actively.

WILKIE: My question is to Stuart Taylor. I recall that the work you published with Reinhardt Rüdel suggested that the left-hand end of the force-length curve is at least in part due to imperfect penetration of excitation, used in a rather loose sense, to the center of the fiber. I wanted to ask Stuart whether I had got the message correctly from your presentation, that you were saying, in addition, not only does the excitation proceed imperfectly, but that having done so, it releases less calcium than it would do at a longer sarcomere length.

TAYLOR: Yes, that's true (e.g., see Figure 11 of Blinks et al., J. Physiol. 277: 291-323, 1978).

NOBLE: And do you believe that in Dr. Sugi's experiment, where he has a skinned fiber, this is overcome?

TAYLOR: Yes, I believe it can be overcome to a great degree. I'd also like to support Dr. Sugi's conclusion with some experiments performed on intact single fibers (Lopez et al., Science 214: 79-82, 1981; Taylor et al., Can. J. Physiol. Pharmacol. 60: 489-502, 1982). These results are similar to those shown by Dr Sugi. However, they were obtained on single fibers from tibialis anterior muscles of *Rana temporaria* studied at 15°C. Twitch potentiating agents markedly increased the plateau value of fused tetanic force in fibers stimulated to shorten below the striation spacing at which filament overlap is optimal. The effect is greatest in fibers studied in the warm and allowed to rest for a sufficient interval between contractions. Striation spacings were measured optically before each tetanic aequorin response and force response were recorded in the dark, and in separate experiments the period during contraction was photographed via a microscope to eliminate the possibility that the effects on force were associated with any changes in the uniformity of the striation spacing.

The conclusions of other investigators may differ from mine. But their results are not necessarily different. Julian and Morgan (1981), for example, routinely elicited tetani in their intact fiber studies with only 100 seconds of rest for each second of stimulation. This is less rest than is needed to obtain reproducible aequorin responses from muscle fibers (Blinks et al., 1978). More than 100 seconds of rest is required after only one twitch at 15°C to elicit rested-state responses. At 5°C the required interval of rest is even greater (Blinks et al., 1978). Hence if fibers are already potentiated by frequent contractions the addition of a potentiating agent increases force by very little (Julian and Morgan, 1981). Now can I have slide #75?....no, that's #175.

NOBLE: Can I interject for the moment and ask Dr. Gordon whether he's worried about there being no ascending limb, or whether this is O.K. according to theory.

GORDON: No, I'm really not worried about that. I think one of the things we proposed was that it might be some inhibition to shortening, because of steric hindrance of the filaments, which we hypothesized because the maximum unloaded shortening velocity decreased substantially at sarcomere length of 1.6 μm, where the thick filaments would hit the Z-line. But presumably this is a retarding force to shortening so if the active force could overcome this, you could have some higher forces; I don't see any real problems with that. What we proposed were really speculations and not based on a lot of evidence.

TAYLOR: These figures show the effects of chemical potentiators on single fibers at 15°C (Figs. 1B and 2B of Lopez et al., 1981; Fig 4 of Taylor et al., 1982). I don't know if the effects would be greater if these particular agents were used together rather than separately. I'd also like to comment that I don't see a conflict between these results and the results of Gordon et al. (1966), which were obtained in the cold. Our results indicate that Ca^{2+}-induced activation is normally a factor even in the cold when fibers have shortened below the length at which filament overlap is optimal, and it becomes progressively a more important factor as a muscle is warmed (Lopez et al. 1981). The results of Godt and Lindley (J. Gen. Physiol. 80: 279-297, 1982), for example, support the idea that temperature markedly influences the nature of Ca^{2+}-induced activation.

EDMAN: Well, inadequate activation may very well influence the ascending limb of the length-tension relation in intact fibers, even within the range 2.0-1.65 μm sarcomere length. But I want to point to a paper by Moss (J. Physiol. 292: 177-192, 1979) where he described the same kind of experiments that Dr. Sugi now presented. Moss made a point of the fact that he had very uniform sarcomere patterns during contraction. At any rate, his length-tension curve was very much the same as that obtained in intact fibers during tetanic contraction.

SUGI: In my experience all sarcomeres are visible during half maximal activation at pCa 6.5, while some sarcomeres become invisible during the maximum activation at pCa 5.5. The actual values of pCa differ from investigator to investigator according to the equation used to calculate pCa values, and I suspect that in Moss' case, his activation solution produced substantially submaximal activation, since his preparation exhibits very clear sarcomere spacings from end to end. We have some data that the ascending limb is present in half-maximally activated preparations.

EDMAN: It is a real problem to get uniform sarcomere behavior during contraction of skinned muscle fibers. In Dr. Sugi's experiments I suggest that one tries to establish whether the very short sarcomeres are able to shorten actively during the *entire* rising phase of the auxotonic contraction.

WINEGRAD: If you don't restrain myofibrils you don't get this dispersion of sarcomere lengths, even when the myofibrils shorten down to less than 1.5 μm. We did an experiment trying to estimate what the resistance to shortening was at those lengths by gradually raising the concentration of calcium and measuring the sarcomere length in these freely suspendable myofibrils. From the tension-pCa curve of a mounted myofibril, we inferred what the resistance to shortening should be at any given sarcomere length in the freely suspended fibers based on the particular calcium concentration that was necessary to produce that degree of shortening. In order to get shortening down to the equivalent of about 1.4 or 1.3 μm, no more than 10-15% of the total maximum calcium-activated force is required in the absence of restraint. There may be an error from problems about activation of calcium at very short lengths, but the likelihood is that error would result in an *overestimation* of resistance to shortening rather than an underestimation. I think that what Professor Sugi is showing us can be explained by the fact that there isn't a great deal of resistance to shortening when you have good activation and have bypassed any problems in the function of the tubular systems.

GORDON: A comment about the nonuniformity. I think in the case of the Fabiatos' preparation, a very small cardiac cell, where they can visualize the whole cell, compared to say Moss's study, where he's using a much larger skeletal fiber, the Fabiatos can show that they have very uniform striations. They can activate with a high enough calcium concentration to obtain maximum activation. Moss stayed away from using high calciums, and because of the shift in the calcium sensitivity with sarcomere length, I suspect he wasn't maximally activating his fibers. Thus this results in less force than you probably could get if you upped the calcium.

TAYLOR: I would like to make a comment on Dr. Edman's remark about Moss's paper (J. Physiol. 292: 177-192, 1979). In my talk I focused on the differences between what we did and what Julian and Morgan (J. Physiol. 319: 205-217, 1981) did as they attempted to reproduce our experiments (Lopez et al., 1981). But I didn't mention that Moss limited his studies to 5°C, which is relevant. All the effects that we observe (Lopez et al., 1981) decrease progressively as the temperature is decreased. They tend to become much less significant when one gets down to 0°C. Moss, himself, offers the suggestion in his paper that the seemingly contrary results might be due to the fact that he looked only at low temperature. "A" length-tension relation is not the same as "the" length-tension relation, and I see no reason why we should expect a single factor such as thick filaments colliding with Z-lines to explain the relation under all conditions. Those who described the length-tension relation that has been our basis for comparison maintained from the start that there are several possible reasons for an ascending limb (Huxley, A.F., 1965. Proc. XXIII Rd. Inter. Congr. Physiol. Sci., Excerpta Medical Inter. Congr. Series No. 87, 383-387).

POLLACK: I'd like to return to Al Gordon's earlier comment. Haruo Sugi showed that his tension was flat down to about 1.2 μm, and a number of others have also shown that tension doesn't decrease down to lengths shorter than the length of the thick filament. Now, if we assume that thick filaments *don't* shorten, wouldn't you expect at least some decrease of force below 1.6 to 1.7 μm? Would the implication be that the thick filament doesn't provide any resistance to shortening?

GORDON: Or not a lot of resistance to shortening. If you're dealing with 10% of maximum force, I think that would be consistent with the data.

POLLACK: So, in other words, the resistive force of the thick filament would have to be very small, relative to 2 or 3 kg/cm^2.

GORDON: Could be. Could be consistent with that.

SARCOMERE LENGTH CHANGES IN SINGLE FROG MUSCLE FIBRES DURING TETANI AT LONG SARCOMERE LENGTHS

John D. Altringham and G.H. Pollack

Department of Anesthesiology and Division of Bioengineering RN-10, University of Washington, Seattle, Washington 98195, U.S.A.

ABSTRACT

Laser diffraction and photomicrography have been used to monitor sarcomere length changes in single muscle fibres of the frog, at long sarcomere lengths, during fixed end tetani.

In the central 90% of all fibres, changes in sarcomere length were consistently less than 0.25 μm. Sarcomere length showed an initial rapid change, followed by a progressively slower increase, which persisted throughout a 4s tetanus. Sarcomere length in the terminal 200-400 μm segment at each end of a fibre decreased rapidly by up to 1 μm in the first second of a tetanus. This shortening was accompanied by a marked increase in disorder of the striation pattern.

Maximum isometric tensions in fixed end tetani were much greater than those predicted by crossbridge theory over the entire range of sarcomere lengths studied. An analysis of the intersarcomere dynamics suggests that this extra tension may be explained by known phenomena on the basis of a progressive increase in sarcomere length dispersion along the fibre.

INTRODUCTION

The cross bridge theory of muscle contraction predicts that cross bridges act as independent force generators (Huxley 1957), and that tension is therefore proportional to the overlap between actin and myosin filaments. Tension should thus decrease linearly from a maximum at a sarcomere length of 2.2 μm, to zero at around 3.65 μm. Ramsey and Street (1940) obtained the first full length tension curve for isolated single muscle fibres. *Maximum* tetanic tension was found to be inversely proportional to muscle length, but did not fall to zero until the fibre was stretched to 200% of its resting length, equivalent to a sarcomere length of approximately 4.0 μm.

Tension development during isometric tetanus in stretched single fibres is initially rapid, but then slows progressively before reaching a plateau which may last for many seconds. As sarcomere length increases beyond approx. 2.4 μm, the rate of rise of tension decreases. This slow rise of tension has been attributed to a progressive increase in the range or "dispersion" of sarcomere lengths along the length of the fibre (Gordon, Huxley & Julian, 1966a,b; Julian and Morgan, 1979). As sarcomeres at the ends of the fibre shorten (thereby increasing actin-myosin overlap) and generate more tension, the central sarcomeres lengthen, and in some way bear the extra tension. On this assumption, tension development was divided into two phases, an early rapid phase, representing the true force-generating capacity of the sarcomeres at a given length, and a slow "creep" phase caused by the progressive increase in sarcomere length dispersion. In an attempt to eliminate creep, Gordon et al. (1966 a,b) used a "spot follower" technique to hold sarcomere length constant during tetanus. Creep was not entirely abolished, and an extrapolation technique was used to estimate the pre-creep tension. In plotting their length-tension relation, Gordon et al. used pre-creep tension rather than maximal isometric tension to obtain a length-tension curve which correlates well with that predicted from filament dimensions.

Since that time, a number of workers have studied sarcomere dynamics in an attempt to confirm the existence of increasing sarcomere length dispersion during tetani at long sarcomere lengths. Using a spot follower technique and photomicrography, Julian, Sollins and Moss (1978) and Julian and Morgan (1979) demonstrated that sarcomeres close to the ends of the fibres did in fact undergo considerable shortening, stretching sarcomeres in the central region. It was concluded that the changes were sufficient to explain the extra "creep" tension. Ter Keurs, Iwazumi and Pollack (1978) used a focused laser beam to monitor sarcomere length changes during fixed end tetani, in fibres chosen for their sarcomere length uniformity at rest. Sarcomere length dispersion was consistently <4% of mean sarcomere length during all phases of a tetanus, and was therefore unable to account for the extra tension above that predicted by the length-tension curve of Gordon, et al.. It has been suggested (Huxley, 1980) that laser diffraction is not an appropriate method with which to investigate the problem, since it detects only those striations which are regular over a large number of sarcomeres, and may miss small populations of non-uniform sarcomeres. We have used both laser diffraction and photomicrography to study sarcomere length changes during long tetani, in an attempt to explain this apparent inconsistency between the two methods.

METHODS

Single fibres from the anterior tibialis muscle of *Rana temporaria* were used in all experiments. Care was taken during dissection to display the tapered end regions. Fibres were subjected to fixed end tetani of 4s duration at randomly chosen sarcomere lengths, usually >2.8 μm (a number of control experiments were also performed around 2.2 μm).

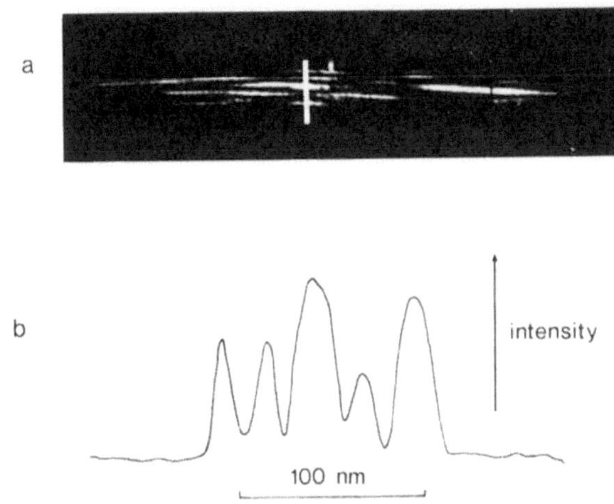

Figure 1: a) Microstructure of 1st order diffraction line, showing fine, parallel lines. b) Densitometric scan of the above diffraction line, through the section indicated by white line.

During the course of a tetanus, the first order laser diffraction pattern, projected onto a frosted screen, was photographed at 5.6 frames s^{-1}, or photomicrographs were taken at 4 frames s^{-1}. Diffraction patterns were analysed on a microdensitometer (Fig. 1), a change in sarcomere length of 1 nm could be detected. Photomicrographs were projected onto a film reader, and mean sarcomere length obtained by averaging over 3 parallel lines of 20 sarcomeres at 3 points in each frame. Force was measured with a silicon beam strain gauge (AME, Horten, Norway). All experiments were performed at 5-10°C.

RESULTS

Length Tension Data

Maximum isometric tensions were typically around 25 to 30 Ncm^{-2}. Representative tension traces are shown in Fig. 2. As sarcomere length was increased, the rate of rise of tension decreased, and maximum tension was reached at a progressively later time after the onset of contraction. In all cases, maximum tension was reached within 2s of the start of stimulation. At sarcomere lengths less than around 3.4 μm, there was often a small overshoot after which tension declined to a steady level, or continued to decline at a very much slower rate. This overshoot was most prevalent at the shortest sarcomere lengths. At sarcomere lengths less than around 3.4 μm, the rate of rise of tension usually decreased smoothly. It was therefore not possible to plot "pre-creep" tension by the extrapolation method of Gordon et al., where segment length clamping resulted in a more obvious 2 phase tension curve. For this reason maximum isometric tension was plotted against sarcomere length, as

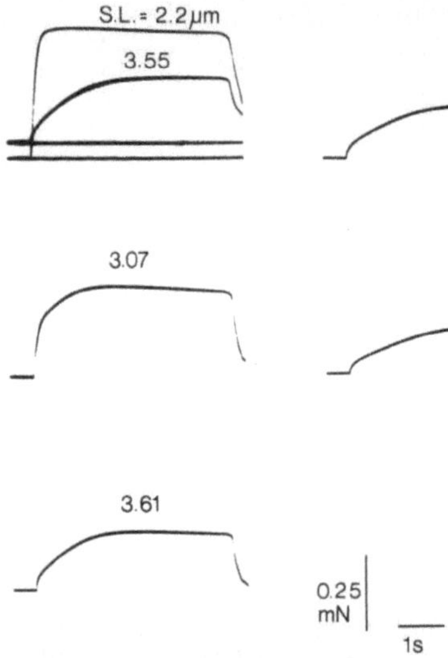

Figure 2: Representative tension records from fixed end tetani. Figures refer to sarcomere length at peak of contraction, measured by laser diffraction.

measured in the central portion of the fibre by diffraction. It is this component of the maximum tension, above that prediced by the cross-bridge theory, which must be explained by any hypothesis based on creep. Fig. 3 shows length tension data obtained from 6 fibres. Active tension declines slowly with increasing sarcomere length to around 0.5 P_0 at a sarcomere length of 3.65 μm where thick/thin filament overlap is presumed to be zero. Resting tensions were small or absent below a sarcomere length of 3.0 μm, rising to around 0.15 P_0 at 3.65 μm.

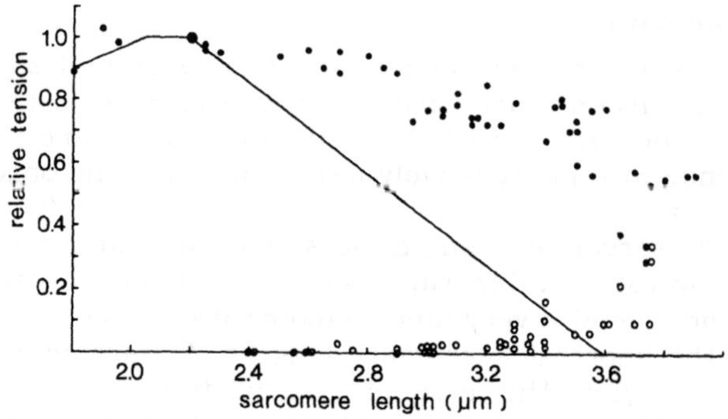

Figure 3: Maximum isometric tension plotted against sarcomere length (measured in the central segment of fibres). Data from 6 fibres. Solid line from Gordon et al., (1966b). Normalised to tension at 2.2 μm. Closed circles = active tension. Open circles = resting tension.

Resting Sarcomere Length Distribution

Resting sarcomere length distribution was measured by both direct photography of the striation pattern, and with the phase locked loop device. Results from the two techniques were comparable. A representative scan is shown in Fig. 4 from a fibre after multiple contractions at sarcomere lengths between 2.2 and 3.7 μm. Sarcomere length always decreased towards the ends of the fibres at sarcomere lengths >2.6 μm, but total dispersion at all sarcomere lengths was usually 0.1-0.2 μm, and only on one or two occasions, >0.25 μm. No significant increase in sarcomere length dispersion was detected during an experiment in fibres which showed steady, reproducible tension records.

Sarcomere Length Changes During Tetanus

Changes in the central 90% of the fibre length. Due to the large diameter of the laser beam (1 mm), diffraction measurements were necessarily restricted to this region. The following results were obtained from 18 fibres, 8 using laser diffraction, and 10 using photomicrography.

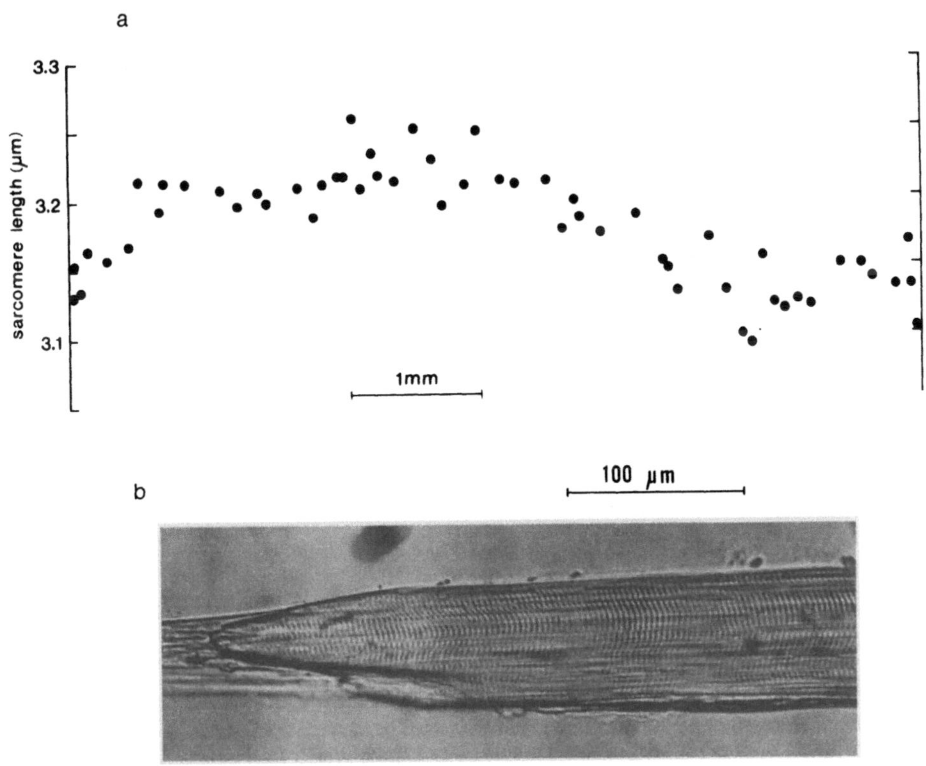

Figure 4: a)Sarcomere length measurements, along the entire length of a typical fibre (2-4-82) after multiple contractions at various sarcomere lengths. b) Micrograph of the same fibre, taken during the above measurements.

Comparable results were obtained from the two techniques. The maximum sarcomere length change over an entire contraction was <0.2 μm from diffraction (median) and <0.25 from photomicrography (mean). The initial change in sarcomere length was a rapid increase, followed by a progressively slower increase. The rapid lengthening was occasionally sarcomere^{-1}s^{-1}. In a small number of cases local slow shortening was observed in micrographs, not observed by diffraction, due to the summing of large sarcomere populations by this technique.

An example of the fine parallel microstructural lines present in the first order diffraction line is shown in Fig. 1, together with a densitometric scan of the lines, to illustrate the method of analysis. Typical diffraction results obtained during tetanic stimulation are illustrated in Fig. 5. The maximum width or dispersion of the diffraction line recorded during rest or tetanus was <0.25 μm, and commonly 0.1-0.2 μm. The early rapid change in sarcomere length was usually associated with major changes in the microstructure.

Photographs from a representative contraction are shown in Fig. 6. Throughout the course of a tetanus, the striation pattern remained very clear, with sarcomeres retaining excellent register, explaining the close agreement between results from diffraction and photomicrography.

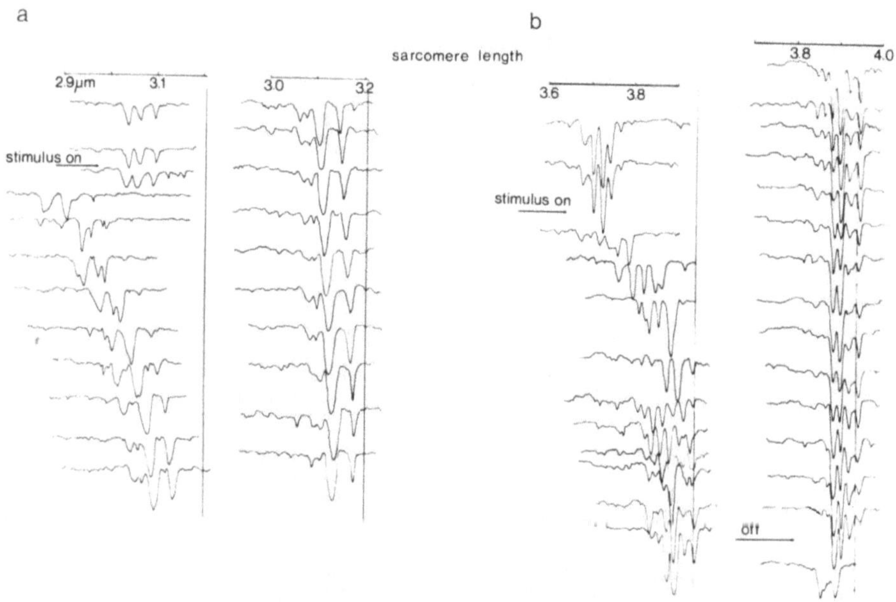

Figure 5: a) Densitometric scans of contraction at a resting sarcomere length of 3.05 μm, taken from preceeded by a transient rapid decrease in sarcomere length. Local changes in median/mean sarcomere length were moderately repeatable from contraction to contraction, as noted by Cleworth and Edman (1972). During the plateau phase of a tetanus, diffraction always showed an increase in sarcomere length, ranging from 5-35 nm sarcomere^{-1}s^{-1}. Photomicrography gave lengthening velocities consistently <0.1 μm the middle of a fibre (20-11-81). b) As above, fibre of (1-12-81), resting sarcomere length in middle of fibre = 3.70 μm.

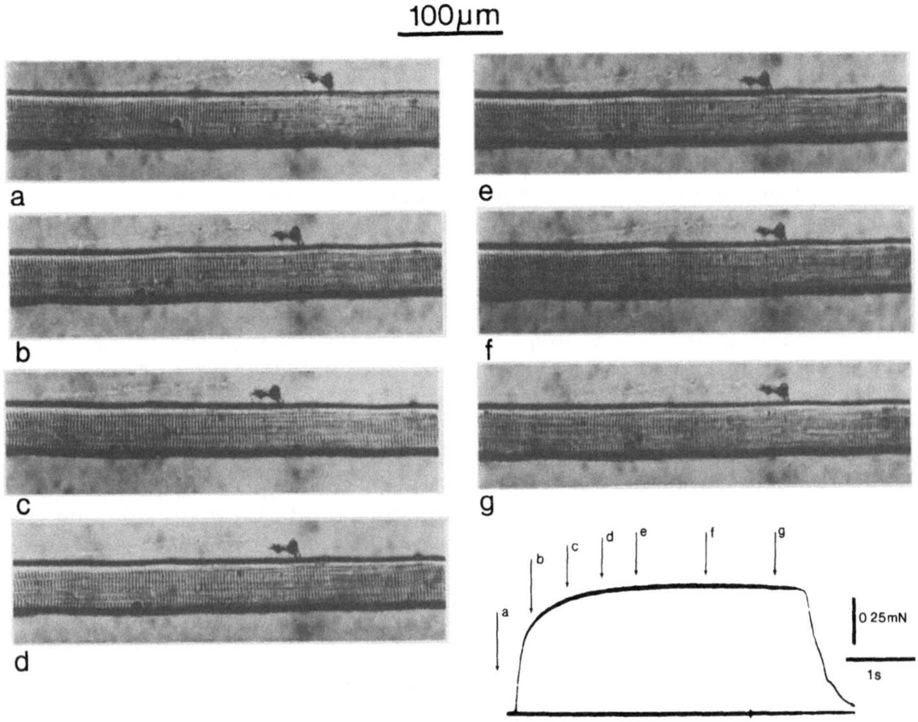

Figure 6: Series of micrographs from the middle segment of a fibre (30-3-82) during tetanus. Mean sarcomere lengths in frames a,b,d and f respectively were 3.30, 3.27, 3.38 and 3.40 μm. Letters indicate time during tetanus at which frame was taken (see inset).

Sarcomere length changes at the ends of fibres. The changes in sarcomere length over the terminal 200-400 μm were more complex. To study this region as a whole, low power micrographs were taken of the entire region, or overlapping high power micrographs taken in successive contractions. At sarcomere lengths around 2.2 μm, changes were of a similar magnitude to those observed in the central region of the fibre.

All other experiments were performed at sarcomere lengths >2.8 μm. At these lengths, there was very little translation of the fibres due to slightly eccentric attachment to the apparatus, or to tendon compliance. Since fibres were dissected with very fine tendons, the latter could prove troublesome at shorter lengths. It was therefore easier to obtain photographs of a specific area of an end region throughout each contraction. In 9 of the 10 fibres studied, contraction was accompanied by extensive shortening over the last 100-300 μm at both ends of the fibre. One fibre showed shortening at only one end. The magnitude of this shortening, and its extent along the length of the fibre were dependent upon the initial sarcomere length. The relation can be best described with an example. Fig. 7 shows the mean sarcomere length changes measured over 3 adjacent regions of a fibre at 3 initial sarcomere lengths (see fig. legend for details).

In general terms, as sarcomere length increased from 3.0 μm to 3.7 μm, the magnitude of the initial fall in sarcomere length tended to decrease, and the length of the fibre over which it occurred also decreased. In all 3 cases, the terminal zone shortened, but to a lesser extent at 3.7 μm than at 3.0 μm. The adjacent zone undergoes a transition from shortening to lengthening as sarcomere length is increased, as does the most central zone, but at a shorter sarcomere length. This pattern was observed in all fibres studied at widely different sarcomere lengths using photomicrography. In the 8 fibres studied using laser diffraction, sarcomere lengthening in the central region was, despite some variation along the length of a fibre, generally more rapid at shorter sarcomere lengths than at long lengths, consistent with the more extensive shortening at the ends observed by .photomicrography. An example of the behaviour of the terminal sarcomeres is illustrated in Fig. 8. A consistent feature of all contractions was an increase in irregularity of the striation pattern with shortening. This eventually prevented sarcomere length measurement over large areas in the end regions. The final frame in Fig. 8 illustrates the rapid reversal of this condition upon relaxing the fibre.

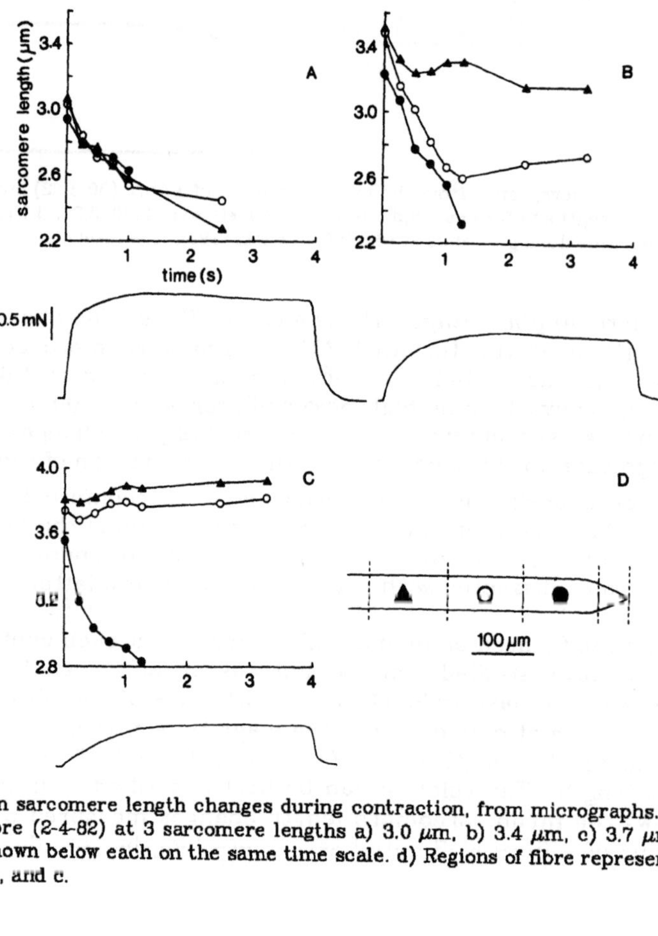

Figure 7: Mean sarcomere length changes during contraction, from micrographs. In the end region of a fibre (2-4-82) at 3 sarcomere lengths a) 3.0 μm, b) 3.4 μm, c) 3.7 μm. Tension records are shown below each on the same time scale. d) Regions of fibre represented by the symbols in a,b, and c.

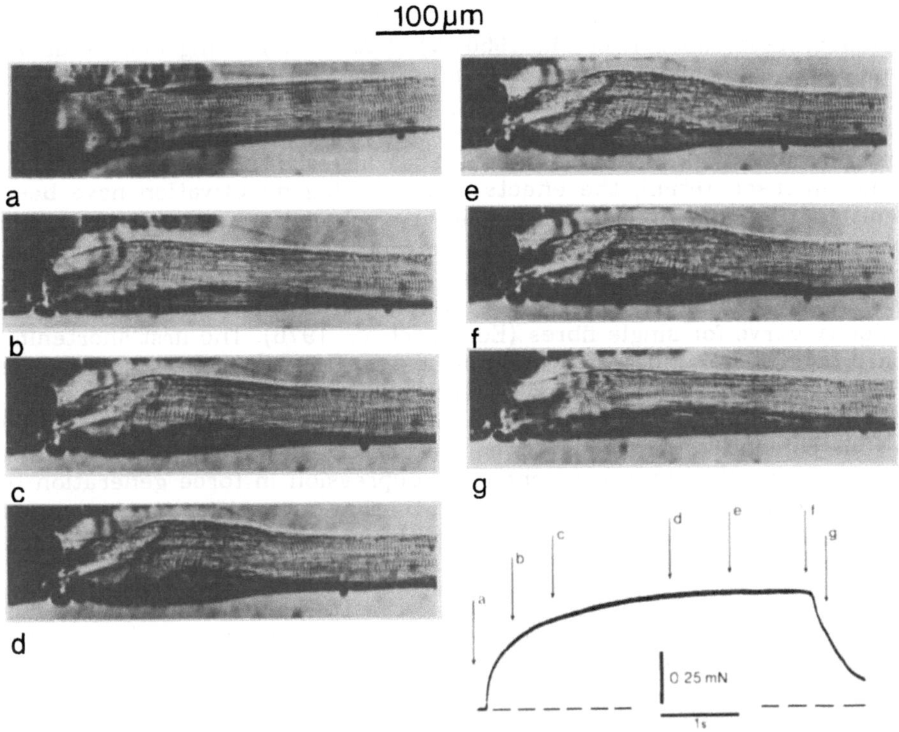

Figure 8: Series of micrographs from the end region of the fibre shown in Fig. 6 (30-3-82) during tetanus. Letters indicate time during tetanus at which frame was taken (inset). Note the localised nature of the sarcomere shortening, concomitant increase in fibre diameter, and the rapid loss of sarcomere register.

DISCUSSION

The initial question we asked was: do the terminal sarcomeres of stretched fibres shorten during tetanus? The results clearly indicate that they do, and are in contradiction to the results from the diffraction studies of ter Keurs et al. (1978), using semitendinosus fibres of *Rana pipiens*. The differences may be due to species/muscle differences, or may indeed be due to their use of diffraction to monitor sarcomere length.

Can the observed changes in sarcomere length explain the extra tension above that predicted by the length-tension relation of Gordon et al.? If this tension results from some form of instability, as proposed by Gordon et al.(1966a,b), the problem can be thought of as two separate processes: the terminal sarcomeres must be able to generate the extra force, and the sarcomeres in the middle of the fibre must be able to support this tension. We will consider the end region first. Along with the length-tension curve of Gordon et al., which is assumed, any calculations on force generation must take into account possible effects of the force-velocity relation, and of shortening deactivation. It has been shown (Edman, 1966; Gordon et al., 1966a,b) that if shortening occurs under moderate loads, and immediately after the onset of stimulation, then force generation is independent of past history, and is not significantly depressed. Since most of the internal shortening occurs early in the contraction (see Fig. 7), and Edman (1980) has shown deactivation to be small in fused tetani, the effects of shortening deactivation have been omitted from the analysis. The following calculations are based on the observed sarcomere length, an estimate of the shórtening velocity from plots similar to those shown in Fig. 7, the predicted tension from the length-tension curve and its depression based on a published force-velocity curve for single fibres (Edman et al., 1976). The first shortening pattern in Fig. 7 illustrates a simple case. There is little dispersion of resting sarcomere length, and all sarcomeres shorten at similar velocities, to a similar extent. As a first approximation, they can be considered as one population. Assuming a mean sarcomere shortening velocity of 0.1 sarcomeres s^{-1}, with a corresponding depression in force generation to 0.84 of that predicted from the length-tension curve, then the estimated tension can be plotted and compared with the measured tension. This has been done in Fig. 9a.

Despite this simplification, the fit is reasonably good. A second analysis is shown in Fig. 9b,c, the calculations being based on the longest sarcomere population in Fig. 9b, on the basis that they represent the weakest link. Again, the estimated tensions are very close to the measured values. This method of analysis is an obvious oversimplification,

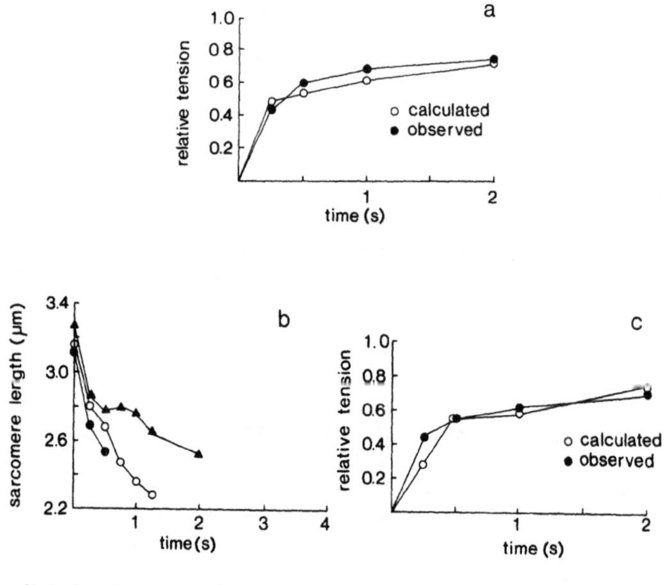

Figure 9: a) Predicted and measured tensions during contraction for records shown in Fig. 7a (see text for details). b) Sarcomere length changes during contraction, from micrographs, in the end region of a fibre (30-3-82). c) Predicted and measured tensions for contraction illustrated in b.

since it takes into account only one end of the fibre. Experiment shows however that both ends usually behave similarly, if the dissection results in a clean tendonous insertion at both ends (excessive connective tissue may introduce more gross sarcomere length dispersions). Also, sarcomere length, and changes in sarcomere length, change continuously, but not necessarily uniformly, along the end of the fibre. Any analysis which divides this into essentially arbitrary regions is clearly only an approximation. Further, more complex patterns of lengthening and shortening, as illustrated in Fig. 7b, cannot be subject to such simple treatment. Consider the upper trace of Fig. 7b; mean sarcomere length decreases slowly before becoming relatively stable. This 120 μm segment of the fibre is in a transition zone, between shortening and lengthening sarcomeres. By subdividing this segment it can be shown that sarcomeres close to the ends of the fibre decrease in length faster than the mean, those closer to the middle stretch slowly. The changes illustrated are only a mean, an analysis of these more complex patterns of behavior would require accurate sarcomere length measurements over much shorter segments than those used in the present study. As a test of the creep hypothesis, we analysed the more simple patterns of behavior. In all cases, the fit was good, suggesting that the explanation of tension creep provided is a good working hypothesis.

The extra force generated by the terminal sarcomeres must be borne in some way by the sarcomeres in the central 90% of the fibre. From Fig. 3 it can been seen that this extra tension may be up to 60% of the maximum isometric tension at very long sarcomere lengths. A common feature of all contractions was an initially rapid increase in sarcomere length (Fig 5) (in a few cases preceeded by a rapid shortening), followed by a progressively slower increase. This early sarcomere stretch may explain their ability to sustain greatly elevated tensions. The effects of stretches of different amplitude and velocity on force enhancement have been studied by Edman et al.(1981). A critical amplitude of stretch of only 16.6 nm per half sarcomere was required to increase tension markedly to a "break" point, after which tension remained constant as long as the stretch was maintained. The degree of enhancement was variable from fibre to fibre, in the range of 40 to >200% pre-stretch tension. The critical amplitude was essentially independent of velocity of stretch and sarcomere length. After stretch, the force decayed to pre-stretch levels over approximately 2s at short sarcomere lengths. At lengths >2.7 μm, a considerable residual enhancement, which did not decay, remained after stretch (approximately 50% pre-stretch tension, see Edman et al,. 1978, Fig. 6; 1981, Fig. 3). Thus, a sarcomere length change of only 33 nm is required to recruit a large force component, which remains elevated after stretch, in the range of sarcomere lengths studied in the present work. The sarcomere length changes observed in the present study are of sufficient magnitude to produce these extra tensions. Stretch enhancement may therefore explain the ability of fibres to maintain elevated tensions during tetani at long sarcomere lengths.

Considerable tensions were obtained even when the sarcomeres in the central portion of the fibre were stretched beyond overlap. As shown in Fig. 7c, terminal sarcomeres were still shorter than 3.65 μm, and capable of force generation. The large resting tensions of the central sarcomeres at these lengths were sufficient to support the active tension of these short sarcomeres. There is increasing evidence (e.g. Locker and Leet, 1976) for the existence of a third filament, connecting the myosin filaments to the Z line - the "gap" filament first proposed by Huxley and Peachey in 1961. This third filament, and possibly others presently being studied (e.g. Wang & Williamson, 1980; Dos Remedios & Gilmour, 1978; Magid, 1981, see also this symposium) may be responsible for the high resting tensions at long sarcomere lengths. The existence of some continuity between the A band and the Z line is strongly suggested by the ability of the central sarcomeres to be reversibly stretched beyond overlap, a number of times during an experiment, without impairment of force generation. A third filament would act as a stabiliser preventing myosin disorientation beyond overlap, and would guide the myosin filaments back between the actin filaments with shortening. The mechanical constraints placed on very long sarcomeres by a stretched third filament may explain the stabilising effect noted under these conditions. As sarcomere length was increased beyond 3.0 μm, the terminal zone of sarcomere shortening became increasingly more restricted.

Are the small differences in resting sarcomere length along the fibre sufficient to trigger sarcomere creep? With a steep descending limb to the length-tension relation, any small difference will result in instability, and it could occur at any point along the fibre. However the large decreases in sarcomere length associated with tension creep always occurred at the ends. Despite the local changes in sarcomere length illustrated in Fig. 4, mean sarcomere length showed a consistent decrease towards the ends in all fibres. It may be that these larger zones of short sarcomeres initiate creep before instability becomes significant in the central region, and stretching of these central sarcomeres will greatly increase their stability (Edman, Elzinga and Noble, 1983). It should be noted, that force enhancement by stretch, shortening deactivation, and the presence of passive, non actin/myosin filaments will all tend to stabilise the intrinsic instability of the descending limb. The presence at rest of short terminal sarcomeres may be due to thicker connective tissue layers around the insertions. These layers may resist stretching to a slightly greater extent, and support a little more of the resting tension. In support of this idea, a zone of shorter sarcomeres was found around the end plate of a number of rejected fibres, where connective tissue and other debris could not be removed. Activation led to shortening in this zone.

In conclusion, the results presented in this paper lend validity to the original hypothesis of Gordon et al. (1966a,b), that tension creep at long sarcomere lengths is caused by increasing sarcomere length inhomogeneity.

ACKNOWLEDGEMENTS

Grateful thanks to Dr. Tsukasa Tameyasu for many stimulating and fruitful discussions. John Altringham was supported by a N.A.T.O./S.R.C. Postdoctoral Fellowship during the course of this work.

REFERENCES

Cleworth, D.R. & Edman, K.A.P. (1972) Changes in sarcomere length during isometric tension development in frog skeletal muscle. J. Physiol. 227: 1-17.

dos Remedios, C.G. & Gilmour, D.J. (1978) Is there a third type of filament in striated muscles? J. Biochem. 84: 235-238.

Edman, K.A.P. (1966) The relation between sarcomere length and active tension in isolated semitendinosus fibres of the frog. J. Physiol. 183: 407-417.

Edman, K.A.P. (1980) Depression of mechanical performance by active shortening during twitch and tetanus of vertebrate muscle fibres. Acta Physiol. Scand. 109: 15-26.

Edman, K.A.P., Elzinga, G. and Noble, M.I.M. (1978) Enhancement of mechanical performance by stretch during tetanic contractions of vertebrate skeletal muscle fibres. J. Physiol. 281: 139-155.

Edman, K.A.P., Elzinga, G. and Noble, M.I.M. (1981) Critical sarcomere extension required to recruit a decaying component of extra force during stretch in tetanic contractions of frog skeletal muscle fibres. J. Gen. Physiol. 78: 365-382.

Edman, K.A.P., Elzinga, G. and Noble, M.I.M. (1983) Stretch of contracting muscle fibres. Evidence for regularly spaced active sites along the filaments and enhanced mechanical performance. This volume.

Edman, K.A.P., Mulieri, L.A. and Scubon-Mulieri, B.C. (1976) Non-hyperbolic force-velocity relationship in single muscle fibres. Acta Physiol. Scand. 98: 143-156.

Gordon, A.M., Huxley, A.F. and Julian, F.J. (1966a) Tension development in highly stretched vertebrate muscle fibres. J. Physiol. 184: 143-169.

Gordon, A.M., Huxley, A.F. and Julian, F.J. (1966b) The variation in isometric tension with sarcomere length in vertebrate muscle fibres. J. Physiol. 184: 170-192.

Huxley, A.F. (1957) Muscle structure and theories of contraction. Prog. Biophys. Biophys. Chem. 7: 255-318.

Huxley, A.F. (1980) Reflections on muscle. Publ. Liverpool University Press.

Huxley, A.F. & Peachey, L.D. (1961) The maximum length for contraction in vertebrate striated muscle. J. Physiol. 156: 150-165.

Julian, F.J. & Morgan, D.C. (1979) Intersarcomere dynamics during fixed end tetanic contractions of frog muscle fibres. J. Physiol. 293: 365-378.

Julian, F.J., Sollins, M.R. & Moss, R.L. (1978) Sarcomere length non-uniformity in relation to tetanic responses of stretched skeletal muscle fibres. Proc. Roy. Soc. Lond. B. 200: 109-116.

Locker, R.H. & Leet, N.G. (1976) Histology of highly stretched beef muscle. IV. Evidence for movement of Gap filaments through the Z-line, using the N_z-line and M-line as markers. J. Ultrastruct. Res. 56: 31-38.

Magid, A. (1981) Interfilamentary forces in relaxed detergent-skinned frog skeletal muscle. Biophys. J. 33: 226a.

Myers, J., Tirosh, R., Jacobson, R.C. & Pollack, G.H. (1982) Phase locked loop measurement of sarcomere length with high time resolution. IEEE Trans. B.M.E. 29: 463-466

Ramsey, R.W. & Street, S.F. (1940) The isometric length-tension diagram of isolated skeletal muscle fibres of the frog. J. Cell. Comp. Physiol. 15: 11-34.

Tameyasu, T., Ishide, N. and Pollack, G.H. (1982) Discrete sarcomere length distribution in skeletal muscle. Biophys. J. 37: 489-492.

ter Keurs, H.E.D.J., Iwazumi, T. & Pollack, G.H. (1978) The sarcomere length-tension relation in skeletal muscle. J. Gen. Physiol. 72: 565-592.

Wang, K. & Williamson, C.L. (1980) Identification of an N_2-line protein of striated muscle. Proc. Natl. Acad. Sci. U.S.A. 77: 3254-3258.

DISCUSSION

POLLACK: Since John and I differ in our interpretations, we agreed in advance that the best way to proceed was to let John present and then for me to say a few words about why my view of the interpretation is different from his. We differ in that John believes the results fit the "classical" length-tension diagram, while I do not.

The central issue is whether sarcomeres can or cannot "climb up" the descending limb. Figure D-1 shows the "traditional" length-tension curve along with a schematic drawing of the time course of the rise in tension observed by Gordon, Huxley, and Julian (1966). Their idea was that the slow-rise phase of tension development is actually caused by sarcomeres "creeping up" the descending limb by virtue of instability. As they creep up, it was theorized, their tension should increase, thereby conferring still more strength, and further tendency to shorten and develop additional tension. So the central question is, *can* the sarcomeres actually creep up the descending limb?

Any standard length-tension diagram, whatever its shape, is derived from a series of static measurements, each one taken at a different sarcomere length; you get a series of data points, and then connect them with a line. But, since these are *static* measurements, they say nothing of the dynamics of going, say, from one point to another point *during* contraction. I'd like to draw an analogy to illustrate how failure to make the distinction can lead to erroneous conclusions.

Figure D-2 shows a series of mountain peaks at the right, similar to what I see from the window of my office. The height of each peak is plotted at the left; there is a temptation to interconnect the points, forming a smooth curve. One then might be further tempted to say that you could move from one point to another along the curve, but in reality such a feat

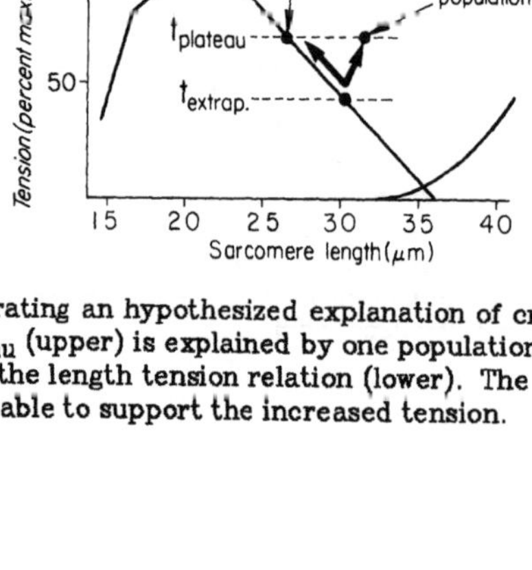

Figure D-1: Diagram illustrating an hypothesized explanation of creep. The slow rise of tension from t_{extrap} to $t_{plateau}$ (upper) is explained by one population of sarcomeres shortening up the descending limb of the length tension relation (lower). The remaining sarcomeres are stretched and are thereby able to support the increased tension.

Figure D-2: Static vs. dynamic aspects of the length tension relation. Points in the length tension relation (left) are static; they are obtained in separate contractions, each at a different sarcomere length. Connecting the dots to form a smooth curve implies tht dynamic behavior might also follow the curve, but the analogy at the right shows such an implication may be misleading: A curve drawn to interconnect the mountain peaks does not imply that a climber ought to be able to traverse the range along such a route.

is evidently impossible. I believe this situation is parallel in the length-tension measurements. One is (I believe incorrectly) tempted to ascribe dynamic properties to a curve devised from a series of static measurements.

Despite this theoretical objection, it still *might* be possible, fortuitously, that the "static" length-tension curve does have application to dynamic events; perhaps under some conditions, sarcomeres could move up or down this curve. Let's consider going down first, since it's easier. Theory predicts that there ought to be at least *some* condition in which stretch produces a decrease of tension; but I have never seen it. The literature is filled with examples of stretch -- either during or after -- giving an *increase* of tension. Of course we can invoke rationalizations as to why the experiment fails to fit the expectation; but, in essence stretched sarcomeres *do not* follow the prediction of the length-tension curve.

How about shortening sarcomeres? When you shorten the fiber, can you actually climb *up* the descending limb? The clear experiment, I think is to tetanize the fiber, let it develop steady tension, and then let it shorten very slowly. If the hypothesis is correct, the sarcomeres should creep up the descending limb and thereby increase the fiber tension. The experiment is shown in Figure D-3. This is a tetanic tension record taken at 3.2 μm sarcomere length. The preparation was in all ways comparable to the ones John Altringham showed. With the slowest release, it can be seen that there's no discernible effect on tension. The next release is just a bit faster, and you can see that the tension begins to *decrease*, and if the release is still faster the tension decreases further and so on. This experiment has been repeated in John's hands at a number of different sarcomere lengths, and at a number of different times after the onset of stimulation. In all records of this type, the tension *decreases* during shortening; it never increases.

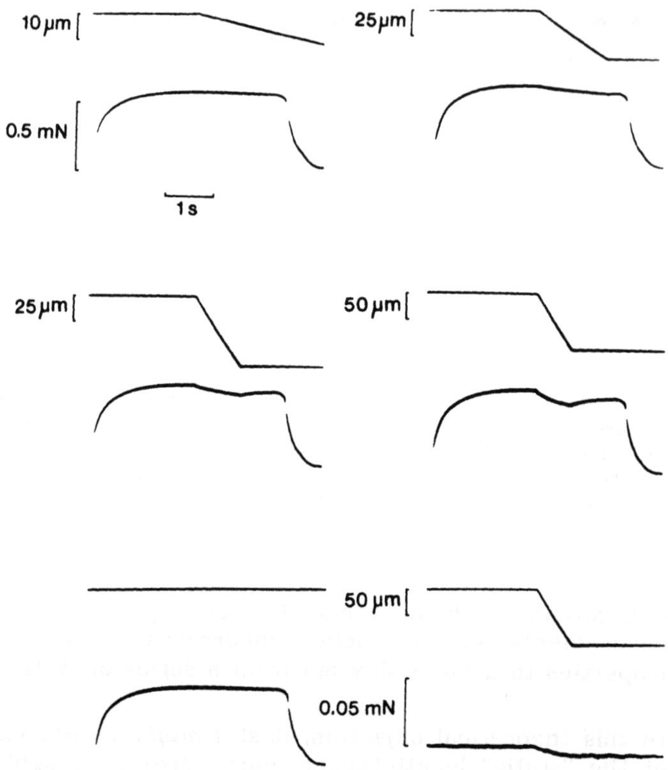

Figure D-3: The effect of shortening at low velocities in tibialis anterior fibers. The hypothesized explanation of creep implies that slow shortening up the descending limb should cause an *increase* of tension from isometric, while the records show a *decrease*. In all six panels length is shown on upper and tension on lower records. Lower left panel shows control isometric record, while lower right shows a release of the unstimulated fiber. The results of the upper four panels show that there is no velocity at which the tension during shortening creeps to a higher than isometric value.

So I take this -- and John disagrees -- I take this as evidence that the sarcomeres cannot creep up the descending limb. When they shorten, no matter how slowly, they actually *decrease*, not increase, their tension. And, when they are stretched they actually *increase*, not decrease their tension. Both are *opposite* the expectation derived from the static length-tension curve. Stability is assured.

Now, a completely independent argument as to why I think the instability-creep hypothesis is incorrect is that there are violations. John showed that the borderline between shortening and stretch near the end of the fiber is rather sharp. Just on either side of that border the difference in initial sarcomere length may be smaller than 0.1 μm. So this hypothesis, then, suggests that instability can develop with only a *very* small difference of degree of overlap, translating to differences of contractile strength on the order of perhaps only 5% or so.

If that were *really* the case, then very small differences of contractile strength along the full length of the fiber, arising, for example, out of differences of cross-section or of activation level, should trigger the same kind of "instability" that occurs near the ends. Once triggered, since it is an instability, you should get a progressively developing increase of inhomogeneity, winding up with a series of long populations and short populations. Yet (apart from the myotendinous regions) this does not happen. The majority of the fiber remains homogeneous.

This point is well illustrated in a study by Julian and Moss (J. Physiol. 304: 529-539, 1980) in which they used skinned fiber segments. Here the range of initial sarcomere length inhomogeneity along the specimen was 0.24 μm, at the upper limit of the difference between the central region and the ends in our studies. Despite this initial degree of inhomogeneity of sarcomere length, as you can see (Figure D-4), during activation, this striation pattern is well-maintained, and the degree of inhomogeneity increases only slightly. There is no apparent instability despite the relatively large initial SL inhomogeneity. This, it seems to me, violates the instability hypothesis.

So my point is that there's certainly shortening at the ends in most cases in the present experiments -- though apparently not in earlier experiments with semitendinosus fibers (ter Keurs et al., J. Gen. Physiol. 72: 565-592, 1978). But why the ends shorten is *not* due to instability. I

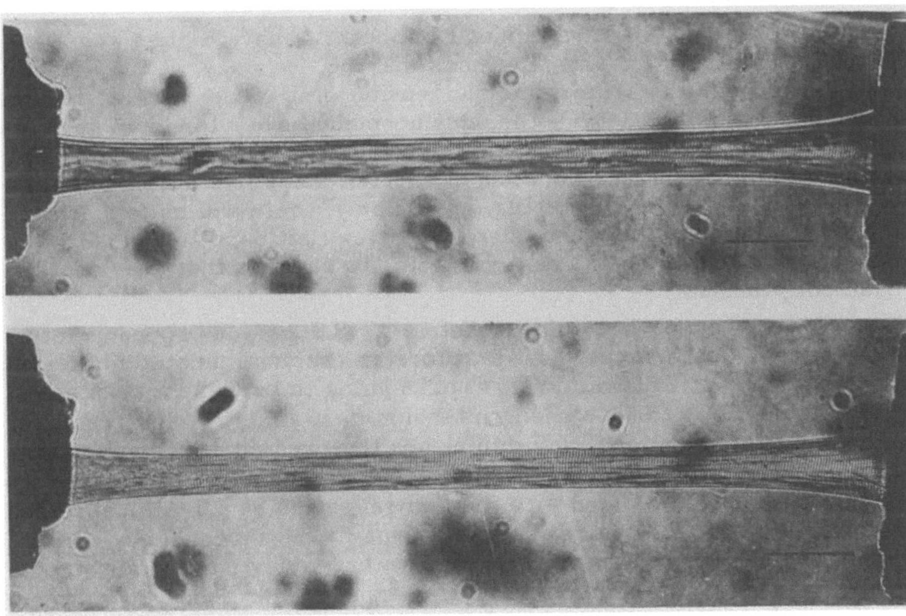

Figure D-4: Skinned fiber segments obtained from anterior tibialis muscles of the frog, taken from Julian and Moss (1980). Calibration bar, 100 μm. Despite a measured sarcomere heterogeneity amounting to 0.24 μm along the fiber in the relaxed state (upper), the activated fiber (lower) fails to develop the gross heterogeneities predicted by the instability-creep hypothesis.

think the reason may be that the ends are naturally stronger than the central region. Why do I say so? Well, for a couple of reasons. First of all, there are some differences in the terminal 200 or 300 μm in membrane properties. For example, Almers (in *Disorders of the Motor Unit.* Ed. D.L. Schotland, pp 349-366, 1982) has shown recently that the number of sodium channels differs in the terminal several hundred μm. Katz and Miledi (J. Physiol. 170: 379-388, 1964) showed some time ago that in the terminal 200 or 300 μm -- the same as the region that shortens -- the sensitivity to acetylcholine is approximately ten times higher than in the remainder of the fiber. And, further, there are some differences of resting stiffness at the ends that cause the sarcomere length inhomogeneity to begin with. It's pure speculation that any of these might necessarily give rise to any difference in contractile strength at both ends. But I just wanted to point out that there are some documented differences in properties between the very ends and the rest of the fiber that *could* predispose the terminal 50-100 sarcomeres to behave in a different way.

The following argument, though convinces me that the ends are indeed likely to be stronger, possibly much stronger. The reasoning requires a little detail, but is straightforward. Suppose we consider the true isometric tension of the central region versus the true isometric tension at the end region, both at the same sarcomere length, let's say 2.8 μm. What I wish to demonstrate is that the isometric tension developed by the end sarcomeres is substantially higher than by the central ones, at the *same* sarcomere length.

Figure D-5 shows the reasoning. Suppose we have a fiber where the central region is at 2.8 μm. The ends may be, let's say, at 2.7 μm. We want to consider the true isometric tension first of the central region. During fixed-end contractions the ends normally have a tendency to shorten, while the central region is lengthened. So, in order to maintain the central region isometric you need to shorten the whole fiber somewhat. So long as the fiber is shortening, even slowly, there will be *less* tension than when you don't allow the fiber to shorten (left panel). So at 2.8 μm, the isometric tension at the center is going to be lower than the "control" (fixed-end) tension.

Now let's keep the *ends* isometric at 2.8 μm during contraction (right panel). To do so, we must first stretch the unstimulated fiber a bit to get the ends at 2.8 μm and then we're going to hold them constant at 2.8 μm. According to John's length-tension diagram, stretching the fiber a bit to get the ends to 2.8 μm diminishes the fixed-end tension by just a few percent. So the "control" tension record is going to be very similar to the control record obtained with the central region at 2.8 μm, depressed by just a few percent. Since the ends normally have a tendency to shorten, in order to keep the *ends* at 2.8 μm what you need to do is stretch the whole fiber, and when you do that, obviously you're going to get *more* tension than in the control record -- probably much more since the end shortening is ordinarily vigorous. In other words, the "end isometric" tension at 2.8 μm is going to be substantially higher than the control.

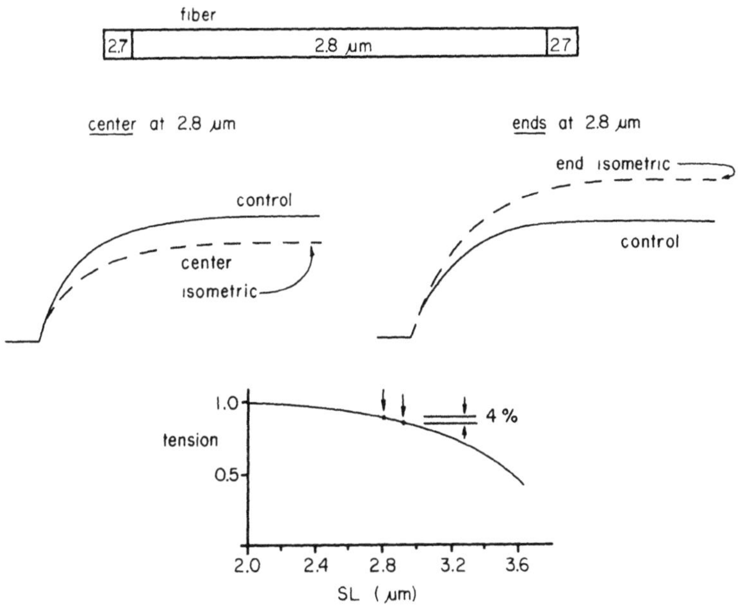

Figure D-5: Hypothetical experiment in which the isometric tension developed by the ends relative to that of the center is measured. The 'result' implies that the ends may be naturally stronger than the central region, accounting for the shortening often, but not always, found near the tendinous insertions. See text for explanation of panels.

So what we have here -- it's hypothetical, and needs to be tested -- is that the isometric tension of the ends is likely to be substantially higher than that of the middle, perhaps by a large margin; this could stem from the differences in membrane properties mentioned above. It offers a natural explanation for why the ends generally shorten and stretch the middle without having to invoke instability.

To recapitulate, the isometric length-tension diagram is first of all, derived from a series of *static* measurements. It implies *nothing* of the dynamics that occur during shortening or stretch from one length to another. If the static length-tension diagram were found to apply *during* contraction, it could only be fortuitous. However, experiments of releasing the fiber very slowly from a tetanic plateau show that it does not apply. Tension does not creep up during slow shortening; tension goes *down*. The sarcomeric behavior is stable. This explains also why moderate initial sarcomere length inhomogenetities or activation variations along the central region of the fiber generally fail to trigger gross inhomogeneities. I think the reason ends generally (but not always) shorten is simply that they may be considerably stronger. This explains also why moderate sarcomere length inhomogeneities or activation variations within the central region generally fail to trigger gross inhomogeneities.

ALTRINGHAM: Gerry and I have argued these points for some time, I think it's only fair to let somebody else join in. However, I would like to make a couple of points. The first is that if you assume there is only the static length-tension curve operating at sarcomere lengths on the descending limb, then yes, the instability should cascade and go to an endpoint. But there may be stabilizing factors like stretch enhancement, and with other factors like this it's not necessary for the instability to go to its endpoint.

One of the points -- if I could have my slides back again -- is raised by other release experiments and I think personally that these may be more appropriate. Since I measured, to a first approximation, an exponential decrease in sarcomere length during the creep phase early in contraction, I thought it was maybe more relevant in a release experiment to mimic this situation. Just take this diagram first (Fig. D-6, panel B). We see that at 2.2 μm there's no creep; isometric tension rises rapidly to a plateau. Now assuming there is no creep, then at 3 μm if there are no sarcomere length dependent activation effects, then tension rise should follow the same line and simply plateau at a new point. Now, if creep is present, then it can be seen experimentally that initially, tension is depressed (presumably due to the force-velocity relation), but as the velocity decreases, then tension rises to a new level above that predicted by the isometric length-tension curve. We can see that in the course of a release superimposed on the creep, all you do is essentially depress the initial phase even further, but again tension rises to a new higher level. This is illustrated in an actual experiment (A panel); starting at a sarcomere length of 3.6 μm there is an extended creep phase rising to a plateau. If sarcomere length is decreased to 3.3 μm exponentially and then held at a relatively slow velocity, then tension is initially depressed but it rises to a new, higher level. So I see this as being perfectly compatible with the idea that sarcomeres have the ability to creep up the descending limb. I think the difference in Gerry's interpretation is that sarcomeres should follow the static descending limb; I think it's a more complicated pathway.

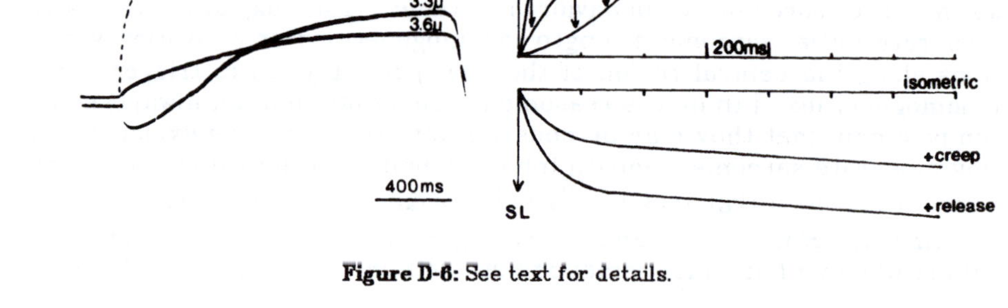

Figure D-6: See text for details.

TER KEURS: In the eloquent discussion of the discrepancies between your interpretations, I missed one thing, and that is that a possible explanation might be that activation of the ends of the fiber rises more rapidly and, therefore, causes the terminal sarcomeres to shorten more quickly than the sarcomeres in the center of the preparation. That might set up such a behavior, if there's a difference in speed of activation of the various areas. More important than adding another possibility is the question I would like you to answer. Did you, instead of releasing a fiber, try to control sarcomere length?

ALTRINGHAM: No, we don't yet have the facilities but we're planning to do it.

TER KEURS: I think that would settle the matter, then. No doubt that would settle it.

HOMSHER: I think Gerry has it wrong about climbing up the length-tension curve, because if you take, say, a semitendinosus muscle and bring it up to 3.6 μm, and begin shortening within 100 milliseconds or so of stimulation, the force will rise up as you shorten.

ALTRINGHAM: Even in some of the constant velocity release experiments, where you see the tension initially declines, it can begin to increase again during shortening if the release is large.

EDMAN: Could I just comment here? It doesn't climb up the static length-tension diagram actually. I'm commenting on Gerry's point. You *have* to get this decrease in force, certainly, of that segment which is decreasing in length. But still the force will increase eventually, and that can be shown in the segments I'm going to present here.

POLLACK: If you *require* a decrease of force in a shortening segment before the eventual increase of force, then the creep argument cannot be correct. The shortening segment in a fixed-end tetanus is the ends. During the time the ends are shortening, the tension is always rising - rapidly at first and then more slowly toward a plateau. The tension does not decrease before it increases. Therefore, any argument that requires a shortening segment (or, for that matter, any segment along the fiber) to *decrease* its force transiently, must, it seems to me, be invalid.

LENGTH-TENSION-VELOCITY RELATIONSHIPS STUDIED IN SHORT CONSECUTIVE SEGMENTS OF INTACT MUSCLE FIBRES OF THE FROG

K.A.P. Edman and C. Reggiani

Department of Pharmacology, University of Lund, S-223 62 Lund, Sweden

ABSTRACT

Length changes of consecutive, 0.5-0.8 mm long segments of frog single muscle fibres were studied by photoelectric recording of opaque markers placed on the fibre surface. There was a marked redistribution of segment length during an ordinary isometric contraction (fixed fibre ends) at both 2.15 and 2.6-2.8 μm sarcomere length. This length redistribution can explain the tension 'creep' that occurs during standard isometric contractions on the descending limb of the length-tension relation. Length clamp of individual segments eliminated tension creep completely. Active force of length-clamped segments was investigated within the range 2.20-3.65 μm sarcomere length. The descending limb of the length-tension relation (determined in segments where no tension creep occurred) was not strictly linear but had a slightly sigmoid shape. Active force was reduced to zero at a sarcomere length close to 3.65 μm. While isometric force varied only moderately between different segments, the velocity of unloaded shortening (V_0) was found to vary greatly (by 22-50%) along the length of a fibre. V_0 did not correlate with the passive resistance to a length change, the isometric force or the cross-sectional area of the individual segments. Local differences of the internal milieu and/or coexistence of myosins of different kinetic properties within a single fibre may account for the observed differences in V_0.

INTRODUCTION

Force production during a tetanus recorded in whole muscle (Abbott and Aubert, 1952; Joyce, Rack and Westbury, 1969), isolated muscle fibres (Edman, 1966) or even in a length-clamped segment of a single fibre (Gordon, Huxley and Julian, 1966a,b), usually involves a certain amount of tension 'creep'. That is, after an initial rapid rise of tension there is a relatively slow increase in force until a maximum level is finally reached. So far there has been no general agreement about the origin of tension creep, that is whether it is due to instability of sarcomere length or whether in fact it represents the true mechanical behaviour at sarcomere level (Gordon et al., 1966a; Noble and Pollack, 1977; Julian, Moss and Sollins, 1978; Sugi and Pollack, 1979). A clear answer to this question is needed before we can establish the exact shape of the length-tension curve above slack length where tension creep is pronounced. In the present experiments we have investigated in more detail how the mechanical performance differs along the length of a muscle fibre, with respect to both force and maximum velocity of shortening. Furthermore, we have tried to eliminate the problem of tension creep by using a technique which enables us to length-clamp very short segments of the fibre.

METHODS

The essentials of the experimental arrangement are illustrated in Fig. 1. Single fibres from the tibialis anterior muscle of *Rana temporaria* were mounted horizontally in a thermostatically controlled bath at 1-2°C between a force transducer and the shaft of an electromagnetic puller. Pre-cooled Ringer solution (composition, see below) was perfused through the chamber (volume, 2.5 ml) at a rate of 1.7 ml/min. The bath tempera-

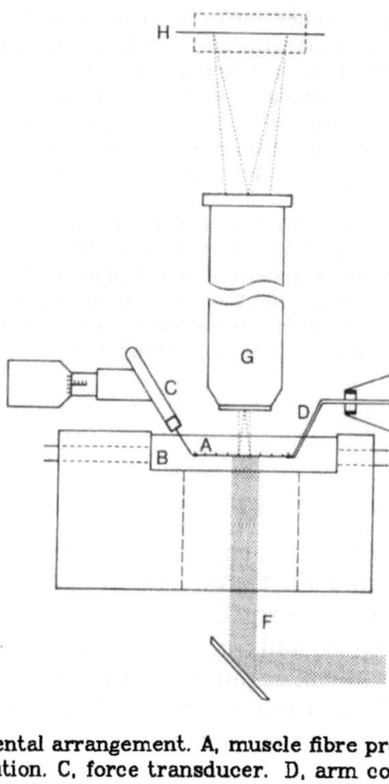

Figure 1: Experimental arrangement. A, muscle fibre provided with markers. B, trough containing Ringer solution. C, force transducer. D, arm connected with puller. E, electromagnetic puller. F, expanded laser beam. G, microscope. H, stage for Reticon CCPD 1024.

ture was measured at many different sites along the fibre by means of a thermistor probe that was operated by a micromanipulator. The difference in temperature along the fibre was less than 0.2°C. The bath temperature was kept constant to ±0.2°C during any given experiment.

Small clips of aluminium foil were attached to the tendons close to the insertion of the fibre. The clips had a hole which fitted the hook on the force transducer and the puller arm, and the free portion of the clip was folded around the hook to prevent any change in position during the experiment. It was possible to adjust the attachment of the fibre in such a way that the sideway movement of the fibre during contraction was negligible (< 15 μm). By twisting the aluminium clips appropriately it was also possible to minimize any tendency of the fibre to twist during contraction.

The fibre was stimulated supramaximally by means of two platinum plate electrodes placed on either side of the preparation. Fused tetani of 1-3 s duration were studied.

The electromagnetic puller was similar to that previously described (Edman, 1975) but it could produce faster movement. The rise time of a 100 μm step was 0.3 ms.

The contractile behaviour of individual segments along the fibre was studied using the photoelectric recording technique described by Edman and Höglund (1981). For this purpose opaque markers of black dog's hair (approximately 50 μm wide and 200 μm long) were placed on the upper surface of the fibre perpendicularly to its long axis. The markers were spaced at 0.5-0.8 mm intervals along the entire length of the fibre, the markers next to the tendons being placed 0.2-0.4 mm from the fibre-tendon junction. Care was taken to have the fibre free of connective tissue. Under these conditions the markers adhered well to the fibre surface. They maintained their position on the fibre over many hours of experimentation even after numerous release recordings. The fibre was illuminated from below by a He-Ne laser and a real image of the fibre was projected through a microscope onto a photodiode array (Reticon 1024) that was placed on a stage above the microscope. The position of two adjacent markers was sensed in this way by the photodiode array (Fig. 1), which was scanned at 4 kHz. An analogue computer calculated the actual distance between the two markers from each scan. A signal was displayed on an oscilloscope screen showing the per cent change in length between the two markers, i.e. of one segment. The accuracy of this measurement was 0.1-0.2% of the segment length. The laser beam, microscope and photodiode array could be translated together to enable measurement from any pair of markers, i.e. from any segment along the fibre.

Length changes within the segments located between the outermost markers (specified above) and the fibre-tendon junction were studied by cinephotographic recording (64 f.p.s) of fine Nylon filament markers (ca 13 μm wide) that were placed 0.1-0.2 mm apart within this region of the fibre. One of these markers was placed on the fibre-tendon junction. The inter-marker distances could be measured with an accuracy of ± 0.5% by this recording.

The resting sarcomere length was measured 1. by laser diffraction technique (Cleworth and Edman, 1972) and 2. by direct microscopy at 800× magnification using a 40× water immersion objective and an ocular micrometer (Edman and Kiessling, 1971). In experiments where the length-tension relationship was studied in individual segments (length clamp) the resting sarcomere length was measured in each segment using both of the above methods. The two measurements agreed to within 0.02 μm at the various sarcomere lengths considered between 2.20 and 3.65 μm. A mean value of these measurements was used for the determination of the length-tension relation.

The cross-sectional area of the fibre was measured in individual segments midway between two markers. This measurement was performed in a special chamber after adjusting the sarcomere length to approximately 2.25 μm, as described in detail previously (Edman, 1975).

The signals from the force and displacement transducers and the signals from the photodiode array were displayed on Tektronix 5113 storage oscilloscopes and photographed on 35 mm film. Measurements from the film records were performed in a Nikon model 6C profile projector (Edman, 1979).

The Ringer solution had the following composition (mM): NaCl 115.5, KCl 2.0, $CaCl_2$ 1.8, NaH_2PO_4–Na_2HPO_4 2.0, pH 7.0.

RESULTS AND DISCUSSION

1. Segment Length Changes During Tetanus at Fixed Fibre Ends

Fig. 2 shows examples of segment length changes during 3 sec isometric tetani performed at 2.15 μm and 2.60 μm sarcomere lengths. These records have been collected from different contractions induced at 4 min intervals. The superimposed records in the right hand panel have been repeated after 1.5-2.5 hours experimentation and it is evident that the mechanical behaviour of each individual segment is remarkably consistent over this time. A series of records like this therefore depicts the complete pattern of length changes during any *given* contraction. In the analysis of the records distinction was made between the initial length change which coincides with the steep rise in tension and the subsequent slow change which coincides with the 'plateau' phase of the tetanus.

The upper diagrams of Fig. 3 show the amplitude of the initial length change in different segments of two representative fibres (A,B) that were stimulated to produce a 3 sec isometric tetanus of 2.15 μm (filled circles) and 2.60 μm (open circles) sarcomere lengths, respectively. The segments studied cover the entire length of the fibre except for a portion, 4-5% of the fibre length, located between the fibre-tendon junction and the marker next to the tendon. (Length changes within the end portions have been studied separately, see further below). As can be seen in Fig. 3, there was shortening, although to different degrees, in nearly all segments during the initial phase, and this applies also to contractions at long length. The vertical bars show the amplitude and direction of the

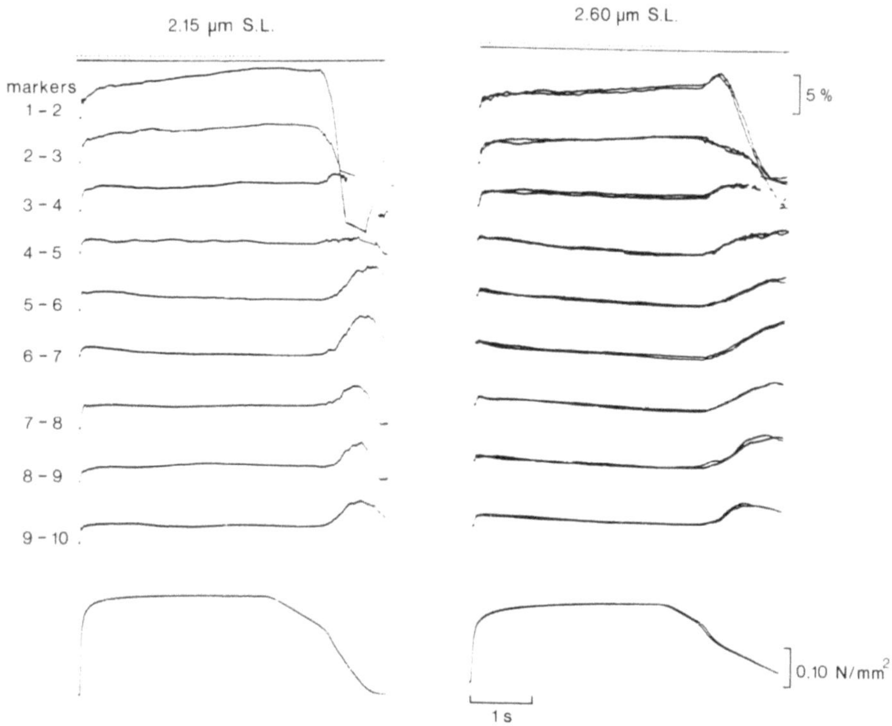

Figure 2: Oscilloscope records of segment length changes during 3-s isometric tetanus of single muscle fibre at 2.15 μm (left) and 2.60 μm (right) sarcomere length. Segments identified by two adjacent markers. Each record refers to a separate contraction performed at 4 min intervals. Bottom traces: isometric force. Superimposed traces in right panel are separated in time by 1.5-2.5 h.

subsequent slow segment length change. It is evident that this second, creeping length change may appear as both shortening and elongation, and it is not possible to predict from the first phase how the subsequent length change will come out.

In the lower diagrams of Fig. 3 the length creep (following the initial length change) has been replotted for the two fibres, and some interesting features become apparent. In the first place there was a substantial length creep, both in the form of shortening and elongation, at 2.15 μm sarcomere length, where no tension creep is generally seen. Furthermore, there was a *net shortening* if all the segments are taken together. This would seem to imply that the very tips of the fibre were actually stretched under these conditions. By contrast, at 2.60 μm sarcomere length, the majority of the segments were found to elongate to various degrees resulting in a *net increase in length* of the segments. This means that the tips of the fibre must have been shortening to a corresponding degree. That the end regions do behave as predicted from these results was confirmed in separate experiments in which length changes within

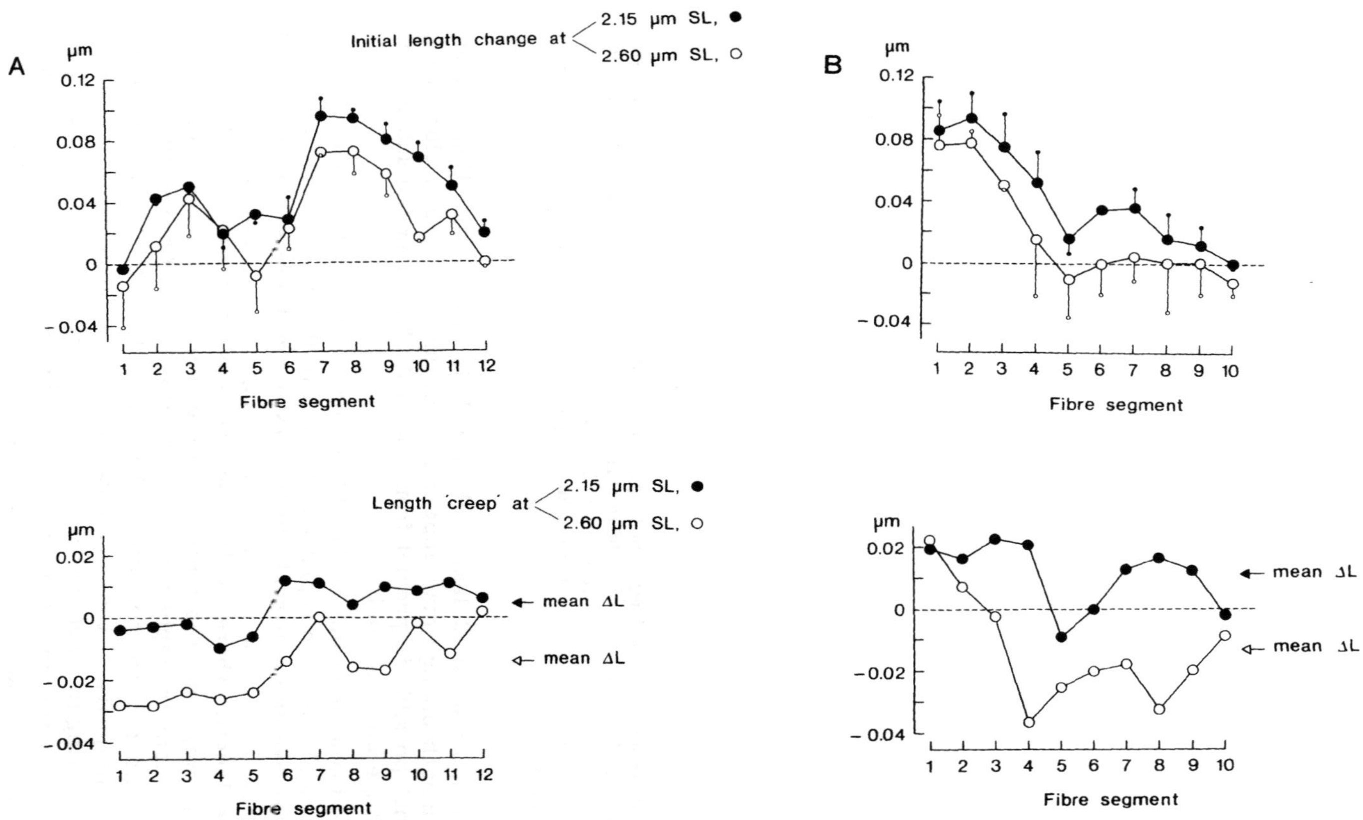

Figure 3: *Upper diagrams:* Amplitude of initial length change of individual segments during 3 s isometric tetanus in two different fibres (A and B) at 2.15 μm and 2.60 μm sarcomere length. Length change expressed as μm/sarcomere. Vertical bars show amplitude and direction of the subsequent slow segment length change. *Lower diagrams:* Amplitude of slow segment length change replotted for the two fibres. Arrows indicate mean amplitude of slow length change at the two sarcomere lengths.

the end regions of the fibre were studied by means of cinephotographic recording of thin Nylon markers (see Methods).

The data presented in Fig. 2 and Fig. 3 clearly show that there is a considerable length redistribution between different segments during isometric contraction, at both slack length and more extended fibre lengths. Tension creep will occur at a stretched length, because the stronger segments, which shorten, will steadily improve their strength as they approach the plateau of the length-tension relation. These segments will cause a corresponding force enhancement of the weaker segments which are stretched. Tension will therefore rise slowly. At slack length, on the other hand, no tension creep will occur, because the stronger segments do not gain strength as they shorten over the plateau of the length-tension relation. However, due to the intrinsic movement along the fibre the measured tension at slack length can be estimated to be ca. 2% lower than the true isometric force of the strongest segments and correspondingly higher than the maximum isometric force of the weaker, elongating segments.

As illustrated in Fig. 3 (lower left diagram) some segment may be found to stay constant in length while tension creep occurs. This finding would seem to argue against the idea that tension creep is entirely due to sarcomere instability (also see ter Keurs, Iwazumi and Pollack, 1978, 1979). The results presented in the following section strongly suggest, however, that tension creep does not represent the true mechanical behaviour at sarcomere level. It is therefore probable that a redistribution of length does occur during tension creep within a segment whose overall length remains virtually constant during the creep phase.

2. Force Development During Length Clamp of Individual Segments

If tension creep were solely due to sarcomere instability, one might expect that this phenomenon would disappear by length-clamping very short segments of the fibre. The following experiments were performed to test this point. Fig. 4 compares tetanic contractions at four different lengths. In the left hand panel ordinary isometric tetani are shown, i.e. recordings at fixed fibre ends. The right hand panel shows length clamp recordings at the same resting sarcomere lengths.

The ordinary 'isometric' tension myogram can be seen to include some tension creep already at 2.20 μm sarcomere length, but the creep phase becomes more and more pronounced in the records at longer sarcomere lengths. By contrast, the length-clamp records exhibit no substantial tension creep at any sarcomere length. It is furthermore of interest to note that virtually no force is produced near 3.65 μm sarcomere lengths where thick and thin filaments are in end-to-end position.

These results clearly show that tension creep can be eliminated if recording is made from a short segment where the sarcomere population is likely to be fairly uniform. The tension derived by this measurement is probably very close to the true sarcomere isometric force.

It is essential to point out that occasional segments do show some creep, probably because there is enough sarcomere instability in such a

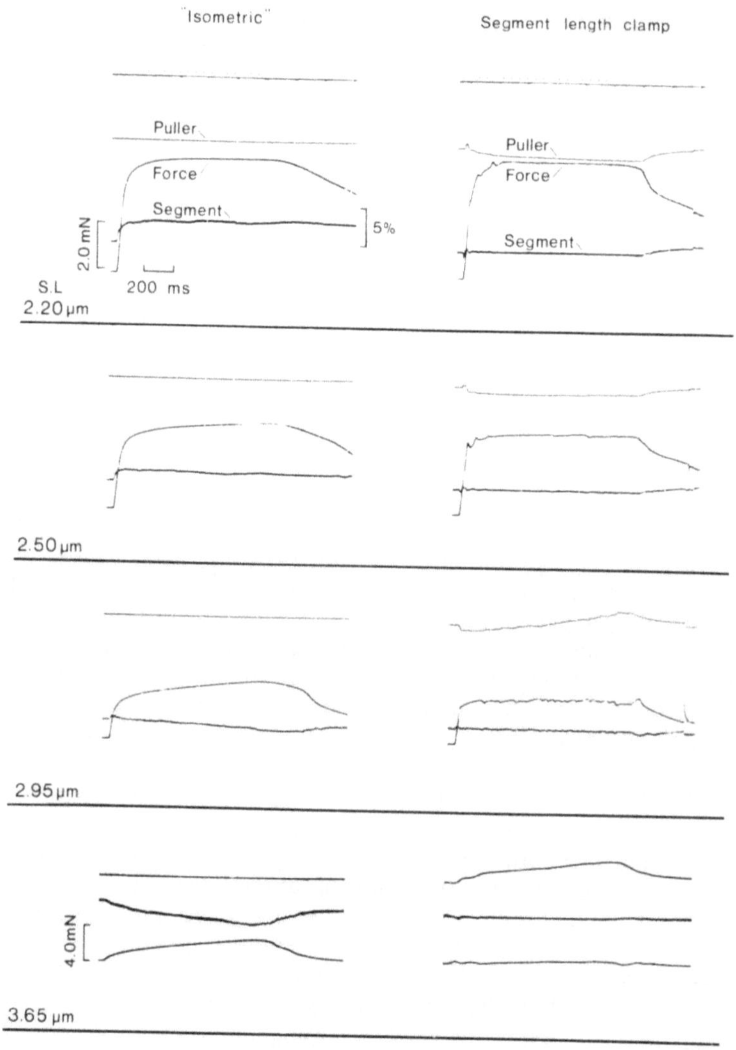

Figure 4: Tetanic contractions of single muscle fibre at four different sarcomere lengths. Left panel: ordinary isometric recording (fixed fibre ends) showing length change of one segment. Right panel: force production during segment length clamp.

segment to produce a redistribution of sarcomere length. Furthermore, tension creep regularly develops in segments that have previously produced creep-free tension, when the fibre starts to deteriorate after long usage. This observation accords with the view that tension creep is due to sarcomere instability.

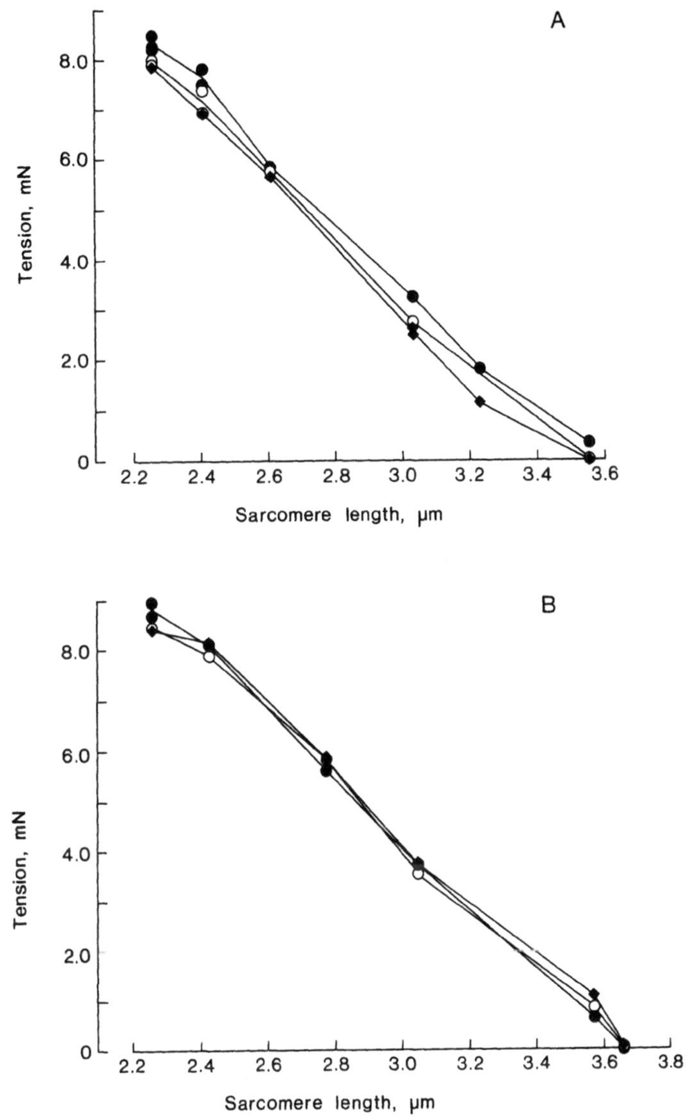

Figure 5: Relation between tetanic force and sarcomere length recorded during length clamp of three different segments. A and B show data from two different fibres.

3. Descending Limb of Length-Tension Curve

Fig. 5A, B shows results from length clamp recordings of the kind described in the preceeding section. Illustrated are force values of length-clamped segments (showing no tension creep) plotted against sarcomere length. It can be seen that the length-tension relationship is not strictly linear (c.f. Gordon et al., 1966b) but slightly sigmoid. Thus, extrapolation from the middle portion of this relationship intersects the abscissa at 3.3-3.4 μm. It is of interest to note, however, that the *measured* tension does reach zero level at a point close to 3.65 μm sarcomere length where the A and I filaments are end-to-end. These findings support the view that force production is related to the area of overlap between the thick and thin filaments. The results do seem to indicate, however, that there is a superimposed effect of some other factor than the area of overlap between the A and I filaments that modulates the length-tension relation. The nature of this second factor is unclear at the present time. The non-linear shape of the length-tension relation may reflect a length dependent activation of the contractile system.

4. Differences in Velocity of Unloaded Shortening Along Single Muscle Fibres

The velocity of shortening at zero load, V_0, is generally held to be an indication of the turnover rate of the myosin cross-bridges. Previous studies (Edman, 1979) of frog intact muscle fibres have shown that V_0 is independent, over a wide range, of the sarcomere length and of the fibre's capacity to produce force. Recent evidence has shown that V_0 varies substantially along the length of a muscle fibre (Edman and Reggiani, 1982).

Fig. 6 shows the approach used for recording the velocity of "unloaded" shortening in individual segments. At a given time during the tetanus plateau the puller was quickly released to produce a drop in force to very nearly zero level. The puller movement was thereafter adjusted appropriately to keep the tension in the fibre just above zero (about 2% of maximum force). The fact that some force was allowed ensures that the fibre did not bend or get slack at any time during the motion. This point was also confirmed by cinephotographic recording of the fibre during the release movement using a 16 mm Bolex camera (64 f.p.s.) fitted on a Zeiss Stereo microscope. The shortening of one segment, i.e. the decrease in distance between two adjacent markers was recorded (upper trace, Fig. 6). The length change of the segment exhibited an initial rapid phase followed by a straight ramp. The slope of this straight ramp

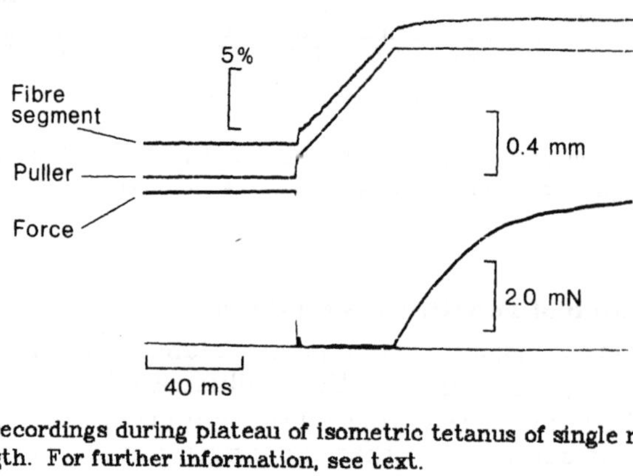

Figure 6: Release recordings during plateau of isometric tetanus of single muscle fibre at 2.2 μm sarcomere length. For further information, see text.

provided a measure of the velocity of shortening at nearly zero load in that particular segment.

Fig. 7 shows velocity data from different segments of one single fibre. The individual data are measurements from repeated tetani performed at two minute intervals. It can be seen that the values were quite consistent for repeated measurements in each respective segment, but it is evident that there was a considerable variation of V_0 *along* the fibre, ranging between 1.85 and 2.45 lengths/sec, i.e. a difference of about 30%.

It was of interest to determine the mean velocity of shortening of the different segments, for if the observed distribution of shortening velocity is correct, then the calculated mean should agree with the speed of shortening recorded for the fibre as a whole. In this computation account was taken of the actual length of each segment. The velocity of shortening of the whole fibre was determined from the speed of puller movement and the overall length of the fibre. The values of shortening velocity derived by these two completely independent measurements were in very good agreement. The difference was less than 1% in the example given in Fig. 7, and in no case did the difference exceed 5%.

The velocity pattern along the fibre differed from one preparation to another but remained constant in any given fibre throughout an experiment (Edman and Reggiani, 1982). In eight preparations the difference in velocity along the fibre varied between 22 and 50% of the mean value of V_0 in each fibre. There was no correlation between V_0 and the segments' ability to produce force (measured by length clamp). There is also evidence (Edman and Reggiani, 1982) that the differences in V_0 were *not* due to regional differences in the passive resistance to shortening.

What is then the mechanism underlying the V_0 differences along the fibre? We know from our previous experiments (Edman and Mattiazzi, 1981) that V_0 is sensitive to the intracellular pH and probably also to the

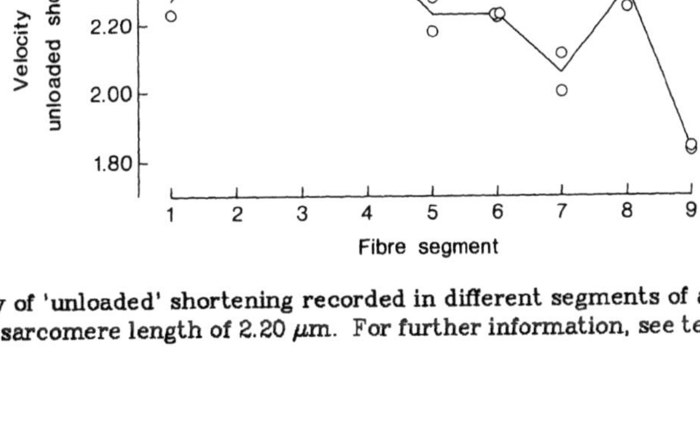

Figure 7: Velocity of 'unloaded' shortening recorded in different segments of a single muscle fibre at a resting sarcomere length of 2.20 μm. For further information, see text.

concentration of breakdown products of ATP. Local differences of the intracellular milieu may therefore be considered as one possible cause of the variation in V_0 within the fibre. Another, more intriguing possibilty would be that *myosins of different kinetic properties* exist along the fibre. Such a non-uniformity with respect to the kinetic properties of myosin is not unreasonable in view of the fact that fast and slow types of myosin have been found to coexist in individual fibres of adult mammalian skeletal muscle (Gauthier and Lowey, 1979; Lutz, Weber, Billeter and Jenny, 1979).

REFERENCES

Abbott, B.C. and Aubert, X.M. (1952). The force exerted by active striated muscle during and after change of length. J. Physiol. 117: 77-86.

Cleworth, D.R. and Edman, K.A.P. (1972). Changes in sarcomere length during isometric tension development in frog skeletal muscle. J. Physiol. 227: 1-17.

Edman, K.A.P. (1966). The relation between length and active tension in isolated semitendinosus fibres of the frog. J. Physiol. 183: 407-417.

Edman, K.A.P. (1975). Mechanical deactivation induced by active shortening in isolated muscle fibres of the frog. J. Physiol. 246: 255-275.

Edman, K.A.P. (1979). The velocity of unloaded shortening and its relation to sarcomere length and isometric force in vertebrate muscle fibres. J. Physiol. 291: 143-159.

Edman, K.A.P. and Höglund, O. (1981). A technique for measuring length changes of individual segments of an isolated muscle fibre. J. Physiol. 317: 8-9P.

Edman, K.A.P. and Kiessling, A. (1971). The time course of the active state in relation to sarcomere length and movement studied in single skeletal muscle fibres of the frog. Acta Physiol. Scand. 81: 182-196.

Edman, K.A.P. and Mattiazzi, A. (1981). Effects of fatigue and altered pH on isometric force and velocity of shortening at zero load in frog muscle fibres. J. Muscle Res. Cell Motil. 2: 321-334.

Edman, K.A.P. and Reggiani, C. (1982). Differences in maximum velocity of shortening along frog muscle fibres. J. Physiol. 329: 47-48P.

Gauthier, G.F. and Lowey, S. (1979). Distribution of myosin isoenzymes among skeletal muscle fiber types. J. Cell Biol. 81: 10-25.

Gordon, A.M., Huxley, A.F. and Julian, F.J. (1966a). Tension development in highly stretched vertebrate muscle fibres. J. Physiol. 184: 143-169.

Gordon, A.M., Huxley, A.F. and Julian, F.J. (1966b). The variation in isometric tension with sarcomere length in vertebrate muscle fibres. J. Physiol. 184: 170-192.

Joyce, G.C., Rack, P.M.H. and Westbury, D.R. (1969). The mechanical properties of cat soleus muscle during controlled lengthening and shortening movements. J. Physiol. 204: 461-474.

Julian, F.J., Moss, R.L. and Sollins, M.R. (1978). The mechanism for vertebrate striated muscle contraction. Circulation Res. 42: 2-14.

Lutz, H., Weber, H., Billeter, R. and Jenny, E. (1979). Fast and slow myosin within single skeletal muscle fibres of adult rabbits. Nature 281: 143-144.

Noble, M.I.M. and Pollack, G.H. (1977). Molecular mechanisms of contraction. Circulation Res. 40: 333-342.

Sugi, H. and Pollack, G.H. (1979). *Cross-Bridge Mechanism in Muscle Contraction*. Univ. Tokyo Press.

ter Keurs, H.E.D.J., Iwazumi, T. and Pollack, G.H. (1978). The sarcomere length tension relation in skeletal muscle. J. Gen. Physiol. 72: 565-592.

ter Keurs, H.E.D.J., Iwazumi, T. and Pollack, G.H. (1979). The length tension relation in skeletal muscle: revisited. In: *Cross-Bridge Mechanism in Muscle Contraction*. Ed. Sugi, H. and Pollack, G.H. pp. 277-295. Tokyo Univ. Press.

DISCUSSION

PODOLSKY: I was wondering whether the variation in speed was as great if you loaded the fiber and measured the speed of the different segments as a given fraction of load.

EDMAN: I haven't yet made a complete study of the force-velocity relation in individual segments. However, I have recorded the maximum isometric force (P_0) in the different segments along the fiber by means of length clamp technique, and in some experiments recorded V_0 in the same segments. The results show that there are small differences in P_0 among different segments but these differences in P_0 do not correlate with the variation in V_0 along the fiber. For instance, a segment having a high V_0 may have a relatively low P_0, or vice versa. However, I cannot yet tell you if the shape of the force-velocity relation varies in any fundamental way in the different segments along the fiber.

HOUSMANS: How confident are you that the surface marker does not misrepresent to some extent the movement of the core of the fiber itself?

EDMAN: You are actually asking if the membrane is doing something different from what the layers underneath are doing. Remember that there are T-tubules in each sarcomere which tie down the membrane into the inner structures of the fiber. The M-lines are also connected with the plasma membrane (Brown & Hill, J. Microscopy 125: 319-336, 1982). A change in distance between surface markers will therefore represent a change in length of the myofibrils. As expected, there is a very good agreement between sarcomere length measurements made by the laser diffraction technique and segment length measurements performed with the marker technique. The markers stick onto the fiber surface very well, provided the fiber is freed from connective tissue. Under these conditions I am confident that any change in distance that one records between two adjacent markers does represent a change in mean sarcomere length in that particular segment.

HOUSMANS: Does dispersion in precontractile sarcomere lengths and segment length run more parallel at longer sarcomere lengths than at shorter sarcomere lengths?

EDMAN: Well, we haven't studied the dispersion of sarcomere length in the various segments in detail. However, the same good correlation exists between segment length and sarcomere length over the entire length-tension relation. This, of course, is not to say that the sarcomere spacing is the same in all segments along the fiber.

Actually, there is quite a big difference in resting sarcomere length between the ends and the middle region of the fiber, and I should like to come back to that problem for just a second, because our measurements do not seem to agree with one of Dr. Pollack's statements. The interesting thing is that the sarcomere length at the very tips of the fiber is not significantly different from that in the middle region of the fiber when we are at slack length (approximately 2.1 μm sarcomere spacing). Indeed, in some fiber ends the sarcomere spacing may actually be somewhat larger than that in the middle region but still within the plateau of the length-tension relation. During isometric contraction at *slack length* the end segments are found to elongate, not shorten. This may be due to the fact that the tips of the fiber are tapering, and this would make them inherently weaker, since the cross-sectional area would be smaller than in the central segments. On the other hand, when the fiber is extended to a resting sarcomere length of, say, 2.6 μm in the middle region, the sarcomeres at the tips will be approximately 0.25 μm shorter. This more favorable position of the end sarcomeres on the length-tension curve will more than compensate for their inherent weakness and they will now become stronger than the central segments. The tips will therefore always shorten during fixed-end contractions above slack length.

POLLACK: You mentioned earlier that the difference of isometric tension of different segment along the fiber was up to ±5-7%.

EDMAN: Yes, about ±4%.

POLLACK: The question I have is, if you believe in the possibility of instability, in climbing up the descending limb, shouldn't this difference of strength along the fiber, especially at stretched lengths, where you showed it can be quite large (Fig. 5), be enough to trigger this sort of instability? Shouldn't the inhomogeneity develop progressively, so that you'd see some regions that wound up ultimately with very short sarcomere length and some with very long sarcomere length?

EDMAN: Yes, you do. I showed that. In the diagrams I showed (Fig.3) you could see some segments shorten greatly whereas others elongated, and there were great differences along the entire length of the fiber.

POLLACK: Those changes were relatively modest. I was talking about the gross changes of sarcomere length, for example, shortening of a micron or so that John Altringham reported to occur at the ends in most specimens, the kind of shortening that some like to attribute to instability. Why shouldn't the same phenomenon occur in the middle?

EDMAN: Well, the same phenomenon does occur in the middle of the fiber. The point is, however, that at a prestretched fiber length the ends are by far the strongest segments. They will therefore shorten at a greater speed than any of the central segments. Note, however, that there is a *continuous* redistribution of length between the central segments during the tetanus period (Fig. 2). This redistribution of length is quite substantial during a 3 sec tetanus and becomes more and more pronounced if the fiber is activated during progressively longer times.

CECCHI: Have you measured the elasticity with a segment clamp?

EDMAN: Yes, as you may have noticed from the shortening record where I produced a release (Fig. 6) there was an initial step of the segment length change. That is an elastic step which, of course, does not involve the tendons. So if we extrapolate backwards to the onset of release we would get a value of the compliance in that segment.

CECCHI: How much was it?

EDMAN: I haven't pursued that in detail. We are trying to computerize these things now. But preliminary measurements have shown that the compliance is about six nanometers per half sarcomere, but this will be investigated further.

GULATI: Coming back to the differences in speed of shortening at zero load along the segment, can you completely rule out the possibility of small differences in passive load along different segments, for instance, from differences in the distribution of internal membrane system or even the differences in connective tissue along the fiber?

EDMAN: Yes, I think I can do that. When we first observed these differences in velocity, it was essential to rule out a number of trivial causes. Temperature differences along the fiber could be one which we can safely rule out in our set-up. However, another possibility would be that there are differences in the passive resistance to shortening along the fiber. We approached this problem by applying a stretch to the resting fiber using a speed that corresponded with the shortening velocity. The rationale of this experiment was to see if the rate of stretch of the segments differed along the fiber which would tell us if there was any difference in the passive resistance to a length change. Slight differences were recorded, but they did not correlate with the observed variation in V_0 along the fiber, as I presented. We have therefore reasons to believe that the observed variation in V_0 along the fiber is not due to local differences in the passive visco-elastic properties of the fiber.

COLOMO: Have you seen any difference between different segments concerning speed of force development during a tetanus in length clamped segments?

EDMAN: This is rather difficult to establish, and we have not pursued this problem yet. There may well be measurable differences in the speed of onset of force development in different segments.

FORCE - SARCOMERE-LENGTH RELATION AND FILAMENT LENGTH IN RAT EXTENSOR DIGITORUM MUSCLE

H.E.D.J. ter Keurs, A.R. Luff* and Susan E. Luff+

Experimental Cardiology, Academisch Ziekenhuis Leiden, 2333AA Leiden, The Netherlands
**Department of Physiology, Monash University, Clayton, Victoria 3168, Australia*
(to whom correspondence should be sent)
+Department of Botany, Monash University etc.

ABSTRACT

Relations between sarcomere length (SL) and force (F) were studied in ten fiber bundles (six to twenty fibers) from rat extensor digitorum muscles. A bundle (60 μm by 200-300 μm) was mounted in a glass covered perfusion chamber containing modified Krebs Henseleit buffer at 25°C, oxygenated with 95% O_2, 5% CO_2 and pancuronium bromide (8 mg/1). F (Disa 51E 01 transducer) and SL (laser diffraction and light microscopy) were measured; the latter could be controlled by a servomotor system. 200-500 ms tetanic stimulus trains were applied via platinum electrodes parallel to the muscle with 20% above maximal intensity, 160 Hz frequency and 1 ms duration of pulses. Tetani were at 2 min intervals. F attained a steady value 100 ms after the start of the tetanus at 2.0 - 2.5 μm SL and 350 ms at 3.5 μm SL.

Active force, measured during tetani in which sarcomere length was held constant, was maximal between SL = 2.15 μm and 2.65 μm and declined in linear fashion with SL to zero at SL = 3.90 μm. Active force at SL = 2.00 μm was 95% of maximal force. Passive force was manifest above SL = 3.10 μm and was 10% of maximal force at 3.80 μm.

Eight similar bundles were processed conventionally for electron microscopy (Philips EM 201A) while SL was measured during the processing steps. Measurements were made from micrographs of longitudinal sections. SL measured from the micrographs were consistent with the observed shrinkage (5%). Actin periodicity was 41.5 ± 0.19 nm; twenty-seven periods per actin filament were found. Filament lengths were corrected for an assumed actin periodicity of 39 nm. Actin length was 1.13 ± 0.013 μm; myosin length was 1.53 ± 0.015 μm. Bare zone was 0.17 μm ± 0.01 μm.

These filament lengths would give optimum overlap at SL between 2.26 and 2.43 μm and a linear decrease to zero with increasing SL from 2.43 μm to 3.79 μm. Actual force was consistently higher than predicted by overlap and force was maintained to both the left and the right of the predicted plateau.

INTRODUCTION

Since the comprehensive study of Gordon, Huxley and Julian (1966) the force - sarcomere length relation of frog skeletal muscle has been considered as a corner stone of the cross-bridge mechanism of contraction. Force decreased in linear fashion with increase of sarcomere length from a maximum at 2.25 μm to zero at 3.65 μm. Force decrease seemed to correlate well with diminishing overlap between actin and myosin. Recently ter Keurs, Iwazumi and Pollack (1978) have presented results from frog single fibers in which force did not decrease in a linear fashion on the descending limb and in which maximum force was developed up to a sarcomere length of 2.5 μm. Later, ter Keurs and Elzinga (1981) have shown that force during tetani at constant sarcomere length decreased linearly with sarcomere length from maximum at 2.25 μm to zero at 3.65 μm. These fibers exhibited substantially larger passive force, though, than the fibers used in the former study (ter Keurs et al., 1978). There is, therefore, some doubt about the exact form of the descending limb of the force-sarcomere length relation.

It has generally been assumed that the force-sarcomere length relation in mammalian skeletal muscle is similar to that in frog. Both direct and indirect measurements suggest otherwise. Close (1972) suggested an optimum of sarcomere length at 2.5 μm in rat, Buller and Lewis (1965) estimated optimal initial sarcomere length of 3.1 μm in soleus and 2.7 μm in flexor hallucis of cat; Rack and Westbury (1969) derived an optimal initial sarcomere length of 2.8 - 3.0 μm in cat soleus. McCarter et al. (1977) measured initial sarcomere length in lateral omohyoideus of rat by means of laser diffraction techniques and found that maximal force is developed at an initial sarcomere length of 3.2 μm.

Studies on mammalian skeletal muscle indicate that during tetanic contractions in which muscle length is kept constant maximal force is developed at longer sarcomere length than in frog muscle. Data available on the ultrastructure of mammalian skeletal muscle (Page and Huxley, 1963; Walker and Schrodt, 1973) suggest that thin filament lengths are longer than in frog.

It was the purpose of this study to determine the force-sarcomere length relation in thin bundles of rat extensor digitorum muscle at constant sarcomere length and to determine filament lengths in the fibers of this muscle in order to correlate isometric tetanic force with overlap between actin and myosin.

METHODS

Preparations

Female Wistar rats were anesthetized with ether. The abdominal aorta was cannulated and perfused with oxygenated physiological saline. The extensor digitorum longus muscle was removed and transferred to a dissecting disk. Oxygenated physiological saline flowed through the chamber at a rate of 5 $ml.min^{-1}$. Bundles of 6-20 fibers, 100 μm by 300

μm in cross section, were dissected from the extensor of the second digit. The bundles were about 11 mm long. After trimming the tendons the bundle was transported to the experimental chamber mounted on the stage of an inverted Zeiss microscope. The tendons were clamped by springloaded stainless steel clips as close as possible to the origin of the fibers.

Solution used throughout dissection and experiments consisted of: NaCl 115 mM; KCl 5mM; $MgCl_2$ 1.2 mM; NaH_2PO_4 2 mM; Na_2SO_4 1.2 mM; $CaCl_2$ 2mM; Na acetate 10mM; $NaHCO_3$ 28 mM; glucose 11 mM; EGTA 0.5 mM equilibrated at 25.5° ± 0.2°C with 95% O_2 and 5% CO_2; pH was 7.4. Pancuronium bromide 8 $mg.l^{-1}$ and insulin 4 $IU.l^{-1}$ were added. The solution flowed through the experimental chamber at a rate of 2 $ml.min^{-1}$.

The method of measurement and control of sarcomere length, muscle length and force, that was used has been described previously (van Heuningen et al., 1982). In short, force was measured with a capacitive force transducer connected to a reactance converter (DISA 51 E01); sensitivity 0.1 $V.mN^{-1}$; linearity 5% up to 10 mN; drift 0.2 $mN.hr^{-1}$; resonant frequency 460 Hz; compliance 0.8 $\mu m.mN^{-1}$. Muscle length was measured by means of a variable mutual inductance displacement transducer (Metrisite) incorporated in a conventional servomotor (Gould Brush 869223).

A laser beam converged to a diameter of 300 μm was projected onto the bundle. The emanating first or second order bands of the diffraction pattern were deflected by mirrors onto two photodetector systems. One of the photodetector systems was a photodiode array (Reticon RL 512 G) which was scanned 4× $msec^{-1}$. The intensity of the zero order close to the first (or second) order was measured during each scan and its distribution under the selected order was approximated by an exponential decay. The simulated zero order was subtracted from the intensity distribution of the selected order. The median position of the intensity distribution of the selected order was calculated.

The second photodetector was a lateral effect photodiode (United Detector Technology 220 DP; Rijnsburger et al., in preparation). The difference and sum of the detector output currents are directly related to first and zero moment of the distribution of light impinging on the detector. Zero moment and first moment of the zero order intensity distribution were calculated with the aid of i) photodiodes at both ends of the detector and ii) network which similated an adjustable parabolic decay (van Heuningen et al., 1982). The position of the centre of gravity of the selected order was calculated from the ratio of its first- and zero moment after subtraction of the simulated zero order contribution.

The voltages proportional to position of the median of the selected order on the photodiode array and to the position of the centre of gravity on the lateral effect photodiode were converted to voltages proportional to sarcomere length by means of a non linear amplifier which had the same transfer function as the Bragg equation. The circuitry was calibrated before every experiment with test gratings (calibration error was smaller than 0.07% for the lateral effect diode; <0.4% for the photodiode array).

Sarcomere length control was obtained by a servo motor system which allowed potentiometer selection of either muscle length or sarcomere length for feedback. Signal levels were adjusted to attain critical damping of the control loop prior to switching to sarcomere length control. Characteristics of the system were: step response time 3.3 msec; frequency response 190 Hz (-6 dB); overshoot 3%.

A clear region of the bundle close to the force transducer was selected for study. Consequently minimal longitudinal translation of the region occurred during sarcomere length control. Small tissue remnants provided natural markers which were used to monitor translation. Direct stimulation was achieved with platinum wires embedded in the chamber wall parallel to the bundle. Stimulus pulses were 2 msec in duration; their amplitude was 20% supramaximal for isometric twitches. Stimulus rate was 160 Hz for tetani applied for 0.2 to 0.5 sec, every 100 sec. Initial muscle length was set slightly beyond slack length. Sarcomere length was then always 2.55 μm to 2.60 μm. Length changes were made in steps of 0.1 μm to 0.2 μm. At the new sarcomere length three tetani were recorded. It was common that sarcomere length during the first tetanus at any new length slightly differed from the second, third or any subsequent tetani.

Bundles were discarded when substantial sarcomere length change at the onset of contraction was observed, even if length remained constant during the tetanus. They were also discarded if sarcomere length changed substantially during the plateau of the tetanus.

Electron Microscopy

Similar bundles, dissected identically, were mounted in aluminum frames and fixed and mounted at various sarcomere lengths. Sarcomere length was monitored by a diffraction system during the fixation process. One procedure used consisted of fixation in 1% glutaraldehyde and 3% formaldehyde in physiological saline for 1 hour, then rinsed several times in physiological saline for 20 minutes, followed by fixation in 1% OsO_4 in 0.05 M potassium ferrocyanide solution overnight. Next the specimen was rinsed in physiological saline, dehydrated in ethanol, soaked in epoxy propane for 20 minutes, transferred to a 1:1 mixture of epoxy propane and Epon for 120 minutes and then left overnight in pure Epon. The following day the specimen was embedded in fresh Epon and polymerized for 2 days at 60°C.

The other procedure was similar to that of Page (1974). The bundles were fixed in 6% glutaraldehyde in 0.1 M cacodylate buffer (pH 7.2 - 7.4) for 120 minutes washed 3 times in buffer for 20 minutes each time and then fixed in 1% OsO_4 in 0.1 M cacodylate buffer for 30 minutes. Following a brief wash in buffer they were dehydrated in alcohol. Subsequent procedures were identical to the first procedure.

Many of the precautions and checks mentioned by Page and Huxley (1963) were used in the section cutting and photographing of the fibers. The sections were cut on an LKB microtome with the long axis of the bundle parallel to the edge of the knife. Once the block was trimmed, thick sections (1 μm) were cut on glass knives to check the alignment of the block. Adjustments were made until the fibers in the section showed no significant deviation into or out of the plane of section. Silver-grey thin sections were then cut with diamond knife and mounted on uncoated 200 mesh grids. These were stained with uranyl acetate and lead citrate. Sections were examined in a Philips 201A electron microscope, at a magnification of 20,000×. Firstly, a calibration grid was photographed, then several grid areas from several fibers and finally the calibration grid was rephotographed, on 35 mm film. The difference in the calibration from the end of the film to the other was never greater than 1%, and the mean of the two measurements was used for that film. The calibration from one film to another did not change by more than ± 0.8%. All prints were made with the same degree of enlargement and no distortion of the print itself was detected during processing. Measurements were made directly from a total of about 300 photographic prints.

RESULTS

Stability

It was possible to obtain up to 60 tetani with only a few percent decrease of maximal force development. Consistent with good viability of the bundles it was found that only little variation of sarcomere length was observed when the fibers were scanned from tendon to tendon. Sarcomere length dispersion of less than ± 0.10 μm was found at rest at all sarcomere lengths studied. Fiber bundles that showed larger sarcomere dispersion rapidly deteriorated and were discarded.

Force Development at Constant Muscle Length

Figure 1 shows typical records obtained from the same fiber bundle at two different initial sarcomere lengths. Force rose rapidly and reached a peak within 100 ms of onset of stimulation. Thereafter a slight decline of 1-2% in force was observed. This decline in force during an isometric tetanus has repeatedly been observed in vivo in mammalian muscle and appears to be characteristic of fast twitch muscle. Maximum force produced by these preparations was usually between 0.25 and 0.3 $N.mm^{-2}$, which is consistent with active stress developed in intact extensor digitorum longus muscle of rat in vivo.

Fig 1A shows some initial shortening of sarcomeres at the onset of the tetanus, thereafter sarcomere length remained constant until well into the relaxation phase. At a longer sarcomere length (Fig 1B, Sl = 2.70 μm) sarcomere length also measurably changed during the plateau of tetanus following initial shortening. Most preparations showed this phenomenon at sarcomere lengths above 2.6 μm.

In view of the fact that sarcomere length did not remain constant at sarcomere lengths beyond 2.6 μm in tetani at constant muscle length it was necessary to use sarcomere control as described in the methods section.

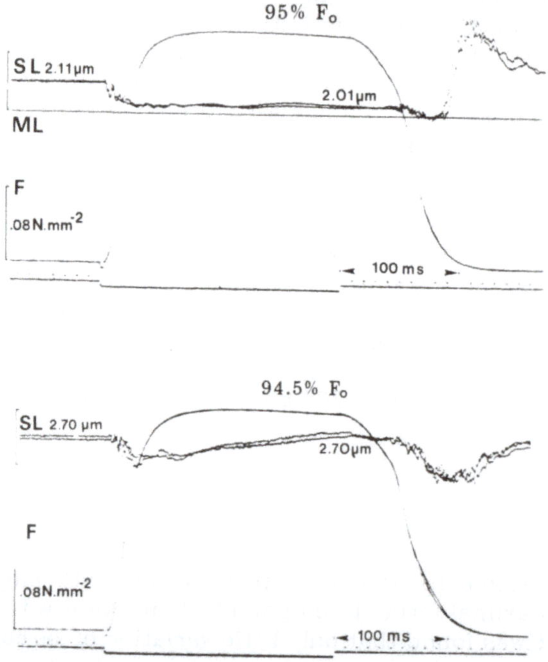

Figure 1: Typical records of force development and sarcomere behaviour during tetani at 25.5°C in a muscle extensor digitorium longus bundle of rat. Upper traces show sarcomere length, second traces show muscle length, third traces show tetanic forces upon 160 Hz stimulation indicated by the block pulse in the timing signal of the bottom trace. Calibrations indicated: calibration bar of sarcomere length ± 0.1 μm. Initial sarcomere lengths 2.01 μm in panel A and 2.70 μm in panel B respectively.

Force Development at Constant Sarcomere Length

We used sarcomere length control in contractions which started at sarcomere length above slack. Fig 2 shows typical records obtained from a bundle in which sarcomere length was held constant at initial length, for three initial lengths (3.33 μm, 3.48 μ and 3.64 μm, in A, B and C respectively). It is clear from Fig. 2 that sarcomere length was constant to within 0.01 μm during the tetani except for a small transient at the onset of contraction. Force rose to plateau level with a sarcomere length dependent rate and then remained constant until well into relaxation. Gross sarcomere nonuniformity is known to occur during relaxation (Edman, 1980) and probably caused these transient sarcomere length changes. Force and sarcomere behaviour during relaxation were neglected in this study. Translation of the region under study always occurred upon sarcomere control. The amount of translation was always (5 experiments) less than 200 μm, however, and was minimally 40 μm. From tetani during which sarcomere length remained constant we plotted the relation between force and sarcomere length. In each of 5 experiments at constant sarcomere length a linear relation between force and

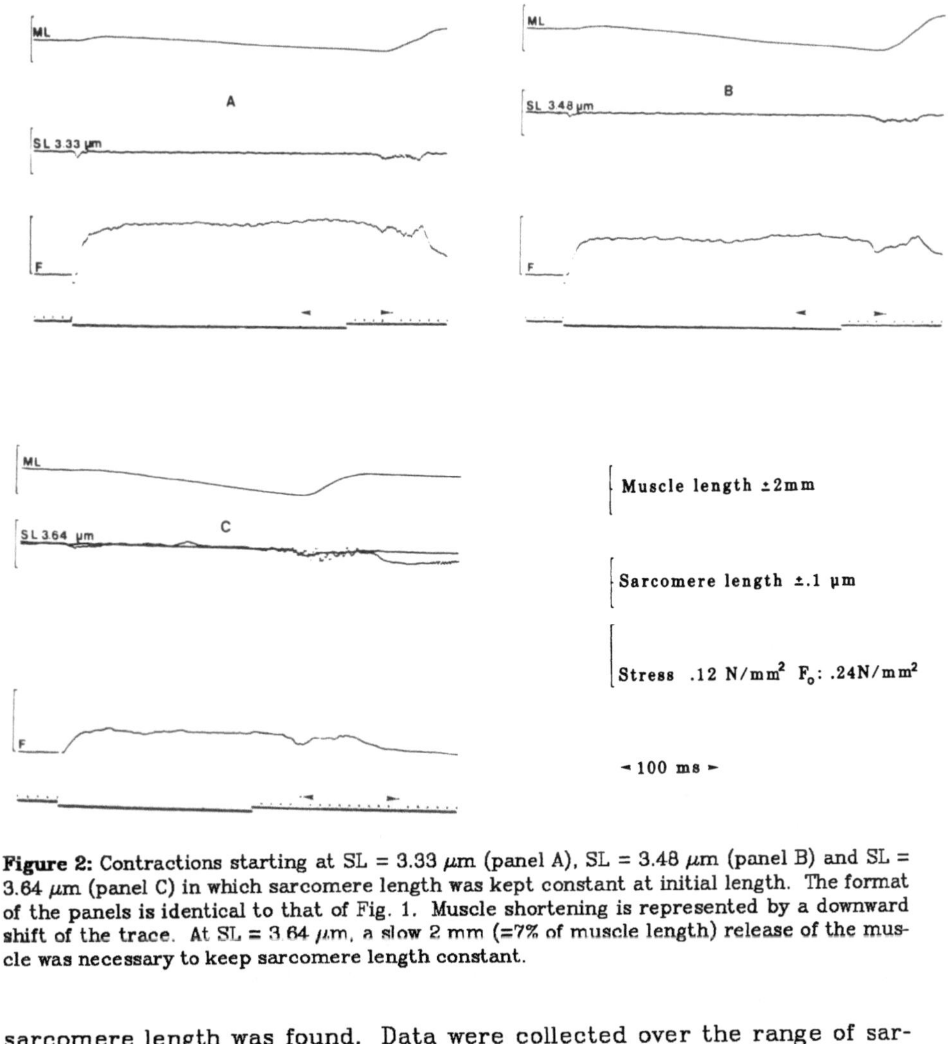

Figure 2: Contractions starting at SL = 3.33 μm (panel A), SL = 3.48 μm (panel B) and SL = 3.64 μm (panel C) in which sarcomere length was kept constant at initial length. The format of the panels is identical to that of Fig. 1. Muscle shortening is represented by a downward shift of the trace. At SL = 3.64 μm, a slow 2 mm (=7% of muscle length) release of the muscle was necessary to keep sarcomere length constant.

sarcomere length was found. Data were collected over the range of sarcomere lengths from 2.65 μm to 3.50-3.80 μm (Fig 3). It is clear that force starts to fall above a sarcomere length of 2.65 μm. Linear regression of force vs sarcomere length (Fig 3, solid line) indicated a force intercept at about 3.9 μm. The plateau of the force - sarcomere length relation extended from 2.15 to 2.65 μm (11 experiments). Significant passive force was exerted by these bundles only above 3.5 μm.

Filament Lengths

Measurements were made from enlarged micrographs (see Fig. 4). Five bundles, that were fixed at different sarcomere lengths were accepted as it was possible to measure their actin periodicity.

Furthermore measurements were only accepted from sarcomeres in which i) individual myosin filaments were visible, ii) no region of low

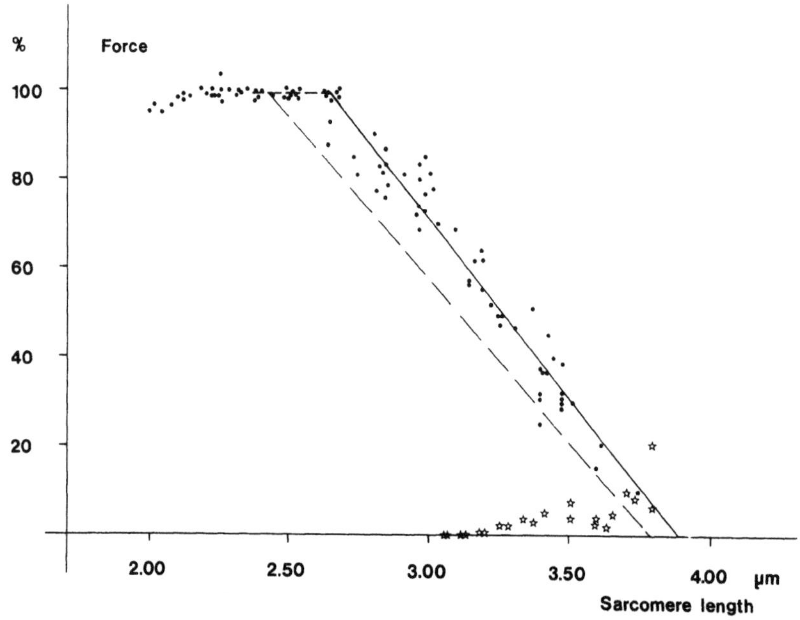

Figure 3: The relation between force (dots) and sarcomere length derived from contractions at constant sarcomere length of m. extensor digitorum of rat. Drawn line is the regression line through force-sarcomere coordinates above SL = 2.65 μm. Dashed line: overlap predicted on the basis of filament lengths. Passive force development is indicated by asterisks.

electron density at the end of the myosin filaments was encountered, iii) the border between the I-band and A-band, iv) the myosin filament was symmetrical with respect to Z and M lines. Actin periodicity was used as an internal calibration in order to correct for filament changes which occur during process of the bundles for electron micrography. Actin periodicity was calculated from the regression coefficient of the line between the length of strings of actin periods and the number of actin spacings in the string.

The results for individual fiber bundles are summarized in Table 1. All the filament lengths have been corrected on the basis that the true actin periodicity in living muscle is 39 nm. It has been convincingly argued (Huxley and Brown, 1967; Huxley, 1972) as a result of X-ray diffraction experiments that the commonly observed meridional reflection at about 38.5 nm in living frog muscle is attributable to the troponin complexes spaced along the actin filaments. For the purposes of these calculations it has been assumed that the value obtained for frog muscle is applicable to mammalian skeletal muscle. This may not necessarily be true as Vibert, Hazelgrove, Lowy and Poulsen (1972) consider that real species differences exist for the actin repeat and therefore, possibly for the spacing of troponin on the actin filament.

The overall mean value for the length of the actin filament was 1.13 $\mu m \pm 0.013$ μm (SEM) and for the myosin filament 1.53 $\mu m \pm 0.15$ μm

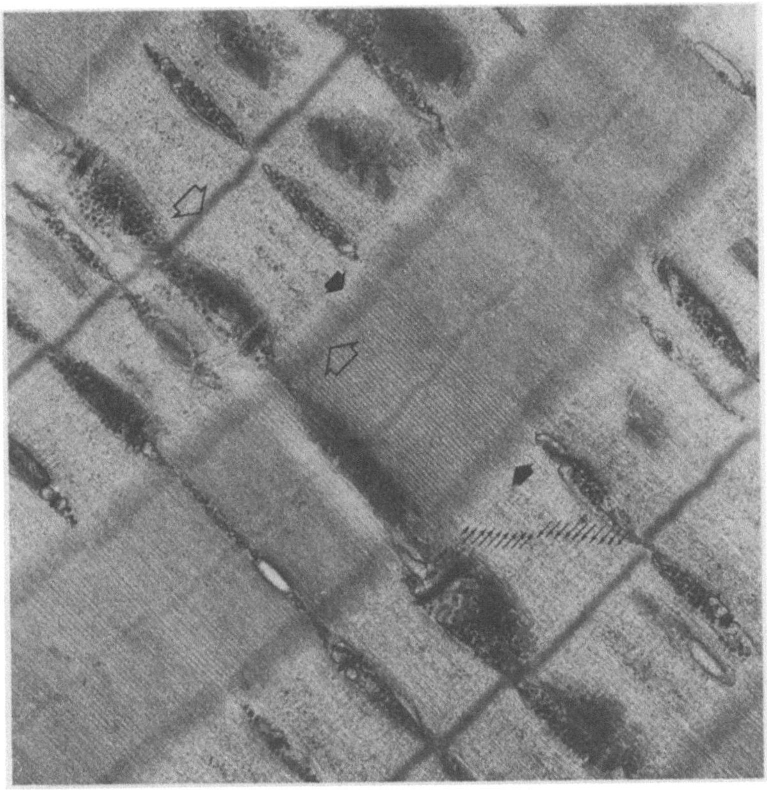

Figure 4: Electron micrograph of fiber bundle no. 4, see Table 1, showing actin periodicity (arrows) and filament properties (see text) that served as criteria to accept results. Actin filament length (open arrows) and myosin filament length (closed arrows) are indicated. Procedure one of fixation (see methods) Actin periodicity 41.43 nm in this bundle was calculated by linear regression of the length of strings of actin periods on the number of actin spacings in the string (r=0.99).

(SEM), the bare zone in the middle of the thick filament was 0.17 μm ± 0.01 μm. This latter value is in good agreement with the value of 0.15-0.16 μm obtained by Craig and Offer (1976) for the bare zone of rabbit psoas muscle determined by antibody labeling.

Figure 3 shows percentage filament overlap calculated from the filament lengths above against sarcomere length, and superimposed on the graph of force against sarcomere length. On the basis of filament overlap the plateau should extend from 2.26 μm to 2.43 μm and the zero overlap intercept on the descending limb should occur at 3.79 μm.

Table 1: Actin periodicity, actin filament length and myosin length measured from 5 bundles of extensor digitorum longus muscle of rat.

Muscle bundle no.	Actin period nm	Actin μm	Myosin μm
1	42.83 ± 0.38	1.229 ± 0.006	1.660 ± 0.007
	39	1.119 ± 0.02	1.512 ± 0.03
4	41.77 ± 0.16	1.138 ± 0.009	1.638 ± 0.003
	39	1.098 ± 0.03	1.530 ± 0.01
7	41.60 ± 0.21	1.239 ± 0.004	1.643 ± 0.005
	39	1.162 ± 0.01	1.540 0.01
7	41.06 ± 0.15	1.240 ± 0.008	1.601 ± 0.018
	39	1.178 ± 0.01	1.521 ± 0.02
8	41.72 ± 0.125	1.176 ± 0.003	1.638 ± 0.002
	39	1.099 ± 0.01	1.531 ± 0.007
11	40.25 ± 0.13	1.157 ± 0.005	1.581 ± 0.007
	39	1.121 ± 0.01	1.531 ± 0.01

When lengths are corrected for actin periodicity of 39 nm, average actin length from these data is 1.13 μm ± 0.013 μm; average myosin length: 1.53 μm ± 0.015 μm. The bare zone of myosin was 0.17 μm ± 0.01 μm.

DISCUSSION

Although a linear descending limb was obtained from fiber bundles in which sarcomere length control was applied there was a lack of correlation between force and filament overlap. Actual force development was consistently higher than would have been predicted from filament overlap and force was maintained to both the left and right of the predicted plateau. The experimental procedure used in these experiments was basically similar to the earlier experiments of Gordon et al. (1966). Their experiments involved the use of a spotfollower device which served to control the length of a predetermined segment of a single muscle fiber. In our experiments a short length of fibers within the bundle was also controlled to the extent that the average sarcomere length was held constant. The sarcomeres subject to control were not always exactly the same, in any one contraction there was always some longitudinal translation, albeit small, which caused sarcomeres to be moved into and out of the observed region.

The sarcomere length control techniques used in these experiments would seem to be a valid method. Both average sarcomere length and force remained reasonably constant during control. This method is probably at least as good as the segment length control methods used by other workers and has certain advantages. The main one being that sarcomere length itself is the control signal rather than a fixed length of fiber within which an unknown amount of sarcomere length dispersion can occur.

The results on the force - sarcomere length relation are at variance with those of McCarter et al. (1977) on lateral omohyoideus muscle. They obtained a maximum of force development at SL = 3.2 μm in tetani at constant muscle length. They failed to find a force plateau of any significance. Passive force depended on sarcomere length in a similar manner to this study. It is possible that the difference between their and our studies resides in our use of sarcomere length control in small bundles of fibers.

The measured values for filament length in this study were corrected on the basis that the I-filament periodicity was 39 nm rather than the values of 40-43 nm obtained by direct measurement. This was based on the X-ray diffraction measurements of Huxley and Brown (1967) and confirmed by more recent observations of for example, Vibert et al. (1972) (of Wray and Holmes, 1981). Huxley (1972) showed that the 38.5 nm meridional reflection was unchanged between rest and contraction and specifically attributed the reflection to the troponin. Correction for actin periodicity had the result that the corrected filament lengths were all shorter than the measured lengths. This might be regarded as a slightly anomalous result as the usual effect of fixation and processing material for electron microscopy is to cause shrinkage. However, it is possible that within the I-band region where the measurements of periodicity were usually made, the filaments could have been stretched by shrinkage within the A-band region. Alternatively both actin and myosin could have been stretched because the bundles were cut with the knife edge parallel to the bundle causing compression of the blocks containing the bundle in the direction of motion of the knife and expansion of in longitudinal direction. Finally it should be pointed out that the value of 39 nm for the I-filament periodicity was derived from studies on frog sartorius and rabbit psoas muscle. It is possible, though perhaps unlikely that this periodicity is different in rat m extensor digitorum longus. The number of actin periods (n=27) that we observed is similar to the number that has been described before in rat (Walker and Schrodt, 1973).

That force is maintained at a maximum to the left of the region where overlap is maximal need not necessarily cause concern. In this region, although the tips of the opposing thin filaments are overlapping, this is occurring within the thick filament bare zone, therefore there is no real interference in the association between actin and myosin, assuming there is no active association in the region of the bare zone. However, the finding that plateau force extends further to the right and is maintained at a consistently higher level down the descending limb than would be predicted by filament overlap alone, deserves comment. In some ways the lack of correlation between filament overlap and force development over the descending limb is small, i.e. of the order of 5%. But Fig. 3 clearly shows that the scatter in the force data points is reasonably narrow and the vast majority are clearly above the filament overlap line. It is also possible that the value of I-filament periodicity used to correct the filament length data is inappropriate, however this seems unlikely. This leaves us with the proposition that force development in the descending limb is not wholly determined by the degree of filament overlap.

An alternative explanation could be along the lines proposed by ter Keurs et al. (1978) that the cross-bridges located immediately either side of the bare zone contribute little if any force. The apparent linear fall of force over the descending limb then could result from simultaneous length dependent activation and length dependent sensitivity of the myofilaments to calcium (Fabiato, 1978).

REFERENCES

Buller, A.J., Lewis, D.M. (1965). The rate of tension development in isometric tetanic contractions of mammalian fast and slow skeletal muscle. J. Physiol. 176: 337-354

Close, R.I. (1972). Dynamic properties of mammalian skeletal muscle. Physiol. Rev. 52: 129-197

Craig, R., Offer, G. (1976). Axial arrangement of cross bridges in thick filaments of vertebrate skeletal muscle. J. Mol. Biol. 102: 325-332

Edman, K.A.P. (1980). The role of non-uniform sarcomere behaviour during relaxation of striated muscle. Eur. Heart J. 1/suppl A: 49-57

Fabiato, A. & Fabiato, F. (1978). Myofilament-generated tension oscillations during partial calcium activation and activation dependence of the sarcomere length tension relation of skinned cardiac cells. J. Gen. Physiol. 72: 667-669

Gordon, A.M., Huxley, A.F., Julian, F.J. (1966). Tension development in highly stretched vertebrate muscle fibers. J. Physiol. 184: 143-169

van Heuningen, R., Rijnsburger, W.H., ter Keurs, H.E.D.J. (1982). Sarcomere length control in striated muscle. Am. J. Physiol. 242: H411-H420

Huxley, H.E. (1973). Structural changes in the actin- and myosin-containing filaments during contraction. Cold Spring Harbor Symp. Quant. Biol. 37: 361-377.

Huxley, H.E., Brown, W. (1967). The low angle X-ray diagram of vertebrate striated muscle and its behavior during contraction and rigor. J. Mol. Biol. 30: 383-434.

ter Keurs, H.E.D.J., Iwazumi, T., Pollack, G.H. (1978). The sarcomere length-tension relation in skeletal muscle. J. Gen. Physiol. 72: 565-592.

ter Keurs, H.E.D.J., Elzinga G. (1981). The sarcomere length-force relation of frog muscle; effects of sarcomere motion and species. VII Intern. Biophys. Congress & III Pan-Am. Biochem. Congress (in the press)

McCarter, R., Radicke, D., Yu, B.P. (1977). A model preparation for studying fast mammalian skeletal muscles. Proc. Soc. Exp. Biol. & Med. 156: 40-45 (39871)

Page, S.G., Huxley, H.E. (1963). Filament lengths in striated muscle. J. Cell. Biol. 19: 369-391

Page, S.G. (1974). Measurements of structural parameters in cardiac muscle. In: *The Physiological Basis of Starling's Law of the Heart.* Ciba Symp. 24, Elsevier Excerpta Medica Ned North-Holland, A'dam, London, New York

Rack, P.H.M., Westbury, D.R. (1965). The effects of length and stimulus rate on tension in the isometric cat soleus muscle. J. Physiol. 204: 443-460

Vibert, P.J., Hazelgrove, C.J., Lowy, J. (1972). Structural changes in actin containing filaments of muscle. J. Mol. Biol. 71: 757-767

Walker, S.M., Schrodt, G.R. (1973). Segment lengths and thin filament periods in skeletal muscle fibers of the rhesus monkey and the human. Anat. Records 178: 63-82

Wray, J.S., Holmes, K.C. (1981). X-ray diffraction studies of muscle. Ann. Rev. Physiol. 43: 553-565

DISCUSSION

NOBLE: In the case of your rat data, one could argue that the curve would fit the prediction from filament overlap if you use your measured filament lengths, and I wondered if this would also be the case for your data on frog?

TER KEURS: I think the original overlap relation that was suggested in the paper of Gordon, Huxley and Julian was based upon assumed filament lengths in the frog which were slightly larger than later verified. So there's a similar disparity between the mechanical data and the overlap data based on electron microscopic measurements in frog as well as in the rat. We found it troublesome enough to get good electron microscopy from rat extensor digitorum, but did not do these measurements for frog.

HUXLEY: I think the length that's perhaps more difficult to arrive at is the A-band length, isn't it? It's not as easy to get an internal calibration on that, but I wondered whether you could measure the 430 or 440 Å periodicity in the central region of the A-band and use that as a possible calibration for longitudinal shrinkage of A-band length. Can you see that in the pictures?

TER KEURS: We are studying the A-band region on the electron micrographs. We also study laser diffraction patterns of the A-bands to obtain data for myosin periodicity. I think it is necessary to follow up that point to exclude the possibility that, while we process the fibers myosin length changes. I should mention the fact that the fiber was studied with laser diffraction during the whole fixation procedure. We found no appreciable shrinkage except during osmification of the fiber. Then elongation of the fiber was seen. In the epon stage, two things possibly could happen. Myosin could shorten at the expense of actin; this would both tilt and shift the predicted overlap relationship slightly. Secondly, we did cut the fiber with the knife edge parallel to its longitudinal axis. This, in principle, could cause lateral compression of the sections and, therefore, longitudinal expansion of the filaments. Assuming this happened, we corrected for the 39 nm periodicity. Whether that corrected value is right I don't know, because I don't know what the actual actin repeat would be in X-ray data.

HUXLEY: Well, probably 38.5 nm might be better

TER KEURS: It's easy to correct for that.

TREGEAR: Is it known for the rat? Is the actin spacing known?

HUXLEY: It's got to be seven times the actin periodicity.

GORDON: When does the force start to fall off at short sarcomere lengths?

TER KEURS: We tried to assess this but the fibers were damaged by shortening below SL = 1.8 μm. When we allowed the muscle to shorten from slack length to below 1.8 μm, and then returned to plateau length, a force loss of 10% was a consistent finding. So I don't have an exact answer.

POLLACK: Henk, if I understood your data correctly, it appeared that the difference between the muscle length isometric and the sarcomere length isometric was not all that great, perhaps a shift of the length-tension relation of a tenth of a micron or so. Is that observation correct?

TER KEURS: For pipiens and temporaria, that was true. For rat, as you recall, the curve measured at constant muscle length went down to zero at 4.5 μm, so imposing the clamp there had more of an effect. I don't know whether that may be related to the fact that our pipiens and temporaria fibers had a very stiff parallel elastic element and the extensor digitorum had a very compliant parallel elastic element, just like the J. Gen. Physiol. data.

NOBLE: I didn't hear it, but did you eliminate the creep with your sarcomere length clamp in the case of the frog? You said it for the rat, but --

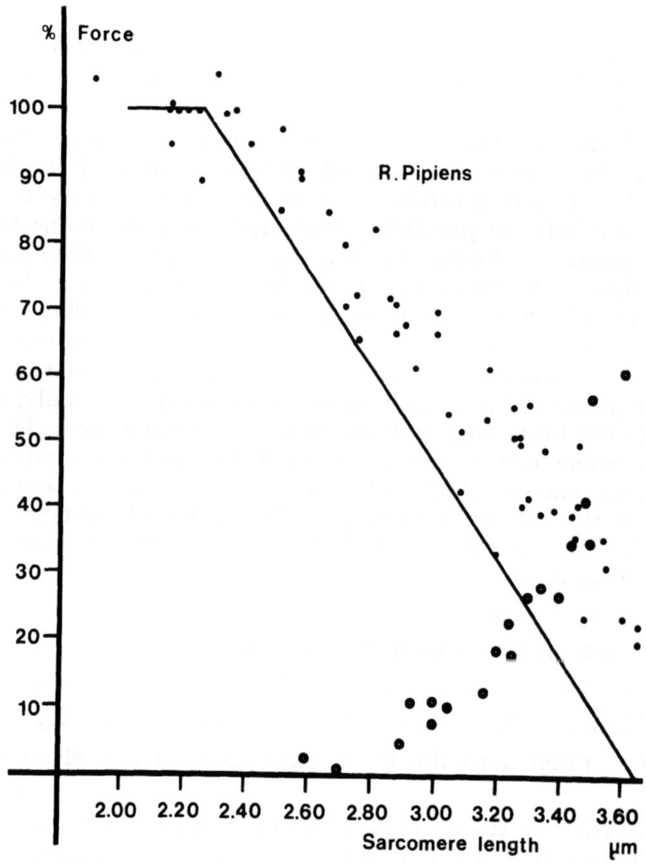

Figure D-1: Length-tension relation obtained under sarcomere length clamp using laser diffraction.

TER KEURS: Yes, what I said for the rat held true for the frog as well.

POLLACK: Could you draw the two length-tension curves where you have measured both the muscle length isometric and sarcomere length isometric?

TER KEURS: Yes. Let me talk about pipiens and temporaria. The passive force (Fig. D-1, heavy dots) started to develop in both species exponentially above 2.6 μm sarcomere length and attained 50% of maximal force at 3.50 μm. Plateau force during tetani at constant *muscle* length was maximal between 2.0 μm and 2.25 μm and decreased linearly with sarcomere length above 2.25 μm in both species. Force was zero at 3.75 μm in *R. temporaria* and at 3.95 μm in *R. pipiens*.

When sarcomere length during tetani was held constant the slow phase of force increase disappeared and segment length was observed to remain constant. Force then was maximal between 2.0 and 2.25 μm as well, but decreased linearly with sarcomere length over the descending limb to zero at 3.65 μm in *R. temporaria* and to zero at somewhat above 3.80 μm in *R. pipiens* (Fig. D-1, light dots). So, you are correct in saying that control introduced only small differences in force of these fibers compared to muscle isometric contractions.

POLLACK: So, in the frog, the difference between actually doing a sarcomere length clamp and doing a simple muscle isometric contraction gives you only a modest shift in the length tension curve, not an extraordinary shift. I think Paul (Edman) found the same. If this is also true of our fibers (Altringham and Pollack, this symposium), then our "clamped" length-tension relation will remain rather flat, unlike what you and Paul found but similar to our earlier results with semitendinosus fibers (ter Keurs, Iwazumi, and Pollack, 1978). We'll need to check that.

SOME SPECIFIC PREDICTIONS AND EXPERIMENTS ON SINGLE MYOFIBRILLAR MECHANICS

Tatsuo Iwazumi

Department of Physiology and Biophysics, University of Texas, Galveston, TX

ABSTRACT

The length-tension relationship of myofibrils approximately 50 sarcomeres long from skinned single atrial cells was measured. All sarcomeres were observable throughout the experiments. The tension developed at submaximal Ca^{2+} concentrations consistently increased with the sarcomere length.

Single myofibrils are the most desirable preparations for investigating sarcomere mechanics. However, due to their extremely small size, difficulties associated with handling of the preparation and the measurements of mechanical properties have prevented their practical use. These difficulties have recently been overcome by the development of a new instrument (Iwazumi, 1982). Briefly, the new instrument is able to measure tensions from 0.1 μg to 1 mg with a bandwith of DC to 30 KHz and to control the length with a resolution of 2 nm at a speed of 20 μs at 1 μm step length change (about 1% of myofibril length). The sarcomere length can be measured directly from the image of the myofibril projected onto a 1024 element linear photodiode array.

The preparations used in the experiments were detergent skinned atrial single cells from the bull frog. These cells were enzymatically dispersed and had usually only two myofibrils in their extremities. These cardiac myofibrils were preferred to skeletal myofibrils mechanically torn from the large bundle since the tearing caused excessive stretch of myofibrils thus resulting in the loss of sarcomere uniformity. The length of myofibril mounted between the force and length transducers was about 100 μm at slack length containing about 50 sarcomeres.

The purpose of the experiment was to test a set of predictions from a field theory of muscle contraction (Iwazumi, 1978): 1. At submaximal Ca^{2+} concentrations myofibril sarcomere length-tension curves will exhibit positive slopes (i.e., active tension increases with sarcomere length) up to the limit of the overlap; 2. At supramaximal Ca^{2+} concentrations

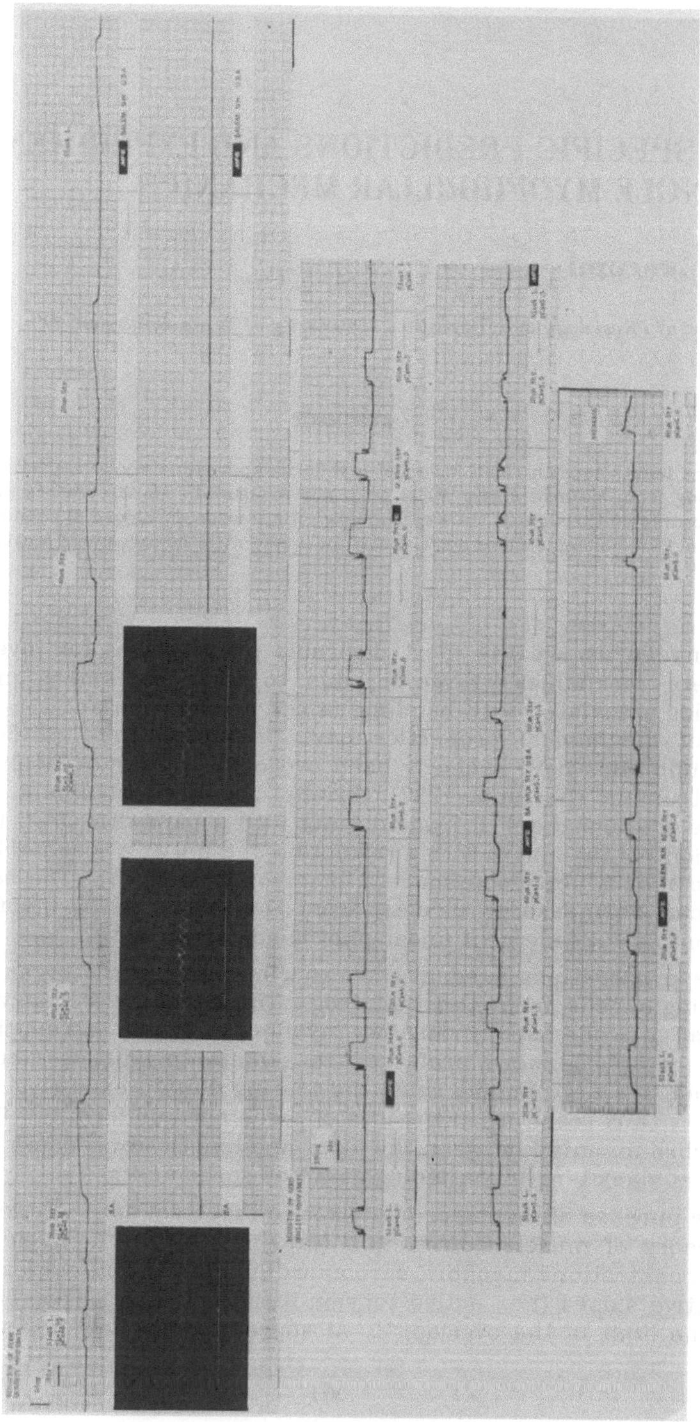

the curves will become flat, the sarcomeres will become unstable, and some will eventually fail to maintain the overlap of thick and thin filaments, resulting in a sudden drop of tension. This phenomenon will not be clearly observable in large multifibrillar preparations.

Two typical behaviors of atrial myofibrils are shown in Fig. 1. The first trace is a continuous tension record of a poor quality myofibril activated by pCa=6.0 solution (for the solution compositions, see Iwazumi & Pollack, 1981). Observe that the tension rose rather slowly and the plateau value increased with the sarcomere length. As the extent of the stretch is reduced, the tension rise became longer and the plateau tension was much smaller than that of the previous contraction at the same length. This rapid deterioration of tension is very common in the detergent skinned atrial myofibrils, and it appears to be due to a trace amount of proteinase remaining with the preparation, despite thorough rinsing before skinning. Three photographs are the image intensity of sarcomeres at respective stretches. Note that even in very short segments the sarcomere length inhomogeneity is evident. The second trace is also a continuous tension record of a good quality myofibril activated by solutions with pCa=6.0 (first row), pCa=5.5 (second row), and pCa=5.0 (third row). It is apparent that this myofibril showed some different behavior compared with the previous one. First, the tension rose much faster. The twitch-like tension spikes are due to the ringing of solution flow control servo system because of my negligence of proper adjustment; the myofibril was responding to each bolus of contracting solution. Second, the rate of tension deterioration was much lower than the previous myofibril, although it is still much greater than that found in good skinned fibers (Iwazumi and Pollack, 1981). An important common finding in these two myofibrils and also in uniform skinned fibers is that the active tension increases with sarcomere length at submaximal Ca^{2+} concentrations.

Testing the second prediction in the previous paragraph turned out to be rather difficult because atrial myofibrils deteriorate significantly during submaximal Ca^{2+} activations, and the sarcomere uniformity is also

Figure 1: *Top*: Raw tension data from a poor quality atrial myofibril that suffered a minor proteinase attack. Calibrations are 25 μg/large div. and 5 sec/ div. and 5 sec/large div.. The myofibril was activated by pCa=6.0 solution at slack length (SL=1.98 μm), 20 μm stretch (2.38 μm), 40 μm stretch (2.76 μm), 60 μm stretch (3.02 μm), 40 μm, 20 μm and slack length. Note that each time resting myofibril was stretched the resting tension increased then considerably stress relaxed. Also note that active tension rose slowly. These are typical responses of damaged myofibrils. Inset photos are sarcomere image video signals at slack, 40 μm, and 60 μm stretches. Substantial sarcomere length non-uniformity is evident even in very short segments. Second, third, and fourth traces are from a good quality atrial myofibril. Calibrations are 50 μg/large div. and 5 sec/large div.. Second trace was obtained at pCa=6.0, third pCa=5.5, and fourth pCa=5.0. Note that passive stretches caused very little stress relaxation in the second trace but noticeably more in the fourth indicating progressive deterioration of the myofibril. In good myofibrils tension development after solution switching was very rapid. Regardless of myofibril quality active tension increased with stretch up to the limit of myofilament overlap. When the limit was exceeded the myofibril suddenly failed at the moment of activation as seen at the very end of the fourth trace.

impaired. As seen at the end of the last trace, such a myofibril failed at the moment of activation; therefore, the expected phenomenon could not be demonstrated without ambiguity. An alternative procedure is to start the experiment with supramaximal Ca^{2+}, although this will destroy the myofibril after a few activations.

REFERENCES

Iwazumi, T. (1982). High performance instrument for myofibrillar mechanics. Biophysical Society Abstract 37: 357a.

Iwazumi, T. and Pollack, G.H. (1981). The effect of sarcomere non-uniformity on the sarcomere length-tension relationship of skinned fibers. J. Cell. Physiol. 106: 321-337.

Iwazumi, T. (1978). A new field theory of muscle contraction. In: *Cross-bridge mechanism in muscle contraction*. G.H. Pollack & H. Sugi, eds. Univ. of Tokyo Press, 611-632.

DISCUSSION

A comment was made about the outstanding stability of the the force transducer, and the method used to achieve it. The baseline stability depends primarily on temperature and solution flow. The flow is highly regulated by a servo mechanism combined with an unusual upside-down chamber in which water is contained only by its surface tension. This trick is possible because the chamber volume is less than 50 μl.

Another conferee suggested that is was a good idea to show continuous tension records because they clearly demonstrate that the active tension rose with sarcomere length. This was the reason raw data were shown instead of length-tension diagrams.

The question was raised if there was a correlation between the active tension and the resting tension, both of which increased with sarcomere length in these experiments. Both tension curves showed similarity only at low Ca^{2+} concentrations (pCa 5.5) but not at high concentrations.

The final question asked was how does my field theory explain increasing active tension with the sarcomere length. The theory predicts that the tension is proportional to the Ca^{2+} concentration (provided it is less than saturating value) at the tip of the thin filament, and that the distribution of Ca^{2+} concentration within the A-band is concave upward. Combining these with the structure of sliding filament, one can readily see that active tension increases with sarcomere length at constant Ca^{2+} concentration outside the sarcomeres. The reason for concave Ca^{2+} distribution within the A-band is that the thick filament has a bipolar structure, and a potential gradient is established along the filament when its center is positively charged. Ca ions see this positive gradient when they diffuse into the A-band; therefore, a steady state concentration distribution is established which is lowest at the center and equal to external concentration at both ends. Experimental evidence for the concave distribution of Ca^{2+} within the A-band can be found in the autoradiographic data by Winegrad (Fed. Proc. 24(1): 1140, 1965).

$1\ kgf/cm^2$ - THE ISOMETRIC TENSION OF MUSCLE CONTRACTION: IMPLICATIONS TO CROSS-BRIDGE AND HYDRAULIC MECHANISMS

Reuven Tirosh

Department of Cell Biology, The Weizmann Institute of Science, P.O. Box 26, Rehovot, Israel

ABSTRACT

Attention is drawn to experimental results from many laboratories which indicate that the isometric force (F) in the contraction of striated muscle fibers is linearly proportional to their variable cross-section area (A). Reversible swelling of intact, skinned, or glycinerated fibers can be induced by changes in tonicity, ionic strength or pH. In all cases where careful measurements of F and A are reported, the maximal isometric tension namely,

$T = F/A$, is found around $1\ kgf/cm^2$,

even though F and A may change more than threefold for a given fiber at a certain length. These results seem to be independent of the fiber length or temperature. Thus, the isometric tension T in striated muscle does not depend on the number or the rate of the interacting cross-bridges.

This result of constant isometric tension, which has so far received little attention, is however, a simple prediction of the hydraulic mechanism which is proposed for muscle contraction. Therefore, the hydraulic model, which is based on the hypothesis of vectorial flux of energetic protons deserves serious consideration.

INTRODUCTION

The idea that force generation during muscular contraction is a result of mechanical involvement of the contractile proteins had prevailed long before the cross-bridge mechanism was proposed. The refinement of that basic concept of force generation was possible after the more detailed macromolecular architecture was revealed by optical and electron-microscopic observations (Huxley, 1972). The biochemical identification of actin and myosin as the enzymatic system hydrolysing ATP, has led to various proposals of conformational or electrical changes during the cyclic interaction of the cross-bridges as responsible for force generation. Most important, however, have been those experiments that aimed at verifying more directly the basic assumption about the mechanical involvement of the cross-bridges as force generators during contraction. Two types of experiments were most suitable for that purpose: those checking the need for a *continuous structure* of the contractile proteins, and those checking a *linear force-length relation.* In the first type, it was demonstrated that extraction of the myosin filaments from myofibrils left a ghost incapable of contracting, whereas reincorporation of myosin enabled a contraction in the presence of MgATP. On the other hand, a similar incorporation of heavy meromyosin (HMM) - the soluble chemically active fragment of myosin - had failed to show any contraction. In the second type of experiments, quantitative observation of a linear decline of the force-length relation in correlation with the degree of cross-bridge interaction, further supported the general consensus that force generation was due to mechanical involvement of the proteins (Gordon, Huxley and Julian, 1966).

Being interested in the basic question of how chemical energy is transformed into mechanical energy and vice-versa, I have collaborated in a research project in the laboratory of Professor A. Oplatka who pursued theoretical and experimental research of the above problems in muscle contraction (Oplatka, 1972), extending the interest of the late Professor Aharon Katchalsky in reversible mechano-chemical conversion (Steinberg, Oplatka and Katchalsky, 1966). With an aim to achieve reconstitution of a mechano-chemical effect in the acto myosin system, I focused on the phenomena of cytoplasmic streaming rather than muscle contraction. After demonstration of active streaming in microcapillaries (Tirosh, Oplatka and Chet, 1973), a hydraulic mechanism for muscle contraction was proposed (Tirosh, 1978; Tirosh, Liron and Oplatka, 1979 a,b). In this model force is generated as a result of pressure gradients that are developed within the water phase due to ATP hydrolysis. The enzymatic proteins are considered to be responsible for vectorial orientation of the chemical reaction but they need not directly participate in force generation.

With this idea we returned to the two types of experiments mentioned above, but with the following predictions by the hydraulic mechanism of force generation: (i) There is no need for a continuous structure of the proteins; (ii) The maximal isometric tension depends on the cohesive forces of the water phase.

Tension With No Continuous Structure of the Proteins

The first prediction was verified by re-examination of contraction of myofibril ghosts under reinsertion of the soluble fragment HMM (Oplatka, Gadasi, Tirosh, Lamed, Muhlrad and Liron, 1974). This time, however, HMM was included in the irrigating solution of MgATP which is known to induce dissociation of the myosin heads from actin. Such inclusion is not necessary when reincorporating the filamentous myosin within the sarcomere because it cannot diffuse out on addition of MgATP.

Further verification of the first prediction was demonstrated by measuring active tension in skinned glycerinated fibers in the presence of HMM and after inactivation of the intact myosin (Borejdo and Oplatka, 1976). The force developed was smaller than that obtained with myosin insertion and in both cases the force was rather small relative to that obtained by the same fiber before inactivation of the intact myosin. However, as will be shortly clarified, the mechanical potential of the reconstituted fiber is probably underestimated by taking into account the value of the force rather than that of the actual tension developed.

1 kgf/cm^2 - The Maximal Isometric Tension

The idea of active streaming has been found to be widely useful in explaining various features of structure, function and energetics of muscle (Tirosh, 1978; Tirosh, Liron and Oplatka, 1979 a,b). Here, only one interesting aspect is briefly discussed, namely the contractile force F as the lengthwise compressive force that acts on each single sarcomere. It is therefore given by $F = A.\Delta P$, where ΔP is the pressure difference which is developed on the opposing surfaces of the I regions in each sarcomere. The development of such a pressure difference is assumed to be created along the overlap regions between actin and myosin filaments as a result of vectorial movement of energetic protons towards the center of each sarcomere. Therefore the maximum value of ΔP which can be developed is physically dependent on the cohesive forces of the water phase. If the structure of a single fiber is homogeneous, then the tension T is well defined and is given by: $T = F/A$. Comparing the two equations above we get $T = \Delta P$. Thus, the maximal *compressive* tension that can be generated is equal to the maximal pressure difference that can be developed *within* the continuous water phase. It is therefore anticipated that in isometric contraction, above a certain threshold of activity, the contractile tension may reach a maximal value, which is a measure of the cohesive forces of the fluid phase. If so, then the maximal isometric tension should be independent of length or width of the fiber, or of its ATPase activity down to a critical level.

A linear relation between maximum value of F and A is indeed found in isometric contraction of intact, skinned or glycerinated fibers, where cross section area is changed by various conditions of the bathing solution like: tonicity, ionic strength or pH (Edman and Anderson, 1968; Hellam and Podolsky, 1969; April and Brandt, 1973; Gordon, Godt, Donaldson and Harris, 1973; Godt and Maughan, 1977, Robertson, 1977; Maughan, 1981). Moreover, if the *actual* tension is calculated it is found to be around 1 kgf/cm^2 (Gordon et al., 1973, p. 554; Hellam and Podolsky, 1969, p. 815 (after dividing by 1.44; see Godt and Maughan, 1977, p. 110); Robertson, 1977, p. 27; Filo, Bohr and Rüegg, 1965; Csapo, 1973 II, p. 16; Heinl, Kuhn and Ruegg, 1974; Borejdo and Oplatka, 1976, p. 244). Through critical reading of the references, it becomes clear that in the convention of the cross-bridge mechanism one might discard this unique result, because any effect on the value of the isometric force is considered to reflect some direct effect on the mechanical potential of the individual cross-bridge. The isometric force is therefore generally reported relative to a reference value at some length; but then it is important not to refer to it as tension. This remark applies also to the "length-tension relation" which can be called length- force relation, while it should be noted that the actual isometric tension of 1 kgf/cm^2 seems to be independent of length (Tirosh, Liron and Oplatka, 1978a).

Intact muscles have generally variable cross section area; the average value, A , is therefore smaller than the maximal value of A that is considered in the hydraulic model. Thus, the calculated value of an *effective* tension T*, can become reasonably bigger than the actual tension. Thus, it can be explained how the values reported for the *effective* tension in isometric contraction of the various muscles are in the range: $1 \leq T^* \leq 12$ kgf/cm^2 (Zachar and Zacharova, 1966), while the actual tension is 1 kgf/cm^2. It is also remarkable that about the same effective tension is obtained for muscles (including smooth and striated muscles) in a domain where the specific ATPase activity varies more than two orders of magnitude. This independence of isometric tension on the rate of chemical activity in a wide range is found also for the same muscle by changing the temperature. All of these results are simply explained by the proposal that the isometric tension of 1 kgf/cm^2 is a measure of the cohesive forces of the water phase.

DISCUSSION

It is clear that the proteins are enzymatically involved in ATPase activity; however there is no convincing evidence to verify the general assumption that a useful force is either generated or transmitted by the catalytic proteins themselves. This assumption has been established for contraction of striated muscle on structural basis, but its mechanical verification by force-length measurements is a subject of debate (ter Keurs, Iwazume and Pollack, 1978; Pollack, 1982). The controversial wide domain of the related experimental results, however, was predicted by the hydraulic model of muscle contraction (Tirosh et al., 1979a). In various experimental setups it was clearly verified that soluble myosin species can participate in generation of streaming, contraction and tension (Tirosh and Oplatka, 1982). It was therfore concluded that the filamentous structure of myosin in striated muscle is presumably needed for optimal packing and operation, but is not obligatory for force generation. Moreover, a bipolar structure of the myosin filament may actually be an obstacle to long range streaming, since it can interfere with unidirectional polarization of the actin filaments. This may explain why it has been difficult to find myosin filaments in smooth muscles and in non-muscle cells under physiological conditions (Pollard and Weihing, 1974).

REFERENCES

April, E.W. and Brandt, P.W. (1973). The myofilament lattice: studies on isolated fibers. J. Gen. Phsiol. 61: 490-508.

Borejdo, J. and Oplatka, A. (1976). Tension development in skinned glycerinated rabbit psoas fiber segments irrigated with soluble myosin fragments. Biochim. Biophys. Acta. 440: 241-258.

Csapo, A.I. (1972). The uterus-model experiments and clinical trials. In: *The Structure and Function of Muscle*, vol. 2, p. 16, ed. Bourne, G.H., New York and London: Academic Press.

Edman, K.A.P., Anderson, K.E. (1968). The variation in active tension with sarcomere length in vertebrate skeletal muscle and its relation to fiber width. Experientia (Basel) 24: 134-136.

Filo, R.S., Bohr, D.R. and Ruegg, J.C. (1965). Glycerinated skeletal and smooth muscle: calcium and magnesium dependence. Science 147: 1581-1583.

Godt, R.E. and Maughan, D.W. (1977). Swelling of skinned muscle fibers of the frog. Biophys. J. 19: 103-116.

Gordon, A.M., Huxley, A.F. and Julian, F.J. (1966). The variation of isometric tension with sarcomere length in vertebrate muscle fibers. J. Physiol. 184: 170-192.

Gordon, A.M., Godt, R.E., Donaldson, S.K.B., and Harris, C.E. (1973). Tension in skinned frog muscle fibers in solutions of varying ionic strength and neutral salt composition. J. Gen. Physiol. 62: 550-574.

Heinl, P., Kuhn, H.J. and Ruegg, J.C. (1974). Tension responses to quick length changes of glycerinated skeletal muscle fibers from the frog and tortoise. J. Physiol. 237: 243-258 (see p. 254).

Hellam, D.C. and Podolsky, R.J. (1969). Force measurements in skinned muscle fibers. J. Physiol. 200: 807-819.

Huxley, H.E. (1972). Molecular basis of contraction in cross-striated muscles. In: *The Structure and Function of Muscle*. vol. 1, pp. 301-387, ed. Bourne, G.H. New York and London: Academic Press.

Krasner, B. and Maughan, D.W. (1981). Dextran T500 decreases skinned fiber width, tension and ATPase. Biophys. J. 33: 27a.

Oplatka, A. (1972). On the mechanochemistry of muscular contraction. J. Theor. Biol. 34: 379-403.

Oplatka, A., Gadasi, H., Tirosh, R., Lamed, Y., Muhlrad, A. & Liron, N. (1979). Demonstration of mechanochemical coupling in systems containing actin, ATP and non-aggregating active myosin derivatives. J. Mechanochem. Cell Motility. 2: 295-306.

Pollack, G.H. (1982). The sliding filament/cross-bridge theory: a critical review. Submitted to Physiol. Rev.

Pollard, T.D. and Weihving, R.R. (1974). CRC Crit. Rev. Biochem. 2: 1

Robertson, S.P. (1977). pH dependence of calcium-activated tension of skinned skeletal muscle fibers. Ph.D. Thesis, pp. 26-27, 56-60. University of Washington, Seattle.

Steinberg, I., Oplatka, A. and Katchalsky, A. (1966). Mechanochemical engines. Nature 210: 568-571.

ter Keurs, H.E.D.J., Iwazumi, T. and Pollack, G.H. (1978). The sarcomere length tension relation in skeletal muscle. J. Gen. Physiol. 72: 565-592.

Tirosh, R. Oplatka, A. and Chet, I. (1973). Motility in a "cell sap" of the slime mold physarum polycephalum. FEBS Letters, 34: 40-42.

Tirosh, R. (1978). Elementary Aspects of the Mechanochemical Coupling in the Actomyosin System. Ph.D. Thesis, The Weizmann Institute of Science, Rehovot, Israel (hebrew text).

Tirosh, R., Liron, N. and Oplatka, A. (1979a). A hydrodynamic mechanism for muscular contraction. In: *Cross-Bridge Mechanism in Muscle Contraction*. pp. 593-609, ed. Sugi, H. and Pollack, G.H. Tokyo: University of Tokyo Press.

Tirosh, R., Liron, N. and Oplatka, A. (1979b). A proposal for the molecular basis of cytoplasmic streaming. In: *Cell Motility: Molecules and Organization*. pp. 133-145, ed. Hatano, S., Ishikawa, H. and Sato, H. Tokyo: University of Tokyo Press.

Tirosh, R. and Oplatka, A. (1982). Active streaming against gravity in glass microcapillaries of solutions containing acto-heavy-meromyosin and native tropomyosin. J. Biochem. (Tokyo), 91: 1435-1440.

Zachar, T. and Zacharova, D. (1966). Potassium contractures in single muscle fibers of the crayfish. J. Physiol. 186: 596-618.

DISCUSSION

WILKIE: You have spoken very clearly about a hypothetical mechanism for generating this force, one kilogram force per square centimeter, but the other aspect of muscular contraction that is at least as important, which is the production of work or mechanical power, is something that of course is very different in different muscles, is different at different temperatures, and is of the essence of the usefulness of muscles in the body. Now would you like to say something about the way in which your proposal relates, not really to the force development, but to work and mechanical power?

TIROSH: Thank you very much for this question. I'd like to say that my main interest was, indeed, not in force generation but rather in energy conversion. My first steps were in considering how biochemical energy is transformed into mechanical energy. Indeed, my conclusion about force generation came out as the result of considering mechanochemical transformation. Thus, indeed, my theory for muscle contraction takes into account the chemical input and the work and heat output in various muscles. I don't want to praise myself, but ...

WILKIE: Why not?

TIROSH: Thank you. But the mathematical formulation of the hydronamic approach to muscle contraction has encountered unexplained difficulties of publication despite the very good agreement with experimental results, which relates to the force-velocity relation at different temperatures, in different muscles. I think that these theoretical results really need the attention of other people working in this field.

WILKIE: Can you attack the problem of transduction of chemical energy into mechanical power?

TIROSH: Yes!! This is primary. What I want to say is that the prediction about isometric force is a consequence, and is secondary about isometric force are secondary.

TER KEURS: You refer to smooth muscle as well. I would like to know whether the structural conditions of your model are met in smooth muscle.

TIROSH: The whole idea of my alternative approach to this problem is that you have to consider force generation not by the protein system but rather by the water phase. Thus, my theoretical approach to muscle contraction was developed after I had demonstrated active streaming in a soluble system of actomyosin and, therefore, it includes an explanation of active streaming, and of force generation in smooth muscle as well as in striated muscle. But I understand that it's not easy to consider this kind of idea. So I won't be able to present it now. I have six years of very good experience which has proven to me that despite the fact that I consider the basic ideas to be simple, it's not so easy to transfer them.

But you all know that the physiological function of the brain is dependent on the function of the heart. Therefore, in order to be able to transfer new ideas reasonably, first I'd have to generate some empathy for these ideas. Therefore, instead of bringing the elementary aspects of the model, I'd like to mention four historical demonstrations that maybe will raise your empathy for these ideas.

But before I try to build these arguments, I want you to understand my philosophical approach to this problem. As scientists, I think that it's worthwhile for us to admit that we are limited. It's difficult to reach the absolute truth. My idea of reaching closer to the truth is this: it's worthwhile for us as scientists in the research of certain problems to seek two hypotheses; not more, but not less. The criteria for these two hypotheses is that we should strive to reach two hypotheses that may be alternatives, exclusively alternatives. For example, in the research of muscle contraction there is a reasonable hypothesis that force is generated by the contractile proteins. I don't deny it. On the other hand, I'm asking myself, "is there an alternative to this?" There is. This is that the force is *not* generated by the contractile proteins.

Anyhow, I want to stress that I am considering it very important to remain faithful to the ideals of force generation by the contractile proteins like you remain faithful to a woman. It's very easy to remain faithful where you have only one woman; the real challenge is to have two women and remain faithful to both!

But what is an alternative hypothesis? My proposal is the water phase; the hydraulic mode of force generation is indeed not a mystical possibility.

NOBLE: When you refer to contractile proteins, do you mean contractile filaments?

TIROSH: Contractile filaments and/or contractile proteins.

NOBLE: All right. Then can I ask you another philosophical question. I'm a great believer in evolution, and I think a very striking feature of muscle is this orderly array of filaments. I wonder how this evolved through natural selection if it wasn't really necessary.

TIROSH: Yes. I'd like to bring a simple answer to this question. I indeed think, as I presented in the theory, that this highly structured order in muscle is really most important. But the main goal of the highly-structured system, according to my model, is related to the vectorial function of the enzymatic proteins, namely: to induce molecular orientation of MgATP hydrolysis so as to obtain vectorial fluxes of energetic protons. Beyond this role of vectorial catalysis there is no need for the proteins to be involved mechanically in force generation. Now it is well-known that the enzymatic proteins and the stereo-specific interaction of an enzyme with a substrate adapt very well to this proposal. So I don't deny what appears quite clearly in muscle structure.

But I just grant, as was presented very clearly by Hugh Huxley, himself, in a lecture he gave at the University of Washington, that there is *still* not enough evidence for a mechanical participation of the proteins in force generation. There are more experiments needed for that matter. I don't deny he importance of these experiments. But I'm trying to think in terms of an alternative concept as well.

INGELS: I'd just like to make a statement. You mentioned the evolutionary question. One of the earliest life forms we know about is the bacteria and, as most of you know, Dr. Berg at Cal Tech now and Dr. Adler have shown that the flagella of the bacteria, the earliest life form we have, rotates as a result of a proton flux.

POLLACK: Reuven, you didn't finish your point about the four historical demonstrations.

TIROSH: Yes, you already know my philosophical attitude towards working with alternative hypotheses; thus I'm in favor of bigamy in research -- looking for and treating seriously two complementary approaches. In this context I mention four historical demonstrations of force generation which have revealed a real need to consider cross-bridge as well as, for example, hydraulic mechanisms.

The first one is in relation to Archimedes. You know he was interested in the lever mechanism and even had the idea of using a lever to move the world, only give him a fixed rigid point in space. This is clearly a cross-bridge mechanism. On the other hand, you are all familiar with his well-known law of buoyant force, which has to do with hydrodynamic aspects. Now you should realize that it was hard for him to reach this law because, as is well documented, this distinguished and enthusiastic scientist found himself running out into the streets naked, shouting with excitement of what he had just realized while deep in his bath.

The second historical confrontation of the cross-bridge mechanism has to do with the Magdeburg demonstration. All people acknowledge the force of horses and chains and so on, but it was unbelievable that there was also a force or pressure of the atmosphere. So the way it was demonstrated was by pairing two matched halves of a rigid hollow sphere, connecting both sides to strong horses by chains, and then evacuating it. In that way, by watching the powerful stretch of the horses, people realized one should consider cross-bridge mechanism vs. hydraulic mechanism!

Another demonstration is that of Newton lying under an apple tree, so we are told, and wondering about an apple that fell forcefully on his head. He wondered, "What is the source of this force?" I would like to say that, would he be bad tempered, he could get very good evidence for cross-bridge mechanism for this fall and thereby ignore his interpretation about gravitation. Suppose he thought that the tree had *thrown* the apple on him ..... The whole point is that he could get evidence for this suspicion just by hitting the tree and getting a flux of apples on his head. This would be a striking demonstration of a crossbridge-like throwing mechanism.

The last historic anecdote has to do with the modern industrial revolution which, as we know, was brought about by the locomotive. If you watch the movement of the wheels and tie rods of an old locomotive, without knowing anything about the steam engine, you could envision the cross-bridge mechanism very nicely. The tie rods obviously propel the locomotive in a crossbridge-like manner. Then the question would arise, "Why do they call it a steam engine?" In a short while you could see and hear the answer, realizing that steam is needed to blow the whistle!

GENERAL DISCUSSION

The Length-Tension Relation

NOBLE: Perhaps we should focus on the descending limb since there has already been adequate discussion of the ascending limb after Stuart Taylor's presentation. Clearly, the cross-bridge theory demands a certain kind of length-force curve. This needs to be followed under *all* circumstances. We are interested in any kind of variability such as that discussed in previous presentations.

I would like to start by putting out the suggestion that the clamped experiments that we've seen appear to have dispensed with the problem of creep. Does anybody propose to argue with that?... If not, should we have some discussion about reported deviations away from the predictions of the descending limb.

POLLACK: The experiments that Henk ter Keurs and Paul Edman reported show good agreement with the classical prediction. On the other hand, it is very interesting that in both sets of experiments with frogs, the difference between the segment length isometric, and muscle length isometric length-tension curves was relatively modest. So, in their muscle length isometric cases, there is also little deviation from the classical prediction.

On the other hand, there are other results -- for example our Seattle, results (ter Keurs et al., 1978) and the ones presented here by John Altringham -- in which the muscle length isometric curve deviates *markedly* from the classical prediction. So I am just wondering whether there may be some intrinsic differences in fibers that give rise to the large difference in the muscle length isometric contractions in two cases.

The only difference I could see, as Henk pointed out, is the difference in resting tension. In the Seattle experiments, both ter Keurs et al., and the ones presented here by Altringham, the resting tension was very low up to beyond 3 μm sarcomere length, but in the new experiments reported by ter Keurs and I believe also by Edman, there was considerable resting tension. This raises the question of whether the differences of resting tension can somehow be involved in the differences of length-tension relations. This relates to the issue of whether the resting tension is really all "resting."

If you assume the resting tension is some sort of parallel elastic element that should be subtracted from total tension, then clearly the analysis was done properly, i.e., you should compute the active component of tension by subtracting the resting tension from the total. On the other hand, if the resting tension is in part due to some active processes, then it is not clear that you should be subtracting all of it from the total tension. Should you be subtracting part of it, all of it, or none of it? That's the question I raise. Should we be considering the sum of so-called active tension and part of the resting tension, which would give you a length-tension curve that has somewhat of a tendency to be a bit flatter, or not?

NOBLE: Well, Henk ter Keurs was involved in both experiments in Seattle and in Leiden where there was a tremendous difference in the muscle isometric length-tension curves obtained. Would you like to comment?

TER KEURS: Yes, first of all I don't have an explanation for the differences in passive tension, and I think it is very worthwhile to study passive tension next to active tension development. Because of the problem Jerry raised, we did our measurements mostly in the range of sarcomere length in the frog between 2.2 and 3.2 μm, and if we were to add passive tension to active tension, that would cause only changes of the results between 2.85 and 3.2 μm of any significance. That would mean that we would obtain a descending limb in the experiments which I have described, on *Rana temporaria*, which would go down to 2.85 μm and then would go up. But I think such a discontinuity is not very likely to expect as a property of the force generators. Experimentally, we took care to stay in a range which might be considered safe.

GORDON: A comment on how you treat passive tension. I think the best experiments are the ones that Richard Podolsky did in skinning part of the fiber and looking at the sarcomere length in the skinned versus the unskinned as you stretch the fiber. Up to the sarcomere length of about 3.2 μm he got equivalent stretches in sarcomere length in both skinned and unskinned, but beyond 3.2 the unskinned region was very stiff and the skinned region then stretched a lot. 3.2 μm was the length at which passive tension increased dramatically in the intact fiber, which seemed to imply that the major portion of the passive tension was borne by the elements you remove when you mechanically skin the fiber. I think that's a reason why you might subtract it if you're looking at what's happened inside.

POLLACK: I think that's reasonable. I'm just wondering then what another possibility is for the differences in the two sets of results if it's not passive tension.

GORDON: Another suggestion for the difference is that we found that you could enhance the creep a lot by the stimulation conditions. If we stimulated the fibers strongly, with a substantial amount of overlap, we could make the muscle-isometric case creep a lot more than occurred originally in the fiber.

SCHOENBERG: I'd like to go back a moment to the question of resting tension. Our observation was that in preparations with a lot of resting tension, there seems to be not much difference when you went to laser clamp and muscle clamp. We haven't studied this problem in great detail, but it's been my impression at least that preparations which have a large amount of resting tension have better stability of the striation pattern, so one possible reconciliation of these results is that when you do have a large amount of *passive* resting tension, it keeps the sarcomeres in better registration and you get closer to the answer than you would get if you indeed clamp the sarcomeres in another way.

WINEGRAD: Years ago Joy Frank and I looked at the conformation of the sarcoplasmic reticulum as a function of sarcomere length. When the fiber was stretched to approximately 3.2 μm sarcomere length, all of the extension of the SR took place in the tubular portion of the reticulum; the terminal cisternae maintained their same dimensions. But when the sarcomeres began to stretch beyond 3.2 μm the terminal cisternae became distorted and the calcium distribution within the terminal cisternae was distorted as well. So at long sarcomere lengths I think you have, in addition to effects of the reticulum as an elasticity, the possibility of altered EC coupling.

SCHOENBERG: Saul, the experiments of which I am speaking were done on skinned fibers, so they were fully calcium activated and it's been my impression with skinned fibers that those which have the largest amount of resting tension preserve the striation pattern the best. I don't know what you make of that, but I think it does offer an explanation.

WINEGRAD: I was referring more to Al Gordon's description of the Podolsky experiment, where only part of it was skinned.

TIROSH: I'd like to make a practical comment in respect to measurements of the force-length relation. It has to do with the fact that I was involved in dissection of fibers. I want to say that I was really lucky because after some several months of devoted work, I was found to be incompatible with this task. We very often found that the fibers I dissected showed very uneven striations. I would consider such preparations as improper for taking measurements. On the other hand, I realize that my more skillful colleagues could dissect fibers that fulfill very important criteria of minimal structural damage, which could be checked by scanning the fiber with the laser to check uniformity. So I suggest these criteria be adopted by all of those doing length-tension experiments.

NOBLE: Can I ask one of your colleagues who has superior manual dexterity to respond to your point. ter Keurs?

TER KEURS: Well it's a standard protocol to measure sarcomere length distribution, both in single fibers and fiber bundles. Fibers that have nonuniformities larger than $\pm$ 0.1 μm away from average sarcomere length along the fiber are rejected in our lab.

TIROSH: But not always in other labs.

NOBLE: Personally, I would have thought that the presentations we had this afternoon fulfilled those criteria.

IWAZUMI: I was the person responsible for most of the fiber dissections in our length-tension experiments (ter Keurs et al., 1978). I dissected literally thousands of fibers and selected very few. Most fibers met the criterion of sarcomere length variation within ±0.05 μm tendon to tendon. I think this was more restrictive than in the other studies.

WILKIE: Just before the session, I was having a discussion with another scoundrel whose name I won't disclose, and the question arose as to whether single isolated fibers really were particularly suitable objects for making high-grade mechanical measurements. When they are assembled in an anatomical muscle with their connective tissue, the connective tissue isn't there at random -- nothing is. The microanatomy of bone provides the most obvious example because the trabeculae of the bone exactly mimic the lines of tension and compression to which the bone is subject. One really is driven to wonder whether the arrangement of collagen and of other connective tissue in the anatomical muscle isn't such that when you remove it, you get a very distorted impression of what the muscle fibers really were doing before you had removed their connective tissue arrangements. They are there for a purpose..

NOBLE: I'm glad you said that, because in my field I have to defend myself for studying whole hearts, and on the other hand we have people like Fabiato down to single cells, and also a poster outside by Iwazumi about single myofibrils, so the opposite of your approach would be to go to an even smaller component.

WILKIE: I would say that of course you should use isolated single fibers for some purposes. Think of aequorin experiments, for example. But for making mechanical measurements, with a few clear exceptions, they may not be the ideal objects which *most people* now seem to assume they are. This is especially important where the interpretation depends critically on small differences in the experimental results which is often the case. You're hardly considered to be respectable unless you are working on isolated single fibers. We two scoundrels were discussing the validity of this point of view. I certainly have doubts about it!

NOBLE: I'm quite happy to join the scoundrels!

BRESSLER: First, I would like to invite Dr. Wilkie to join our MRC Committee when I get my grant reviewed next year. But I might just comment that of course in the early '70's we were doing whole muscle work exclusively, and our findings paralleled the single muscle fiber work in a qualitative way. We saw then that stiffness vs. tension was length-dependent on both the ascending limb and the descending limb; and the Huxley-Simmons group is finding the same thing. So in a qualitative way, you are absolutely right. But the problem is when you want to start to quantify the information, to consider subtle differences, then you're dealing with potential artifacts of connective tissue and tendons. So then you're forced into the single fiber stuff. I've spent many years tooling up for that, and I'm just about there .... and now you say that it's all unnecessary!!

EDMAN: I would like to defend single fibers. They are certainly quite suitable for mechanical measurements, if you dissect them properly, and particularly if you use length clamps and things like that. They have the obvious advantage that you can be sure that you have a physiological internal environment, e.g. ionic strength, which I think is rather essential for discussion of mechanics.

NOBLE: But you can argue that it is unknown and uncontrolled, and I guess that is the reason some people are going even to a single myofibril.

IWAZUMI: I decided to use single myofibrils, actually, to avoid striation inhomogeneity. I found the nonuniformity to be a crucial problem not only in single fibers but in skinned fibers as well. I finally constructed the "dream" machine to do single myofibril mechanics, and to my surprise, even down to single myofibrils 100 μm long, I *still* see nonuniformity in poor quality specimens. That means I haven't eliminated the selection problem. I still have to do lots and lots of selection to obtain uniform myofibrils. However, the single myofibril lets you observe *all* sarcomeres, which is a definite advantage over the large bundle myofibril preparation. So what I really urge is that the protocol has to include quantitative data about the uniformity of sarcomeres. Otherwise, we can't compare the results.

INGELS: I want to make a quick statement backing that concept. When we do spot follower experiments, we're getting rid of the problems at the ends, but we're introducing new "ends." That has always bothered me. It seems like a tacit assumption that perhaps now we've solved the problem, but we always have these two ends and everything in between.

EDMAN: Yes, but as I demonstrated yesterday (Fig. 4, Edman & Reggiani, this volume) one can get creep-free tension by recording from very short segments of a single muscle fiber. So it is possible to get rid of the "problem at the ends" about which you are concerned, at least on single fiber level.

NOBLE: Dr. Tirosh, would your model predict a length-tension curve similar to the Gordon, Huxley, Julian one (J. Physiol. 184: 170-192, 1966), or something different from that?

TIROSH: Thank you for this question. In fact, a theoretical treatment of force-length relation (Tirosh, Lizon & Oplatka 1978, in *Cross Bridge Mechanism in Muscle Contraction*, eds. Sugi & Pollack, pp 593-609), in the hydrodynamic model was presented at the previous congress. What comes out I had expected before I knew the experimental results that the group from Pollack's lab reported for the first time in an abstract at the Biophysical Meeting in 1977 (Biophys. J. 17: 199a, 1977). I expected their results of constant force-length relation as well as the Gordon, Huxley and Julian results. That is, the model can explain the wide domain of experimental results that are being obtained. For everyone who is interested, I would be very happy to explain this wide domain of experimental results and what I believe is needed for the achievement of either the declining relation of Gordon, Huxley & Julian, or the wider constant relation of Pollack's group.

NOBLE: Maybe we should finish by suggesting that any of us who haven't got that book should get a copy!

ACTIVATION OF THE MYOFILAMENTS

INTRODUCTION

Calcium activates muscle contraction in all muscles investigated (Ebashi and Endo, Prog. Biophys. Molec. Biol. 18: 123-183, 1968). The central questions for each muscle are how myoplasmic calcium concentration is regulated and how elevated intracellular calcium brings about cross-bridge interaction and muscle contraction. In most muscles, intracellular calcium is regulated by its release from and active transport into the sarcoplasmic reticulum (SR), the smooth endoplasmic reticulum of muscle. In addition, cardiac, smooth, and some skeletal muscles have voltage-dependent calcium channels in the cell membraane allowing calcium fluxes from the extracellular fluid during activation, but in most cases it is probably the calcium released from the intracellular stores that brings about contraction (Fabiato and Fabiato, Ann. Rev. Physiol., 41: 473-484, 1979).

The link between muscle electrical activation and calcium release from this intracellular store is uncertain with at least two hypothesized mechanisms (Endo, Physiol. Rev. 57: 71-108, 1977). The first is that the depolarization of the transverse tubular or surface membrane is transmitted in some manner to the sarcoplasmic reticulum so that calcium channels in the SR membranes are opened, allowing stored calcium to flow into the sarcoplasm. This was initially hypothesized to be due to a depolarization-induced opening of these channels (Endo, 1977) although this is questioned because electron probe microanalysis of fixed tissues shows no significant concentration differences across the sarcoplasmic reticulum membrane for ions (K and Cl) to which cell membranes are normally permeable (Somlyo et al., J. Cell Biol. 90: 577-594, 1981).

The second hypothesized mechanism for calcium release is that elevation of intracellular calcium concentration causes an increased calcium permeability of the SR membrane, a calcium-induced calcium release (Endo et al., Nature (Lond.) 228: 34-36, 1970; Ford and Podolsky, Science 167: 58-59, 1970). Which mechanism is used physiologically has not been established in all cases. Endo (1977) argues against calcium-induced calcium release being the predominant mechanism in frog skeletal muscle. However, Fabiato (Fed. Proc. 41: 2238-2244, 1982) argues that in mammalian cardiac muscle, calcium-induced calcium release may be the primary mechanism allowing calcium entering through cell membrane calcium channels to induce calcium release from the stored sites in the sarcoplasmic reticulum and that even in skeletal muscle the SR is very sensitive to changes in intracellular calcium.

The relative importance of these two mechanisms can be studied using contractures induced by caffeine and inhibited by procaine, since both are hypothesized to act through the calcium-induced calcium releasing system (Endo, 1977). In the paper by Kurihara, Konishi and Sakai, this technique is used to investigate calcium release during contractures produced by rapid cooling of the muscle in the presence of caffeine.

The uptake of calcium by the sarcoplasmic reticulum has been extensively studied and the ATP requiring calcium transport protein identified and investigated (Tada et al., Physiol. Rev. 58: 1-79, 1978). This active transport can lower the myoplasmic calcium concentration sufficiently to bring about muscle relaxation. In skeletal muscle, there appears to be a spacial separation between the predominant site of calcium release from the terminal cisternae or sacs (which are both located adjacent to the transverse tubules and contain most of the SR volume) and the sites of calcium uptake, the longitudinal tubules (which contain the majority of the SR membrane area). Calcium localization studies by Winegrad (J. Gen. Physiol. 51: 65-84, 1968) and Somlyo et al. (1981) have demonstrated this apparent difference between the sites of release and uptake and also the slow return of calcium from the myofibrils to the site from which it is released. An interesting question has been whether Ca is shuttled as a free ion between the myofilaments and the SR pump sites or as abound complex with a myoplasmic protein. The paper by Gillis, Thomason, LeFevre, and Kretsinger discusses a calculation of Ca binding to myofilament proteins and uptake by the SR to test the possibility that the myoplasmic Ca-binding protein parvalbumin acts as this calcium shuttle. Because of the high calcium affinity of parvalbumin, it must be considered in the analysis of calcium movements in muscle.

Calcium activates muscle contraction through different mechanisms in different muscles (Adelstein and Eisenberg, Ann. Rev. Biochem. 49: 921-956, 1980). Calcium activation of cross-bridge interaction was first demonstrated by Ebashi and coworkers to occur through calcium binding to the myofibrillar protein troponin on the thin filament in rabbit skeletal muscle (Ebashi et al., J. Biochem. 64: 465-477, 1968). Next, it was shown in skeletal muscle from a scallop that the site of calcium control was on the thick filament (Lehman and Szent-Gyorgi, J. Gen. Physiol. 66: 1-30, 1975). Finally, in smooth muscle there have been two hypothesized mechanisms of calcium activation, one a calcium-calmodulin regulated phosphorylation of a myosin light chain (Adelstein and Eisenberg, 1980) and the other a calcium regulation of the thin filament through the protein leiotonin (Ebashi et al., 1982).

The detailed molecular mechanism by which calcium controls contraction has not been worked out for any of these systems. The thin filament regulation through calcium binding to troponin has been the most extensively studied. The steric hinderance model (of H.E. Huxley, J. Haselgrove, D.A.D. Parry, and J. Squire - see J. Squire, *The Structural Basis of Muscular Contraction*, New York: Plenum Press, 1981) through which Ca allows cross-bridge interaction by binding to troponin allowing tropomyosin to move on the thin filament away from its site blocking actin-myosin interaction is attractive but doesn't explain all the data. Although it has been hypothesized in vertebrate skeletal muscle that among the various calcium binding sites (Robertson et al., Biophys. J. 34: 559-569, 1981) the one responsible for regulation of contraction is the calcium specific site on troponin, other factors such as MgATP and Mg affect calcium sensitivity in "skinned" muscle fibers but do not directly compete with calcium for binding to this site. On the basis of biochemical measurements of calcium binding to myofibrils in the presence or absence of ATP, Bremel and Weber (Nature 238: 97-101, 1972) hypothesize that myosin interaction with thin filaments increased calcium binding to troponin so that cross-bridge interaction might affect calcium binding per se. If this were true, in the thin filament regulated system decreases in MgATP should be expected to increase calcium binding to thin filaments because of a decreased dissociation of actin and myosin. This could show up as a changed muscle calcium sensitivity.

The paper by Godt and Morgan shows that even in a scallop muscle where the regulation is presumably through the thick filament, MgATP shifts the calcium sensitivity, which would indicate that cross-bridge interaction per se may change calcium sensitivity of the myofilaments, whether calcium was bound to the thick or thin filament. The paper by Gordon, Ridgway and Martyn addresses this question and shows that in three different kinds of experiments in barnacle muscle fibers (both intact and "skinned") cross-bridge interaction and force affects Ca sensitivity such that increased force produces increased calcium sensitivity. Thus, there is direct physiological evidence that in addition to calcium modifying cross-bridge interaction, cross-bridge interaction can modify calcium sensitivity.

-A.M. Gordon

CALCIUM SENSITIVITY IS MODIFIED BY CONTRACTION

Albert M. Gordon, Ellis B. Ridgway*, and Donald A. Martyn+

Department of Physiology & Biophysics, University of Washington, Seattle, WA 98195
**Department of Physiology, Medical College of Virginia*
+Center for Bioengineering, University of Washington.

ABSTRACT

This paper summarizes three lines of experimental evidence showing that crossbridge interaction affects calcium sensitivity and probably also affects calcium binding. Evidence is presented that this is a true hysteresis, not just a slow approach to equilibrium.

1) In barnacle single muscle fibers injected with aequorin to monitor intracellular Ca, a long duration stimulus under voltage clamp conditions can produce a long duration calcium transient and force record which both approach steady levels. However, if the stimulus is briefly elevated to transiently produce a higher force early in the contraction, the same steady state Ca level can eventually maintain a higher steady force. Thus Ca sensitivity is modified.

2) In "skinned" barnacle muscle activated by Ca in the presence of buffered Ca, MgATP, and pH, force was measured in split, detergent treated fiber while it was transferred consecutively between solutions which were relaxing, submaximal contracting, maximal contracting, the same submaximal contracting, and finally relaxing. The submaximal contracting solution produced more force when stepping down in Ca concentration (as in relaxation) than when stepping up. Extending the time in the initial submaximal contracting solution did not result in more force. Thus, the force-pCa relationship shows marked hysteresis. The same phenomenon was seen in frog and mammalian muscle fibers. These experiments confirm the findings that contraction modifies Ca sensitivity.

3) In the barnacle single muscle fiber preparation (under both voltage clamp and controlled length conditions), phasic (400 msec) depolarization leads to a calcium transient and a twitch contraction. Releasing the muscle to allow it to shorten rapidly during the declining phase of the calcium transient causes the force to fall and leads to extra Ca in the sarcoplasm. Rapidly stretching the muscle produces the opposite effect. The extra Ca probably comes from a myofilament Ca activating site. Thus, a length change (force change) affects Ca binding.

It has been well established that calcium activates muscle contraction (Ebashi & Endo, 1968). This occurs through binding of calcium to myofibrillar or sarcoplasmic proteins which initiates a sequence of events leading to the interaction of actin with myosin and eventually to contraction. The question we address is, does muscle contraction affect calcium binding to the activating sites? The existence of an effect of contraction on calcium sensitivity has been suggested on theoretical grounds (Adelstein & Eisenberg, 1980; Taylor, 1979) and there is evidence for this in myofibrils in the absence of ATP (Bremel & Weber, 1972; Fuchs, 1977). However, it has not been demonstrated either in intact muscle or in skinned fibers under physiological conditions. In a recent paper (Ridgway, Gordon, & Martyn, 1983) we presented three lines of evidence leading to the conclusion that there is hysteresis in the relation of muscle force to intracellular free calcium such that contraction increases the calcium sensitivity probably through an increase in calcium affinity. From these studies arose the question of whether there is a true hysteresis in the calcium-force relationship or just a slow approach to equilibrium. In this paper, we will summarize the evidence previously presented along with additional evidence to show that there is a true hysteresis in the calcium-force relationship.

In intact barnacle single muscle fibers, cannulated and microinjected with the bioluminescent calcium indicator, aequorin (Ridgway & Ashley, 1967), it is possible to produce long lasting calcium and force responses (Figure 1A). A steady depolarization lasting 10 sec produces a calcium response in the fiber which is initially high and then reaches a relatively low but steady value. Near the end of the stimulation, the force and light both achieve a quasi steady-state relationship. When the stimulating conditions are slightly changed (Figure 1B) so that a brief initial higher depolarization is superimposed on the long steady depolarization, both the light (Ca) and force increase transiently before arriving again at steady values which are approximately the same for the light as in Figure 1A, but elevated for the force. Thus, near the end of this pulse, the same "steady" Ca level gives a higher force in 1B than in 1A. Hence, the final force depends not only on the final calcium, but also on the history of the fiber. To investigate this further, we manually adjusted the stimulating voltage to produce three varieties of force waveforms. The first was as indicated in Figure 1B, where the force overshoots the final steady level. The other two are illustrated in Figures 1C and 1D, a fast increase (1C) or a slow increase (1D) to a final steady level with the calcium response producing these force changes. The final steady Ca level is the same in Figures 1C and 1D, but the final force level is substantially reduced in 1D. When this was done for a number of different steady Ca levels achieved for the three stimulating conditions, the force at a given Ca level was always higher if the fiber had previously experienced high Ca and high force. In other words, the muscle requires less Ca to hold a particular force level if the muscle is relaxing from a higher force level than it requires to produce that force if the muscle is contracting up to that force level. That is, there is hysteresis in the steady state relationship between free calcium and muscle force.

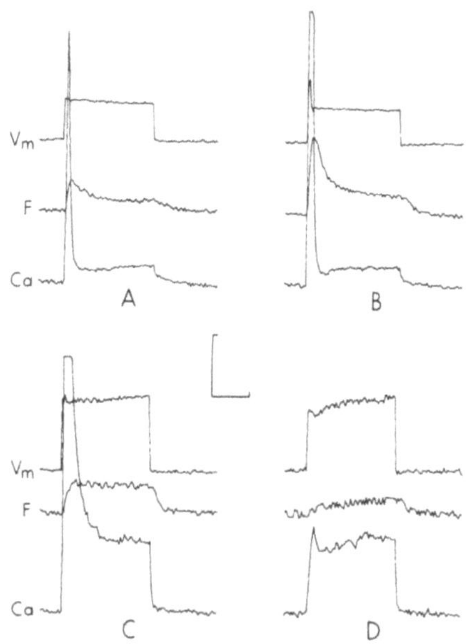

Figure 1: Quasi-steady state hysteresis in the force-Ca relationship. Panel A: long stimulus pulse at constant amplitude. Panel B: long stimulus with brief initial depolarization. Trace V_m: membrane potential at 40 mV/cal. Trace F: isometric force at 46 mN/cal. Trace Ca: aequorin light signal (Ca transient) at 1.0 μamp/cal. Horizontal sweep: 4 sec/cal. Temperature: 7°C. Fiber length: 23 mm. Fiber weight: 46 mg. Panel C: long stimulus adjusted to produce step force response. Panel D: long stimulus adjusted to produce slowly rising force response. Trace V_m: membrane potential at 20 mV/cal. Trace F: isometric force at 46 mN/cal. Trace Ca: aequorin light signal (Ca transient) at 2.0 μamp/cal. Horizontal sweep: 4 sec/cal. Temperature: 7.5°C. Fiber length: 24 mm. Fiber weight: 78 mg.

To extend, quantify, and confirm this result, we turned to the skinned barnacle fiber preparation. We skinned barnacle single muscle fibers mechanically by splitting them and then soaked them for up to an hour in a relaxing solution containing 1% Triton X-100 to disrupt the surface and sarcoplasmic reticulum membranes. It is possible to activate and relax such a fiber while measuring force under controlled ionic conditions set by the constituents in the bathing solution. The force transducer and method of changing solutions was similar to that used by Hellam & Podolsky (1969). The solution was designed to buffer against changes in ATP [using a creatine phosphate (20 mM)-creatine phosphokinase (10 U/ml) ATP regenerating system], pH (using 60 mM pH buffer, TES), and Ca (using 30 mM EGTA as a Ca buffer). The other constituents were (in mM): K propionate 170, Na propionate 30, Mg^{2+} 1.5, and MgATP 4.5, Solution mixing for these experiments involved the technique of Ashley and Moisescu (1977). To activate the fiber, we transferred it from a relaxing solution into a weakly buffered relaxing solution (buffered with HDTA), then to a series of activating solutions of different Ca concentrations and finally back to a relaxing solution. HDTA treatment was used to

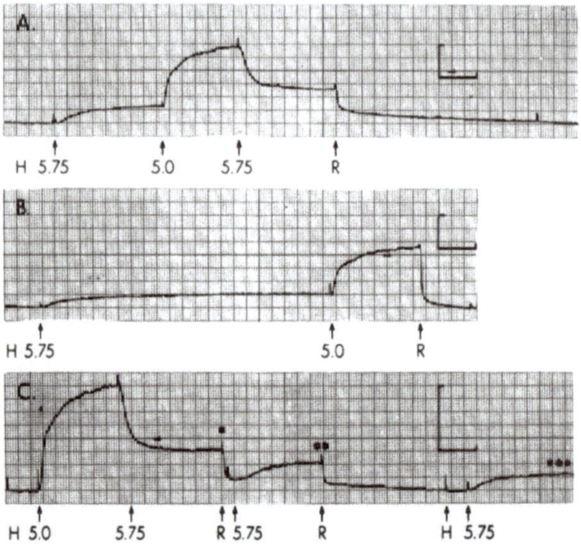

Figure 2: Hysteresis in the force-pCa relationship in skinned muscle fibers. Panel A: at the arrows the fiber is transferred consecutively to a solution with EGTA as the calcium buffer (R)(not shown), a relaxing solution with HDTA as the Ca buffer (H) shown on the left, solutions with pCa as indicated (5.75, 5.0, and 5.75 again with EGTA as the Ca buffer), a relaxing solution with EGTA (R), and finally to HDTA solution (H). Panel B: the same fiber as in Panel A, but transferred consecutively from solutions with HDTA as the Ca buffer, to a solution with EGTA and a pCa of 5.75, 5.0, and then back to a relaxing solution (R) as indicated. The fiber is maintained in pCa 5.75 for a time equivalent to the complete contraction cycle shown in Panel A. Panel C: the same fiber later in the experiment transferred from the relaxing solution (H) to pCa 5.0 to 5.75 and then briefly to the relaxing solution (R) at (*). The fiber is then returned to the pCa 5.75 solution before the force has diminished to the base line. After a steady force is again achieved (at **), the fiber is relaxed in R. It is then transferred from relaxing solution (H) to the pCa 5.75 and a steady force achieved at ***. Horizontal and vertical bars represent 1 min and 0.1 mN respectively. The fiber diameter was 100 μm. Temperature: 22°C.

produce a faster response. The same increased speed could be achieved with an EGTA solution with a pCa of 7, slightly subthreshold. Presumably, these solutions load some high affinity calcium binding sites which are not involved directly in contraction. The sequence of solution changes and resulting forces are illustrated in Figure 2. In each solution, we waited long enough for the force to reach a "steady level." As can be seen, the fiber produces a substantially higher force with the same submaximal Ca activating solution when the Ca concentration was decreased to that level (as in relaxation), compared to an increase to that level (as at the beginning of contraction).

It is important to determine whether the higher force when stepping down in Ca (as in Figure 2A) was due only to the total time in contracting solution rather than to hysteresis. Stated another way, would the force have reached the same level if we had left the fiber in the initial submaximal Ca activating solution long enough? The data shown in Figure 2B show that there is long term stability of the force when stepping up to a submaximal Ca concentration. Similar long term stability of the force occurs when stepping down in Ca and force, as the force will hold at the elevated level for many minutes. It is difficult to experimentally determine how long the force will stay at this elevated level for each fiber because very long, high level contractions are accompanied by declining maximum force and an increase in force baseline in many fibers. However, fibers with relatively stable baselines have shown elevated forces for over five minutes.

The effects of high force and Ca are surprisingly long lasting, persisting even after the fiber is partially relaxed. This point is illustrated in Figure 2C. The fiber was first exposed to a pCa = 5.0 solution in which it produces maximum force and then to a pCa = 5.75 solution in which the contraction is submaximal but elevated over that produced by the same solution when initiated in a fully relaxed fiber (compare 2C * with ***)(hysteresis). At * the fiber was exposed briefly to a relaxing solution just long enough to cause partial relaxation before being re-exposed to the pCa = 5.75 solution. The steady force in this pCa = 5.75 solution at ** is intermediate between that obtained for the same solution when contraction is initiated in a fully relaxed fiber (2B or 2C at ***) and that when contraction is initiated in a fully contracted fiber (2C at *). This particular fiber displayed no change in resting, baseline force.

When different submaximal Ca activating solutions are used in the sequence of contractions shown in Figure 2A, one generates a full pCa-force relationship under the two conditions of increasing Ca from a relaxing solution and decreasing Ca from the maximal activating solution. As shown in Figure 3, these curves are quite different. Presumably the fiber contracts along one curve and relaxes along the other. The curves were obtained for a fiber in solutions with a free Mg^{2+} of 5.5 mM instead of 1.5 mM concentration used in the case of the fiber shown in Figure 2. The muscle is less sensitive to Ca when the Ca is being increased (contraction) than when it is being decreased (relaxation). The lines through these points are fits of the Hill equation [fraction $F = Ca^n / (Ca^n + K^n)$] to the data. The position of the curves (K) shifts by 0.06 pCa units and the slope of the curve (n) changes from 5.3 to 3.8 in going from increasing Ca to decreasing Ca. The comparable data for a free Mg^{2+} of 1.5 mM are 0.13 pCa units and 3.3 to 2.6. Thus, there is an increased sensitivity and possibly a decreased slope in the force-pCa relationship produced by the intervening contraction. Similar changes in Ca sensitivity have been observed in skinned muscle fibers from both the frog and the rat. Thus, this is not unique to barnacle fibers.

The conclusion from these two studies on barnacle muscle is that there is hysteresis in the force-pCa relationship with muscle contraction producing increased Ca sensitivity. The increased Ca sensitivity might be due to an increased Ca binding, increased effectiveness of the calcium already bound, or both. Additional experiments involving length changes on aequorin injected fibers from the giant barnacle, described in preliminary form (Gordon & Ridgway, 1978), bear indirectly on this issue.

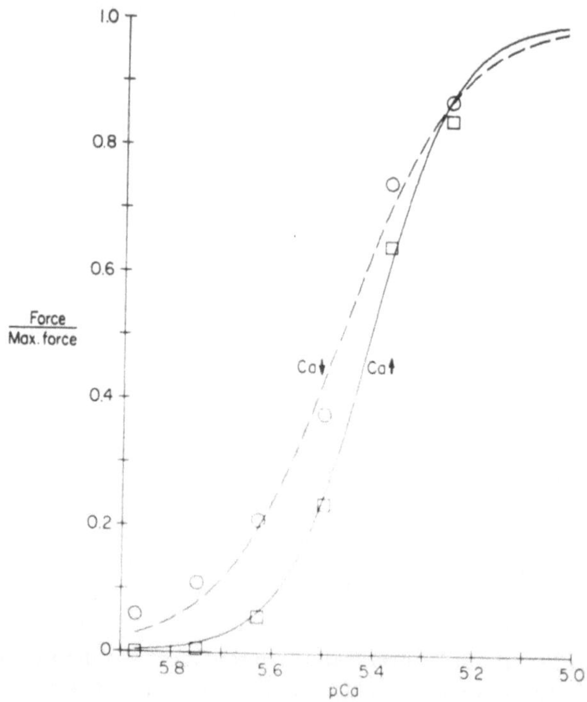

Figure 3: Force-pCa relationship. Force as a fraction of maximum as plotted against pCa for two conditions, stepping up (increasing Ca) and stepping down (decreasing Ca) in concentration. Curves are fitted to the Hill equation, fraction force equals $[Ca]^n/([Ca]^n+K^n)$, with n and pK = 5.30 and 5.41 pCa respectively, for increasing Ca and 3.8 and 5.47 pCa for decreasing Ca. Mg^{2+} = 5.5 mM. For both conditions, maximum force is produced at a pCa of 5.0. These points are not plotted but used in the curve fitting.

Figure 4 shows a contraction in an aequorin-injected, voltage-clamped muscle fiber. If the muscle is allowed to shorten during the declining phase of the Ca transient, the force falls rapidly and there is a burst of myoplasmic Ca seen on the declining phase of the Ca transient (see arrow on figure). Extra light (Ca) is seen in the lower trace, which is the difference between the experimental and a control Ca transient. Thus, allowing the muscle to shorten mobilizes additional Ca from some site within the muscle. The opposite is true if one stretches the muscle during the declining phase of the Ca transient. An increase in force and a decrease in the light signal is seen, implying additional Ca binding to some site. A similar increase of light on shortening has been seen in aequorin-injected cardiac muscle (Allen & Kurihara, 1982).

We believe that the extra Ca seen on release comes from the myofilaments and is related to the Ca activating the contraction. Other possible sources for the extra calcium include the sarcoplasmic reticulum, the extracellular pool and soluble Ca binding sites in muscles, but the evidence against these other sources and in support of the

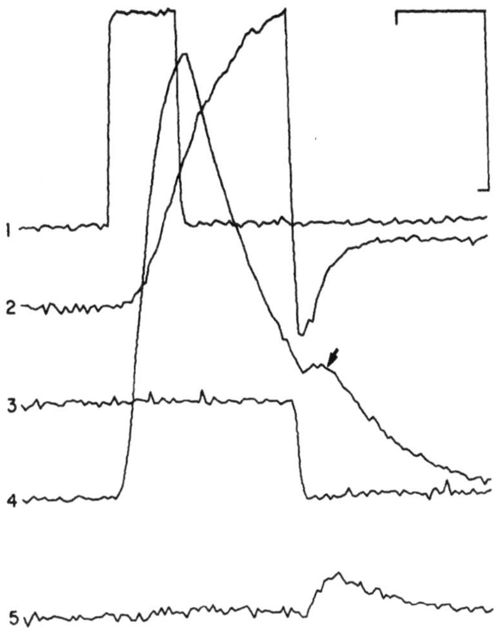

Figure 4: Effect of a quick release on the falling phase of the calcium transient. The record is a single trace and is not averaged. Trace 1: membrane potential at 40 mV/cal. Trace 2: isometric force at 0.1 N/cal. Trace 3: fiber length at 2.1 mm/cal. Trace 4: calcium transient (aequorin light signal) at 0.4 μamp/cal. The arrow indicates the "bump" of extra light accompanying the length change. Trace 5: computer subtraction of a control calcium transient from the calcium transient shown in Trace 2, to show the magnitude and time course of the extra light at 0.4 μamp./cal. Horizontal sweep rate: 400 msec/cal. Temperature: 7°C. Fiber length: 25 mm. Fiber weight: 48 mg.

myofibrillar source is strong. Some of this evidence has been discussed in a previous publication (Gordon & Ridgway, 1978; Ridgway, Gordon & Martyn, 1983). We will summarize here the evidence in favor of the hypothesis that a length change is sampling the Ca bound to a myofibrillar activating site. First, the extra Ca we see for a given decrease in length increases when the muscle force is increased either by increasing the stimulus intensity or through summation of force using double stimuli. In both conditions, increased force is probably produced by increased bound activating Ca. Thus, increased extra light (e.g., calcium) is consistent with the hypothesis that the Ca is from the myofilaments. Second, when the time of the length change is varied through the contraction, changes in muscle length early in contraction (when the light level is high and the force is low) produce much more extra light (extra Ca) and greater redeveloped force. There is a strong correlation between the amount of force redeveloped and the amount of extra light seen on release unless the release is very early in the contraction. This is consistent with the hypothesis that we are sampling the Ca on the myofilaments since there would be more Ca on the myofilaments if more force is redeveloped. Finally, if the extra light seen on muscle shortening is converted to extra calcium and is plotted as occurring at the time of the muscle shortening, its time course is intermediate between that of the free, myoplasmic Ca (the calcium transient) and force as would be expected for calcium bound to a site activating the myofilaments. This suggests that the length change disturbs the binding Ca to the myofibrillar activating sites such that the decrease in muscle length (force) causes a decrease in Ca binding. The decreased binding is consistent with the hypothesis that the hysteresis observed in the force-pCa relationship and the change in Ca sensitivity is due to a force-induced change in Ca binding to activating sites. However, we have not proven that this accounts for the changed sensitivity observed in the long stimulation pulse or skinned fiber experiments. It is possible there may also be changes in how "effective" a given extent of Ca binding is in facilitating force development. For example, in a thin filament regulated system, where tropomyosin displacement regulates force, a considerable amount of calcium binding to troponin may be necessary to accomplish the initiated tropomyosin displacement, but once displaced, myosin binding to actin could provide some energy toward maintaining the displacement (Murray & Weber, 1981) and the calcium requirement could thereby be reduced, or attached myosins may simply block the tropomyosin. At any rate, these possibilities and others would increase the apparent "effectiveness" of a given calcium level, without a change in calcium affinity or binding.

The details of the mechanism for feedback between contraction and Ca binding remain to be worked out. One possibility is that active cycling crossbridges increase Ca binding to thin filaments. Bremel and Weber (1972) have already suggested that rigor crossbridges have this effect. This mechanism cannot be distinguished from alternatives such as Ca binding to the thick filament.

Our conclusion is that the relationship between Ca concentration and force depends upon the history of the muscle. Increased Ca sensitivity is produced by muscle contraction. This is probably due, at least in part, to increased Ca binding.

As pointed out earlier, this has a number of implications (Ridgway, Gordon, & Martyn, 1983). First, the shape of the force-pCa relationship will depend upon crossbridge interaction in such a way that factors that affect crossbridge interaction may also affect this curve. This could be the mechanism by which factors such as MgATP (Godt, 1974) (which is necessary for crossbridge detachment), pH (Fabiato & Fabiato, 1978; Robertson & Kerrick, 1979) (affecting myosin ATPase), or fiber type (Kerrick, Secrist, Coby & Lucas, 1976) (with different ATPase rates) affect Ca sensitivity. Another implication is that the relationship between force and Ca concentration would be expected to show positive cooperativity with increased sensitivity being produced by increased force. This would explain the steep relationship between the force and Ca (Brandt, Cox & Kawai, 1980). Finally, the results imply that relaxation could be prolonged in muscles due to increased Ca sensitivity produced by contraction.

REFERENCES

Adelstein, R.S. and Eisenberg, E. (1980). Regulation and kinetics of the actin-myosin-ATP interaction. Ann. Rev. Biochem. 49: 921-956.

Allen, D.G. and Kurihara, S. (1982). The effects of muscle length on intracellular calcium transients in mammalian cardiac muscle. J. Physiol. 27: 79-94.

Ashley, C.C. and Moisescu, D.G. (1977). Effect of changing the composition of the bathing solutions upon the isometric tension-pCa relationship in bundles of crustacean myofibrils. J. Physiol. 270: 627-652.

Brandt, P.W., Cox, R.N. and Kawai, M. (1980). Can the binding of Ca^{2+} to two regulatory sites on troponin C determine the steep pCa/tension relationship of skeletal muscle? Proc. Natl. Acad. Sci. USA 77: 4717-4720.

Bremel, R.D. and Weber, A. (1972). Cooperation with actin filament in vertebrate skeletal muscle. Nature 238; 97-101.

Ebashi, S. and Endo, M. (1968). Calcium ion and muscle contraction. Prog. Biophys. Mol. Biol. 18: 123-183.

Fabiato, A. and Fabiato, F. (1978). Effects of pH on the myofilaments and the sarcoplasmic reticulum of skinned cells from cardiac and skeletal muscles. J. Physiol. 276: 233-255.

Fuchs, F. (1977). The binding of calcium to glycerinated muscle fibers in rigor. The effect of filament overlap. Biochim. Biophys. Acta 491: 523-531.

Godt, R.E. (1974). Calcium-activated tension of skinned muscle fibers of the frog: dependence on magnesium adenosine triphosphate concentration. J. Gen. Physiol. 63: 722-739.

Gordon, A.M. and Ridgway, E.B. (1978). Calcium transients and relaxation in single muscle fibers. Eur. J. Cardiol. 7: 27-34.

Hellam, D.C. and Podolsky, R.J. (1969). Force measurements in skinned muscle fibres. J. Physiol. 200: 807-819.

Kerrick, W.G.L., Secrist, D., Coby, R. and Lucas, S. (1976). Development of difference between red and white muscles in sensitivity to Ca^{2+} in rabbit from embryo to adult. Nature 260: 440-441.

Murray, J.M. and Weber, A. (1981). Competition between tropomyosin and myosin and cooperativity of the tropomyosin-actin filament. In: *The Regulation of Muscle Contraction*. A.D. Grinnell and M.A.B. Brazier (Eds.), New York: Academic Press.

Ridgway, E.B. and Ashley, C.C. (1967). Calcium transients in single muscle fibers. Biochem. Biophys. Res. Comm. 29: 229-234.

Ridgway, E.B., Gordon, A.M. and Martyn, D.M. (1983) Hysteresis in the force-pCa relationship in muscle. Science, in press.

Robertson, S.P. and Kerrick, W.G.L. (1979). The effects of pH on Ca^{2+} -activated force in frog skeletal muscle fibers. Pflügers Archiv 380: 41-45.

Taylor, E.W. (1979). Mechanism of actomyosin ATPase and the problem of muscle contraction. CRC Crit. Rev. Biochem. 6: 103-164.

DISCUSSION

EDMAN: Your results would seem to give an explanation for the shortening induced deactivation. Our recent experiments on chemically skinned muscle fibers of frog and rat (Ekelend & Edman, Acta Physiol. Scand., in press) have confirmed that active shortening causes a transitory depression (lasting 1-2 s) of the fiber's ability to produce force. This depressant effect can be almost completely eliminated by increasing the free calcium concentration to a high level (about 10 μM). We interpreted this finding to mean that active shortening temporarily decreases the affinity of the binding sites for calcium. Your data seem to be consistent with this view. However, your results differ from those presented by Stuart Taylor. Can you comment on that?

GORDON: We're dealing with very different conditions. His releases were during the plateau of a tetanus, where both calcium release and calcium uptake are going on at the same time. In addition, the myofilaments are fully saturated with calcium. We're working far below saturation, at about 20% of maximum force, so I think the experiments are quite different.

EDMAN: Is this muscle you are studying myosin regulated?

GORDON: There's some question about that. Andrew Szent-Györgyi speculated that it is a dual regulated system (Lehman and Szent-Györgyi, J. Gen. Physiol. 66: 1-30, 1975),but he tells me that the myosin side of that regulation is less well determined because of low measured ATPase rates. Initial experiments on effects of Tn-I that Glenn Kerrick discussed with me imply that there is at least a thin-filament regulated system, and possibly a thick filament one as well, in barnacle muscle.

TER KEURS: You've alluded to the fact that you expect that the extra calcium pulses following a release are not due to calcium released from the membranes. Would you elaborate on that a little bit more, and specifically do you see extra calcium pulses with aequorin in skinned fibers?

GORDON: We've not looked at skinned fibers but it would be a good experiment to do. The arguments against the membranes are as follows: first for the surface membrane, we've published experiments previously on the effects of muscle length on calcium release. These show that muscle length affects calcium release under constant current conditions. When you stretch the fiber, more calcium is released. When you release the muscle there is a smaller calcium transient, implying less calcium release from the sarcoplasmic reticulum. These are opposite to the effects described in my talk. Secondly the currents that we saw accompanying the length change had an equilibrium potential more like that expected for a potassium or chloride current than a calcium current.

As for the sarcoplasmic reticulum membranes as a source of the extra calcium, there are a number of experiments that argue against that: (1) We think we have adequate voltage control of the membrane during the declining phase of the calcium transient. We normally repolarise the membrane to the resting potential during this period, but if we step down to a hyperpolarized level, there is no effect on the calcium transient. Thus, the membrane potential is controlled, at least it's not depolarized. (2) The membrane potential is back down to the control level at the time we do our length changes. (3) If we use paired stimulation with two identical stimuli, the second stimulation, even though it's almost identical in amplitude to the first, will produce a calcium transient which is lower in amplitude than the first. In other words, the SR needs some repriming. If we change muscle length on the second of these two pulses, at a time when the SR's ability to release calcium is compromised but the force is higher due to summation of force, we get more extra light than on the first. These kinds of experiments led us to conclude that the extra calcium is not coming from the sarcoplasmic reticulum.

WINEGRAD: You indicated that even after the tension had come back to the basline for a period of time, you still had some residual effect upon the sensitivity to calcium. That would suggest that the sink that holds your calcium and releases it during the tension transients has not regained all of its calcium by the time you come back to baseline. This might suggest the possibility that in addition to any other sink besides the myofibrils, there may be something like a double calcium switch where both calcium sites must be saturated to produce tension.

GORDON: That is an interesting speculation. In the experiment to which you refer, the residual effect on Ca sensitivity was seen when the fiber was partially relaxed but the tension had not come back completely to the baseline before the fiber was reactivated. In all experiments when we allowed the fiber to relax fully there was little or no residual effect on the calcium sensitivity.

WINEGRAD: The thing that concerns me is the effect continues after tension has returned to the baseline, when presumably no calcium is bound to the regulatory sites of troponin. Could there be important calcium binding sites on the cytoskeleton?

GORDON: It's conceivable that calcium could be bound to other sites, and it could be that there is some effect that has a very long time constant. I'm really surprised, myself, by the fact that when you step down in calcium, the force *holds* at this level for many, many minutes. It's a non-equilibrium situation with the force far above what it would have been had it been stepped up to that level. Whether or not this is due to the same effect described by Bremel and Weber of rigor cross-bridge activation of thin filaments, I don't know.

GILLIS: May I suggest that your increase in calcium sensitivity could be due to a reduced off-rate of calcium from troponin, because the on-rate is so high that it cannot change.

GORDON: That is a good suggestion that we have considered.

CHANGES IN $[Ca^{2+}]_i$ INDUCED BY RAPID COOLING OF SINGLE SKELETAL MUSCLE FIBRES TREATED WITH LOW CONCENTRATION OF CAFFEINE

S. Kurihara, M. Konishi and T. Sakai

Department of Physiology, The Jikei University School of Medicine, 3-25-8 Nishishinbashi Minato-ku, Tokyo, Japan

ABSTRACT

In single skeletal muscle fibres treated with low concentration of caffeine, lowering the bathing solution temperature from 18°C to below 7°C increased $[Ca^{2+}]_i$ in three phases depending on caffeine concentration and temperature. Tension could be fully developed (rapid cooling contracture, RCC) by the second phase of the released Ca^{2+}. The second and third phases were inhibited by low concentrations of procaine, and the first phase was blocked by a higher one. RCC was observed even in Ca^{2+}-free solution. The mechanism of $[Ca^{2+}]_i$ changes and RCC during cooling was discussed.

INTRODUCTION

Ca^{2+} release from SR is an essential step in E-C coupling of skeletal muscle fibres (Ebashi and Endo, 1968). However, the mechanism of Ca^{2+} release from SR has not been fully clarified. It is known that when the temperature of the bathing solution is lowered from room temperature to below 4°C, contracture is initiated in skeletal muscle fibres treated with low concentrations of caffeine (rapid cooling contracture, RCC) (Sakai and Kurihara, 1974).

In the present study, $[Ca^{2+}]_i$ changes in RCC were measured with a Ca^{2+} sensitive photoprotein, aequorin (Aq) to explore the mechanism of Ca^{2+} release from SR.

METHODS

Single muscle fibres were dissected from M. tib. ant. of R. temp. and were mounted in a chamber with 2 mm trough for rapid solution change with ends attached to a hook and a tension transducer. Sarcomere length of each fibre was adjusted to a 2.3 μm by observing laser diffraction pattern. Ringer solution was (mM); NaCl, 115; KCl, 2.5; NaH_2PO_4, 0.85; Na_2HPO_4, 2.15; $CaCl_2$, 1.8; pH 7.0. The temperature of the solution was monitored continuously with thermistor thermometer with 0.1 sec response time constant. For rapid cooling, temperature was lowered from 18°C to the desired temperature within 3 sec by flowing a precooled solution. For measurement of $[Ca^{2+}]_i$, Aq was microinjected into several points of the fibre by applying pressure and the resultant light was detected with a photomultiplier (Blinks et al., 1978).

RESULTS

In single muscle fibres treated with low concentrations of caffeine (0.4-1.2 mM) for 5 min, a slight increase of light (first phase) was recognized when the temperature of the solution was lowered from 18°C to below 7°C. Tension did not develop at low concentrations of caffeine (subthreshold for RCC) and threshold concentrations of caffeine for RCC were different fibre to fibre. Increasing the caffeine concentration slightly increased the first phase and initiated subsequent rise of the light (second phase). Tension slowly started to develop just before the second phase and finally reached maximum level during cooling. However, the second phase increased further and evoked a remarkable light (third phase) at high concentrations of caffeine and lower temperature (lower than 4°C). In single muscle fibres, caffeine concentrations 0.2 mM lower than threshold could not induce RCC. The peak tension of RCC (at about 3°C) was 67% of the tetanus tension at 18°C and 81% at 5°C. $[Ca^{2+}]_i$ level of the first phase calculated from calibration curve of Aq (Allen and Blinks, 1979) was about 1 μM and the second phase was 2 μM. At the peak of the third phase, $[Ca^{2+}]_i$ finally reached 10^{-5}~10^{-4}M. These figures are a very rough estimation and we are currently continuing experiment for calibration. Single muscle fibres in Ca^{2+}-free Ringer solution with 1 mM EGTA and 3 mM $MgCl_2$ still showed tetanus response and RCC without substantial changes of light. Low concentrations of procaine (0.05-0.2 mM) inhibited the second and third phases of the light. Increase of procaine concentration finally abolished the first phase. The rate of RCC development was delayed when the second phase was inhibited by procaine, and RCC was completely abolished by high concentrations of procaine.

DISCUSSION

Present results suggest RCC is caused by the increase of $[Ca^{2+}]_i$ in three phases. The second and third phases of the light were initiated regeneratively when $[Ca^{2+}]_i$ reached about 1~2 μM. These phases are more sensitive to procaine known as a Ca^{2+}-induced Ca^{2+} release inhibitor (Thorens and Endo, 1975) than the first phase. This suggests the second and third phases are probably regenerative Ca^{2+} release processes from SR even though the calculated $[Ca^{2+}]_i$ level of the first and second phases was lower than that obtained in skinned fibres (Endo, 1975). For the generation mechanism of the first phase, Ca^{2+}-induced Ca^{2+} release might be

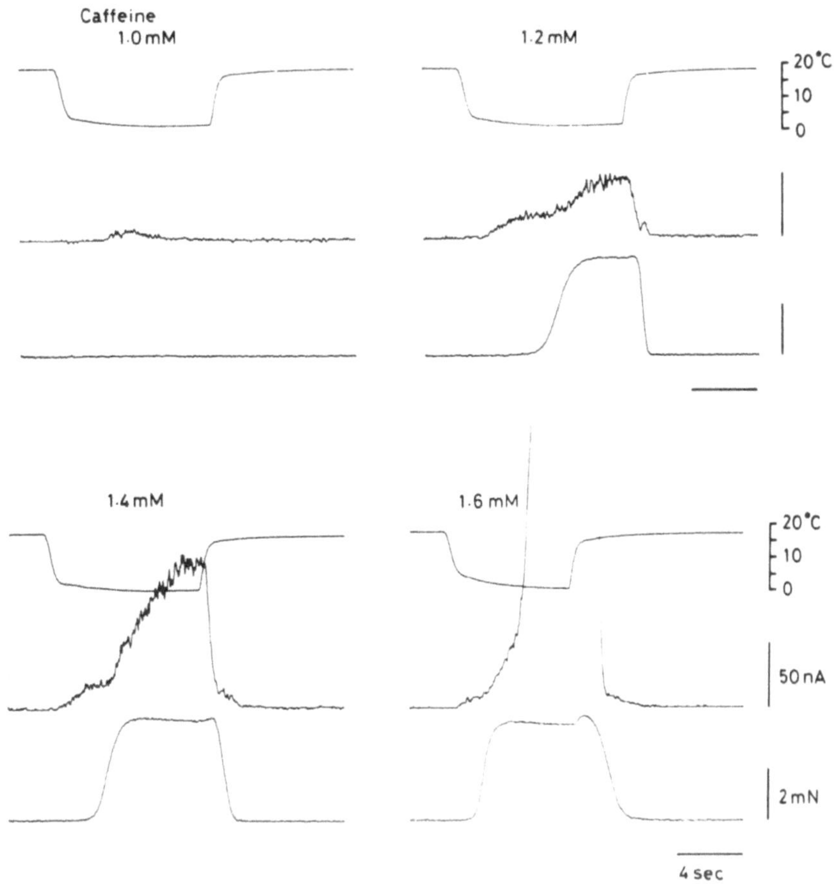

Figure 1: Changes of Aq light signal and tension during cooling. A single muscle fibre (diameter=200 μm, length = 5.2 mm) was treated with each concentration of caffeine for 5 min. Temperature was lowered from 18°C to 4°C (within initial 3 sec) and then to 1°C (creep). At 1 mM, a slight rise of light signal was recognized without tension development. Two phases of light signal were recognized at 1.2 and 1.4 mM, and tension started to rise. The second phase still increased even though tension reached maximum level. At 1.6 mM, the third phase (scaled out) followed the second phase. Light signal was recorded through a 5 HZ filter.

considered. However, $[Ca^{2+}]_i$ lower than 1 μM might be insufficient for Ca^{2+}-induced Ca^{2+} release. On the other hand, since a part of caffeine effects is considered to be due to the action through T-tubules (Lüttgau and Oetliker, 1968), the possibility which the first phase might be Ca^{2+} from T-tubules wall cannot be excluded completely. However, we are still careful for determining the generation mechanism of the first phase. Present result further suggested that tension development was governed by much smaller $[Ca^{2+}]_i$ change than that obtained in skinned fibre, even though the influence of temperature change on the Aq light signal must be carefully considered for $[Ca^{2+}]_i$ calculation.

ACKNOWLEDGEMENTS

The aequorin used in the study was prepared in the laboratory of Prof. J.R. Blinks with support from NIH Grant HL 12186.

REFERENCES

Allen, D.G. and Blinks, J.R. (1979). The interpretation of light signals from aequorin-injected skeletal and cardiac muscle cells: A new method of calibration. In: *Detection and Measurement of Free* Ca^{2+} *in Cells*. pp. 159-174, ed. Ashley, C.C. and Campbell, A.K. Elsevier/North-Holland.

Blinks, J.R. and Rüdel, R. and Taylor, S.R. (1978). Calcium transients in isolated amphibian skeletal muscle fibres: Detection with aequorin. J. Physiol. 277: 291-323.

Ebashi, S. and Endo, M. (1968). Calcium and muscle contraction. Prog. Biophy. Mol. Biol. 18: 123-183.

Endo, M. (1975). Mechanism of action of caffeine on the sarcoplasmic reticulum of skeletal muscle. Proc. Japan Acad. 51: 479-484.

Lüttgau, H.C. and Oetliker, H. (1968). The action of caffeine on the activation of the contractile mechanism in striated muscle fibres. J. Physiol., 194: 51-74.

Sakai, T. and Kurihara, S. (1974). A study on rapid cooling contracture from the view point of excitation-contraction coupling. Jikeikai Med. J. 21: 47-88.

Thorens, S. and Endo, M. (1975). Calcium-induced calcium release and "Depolarization"-induced calcium release: their physiological significance. Proc. Japan Acad. 51: 473-478.

DISCUSSION

WINEGRAD: Have you measured membrane potential during the cooling contracture?

KURIHARA: Yes, the depolarization is less than 10 mv.

WINEGRAD: Have you observed cooling contractures in skinned fibers?

KURIHARA: Yes. Dr. Sakai has observed it.

WINEGRAD: Is the procaine effect antagonised by an increase of caffeine?

KURIHARA: I have not done this experiment, but cooling contractures induced by lowering the caffeine concentration are suppressed by lower concentrations of procaine.

HOUSMANS: Is the conduction of the temperature change fast?

KURIHARA: Yes, conduction of heat to the fiber core is done within 0.1 sec since the fiber diameter is 100-200 μm.

PODOLSKY: Your results show that the caffeine sensitivity is different from fiber to fiber. This is a troublesome aspect of using the RCC technique.

CONTRACTILE RESPONSES TO MgATP AND pH IN A THICK FILAMENT REGULATED MUSCLE: STUDIES WITH SKINNED SCALLOP FIBERS

Robert E. Godt and J.L. Morgan

Dept. of Physiology, Medical College of Georgia, Augusta, GA 30912

ABSTRACT

The striated adductor of the Atlantic deep sea scallop (*Placopecten magellanicus*), a thick filament regulated muscle, contains little or no troponin. We examined the effect on activation of two agents (MgATP and pH) that alter the contractile threshold of thin filament regulated muscle, presumably through effects on troponin, to see if they also alter that of thick filament regulated muscle. We find that decreasing MgATP from 2 to 0.1 mM shifts the force-pCa curve of chemically skinned scallop muscle to the left by about 0.8 log units (i.e. Ca^{2+} sensitivity increases some six-fold). Under similar conditions the force-pCa relation of frog skinned fibers shifts leftward by almost the same amount, 0.7 log units (Godt, 1974). The force-pCa curve of scallop was unaffected by a decrease in pH from 7 to 6.5. It is especially interesting because: A) the force-pCa relation of skinned fibers from frog (Robertson and Kerrick, 1979) and striated adductor of the Pacific scallop (*Chlamys hastata hericia*) (Donaldson, unpublished observations) is shifted to the right by about 0.5 log units over this pH range. Furthermore, B) decreasing pH is reported to decrease the calcium affinity of Placopecten myofibrils (Chantler et al., 1981). Thus the molecular details of thick filament regulation appear to be more complex and varied than hitherto supposed.

INTRODUCTION

For some time we have been investigating the effects of alterations in MgATP and pH on calcium activation. In vertebrate striated muscle these agents have been shown to vary Ca^{2+} sensitivity presumably by affecting Ca^{2+} affinity of troponin (Weber and Murray, 1973; Robertson and Kerrick, 1979). If this be true, it seemed worthwhile to observe their effects on a thick filament regulated muscle, the striated adductor of the Atlantic deep sea scallop (*Placopecten magellanicus*), which is reported to have little or no troponin (Lehman, 1982).

METHODS

Sea scallops (*Placopecten magellanicus*) were obtained from the Marine Biology Laboratory, Woods Hole, MA. Small bundles of fibers (100-200 μm dia.) were taken from the striated adductor muscle. The fibers were chemically "skinned" in a relaxing solution containing 0.5% Triton X-100, a non-ionic detergent, and 50 μg/ml saponin for at least 20 min. Solutions contained (mM): 1 Mg^{2+}, 15 creatine phosphate, 20 Imidazole, KCl so that ionic strength was 0.2M; EGTA, MgATP and pH as indicated; 0.5 mg/ml creatine kinase, 22-23 C (cf. Godt and Lindley, 1982). The fibers were stretched to a sarcomere spacing of 2.6 μm as indicated by laser diffraction. The experimental protocol was similar to that employed by Simmons and Szent-Gyorgyi (1978). Fibers were transferred from contracting solutions (with 5mM total EGTA) to relaxing solution with 40 mM EGTA (pCa>8). To speed up force production, before exposure to high Ca^{2+} solutions the fiber was immersed for at least 1 min in a relaxing solution with 0.1 mM EGTA.

RESULTS

The force-pCa relation at 2 mM MgATP shown in the Figure is nearly identical to that of Simmons & Szent-Gyorgyi presented in Chantler et al. (1981, their Fig. 8) measured in similar solutions. When fibers are transferred from 2 to 0.1 mM MgATP relaxing solution we observed no change in force, i.e. the fibers did not go into rigor. The pCa required for 50% activation is some 0.8 log units higher in 0.1 mM MgATP than in high MgATP (see Figure). At low MgATP, when fibers are transferred from contracting to high EGTA relaxing solutions, force tends to "hang up". Raising creatine phosphate in the relaxing solution to 30 mM had no effect on the hang up. If the fiber was slackened during this phase force redeveloped, indicating that this force is being maintained by actively cycling rather than rigor cross-bridges. The hung up force quickly declined when the fiber was transferred to a high MgATP relaxing solution. Under our conditions, calcium sensitivity and maximal Ca^{2+} activated force (pCa 4) of skinned scallop fibers was the same at pH 7 and 6.5 (see Figure).

DISCUSSION

We find that decreasing MgATP shifts the force-pCa curve of scallop comparably to that of frog. Weber and colleagues (e.g. Bremel and Weber, 1972; Weber & Murray, 1973) first observed an increased Ca^{2+} sensitivity of rabbit myofibrils with decreased MgATP, due to increased binding of Ca^{2+} to the myofibrils. In explanation, Weber hypothesized that as MgATP decreases, an increasing number of ATP-free "rigor" cross-bridges binds to the thin filament. Through cooperative interactions among the thin filament proteins, this in turn increases the affinity of the low affinity Ca^{2+} sites on troponin that supposedly regulate contraction (Potter and Gergely, 1975). Subsequent experiments with skinned fibers from a

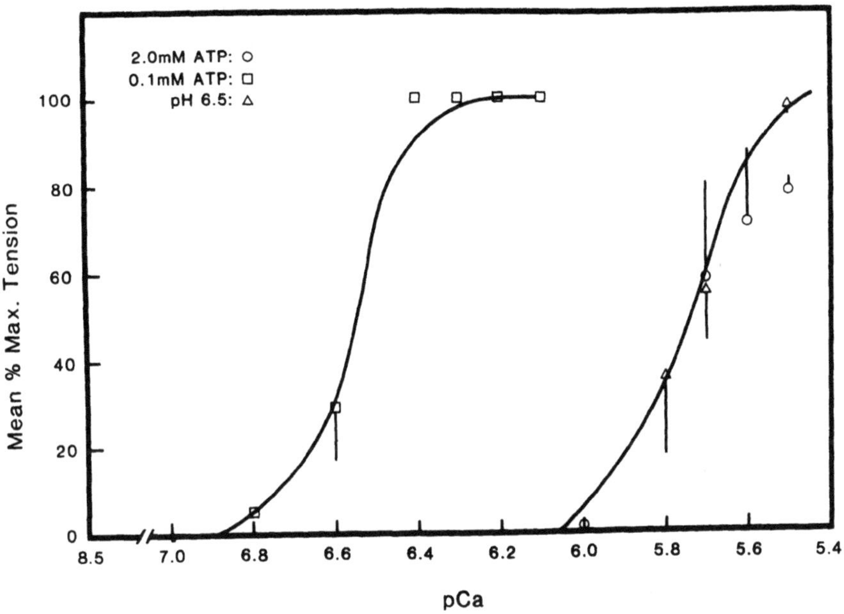

Figure 1: Relation between normalized force and pCa. Force in pCa 4 defined as 100%. Vertical bars represent SEM.

number of thin filament regulated muscles showed that force-pCa curves shifted to the left (i.e. Ca^{2+} sensitivity increased) as MgATP declined. Because scallop myofibrils contain little or no troponin (Lehman, 1982) we did not expect that Ca^{2+} affinity in this muscle would be affected by MgATP. However, Adelstein and Eisenberg (1980) argue that apparent cooperativity within the contractile apparatus is not a novelty but rather is an inherent property of the regulatory system. If binding of Ca^{2+} regulates the attachment of cross-bridges, as few would deny, then considerations of detailed balance require that attachment of cross-bridges must affect the binding of Ca^{2+}. Thus one should observe a leftward shift of the force-pCa relation of scallop fibers with decreased MgATP.

Donaldson et al.(1978) reported that a decrease in pH from 7 to 6.5 decreases the Ca^{2+} sensitivity of skinned fibers from the Pacific coast scallop (*Chlamys hastata hericia*), shifting the force-pCa curve to the right by at least 0.5 log units (Donaldson, personal communication) but having no effect on maximal Ca^{2+} activated force. Chantler et al. (1981) demonstrated that a decrease in pH from 7.5 to 7 decreased the Ca^{2+} binding to Placopecten myofibrils by about an order of magnitude. Under our conditions, both the force-pCa relation and the maximal Ca^{2+} activated force of chemically skinned Placopecten fibers are unaffected by a decrease in pH from 7 to 6.5. Differences in results from Placopecten and Chlamys fibers might be ascribed to species differences between Atlantic and Pacific coast scallops. The apparent discrepancy with the Ca^{2+} binding studies on Placopecten myofibrils is less easily explained, although it may be significant that these binding studies were carried out at lower ionic strength (0.08-0.09M) than that of our solutions (0.2M). At present we can only speculate about these discrepancies. Nevertheless, these differing studies indicate to us that the thick filament regulatory system may be more complex and varied, and therefore even more interesting than hitherto suspected.

REFERENCES

Adelstein, R.S. and Eisenberg, E. (1980). Regulation and kinetics of the actin-myosin-ATP interaction. Ann. Rev. Biochem. 49: 921-956.

Bremel, R.D. and Weber, A. (1972). Cooperation within actin filament in vertebrate skeletal muscle. Nature 238: 97-101.

Chantler, P.D., Sellers, J.R. and Szent-Gyorgyi, A.G. (1981). Cooperativity in scallop myosin. Biochem. 20: 210-216.

Donaldson, S.K.B., Bolles, L., Farrance, M. and Lucas, S. (1978). H^+ and Mg^{2+} effects on scallop skeletal and rabbit left ventricular skinned force. Biophys. J. 21: 88a.

Godt, R.E. (1974). Calcium-activated tension of skinned muscle fibers of the frog: Dependence on MgATP concentration. J. Gen. Physiol. 63: 722-739.

Godt, R.E. and Lindley, B.D. (1982). The influence of temperature upon contractile activation and isometric force production in mechanically skinned muscle fibers of the frog. J. Gen. Physiol. 80: 279-297.

Lehman, W. (1982). Troponin-like components in molluscan muscles. Biophys. J. 37: 40a.

Potter, J.D. and Gergely, J. (1975). The calcium and magnesium binding sites on troponin and their role in the regulation of myofibrillar adenosine triphosphatase. J. Biol. Chem. 250: 4628-4633.

Robertson, S.P. and Kerrick, W.G.L. (1979). The effects of pH on Ca^{2+} activated force in frog skeletal muscle fibers. Pflugers Arch. 380: 41-45.

Simmons, R.M. and Szent-Gyorgyi, A.G. (1978). Reversible loss of calcium control of tension in scallop striated muscle associated with the removal of regulatory light chains. Nature 273: 62-64.

Weber, A. and Murray, J.M. (1973). Molecular control mechanisms in muscle contraction. Physiol. Rev. 53: 612-673.

DISCUSSION

Dr. Rall pointed out that the fibers in the scallop striated adductor are quite small, each containing but a single myofibril. Based on the force records presented in the poster, calculations showed that in 2 mM MgATP solutions this skinned preparation was generating a maximal force/cross-sectional area of about 50 mN/mm^2. This is quite similar to the maximal tetanic tension Dr. Rall observed with the intact scallop muscle (Rall, J. Physiol. 321: 287-295, 1981).

FORMATION OF CALCIUM-PARVALBUMIN COMPLEX DURING CONTRACTION. A SOURCE OF "UNEXPLAINED HEAT"?

J.M. Gillis, D. Thomason*, J. Lefevre and R.H. Kretsinger*

Département de Physiologie, Université Catholique de Louvain, B-1200 Bruxelles, Belgium
**Department of Biology, University of Virginia, Charlottesville, VA, USA 22901*

ABSTRACT

Computer simulation of the kinetics of the distribution of Ca between troponin, parvalbumin and the sarcoplasmic reticulum, during contraction and relaxation shows that parvalbumins can contribute significantly to the rate of relaxation and to the post contractile translocation of calcium. The binding of Ca to parvalbumin is an exothermic process which may account for about 20% of the 'unexplained heat' during contraction.

Fast skeletal muscles from Fishes and Amphibians contain a high concentration of parvalbumin, a cytoplasmic Ca binding protein. Gillis & Gerday (1978) have shown that Ca-free parvalbumins added to a suspension of Ca-activated myofibrils can reduce the ATP-ase activity down to the very low level obtained with addition of EGTA, by binding the calcium of the medium. This Ca bound to the parvalbumin can then be sequestered by isolated vesicles derived from the sarcoplasmic reticulum (SR) so that Ca-free parvalbumins can be regenerated. From these findings it was proposed that parvalbumins can act as a soluble relaxing factor transferring Ca from myofibrils to the SR.

Recently, Gillis et al. (1982) have studied by computer simulation the kinetics of the Ca-troponin and Ca-parvalbumin complexes in response to a Ca pulse meant to mimic maximal activation of the muscle i.e. to obtain 90% saturation of troponin. The model contains three components: (1) the Ca specific, or regulatory sites of troponin ('T-sites'); (2) the high affinity or Ca-Mg sites of troponin and parvalbumins, considered as a single class of sites ('P-sites') and (3) the Ca pump of the sarcoplasmic reticulum (see legend of Fig. 1 for the conditions of simulations). Figure 1 gives the results obtained in the case of a frog muscle: in response to the

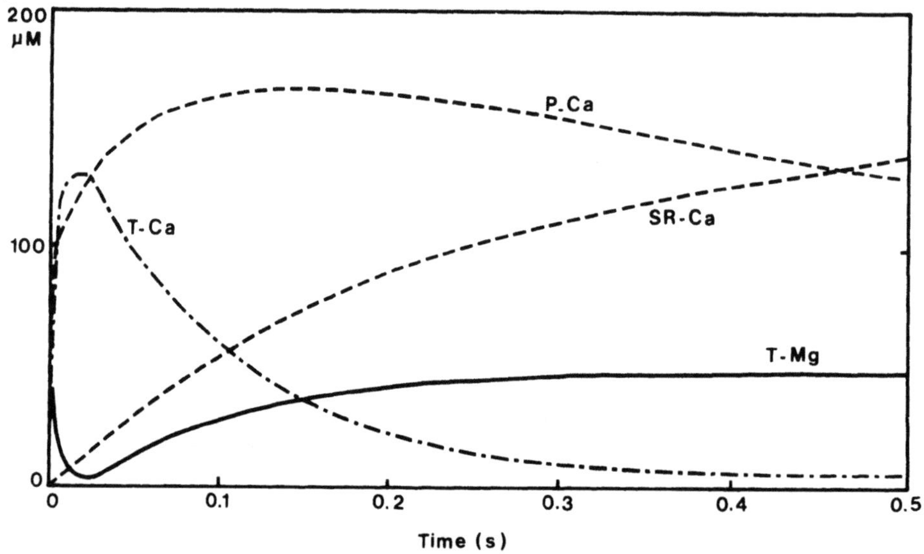

Figure 1: Evolution of the Ca-troponin complex (T-Ca, only the regulatory sites); and of Ca complexes with the high affinity sites (Ca-Mg sites, from parvalbumins and troponin: P-Ca) after a single Ca pulse of 200 μmoles.L^{-1} (delivered in 20 msec). Conditions of simulation:
- Sites concentration: regulatory 140 μM; Ca-Mg 840 μM (700 from parvalbumins + 140 from troponin).
- Rate constants:

	+Ca	+Mg
Regulatory sites		
ON	10^{8} M^{-1} sec^{-1}	$10^{4.8}$ M^{-1} sec^{-1}
OFF	$10^{1.5}$ sec^{-1}	$10^{2.3}$ sec^{-1}
Ca-Mg sites		
ON	10^{8} M^{-1} sec^{-1}	$10^{4.8}$ M^{-1} sec^{-1}
OFF	10^{0} sec^{-1}	$10^{0.5}$ sec^{-1}

SR-Ca: amount of Ca taken up by the SR, functioning according to a Michaelis-Menten kinetics. V_{max}: 700 μmoles Ca.sec^{-1} (25°C); K_m: 10^{-7} M Ca^{2+}. For sources of the concentrations and kinetics data, see references given in Gillis et al. (1982).
- Resting conditions (μM):

		apo	+Ca	+Mg
(In the presence of:				
Ca^{2+}: 2.5×10^{-8} M	regulatory sites	83	5	52
Mg^{2+}: 3×10^{-3} M)	Ca-Mg sites	19	40	778

Ca pulse, regulatory sites of troponin are 90% Ca-saturated within 20 msec, while the formation of Ca-parvalbumin increases at a much slower rate, limited by the slow dissociation of Mg from the Ca-Mg sites (95% occupied by Mg, at rest). The presence of a high concentration of cytoplasmic Ca binding proteins thus does not prevent muscle activation.

Relaxation i.e. dissociation of the Ca-troponin complex starts early, as Ca binds to parvalbumin and is extracted from the system by the SR pump. After 100 msec, [T-Ca] has dropped from 123 μM to 30 μM, while the amount of Ca bound to the Ca-Mg sites has increased from 40 to 135 μM; a gain of 95 μmoles Ca, about 50% of the pulse. Then, decalcification of the Ca-Mg sites proceeds slowly as Ca is removed from the system by the SR activity. When relaxation is complete there is still about 75 μmoles of Ca attached to the Ca-Mg sites, mainly on parvalbumins. The return to the resting level thus proceeds at very different rates for the two types of Ca binding sites. The difference is still more pronounced after repetitive simulations. At the low frequency of 2/sec, [Ca-troponin] completely returns to the resting level before the next pulse. This is not the case for [Ca-Parvalbumin]. As simulations are repeated, there is a progressive build up of [Ca-Parvalbumin] which reaches a steady state value of about 150-200 μM after 10 pulses. This level is critically dependent on the frequency of stimulation and on the SR activity which eventually removes the Ca from the medium (see Gillis et al., 1982 for more details).

These results show that (1) an important fraction of the Ca released for activation is still bound on the Ca-Mg sites (mainly on parvalbumins) at the time troponin relaxation is complete; (2) this Ca is slowly reaccumulated by the SR during the *recovery* period.

These conclusions offer a new view of well established observations.

1. The questions of the 'unexplained heat'. In frog muscle, the energy balance sheet is incomplete: the sum (work + heat) is not completely accounted for by the amount of ATP and PC split during the contraction (Gilbert et al., 1971): there is an excess of about 20-40 $J.L^{-1}$ of 'unexplained heat'. Extensive search for exothermic reactions not accounted for by ATP and PC hydrolysis has been largely unsuccessful (Curtin & Woledge, 1978). When it was realized that parvalbumins could interfere with the Ca movements during contraction, Curtin & Woledge (1978) proposed that Ca binding to parvalbumins could contribute significantly to the heat production, as the heat of binding amounts to 27 KJ/mole of Ca, even in the presence of millimolar concentration of Mg^{2+} (Woledge & Closet, unpublished). Our simulations allow us to estimate quantitatively this contribution: at the end of the relaxation the excess amount of Ca bound to the Ca-Mg sites varies between 100 μmoles.L^{-1} (single twitch) to 200 μmoles (repetitive simulations). The binding of these quantities could have liberated 2.7-5.4 $J.L^{-1}$, accounting for 14% of the 'unexplained heat'. This contribution is significant but not very large. Obviously the 'unexplained heat' reflects exothermic processes of different origins.

   The eventual return of the Ca into the SR should be associated with an ATP hydrolysis of 50-100 μmoles.L^{-1} *during the recovery*.

2. Post contractile translocation of Ca. The autoradiographic study of Winegrad (1967) showed that ^{45}Ca was mainly located in the interfibrillar space during the first few minutes of the recovery period, and slowly reaccumulated in the terminal cisternae. He

proposed that Ca was picked up by the longitudinal elements of the SR and moved to the cisternae by intrareticular translocation. The low resolution of the light microscope did not permit to demonstrate that this interfibrillar Ca was actually inside the SR. Our study suggests on the other hand that after relaxation, a large fraction (~50%) of the Ca has not yet returned into the SR but is bound to soluble Ca binding proteins, in the interfilamentary space. This interpretation fits perfectly well with the direct element localization (electromicroprobe analysis) by Somlyo et al. (1981), who show that the Ca released by the stimulus is not restored into the longitudinal SR.

REFERENCES

Curtin, N.A. and Woledge, R.C., (1978) Energy changes and muscular contraction. Physiol. Rev. 58: 690-761.

Gilbert, C., Kretzschmar, K.M., Wilkie, D.R. and Woledge, R.C. (1971) Chemical change and energy output during muscular contraction. J. Physiol. 218: 163-193.

Gillis, J.M. and Gerday, C. (1977) Calcium movement between myofibrils, parvalbumins and sarcoplasmic reticulum in muscle. In: *Calcium-binding proteins and calcium function* (Wasserman, R.M., Corradino, R.A., Carafoli, E., Kretsinger, R.H., MacLennan, D.H. and Siegel, F.L., eds.) pp. 193-196, New York: North-Holland.

Gillis, J.M., Thomason, D., Lefevre, J. and Kretsinger, R.H. (1982). Parvalbumins and muscle relaxation: a computer simulation study. J. Musc. Res. Cell Mot. (in press).

Somlyo, A.V., Gonzales-Serratos, H., Schuman, H., McCleilan, C. and Somlyo, A.P. (1981) Calcium release and ionic changes in the sarcoplasmic reticulum of tetanized muscle: an electron probe study. J. Cell Biol. 90: 577-594.

Winegrad, S. (1968) Intracellular calcium movements of frog skeletal muscle during recovery from tetanus. J. Gen. Physiol. 51: 65-83.

DISCUSSION

INGELS: I'd like to offer support for your central idea from two different lines. One, that I'm rather intimately familiar with, was my doctoral work in 1966. I modeled the diffusion equation for calcium in a sarcomere and was forced by the nature of the speed of relaxation to postulate just such a mechanism because there simply wasn't time for diffusion to get the calcium which had been released from the SR back into the SR. It was orders of magnitude off. So I had to model this in the computer, and it had just the properties you're describing. It was uniformly distributed throughout the sarcomere and then could take the calcium back to the SR. And the other one is, because of my interest in this, maybe not everyone has seen this, but there was an article just published by Celio and Heizmann in Zurich (Nature 297: 504, 1982) which showed parvalbumin associated with fast-contracting muscle fibers, and they can specify the speed of relaxation or the speed of the muscle into five subgroups which each have different amounts of parvalbumin in them. So the amount of parvalbumin is critical to the speed at which these muscles can relax.

GILLIS: Well in fact I've tabulated the amounts of parvalbumin of the various types of lower vertebrate muscles going from frog, to pike, to carp, and eventually to the swim bladder muscle of the toad fish which contracts very fast. There's a strict correlation between speed of contraction and relaxation and the content of parvalbumin in lower vertebrates.

MAUGHAN: Gupta and Moore (J. Biol. Chem. 255(9): 3987-3993, 1980) report a value around 0.6 mM for intracellular free magnesium. I'm wondering how such a low value would affect the results of your simulations. Perhaps I should rephrase the question and ask, what part of your simulation would be most sensitive to the free magnesium concentration?

GILLIS: The value of 3 mM comes from the NMR work of Cohen & Tyler-Burt (Proc. Nat. Acad. Sci. 74: 4271, 1977). If you take a lower value for free magnesium, the main effect will be to have less parvalbumin occupied by magnesium. And therefore, more apo-parvalbumin (i.e. metal-free) into the system. The interest of having the large fraction of the parvalbumin occupied by magnesium is that it reduces the straight competition from troponin. Otherwise you would never see a muscle being activated. If you have troponin in one side and a lot of parvalbumin on the other side, and they compete for calcium, given the relationship of the concentrations, troponin will never win. But we know it does. So this is due to the fact that parvalbumin is occupied by magnesium and dissociation of magnesium is rather slow.

HOMSHER: I agree with Jean-Marie's calculation, in theory, but I think I disagree quantitatively. The basis of this is that you have to be able to get more enthalpy from calcium binding to the parvalbumin. Based on calculations from the dissociation constant for calcium to parvalbumin and magnesium to parvalbumin, you'd suspect that in a maintained tetanus, where the concentration of calcium would be presumably in excess of 10^{-5} M, it would almost totally load up the parvalbumin completely with calcium. Is that right?

GILLIS: Yes. If you stimulate the muscle at a high frequency to get sustained tetanus.

HOMSHER: Well, it's a relevant consideration for the unexplained enthalpy in a long sustained tetanus.

GILLIS: True.

HOMSHER: And assuming that 10^{-5} M calcium is the maintained level or higher, then one would expect something on the order of 20 millijoules per gram. So in principle I totally agree with you.

GILLIS: Well, if all parvalbumin is completely saturated with calcium, yes I agree.

KRUEGER: The place where speed of relaxation has an evolutionary effect of physiological significance is cardiac muscle.

TER KEURS: Specifically, you find variations of speed of relaxation in cardiac muscle from various species, for example rat and rabbit.

KRUEGER: What is known about parvalbumin in cardiac muscle?

GILLIS: In heart, whatever the animal, there is a very low concentration. There is a very strict distinction between fast and slow muscle. If you take a fish, for example, you have this border of red fibers which are used for slow swimming, and they have very little parvalbumin. The main body, the white fibers, used for quick swimming, have the high content. In mammals, the amount of parvalbumin is something like 50 micromolar compared with 500 here. In heart, whatever the animal, there is a very low concentration.

SHIMIZU: I couldn't understand why you neglected the role of calcium binding proteins inside the SR.

GILLIS: This is taken into account by the equation of SR function which is coming from people who are using SR vesicles. I consider the SR as a whole, as a complete system, with its calsequestrin and other possible Ca binding sites.

WILKIE: At what stage do you think the active transport of calcium by the calcium pump is taking place? Throughout the prolonged contractions? Afterwards? Or when?

GILLIS: In the simulations we show here, it was working all the time from the start. As soon as calcium is available the SR pumps it. But due to the fact that the free calcium level is not very high, the SR pump is probably not working at its maximum rate. As you know, the transient of calcium goes down quickly to micromolar values or even less.

GODT: Can you tell us how your analysis differs from that published recently by Potter and Robertson (Biophys. J. 34: 559, 19--)?

GILLIS: Yes. There are two main differences. The first one is in their model, Robertson and his colleagues were simulating a given function of free calcium concentration to drive their model. I think one should do it the other way around -- we inject a certain amount of calcium and see how it is redistributed according to the ligand's characteristics. That's one point. The second point is that they consider the situation which exists in mammals in which the parvalbumin concentration is very low indeed. So they conclude that there was no physiological role of parvalbumins. They did not consider the case of high concentrations in frogs and fishes.

EBASHI: I'd like to ask a biological question. Why do animals living in water have more parvalbumin?

GILLIS: I must say personally I am not responsible for ths discrimination. It is a rather striking observation that the high concentration of parvalbumin is found, as I say, not only in fast skeletal muscle, but in lower vertebrates, and when I say a high concentration, it means going up to mM concentration, far above troponin. Then what I'd like to propose, but it is purely speculative, is that it's a way to help relaxation when the SR cannot work fast enough. And this may be the case for lower vertebrates. They have to live in conditions where the external temperature may change from cold to lukewarm. The presence of parvalbumins may help to get quick relaxation, in quick movements to escape or to attack prey, even at low temperature. This may confer a higher survival value.

ALTRINGHAM: Just another comment on that. Many fish, when they're swimming to capture prey or escape from predators, have tail beat frequencies of up to 70 Hz. They therefore need to relax the muscle of one side of the body very quickly, before the other side begins to contract, so rapid relaxation, aided by parvalbumins is very advantageous.

INGELS: I should point out that this paper I referred to is a mammalian muscle, and it's widespread in mammalian muscle.

EBASHI: Yes, but the content is different.

INGELS: The content is different, and what they're pointing out is that the content is different depending on the speed of relaxation of the muscle. To model it you can draw an analogy if you were to take a perfume atomizer right here and spray it into the room. Let's say the sarcoplasmic reticulum were on the walls. Just imagine how long it would take that if the perfume were calcium, -- how long it would take that perfume, once that it evened out all its concentration gradients to get stuck to the walls and leave the room. You have to have something *in* the room which takes the calcium, binds it, removes it as a free calcium store, *then* this parvalbumin-calcium complex diffuses around randomly, and finally sticks and transfers its calcium to the wall. That's what I modeled mathematically in 1965, and that's what you had to have, or muscle would never relax.

KRUEGER: The distribution of parvalbumin within the cell .. Is there any way of getting at this? What is the distribution within the contractile lattice?

GILLIS: In fact, parvalbumins are highly soluble. Against that there were some objections: people say that they are soluble because to extract them the SR is destroyed and the parvalbumins are inside the SR. To clarify this point we did an experiment some years ago on a skinned fiber preparation from frog. We kept it in oil in order to prevent anything leaking, tranferred it to a plate for immunodiffusion against antibodies, for parvalbumin, and then titrated how much parvalbumin can diffuse out, without having destroyed the internal structure of the SR. We found that the whole content had diffused out, so I think it's a rather good argument to say that they are probably in solution in the cell (Gillis et al., Biochem. Biophys. Acta 585: 444, 1979).

KRUEGER: Is it in the contractile lattice?

GILLIS: Well that's another point. In some animals it was possible to localize some parvalbumin-like compound with fluorescent antibodies; but it was, in fact, an extremely small fraction of the total content of the fiber. So I think that even if some groups have localized parvalbumin in the A-band, the amount that is localized is very small compared to the total content.

TENSION TRANSIENTS AND STIFFNESS

INTRODUCTION

In the sliding filament model of muscle contraction (A.F. Huxley Prog. Biophys. biophys. Chem. 7: 255-318, 1957), force generation and shortening result from the alternate formation and breaking of cross-links between the thick and thin filaments. It is well known that relaxed muscle is highly extensible, while contracting muscle becomes considerably stiffer. This change is explained as being due to the formation of cross-links between the filaments. Muscle stiffness is normally expressed as $\Delta P/\Delta L$, where ΔL is a rapid change in muscle length, while ΔP is the coincident change in tension, and is generally believed to be proportional to the number of cross-links in each sarcomere. Bressler's contribution examines stiffness using both stretches and releases, while Cecchi's paper is concerned with measurement of muscle stiffness with sinusoidal length changes in the kHz region.

When a small rapid length change is applied to an isometrically contracting muscle fiber, the resulting rapid tension change is followed by the recovery of tension towards the initial isometric level. The tension recovery is divided into two phases; the initial quick recovery within a few msec and the subsequent much slower recovery (Huxley & Simmons, Nature 233: 533-538, 1971). The initial quick tension recovery has been interpreted as being due to the tendency of the myosin head to rotate along the thin filaments, and a hypothesis has been put forward that myosin head rotation is responsible for force generation and shortening in striated muscle. Thus, the tension transients following rapid length changes are now being studied under a variety of conditions to provide information about the kinetic properties of the cross-bridges. The papers of Shimizu, Fay and Rüegg are related to this subject.

Since a single muscle fiber consists of thousands of sarcomeres connected in series, a rapid change applied to one end of the fiber may not necessarily produce uniform sarcomere length changes along the entire fiber length, while the tension transients are being recorded at the other end. Sugi's paper is concerned with the detection of such a longitudinal nonuniformity of sarcomere length changes following rapid changes in fiber length, suggesting that the current interpretation of the tension transients should be reconsidered. Nishiyama then provides a multisegmental model to analyze their distributed behavior.

Another way of studying cross-bridge kinetics is so-called "sinusoidal analysis", in which sinusoidal vibrations of varying frequency are applied to the preparation and the resulting sinusoidal tension changes are recorded to give the elastic and viscous muduli as a function of the frequency of the applied vibration. Kawai uses this technique to study cross-bridge behavior.

In all the methods described above, the applied length perturbations are relatively large, and cannot be free of the possibility that the state of the contractile system is altered by the experimental procedure *per se.* Hatta reports that the ultrasonic waves in the MHz region can be successfully used to measure muscle stiffness changes with negligibly small perturbations and extremely high time resolution.

SYMMETRIC AND ASYMMETRIC PROCESSES IN THE MECHANO-CHEMICAL CONVERSION IN THE CROSS-BRIDGE MECHANISM STUDIED BY ISOMETRIC TENSION TRANSIENTS

H. Shimizu and H. Tanaka

Faculty of Pharmaceutical Sciences, The University of Tokyo, Hongo, Bunkyo-ku, Tokyo 113, Japan

ABSTRACT

The substrate concentration dependence of isometric tension transients in response to a quick stretch and release of glycerinated muscle fibers was studied during the first few seconds with a time resolution of sub-milliseconds. Observed tension transients apparently comprise five rate processes, which are named process (1), (1'), (2), (3) and (4) in order of fast to slow. It was found that they might be classified into two groups: one is composed of processes (1), (1') and (4) and the other of processes (2) and (3). The processes belonging to the first group give a symmetric response with respect to the direction of the length change such that not only the rate but also the fraction of the processes are about the same in magnitude for stretch and release. However, this is not the case for the processes belonging to the second group, which give an asymmetric response. In addition, the rate of the processes belonging to the second group depends on substrate concentration while the rate of those belonging to the first group does not when substrate concentration is higher than a few tens of micromolar. The symmetric processes are attributed to mechanical changes in the property of cross-bridge whereas the asymmetric ones to those accompanying chemical changes. The early tension recovery phase is comprised not only of processes (1) and (1') but also of (2) whose rate is proportional to substrate concentration and the square of substrate concentration for stretch and release, respectively.

INTRODUCTION

Chemo-mechanical energy transformation in skeletal muscle fibers takes place at cross-bridges attaching to thin filaments (Huxley, 1969). However, the molecular mechanism of coupling between the performance of mechanical work and the splitting of ATP has not been clarified yet. To investigate elementary mechanical processes of muscle contraction, Huxley and Simmons (1971) applied a step length change to activated intact skeletal muscle. They observed four phases in the tension transient following the step length change. The four phases can be assigned to the elementary physical changes in the state of cross-bridges (Huxley, 1974). To quantitatively explain observed data in the transient states as well as in the steady state by using a single model, the presence of one detached state and, at least, two attached states (preactive and active states) must be assumed for cross-bridges (Huxley, 1973; Julian et al., 1974; Nishiyama et al., 1977; Eisenberg et al., 1980). It is speculated from the three-state models that the elementary processes of the energy transformation are comprised of the attachment, conformational change and detachment of cross-bridges.

Tension transients in activated skinned muscle fibers at various substrate concentrations carry important information on physiochemical processes occurring in the chemo-mechanical transformation. Skinned muscle fibers are useful in obtaining such information because mechanical properties of the fibers can be measured under various chemical conditions (Arata et al., 1977; Cooke & Bialek, 1979; Ferenczi, 1979; Chaen et al., 1981). In this study we examined the substrate dependence of tension transients of glycerinated rabbit psoas fibers in various concentrations of substrate not only against quick release but also against quick stretch. We were also interested in examining whether the chemo-mechanical change in muscle is unidirectional or not. Not only asymmetric but also symmetric tension recovery processes in relation to the direction of length change were observed. It was found that the symmetric processes were due to mechanical changes in the property of cross-bridge, whereas the asymmetric processes are influenced by substrate concentration suggesting that chemical changes are involved. In short, the unidirectional property of the chemo-mechanical process in muscle is caused by the presence of the asymmetric processes.

MATERIALS AND METHODS

Materials

Glycerol-treated muscle fibers were prepared by the method described below. A strip of rabbit psoas fibers (1-2 mm in diameter) which had been dissected and tied to a glass stick was immersed into a stock solution containing 100 mM KCl, 20 mM MOPS, 4 mM $MgCl_2$, 4 mM EGTA, 6 M glycerol, (pH 7.0) and incubated for 24 hours at 0°C. Then the solution was exchanged with a fresh one in which the strips were stored for 1-3 months at -20°C.

A bundle of 4-6 single fibers was dissected before use from the strip of the above-mentioned glycerinated muscle in the stock solution. The sarcomere length of the fibers was measured as 2.3-2.5 μm by He-Ne-laser light diffraction. The bundle was mounted on the apparatus described below after adjusting its length to 4 mm. Then both ends of the bundle were fixed with collodion to the extension from a servo motor and to a tension transducer, respectively.

ATP and phosphocreatine were purchased from Kyowa-Hakko Co. and Sigma Chemical Co., respectively. All other chemicals were of the highest grade commercially available and used without further purification. Creatine kinase (EC 2.7.3.2) was prepared from rabbit skeletal muscle by the procedure of Node et al. (1955). Its activity was about 60 μmol/mg/min in 4 mM $MgCl_2$, 4 mM ADP, 4 mM phophocreatine, 20 mM MOPS at pH 7.0 and 25°C.

Solutions

Relaxation solutions (solution R) were composed of 4 mM $MgCl_2$, 4 mM EGTA, 20 mM MOPS, 70 mM KCl, 20 mM phosphocreatine and 3 mg/ml creatine kinase with various amounts of ATP. Contraction solutions (solution C) contained 4 mM $CaCl_2$ in addition. The ionic strength of solutions C and R ranged from 0.163 to 0.178. The pH of the solutions was adjusted by adding a proper amount of KOH. The concentrations of ionic species and substrate (MgATP) were calculated by a computer iteration procedure according to the method of Perrin and Sayce (1967).

The ability of the ATP feeder system, creatine kinase-phosphocreatine, which was added to eliminate the effect of diffusion of substrate into the fiber, was checked by observing the velocity of unloaded shortening at 34 μM MgATP. It was found that the shortening velocity increased as both the concentrations of creatine kinase and phosphocreatine were increased and that more than 1 mg/ml of creatine kinase and 20 mM phosphocreatine saturated the velocity. Therefore, subsequent experiments were performed with amounts of creatine kinase and phosphocreatine sufficient to eliminate the effect of diffusion.

Measurement of Tension Transients

A block diagram of our data acquisition system for recording tension transients in response to a quick length change is illustrated in Fig. 1. A microcomputer with Z-80 (Zilog) was used. It had a 12 kbyte read only memory, 48 kbyte random access memory, cassette magnetic tape unit, floppy disk unit, video display, thermal printer, 8 bit A/D converter (Burr Brown, ADC80), 8 bit D/A converter (Burr Brown, DAC80), and an interface to fetch data from an 8 bit digital recorder (Hitachi, VC-800). The length of the fiber bundle was controlled by the servo motor (General Scanning, G-100PD with the rise time of 320 μs) on the extension from which a muscle bundle was fixed. The motor was controlled by the D/A converter output signal. The angle of rotation of the motor was measured by the output signal of an internal capacitive transducer. The tension response of the fiber bundle was signalled by a universal integrated sensor (Kulite Semiconductor, SB-45-350-200(64)); natural frequency and time constant for damping were about 7.5 kHz and 0.15 ms, respectively, when the fiber bundle was attached. Its sensitivity and compliance were 2 mV/mM and 0.035 μm/mN, respectively.

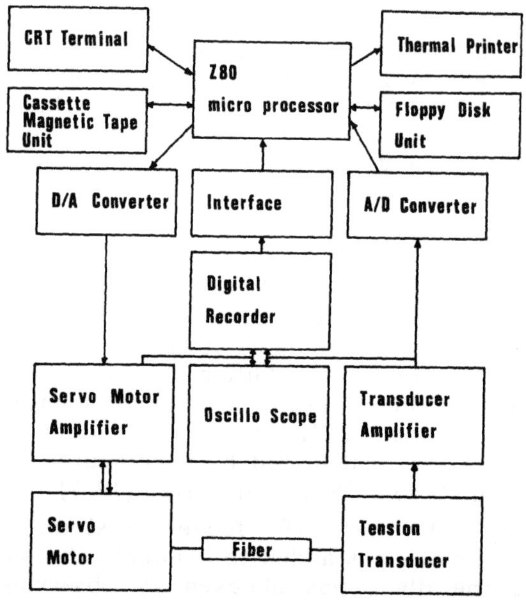

Figure 1: A block diagram of the data acquisition system for recording tension transients.

The tension sensor output signal was amplified and sent to the digital recorder as well as to the A/D converter. The most rapid part of the transient signal against a quick length change, the first 10 ms, was recorded with a digital recorder with a sampling rate of 10 μs. The other part of the signal was recorded with the microcomputer with a sampling rate of 104 μs to 31.25 ms, which was adjusted by a microcomputer program. Recorded data were transferred to a cassette magnetic tape as well as to a floppy disk, and stored . The recorded trace of tension was transferred to a high speed large scale computer of the University of Tokyo (Hitachi, M200H) via the floppy disk. Most of data processing was done by this large computer.

For the rate analysis of the tension recovery, the "SALS" program (Nakagawa and Oyanagi, 1980) provided by the computer center of the University of Tokyo was utilized. Data points used for the calculation were taken from the stored data according to the following principle: the datum at the time when the tension had the extreme (maximal or minimal) value just after the length change was taken as the first point, then successive 79, 67, 63, 64, 64, 64, 63 and 32 points were sampled up to 2 s at time intervals of 0.01, 0.03, 0.104, 0.313, 1.042, 3.125, 10.417 and 31.25 ms, respectively.

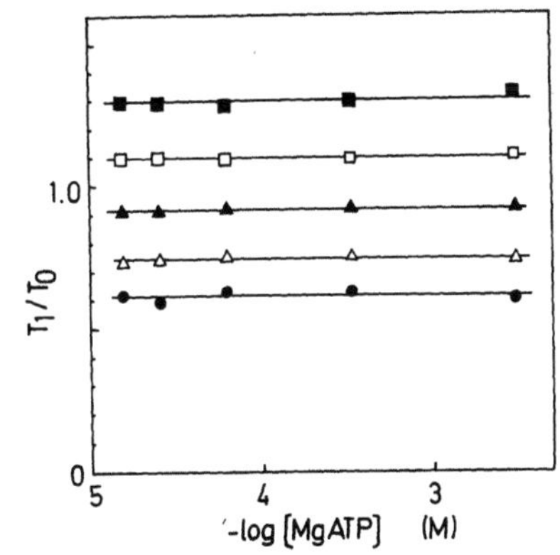

Figure 2: T_1/T_0 as a function of substrate concentration; ■, □, ▲, △ and ● denote the step amplitudes of +3.0, +1.0, -1.0, -3.0 and -5.0 nm/half-sarcomere respectively. Each data point denotes the average of 2-4 observations at pH 7.27 and 2-4°C.

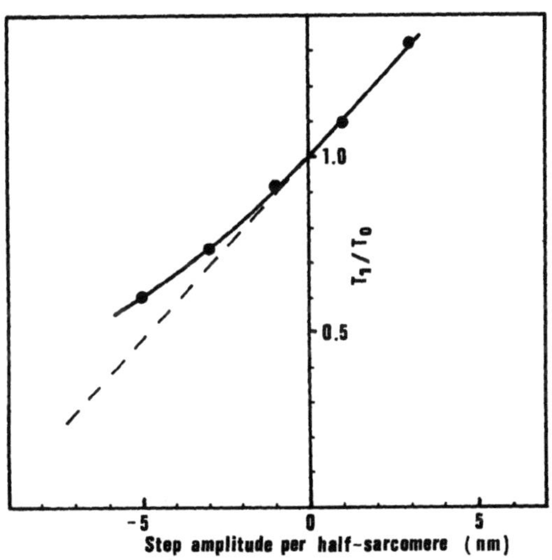

Figure 3: T_1/T_0 against the step amplitude of a length change. Substrate concentration is 3.1 mM. Each data point is the average of 4 observations at pH 7.27 and 2-4°C.

RESULTS

In Fig. 2, T_1/T_0 is plotted as a function of substrate concentration. This ratio is not affected by substrate concentration, [S], at least in the range of our observation which is higher than 15 μM. The ratio T_1/T_2 against various amplitude of a step length change plotted in Fig. 3 is concave. Extrapolation of the linear part of the T_1/T_0 curve intersects the horizontal axis at -9.5 nm/half-sarcomere. This value is rather larger than the one reported by Ford et al., (1977). One reason for this may be attributed to the difference in the elasticity of a cross-bridge between

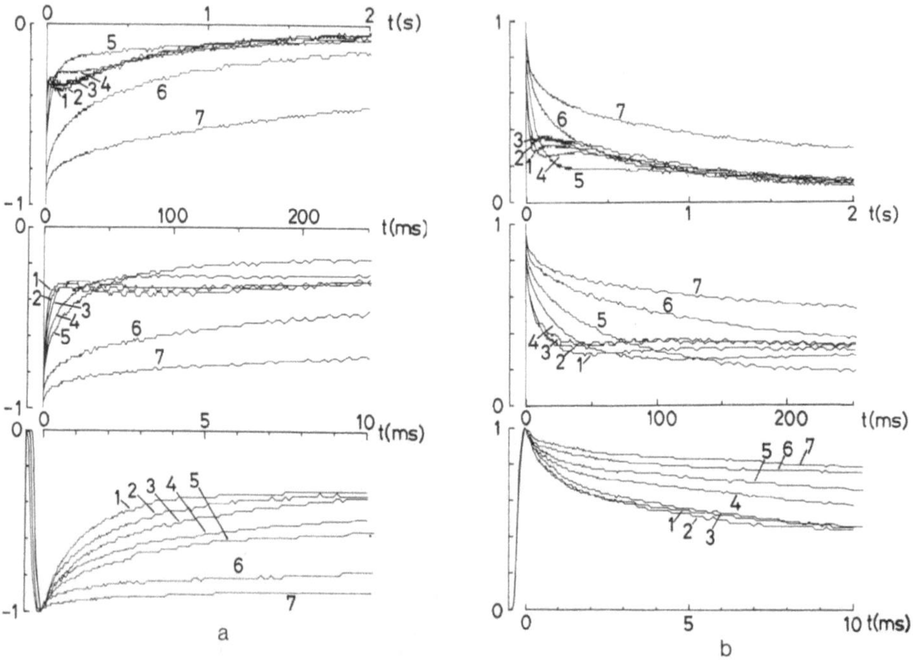

Figure 4: Superimposition of typical tension traces at various substrate concentrations for a quick release, (a), and for a quick stretch, (b), at pH 7.10 and 2-4°C. Data are normalized to each extreme value. Ordinate shows $(T(t) - T_0)/|T_1 - T_0|$. Concentrations of substrate for the curves 1, 2, 3, 4, 5, 6, and 7 are 3.0 mM, 860, 330, 120, 60, 25 and 15 μM. Overall time scales in the three kinds of figures are taken from the bottom trace to the top at 10, 250 ms and 2 s, respectively.

frog fast muscle and rabbit psoas muscle rather than to that in the rise time of the servo motor between ours and theirs; our rise time is 320 μs, which is 1.6 times slower than theirs.

As is seen in Fig. 3, an almost linear response was obtained in T_1/T_0 against length changes where the step amplitude was less than 3 nm/half-sarcomere. Therefore, we used a step amplitude of 3 nm/half-sarcomere in the following experiments on tension transients. The average values of T_1/T_0 for a quick stretch or release of 3 nm/half-sarcomere are 1.29 and 0.71 respectively.

The tension traces for a quick stretch or release of 3 nm/half-sarcomere are normalized to each extreme value and are plotted in Figs. 4(a) and (b), respectively. In these figures, the same tension traces are plotted in three different time scales. From the bottom to the top, the overall time scales are taken as 10 ms, 250 ms and 2 s, respectively. The origin of the time t is taken at the time when tensions begin to recover.

The time course of the tension recoveries may be expressed as a superimposition of exponential processes as reported by Abbot and Steiger (1977):

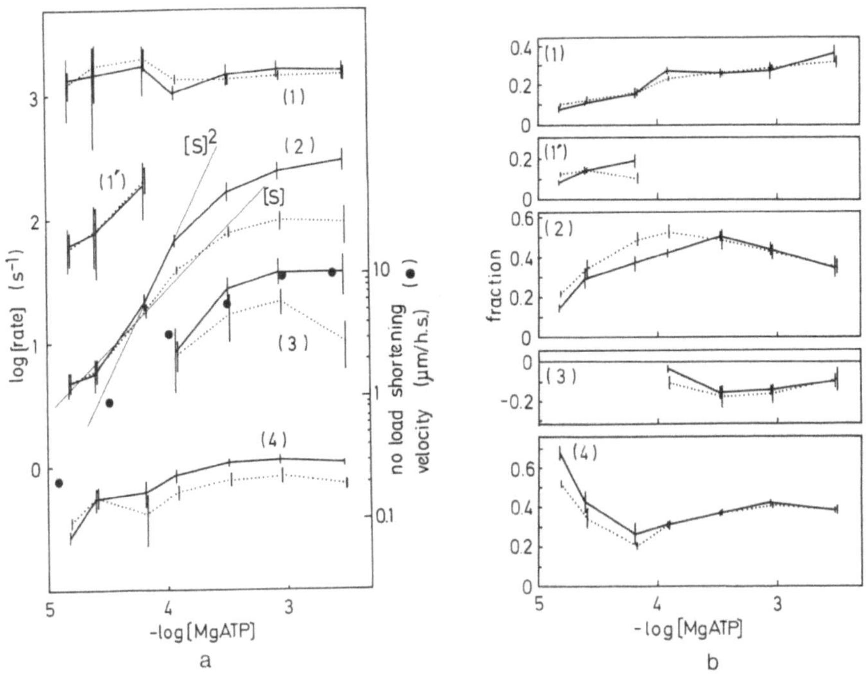

Figure 5: Results of rate analyses plotted as a function of substrate concentration in a logarithimic scale. Each data point is the average of 4-9 observations. Rate constants (K_i) on a logarithmic scale are shown in (a) and fractions (A_i) in (b). Solid and dotted lines respectively indicate the data for a quick release and stretch. Small solid circles show the velocity of unloaded shortening observed by Chaen et al. (1981). Thin linear lines respectively represent a quantity porportional to substrate concentration, ([S]), and to the square of substrate concentration, ($[S]^2$).

$$(T(t) - T_0)/(T_1 - T_0) = \sum A_i \exp(-K_i t) \quad (1)$$

where T(t) represents the tension at time t, T_0 isometric tension, T_1 the extreme value of tension in response to a quick length change and A_i the size of the fraction of the ith process of which the rate constant is K_i in the total process.

In Figs. 5 (a), (b) are respectively shown rate constants (K_i) and fractions (A_i) obtained by analysis (1) applied to tension traces in various [S]. Each point stands for the average of 4-9 data. The tension recovery is comprised of five rate processes, (1), (1'), (2), (3) and (4) in order of the magnitude of their rates. Process (1') is observable only when [S] is below 66 μM while in process (3) with a negative fraction only above 120 μM.

In Fig. 5 the most clear dependence on [S] is seen in K_2. For quick stretch, K_2 is a linearly increasing function of [S] and reaches a plateau of about 100 s^{-1}. On the other hand K_2 for quick release can be approximated by a second order increasing function with a plateau at 300 s^{-1}. For quick release, K_3 is also a linearly increasing function of [S] with a plateau of about 40 s^{-1}. However, K_3 for quick stretch does not have a simple dependence on [S]. When [S] is above 66 μM, K_4 is not so clearly affected by the concentration of substrate and it seems slightly larger for release than for stretch. However, it becomes an increasing function of [S] below 25 μM where A_4 is decreasing.

DISCUSSION

As was reported by Huxley (1974), the tension response of an intact frog muscle fiber to a step length release can be divided into four phases, 1 to 4. According to Huxley, phases 1, 2, 3 and 4 are assigned to the simultaneous drop of tension, rapid early tension recovery, extreme reduction or even reversal of rate of tension recovery and gradual recovery of tension towards the isometric tension, respectively. As is shown in Figs. 4, these four phases are observable in the tension response of glycerinated rabbit psoas muscle fibers when substrate concentration is higher than 100 μM. A similar tension response is also observed in the opposite direction for a quick length stretch. According to Huxley and Simmons (1971), Huxley (1974) and Ford et al. (1977), phase 1 is attributable to the instantaneous elasticity of cross-bridges and the other phases to changes in a mechanical state of the cross-bridges. Fig. 2 shows that the amplitude of phase 1 which is normalized to isometric tension, $(1 - T_1/T_0) \equiv X$, is not affected by [S], when [S] is higher than 15 μM. Furthermore, if we take into account of the effect of phase 2, X becomes a linear function of the amplitude of the length step (Fig. 3). Considering both Fig. 2 and Fig. 3, the coefficient between X and the step amplitude is concluded to be independent of [S]. Thus, it can be said that the instantaneous elasticity of cross-bridges does not depend on [S].

Kawai (1978) applied a sinusoidal analysis to isometrically contracting muscle fibers. He found that the tension change in response to a length change is approximately described by a transfer function which is the sum of three independent exponential rate processes. The three processes, process (A), (B) and (C) in order of slow to fast, have a close correlation to phase 4, 3 and 2, respectively (Kawai and Brandt, 1980). Moreover, comparing the rate constants mutually, their processes (A), (B) and (C) correlate with our processes (4), (3), and (2), respectively. Since Kawai and his collaborators did not observe rate processes which were faster than process (C), our process (1) and (1') might not be discriminated by their sinusoidal analysis. Cox and Kawai (1981) observed the [S]-dependence of three rate processes. Their results do not contradict our observations of [S]-dependence of rates and fractions of the processes shown in Fig. 5.

Figs. 5(a,b) show that the five rate processes comprising the time course of tension recovery may be classified into two groups: one is composed of processes (1), (1') and (4) and the other is composed of processes (2) and (3). In the first group, not only rate but also fraction are almost the same in magnitude for stretch and release. As a result, the processes belonging to this group give a symmetrical response with respect to the direction of the length change. The mechanical changes in cross-bridges coupling to these paired symmetric processes might have almost the same rate constant magnitude for quick release and stretch. However, this is not the case for the processes belonging to the second group, which give an asymmetric response with respect to the direction of the length change. It should be noted that only the rates of the processes belonging to the second group depend on substrate concentration. These facts seem to indicate that the second processes, (2) and (3), are related to a chemical change but the first ones, (1), (1') and (4), are not. When [S] is lower than 66 μM, K_1, and K_4 also have [S] dependence. However K_1 and K_4 are almost the same in magnitude between stretch and release. This suggests that they would not couple with chemical changes directly.

It has been considered that phase 2 for release is due to a conformational change in attached cross-bridges from preactive to active state (Huxley and Simmons, 1971; Nishiyama et al., 1977; Eisenberg et al., 1980). This will also be the case for phase 2 for stretch: a conformational change from active to preactive state. Thus, process (1) and (1') which are observed at the earlier part of phase 2 can be assigned to the conformational change in the attached cross-bridges without associating with the chemical change. This is in contrast to process (2) observed at the latter part of phase 2; K_2 for release and stretch seem to be proportional to $[S]^2$ and [S], respectively. Shimizu and Yamada (1975) discussed that the transition from the preactive to active state, would be accelerated by using chemical energy from ATP decomposition. As phase 2 is not coupled with a detachment process (Ford et al., 1974; Julian & Morgan, 1981), the presence of substrate concentration dependence in the rate of process (2) will be interesting in relation to so-called direct decomposition of substrate without detachment of cross-bridge (Inoue et al., 1979; Stein et al., 1979).

As shown in Fig. 5(a), the substrate dependence of K_3 for release has a close correlation with that of the velocity of unloaded shortening observed by Chaen et al., (1981) with the same kind of glycerinated muscle fiber. As they concluded that the velocity of unloaded shortening is determined by a rate of cross-bridge detachment, we may assign process (3) for relase to the detachment of cross-bridges. Process (3) for release is a tension decreasing process, suggesting this process is due to the detachment of force-producing (i.e. active) cross-bridges (Huxley, 1974). As the substrate concentration dependence of unloaded shortening velocity seems to be proportional to [S], we would say a single ATP molecule is decomposed when a cross-bridge is detached. Process (3) for stretch is a tension increasing process, suggesting this process is due to the attachment of cross-bridges to the active state. But this will not be the exact reverse process, because direct attachment of cross-bridges to active state has a very small probability. It would be a double step-change via preactive state (Nishiyama et al., 1977; Eisenberg et al., 1980).

Process (4) for release is an attachment process of cross-bridges (Huxley, 1974). As the cross-bridge attachment is thought to take place in two stages (Huxley, 1973; Nishiyama et al, 1977; Eisenberg et al., 1980), process (4) for release may be assigned to attachment of detached cross-bridges to the preactive state. Thus, process (4) for stretch may be assigned to detachment from the preactive state, which is the reverse process. From our observations, K_4 is not affected by [S] when [S] is higher than 66 μM. Moreover K_4 for release has almost the same magnitude as that for stretch. These observations are consistent with the view that the attachment to the preactive state can be considered as a physical reversible process (Huxley, 1973; Nishiyama et al., 1977; Eisenberg et al., 1980).

When [S] is decreased below 66 μM, K_4 decreases. Especially below 25 μM, K_4 seems to be almost proportional to [S], and A_4 rapidly increases as [S] is decreased. When [S] is further decreased from 15 μM, K_4 decreases and A_4 greatly increases. Consequently, tension recovery phases other than process (4) disappear (Heinl et al., 1974). This seems to mean that the number of cross-bridges which substantially contribute to tension recovery would decrease. Marston & Tregear (1974) reported that when [S] is decreased below 30 μM, the amount of nucleotide bound to myosin head decreases in calcium activated myofibrils. Myosin heads without bound nucleotide would not contribute to tension recovery. Thus, the increase in A_4 in low [S] seems to be related to an increase in the number of myosin heads without attached nucleotide. As K_4 has an [S]-dependence when [S] is lower than 30 μM, we can speculate that process (4) at these substrate concentrations is a process of bringing myosin heads to a state contributing to tension recovery by binding substrate. Process (1') appears only where [S] is lower than 66 μM. As the sum of A_1 and A_1' is almost independent of [S], process (1') would be a part of process (1). The splitting of process (1) into (1) and (1') might be caused by the existence of cross-bridges without bound nucleotide, which might affect the motions of active and preactive cross-bridges.

REFERENCES

Abbott, R.H. and Steiger, G.J. (1977). Temperature and amplitude dependence of tension transients in glycerinated skeletal and insect fibrillar muscle. J. Physiol. 266: 13-42.

Arata, T., Mukohata, Y. and Tonomura, Y. (1977). Structure and function of the two heads of the myosin molecule. VI. ATP hydrolysis, shortening and tension development of myofibrils. J. Biochem. 82: 801-812.

Chaen, S., Kometani, K., Yamada, T. and Shimizu, H. (1981). Substrate-concentration dependences of tension, shortening velocity and ATPase activity of glycerinated single muscle fibers. J. Biochem. 90: 1611-1621.

Cooke, R. and Bialek, W. (1979). Contraction of glycerinated muscle fibers as a function of the ATP concentration. Biophys. J. 28: 241-258.

Cox, R.N. and Kawai, M. (1981). Alternate energy transduction routes in chemically skinned rabbit psoas muscle fibers: a further study of the effect of MgATP over a wide concentration range. J. Mus. Res. Cell Motil. 2: 203-214.

Eisenberg, E., Hill, T.L. and Chen, Y. (1980). Cross-bridge model of muscle contraction: Quantitative analysis. Biophys. J. 29: 195-227.

Ferenczi, M.A., Goldman, Y.E. and Simmons, R.M. (1979). The relation between maximum shortening velocity and the magnesium adenosine triphosphate concentration in frog skinned muscle fibres. J. Physiol. 292: 71-72P.

Ford, L.E., Huxley, A.F. and Simmons, R.M. (1977) Tension responses to sudden length change in stimulated frog muscle fibres near slack length. J. Physiol. 269: 441-515.
Heinl, P., Kuhn, H.J. and Ruegg, J.C. (1974). Tension responses to quick length changes of glycerinated skeletal muscle fibers from the frog and tortoise. J. Physiol. 237: 243-258.
Huxley, A.F. (1973). A note suggesting that the cross-bridge attachment during muscle contraction may take place in two stages. Proc. R. Soc. B 183: 83-86.
Huxley, A.F. (1974). Muscular contraction (review lecture). J. Physiol. 243: 1-43.
Huxley, A.F. and Simmons, R.M. (1971). Proposed mechanism of force generation in striated muscle. Nature, Lond. 233: 533-538.
Huxley, H. (1969). The mechanism of muscular contraction. Science, N.Y 164: 1356-1366.
Inoue, A., Takenaka, H., Arata, T. and Tonomura, Y. (1970.) Functional implications of the two-headed structure of myosin. Adv. Biophys. 13: 1-194.
Julian, F.J. and Morgan, D.L. (1981). Variation of muscle stiffness with tension during tension transients and constant velocity shortening in the frog. J. Physiol. 319: 193-203.
Julian, F.J., Sollins, K.R. and Sollins, M.R. (1974). A model for the transient and steady-state mechanical behavior of contracting muscle. Biophys. J. 14: 546-562.
Kawai, M. (1978). Head rotation or dissociation? A study of exponential rate processes in chemically skinned rabbit muscle fibers when MgATP concentration is changed. Biophys. J. 22: 97-103.
Kawai, M. and Brandt, P.W. (1980). Sinusoid analysis: a high resolution method for correlating biochemical reactions with physiological processes in activated skeletal muscles of rabbit, frog and crayfish. J. Mus. Res. Cell Motil. 1: 279-303.
Marston, S.B. and Tregear, R.T. (1974). Nucleotide binding to myosin in calcium activated muscle. Biochem. Biophys. Acta 333: 581-584.
Nakagawa, T. and Oyanagi, Y. (1980). Program system experimental sciences. In: *Recent Developments in Statistical Inference and Data Analysis*, pp. 221-225, ed. Matusita, K. Amsterdam, New York and Oxford: North-Holland Publishing Company.
Nishiyama, K., Shimizu, H., Kometani, K. and Chaen, S. (1977). The three-state model for the elementary process of energy conversion in muscle. Biochem. Biophys. Acta. 460: 523-536.
Noda, L., Kuby, S. and Lardy, H. (1955). ATP-creatine transphosphorylase. In: *Methods in Enzymology*, vol. II, pp. 605-616, ed. Colowick, S.P. and Kaplan, N.O. New York and London: Academic Press.
Perrin, D.D. and Sayce, I.G. (1967). Computer calculation of equilibrium concentrations in mixtures of metal ions and complexing species. Tantala 14: 833-842.
Shimizu, H. and Yamada, T. (1975). The synergetic enzyme theory of muscular contraction: a two-headed myosin model. J. Theor. Biol. 49: 89-109.
Stein, L.A., Schwarz, R.P., Jr., Chock, P.B. and Eisenberg. E. (1979). Mechanism of actomyosin adenosine triphosphatase. Evidence that adenosine 5'-triphosphate hydrolysis can occur without dissociation of the actomyosin complex. Biochemistry 18: 3895-3909.

DISCUSSION

KAWAI: I am very pleased to see your results from experiments which used the step length change technique. They are very comparable to what I have been observing using sinusoidal length changes, especially in terms of the effect of MgATP on rate constants and associated K_m values. We both agree that processes going on during phases 2 and 3 are involved in MgATP binding because of their MgATP sensitivity. This is solid experimental evidence, which should be incorporated in the cross-bridge modeling.

However, our interpretations of the reactions involved in the rate limiting steps of the two processes at saturating MgATP concentrations are opposite. We reasoned (Cox & Kawai, J. Mus. Res. & Cell Mot. 2:203-214, 1981) that the rate limiting step for process B (phase 3) is a pathway in which detachment is not mandatory (Inoue et al., J. Biochem. (Tokyo) 74: 923-934, 1973; Stein et al., Biochem. 18: 3895-3909, 1979), because a myosin cross-bridge does not necessarily have to dissociate from actin in order to produce oscillatory work on the forcing apparatus. We also reasoned that the rate limiting step for process C (phase 2) is the actomyosin dissociation reaction, because the muscle can shorten even when process B (oscillatory work) is absent at MgATP concentrations around 0.1-0.3 mM. It is my conviction that muscle shortening requires the detachment reaction. What is your opinion?

TANAKA: Your interpretation would be one of the possibilities but I cannot agree to it for the following two reasons: One is that in the early tension recovery phase no detachment occurs as shown by the double step length change experiments of Ford et al., (J. Physiol. 240: 42-43, 1974). We have verified their experiments and found that no remarkable change occurs in stiffness at the end of the early phase, which indicates process 2 does not correlate to detachment. The second reason is that if process 3, which is related to the oscillatory work you saw, would have a correlation to the undissociating pathway of the actomyosin ATPase reaction, its fraction should be increased under conditions to enhance the pathway (Inoue et al., 1979). However, from our preliminary experiments, such conditions do not necessarily enhance the fraction. As you have reported previously (Kawai & Orentlicher, Biophys. J. 16: 152a, 1976), and we have also verified, inorganic phosphate greatly enhances the fraction. So, I don't think there is evidence to suggest a relation between the process 3 and the pathway of the ATPase reaction.

WARSHAW: What was the extent of your length step?

TANAKA: Three nm per half sarcomere.

WARSHAW: If you went to larger stretches and releases, could you still assign which processes are symmetric and ATP-dependent. Are they still the same?

TANAKA: I cannot answer your question because we have not performed enough experiments with length changes other than 3 nm.

HOMSHER: I'd like to ask three questions about the interpretability of these kinds of experiments. First of all, when you have ATP in there -- and the ATP is coming from the outside -- it seems to me you are very likely to have radial gradients across the width of muscle fiber, and if you have rate constants that are functions of the ATP concentration, then you're going to have different cylinders contracting at different rates or undergoing their transitions at different rates, which will actually confound your tension records.

Secondly, from a set of records your curve stripping looks to be about five different rate constants. Andrew Huxley pointed out over 20 years ago that even when you try to find rate constants and amplitudes for a two-compartment analysis, the rate constants and amplitudes need to differ by a large amount in order to really get accurate curve stripping. It seems to me that when you have five it really compounds the situation.

I also wonder about the extent to which, by releasing at this one distance, you're making an assumption that the rate constants governing transitions between different cross-bridge states are independent of the force that's exerted on them, and how this, too, might affect your results. In other words, it seems that there are a lot of factors here interacting with one another, which really render this kind of interpretation difficult --and it isn't only the case of your interpretation but maybe even that of Huxley and Simmons that cannot really be interpreted in terms of rate constants of processes going on in the cross-bridges.

TANAKA: For the first question, we used a large enough amount of ATP regenerating system to eliminate the radial gradient of the concentration of ATP. The ability of the back-up system was checked even at concentrations of MgATP as low as 34 μM by measuring the no-load shortening velocity at various concentrations of the backup system. A sufficient concentration of the backup system was used in the experiments, at which no-load shortening velocity was fully saturated.

For your second question, I think our rate analysis is reasonable for the following two reasons. First, all the rate constants of the neighboring processes differ by almost an order of magnitude. Owing to this, there is only little uncertainty in our classification of rate constants and fractions. Errors in our rate analysis were so small that they were included in the error bars of the figures just shown. Thus, I am sure our analysis has little arbitrariness. Second, the tendencies of the substrate dependence of the rates and the fractions of all the processes differ one from another. So the processes have no correlation with one another.

My answer to the third question is as follows: As in Huxley's two-state model in 1957, successful cross-bridge models such as three-state models are based on a common assumption that transition probabilities between different cross-bridge states and the magnitude of force exerted at the cross-bridge were functions of the distance between the cross-bridge and the interacting actin sites. Once we have accepted this assumption, we have the situation that the specific distribution of attached cross-bridges depends on the sliding velocity. Consequently, in our assumption, the transition would depend on the extent of average force exerted by cross-bridges.

RITZ-GOLD: I'd be interested in knowing whether you have tried using a non-hydrolyzable substrate analog like MgAMP-PNP or MgPPi and, if so, what the kinetic transient in response to a concentration jump would look like under your experimental conditions. I would expect that such analogs might act as allosteric effectors to produce a release internally (i.e., mechanochemically). As such, their primary effect would be to destabilize rigor complexes by forming the ternary complex intermediate. In the case of MgATP, the rate of dissociation of ATP-myosin from actin would be much faster than the rate of conformational change of myosin from the rigor (45°) to the high affinity relaxed (90°) conformation. On the other hand, if the rate of dissociation were to be significantly slower than the rate of conformational change for an analog such as MgPPi, then one might anticipate a transient formation of significant quantities of the non-equilibrium ternary complex intermediate.

TANAKA: I haven't done this type of experiment with ATP analogs, but plan to do so in the future.

KAWAI: A comment to that question. I tried AMP-PNP but found no effect when both MgATP and MgAMP-PNP were present in equi-molar bases.

RITZ-GOLD: Did you do your experiments with MgAMP-PNP only in the absence of MgATP?

KAWAI: Yes I did. In the absence of MgATP, a cross-bridge assumes a rigor configuration with a typical elastic response to sinusoidal length changes, i.e., very little viscosity in the frequency range 0.25-167 Hz (Kawai & Brandt, J. Mus. Res. Cell Mot. 1: 279-303, 1980). This characteristic does not change even after MgAMP-PNP is introduced in millimolar concentrations, although the rigor tension is reduced and stiffness diminishes. This experiment shows that the length change does not modify the rate constants involved in MgAMP-PNP reactions.

RITZ-GOLD: What if you use pyrophosphate (PPi)?

KAWAI: Just the same as AMP-PNP. There is very little change in the frequency profile of elasticity, and there is very little viscosity.

RITZ-GOLD: In experiments with maleimide spin-labeled rabbit psoas fibers, I've observed transient changes in cross-bridge mobility state in response to MgPPi concentration jumps. Since these changes in cross-bridge state might be equivalent to a relaxation transient, I was wondering whether you have observed transient changes in tension in response to a MgPPi jump in your system.

KAWAI: Yes, my result agrees with yours. 5-10 mM MgPPi is needed to have any effect in rabbit psoas or crayfish walking leg muscles. As in the case of the MgAMP-PNP experiment, rigor tension is reduced and stiffness declines, but there is very little change in the frequency profile of the complex stiffness. The point I have to make clear is that these transients (processes 2-4) are present in a fully activated muscle, but they are absent in a rigor condition: hence the processes reflect kinetics of actively cycling cross-bridges. If you do a model analysis, you will find that any apparent (experimentally observed) rate constant is a complicated function of many intrinsic rate constants involved in the cross-bridge cycle scheme. But you will find that only one or two of them mostly affect the apparent rate constant, hence you can establish a mapping relationship. We have been doing this experimentally.

TREGEAR: Going back to Earl Homsher's criticisms which are very interesting, can I first ask one short question? At low concentrations of substrate, Dr. Tanaka, did you change the concentration of the backup system and find that it had no effect on your response? That is the easiest control. Have you done that?

TANAKA: Yes, I have done it. The concentration of our backup system was sufficiently high that the change of it gave no effect on the tension response.

TREGEAR: The second question Homsher asked, if I understood you correctly, you did not answer directly because you gave a theoretical answer where what he was saying was empirically, it is difficult to distinguish many different rate constants. But I just wanted to point out that the third rate constant in this system is the other way -- the process is going the other way -- and that actually makes the analysis a good deal easier, much easier to separate out the processes.

A COMPARISON OF MUSCLE STIFFNESS MEASUREMENTS OBTAINED WITH RAPID RELEASES OR STRETCHES OF FROG SEMITENDINOSUS FIBERS

Bernard H. Bressler and Laureen A. Dusik

Department of Anatomy, University of British Columbia, Vancouver, B.C., Canada V6T 1W5

ABSTRACT

The stiffness-tension relationship has been studied during the isometric twitch. All experiments were carried out at 0°C using small fiber bundles from frog semitendinosus. Rapid shortening and lengthening steps of 3-6 nm/hs, complete in 500 μsec were given at various times during the twitch. Instantaneous stiffness was measured as the ratio of ΔP to ΔL and expressed as a relative change to the maximum values recorded at the plateau of an isometric tetanus. The time course of the stiffness changes followed the tension for both releases and stretches. However, the stiffness measured with a rapid stretch was higher than when measured with a rapid release. This raises the possibility that the instantaneous elasticity is non-linear in stretches.

INTRODUCTION

The correlation of muscle stiffness (Huxley and Simmons, 1971; Bressler and Clinch, 1974, 1975) as well as muscle force (Gordon, Huxley and Julian, 1966) to the degree of overlap of the actin and myosin filaments has led to the concept that a major portion of a contracting muscle's instantaneous elasticity resides in the cross-bridges. An essential element in our current model of cross-bridge action (Huxley, 1974; Eisenberg and Hill, 1978) is that the instantaneous elasticity is characterized as being linear. The purpose of the present study was to verify this concept experimentally by comparing the properties of the instantaneous stiffness of contracting skeletal muscle measured with both rapid lengthening and shortening steps.

METHODS

Experiments were carried out on isolated fibre bundles from the frog (Rana pipiens) semitendinosus muscle maintained in Ringer solution at 0.5°C. Rapid step changes of length, of constant amplitude, were given at various times during the isometric twitch. Instantaneous stiffness values were measured as the ratio of ΔP to ΔL and expressed as a relative change to the maximum stiffness recorded at the plateau of an isometric tetanus. A laser diffraction technique, modified from Iwazumi and Pollack (1979) was used to measure sarcomere length at various points along the fibres.

RESULTS

Fig. 1 shows representative records of the tension transients produced with a rapid release or rapid stretch given at 50 msec after the beginning of an isometric twitch. In both panels the record with a length change is superimposed on the record of an isometric twitch which had been given a pre-stretch or a pre-release 10 msec before stimulation began. The fact that the tension prior to the length change superimposed on the record from a twitch with a pre-release or pre-stretch indicated that the length step did not shift the fibres along the tension-length curve. In these records a 2.9 nm/hs length change produced a tension change of 0.17 P/P_o and 0.22 P/P_o for the release and stretch respectively. The relationship between instantaneous stiffness and the twitch tension at which the length perturbation occurred, for seven experiments is shown in Fig. 2. The solid line in this figure is the predicted relationship if the stiffness was directly correlated to the tension. It may be seen that the stiffness values obtained with shortening steps are more closely correlated with the tension than those obtained with lengthening steps.

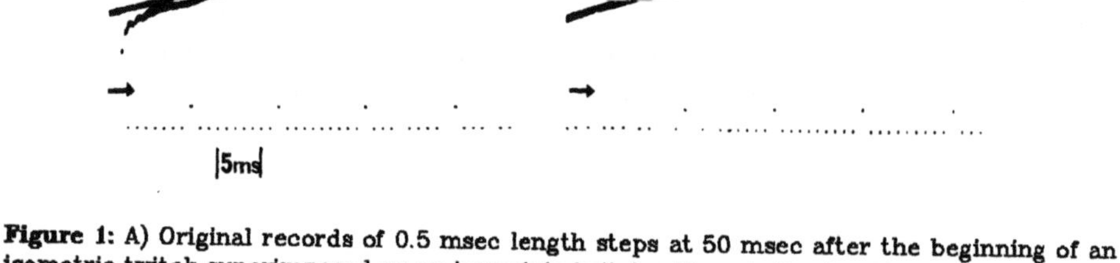

Figure 1: A) Original records of 0.5 msec length steps at 50 msec after the beginning of an isometric twitch superimposed on an isometric twitch with a similar length step occurring 10 msec before stimulation began. Arrows indicate the original tension baseline. Length changes expressed in nm/half-sarcomere. Oscilloscope was triggered 5 msec before length change occurred.

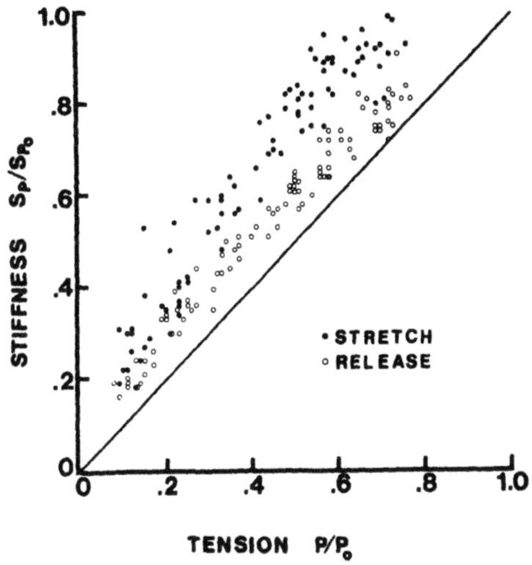

Figure 2: A comparison of twitch tension to stiffness, measured with rapid stretch or release, for seven experiments.

Moreover, the higher stiffness values recorded with rapid stretch were more pronounced at higher tension levels, which corresponded to the late stages of the initial development of tension, the peak, and the early relaxation phase of the isometric twitch.

DISCUSSION

Previous reports by Bressler and Clinch (1974; 1975) and Ford, Huxley and Simmons (1981) have suggested that stiffness measurements in a contracting muscle provide an index of the total number of cross-bridges that are formed. This conclusion was based on stiffness values that were obtained with rapid releases or a combination of releases and stretches to generate the T_1 curve. An additional constraint on the use of stiffness measurements as an index of the total number of cross-links that are formed at a given tension, is that the instantaneous elasticity is linear. Our current results do not appear to satisfy this condition. The fact that the instantaneous stiffness is higher when measured with a rapid stretch than when measured with a rapid release suggests that the instantaneous elasticity of the cross-bridges is non-linear (cf. Ford, Huxley and Simmons, 1981).

ACKNOWLEDGEMENTS

This work was supported by the Medical Research Council of Canada and a Pre-doctoral Fellowship from the Muscular Dystrophy Association of Canada to L.A. Dusik.

REFERENCES

Bressler, B.H. and Clinch, N.F. (1974). The compliance of contracting skeletal muscle. J. Physiol. 237: 477-493.

Bressler, B.H. and Clinch, N.F. (1975). Cross-bridges as the major source of compliance in contracting skeletal muscle. Nature 256: 221-222.

Eisenberg, E. and Hill, T.L. (1978). A cross-bridge model of muscle contraction. Prog. Biophys. Molec. Biol. 33: 55-82.

Ford, L.E., Huxley, A.F. and Simmons, R.M. (1981). The relation between stiffness and filament overlap in stimulated frog muscle fibres. J. Physiol. 311: 219-249.

Gordon, A.M., Huxley, A.F. and Julian, F.T. (1966). The variation in isometric tension with sarcomere length in vertebrate muscle fibres. J. Physiol. 184: 170-192.

Huxley, A.F. (1974). Muscular contraction. J. Physiol. 243: 1-43.

Huxley, A.F. and Simmons, R.M. (1971). Mechanical properties of the cross-bridges of frog striated muscle. J. Physiol. 217: 60P-61P.

Iwazumi, T. and Pollack, G.H. (1979). On-line measurement of sarcomere length from diffraction patterns in muscle. IEEE Trans. Biomed. Engineering 26: 86-93.

DISCUSSION

Dr. Gordon asked whether the possibility had been considered that the rate constants for the fast initial redevelopment phase after the step change of length might be different in stretch versus release. This would lead to an underestimate of the actual tension drop during release if the rate constant for release were faster than for stretch. In fact, that point had not been considered, but the rate constants will be measured to see whether this effect could significantly influence the data. There has been some effort to keep the length steps small where the rate constant of the initial redevelopment phase would be slow compared to larger length steps.

A lengthy discussion between Drs. Bressler and Cecchi served to confirm that the two sets of results were similar on the following points:

(1) the stiffness leads the tension during the initial development of tension in an isometric tetanus.

(2) the stiffness during the initial phase of relaxation, up to the shoulder, lags the tension decline. After the shoulder the tension and stiffness are once again more closely related.

(3) in spite of the fact that Cecchi used 9 kHz sine wave oscillations and 500 μsec steps were used here, the measured stiffness was nearly the same for both preparations, i.e. between 5-7 nm/half sarcomere. It appears that the limiting factor may not be speed, but rather, without some form of sarcomere-length control, it becomes difficult to eliminate the tendon compliance. Moreover, in a toe fiber (Cecchi's preparation) there is a greater proportion of tendon to muscle tissue than in a semitendinous fiber (Bressler's preparation.)

TENSION TRANSIENTS IN SKINNED MUSCLE FIBRES OF INSECT FLIGHT MUSCLE AND MAMMALIAN CARDIAC MUSCLE: EFFECT OF SUBSTRATE CONCENTRATION AND TREATMENT WITH MYOSIN LIGHT CHAIN KINASE

J.C. Rüegg, H.J. Kuhn*, K. Güth, G. Pfitzer and F. Hofmann

II. Physiologisches Institut und Pharmakologisches Institut der Universität Heidelberg Im Neuenheimer Feld 326
**Abteilung für Physiologie der Universität Ulm, FRG*

ABSTRACT

Glycerinated single fibres from the dorsal longitudinal muscle of Lethocerus maximus were isometrically contracted in MgATP-salines (10 μM Ca^{2+}; 1.5 mM Mg^{2+}; pH 6.7; 22°C and 20 mM PEP; 100 U/ml pyruvate kinase). The ratio of ATPase activity to tension decreased by a factor of 2 after reducing the ATP-concentration from 15 to 0.5 mM. At all ATP-concentrations (0.5 - 15 mM), the fibres showed tension adjustments in response to small step changes in length characteristic to an actively contracting muscle: i) an elastic phase which did not depend on ATP-concentration ii) a quick phase of stress relaxation with at least two exponential components; iii) a phase of delayed tension generation. An increase in size of the length step and/or a decrease of ATP-concentration slowed the quick phase and the delayed phase. Similar results have been obtained with skinned cardiac muscle (pig right ventricle). To see, how the isolated contractile system is affected by an increase in the light chain phosphorylation, tension transients were studied in skinned right ventricular muscle fibres before and after incubation with ATPγS (2 mM), pure myosin light chain kinase (9 μg/ml), Calmodulin (1 μM) and Ca^{2+} (0.8 μM). While isometric tension development elicited by 20 μM Ca^{2+} in the ATP salt solution was barely affected in presence of the enzyme, the ATPase activity was decreased by about 25% of the control. There was also a marked decrease (about 50%) in the contraction velocity as determined by the recovery of tension following a quick release. Quick stretches cause an immediate increase in tension followed by a rapid fall and a subsequent rise in tension. The velocity of this tension rise decreased by approximately 30% after incubation with myosin light chain kinase.

INTRODUCTION

It is usually assumed that force of muscle fibres is regulated by the number of crossbridges which are switched on and which are attached to actin at any one moment. Recently, evidence has accumulated that, in addition, the cycle time of crossbridges may also be regulated: For instance, Cooke et al. (1981) reported that in skinned skeletal muscle fibres, there was no loss in force after thiophosphorylation of the P-light chain of myosin (M_r 19,000), whereas the ATPase activity was decreased by as much as 50%. Obviously, the number of crossbridges acting was not reduced, but there was a reduction in speed, at which they cycle. It was also reported that a reduction in substrate concentration would slow the contractile response of skinned muscle fibres (e.g. Cox and Kawai, 1981). In this paper, we report the effects of changes in substrate concentrations on skinned insect flight muscle and cardiac muscle fibres as well as the effects of thiophosphorylation on the latter.

METHODS

Fibres from dorsal longitudinal muscle of Lethocerus were glycerinated according to Jewell and Rüegg (1966). Chemically skinned right ventricular muscle fibres were obtained as described by Herzig et al. (1981). The fibres were mounted horizontally between the carbon rod extension of a force transducer (Aksjeselskapet AME 801) and a length step generator, and force and stiffness were measured as described by Güth (1981). The resonance frequency of the transducer was more than 5 kHz, base line stability ± 10 μN at room temperature, system compliance was less than 2 N/mm. The preparations were relaxed in a solution ("relaxing solution") containing imidazole 20 mM, ATP 10 mM, $MgCl_2$ 12.5 mM, NaN_3 5 mM, EGTA 5 mM, pH 6.7, 24°C. Raising the free $[Ca^{2+}]$ to 20 μM by replacement of EGTA by Ca-EGTA induced a maximal contraction. All solutions contained an ATP regenerating system: Either creatinphosphate 10 mM and creatinekinase 1 mg/ml or phosphoenolpyruvate 5 mM and pyruvatekinase 0.11 mg/ml for experiments, in which the ATPase activity was measured. The latter solutions contained in addition NADH 40 mM and lactic dehydrogenase 0.125 mg/ml. Thiophosphorylation of the M_r 19.000 myosin light chain was induced by incubation of the fibres in an ATP free solution containing imidazole 20 mM, $MgCl_2$ 12.5 mM, NaN_3 5 mM, EGTA 5 mM, $CaCl_2$ 2.5 mM, Calmodulin 1 μM, pure cardiac myosin light chain kinase 7.5 μg/ml, ATPγ-S 2 mM for 30 min. The ATPase activity of the preparations was estimated with a coupled enzymatic assay from the decrease in [NADH] (c.f. Peterson, 1980). Cardiac myosin light chain kinase was prepared according to Wolf and Hofmann (1980).

RESULTS AND DISCUSSION

Figure 1 shows the substrate concentration dependence of the ATP splitting rate, immediate active stiffness and active force in Ca-activated glycerinated insect flight muscle. When the ATP concentration was

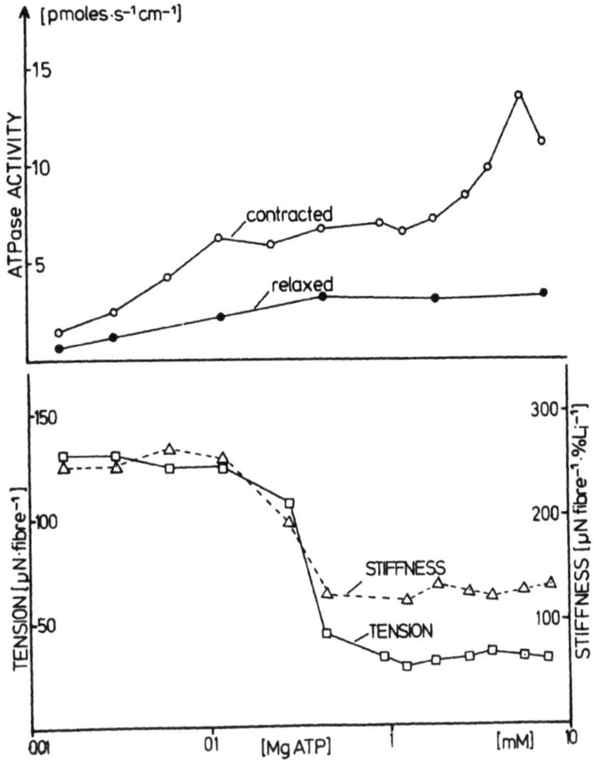

Figure 1: The effect of substrate concentration on ATP splitting rate and contraction of single glycerinated insect flight muscle fibres (Lethocerus maximus, dorsal longitudinal muscle). Abscissa Mg-ATP concentration (at pMg^{2+} 2.8). Open circles: ATP splitting rate at 10^{-5} M Ca^{2+}. Filled circles: at 10^{-8} M Ca^{2+}. Triangle: Active stiffness (total immediate stiffness-passive stiffness in relaxing solution). Open squares: Active force (total force-force in relaxing solution). The ATP saline contained an ATP regenerating system containing phosphoenolpyruvate (PEP) and phosphoenolpyruvate kinase to keep the concentration of high energy phosphate (PEP + ATP) at 20 mM at all ATP concentrations. pCa 5, pH 6.7, T 21°C.

lowered from 10 to 0.5 mM at constant free Mg-ion concentration (in presence of an ATP regenerating system), the Ca-activated Mg-dependent ATPase activity decreased; but both active tension and stiffness increased: The fibres enter the high tension state (c.f. Jewell and Rüegg, 1966), in which tension is maintained more economically than at high concentration of MgATP. Although there are more crossbridges attached at any one time (immediate stiffness is increased), these crossbridges split less ATP. If each cyclic interaction of crossbridges involves the splitting of one ATP molecule, this result, presumably, signifies that the cycle time of crossbridges increased. Note also that the tension to stiffness ratio increases as well, when the ATP concentration is lowered, suggesting a higher force per crossbridge.

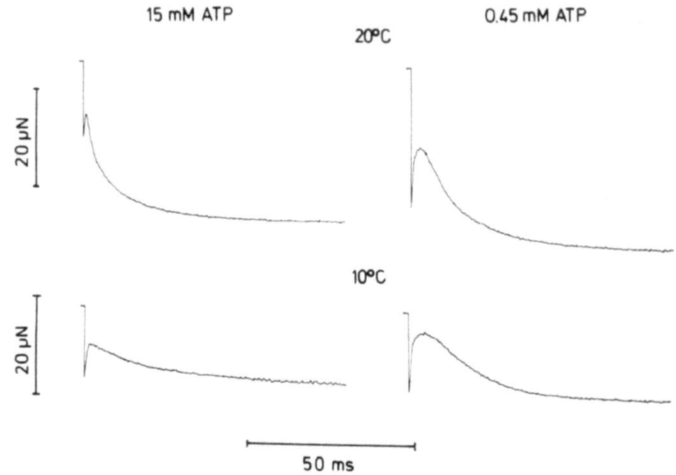

Figure 2: Delayed tension fall following quick releases (0.13% L_0) of single glycerinated fibre from dorsal longitudinal muscle of Lethocerus maximus. Effect of ATP concentration and temperature. For conditions, see Fig. 1.

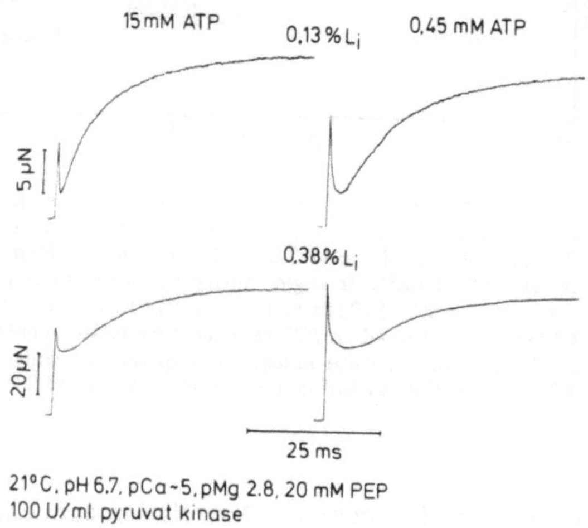

Figure 3: Tension transients of single glycerinated dorsal longitudinal muscle from Lethocerus maximus in response to 0.13% or 0..38% stretch. Note effect of substrate concentration. For conditions, see Fig. 1.

The question arises, whether the increased holding economy observed at low ATP concentration is due to a decrease in the rate constant of crossbridge detachment, since such a relationship between detachment rate constant and holding economy has been already suggested by Pybus and Tregear (1972). Also, an increase in substrate concentration is known to increase the rate of actomyosin dissociation by

ATP (Eccleston et al., 1976). Güth et al. (1981) showed that the rate of net detachment of crossbridges in insect flight muscle can be determined from the rate of delayed tension fall following a quick release which also causes a delayed fall in immediate stiffness. Note (c.f. Fig. 2) that the rate of tension decay is high in presence of 15 mM ATP, but decreases both after reduction of the ATP concentration to 0.45 mM or after lowering the temperature from 20 to 10°C. We suggest that the crossbridge detachment rate is decreased by lowering the ATP concentration and by lowering the temperature. The tension responses obtained following a quick stretch are more difficult to interpret. Immediate tension increase is followed by a quick tension fall consisting of at least 2 and possibly more components, and this is followed by a delayed tension rise. The reduction in ATP concentration slows these transients.

The processes are also slowed, when the amplitude of stretch is increased (Fig. 3).

Like insect flight muscle, skinned cardiac muscle (Steiger, 1971) also respond to quick stretch with a rapid increase in tension followed by a quick tension fall and then by a delayed tension rise. The quick phase as well as the delayed tension rise become considerably slower after reduction in ATP concentration (Fig. 4). Since the later phase of the quick

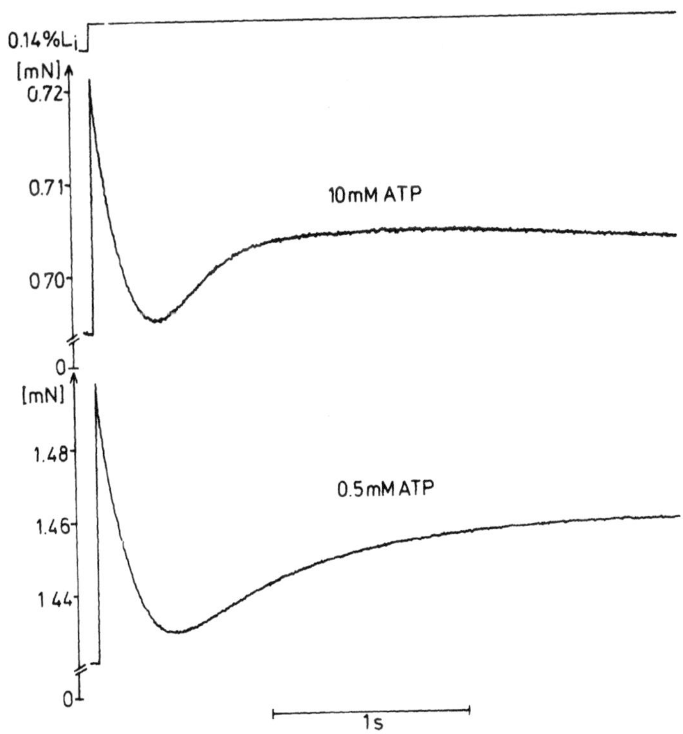

Figure 4: Tension transients in response to quick stretch (0.14% length change) of glycerinated pig right ventricle trabecular fibre bundle. Effect of substrate concentration.

phase as well as the delayed tension rise can be described by a single exponential, it is interesting to see, how the rate constants of these processes change with a change in ATP concentration. In Fig. 5, the reciprocal rate constant of the quick phase and of the delayed tension rise are plotted versus the reciprocal ATP concentration. In this way, a linear relationship is obtained. The abscissa intercept indicating a K_M of less than 100 μM.

In an attempt to thiophosphorylate the 19.000 Dalton light chain of cardiac myosin, skinned fibres were incubated with myosin light chain kinase (7.5 μg/ml), 0.8 μM Ca^{2+} and 0.02 mg/ml calmodulin and 2 mM ATPγS. After 30 min incubation, the preparations were incubated into relaxing solution, and an isometric contraction was elicited by increasing Ca^{2+} concentration to 20 μM. The lower curve of Fig. 6 shows the tension changes following quick release and restretch after treatment with myosin light chain kinase. These transients may be compared with the transient in control experiments, where incubation occurred without myosin light chain kinase. Compared with the control, the recovery following quick release is slower as are the transient tension responses following quick stretch. No decrease in the rate constants of tension responses has been observed, when the preparation was incubated with c-AMP dependent protein kinase and ATP.

The slow tension recovery suggests that the contraction velocity might be reduced after thiophosphorylation. Effects of contraction speed could be tested more directly by measuring the unloaded shortening velocity using Edman's procedure (c.f. Herzig and Rüegg, 1980). When the isometrically contracted fibre was suddenly released by about 12%, tension immediately dropped to zero, and the fibre bundle became slack; it

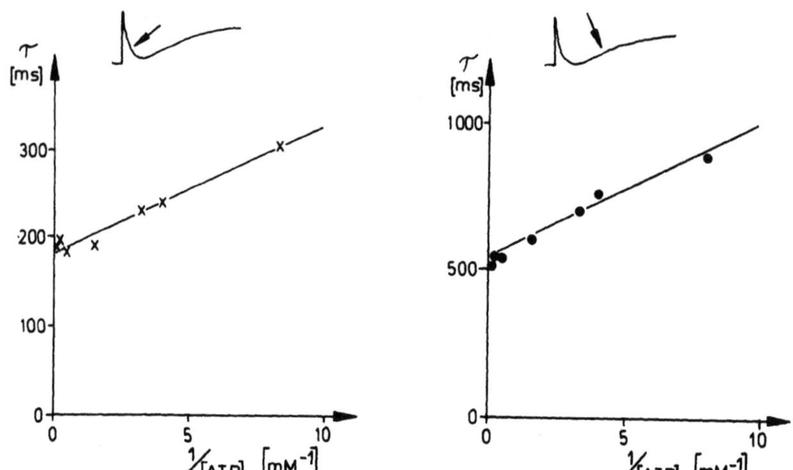

Figure 5: Effect of substrate concentration on tension transients of pig right ventricular trabecula. Plot of time constant of exponential tension changes versus reciprocal ATP concentration. Left: Late quick phase; right: Delayed tension rise following a quick stretch by 0.14%.

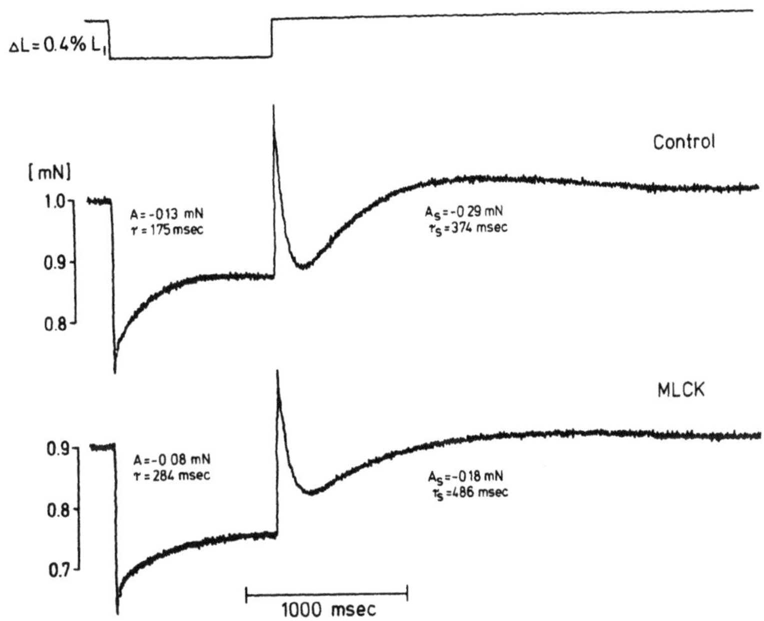

Figure 6: Effect of quick release (by 0.4% L_0) and restretch on bundles of glycerinated pig right ventricle fibre bundle. Above: Control experiments; below: response to length change following incubation with myosin light chain kinase (MLCK) and ATPγS.

then shortened under no load for the next 100 to 200 ms, until it was no longer slack, and tension started to recover. The lag time lasting from tension release until the onset of tension recovery increased by about 50% after treatment with myosin light chain kinase, suggesting that the velocity of unloaded shortening decreased (Fig. 7).

The reduction of shortening velocity following incubation with myosin light chain kinase is paralleled by a decrease in ATP splitting rate of the fibre bundles as measured in a NADH-coupled system. The NADH concentration was measured at 3 min intervals. Fig. 8 shows that the NADH concentration falls linearly with time suggesting a constant rate of ATP splitting. Following incubation with ATPγS, the rate of ATP splitting is reduced by about 20%, but no reduction is observed in control experiments.

Since tension and immediate stiffness are unaltered after treatment with MLCK, the reduction in ATP splitting rate presumably suggests that the crossbridge cycles have become slower. In principle, these results are very similar to those observed by Cooke et al. (1981) in the case of glycerinated psoas fibres.

In conclusion then, the crossbridge cycles become slower in cardiac muscle after reduction of substrate concentration or after treatment with myosin light chain kinase and ATPγS, a procedure which has been shown to thiophosphorylate the light chains of sceletal muscle. Reduction in cycling rate of crossbridges apparently renders tension maintenance

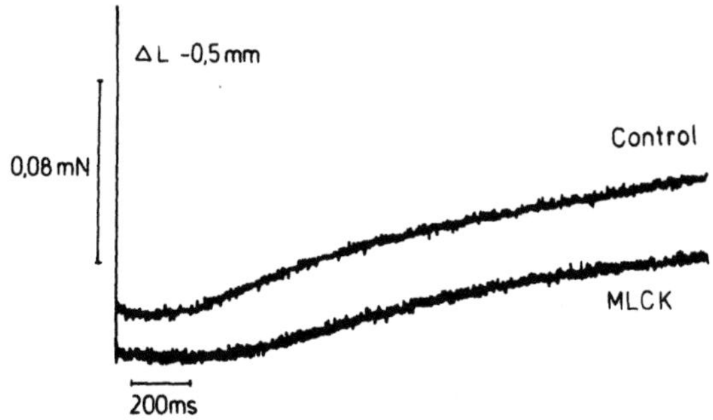

Figure 7: Lag time and tension recovery following a quick release (by 12%) of skinned cardiac right ventricle fibre bundle. Upper tracing: Control experiment. Lower tracing: Experiment following incubation with MLCK and ATPγS. Increase in lag time indicates reduction in maximum shortening velocity.

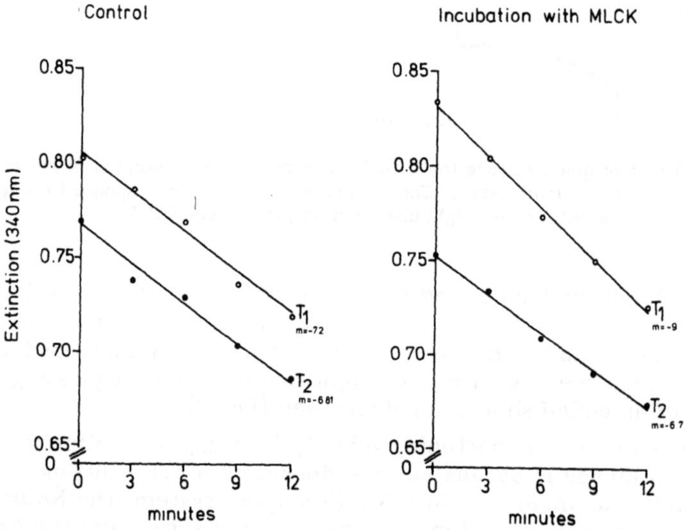

Figure 8: ATP splitting by glycerinated cardiac right ventricle fibre bundles measured as decrease in NADH absorption as function of incubation time. Left: Control measurements of ATPase activity between first contraction (T_1) and second contraction (T_2) changed less than 3.3 ± 2.4% (n=7), whereas force decreased by about 10%. Right: Measurement of ATP splitting rate in first contraction before (T_1) and in second contraction (T_2) after incubation with myosin light chain kinase and ATPγS. In 6 experiments, the ATPase activity following ATPγS treatment was decreased by 20.8 ± 1.4% ($\bar{x}$ ± SE) of control values before thiophosphorylation, while force was similar to that in controls.

more economical (c.f. Rüegg, 1971). This is particularly true in the case of living smooth muscle, where during relaxation, tension can be maintained almost without energy consumption (Siegman et al., 1981) or in skinned smooth muscle (guinea pig taenia coli). In these muscles, contraction can be induced by rising the Ca-ion concentration; after lowering the Ca-ion concentration by EGTA, relaxation is slow, but tension is maintained without active state (Schneider et al., 1981) and at a very low ATP splitting rate (Güth and Junge, 1981). Unlike skeletal muscle, phosphorylation of the M_r 20.000 light chain is said to increase cycling frequency and to decrease the holding economy (Dillon et al. 1981).

ACKNOWLEDGEMENT

This work was supported by the Deutsche Forschungsgemeinschaft.

REFERENCES

Cooke, R., Franks, K., Ritz-Gold, C.J., Toste, T., Blumenthal, D.K. and Stull, J.T. (1981). The function of myosin light chain phosphorylation in skeletal muscle. Biophysical J. 33: 235a.

Cox, R.N. and Kawai, M. (1981). Alternate energy transduction routes in chemically skinned rabbit psoas muscle fibres: a further study of the effect of MgATP over a wide concentration range. J. Muscle Res. and Cell Mot. 2: 203-214.

Dillon, P.F., Aksoy, M.O., Driska, S.P. and Murphy, R.A. (1981). Myosin phosphorylation and the crossbridge cycle in arterial smooth muscle. Science 211: 495-497.

Eccleston, J.F., Geeves, M.A., Trentham, D.R., Bagshaw, C.R. and Mrwa, U. (1976). The binding and cleavage of ATP in the myosin and actomyosin ATPase mechanism. In: *Molecular Basis of Motility*, pp. 42-52, ed. Heilmeyer, L., Rüegg, J.C., Wieland, T., Heidelberg: Springer Verlag.

Güth, K., Kuhn, H.J., Tsuchiya, T., Rüegg, J.C. (1981). Length dependent state of activation-length change dependent kinetics of cross bridges in skinned insect flight muscle. Biophys. Struct. Mech. 7: 139-169,.

Herzig, J.W., Rüegg, J.C. (1980). Investigations on glycerinated cardiac muscle fibres in relation to the problem of regulation of cardiac contractility - effects of Ca^{2+} and cAMP. Basic Res. Cardiol. 75: 26-33.

Herzig, J.W., Köhler, G., Pfitzer, G., Rüegg, J.C. and Wölffle, G. (1981). Cyclic AMP inhibits contractility of detergent treated glycerol extracted cardiac muscle. Pflügers Arch. 391: 208-212.

Jewell, B.R. and Rüegg, J.C. (1966). Oscillatory contraction of insect fibrillar muscle after glycerol extraction. Proc. Roy. Soc. B. 164: 428-459.

Junge, J., Güth, K. (1981). High force-low ATPase activity observed immediately after withdrawing Ca from the contracting smooth muscle (skinned Taenia coli). Pflügers Arch. 391: R 38.

Peterson, J.W. (1980). Vanadate ion inhibits actomyosin interaction in chemically skinned vascular smooth muscle. Biochem. Biophys. Res. Comm. 95: 1846-1853.

Pybus, J. and Tregear, R. (1972). Estimates of force and time of actomyosin interaction in an active muscle and of the number interacting at any one time. Cold Spring Harbor Symp. Quant. Biol. 37: 655-668.

Rüegg, J.C. (1971). Smooth muscle tone. Physiol. Rev. 51: 201-258.

Schneider, M., Sparrow, M.P. and Rüegg, J.C. (1981). Inorganic phosphate promotes relaxation of chemically skinned smooth muscle of Guinea pig Taenia coli. Experientia 37: 980-982.

Siegman, M.J., Butler, Th. M., Moers, S.U. and Davies, R.E. (1980). Chemical energetics of force development, force maintenance and relaxation in mammalian smooth muscle. J. Gen. Physiol. 76: 609-629.

Steiger, G.J. (1971). Stretch activation and myogenic oscillation of isolated contractile structures of heart muscle. Pflügers Arch. 330: 347.

Wolf, H., Hofmann, F. (1980). Purification of myosin light chain kinase from bovine cardiac muscle. Proc. Natl. Acad. Sci. 77: 5852-5855.

DISCUSSION

COOKE: We've shown that both phosphorylation and thiophosphorylation of myofibrils can control their ATPase activity. So I think that thiophosphorylation may be doing the same thing as phosphorylation.

RUEGG: Have you some experience with cardiac muscle?

COOKE: Yes, we looked at cardiac myofibrils. And this is a preparation in which myofibrils are very slightly cross-linked with glutaraldehyde before addition of ATP. This prevents contraction and maintains an ordered sarcomere array. And, under those conditions, one gets an effect of phosphorylation, a decrease of ATPase activity, similar in both cardiac and skeletal myofibrils. However, if you don't crosslink the myofibrils, if you allow them to contract or if you look at the ATPase of the purified proteins, we see no effect of phosphorylation. So I think it's an effect which one only finds in the organized filament array.

BRESSLER: Am I correct in assuming that you showed no change due to the added myosin light chain kinase following the quick release? Did it have no effect on the recovery process following the quick release, but only after a quick stretch?

RUEGG: No, the recovery process is slowed down both after quick release as well as after stretch.

GORDON: I want to mention also that at the Biophysical Society meeting, Rick Moss reported that in skeletal muscle, if he removed the DTNB light chain, he got a lower shortening velocity, implying a role for that particular light chain in controlling shortening velocity. How sure are you that you're not phosphorylating other proteins within the muscle, for example TN-I, or proteins like that?

RUEGG: Well, that we don't know, since quantitative estimates of phosphorylation are still lacking. On the other hand, addition of c-AMP and c-AMP-dependent protein kinase does cause phosphorylation of TN-I, but not the mechanical effects mentioned above.

TER KEURS: How did shortening velocity in these experiments compare with shortening velocity in intact preparations of the same kind?

RUEGG: I don't know the shortening velocity of living pig ventricular muscle. We used skinned pig cardiac muscle fibers and the unloaded shortening velocity is one length per second or less at 20°C.

NOBLE: I wonder how important this is in physiological conditions in heart muscle, because England (in *Cardiac Metabolism*, Ed. Drake-Holland and Noble, John Wiley, New York, 1983) has shown that the amount of phosphorylation of the myosin light chain is almost constant, however much you alter the conditions, such as, for instance, catecholamine stimulation.

RUEGG: Yes, I'm very much aware that so far most attempts to show changes in the extent of myosin light chain phosphorylation under physiological conditions failed to find these changes.

CECCHI: When you changed the ATP concentration, there was a change in the relation between stiffness and tension. Do you have an explanation for that?

RUEGG: No, we don't have any explanation for it, but it presumably means that the force per cross-bridge is not constant under these conditions. It could be that the population of cross-bridges attached at various attitudes changes. It's a very interesting point, one which may be worthwhile to follow up.

TREGEAR: I would just like to report that Loxdale found the same thing on insect flight muscle in his thesis about three years ago.

GORDON: On that particular point, I wish Bob Simmons were here, since they've done measurements on the stiffness-force ratio in frog skeletal fibers. My impression was that at low ATP concentration, the ratio of stiffness to force actually went *up*. And you showed it going down. Am I correct in that?

SCHOENBERG: I think that is what they reported at the last Biophysical Society meeting (Hamrell et al., Biophys J. 37: 356a, 1982).

RUEGG: We also find that in skeletal skinned muscle in rigor there is more stiffness per force than in normal contraction (Herzig et al., Pflügers Arch. 389: 97-103, 1981), so that in this respect we would agree with Simmons. In skinned insect flight muscle the stiffness-tension ratio increases when the magnesium concentration is lowered at constant ATP concentration (Griffiths et al., Pflügers Arch. 382: 155-163, 1979). At very low concentration of both magnesium and ATP and in the presence of an ATP regenerating system we found (cf. Fig. 1) that insect flight muscle enters a "high tension state" which is characterized by a lower stiffness to tension ratio.

WINEGRAD: Have you looked at the stiffness- tension ratio in your cardiac fibers?

RUEGG: Yes, in cardiac muscle fibers the stiffness to tension ratio seems to increase somewhat.

WINEGRAD: Over what range of ATP concentration?

RUEGG: Well, from 0.1 to 10 mM.

EDMAN: Do your ATPase value and S_0 value depend on the calcium concentration?

RUEGG: The ATP splitting rate is very much calcium dependent and the S_0 value is also clacium dependent (Herzig and Ruegg, Basic Res. Cardiol. 75: 26-33, 1980). Maybe I should add that we agree that there is an effect of ATP concentration on phase 2, and we know that phase 2 is not associated with a change in stiffness, suggesting that there is no net detachment of cross-bridges. I should stress no *net* detachment, because there could be also detachment and reattachment of bridges during phase 2.

TENSION TRANSIENTS IN SINGLE ISOLATED SMOOTH MUSCLE CELLS

David M. Warshaw and Fredric S. Fay

Department of Physiology, University of Massachusetts Medical School, 55 Lake Avenue N., Worcester, Massachusetts 01605

ABSTRACT

Tension transients have been recorded for the first time in a single smooth muscle cell. The transient contains a linear elastic response and a biphasic recovery which appear to originate from the cross-bridges. A comparison of transients in smooth and fast skeletal muscle fibers suggests that the cross-bridge in smooth muscle is more compliant than in striated muscle and that transitions between several cross-bridge states occur more slowly.

The cyclic interaction of cross-bridges in smooth muscle is believed to be responsible for force generation or active cell shortening (Fay, Rees and Warshaw, 1981). In order to obtain insight into the cross-bridge mechanism in smooth muscle, we have undertaken a study of the response of single isolated smooth muscle cells (SMC) to rapid changes in length during isometric force production. For this purpose, we have attached a single SMC isolated by enzymatic disaggregation (Fay, Hoffman, Leclair and Merriam, 1982) of the stomach muscularis of the toad, *Bufo marinus*, around special probes, which are in turn connected to a length driver and newly designed force transducer (Fig. 1A).

Tension transients recorded in a single SMC by this method in response to either a rapid release or stretch are illustrated in Figure 1B. Three general features are apparent: 1) an immediate tension change that coincides with the applied length step; 2) a rapid partial recovery of tension that occurs during the length step, and; 3) upon completion of the length step, a further recovery of tension that is resolvable into two components and results in virtually complete recovery in 1 second. Strobe photographs of the distribution of marker beads on the surface of the cell at various times after completion of the length step revealed that the length change was uniformly distributed over the cell (Fig. 1A). Thus, the tension transients appear to reflect the response of the cell as a whole,

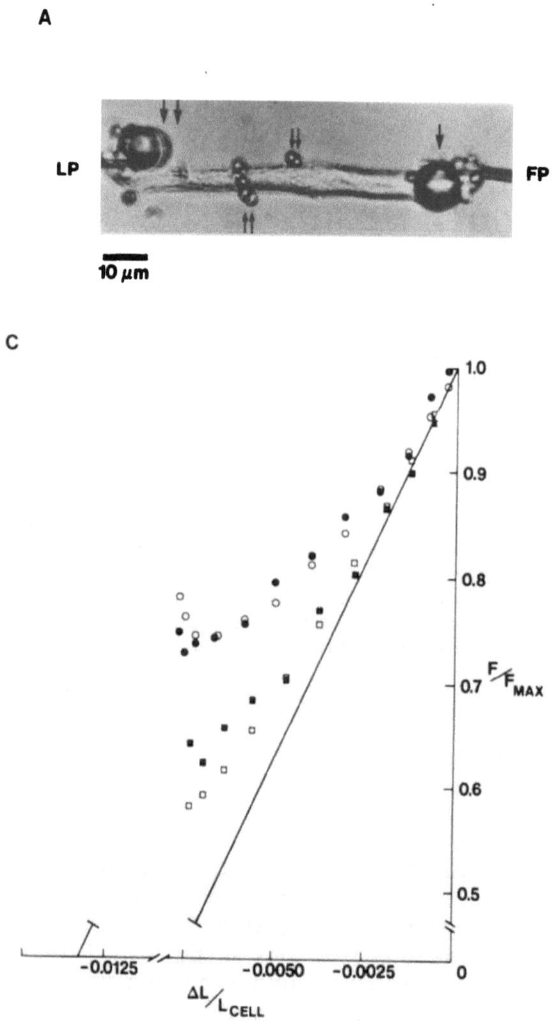

Figure 1: A) Single smooth muscle cell tied between two microprobes. A single cell was attached by anionic exchange resin beads to two microprobes allowing knots to be tied by micromanipulation (Fay, 1979). The microprobes, FP and LP, were in turn connected to a force transducer (natural frequency = 425 Hz, resolution = 1 μg) and piezoelectric displacement device (± 2.5% L_{cell} in 1.0 ms) respectively. The position of the displacement device was detected by an eddy current sensor (frequency response = 5 KHz; resolution = 0.03 μm) capable of recording the small length steps imposed. Cells were maximally activated to contract by transcellular electrical stimulation through platinum paddles. Tiny resin beads that appear on the cell surface act as length markers. By using double exposure flash photographs, changes in bead spacing (arrows) confirm that the displacement sensor provides a faithful electronic record of the length step. In addition, the length step is distributed uniformly throughout the cell.

B) A typical pair of tension transients in response to a release and stretch of 0.75% cell length complete in 5 ms were obtained at the peak of contraction. Tension records are computer traces after digitally correcting for the force transducer's mechanical properties

B

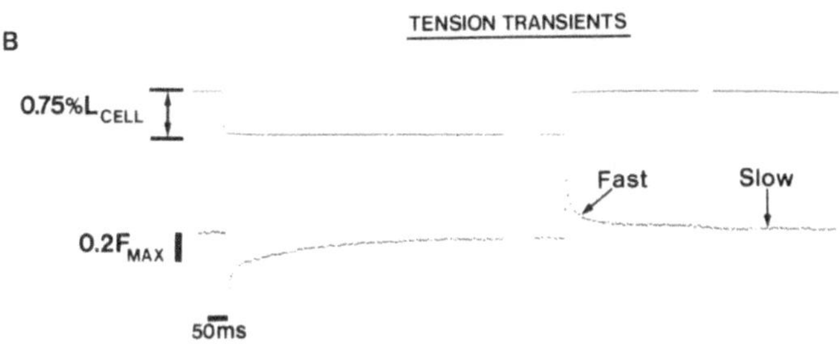

D

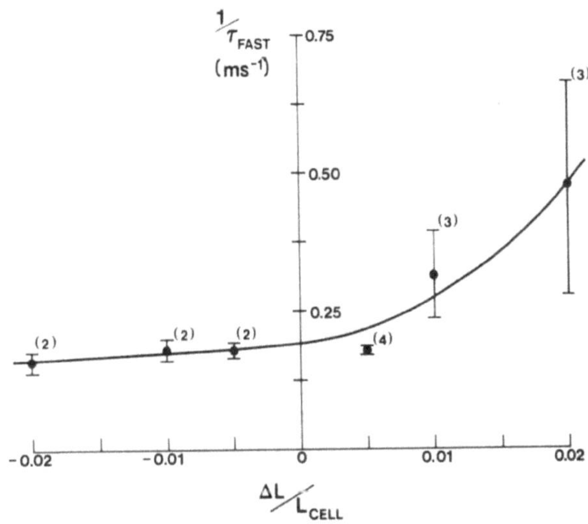

(Ford et al., 1977). The tension recovery after completion of the length step is described mathematically by a fast and slow exponential process. The usual method for determining the magnitude and rate constant of the two exponential processes utilized a nonlinear regression by least squares analysis.

C) Length:force characteristic for a single cell. Plotted are observed data (closed symbols) and a model response (open symbols) for releases of varying speeds (■, □ = 5 ms; ●, ○ = 25 ms). Model response was determined by first assuming that the recovery process during the length step was similar to the recovery observed upon completion of the length step and secondly superimposing this recovery upon a linear elasticity (solid line) comparable in slope to the initial portion of the length:force relationship. The linear elastic response is extrapolated to zero force with an intersection of 1.3% cell length. The mean ± standard error for nine cells equaled 1.6 ± 0.2%.

D) The amplitude of the length step versus the rate constant ($1/\tau_{fast}$) of the fast recovery process. Data from three cells are shown with the curve of the best fit ($1/\tau_{fast}$) = 0.19 + 4.43 $\Delta L/L_{cell}$ + 323.33 $(\Delta L/L_{cell})^2$ + 8627.45 $(\Delta L/L_{cell})^3$; $r^2 = 0.95$; drawn through the means and standard errors. Numbers in parentheses indicate the total number of trials obtained from these cells for a given amplitude step.

and the knotted portions do not appear to represent areas of high compliance which might seriously distort the tension transients.

In a single striated muscle fiber, the relationship between length and force during the step change in length is believed to reflect the properties of a linear elastic element that resides within the cross-bridge (Ford, Huxley and Simmons, 1977). When the relationship between length and force during either releases or stretches in a single SMC at the peak of force production was examined (see Fig. 1C), force was not linearly related to length as in striated muscle. The nonlinear nature of the length: force relationship (L:F) in the isolated cell does not appear to reflect the fiber's true elastic properties. Instead, the nonlinearity appears to reflect superposition of recovery processes upon the elastic response, as is evident by comparing the L:F obtained on the same fiber utilizing releases of two different speeds (Fig. 1C). In this case, the deviation from linearity in the slower release (25 ms) is minimized when a faster release (5 ms) is applied since the release occurs at a rate comparable to that for the recovery process. Therefore, the nonlinearity of the L:F in the single cell clearly to some extent reflects recovery processes that occur during the step as mentioned above. In order to characterize the true underlying elastic response of the single cell, tension recovery during the step has been removed mathematically (Ford et al., 1977) by assuming that the observed response during the length change reflects an elastic response and recovery identical in nature to the recovery processes observed after completion of the step (Figs. 1B,D). As seen in Figure 1C, the observed nonlinear responses during a length change of 0.75% L_{cell} are accounted for if we assume a linear fiber elasticity of 77 $F_{max}/\Delta L_{cell}$ upon which is superimposed recovery processes identical to those following completion of the length step. These results also indicate that a good estimate of the true elastic response may be obtained from releases of less than 0.5% L_{cell} completed in under 5.0 ms. In order to determine if the elastic response originates in the cross-bridge as in striated muscle, stiffness was determined during the onset of force development. If stiffness resides in the cross-bridges then upon activation as the number of cross-bridges acting in parallel increases and force rises, a proportional increase in fiber stiffness (i.e., the sum of cross-bridge stiffnesses) should be observed. When stiffness was determined in a single SMC at several points during the development of force, fiber stiffness varied in direct proportion to the force. This result definitely supports the view that the elastic responses recorded originate within the cross-bridges. By extrapolating the linear relationship for the elastic response of the cell (Fig. 1C), we find that a release of 1.3% L_{cell} is required to drop force within the cross-bridges to zero. This represents a considerably larger release than the 0.5% per half-sarcomere required to drop force to zero in fast skeletal muscle (Ford et al., 1977). If we assume that a mechanically similar cross-bridge exists in smooth muscle, then the effective half-sarcomere should be 0.4 μm. This value does not agree with structural data for smooth muscle from these amphibia (Fay, Fogarty, Fujiwara and Tuft, 1982) as well as those from other species (Ashton, Somlyo and Somlyo, 1975) which suggest that the effective half-sarcomere length in smooth muscle is probably longer rather than shorter than in striated muscle. Thus the mechanical data coupled with structural data for smooth muscle suggest that differences in the elastic properties of smooth versus skeletal muscle fibers are most likely explained by some difference in the cross-bridge whose properties are being probed.

As shown in Figure 1B, the recovery of tension upon completion of the step is biphasic and can be adequately described mathematically by two exponential processes (τ_{fast} = 5−20 ms; τ_{slow} = 50−300 ms). We believe these recoveries reflect cross-bridge activity since this recovery is lost in rigor-like cells that can no longer shorten following prolonged activation (> 2 min.) although both maximal force and stiffness remain elevated. In comparison to fast striated muscle at 4°C (Ford et al., 1977), the rates of both the fast and slow phases of tension recovery in a single smooth muscle cell at 20°C are at least an order of magnitude slower. This is in accord with recent energetic (Paul, Glück, Rüegg, 1976; Siegman, Butler, Mooers and Davies, 1980) and biochemical studies (Marston and Taylor, 1980) which suggest that the overall actomyosin cycle as well as a number of specific steps in the cycle are slowed in smooth muscle relative to striated muscle. If we can assume that the fast phase of tension recovery in smooth is similar to that in fast skeletal muscle fibers and reflects the transition of a portion of the attached cross-bridge population from an attached low force state to an attached high force state, then this transition may be much slower in smooth muscle as predicted by our data.

The length dependence of the fastest phase of tension recovery is shown in Figure 1D. This length dependence of the rate constant for rapid tension recovery in single smooth muscle fibers is opposite to that observed in fast skeletal muscle fibers (Ford et al., 1977). It is rather interesting that tension transients in slow skeletal muscle fibers of the tortoise (Heinl, Kuhn and Rüegg, 1974) exhibited a similar length dependence for the fast phase of tension recovery to that in our SMC. While tortoise skeletal muscle is structurally similar to fast skeletal muscle (Page, 1968), it is more like smooth muscle in terms of the high economy of its contraction (Woledge, 1968). Thus the similarity in the length dependence for the rate constant in slow skeletal muscle of the tortoise and SMC may indicate that differences in the inherent length dependence of step(s) in the cross-bridge cycle may be a common feature of slow economical muscles.

The tension transients observed for the first time in a single SMC suggest that cross-bridges within smooth muscle appear to be characterized by a linear but more compliant elasticity than in fast skeletal muscle, and that transitions between attached cross-bridge states occur at least an order of magnitude more slowly. Further analysis of the tension transient in single smooth muscle cells promises to aid in our understanding of smooth muscle's characteristic ability to produce equivalent forces to striated muscle while using much less energy and having considerably less myosin.

ACKNOWLEDGEMENTS

DMW supported by postdoctoral fellowships from the MDA and NIH (HL 05770) and FSF by grants from the NIH (HL 14523) and MDA.

REFERENCES

Ashton, F.T., Somlyo, A.V. and Somlyo, A.P. (1975). The contractile apparatus of vascular smooth muscle: intermediate high voltage stereo electron microscopy. J. Mol. Biol. 98: 17-29.

Fay, F.S. (1976). Mechanical properties of single isolated smooth muscle cells. I.N.S.E.R.M. Symposium 50: 327-342.

Fay, F.S., Rees, D.D. and Warshaw, D.M. (1981). The contractile mechanism in smooth muscle. In: *Membrane Structure and Function*, Vol. 4, pp. 79-130, ed. Bittar, E.E., New York: John Wiley and Sons.

Fay, F.S., Hoffman, R., Leclair, S. and Merriam (1982). Preparation of individual smooth muscle cells from the stomach of *Bufo marinus*. In: *Methods in Enzymology*, Vol. 85, pp. 284-292, ed. Cunningham, L.W. and Frederiksen, D.W. New York: Academic Press.

Fay, F.S., Fogarty, K., Fujiwara, K. and Tuft, R. (1982). Contractile mechanism of single isolated smooth muscle cells. In: *Basic Biology of Muscles*, pp. 143-157, ed. Dewey, M., Levine, R. and Twarog, B. New York: Raven Press.

Ford, L.E., Huxley, A.F. and Simmons, R.M. (1977). Tension responses to sudden length change in stimulated frog muscle fibers near slack length. J. Physiol. 269: 441-515.

Heinl, P., Kuhn, H.J. and Ruegg, J.C. (1974). Tension responses to quick length changes of glycerinated skeletal fibers from the frog and tortoise. J. Physiol. 237: 243-258.

Marston, S.B. and Taylor, E.W. (1980). Comparison of the myosin and actomyosin ATPase mechanisms of the four types of vertebrate muscles. J. Mol. Biol. 139: 573-600.

Page, S.G. (1968). Fine structure of tortoise skeletal muscle. J. Physiol. 197: 709-715.

Paul, R.J., Glück, E. and Rüegg, J.C. (1976). Cross-bridge ATP utilization in arterial smooth muscle. Pfluegers Arch. 361: 297-299.

Siegman, M.J., Butler, T.M., Mooers, S.V. and Davies, R.E. (1980). Chemical energetics of force development, force maintenance, and relaxation in mammalian smooth muscle. J. Gen. Physiol. 76: 609-629.

Woledge, R.C. (1968). The energetics of tortoise muscle. J. Physiol. 197: 685-707.

SARCOMERE LENGTH AND FORCE CHANGES IN SINGLE TETANIZED FROG MUSCLE FIBERS FOLLOWING QUICK CHANGES IN FIBER LENGTH

Haruo Sugi and Takakazu Kobayashi

Department of Physiology, School of Medicine, Teikyo University, Itabashi-ku, Tokyo 173, Japan

ABSTRACT

By use of an optical system, with which the beam of the first-order diffraction line of He-Ne laser light from a single frog skeletal muscle fiber was split by the wedge-shaped mirror to be focused on two photodiodes (Haugen & Sten-Knudsen, 1976), small changes in sarcomere length (less than 1 Å) could be recorded during quick fiber length changes (up to 1.2% of L_0, complete within 0.2-0.4 msec) applied at the plateau of isometric tetanus. Data were only obtained on fibers which showed typical sinusoidal sarcomere length changes in response to sinusoidal fiber length changes during tetanus with a linear relation between their magnitudes. Measurements of sarcomere length changes were made at various points along the fiber length. The interval between the onset of fiber length changes at one fiber end and that of force change recorded at the fixed fiber end was explained by the propagation of mechanical impulse at about 180 m/sec. In the case of quick releases, the onset of sarcomere shortening near the fixed fiber end tended to take place after that of force change, especially with long fibers, indicating that the drop in force during a quick release may not always be associated with sarcomere shortening along the entire fiber length. This implies that the force changes in response to rapid length changes may not give correct information about the cross-bridge properties.

Irrespective of the point at which sarcomere shortening was recorded, it was always observed that the onset of quick force recovery occurred while sarcomere shortening was still in progress. Such a phenomenon can be simulated by a viscoelastic multi-segment model with series elasticity located in each segment.

INTRODUCTION

It is generally believed that contraction in striated muscle results from the alternate formation and breaking of cross-links between the projections on the thick filaments, i.e. the cross-bridges, and the sites on the thin filaments (A.F. Huxley, 1957; H.E. Huxley, 1960). In the contraction model of Huxley & Simmons (1971), each cross-bridge consists of a myosin head and an elastic link extending from the thick filament. According to them, the instantaneous elasticity in frog muscle fibers, as determined by applying quick length changes to isometrically contracting fibers and measuring the resulting force changes (Huxley & Simmons, 1971, 1973; Ford, Huxley & Simmons, 1977), may largely reside in the elastic link of the cross-bridges, while the straight length-force relation of the cross-bridge elasticity is truncated by the early force recovery due to the rotation of the myosin head.

Since a muscle fiber is composed of thousands of sarcomeres connected in series, the above contraction model rests on the assumption that the cross-bridges in every sarcomere behave in a similar manner in response to length changes applied to one end of the fiber. Sugi & Tameyasu (1979) studied the segmental length changes of a tetanized frog muscle fiber in response to a quick release with an ultra high-speed cinecamera, and found that the initial segmental shortening was localized near the released fiber end, suggesting that the initial force drop in response to a quick release is not due to the cross-bridge elasticity. The accuracy of measurement of the fiber segmental length changes was, however, about 1%, and could not exclude the possibility that a quick sarcomere shortening of about 0.5% had taken place along the entire fiber length before the local segmental shortening became obvious, since the experiments of Ford et al., (1977) indicate that the minimum amount of quick release required to drop the isometric force to zero is about 0.5% of the fiber length.

The present experiments were undertaken to examine the nonuniformity of sarcomere length changes in response to rapid fiber length changes along the fiber length by recording local sarcomere length changes by optical diffraction with He-Ne laser light.

METHODS

Single fast muscle fibers (50-120 μm in diameter) were isolated from the semitendinosus or tibialis anterior muscles of the frog *(Rana japonica)*. The fiber was mounted horizontally at its slack length (L_0, 0.4-1 cm excluding tendons) by tying the tendons to the shaft of a servo motor (General Scanning, G-100PD) and a force transducer (Aksjelskapet, AME-80, natural frequency of oscillation, 5 kHz) respectively with braided silk thread (Fig. 1). Precooled Ringer solution (2-4°C) was continuously circulated through the experimental chamber.

As shown in Fig. 1, the sarcomere length changes were measured by use of optical diffraction with He-Ne laser light (0.1 mm in diameter); the beam of the first-order diffraction line was split by a wedge-shaped mirror to be focused on two photodiodes at both sides of the mirror (Haugen & Sten-Knudsen, 1976). The distance between the fiber and the mirror edge along the first-order diffraction line was fixed to 6.5 cm. The angle of rotation of the mirror edge from the incident beam was initially adjusted in such a way that the edge of the mirror split the beam into two halves of equal intensities. Thus, when a small sarcomere length change took

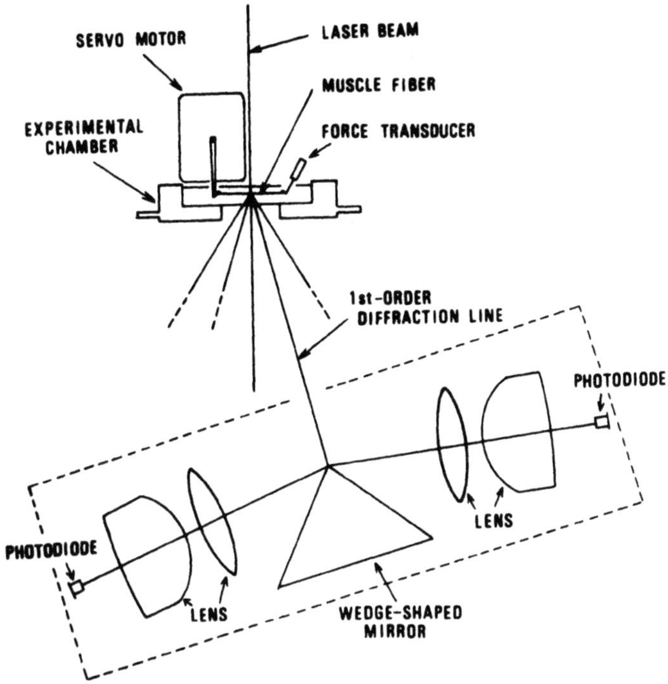

Figure 1: Diagram of the experimental arrangement. Single muscle fibers were mounted horizontally in the experimental chamber between the servo-motor and the force transducer. The sarcomere length changes were measured by projecting the beam of the first-order diffraction line on the wedge shaped mirror, so that the beam was initially split into two halves of equal intensity. For further explanations, see text.

place, it caused a small shift of the first-order diffraction line to produce a difference in voltage signal between the two photodiodes. Thus, small sarcomere length changes were linearly related to the changes in difference signal between the photodiodes.

When resting muscle fibers were subjected to a series of stretch-release cycles with the servo motor (20 Hz), the amplitude of the difference signal (V_d) was proportional to that of fiber length changes (Fig. 2). The sarcomere length change (ΔS) was estimated from the relation: $\Delta S = S \cdot \Delta L / L$ where S, L and ΔL are the sarcomere length, the fiber length and the fiber length change respectively (Haugen & Sten-Knudsen, 1976). As shown in Fig. 2, the V_{d_0} versus ΔS relation was linear at least within the range of ΔS up to 300 Å . These results may be taken to indicate that V_d serves as a valid measure of ΔS up to 300 Å . Similar results were obtained when resting fibers were subjected to sinusoidal length changes. A series of preliminary experiments indicated that, with the above method, the sarcomere length changes as small as 1 Å could be detected with a time resolution of less than 10 μsec.

The fiber was initially kept isometric, and tetanized maximally with 1 msec rectangular current pulses at 10-20 Hz given through a multi-

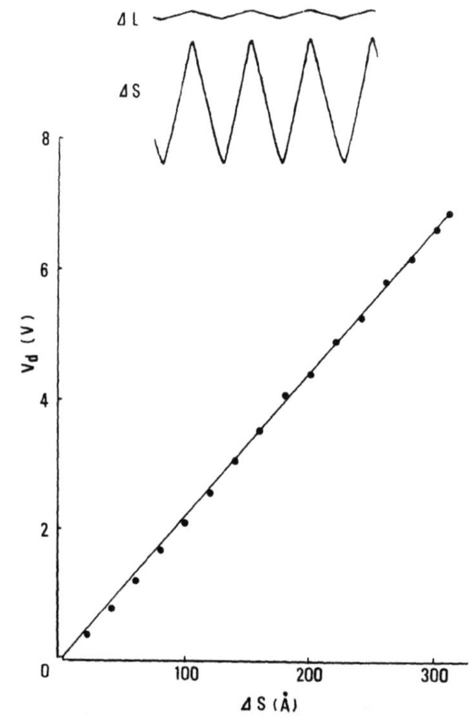

Figure 2: Relation between the difference signal (V_d) and the sarcomere length change (ΔS). Inset shows an example of sarcomere length changes in response to a series of stretch-release cycles at 20 Hz.

electrode assembly. When the full isometric force was developed, the fiber length was changed by up to 1.2% of L_0 with variable velocities by means of the servo motor system, the fiber length changes being sensed with a differential capacitor (demodulation frequency, 2 MHz) incorporated in the servor motor system. The sarcomere length of the fibers at their slack length was 2.1-2.2 μm, and during the isometric force development, the sarcomere length shortened by 50-100 nm due to the internal shortening of the fiber by stretching the tendinous ends. Since the internal shortening occurring during each tetanus was reproducible, it was possible to adjust the mirror position in such a way that, after the completion of initial sarcomere shortening due to the internal shortening, the beam of the first-order diffraction line was split by the mirror wedge into two halves of almost equal intensity.

RESULTS

Fiber length versus sarcomere length relation in tetanized fibers. The linear relation between V_d and ΔS in resting fibers (Fig. 2) does not, however, necessarily mean that the linear V_d versus ΔS relation also holds in tetanized fibers. As a matter of fact, if sinusoidal fiber length

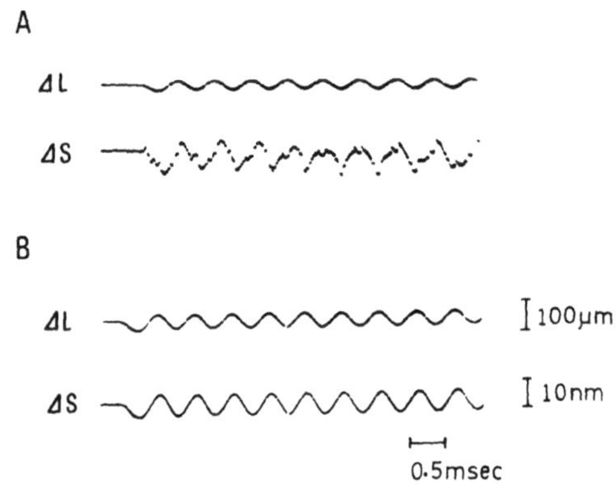

Figure 3: Examples of sarcomere length changes in response to sinusoidal fiber length changes in tetanized fibers. The sarcomere length signals are complex in A, and typically sinusoidal in B.

changes (2 kHz) were applied to the fiber generating the full isometric force (P_0), the resulting sarcomere length signals were in many cases not sinusoidal in shape, but showed various complex patterns (Fig. 3A). According to Rüdel & Zite-Ferenczy (1979), this is due to different myofibrillar planes of slightly different sarcomere length and different orientation with respect to the incident beam; different myofibrillar planes will happen to fulfill the Bragg condition during the sinusoidal fiber length changes to result in complex sarcomere length signals.

In the present study, however, we could obtain fibers which showed sinusoidal sarcomere length signals in response to sinusoidal changes in fiber length (Fig. 3B) irrespective of the position illuminated by the laser beam, and the V_d versus ΔL relation was linear over the range of V_d up to 180 Å, indicating that these fibers may consist of myofibrils with reasonably uniform sarcomere length and orientation with respect to the incident beam. Such fibers could be obtained only one out of 3-5 fibers, and the experimental data were obtained only from these fibers.

Sarcomere length and force changes in response to fiber length changes at various velocities. Fig. 4 shows examples of the changes in fiber length, sarcomere length and force when a tetanized fiber was stretched by about 0.7% of L_0 at four different velocities on a relatively slow time base, the sarcomere length being measured at the middle of the fiber. The time course of sarcomere lengthening was almost similar to that of fiber lengthening, while the force continued to rise during the sarcomere lengthening and started to decay at the completion of the sarcomere lengthening. The increase in force above the isometric level during a given amount of stretch was larger, the larger the stretch velocity.

Fig. 5 shows examples of the changes in fiber length, sarcomere length and force when a tetanized fiber was released by about 0.7% at four different velocities. With slow release velocities (Fig. 5A, B), the time course of sarcomere shortening took place almost in parallel with that of fiber shortening, while the force continued to fall during the sarcomere

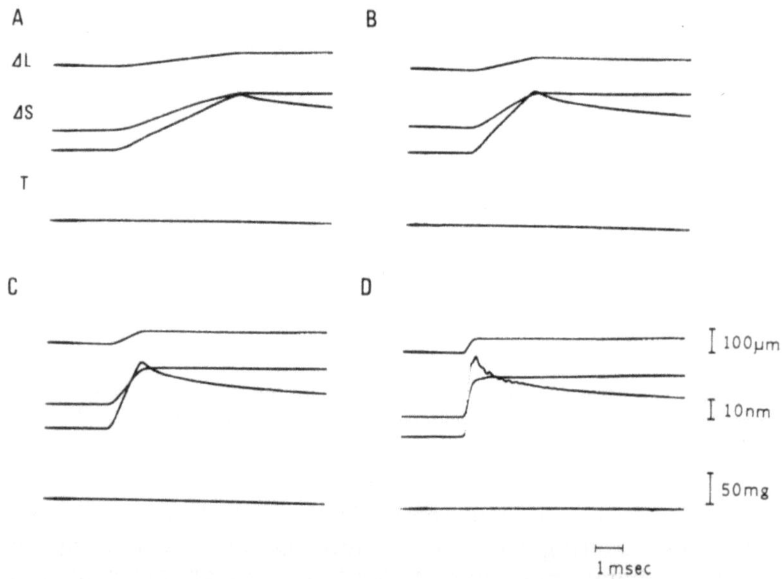

Figure 4: Examples of changes in fiber length (upper traces), sarcomere length (middle traces) and force (lower traces) when a tetanized fiber was stretched by about 0.7%. The stretch velocity was 1.5 L_0/sec in A, 3.0 L_0/sec in B, 6.8 L_0/sec in C and 19.5 L_0/sec in D. Records A-D were obtained from one and the same fiber.

shortening and started to rise at the completion of the sarcomere shortening. With faster release velocities, on the other hand, the sarcomere length tended to exhibit oscillatory changes of small amplitudes after the completion of fiber shortening (Fig. 5C, D). The decrease in force below the isometric level increased with increasing release velocity. It was noticed that, with a release velocity of about 20 L_0/sec, the force started to rise while the fiber and the sarcomere still continued to shorten (Fig. 5D). This phenomenon will be later described in detail.

Relation between sarcomere length and force changes in response to quick fiber length changes. Fig. 6 shows the changes in fiber length, sarcomere length and force in response to a quick stretch (A,B) and a quick release (C,D) applied to a tetanized fiber on a fast time base; the sarcomere length changes were recorded near the moving end of the fiber in A and C, and near the fixed end of the fiber in B and D. The velocity of quick fiber length changes was about 20 L_0/sec, and the fiber length changes were complete within 0.2-0.4 msec. Near the moving fiber end, the fiber length and the sarcomere length started to change simultaneously, while the force began to change with a definite delay (Fig. 6A, C). The delay of the onset of force changes in response to applied fiber length changes can be explained as being due to the time required for transmission of longitudinal mechanical wave along the entire fiber length to be sensed with the force transducer at the fixed fiber end. The transmission velocity along the fiber was estimated from the interval between the

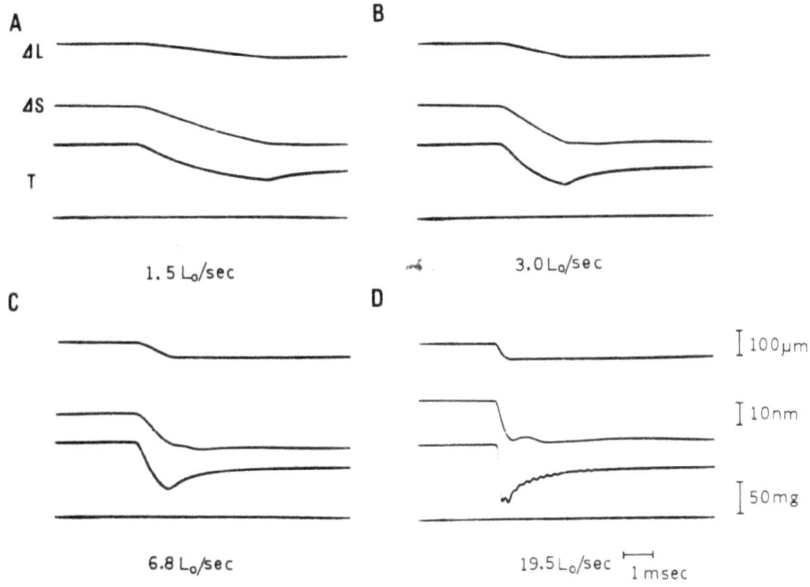

Figure 5: Examples of changes in fiber length, sarcomere length and force when a tetanized fiber was relesed by about 0.7%. The release velocity was 1.5 L_0/sec in A, 3.0 L_0/sec in B, 0.8 L_0/sec in C and 19.5 L_0/sec in D. Records A-D were obtained from one and the same fiber.

onset of fiber length change and that of force change and the fiber length, and was about 180 m/sec in agreement with the results of Schoenberg, Wells & Podolsky (1974).

Near the fixed fiber end, on the other hand, the sarcomere length started to change with a definite delay in response to applied fiber length changes (Fig. 6B, D). In the case of muscle fibers of less than 0.7 cm in length, the delay of the onset of sarcomere length changes was as long as that of the onset of force change (Fig. 6B, D). In the case of longer muscle fibers, however, it was sometimes observed that the sarcomere length started to change after the onset of force changes, as shown in Fig. 7. This may be taken to indicate that the transmission velocity of mechanical wave is not necessarily identical with that of sarcomere length changes along the fiber length.

In the case of quick stretches, the fiber length and the sarcomere length increased almost in parallel with each other, and the force started to decay as soon as the fiber and sarcomere lengthening was completed (Figs. 4D, 6A, B and 7A). In the case of quick releases, on the other hand, the sarcomere length changes tended to persist after the completion of fiber shortening, and it was always observed that, irrespective of the point where the sarcomere length changes were recorded, the initial rapid fall in force in response to the quick release was followed by the subsequent rise in force which started while the sarcomere shortening was still in progress (Figs. 5D, 6C, D and 7B).

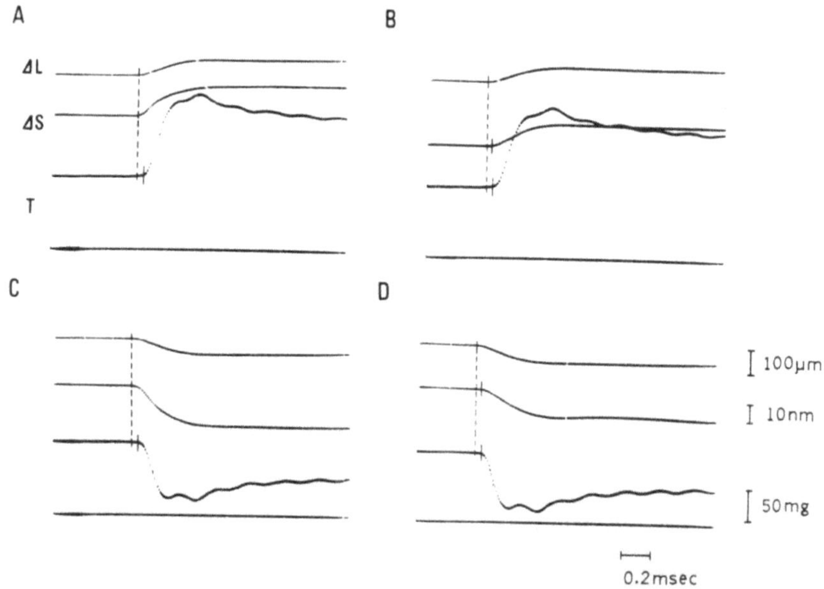

Figure 6: Examples of changes in fiber length, sarcomere length and force when a tetanized fiber was quickly stretched (A, B) or released (C, D) by about 0.7%. The sarcomere length changes were recorded near the moving fiber end in A and C, and near the fixed fiber end in B and D. Vertical broken line indicates the time of onset of fiber length change, while the onsets of sarcomere length and force changes are indicated by short vertical lines.

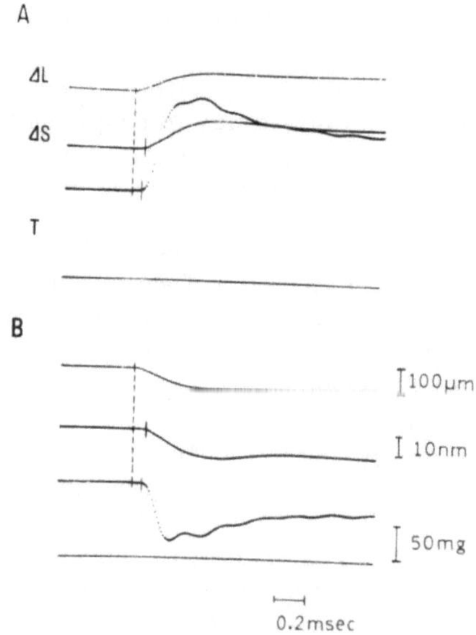

Figure 7: Examples of records showing the onset of sarcomere length changes near the fixed fiber end after that of force changes.

DISCUSSION

In the present experiments, it was possible to select muscle fibers which showed a linear fiber length versus sarcomere length relation both at rest and during activity (Figs. 2 and 3B), and the sarcomere length changes in response to applied fiber length changes could be studied in detail under various conditions. In this paper, we will discuss only the sarcomere length changes in response to quick changes in fiber length, since the contraction model of Huxley & Simmons (1971), which has been central in the field of muscle mechanics, has been constructed from the isometric force transients following quick fiber length changes.

The onset of sarcomere length changes in response to quick fiber length changes showed a definite delay when the sarcomere length was monitored near the fixed fiber end (Fig. 6B, D). The delay in the onset of sarcomere length changes near the fixed fiber end was almost the same as that in the onset of force changes in the case of short muscle fibers. In the case of long fibers, however, the sarcomere length changes tended to take place after the onset of force changes (Fig. 7). On the other hand, there were no appreciable changes in the mode of force changes in response to quick fiber length changes between short and long fibers.

These results may be taken to indicate that the longitudinal mechanical wave, which starts at the moving fiber end, propagates along the fiber length, and reaches to the fixed fiber end to cause the force changes sensed by the force transducer, is not identical with the wave of sarcomere length changes, which also starts at the moving fiber end and propagates along the fiber length. The result that the force change can start before the onset of any detectable sarcomere length changes (Fig. 7) implies that the longitudinal mechanical wave giving rise to the force changes propagates along the fiber length without any appreciable sarcomere length changes. This view is in accord with the results of Sugi & Tameyasu (1979) that, when a tetanized fiber was subjected to a quick release, the shortening was confined to a fiber segment nearest the moving fiber end, while the whole fiber shortened by more than 1%, i.e. the amount required to drop the isometric force from P_0 to zero.

Another interesting result obtained in the present study is that, in response to a quick release, the initial drop in force was followed by the subsequent force redevelopment which started while the sarcomere shortening was still in progress (Figs. 5D, 6D and 7B). This phenomenon was always observed provided the point where the sarcomere length changes were recorded was distant from the moving fiber end. It contradicts the idea that the quick force recovery following a quick release results from the myosin head rotation which starts after the sarcomere shortening is over (Huxley & Simmons, 1971).

Fig. 8 shows a viscoelastic multi-segment model developed for simulating the sarcomere length and force changes observed in the present exeriments (Nishiyama, 1982). The zone of overlap between the filaments in sarcomere is represented by a viscous element and an elastic element, both of which are connected in parallel with the contractile component, while the non-overlap zone is represented by another elastic element,

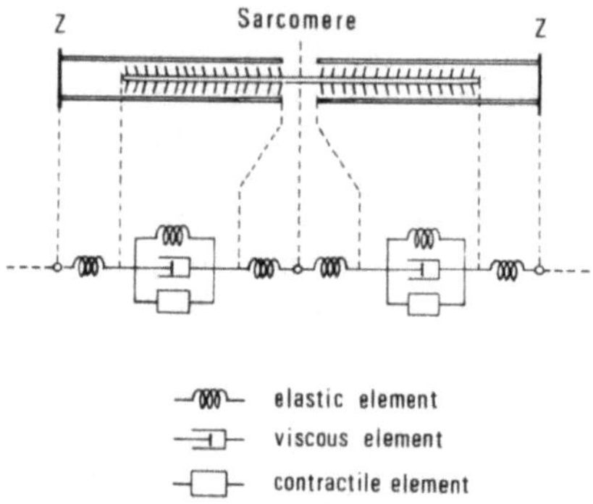

Figure 8: A viscoelastic multi-segment model used to simulate the present results.

which constitutes a series elasticity in every non-overlap zone. The elastic element at the non-overlap zone is based on the electron microscopic observation that the thick filaments are extensible only at the H-zone, while the thin filaments are extensible everywhere except for the zone of overlap (Sugi & Suzuki, 1980). The contractile component is a two-state model described elsewhere (Tsuchiya, Sugi & Kometani, 1979; Sugi & Tsuchiya, 1981). By considering a muscle fiber consisting of many sarcomeres connected in series, the onset of quick force recovery before the completion of sarcomere shortening can well be simulated, suggesting that the force changes following a quick release results from the viscoelastic and multi-segmental nature of muscle fibers, but not from the response of the cross-bridges taking place uniformly along the fiber length.

REFERENCES

Ford, L.E., Huxley, A.F. and Simmons, R.M. (1977) Tension responses to sudden length change in stimulated frog muscle fibers near slack length. J. Physiol. 269: 441-515.

Haugen, P. and Sten-Knudsen, O. (1976) Sarcomere lengthening and tension drop in the latent period of isolated frog skeletal muscle fibers. J. Gen. Physiol. 68: 247-265.

Huxley, A.F. (1957). Muscle structure and theories of contraction. Prog. Biophys. Biophys. Chem. 7: 255-318.

Huxley, A.F. and Simmons, R.M. (1971). Proposed mechanism of force generation in striated muscle. Nature, Lond. 233: 533-538.

Huxley, A.F. and Simmons, R.M. (1973). Mechanical transients and the origin of muscular force. Cold Spring. Harb. Symp. Quant. Biol. 37: 669-680.

Huxley, H.E. (1960). Muscle cells. In: *The Cell*, vol. 4, ed. Brachet, J. and Mirsky, A.E., pp.365-481, New York and London: Academic Press

Nishiyama, K. (1982). Analysis of mechanical behavior of muscle by a multi-sarcomere model. (This volume)

Rüdel, R. and Zite-Ferenczy, F. (1979). Do laser diffraction studies on striated muscle indicate stepwise sarcomere shortening? Nature, Lond. 278: 573-575.

Schoenberg, M., Wells, J.B. and Podolsky, R.J. (1974). Muscle compliance and the longitudinal transmission of mechanical impulse. J. Gen. Physiol. 64: 623-642.

Sugi, H. and Suzuki, S. (1980). Extensibility of the myofilaments in vertebrate skeletal muscle as studied by stretching rigor muscle fibers. Proc. Jap. Acad. B56: 290-293.

Sugi, H. and Tameyasu, T. (1979). The origin of the instantaneous elasticity in single frog muscle fibers. Experientia, 35: 227-228.
Sugi, H. and Tsuchiya, T. (1981). Isotonic velocity transients in frog muscle fibers following quick changes in load. J. Physiol. 319: 219-238.
Tsuchiya, T., Sugi, H. and Kometani, K. (1979). Isotonic velocity transients and enhancement of mechanical performance in frog skeletal muscle fibers after quick increases in load. In: *Cross-bridge Mechanism in Muscle Contraction*, ed. Sugi, H. and Pollack, G.H., pp.225-249. Tokyo: University of Tokyo Press and Baltimore: University Park Press.

DISCUSSION

POLLACK: In your early experiments with the cine records, it seemed that not only was there a time difference but there was also a difference of magnitude of the amount of release between the moving and the fixed end. In other words, the sarcomere length change was larger at the moving end than at the fixed end. Was that confirmed in your later results, or not?

SUGI: Yes. In the present experiments, the sarcomere length changes tended to be smaller at the fixed end than at the moving end.

CECCHI: I would like to know how long was the fiber you used for these experiments.

SUGI: Four to six millimeters in the case of tibialis anterior, and eight to ten millimeters in the case of semitendinosus..

CECCHI: Have you tried to measure the amount of the length step necessary to discharge the isometric force completely?

SUGI: A little less than 1%, about 0.8% or so.

HOUSMANS: I would like to know, Professor Sugi, if this difference in behavior of the fixed end also holds in the case of load step experiments?

SUGI: I have never examined sarcomere length changes during load steps. But, the completion of the load step requires longer time than the rapid length step by the present technique, and I think that the degree of non-uniformity of sarcomere length change may be much smaller in the case of load step experiments.

INGELS: I'd like to point that Alan Gott has been working, as of course we know, in this field for about ten years now. He's been warning me on the phone from time to time that the quick release type experiments would be hazardous because of the shock wave effect that would transmit through the muscle. You can imagine all kinds of constructive and destructive interference, reflections and whatever. My question would be: how do your results and your models compare with the things that he's been saying?

SUGI: Our experimental results are, I think, compatible with Gott's multisegment muscle model (Gott in *Cross-bridge Mechanism in Muscle Contraction*, Ed. Sugi and Pollack, 1979).

KRUEGER: Haugen and Sten Knudsen (J. Gen. Physiol. 68: 247-265, 1976) reported an anamolous findin · with diffraction methods: that sometimes the small sarcomere length cn ,nge occurring in the latency period might be in a direction opposite to what they predicted. That might account for Rüdel and Zite-Ferenczy's model (Nature, 278: 273, 1979). Have you ever seen any, let's say, slow lengthening of the sarcomeres?

SUGI: I observed such phenomena in "bad" fibers which showed complex sarcomere length changes in response to sinusoidal changes in fiber length (Fig. 3A). But I never saw such phenomena in "good" fibers which showed typical sinusoidal sarcomere length changes in response to sinusoidal changes in fiber length (Fig. 3B).

SCHOENBERG: I would just like to amplify what Professor Sugi has said. We also see "good" fibers and "bad" fibers in this regard, and we look at them in the microscope and our feeling is that Rüdel and Zite-Ferenczy are right. You see these funny stepwise effects if the striation patterns are not perfectly aligned. I'm very pleased that Professor Sugi has found such nice fibers in the frog, and we may go back and look at frogs. The rabbit is at least initially a lot better in that when you get nice uniform striations across the fiber, you don't see these "bad" fibers.

KRUEGER: I'm not challenging Rüdel's model at all. What I would like to comment on is that with extremely rapid length changes, and with extremely fine and precise sarcomere length measurements, I wonder whether there's any possibility that the zero order might shift a little bit, causing a slight rocking in the diffraction axis. I don't think this will affect the results but it might explain the small "humps" that you find in the sarcomere length trace in your "good" fibers.

SCHOENBERG: I looked for it in our "bad" fibers and I didn't see it, but maybe it should be repeated in "good" fibers.

SUGI (added after the discussion): We checked Krueger's point on "good" fibers, and did not find any detectable shift of zero order.

GORDON: I was interested that there were some other differences between your "good" fibers and "bad" fibers. Are there some other properties that you could measure that showed a difference?

SUGI: I have just started this kind of study, and I do not see any appreciable difference at least concerning the tension transients following fiber length changes.

POLLACK: With regard to the "bad" fibers that show stepwise sarcomere length changes, I'm pre-empting what I'd like to say this afternoon, but I'd like to mention that almost all of our fibers must be "bad" ones. I'm going to report that we've used methods other than optical diffraction to look at sarcomere shortening. We've been using a new, on-line method of imaging of the striation pattern with incoherent light, and I just want to say that we can observe step-wise phenomena comparable to that found using optical diffraction in virtually every fiber. The fibers' "bad" behavior is not restricted to optical diffraction.

HUXLEY: Could you say anything yet about the distribution of the total extent of elasticity in the three elastic elements in your model? I mean the one in the I-band, the H-zone, and the one in the zone of overlap.

SUGI: Based on our experiments, in which we stretched rigor muscle fibers, the I-band contribution to the series elasticity appears to be twice as large as the H-zone contribution.

HUXLEY: I didn't mean on the rigor fibers, but where you're using that model on the quick-release experiments.

SUGI: At present, I cannot answer your question exactly. Dr. Nishiyama is trying to simulate the present results by changing model parameters.

EDMAN: One implication of your finding would seem to be that the stiffness value one arrives at in a quick-release or quick-stretch measurement will be dependent on the length of the fiber (or fiber segment), from which one records.

SUGI: I have never studied the relation between fiber length and stiffness. Stiffness values are also dependent on the speed of length perturbations.

EDMAN: But according to your results the length of the preparation will have an influence on the stiffness measurement. In order to drop the force to zero you will need a proportionally smaller release step (in terms of fiber length) when the measurement is made on a longer fiber. The stiffness of the longer fiber will therefore be overestimated. Increasing the speed of release would have a similar effect; it would lead to overestimation of the fiber stiffness.

SUGI: Yes, I agree. Probably longer fibers may tend to yield larger values of stiffness.

POLLACK: Have you tried to analyze your results in terms of the sort of method that Ford, Huxley and Simmons (J. Physiol. 269: 441-515, 1977) have used?

SUGI: I think that, in their experiments, tension changes are largely viscoelastic phenomena; the cross-bridges may have no time to move at the early phase of tension transients.

POLLACK: But if you *tried* to analyze it, what implications do you think you might draw?

SUGI: I think that the story of cross-bridge elasticity and cross-bridge head rotation should be re-examined.

ANALYSIS OF MECHANICAL BEHAVIOR OF MUSCLE BY A MULTI-SARCOMERE MODEL

Ken'ichi Nishiyama

Laboratory of Statistics, Faculty of Economics, Teikyo University, 359 Ootsuka, Hachioji, Tokyo 192-03, Japan.

ABSTRACT

We have made a multi-sarcomere model and analysed the rapid transient process of active muscle, using the model. We found that it was essential to assume an elastic component in the non-overlap region of each sarcomere in simulating this process.

It is controversial whether the rapid transient process of muscle can be understood only by cross-bridge properties. To clarify this point, we made a multi-sarcomere model and analysed the recent observation of the transient process by Sugi and Kobayashi (1982). As shown in Figure 1, our model consists of a series of many identical sarcomeres, each sarcomere

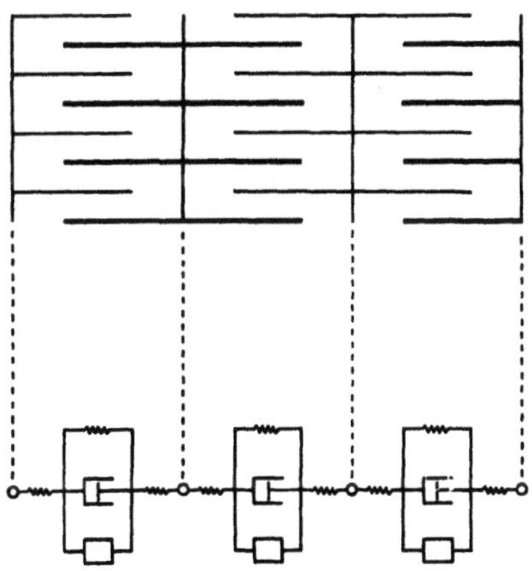

Figure 1: A simplified multi-sarcomere model.

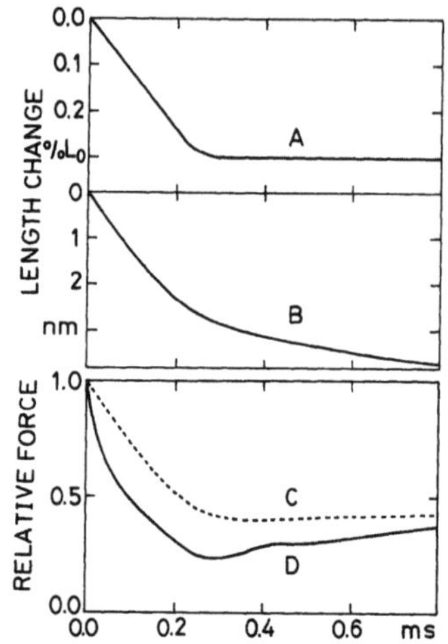

Figure 2: Length and force changes in our model, following a quick release. A: quick release, B: sarcomere length change near the fixed end, C: force from cross-bridge cycling, D: force at the fixed end.

being divided into two regions connected in series: one is the overlap region between the thick and thin filaments, and the other the non-overlap region. The non-overlap region is represented by an elastic component, based on the observation by Sugi and Suzuki (1980) that both filaments in this region are extensible. The overlap region consists of two parallel components: a passive viscoelastic component of Voigt type and an active component producing contractile force by cross-bridge cycling.

We made a detailed computer simulation and found that our model can explain the observation of Sugi and Kobayashi: when a tetanized muscle fiber is quickly released by less than 1% and then held at constant length, the force first drops nearly in phase with the muscle length and then starts to recover, whereas the shortening of sarcomere spacings still continues after the beginning of force recovery. A result of our calculation is given in Figure 2. It was essential in simulating this result to assume an elastic component in the non-overlap region. It is also shown that the force exerted by cross-bridge cycling is not always identical with that measured at one end of the model, especially during the early phase of quick force recovery. This suggests that the force measured at one end of a muscle fiber during a quick release does not necessarily reflect the state of the cross-bridges, implying that the quick force recovery after quick release results from the viscoelastic, multisegmental nature of the muscle fiber, and not from cross-bridge head rotation.

ACKNOWLEDGEMENTS

I would like to thank Professor Sugi for helpful suggestions and discussions.

REFERENCES

Sugi, H. and Kobayashi, T. (1982). Sarcomere length and force changes in single tetanized frog muscle fibers following quick changes in fiber length. (This volume).

Sugi, H. and Suzuki, S. (1980). Extensibility of the myofilaments in vertebrate skeletal muscle as studied by stretching rigor muscle fibers. Proc. Japan Acad. 56 Ser. B: 290-293.

THE KINETICS OF CROSS-BRIDGE ATTACHMENT AND DETACHMENT STUDIED BY HIGH FREQUENCY STIFFNESS MEASUREMENTS

Giovanni Cecchi, Peter J. Griffiths* and Stuart Taylor+

Istituto di Fisiologia Umana, Viale Morgagni 63, Florence, Italy
**University Laboratory of Physiology, Oxford, OX1 3PT, England*
+The Department of Pharmacology, Mayo Foundation, Rochester, MN 55905 USA

ABSTRACT

Muscle fiber stiffness, supposedly an indication of attached cross-bridges, was measured throughout tetanic contraction and subsequent relaxation. Stiffness increased at a rate faster than the development of force during the rise of tetanic contraction and decreased more slowly than force during relaxation. One explanation for these results is that long-lived cross-bridge states may exist between attachment, force generation and detachment.

INTRODUCTION

The application of quick length changes to muscle has long been used as a means to investigate properties such as stiffness and to deduce the time course of the force-generating molecular events of the contractile apparatus. This approach is of greatest value when the compliance in series with the force generators is small and the speed of the length change is great. Therefore, we have used preparations with relatively little compliance in the end attachments and apparatus with a high natural frequency to further study an idealized model of muscle contraction. This inquiry arose during the course of our attempts to examine the relations among myoplasmic Ca^{2+} levels, length changes, and the time course of relaxation which are outlined elsewhere in this volume (Cecchi, Griffiths and Taylor, 1983).

Most current models for the molecular mechanism of striated muscle contraction suppose that force generation occurs at identical sites of interaction uniformly distributed in the zone of each half sarcomere where thick and thin filaments overlap. In the model where each site of interaction is associated with an independent force generator, it is also supposed that the lateral projections from the thick filaments (i.e., the cross-bridges) attach to the thin filaments, thereby formimg extensible links prior to force production (A.F. Huxley, 1980; H.E. Huxley, 1982). Hence, the measurement of instantaneous stiffness at various times during a contraction cycle makes it possible to compute the relative number of attached cross-bridges.

In isometric contractions of intact skeletal muscle fibers at different sarcomere lengths, stiffness varies in direct proportion to the degree of overlap between thick and thin filaments, and the ratio between stiffness and force is essentially constant (A.F. Huxley and Simmons, 1973). However, in muscle fibers with the surface membrane removed or destroyed, the ratio between stiffness and force can be increased by replacing ATP with an analog that is hydrolyzed much more slowly; presumably cross-bridges are attached but they are in a state generating less than maximum force (Kuhn, 1981). We measured the stiffness of intact fibers during the tetanus rise and during relaxation and found that the ratio between stiffness and force was not constant during those periods. This leads us to suggest that cross-bridges may exist in a state generating relatively less force during the tetanus rise and during relaxation than during the plateau of a tetanic contraction.

METHODS

Single fibers were isolated intact from the tibialis anterior and lumbricalis" digiti IV muscles of the frog *Rana temporaria*. There are several lumbricalis muscles on the plantar surface of the frog's foot. But the one we used arises from the tendons of the flexor longus, runs along the fourth toe on the side nearest the fifth toe, and inserts into a phalanx of the fourth toe (Ecker, 1971). We used these fibers because they were relatively short, and it was possible to impose very rapid length changes without exciting a natural mode of vibration of the fibers; hence a delay in transmission of a displacement from one end of a fiber to the other was not a factor (e.g., see the Figure in the discussion following this paper). Fibers were stimulated tetanically for about 0.5 s at regular 5 min intervals by way of platinum plate electrodes, which extended parallel to the long axis of a fiber, along the whole length of the chamber. The stimulus frequency was adjusted between 10 and 20 Hz at 4°C to give a just fused tetanus. Length changes were produced with a low-inertia moving coil system, and steps could be completed in less than 130 μs. Force measurements were made with a capacitance force transducer with a natural frequency of 50 kHz and a compliance of 0.05 μm/mN (Cecchi, Griffiths and Taylor, 1982b). We applied high frequency, small, sinusoidal length changes to one end of a fiber, recorded the changes in force at the other end, and calculated stiffness as a ratio between the force and length sinusoids during the following periods: i) during the onset of tetanic contraction; ii) during the tetanic plateau; iii) during the periods of recovery from a rapid step release imposed on the plateau; and iv) during relaxation. We inferred that the influence of end attachments was relatively small and that the elastic properties could be attributed mostly to a muscle fiber itself, by measuring the extrapolated amount of instantaneous shortening necessary to just discharge the maximum tetanic force developed by a fiber (e.g., Cecchi et al., 1982b). Responses were displayed on a Nicolet Explorer digital oscilloscope and then stored on magnetic discs. Because of the limited memory of the oscilloscope and our wish to obtain the highest possible resolution, each experimental point shown in the figures was determined from a different tetanic response of the same fiber (e.g., there were 19 responses in Figure 1). The time base of the Nicolet was set between 1 and 5 μs per point. We used data from those experiments in which the difference in maximum tetanic force among the responses was less than 5%.

RESULTS

Figure 1 shows the time course of the changes in stiffness (X) and force (●) during the recovery from a relatively big (1% of fiber length) step release imposed on the tetanus plateau. As already described, the release produced an initial drop of force followed by its rapid early recovery. Then, during the next 10 ms or so, there was an extreme reduction in the rate of recovery of force followed by a gradual recovery which approached the value during the tetanus plateau asymptotically

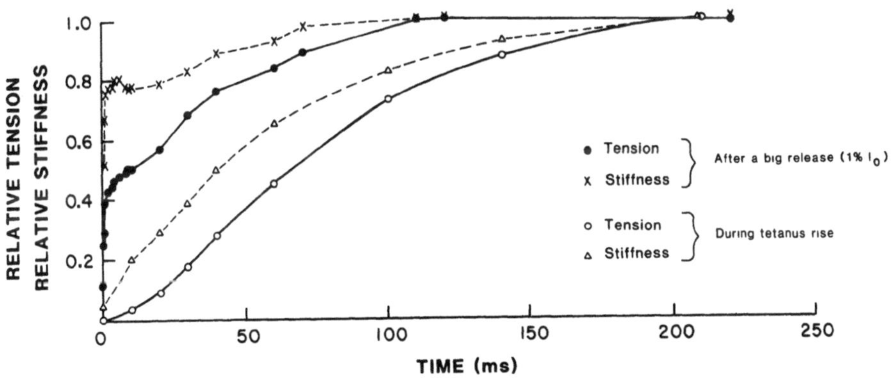

Figure 1: The relative stiffness and relative tension of a single fiber (10.xi.81) after a quick release imposed on the tetanus plateau and during the rising phase of tetanic contraction. Small amplitude high frequency oscillations were applied to one end of the fiber and the first determination of stiffness after the quick release was made when force had recovered to 0.12 times the value on the tetanus plateau. The length of each fiber used limited the maximum frequency of oscillation we applied to determine stiffness because we wanted to avoid any significant contribution from fiber inertia (e.g., see the Figure in the discussion following this paper). The fiber used in this particular experiment was from the lumbricalis" digiti IV muscle of the frog *Rana temporaria* and was relatively short (2.3 mm long). The extrapolated amount of instantaneous shortening necessary to just discharge the maximum tetanic force developed by this fiber was -6.7 nm per half sarcomere at 4°C. The resting striation spacing was set to 2.25 μm while we observed the laser diffraction pattern from the fiber. The amplitude of the applied oscillations was 1.2 μm peak to peak and the frequency was 4 kHz. As we previously determined, no phase shift occurred between the force and length sinusoids at this frequency, except at the point measured prior to stimulation (Cecchi, Griffiths and Taylor, 1982a).

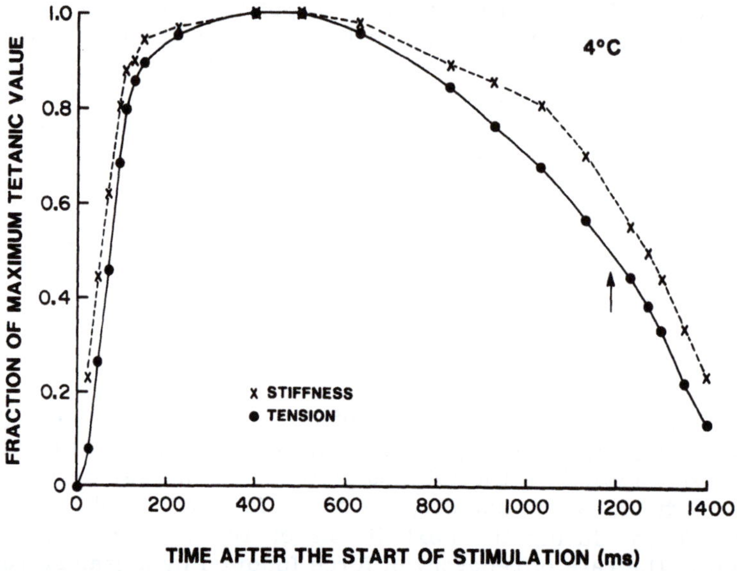

Figure 2: The relative stiffness and relative tension of a single fiber (29.x.81) during the rising phase and during relaxation of tetanic contractions. Each point was determined from a separate contraction of the same fiber, as described in the Methods and the legend to Figure 1. However, this fiber was isolated from the tibialis anterior muscle, was 6.8 mm long, and the maximum frequency of oscillation was necessarily lower (2.5 kHz). The resting striation spacing was 2.2 μm. The amount of instantaneous shortening necessary to just discharge the maximum tetanic force was determined by extrapolating the linear portion of a "T_1 curve" (Cecchi, Griffiths and Taylor, 1982b), and the value for this fiber was -5.39 nm per half sarcomere. All points were scaled relative to the maxima of the tetani (i.e., 1.0) which occurred 400 ms after the start of stimulation. Other details were as described in the legend to Figure 1. The arrow indicates the point at which segment length began to deviate from the constant length maintained during the tetanic plateau, as described in the legend of Figure 3.

(A.F. Huxley and Simmons, 1973).

The release also produced an immediate drop in stiffness, but the relative decrease in stiffness was much smaller than that of force. Hence the ratio of stiffness to force during the initial phase of the recovery was higher than at the tetanic plateau. The initial decrease in stiffness was followed by partial recovery and a slight second fall approximately 10-15 ms after the release. The second drop in stiffness occurred although force was rising at the time and, therefore, could not have been due to compliance in the tendons. A similar change has been reported by others and its origin is not yet clear (Stienen and Blangé, 1981). Then stiffness gradually approached its value just prior to the release, but this phase of recovery preceded the recovery of force.

Figure 1 also shows the time course of the changes in relative stiffness (Δ) and relative force (O) during the rise of a series of tetani from the same fiber. We found that fiber stiffness increased faster than force and continued to lead the development of force by about 15 ms throughout most of the rise.

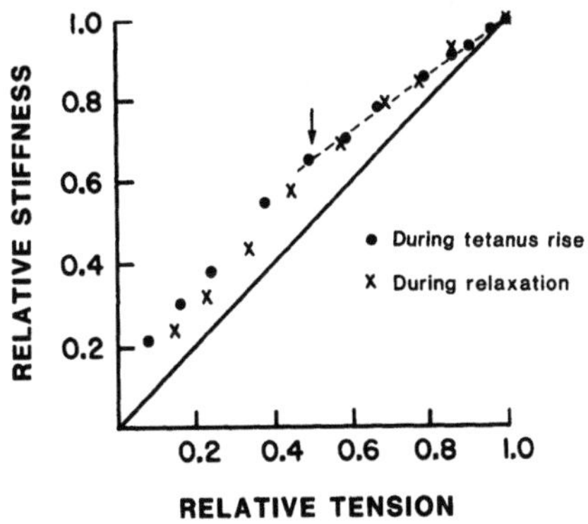

Figure 3: The same data shown in Figure 2 are plotted here independent of time. The solid line shows the relationship expected if the ratio of stiffness to tension were constant at all instants on the tetanus rise and during relaxation. The dashed line is drawn coincident with the phases of a tetanus during which the ratio of stiffness to tension is the same, but not constant. The arrow indicates the end of the isometric phase of relaxation, that is the "shoulder" (A.F. Huxley and Simmons, 1970; 1973). To the left of the arrow the ratio of stiffness to tension during relaxation falls below the ratio during the tetanus rise. The length of various segments of the fiber was monitored throughout contraction by a solid-state line scanner (Reticon RL-128 G). Deknatel surgical silk (number 5-0; black braided; non-capillary; Type B) was unbraided into small filaments, and short strands were tied in single overhand knots and tightened around the fiber until constriction of the fiber was first detectable. These served as firmly attached, non-injurious opaque markers when their image was focused by way of a dissecting microscope onto the photodiode array, because the length changes and radial expansion of a fiber during these contractions were relatively small. The array consisted of 128 elements, had a scanning frequency of 1 Mhz; and could resolve movements of less than 2 μm.

The results from another experiment in which the time course of stiffness and force was also observed during relaxation are shown in Figure 2. The stiffness (X) and force (•) values plotted were measured at selected instants on the rising and falling phase of a series of tetani. Stiffness not only had higher relative values than force all during the rise, but relative stiffness remained higher than relative force during the relaxation phase as well.

Figure 3 shows the same data with the time parameter removed. Stiffness at selected instants was plotted against force at the same instant, and fiber stiffness was greater than predicted by a constant ratio between stiffness and force on both the rising (•) and falling (X) phase of the tetanus. Another feature that becomes apparent when the data are plotted in this way is the difference that begins at the arrow, that is, at the end of the isometric phase of relaxation. The arrow in Figure 2 is placed at the same moment. The ratio of stiffness to force in Figure 3 is the same during relaxation with no shortening, as it is during the rise where there is shortening (A.F. Huxley and Simmons, 1970). But after the point on the relaxation phase where segment length begins to deviate from the constant length maintained during the tetanic plateau, the ratio becomes less. This is probably another indication of the longitudinal inhomogeneity in the duration of activity - known to occur during relaxation from tetani (A.F. Huxley and Simmons, 1973). One implication of this finding in regard to the effect of stretch on the apparent rate of relaxation is discussed elsewhere in this volume (Cecchi et al., 1983).

DISCUSSION

Our data indicate that stiffness leads force during the tetanus rise. One way to account for this observation would be to assume there is a relatively long-lived cross-bridge state between attachment and force generation, which is the idea we find most attractive. Such a state has been proposed by H.E. Huxley (1979a; 1979b) as a possible explanation for the observation that the equatorial X-ray diffraction pattern from contracting frog muscle changes more rapidly than force during the tetanus rise. Other explanations have been suggested for the X-ray events during the rise (H.E. Huxley and Haselgrove, 1976), but not all could also account for our stiffness measurements. Biochemical experiments over the last decade have also demonstrated a rate-limiting step in the early stages of a cross-bridge cycle, one that occurs not only when cross-bridges are unattached, but also when they are attached (Stein, Schwarz, Chock and Eisenberg, 1979). However, recent evidence suggests that this state probably cannot account for our observations either (Chalovich, Chock and Eisenberg, 1981).

The second fall in stiffness that we observed about 10 to 15 ms after the release, which corresponds to phase 3 of the tension transient described by A.F. Huxley and Simmons (1973), cannot be attributed to tendon compliance because the force is rising during this period. Therefore, it represents a real phenomenon associated with cross-bridge activity. It has already been suggested (A.F. Huxley, 1974) that this phase could be due to a slow detachment and rapid re-attachment of cross-bridges. The reduction in the recovery of force about 10 to 15 ms after the release (Figure 1) is best explained by assuming a change in the cross-bridge distribution to one favoring cross-bridges generating more force. Alternatively, one could assume there is an increased rate of breaking of the cross-bridges that are not generating force.

Our data also indicate that during the final recovery of force to the tetanic value, which corresponds to phase 4 of the tension transient (A.F. Huxley and Simmons, 1973), stiffness is always higher than force. This result seems consistent with the results obtained recently by H.E. Huxley, Simmons, Faruqi, Kress, Bordas and Koch (1981), where changes in X-ray diffraction patterns suggest that cross-bridges could re-attach in a more perpendicular orientation during this phase, and for a period generate little or no force.

In addition, we observed that force declined during relaxation more precipitously than stiffness. During relaxation the change in instensity of the equatorial X-ray diffraction pattern and the X-ray layer lines sometimes lags behind force (Matsubara and Yagi, 1978; H.E. Huxley, 1980; 1982). These observations could be accounted for in a number of ways.

The process of relaxation might involve a reduction in the ability of attached but non-force generating cross-bridges to make the transition into a force generating state. Alternatively, (i) detachment of force generating cross-bridges may be accelerated during relaxation, (ii) force generating bridges may actually rotate to a non-force generating conformation or (iii) detachment of some bridges may be followed by very rapid re-attachment at new locations (Tregear and Marston, 1979; Griffiths, Kuhn and Rüegg, 1979; Kuhn, 1981). Further evidence is needed to decide conclusively among these and other explanations.

ACKNOWLEDGEMENTS

This work was supported in part by fellowships from the Muscular Dystrophy Association of America, Inc. and the Minnesota Heart Association (to G.C. and P.J.G.), and by grants-in-aid from the Minnesota Heart Association and U.S. Public Health Service (NS 14268). We are indebted to L.A. Wanek for assistance with the experiments, to L.F. Wussow for preparation of the figures and the manuscript, and to R. Rüdel, S. Winegrad and M.K.D. Pagala for helpful criticism. We thank the Muscular Dystrophy Association for a service after this symposium that made us aware of the similar conclusions already drawn from different evidence (P. Mason and H. Hasan, 1980. Experientia 36: 949).

REFERENCES

Cecchi, G., Griffiths, P.J. and Taylor, S. (1982a). Mechanical resonance of single muscle fibers studied by high-frequency sinusoidal vibrations. Biophys. J. 37: 120a

Cecchi, G, Griffiths, P.J. and Taylor, S. (1982b). Muscular contraction: Kinetics of cross-bridge attachment studied by high frequency stiffness measurements. Science 217: 70-72.

Cecchi, G., Griffiths, P.J. and Taylor, S. (1983). Changes in intracellular Ca^{2+} induced by shortening imposed during tetanic contractions. This volume.

Chalovich, J.M., Chock, P.B. and Eisenberg, E. (1981). Mechanism of action of troponin·tropomyosin. J. Biol. Chem. 256: 575-578.

Ecker, A. (1971). *The Anatomy of the Frog.* Translated by G. Haslam. Amsterdam, A. Asher & Co., N.V.

Griffiths, P.J., Kuhn, H.J. and Rüegg, J.C. (1979). Activation of the contractile system of insect fibrillar muscle at very low concentrations of Mg^{2+} and Ca^{2+}. Pflugers Arch. 382: 155-163.

Huxley, A.F. (1974). Muscular contraction (Review Lecture). J. Physiol. 243: 1-43.

Huxley, A.F. (1980). Reflections on Muscle. In: *The Sherrington Lectures,* vol XIV, 111 pp, Princeton Univ. Press, Princeton, NJ.

Huxley, A.F. and Simmons, R.M. (1970). Rapid "give" and the tension "shoulder" in the relaxation of frog muscle fibres. J. Physiol. 210: 32-33P.

Huxley, A.F. and Simmons, R.M. (1973). Mechanical transients and the origin of muscular force. Cold Spring Harbor Symp. Quant. Biol. 37: 669-680.

Huxley, H.E. (1979a). In: *Structure in the Molecular Basis of Force Development in Muscle,* pp. 1-13, ed. Ingels, N.B. Jr., Palo Alto Medical Research Foundation, Palo Alto, CA.

Huxley, H.E. (1979b). Time resolved X-ray diffraction studies on muscle. In: *Cross-Bridge Mechanism in Muscle Contraction,* pp. 391-405, ed. Sugi, H. and Pollack, G.H., University Park Press, Baltimore.

Huxley, H.E. (1980). The movement of myosin cross-bridges during contraction. In: *Muscle Contraction: Its Regulatory Mechanisms,* pp 33-43, ed. Ebashi, S., Maruyama, K. and

1982a M., Japan Scientific Societies Press, Tokyo.

Huxley, H.E. (1982). Guest Lecture: The Mechanism of Force Production in Muscle. In: *Disorders of the Motor Unit*, pp. 1-11, ed. Schotland, D.L., John Wiley and Sons, NY.

Huxley, H.E. and Haselgrove, J.C. (1976). The structural basis of contraction in muscle and its study by rapid X-ray defraction methods. In: *Myocardial Failure*, pp. 4-15, ed. Riecker, G., Weber, A. and Goodwin, J., Springer, Berlin.

Huxley, H.E., Simmons, R.M., Faruqi, A.R., Kress, M., Bordas, J. and Koch, M.H.J. (1981). Millisecond time-resolved changes in X-ray reflections from contracting muscle during rapid mechanical transients, recorded using synchrotron radiation. Proc. Natl. Acad. Sci. 78: 2297-2301.

Kuhn, H.J. (1981). The mechanochemistry of force production in muscle. J. Muscle Res. & Cell Motil. 2: 7-44.

Matsubara, I. and Yagi, N. (1978). A time-resolved X-ray diffraction study of muscle during twitch. J. Physiol. 278: 297-307.

Stein, L.A., Schwarz, R., Chock, P.B. and Eisenberg, E. (1979). The mechanism of the actomyosin ATPase: evidence that ATP hydrolysis can occur without dissociation of the actomyosin complex. Biochemistry 18: 3895-3909.

Stienen, G.J.M. and Blangé, T. (1981). Local movement in stimulated frog sartorius muscle. J. Gen. Physiol. 78: 151-170.

Tregear, R.T. and Marston, S.B. (1979). The cross-bridge theory. Ann. Rev. Physiol. 41: 723-736.

DISCUSSION

KAWAI: You said that the decrease in stiffness found during phase 1 could be due to non-Hookean tendon compliance. It could also be due to a non-linear stiffness in crossbridges unless you can tell the difference. To distinguish between the two possibilities you should measure the length changes at the sarcomere level and correlate those with the changes in force to calculate the stiffness.

SUGI: I had the impression that with slow releases the force and the stiffness went in parallel as some parameters were changed. We got the same results with ultrasonic waves during force development. But your stiffness vs. force relation was complex when you used the quick release method. I think this is also consistent with my results because there is non-uniformity when a fast length change is imposed.

CECCHI: I must remind you that the transmission time in our preparation was much shorter than the release time. Therefore, We believe that the differences we found in the stiffness to tension ratio measured during the tetanus rise compared to that measured after or during a quick release, are real differences due to crossbridge activity and are not due to nonuniformity.

KRUEGER: A simple question about events on the shoulder in relaxation. From what you said before, I inferred that there was no phase shift between the length and the tension of that part of contraction as well.

CECCHI: Yes, in *all* the experiments I presented here there was no phase shift between force and length sinusoids. We were worried about this point and we checked the phase shift carefully. If you use length oscillations at a frequency of 1 kHz you get a phase shift with the force preceeding the length. Presumably this is due to the intrinsic recovery mechanism of the crossbridges and to some viscosity in the fiber, which is small but significant. But in the range of 4 to 7 kHz there was no phase shift. At a freqeuncy higher than 9 kHz we again got a phase shift. This effect is presumably due to the inertia of the fiber.

POLLACK: I take it from your comments, then, that you are assuming that the length change, both the step and the sinusoidal oscillation, are imposed uniformly, both spatially and temporally. Do you have any evidence to support that?

CECCHI: Well, first I would like you to remember that we did the experiments on very short fibers, about 2 to 2.5 mm long. The transmission time in these fibers was around 12 μsec. Considering the viscosity of a fiber and of the surrounding medium, the effects of the transmission time on the distribution of a length perturbation have completely disappeared at the end of the step and at the end of the first or second period of the oscillations we imposed. Another point to consider is that the shape of the tension transients was exactly the same when we used fibers 5 to 7 mm long. Also the absence of a phase shift is evidence against a significant effect due to the transmission time.

BRESSLER: The fact that the stiffness stayed up whereas tension fell, during a rapid release, would you consider an interpretation of that as that there is really not much detachment of cross-bridges during the actual step itself? Because if the stiffness were linear, then as long as the bridge were attached, irrespective of its force, it would contribute to the total stiffness.

CECCHI: That is also our interpretation. But we still have to explain why there is a drop in stiffness at the end of the release. We think that this drop could be due to non-Hookean compliance of the tendons, but we cannot exclude the possibility that the non-linear compliance is in the cross-bridges. Our results are in general agreement with the Huxley and Simmons interpretation of the tension transients, that during the quick release and during the fast recovery phase there is not a significant change in the number of attached crossbridges but only a change in the crossbridge distribution among different states, for example, force generating or non-force generating.

TANAKA: Have you done any measurements of stiffness during stretch?

CECCHI: Yes, we did a few experiments of this type but we have not yet completely analyzed the results. What I can say is that during the stretch, when the tension rises, there is an increase in stiffness, but the increase in stiffness is much smaller than the increase in tension. The records are symmetrical with the records we obtained during the quick release. I cannot claim the numbers are exact, but in experiments with a stretch that produced an increase in tension of about 15%, we found an increase in stiffness of less than about 5%.

HUXLEY: I wonder if you have done any experiments where you do a release sufficient to drop the stiffness, and then immediately or within a millisecond you restretch back to the original length?

CECCHI: No, we did not. We did some experiments with rather big releases in order to completely detach the crossbridges and we observed the complete time course of the stiffness during the recovery of tension. The time course of the stiffness was very similar to the time course we found during the tetanus rise, indicating a complete detachment of the crossbridges due to the big release, but we never tried to restretch.

RUEGG: Dr. Güth and his colleagues have done experiments of the kind Dr. Huxley suggested. They released skinned insect flight muscle or frog skeletal muscle and then restretched them after about 2 msec. Tension and stiffness were the same before the release and after restretching the fiber.

HUXLEY: But the stiffness decreased during release and was restored again after restretch, wasn't it?

RUEGG: Yes, if the fiber is released by a large amount, more than 1% L_0, stiffness decreases during release, i.e., the slope of tension-length curve is bent during the release. But during restretch to the initial length (after 2 ms) the slope of the tension-length curve becomes similar to that at the beginning of the release, which means that stiffness as well as the tension is restored by restretching 2-3 msec after the release. (cf. P.J. Griffiths et al., Biophys. Struc. Mech. 7: 107-124, 1980).

TER KEURS: I would like to go back to your initial hypothesis, where you assume that the S-1 head goes through states 2 to 7 or so. You assume that the S-1 head is rigid in all those situations. However, if you assume that the S-1 head is more flexible sideways than extensible you would not expect differences in stiffness of the system depending on the state of activity.

CECCHI: That is what we assumed. The stiffness of a crossbridge is assumed to be constant, independent of the state while the tension developed is different. On the other hand, during phase 3 of recovery, the stiffness goes down while the tension goes up. You can explain this result by assuming there is a reduction in the number of attached crossbridges, as indicated by the decrease in stiffness, and that a single crossbridge develops more force than previously.

RALL: You said that you thought part of the 20% drop in stiffness may be due to a residual tendon effect. Can you make that drop greater by having more tendon? In other words, can you make the situation worse?

CECCHI: I suppose this is possible but we never tried. We wanted to reduce the tendons as much as possible.

RALL: I believe if you increase the tendons it wouldn't change.

CECCHI: It depends very much on the shape of the tension-extension curve of the tendons. If it is approximately exponential I do not know how much the situation might be changed by having more or less tendon.

KAWAI: I have a suggestion. Why not just take the tendon out and make your measurements for comparative purposes?

CECCHI: We made measurements on pieces of tendon and we found that the tendon compliance is non-Hookean, but in order to make quantitative measurements you should be able to dissect a piece of tendon exactly equal to tendon inserted on the fiber you used for the experiment. On the other hand, the tendon attachments in our single fiber preparations are so small that it is practically impossible to dissect them out and make the measurements.

HOUSMANS: I would like to show results that are relevant to Dr. Cecchi's communication and also to the papers preceding his. We obtained, some years ago, similar results in tetanized mammalian cardiac muscle, where we did not impose changes in *length*, but rather changes in *load* (Brutsaert and Housmans, in *Crossbridge Mechanism in Muscle Contraction*. Ed. H. Sugi and G.H. Pollack, pp241-258, 1979).

Figure D-1A shows, on a very slow time scale, records of force and shortening of an afterloaded tetanic contraction in cat papillary muscle. During steady shortening, load was abruptly increased and this induced a rapid lengthening of the muscle at constant load. The details of such a

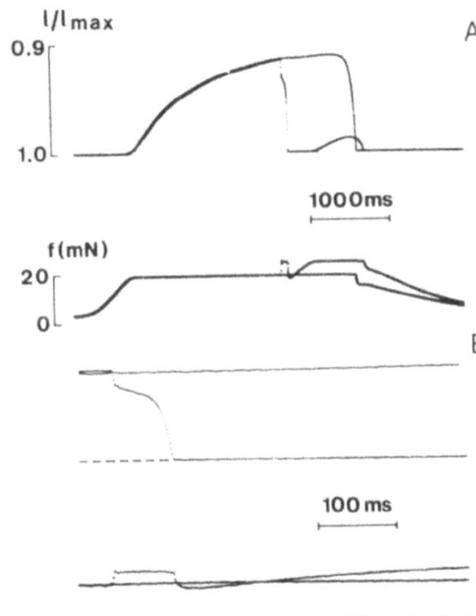

Figure D-1: Length responses to sudden increases of load during tetanic contraction of cat papillary muscle. Length and force calibrations are the same for panels A and B. *Panel A* (slow sweep): Load was rapidly increased during steady shortening of an afterloaded tetanic contraction. This maneuver provoked rapid yielding of the muscle, followed by a transient fall in isometric force, although the imposed higher load was maintained throughout the remainder of the contraction. The control afterloaded tetanus was superimposed. Peak shortening of the control contraction coincides with the end of the stimulation train. Tetanization of the muscle was as described in Brutsaert & Housmans (1979). *Panel B* (fast sweep): Same contractions as in panel A (sweep was triggered shortly before the increase in load). The train of electrical stimulation was continued throughout the entire sweep. Muscle characteristics: length at l_{max}, 5.25 mm; mean cross-sectional area 0.62 mm^2; ratio of resting to total isometric tension at l_{max} 8.9%; temperature 19°C.

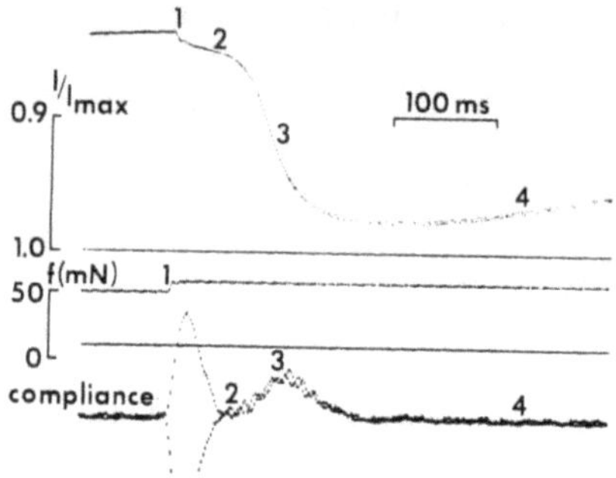

Figure D-2: Changes in muscle compliance during load-induced length transient in tetanic contraction of cat papillary muscle. From top downwards: records of length, force, and mechnical compliance as a function of time of two superimposed load-clamped tetanic contractions, similar to that in Figure 1-B. The initial transient in the compliance tracing, at the time of the load step, is a consequence of the settling time of electronic correlators used to decompose total muscle stiffness on-line into inertial, viscous and elastic (compliance) components. Compliance signals of superimposed contractions therefore do not superimpose immediately after a rapid change in load or length. During the second half of phase 2 and thenceforth, however, a unique measurement of compliance was obtained. See text. Muscle characteristics: length at l_{max}, 8.5 mm; mean cross-sectional area 0.88 mm^2; ratio of resting to total isometric tension at l_{max} 12.0%; temperature 19°C.

load-induced length transient are apparent in Figure D-1B. The sudden increase of load induced an initial elastic extension of the muscle (phase 1). Thereafter, at the constant higher load, where series elastic components do not contribute to length changes, there is a slow, constant-velocity lengthening (phase 2), followed by a much more rapid "yielding" (phase 3). In tetanus, this load-induced three-step isotonic yielding is eventually followed by a slow recovery of force (Fig. D-1B) or shortening (Fig. D-2) at the higher load. There are striking similarities between load-induced length transients in cardiac muscle and force transients after quick release experiments in skeletal muscle. We have measured, on-line, the changes of mechanical compliance (1/stiffness) that occur during these load-induced length transients, using 30-60 Hz sinusoidal load oscillations. The compliance signal (bottom trace in Figure D-2) is the elastic component of total stiffness, whereby corrections were made for the small contributions of viscosity and inertia at any given oscillation frequency (Lewis et al., Cardiovasc. Res. 14: 339-344, 1980).

In Figure D-2, two load-induced length transients, similar to that in Figure D-1, were superimposed. It did not prove possible to obtain a reliable measurement of compliance during phase 1 extension and the first part of phase 2 lengthening. However, during the second part of phase 2, compliance reached a steady value, only to increase at the transition of phase 2 slow lengthening and phase 3 yielding. Compliance reached a peak value halfway through phase 3, and returned to control levels during the slow recovery (phase 4). The changes of compliance under consideration are consistent with the hypothesis that phase 2 represents back rotation of crossbridges over their range of attachment, with subsequent detachment during phase 3, and followed by eventual reattchment in phase 4 (Brutsaert and Housmans, 1979).

I wish to indicate parallelism with Dr. Cecchi's results: we see no change in compliance or stiffness during phase 2, in agreement with his data that indicate a relatively long-lived cross-bride state; his fall in stiffness during tension recovery in skeletal muscle (phase 3) and the increased compliance during phase three yielding in cardiac muscle may reflect an increased rate of cross-bridges that are not generating force or not sustaining load.

GORDON: Dr. Cecchi, you showed a decline in stiffness during a tension transient that, I gather, you would like to attribute to the tendon. And other people have talked about measurements showing no change in stiffness during this state. What's your conclusion now as to whether stiffness undergoes changes during, say, phase 2 of the force transient?

CECCHI: Somebody said earlier that during the quick release there are no changes in stiffness. But I think he was referring to the theoretical situation according to Huxley and Simmons more than to an experimental result. Huxley and Simmons found that at the end of a quick release there was a smaller stiffness than at the tetanic plateau. They measured the stiffness from the slope of the T_1 curve and the slope was dependent on the velocity of release. This is because the crossbridge recovery is very fast and it starts during the release itself, truncating the tension transient. In our experiments we used sinusoidal length changes at a frequency between 4 and 9 kHz. At 9 kHz there is practically no recovery, so that the fall in stiffness we found cannot be attributed to the crossbridge recovery. It is a real fall in stiffness but we do not know if it is due to the tendon compliance or crossbridge compliance. We have to do the same experiments without the influence of the tendons.

GORDON: I brought the issue up because in terms of the people who are looking at the effects of ATP on rate constants, some of the interpretations are based on whether crossbridges are detached or not. So those depend on the stiffness measurements.

CECCHI: I suppose the situation could be different because in skinned fibers you have some compliance from the fiber attachments, and for this type of measurement every contribution to the total fiber compliance is very important.

COOKE: Dr. Cecchi, if one interprets changes in stiffness as indicating changes in attachment and detachment of crossbridges, then when you stretch the muscle do you interpret that as more crossbridges attaching, because the stiffness goes up?

CECCHI: No, as I said before we did only a few experiments with stretches, and I cannot be very exact but I am confident that the increase in stiffness we observed during the stretch was much smaller than the increase in tension. As an indicative value I would say that for an increase in tension of about 15 to 20% we measured an increase in stiffness of only about 5%. This increase in stiffness could be due just to the stretching of the tendons.

COOKE: I do not understand why.

CECCHI: I would say that our result suggests that only a small portion of the increase in force could be explained by an increase in the attached crossbridges.

COOKE: So you are not interpreting that as a change in crossbridge number.

CECCHI: Right, we are using the stiffness to tension ratio as an indication of change in crossbridge number or a change in crossbridge distribution among different states generating more or less force. According to our results we think that the increase in force produced by the stretch is not due to an increase in the crossbridge number.

COOKE: How fast does the change in stiffness occur? What is your limit of resolution on the initial rise or fall in stiffness?

CECCHI: The fastest change in stiffness appears during the quick release, and how fast it occurs depends directly on the velocity of release used. During the tetanus rise or relaxation the stiffness changes are much slower, as I showed previously (Fig. 1 and 2). For the second question, using oscillations at a frequency of 7 kHz, we have been able to measure the time course of the change in stiffness during a length step of 1 millisecond, making one measure every 142 μs.

NOBLE: If you oscillate one end of the fiber, would you not expect the waves to be transmitted to the other end where it's fixed, and then reflected back so you would have an attenuated wave coming back, which would then interfere and give you apparent changes in stiffness which aren't actually real?

CECCHI: Yes, this is what we are expecting. As you said, if we oscillate one end of the fiber we produce a wave which travels towards the transducer end and is reflected back to the other end, interfering with the direct wave. The degree of interaction depends on both the amplitude and the phase of the reflected wave. If you apply length oscillations at a frequency equal to half the transmission time, the reflected wave should be perfectly in phase with the direct wave and the fiber should start to resonate.

In order to measure how big this interaction is, we did a few experiments on long fibers isolated from iliofibularis muscles and measured the force at one end of a fiber as a function of the oscillation frequency of the length changes applied to the other end. As you can see (Fig D-3) the fiber is able to resonate and as the frequency applied approaches the fiber's resonance frequency, the force recorded (and therefore the apparent stiffness) increases and starts to show a phase delay with respect to the length oscillations. At the resonance frequency (about 12 kHz in the fiber shown here) the force reaches a maximum amplitude and is about 5 times higher than the force recorded at 1 kHz and it lags the length oscillations by about 90°. The behavior of both the amplitude and the phase shift of the force oscillations indicate the presence of a significant amount of damping of the longitudinal oscillations produced by viscous forces in the fiber and the surrounding fluid. The increase in

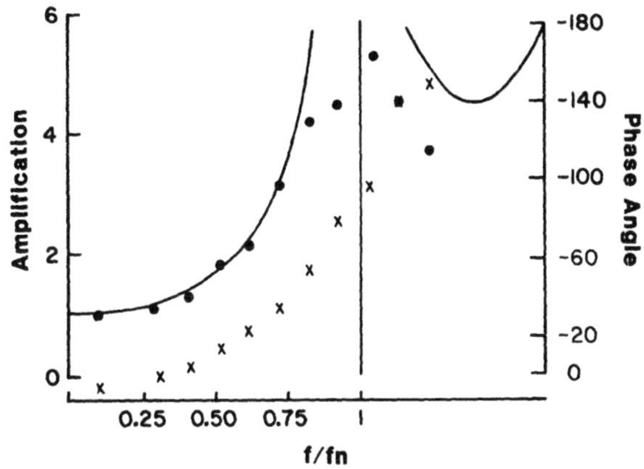

Figure D-3: The relationship between the frequency of small, sinusoidal length changes applied to one end of a fiber during the plateau of a tetanus, and the change in force recorded at the other end. The frequency of the length changes (f) is plotted on the abscissa as a fraction of the frequency (f_n) at which a 90° phase shift between force records and length changes was detectable. The amplitude (●) and phase angle (X) of the force responses are the ordinates shown on the left and right side of the figure respectively. The amplification of force is plotted as a ratio between the peak response at a given frequency and the response at 1 kHz. The continuous lines represent the theoretical behavior of an elastic rod with negligible viscous forces, along which quick length changes are uniformly distributed. The lines were calculated from the model of Schoenberg, Wells and Podolsky (Muscle compliance and the longitudinal transmission of mechanical impulses. J. Gen. Physiol. 64: 623-642, 1974). We deliberately used relatively long fibers for these experiments in order to excite natural modes of vibration in a fiber. The data shown were obtained from an iliofibularis fiber (22.ix.81) that was 11.3 mm long when its striation spacing was 2.20 μm (at 4°C). We also determined the range of frequencies where force did *not* undergo mechanical amplification in experiments on shorter fibers, and used only data obtained at these frequencies of oscillation to determine stiffness (Cecchi, Griffiths and Taylor, 1982b; Cecchi, Griffiths and Taylor, this volume). Hence the relatively high stiffness we observed during the rise of tetanic contraction and during relaxation was not due to a delay in transmission of mechanical perturbations from one end of a fiber to the other.

force as compared to the force measured at 1 kHz (where there should not be a mechanical amplification) is very small up to a frequency of 0.2-0.25 of the resonance frequency of the fiber. Therefore, to avoid resonance artifacts it is necessary to use a maximum oscillation frequency no higher than about a fourth the resonance frequency. This is exactly what we did. The resonance frequency (as calculated from the transmission time) of our short fibers was around 35-40 kHz and with these fibers we used a maximum oscillation frequency of 7-9 kHz. With fibers isolated from semitendinosus muscles, which have a much lower resonance frequency, we used a maximum frequency of 2.5 kHz (Fig. 2 and 3). Under these conditions the amplitude of the reflected wave should be too small to produce a significant effect on the stiffness measurements.

THE ROLE OF Ca^{2+} IN CROSS-BRIDGE KINETICS IN CHEMICALLY SKINNED RABBIT PSOAS FIBERS

Masataka Kawai, Philip W. Brandt, and Robert N. Cox

Muscle Physiology Division, Departments of Neurology and Anatomy, College of Physicians and Surgeons of Columbia University, 630 West 168th St., New York, N.Y. 10032, U.S.A.

ABSTRACT

The sinusoidal analysis method is applied to chemically skinned rabbit psoas preparations, and the effect of Ca ions on exponential processes is determined. The results are compared to the predictions of two classes of models, those in which the binding of Ca^{2+} to the thin filaments acts as a switch to fully activate a segment of the filament (switch hypothesis), and those in which the binding of Ca^{2+} results in a graded increase in activity in proportion to the fraction of sites occupied (graded hypothesis). Although qualitative features of our results are in accord with the switch hypothesis, quantitative examination of the results shows that there is an extra process (B'), which produces a W-shaped phase-frequency plot during partial activation. This result is consistent with the assumption that there are two activated states available to the thin filaments of structured muscle.

INTRODUCTION

Two apparently different mechanisms for the regulation of skeletal muscle function by Ca^{2+} have been proposed, and evidence has been adduced in support of both. The difference between the two lies in the proposed effect of Ca binding on the attachment rate of myosin to actin. In one mechanism the binding of Ca^{2+} causes the apparent attachment rate to undergo an all-or-none transition from zero to a fixed value; corresponding actin segments are described as going from the "off" to the "on" state (Podolsky & Teichholz, 1970). In the other mechanism the apparent attachment rate constant is assumed to increase in a graded way, i.e., in proportion to the fraction of regulatory sites occupied by Ca (Julian, 1969).

Podolsky & Teichholz found identical maximum velocities of shortening (V_{max}) in fully and partially activated skinned frog fibers (see also

Gulati & Podolsky, 1981). Because V_{max} is believed to reflect attachment and/or detachment rate constants of myosin cross-bridges, their result supported the argument that the binding of Ca^{2+} to the regulatory sites does not modify cross-bridge kinetics in a graded way, but rather opens active sites to myosin. In contrast, Julian (1971) reported that V_{max} varied with the degree of activation (see also Wise, Rondinone & Briggs, 1971; Julian & Moss, 1981).

That the rate constant(s) of the cross-bridge cycle may be graded in proportion to the amount of Ca bound to the regulatory sites on troponin-C (TnC) raises a challenging question about the mechanism of muscle activation by Ca^{2+}. It is essential therefore to determine the action of Ca^{2+} by an independent method. We have approached this problem by using a method which is called sinusoidal analysis. With this method we detect three or more exponential processes, each of which is assumed to be a composite of many intrinsic rate constants of the cross-bridge cycle.

PREDICTIONS

In vertebrate skeletal muscles, actomyosin ATPase is regulated via a control mechanism on the thin filaments (Ebashi & Endo, 1968; Weber & Murray, 1973; Fuchs, 1974; Szent-Gyorgyi, 1975). Ca binding to a TnC (Troponin C) molecule turns on a segment of thin filament, consisting of 7 monomeric actins, via the tropomyosin-mediated control system; increasing Ca^{2+} concentration results in an increased number of on-state actin units. Myosin heads charged with the hydrolysis product readily react with actin units in the on-state to form cross-bridges. In this scheme, 7 actins, 1 troponin, and 1 tropomyosin work as a basic cooperative regulatory unit.

An additional approach to the Ca regulatory mechanism can be constructed from the report of Stein et al. (1979). They concluded that the "dissociated" actin and myosin species are in fact in rapid equilibrium with a "weakly associated" species (earlier work on this was carried out by Inoue, Shigekawa & Tonomura, 1973). Furthermore, Chalovich, Chock & Eisenberg (1981) reported that Ca binding to TnC controls the phosphate release reaction, which is subsequent to attachment, and which may accompany the power stroke. Thus the attachment reaction is fully reversible and does not require Ca, whereas a subsequent release reaction is almost irreversible and requires Ca-activated actin. If this view is taken, the classical Ca-controlled attachment reaction must include the two sequential reactions described above. For simplicity we shall henceforth lump them together in the "attachment" reaction.

The switch hypothesis for the regulatory action of Ca binding is compatible with the above scheme established by biochemists and X-ray diffractionists. In this mechanism a segment of the actin filament is activated as Ca^{2+} binds to TnC, the number of active segments increasing with increase in Ca binding. In contrast to the behavior of actomyosin in solution, steric limitations on myosin in structured systems allow it to

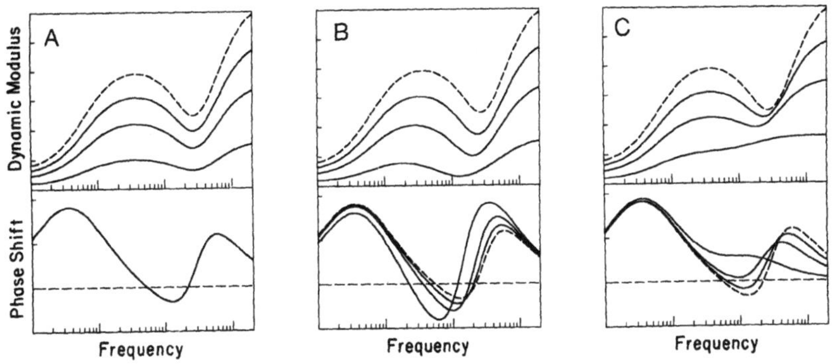

Figure 1: Predictions based on switch hypothesis in *A*, and on graded hypothesis in *B* and *C*. In lower panel of *A*, all 4 plots are superimposed. Broken lines correspond to full activation. Parameters are: H=16.9, A=88.7, B=-88.2, C=141.1, Y_∞=158.5, a=0.9, b=17.5, c=71.6. Values of complex moduli are calculated from an equation with three exponential processes, which lacks the (B') term in Eqn. 1. In all cases Y_∞ is scaled to match the values shown in Fig. 2. In *B*, the characteristic frequency *b* is also scaled to change with Y_∞. Likewise in *C*, *c* is scaled to change with Y_∞. Scaling factors for partial activation are: 0.25, 0.56, 0.79.

interact only with those actins which are in the immediate vicinity. The reaction takes place or not depending on the condition of the actin; hence it is often described as an all-or-none mechanism. The expected reaction will follow first order kinetics, i.e., there should be no concentration terms involved (assuming that myosin is always supplied). This situation is more easily visualized if the Ca-controlled step is the phosphate release reaction (above), which is clearly first order. If the reaction is first order, we expect to see no shift along the frequency axis in the plot of the dynamic modulus and no shift in the phase-frequency plot as Ca^{2+} concentration is elevated (Fig. 1A). This is because the frequencies corresponding to maxima and minima of these plots represent apparent rate constants that in turn have direct functional relationships to the intrinsic rate constants of the underlying chemical reactions. What does change as a function of Ca binding is the number of active cross-bridges, and this is reflected as a proportionate increase in the dynamic modulus (stiffness) and tension (Fig. 1A). The switch hypothesis, as put forward by Podolsky & Teichholz (1970), is a simple modification of A.F. Huxley's (1957) contraction model; Podolsky & Teichholz assumed that the number of active sites increases as the Ca^{2+} concentration is elevated.

In the graded hypothesis the apparent attachment rate constant increases in proportion to the degree of activation. This hypothesis, which is another modification of A.F. Huxley's model, was initially put forward by Julian (1969). This mechanism assumes that, as in a solution of actin and myosin, more actin units become available for interaction with myosin as the Ca^{2+} concentration increases because the number of the active actin units increases with Ca^{2+}. The reaction is bi-molecular and hence follows second order kinetics. If this is the case, we expect to find a continuous shift of the plots of dynamic modulus or phase shift along the frequency axis (Fig. 1B,C), assuming our technique is sensitive to the reaction in question. The situation may be analogous to the substrate effect, in which we found graded changes in apparent rate constants as the substrate concentration was elevated (Kawai, 1978, 1982; Cox & Kawai, 1981). It is beyond doubt that the substrate binding reaction follows second order kinetics.

In either mechanism, the overall rate of attachment increases with Ca^{2+}, and this is followed by increased tension or hydrolysis rate. Hence a measurement of tension or ATPase does not allow us to discriminate between the two possibilities. But a kinetic analysis using sinusoidal waveforms does provide a potential method for differentiating between the two mechanisms.

RESULTS

Small bundles (1-3 fibers) of chemically skinned rabbit psoas muscle are soaked first in relaxing solution which contains EGTA and MgATP, then are activated with solutions of the desired pCa. As soon as a steady tension is developed, the complex stiffness/modulus is collected at 17 frequencies (0.125-167 Hz), and the muscle is relaxed again. Peak-to-peak length oscillation is kept constant at ≦0.2% L_o. This sequence is repeated for different pCa values. The method of sinusoidal analysis was described earlier (Kawai & Brandt, 1980). Complex stiffness and modulus are the same quantity, except that the latter is normalized to the physical dimension (cross-sectional area and reciprocal length) of the muscle. Complex modulus has 2 components: amplitude and phase shift. We call the amplitude component the dynamic modulus. It is analogous to stiffness and hence represents the resistance to stretch.

Our standard activating solution contains (Na salts in mM): 6 total EGTA, 5 MgATP, 5 free ATP, 7.5 phosphate, 16 phosphocreatine, 80 unit/ml phosphocreatine kinase, 4 sulfate, 42 propionate, 10 MOPS, and Ca to achieve a desired pCa; pH is adjusted to 7.00±0.01, and the ionic strength is maintained at 201 mM. The experiments were carried out at 20°C. Under these conditions tension threshold is observed at pCa 6.0, and full tension is observed by pCa 5.0-5.5; the pCa-tension relation is very steep with a Hill coefficient ≧4 (cf. Brandt, Cox, Kawai & Robinson, 1982).

Fig. 2 shows complex modulus data obtained at various pCa values. It is seen from Fig. 2A that the dynamic modulus progressively increases with increase in Ca^{2+} concentration. More importantly, its frequency profile does not change significantly, and the characteristic dip (a minimum) is always observed at the same frequency (25 Hz). In the phase-frequency plot (Fig. 2B), the minimum at 17 Hz and the maximum at 50 Hz do not shift noticeably, indicating that there is no change in rate constants $2\pi b$ and $2\pi c$ with Ca concentration (these correspond to phases 3 and 2, respectively, of Huxley & Simmons, 1971). The change of the slowest rate constant $2\pi a$ (corresponding to phase 4) at around 1 Hz appears to be minimal. This result is consistent with the switch hypothesis.

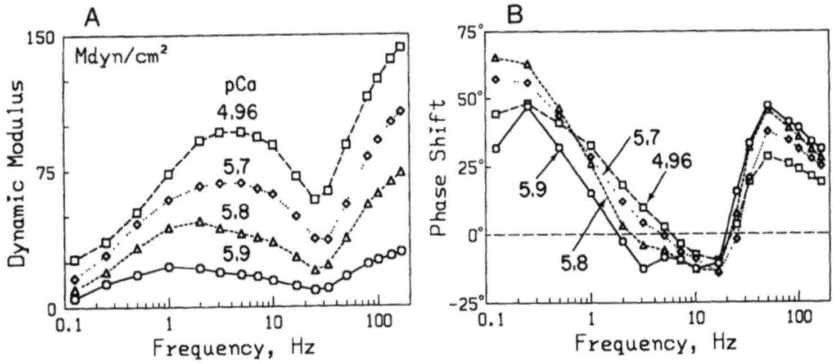

Figure 2: Complex modulus data obtained at 4 pCa values (indicated) are plotted against frequency. Fig. 2B was reproduced from Kawai et al. (1981).

A close examination of Fig. 2A, however, reveals that there is a subtle shape change in the frequency range 2-7 Hz during partial activation. The effect is clearer in Fig. 2B where an extra minimum is observed at pCa 5.9. Although the extra minimum becomes less pronounced as the activation progresses, its position appears to stay at the same frequency (3 Hz). Hence the phase-frequency plot is W-shaped at partial activation, while it is V-shaped at full activation (Kawai, Cox & Brandt, 1981). We have designated the extra minimum process (B'). It is related to process (B) (whose minimum is at 17 Hz in phase-frequency plot) in the sense that both are phase delays (phase shift is negative: Fig. 2B) hence both produce oscillatory work on the forcing apparatus.

In order to obtain population parameters which we call "magnitudes," we fitted the complex modulus data to the following equation, which consists of 4 exponential processes, by the least squares' method.

$$\begin{array}{ccccc} & \text{Process (A)} & \text{Process (B')} & \text{Process (B)} & \text{Process (C)} \\ Y(f) = H + & Afi/(a+fi) & +\ B'fi/(b'+fi) & +\ Bfi/(b+fi) & +\ Cfi/(c+fi) \end{array} \quad (1)$$

Each term is called an exponential process, because its inverse transform (in the time domain) consists of an exponential term (Kawai & Brandt, 1980). Here f is the frequency of oscillation; $i = \sqrt{-1}$; a , b' , b , c are characteristic frequencies; A, B', B, C are magnitudes of processes (A), (B'), (B), (C), respectively. The signs of the magnitude parameters are self-contained; we expect B and B' to be negative, because (B) and (B') are phase delays. In order to carry out fittings, we assumed that all characteristic frequencies are independent of Ca^{2+} concentration: a = 0.9 Hz, b' = 3 Hz, b = 19 Hz, c = 80 Hz (Fig. 3A). This is justified because of the above observation that the positions of the minima and maxima in the frequency plots do not change with Ca^{2+} concentration (Fig. 2). The values chosen for a, b, c are the best-fit parameters for full activation (pCa 4.96) in the three time constant model, which lacks the B' term in Eqn. 1; the value chosen for b' is based on the position of the extra minimum in the phase-frequency plot at partial activation (Fig. 2B). The above assumptions are necessary because of the small magnitude of

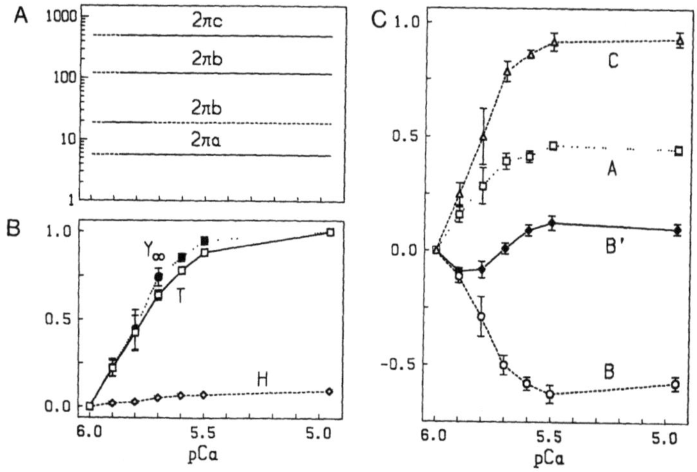

Figure 3: *A* shows the values (constants) of the 4 rate constants. Sections shown in broken lines indicate that respective magnitudes are small and therefore not well determined. *B* and *C* are magnitude parameters (as indicated) derived from fitting the complex modulus data to Eqn. 1. Also included is tension (T) in *B*. Average from 9 experiments with SEM bars (those smaller than the symbol size are not shown).

process (B'), which makes the fitting more difficult, and because of the complexity of the 4-exponential fitting procedures. The value of a may not be entirely constant, and in our experience it is slightly less at a partial activation.

The above assumptions leave only the magnitude parameters (H, A, B', B, C) to be fit. This fitting is an easy one since Eqn. 1 is linear with respect to these parameters. The fit, which includes complex variables, was performed by a method similar to that described in Appendix 2, step 2, of Kawai & Brandt (1980); the results are averaged and plotted in Fig. 3B & C. Also included are the tension and Y_∞ (Fig. 3B). The latter quantity, which corresponds to instantaneous stiffness, is calculated by extrapolation to infinite frequency in Eqn. 1:

$$Y_\infty = H + A + B' + B + C \qquad (2)$$

Y_∞ is often used as an indicator of the number of attached cross-bridges, which is true only if cross-bridge stiffness is the same everywhere in each half sarcomere, and the thick and thin filaments are rigid. The actual value of Y_∞ may be slightly higher than that calculated by Eqn. 2 because of our limitation in band-width (167 Hz) and because the expression for process (C) in Eqn. 1 is only an approximate description of the data. Before averaging, the tension is normalized to P_o (tension developed at pCa 4.96) and magnitude parameters to Y_∞ at pCa 4.96. Their absolute values are: $P_o = 2.03 \pm 0.14$ Mdyn/cm^2; $Y_\infty = 183 \pm 115$ Mdyn/cm^2 (N=9, SEM are shown).

From Fig. 3B it is clear that Y_∞ remains approximately proportional to tension as the Ca^{2+} concentration is increased. Similarly, we find that

H (corresponding to DC stiffness) is scaled with tension and positive at all times. Fig. 3C shows that magnitude parameters *A* and *C* also increase monotonically with tension. *B* is negative, as expected, but deviates slightly from linearity in the range pCa 5.8-5.9. *B'* is initially negative at partial activation, but becomes positive toward full activation. The reason why *B'* becomes positive is probably fortuitous and may be simply that an equation with 4 rate constants describes the data better than one with 3 rate constants.

As discussed above, the negative value of *B'* in the range pCa 5.8-5.9 is significant and is important for our current interest. Fig. 3C shows that substantial numbers of bridges contribute to both processes (B) and (B') at partial activation. Then, as the level of activation increases, the *B'* population diminishes while the *B* population increases. An implication of this observation is as follows. At partial activation active cross-bridges are divided into two populations; those attaching with a lower rate constant and those attaching with a higher rate constant. As the degree of activation increases, the number of cross-bridges attaching with the higher rate constant is increased, and the number attaching with the lower rate constant is reduced.

DISCUSSION

The most important observation in the present report is that the dynamic modulus at all frequencies is approximately scaled with tension as the Ca^{2+} concentration is changed (Fig. 2A). It is also important to note that the characteristic frequencies (maxima and minima) of the frequency plots do not appear to shift progressively along the frequency axis (Fig. 2A,B). This observation is consistent with the assumption that the apparent rate constants are not affected by Ca^{2+} concentration and that whatever actomyosin reaction is controlled by Ca follows first order kinetics. If the Ca-controlled reaction is the actual attachment transition, this observation (first order reaction) implies that the extensibility of a myosin head is sufficiently limited that it can only reach actins within its immediate vicinity. If the Ca-controlled reaction is the phosphate release step (subsequent to attachment), as claimed in the report of Chalovich et al. (1981), then the above results can be readily explained: cross-bridges are already made, and activation of an actin unit by Ca promotes this subsequent reaction, which is first order. Thus our observations seem to be qualitatively consistent with the variation of the switch hypothesis.

To fit our data fully, a modification is required to the above simplified mechanism. This is because (as is clear upon careful examination of Fig. 2B) the phase-frequency plot is W-shaped at partial activation, while it is V-shaped at full activation. The component responsible for the extra minimum at partial activation is small but consistently present. We named this process (B'). As usual, process (B) is the major oscillatory work component and is responsible for the conventional V-shaped appearance. As the degree of activation increases with increased binding of Ca^{2+} to TnC, the magnitude of process (B) grows at the expense of process (B'), and presumably at the expense of the number of inactive segments of actin. The increase in the total number of cross-bridges can be quantified by the increase in Y_∞.

Cross-bridges responsible for both processes (B) and (B') must be arranged in parallel within a half sarcomere. Were they arranged in series, sinusoidal analysis would detect only one exponential process the rate constant of which would be somewhere between those of processes (B) and (B'). We can therefore rule out the undesirable possibility that the slow rate constant reflects the compliance of a damaged end segment of a fiber, if such existed. The parallel arrangement implies that two reactions corresponding to processes (B) and (B') take place side by side. The two reactions must be equivalent steps in the cross-bridge cycle because the muscle can function without process (B'), because $B + B'$ appears to be scaled with Y_∞ (Fig. 3), and because both processes generate oscillatory work.

It is generally believed that Ca control in rabbit psoas takes place via the thin filament control mechanism (Szent-Gyorgyi, 1975). The presence of two rate constants in the same step may mean that there are two active conditions of the thin filament. Because the relative magnitudes of the two processes are controlled by Ca, these two activated conditions must also be controlled by Ca. It is interesting to consider how this could come about.

One possibility is that the cooperative regulatory unit is not limited to the basic unit consisting of 1 troponin, 1 tropomyosin, and 7 actins. Recently, Nagashima & Asakura (1982) reported that up to 16 regulated actins cooperate and suggested that there may be attractive interactions between the basic units. As they suggested, it is possible that 2 tropomyosins across the actin helix interact. In this case, full activation may require binding of Ca^{2+} to two juxtaposed cooperative units: an activation of one basic unit may allow the "attachment" reaction to proceed at a low rate (corresponding to process (B')), while activation of both units may allow the reaction to proceed at the higher rate (corresponding to process (B)). Since there are two Ca binding sites of physiological significance per troponin (Potter & Gergely, 1975), cooperation between more than one basic unit could explain the high Hill coefficient such as observed in the present work.

Another possibility is that the two active states of the thin filament correspond to an "on" state and a "potentiated" state, such as proposed by Weber & Murray (1973). In this mechanism the number of myosin heads attached to the basic cooperative unit also determines its activity. When it is activated only by Ca binding (no myosin heads bound), then the corresponding unit exhibits an "on" state, which allows a low "attachment" rate, and which corresponds to process (B'). When 30% or more of the cross-bridges are attached to an "on" state unit, the unit enters a "potentiated" state, which allows a high "attachment" rate, and which corresponds to process (B). Nagashima & Asakura (1982) used NEM-S1 (N-ethylmaleimide-treated myosin subfragment 1) to populate actin in an irreversible way and measured ATPase activity with a small amount of normal S1. By showing that the presence of NEM-S1 progressively potentiated the ATPase, they provided remarkably clear support for the Weber-Murray mechanism. Such continuous effect, however, is not in accord with our observation of two discrete rate constants at partial activation; hence we may dismiss this mechanism as a possible explanation of the present observations.

Although we have not considered a myosin-linked Ca control mechanism in rabbit psoas, such a mechanism is not entirely impossible. Two possible mechanisms involve isoenzymes (Weed, Hall & Spurway, 1975) or light chain phosphorylation (Barany, Barany, Gillis & Kushmerick, 1979). Both produce more than one kind of myosin that may differ in ATPase activity. For instance, the presence of isoenzymes may explain the W-shaped appearance of the phase-frequency plot (Fig. 2B) during partial activation. However, it could not explain our failure to observe a W-shaped appearance at full activation. The role of phosphorylation steps needs further clarification, although its significance in vertebrate skeletal muscles is questionable (Barsotti & Buttler, 1981; Cooke, Franks, Ritz-Gold, Toste, Blumenthal & Stull, 1981; Crow & Kushmerick, 1981).

Another important result of the present study is the observation that Ca ions affect the middle frequency range, in which oscillatory work is prominent (i.e., phase 3 of Huxley & Simmons, 1971). When this fact is combined with a recent report by Chalovich and others (1981), in which it was shown that the reaction controlled by Ca is the phosphate release step that presumably accompanies the power stroke, then it follows that the power stroke is involved in process (B) (phase 3). This conclusion is in contrast to the proposal of Huxley & Simmons (1971) that the power stroke is involved in phase 2.

Our observation that there is an additional slow process at partial activation implies that V_{max} would be slower if force-velocity experiments were carried out at low Ca^{2+} concentrations, as observed by Julian (1971) and Wise et al. (1971). This is because V_{max} depends on both slowly and rapidly cycling cross-bridges. Our finding, however, that none of the rate constants changes with Ca^{2+} concentration is consistent with a modified switch mechanism as discussed above. A new addition is that there are two qualitatively different switches to account for the data. We are aware of the argument that V_{max} and rate constants may not correlate, due to the difference in strain on a cross-bridge in the two experimental conditions. The significance of this argument depends on the degree of strain sensitivity of the rate constants, and thus the relationship between the two measurements is model dependent.

ACKNOWLEDGEMENTS

The present work is supported by grants from NSF (PCM 80-14527), NIH (NS11766), and MDA.

REFERENCES

Barany, K., Barany, M., Gillis, J.M., and Kushmerick, M.J. (1979). Phosphorylation-dephosphorylation of the 18,000-dalton light chain of myosin during the contraction-relaxation cycle of frog muscle. J. Biol. Chem. 254: 3617-3623.

Barsotti, R.J., and Butler, T.M. (1981). Effect of myosin LC_2 phosphorylation on high-energy phosphate usage in contracting skeletal muscle. Biophys. J. 33: 234a (Abstr.)

Brandt, P.W., Cox, R.N., Kawai, M., and Robinson, T. (1982). Regulation of tension in skinned muscle fibers: effect of cross-bridge kinetics on apparent Ca sensitivity. J. Gen. Physiol. 79: 997-1016.

Chalovich, J.M., Chock, P.B., and Eisenberg, E. (1981). Mechanism of action of troponin-tropomyosin. J. Biol. Chem. 256: 575-578.

Cooke, R., Franks, K., Ritz-Gold, C.J., Toste, T., Blumenthal, D.K., and Stull, J.T. (1981). The function of myosin light chain phosphorylation in skeletal muscle. Biophys. J. 33: 235a (Abstr.)

Cox, R.N., and Kawai, M. (1981). Alternate energy transduction routes in chemically skinned rabbit psoas muscle fibers: a further study of the effect of MgATP over a wide concentration range. J. Muscle Res. Cell Mot. 2: 203-214.

Crow, M., and Kushmerick, M.J. (1981). Light chain phosphorylation and muscle energetics. Biophys. J. 33: 236a (Abstr.)

Ebashi, S., and Endo, M. (1968). Calcium ion and muscle contraction. Prog. Biophys. Mol. Biol. 18: 123-183.

Fuchs, F. (1974). Striated muscle. Ann. Rev. Physiol. 36: 461-502.

Gulati, J., and Podolsky, R.J. (1981). Isotonic contraction of skinned muscle fibers on a slow time base. Effect of ionic strength and calcium. J. Gen. Physiol. 78: 233-257.

Huxley, A.F. (1957). Muscle structure and theories of contraction. Prog. Biophys. biophys. Chem. 7: 255-318.

Huxley, A.F., and Simmons, R.M. (1971). Proposed mechanism of force generation in striated muscle. Nature 233: 533-538.

Inoue, A., Shigekawa, M., and Tonomura, Y. (1973). Direct evidence for the two route mechanism of the acto-H-meromyosin-ATPase reaction. J. Biochem. (Tokyo). 74: 923-934.

Julian, F.J. (1969). Activation in a skeletal muscle contraction model with a modification for insect fibrillar muscle. Biophys. J. 9: 547-570.

Julian, F.J. (1971). The effect of calcium on the force-velocity relation of briefly glycerinated frog muscle fibers. J. Physiol. (Lond.). 281: 117-145.

Julian, F.J., and Moss, R.L. (1981). Effects of calcium and ionic strength on shortening velocity and tension development in frog skinned muscle fibres. J. Physiol. 311: 179-199.

Kawai, M. (1978). Head rotation or dissociation? A study of exponential rate processes in chemically skinned rabbit muscle fibers when MgATP concentration is changed. Biophys. J. 22: 97-103.

Kawai, M. (1982). Correlation between exponential processes and cross-bridge kinetics. In: *Basic Biology of Muscle: A comparative approach.* pp. 109-130. eds. B.M. Twarog, R.J.C. Levine, M.M. Dewey, Raven Press, New York.

Kawai, M. and Brandt, P.W. (1980). Sinusoidal analysis: a high resolution method for correlating biochemical reactions with physiological processes in activated skeletal muscles of rabbit, frog and crayfish. J. Muscle Res. Cell Mot. 1: 279-303.

Kawai, M., Cox, R.N., and Brandt, P.W. (1981). Effect of Ca ion concentration on cross-bridge kinetics in rabbit psoas fibers. Evidence for the presence of two Ca-activated states of thin filament. Biophys. J. 35: 375-384.

Nagashima, H., and Asakura, S. (1982). Studies on co-operative properties of tropomyosin-actin and tropomyosin-troponin-actin complexes by the use of N-ethylmaleimide-treated and untreated species of myosin subfragment 1. J. Mol. Biol. 155: 409-428.

Podolsky, R.J., and Teichholz, L.E. (1970). The relation between calcium and contraction kinetics in skinned muscle fibres. J. Physiol. (Lond.). 211: 19-35.

Potter, J.D., and Gergely, J. (1975). The calcium and magnesium binding sites on troponin and their role in the regulation of myofibrillar adenosine triphosphatase. J. Biol. Chem. 250: 4628-4633.

Stein, L.A., Schwarz, R.P., Chock, P.B., and Eisenberg, E. (1979). Mechanism of actomyosin adenosine triphosphatase. Evidence that adenosine 5' triphosphate hydrolysis can occur

without dissociation of the actomyosin complex. Biochemistry 18: 3895-3909.
Szent-Gyorgyi, A. (1975). Calcium regulation of muscle contraction. Biophys. J. 15: 707-723.
Weber, A., and Murray, J.M. (1973). Molecular control mechanisms in muscle contraction. Physiol. Rev. 53: 612-673.
Weeds, A.G., Hall, R., and Spurway, N.C. (1975). Characterization of myosin light chains from histochemically identified fibers of rabbit psoas muscle. Febs. Letters 49: 320-324.
Wise, R.M., Rondinone, J.F., and Briggs, F.N. (1971). Effect of calcium on force-velocity characteristics of glycerinated skeletal muscle. Am. J. Physiol. 221: 973-979.

DISCUSSION

GULATI: Is it possible to explain your results obtained on partially activated fibers by heterogeneity in the the sarcomeres?

KAWAI: I do not believe sarcomere heterogeneity underlies our data. This is because when you look at the fiber under the light microscope, the striation pattern is very regular in fibers activated to between 0 and 70% P_0, whereas the W-shaped phase-frequency plot is prominent at 20% P_0 and only a shoulder is detectable at 40% P_0 (Fig. 2B). Between 70% and 100% P_0 you observe a typical V-shaped appearance in the phase-frequency plot. The resolution of the sarcomere pattern at P_0 is poor and the fiber appears milky in ordinary optics. I believe this is related to a change in optical density combined with the large focal depth of the ordinary optics. The milky appearance does not necessarily mean that the sarcomere lengths are heterogeneous. An important point here is that, except for the scaling factor, the Nyquist plot is almost identical at 70% P_0 (very regular sarcomere pattern) and at 100% P_0 (milky appearance). This is also applicable to the phase-frequency plot (Fig. 2B). Therefore the milky appearance under the ordinary optics does not qualitatively affect our measurements.

GULATI: A number of previous studies on skinned fibers (Thames et al., J. Gen Physiol. 63: 509-530, 1974; Gulati and Podolsky, J. Gen Physiol. 78: 233-257, 1981) have suggested the presence of cross-bridge heterogeneity as a possible explanation for the observed decrease in speed in partially activated fibers. Since the conditions of these experiments had not been physiological, we re-studied the isotonic contraction properties of temperature-step activated intact fibers under partial activation. The slide (Fig. D-1) shows the speed of unloaded shortening (normalized to the speed of activation level of 0.8 P_0) as a function of the degree of activation (β). In these experiments β was varied from 1.0 down to 0.2. Note that there is a small but progressive tendency for the speed to decrease with lowering the degree of activation. The speed may be reduced by up to 20% at the lowest β. If the assumption is made that a fixed internal load is present in the fiber throughout, its effect can be predicted quantitatively by using Hill's force-velocity equation. This prediction for a constant internal load amounting to 2% P_0 is shown by the solid line, in this figure. So the take-home point from our data is that a very small (but constant) internal load, if present, can have substantial effects on the contraction properties of the fiber in partially activated conditions.

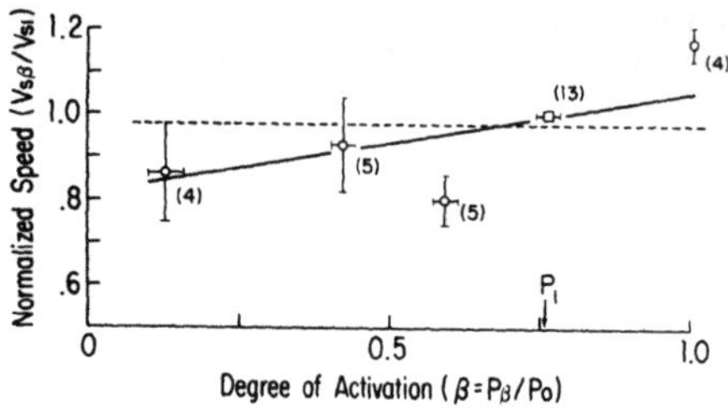

Figure D-1: Effect of the degree of activation on the unloaded speed of shortening of isolated tibialis fibers from the frog. P_1=maximal isometric force with temp step activation (Gulati and Babu, Science 215: 1109-1112, 1982); P_0=isometric tetanic force with electrical stimulation; P_β=submaximal isometric force with temp step activation with varying caffeine; V_{S1}=unloaded speed with slack test, fiber maximally activated with the temp step; $V_{S\beta}$=unloaded speed for submaximally activated fiber. Dashed line shows the mean value of all the data points. Solid line shows the theoretical relation, assuring the existence of an heterogeneous population of cross-bridges resulting in a fixed internal load of 2% P_0 at all β values. The numeric within parentheses shows the number of fibers (Gulati and Babu, unpublished).

KAWAI: There is a profound difference in anticipated results of the two techniques. In small length perturbation experiments such as in our methods of sinusoidal analysis, the internal load (if present) does not affect the apparent rate constants. This is because we can assume, as Dr. Gulati did, that cross-bridges and the internal load are arranged in parallel. In our sinusoidal analysis method the length of a half sarcomere is oscillated only by ±1 nm, which we believe is a lot less than the distance for a cross-bridge stroke. If an extra spring (internal load) is added, it cannot interfere with cross-bridge's motion when the length of the half sarcomere is externally controlled. Transient forces from the cross-bridge and the spring are simply summated to register tension. Hence we can separate the effect of the internal load from that of the cross-bridge by sinusoidal analysis. Note that the internal load is included in the H-parameter (DC stiffness) in our measurements. As you see in Fig. 3B we find the magnitude of the H-parameter to be very small, hence only a very small internal load is possible. The effect of internal load will be considerably larger in experiments which grossly alter the length of muscle, such as in force-velocity experiments, because an internal load will interfere significantly with the speed of shortening.

KRUEGER: Your response to the question on heterogeneity was based on looking at the overall pattern of striations. You said some look better than others, and so on. Is there any mechanical evidence which would also support this because it's not very easy to measure uniformity or to develop an algorithm to determine how a "defined" state of nonuniformity would specifically affect these results?

KAWAI: No, it is not easy. Yet you can suppose two simple cases. One is a parallel arrangement, the other a series arrangement. If you imagine two sarcomeres (having grossly different rate constants) arranged in series, then you would observe a single exponential process with a rate constant between the two. Because the rate constants change very little with the Ca^{2+} concentration, this kind of heterogeneity is apparently not the case. On the other hand, if you imagine that two heterogeneous sarcomeres are arranged in parallel, then our sinusoidal analysis would resolve the two rate constants. In this case you would see this heterogeneity more clearly at higher tension where heterogeneity involvement is presumably more. This is opposite to our observation. Thus I do not believe that this kind of heterogeneity is the source of our measurements of the W-shaped phase-frequency plot at partial activation.

TREGEAR: With regard to Dr. Krueger's question about microscopic heterogeneity, have you done the experiments with different amplitudes of oscillation?

KAWAI: Yes. We have done some experiments at maximal activation. Our results show that the rate constants are almost unchangeable if the peak-to-peak oscillation is kept below 0.5%. This does not mean that rate constants are length independent. In fact, the sinusoidal analysis averages any length-sensitive rate constant within its excusion range.

TREGEAR: It's not the rate constant I was interested in. When you lowered the amplitude, I wanted to know whether the phase shift with frequency response became sharper.

KAWAI: The sharpening is significant if the osciallation amplitude is reduced in the range 2% to 0.5%. The sharpening is much less if the amplitude is below 0.5%. There is very little sharpening if the amplitude is reduced below 0.2%. The amplitude of the experiment I reported today is 0.17%.

TREGEAR: That is not a low amplitude in terms of how this has been done in the past.

KAWAI: That is not a low amplitude with respect to results on insect muscle fibers in which you see a continuous effect of the amplitude reduction (Cumenetti and Rossmanith, J. Muscle Res. Cell Mot. 1: 345-356, 1980). 0.17% is low amplitude for rabbit psoas in which you see a much smaller effect of amplitude change. It appears that nonlinearity at the optimal frequency of oscillatory work is much greater in insect muscles than in rabbit muscles.

TREGEAR: Never mind whether it is insect or rabbit work, you have got some evidence that might possibly be due to microscopic heterogeneity. And I thought if you had done it at lower amplitude the response might have sharpened up. That was what I wanted to know.

KAWAI: As I said, the sharpening of the phase-frequency plot is very little. On the contrary, the data necessarily become noisier at lower amplitudes, hence the resolution decreases. If you are asking if I changed Ca concentration at very low amplitude, the answer is no. Actually this experiment is almost impossible at present, because the signal-to-noise ratio is proportional to the product of tension and oscillation amplitude, and we are already very close to the limit.

TIROSH: I would like to comment on the effect of ionic strength on V_{max} under different concentrations of Ca. As far as I know, there were differences of ionic strength in the experiments of Podolsky & Teichholz (J. Physiol. 211: 19-35, 1970) and Julian (J. Physiol. 218: 117-145, 1971) relative to yours. We have heard here the presentation of Dr. Schoenberg about the effect of ionic strength on the interaction of cross-bridges with actin at rest. Could you please comment?

KAWAI: Yes. The ionic strength of the experiment I reported today is 201 mM. We also made measurements at 128 mM and 265 mM. The results were not much different from those at 201 mM, except that the magnitude parameters (tension included) decreased linearly with increase in the ionic strength in the above range. We could not go lower in ionic strength than 128 mM because of the presence of essential multivalent ions, most of which are used to buffer MgATP. One subtle difference in the low ionic strength experiments is that the frequency of the main valley (at ~13Hz) in the phase-frequency plot increased slightly with increase in Ca concentration. These results were published recently (Kawai et al., Biophys. J. 35: 375-384, 1981).

GORDON: Did the data depend on the order in which you did the experiment, that is, whether you used low Ca first or not?

KAWAI: They didn't. I have to remind you, however, that the Ca concentration was not continuously increased in a single activation. The muscle fiber was relaxed after each activating pCa solution.

HOMSHER: What were your exact experimental conditions?

KAWAI: The experiment was done at 20°C, pH 7.00, and ionic strength at 201 mM. Detailed solution compositions are (Na salts in mM): 6 CaEGTA/EGTA, 5 MgATP, 5 free ATP, 7.5 inorganic phosphate, 4 sulfate, 42 propionate, 10 MOPS, 16 creatine phosphate, and 80 U/ml kinase.

TIROSH: In this respect, I want to mention that, as far as my experience is concerned, it's worthwhile to keep the Mg concentration higher than that you have used. The basis is that, in muscle, the concentration of Mg is about 8 mM, and ATP is 3 to 4 mM. From my experience with reconstitution of mechanochemical effects with actomyosin system this is an essential necessity.

DREIZEN: Have you looked at any other fiber besides rabbit psoas, for example soleus slow muscle? The reason is the amount of heterogeneity that you find in soleus might be less in the sense that phosphorylation is diminished; in addition there would be only one myosin isozyme.

KAWAI: I haven't done calcium experiments on soleus, but I have done the maximal activation experiments.

PODOLSKY: I was wondering whether you looked at frog muscle, because the experiments of ours you've quoted were done on frog muscle.

KAWAI: Again, I looked at frog muscle at full activation, but not with calcium change.

PODOLSKY: Did it behave the same as rabbit?

KAWAI: In the case of frog semitendinosus muscle, oscillatory work was slightly less, but very comparable (Kawai & Brandt, J. Muscle Res. Cell Mot. 1: 279-303, 1980). In the case of rabbit soleus muscle, which is a slow muscle and one would expect a different mechanical response, oscillatory work is almost unidentifiable (Kawai et al., Biophys. J. 41: 341, 1983).

HOMSHER: Again to Dr. Kawai. Have you looked at living muscle fibers, like living frog fibers, and compared it to skinned fibers?

KAWAI: I have looked at intact and skinned crayfish walking leg fibers, intact frog semitendinosus muscles and skinned semitendinosus bundles, and skinned rabbit psoas. These are all fast twitch fibers and they generally yield typical 3 exponential processes and typical Nyquist plots of bifoliate shape (data are published in Kawai & Brandt, J. Muscle Res. Cell Mot. 1: 279-303, 1980). In skinned fibers, the magnitudes and the rate constants of processes are functions of solution parameters, hence they depend on the experimental condition.

GORDON: With respect to measurements of rate constants and the effect of MgATP on them, where do we stand on the issue of whether on the one hand cross-bridges remain attached with their rates being affected by MgATP in an attached state, or attaching and detaching?

KAWAI: The solid experimental evidence is that rate constants of processes C & B (phase 2 & 3 of Huxley, respectively) change proportionately to MgATP concentration in the low mM range (Kawai, in *Cross-bridge Mechanisms in Muscle Contraction*, pp. 149-169. Ed. H. Sugi and G.H. Pollack, Univ. Tokyo Press, Tokyo, 1979). We infer from this observation that the rate constant of MgATP binding is measured by processes C & B, because such binding is second order in nature, hence associated apparent rat constant changes proportionate to the MgATP concentration. There is no disagreement among the speakers in this respect. In fact my measurement shows that the second order binding constant is 3×10^6 $M^{-1}sec^{-1}$ by using the rate constant of process C (Kawai, in *Basic Biology of Muscles: A Comparative Approach*, pp. 109-130. Ed. B.M. Twarog, R.J.C. Levine and M.M. Dewey, Raven Press NY, 1982). This is in good agreement with the value $1-3\times10^6$ $M^{-1}sec^{-1}$ of White & Taylor (Biochem. 15: 5818-5826, 1976) who used extracted protein systems from rabbit psoas. Thus, at low mM concentration we can safely conclude that the MgATP binding rate limits phases 2 and 3.

Disagreement arises in terms of the rate limiting step at higher MgATP concentration where rate constants of processes B & C saturate. I assumed (Kawai, Biophys. J. 22: 97-103, 1978) that it is the dissociation step which ensues after MgATP binding. This assumption was based on the available biochemical scheme (Lymn and Taylor, Biochem. 10: 4617-4623, 1971) at the time, together with the fact that the rate constant of process C is about 600 sec^{-1} on saturation (White & Taylor showed that the rate constant of dissociation was >500 sec^{-1}). We have heard presentations of Drs. Tanaka, Rüegg, and Cecchi, who concluded that there is no change in the gross numbers of attached cross-bridges during rapid phase 2 by double step experiments. These experiments suggest that dissociation or back association does not take place during phase 2, although rapid fall of stiffness which is simultaneous with the length release is not accounted for. This may be due to compliant and nonlinear tendon end, and in this respect there is room for technical improvement in the experiments. With the increased number of intermediary states of the enzyme-substrate complex in the cross-bridge cycle, and with the incorporation of branch pathway(s), we may have to reconsider reactions involved in the rate limiting step.

MUSCLE STIFFNESS CHANGES DURING ISOMETRIC CONTRACTION IN FROG SKELETAL MUSCLE AS STUDIED BY THE USE OF ULTRASONIC WAVES

I. Hatta, Y. Tamura, T. Matsuda, H. Sugi* and T. Tsuchiya*

Department of Applied Physics, Faculty of Engineering, Nagoya University, Chikusa-ku, Nagoya 464, Japan
**Department of Physiology, School of Medicine, Teikyo University, Itabashi-ku, Tokyo 173, Japan*

ABSTRACT

In order to measure muscle stiffness changes with a high time resolution and with minimal disturbance to the contractile mechanism *per se*, we constructed an apparatus with which the propagation velocity of ultrasonic waves (MHz region) in the longitudinal or transverse direction was measured to serve as a measure of muscle stiffness. The longitudinal muscle stiffness started to increase on stimulation before the onset of isometric force, and reached a maximum before the peak twitch force. Analysis of experimental data indicated that, during an isometric tetanus, the increment of muscle longitudinal stiffness was about $6\times10^7 N/m^2$, a value similar to those obtained by Truong (1974) and Ford et al. (1981) with sinusoidal vibrations (3 kHz) and length steps respectively. This suggests that the increment of muscle longitudinal stiffness during the activation of the contractile system results from the recruitment of an almost non-dispersive elastic component.

In the case of transverse muscle stiffness, on the other hand, it started to decrease on stimulation before the onset of isometric force, and reached a minimum before the peak twitch force. Possible causes of this unexpected result is discussed in relation to the molecular mechanism of muscle contraction.

INTRODUCTION

Although it is generally believed that contraction in striated muscle results from the alternate formation and breaking between the projections on the thick filaments, i.e. the cross-bridges, and the sites on the thin filaments (A.F. Huxley, 1957; H.E. Huxley, 1960), the mechanism of the cross-bridge operation producing force and motion is still a matter for debate and speculation. To give information about the number and the state of the cross-links at various stages of muscle contraction, muscle stiffness changes have been studied by applying step or sinusoidal length changes and measuring the resulting force changes (e.g., Julian and Sollins, 1975; Ford, Huxley and Simmons, 1977; Rüegg, Güth, Huhn, Herzig, Griffiths and Yamamoto, 1979) or the resulting longitudinal propagation of mechanical waves (Schoenberg and Podolsky, 1974; Truong, 1974). These methods utilize, however, relatively large perturbations to muscles or muscle fibers, and may not be free from the possibility that the state of the contractile system is altered by the measurement procedure, *per se.*

The present experiments were undertaken to record muscle stiffness changes with negligibly small perturbations and an extremely high time resolution, by measuring the propagation velocity of ultrasonic waves in the longitudinal or the transverse direction in frog skeletal muscle. It will be shown that the longitudinal muscle stiffness increases during isometric force development, while the transverse stiffness decreases.

METHODS

Two different methods were used to study muscle stiffness changes. For measuring muscle stiffness in the longitudinal direction (longitudinal muscle stiffness), the semitendinosus muscle of the bullfrog (*Rana catesbeiana*) was mounted in a lucite chamber in a manner shown in Fig. 1A; the muscle was bent at right angles near both ends, the proximal end being fixed in position while the distal end being connected to a strain gauge (Shinko, type UT) to record isometric force. A ceramic piezoelectric transducer, which was a circular plate (diameter, 8 mm) of $PbZrO_3$-$PbTiO_3$ (Berlincourt, Curran and Jaffe, 1964), was made in contact with the muscle surface at each bent end portion in such a way that the plane of both ceramic plates was perpendicular to the long axis at the middle portion of the muscle. The distance between the two ceramic plates was 2-2.5 cm.

For measuring muscle stiffness in the transverse direction (transverse muscle stiffness), the sartorius muscle of the bullfrog was mounted in a lucite chamber with its lower surface in contact with the bottom of the chamber; the pelvic end was fixed in position while the tibial end was connected to the strain gauge (Fig. 1B). A ceramic transducer (diameter, 6mm) was brought into contact with the upper muscle surface at the middle portion, so that the place of the ceramic plate was perpendicular to the muscle long axis and in parallel with the bottom of the chamber. The distance between the ceramic plate and the bottom of the chamber was 1.5-2 mm.

The ceramic piezoelectric transducer not only produces ultrasonic waves by applying sinusoidal voltage of ultrasonic frequency, but also senses ultrasonic waves to convert them into electric voltage. In the case of longitudinal stiffness measurements, one ceramic transducer was repetitively energized with sinusoidal voltage from a function generator (Wavetek, model 162) to transmit brief trains of ultrasonic waves to the

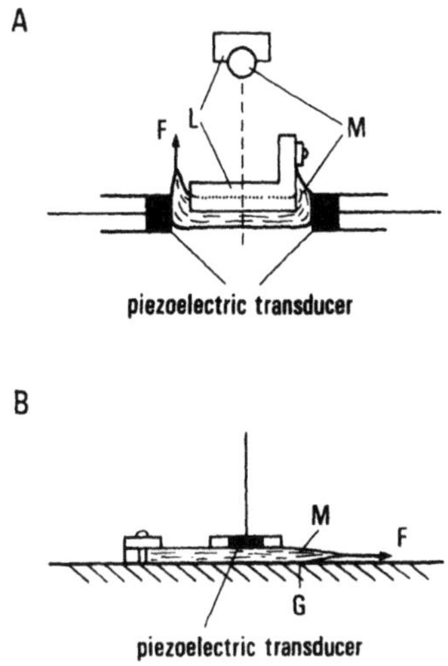

Figure 1: Methods of recording muscle stiffness changes during isometric contraction in frog skeletal muscle. A: Experimental arrangement for measuring the longitudinal stiffness. The semitendinosus muscle of the bullfrog was held by a lucite block (L) with a cylindrical groove of about 5 mm diameter, and bent around the rectangular corner of the block near both ends. Two ceramic piezoelectric transducers were then made in contact with the muscle surface in such a way that the plane of the transducers was perpendicular to the long axis of the muscle. One transducer was used to transmit ultrasonic waves to the muscle, while the other was used to sense the waves propagated through the muscle. B: Arrangement for measuring transverse muscle stiffness. The sartorius muscle of the bullfrog was mounted horizontally with its lower surface in contact with the bottom of the chamber (G) which serves as a reflecting plane of ultrasonic waves. The piezoelectric transducer was in contact with the upper muscle surface so that its plane was in parallel with the bottom of the chamber. Ultrasonic waves from the transducer propagated across the muscle, and were reflected by the bottom of the chamber to be sensed by the same transducer. In both A and B one end of the muscle was fixed in position, while the other end was connected to the strain gauge (F). In both A and B, the muscle was stimulated with a pair of Pt plate electrodes parallel to its long axis (not shown). From Tamura et al. (1982).

muscle at a constant interval. The wave trains propagated longitudinally through the middle portion of the muscle to be sensed by another ceramic transducer. In the case of transverse stiffness measurements, on the other hand, the trains of ultrasonic waves, which were transmitted to the muscle by the ceramic transducer, propagated across the muscle and were reflected by the bottom of the chamber to be sensed by the same transducer as trains of echo waves.

For accurate determination of the change in propagation velocity of ultrasonic waves in the muscle, the interval between the original and the propagated ultrasonic signals was divided into two parts, T and ΔT (Fig. 2). A rectangular pulse of a constant duration T with an accuracy of 9.5 figures was generated by a synthesizer (Rockland, type 5100). The duration ΔT was equal to the interval between the end of the pulse T and the beginning of the first propagated signals. The pulse with a duration ΔT was integrated by a time-to-amplitude converter to obtain an amplitude

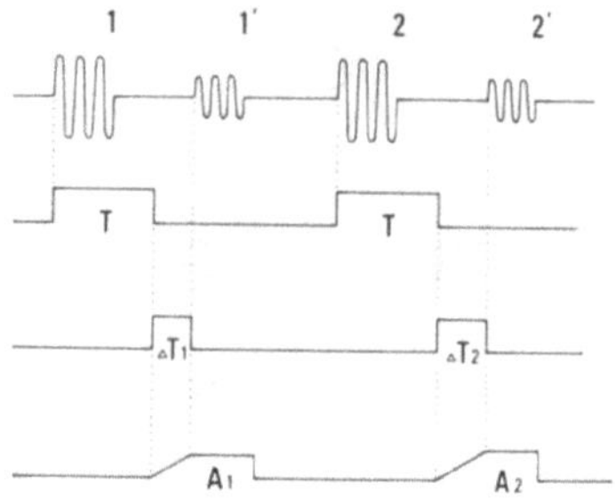

Figure 2: Diagram illustrating the method for measuring propagation velocity changes. Top record shows two original wave trains (1 and 2) and the corresponding propagated wave trains (1' and 2'). As shown in two middle records, the interval between the original and the propagated trains was divided into two parts: a rectangular pulse of constant duration (T) and an additional brief pulse of variable duration ΔT. The brief additional pulses (ΔT_1 and ΔT_2) were integrated by the time-to-amplitude converter to give the amplitude signals A_1 and A_2 in the bottom record. From Tamura et al. (1982).

signal A proportional to the time ΔT, (Odru, Riou, Vacher, Deterre, Peguin and Vanoni, 1978; Nakajima, Tanaka, Shimazaki, Yamanaka, Kinoshita and Wada, 1979). As the propagation velocity of ultrasonic waves changed with time to give values ΔT_1, ΔT_2, etc., the corresponding amplitude signal A_1, A_2, etc., were successfully recorded in a digital-memoryscope (Hitachi, type VC 801). The time resolution of muscle stiffness measurements was limited by the frequency of repetition of the original ultrasonic wave trains, being 250 μsec for the longitudinal stiffness and 30 μsec for the transverse stiffness. The amplitude of perturbations produced in the muscle during the application of ultrasonic waves (~1 Å) was negligibly small.

The muscles were soaked with Ringer solution (NaCl 115 mM, KCl 2.5 mM, $CaCl_2$ 1.8 mM, pH 7.3 by Na-phosphate buffer), and at each time of stiffness measurements the chamber was drained and the muscles were stimulated maximally with single or repetitive (50 Hz) 0.5 msec current pulses through a pair of Pt plate electrodes in contact with the muscle surface along its entire length. The resulting isometric force development was also recorded in the memoryscope, and its time course was then displayed on a chart recorder together with that of stiffness changes (see Figs. 4 and 6). To minimize the muscle movement during the period of mechanical response, the initial muscle length was made at about 1.2 times the slack length. The experiments were performed at room temperature (20-25°C) unless otherwise stated.

RESULTS

Frequency Dependence of the Longitudinal Propogation Velocity in Resting Muscle

Fig. 3 shows the dependence of the longitudinal propagation velocity

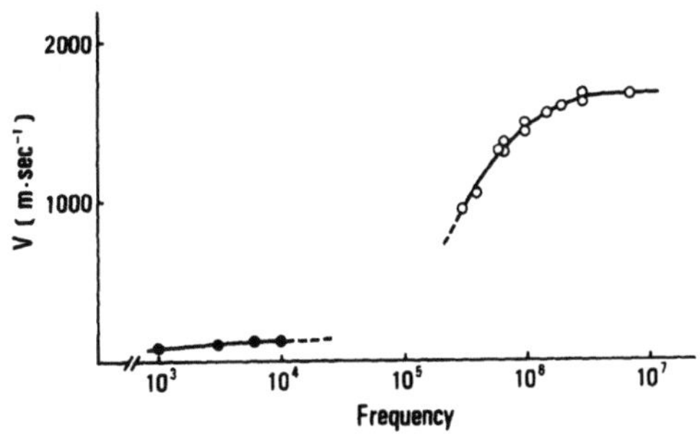

Figure 3: Dependence of the longitudinal propagation velocity in resting frog skeletal muscle on the frequency of ultrasonic waves applied. The measured values of propagation velocity (V) are plotted against the wave frequency in logarithmic scale (open circles). The data reported by Truong (1974) with kHz region waves are also shown (filled circles).

in the resting muscle on the ultrasonic wave frequency (400 kHz-7 MHz). For the sake of comparison, the data points obtained by Truong (1974) with sinusoidal vibrations (1-10 kHz) are also shown. The longitudinal propagation velocity in resting muscle increased with increasing ultrasonic wave frequency, reflecting the viscoelastic nature of muscle structures. At 3 and 7 MHz, the propagation velocity was the same within the accuracy of measurements, being $1.60 \pm 0.05 \times 10^3$ m/sec (mean ±S.D., n=20).

When mechanical waves at a constant frequency ω propagate through the resting muscle, their phase velocity V_p is related to the resting muscle stiffness $E_o(\omega)$ as:

$$E_o(\omega) = \rho V_p^2 \tag{1}$$

where ρ is the density of muscle. The resting muscle stiffness can be determined from equation (1), if the measured propagation velocities can be regarded as the phase velocities but not as the group velocities. This condition is satisfied in the range of wave frequency where the frequency dependence of the propagation velocity is not observable. The similarity of the propagation velocity between 3 and 7 MHz (Fig. 3) indicates that the measured velocities with ultrasonic waves of 3-7 MHz are equal to the phase velocities.

In addition, in the case of the longitudinal propagation, the wavelength λ should satisfy the condition, $a/\lambda > 2.5$, where a is the radius (Tu, Brennan and Sauer, 1955); otherwise the wave propagation along the muscle long axis takes place with mixed modes to complicate the condition. In the present study, the muscle cross-section was nearly circular (Fig. 1) with a radius of about 2.5 mm, while the wavelengths of 3 and 7 MHz waves are about 0.5 and 0.2 mm respectively. Thus, the propagation velocity with 3-7 MHz ultrasonic waves can serve as a valid measure of the resting muscle stiffness, which can be calculated from equation (1) to be about $2.56 \times 10^9 N/m^2$ assuming that ρ is 1 g/cm^3.

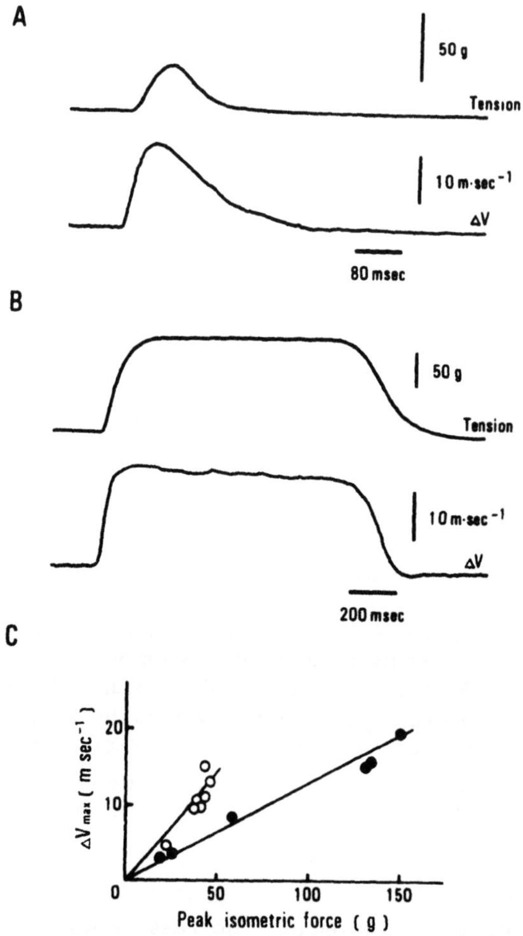

Figure 4: Changes in the longitudinal propagation velocity of ultrasonic waves during isometric contraction. A: Change in the longitudinal propagation velocity (ΔV, lower trace) during an isometric twitch (upper trace). B: Change in the longitudinal propagation velocity during an isometric tetanus. In both A and B, 3 MHz waves were used. In this and subsequent figures, the upward and downward deflection of velocity trace indicate an increase and a decrease in propagation velocity respectively. C: Relation between the maximum increment of the longitudinal propagation velocity (ΔV_{max}) and the peak switch (open circles) or tetanic force (filled circles). Peak forces were varied by applying submaximal stimuli of various strengths.

Changes in the Longitudinal Muscle Stiffness During Isometric Contraction

Fig. 4A is a typical example showing the changes in propagation velocity of 3 or 7 MHz ultrasonic waves along the longitudinal muscle axis during an isometric twitch. On stimulation, the longitudinal propagation velocity started to increase 10-20 msec before the onset of isometric force development, reached a maximum while the isometric force still continued to rise, and then began to decrease to the initial value. The return of the propagation velocity to the initial value was complete before the completion of relaxation of the twitch force. During an isometric tetanus, the propagation velocity increased to a steady value during the rising phase of tetanic force, and the value was maintained until the beginning of relaxation (Fig. 4B).

Fig. 4C shows the relation between the peak twitch or tetanic force and the maximum increment of the longitudinal propagation velocity above the resting value; the twitch and tetanic forces being varied by applying single or repetitive pulses of various submaximal strengths. The amount of the maximum increment of the propagation velocity was nearly proportional to both the twitch and the tetanic forces. However, the slope of the regression line for the twitch force was definitely steeper than that for the tetanic force.

Fig. 5 shows the relation between the increment of propagation velocity and the isometric force during the course of a twitch and a tetanus (0.2-1 sec duration). It can be seen that for a given amount of isometric force, the increment of propagation velocity of isometric force, the increment of propagation velocity above the resting value is much larger during the rising phase of isometric force than during the relaxation phase. The rate of the increase in propagation velocity with isometric force was maximum at the beginning of isometric force development, and decreased gradually while the force continued to rise towards the maximum steady tetanic force. Thus, the amount of the increment of propagation velocity during a tetanus with a peak force of about 130 g was only 40% larger than that during a twitch with a peak force of about 50 g (Fig. 5). This results in the difference in slope of the propagation velocity versus peak force relation between twitches and tetani (Fig. 4C).

The increment of the longitudinal propagation velocity above the resting value was 0.8-1% during isometric twitches, and 1.0-1.3% during isometric tetani with ultrasonic waves of 3 or 7 MHz. The above increment increased with decreasing frequency of ultrasonic waves below 3 MHz, though the data were not used for the reasons described previously.

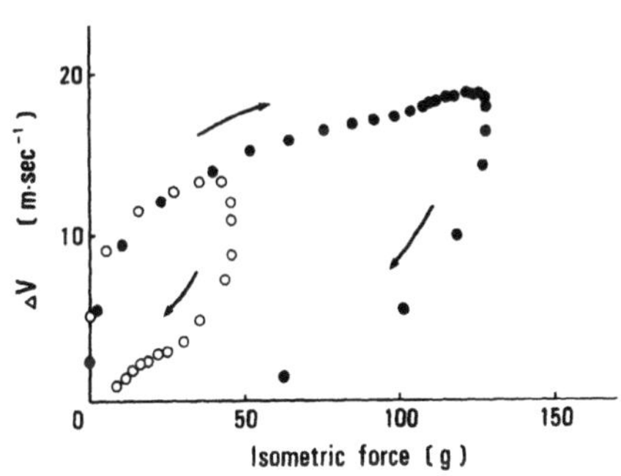

Figure 5: Relation between the increment of the longitudinal propagation velocity and the isometric force during the course of an isometric twitch (open circles) or tetanus (closed circles). The data points were plotted every 10 msec, their sequence being indicated by arrows.

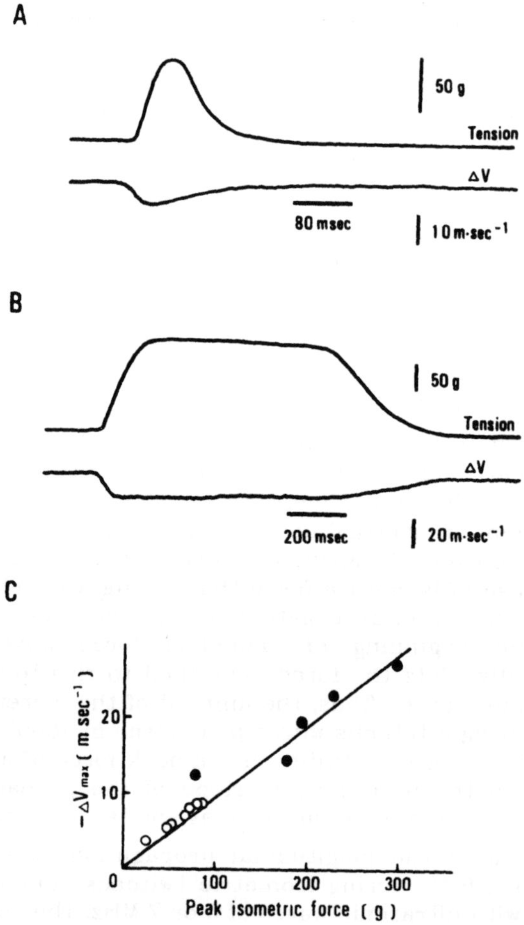

Figure 6: Changes in the transverse propagation velocity of ultrasonic waves during isometric contraction. A: Change in the transverse propagation velocity during an isometric twitch. B: Change in the transverse propagation velocity during an isometric tetanus. In both A and B, 7 MHz waves were used. C: Relation between the maximum decrement of the transverse propagation velocity and the peak twitch (open circles) or tetanic force (filled circles). Peak forces were varied by applying subthreshold stimuli of various strengths.

Assuming that the muscle density ρ does not change during isometric contraction, the increment of the longitudinal muscle stiffness above the resting value $\Delta E(\omega)$ can be calculated as:

$$\Delta E(\omega) = \rho V'^2_p - E_o(\omega) \qquad (2)$$

where V'_p is the longitudinal propagation velocity reached during isometric contraction. During isometric tetani (0.2-1 sec duration), the value of $\Delta E(\omega)$ is calculated from equation (2) to be $5\text{-}7 \times 10^7 N/m^2$.

Changes in the Transverse Muscle Stiffness During Isometric Contraction

A typical example showing the change in propagation velocity of ultrasonic waves in the direction perpendicular to the longitudinal muscle axis during an isometric twitch is presented in Fig. 6A. In contrast with the longitudinal propagation velocity, the transverse propagation velocity started to decrease before the onset of isometric force development, and reached a minimum prior to the peak twitch force. With 7 MHz waves, the transverse propagation velocity in the resting muscle was $1.5 \pm 0.1 \times 10^3$ m/sec (mean ± S.D., n=20), and it decreased by about 1% during a twitch. During an isometric tetanus, the propagation velocity decreased to a steady value during the development of tetanic force, and this value was maintained until the beginning of relaxation (Fig. 6B).

The relation between the peak twitch or tetanic force and the decrement of the transverse propagation velocity is shown in Fig. 6C, when the twitch and tetanic forces were varied by applying single or repetitive pulses of submaximal strengths. Unlike the longitudinal propagation velocity (Fig. 4C), the data points of both twitch and tetanic forces distributed along a single regression line, indicating that the decrement of propagation velocity is proportional to the isometric force attained irrespective of the type of stimulation.

Fig. 7 shows the relation between the decrement of propagation velocity and the isometric force during the course of a twitch and a tetanus (0.2-1 sec duration). For a given amount of isometric force, the decrement of propagation velocity below the resting value was much larger during the rising phase of isometric force than during the relaxation phase.

In a few cases, the peak twitch force was reduced by immersing the muscle in Ringer solutions made 1.5-3 times hypertonic (1.5-3 R) by the addition of NaCl. The decrement of propagation velocity was also proportional to the isometric force in hypertonic solutions, and the data points also distributed along the same regression line in Fig. 6C. In the 3R

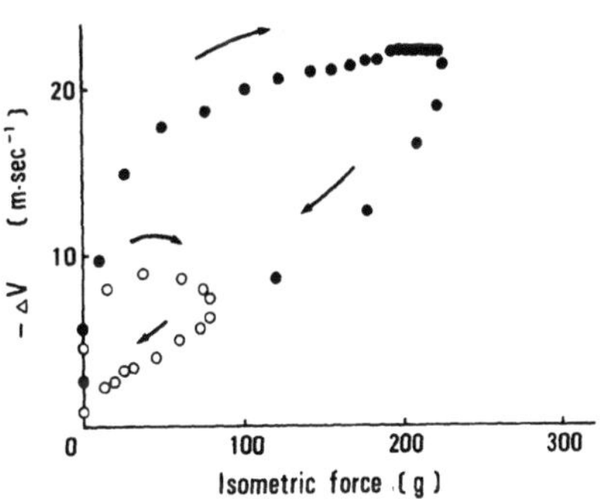

Figure 7: Relation between the decrement of the transverse propagation velocity and the isometric force during the course of an isometric twitch (open circles) or tetanus (filled circles). The data points were plotted every 10 msec, their sequence being indicated by arrows.

solution, the transverse propagation velocity exhibited no significant change in response to stimulation after the disappearance of twitch force. Since the hypertonic solutions are known to reduce or eliminate twitch force without affecting the action potential (Hodgkin and Horowicz, 1957), these results indicate that the decrease in the transverse muscle stiffness, as indicated by the decrease in the transverse propagation velocity of ultrasonic waves, is related with the isometric force development but not with the generation of action potential.

DISCUSSION

Increase in the Longitudinal Muscle Stiffness During Isometric Contraction

In the present experiments, we have succeeded in continuously recording the time course of muscle stiffness changes during a single isometric response with extremely small disturbances to the contractile mechanism (Figs. 4A, B and 6A, B). The longitudinal muscle stiffness started to increase 10-20 msec before the onset of isometric force development, and reached a maximum during the rising phase of isometric force (Fig. 4A, B). The time course of change in the longitudinal muscle stiffness during isometric contraction resembles that of the intensity ratio of two equatorial reflections, $I_{1,0}/I_{1,1}$ (Haselgrove and Huxley, 1973; Matsubara and Yagi, 1978; Amemiya, Sugi and Hashizume, 1979), which has been regarded to reflect the radial movement of the cross-bridges towards the thin filaments. This may be taken to indicate that the longitudinal muscle stiffness changes are associated with the events responsible for force generation in muscle.

The increment of the longitudinal stiffness above the resting value during an isometric tetanus was about 6×10^7 N/m^2 with ultrasonic waves of 3 or 7 MHz. Since Truong (1974) has measured the longitudinal propagation velocity of 3 kHz mechanical waves in tetanized frog muscle, we calculated the increment of muscle stiffness during a tetanus from his data, and obtained a value also of the order of 10^7 N/m^2. A similar value of muscle stiffness can also be obtained from the slope of T_1 curve obtained from length step experiments of Ford et al. (1981). The above comparison leads to the idea that the increase in muscle stiffness, which has been observed by many investigators by applying length perturbations with different techiques, may result from the formation of an almost non-dispersive elastic element exhibiting almost constant stiffness values over a wide frequency range (from kHz to MHz region) of applied perturbations.

Decrease in the Transverse Muscle Stiffness During Isometric Contraction

In the present study, it was found that the transverse muscle stiffness decreased below the resting value during isometric contraction (Fig. 6A, B). In calculating the muscle stiffness from the propagation velocity, it is assumed that the muscle density ρ remains unchanged during isometric contraction; if ρ increases during contraction, this may contribute to the decrease in propagation velocity. The possibility that ρ changes during isometric contraction may be, however, excluded by the following reasons: (1) the myofilament lattice spacing does not change appreciably during isometric contraction; (2) the muscle volume decreases very slightly during contraction (Baskin and Paolini, 1966); and (3) the decrease in the transverse stiffness could still be observed in hypertonic solutions in which the ionic strength of the myoplasm increases considerably.

It may be argued that, due to local segmental shortening taking place within an isometrically contracting muscle, the overlap between the filaments might increase at the middle part of the muscle to result in an apparent increase in protein density underneath the ceramic transducer. The muscle segmental length changes differs, however, from preparation to preparation, and the local shortening does not necessarily take place at the middle part of the preparation (Kobayashi and Sugi, unpublished). As a matter of fact, we always observed the stiffness decrease when the position of the transducer was altered along the muscle length. Moreover, our tentative calculations indicate that an increase in protein density loads to an increase in propagation velocity; an effect opposite to the decrease in propagation velocity observed. Thus, it may be safe to conclude that the decrease in the transverse propagation velocity results from the decrease in the transverse muscle stiffness, but not from the increase in protein density.

Concerning the underlying mechanism for the decrease in the transverse muscle stiffness during contraction, one possibility may be that the nature of the cross-links between the cross-bridges and the thin filaments is such that the strain in the transverse direction is largely taken up by the deformation of the cross-links to result in softening of the muscle in the transverse direction. Much more experimental work is needed to clarify how the molecular mechanism of force generation is related to the stiffness changes observed in the present study.

REFERENCES

Amemiya, Y., Sugi, H. and Hashizume, H. (1979). X-ray diffraction studies on the dynamic properties of cross-bridges in skeletal muscle. In: *Cross-bridge Mechanism in Muscle Contraction*, ed. Sugi, H. and Pollack, G.H., pp. 425-443. Baltimore: University Park Press and Tokyo: University of Tokyo Press.

Baskin, R.J. and Paolini, P.J. (1966). Muscle volume changes. J. Gen. Physiol. 49: 387-404.

Berlincourt, D.A., Curran, D.R. and Jaffe, H. (1964). Piezoelectric and piezomagnetic materials and their function in transducers. In: *Physical Acoustics*, Vol. I, Part A, ed. Mason, W.P., pp. 169-270, New York and London: Academic Press.

Ford, L.E., Huxley, A.F. and Simmons, R.M. (1977). Tension responses to sudden length changes in stimulated frog muscle fibres near slack length. J. Physiol. 269: 441-515.

Ford, L.E., Huxley, A.F. and Simmons, R.M. (1981). The relation between stiffness and filament overlap in stimulated frog muscle fibres. J. Physiol. 311: 219-249.

Haselgrove, J.C. and Huxley, H.E. (1973). X-ray evidence for radial cross-bridge movement and for the sliding filament model in actively contracting skeletal muscle. J. Molec. Biol.

77: 549-568.
Hodgkin, A.L. and Horowicz, P. (1957). The differential action of hypertonic solutions on the twitch and action potential of a muscle fibre. J. Physiol. 136: 17P.
Huxley, A.F. (1957). Muscle structure and theories of contraction. Prog. Biophys. Biophys. Chem. 7: 255-318.
Huxley, H.E. (1960). Muscle cells. *The Cell*, Vol. 4, ed. Brachet, J. and Mirsky, A.E., pp. 365-481. New York and London: Academic Press.
Julian, F.J. and Sollins, M.R. (1975). Variation of muscle stiffness with force at increasing speeds of shortening. J. Gen. Physiol. 66: 287-302.
Matsubara, I. and Yagi, N. (1978). A time-resolved X-ray diffraction study of muscle during twitch. J. Physiol. 278: 297-307.
Nakajima, H., Tanaka, H., Shimazaki, O., Yamanaka, K., Konishita, T. and Wada, Y. (1979). New instrument for rapid and accurate measurement of ultrasonic velocity and attenuation using a microcomputer system. Jpn. J. Appl. Phys. 18: 1379-1385.
Odru, R., Riou, C., Vacher, J., Deterre, Ph., Peguin, P. and Vanoni, J. (1978). New instrument for continuous and simultaneous recording of changes in ultrasonic attenuation and velocity. Rev. Sci. Instrum. 49: 238-241.
Rüegg, J.C., Güth, K., Kuhn, H.J., Herzig, J.W., Griffiths, P.J. and Yamamoto, T. (1979). Muscle stiffness in relation to tension development of skinned striated muscle fibers. In: *Cross-bridge Mechanism in Muscle Contraction*, ed. Sugi, H. and Pollack, G.H., pp. 125-148. Baltimore: University Park Press and Tokyo: University of Tokyo Press.
Schoenberg, M., Wells, J.B. and Podolsky, R.J. (1974). Muscle compliance and the longitudinal transmission of mechanical impulses. J. Gen. Physiol. 64: 623-642.
Tamura, Y., Hatta, I., Matsuda, T., Sugi, H. and Tsuchiya, T. (1982). Changes in muscle stiffness during contraction recorded using ultrasonic waves. Nature 299: 631-633.
Truong, X.T. (1974). Viscoelastic wave propagation and rheologic properties of skeletal muscle. Am. J. Physiol. 226: 256-264.
Tu, L.Y., Brennan, J.N. and Sauer, J.A. (1955). Dispersion of ultrasonic pulse velocity in cylindrical rods. J. Acoust. Soc. Am. 27: 550-555.

DISCUSSION

POLLACK: I think your observation that the transverse stiffness decreases during isometric contraction is a most interesting and potentially important one. I'm wondering, though, whether there might be an interpretation other than the one you offer. For example, suppose that there were transverse elements that interconnected thick filaments, be they struts or bridges or whatever, and suppose that during isometric contraction these elements changed their stiffness. Could this possibly account for the change of stiffness that you measured during contraction?

HATTA: Yes. But it is very difficult to distinguish among mechanisms. Our interpretation is not a fixed one.

FAY: I would like to follow up Jerry Pollack's question. If you stretch the muscle into non-overlap and do the same experiment, do you still see a drop in radial stiffness?

SUGI: Many people say that we should change the amount of overlap, but it is not so simple. The problem is that there is considerable development of resting tension, and the measured velocity is dependent on the resting tension and it is difficult to separate the effect from that of a change in the amount of overlap. But anyway, we were able to change isometric force in many ways, for example, by using submaximal stimulation (Figs. 4C and 6C). Also we used hypertonic solutions to reduce isometric force, as suggested by Dr. Huxley. In this case too, we obtained a similar relationship between the maximum of isometric force and the amount of increment of stiffness. But we have never done the experiment without overlap.

GILLIS: May I suggest you to check your basic equation $c = \rho V^2$, by soaking the muscle in D_2O Ringer. In that case you would increase the density of the muscle, and see whether you are still satisfied with the equation.

HATTA: Many people think that the change of the density also gives the velocity change, but it is wrong. In our case, the density is almost always one. But if we suppose that the muscle is composed of protein and water and the protein content increases during contraction, this gives an increase of velocity in the transverse direction, contrary to the experimental finding.

MATSUBARA: About the change in the transverse stiffness in your presentation, it is known from Oosawa's work that the flexibility of the thin filament increases significantly during activation. I wondered whether this quantitatively would account for your change in stiffness?

HATTA: To check this, we would need to determine the magnitude of the coupling constant between strain and cross-bridge motion.

HUXLEY: Have you looked in detail at the time course of the stiffness change, relative to tension? Does the change in velocity happen significantly ahead of the change in force? I'm wondering whether you could distinguish mechanisms where the velocity change depends either on force or on activation, for example on the time course of bound calcium.

HATTA: May I show a slide (cf. Fig. 4A)? The upper is a twitch curve and the lower is the velocity change. In my opinion, if cross-bridge attachment happens, the velocity should be changed. But there are many states in the cross-linking process. Some of these states are associated with the contraction. So the velocity change includes all things which happen in a short time.

HUXLEY: Well, look, I'm not asking about interpretation, I'm asking about the experimental facts first. If you take those two top curves, the tension and the change of velocity, and plot them together as a percentage of the maximum change, do they coincide? If they don't coincide, how much difference is there between them? Because one can't see on that curve.

GORDON: Many of us are not used to looking at the kind of time course plotted as in Fig. 5. they went up and down together, there'd be a straight line. Is that right?

HATTA: Yes. But, our results are different from that. As seen in Fig. 5, the velocity change precedes that of isometric force in both the force development and the relaxing phase.

GORDON;: So the stiffness doesn't hang on longer, as in the results of Dr. Cecchi? Is there a discrepancy there?.

CECCHI: Yes. We did have just the opposite result.

TAYLOR: I'd like to attempt to clarify some discrepancy between your results and some of the data that we presented this morning. Specifically, why there are these apparent differences in the relation between stiffness and force on the descending phase of the tetanus between your data and that of Dr. Cecchi? Your stiffness outlasted force on falling phase of tetanus.

SUGI: The methods are completely different. In Cecchi's experiments with vibrations of kHz range, the resting muscle stiffness is negligibly small compared to the active muscle stiffness. In our experiments with MHz waves, on the other hand, the resting stiffness is very large, reflecting the viscoelastic nature of muscle structures. The active stiffness can be measured as an increment which is only a small fraction of the resting stiffness. Therefore, in our stiffness measurements, we are very much concerned about viscosity. So, I think it is natural that different results can be obtained by the two completely different methods.

BRESSLER: We too have shown a very nice relationship, with the very fast steps, of the stiffness staying up during the early part of the relaxation phase and then finally falling and following the shoulder in the tetanus (cf. Bressler and Dusik, this symposium). So during the relaxation phase your results differ from Cecchi's and ours. It needs to be explained.

SUGI: I would like to state that, unlike ordinary stiffness measurements including yours, our stiffness measurements do not include tendon compliance which may greatly mask stiffness changes in the contractile component *per se*.

GENERAL DISCUSSION

Interpretation of Stiffness Measurements

GORDON: One issue that was brought up in the first few papers of this session had to do with the time course of stiffness change during a tetanus. There seemed to be some agreement that stiffness increased before tension at the beginning of the tetanus, and the interpretation of at least one speaker was that there was an attached state that wasn't generating force. Are there other interpretations?

NOBLE: I wonder why you assume that there has to be an attachment, because I haven't heard any firm evidence that there *is* an attachment. It would seem to me that if the system is activated, whatever the mechanism of force generation between the filaments, you would get the changes in stiffness that you describe.

GORDON: Dr. Huxley presented evidence from the X-ray data that the changes in the X-ray pattern are consistent with the movement of the heads.

NOBLE: Yes, I think that's right. Movement. And with the fluorescent and paramagnetic probes we've heard about changes in orientation. But you're talking about mechanical fixation. What's the evidence for that?

TAYLOR: The stiffness.

NOBLE: Thank you. I'm saying that you'll get a change in stiffness if the system is tense and the filaments are tense, whatever the mechanism by which that comes about.

SCHOENBERG: I think is was actually Dr. Hugh Huxley, a few years ago, who pointed out that it's very difficult to draw conclusions about there being an attached state which doesn't produce force from experiments which are not done under perfectly isometric conditions. This is because, as we know from A. V. Hill's work, sarcomere shortening against any compliance will delay the appearance of force. Shortening will also likely change the rate at which cross-bridges attach. Unless we know the precise effect of shortening upon both the rate of force development and the rate of cross-bridge attachment, it is very difficult to infer how the force would compare with the stiffness under isometric conditions.

TAYLOR: Dr. Cecchi showed that this deviation also occurred on the isometric phase of relaxation, where there is no shortening of the sort you describe.

SCHOENBERG: Yes, I think that's true, and my comments were just directed toward the rising phase of the tetanus.

GORDON: In the phase of relaxation of isometric force after a tetanus, you (Dr. Cecchi) showed there that the stiffness declined less rapidly than the force. Are you implying then that there is an attached state which is not generating force that occurs during relaxation?

CECCHI: I will just say that the more likely explanation we can give for this finding is a change of the cross-bridge distribution among different states, generating more or less force. But, could I comment on what Mark Schoenberg said? I think he raised a very important point. As I said during my presentation the result that during the tetanus rise the stiffness precedes the tension could be explained in many different ways. I do not intend to talk about all the possible explanations now, but I would like to comment about the possibility raised by Mark Schoenberg. I agree with him, if the situation is not perfectly isometric there is a shortening of the contractile elements against the tendons, and the relationship between stiffness and tension is altered. According to previous data (A.F. Huxley and R.M. Simmons, Cold Spr. Harbor Symp. Quant. Biol. 37: 669-680, 1972; F.J. Julian and M.R. Sollins, J. Gen. Physiol. 66: 287-302, 1979 and to the A.F. Huxley model, Prog. Biophys. and Mol. Biol. 7: 255-318, 1957) shortening would tend to increase the stiffness to tension ratio. However, a rough calculation shows that in order to explain our results, an improbably large amount of shortening would have to occur during the tetanus rise. For example, in a fiber we measured a stiffness to tension ratio of 1.9 during the tetanus rise when the tension was 0.22 P_0. According to the Huxley model this ratio is in agreement with a velocity of shortening of $0.37V_{max}$. At a tension of 0.27 P_0 (4.5 ms after the first measurement) the ratio was found consistent with a velocity of $0.30V_{max}$. Even taking the mean velocity equal to the lowest velocity, the tendons should have been extended by about 8 μm as tension increased from 0.22 to 0.27 P_0. However the T_1 curve of this fiber shows that the extension of the total fiber compliance from 0.22 to 0.27 P_0 was less than 1.8 μm. Therefore the amount of shortening present during the tetanus rise should be too small to explain our result.

SUGI: I want to mention something on the positive side. We have been measuring the equatorial intensity ratio, I(1,0)/I(1,1), using X-ray diffraction, as Dr. Hugh Huxley has done, and when I began the ultrasonic wave experiments with Dr. Hatta I first noticed that the time course of rising stiffness as measured by ultrasonic waves was very similar to the time course of the equatorial intensity ratio. So, I think the stiffness increase, as measured by ultrasonic waves, may actually represent the formation of cross-links between the filaments.

I also want to mention the influence of shortening. We examined the changes in I(1,0)/I(1,1) in frog skeletal muscle during shortening against an inertial load (Amemiya et al. Proc. Japan Acad. 56(b): 235-240, 1980).

In this case, both muscle length and force changed with time independently of each other. We found that I(1,0)/I(1,1) followed the force change very closely as the muscle length was changing. This indicates that I(1,0)/I(1,1) is sensitive to the force change but not to the length change.

Finally, I want to point out that during the relaxation phase of the isometric mechanical response some part of the muscle is still shortening while another part is being stretched, resulting in a considerable segmental nonuniformity along the muscle length. Such a nonuniformity is believed to be much less during the rising phase, and our ultrasonic wave data, showing that the stiffness for a given isometric force is much smaller during the relaxation phase than during the rising phase, may be compatible with this.

SCHOENBERG: Had I known this was going to be such a burning issue I would have brought a 15 minute talk, but I'll try to be very brief. Jay Wells and I have performed very similar experiments using the transmission time technique as a measure of stiffness. We used the semitendinosis muscle bundle preparation, so I believe it probably has much more compliance than the preparation that Dr. Cecchi has spoken of. But our results are in qualitative agreement with the findings of Dr. Cecchi, both during the rising phase as well as during the falling phase. But we did one additional experiment where we shone the laser through the preparation in order to get an idea of how much shortening we had. Just as Dr. Cecchi did, we used the model of A.F. Huxley (Prog. Biophys. and Mol. Biol. 7: 255-318, 1957) to see how much of the lag we saw on the the rising phase might be due to the shortening. In our preparation with quite a bit of compliance, we found about half the lag could be accounted for by that model. The problem is, that the calculation is very model sensitive, because if we performed the same computation using the model of Podolsky and Nolan (Cold Spr. Harbor Symp. Quant. Biol. 37: 661-668, 1973) and we obtained a different answer.

We also showed, as was pointedout by A.F. Huxley and R.M. Simmons (Cold Spr. Harbor Symp. Quant. Biol. 37: 669-680, 1972) that relaxation seems to have two phases that are separated by a shoulder. If you look at the force trace, you see a slow decrease in force followed by a shoulder and a more rapid decrease in force, and as pointed out by Professor Huxley (reference above) and Cleworth and Edman (J. Physiol. 227: 1-17, 1972) and we confirmed in our own experiments using the laser, that up until the shoulder, sarcomere lengths are pretty uniform. In agreement with Dr. Cecchi, we found that during the phase up to the shoulder there's very little change in stiffness, although the force has already started to drop.

TIROSH: I'd like to say that we should keep some reservation in respect to the interpretation of the measurement of the elastic properties of the muscle. As was clearly indicated, mainly by Professor Sugi, but I want to add to this; we have other elastic elements besides the cross-bridges. We have heard here, that the M-band is stretched, the Z-band is also amenable to stretch, and the connecting filaments might be stretched. These should be kept in mind when trying to interpret stiffness changes.

INGELS: I just wanted to make one point. That is, Max Perutz a few years ago talked about protein interactions as very flexible objects, and the electrostatic effects dominating in these interactions. I think when we think of the word "attachment," we kind of think of a bowling ball going back into a wooden shelf and "clunk." But I think if we were riding a myosin molecule, when we got about 10 Å away from the actin filament or actin monomer and someone asked us if we were attached or not, we might wonder what the hell was going on. And when we hit, collided, rebounded, electric fields interacted and so forth, we'd better be careful about this term "attachment" because it is probably richer at the molecular level, at which we're trying to focus down on, than perhaps we're used to dealing with in our own ordinary system.

INFLUENCE OF MYOFILAMENT LATTICE DIMENSIONS ON CONTRACTILE FUNCTION

INTRODUCTION

In the widely accepted sliding filament theory of muscle contraction, force is assumed to arise from interactions between neighboring thick and thin filaments. It is conventionally held that these interactions are, in fact, the physical attachment of angled cross-bridges which are able to stretch across the rather large distance between the surfaces of the myofilaments (about 13 nm in muscle at rest length). Inasmuch as intact muscle fibers maintain a constant volume, the distance between the myofilaments is not fixed but varies with muscle length. The variation in interfilament distance is not trivial. For example, surface-to-surface distance decreases some 40% when a fiber is stretched from 2.1 μm to 3.4 μm (cf. Schoenberg, Biophys. J. 30: 51-68, 1980). A priori, it seems likely that such changes in working distance would influence cross-bridge performance. The prevailing view is, however, that cross-bridges accomodate these changes in working distance and maintain constant force per bridge over a wide range of muscle lengths by dint of their doubly hinged configuration (Huxley, Science 164: 1356-1366, 1969).

The bulk of experimental evidence from intact muscle appears to indicate that interfilament spacing has little or no influence on contractile performance. For example, the linearity of the descending limb of the active force-length relation is consistent with a model where force per bridge is constant over this entire range of lengths (Gordon et al., J. Physiol. 184: 170-192, 1966). The velocity of unloaded shortening is independent of sarcomere length over a range of lengths above optimum (Gordon et al., 1966; Edman & Hwang, J. Physiol. 269: 255-272, 1977). Furthermore, the general shape of the length-tension relation and the velocity of unloaded shortening are reportedly unchanged when fibers are exposed to moderately hyper- or hypotonic solutions which respectively shrink or swell the fiber (and hence the myofilament lattice) (Edman & Andersson, Experientia 24: 134-136, 1968: Edman & Hwang, 1977; but see Gulati & Babu, this volume, for contrary evidence). Changes in tonicity will also vary intracellular ionic strength which is known to affect cross-bridge properties. If, however, fibers are exposed to isosmotic (though hypotonic) solutions with high KCl, the fibers swell but the internal ionic strength should in principle be maintained. Under these conditions the force of swollen fibers is similar to that of normal fibers arguing that marked changes in interfilament spacing per se have little influence on force production (April & Brandt, J. Gen. Physiol. 61: 490-508, 1973; see also Gulati & Babu, Science 215: 1109-1112, 1982).

With intact fibers it is difficult to precisely control length, interfilament spacing, and internal ionic strength independently of one another. In this regard, the skinned fiber preparation offers significant advantages in the investigation of possible influences of interfilament spacing on contraction. With the membrane removed or chemically breached, the solutions bathing the contractile apparatus are under precise experimental control. The interfilament spacing can be varied independent of fiber length at constant ionic strength by including long-chain neutral polymers (such as Dextran or polyvinylpyrrolidone) in the bathing media. If these polymers are large enough to be effectively excluded from the myofilament lattice the resultant osmotic force shrinks the lattice reversibly (Godt & Maughan, Biophys. J. 19: 103-116, 1977; Maughan & Godt, Biophys. J. 28: 391-402, 1979; Goldman et al., J. Physiol. 295: 81P, 1979; Magid and Reedy, Biophys. J. 30: 27-40, 1980). Since lattice spacing tends to swell when fibers are skinned in polymer-free medium, one is able with this technique to study the influence of increased as well as decreased interfilament spacing on contractile properties.

Skinned fiber experiments of this sort show that changes in interfilament spacing appear to influence a number of contractile properties. For example, osmotic compression of skinned fibers markedly reduces isometric force production (Maughan & Godt, Pfluegers Arch. 390: 161-163, 1981a) and the attendant ATPase activity (Krasner & Maughan, Biophys. J. 33: 27a, 1981), as well as the isotonic velocity of shortening (Tsuchiya, this volume). Marked compression leads to an increase in axial elastic modulus and to changes in force transients which might well reflect changes in cross-bridge behavior (Berman & Maughan, Pfluegers Arch. 393: 99-103, 1982; Maughan & Berman, this volume; Tsuchiya, this volume). Such changes might be expected since X-ray diffraction studies demonstrate that compression of the lattice affects cross-bridge configuration (Maeda, this volume). Compression with polymers can influence the calcium activation of skinned fibers as well. When swollen fibers are compressed back to near their in vivo size their calcium sensitivity increases (i.e. activation occurs at lower Ca^{2+} concentrations). Further compression decreases the calcium sensivity (Godt & Maughan, Pfluegers Arch. 391: 334-337, 1981). We therefore have evidence that, in contrast to much of the evidence from intact fibers (see however Gulati & Babu, this volume), changes in interfilament spacing can indeed affect contractile performance. It is not clear at present how to reconcile these apparent discrepancies.

Conversely, one can utilize the skinned fiber preparation to investigate the influence of cross-bridge interaction on interfilament spacing. When an angled cross-bridge binds and exerts an axial force there should also be a radial component of force which would tend to compress the myofilament lattice. Such radial compression is not observed in intact fibers. However, one might not expect to observe significant changes in interfilament spacing in intact fibers since any decrease in fiber volume would be countered by large osmotic forces across the sarcolemma which would tend to maintain constant cell volume. Mechanical removal of the membrane or its compromise with chemical treatment removes this constraint and permits observation of any volume change associated with cross-bridge attachment. As expected, the attachment of cross-bridges either with rigor produced by the withdrawal of ATP or during calcium activation in ATP-containing solutions produces changes in fiber diameter and myofilament spacing consistent with the angled cross-bridge model (Goldman et al., 1977; Maughan & Godt, J. Gen. Physiol. 77: 49-64, 1981b; Matsubara et al., this volume).

-Robert Godt

FORCE RESPONSE TO WIDTH AND LENGTH PETURBATIONS IN COMPRESSED SKINNED SKELETAL MUSCLE FIBERS

David W. Maughan and Michael R. Berman

Department of Physiology and Biophysics, University of Vermont School of Medicine Burlington, VT 05405

ABSTRACT

Previous studies have shown that radial compression of calcium-activated skinned skeletal muscle fibers, with attendant reduction of filament lattice spacing, reduces isometric force generation. In relaxed skinned fibers, radial compression produces a marked increase in axial elastic modulus, and the response to a small amplitude length perturbation resembles that of a muscle in rigor. We interpret these results as indicating that radial compression of the myofilament lattice produces "hindered" cross-bridges which are load bearing but not force generating.

The experiments reported here were designed to study the effect(s) of "hindered" cross-bridges on both the time course of isometric force responses following Ca^{2+} activation and fiber width and length perturbations. The experiments were carried out at room temperature on radially compressed skinned single rabbit soleus fibers. Force development following step-wise Ca^{2+} activation and step-wise changes of fiber width was "slow" (60-90 sec) compared to that in normal width fibers (~1 sec), and could be approximated by a single exponential curve. Force redevelopment following a length release in compressed fibers was both more rapid and more complicated than force development following activation and width steps, and required a double exponential curve for an adequate description. The results are consistent with the notion that hindered cross-bridges form as a result of lattice compression, and that the hindered bridges affect the force responses following width and length perturbations.

INTRODUCTION

Calcium activated force is reduced (or even abolished) when the width of a skinned skeletal muscle fiber is decreased by osmotic compression (Maughan & Godt, 1981 and April & Maughan, 1982). X-ray diffraction studies show that osmotic compression of skinned fibers results not only in a reduction of fiber width but also in a reduction of myofilament lattice spacing (April, Farrell & Schreder, 1977; Goldman, et al. 1979; Magid & Reedy, 1980; Millman & Nickel, 1980). One possible explanation for the reduction in Ca^{2+} activated force is that "hindered" (load-bearing, but not force generating) cross-bridges are formed when the lattice spacing is reduced by osmotic compression. Two recent sets of experiments lend support to this hypothesis of hindered bridge formation. First, actomyosin ATPase measurements in Ca^{2+}-activated compressed skinned fibers show that the rate of ATP hydrolysis is reduced in roughly the same proportion as is the Ca^{2+}-activated force. (Krasner & Maughan, 1981). These results can be interpreted as indicating that either fewer cross-bridges are cycling or that the cross-bridge cycling rate has been reduced. Second, experiments on relaxed skinned fibers show that the axial elastic modulus ("instantaneous" stiffness) increases considerably as fiber width is reduced (Berman & Maughan, 1982). Because no active force was generated in these latter experiments, the observed increased elastic modulus supports the notion of hindered cross-bridge formation.

In performing the experiments on Ca^{2+} activated fibers we noticed that when compressed fibers were switched from a relaxing to an activating solution, the force took an inordinately long time (as compared with *in situ* width fibers) to reach its plateau value. Such a reduction in rate of force development could be attributed either to a direct effect of lattice compression on cross-bridge kinetics or, alternatively, to an indirect effect on the rate of force generation due to an increase in internal load originating from the postulated hindered bridges. To further investigate the effect of fiber (lattice) compression on the rate of force generation, we performed a series of experiments in which we measured and analyzed the force responses to width and length perturbations in compressed fibers.

MATERIALS AND METHODS

Single fiber segments from rabbit soleus muscles were isolated under relaxing solution (pCa<8) containing 62.9 mM KCl, 15 mM creatine phosphate, 4.75 mM $MgCl_2$, 5mM EGTA, 3.16 mM Na_2ATP, 20 mM imidazole (pH 7, ionic strength 0.15 M, 22°C) yielding 1 mM Mg^{2+} and 3 mM $MgATP^{2-}$ (Berman & Maughan, 1982). Fiber segments (0.5-2.0 mm long) were clamped between a force transducer and a displacement generator. Once clamped, fibers were chemically skinned by soaking for 20-30 min in relaxing solution containing 0.5% v/v Triton X-100 and 0.5% w/v Brij 58. The activating solution (pCa 4.0) had the same characteristics as the relaxing solution except that 5.186 mM $CaCl_2$ was added and other constituents modifed slightly to maintain a constant ionic strength. Fiber segment width was reduced by transferring the preparation to solutions containing Dextran T500 (0-0.26 g/ml, Godt & Maughan, 1981). Fiber width and average striation spacing were measured from photomicrographs. Acceptable fibers exhibited no obvious tapering or twisting and had a uniform average striation spacing (± 0.05 μm) from end to end when viewed in relaxing solution. On activation, fibers which exhibited more than 0.2

μm reduction in average striation spacing or any obvious heterogeneity were discarded. Force was measured at one end of the fiber and, in some experiments, small amplitude rapid displacements were applied and measured at the other end of the fiber as described in detail elsewhere (Berman & Maughan, 1982).

RESULTS

Force Development and Redevelopment After Release in Compressed Fibers

Fig. 1 shows the time course of force development in a fiber which was compressed while relaxed, then activated while the width was held constant, and finally subjected to a small amplitude release and restretch once the force reached a steady value. Force development following activation was rather slow, whereas the time course of force redevelopment after release was comparatively quick. Also, the time course of force development following the release was more complicated than was the time course of force development following activation.

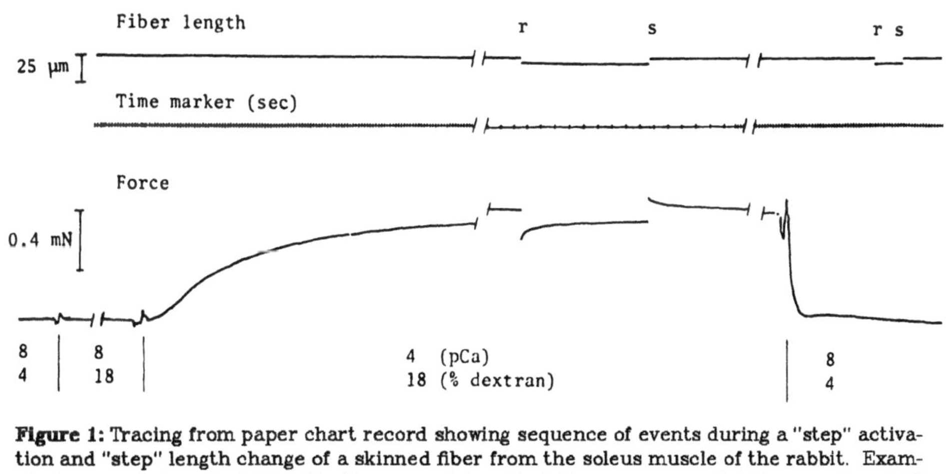

Figure 1: Tracing from paper chart record showing sequence of events during a "step" activation and "step" length change of a skinned fiber from the soleus muscle of the rabbit. Example from experiment of 4-29-82, #3; segment length, 913 μm; average striation spacing, 3.07 μm; segment width in pCa 8, 4% (0.04 g/ml) dextran-containing relaxing solution, 43.8 μm; temperature 22°C. Initially the fiber was bathed in the pCa 8, 4% dextran relaxing solution, then transferred to pCa 8, 18% dextran relaxing solution. After the fiber width was reduced to its final equilibrium value (32.9 μm) the fiber was transferred to pCa 4, 18% dextran activating solution. Force developed slowly, and after it reached its plateau the fiber was rapidly shortened (10 msec) by 6.7μm (0.73% of the fiber segment length), as indicated by the letter *r* next to the length trace. The fiber was rapidly restretched (letter *s)* and then returned to the pCa 8, 4% relaxing solution. Transfers to varius solutions are marked by vertical bars below the force record. Note that release *(r)* and stretch *(s)* in pCa 8, 4% dextran relaxing solution produced no measurable response in force, indicating that stiffness is negligible compared to that observed in pCa 4, 18% dextran activating solution.

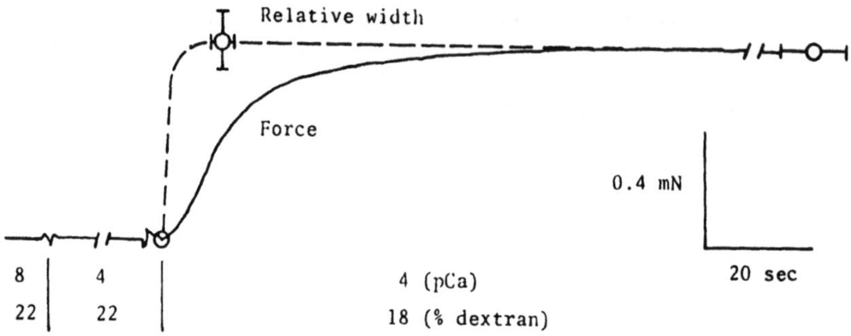

Figure 2: Record of force development following width "step." Superimposed on tracing is a composite plot of relative width versus time following transfer from pCa 4, 22% dextran activating solution to pCa 4, 18% dextran activating solution; relative width amplitudes and elapsed times given (means ± S.E.M., 5 fibers). For comparison, the width relative to that in pCa 8, 4% dextran relaxing solution has been normalized such that the (relative) equilibrium width coincides with the force plateau. Dashed line based on visual inspection of fiber width changes in similar experiments. Note that the normalized relative width reaches its equilibrium value sooner than does the force. Force record from experiment 3-10-82, #1; segment length, 2380 μm; segment width in pCa 8, 4% dextran solutions, 50.5 μm; relative width in pCa 4, 22% dextran solution, 0.71; relative width in pCa 4, 18% dextran solution, 0.77; average striation spacing, 2.98 μm.

Force Development Following Width "Step"

When a relaxed fiber is compressed and then activated by switching to a high calcium solution while compressed it is possible that the subsequent time course of force development is influenced by the diffusion of activator calcium (both as free calcium and as calcium bound to EGTA) into the interior of the fiber (Orentlicher, Reuben, Grundfest & Brandt, 1974; Moisescu & Thieleczek, 1978). To eliminate the possible effect(s) of diffusion of calcium on the observed time course of force development we carried out a series of experiments (Fig. 2) in which fibers were greatly compressed with dextran while relaxed--in fact so compressed that when activated by switching to a high calcium solution containing the same dextran concentration, no force developed (Maughan & Godt, 1981). After allowing for a period of complete equilibration of Ca^{2+} throughout the fiber, the fibers were then transferred to a solution of the same calcium concentration but containing less dextran, so that the fiber rapidly expanded to a less compressed state. Under these conditions the time course of force development was not affected by diffusion of calcium since calcium concentration was constant throughout the fiber during the final transfer.

It is possible, however, that in the width step experiments the time course of width equilibration may influence the observed time course of force development. This is unlikely since Fig. 2 shows that the width reaches its final equilibrium value before force. Thus the observed slow time course of force development is most likely a reflection of reduced lattice spacing and is little, if at all, affected by the width transient.

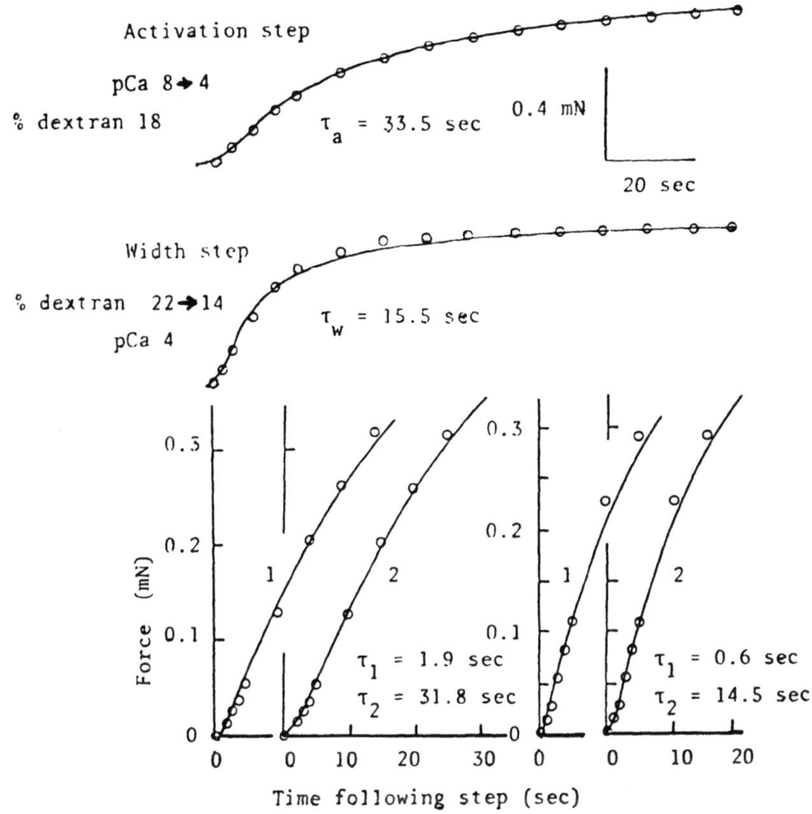

Figure 3: Top traces: comparison of force record following activation "step" with force record following width "step". The open circles superimposed on the force records are from a single exponential fit to the data points. For the activation step the time constant is 33.5 sec; for the width step time constant is 15.5 sec. Activation step from experiment of Fig. 1; width step from experiment 3-23-82, #1; segment length, 1243 μm; segment width in pCa 8, 4% dextran solution, 62.8 μm; relative width in pCa 4, 22% dextran solution, 0.72; relative width in pCa 4, 14% dextran solution, 0.79; average striation spacing, 3.29 μm. Bottom set of traces: comparison of single and double exponential fits to the force records for both activation and width steps of top traces. Experimental points given by open circles, exponential fits shown by solid curves; number of time constants indicated by number next to each curve. Curves on left, activation step; curves on right, width step. Note that except for the first few seconds a single exponential function describes the data nearly as well as does a double exponential function.

Describing Force Development

The time course of force development following either an activation step or a width step can, as a first approximation, be fit by a single exponential curve of the form Force = A + Bexp $(-t/\tau)$, where A and B are constants and τ is the time constant associated with force development. Fig. 3 (bottom) shows force records fit by both a single and a double exponential curve. Notice that the single exponential curve fits the data nearly as well as does the double exponential and, further, that the longer

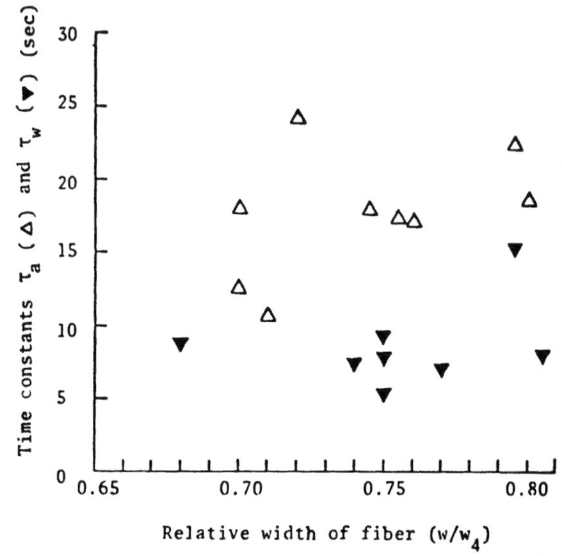

Figure 4: Time constants of force development following activation and width steps as a function of fiber width relative to width in pCa 8, 4% dextran relaxing solution (w/w_4). Open upright triangles, time constants from activation steps; filled inverted triangles, time constants from width steps. Pooled data from 16 fibers. Mean ± S.D., τ_a = 19.5 ± 7.9 sec; τ_w = 8.5 ± 3.1 sec. Time constants from single exponential expressions fit to the time courses of force development. Activation steps produced by transferring compressed skinned fibers from a pCa 8, 18 or 22% dextran relaxing solution to a pCa 4, 18 or 22% dextran activating solution. Width steps produced by switching compressed skinned fibers from a pCa 4, 22% dextran activating solution to a pCa 4, 18 or 14% dextran activating solution.

time constant derived from the double exponential fit is close to that derived from the single exponential fit. It is possible, however, that the short (initial) time constant derived from the double exponential fit reflects the initial activation or width transients referred to earlier. Since we are primarily interested in the slow phase of force development, we have chosen for simplicity to use a single exponential fit.

Effect of Relative Width on Force Development

As mentioned in the Introduction, previous experiments have demonstrated that Ca^{2+} activated force falls when fibers are compressed and that the axial elastic modulus of relaxed fibers increases with compression. Based on these earlier results, one would expect that if the postulated hindered cross-bridges formed by compression retard force development, then force development will be slower in fibers of smaller relative width. Unfortunately the results to date are inconclusive. Fig. 4 is a plot of time constants derived from single exponential fits to the time course of force development from activation and width step experiments as a function of relative width. As can be seen there is no relation ($p > 0.05$) between the time constants and relative width, at least over the limited range of relative widths examined. Similarly, no obvious relation

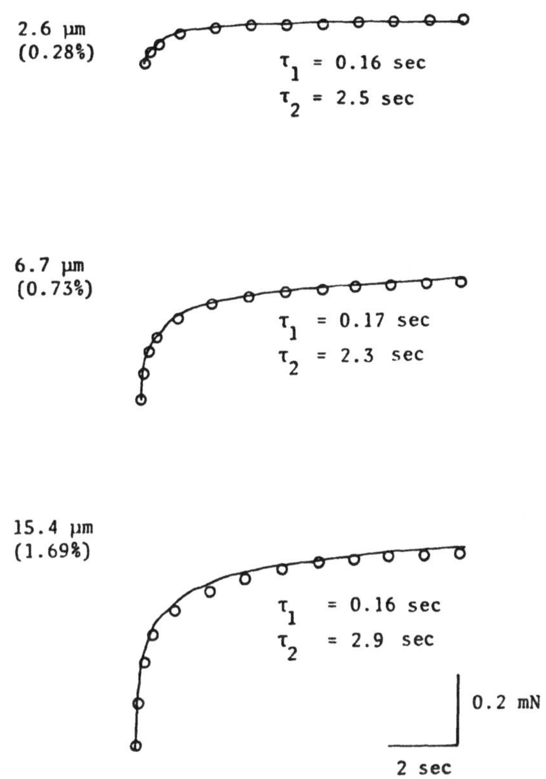

Figure 5: Record of force redevelopment following rapid release. Amplitude of release (as absolute value and as percent of fiber length) indicated to the left of each tracing. Double exponential fits to the records indicate by open circles, where the fast (τ_1) and slow (τ_2) time constants are given near each curve. Note that the time constants appear to be relatively insensitive to release amplitude. Example from experiment of Fig. 1; fiber in pCa 4, 18% dextran activating solution. All releases at force plateau (0.72 mN, indicated by horizontal line at upper left of each record). Average striation spacing 3.02 μm; fiber width 32.9 μm; fiber length 913 μm.

existed between the time constants and the absolute width of the fiber segments or between either relative or absolute widths and the rise-time to one-half plateau force.

Describing Force Redevelopment Following Release

In contrast to the single exponential fit used to describe force development following activation steps and width steps, force redevelopment following a small amplitude release requires at least a two exponential fit of the form Force = A + Bexp (t/τ_1) + Cexp (t/τ_2), where A,B and C are constants and τ_1 and τ_2 are the time constants associated with force redevelopment. Fig. 5 illustrates that, in most cases, two exponentials fit the data well. Three exponentials (not shown) did not improve the fit

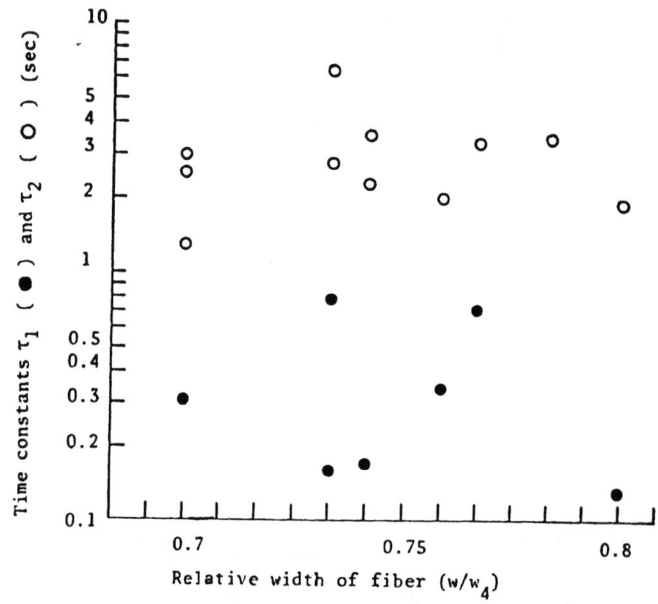

Figure 6: Time constants of force redevelopment following a rapid release as a function of relative width. Note that the fast (τ_1) time constant, indicated by filled circles, and the slow (τ_2) time constant, indicated by open circles, differ by an order of magnitude. Pooled data from 8 fibers. Mean ± S.D. of τ_1, 0.31 ± 0.23 sec; of τ_2 2.98 ± 1.37 sec. Release amplitude in all cases was 6.7 μm (0.36-0.73% of fiber length). Fibers in pCa 4, 14 or 18% dextran activating solution; releases at force plateau.

significantly. Therefore, for simplicity, we chose to fit the force response with a double exponential curve. We would expect that if the time course of force redevelopment following release reflects the effects of the postulated hindered cross-bridges, one or both of the time constants would depend upon the relative fiber width. As shown in Fig. 6 there is no relation ($p > 0.05$) between time constants and relative width over the limited range of relative widths investigated.

DISCUSSION

The rate of force development following Ca^{2+} activation in osmotically compressed skinned fibers is substantially slower than the rate of force development in uncompressed fibers. For example, the time constant of force development in rabbit soleus fibers at *in situ* width (between 0.02 and 0.05 g/ml Dextran T500; Godt & Maughan, 1981) is approximately three seconds (Krasner & Maughan, 1981 and unpublished), whereas in compressed fibers we observe values in the range of 10-20 seconds (Fig. 4). Similarly, skinned soleus muscle fibers from the rat (room temperature) develop force in less than one second (Stephenson & Williams, 1981).

Orentlicher, et al. (1974) discuss the effect of calcium equilibration on the rate of rise of force following an activation step in skinned fibers. From their analysis, we conclude that, though full calcium equilibration may take up to several seconds (thus influencing the beginning of force development), it is unlikely that calcium equilibration contributes appreciably to the very long time constants we observe. For experimental verification, we plotted the exponential time constants (and the half-rise times) of force development as functions of the absolute width of the skinned fiber segments. If the rise times are calcium diffusion limited, then the time constants (and half-rise times) ought to increase with fiber absolute width; in fact, neither parameter correlated with fiber width, lending support to the idea that the rise of force observed with calcium activation is limited by a process other than calcium diffusion. Furthermore, it appears that it is not deficiency of calcium *per se* that causes the sluggish behavior, since activation with submaximal levels of calcium (pCa 6.00-6.25) at near *in situ* width (0.04 g/ml Dextran T500) produced a comparatively rapid development of force, with half-rise times of less than a second (unpublished results). One must, however, approach this data cautiously inasmuch as fibers activated near *in situ* width, under our experimental conditions, do not maintain striation registration or homogeneity of spacing long enough after reaching a force plateau to verify structural integrity during or immediately after the relatively rapid force development.

We performed "width step" experiments in an attempt to observe force development under the simpler conditions of constant activation. In these experiments, fibers were activated while extremely compressed and then quickly expanded at full activation. Generally, force development following width steps was 2-3 times faster than following activation steps. Since the new equilibrium width is reached quickly in comparison to the force, and since the width step is presumably carried out under conditions of constant activation, the time course of isometric force development is simpler to interpret. Two simple interpretations are, first, it is possible that the width step permits a population of active cross-bridges to undergo movement, to cycle, and thus to produce force, while at the same time the population of hindered cross-bridges retards the rate of force development. Alternatively, it is also possible that, while the width step allows active cross-bridges to undergo movement and cycle, it does so while it allows (or forces) some hindered bridges to detach (and possibly to enter into the cycling population). Thus the time course of force development may reflect the detachment rate of hindered bridges if the cycling rate of the active cross-bridges and the effects of the internal load inflicted by the remaining hindered bridges are not limiting factors. At present, our experiments are not able to distinguish between these two possibilities.

Inasmuch as factors other than actin-myosin interaction may affect the time course of force development following width steps (such as the diffusion of water into the fiber), we studied force redevelopment following a length release at a force plateau, under conditions of constant width and activation. As pointed out by Griffiths, Kuhn, Güth & Rüegg (1979), results obtained under these conditions should reflect only actin-myosin interactions. Comparing the records of Fig. 5 to those of Ford, Huxley, and Simmons (1977), it would seem that the time course of force development following a release is different than that observed in living single semitendinosus fibers of the frog. Specifically, the responses reported here have only three observable phases rather than four, and both the "fast" and "slow" recovery phases are far slower than the recovery phases reported by Ford et al. (1977). However, even though the time scales are different we think that our responses can be interpreted in a manner similar to those of Ford et al. (1977); i.e., one can distinguish a simultaneous drop of force during the applied length change (Phase 1), a "rapid" early force recovery following the length changes (Phase 2) and a reduction in the rate of force recovery (Phase 3). In compressed fibers we have never observed a reversal of the rate as seen in some cases by Ford et al. (1977). Our releases were probably not held long enough for us to observe the gradual recovery of force, with asymptotic approach to isometric force (Phase 4), if it exists at all in compressed fibers. Given this analogy, our time constants appear to be two orders of magnitude greater than those reported by Ford et al. (1977). Certainly some of this difference is attributable to differences between amphibian fast muscle and mammalian slow muscle and to the different temperatures involved. Nonetheless, we feel that the extremely slow time constants reported here are at least attributable in part to the compressed state of the myofilament lattice.

Fig. 7 shows a composite diagram of the data which we believe supports our hypothesis that hindered cross-bridges form as a result of lattice compression. The figure depicts a) the roughly linear reduction in force produced when an activated skinned fiber is compressed (Maughan & Godt, 1981; Krasner & Maughan, 1981; April & Maughan, 1982; Maughan & Berman, 1982) b) the rather steep increase in axial elastic modulus with compression found in relaxed fibers (Berman & Maughan, 1982), and c) a fall in actomyosin ATPase activity in compressed activated fibers (Krasner & Maughan, 1981). The inverse relation between force and axial elastic modulus suggests that the elastic modulus increases because some of the active, normally cycling cross-bridges have converted to hindered bridges as a result of compression. This notion is supported by elastic modulus measurements in activated compressed fibers (unpublished), which suggest that elastic modulus does not change appreciably as force is reduced by compression. This points to an exchange between the hindered and non-hindered population with the total population of attached cross-bridges being roughly constant in number. The fall in actomyosin ATPase activity is consistent with what would occur if some of the active bridges were taken out of the cycling population.

Note that the ATPase activity falls more sharply with compression than does isometric force. This suggests that compression is more effective in reducing the cycling rate (and thus the hydrolytic activity) of the remaining bridges than in reducing the average force per cross-bridge under isometric conditions. A change in the kinetic parameters could occur if compression alters a cross-bridge attachment or

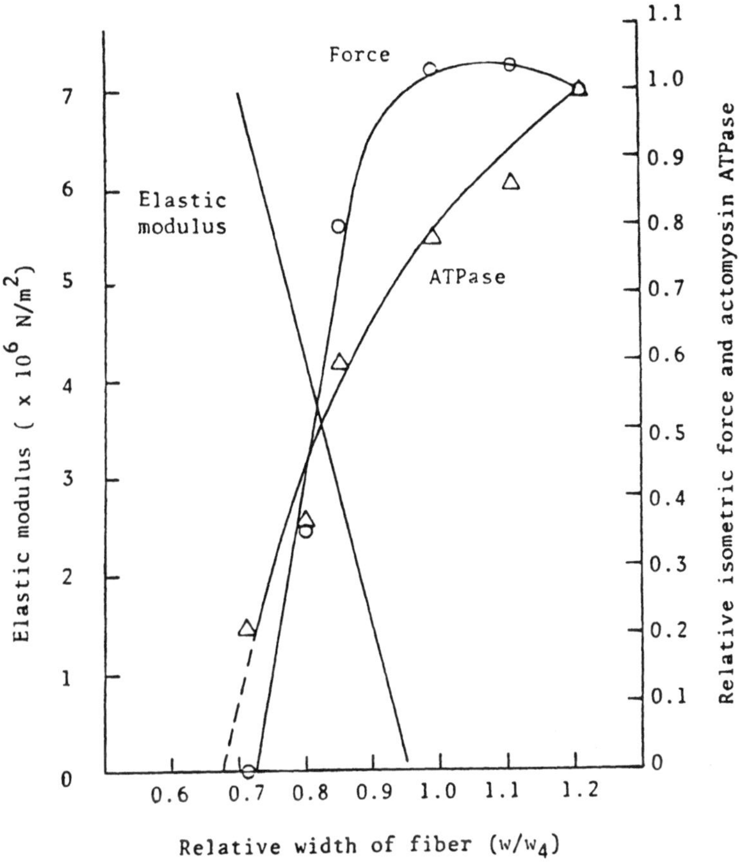

Figure 7: Composite figure showing the effect of relative width on isometric force and actomyosin ATPase activity in activated skinned fibers and axial elastic modulus in relaxed skinned fibers. Force and ATPase data for rabbit soleus muscles are from Krasner & Maughan (1981 and unpublished). Active ATPase of the fully activated skinned fibers at each concentration of Dextran T500 was calculated from the total ADP (minus that from non-actomyosin sources) produced as an end product of ATP hydrolysis and measured using high pressure liquid chromatography. Elastic modulus data (for frog semitendinosus muscle) adapted from Berman & Maughan (1982). This force vs. width data plotted is similar to force vs. relative width data from frog semitendinosus mechanically-skinned fibers (Maughan & Godt, 1981) and functionally-skinned multi-cellular bundles from guinea pig ventricle (Maughan & Berman, 1982). Insofar as the frog skeletal and guinea pig cardiac data were taken from experiments carried out with different animal species and solution compositions, the similarity of results suggests a common underlying mechanism of force inhibition with lattice compression.

detachment rate constant, or perhaps a formation or dissolution rate constant of an important intermediate. For example, a forced reduction in the axial tilt angle with which the subfragment-1 moiety of the cross-bridge attaches to actin may, in compressed fibers, alter a rate constant which leads to a slower time course of force development (Fig. 1-4) and a reduced force amplitude and the ATPase activity (Fig. 7). In the cross-bridge model of Eisenberg & Greene (1980), the free energy change associated with the binding of ATP to actomyosin is diminished at smaller angles of attachment of the cross-bridge, so that the probability that ATP binds and eventually is hydrolyzed is lessened. One could speculate that this feature of the Eisenberg & Greene (1980) model accounts for the decline in force and ATPase activity observed with compression. However, one cannot readily distinguish on the basis of the present results an effect of a direct change in some rate constant from an effect of rigor-like "hindered" bridges which form as a result of compression and which inflict pronounced internal loads upon the normally cycling cross-bridges.

ACKNOWLEDGMENTS

This work was supported by an NIH postdoctoral fellowship (PHS IF32-HL06543-01) to MRB and by grants from the American Heart Association (78-634 and 79-165). DWM is an Established Investigator of the American Heart Association.

REFERENCES

April, E.W., Farrell, M. and Schreder, J. (1977). Osmotically induced changes in the filament lattice of skinned striated muscle fibers. Biophys. J. 17: 174a.

April, E.W. and Maughan, D.W. (1982). Correlation between interfilament spacing and force generation in striated muscle. Biophys. J. 37(2): 129a.

Berman, M.R. and Maughan, D.W. (1982). Axial elastic modulus as a function of relative fiber width in relaxed skinned muscle fibers. Pflugers Arch. 393: 99-103.

Eisenberg, E. and Greene, L.E. (1980). The relation of muscle biochemistry to muscle physiology. Ann. Rev. Physiol. 42: 293-309.

Fabiato, A. and Fabiato, F. (1978). Myofilament-generated tension oscillations during partial calcium activation and activation dependence of the sarcomere length-tension relation of skinned cardiac cells. J. Gen. Physiol. 72: 667-699.

Ford, L.E., Huxley, A.F. and Simmons, R.M. (1977). Tension responses to sudden length change in stimulated frog muscle fibres near slack length. J. Physiol. 269: 441-515.

Godt, R.E. and Maughan, D.W. (1981). Influence of osmotic compression on calcium activation and tension in skinned muscle fibers of the rabbit. Pflug. Arch. 391: 334-337.

Goldman, Y.E., Matsubara, I. and Simmons, R.M. (1979). Lateral filamentary spacing in frog skinned muscle fibers in the relaxed and rigor states. J. Physiol. (Lond.) 295: 81p.

Griffiths, P.G., Kuhn, H.J., Güth, K. and Rüegg, J.C. (1979). Rate of isometric tension development in relation to calcium binding of skinned muscle fibers. Pflugers Arch. 382: 165-170.

Krasner, B. and Maughan, D. (1981). Dextran T500 decreases skinned fiber width, tension and ATPase. Biophys. J. 33: 27a.

Magid, A. and Reedy, M.K. (1980). X-ray diffraction observations of chemically skinned frog skeletal muscle processed by an improved method. Biophys. J. 30: 27-40.

Maughan, D. and Berman, M. (1982). Radial compression of functionally-skinned cardiac bundles abolishes calcium activated force. Biophys. J. 37: 363a.

Maughan, D.W. and Godt, R.E. (1981). Inhibition of force production in compressed skinned muscle fibers of the frog. Pflugers Arch. 390: 161-163.

Millman, B.M. and Nickel, B.G. (1980). Electrostatic forces in muscle and cylindrical gel systems. Biophys. J. 32(1): 49-63.

Moisecu, D.G. and Thieleczek, R. (1978). Calcium and strontium concentration changes within skinned muscle preparations following a change in the external bathing solution. J. Physiol. 275: 241-262.

Orentlicher, M., Reuben, J.P., Grundfest, H. and Brandt, P.W. (1974). Calcium binding and tension development in detergent-treated muscle fibers. J. Gen. Physiol. 63: 168-186.
Stephenson, D.G. and Williams, D.A. (1981). Calcium-activated force responses in fast- and slow-twitch skinned muscle fibres of the rat at different temperatures. J. Physiol. 317: 281-302.

DISCUSSION

GODT: I think it's really very interesting that in the last slide (Fig. 7), the relationship between force and ATPase falls off in such a way that the tension cost is least at the interfilament spacing of the normal fiber lattice. That's interesting in light of the observation David and I published last year that, in fact, the calcium sensitivity of the contractile apparatus is also highest at the filament spacing normally found in the intact lattice. If the lattice spacing is either swollen or shrunken, then the contractile apparatus is less sensitive to calcium. So it's a similar thing with tension cost.

SCHOENBERG: Have you been able to examine the phenomenon as a function of the sarcomere length?

MAUGHAN: This set of experiments was carried out at one sarcomere length. Further experiments should be done at various sarcomere lengths, using X-ray methods to monitor filament lateral separation, because I believe that the underlying phenomenon is one of lattice compression rather than one of filament overlap.

HUXLEY: Have you got any measurements on ATPase on relaxed fibers which have been compressed?

MAUGHAN: Yes. These experiments which I've described were done with the ATPase measured in both the active state and the relaxed state. What you saw was the total ATPase measured in the active fiber minus the ATPase measured in the relaxed fiber. There was no change in basal ATPase in relaxed fibers over a wide range of compressions.

PODOLSKY: You showed that the elastic modulus for your compressed, relaxed fiber was about the same as for the rigor fiber. I was wondering whether if you squeeze the fiber even further you can get it much stiffer than the rigor fiber, or whether the rigor stiffnesss is sort of an upper limit.

MAUGHAN: The data that you're referring to show that the elastic modulus of the compressed fiber is about two-thirds that of the rigor fiber. The elastic modulus increases with compression but, in fact, we've never really been able to achieve the same stiffness in compressed fibers that is found in rigor fibers, but it's very close. In active fibers we do not see a very great tendency of the elastic modulus to increase with compression; this tends to support the idea that there's an exchange of population of crossbridges from an active, cycling, hydrolyzing population to a non-active but still-attached population.

LATERAL SHRINKAGE OF THE MYOFILAMENT LATTICE IN CHEMICALLY SKINNED MUSCLES DURING CONTRACTION

I. Matsubara, Y. Umazume and N. Yagi

Department of Pharmacology, Tohoku University School of Medicine, Seiryo-machi, Sendai, Japan 980

ABSTRACT

A toe muscle was isolated from a hind limb of the mouse and treated with saponin to make the sarcolemma more permeable to the solutes of the bathing medium. The equatorial X-ray diffraction pattern was recorded to determine the 1,0 spacing of the hexagonal myofilament lattice. The spacing in relaxed muscle at a sarcomere length of 2.5 μm was 408 Å. When the muscle was maximally activated at pCa 4.4, a steady isometric tension of 1.3 kg/cm^2 was produced and the spacing decreased to 384 Å. A decrease in spacing of the same magnitude was observed when a relaxed muscle went into rigor, although the rigor tension was only 0.1 kg/cm^2, 8% of the maximum contractile tension.

From the intensity ratio of the 1,0 and 1,1 reflections the number of myosin heads transferred radially to the vicinity of the thin filament was calculated. During the maximum activation at pCa 4.4 the amount of the radial transfer was 96% of that in rigor. When the muscle was activated at a lower calcium concentration, the lattice shrinkage was smaller and the radial transfer was also smaller.

These findings suggest that the lateral force underlying the lattice shrinkage may be due to lateral elasticity of cross-bridges.

INTRODUCTION

When the sarcolemma of a muscle fibre is removed or disrupted the lateral spacing between the thick and the thin filaments increases. This increase has been attributed to removal of the osmotic constraint upon the fibre volume (Matsubara & Elliott, 1972). The myofilament lattice of such "skinned" fibres, compared with that of intact fibres, responds quite differently to various changes in physiological conditions.

When a skinned fibre is put into rigor at a sarcomere length where the filaments overlap maximally, the spacing decreases by about 10%.

Goldman, Matsubara and Simmons (1979) studied this phenomenon at various sarcomere lengths and found that the decrease in the spacing was roughly proportional to the amount of the overlap. This finding led them to conclude that the lateral force underlying the lattice shrinkage may be produced by cross-bridges. They estimated the magnitude of the lateral force from the osmotic pressure which shrunk the lattice to the same extent; the result suggested that the lateral force produced by each thick filament is surprisingly large, and comparable to the sliding force it produces.

Also during contraction the interfilament spacing in "skinned" muscle preparations decreases; this was first discovered by Shapiro, Tawada & Podolsky (1979) using a glycerol extracted muscle. The present study was carried out in an attempt to clarify the nature of the lateral force produced during contraction.

METHODS

Specimen Technique

A toe muscle (m. extensor digitorum longus) was dissected from a hind limb of a 3-5 week old mouse anaesthetized with sodium pentobarbital (50 mg/kg). The muscle was treated with saponin (50 μg/ml) for 1 hr using the method described by Endo and Iino (1980); this made the sarcolemma permeable to solutes of the bathing medium. Then the muscle was held isometrically in a specimen chamber by connecting the proximal tendon to a force transducer (RCA 5734) and the distal tendon to a hook mounted on a micrometer. The chamber had mylar windows to pass X-rays through the middle of the muscle. The sarcomere length was adjusted to 2.5 μm using the light diffraction method. The cross-sectional area of the muscle at this length was obtained by measuring the thickness and the width in the middle of the muscle.

Bathing Solutions

The specimen chamber was filled with a relaxing solution of the following composition (in mM/l): KMs (potassium methanesulphonate), 70; $Mg(Ms)_2$, 11.4; ATP, 10.9; EGTA, 10; PIPES, 20 (pH=6.8 at 20°C). This composition gave a MgATP concentration of 10 mM/l, a free Mg concentration of 1 mM/l and an ionic strength of 0.20. In certain experiments (specified in the text) a relaxing solution of a lower ionic strength of 0.15 was used; the KMs concentration of this solution was 20.5 mM/l.

The muscle was activated by replacing the relaxing solution with a calcium containing solution. Different calcium concentrations ranging from pCa 6.0 to 4.4 were used to obtain different degrees of activation. For example, the composition of the maximally activating solution was: KMs, 50.5; $Mg(Ms)_2$, 11.0; ATP, 11.0; EGTA, 10; $Ca(Ms)_2$, 9.8 (pCa=4.4); PIPES, 20 (pH=6.8). All the activating solutions had a MgATP concentration of 10 mM/l, a free Mg concentration of 1 mM/l and an ionic strength of 0.20.

To put the muscle into rigor a solution of the following composition was used: KMs, 126.6; $Mg(Ms)_2$, 1.4; EGTA, 10; PIPES, 20(pH=6.8). This solution had a free Mg concentration of 1 mM/l and an ionic strength of 0.20.

X-ray Diffraction

The equatorial X-ray diffraction patterns were recorded using a double-mirror Franks camera (Franks, 1955; Elliott & Worthington, 1963) combined with a fine-focusing rotating-anode generator (Rigaku, FR). The muscles were exposed at room temperature (17-21°C) for 5-10 min with a specimen-to-film distance of 22 cm.

The positions of the 1,0 and the 1,1 reflections from the hexagonal myofilament lattice were measured with a comparator (Nikon, model 6C) to obtain the 1,0 lattice spacing. The integrated intensities of these reflections were measured with a densitometer (Joyce-Loeble, scandig 3). From the intensity ratio of the two reflections, I_{10}/I_{11} the number of myosin heads associated with the thin filament was calculated according to the method described by Huxley (1968) and Haselgrove and Huxley (1973). A possible effect of variation in the spacing on the intensity ratio was assumed to be negligible.

Experimental Protocol

At the beginning of each experiment a diffraction pattern was recorded from the muscle at the relaxed state. This was followed by exposures during contractions in two different calcium concentrations; each diffraction pattern was recorded while the muscle was producing a steady tension. When the two contractions did not include the maximum contraction at pCa 4.4, a brief contraction at this pCa was carried out in order to measure the maximum tension. An interval of at least 15 min was left between contractions.

After the contractions the muscle was put into rigor; the diffraction pattern was recorded while a steady rigor tension was produced. In some experiments the rigor pattern was recorded before the contractions.

RESULTS

Tension Production

The threshold of tension production was pCa 5.8. The maximum tension observed at pCa 4.4 was 1.3 kg/cm^2. A half value of the maximum tension was produced at pCa 5.6-5.4.

In rigor, these preparations gave a steady tension of about 0.1 kg/cm^2.

Changes in the Lattice Spacing

Figure 1 shows the 1,0 lattice spacing obtained in the relaxed state and during contractions at various calcium concentrations. The centre-

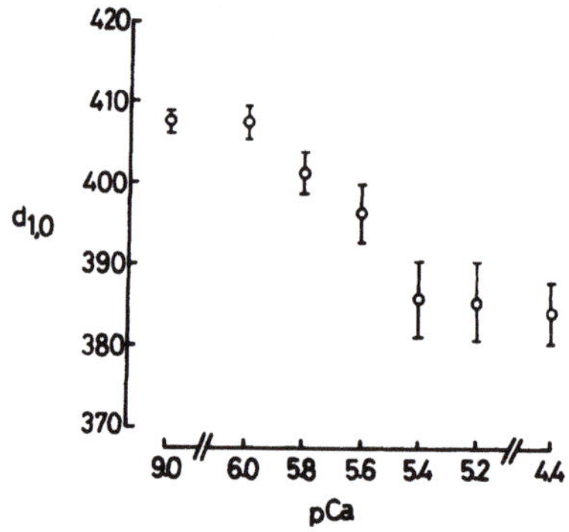

Figure 1: The 1,0 lattice spacing (d_{10}; in angstrom) plotted against the calcium concentration (pCa) of the bathing solution. Each point represents a mean value (±S.E. of the mean) for 6-10 muscles.

to-centre distance between the thick and the thin filaments corresponds to 2/3 of this spacing.

The mean spacing in the relaxing solution was 408 Å. This was greater than the spacing (375 Å) in the intact muscle suspended in Tyrode solution at the same sarcomere length (2.5 μm). The increase in the spacing caused by saponin treatment has already been reported by Endo & Iino (1980) using a Xenopus muscle.

When the calcium concentration was increased stepwise from the relaxing level, the spacing started decreasing at pCa 5.8 which was the same as the threshold for tension development. A further increase in the calcium concentration caused a greater decrease in the spacing. During the maximum contraction at pCa 4.4 the spacing was 384 Å, 6% smaller than that in the relaxed state.

In muscle in rigor the spacing was 386 Å, almost the same as that of maximally contracting muscle.

The above experiments were carried out at an ionic strength of 0.20. When the same experiments were repeated at a lower ionic strength of 0.15, the spacing behaved differently. The spacing in relaxed muscle at this low ionic strength was 390 Å. When the muscle was activated at pCa 4.4 or put into rigor, keeping the ionic strength low, no detectable change occurred in the spacing. The maximum tension at this ionic strength was 1.1 kg/cm^2.

Number of Myosin Heads Associated With Actin

The intensity ratio of the 1,0 and the 1,1 equatorial reflections (I_{10}/I_{11}) in relaxed muscle was 1.78. The ratio decreased to 0.32 during the maximum contraction at pCa 4.4. Contractions at lower calcium

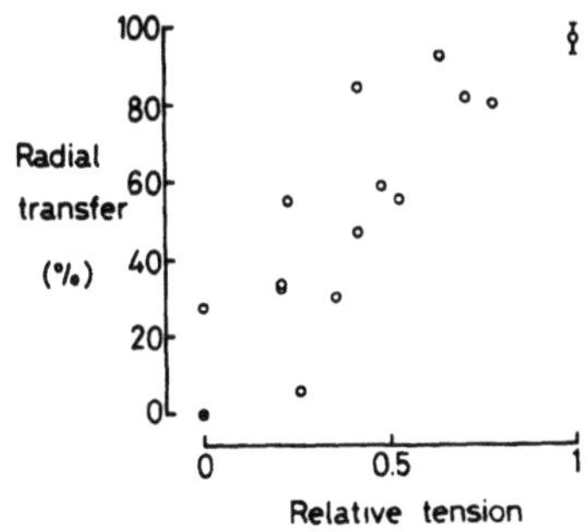

Figure 2: The fraction of myosin heads transferred radially to the vicinity of thin filaments plotted against the tension relative to the maximum tension. The filled circle represents the relaxed state. Each open circle represents a measurement during a contraction of a muscle, except at the maximum tension where the open circle represents a mean value (±S.E. of the mean) for 5 muscles.

concentrations gave intermediate ratios. In muscle in rigor the ratio was 0.29.

From these intensity ratios the fraction of myosin heads which were transferred to the vicinity of the thin filaments during contraction was calculated. The calculation was based on an assumption that the mass transferred from the thick to the thin filaments when a relaxed muscle goes into rigor (i.e. the "rigor transfer") represents the total mass of myosin heads in the double hexagonal filament lattice (Huxley, 1968). In Fig. 2 the fraction of myosin heads thus calculated to be transferred during contraction is plotted against the tension relative to the maximum tension. The average transfer occurring during the maximum contraction at pCa 4.4 was 96%, very close to the rigor transfer. During weaker contractions at lower calcium concentrations the transfer was less; the fractional transfer seemed roughly proportional to the relative tension.

DISCUSSION

Different Behaviour of the Lattice at Different Ionic Strengths

The present results showed that the lattice spacing decreased when a muscle contracted or went into rigor at an ionic strength of 0.20, but the decrease did not occur at a lower ionic strength of 0.15. A possible explanation of this difference might be as follows. Based upon experiments on skinned fibers in rigor, Goldman et al. (1979) suggested that there may be an optimum interfilament spacing for actin-myosin interaction and that the interaction tends to bring the spacing to the optimum value. This idea can be extended to the present case. Suppose that the optimum spacing in our "skinned" muscle is 380-390 Å. Then, at an ionic strength of 0.20 where the spacing at the relaxed condition is markedly greater than the optimum, a significant decrease in the spacing would be observed on activation and on going into rigor. On the other hand, at an ionic strength of 0.15 where the spacing at the relaxed condition is already close to the optimum, no significant change would occur to the spacing. Thus the present results are compatible with the idea of optimum spacing.

Nature of the Lateral Force

Two observations suggested strongly that the lateral force underlying the spacing changes on going into rigor is produced by cross-bridges (Goldman et al., 1979); (1) the magnitude of the spacing decrease is roughly proportional to the amount of the overlap between the thick and the thin filaments; (2) when an osmotically shrunk fibre goes into rigor the lattice spacing increases (such a change in the direction is hard to explain in terms of electrostatic forces). The following observations suggest that the lateral force operating during contraction may also be produced by cross-bridges. (1) The magnitude of the lattice shrinkage during contraction is similar to that observed on going into rigor. (2) The shrinkage is greater at calcium concentrations at which a greater number of cross-bridges are supposed to be active. (3) The shrinkage is smaller in stretched muscles (Shapiro et al., 1979).

If the lateral force is to be attributed to cross-bridges, then two possibilities must be considered concerning the mechanism of the lateral force production. One possibility is that the force is the lateral component of the cross-bridge force produced by the actin-myosin interaction. That is to consider the cross-bridge force to be transmitted to the shaft of the thick filament at a certain angle; then the cross-bridge force will have a component parallel to the filament, which is the sliding force, and a component perpendicular to the filament, which is the lateral force. In this case the lateral force may be expected to be proportional to the sliding force as long as the angle stays the same. However, this expectation was not fulfilled in muscles going into rigor. Namely the rigor tension was less than 8% of the maximum contractile tension, yet the spacing decreased to almost the same extent as during the maximum contraction. To overcome this difficulty one could argue that the angle of the force transmission may be greater in rigor, making the lateral force very large compared to the rigor tension. However, the following observation (unpublished) suggests that this is unlikely to be the case. When a "skinned" muscle in rigor was released by 1% of its length, the rigor tension disappeared and there was no tension recovery. We compared the lattice spacings before and after the release. If the lateral force were a component of the cross-bridge force, then the spacing would be expected to increase after release. However, in practice, there was no such increase in the spacing; the lateral force seemed almost the same whether or not the rigor tension was present. Thus, as long as we assume that the origin of the lateral force during contraction is the same as that in rigor, we have to conclude that the lateral force is unlikely to be a component of the cross-bridge force produced by the actin-myosin interaction.

Another possibility is that the lateral force is produced by a different mechanism from that of the tension production. According to the idea proposed by Goldman et al. (1979), cross-bridges displaced from their optimum lateral positions produce an elastic lateral force. If this is the real mechanism, the lateral force may be expected to be proportional to the number of myosin heads transferred to the vicinity of the thin filaments. This agrees with the present observations, in which muscles during maximum contraction and in rigor, both of which had roughly the same number of myosin heads in the vicinity of the thin filaments, showed an identical spacing change.

There are various possible mechanisms by which the cross-bridge can produce an elastic lateral force. For example, suppose that S-2 and light meromyosin are linked by many hydrogen bonds. Then a displacement of the head from its optimum lateral position would break a certain number of the bonds. Such a molecular process would produce an effect which can be regarded as elasticity.

In conclusion, the interfilament spacing decreases when "skinned" muscles contract, and our results are compatible with the view that this decrease is caused by lateral elasticity of cross-bridges.

REFERENCES

Elliott, G.F. and Worthington, C.R. (1963). A small-angle optically focusing X-ray diffraction camera in biological research. Part I. J. Ultrastruct. Res. 9: 166-170.

Endo, M. and Iino, M. (1980). Specific perforation of muscle cell membranes with preserved SR functions by saponin treatment. J. Muscle Res. Cell Motil. 1: 89-100.

Franks, A. (1955). An optically focusing X-ray diffraction camera. Proc. phys. Soc. B68: 1054-1064.

Goldman, Y.E., Matsubara, I. and Simmons, R.M. (1979). Lateral filamentary spacing in frog skinned muscle fibres in the relaxed and rigor states. J. Physiol. 295: 80-81P.

Haselgrove, J.C. and Huxley, H.E. (1973). X-ray evidence for radial cross-bridge movement and for the sliding filament model in actively contracting skeletal muscle. J. molec. Biol. 77: 549-568.

Huxley, H.E. (1968). Structural difference between resting and rigor muscle: evidence from intensity changes in the low-angle equatorial X-ray diagram. J. Molec. Biol. 37: 507-520.

Matsubara, I. and Elliott, G.F. (1972). X-ray diffraction studies on skinned single fibres of frog skeletal muscle. J. Molec. Biol. 72: 657-669.

Shapiro, P.J., Tawada, K. and Podolsky, R.J. (1979). X-ray diffraction of skinned muscle fibres. Biophys. J. 25: 18a.

DISCUSSION

POLLACK: Having such bonds between the S-2 and light meromyosin such as you suppose -- would that tend to preclude the possibility that there's any elasticity in the S-2? Have you given up that idea?

MATSUBARA: No, our results do not rule out the possibility you mentioned. The hydrogen bonds, which we postulated in one of the slides as a possible explanation, do not need to cover the whole length of S-2.

ROWE: Am I right in saying that the argument against the radial force being an effective component of the contracting force, because of the rigor experiment, rests upon the assumption that rigor tension arises from at least partial distortion of the S-1 component of the cross-bridge? But if it arose from some other mechanism, something in the S-2, then this argument would not hold good. Is that correct?

MATSUBARA: What you said is correct. In our argument we assume that both contractile and rigor tensions are produced by S-1 reacting with actin and that the forces are transmitted to the shaft of the thick filament through S-2. If you assume differently in these respects, our argument may not hold.

ROWE: If you say rigor is locked at 45°, and rigor tension is generated simply by a bit of stretching of S-2 or something, then you wouldn't expect a radial component. Would you?

HUXLEY: Well, it would depend on the angle of the S-2. I mean if the S-2 was not parallel to the axis, you would expect there to be a radial force component.

ROWE: Yes but it's a "cosine theta" factor and the S-2 is relatively long.

MATSUBARA: We are not ruling out the presence of the radial component. What we are saying is that the main force responsible for the shrinkage is unlikely to be the radial component of the contractile force. This has led us to postulate a radial force which is independent of the contractile force, as the main force.

TREGEAR: What happened at an ionic strength of 0.15 M when you activate? Did you say what happened to the equatorial intensity ratio? Does it also change?

MATSUBARA: Yes, the equatorial intensity ratio changed on activation. But the spacing did not change at this ionic strength; our interpretation is that the spacing at the resting state is already close to the optimum spacing for cross-bridges. The maximum tension at this ionic strength was not significantly different from that obtained at 0.20.

TREGEAR: Was the equatorial intensity change towards rigor? Did it go the whole way towards rigor?

MATSUBARA: Not quite the whole way; we observed approximately 90% of the rigor transfer (i.e. the radial transfer of myosin heads occurring on going into rigor) during contraction.

TREGEAR: So there isn't a one-to-one correlation between the motion of the cross-bridge and the tension, is there?

MATSUBARA: At a fixed ionic strength there is a good correlation between tension and the amount of radial transfer, as we have shown in one of the slides. However, we do not know whether this correlation would hold when the tension is varied by changing the ionic strength. If the ionic strength affects the contractile force produced by each myosin head reacting with actin, then I expect that the correlation, such as we have shown in the slide, will not hold.

MAGID: I reported some results from experiments where I looked at the lattice spacing as a function of PVP concentration last year at the Biophysical Society meeting (Biophys. J. 33: 226a, 1981). These were experiments with Triton-skinned muscle. I just wanted to say that at around the lattice spacing which corresponds to the *in vivo* spacing, the lattice is very compliant. It takes a very small amount of PVP, that is, osmotic pressure, to produce large changes in filament spacing. Whereas at higher PVP concentrations the lattice becomes progressively stiffer. However, I think that my data indicate that, under the conditions you are studying, very small forces would suffice to move the filaments closer together. They do not need to be massive forces. What I am trying to emphasize is that resting muscle possesses a radial stiffness. Millman and Nickel reported at the Biophysical Discussion (Biophys. J. 32: 49, 19--) that the rigor and relaxed radial force curves more or less parallel each other with an offset between them, as if the rigor conditions impose an additional radial stiffness component upon that which seems to be present in resting muscle.

MAUGHAN: I should note that in the highly compressed fiber, radial forces can be very large. Bob Godt and I looked at changes in fiber width attending contraction at various degrees of initial compression (Maughan and Godt, J. Gen Physiol. 77: 49-64, 1981). As we switched a fiber from a relaxed state to a rigor state, we measured a radial force on the order of the normal axial cross-bridge force, or even much greater. Ernest April and I also noted very large radial forces in compressed skinned crayfish fibers. The large radial forces seem to be of non-cross-bridge origin, or at least they are not related to head rotation.

SCHOENBERG: If I understood you correctly, you took a rigor fiber which was exerting some tension and then you released it to zero tension, and you didn't notice any lateral spacing change. what was the overall tension change compared to P_0 in that experiment?

MATSUBARA: The overall tension change was approximately 100 grams/cm^2, which corresponded to 8% of the maximum contractile tension.

SCHOENBERG: How large was it when you started?

MATSUBARA: The initial rigor tension was about 8% of the maximum tension; it decreased to zero when we released the muscle by 1% of its length.

SCHOENBERG: So with a change of 8% of P_0 you aren't able to see any change in the lateral spacing?

MATSUBARA: That is correct. Our conclusion is that the lateral force is unlikely to be a component of the contractile force.

FAY: How do you reconcile your results indicating an increase in the radial rces associated with cross-bridge action with Dr. Hatta's results this morning where he showed results indicating that there was a decrease in radial stiffness associated with muscle contraction?

MATSUBARA: We have not made any stiffness measurement on our preparation. However, I would like to point out that the radial stiffness, measured with the method described by Dr. Hatta, is affected not only by cross-bridges but also by other factors. For example, a change in lateral flexibility of the thin filament, which is known to occur on activation, will affect the stiffness.

POLLACK: In your model, how would the cross-bridge reach out to find its attachment side? It seems to me that there's always a force that prevents the S-2 from swinging out, so, for example, if you consider cases, either at short sarcomere length length or at low ionic strength, where the interfilament spacing is relatively large, how would the cross-bridge be able to reach out?

MATSUBARA: Perhaps the thin filament can fluctuate. I am referring to the work by Professor Oosawa's group. They have shown that the lateral flexibility of the thin filaments increases on activation. Secondly, there might still be some vibration of bridges.

EFFECT OF LATTICE SPACING ON CROSS-BRIDGE ORIENTATIONS IN RELAXED CRAB MUSCLE

Yuichiro Maéda

Max-Planck-Institute, Department of Biophysics, Heidelberg, W. Germany D-6900

I would like to make a contribution which is relevant to Dr. Matsubara's work as well as to Dr. Maughan's work and to Dr. Robinson's finding, presented earlier at this symposium.

I recorded X-ray diffraction patterns, not only the equatorial but also the axial patterns, from skinned crab single fibers, the lattice spacing of which was changed by osmotic pressure which was controlled by the amount of dextran added to the relaxing solution.

By analyzing the myosin reflections, which are not sampled by the lattice, the distance (r_0) between the center of the heads and the thick filament axis, the tilt angle of the heads, as well as dimensions of the heads could be estimated.

The main results are the following (Fig. 1):

(1) By chemical skinning, the lattice was swollen from a = 60 nm to 72 nm.

(2) The more dextran was added, the smaller the lattice spacing and the shorter the distance r_0 became. The relation between r_0 and the spacing a consists of 3 phases. In the first phase, from a = 72 to 63 nm, reduction in the spacing was not accompanied by a decrease in the distance r_0, and the heads remained almost perpendicular to the filament axis. In the second phase, from 63 to 50 nm, compression of the lattice was accompanied by a linear decrease of the distance r_0, and by an increase of the tilt angle as well as slew angle of the heads. In the third phase, the filaments considerably resisted against the osmotic pressure. Although the lattice could be compressed to 45 nm, no further decrease in r_0 was observed.

The simplest explanations of the results are as follows:

(1) The myosin heads can be pushed back by the surrounding thin filaments, without specific interaction with the thin filaments, as indicated by the absence of the rigor diffraction pattern. There appears to be simple steric hindrance between the heads and the thin filaments.

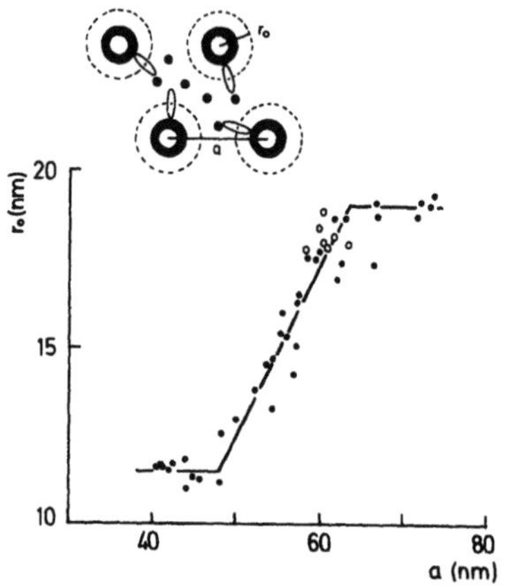

Figure 1: Relation between the distance (r_0) of the centre of gravity of myosin head from the thick filament axis and the lattice spacing (a). Closed circles are from skinned fibres in the relaxing solution, open circles are from live relaxed muscles, both at normal sarcomere lengths, 5.5-6.0 μm.

(2) For some unknown reasons, in the fibers swollen beyond the normal dimension, heads will not move out beyond the upper limit of r_0 = 19 nm.

(3) On the other hand, if heads have already been pushed back and tilted down on the surface of the thick filaments, further compression results solely in further reduction in the average distance between the heads and the thin filaments.

(4) Live, relaxed muscle gives approximately a = 60 nm and r_0 = 18 nm, where heads are slightly tilted relative to the filament axis and the tips of heads must be quite close to the surface of thin filament.

(5) Elongated heads, being 16-18 nm long and 3-4 nm thick, which are arranged in a helical manner with 4-fold rotational symmetry could explain main features of the observed intensity profiles.

I am making three comments:

(1) Concerning Dr. Matsubara's talk, it is interesting to correlate my finding with his finding that, at maximum activation, skinned mammalian muscle tends to assume a lattice spacing which is slightly larger than that of live muscle. Results obtained from crab muscle show that, in the living resting state, the distance r_0 is slightly smaller than that for skinned fibers to which no osmotic pressure is applied, and that the elongated heads are slightly tilted relative to the filament axis. In addition, the tip of the head is situated very close to the surface of the thin filament. The configuration of the heads appears to be restrained in live relaxed muscle. Therefore, if the heads take on a configuration which fits only to a more expanded lattice during contraction, activated muscle fibers would have a larger lattice spacing. It is also interesting to know how the tension generating process is affected by reduction in spacing between thin and thick filaments.

(2) Concerning Dr. Maughan's presentation, according to his interpretation, the heads might make a rigor complex with the thin filaments in compressed vertebrate fibers in relaxing solution. However this seems to be very unlikely. According to my observations, no rigor patterns were recorded from compressed fibers if the concentration of ATP was high enough. At ATP concentration of lower than 4 mM, on the other hand, the rigor patterns were observed. This can be explained by ATP depletion inside the fibers.

(3) Concerning Dr. Robinson's presentation, I think it is very unlikely that the filaments are so bulky. If the filaments are so thick, the space between thick and thin filaments would be so narrow that the heads could not behave as shown in the results I have just presented.

Finally, I would like to make a point. The system used here offers two advantages, namely that the myosin reflections are unsampled and that the configuration of the heads can be altered. Crab muscle should also allow detection of configurational changes of the heads during contraction.

ISOTONIC CONTRACTION OF TEMP-STEP ACTIVATED MUSCLE FIBERS WITH VARIED TONICITY: EFFECTS OF CELL VOLUME AND THE DEGREE OF ACTIVATION

Jagdish Gulati and Aravind Babu

Department of Medicine & Cardiovascular Research Center, Albert Einstein College of Medicine, Bronx, New York 10461

ABSTRACT

These studies on intact fibers describe the effects of calcium, ionic strength and volume on the contraction properties. The results provide firm evidence that cell volume affects the speed but not the force. On the other hand, sarcoplasmic ionic strength affects the force development, with no effect on unloaded speed of shortening. These results suggest that there are essential differences in the rate limiting steps for isometric and isotonic properties of the cross-bridge mechanism. The studies at various degrees of activation indicate that Ca acts as a simple "on-off" switch for cross-bridge activation, in intact fibers.

Studies of the effects of calcium and cell volume on the steady-state contraction properties are described. The studies of the volume effects provide insights into the relations between the structural and kinetic properties of the cross-bridge mechanism. Similarly, the Ca effects give the information on the cross-bridge activation mechanism and on the kinetics.

Isolated intact single fibers from frog skeletal muscles were activated by an *instantaneous* temperature step from 25° to 0°, in the presence of 0.2-3 mM caffeine (Gulati & Babu, 1982). The volume effects were studied with maximal activation in hypertonic Ringer solution. The osmolarity was controlled with a) impermeant sucrose, which decreases the cell volume thereby increasing the sarcoplasmic ionic strength, and b) permeant KCl, which changes the ionic strength at constant volume. The calcium effects were studied by varying caffeine, yielding various degrees of activation (beta). The speed of unloaded shortening (V_s) was measured in each case at 0°C by the slack-test method (Fig. 1).

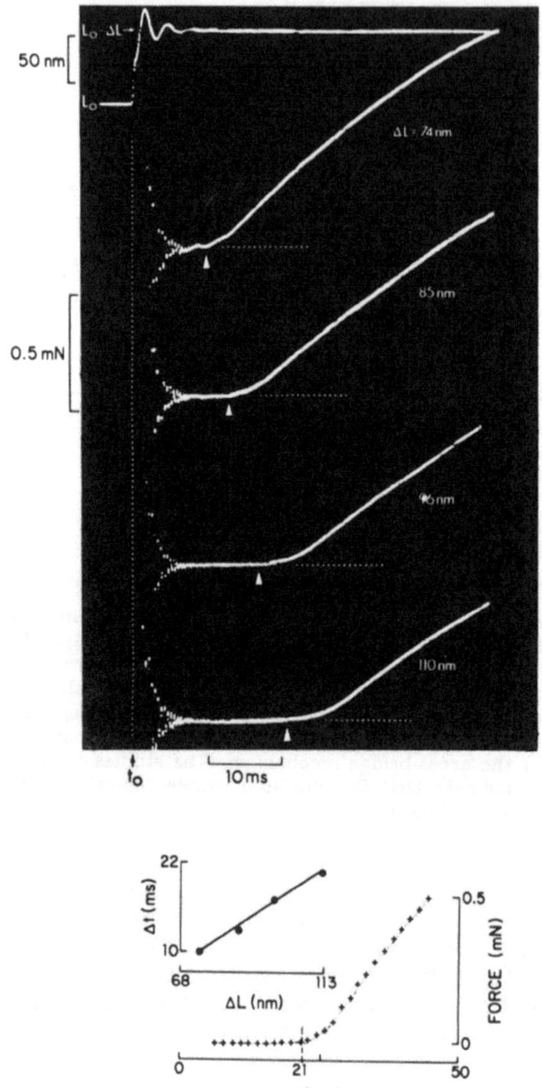

Figure 1: Slack test on a fiber activated by applying instantaneous temperature step from 25° to 0°C in standard solution with 2 mM caffeine. The shortening steps of the indicated magnitude were applied at the force plateau (not shown) on maximally activated fibers. The top trace shows a typical slack step of 74 nm per half sarcomere at time t_o. The second trace shows the actual force response to this step. At t_o (vertical dashed line) force drops to the zero level (horizontal dashed line) and rises again after a time period Δt. Note that Δt increases with larger steps. The method of computer analysis of the data is shown in drawings at the bottom. The force traces were digitized (x) and fitted with a polynomial function. The time at which force deviated from the base line was computed, as shown in the bottom drawing that corresponds to the actual force trace above for $\Delta L = 110$ nm. The slope of the ΔL vs Δt plot gives the inverse of the speed of shortening (V_s) at zero load. The mean value of V_s at 0°C in Ringer solution (1.0T) on 33 fibers was 2.7 μm/half sarcomere per sec.

Table 1: Effects of hypertonicity with sucrose and KCl, and the degree of activation on contraction parameters (0°C).

A. Osmolarity (T)

	Impermeant sucrose		Permeant KCl	
	1.4T	1.6T	1.4T	1.6T
Relative speed[1]	0.60 ± 0.08(6)	0.48 ± 0.06(4)	0.95 ± 0.14(8)	0.98 ± 0.06(5)
Relative force	0.72 ± 0.01(11)	0.58 ± 0.01(5)	0.68 ± 0.02(13)	0.53 ± 0.02(7)
Relative volume	0.76 ± 0.01(25)	0.66 ± 0.02(4)	1.00 ± 0.01(13)	1.00 ± 0.01(7)
Estimated filament[2] separation (nm)	9.6	8.7	13.2	13.2

(1) All the measured parameters are given relative to their values in standard Ringer solution. In each case, numerics within parenthesis indicate the number of single fibers used for the data points.

(2) The radial separation between the surface of thick to that of thin myofilament is taken from Fig. 2 and Table 1 of Gulati and Babu (82).

B. Degree of Activation (β)

	$\beta = 1$	$\beta = 0.76$ (range: 0.84-0.64)	$\beta = 0.36$ (range: 0.46-0.18)
Speed of shortening, Vs (μm/½ sarc/sec)	3.32 ± 0.33(4)	2.82 ± 0.16(10)	2.55 ± 0.15(6)
Normalised speed	1.0	0.85	0.77
Estimated internal load (fraction of β)	–	0.03	0.06

(β = Force level fraction of electrical tetanic stimulation).

The results in hyperosmotic solutions with sucrose and KCl are summarized in Table 1A. Force is decreased in 1.4T and 1.6T solutions respectively to 0.7 and 0.5 times the force in 1.0T, with both solutes. On the other hand, V_s is decreased only in sucrose but remains relatively unaffected in KCl. Conclusions from these studies are: 1) Increase in cell ionic strength decreases the force development, but there is no effect of ionic strength on the speed of shortening. 2) Force is independent of cell volume; therefore, force per cross-bridge is constant with variation in thin-thick filament separation from 6 nm to 18 nm. 3) In contrast, V_s and the kinetics of cross-bridge turnover are affected by cell volume. (Cell chloride is presumably high with KCl and low with sucrose, suggesting that this influence of volume on V_s might be a Cl effect. Such effect is unlikely, however, because V_s in skinned fibers is similar to that in intact fibers, in spite of large differences in Cl concentrations in bathing milieus of the cross-bridges in the two fiber preparations). The present evidence, that V_s is independent of ionic strength, is opposed to the conclusion of Edman & Hwang (1977) that was based on data of impermeant solutes only. Our data demonstrate that the decrease in speed with sucrose is a volume effect. This effect may be produced by an increased internal load, possibly due to filamentary interaction at compressed distances of radial actin-myosin separation (Hill, 1968), and/or increased cell viscosity.

The volume effect in this study also raises a question regarding the influence of filament separation on the kinetics of cross-bridge turnover, at a fixed (2.2-2.3 μ) sarcomere length. The above explanation that the effect on speed is by a separate internal load, assumes that the intrinsic turnover kinetics of acto-myosin cross-bridges are constant with compressed filament separations. This suggests that, during contraction, movements of a cross-bridge may occur within restricted domains of the S-1 (or the S-2) moiety, so that the possible effects due to steric hindrance are avoided at the compressed spacings. Therefore, it could be that the 45° rigor conformation of a bridge may not be part of its normal working stroke (Gulati and Babu, 1982). It should be noted that this may also be an explanation for the recent interesting observation, in fast-time-resolved X-ray studies (Huxley et al., 1981), of lack of rigor patterns in the equatorial reflections and in the off-meridional layer-line reflections, when quick shortening steps were applied to contracting muscles.

Regarding the Ca effect, the steady force and the slack test records were obtained in a wide β-range (0.18 P_o to P_o). The unloaded speed of shortening is found to be relatively constant, although it may be noted (Table 1B) that there is a tendency for V_s to decrease slightly with β. This result is also explained by the existence of a constant internal load at all β, amounting to less than 2% of P_o (last row in Table 1B, estimated by using A.V. Hill's force-velocity equation). The degrees of activation very likely reflect the steady-state free Ca levels within the cell. The results show therefore that the cross-bridge kinetics for V_s are relatively insensitive to Ca ions. These findings provide support for a simple "on-off" switch-like action of Ca on cross-bridge mechanism, in intact muscle.

ACKNOWLEDGEMENTS

These studies were supported by grants from NIH (AM-26632) and the Muscular Dystrophy Association.

REFERENCES

Edman K.A.P. and Hwang, J.C. (1977). The force-velocity relationship in vertebrate muscle fibers at varied tonicity of the external medium. J. Physiol. 269: 255-272.

Gulati, J. and Babu, A. (1982). Tonicity effects on intact single muscle fibers: relation

between force and cell volume. Science 215: 1109-1112.
Hill, D.K. (1968). Tension due to interaction between the sliding filaments in resting striated muscle. The effect of stimulation. J. Physiol. 199: 637-684.
Huxley, H.E., Simmons, R.M., Faruqi, A.R., Kress, M., Bordas, J. and Koch, M.H.J. (1981). Millisecond time-resolved changes in X-ray reflections from contracting muscle during rapid mechanical transients, recorded using synchrotron radiation. Proc. Natl. Acad. Sci. 78: 2297-2301.

DISCUSSION

A common question asked by many was: could the effects of sucrose and KCl hypertonicities on the speed of shortening (V_s) be explained by possible influences of these solutes on the mechanisms of Ca release? The answer is no, because maximal force in these experiments was strictly related to tonicity (and to intracellular ionic strength), suggesting that, at a given tonicity, the degrees of activation are quite likely to be similar with the two solutes. Another reason is that, when tested in a solution of normal tonicity, the speed is independent of the degree of activation. Therefore even if slightly unequal amounts of calcium were released during activations in sucrose and KCl, this inequality would not be an important factor for the effects of tonicity on V_s.

Another question was on the conclusion that the decrease in speed in the sucrose solution is a volume effect. Is this conclusion consistent with the fact that the speed is constant with increasing sarcomere length? This question is important because myofilaments pack more densely in shrunken fibers, and the volume effect may therefore indicate enhanced interaction by increasing the internal load. The filament crowding occurs also with increasing sarcomere length (in normal tonicity solution), but the sarcomere length-velocity relation indicates that internal load is substantially constant in this case. One explanation is that the filament interaction may be quite negligible in a solution of normal tonicity. Another possibility is that the interaction due to filament crowding at long sarcomere lengths is offset by the decrease in overlap.

CHANGES IN MECHANICAL PROPERTIES IN OSMOTICALLY COMPRESSED SKINNED MUSCLE FIBERS OF FROG

Teizo Tsuchiya

Department of Physiology, School of Medicine, Teikyo University, Kaga, Itabashi-ku, Tokyo, 173 Japan

ABSTRACT

The mechanical properties of frog skinned muscle fibers were examined in PVP solution. In a high concentration of PVP, the velocity of isotonic shortening and lengthening decreased remarkably and the stiffness increased not only in the activation solution but also in the relaxing solution.

In skinned muscle fibers, the diameter of a fiber can be changed osmotically by using a long chain polymer (Maughan and Godt, 1979). Several mechanical properties of frog muscle fibers osmotically compressed by PVP (polyvinylpyrrolidone) were investigated in the present experiment. Ca-activated tension increased by 30% in 4% PVP above that in 0% PVP but was depressed remarkably in 12% PVP. When the activating solution containing 4% PVP was changed to one containing 12% PVP, the higher tension was maintained for several minutes. The shortening and lengthening velocities measured by an electronic servo system decreased with the concentration of PVP. In the high tension state in 12% PVP, a fiber did not show appreciable isotonic shortening and lengthening (Fig. 1). The force-velocity relation was examined by measuring the tension change during constant velocity shortening and lengthening. The tension dropped more rapidly and reached a lower level during the constant velocity shortening as the concentration of PVP was increased. During constant velocity lengthening, the tension rose to a higher level with increasing concentration of PVP, especially in the high tension state in 12% PVP. The tension rose to more than three times the isometric tension without "give" phenomenon.

These results in constant velocity experiment were qualitatively the same as those in isotonic experiments. The tension redevelopment after a large release to zero tension was observed. When a fiber was quickly

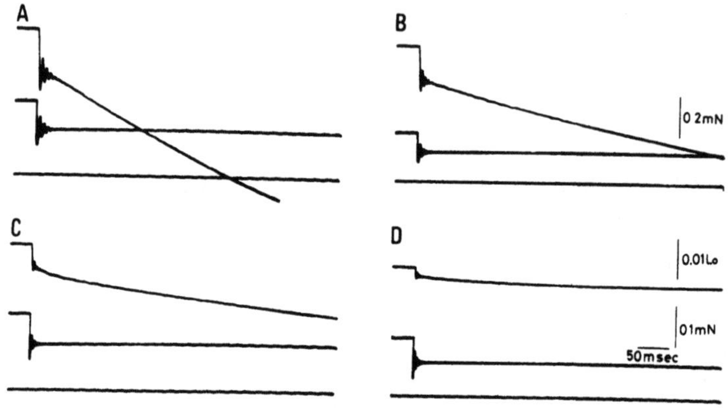

Figure 1: The decrease in shortening velocity of a skinned fiber in the activating solution containing 0% (A), 4% (B), 8% (C) and 12% (D) PVP at the same relative load. In D, the tension is held at the high level by the successive treatment of 4% and 12% PVP. Note no shortening at the end of the length record in D. The tension in A and C is recorded at the same sensitivity as in D.

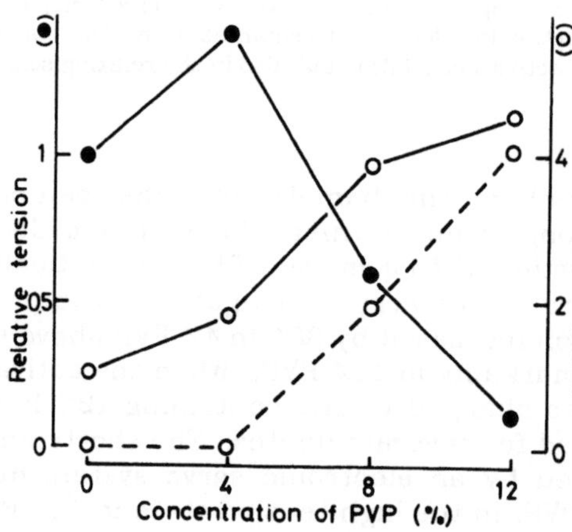

Figure 2: Variations of isometric tension (solid circle) and stiffness in activating (hollow circle with continuous line) and relaxing solution (hollow circle with interrupted line) with PVP concentration. Note the remarkable increase in stiffness in relaxing solution containing 8 and 12% PVP.

released, the tension dropped and stayed at zero for a longer time and redeveloped more slowly with increasing concentration of PVP.

A short length step (0.5% L_0) was imposed to measure fiber stiffness. In 0 and 4% PVP, the stiffness was zero in the relaxing solution but was higher in the activating solution. In relaxing solution, the stiffness in 8 and 12% PVP increased remarkably and a similar increase in stiffness was observed with increasing PVP concentration in activating solution. Stiffness was higher in activating solution than relaxing solution at any concentration of PVP (Fig. 2). It was reported that a high viscosity medium containing sugar, which could penetrate into a fiber, inhibited tension development (Endo et al., 1979). The effects of increased viscosity by sucrose on stiffness and tension redevelopment were examined. These effects, however, were quite different from those of PVP solution. This fact seems to show that most of the above-mentioned effects of PVP were induced by osmotic compression.

REFERENCES

Endo, M., Kitazawa, T., Iino, M. and Kakuta, Y. (1979). Effect of "viscosity" of the medium on mechanical properties of skinned skeletal muscle fibers. In: *Cross-bridge Mechanism in Muscle Contraction*, pp. 365-376, ed. Sugi, H. and Pollack, G.H. University of Tokyo Press.

Maughan, D.W. and Godt, R.E. (1979). Stretch and radial compression studies on relaxed skinned muscle fibers of the frog. Biophys. J. 28: 391-402.

DISCUSSION

MAUGHAN: I recommend the use of Dextran T 500, since the average molecular weight of Dextran T 500 is larger than that of PVP-40, and it is more effectively excluded from the interior of a skinned fiber. Further, the bonding of Dextran T 500 to metal ions is weaker than that of PVP-40.

SAEKI: You measured the shortening velocity at the same relative load. How did you adjust it?

TSUCHIYA: I used a servo system to control isotonic load. The isometric tension was memorized in sample-hold circuit and the load was adjusted so that the relative value would be constant.

CONTRACTION DYNAMICS

INTRODUCTION

This section comprises a number of contributions that are off the beaten path; i.e., some of the results are ones which do not necessarily find easy interpretation within the framework of the cross-bridge model.

The first paper, offered by M.I.M. Noble, presents the results of a long line of experimentation on the effects of stretch on single fibers of frog skeletal muscle fibers. When a fiber sitting at a particular length along the descending limb of the length-tension relation is stimulated to a steady tetanic tension, then stretched during stimulation to a longer length, the resultant tension is not the lower tension predicted by the shape of the length-tension relation; it is higher. The "extra tension" that is elicited by stretch has several separable components that are considered in detail. The investigators are divided on their interpretation. Edman feels that the results can be fitted into the framework of the cross-bridge theory if a parallel elastic element that is not present during isometric contraction is recruited during the stretch. Noble, on the other hand, claims their experiments rule out such a possibility, and suggests that no plausible hypothesis lying within the framework of the theory has yet been proposed.

The effects of stretch are also studied in the short presentation by Nichols. Muscle stretch causes an increase of tension; but when the stretch is sufficiently large, the fiber "yields." Nichols investigates the effect of velocity of stretch on the yielding phenomenon.

The paper presented by Colomo describes the characteristics of the early phase of activation. Since the early work of A.V. Hill in which the time course of the development of the "active state" had been of central interest, there has been an ongoing effort to pinpoint in a more definitive way the manner in which contractility develops early in contraction. Colomo finds, interestingly enough, that although tension, stiffness, and the force-velocity relation all require some time to attain their full value, the maximum, unloaded velocity of contraction is not at all time dependent. The insensitivity of maximum shortening velocity to activation level, and also to sarcomere length (Edman, Tokyo) is a feature that seems worthy of attention.

The presentation by Pollack reviews the various methods that have been used to observe stepwise sarcomere shortening -- laser diffraction, high-speed cinemicrography, and on-line imaging of the striation pattern.

These methods appear to show self-consistency. In addition, he describes the sounds that are generated as the muscle shortens. These are discrete, not continuous, thereby supporting the notion that the contractile process is discrete and synchronized over a large region of the muscle.

In a second communication, Pollack presents a preliminary model to account for this unexpected phenomenon. The central point in the model is that each shortening step is caused by a local shortening of a region along the thick filament. The model invokes a number of controversial views, but is capable of explaining not only the steps and pauses in a quantitative way, but also the fact that this discrete behavior is synchronous over a large region of space.

Stepwise behavior is also considered by Ritz-Gold. Using a spin label, she shows that the electron spin resonance signal is consistent with stepwise changes in cross-bridge state. She relates these stepwise cross-bridge states to the phenomenon of stepwise sarcomere shortening using a model that invokes catastrophe theory.

Finally, Natori, who triggered a revolution in approach by inventing the skinned fiber, offers the final paper. When immersed in distilled water, both dehydrated fibers and skinned fibers have the ability to develop tension and shorten. Natori suggests that distilled water-induced contraction may constitute a simple model for the study of the contractile mechanisms, in much the same way that skinned fibers have, heretofore.

STRETCH OF CONTRACTING MUSCLE FIBRES: EVIDENCE FOR REGULARLY SPACED ACTIVE SITES ALONG THE FILAMENTS AND ENHANCED MECHANICAL PERFORMANCE

K.A.P. Edman, G. Elzinga* and M.I.M. Noble+

Department of Pharmacology, University of Lund, Sweden
**Department of Physiology, Free University, Amsterdam, The Netherlands*
+The Midhurst Medical Research Institute, Midhurst, West Sussex, England

ABSTRACT

Single frog skeletal muscle fibres were stretched during fused tetanic contractions. The force increase during stretch exhibited a breakpoint at a mean critical length change of 16.6 nm per half sarcomere that was independent of stretch velocity and sarcomere length. The early decaying extra force after stretch (component 2) was removed by a small quick release, leaving a longer lasting component (component 3). The amplitude of release required increased with time up to the angle in the force record during stretch, was constant for the remainder of the stretch and decreased with time after the end of stretch; it was consistently less than the critical amplitude of stretch (above). Component 3 occurred at sarcomere lengths above 2.3 μm and was amplitude dependent. The final force after stretch was usually higher than the isometric force at the starting length of the stretch. Non-uniformity as a cause of this component was examined by (a) laser diffraction studies which showed sarcomere stretch at all locations and (b) 0.6-0.7 mm long segments along the entire fibre which all elongated during stretch. After stretch the sarcomeres and segments were significantly more stable than during control isometric tetani. Segments which were clamped by a servo system demonstrated component 3. Shortening during contraction followed by stretch back to the starting length led to nearly as much force enhancement as stretch alone, suggesting that component 3 is not due to a passive elastic element recruited during activation. An increase in temperature decreased components 1 (velocity dependent force during stretch) and 2 but increased component 3. The critical length features of component 2 suggest a cross-bridge mechanism. However, the sarcomere length dependence of all components differs from that of isometric force and from predictions based on filament overlap.

INTRODUCTION

In the previous symposium we separated the response to stretch of contracting single frog skeletal muscle fibres into three components: (1) a velocity dependent force present only during stretch, (2) a component which was independent of velocity but was dependent on amplitude of stretch up to 18 nm per half-sarcomere during stretch; at greater amplitudes of stretch, no further rise in force was found. This component decayed away completely about 2 seconds after the end of stretch; (3) a component only present at sarcomere lengths above 2.3 μm. This component was dependent on the amplitude of stretch over the range studied and its force-extension curve depended on the starting sarcomere length. This component decayed to a value which remained higher than control isometric tension at the stretched length in tetani lasting up to 8 sec (Edman, Elzinga and Noble, 1979).

Julian and Morgan (1979) subsequently published evidence suggesting that non-uniform distribution of length changes accounted for some of these effects. We have therefore studied segment length changes in considerable detail (Edman, Elzinga and Noble, 1982) and explored the detailed characteristics of component 2 (Edman, Elzinga and Noble, 1981).

METHODS

Single fibres from the tibialis anterior muscles of Rana temporaria were studied with methods previously described in detail (Edman, Elzinga and Noble, 1978, 1981 and 1982). Changes in segment lengths along the fibre were studied by placing about 10 markers at roughly equal distances (less than 1 mm apart) along the fibre. Segment length was measured continuously by the method of Edman and Höglund (1981).

RESULTS AND DISCUSSION

Component 2 - Decaying Force Enhancement Associated with a Critical Amount of Filament Sliding

The force increase during stretch exhibited a breakpoint at a critical length change (average: 16.6 nm per half sarcomere) that was independent of velocity of stretch and of sarcomere length between 1.8 and 2.8 μm. (Edman, Elzinga and Noble, 1979, 1981). After stretch there was an early decaying force component with a force extension curve similar to that during stretch which disappeared over about 2 seconds. This component was removed by a small quick release leaving a longer lasting component (Fig. 1). The critical amplitude of release required to produce this result was found by clamping the fibre to a load at which there was zero velocity of shortening (Fig. 2). This amplitude increased with time up to the angle in the force record during stretch, was constant for the remainder of the stretch, and decreased with time after the end of stretch (Fig. 3); it was constantly less than the critical amplitude of

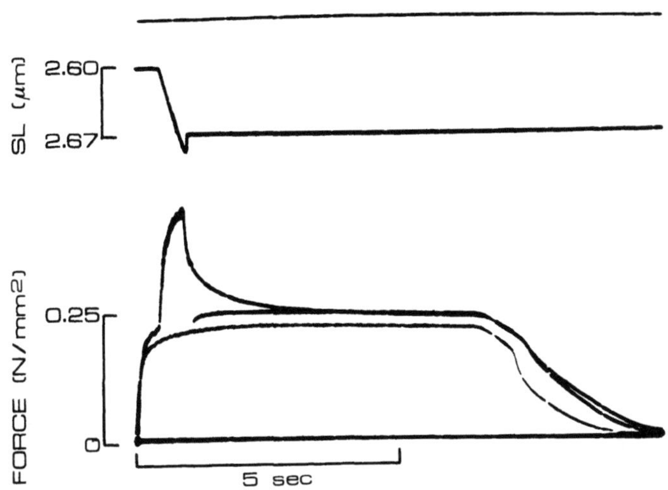

Figure 1: Bottom: 3 superimposed myograms - isometric at the final length, stretch and stretch plus release. Top: fibre length changes for stretch plus release. Scale indicates sarcomere length changes.

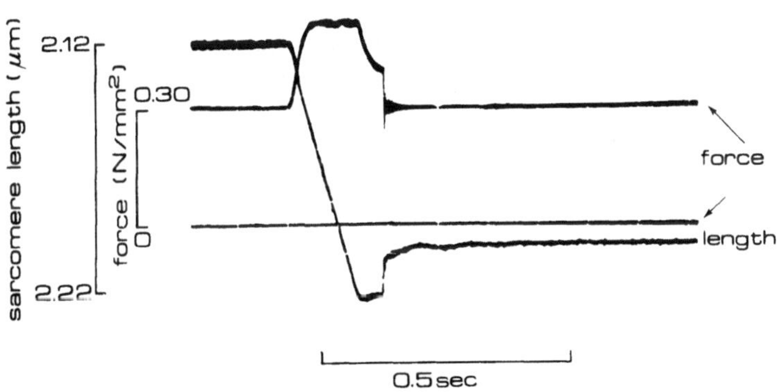

Figure 2: Stretch followed by release to a load clamp producing zero velocity of length change.

stretch required to reach the break point of enhancement during stretch (Fig. 3) but was also independent of sarcomere length (Fig. 4a).

The finding of critical amplitudes of sarcomere length change which are independent of sarcomere length (eg. Fig. 4) is perhaps amongst the strongest evidence available that the generation of force in a muscle fibre results from an interaction of fixed periodic structures along thick and thin filaments. The phenomenon cannot be explained by theories depending on generalised forces acting between filaments or by contraction of continuous filaments attached to the Z discs (Ingels and Thompson, 1966; Elliott, Rome and Spencer, 1970; Hoyle, 1966). However, the idea that the

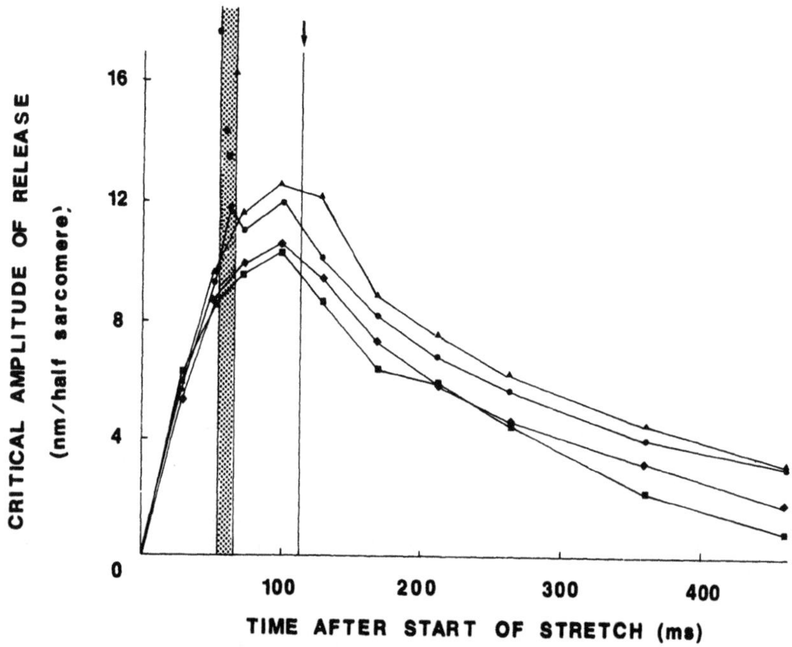

Figure 3: Releases according to the method in Figure 2 were applied at different times after the start of stretch. Arrow indicates the end of stretch. Shaded area indicates time of angle in force record; symbols within this area indicate corresponding amplitudes of stretch.

extra force during stretch is a physical distortion of the cross-bridges does not explain the discrepancy between the critical amplitude of stretch and release (Fig. 3). Nor does the idea that fibre force depends on cross-bridge number explain the fact that the force drop accompanying the critical release showed a small increase up to an optimum magnitude at 2.4-2.7 μm sarcomere length, with a decrease at longer length (Fig. 4b); i.e., the optimum sarcomere length for component 2 was clearly higher than the 2.2 μm for the length tension curve, where optimum overlap of thick and thin filaments occur.

Component 3 - Long-Lasting Residual Force Enhancement After Stretch

All three components of the stretch response were elicited when a given segment was examined (Fig. 5, Edman, Elzinga and Noble, 1982). If non-uniformities are responsible for component 3 (Julian and Morgan, 1979), they must be occurring within a 0.5-1.0 mm long segment and not *between* such segments. Such non-uniformities were not detected by laser diffraction techniques.

Further features of component 3 exclude non-uniformities as its cause: (1) force after stretch is usually higher than that recorded isometrically at the *starting* length (Fig. 5,6) although all measurable sarcomere lengths and segments lengthen during stretch (Hill, 1978; Edman, Elzinga and Noble, 1978, 1979, 1981, 1982). (2) At any finite load, the fibre

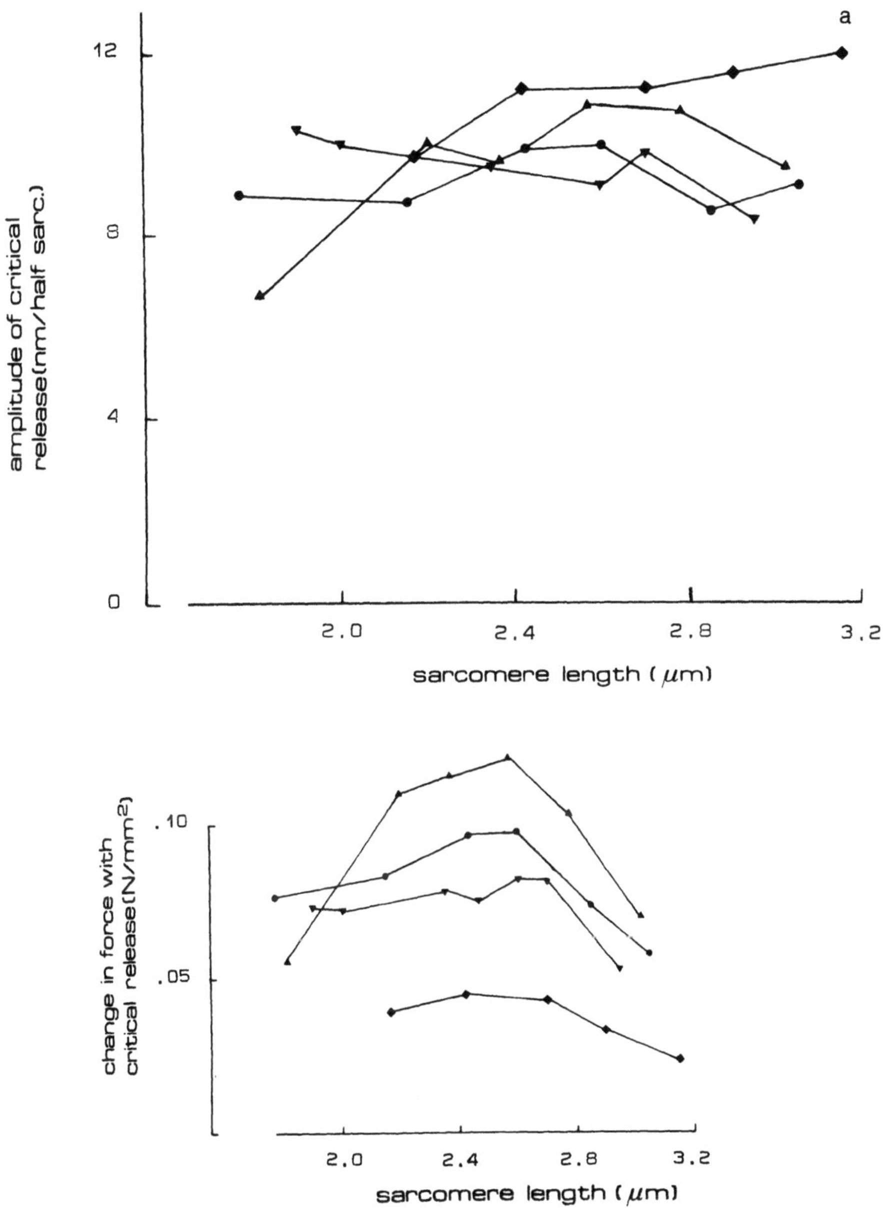

Figure 4: Relationship to sarcomere length of amplitudes of (a) length change and (b) force change accompanying releases performed as in Figure 2.

shortens faster after stretch (Cavagna and Citterio, 1974; Edman, Elzinga and Noble, 1978, 1979; Sugi and Tsuchiya, 1981). (3) Segment length changes and sarcomere length changes (laser diffraction) are much smaller following stretch than during control isometric tetani (Edman,

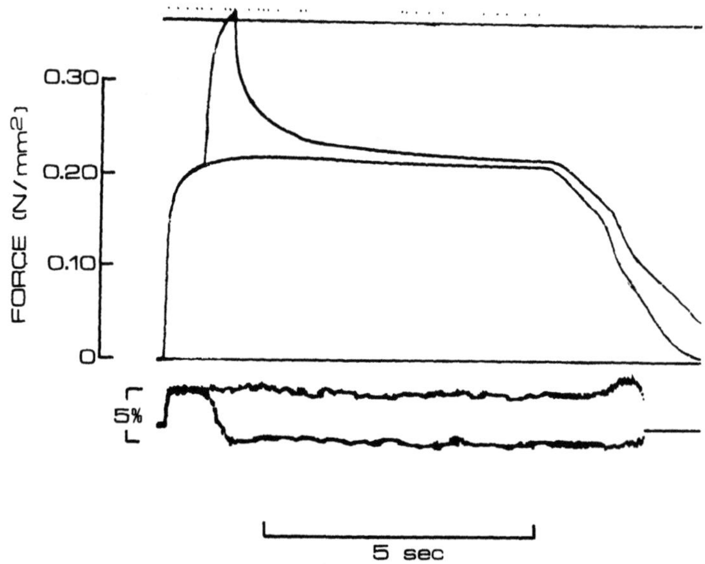

Figure 5: Force and segment length (below) recorded during stretch and during isometric contraction at the short (starting) length.

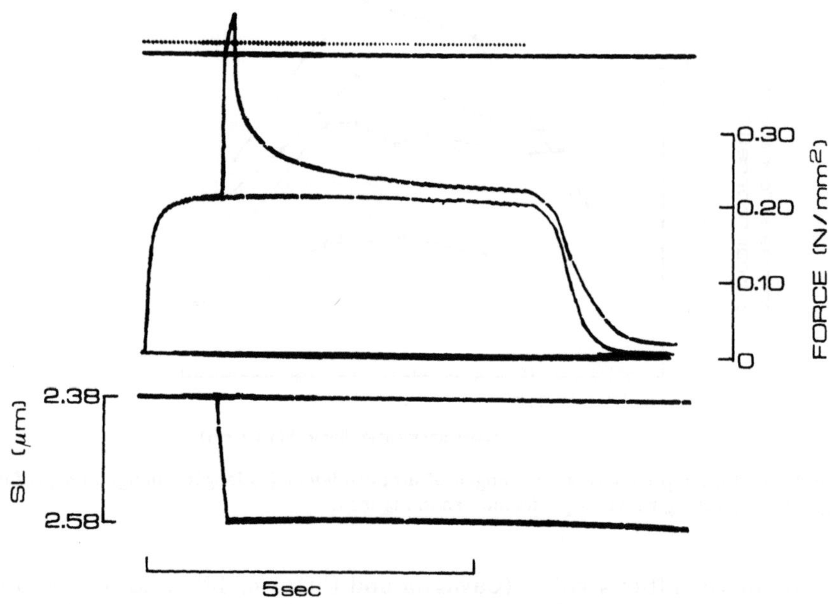

Figure 6: Force and fibre length recorded during stretch and during isometric contraction at the short (starting) length. Resting force is superimposed. Length scale indicates sarcomere length changes.

Elzinga and Noble, 1982). (4) Component 3 does not develop with time as non-uniformities develop. It can be revealed at full amplitude immediately after stretch by applying a small critical release to remove component 2 (Fig. 1). (5) *Isotonic* lengthening under a load greater than isometric tension (P_0) causes the fibre to *shorten* against a load greater than P_0 (Sugi and Tsuchiya, 1981). (6) A further increase in load following such isotonic lengthening is accompanied by a lower initial velocity of lengthening than occurs without such lengthening (Sugi and Tsuchiya, 1981).

What could be the mechanism underlying component 3 - is it an active or a passive phenomenon? The idea that a passive elastic element is recruited during the stretch is suggested by the proportionality between the magnitude of the effect and the increase in force during the plateau phase of the myogram during stretch (Fig. 7). The possibility that this might be a tightening of the sarcolemma was tested by swelling the fibre by immersion in hypotonic solution. No exaggeration of the component was observed. We further argued that if recruitment of any

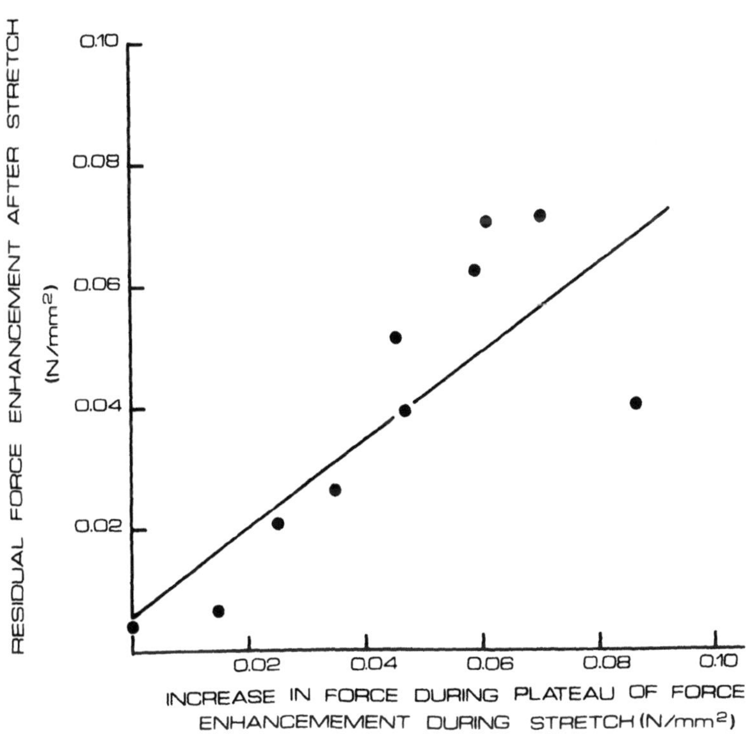

Figure 7: Relationship of component 3 to the increase in force during stretch after the angle in the record.

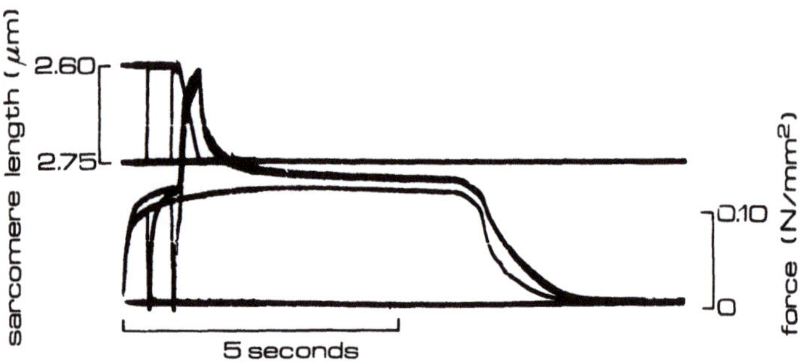

Figure 8: Shortening at two different times before stretch superimposed on stretch alone and an isometric contraction at the long (final length). Note that prior shortening does not influence force enhancement after stretch.

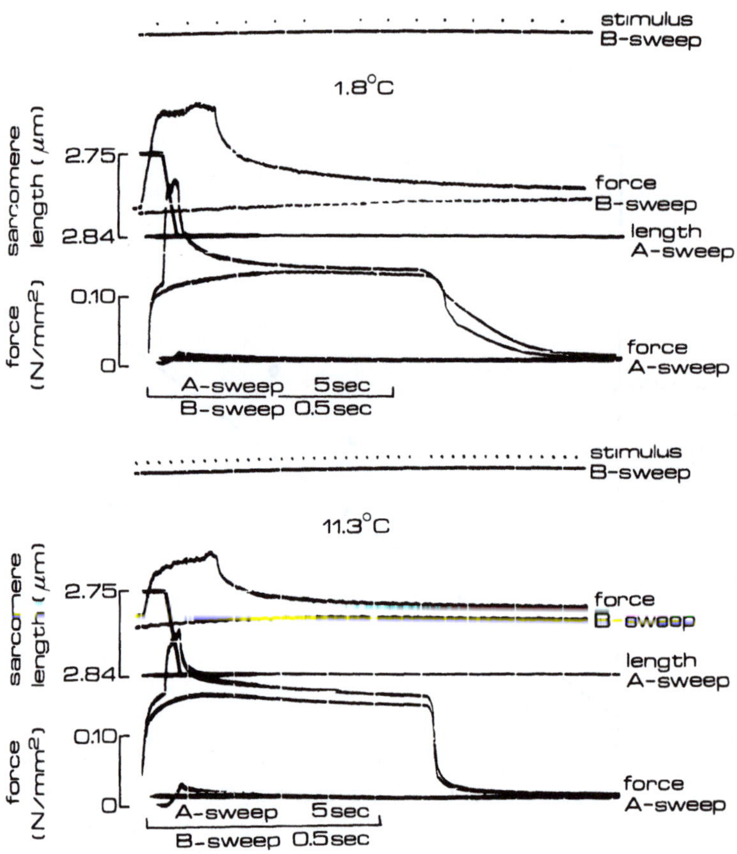

Figure 9: Records at two different oscilloscope sweep speeds and two different temperatures.

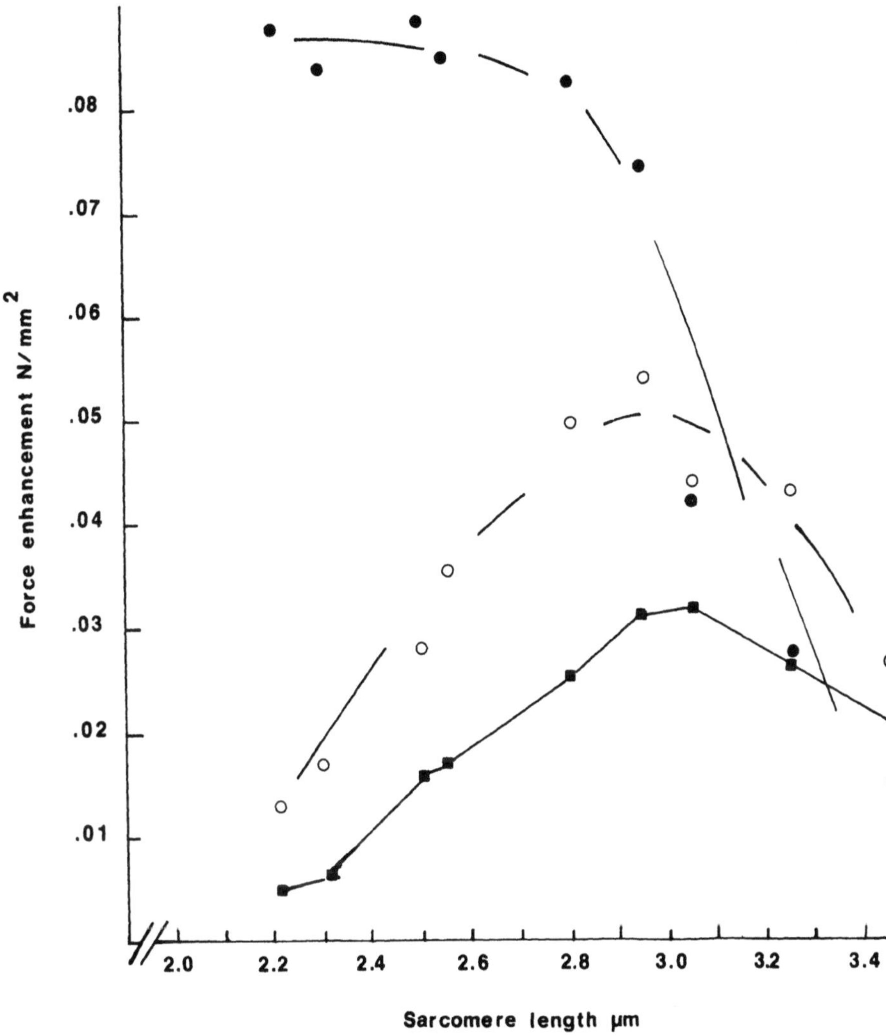

Figure 10: Relationship to sarcomere length of components of the response to stretch. ● component 1, ○ component 2, ■ component 3.

passive structure was involved, activation of the fibre at the final length, release to the short length and finally stretch should result in no force enhancement. However, full force enhancement resulted from this maneuver (Fig. 8).

Relationship of Components 2 and 3 to Cross-Bridge Mechanisms

Differentiation between active and passive processes should be given by studying the effect of an increase in temperature. This produced a decrease in component 2 and an increase in component 3 (Fig. 9). The evidence presented in the section concerning component 2 (above), suggested that this feature resides in periodic structures along the filaments. The cross-bridges are the obvious candidates for this, and their enzymatic activity (rate of turnover) is increased by increase in temperature. This results in a decrease in the force generated during and early after stretch (Fig. 9).

In the case of component 3, temperature increases the magnitude of the response but not in the manner predicted by an enzymatic process. Moreover, the velocity of shortening at zero load is not increased by stretch as would be expected if stretch caused an activation of cross-bridge enzymatic activity. The failure of stretch to produce residual enhancement of heat production argues the same case (Curtin and Woledge, 1979). On the other hand, evidence from X-ray diffraction studies (Sugi, Amemiya and Hashizume, 1977) suggests that stretch does not induce any change in the number of cross-bridges interacting with thin filaments. Clearly it is not possible at this time to offer explanations for these effects in terms of cross-bridges. A similar view emerges from the discussion of the subject given by Sugi and Tsuchiya (1981).

A major difficulty confronting attempts to make such explanations is the requirement that cross-bridge number be proportional to overlap of thick and thin filaments. The manifestation of this in terms of the isometric length-tension relationship (Gordon, Huxley and Julian, 1966; Edman, 1966) was consistently met in our experiments with long tetani. However, the components of the response to stretch consistently fail to meet this requirement (Fig. 10).

ACKNOWLEDGEMENTS

Supported by a Wellcome Trust European Collaborative Grant and by a grant from the Swedish Medical Research Council (project No. 14X-184).

REFERENCES

Cavagna, G.A. and Citterio, G. (1974). Effect of stretching on the elastic characteristics and the contractile component of frog striated muscle. J. Physiol. 239: 1-14.

Curtin, N.A. and Woledge, R.C. (1979). Chemical change, production of tension and energy following stretch of active muscle of frog. J. Physiol. 297: 539-550.

Edman, K.A.P. (1966). The relation between length and active tension in isolated semitendinosus fibres of the frog. J. Physiol. 183: 407-417.

Edman, K.A.P., Elzinga, G. and Noble, M.I.M. (1978). Enhancement of mechanical performance by stretch during tetanic contractions of vertebrate skeletal muscle fibres. J. Physiol. 281: 139-155.

Edman, K.A.P., Elzinga, G. and Noble, M.I.M. (1979). The effect of stretch of contracting skeletal muscle fibres. In: *Cross-bridge Mechanism in Muscle Contraction*, Ed: Sugi, H. and Pollack, G.H. University of Tokyo Press, Tokyo.

Edman, K.A.P., Elzinga, G. and Noble, M.I.M. (1981). Critical sarcomere extension required to recruit a decaying component of extra force during stretch in tetanic contractions of frog skeletal muscle fibres. J. Gen. Physiol. 78: 365-382.

Edman, K.A.P., Elzinga, G. and Noble, M.I.M. (1982). Residual force enhancement after stretch of contracting frog single muscle fibres. J. Gen. Physiol. 80: 769-784.

Edman, K.A.P. and Höglund, O. (1981). A technique for measuring length changes of individual

segments of an isolated muscle fibre. J. Physiol. 317: 8-9P.
Elliott, G.F., Rome, E.M. and Spencer, M. (1970). A type of contraction hypothesis applicable to all muscles. Nature 226: 417-420.
Gordon, A.M., Huxley, A.F. and Julian, F.J. (1966). The variation in isometric tension with sarcomere length in vertebrate muscle fibres. J. Physiol. 84: 170-182.
Hill, L. (1977). A-band length, striation spacing and tension change on stretch of active muscle. J. Physiol. 266: 667-685.
Hoyle, G. (1966). A new formulation of the mechanism of muscular contraction. Fed. Proc. 25: 465.
Ingels, N.B. and Thompson, N.P. (1966). An electrokinematic theory of muscle contraction. Nature 211: 1032-1035.
Julian, F.J. and Morgan, D.L. (1979). The effect on tension of non-uniform distribution of length changes applied to frog muscle fibres. J. Physiol. 293: 379-392.
Sugi, H., Amemiya, Y. and Hashizume, H. (1977). X-ray diffraction of active frog skeletal muscle before and after a slow stretch. Proc. Japan Acad. 53B: 178-182.
Sugi, H. and Tsuchiya, T. (1981). Enhancement of mechanical performance in frog muscle fibres after quick increases in load. J. Physiol. 319: 239-252.

DISCUSSION

RUEGG: You stretched comparatively slowly. If you made a rapid stretch, would you see a delayed rise in tension similar to that observed in stretch-activation of many vertebrate skinned fibers?

NOBLE: We haven't done that very often with the quite large amplitudes of stretch used in our study. If we stretch very fast, we get a very high peak of force during stretch. The force enhancement after stretch is then slightly less (for components 2 and 3) than that obtained at more reasonable velocities of stretch.

RUEGG: No delayed rise?

NOBLE: No.

KAWAI: Do you think this extra mechanism to stabilize the sarcomere structure is based on the cross-bridges, or something else?

NOBLE: Well that's what I'm asking you.

KAWAI: I think it's in the cross-bridge. And I think it has to do with a dissociation reaction: if the cross-bridge is strained, then the dissociation becomes impossible. Or its rate constant becomes very close to zero, and then stops dissociating. This makes for a protective device, as you mentioned.

NOBLE: Do you imagine that the cross-bridges are just hanging on there to sustain this force?

KAWAI: Yes.

NOBLE: Well, okay. Supposing that is so, you're stretching, they take up strain, they're hanging on there, then you reach the critical angle. and presumably they then come off, do they?

KAWAI: Well, this assumption requires that the bridges don't come off.

NOBLE: Well, how do you get sliding then?

HOMSHER: Curtin and Woledge, I think, did the critical experiment that shows that Kawai's explanation won't work.

NOBLE: Yes, Curtin and Woledge found that the net energy release is not any different after stretch. Isn't that so?

CURTIN: That's right.

EDMAN: The residual force enhancement after stretch (component 3) may be recruited from strained passive structures between the myofibrils. When the fiber goes into isometric contraction, there is some sliding between the myofibrils and the interconnecting structures will be strained. When now the fiber is stretched, force will be produced in these passive structures, and it will be added to the active force. The effect of this stretch may well increase at longer sarcomere lengths, as is actually found experimentally (Fig. 12).

NOBLE: Well, Dr. Elzinga, our absent co-author, did the experiment to test this hypothesis. He imposed a release prior to the stretch, and found as much enhancement as when the release had not been applied (cf. Fig. 8). I think that disproves it.

EDMAN: No, I don't think the explanation I propose is ruled out by this experiment. The passive structure may just assume the same state at a given sarcomere length irrespective of the prehistory during the contraction.

NOBLE: No, I think the argument is that the element is there, recruited at the beginning of stretch. So, if you now relax the muscle and bring it back to that length at which it was recruited, i.e. not stretching the element beyond its original length (at which it was recruited), then there should be no force enhancement. But there is!

HUXLEY: Going back to your previous discussion with Dr. Kawai, I don't see why you should rule out cross-bridges doing it. Supposing it ...

NOBLE: Well, don't get me wrong. I haven't ruled it out. But I haven't yet heard a version of the cross-bridge theory which will mathematically predict this response.

HUXLEY: Well, I don't know about mathematically predicting it, but supposing that there were a region of cross-bridge distortions where the detachment rate was extremely low -- and at the same time supposing that if the cross-bridges were forced even further, I mean an even greater force was applied to them -- they did begin to detach. The ones that were in this intermediate range would detach extremely slowly and give rise to your long-lasting force. Then the extensive release necessary to release that force would probably correspond to the length over which that low detachment rate took place. There would be no reason why that should necessarily be the same distance as you have to stretch the muscle in the first place, before you came to the break in the tension curve.

NOBLE: Well, what do you think happens after the break in the tension curve?

HUXLEY: Well, I should imagine if you wanted to construct a model you could say first of all that you started moving cross-bridges beyond their normal attachment point. And most of them were continuously detaching. And as the extent to which they were distorted increased, their detachment rate increased. So you got to a point where the detachment rate was very high, and at that point most of the bridges kept coming off, so that fixed the initial tension you reach. But that there were always some bridges which wouldn't have come off by then, and they then got carried into the region where they detached extremely slowly.

EDMAN: We have considered a mechanism like the one you propose, but there are a number of observations which are difficult to fit with it. For instance, the residual force enhancement after stretch increases steadily with the amplitude of stretch, and the stretch I refer to is *much* larger than the presumed working range of a cross-bridge, also there is no extra heat production.

HUXLEY: Well if the bridges are on, there is no reason why there should be extra heat.

EDMAN: That's true, but the residual force enhancement after stretch increases with the sarcomere length at which the stretch is performed. A maximum is obtained at about 3.0 μm sarcomere length. That is, the amount of extra force produced by stretch increases with *decreasing* area of overlap.

GULATI: What is the shortest stretch you need to see the steady enhancement of tension?

NOBLE: It is a continuous function of amplitude of stretch, but we've usually used fairly big stretches to see it.

HUXLEY: I was going to ask whether the change in behavior with sarcomere length is sensitive to interfilament spacing.

NOBLE: Well, we did some experiments with hypotonic solutions to swell the fibers. All that seemed to happen was that everything was in proportion. There was a higher tetanic force, and all these components were increased proportionally.

VELOCITY SENSITIVITY OF YIELDING DURING STRETCH IN THE CAT SOLEUS MUSCLE

T.R. Nichols

Department of Kinesiology, University of Washington, Seattle, Washington 98195

ABSTRACT

Yielding of muscle force occurs when a muscle is stretched by a sufficiently large amount. A higher velocity of stretch requires a larger stretch to cause yielding, and the preyield stiffness is constant. The yield is attributed to cross-bridge disengagement and the velocity sensitivity to a mechanical filtering of this event.

During ramp stretch the stiffness of an activated muscle will eventually fall abruptly (Rack and Westbury, 1974; Sugi, 1972). At unfused activation rates, this yielding is accentuated and may result in a decrease in force as well as stiffness and a further drop in force after the length is held at the new value. (Joyce, et al., 1969; Nichols and Houk, 1976; Flitney and Hirst, 1978). Both the force change at the time of the first yield (FR) and the stretch required to produce it (LR) increase with velocity (Figure 1; Nichols, 1974; cf. Flitney & Hirst, 1978). Moreover, the drop in force occurring after the ramp is held increases in parallel with the first yield. The preyield stiffness, calculated by the ratio yield force change/yield length change, remains relatively constant over a range of velocities and stimulation rates.

If it is assumed that the cross bridges are forcibly detached by a fixed amount of stretch, that they have viscoelastic properties and that there is an undamped series elasticity, then the velocity sensitivity of the yielding observed externally can be predicted reasonably well as shown in Figure 1. The solid lines represent the behavior of a simple viscoelastic system subjected to ramp stretch. The lines show the overall length and force changes of the system when the internal length (XR) reaches a fixed critical value at each velocity. The symbols represent experimental points. For the experiment shown, the model predicts that the critical internal stretch is 75 μm, which works out to be about 3 nm per half sarcomere according to histological measurements for this muscle made by

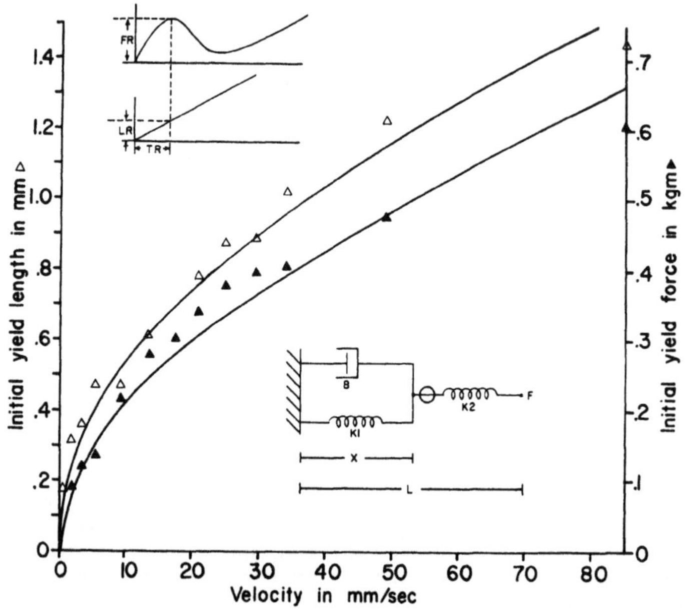

Figure 1: Velocity sensitivity of initial yield length and force change in cat soleus muscle. Stimulation: 8 pps, distributed (Rack and Westbury, 1969). Temperature: 37°C. Input: ramp stretch, 3.4 mm, velocity variable. Upper inset illustrates measurement of force change (FR, ▲) and length change (LR, △) at initial yield. These values are plotted versus velocity of the ramp. Solid lines represent the change in force and length of the viscoelastic system shown in the lower inset at a point where the internal length, X, reaches a fixed value, XR, during each ramp stretch.

Rack and Westbury (1969). This value is in the range of the stretch needed to produce the "give" in frog muscle (7.5-8.5 nm phs, Flitney & Hirst, 1978).

The value of the preyield stiffness was found not to vary systematically with stimulation rate. If stimulation rate influences the number of cross bridges attached at any given time, then the invariance of preyield stiffness with rate suggests that the series elasticity lies predominantly in a common structure rather than largely in the cross bridges themselves. About one quarter of the preyield stiffness can be accounted for by the stiffness of the tendon.

This study suggests that yielding is due to the mechanical detachment of cross bridges. The increase in stretch required to produce the yield with velocity may be due to a mechanical filtering effect of the viscoelastic properties of the cross bridges interacting with the series elasticity.

REFERENCES

Flitney, F.W. & Hirst, D.G. (1978). Cross-bridge detachment and sarcomere 'give' during

stretch of active frog's muscle. J. Physiol. 276: 449-465.
Joyce, G.C., Rack, P.M.H. & Westbury, D.R. (1969). The mechanical properties of cat soleus muscle during controlled lengthening and shortening movements. J. Physiol. 204: 461-474.
Nichols, T.R. (1974). Soleus Muscle Stiffness and Its Reflex Control (Ph.D. Thesis). Cambridge: Harvard University.
Nichols, T.R. & Houk, J.C. (1976). Improvement in Linearity and regulation of stiffness that results from actions of stretch reflex. J. Neurophysiol., 39: 119-142.
Rack, P.M.H. & Westbury, D.R. (1969). The effects of length and stimulus rate on tension in the isometric cat soleus muscle. J. Physiol. 204: 443-460.
Sugi, H. (1972). Tension changes during and after stretch in frog muscle fibers. J. Physiol. 225: 237-253.

FORCE-VELOCITY RELATION AND STIFFNESS IN FROG SINGLE MUSCLE FIBRES DURING THE RISE OF TENSION IN AN ISOMETRIC TETANUS

Carlo Ambrogi Lorenzini, Francesco Colomo[§] and Vincenzo Lombardi

Instituto di Fisiologia, Università degli Studi di Firenze, Viale G.B. Morgagni 63, I-50134 Firenze, Italy
[§]to whom correspondence should be sent.

ABSTRACT

The force-velocity *(T-V)* relation and the force-extension (T_1) relation are determined at preset times and at increasing isometric tensions during a tetanic contraction in frog single muscle fibres in which the passive compliance in series with the sarcomeres was made very small.

The slope of the instantaneous T_1 relation, the fibre stiffness, increases roughly in proportion to the level of the rising isometric tension at which the measurements were made.

The value of V_0 (the velocity of shortening under zero load) is time-independent, whereas the force T exerted during shortening at any velocity V lower than V_0 increases gradually with time after the beginning of the tetanus volley and attains its steady state level before the isometric tension has attained the tetanus plateau and the fibre stiffness its final value.

It is concluded that the delay of the development of the isometric tension and of the fibre stiffness with respect to the development of the *T-V* relation is determined by a specific factor of the contractile process. It is interesting to note that in a cross-bridge model of contraction, in which the value of the rate constant for cross-bridge formation is moderate, the recruitment of actin sites which is measured by the characteristics of the instantaneous *T-V* relation, is expected to lead significantly the actual cross-bridge formation, which is measured both by the instantaneous isometric tension and by the instantaneous stiffness.

INTRODUCTION

The present report deals with the development of the contractile process in intact skeletal muscle fibres of the frog. In terms of cross-bridge models of contraction the force-velocity (*T-V*) relation and the force-extension (T_1) relation are used here to obtain information about the mechanisms of activation and of force production (Huxley, 1957, 1974; Julian, 1969; Julian and Sollins, 1973; Podolsky and Nolan, 1973).

Previous work on this subject (Cecchi, Colomo and Lombardi, 1978, 1981; Cecchi, Colomo, Lombardi and Piazzesi, 1979) has shown that during the rise of tension in an isometric tetanus the force- velocity relation attains gradually its steady state characteristics and that at any time there is a significant delay of the development of isometric tension with respect to the development of the T-V relation. The experiments described here were performed in order to investigate the nature of this delay. The results obtained, while excluding that the delay between the development of the T-V relation and the development of the isometric tension may be attributed to the presence in the fibres of a significant passive compliance in series with the sarcomeres, indicate that it is determined by a specific event of the contractile process. This view agrees with the predictions of a cross-bridge model of contraction in which the value of the rate constant for cross-bridge formation is moderate (Huxley, 1957).

METHODS

Techniques and methods of procedure are in great part similar to those described in a previous paper (Cecchi et al., 1978). Experiments were performed at about 4.5°C on over-all directly stimulated single fibres isolated from anterior tibialis muscle of the frog (*Rana esculenta*). Stimuli of alternating polarity, 0.5 ms duration and 1.5 times the threshold strength were used. In order to minimize the amount of compliance in series with the sarcomeres special care was taken in dissecting and mounting the muscle fibres to make the length of the tendon attachments as short as possible. In all fibres used here the length of both tendons was not greater than about 350 μm. Moreover, the connections between fibre tendons and tension or length transducers were made by means of the aluminum foil clips described by Ford, Huxley and Simmons (1977). The T-V relation and the T_1 relation were determined at preset times during an isometric tetanus either at low initial tensions or at the plateau. Controlled-velocity releases for determining the T-V relation and step-length changes for determining the T_1 relation were imposed on the muscle fibres by means of a loudspeaker-length transducer servo system derived from that described previously by Cecchi, Colomo and Lombardi (1976). Step-length changes of 150 μm were complete in about 150 μs. Tensions were measured by means of a capacitance-gauge transducer similar to that described by Huxley and Lombardi (1980). The resonance frequency of the tension transducers used here ranged from 40 to 50 kHz. In all the experiments the average sarcomere length of the resting muscle fibres was about 2.25 μm. The outputs from tension and length transducers were recorded on a digital oscilloscope (Nicolet, Explorer III) and stored in its floppy disc memory. Measurements of the responses were made directly by means of the internal recording system of the Explorer oscilloscope.

V_0 (μm/s per half-sarcomere) is the smallest velocity of release required to drop and to maintain at zero the isometric tension or, conversely, the velocity of shortening under zero load. Δl (nm per half-sarcomere) is the amount of step-release required to drop to zero the isometric tension developed during a tetanic contraction. T_r is the isometric tension developed during a tetanic contraction at the times when releases were imposed to the fibre. T_1 is the instantaneous tension attained in response to a step-length change. T_0 is the observed value for

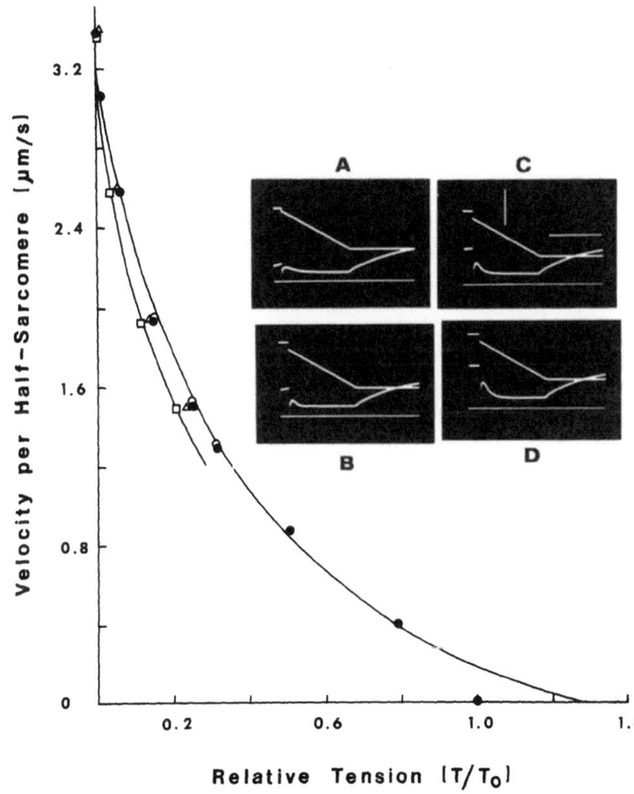

Figure 1: Relative *T-V* relations determined at different times during the rise of tension and the plateau in an isometric tetanus in a muscle fibre at 4°C. Squares, triangles, empty circles and filled circles refer to *T-V* data-points obtained respectively 40 ms (T_r=0.41T_0), 55 ms (T_r=0.59T_0), 80 ms (T_r=0.82T_0) and 220 ms (T_r=T_0) after the beginning of the stimulus volley. The curve fitted to filled circle data-points was drawn by means of Hill's hyperbolic equation and then scaled on the T/T_0 axis by a factor of 0.76 to fit square data-points. The large discrepancy between T_0 and T_0^* is due to the non-hyperbolic behaviour of the plateau *T-V* relation in the high load region (Edman et al. 1976). The inset shows sample responses (middle traces) to releases imposed 40 ms (*A*), 55 ms (*B*), 80 ms (*C*) and 220 ms (*D*) after the beginning of the stimulus volley. Top traces are length changes. Bottom traces are resting tensions. The velocity of the shortening was 1.49 μm/s per half-sarcomere. T_0: 220 kN/m^2; stimulation frequency: 14/s; L_0: 6.98 mm; sarcomere length: 2.25 μm; cross-sectional area: 7,677 μm^2. Vertical calibration: 180 kN/m^2 or 35 nm per half-sarcomere. Time calibration: 20 ms.

plateau tetanic tension. T_0^* is the intercept on the load axis for Hill's hyperbolic equation (Hill, 1938) which was calculated using a computer program (Minuit, CERN computer, series 6000) to find T_0^* and the values for the constants a and b by direct searching.

RESULTS

Figs. 1,2, and 3 illustrate the results of an experiment performed at 4°C, in which T-V and T_1 relations were determined at various times during an isometric tetanus. Comparable results were obtained in all the six fibres used during the course of this work.

The T-V relations (Fig. 1) were determined 40, 55, 80 and 220 ms after the beginning of the tetanus volley. At these times the isometric tension had risen, respectively, to 0.41 T_0, 0.59 T_0, 0.82 T_0 and T_0. It was confirmed (Cecchi, Colomo and Lombardi, 1978) that the value of V_0 was

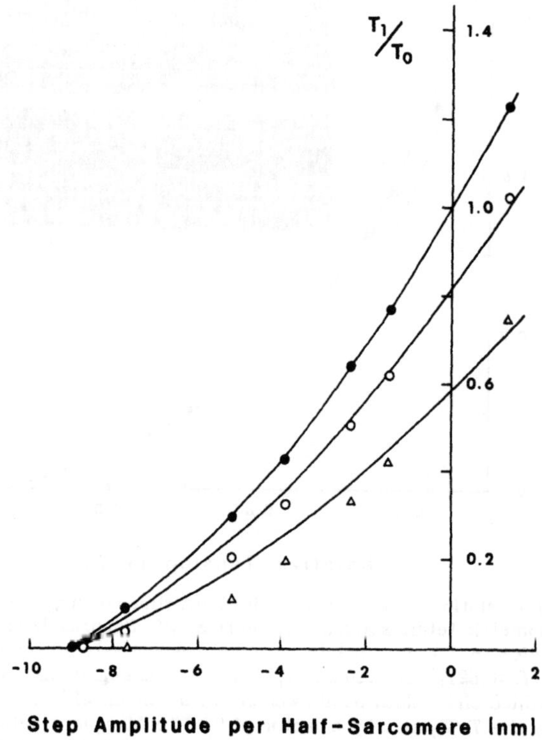

Figure 2: T_1 relations determined for the same muscle fibre as in Fig. 1 at different isometric tensions during the development and the plateau of a tetanic contraction. Triangles, empty circles and filled circles refer to measurements made respectively at 0.59 T_0, at 0.82 T_0 and at T_0. The curve fitted to filled circle data-points is a parabola, calculated by means of the same computer program as that used for Hill's hyperbola, and scaled down on the ordinate according to the isometric tension developed at the times when step-length changes were imposed to the fibre.

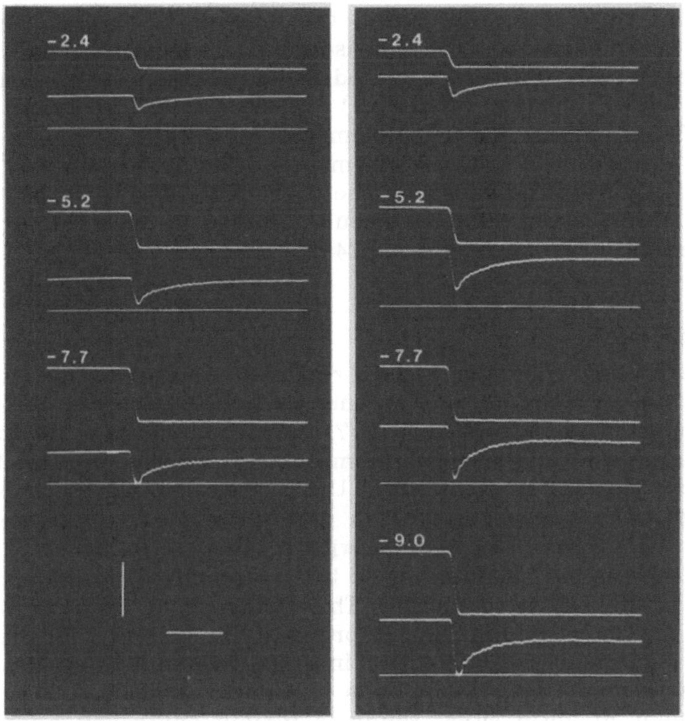

Figure 3: T_1 records obtained with different amount of step-release at 0.59 T_0 or 55 ms (left-hand column) and at T_0 or 220 ms (right-hand column) from the same experiment as in Figs. 1 and 2. In each frame the upper trace measures the length change, the middle trace represents the tension response and the lower trace is the resting tension. Figures in each frame indicate the amount (nm per half-sarcomere) of step-release imposed to the fibre. Vertical calibration: 220 kN/m^2 or 8 nm per half-sarcomere. Time calibration: 1 ms.

time-independent, whereas the steady force T exerted at any velocity of shortening V lower than V_0 increased with time after the beginning of stimulus volley and attained its steady state value before the isometric tension had attained the plateau. Since T-V data-points obtained at 55 ms (T_r=0.59T_0) were practically indistinguishable from those obtained later at 80 ms (T_r=0.82T_0) or at 220 ms T_r=T_0 it was concluded that in this fibre at 4°C the T-V relation attained its steady state characteristics at about 55 ms, well before the isometric tetanus tension had attained the plateau.

The T_1 relations (Fig. 3) were determined at 55 ms (T_r=0.59T_0), at 80 ms (T_r=0.82T_0) and at 220 ms (T_r=T_0). It is clear from Fig. 3 that the slope of the T_1 relations (the fibre stiffness) increased roughly in proportion to the level of the rising isometric tension at which step-releases were imposed to the fibre even beyond 55 ms after the beginning of the stimulus volley, or 0.59 T_0, when the T-V relation attained its steady state characteristics.

In the muscle fibre of Fig. 3 the amount of step-release per half-sarcomere required to drop the isometric tension to zero was 7.7 nm (observed) or 5.1 nm (extrapolated from the linear part of the T_1 relation) at 0.59 T_0, 8.8 nm (observed) or 5.66 nm (extrapolated) at 0.82 T_0 and 9.05 nm (observed) or 6.12 nm (extrapolated) at T_0. The observed and the extrapolated values of Δl (means and standard errors) for the six fibres used during the course of the present work were respectively: 7.00 nm ± 0.41 or 4.72 nm ± 0.27 at about 0.38 T_0; 8.22 nm ± 0.16 or 5.84 nm ± 0.10 at about 0.76 T_0; 8.62 nm ± 0.24 or 6.16 nm ± 0.09 at T_0.

DISCUSSION

The finding that the values reported above for the step-release required to drop the plateau tetanic tension to zero are comparable to those reported by Ford et al. (1977) for muscle fibres at about the same temperature and under length clamp conditions should imply that in the fibres used for the present work, the passive compliance in series with the sarcomere was negligible. In other words, the large delay of the development of isometric tension with respect to the development of the *T-V* relations is not attributable to the presence of a passive compliance in series with the sarcomeres. This delay could be determined by a specific factor of the contractile process. For instance, in the case of a cross-bridge model of contraction in which the value of the rate constant for cross-bridge formation (f_1) is moderate (Huxley, 1957, 1974) the recruitment of actin sites, which is mainly controlled by the rate of release of Ca^{2+} and is measured by the characteristics of the instantaneous *T-V* relation (Cecchi et al., 1978, 1981), is expected to lead significantly the actual cross-bridge formation, which is measured by the instantaneous isometric tension or by the instantaneous fibre stiffness (Huxley, 1957, 1974). In this connection, evidence that the value of f_1 is moderate and therefore responsible for the slower development of isometric tension with respect to that of the *T-V* relation is given by the observation that the fibre stiffness, i.e. the actual number of attached cross-bridges, increases roughly in proportion to the rising tetanic tension.

It must be noted however that in accordance with the results of previous work of Cecchi, Griffiths and Taylor (1982) the value of Δl was significantly greater the higher the level of tension at which step-releases were imposed, as if a small series compliance was still present in the fibres and therefore during the tetanus rise the fibre stiffness was increasing more steeply than the isometric tension. These findings disagree with the results of previous work of Huxley and Simmons (1973) showing that under length clamp conditions the fibre stiffness increases proportionally to the tension developed. The reasons for this discrepancy are not clear. It is possible that the faster rise of fibre stiffness with respect to that of isometric tension may be due to a small series compliance still present in the fibres, but it can not be excluded that this lag may be determined by other events of the contractile process (Cecchi et al., 1982).

ACKNOWLEDGEMENTS

The authors wish to thank Mrs. C. Lelmi for typing the manuscript and Mr.A. Vannucchi for the preparation of the illustrations. This work was supported by grants from the Consiglio Nazionale delle Ricerche, from the Council Board of Florence University and from the Ministero della Pubblica Istruzione.

REFERENCES

Cecchi, G., Colomo, F. and Lombardi, V. (1976). A loudspeaker servo system for determination of mechanical characteristics of isolated muscle fibres. Boll. Soc. Ital. Biol. Sper. 52: 733-736.

Cecchi, G., Colomo, F. and Lombardi, V. (1978). Force-velocity relation in normal and nitrate-treated frog single muscle fibres during rise of tension in an isometric tetanus. J. Physiol. 285: 257-273.

Cecchi, G., Colomo, F. and Lombardi, V. (1981). Force-velocity relation in deuterium oxide-treated frog single muscle fibres during the rise of tension in an isometric tetanus. J. Physiol. 317: 207-221.

Cecchi, G., Colomo, F., Lombardi, V. and Piazzesi, G. (1979). Development of activation and rise of tension in an isometric tetanus. Pflüger's Arch. 381: 71-74.

Cecchi, G., Griffiths, P.J. and Taylor, S.R. (1982). Muscular contraction: the kinetics of cross-bridge attachment studied by high frequency stiffness measurements. Science, N.Y. 217: 70-72.

Edman, K.A.P., Mulieri, L.A. and Scubon-Mulieri, B. (1976). Non-hyperbolic force-velocity relationship in single muscle fibres. Acta Physiol. Scand. 98: 143-156.

Ford, L.E., Huxley, A.F. and Simmons, R.M. (1977). Tension responses to sudden length change in stimulated frog muscle fibres near slack length. J. Physiol. 269: 441-515.

Hill, A.V. (1938). The heat of shortening and the dynamic constants of muscle. Proc. Roy. Soc., B. 126: 136-195.

Huxley, A.F. (1957). Muscle structure and the theories of contraction. Progr. Biophys. biophys. Chem. 7: 255-318.

Huxley, A.F. (1974). Muscular contraction (Review lecture). J. Physiol. 243: 1-43.

Huxley, A.F. and Lombardi, V. (1980). A sensitive force-transducer with resonant frequency 50 kHz. J. Physiol. 305: 15-16P.

Huxley, A.F. and Simmons, R.M. (1973). Mechanical transients and the origin of muscular force. Cold Spring Harbor Symp. Quant. Biol. 37: 669-680.

Julian, F.J. (1969) Activation in a skeletal muscle contraction model with modification for insect fibrillar muscle. Biophys. J. 9: 547-570.

Julian, F.J. and Sollins, M.R. (1973). Regulation of force and speed of shortening in muscle contraction. Cold Spring Harbor Symp. Quant. Biol. 37: 635-646.

Podolsky, R.J. and Nolan, A.C. (1973). Muscle contraction transients, cross-bridge kinetics and the Fenn effect. Cold Spring Harbor Symp. Quant. Biol. 37: 661-668.

DISCUSSION

TIROSH: As we have heard, it is possible to enhance tension generation by stretch. Did you check the force-velocity relation during force enhancement after stretch?

COLOMO: No, I didn't. But previous experiments performed using whole muscles (G. Cavagna and G. Citterio: J. Physiol. 239, 1-14, 1974) or single fibers (Edman et al., J. Physiol. 281: 139-155, 1978) have shown that potentiation by stretch is associated with shift towards higher force values of the force-velocity relation. The velocity of shortening at zero load remained unchanged.

TREGEAR: I was just wondering, are you actually suggesting that the crossbridges have a different property during the beginning of contraction, that is, the average crossbridge property is different during the rising phase of a tetanus?

COLOMO: No, I am not suggesting that. We have shown that the tendon compliance cannot be responsible for the delay of the development of the isometric tension with respect to that of the force-velocity relation. This delay should be determined at least in part by a specific activation factor. In Huxley's crossbridge model, in which the rate constant controlling the crossbridge formation (f_1) is taken to be relatively small, the development of the force-velocity relation, which measures the recruitment of actin sites for crossbridge formation, is expected to lead the development of isometric tension, which measure the actual number of force-generating crossbridges. In this way the development of the force-velocity relation and in turn that of fiber stiffness and of isometric tension may be attributed to a time-dependent increase in the number of actin sites available for crossbridge formation.

WARSHAW: Am I right that your data is opposite to that of Cecchi?

COLOMO: If you look carefully at our data you can see that during the tetanus rise, fiber stiffness and isometric tension increase together. However, there is no doubt that the development of fiber stiffness leads that of isometric tension. This result is the same, not the opposite, as that communicated here by Cecchi et al. The point to note is that while our experiments were designed to investigate the time relations between the development of activation and that of stiffness and isometric tension, the experiments by Cecchi et al. were designed to study the mechanics of crossbridges.

STEPWISE SHORTENING: EVIDENCE AND IMPLICATIONS

Gerald H. Pollack, Reuven Tirosh, Frank V. Brozovich, Joan W. Lacktis, Robert C. Jacobson, and Tsukasa Tameyasu

Division of Bioengineering & Department of Anesthesiology WD-12, University of Washington, Seattle, Washington 98195

ABSTRACT

The observation that sarcomeres shorten in steps has proved controversial. On the one hand, the phenomenon implies that the contractile process cannot be based on a molecular mechanism that behaves in a random manner: The fact that the steps and pauses characterize the kinetics of large volumes of tissue implies that the elements comprising such volumes must stop and pause synchronously. On the other hand, since current contractile models do not anticipate synchronized behavior, there has been considerable speculation that the phenomenon might not be a genuine feature of contraction, but an instrument-based artifact. We present here a review of observations made with four methods that have been brought to bear on the question. All four show discrete, synchronized contractile behavior. The observation of steps with multiple independent methods implies either that each technique harbors its own "gremlin" that generates spurious steps and pauses of a similar nature, or that the phenomenon is genuine. Finally, some consistent properties of the distribution of step size are considered with respect to possible molecular models.

The waveform of sarcomere shortening is seldom smooth. We first noticed several years ago, using a high resolution diffractometer, that the shortening pattern often consisted of a series of "steps" interspersed with periods of "pause" during which there was little or no shortening (Pollack et al., 1977). The waveform resembled a staircase.

The finding was serendipitous and, at first, puzzling. Because the incident laser beam encompassed thousands of myofibrillar sarcomeres, the stepwise behavior implied extraordinary synchrony; for otherwise any locally discrete behavior would have been smoothed out when measurements were made over as large a number of sarcomeres as they were. Such synchronous behavior is not necessarily anticipated by current models of contraction (A.F. Huxley, 1957; H.E. Huxley, 1969), as cross-bridges are presumed to behave in an independent, random manner. Thus, the observation raised many questions.

Our thrust during the past several years has been in two directions: first, to verify that the phenomenon was not an experimental artifact, as had been suggested earlier (Rüdel and Zite-Ferenczy, 1979); this was accomplished by interrogating the muscle using a variety of methods, each confirming discrete, synchronous behavior. The second goal has been to establish the basic features of the phenomenon, particularly the size of the step and the factors that affect it. This paper is intended to serve as a review of these developments.

Early Work

Early observations were made in both cardiac and skeletal muscle (Pollack et al., 1977; 1979). The specimen was transilluminated with a variably compressed laser beam. A single first order of the diffraction pattern was projected onto either a Schottky barrier photodiode, kymographic film, or, in most experiments, a linear photodiode array from which the centroid of the first order could be located with high time resolution (Iwazumi and Pollack, 1979). In these early experiments, contraction was auxotonic, i.e., sarcomere shortening occurred by virtue of abundant compliant tissue adjacent to fixed clamps. Examples of records of sarcomere length changes obtained using the diffraction method are shown in Figure 1.

The suggestion by Rüdel and Zite-Ferenczy (1979) that the discontinuous feature of such records might arise out of reflections off fortuitiously tilted "Bragg planes" rather than as genuine behavior of the contractile mechanism prompted additional work. A normally incident laser beam gives rise to two first orders whose angular separation is inversely related to the striation spacing. The Rüdel - Zite-Ferenczy hypothesis predicts that each first order arises from reflections off appropriately tilted planes of striations, and that certain fortuitous arrangements of these planes could give apparent stepwise behavior. Since each first order necessarily derives from a different set of planes, any fortuitous stepwise behavior in one first order would not ordinarily be predicted to appear in the complementary first order. We checked this by measuring the dynamic behavior of both first orders using a dual diffractometer.

Dual Diffractometry

Single fibers of frog toe muscle were illuminated with a laser beam compressed to the diameter of the fiber. The diffracted light was collected with a 40× water immersion objective. The diffraction pattern emerging from the vertical tube of the microscope was split symetrically with a prism, such that one first order was directed rightward, the other leftward, each onto a separate photodiode array. With this method, we could measure the dynamic behavior of both first orders.

The results were mixed, and therefore inconclusive. On the one hand, occasional records showed stepwise behavior in one first order and smooth behavior in the other, as observed by Rudel and Zite-Ferenczy

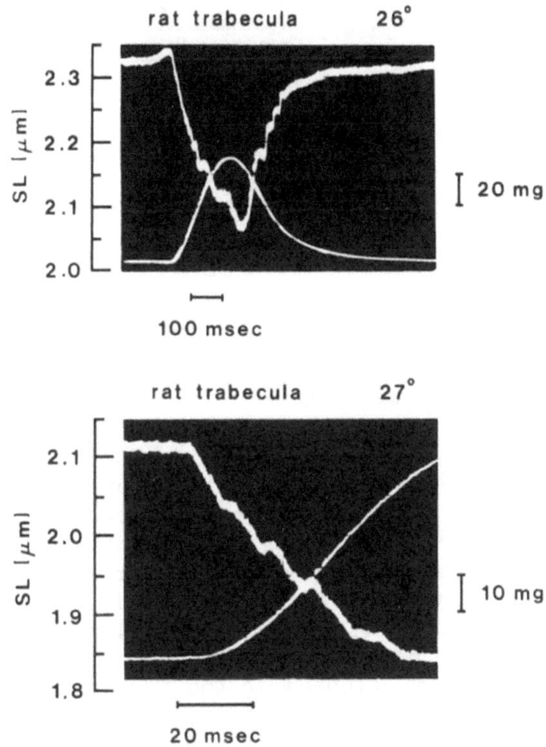

Figure 1: Early records of sarcomere shortening in cardiac muscle obtained using laser diffraction. Sarcomere shortening ocurred at the expense of the compliant ends of the specimen. Note the sharp transitions between steps and pauses.

(1979). The majority showed stepwise behavior in both, but the coincidence of pauses was not always evident. For example, in Fig. 2, some pauses in one of the first orders had counterparts in the other, while other pauses had no counterpart. Quantitative analysis of the degree of coincidence of pauses in the two orders indicated that, within the constraints of the criterion of coincidence (sarcomere length coincidence - within 0.025 μm; time coincidence - over at least 80% of the duration of the pause), 19% (N=219) of the pauses in one first order had counterparts in the other first order.

One interpretation of these results is that there is no general correlation of steps and pauses between the two first orders; when correlation does occur, it is by chance. This interpretation does not seem likely for the following reason. In order to produce a single pause, the Bragg hypothesis requires a coincidence of presumably random events (cf. Rüdel and Zite-Ferenczy, 1979); to produce a cascade of pauses and steps in a single first order requires a string of such coincidences; and to observe counterparts in the two first orders, each arising out of *independent* Bragg planes, demands, in 19% of all cases, what would appear to be an unrealistic degree of coincidence.

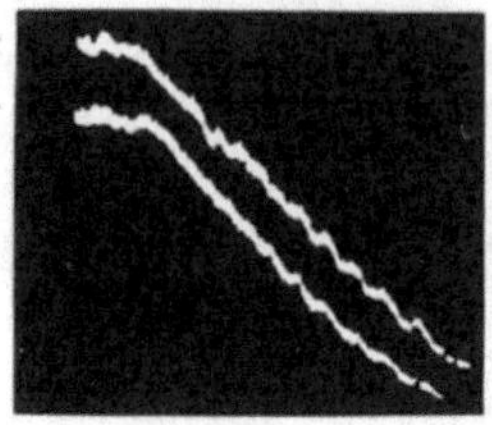

Figure 2: Time course of sarcomere shortening in a single fiber of frog toe muscle as measured by laser diffraction. The records show the waveforms obtained from each of the two first orders in the same region of the fiber. Note that some, but not all, pauses appearing in one waveform appear in the other.

An alternative interpretation is that the observed differences of behavior of the two first orders arise as a natural consequence of striation skew, though not necessarily involving Bragg reflections. If the striations lie normal to the fiber axis, the dynamic behavior of the two first orders should be identical. However, given the natural helicoid fiber structure (Peachey and Eisenberg, 1978) and the unavoidably imperfect mounting of the fiber, some striation skew is inevitable. Striations skewed rightward will contribute more intensity to the right than to the left first order, and those skewed leftward will behave oppositely (Yeh et al., 1980). When the skew is severe, the degree of preferential contribution will increase. Given an abundance of such regions and a distribution of skew angles within the illuminated field, it is evident how the dynamic behaviors of the two first orders might sometimes differ in fine detail. Such differences do not necessarily imply that the information contained in either of the two orders is artifactual.

High Speed Cinemicrography

A more definitive test of the genuine nature of the stepwise shortening phenomenon lies in the analysis of the dynamics of the striation pattern, itself. This work has proceeded two ways, first, by analysis of the striation pattern during shortening, obtained by high-speed cinemicrography (Delay et al., 1981), and later by striation image analysis with an on-line method (Myers et al., 1982).

For the high speed film analysis, single fibers of frog semitendinosus muscle were illuminated with a laser beam. The image of the striations was projected onto the film plane of a high-speed rotating prism camera operating at 4000 frames per second. The optical diffraction pattern, obtained during the same contraction for comparison, was projected through the vertical tube of the microscope onto a photodiode array, from which sarcomere length was obtained as described above.

Each film frame was analyzed using multple scan microdensitometry to obtain the mean striation spacing. Frame by frame analysis then gave the time course of sarcomere shortening. Figure 3 shows one such result.

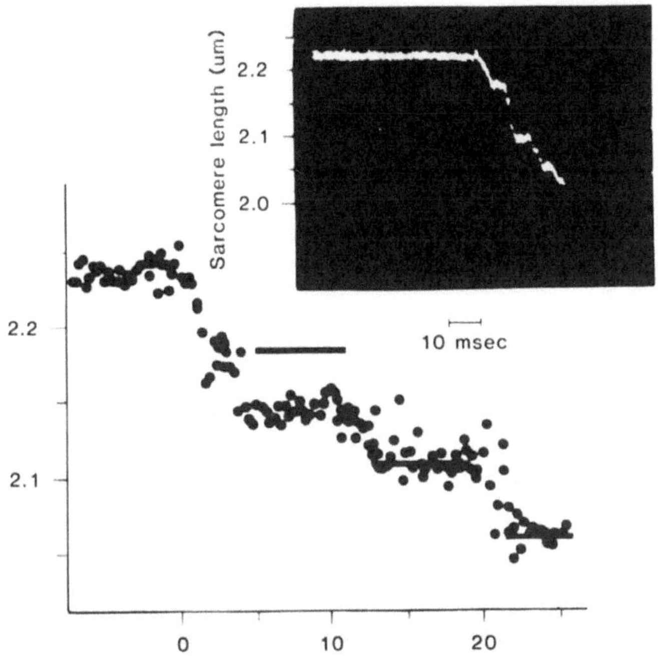

Figure 3: Time course of sarcomere shortening in a single fiber of frog semitendinosus muscle. Records were obtained during the same contraction using different methods. Inset shows shortening waveform obtained by diffraction. Points on the graph were obtained from analysis of successive frames of high-speed cine records of the striation pattern. Bars represent the locations of pauses on the diffraction record.

Pauses and steps are apparent, though the steps obtained with the cine method are not identical to those measured using optical diffraction. The differences presumably follow from what has been discussed above: namely, the laser may not sample the illuminated volume uniformly. Similarly, the volume sampled by the imaging method includes only an optical section through the fiber; thus, the regions sampled by the two methods are unlikely to be identical, and some differences of results are anticipated. Such differences notwithstanding, both methods show qualitatively similar stepwise shortening patterns.

On-Line Analysis

The time demands imposed by cine analysis prompted the development of an on-line approach. The image of the striation pattern, obtained using an incoherent, Xe light source, was projected onto a photodiode array. The array was scanned across the striations every 256 μsec, each scan giving a waveform whose frequency was equal to the spatial frequency of the striated image. This signal was then passed into a phase-locked loop circuit, which extracted the principal frequency and converted it to sarcomere length. The phase-locked loop method is a well

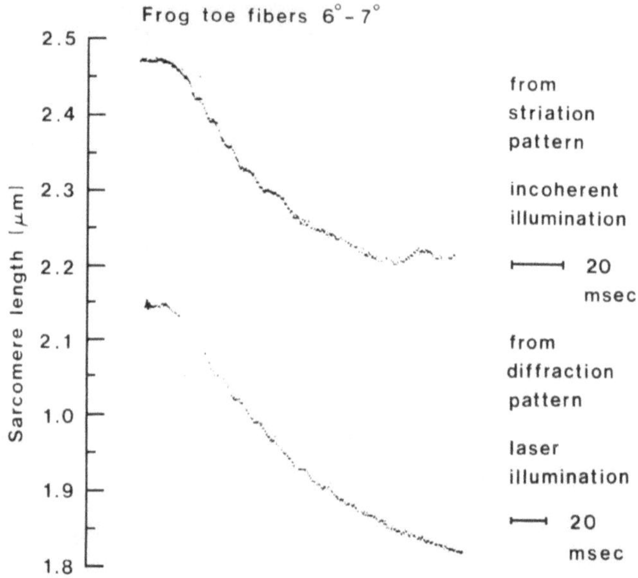

Figure 4: Time course of sarcomere shortening obtained using imaging (top) and diffraction (bottom) methods, both on-line. Comparable features are seen in the two traces. Note slightly different time scales.

known means of extracting the frequency of quasi-periodic signals, remaining effective even when the noise level is relatively high. The method was demonstrated to have excellent linearity, and could resolve sarcomere length changes as small as 1 nm (Myers et al., 1982).

Figure 4 shows an example of a record obtained with this method (top). The patterns of stepwise shortening are qualitatively similar to those obtained with optical diffraction under comparable conditions (Fig. 4, bottom).

Stepwise shortening has now been examined by this method under two modes of contraction: auxotonic contractions, as described above, and lightly loaded isotonic contractions obtained using a lightweight lever. The patterns of shortening were qualitatively similar, but the yield was different. With the auxotonic contractions, stepwise phenomena could be observed in only about half the fibers, while in the lightly loaded isotonic contractions, 80-90% of fibers studied gave abundant stepwise phenomena.

A possible reason for the difference in yield may lie in differences in the influence of "end effects." Since the pattern of steps in any one myofibril depends critically on the load (Vassallo and Pollack, 1982), conditions which exacerbate load nonuniformity across the fiber will tend to give different step patterns in different myofibrils, thereby smoothing the waveform detected from an ensemble of myofibrils. Tension is least likely to be uniformly distributed across the fiber near the ends of the specimen, where asymmetric tendinous insertions and skewed mounting of the tendon will provide nonaxisymmetric load distribution. At an axial distance of perhaps 10 diameters from the attachment point, however, such "end effects" should give way, and tension distribution across the fiber is likely to be considerably more uniform. However, the fibers studied here were only 10-20 diameters in length -- short fibers were specifically chosen to minimize the effects of axial translation during contraction -- so that end effects are unlikely to have ever given way completely.

As a consequence, in experiments in which substantial tension was allowed to develop, thereby placing stress on the tendinous ends, such end effects would be accentuated. In such instances, nonuniform load distribution would be promoted and stepwise phenomena would be least detectable, as was indeed found in the auxotonic contractions. On the other hand, when little or no tension was allowed to develop, so that the ends were barely stressed, load distribution should have tended to remain uniform. This may explain why in fibers shortening under light load stepwise phenomena could be observed so much more regularly.

The same argument may also explain why stepwise shortening can be observed most clearly and most frequently in unstimulated fibers. Figure 5 shows examples of the results of trapezoidal length changes imposed on an unstimulated fiber. Cascades of steps of moderately regular size are seen both during the release and the stretch phase. Figure 6 shows details of a release on an expanded time scale.

Why should steps be present in unstimulated fibers? Some filamentary interaction apparently persists between contractions (D.K. Hill, 1968). Our recent observations indicate that when presented with zero load, unstimulated fibers can shorten about as fast as stimulated fibers

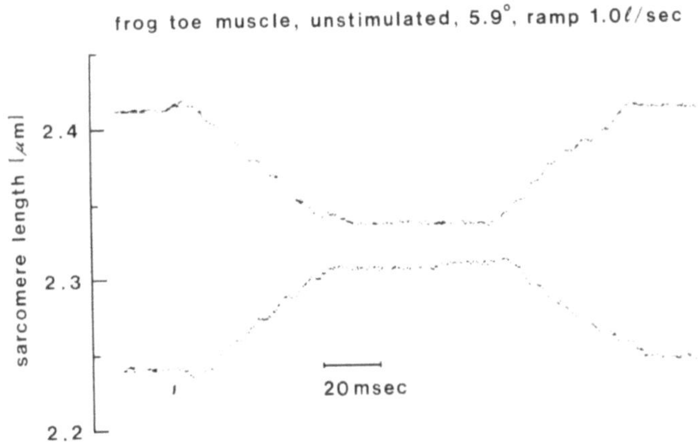

Figure 5: Response to imposition of a trapezoidal change of muscle length in a relaxed fiber. Phase-locked loop sarcomere length detection system was used. In the upper record, release was followed by stretch, while in the lower record stretch was followed by release. Note pauses during shortening and stretch.

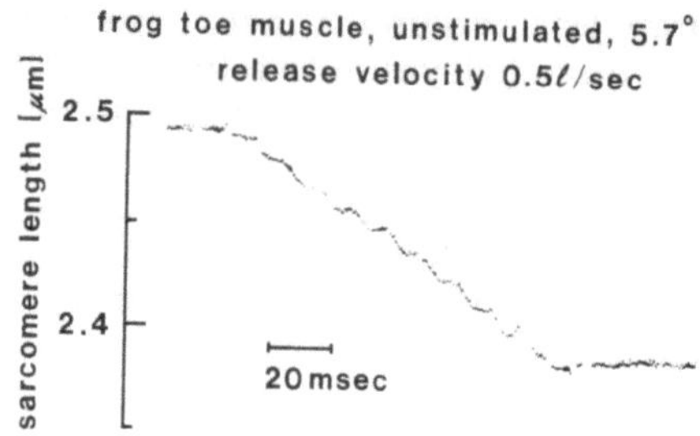

Figure 6: Expanded record of sarcomere length change in response to a trapezoidal change in muscle length. Only the shortening and a segment of the flat portion of the waveform are shown. Note the regularity of the steps and the sharp transitions between pauses and steps.

shortening under similarly unloaded conditions. The observation that V_{max} is comparable in stimulated and unstimulated fibers adds strength to the view that whatever mechanism is responsible for shortening during "contraction" may also be at play during shortening in unstimulated fibers. It is not unreasonable, therefore, to expect stepwise phenomena in unstimulated fibers.

A special advantage in the unstimulated fibers is that axial translation can be almost completely circumvented because imposed length changes are uniformly distributed along the fiber. By peering into the microscope, it has been possible to select regions (near the fixed end) that translate by no more than two striation spacings during the course of a ramp release; essentially, the same region is thus followed throughout the imposed length change. Such regions show stepwise phenomena comparable to those shown in Figures 5 and 6. Details of the shortening pattern are minimally contaminated by the influence of translation of the tissue across the field of view.

Myofibrillar Preparations

We have recently developed an apparatus to measure the dynamics of myofibrillar bundles shortening under light load. Both frog and fish (trout) fibers have been studied. Preparations were glycerinated and bundles were manually teased to obtain preparations typically 1-15 μm wide and 100-200 μm long. Each end of the specimen was wrapped around the tip of a glass micropipette. One pipette was fixed, while the other served as a lever, pivoted at its center with a jewel, and balanced with a light watch spring. The image of the striations was projected onto a photodiode array, and the time course of the sarcomere length was measured using the phase-locked loop method described above. The optics were arranged so that approximately the full width of the preparation filled the narrow dimension of the array. Further, the striated image near the fixed end of the specimen was used. These features minimized the effects of translation.

Although this project is still in developmental stages, a preliminary result is shown in Fig. 7. Upon activation with a solution containing ATP,

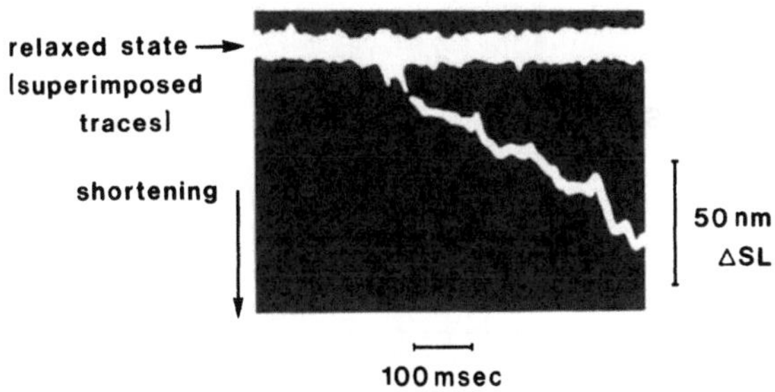

Figure 7: Sarcomere shortening in a thin myofibrillar preparation of frog tibialis anterior muscle. Preliminary data. Although such records are relatively noisy, steps of 10-25 nm, similar to step size found in the intact preparations, seem to appear.

shortening took place. The signals are still sufficiently noisy that definitive conclusions cannot be reached at this stage; however shortening patterns were generally not smooth, but showed one or more discontinuities such as those of Fig. 7.

Sounds

In pursuit of an approach that did not depend on the optical properties of the muscle, we considered measurements of sound. Since fiber shortening is associated with lateral expansion of the filament lattice, we reasoned that each rapid shortening step might be associated with an equally rapid expansion of the lattice. Such a stepwise increase of fiber cross-section would generate a shock wave in the surrounding bath, which ought then to be detectable with a submersed microphone.

Using frog sartorii shortening under light load, we confirmed that shortening was associated with the generation of a series of discrete sound bursts (Brozovich and Pollack, 1983). An example is shown in Figure 8. Various control experiments were carried out to check for artifacts. For example, the Q_{10} of the interval between sounds was measured and found to be approximately 2.0, consistent with a physiological genesis. Because the sounds were not continuous, but were impulsive, the results added support to the conclusion that the shortening process is both discontinuous and synchronous.

Step Size and Molecular Mechanism

One of the interesting questions is whether the step size relates directly to some significant molecular dimension. Initial results were discouraging in that the step sizes were not uniquely equal to either 14.3 nm or 5.5 nm. On the other hand, it seemed that synchronized, repeatable behavior such as observed, ought necessarily to be dimensionally related to some important molecular structure.

Final results still are not yet at hand, but histograms of step size have turned up potentially interesting features. Of the many histograms generated under a variety of conditions, the most consistent feature has

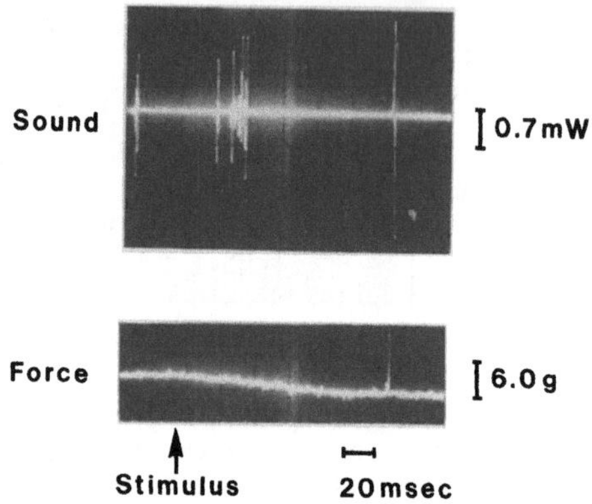

Figure 8: Sounds recorded from a microphone placed beside a sartorius muscle shortening under light load. Note the series of sharp, discrete sounds, indicating synchronous, discrete, contractile behavior.

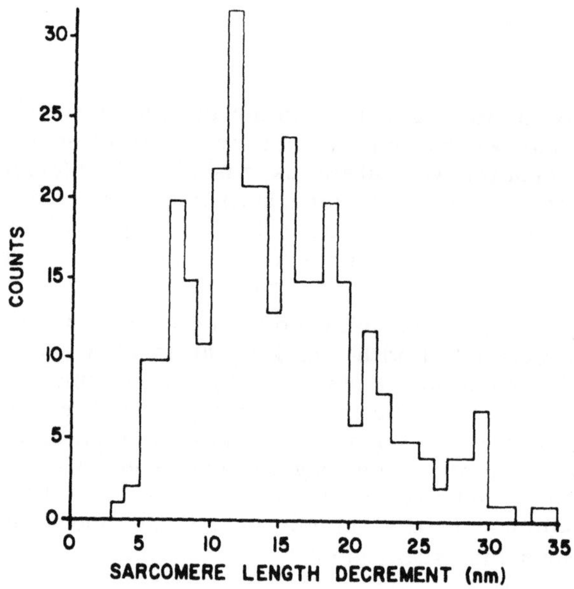

Figure 9: Distribution of step sizes, i.e. of the sarcomere length differences between two consecutive pauses. Data obtained from auxotonic contractions of frog toe muscle. Note the apparent discreteness of the histogram, indicating a series of preferred step sizes. Total count, 328.

been that virtually all histograms are multipeaked; the distribution is rarely smooth. Figure 9 shows an example of a histogram obtained from frog toe fibers contracting auxotonically. Several peaks are apparent. In

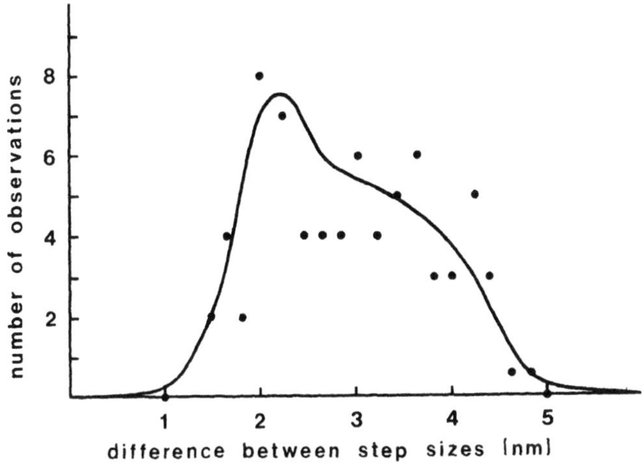

Figure 10: Representative plot of the distribution of the difference between peaks of the histograms such as those of Figure 9, except that data were obtained from unstimulated fibers. Note that the step size differences fall in the range of 1 to 5 nm, with predominance just below 3 nm. This suggests the hypothesis that step sizes may be multiples of a quantal value just under 3 nm.

this case the main peak is located at 11 nm, approximately. Depending on the loading conditions, however, and to some extent on the particular fiber, the location of the main peak may shift considerably. A feature that is emerging consistently is that the *difference* between peaks is generally on the order of two to four nanometers.

Higher-resolution analysis has helped to define this feature more clearly. Figure 10, for example, shows a plot of the distribution of the *differences* between histogram peaks, the data in this case obtained during ramp release of unstimulated fibers. Figure 10 shows some weak tendency for a peak to appear just above 2.0 nm. with a shoulder just above 3.0 nm. In the experiments with auxotonically contracting fibers, we have also found that comparable "peaks" tend to appear, namely, at around 2.2 nm and 3.3 nm (Jacobson et al., submitted). It is interesting that the sum of these is 5.5 nm, the actin repeat along the thin filaments; thus, step sizes could in some way reflect the kinetics of actomyosin interaction during sliding.

An alternative possibility is that these minor peaks are not genuine, and that there is but one main peak whose centroid is just below 3 nm. This would imply that there is a single "quantum" just under 3 nm per sarcomere, and step sizes are integer multiples of this. The breadth of the distribution in Figure 10 would then be the result of random noise. This quantum value (2.8 nm) has particular significance in a new model being developed to understand the molecular mechanism underlying the generation of steps (cf. Pollack, this symposium).

At present we are unable to distinguish between the two hypotheses. The "quantum" step size, if indeed there is one (or two), seems to lie somewhere between 2.0 and 3.5 nm, but we are not yet in a position to identify it with certainty.

CONCLUSION

We have interrogated the contractile mechanism using a variety of methods. Each result is either consistent with, or has demonstrated clearly the presence of, stepwise shortening. This phenomenon is likely to be related to the phenomenon of stepwise bending observed in cilia (Baba, 1979). In the case of muscle, however, the steps appear somehow to be synchronized over regions which by molecular standards, are vast. This is a remarkable feature, which requires explanation at the molecular level.

On the other hand the synchronous behavior may be taken as a "gift" of nature. Its consequence is that any shortening step measured at the macroscopic level must reflect a comparable shortening step at the molecular level. Similarly, a measured pause must represent a molecular pause. Thus, stepwise shortening provides a tool by which molecular dynamics may be probed with macroscopic methods, an exciting prospect.

ACKNOWLEDGEMENT

Frank Brozovich was supported by a stipend from the Medical Student Research Training Program of the University of Washington.

REFERENCES

Baba, S.A. (1979). Regular steps in bending cilia during the effective stroke. Nature 282: 717-720.

Brozovich, F. and Pollack, G.H. (1983). Muscle contraction generates discrete sound bursts. Biophys. J. 41: 35-40.

Delay, M.J., Ishide, N., Jacobson, R.C., Pollack, G.H., Tirosh, R. (1981). Stepwise sarcomere shortening: Analysis by high-speed cinemicrography. Science 213: 1523-1525.

Hill, D.K. (1968). Tension due to interaction between the sliding filaments in resting striated muscle. The effect of stimulation. J. Physiol. 199: 637-684.

Huxley, A.F. (1957). Muscle structure and theories of contraction. Prog. Biophys. Biophys. Chem. 7: 255-318.

Huxley H.E. (1969) The mechanism of muscular contraction. Science 164: 1356-1366.

Iwazumi, T. and Pollack, G.H. (1979). On line measurement of sarcomere length from diffraction patterns in cardiac and skeletal muscle. IEEE Trans. on Biomed. Eng. 26(2): 86-93.

Jacobson, R.C., Tirosh, R., Delay, M.J. and Pollack, G.H. Quantized nature of sarcomere shortening steps. (submitted).

Myers, J., Tirosh, R., Jacobson, R.C. and Pollack, G.H. (1982). Phase locked loop measurement of sarcomere length with high time resolution. IEEE Trans. on Biomed. Eng. 29(6): 463-466.

Peachey, L.D. and Eisenberg, B.R. (1978). Helicoids in the T system and striations of frog skeletal muscle fibers seen by high voltage electron microscopy. Biophys. J. 22: 145-154.

Pollack, G.H., Iwazumi, T., ter Keurs, H.E.D.J., and Shibata, E.F. (1977). Sarcomere shortening in striated muscle occurs in stepwise fashion. Nature 268: 757-759.

Pollack, G.H., Vassallo, D.V., Jacobson, R.C., Iwazumi, T., and Delay, M.J. (1979). Discrete Nature of Sarcomere Shortening in Striated Muscle. In: *Cross-bridge Mechanism in Muscle Contraction.* ed. H. Sugi and G.H. Pollack. Univ. of Tokyo Press/ Univ. Park Press. pp. 23-40.

Rüdel, R. and Zite-Ferenczy, F. (1979). Do laser diffraction studies on striated muscle indicate stepwise sarcomere shortening? Nature 278: 573-575.

Vassallo, D.V. and Pollack, G.H. (1982). The force-velocity relation and stepwise shortening in cardiac muscle. Circ. Res. 51: 37-42.

Yeh, Y., Baskin, R.J., Lieber, R.L., and Roos, K.P. (1980). Theory of light diffraction by single skeletal muscle fibers. Biophys. J. 29: 509-522.

DISCUSSION

TAYLOR: I don't understand how your technique is capable of detecting things on the order of nanometers, but yet I don't see in your records any sign of something relatively as big as the latency relaxation.

POLLACK: Occasionally we can see a bit of lengthening prior to shortening, but not in the particular records I've presented here. One of the reasons we don't consistently see a drop in tension prior to shortening may be that we used floppy tendons. We do this routinely to permit internal shortening when we use isometric contractions, so that we can observe steps. We also do this to provide a buffer so that smooth shortening imposed by the position of a lever doesn't necessarily constrain the sarcomeres to shorten smoothly. In order to see a sizeable latency relaxation you may need to have short, noncompliant tendinous ends; otherwise sarcomere lengthening may not produce much of a drop in tension.

HUXLEY: Do you ever see any kind of steps in tension records?

POLLACK: We haven't looked carefully. We did publish one record (cf. Sugi & Pollack, 1979) which appeared to show a step in the tension record. At that time we weren't using a high-frequency tension transducer. We have one now, and will need to pursue this question further.

MATSUBARA: You mentioned several methods of avoiding the criticism of Bragg angle effect. But in your first method -- imaging on film with analysis by Japanese students -- I wonder whether by such imaging you can avoid the Bragg angle effect as long as you use laser light.

POLLACK: That's a very good point. There are two answers that I'd like to give. So long as a laser is used, Bragg angle artifacts could arise and may then give rise to steps -- an hypothesis which I, frankly, have not found plausible. Anyway, despite my skepticism, we did develop an alternative method that used incoherent, polychromatic illumination, where Bragg angle effects are not possible. Since that method also showed steps we were able to rule Bragg artifacts.

However, even with the laser illumination method we used earlier, I don't think there was a danger of Bragg artifact since we did not look at one diffraction order alone; in order to obtain the striated image we combined the *two* first orders (as well as higher orders). In the Bragg hypothesis one expects pauses or steps only in *one* first order, not the other. Therefore, if we combinc the two first and higher orders to reconstruct the image, if, perchance, there's a step or pause in only one of those orders, recombination should make it disappear. But it doesn't disappear. So I think that method of using laser illumination for imaging is still valid.

But if one still has reservations about that method, I think those were resolved by the subsequent use of incoherent illumination, where Bragg angle effects are not an issue.

KRUEGER: Jerry, since I'm no longer your postdoctoral fellow, I'm less inclined to disagree with you publicly. Finding steps in passively shortened muscle means that they can no longer be uniquely attributed to a contractile process, or at least the same contractile processes that you see in contracting muscles.

POLLACK: The two shortenings seem to me likely to be generated by the very same process. There has been evidence presented at this meeting and elsewhere (D.K. Hill, J. Physiol. 199: 137-184, 1968) that in the resting state there may be some actomyosin interaction. If actomyosin interaction is producing steps in activated muscle, I don't see any reason why it couldn't do so in the "resting" fiber as well. It's very interesting, in fact, that the unloaded velocity of shortening in these unstimulated fibers is comparable to the unloaded velocity of shortening in the stimulated fibers. They may differ by a factor of only 1.5 or so, supporting the view that there are active-like interactions going on in the unstimulated fibers. These arguments would seem to underline the plausibility that steps in active and unstimulated muscle are caused by the same mechanism.

HOMSHER: If you take the passive muscle and stretch it beyond overlap, do you still see the steps?

POLLACK: We haven't tried that yet. Good experiment.

COOKE: Is there any pattern to the velocity between the steps?

POLLACK: We just published a paper on this subject using heart muscle (Vassallo and Pollack, Circ. Res. – 1982). The velocity between the steps seems to change somewhat with load, though not dramatically. When load is lowest the velocity seems to be highest and vice versa. But the influence is not so strong, and the scatter is large enough to make us unsure whether the dependence is genuine or not. A stronger dependency was found with pauses. There was a linear relationship between load and the duration of the pause. These two relations can account for the force velocity relation. With a higher load, you have longer pauses, possibly a lower step velocity (and also small steps) giving a low "overall" velocity of shortening. At low load, the opposite holds.

SHIMIZU: Have you ever used glycerinated muscle for measuring stepwise shortening?

POLLACK: Yes, the fibrils that I showed you, prepared by Dr. Tameyasu were done that way. They appear to give steps.

EDMAN: Wouldn't you expect to see some kind of steps during the plateau of the tetanus?

POLLACK: Well, if there's no shortening, then there can be no steps.

EDMAN: But I mean the sarcomeres would have a tendency to go one step forward or one step back, even during a steady isometric tetanus.

POLLACK: You're suggesting is that there may be a certain stable sarcomere length and sarcomeres then want you to occupy one or another of these lengths.

EDMAN: Yes, you should assume something like that.

POLLACK: It's a good point. We published a paper last year (Tameyasu and Pollack, Biophys. J. 1982) that appears to confirm that something like that may happen. We looked at the substructure in the optical diffraction pattern and found discrete values of sarcomere length at any one instant in time. It appears that certain discrete sarcomere lengths are preferred.

EDMAN: Well this was actually our study (Cheworth and Edman, 1972). In fact, there are changes in the microstructure during tetanus but they are very slow and continuous.

POLLACK: Yes, we have seen that too; minor elements of the substructure can change gradually, but like your observation, Tameyasu and I found that in good, non-translating fibers, the major elements of the microstructural pattern are solidly "locked in." There seems to be little or no jumping back and forth. With a stiff isometric constraint, it would be difficult for any large number of sarcomeres to step synchronously; under such conditions I would expect only *very* small regions to step. If various regions do this in sequence, this will give apparently smooth microstructural movement. But these always represent minor elements of the partern, and therefore reflect a very small population.

TER KEURS: Jerry, you suggest that the best of fibers, the steps are very consistent. Are the acoustic signals equally consistent, and synchronous with the steps?

POLLACK: We're now setting up to compare the sarcomere length changes with the sounds to see if they're synchronous. The sounds themselves are moderately repeatable from contraction to contraction.

WINEGRAD: If you say you see steps during the tetanus of frog muscle, it might be worthwhile looking during maintained tension in catch muscle, where there presumably is no movement.

POLLACK: I didn't quite say that we saw steps during an isometric tetanus. In fact this is probably the very worst condition to look for steps because there is little or no shortening going on. You may have small regions that are changing their lengths slightly, as Paul Edman intimated, but not any large scale shortening that might occur synchronously over a large area. On the other hand, a tetanus is probably ideal, provided you release the fiber. In that case a uniform condition of activation should exist just prior to release, perhaps promoting broad-based synchrony.

SCHOENBERG: Over what number of sarcomeres are the steps going to synchronous? This information would help those of us who could look for it to know what size beams to use. Clearly the entire fiber can't be stepping synchronously because you're controlling the movement of the ends to be smooth. So over what length of the fiber is the stepwise shortening synchronous?

POLLACK: I wish I could tell you that the region of synchrony is "x microns", but I can't. It's variable. We've done some experiments looking region by region, moving 100 μm at a time. Sometimes the same pause is maintained from one region to another over several hundred microns, and sometimes it's not. You haven't asked, but I assume you're asking me to speculate as to why there are different degrees of synchrony.

SCHOENBERG: Not really. At this point I want to ask you a question. We use an 800 μm laser beam and don't see any steps. Is it possible that our laser beam is just covering too wide an area to see the steps? Should we perhaps shorten the diameter of our beam to reduce the region of the fiber illuminated?

POLLACK: Well, first of all, you're sure that you don't see steps? I'm somewhat surprised, since Bill Halpern at the University of Vermont has shown me many records of steps using a comparable laser beam, nominally 1 mm diameter. What are the experimental conditions?

SCHOENBERG: These are relaxed fibers, and generally we've not looked at shortening, only stretch.

POLLACK: And the stretch is absolutely smooth all the time?

SCHOENBERG: No, it's not absolutely smooth all the time, but it is absolutely smooth 95% of the time. I didn't mean to imply that there might not be stepping. I just wanted to stick some numbers on to say we don't think we see steps, and one possible reason is that we're looking at too large an area of the fiber at one time. In other words, if only 100 μm were stepping and the next 100 μm were stepping a little differently, if you illuminated an area of the fiber as large as we had then you might not see it. So I'm asking you, should I expect to see steps illuminating about 800 μm of the fiber at a time?

POLLACK: I'm not sure. One important factor is the boundary condition. If you use short tendons, then the smooth waveform imposed by the lever is imposed on the fiber itself; this must inhibit synchronous stepping over large regions. I would say that certainly the steps should be more prevalent if you look at smaller regions. In our hands a typical region of synchrony may be on the order of several hundred microns. But as I mentioned, Bill Halpern has been using a 1 mm laser beam, and he sees steps regularly.

SCHOENBERG: He has looked generally at frog fibers and, at least in our hands, the striation patterns in frog fibers are not terribly well aligned. We use rabbit fibers, where you can practically draw a straight line across the striations. We find that when we do terrible things to our fibers, like really pull on them a lot or after they're getting very old and the striation pattern is no longer nicely aligned, then we have more of a tendency to see steps. Now I'm not saying that's the explanation for the steps, but I think the reason they're easy to see in frog fibers is because the striation patterns are not aligned.

POLLACK: I'm not so familiar with rabbit fibers, but I've marvelled at how regular the striations can be in many of the frog single fibers that we've used. In others the striations are less regular, for sure. We haven't attempted any quantitative correlation between the quality of the pattern and the frequency of steps, but we have a well-developed qualitative view. Perhaps Dr. Tirosh would like to say something about that, because he's been involved in these studies and has been concerned with that sort of question.

TIROSH: Yes. Indeed your question has a qualitative answer. Namely, in our hands we usually got clearer stepwise shortening patterns while working with a narrow beam, down to 50 μm. However, one should say that, at the same time, the best steps were obtained while you had a high quality of striation homogeneity, minimal translation of the observed region across the beam during contraction, and low tension. When these conditions were not well fulfilled, you could often get not steps but smooth shortening or noise; but not steps. The clearest steps were associated with the highest quality regions.

MAUGHAN: I was just going to say for the record, before I left Vermont Bill Halpern asked me if his name was referred to with regard to the steps, that steps be referred to as "hesitations."

MAGID: Since you showed some results in resting fibers and that's the only sort of thing I like to study, I want to make a point about the structure. I suggest that the thick filaments are bound together sideways by struts or some kind of bracing structure. This gives a structural basis for averaging out the action of cross-bridges, however independent their action is. This point is relevant in understanding the basis for synchrony. The other thing I wanted to say is that during the project on the isolation of A segments of frog, we very often noted that the A segments would come out of the muscle not as just single, myofibril-wide structures, but as great rafts of A segments, laterally bounded together, very often not in perfect register, but "stepping" along, if you will, in small amounts. These rafts were up to 10, 20 or even 30 myofibrils wide. So I think there is a structural basis for averaging out the contractile behavior.

POLLACK: I support your idea. In fact my poster (cf. Pollack, this symposium) describes a possible molecular mechanism for the steps, and relies to some extent on these struts, or lateral interconnections, that you're talking about; except I think they may be the bridges, themselves.

HUXLEY: Just a technical point. If you're imaging a myofibril with a number of sharp lines or bands onto a photodiode array with perhaps 128 elements, there must be some tendency to get some kind of quantization. What limit can you resolve in the face of that quantization?

JACOBSON: I've worked with this apparatus, and I am familiar with the technical details. First of all, the discrete signal coming out of the photodiode array is filtered so that the discreteness is smoothed out. Secondly, there *is* a certain degree of quantization, you might say, due to the fact that you have an oscillation sinusoid which is averaged over a finite window. And averaging over that window does produce a degree of jitter in the result. The jitter is on the order of 2 nm. We measured that by sliding the window across the striations. And we also calculated it by taking the average of the sinusoid over 20 cycles, which is what we had. It came out 2 nm in both cases.

HOUSMANS: I would like to come back to the question of alternate methods of resolving problems of translation when looking at shortening steps, as in Dr. Pollack's experiments. We have studied a multicellular preparation, the mammalian cardiac muscle that is mechanically somewhat more complex than single skeletal muscle fiber (Housmans et al., this symposium). In *some* experiments, with very good fibers, I've seen "hesitations" in the shortening pattern of segments in the central region of the cardiac fiber (these were segments of 500 μm length). As segment markers we used the tips of glass microelectrodes inserted into the core of the fibers. A Reticon array (CCPD-1024) was used to scan segment length. I would like to show an example of such "hesitations" in the segment shortening pattern. Figure D-1 shows the records of *muscle* shortening and velocity of shortening, and of segment shortening (bottom trace) in an isotonic twitch contraction at preload only (cat papillary muscle). Early in the contraction, there are several "hesitations" or "steps" (arrows in Fig. D-1) in *segment* shortening, that are not seen in the trace of *muscle* shortening (middle trace). These "hesitations" or "steps" are of a magnitude well beyond the resolution of the technique and are different from noise.

In these segments there is no problem related to translation because the length of the segment remains defined as one measures the segment length always *between* the markers. These "steps" were seen exclusively during segment shortening at *low* loads and were reproducible in a given segment in the same experimental conditions. This observation is not the result of any of the obvious mechanical artifacts, such as rotational artifacts, or translation of the markers in a direction other than in the focal plane of observation.

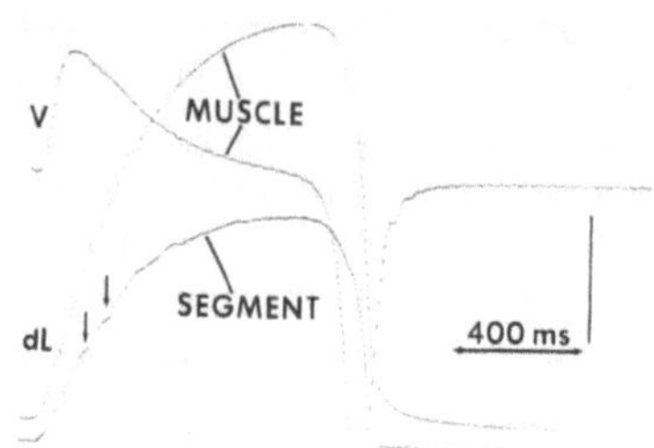

Figure D-1: From top downwards: tracings of velocity of muscle shortening, shortening of muscle, and segment shortening as a function of time in a freeloaded isotonic twitch contraction at preload. The arrows point to hesitations in segment shortening, referred to as "steps" in the text. The vertical calibration bar represents 5% of muscle shortening, 10% of segment shortening and 0.5 l_{max}/sec velocity of muscle shortening. Muscle characteristics: cat papillary muscle, length at l_{max}, 5.75 mm; mean cross-sectional area, 0.42 mm^2; ratio of resting to total isometric force at l_{max} 14.1%; the centrally located segment had a precontractile length of 1475 μm; temperature 26°C.

SCHOENBERG: I'm not sure it applies to the slide you just showed, but there were a few times when we thought we were seeing steps -- the procedure was to use a ramp to stretch out a muscle. But then what we did, is we subtracted our ramp from our sarcomere length measurement. What we found is that things that looked like steps when they're going up at a 45° angle looked just like noise. I don't know if that would explain that result, but when you take noise and you start projecting it at 45° it does tend to look a little bit step-like.

HOUSMANS: Can I answer that? When looking at different segments in the same preparation, in some segments *no* "hesitations" or "steps" were seen, whereas in other segments, "steps" did occur, and were reproducible in those segments. Concerning you question on noise, the "steps" in segment shortenings are well beyond the level of noise, which can be estimated from the baseline of the segment length signal after the contraction. Also, if you look at the trace of shortening of the muscle (middle trace) there is *no* hesitation in *muscle* shortening itself, whereas there is in the segment. As I said, this is a preliminary observation, I am not aware of the particular conditions that tend to induce or provoke this phenomenon in a central segment.

WINEGRAD: In all tracings that we've seen -- tell me whether this is a correct generalization -- one tends to see in the activated muscle, steps during the period of development of tension but not during the decline of tension. Is that an accurate generalization?

POLLACK: It's a fair generalization based on what I showed, but I didn't show all the slides. We do see steps during the decline of tension as well as during development of tension. In fact, we have histograms of step size recorded during the tension decline. These steps do not show up quite as readily as the ones during the contraction phase, but they do show up.

WINEGRAD: Do you get the same histograms for the same fiber?

POLLACK: Several of the same step sizes are seen, but there are some new sizes that tend to appear too.

TIROSH: I'd like to make a general comment on stepwise shortening, that, indeed, if it is real, it's difficult to understand this phenomenon in concepts of cross-bridge mechanisms. However, if one considers other kinds of mechanisms, for example the one that I'm proposing, then stepwise shortening may be understood simply as a *response* of the system. When we found stepwise shortening in the resting condition, I realized that this could represent a viscoelastic response to the system, and therefore was compatible with my model. But at first I thought it was quite a problem, and therefore I felt quite happy to do this work. Jerry, of course, has his own interpretation. That's fine. This brought much more pleasure to the work.

CECCHI: I have a question for Dr. Housmans. You showed that there is a kind of step-like shortening. How can you be sure that this kind of step-wise shortening is not due to the oscillation of the tip of the microelectrode? And secondly, have you tried twice to repeat the records, and have you found the same behavior -- the same step at the same point?

HOUSMANS: The first time I saw this phenomenon, I thought it was some kind of artifact. But it was possible to eliminate this possibility when repeating the same contractions at the regular interval of stimulation and comparing superimposed records of segment shortening. The patterns of stepwise segment shortening, repeated over and over again superimposed perfectly. Then secondly, observation of the microelectrode tips through the microscope did not reveal rotational artifact, nor "swinging" or oscillations of the microelectrode tips; and when we moved the Reticon array parallel to, but farther away from, the muscle so that oscillation artifact, if any, would either appear or disappear, there was no change in the "step" phenomenon that was observed in segment shortening. And again, *not all segments* showed this behavior, but whenever a segment showed stepwise shortening it was reproducible and it was repeatable under the same conditions.

WILKIE: I have a general question. We've heard a great deal of argument back and forth about instrumental questions and experimental differences. Am I right in thinking that the importance of these steps, assuming that they are genuine and they exist, is that there must be some mechanism not yet understood for synchronizing the activity of the cross-bridges? I mean, such phenomena are well known elsewhere. I'm thinking of the ciliated membrane of the esophagus in the frog, where the ciliary beat is sunchronized by means yet unknown to man. But am I right in thinking that's what all this argument back and forth is all about? It has my head spinning.

POLLACK: Yes, thank you. That is precisely what it's about. How is synchrony achieved? Your comment about the synchrony among adjacent cilia reminds me of a paper on *individual* cilia published by a Japanese chap named Baba (Nature 282: 717-720, 1979) several years ago. He found that during the beat, the cilium undergoes stepwise displacements of its angular position. In other words, the bending of the cilium is not smooth, it occurs in steps.

VERDUGO: Jerry, I think Baba's studies were done on the compound cilia.

POLLACK: Yes, that's true of his *Nature* paper. Recently he told me about his new experiments in which single giant cilia showed the same phenomenon even more clearly.

WILKIE: Seemingly, the phenomenon is genuine. What we then have to look for is some mode of transmission along the thick filament that synchronizes the action of the cross-bridges. Isn't that what it's about?

POLLACK: Yes, if one works within the constraints of the cross-bridge theory, that is indeed correct.

WILKIE: Yes, I would assume that most people here would tend to do so (laughter).

POLLACK: It's also necessary to find a mechanism by which the bridges in neighboring myofibrillar sarcomeres are synchronized.

SQUIRE: I would just like to make a very simple structural observation, which is that the A-band and the I-band are not linear arrays; they are actually helical arrangements of things. As you change the overlap it's going to become easier and then more difficult for cross-bridges to interact with actin. So if you had half a sarcomere, and you gradually shortened it, you might expect to see a step-wise shortening. Okay? If you had a myofibril with all the half sarcomeres along it absolutely in register, once again I think you'd expect to see steps. But, if you have sarcomeres which are not exactly in register, then this would tend to be smoothed out. But occasionally, as they're shortening, you might suddenly get back into register at some point and see a step. Okay? So it doesn't seem terribly surprising in terms of a normal cross-bridge model.

EDMAN: But wouldn't you expect to have very large regions that are in synchrony?

SQUIRE: Why?

EDMAN: Dr. Pollack is not looking at one sarcomere. He is looking at a rather long and broad region. So you would have to demand that there is a synchrony over a very wide area.

SQUIRE: What I'm saying is that this sort of thing will be going on all the time in all the sarcomeres, but only occasionally will it happen in phase in all of them. So you suddenly see it. And then it will suddenly disappear again.

EDMAN: Well, wouldn't you expect to see some kind of stepwise shortening or elongation even during maintained sustained tetanic contraction?

SQUIRE: If the amount of overlap between thick and thin filaments was maintained in that situation, then you wouldn't see a step. It's only by changing the overlap that you change the possibility of bridges to interact with actin, so you have to change the overlap. You'd have to assume that there was continuous change of sarcomere length.

POLLACK: John, your comment about the vernier-type effect giving steps is well taken, but I think the probability of achieving synchronization over thousands of myofibrillar sarcomeres would be very low, indeed, unless there were perfect register among sarcomeres. We see cascades of steps very often, especially in the unstimulated fibers, and under unloaded conditions in the stimulated fibers. Secondly, the size of the step varies depending on the conditions. It may range from about 5 nm up to 25 nm. I believe that with this kind of vernier effect the step size ought to be unique.

SQUIRE: Well, I don't think it necessarily would if the sarcomeres are not strictly in phase, but your observation of something about 2 or 3 nm as the basic unit seems to fit in exactly with the actin repeat of 27 Å.

WANG: On the question of the mechanism of synchrony, I'd like to point out that in striated muscle cells there is an extensive network of intermediate filaments which connect all myofibrils together. These filaments are associated intimately with sarcomeres but are distinct from thin filaments, thick filaments, and the third filaments that we had discussed the other day. As you can see from this electron micrograph (Fig. 8), when the majority of thick and thin filaments of a small bundle of myofibrils were extracted away by 0.6 M KI, a network of residual filaments appeared. Note that the very dark Z structures of adjacent myofibrils were linked transversely by short filaments. In addition, Z structures of the same myofibrils were also linked together by many *continuous*, longitudinal filaments! We have found that these longitudinal filaments are actually not the "third filaments," as has been proposed by several investigators. Instead they appear to be intermediate filaments which surround each sarcomere by connecting the peripheries of each Z disc. I think that this filamentous cytoskeletal network, especially the continuous longitudinal filaments which connect sarcomeres in series is well worth considering in understanding the basis for synchronous mechanical properties.

A PROPOSED MECHANISM OF CONTRACTION IN WHICH STEPWISE SHORTENING IS A BASIC FEATURE

Gerald H. Pollack

Division of Bioengineering and Department of Anesthesiology, WD-12, University of Washington, Seattle, WA. 98195

ABSTRACT

A simple model is put forth as an initial attempt to account for the observation that the contractile process is discrete and sychronized over a large region of space.

The basic principle is shown in Figure 1. The driving force for the contractile event is postulated to lie in a shortening of the myosin rod; this causes the bridge to translate along the fiber axis. If the thin filament is bound to the cross-bridge during this shortening event, the thin filament will translate toward the center of the sarcomere, giving a single

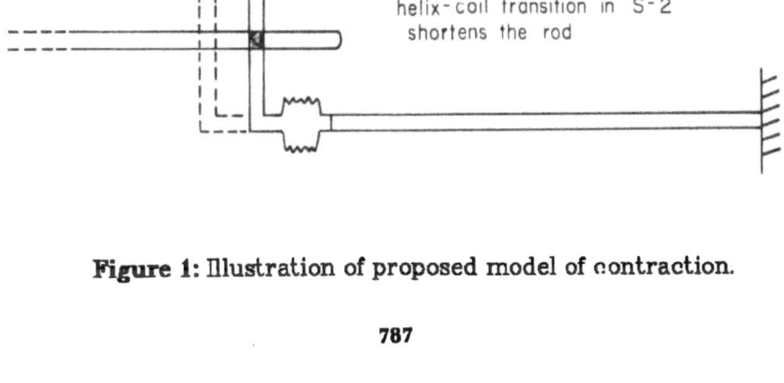

Figure 1: Illustration of proposed model of contraction.

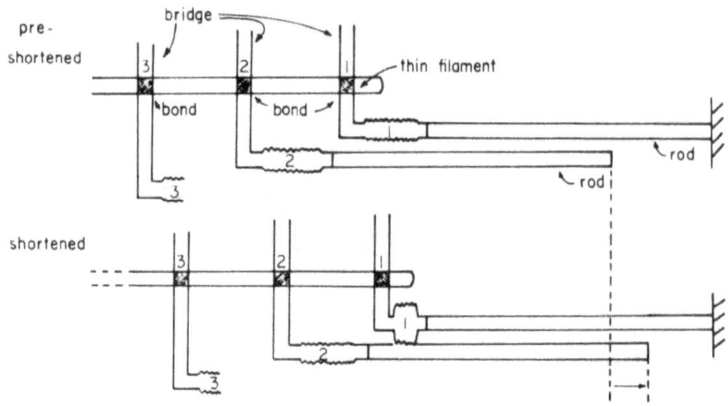

Figure 2: Thick filament structure, comprising a series of overlapping myosin molecules. Shortening of the S-2 of myosin 1 involves sliding of rod 2 over rod 1.

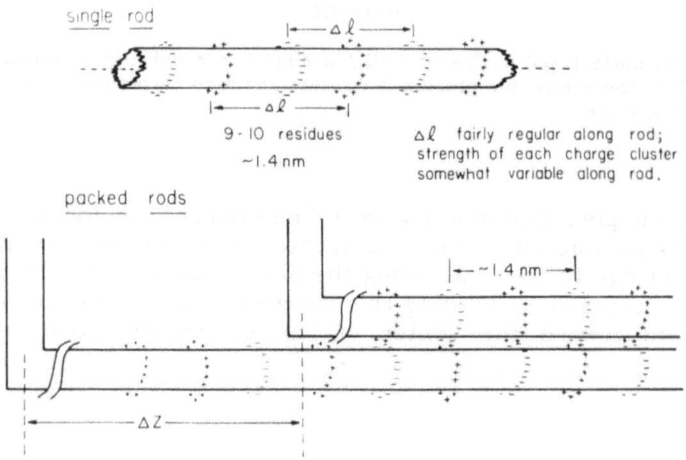

Figure 3: Charge distribution along myosin rod (top). Rods will tend to pack whenever opposite charges are closely apposed (bottom). There is a series of such quasi-stable locations.

sarcomere shortening "step." The ability of the S-2 region of the myosin rod to shorten in the millisecond time range by a helix-coil transition has been demonstrated by Harrington and colleagues (Tsong et al., 1979). Helix-coil transitions are well-known in biology, and the mechanics of such phenomena are understood (Flory, 1956).

In the proposed model the thin filament remains bound to bridges during contraction. It follows that a shortening event is necessarily asso-

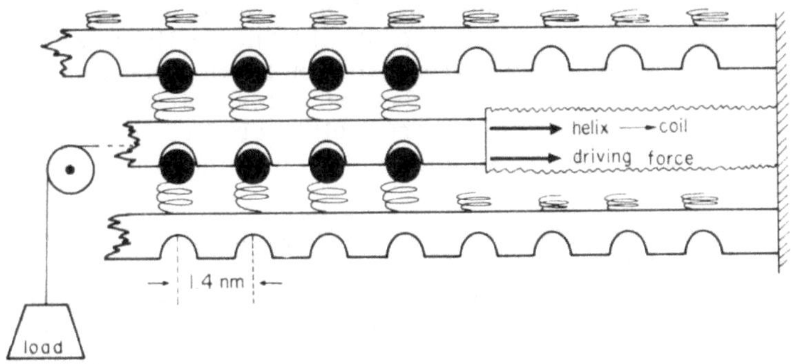

Figure 4: Mechanical analog of proposed contractile mechanism. The balls in sockets correspond to quasi-stable locations where opposite charges are closely apposed.

ciated with a sliding of adjacent rods: Figure 2 shows that as generator 1 shortens, rod 2 slides by rod 1. In order to understand the dynamics of shortening, therefore, it is necessary to understand the nature of rod-rod interactions. This has been studied by McLachlan and Karn (1982). Based on the amino acid sequence along the rod, they showed a remarkably regular charge distribution along the rod. This is summarized in the top section of Figure 3.

The rods will tend to pack whenever opposite charges are closely apposed (Fig. 3, bottom). Because the charge clusters repeat regularly along the rod, a series of such quasi-stable packing configurations is anticipated. Indeed, McLachlan and Karn showed that as one rod was slid (theoretically) by another, a series of sharp potential wells was generated. The deepest well occurred at a relative displacement, ΔZ, of 14.5 nm, as anticipated from the manner in which the rods pack naturally to form the thick filament. The implication of this model, however, is that if rods are made to slide from their natural packing configuration, they will pass through quasi-stable points every 1.4 nm approximately, where they *may* stop.

The helix-coil driver, combined with the above-described packing features, gives rise to the mechanical analog shown in Fig. 4. The spring-loaded balls-in-sockets correspond to the apposed charges: a series of quasi-stable locations exists.

A step is explained as follows: Initially, balls are in sockets; this is the "pause" state. A helix-coil transition in the S-2 region of the rod causes the buildup of force. No rod translation can occur until this driving force rises sufficiently to overcome the resisting force, i.e. the sum of the load and the force required to remove the balls from their sockets. Thus, the onset of shortening is a threshold event. Once this threshold has been reached, the rod will suddenly begin translating rightward, thereby initiating the step. The rod may then stop abruptly at the next notch (1.4 nm), thereby terminating the step and beginning the next pause. However, the rod need not stop there. If the helix-coil generator continues to increase its force as the rod translates, some notches may be skipped and the rod may translate by several notches before stopping. Such relatively large steps would be anticipated when the load is relatively low, or when it is decreasing.

The above model predicts step size to be integral multiples of 1.4 nm in each half sarcomere, or 2.8 nm per sarcomere. Although absolute values of step size are variable, we have indeed found that histograms of step size show multiple peaks, separated predominately, but not exclusively by 2.5-3 nm (see paper by Pollack et al., this symposium). This is close to the predicted quantum separation of 2.8 nm.

The role of ATP in this model is a simple one. It is obvious from Fig. 2, bottom, that at this stage generator #2 could not shorten unless the segment of thin filament between bridges 1 and 2 collapses. Alternatively, if actin dissociates from bridge #1 *after its S-2 has shortened*, then generator #2 could mediate the subsequent shortening step, and so on. This is the presumed role of ATP: to dissociate actin from shortened myosins, so that shortening of other myosins may proceed in sequence. Large scale sarcomere shortening, therefore, involves sequential shortening of myosin rods, beginning near the thin filament tip, and progressing one by one along the thick filament toward the Z line; each "step" is preceded by the splitting of an ATP.

The activation scheme for this model has not yet been worked out. This could involve the phosphorylation of a myosin light chain, a load reduction (latency relaxation), or the binding of a proton generated by ATP hydrolysis; such factors are known to lower the threshold for helix-coil transition in other systems.

A further assumption in this model is that adjacent thick filaments are permanently interconnected along their entire length either by cross-bridges or by other structural units in a ladder-like manner (similar to the M region). Such structural units are apparent in many published and unpublished electron micrographs. A detailed consideration of this possibility may be found in a forthcoming review article (Pollack, 1983).

The effect of such lateral interconnections is to promote synchrony among neighboring thick filaments: one segment (i.e., one myosin rod) along a thick filament cannot undergo substantial shortening unless the respective segment in the neighboring thick filament does so simultaneously. Thus, shortening of thick filament segments are constrained to occur in an all-or-none fashion across a single myofibril. The myofibrillar half-sarcomere behaves as a unit. Local synchrony is thereby assured. But why should the various myofibrillar half-sarcomeres in parallel and series, widely scattered in space, begin their step at the same time? How could such global synchrony be established?

The critical point is that the initiation of the local shortening step is a threshold event. If the time required to reach threshold is everywhere similar, synchrony will be achieved. Recall that a myofibrillar half sarcomere, *behaving as a unit*, cannot begin shortening until the number of activated force generators in parallel across the unit increases to the point where the summed force just exceeds the total resisting force (balls in sockets plus load) on the myofibril. Then, suddenly, local myofibrillar shortening begins. Since the threshold for initiation of shortening involves the summed force of hundreds of generators, modest variations in individual reaction rates will be averaged out, and threshold will be reached essentially simultaneously in widely distributed half-sarcomeres. Thus, the observed global synchrony will be achieved, and the essential features of stepwise shortening will be accounted for.

It is further shown on the poster that this contractile mechanism can explain a variety of additional features of contraction in a natural manner. A full discussion of these must necessarily be reserved for a more detailed communication.

REFERENCES

Flory, P.J. (1956). Role of Crystallization in Polymers and Proteins. Science 124: 53-60.

McLachlan, A.D., and Karn, J. (1982). Periodic charge distributions in the myosin rod amino acid sequence match cross-bridge spacings in muscle. Nature 299: 226-231.

Pollack, G.H. (1983) The sliding filament/cross-bridge theory: A critical review. Physiological Reviews, in press.

Tsong, T.Y., Karr, T. and Harrington, W.F. (1979). Rapid helix-coil transitions in the S-2 region of myosin. Proc. Natl. Acad. Sci. USA 76(3): 1109-1113.

DISCUSSION

Harrington commented that in order for the S-2 to undergo to a helix-coil transition it would probably need to be removed from the surface of the thick filament, unlike the mechanism postulated in the model.

Edman noted that the velocity of contraction under constant load would be predicted by the model to remain constant during steady shortening, whereas he thought the experimental results indicated a variation of loaded velocity with overlap. Pollack responded by indicating that Edman was probably referring to experiments in which quick releases to a fixed load were made at each of a series of sarcomere lengths. However, if contraction is initiated at a stretched length and the fiber is allowed to shorten under constant load, the velocity remains strikingly constant over a long course of shortening (Buchthal and Kaiser, Dan Biol. Medd. 21(7): 1-318, 1951). This is in agreement with the model's prediction.

Both Edman and Hunstman wondered about the predicted shape of the length-tension relation. In fact, the length-tension relation does not emerge as a natural prediction, as in the crossbridge model. Since isometric tension development in this model is a result of thick filament shortening with consequent extension of the connecting filaments (cf.

Magid, this symposium), the tension level depends on a series of factors, not the least of which is the stiffness of the connecting filaments. thus, a uniquely shaped length-tension relation is not predicted. It would be interesting to test, for example, whether the differences in shape of published length-tension curves correlate with differences in the stiffness of the connecting filaments (which would, in turn relate to the steepness of the resting length-tension relation).

Iwazumi commented that the restoring force required to reestablish the initial length after thick filament shortening would be energetically costly.

Regarding the possible activation scheme, Tirosh suggested that proton release following ATP splitting could serve as a trigger of the onset of contraction. Phosphorylation would be too slow. Also, he commented that the model seems to offer no explanation for post-tetanic potentiation.

Huxley wondered whether the model, which predicts thick filament shortening, would be consistent with the X-ray patterns, which show preservation of the 143 Å and 430 Å spacings along the thick filaments during contraction. It is certainly true that if the *entire* thick filament were to shorten uniformly, those two spacings ought to decrease; or disappear if shortening along the length of the filament were non-uniform. However, for a relatively modest amount of sarcomere shortening, only a small segment along the thick filament in each half sarcomere would shorten, leaving the major region of the thick filament unchanged. Under such (typical) experimental conditions, the two spacings under consideration should therefore be preserved, though a diminution of intensity of the 143 Å and 430 Å layer lines would be predicted. This is consistent with what has been observed.

The model also predicts that a new axial repeat spacing corresponding to the regions along the thick filament that have shortened ought to be detectable under certain conditions. If the load is constant during shortening, the steps should be of constant size, and the magnitude of each incremental unit shortening along the thick filament should be consistent. Thus, a region along the thick filament of uniform but shortened spacing is predicted and should be detectable on the meridional X-ray diffraction pattern.

STEPWISE CHANGES IN CROSSBRIDGE STATE AND SARCOMERE LENGTH: DO LATTICE CONSTRAINTS PLAY A CRITICAL ROLE?

C.J. Ritz-Gold and C.M. Gold

841 HSW, Cardiovascular Research Institute, University of California, San Francisco CA 94143

ABSTRACT

Observed abrupt and stepwise changes in crossbridge state induced by MgPPi in skeletal fibers using spin label techniques as well as stepwise changes in sarcomere length in actively contracting fibers are considered in terms of a model based on the principles of catastrophe theory and nucleation of structural transitions in mechanically constrained polymeric assemblies.

We have observed abrupt and stepwise changes in the conformational state of crossbridges in bundles of skeletal muscle fibers induced by the substrate analog MgPPi. A maleimide spin probe was attached to myosin heads to monitor the extent of transition from rigor (rigid) to pseudorelaxed (mobile) state. When different fibers were exposed to increasing concentrations of MgPPi at regular time intervals, response curves as a function of log MgPPi were obtained generally showing one or more large amplitude step increases in extent of pseudorelaxation around 1 mM MgPPi. When exposed to a single maintained dose of MgPPi of different concentrations, fiber bundles showed an initial rapid rise in the mobile state followed by an immediate decay for concentrations below 1 mM and by a lag period and then a decay for higher values.

These patterns of change in crossbridge state as well as similar stepwise patterns observed in sarcomere length during active contraction under isometric constraints (Delay et al., 1981) are being analyzed in terms of a model based on the principles of catastrophe theory. In this model, the potential energy surface is treated as a property of a segment of the fiber, and the segment is considered as a large ordered array of interacting mechanochemical attached-state crossbridge elements. Each element in the myofilament lattice is treated as a two-state protein complex subject to allosteric control by chemical agents (ligands, product) or by mechanical loads that selectively stabilize one of the two states. In addition, it is assumed that mechanical coupling between elements within the lattice gives rise to an excess boundary free energy when two adjacent elements exist in different structural states. The magnitude of this excess free energy will depend on the degree to which the structural change of a given element induces a distortion in its environment.

The abruptness of ligand-induced transitions of the myofilament lattice from one crossbridge conformational state to another may thus be related to the necessity for a certain degree of ligand supersaturation in order to nucleate a structural change in a polymeric lattice that is subject to strong mechanical constraints (Oosawa and Higashi, 1967).

REFERENCES

Delay, M.J., Ishide, N., Jacobson, R.C., Pollack, G.H. and Tirosh, R. (1981). Stepwise sarcomere shortening: analysis by high-speed cinemicrography. Science 213: 1523-1525.
Oosawa, F. and Higashi, S. (1967). Statistical thermodynamics of polymerization and polymorphism of protein. Prog. Theor. Biol. 1: 79-165.

DISCUSSION

Discussion with Dr. K. Nishiyama focused on possible application of his multi-half-sarcomere model to observed abrupt macroscopic changes in sarcomere length and crossbridge state. Since the observed changes are of macroscopic magnitude, and the crossbridge changes occur on a relatively slow time scale, this implies involvement of a macroscopic volume element of the fiber. A model was suggested in which such a macroscopic fiber volume, N, (possibly a cross-section segment) could be taken as an array of many subpopulations [half sarcomeres, N_i] such that

$$N = \Sigma [N_1 , N_2 , N_3 , \ldots N_n].$$

Under certain conditions the entire volume, starting from, e.g., the rigor structural state, could become destabilized by, e.g., random saturation in the presence of a given medium concentration (chemical potential energy) of substrate or analog if lattice structural constraints (internal or external) existed to hinder cross-bridge state change.

It was agreed that, under these conditions, the fiber volume could be considered as a system that is taken to a far-from-equilibrium state. In such a state, the rigor structural state could become metastable with respect to an alternative relaxed structural state of the system if lattice constraints preventing the structural transition were of sufficient magnitude (sufficiently large value of transition activation energy). While the system existed in such a metastable state, fluctuations of individual crossbridges between relaxed and rigor conformations would be occurring. If such a fluctuation in one of the subpopulations (e.g., N_1) were to achieve an amplitude exceeding a critical value at some random point in time, it could supply the small, additional activation energy needed to trigger a transition of the system from the destabilized rigor structural state to a stable relaxed state. This triggered transition event would be initiated in one subpopulation (e.g., N_1) by achievement of a stable nucleus of relaxed-state tropomyosin. Once nucleated, the rigor to relaxed structural change could then propagate throughout the macroscopic volume via kinetic (i.e. dynamic) cooperative interactions between nearest-neighbor crossbridges (spontaneous growth). This process would have the energetic properties of a solid-state nucleated structural phase transition -- in particular, the requirement for a critical extent of destabilization (supersaturation).

Discussion with Dr. D. Martyn concerned possible use of a catastrophe cusp potential energy surface model to account for the occurrence of definite latency periods preceding the onset of contraction or relaxation. In particular, discussions revolved around observations obtained with ferret papillary muscle in which the event representing the onset of relaxation following a tetanic contraction period was confined to points along a contour line in the segment length-time plane. This type of model would be consistent with the independence of onset time with respect to load or initial length. The contour line might then be considered as the line of critical points in phase space at which an abrupt switch could occur in the state of tropomyosin from the potentiated state (T_{M2}) to the blocking state (T_{M1}). This abrupt change would be consistent with known critical properties of the relaxing system.

In addition, a cusp catastrophe potential energy surface representing the state of tropomyosin segments would be consistent with the observed appearance of hysteresis in relative tension - pCa plots. It would predict a delayed loss of tension (i.e., force-generating crossbridges) when the system was initially in an activated state (with T_M in a potentiated state) and then calcium concentration was gradually reduced. Conversely, it would predict a delayed onset of tension when the system was initially in the fully relaxed state (with T_M in the blocking state) and then calcium was gradually increased.

We concluded that it would be important to conduct studies in which the state of crossbridges and the state of tropomyosin could be both monitored to obtain direct evidence for the proposed non-linear relationship between crossbridge state and tropomyosin-troponin complex state.

DISTILLED WATER-INDUCED CONTRACTIONS IN DEHYDRATED AND SKINNED MUSCLE FIBERS

Reiji Natori

Department of Physiology, Jikei University School of Medicine, Minato-ku, Tokyo 105, Japan

ABSTRACT

Dehydrated frog skeletal muscle fibers, prepared by simply immersing the muscle fibers in pure glycerol for 2-3 min, showed a marked sustained tension development in response to distilled water (DW). Similar DW-induced tension responses were also seen in mechanically skinned muscle fibers. The DW-induced mechanical responses were rapidly relaxed by a conventional relaxing solution. The marked reproducibility of the DW-induced responses, together with the simplicity of the dehydration procedure, indicates that the dehydrated fibers can be used as a substitute for the skinned fibers in studying the mechanism of contraction.

In our laboratory, it has been known for many years that, when an air-dried frog skeletal muscle fiber is immersed in distilled water (DW), it shortens to less than 20% of the initial length (Natori & Shibuya, 1954; Natori, 1956). To study the above DW-induced contraction in more detail, we dehydrated single muscle fibers isolated from the sartorius muscles of the bullfrog (*Rana catesbeiana*) by immersing them in pure glycerol for 2-3 min (Fig. 1a, b). The dehydrated muscle fiber with both ends free exhibited transient shortening in response to a relaxing solution containing 120 mM KC1, 4 mM $MgCl_2$, 4 mM ATP and 4mM EGTA (pH 7.0 by MOPS) (Fig. 1c), and marked sustained shortening in response to DW (Fig. 1d).

More quantitative experiments were further performed in the isometric condition. As shown in Fig. 2a, the tension in the fiber rose slightly during the course of dehydration in glycerol, and when the fiber was transferred to the relaxing solution, a transient tension development followed by relaxation took place; the subsequent application of DW produced a marked sustained tension development, which relaxed rapidly on returning the fiber to the relaxing solution. Similar marked tension development by DW was also observed in mechanically skinned fibers (Fig. 2b), though the rate of rise of tension was appreciably smaller than that in the dehydrated fibers. On the other hand, the transient tension

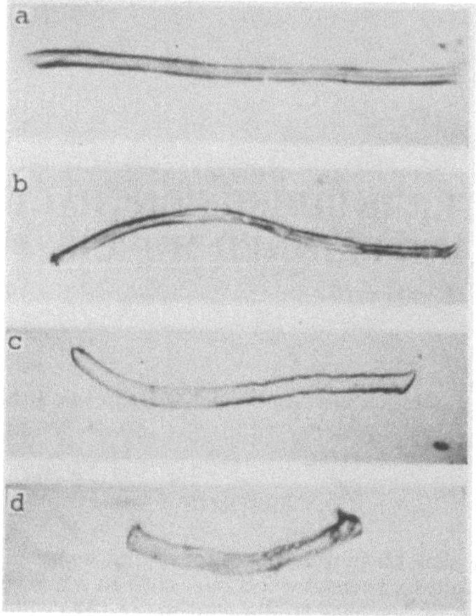

Figure 1: Photomicrographs showing free-ended shortening of dehydrated muscle fiber preparation. a: Isolated single muscle fiber in paraffin oil. b: The same fiber immersed in glycerol for dehydration. c: Transient shortening of the dehydrated fiber in the relaxing solution. d: Marked sustained shortening in DW.

development was hardly observable when skinned fibers were transferred from paraffin oil to the relaxing solution. When the dehydrated fibers were maximally activated in a contracting solution containing 2 mM $CaCl_2$, the peak height of the resulting tension development was much larger than that in response to DW (Fig. 2c).

The DW-induced mechanical response of the dehydrated fiber preparation was reproducible; the application of relaxing solution during the DW-induced mechanical response produced rapid relaxation, and reapplication of DW produced immediate tension redevelopment (Fig. 3a, b). Though the peak height of the response was variable to some extent, and sometimes tended to decrease each time of DW application, the degree of variability was relatively small (Fig. 3c). The tension development to DW was no longer observable if the fiber was previously relaxed to 50% of its *in situ* length.

When the length of the dehydrated fiber was suddenly decreased to a variable extent, the tension drop coincident with the sudden decrease in fiber length was followed by tension redevelopment, which was not so marked that the tension level attained was much lower than that of the DW-induced response at the same fiber length (Fig. 4). The above features of the DW-induced mechanical response was also seen in skinned fibers.

Although the mechanism of the DW-induced response remains to be investigated, its initiation seems to be associated with the release of Ca ions from the sarcoplasmic reticulum (SR) following rapid water intake of

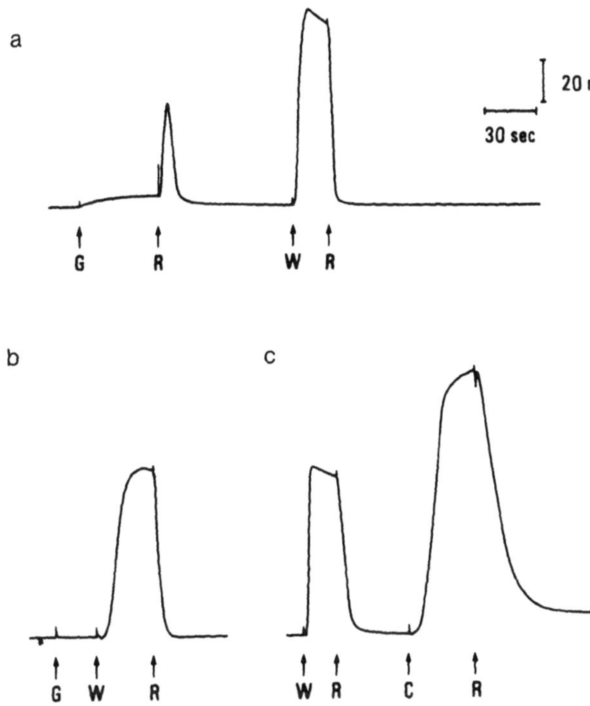

Figure 2: Tension responses in dehydrated and mechanically skinned muscle fibers. a: A muscle fiber was immersed in glycerol (G) for dehydration, transferred to the relaxing solution (R) which produced a transient tension development followed by relaxation, and then placed in DW (W) to result in a marked sustained tension response which relaxed rapidly in the relaxing solution (R). The time base between G and R is three times slower than the rest of the record. b: A mechanically skinned muscle fiber was transferred from paraffin oil to the relaxing solution (R) with no appreciable tension development, and then immersed in DW (W) to result in a marked sustained tension response which also relaxed rapidly in the relaxing solution (R). c: A dehydrated muscle fiber was first made to contract by DW (W), relaxed in the relaxing solution (R), and then maximally activated with a contracting solution containing 2 mM $CaCl_2$ (C).

the preparation. This idea is based on the result that, if $CaCl_2$ was applied to the fiber in which the height of the DW-induced response had been reduced appreciably after repeated application of DW, a large mechanical response to DW developed again, suggesting that the decrease of the DW-induced response results from a gradual decrease in the amount of Ca ions in the SR due to repeated application of DW and the relaxing solution,

Finally, it should be noted that the dehydrated fiber, which can be prepared by a much simpler method than that for making mechnically skinned fibers, may provide a good material for studying the mechanism of muscle contraction.

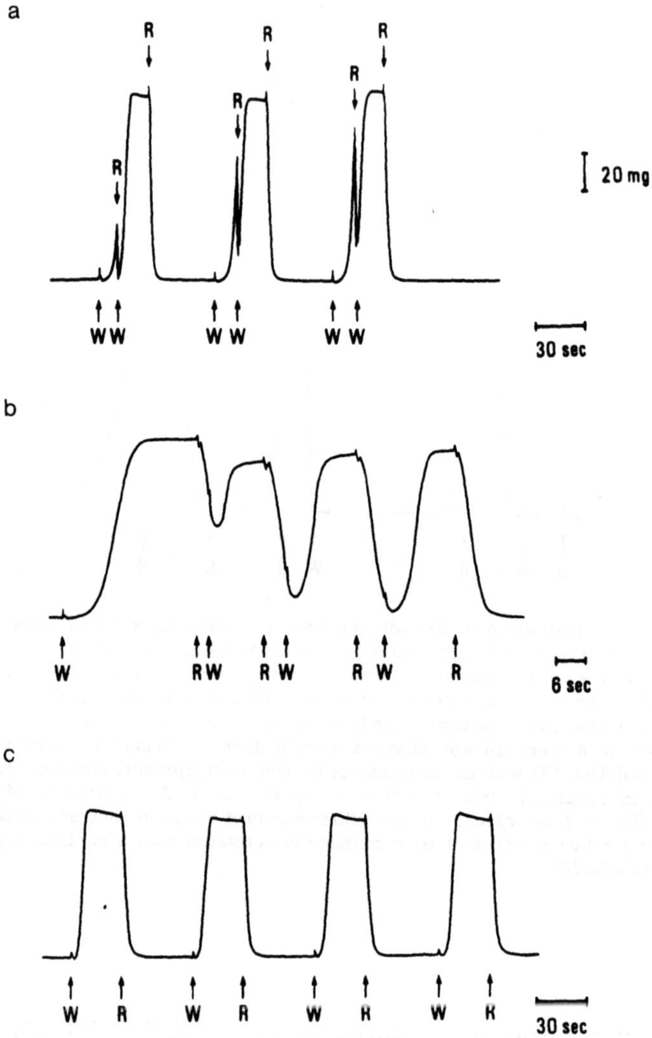

Figure 3: Reproducibility of tension development of the dehydrated fiber in DW (W) and its relaxation in the relaxing solution (R). In a and b, DW and the relaxing solution were applied alternately at various intervals to produce alternate tension development and relaxation. In c, a series of DW-induced responses of almost constant magnitude were produced by alternate application of DW and the relaxing solution at an appropriated interval.

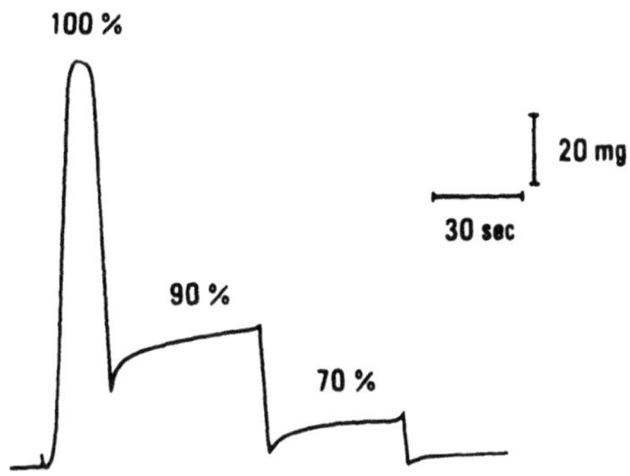

Figure 4: Effect of sudden decreases in fiber length on the DW-induced tension response. The fiber length relative to the initial length is indicated after each sudden decrease in fiber length. Note that the redevelopment of tension after each decrease in fiber length is not marked.

REFERENCES

Natori, R. and Shibuya, M. (1954). Contraction process of dried muscle fiber (in Japanese). Tokyo Jikeikai Ika Daigaku Zasshi 68: 930-932.

Natori, R. (1956). Differences in physiological properties of myofibrils. Small and large muscle fibers. Jikeikai Med. 3: 36-42.

DISCUSSION

Gillis inquired whether or not the distilled water-induced contraction in the dehydrated fibers was due to an effect of varying the ionic strength, since it has been reported at this meeting that muscles can be activated by reducing the ionic strength. In this connection, it was suggested to examine whether D_2O could cause similar contractions. Podolsky asked whether the fibers could shorten actively in distilled water. The answer was that, though the fibers can actually indeed shorten (Fig. 1), the shortening appeared to be not fully reversible, and that the distilled water-induced contraction was somewhat rigor-like in nature. Dr. Magid emphasized the fact that when rabbit or frog myofibrils are washed with distilled water they swell about 5 times, the distance between the myofilaments approaching 1,000 Å, and stressed that it would be quite remarkable if actin-myosin interaction still took place in such a condition.

CARDIAC MUSCLE MECHANICS

INTRODUCTION

The modern era of research in cardiac muscle began in the late fifties, when Abbott and Mommaerts (J. Gen. Physiol. 42: 533, 1959) began experiments with isolated strips of cardiac muscle. Unlike the situation with skeletal muscles, it is not possible to isolate strips of tissue from the heart with tendons at either end, the kind of specimen one prefers for mechanical experiments. The most useful preparation has been the papillary muscle, which interconnects the ventricular endocardial wall with the atrioventricular valves. This tissue is easily isolated, has one tendon, and at least in rare specimens may have a relatively uniform cylindrical shape.

This type of specimen has been used by various groups in rather futile attempts to elucidate the mechanics of heart muscle, until it became clear that an artifact of major proportions existed. Through use of microsphere markers embedded in the capillaries of the tissue and followed during contraction with video microscopy, Krueger and Pollack (J. Physiol. 251: 627, 1975) found that the central region of the specimen shortened grossly, even during isometric contraction, and stretched the ends, which in later studies had been confirmed to be damaged by clamping. Thus, many of the properties -- more often than not complicated and confusing -- that had been attributed to the contractile apparatus, actually reflected the series combination of undamaged and damaged tissue.

The observation of damaged ends made it abundantly clear that reliable information about cardiac mechanics could only be obtained from specimens in which the influence of the damaged ends was minimized or eliminated. This stimulated a number of approaches in which the properties of the presumably undamaged central segment were measured. Three such approaches are employed in the studies presented here. The presentation by Lee Huntsman, exploits the observation that muscle is an approximately constant volume system; i.e., that shortening of a segment is accompanied by a unique and predictable increase of its cross-sectional area. Huntsman and colleagues have developed an interesting method of measuring cross-sectional area continuously. By wrapping a donut-shaped coil around the specimen and placing it in an alternating magnetic field, they are able to monitor changes in local cross-section during contraction.

Two alternative methods are used by Housmans. First, two glass microelectrode tips are plunged into the belly of the specimen along its length. The image of these markers is projected onto a linear photodiode array, and the spacing between the two markers is tracked continuously. A supplementary method is laser diffraction, where the striation spacing of the central segment can be followed during contraction.

The questions of interest in the field of cardiac mechanics have, to a large extent, followed those of skeletal muscle mechanics. Several of these are approached in the presentations included in this section.

Mashima considers the time course of activation. In a modeling study, he considers the effect of the concentration of myoplasmic calcium on the time course of tension development.

A related question is considered by Huntsman. What is the mechanism of falloff of isometric tension at short sarcomere lengths? In cardiac muscle the ascending limb of the length tension relation is steeper than in skeletal muscle. It now appears that this is not due to a diminished release of calcium at short sarcomere lengths (Allen and Kurihara, J. Physiol. 327: 79). Huntsman and colleagues interpret the results of their mechanical experiments as indicating that the mechanism lies in a reduced sensitivity of the myofilaments to calcium at the shorter sarcomere lengths.

A third question concerns the degree of homogeneity along the specimen. Clearly, gross inhomogeneity is present at the ends. But, apart from that, there is the question of whether inhomogeneity exists in the central, undamaged region as well. The experiments presented by Housmans focus on the timing of the onset of relaxation, and indicate that this varies considerably along the length of the preparation.

Finally, some effort is being expended to investigate the kinetics of actomyosin interaction in cardiac muscle. Using barium contractures, Saeki and colleagues present the results of imposing quick changes of load on papillary muscles. The ensuing length transients show several phases, and these phases are interpreted in terms of cross-bridge mehanisms.

MODELING OF CARDIAC MUSCLE CONTRACTION BASED ON THE CROSS-BRIDGE MECHANISM

Hidenobu Mashima and Kazuyuki Kabasawa

Department of Physiology, School of Medicine, Juntendo University, Tokyo, Japan

ABSTRACT

A mathematical model was developed for the cardiac muscle contraction, assuming that the attachment and detachment cycle of the cross-bridge is activated by the internal calcium concentration and the rate constant of the cycle depends on the sliding velocity of myofilaments. The inputs of the model are the rates of calcium release and uptake, while the output is the tension curve of the muscle. The variables are factored into a series of realizable functions and most constants were determined from the dynamic constants for the tetanic contraction of frog ventricular muscle at 20° C. Using this model, the calcium transient curve as well as the change in the number of cross-bridges in each state of the cycle during a given experimental twitch tension curve was calculated with a PDP 11/60 computer, by selecting the input parameters so that the output curve fit the experimental curve. When the twitch tension was increased by increasing initial muscle length, the rate of calcium release increased and that of uptake decreased. At higher external calcium concentrations, the similar changes in the input parameters were observed. In the presence of 5×10^{-8} g/ml adrenalin the duration of activation was markedly prolonged, while the rates of calcium release and uptake show little change.

Since the sliding filament concept was proposed by A.F.Huxley (1957) and H.E.Huxley (1957), various kinetic models have been developed concerning the movement of cross-bridges between thick and thin filaments (Deshcherevskii, 1968; Volkenstein, 1969; Chaplain and Frommelt, 1971; Huxley and Simmons, 1971; Julian et al., 1974). Most of them, however, have dealt with steady state characteristics on the force response on sudden length changes during tetanic contraction. An attempt to develop a mechanochemical model which explained the twitch response as well as the steady contraction was made by Akazawa et al., (1976) for frog skeletal muscle. Although a similar model seemed to be applicable to cardiac muscle, the constants of the model could not be determined in an untetanizable muscle. In 1977, Mashima (1977a,b) succeeded in eliciting a steady tetanic contraction in frog ventricular muscle, and determined the dynamic constants and the internal load from an analysis of the force-load-velocity relation. Now we can start the modelling of frog cardiac muscle based on the cross-bridge mechanism. The degree to which the calculated data fit the experimental curves may help in evaluating the essential points of the mechanism.

Description of the Model

The model consists of three sub-systems: the first (sub-system I) is the Ca regulation system of the cross-bridge, the second (sub-system II) the kinetics of the cross-bridge cycle in relation to actin-myosin-ATP interaction and the third (sub-system III) the dynamics of tension output involving the contractile and series elastic components. The input to the whole system is the internal Ca concentration and the outputs are the tension and shortening of muscle. All of them are functions of time.

(1) Cross-bridge cycle (sub-system II)

The main part of the model is the cross-bridge cycle. Essentials of the molecular mechanism of the cross-bridge cycle can be illustrated as shown in Fig. 1. State R) represents the resting state, state 1) the activation phase of a site on the thin filaments through the binding of the Ca^{2+} to troponin, state 2) the force generating phase by the attachment of the

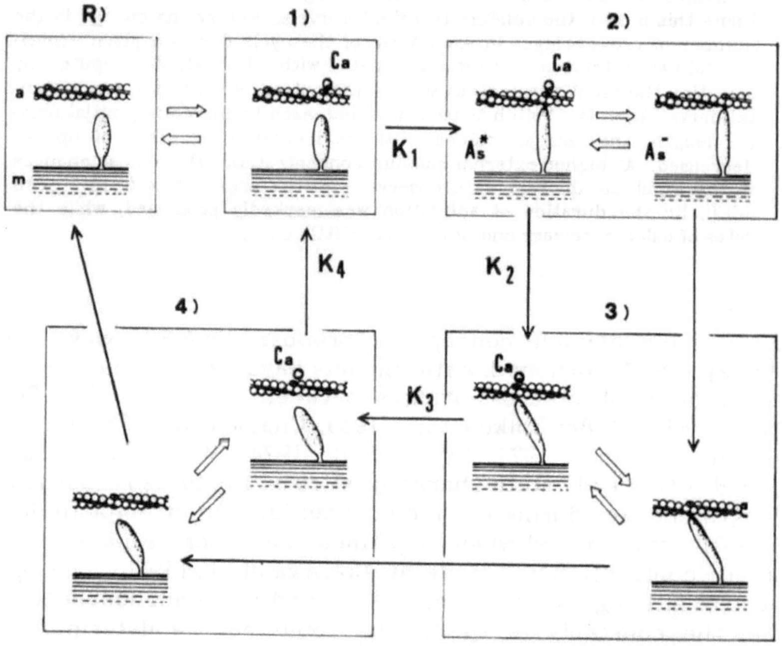

Figure 1: Schematic illustration of the cross-bridge cycle. The arrow → indicates the time order of reactions in the cycle; the arrow ⇒ indicates transition of the active site between active site with Ca(A_s^*) and that without Ca(A_s^-). a, thin filament; m, thick filament.

cross-bridge to the thin filament, state 3) the sliding phase by the movement of the cross-bridge and state 4) the detachment phase of the cross-bridge. When the site is activated by Ca^{2+}, its state immediately changes from R) to 1) and then 2) → 3) → 4) with rate constants of K_1, K_2, and K_3. If the Ca^{2+} concentration is sufficiently high, state 4) turns into state 1) with the rate constant of K_4, and the cyclic reactions will be repeated. One molecule of ATP is split during each cycle of reaction. However, if the Ca^{2+}concentration decreases, the bridges in state 4) or 1) will return to state R).

Assumption (1): The binding of Ca^{2+} to troponin and its dissociation take place uniformly over the whole length of the thin filament without a time lag, and after state 2) has been acquired, the reaction proceeds automatically to state 4) through state 3) even if Ca ions are removed thereafter. Therefore there must be two kinds of active site in states 2), 3), 4) and R), that is, active site with Ca (A_s^*) and without Ca (A_s^-).

Assumption (2): The rate constant K_2 increases proportionally with the shortening velocity, $v\,(=\dot{x})$,

$$K_2 = \begin{cases} \alpha_0+\alpha_1\dot{x} & \text{for shortening} \quad (\dot{x}\geq 0) \\ \alpha_0-\alpha'_1\dot{x} & \text{for lengthening} \quad (\dot{x}<0) \end{cases} \tag{1}$$

where α_1 and α'_1 are constants and α_0 represents the rate of attachment and detachment of cross-bridges in the state of no relative sliding movement.

Other symbols necessary for the present kinetics are as follows:

N : number of available active sites in states 1), 2), 3) and 4)
N^*: number of A_s^*
N^-: number of A_s^- in states 2), 3) and 4)
N^R: number of A_s^- in state R)
$\dot{N}_r$: rate of increase in N^*
$\dot{N}_u$: rate of decrease in N^* — input from sub-system I

n_1, n_2, n_3 and n_4: number of active sites in state 1), 2), 3) and 4), respectively. The number of force generating cross-bridges is n_2

n_1^*, n_2^*, n_3^* and n_4^*: number of A_s^* in state 1), 2), 3) and 4), respectively

n_1^-, n_2^-, n_3^- and n_4^- number of A_s^- in state 1), 2), 3) and 4), respectively (but $n_1^- = 0$).

$$N = n_1+n_2+n_3+n_4 \tag{2}$$

$$N^* = n_1^*+n_2^*+n_3^*+n_4^* \tag{3}$$

$$N^- = n_2^-+n_3^-+n_4^- \tag{4}$$

$$n_j = n_j^*+n_j^- \quad (j=1,2,3,4) \tag{5}$$

The rate of change in the number of activated active sites in the shift from state R) to 1) is $\dot{N}_r N^R/(N^R+N^-)$, where (N^R+N^-) is the total number of A_s^-. Further, when the active site at state j) changes from A_s^-

to A_s^*, the rate of change is $\dot{N}_r n_j/(N^R+N^-)$, $(j=2,3,4)$. Similarly, the rate of change from A_s^* to A_s^- is $\dot{N}_u n_j^*/N^*$, (j=1,2,3,4). Hence, the kinetic equation for A_s^* is

$$\left.\begin{aligned} \dot{n}_1^* &= \dot{N}_r \frac{N^R}{N^R+N^-} + K_4 n_4^* - K_1 n_1^* - \dot{N}_u \frac{n_1^*}{N^*} \\ \dot{n}_j^* &= \dot{N}_r \frac{n_j^-}{N^R+N^-} + K_{j-1} n_{j-1}^* - K_j n_j^* - \dot{N}_u \frac{n_j^-}{N^*} \qquad (j=2,3,4). \end{aligned}\right\} \quad (6)$$

Similarly, the equation for A_s^- is

$$\dot{n}_j^- = \dot{N}_u \frac{n_j^*}{N^*} + K_{j-1} n_{j-1}^- - K_j n_j^- - \dot{N}_r \frac{n_j^-}{N^R+N^-} \qquad (j=2,3,4). \quad (7)$$

From Eqs. (5)-(7), n_j is expressed by a simple equation,

$$\dot{n}_j = K_{j-1} n_{j-1} - K_j n_j \qquad ((j=2,3,4). \quad (8)$$

(2) Tension output (sub-system III)

The dynamic behavior of the muscle can be described by the two component model, which contains a contractile component (CC) and a series elastic component (SEC). But in cardiac muscle, when the initial length is less than 0.9 L_m (L_m is the optimum length at which the maximum force, F_m, can be generated), some internal load (IL) in proportion to the shortening velocity should be introduced (Mashima, 1977b). Therefore, we adopted the contraction model of cardiac muscle shown in Fig. 2. As suggested by Mashima et al. (1972), we assume that each cross-bridge generates a proper force, f, accompanying a proper velocity-dependent force-loss, f_v. Then, a net force produced by a single cross-bridge is $(f - f_v)$. As the number of cross-bridges generating forces is n_2, the total force, P, becomes

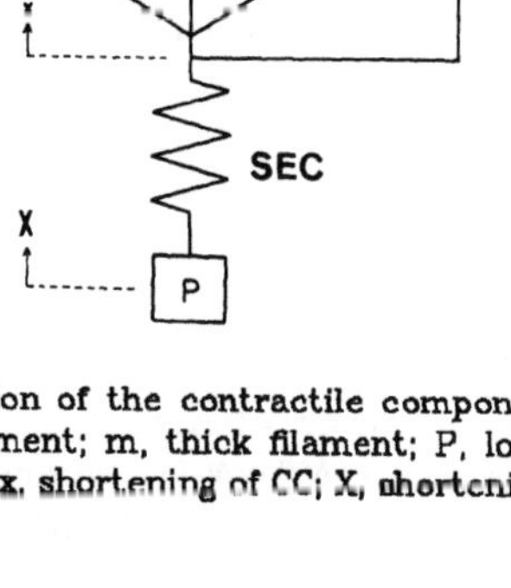

Figure 2: Schematic illustration of the contractile component (CC) and the series elastic component (SEC). a, thin filament; m, thick filament; P, load; f, force; f_v, force-loss in a cross-bridge; IL, internal load; x, shortening of CC; X, shortening of muscle.

$$P = n_2(f - f_v) - \gamma \dot{x} \tag{9}$$

where γ is the constant of IL. In this study, however, only isometric contractions at 0.9 L_m are treated, so that, γ=0. The force-loss, f_v, is assumed simply to be proportional to the velocity,

$$f_v = \begin{cases} \beta \dot{x} & (\dot{x} \geqq 0) \\ \beta' \dot{x} & (\dot{x} < 0) \end{cases} \tag{10}$$

where β and β' are the constants of force-loss.

As for the SEC, the tension-extension relation is

$$P = E_s x \tag{11}$$

where E_s is the elastic coefficient of the SEC. The experimental equation for this relation was obtained from data in frog ventricular muscle at 0.9 L_m (Mashima 1978). That is,

$$E_s = (0.8 + 1.7 \times 10^7 \hat{x}^6) F_m / L_m \tag{12}$$

where $\hat{x} = x / L_m$.

(3) Ca regulation (sub-system I)
The number of activated cross-bridges, N^*, is expressed as follows,

$$N^* = L_p(L) \cdot D([Ca]) \tag{13}$$

where L_p is the effective interaction length, which is a function of muscle length, L, and D is the density of A_s^*, which is a function of Ca concentration. In the isometric contraction, $dL_p/dL = 0$. Therefore, we obtain

$$\dot{N}^* = dD([Ca])/d[Ca] \cdot [\dot{Ca}] \cdot L_p(L) \tag{14}$$

On the other hand, the rates of change in [Ca] and N^* are

$$[\dot{Ca}] = R_{cr} - R_{cu} \tag{15}$$

$$\dot{N}^* = \dot{N}_r - \dot{N}_u \tag{16}$$

where R_{cr} is the rate of release of Ca from the store, R_{cu} the rate of uptake into the store, $\dot{N}_r$ the rate of increase in N^* resulting from Ca release and $\dot{N}_u$ the rate of decrease in N^* caused by Ca uptake. Hence, the outputs to sub-system II are,

$$\begin{cases} \dot{N}_r = \dfrac{dD([Ca])}{d[Ca]} R_{cr} L_p(L) \\ \dot{N}_u = \dfrac{dD([Ca])}{d[Ca]} R_{cu} L_p(L) \end{cases} \tag{17}$$

Thus, the Ca supply to the contractile system can be calculated, if R_{cr} and R_{cu} are given.

According to Endo (1975), R_{cu} is proportional to the [Ca]. Then

$$R_{cu} = K_u([Ca] - Ca_o) \tag{18}$$

where K_u is the constant of Ca uptake, and Ca_o is the threshold or initial concentration. As $D([Ca])$ must be proportional to the tension, we obtain the following equation from the data of Fabiato and Fabiato (1975),

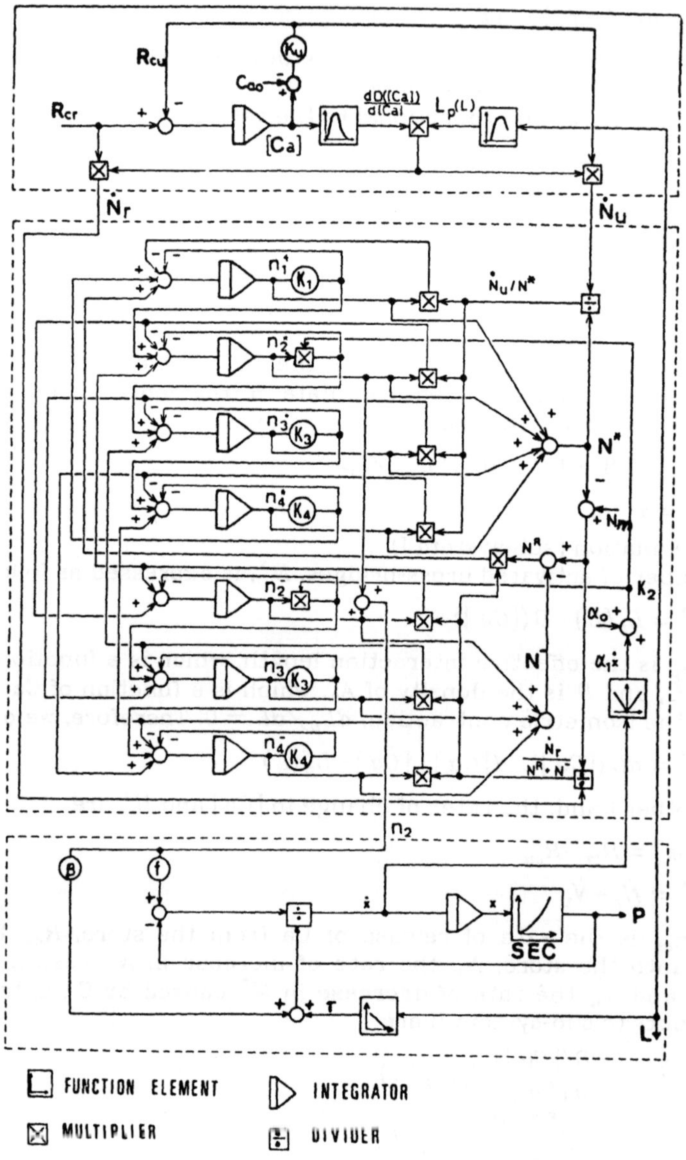

Figure 3: Program for a computer derived from the model. Sub-system I, Ca supply system; Sub-system II, kinetics of the cross-bridge cycle; Sub-system III, dynamics of contraction.

$$\frac{D([Ca])}{D_0} = 1 - \frac{1}{1+K_c[Ca]^2} \tag{19}$$

where K_c is $10^{12}\mathrm{mol}^{-2}$ and D_0 is the value of D([Ca]) at $[Ca] = 10^{-5}$ mol. And Ca_0 was estimated as 0.1×10^{-6} mol. As for $L_p(L)$, the following equation was introduced from the tension-length relation of our preparation

(Mashima 1977a),

$$L_p(L)/L_p(L_m) = -1.7 + 2.7L/L_m \tag{20}$$

(4) The whole model
Finally, the whole model can be constructed by connecting the above-described three sub-systems. The computer program of this model is shown in Fig. 3.

Determination of the Constants

(1) Force-velocity relation
In the steady state contraction, $n_2=n_3=n_4=0$. Hence, from Eqs. (2) and (8), we obtain

$$n_j = \frac{N}{K_j}\frac{KK_2}{K+K_2} \qquad (j=1,2,3,4) \tag{21}$$

where

$$\frac{1}{K} = \frac{1}{K_1}+\frac{1}{K_3}+\frac{1}{K_4} \tag{22}$$

When the muscle is shortening with a constant velocity, v, the theoretical force-velocity relation can be obtained from Eqs. (1), (9), (10) and (21),

$$v\left[\frac{NK\beta}{\alpha_1} + P\right] = \frac{K+\alpha_0}{\alpha_1}\left[\frac{NKf}{K+\alpha_0} - P\right] \tag{23}$$

Comparing Eq. (22) with the force-load-velocity relation for the steady contraction of tetanized frog ventricular muscle (Mashima, 1977b),

$$v(a+P) = b(F_m-P) \tag{24}$$

where a and b are Hill's dynamic constants and F_m is the maximum isometric force, we obtain following relations,

$$a = \frac{N_mK\beta}{\alpha_1}\ ,\ b = \frac{K+\alpha_0}{\alpha_1}\ ,\ F_m = \frac{N_mKf}{K+\alpha_0} \tag{25}$$

where N_m is the value of N at F_m.

(2) Energetic relation
The rate of total energy output, E, of the model is

$$\dot{E} = e_aK_2n_2 \tag{26}$$

where e_a is the splitting energy of one molecule of ATP. Substituting Eqs. (1), (21), (25) into Eq. (26), we obtain,

$$\dot{E} = \frac{e_aN_mK^2/\alpha_1 b}{F_m+a}\left\{\frac{\alpha_0(F_m+a)}{K} + F_m - P\right\} \tag{27}$$

On the other hand,

$$\dot{E} = \dot{W}+\dot{H}_s+\dot{H}_m \tag{28}$$

where W is the external work, H_s the shortening heat and H_m the maintenance heat. From Eq. (24), $\dot{W} = Pv = Pb(F_m-P)/(P+a)$. Assuming

$\dot{H}_s = n_2 f_v v$ and substituting Eqs. (21), (10), (24), we obtain

$$\dot{H}_s = \frac{ab(F_m - P)^2}{(F_m + a)(P + a)}$$

According to Hill and Woledge (1962), the maintenance heat, $\dot{H}_m$, is estimated as λab and for frog skeletal muscle $\lambda = 1$. For cardiac muscle λ was estimated as about 0.5 at 20°C from the data of Gibbs and Loiselle (1978). After all,

$$\dot{E} = \frac{bF_m}{F_m + a}\left\{\frac{\lambda a}{F_m}(F_m + a) + F_m - P\right\} \tag{29}$$

Comparing Eq. (29) with Eq. (27), we obtain

$$bF_m = \frac{e_a N_m K^2}{\alpha_1 b}, \qquad \frac{a}{F_m}\lambda = \frac{\alpha_0}{K} \tag{30}$$

(3) Determination of the constants

Using Eqs. (25) and (30), the constants of the model can be determined by measuring the values of a, b, F_m, L_m, N_m and e_a. These values were determined by Mashima (1977a,b) in frog ventricular muscle at 20°C, that is, $a/F_m = 0.51$, $b/L_m = 0.75\text{sec}^{-1}$, $F_m = 0.46\text{Kg/cm}^2$, $L_m = 1.1\mu\text{m}$ (half sarcomere). The value of e_a must be 7.7×10^{-13} erg and N_m was estimated as $0.55\times10^{13}/\text{cm}^2$/half sarcomere by Huxley (1972) in frog skeletal muscle. Thus, the numerical values of α_0, α_1, β, f and K were determined as shown in Table 1. The constants of lengthening, α'_1 and β', were also determined from the force-velocity constants a' and b' measured in the lengthening experiment (Mashima 1977b). Assuming $K_1 = K_3 = K_4 = 3K$, all constants were determined.

Table 1: Constants of the model (frog ventricular muscle at 20°C)

a/F_m	0.51	K	$11\ \text{sec}^{-1}$
a'/F_m	0.39		$(K_1 = K_3 = K_4 = 33\text{sec}^{-1})$
b/L_m	0.75	α_0	$2.8\ \text{sec}^{-1}$
b'/L_m	0.75	α_1	$18.4/L_m$
F_m	$460\ \text{g/cm}^2$	α'_1	$18.4/L_m$
L_m	$1.1\times10^{-4}\text{cm}$	f	$1.25\ F_m/L_m$
N_m	0.55×10^{13}	β	$0.85\ F_m\ \text{sec}/L_m N_m$
e_a	$7.7\times10^{-13}\text{erg}$	β'	$4.0\ F_m\ \text{sec}/L_m N_m$

Constants on the right were calculated from the experimental data on the left.

Calculation by the Model

(1) Input parameters

In order to calculate the twitch contraction, $R_{cr}(t)$ and K_u must be given. Taking account of the suggestion by Niedergerke (1963), the time course of Ca release was estimated as exponential

$$R_{cr} = C(1-e^{-t/T_r}) \qquad (t \leq T_d)$$
$$\quad = C(e^{T_d/T_r}-1)e^{-t/T_r} \qquad (t > T_d) \qquad (31)$$

where T_r is the time constant, T_d the duration and C the maximum rate of Ca release. Hence, the imput parameters of the model were T_r, T_d, C and K_u.

(2) Simulation of twitch curves

Using the program shown in Fig. 3, twitch tension curves were simulated with a PDP-11/60 computer. All tension curves were obtained at 0.9 L_m or less and at 20°C in frog ventricular strip. Selecting the input parameters (T_r, C, T_d and K_u), the best fit curve was found for the experimental tension curve. The fit was satisfactory as shown in Fig. 4. Thus the most probable values of the input parameters were determined. Then, using these parameters, the time courses of R_{cr}, R_{cu}, d[Ca]/dt(= $R_{cr}-R_{cu}$), n_1, n_2, n_3, n_4 and N during the twitch were all depicted as shown in Fig. 5.

Repeating similar procedures, changes in the input parameters were examined for various tension curves.

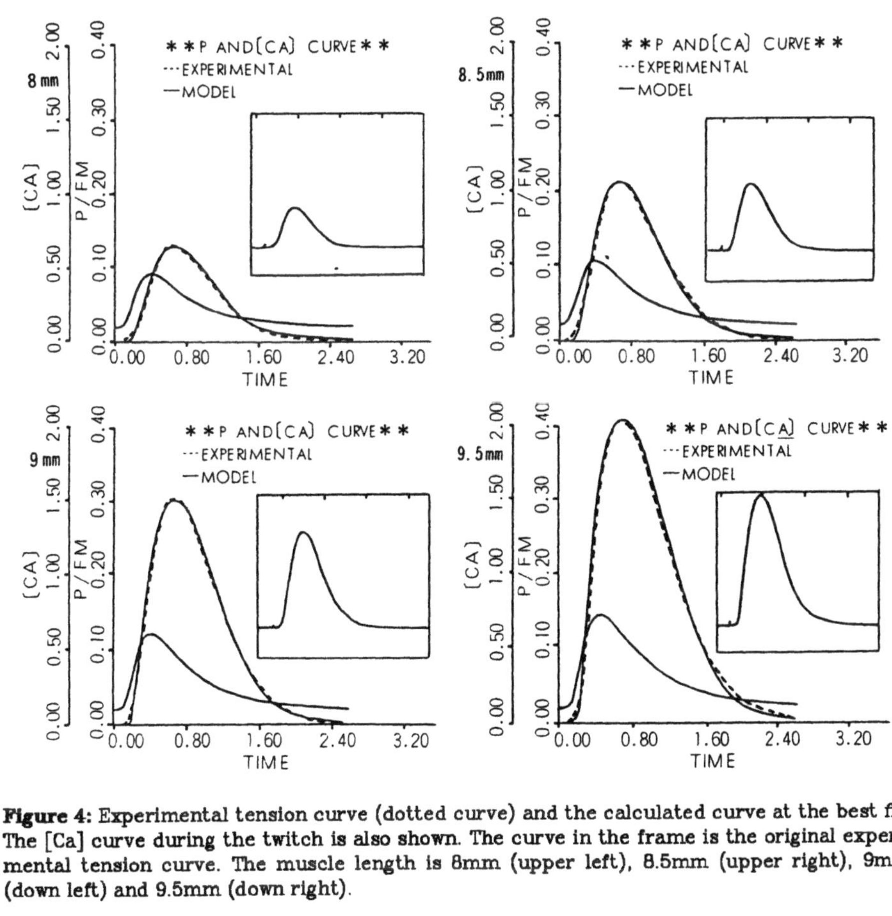

Figure 4: Experimental tension curve (dotted curve) and the calculated curve at the best fit. The [Ca] curve during the twitch is also shown. The curve in the frame is the original experimental tension curve. The muscle length is 8mm (upper left), 8.5mm (upper right), 9mm (down left) and 9.5mm (down right).

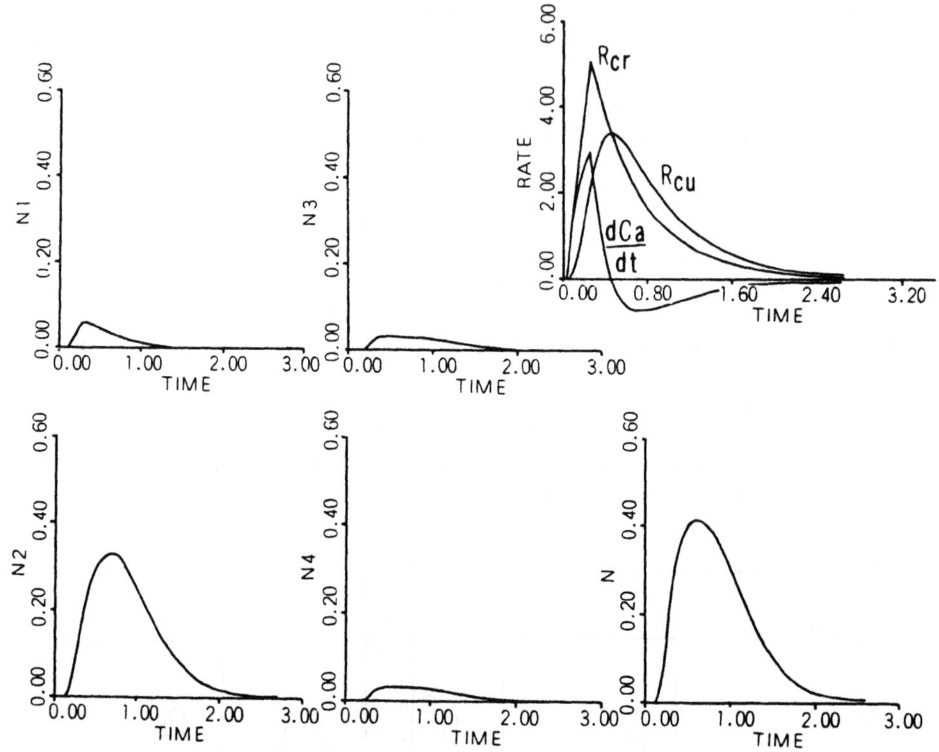

Figure 5: Time courses of n_1, n_2, n_3, n_4, N, R_{cr}, R_{cu} and d([Ca])/dt during the isometric twitch at 9.5mm (=0.9 L_m).

(a) Effect of length

In Fig. 4, the dotted curves are experimental tension curves for frog ventricular muscle at lengths of 9.5 mm (0.90 L_m), 9.0 mm (0.86 L_m), 8.5 mm (0.81 L_m) and 8.0 mm (0.76 L_m). The solid lines are the simulated best fit. The values of input parameters are shown in Table 2. T_r does not alter at all lengths, but C and the effective rate of release, C/T_r, increase and K_u decreases with increasing length. The calculated [Ca] curves are also shown in Fig. 4. The peak of the [Ca] curve clearly precedes the peak of the tension curve conforming to the calcium transient curve reported by Allen and Blinks (1978) in frog atrial muscle.

(b) Effect of external calcium concentration

When the external Ca concentration was changed from 1.8 mM to 3.6 mM or 9 mM, the twitch was potentiated, and the input parameters were estimated as shown in Table 2. The rate of release of Ca, C/T_r, increases and the rate of uptake, K_u, decreases with increasing Ca concentration. The value of Ca_o was usually 0.1×10^{-6} mol but in 1.8 mM Ca the best fit was obtained at $Ca_o = 0.05\times10^{-6}$ mol.

(c) Effect of adrenaline

When 5×10^{-6} g/ml of adrenaline was added to the external solution, the twitch was potentiated and the input parameters were changed as shown in Table 2. The slight increase in the rate of Ca release and the slight decrease in the rate of Ca uptake were seen, but the most remarkable change was a prolongation in the duration of Ca release, T_d. Similar changes were also observed in the aequorin signals.

Table 2: Input parameters

1) Effect of Length (T_d = 0.2 sec, Ca_o = 0.1×10^{-6} mol)

L(mm)	T_r(sec)	C(10^{-6} mol)	C/T_r	K_u(sec^{-1})
9.5	0.48	15.0	31.3	0.370
9.0	0.48	14.8	30.8	0.488
8.5	0.48	14.2	29.6	0.598
8.0	0.48	11.8	24.6	0 725

2) Effect of external Ca concentration

Ca(mM)	T_r(sec)	C(10^{-6} mol)	C/T_r	K_u (sec^{-1})	Ca_o(10^{-6} mol)
9.0	0.15	3.80	25.3	0.165	0.1
3.6	0.25	3.60	14.4	0.350	0.1
1.8	0.45	5.20	11.6	0.670	0.05

3) Effect of adrenaline (5×10^{-8} g/ml)

	T_r (sec)	C(10^{-6} mol)	C/T_r	K_u (sec^{-1})	T_d (sec)
control	0.43	9.10	21.2	0.252	0.2
adrenaline	0.65	13.50	20.8	0.295	0.3

ACKNOWLEDGEMENTS

This work was supported by a Grant-in-Aid for Scientific Research from the Ministry of Education, Science and Culture of Japan and a grant from Takeda Foundation.

REFERENCES

Allen, D.G. and Blinks, J.R. (1978). Calcium transients in aequorin- injected frog cardiac muscle. Nature 273: 509-513.

Akazawa, K., Yamamoto, M., Fujii, K. and Mashima, H. (1976). A mechanochemical model for the steady and transient contractions of the skeletal muscle. Jpn. J. Physiol. 26: 9-28.

Chaplain, R.A. and Frommelt, B. (1971). A mechanochemical model for muscular contraction. I: The rate of energy liberation at steady state velocities of shortening and lengthening. J. Mechanochem. Cell Motility 1: 41-56.

Deshcherevskii, V.I. (1968). Two models of muscular contraction. Biofisika 13: 1093-1101.

Fabiato, A. and Fabiato, F. (1975). Contractions induced by a calcium- triggered release of calcium from the sarcoplasmic reticulum of single skinned cardiac cells. J. Physiol. 249: 469-495.

Gibbs, C. and Loiselle, D. (1978). The energy output of tetanized cardiac muscle: species differences. Pflügers Arch. 373: 31-38.

Hill, A.V. and Woledge, R.C. (1962). An examination of absolute values in myothermic measurements. J. Physiol. 162: 311-333.

Huxley, A.F. (1957). Muscle structure and theories of contraction. Prog. Biophys. Biophys. Chem. 7: 255-318.

Huxley, A.F. and Simmons, R.M. (1971). Proposed mechanism of force generation in striated muscle. Nature 233: 533-538.

Huxley, H.E. (1957). The double array of filaments in cross-striated muscle. J. Biophys. Biochem. Cytol. 3: 631-648.

Huxley, H.E. (1972). The molecular basis of contraction in cross-striated muscle. In: *Structure and Function of Muscle* , 2nd ed., ed. by Bourne, G.H., Academic Press, London, Vol. 1, 301-387.

Julian, F.J., Sollins, K.R. and Sollins, M.R. (1974). A model for the transient and steady state mechanical behavior of contracting muscle. Biophs. J. 14: 546-562.

Mashima, H., Akazawa, K., Kushima, H. and Fujii, K. (1972). The force-load-velocity relation and the viscous-like force in the frog skeletal muscle. Jpn. J. Physiol. 22: 103-120.

Mashima, H. (1977a). Tetanic contraction and tension-length relation of frog ventricular muscle. Jpn. J. Physiol. 27: 321-335.

Mashima, H. (1977b). The force-load-velocity relation and the internal load of tetanized frog cardiac muscle. Jpn. J. Physiol. 27: 485-501.

Mashima, H. (1978). Dynamics of contraction with special reference to calcium. Recent Advances in Studies on Cardiac Structure and Metabolism, 11, Heart Function and Metabolism, 149-157.

Niedergerke, R. (1963). Movements of Ca in beating ventricles of the frog heart. J. Physiol. 167: 551-580.

Volkenstein, M.V. (1969). Muscular contraction. Biochim. Biophys. Acta 180: 562-572.

DISCUSSION

HOUSMANS: Professor Mashima, it seemed to me that that you didn't take into account the transmembrane flux of calcium which is believed to be the main source of activation. In your model it's considered to be coming from calcium stores. Could you clarify that?

MASHIMA: The calcium concentration curve of my model is an input function necessary for obtaining the output curve which simulates the twitch tension curve. The calcium concentration means the concentration which directly activates the myofilaments.

EDMAN: But how do you presume that the outside calcium is influencing the contractility? Does it go in directly and activate or do you assume that it goes into a store and then after that is released by the activation?

MASHIMA: My model does not concern these things. Some calcium may come from outside and work on the stored calcium. And finally, the calcium concentration near the filaments will be raised. My input parameters determine only the shape of this final calcium concentration change. For example, if we raise the external calcium concentration, it is necessary to decrease the rate of calcium release (T_r) and uptake (K_u) to simulate the tension curve (see Table 2).

NOBLE: Well, on the same question, I don't think the problem arises if you solve for the steady state. The internal store problem only comes in if you're talking about transients from one beat to the next.

MASHIMA: Although all constants of the model were determined from the measurable constants of the steady state contraction, the model could even explain the transient twitch contractions. This is the most important point. The calcium curve was calculated as a necessary process to simulate a given twitch curve.

MARTYN: On that point, all the available evidence on the time course of activator calcium indicates that length *per se* doesn't appear to have much of an influence on the rate of uptake of calcium. If anything, it seems to prolong the aequorin transient at short lengths, which is inconsistent with your model.

MASHIMA: The rate of release of calcium (T_r) did not show much change, but the rate of uptake (K_u) slightly decreased as the muscle length increased (Table 2).

TER KEURS: You introduced a length-dependent activation factor, L_p. Could you elaborate on that? The other aspect of the length issue is the introduction of an L_{max}, while, as many measurements on the level of the sarcomere suggest, you cannot find an L_{max} in cardiac muscle.

MASHIMA: I have measured L_{max} in my preparation. L_p means an effective length. On the ascending limb, L_p should be the overlap length, but on the descending limb, it is very difficult to realize the overlap length. As was discussed yesterday, L_p may be an activation factor. But mathematically L_p is expressed by the tension-length curve, or the length-dependent tension development.

TER KEURS: In that equation is L_p just a linear relation?

MASHIMA: Yes. (see Eq (20)), in frog ventricular muscle.

KRUEGER: All of the studies I've seen which have tried to control internal length have shown that the form of the isometric contraction is altered by imposing the control. Do you find, if you stiffen your series elastic component that you can alter the form? I'm not talking about just the amplitude but the actual shape of isometric tension waveform.

MASHIMA: If we alter the stiffness too much, sometimes the computer showed oscillation. So I tried to increase stiffness only a little, in that case, the rising phase of the isometric contraction was steepened but the falling phase did not show much change.

HOUSMANS: One last brief comment on what Dr. Martyn said about the influence of length on the intracellular calcium transients. This has been the work of Drs. Gordon and Ridgway (Europ. J. Cardiol. 7(suppl): 27-34, 1978), of Allen and Blinks (Nature, 273: 509-513, 1978) and also of Allen and Kurihara (J. Physiol. 212: 68-69, 1979; J. Physiol. 305: 29-30P, 1980 and J. Physiol. 310: 75-76P, 1980). From these results it is clear that one must be cautious in applying the interpretation of calcium transients obtained in one species to another species. There are important differences even within the same species between atrial and ventricular cardiac muscle, so that one cannot necessarily apply the conclusions from one species or tissue to the case of frog ventricular muscle.

THE DEPENDENCE OF FORCE AND VELOCITY ON CALCIUM AND LENGTH IN CARDIAC MUSCLE SEGMENTS

Donald A. Martyn, Jeff F. Rondinone and Lee L. Huntsman

Center for Bioengineering WD-12, University of Washington, Seattle, WA 98195

ABSTRACT

The segment length (SL) dependence of force (F) and light load shortening velocity (V_L) was determined for central segments of ferret papillary muscles at different extracellular calcium concentrations. Muscles were maintained at 27°C in a physiological solution which contained in mM: NaCl 140; KCl 5.0; $MgSO_4$ 1.0; NaH_2PO_4 1.0; acetate 20; the pH was 7.4. Calcium concentrations were 1.125, 2.25, 4.5 and 9.0 mM.

Total force-segment length relations were determined from both muscle length isometric (auxotonic) and segment isometric contractions, and were found to be the same for each contraction mode. The peak force generated at a particular segment length was independent of both the amount of shortening during a contraction and the initial SL. Increasing extracellular Ca^{2+} shifted the F-SL relation toward greater force and the SL axis intercept toward shorter SL. Maximum peak twitch tension was achieved in 9.0 mM Ca^{2+} Calcium variations also changed the shape of the total F-SL relation from linear in high Ca^{2}+, to concave in low Ca^{2+}

In order to estimate the active F-SL relations, corrections were made for passive force by two methods. The first assumed that passive force was related to SL, and yielded F-SL relations which were nearly identical to those found for total force. This similarity included the curvature changes observed in different Ca^{2+} concentrations, a finding which is consistent with the hypothesis that length dependent activation is the cause of force decline at short SL. The second method assumed passive force to be related to muscle length, an approach which would be appropriate if, for example, a connective tissue sheath on the muscle dominated passive behavior. These F-SL curves displayed a plateau above 90% SL_{max} and appeared to be vertically shifted versions of each other. Such characteristics are consistent with the possible role of an internal load in causing the decline of force at short SL.

V_L–SL relations were obtained from load clamps to 1 mM, imposed at various times during a segment isometric twitch. The results indicate that 1) V_L declines linearly with SL below 90% SL_{max} and 2) V_L–SL relations are shifted to higher velocity and shorter SL axis intercepts by increasing Ca^{2+}. The slopes of the V_L–SL relations obtained in different calciums are similar. Although an internal load could explain the calcium dependence of V_L, it would not explain the similarity of the slopes of the V_L–SL relations found in different calciums.

INTRODUCTION

The physiological basis of the Frank-Starling law of the heart has been the subject of considerable experimentation and controversy over the past several years. The mechanism which causes force to decline at sarcomere lengths below 2.0 μ is not known for skeletal or cardiac muscle (Gordon et al., 1966) (Taylor, 1974). However, force-sarcomere length relations in cardiac muscle exhibit a steeper decline, than found for skeletal muscle, and may provide a particularly useful preparation for elucidating the mechanism (Pollack et al., 1976) (Julian et al., 1975). Whereas force declines below a sarcomere length of 2.0 μ and reaches zero at about 1.2 μ in frog skeletal fibers, zero force production was obtained at 1.6 μ in cardiac muscle. On the other hand, it has been demonstrated that maximally activated skinned single cardiac cells exhibit a decline of force with decreasing sarcomere length which was much less pronounced than either skeletal or intact cardiac preparations (Fabiato et al., 1975). Beyond the question of mechanism, the observation of considerable contractile inhomogeniety in isolated cardiac muscle (Huntsman et al., 1974) (Krueger et al., 1975) has necessitated reinvestigation of force length relations in sarcomeres or muscle segments.

A number of factors which could influence force-sarcomere length relations have been proposed and can roughly be grouped as those invoking an internal load, which hinders shortening and force production either by myofilament hinderance (Gordon et al., 1966) or by connective tissue on the cell surface (Winegrad, 1980), and those involving length dependent activation (Jewell, 1977). The relatively flat force-length curve obtained by maximally activated skinned cardiac cells (Fabiato et al., 1975) could be explained by the removal of an internal load. On the other hand, evidence has been accumulating that the level of activation declines with decreasing sarcomere length. Data indicate that myofilament sensitivity to calcium declines with sarcomere length in cardiac muscle (Fabiato et al., 1978) (Hibberd et al., 1980) and in barnacle skeletal muscle (Gordon et al., 1978). In addition, the level of activation could be influenced by initial sarcomere length (ter Keurs et al., 1980) or by the amount of shortening done to reach a given sarcomere length (Edman, 1975).

The purpose of the experiments reported here has been to elucidate the influence of length on force and shortening in undamaged segments of isolated ferret papillary muscles. Both force-segment length and lightly loaded velocity-segment length relations have been determined together with the effect of altered extracellular calcium concentration. The results indicate that length dependent myofilament sensitivity is probably the dominant mechanism. Some candidate mechanisms are ruled out by the data directly, while others, if present, are restricted to certain characteristics.

METHODS

Right ventricular papillary muscles were obtained from ferrets anesthetized with sodium pentobarbitol (30 mg per kg). Following dissection, the muscles were mounted in a mechanical testing apparatus which has been described previously (Huntsman et al., 1974). Rapidly flowing bathing solution was provided at 27°C fully aerated with 95% O_2 and 5% CO_2. The solution contained, in mM: NaCl 140, KCl 5, $CaCl_2$ 2.25 (normal); $MgCl_2$ 1.0, $NaHCO_3$ 24; NaH_2PO_4 1.0 and acetate 20. Extracellular calcium was either 1.125 mM, 2.25 mM, 4.5 mM or 9 mM. All other constituents remained the same. Field stimulation at 12 per minute was delivered by 5 millisecond current pulses between platinum electrodes, which were parallel to the muscle.

Measured variables included force (F), segment length (SL), segment velocity (V) and muscle length (ML). Force is normalized to muscle cross sectional area and expressed as mN/mm^2. Muscle length was normalized to the value when developed force was maximal (ML_{max}).

Segment length was determined by measurement of the cross sectional area of the muscle near its center, using a magnetic sense coil (Huntsman et al., 1979). Each muscle was fitted with a coil of appropriate size, with care taken that the coil maintained a constant position in contact with the muscle surface over the physiologic range of muscle lengths. Our experience indicates that segments throughout the central region of the muscle behave identically, so sense coils were positioned conveniently near the center of each preparation. The muscle was allowed to contract under muscle length isometric conditions at ML_{max} for at least an hour to allow full equilibration. Muscle length was maintained at ML_{max} throughout the experiment, except during test perturbations. It was a consistent observation that even though ML could be increased to beyond 110% ML_{max}, SL increased very little beyond that found at ML_{max}; such increases being accompanied by a steep increase in passive force. Segment length was then normalized to this maximal value (SL_{max}). Similar observations of sarcomere length behavior in cardiac muscle have been made by other investigators (Julian et al., 1975) (ter Keurs et al., 1980).

Segment velocity (V) was determined by both analog differentiation and numerical differentiation of the SL signal. Segment velocity is expressed as SL_{max}/sec.

All variables, together with selected timing and control signals, were sampled at 1 or 5 millisecond intervals by an on-line digital computer to ten bit precision. This resulted in least-significant bit resolutions of 0.1 mN force. 10 microns muscle length, and 0.03% segment length.

Except where experimental traces of force, muscle length and segment length are shown, data presented in each figure represent the mean (± standard error) of no fewer than 5 measurements, each from a different muscle. No error bars are shown for data when fewer than 5 measurements were available.

RESULTS

Force-Segment Length Relations

The relationship between force and segment length was determined by two methods; 1) muscle length was held constant, allowing central seg-

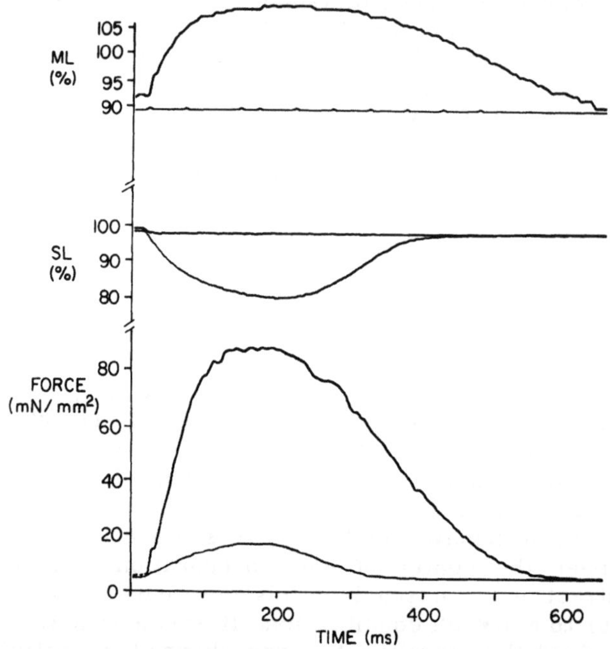

Figure 1: Force (bottom) and segment length (middle) and ML (top) traces for SL auxotonic (lower traces) and SL isometric (upper traces) twitches. Extracellular calcium was 2.25 mM.

ments to shorten auxotonically, and 2) segment isometric contractions, with segment control imposed during diastole or after a brief period of shortening at the beginning of a twitch. In the first case muscle length was stepped from ML_{max} to values from 80 to 100% ML_{max}. Steps to a given ML were of 3 beat duration, the second beat being sampled for data; the muscle was then returned to ML_{max} and allowed to re-equilibrate for at least 10 beats. For segment isometric contractions above slack length it was possible to establish segment control during diastole. Because segment length could not be shortened passively below 90% SL_{max}, segment clamps below that SL had to be initiated after a brief period of active SL shortening. Sample records of F, SL and ML during segment auxotonic and segment isometric contractions are shown in Figure 1. Force segment length data for SL auxotonic and isometric contractions was obtained over a range of SL and are shown in Figure 2, for a representative experiment. The data is plotted in the F-SL plane with SL auxotonic and isometric data superimposed. In the F-SL plane SL auxotonic contractions appear as counterclockwise loops because a low shortening velocity exists at the time of peak force production (Huntsman, et al., 1979). It can be seen that from 97 to 85% SL_{max} the peak total F-SL rela-

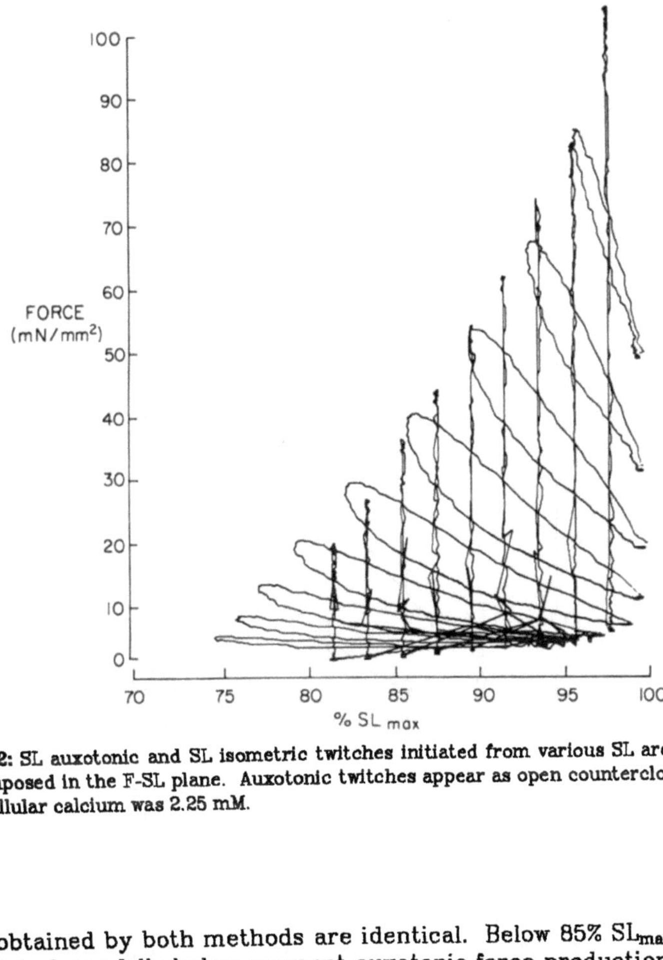

Figure 2: SL auxotonic and SL isometric twitches initiated from various SL are plotted and superimposed in the F-SL plane. Auxotonic twitches appear as open counterclockwise loops. Extracellular calcium was 2.25 mM.

tions obtained by both methods are identical. Below 85% SL_{max}, segment isometric force falls below segment auxotonic force production. At these short lengths the transition from ML control to SL control occurred with unavoidable oscillations, which appeared to diminish peak force production (see Figure 2B). It was not always possible to obtain SL isometric contractions above 97% SL_{max}, as the ML changes necessary to maintain SL control were large and often caused irreversible muscle damage.

Extracellular Calcium Variations

Total peak force-segment length relations from experiments performed with 1.125, 2.25 and 4.5 mM extracellular calcium are presented in Figure 3. Data were obtained from SL auxotonic contractions. Mean values (N=9) and standard error bars are shown. Several experiments (N=5) were performed in 9.0 mM extracellular calcium with no significant increase in force over that found in 4.5 mM calcium. The filled circles and solid line represent the passive F-SL relation. By curve fitting (linear

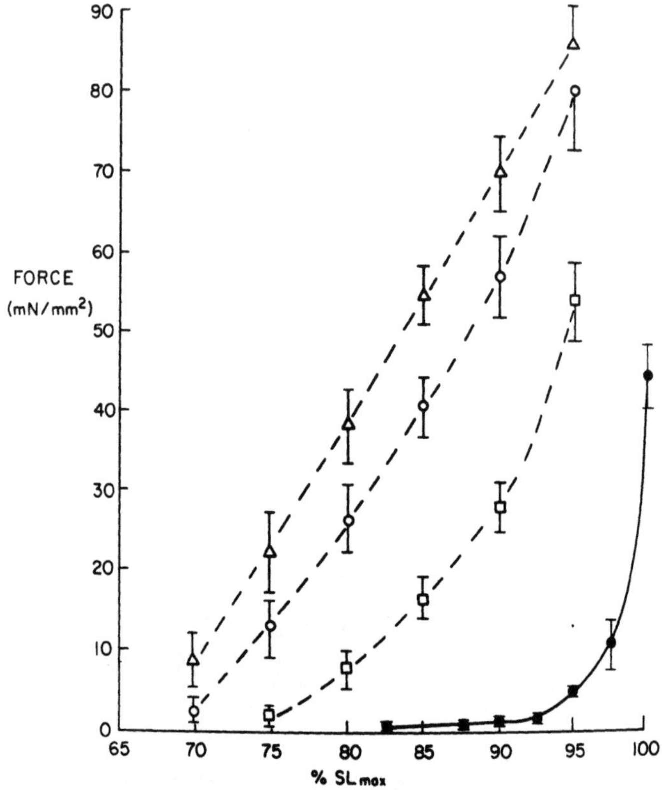

Figure 3: Total force-segment length relations obtained in 4.5 (Δ), 2.25 (○) and 1.125 (□) mM extracellular calcium. Data points are mean values (± SE) for SL auxotonic contractions. The filled circles and solid line represent the passive F-SL relation (N=9).

regression and polynomial least squares fit) the zero force intercepts were found to be 67, 68 and 74% SL_{max} for F-SL relations in 4.5, 2.25 and 1.125 mM extracellular calcium, respectively. SL isometric data, in each calcium, was found to be identical to that obtained from auxotonic twitches.

The data in Figure 3 are uncorrected for passive force. One could imagine that two extreme types of passive correction could be applied. In the first case, passive force might be dependent primarily on segment length. In this case, because of the steep passive F-SL relation, the passive force to be subtracted from total force would be very small for SL below 95% SL_{max}. In fact, the relative shapes and positions of the F-SL relations in the different calciums, corrected in such a way, are little different from the total F-SL curves. On the other hand, if elastic structures which ran from end to end in the preparation, either as internal fibers or as a surface sheath, were important, passive force would be related to muscle length. The magnitude of the correction would then be estimated from the passive F-ML relation. The effects of the two

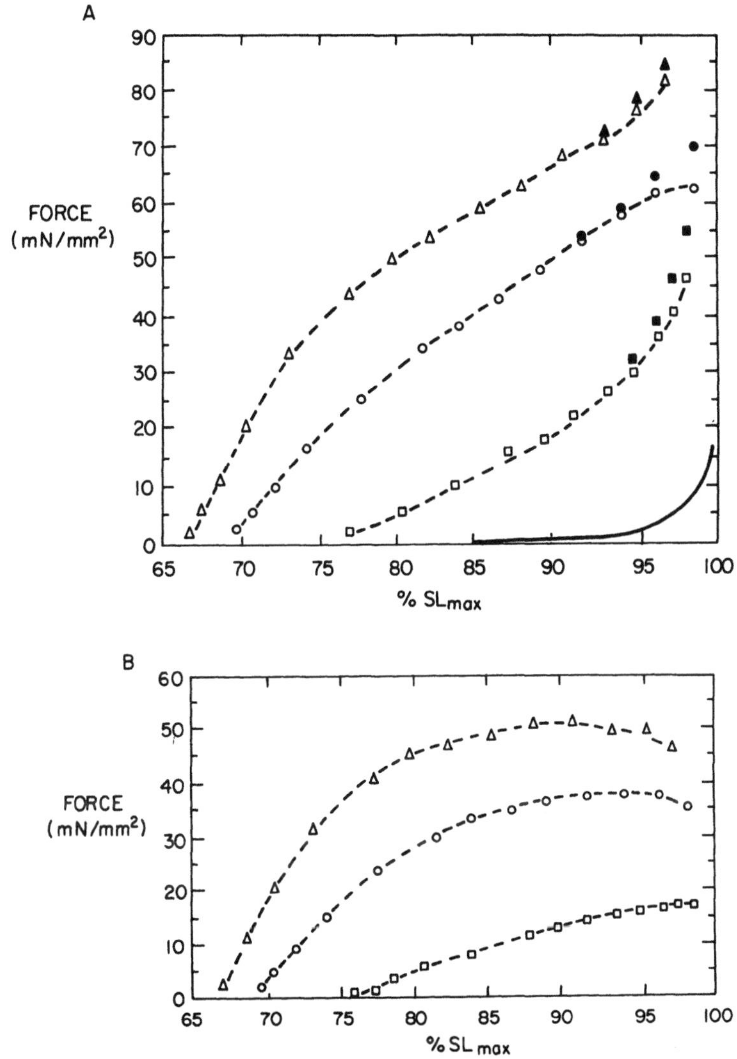

Figure 4: A) Peak SL auxotonic F-SL relations, from a single experiment, which have been corrected for an SL dependent passive force. Data was obtained in 4.5 (Δ), 2.25 (O) and 1.125 (□) mM extracellular calcium. The filled symbols are uncorrected total F-SL data. The passive F-Sl relation is described by the solid line. B) The same data has been corrected; for a muscle length dependent passive force.

corrections on auxotonic F-SL data from a single experiment, done in 1.125, 2.25 and 4.5 mM extracellular calcium are described by Figure 4. With the SL dependent passive force correction (A), F-SL relations changed shape from linear or convex in 4.5 and 2.25 mM calcium to concave in 1.125 mM calcium. When an ML dependent passive force correction (B) is applied F-SL relations in the three calciums all exhibit a

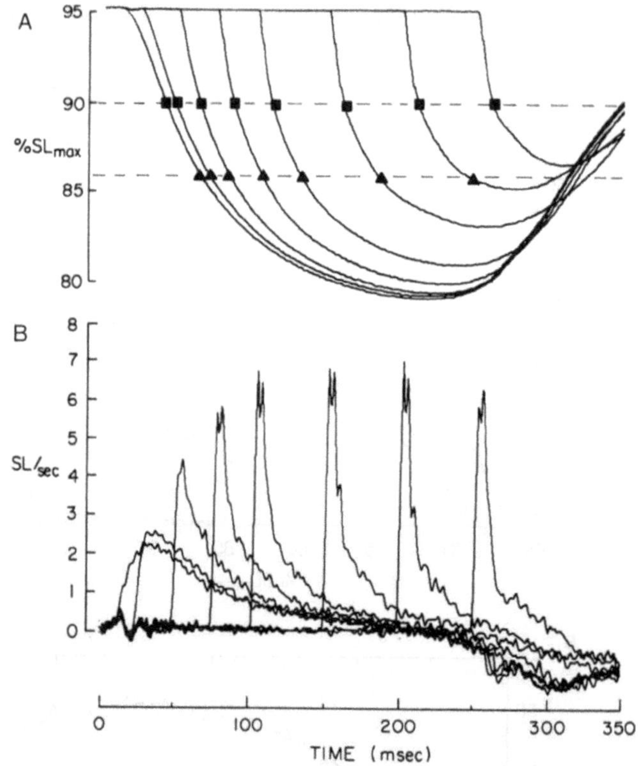

Figure 5: A) Representative traces of segment length for an experiment in which SL was held isometric until a load of 1 mN was imposed at various times during the twitch. The dashed line intersects the segment length traces at 90% and 86% SL_{max} at various times. SL velocity could be determined at each intersect. Extracellular calcium was 2.25 mM. B) Segment shortening velocity (V_L) during 1 mN load releases imposed at various times. Traces were obtained by analog differentiation of the data in Part A.

plateau and appear to be more nearly vertical scaled versions of each other. In each calcium SL isometric F-SL data corrected for passive force by both procedures was identical to the corrected auxotonic data.

V_L – SL Relations

The time and segment length dependence of light load segment velocity (V_L) was determined by releasing the muscle from SL isometric control to force control to 1 mN, at various times during a twitch. The load clamp force of 1 mN corresponds to about 1-2% of maximum isometric force at 95% SL_{max}. Representative SL traces for such a load clamp series are presented in Figure 5A. Corresponding velocity traces are presented in Figure 5B. By choosing specific segment lengths for analysis and using the multiple load clamps initiated at different times, as illustrated in Figure 5A, the time dependence of V_L at a selected segment

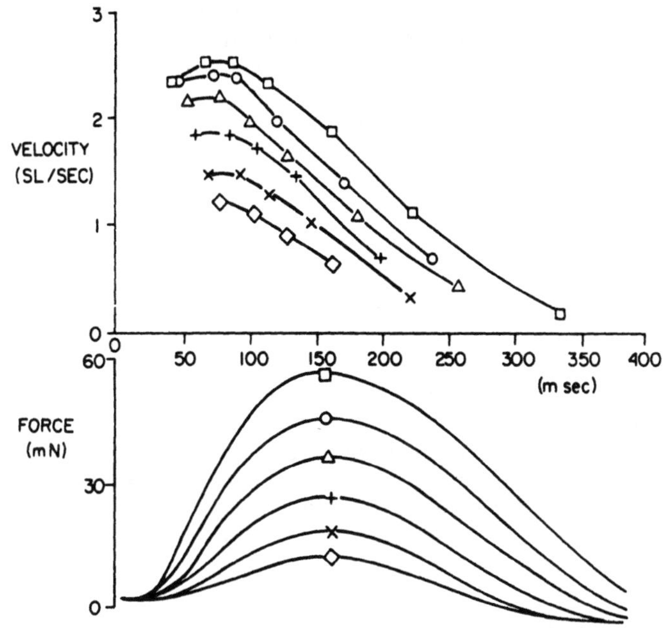

Figure 6: A) V_L as a function of time obtained at a number of segment lengths. Segment lengths corresponding to 90 (□), 88 (○), 86 (Δ), 84 (+), 82 (×) and 80 (◇)% SL_{max}. Data obtained from a single experiment in 2.25 mM extracellular calcium is presented. Initial SL was 95% SL_{max}. B) Segment isometric force production at segment lengths which correspond to those at which V_L-time curves were obtained.

length, was determined. The time course of segment shortening seems to consist of 2 phases, except at early times of load release. The initial phase consists of a rapid rise of V_L to up to 7-10 SL/sec and a decline which is over in 20-30 msec. During the second phase segment velocity declines more slowly with SL and time. The possibility that the early velocity phase, or transient, represents the influence of a passive visco-elastic process on segment velocity (Whalen, et al., 1981) precludes unambiguous interpretation of velocity results obtained at SL much above 90% SL_{max}.

The time dependence of V_L was determined for a number of segment lengths. The data in Figure 6 indicate that V_L rises to a maximum value early during the twitch, plateaus briefly, and then begins to decline before the time of peak segment isometric force production. V_L declines by 20 to 50% of its peak value, by the time of peak force, the relative decline being greatest at shorter SL. The time to peak segment isometric force production was not significantly influenced by SL (Fig. 6B). To determine the SL dependence of V_L, sets of V_L-time curves at various SL (eg: Fig. 6A) were sampled at 100 msec after stimulus. V_L was then plotted against the corresponding SL. At 100 msec V_L was at or near its maximum measured value at all segment lengths below 90% SL_{max}. The load clamp protocol was performed in 1.125, 2.25 and 4.5 mM extracellular calcium. V_L-SL relations obtained in this manner from experiments conducted in the various calcium concentrations are plotted in Figure 7. To facilitate combining data from different muscles, velocity values have been normalized relative to the value obtained at 90% SL_{max} in 4.5 mM extracellular calcium. Each V_L data point is the mean (±SE) value obtained from 5 muscles. The open symbols and dashed lines represent F-SL relations from Figure 3. Force values have also been normalized relative to the force found at 90% SL_{max} in 4.5 mM Ca.

DISCUSSION

The total force-segment relations obtained for various extracellular calcium concentrations in ferret papillary muscle are generally similar to those obtained from cat papillary (Donald et al., 1981) and rat trabecular preparations (ter Keurs et al., 1980) (Gordon et al., 1980). No evidence was found for a plateau range of SL where force production was constant, or declining at long lengths. The changing shape of F-SL relations (Figure 3) at different calcium levels is consistent with the previous results (Gordon et al., 1981) (ter Keurs et al., 1980). The data is supportive of the concept that a length dependent activation process is an underlying mechanism contributing to the fall of force with decreasing lengths (Jewell, 1977).

Length dependent activation could result from a number of possible mechanisms. Initial SL could influence activation, for example, either by affecting calcium release (ter Keurs, 1980) or binding to the filaments. However, the results presented in Figures 1 and 2 indicate that the SL at which a contraction began had no effect on subsequent force production. For example, an SL auxotonic contraction initiated at 100% SL_{max} would shorten to 95% SL_{max} and develop the same force as an SL isometric contraction at 95% SL_{max}. The results indicate that a 5-7% difference in initial SL, as well as a 5-7% difference in SL shortening, has no influence on force production at the final SL. Therefore the results also indicate that shortening induced deactivation (Edman, 1975) is not an important determinant of F-SL relations in cardiac muscle, under the conditions used in this study.

It is unlikely that length dependent activation depends on an influence on released calcium, as it has been shown that myoplasmic calcium levels are similar over a wide range of muscle lengths (Allen et al., 1979). On the other hand evidence has been presented which indicates that myofilament sensitivity to calcium decreases with decreasing sarcomere length in cardiac (Fabiato et al., 1978) (Hibbert, et al., 1980) and barnacle skeletal fibers (Gordon et al., 1978). Our mechanical results indicating that the final shortened length is the chief determinant of force production are consistent with this idea.

However, one must be cautious. When a ML dependent passive force correction is made (Figure 4b), F-SL relations in the three calciums become more nearly vertically scaled versions of each other. The results are then consistent with the influence of an internal load. As a result, data obtained from the measurement of F-SL relations enable one to

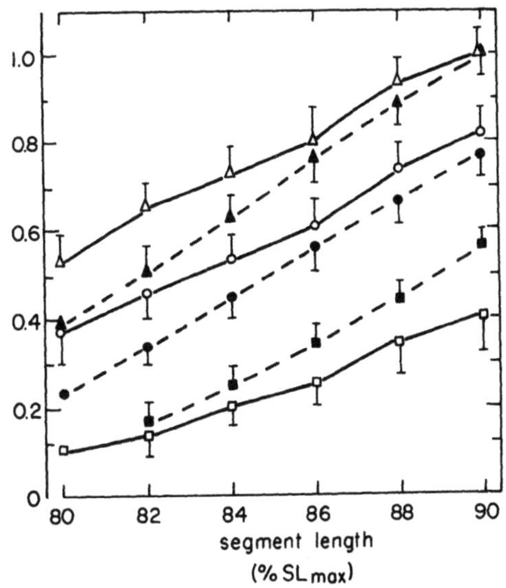

Figure 7: The SL dependence of force (open symbols) and V_L (filled symbols) are compared for 4.5 (Δ), 2.25 (O) and 1.125 (□) mM extracellular calcium. For comparison the mean (±SE) values for force (N=9) and V_L (N=5) are expressed as a fraction of the values obtained at 90% SL_{max} in 4.5 mM calcium.

eliminate the influence of initial length and shortening on activation, but do not enable one to unambiguously chose between length dependent activation and internal loads without complete knowledge about the nature of passive load bearing structures. Though SL dependent passive force seems probable, evidence has been presented that large, axially oriented collagen fibers which span large distances exist in papillary muscles (Caufield et al., 1979). On the other hand, a system of cellular surface fibers, which could restrict sarcomere elongation at long lengths, has been described (Ohrenstein, et al., 1980).

The data in Figure 7 indicate that below 90% SL_{max} peak force and lightly loaded shortening velocity (V_L) exhibit similar dependencies on extracellular calcium and SL. V_L declines somewhat more steeply with SL than does force, particularly in 4.5 and 2.25 mM calcium. However, in order to obtain V_L data down to 80% SL_{max}, it was necessary to sample the V_L-time curves at 100 msec. Had the curves been sampled earlier, V_L at lengths just above 80% SL_{max} would have been slightly higher.

The dependence of maximum shortening velocity (V_{max}) on calcium has been a subject of some controversy in the skeletal muscle literature (Podolsky et al., 1970) (Julian, 1971) (Edman, 1979), whereas in cardiac muscle V_{max} has consistently been shown to depend on extracellular calcium and to decline with sarcomere length below 2.0 μ (Pollack et al., 1976) and muscle length below 92% L_{max} (Brutsaert, 1974). However, both the calcium dependence of V_L and the decline of V_L with SL below 90% SL_{max} (Figure 7) could be caused by a significant internal load (Winegrad,

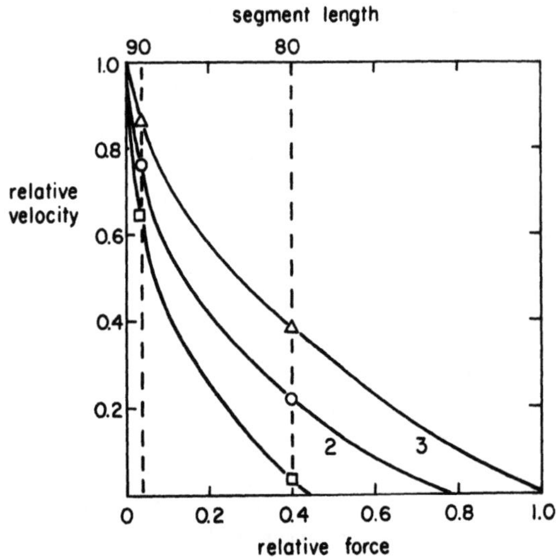

Figure 8: A proposed model in which V_{max} is independent of calcium and P_0 exhibits a calcium dependence. Hypothetical force-velocity curves are presented for 4.5 (3), 2.25 (2) and 1.125 (1) mM extracellular calcium. Force axis intercepts were obtained from data in Figure 3 for 90% SL_{max}. The F-V curves are intersected (dashed lines) at 90 and 80% SL_{max}. The interecepts correspond to velocities measured in 4.5 (Δ), 2.25 (O) and 1.125 (□) mM calcium.

1980). The fact that force declines 30-40% for a 10% decrease of SL indicates that an internal load could be quite large. If this were the case the load experienced by the contractile component during the load clamp would be much greater than the externally applied 1 mN. Under these conditions V_{max} could be quite independent of calcium, while the measured velocity (V_L) would vary with calcium; just as force production would. This notion is graphically presented in Figure 8, where hypothetical force velocity relations in 1.125, 2.25 and 4.5 mM extracellular calciums are described. V_{max} is the same for each F-V curve, while isometric force (derived from Figure 3, at 90% SL_{max}) varies with calcium. It can be seen that at 90% SL_{max} even a small internal load would give an apparent dependence of V_L on calcium. However, as shortening proceeded, the internal load would increase. The apparent difference in V_L, measured in tho 3 oaloiums, should then increase. As a result the V_L-SL relations should diverge from 90 to 80% SL_{max}. The force axis intercept would not appreciably change since Fabiato (1978) has shown that, in skinned cardiac cells with no probable internal load, force declines very little over this range. However, the results in Figure 7 indicate that the V_L-SL relations obtained in 1.125, 2.25 and 4.5 mM calcium do not diverge, below 90% SL_{max}, but are similar in slope, and perhaps slightly convergent.

Thus, a mechanism based purely on internal loads does not appear to account for length dependence of F and V_L. Nevertheless, some internal load may exist. As was mentioned above, SL cannot be passively shortened below 90% SL_{max}, and muscles which shorten extensively under a light load will elongate back to this length. These observations indicate the existence of some restoring force. However, our results suggest that this force may be relatively small. The strong dependence of contractile force and shortening on length is, therefore, probably established principally by length dependent myofilament sensitivity.

REFERENCES

Allen, D.G. and Kurihara, S. (1979). Calcium transients at different muscle lengths in rat ventricular muscle. J. Physiol 292: 680.

Brutsaert, D.L. (1978). The force-velocity-length-time interrelation of cardiac muscle. CIBA Founda. Symp. 24: 155.

Brutsaert, D.L., Dellerk, N., Goethals, M.A. and Housmans, P.R. (1978). Relaxation of ventricular cardiac muscle. J. Physiol. 283: 469-480.

Caulfield, J.B. and Bony, T.K. (1978). Collagen network of myocardium. Circ. II. 58: 240.

Donald, T.C., Reeves, D.N.S., Reeves, R.C., Walter, A.A. and Hefner, L.L. (1980). Effect of damaged ends in papillary muscle preparations. Am. J. Physiol. 238: H14-H23.

Edman, K.A.P. Mechanical deactivation induced by active shortening in isolated muscle fibres of the frog. J. Physiol. (London) 246: 255-275.

Edman, K.A.P. The velocity of unloaded shortening and its relation to sarcomere length and isometric force in vertebrate muscle fibers. J. Physiol. 291: 143-159.

Fabiato, A. and Fabiato, F. (1975). Dependence of contractile activation of skinned cardiac cells on the sarcomere length. Nature 256: 54-56.

Fabiato, A. and Fabiato, F. (1978). Myofilament tension oscillations during partial calcium activation and activation dependence of the sarcomere length tension relation of skinned cardiac cells. J. Gen. Physiol. 72: 667-699.

Gordon, A.M., Huxley, A.F. and Julian, F.S. (1966). Tension development in highly stretched vertebrate muscle fibres. J. Physiol. London 184: 143-169.

Gordon, A.M. and Ridgeway, E.B. (1978). Calcium transients and relaxation in single muscle fibres. Eur. J. Cardiol. Suppl. 7: 27-34.

Gordon, A.M. and Pollack, G.H. (1980). Effects of calcium on sarcomere length-tension relation in rat cardiac muscle: implications for the Starling-Frank mechanism. Circ. Res. 47(4): 610-619.

Hibberd, M.G. and Jewell, B.R. (1979). Length dependence of the sensitivity of the contractile system to calcium in rat ventricular muscle. J. Physiol. 290: 30-31.

Huntsman, L.L., Day, S.R. and Stewart, D.K. (1977). Nonuniform contraction in the isolated cat papillary muscle. Am. J. Physiol. 233(5): H613-H616.

Huntsman, L.L., Joseph, D.S., Oiye, M.Y. and Nichols, G.L. (1979). Auxotonic contractions in cardiac muscle segments. Am. J. Physiol. 237: H131-H139.

Huntsman, L.L. and Stewart, D.K. (1977). Length-dependent calcium inotropism in cat papillary muscle. Circ. Res. 40(4): 336-371.

Jewell, B.R. (1977). A reexamination of the influence of muscle length on myocardian performance. Circ. Res. 40: 221-230.

Julian, F.J. (1971). The effect of calcium on the force-velocity relation of briefly glycerinated frog muscle fibers. J. Physiol. 218: 117-145.

Julian, F.J. and Sollins, M.R. (1975). Sarcomere-length tension relations in living rat papillary muscle. Circ. Res. 37: 299-308.

Julian, F.S. and Moss, R.L. (1976). Absence of a plateau in length-tension relationship of rabbit papillary muscle when internal shortening is prevented. Nature 260: 340-342.

Krueger, J.W. and Pollack, G.H. (1975). Myocardial sarcomere dynamics during isometric contractions. J. Physiol. 25: 627-643.

Lakatta, E.G. and Jewell, B.R. (1977). Length-dependent activation: Its effect on the length-tension relation in cat ventricular muscle. Circ. Res. 40: 251-257.

Lopez, J.R., Wanek, L.A. and Stuart, S.R. (1981). Skeletal muscle: Length dependent effects of potentiating agents. Science 214: 79-82.

Orenstein, J., Hogan, D, and Bloom, S. (1980). Surface cables of cardiac myocytes. J. Mol. Cell. Cardiol. 12: 771-780.

Podolsky, R.S. and Teicholz, L.E. (1970). The relation between calcium and contraction kinetics in skinned fibers. J. Physiol. 211: 19-35.

Pollack, G.H. and Krueger, J.W. (1976). Sarcomere dynamics of intact cardiac muscle. Eur. J. Cardiol. 4 (Suppl): 53-65.

Taylor, S.R. Decreased activation in skeletal muscle fibres at short lengths. In: *The Physiological Basis of Starling's Law of the Heart,* Ciba Foundation Symposium 24, Associated Scientific Publishers, Amsterdam.

ter Keurs, H.E.D.J., Rijnsburger, W.H., Van Heuningen, R. and Nagelsmit, M. (1980). Tension development and sarcomere length in rat cardiac trabeculae. Circ. Res. 46: 703-714.

Whalen, D.A., Martyn, D.A. and Huntsman, L.L. (1981). Passive contributions to load clamp determined segment shortening velocities in isolated ferret papillary muscle. Abstract. Biophys. J. 33: 29a.

Winegrad, S. The importance of passive elements in the contraction of the heart. In: *Cardiac Dynamics,* J. Baan, A.C. Arntzenius and E.L. Yellin, eds., Martinus Nijhoff Publishers, The Hague,

DISCUSSION

EDMAN: I should like to ask you how you explain your finding that force production at a given length is independent of the prehistory during the contraction. Since it is a phasic contraction you are studying, if you allow the muscle to produce some initial shortening, you will measure isometric force at a later time when the activity may have changed a great deal.

HUNTSMAN: I would look at it the other way, Paul. I would say that for a least the first half of the twitch, things are simpler than we thought. It seems that the ability of the muscle to shorten or to generate force is not so history-dependent as we thought. Apparently the dominant factor is the profound length dependence. What happens after the time of peak tension may be another matter though, and these data do not address that. In fact, the data come from a relatively small portion of the total contraction. Nevertheless, we consistently find that isometric and auxotonic contractions yield the same force-sarcomere length relation.

EDMAN: What I want to point out is that the situation in cardiac muscle is quite different from that in skeletal muscle where you are independent of time when you study a tetanic contraction. In skeletal muscle the force produced at a given length is independent of the starting point (Edman, J. Physiol. 183: 407-417, 1966). But in cardiac muscle you are dealing with single cycles of activity, and you are therefore time dependent. Furthermore, both intensity and duration of the activation during a cardiac cycle is likely to vary with the degree of extension of the muscle. In addition, active shortening has a "deactivating" effect in myocardium as well as during a twitch in skeletal muscle (Edman, J. Physiol. 246: 255-275, 1975).

HUNTSMAN: Certainly you are right that the contractile activity is time dependent. But your other points about its dependence on extension and shortening address exactly the surprising aspects of our results. We expected effects of shortening and initial length too, but the data don't show them. One possibility is that there are actually counterbalancing effects involved. However, we've added caffeine to greatly slow the twitches and we've induced tetanus, and still the force-sarcomere length relations are the same. So, the simplest interpretation for now seems to be that the first half of contraction is determined largely by the level of activation and instantaneous length, with little influence by initial length on shortening history.

TER KEURS: Let me start with a question pertinent to your very high velocity peaks. You said they compare very well between active contractions and releases in a passive muscle, i.e., you get very similar shortening patterns. If so, are the shortening patterns in the passive muscle Ca dependent, and are the peaks in the actively contracting muscle also calcium dependent?

HUNTSMAN: Those are excellent questions, but unfortunately we don't yet have all the answers. First, I would emphasize that we've only looked qualitatively at the release transients, particularly those in the resting muscle. The size and duration of the velocity traces appear quite similar to those seen during contraction, but it is important to remember that the passive muscle is released from longer segment lengths than the active muscle. Second, the velocity spikes during contraction do vary with calcium concentration and, in fact, seem to correlate with total force.

TER KEURS: We did a fairly comparable study with trabeculae from rats. We used a procedure in which all measurements started at constant sarcomere length. We studied the effect of time, the effect of initial sarcomere length, and, at constant time and constant sarcomere length, the effect of calcium concentration, on maximal shortening velocity when the load was brought to zero by an isovelocity release. At 25° and external calcium of 2.5 mM, we found that maximal velocity rose in about 20 ms following onset of contraction to a plateau and then remained at a value of the order of magnitude of 12 μm per second, i.e., six muscle lengths per second, up to 120 milliseconds following the start of contraction. Secondly, if one studies maximal velocity as a function of sarcomere length, there's a plateau between 1.85 and 2.3μm. Thirdly, maximal velocity of shortening is a function of external calcium concentration. It increases with external calcium up to 1.2 mM. Above 1.2 mM, maximal velocity of shortening appeared to be constant.

HUNTSMAN: Those are fascinating results and it will be interesting to identify the similarities and differences between our observations. There are, of course, several possibilities to account for differences. Species difference is one. Another is that the velocity transients may be present to some extent in your records but difficult to appreciate without on-line velocity determination. Also, we want to be careful about our velocity determinations. For now, we've chosen the conservative approach and not included the transients in our assessment of active shortening.

TER KEURS: I think then the crux of the matter is the calcium dependence of the transients.

HOUSMANS: I have a few questions. I don't see how you can compare the velocity transients for a passive release in a *resting* muscle with the passive component of the velocity transient of a *contracting* muscle after a quick release, since the stiffness is obviously different in each of those conditions. And, secondly, what method did you use to sort out the active from the passive component of the velocity transient?

HUNTSMAN: First, I agree that it is not at all obvious why the velocity transients from active or passive muscle should be at all comparable. Certainly one would expect the stiffness and any viscous-like properties to be quite different. So far we've only made the initial qualitative observations that the transients appear similar. We have not yet analyzed them, nor tried to separate active and pasive components. Rather, we've just reacted with caution to the observation, and taken care not to include the transients in our assessment of active shortening.

HOUSMANS: My other question is, since you show in ferret papillary muscle that the peak force-length relationship in a twitch is independent of the way in which you reach a particular length -- and, in the rat it seems to be the same, as shown by Strobeck et al. (Fed. Proc. 39: 175-182, 1980) -- what is the indication for that in relaxation? If so, the relaxation would also have to be very uniform when studied at different loads (Brutsaert et al., J. Physiol. 283: 469-481, 1978).

HUNTSMAN: With regard to your last question, I should mention that we do find an interesting observation regarding what appears to be load independent relaxation. We have looked, though, only at what we call the onset of relaxation which is defined to be the time and length in an after-loaded contraction at which shortening ceases: i.e., velocity equals zero. Surprsingly, those points, for a variety of loads and onset times of the isometric-isotonic transition, all fall along a particular line in the segment length-time plane. The data from different loads are intermingled along this line, suggesting that the onset of elongation is not affected by load.

NONUNIFORMITY OF CONTRACTION AND RELAXATION OF MAMMALIAN CARDIAC MUSCLE

Philippe R. Housmans[§], Leonard H.S. Chuck[*], Victor A. Claes and Dirk L. Brutsaert

The Department of Pharmacology, Mayo Foundation, Rochester, MN 55905, U.S.A.
[*]Alberta Heritage Foundation, University of Calgary, Canada
[§]to whom correspondence should be addressed.

ABSTRACT

Nonuniformity of contraction and relaxation in cardiac muscle results from relaxation asynchronies between segments in the central region and their interaction with the prevailing loading conditions, and this may reflect regional differences in activation or force potential at a given time during a twitch.

We have measured the degree of spatial and temporal uniformity during contraction and relaxation of isolated ventricular mammalian myocardium, by examining individual segments in the central region of electrically stimulated cat papillary muscles and trabeculae (cross-sectional area < 0.6 mm^2), preparations where determination of sarcomere length with laser diffraction (Krueger and Pollack, 1975; Nassar, Manring and Johnson, 1974) may be difficult to achieve.

Individual segments (200-800 μm), adjacent to each other, in series along the longitudinal axis of muscles were demarcated by tips of glass microelectrodes inserted into the very core of the muscles. Length changes during contraction and relaxation of each segment were detected by transillumination and projection of the segment marker image onto a 1024-element linear photodiode array.

Two general types of observations were made. In lightly loaded twitch contractions, both the relative extent of shortening and its duration varied from segment to segment, indicating asynchrony in the onset of relaxation (Fig. 1). Delayed relaxation in one area of the muscle may result from regionally prolonged' activation, from inherently "stronger" fibres in that area, or from a combination of both. At higher loads, nonuniformity became much more marked with varied patterns of shortening

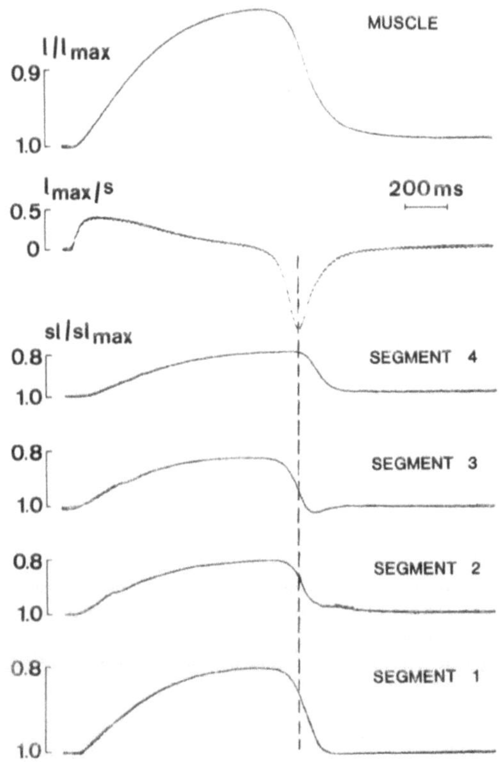

Figure 1: Spatial and temporal nonuniformity of segmental contractions during muscle-isotonic contractions. Muscle length (upper), velocity (second trace from top) and segment shortening in each of four adjacent segments (lower) of an isotonic twitch contraction at preload. All segments located within the central region of the muscle. The vertical dashed line indicates the time of peak relaxation velocity of the muscle to allow comparison of the differences in duration of segment shortening and hence, the asynchrony in onset of relaxation in various segments. Muscle characteristics: L_{max} = 5.0 mm; ratio of resting to total tension at L_{max} = 0.13; cross-sectional area = 0.43 mm^2; initial segment length at muscle length L_{max} for segments 1,2,3 and 4 were 810 μm, 560 μm, 610 μm and 390 μm respectively. Temperature 26°C. Electrical stimulation 10% above threshold, 12/min.

and/or lengthening among segments, but almost exclusively during relaxation. In muscle-isometric twitch contractions (Fig. 2) there was considerable variation among segments in 1) extent of shortening at various times during the twitch, 2) time to peak shortening and 3) onset of segment lengthening. The disparity of shortening-lengthening patterns was most pronounced in late isometric relaxation, with polyphasic shortening-lengthening behaviour in some segments (e.g. segments 2 and 3), whereas prolonged shortening of one segment sometimes occurred at the expense of transient lengthening of another segment beyond its initial length (e.g. segments 4 and 1).

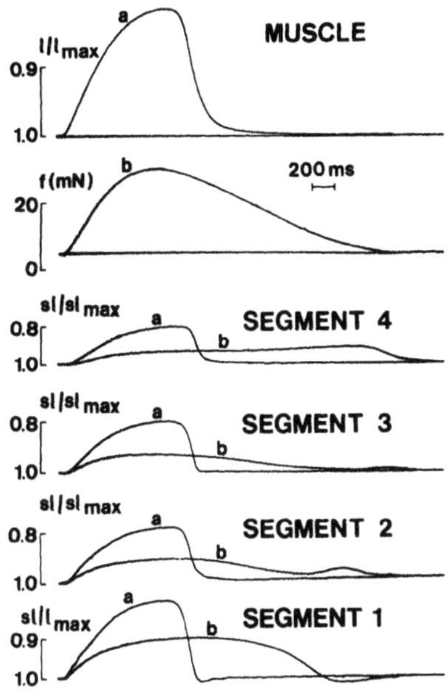

Figure 2: Nonuniform segment shortening during muscle-isometric contractions. Muscle length (upper), force (second trace from top) and segment shortening in each of four adjacent segments in a muscle-isometric twitch (b). The isotonic twitch contraction at preload (a) was superimposed for comparison of time course. Same muscle as in Figure 1.

Peak isometric force was reached at segment lengths shorter than the precontractile length, and more so when precontractile length was shorter. Moreover, in isometric twitch contractions at muscle lengths shorter than L_{max} (i.e. on the ascending limb of the length-tension relation), the dispersion of segment lengths at peak force was larger than for isometric twitches at the muscle length L_{max}. The pronounced nonuniform behavior of segments during isometric relaxation could result from regional differences in activation and force generating capacity or force potential, an effect that manifests itself either at high loads, or, for a given load, at late times after the stimulus, as it was found in skeletal muscle (Stienen and Blangé, 1981) and in single fibres (Edman and Flitney, 1978; Edman, 1982). In conclusion, nonuniformity of contraction and relaxation in cardiac muscle results from asynchrony between segments in the central region and its interaction with the prevailing loading conditions, and, as in skeletal muscle fibres (Edman, 1982), could play a role in the shape of the length-tension relation.

ACKNOWLEDGMENT

Supported by the Cardiovascular Institute (CAVAS-I), University of Antwerp, Belgium, and International Research Fellowship (USPHS) TW 03046.

REFERENCES

Edman, K.A.P. (1982). Length-tension-velocity relationships studied in short consecutive segments of intact muscle fibres of the frog. (This volume).

Edman, K.A.P. and Flitney, F.W. (1978). Non-uniform behavior of sarcomeres during relaxation of skeletal muscle. J. Physiol. 276: 78P-79P.

Krueger, J.W. and Pollack, G.H. (1975). Myocardial sarcomere dynamics during isometric contractions. J. Physiol. 251: 627-643.

Nassar, R., Manring, A. and Johnson, E.A. (1974). Light diffraction of cardiac muscle: sarcomere motion during contraction. In: *The Physiological Basis of Starling's Law of the Heart*, edited by the Ciba Foundation, Elsevier, Excerpta Medica, Amsterdam, pp. 57-82.

Stienen, G.J.M. and Blangé, T. (1981). Local movement in stimulated frog sartorius muscle. J. Gen. Physiol. 78: 151-170.

TRANSIENT LENGTH RESPONSES OF HEART MUSCLE IN Ba^{2+}-CONTRACTURE TO STEP TENSION REDUCTIONS

Yasutake Saeki, Toshimitsu Shibata, Chikako Sato and Keiji Yanagisawa

Department of Physiology, Tsurumi University, School of Dental Medicine, Yokohama 230, Japan

ABSTRACT

The transient length response of the cat papillary muscle in Ba^{2+}-contracture to step tension reduction was found to comprise four different phases. It is tentatively suggested that the transient response is mostly determined by the kinetics of the attachment and detachment of cross-bridges between actin and myosin filaments, which vary appreciably with activation level of the muscle.

To characterize the mechanical properties of activated heart muscle, tension of cat papillary muscle (length: 4.5 - 7.8 mm, cross-sectional area: 0.20 - 0.72 mm^2) in Ba^{2+}-contracture was decreased stepwise (within 4 msec) using a servo-control system and the length response was analyzed. Various amplitudes of step tension reduction from isometric contracture tension (with 4 mM Ba^{2+}) were performed at different initial muscle lengths over the range between L_{max} (the length for maximum developed isometric tension) and 85% L_{max} at 24 - 25°C.

When the tension reductions were less than 70% of the initial tension, the length responses comprised four different phases. The first phase is a rapid shortening during the tension reduction. The second phase is a slow quasi-exponential shortening and the third a slow lengthening. The fourth is an extremely slow and mild shortening, whose velocity does not reach the steady value but decreases with time. With a further increase in amplitude of the tension reduction, the distinction between successive phases of the length transient became less clear (i.e., the first phase of prompt shortening was followed only by the second phase of quasi-exponential shortening). The amplitude of shortening in the second phase increased linearly with increasing amplitude of tension reduction up to 70% of the initial tension. The relative tension (the reduced tension relative to the initial tension)-shortening relationship was found to be independent of the initial muscle length. The amount of lengthening in the third phase increased convexly with increasing amplitude of tension reduction. It reached a maximum value at 40 - 60% tension reduction then decreased with a further increase in amplitude of the tension reduction, and was larger at shorter initial muscle length. Maximum amount of length change in the second and third phase were, respectively, 3.5 ± 0.7% and 1.4 ± 0.3% of the initial muscle length in 9 preparations, The time to peak shortening of the second phase from the end of the first phase (duration of the second phase) was nearly constant regardless of the amplitude of tension reduction but decreased slightly with decreasing initial muscle length. Shortening during the second phase was about three times slower than the corresponding tension transient observed in our previous length-clamp experiment (Saeki, Sagawa and Suga, 1980). The third phase became shorter (from about 500 msec to 200 msec) with increasing amplitude of tension reduction independently of the initial muscle length.

When the level of contracture tension was decreased by decreasing concentration of Ba^{2+} from 4 mM (full activation) to 2 mM (about 70% activation), the amplitude of shortening in the second phase increased with increasing tension reduction up to 50% of the initial tension as in the case of 4 mM Ba^{2+}, but then decreased with a further increase in amplitude of the tension reduction. The maximum amount of lengthening in the third phase was about two times larger than that at 4 mM Ba^{2+}. Duration of the second phase became shorter with increasing amplitude of the tension reduction.

Increasing temperature markedly decreased the amplitude of length response in the second and third phases, and shortened the second and third phases (temperature coefficient Q_{10} of about 4) independently of the initial muscle length. However, the quasi-steady-state length at the end of the fourth phase was fairly insensitive to the change in temperature.

In contrast, the length responses of resting and of rigor muscle treated with iodoacetic-acid were much simpler, changing in a nearly stepwise fashion.

These results suggest that the transient length responses in the second and third phases are mostly determined by the kinetics of the attachment and detachment of cross-bridges, which vary appreciably with activation level and amplitude of the tension reduction. They may also be affected by the passive properties of the muscle in the resting state. One possible explanation for the underlying mechanism of the length transient is that after stepwise tension reduction most of the cross-bridges are brought into the state ready for attachment and in the second phase they start to attach at a finite rate and then in the third phase detach more or less synchronously and finally lapse into the steady-state cycling in a random fashion.

Apart from the cross-bridge mechanism, the major portion of the length transient except for the third phase can be explained in terms of our previous viscoelastic model (Saeki, Sagawa and Suga, 1978) which describes the resting property by a single elastic element and the active property by two elastic elements coupled with one viscous element.

REFERENCES

Saeki, Y., Sagawa, K. and Suga, H. (1978). Dynamic stiffness of cat heart muscle in Ba^{2+}-induced contracture. Circ. Res. 42: 324-333.

Saeki, Y., Sagawa, K. and Suga, H. (1980). Transient tension responses of heart muscle in Ba^{2+} contracture to step length changes. Am. J. Physiol. 238: H340-H347.

DISCUSSION

KAWAI: What do you think about the contribution of damaged ends of the preparation to the length response?

SAEKI: In the case of the tension-clamp experiment such as this, I think the transient length response following the rapid shortening in the first phase seems not be affected seriously by the damaged ends because there can be no change in the length of the damaged end under the condition of constant tension.

WINEGRAD: Do you think that a Ba^{2+} contracture is physiological?

SAEKI: I think the Ba^{2+} contracture is comparable to the K^+ contracture, where Ca^{2+} mediates the activation of the contractile system, since similar length transient responses are obtained from the muscle in K^+ contracture as has been reported by Steiger and his coworkers (Circ. Res. 42, 1978).

We also observed a quite similar length transient from the glycerinated cat papillary muscle which was activated either by Ca^{2+} or by Ba^{2+}

ENERGETICS

INTRODUCTION

When muscle contracts, it converts the energy from chemical reactions into useful mechanical work and into heat. Those of us who study energetics try to learn about contraction by investigating the chemical reactions and heat and work production. The principle that is the basis of our studies is that when a reaction occurs it is accompanied by an energy change.

reactants $\longrightarrow$ products + energy

The ratio of the amount of this energy to the extent of the reaction (in units of joules of energy per mol of reaction) is a characteristic of the reaction; it is called the molar enthalpy change.

$$\frac{\text{energy}}{\text{extent of reaction}} = \Delta H \qquad (1)$$

The important point is that the value of the molar enthalpy change of a rection is a constant for a particular set of conditions of, for example, pH, temperature, ionic strength and magnesium concentration. This is a useful relation because a change in the extent of the reaction is reflected in a change in the amount of energy produced.

To draw quantitative conclusions about energy changes it is usually necessary to know the value of the molar enthalpy change for the reaction in question. It can be determined by *in vitro* experiments in which the reaction occurs under defined conditions so that it is certain that all the energy is being produced by the reaction of interest.

The situation *in vivo*, that is, in the muscle, is more complicated because:

(1) the contractile process can convert energy into heat *and* work and possibly other forms of energy,

(2) a number of reactions occur in muscle, so the energy produced as heat and work is the sum of contributions from them all.

The fact that heat is very non-specific is an important point. It is an advantage if other methods for following the reaction do not exist, and it can indicate that an unsuspected process is occurring. However the non-specific nature of heat does cause problems too. For example, it can be very difficult to be certain that a particular feature of the heat record is due to a particular process.

In this session the papers to be given by K. Yamada, T. Yamada and J. Rall are about experiments on energy production and the mechanical events during contraction of intact muscle or fibres from frog. The energy changes and the mechanical events both reflect the cross-bridge and other reactions. This is the approach used by A.V. Hill in most of the large number of studies he did.

K. Yamada will discuss shortening heat which A.V. Hill discovered and described in his 1938 paper (Proc. Roy. Soc. B. 126: 136-195, 1938) (the force-velocity curve was also described in this paper). Hill's main observation was that extra heat was produced when the muscle was released and allowed to shorten during an otherwise isometric contraction. The characteristics of the extra heat were:

(1) the amount of heat was proportional to the distance shortened. Fairly small releases were used because the initial and final muscle lengths were chosen so that the heat rate during the isometric periods before and after shortening were the same. See also Irving & Woledge (J. Physiol. 321: 411-422, 1981).

(2) the amount of extra heat was independent of the speed of shortening.

(3) the amount of extra heat was independent of the time at which shortening started in the tetanus.

These properties suggest that the reaction producing this heat is probably the same as the one that causes shortening. If it is cross-bridge turnover, then the shortening heat per unit distance shortened should depend on the extent of filament overlap. This point has not really been established yet, although it is obviously an important one. Brief accounts of work on this point have been reported by Lebacq (J. Physiol. (Paris) 65: 440A, 1972) and Irving, Homsher & Lebacq (Fed. Proc. 39: 1730, 1980).

Heat measurements have been done for many years and during this time muscle physiology has changed a great deal. As a result, reasons have arisen for trying to use other preparations besides the sartorius muscle of the frog. A few years ago V. Howarth, R. Woledge and I (J. Physiol. 313: 61-62P, 1981) decided to try single fibres from frog muscle for the following reasons. First, to compare or understand the energy changes that accompany the mechanical events during contraction of single fibres, the experiments had to be done on that preparation. Second, it has been shown that amphibian muscle contains a number of different fibre types (Lännergren, Lindblom & Johansson, Acta Physiol. Scand. 114: 523-535, 1982). It would be interesting to know about their energetic properties. Third, doing and interpreting some pharmacological experiments would be more straight-forward using single fibres. J. Rall has spent the past year in London working on single fibres. Part of this time he was doing heat measurements and as a result many improvements have been made. He will describe our new experiments on single fibres.

The paper to be given by E. Homsher is concerned with energy balance experiments. These experiments are aimed at testing ideas about the identity of the chemical reactions that produce the heat and work during contraction. In these studies a re-arranged form of equation (1) is used:

$$\text{heat} + \text{work} = \sum (\text{extent of reaction} \times \Delta H) \qquad (2)$$

The left side of the equation is the energy observed as heat and work during contraction and the right side is the amount of energy that can be explained by the chemical reactions that are observed to occur. The summation is introduced into the equation because a number of different chemical reactions may be taken into account.

When an energy balance experiment is done a hypothesis is made about the specific reactions that provide the energy and therefore are to be included; it is, of course, necessary to know the ΔH for each reaction. Then measurements are made of the extent of each of these reactions during contraction and a calculation is made of the amount of energy that is due to these reactions. In other words, the right side of equation (2) is evaluated; this quantity is called the explained energy. The amount of heat and work produced by the muscle during contraction is also measured and this gives the left side of the equation, the observed energy. If the original hypothesis about the chemical reactions that are the source of the heat and work, is correct and complete, then the two sides of the equation balance.

The main conclusion of energy balance studies has been that during contraction the heat + work exceeds the amount of energy from simultaneous metabolic reactions (ATP cleavage and the reactions that immediately resynthesize ATP) (Curtin & Woledge, J. Physiol. 288: 353-366, 1979; J. Physiol. 316: 453-468, 1981; Homsher, Kean, Wallner & Garibian-Sarian, J. Gen. Physiol. 73: 553-567, 1979; Homsher, Irving & Wallner, J. Physiol. 321: 423-436, 1981; earlier work reviewed by Curtin & Woledge, Physiol. Rev. 58: 690-761, 1978; Homsher & Kean, Ann. Rev. Physiol. 40: 93-131, 1978). In an isometric tetanus of frog muscle (R. temporaria) at 0°C this excess or unexplained energy amounts to about 200 mJ/g dry weight or 40 mJ/g wet weight (Curtin & Woledge, 1979, see Fig. 1). What is the source of this energy? There is general agreement that it is probably not a metabolic reaction, because these have been taken into account and unidentified metabolic intermediates have not been detected. It seems more likely that the energy comes from a cyclic reaction involving a protein, which might be summarized:

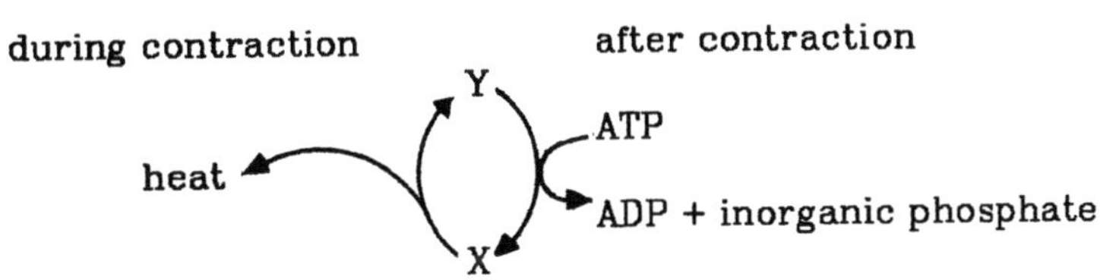

Where X and Y correspond to two different states of a protein or set of proteins. The suggestion is that during contraction part of the cycle occurs and produces heat without ATP cleavage---this "incomplete cycle" would give the energy imbalance results described above. After

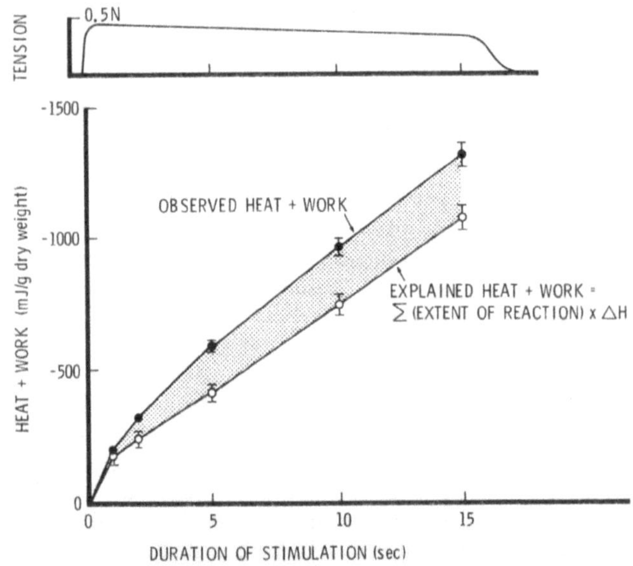

Figure 1: Tension (top) and energy (bottom) produced in a 15 sec isometric tetanus of sartorius muscle from frog (R. temporaria) at l_0 and 0°C. The tension record is a representative example. In the lower part of the figure the filled circles show mean values of energy produced as heat and work and the open circles show the mean values of the amount of energy than can be explained by ATP splitting and the creatine kinase reaction. Bars are ± 1 S.E. of the mean; in two cases the S.E. was too small to be shown. The shaded area indicates the amount of unexplained energy that is to be due to some other reaction. The ratio of wet weight to dry weight was 6.27 ± 0.13 (mean ± S.E., n=10). This figure is based on results of Curtin & Woledge (1979).

contraction the cycle would be completed by return of Y to X coupled directly, or indirectly, to ATP splitting. For the complete cycle the energy output as heat (or heat + work, if work is done) would be balanced by that from ATP cleavage.

If crossbridge reactions produce unexplained energy in this way, then the incomplete cycle may be the shift in the distribution of actin and myosin from the states they occupy in resting muscle (most bridges detached) to the set of states they occupy during contraction (more bridges attached). It is also possible that non-crossbridge reactions participate in "incomplete cycles"; for example, the unexplained energy may be the net energy from the movement of calcium ions from binding sites in the sarcoplasmic reticulum to free solution and binding sites on troponin, parvalbumin, etc.

There is evidence from previous energy balance experiments that both the tension-producing reaction of the crossbridges and also non-crossbridge reactions produce unexplained energy.

The following experiments indicated that crossbridge reactions are involved: (1) the finding of Curtin & Woledge (1981) that the amount of unexplained energy depended on the degree of filament overlap, and (2) the observation of Rall, Homsher, Wallner & Mommaerts (J. Gen. Physiol. 68: 13-27, 1976) and Homsher, Irving & Wallner (1981) that the average rate of production of unexplained energy was greater during rapid shortening (which is surely due to crossbridge turnover) than under isometric conditions. E. Homsher will be describing the energy balance study on rapid shortening and also new results from a study of muscle shortening at a more moderate velocity and performing an appreciable amount of mechanical work.

At least three lines of evidence suggest that unexplained energy is produced by a process other than the tension-generating reactions of the crossbridges. (1) Unexplained energy is produced by muscles at long sarcomere lengths where the overlap between the thick and thin filaments is negligible (Kean, Homsher, Sarian-Garibian & Zemplenyi, Physiologist 19: 250, 1976; Kean, Homsher & Sarian-Garibian, Biophys. J. 17: 202a, 1977; Curtin & Woledge, 1981). (2) The time course of production of tension is different than that for the release of unexplained energy (see Fig.1, Curtin & Woledge, 1979; Homsher, Kean, Wallner, & Garibian-Sarian, 1979). (3) In the second tetanus of a series the tension is almost the same as that produced in the first tetanus whereas there is significantly less unexplained energy in the second than in the first tetanus (Curtin & Woledge, J. Physiol. 270: 455-471, 1977). As mentioned above, it has been suggested that reactions of calcium ions may be sources of unexplained energy. J. M. Gillis, in the section of this volume entitled "Activation of the Myofilaments," describes the results of modeling studies in which the time courses and extents of the reactions of calcium ions are predicted from equilibrium constants and rate constants for these reactions. This information along with the molar enthalpy changes can be used to give values for the heat of these reactions. This information will be very useful in helping to decide, on a quantitative basis, whether it is reasonable to suppose that reactions of calcium ions are the source of an appreciable part of the unexplained energy.

-Nancy Curtin

DEPENDENCE OF THE SHORTENING HEAT ON SARCOMERE LENGTH IN FIBRE BUNDLES FROM FROG SEMITENDINOSUS MUSCLES

Kazuhiro Yamada and Kaoru Kometani

Department of Physiology, Medical College of Oita, Oita 879-56, Japan

ABSTRACT

The relation between shortening heat and sarcomere length was studied using fibre bundles dissected from frog semitendinosus muscles, as well as using whole muscles. The velocity of shortening was at its maximum. The unstimulated muscles showed a large thermoelastic absorption of heat when released at long muscle lengths. However the sarcomere length at which this thermoelasticity started to appear was longer, by at least 0.3 μm per sarcomere, in fibre bundles than in whole muscles. At the same time the amount of heat absorbed was decreased in fibre bundles. The shortening heat in fibre bundles at the sarcomere lengths ranging from 2.17 to 2.74 μm, for which no correction for the thermoelasticity was necessary, decreased linearly with sarcomere length. The shortening heat in fibre bundles at longer lengths and in whole muscles was corrected by subtracting the thermoelastic heat absorption measured separately by releasing unstimulated muscles. After the correction the shortening heat showed an almost similar dependence on sarcomere length in the range from 2.0 to 3.7 μm to that seen in fibre bundles in the sarcomere length range of 2.17 to 2.74 μm.

INTRODUCTION

When active muscle shortens the rate of heat production is greater than it is in an isometric contraction; this excess is described as shortening heat (Hill, 1938). It has been proposed that the shortening heat represents the higher rate of cross-bridge turnover during shortening (Huxley, 1957; Curtin and Woledge, 1978; Homsher and Kean, 1978; Kodama and Yamada, 1979). Then according to the independent force-generator theory (Gordon, Huxley and Julian, 1966), the shortening heat at the maximum velocity of shortening is expected to be directly proportional to the extent of overlap between thick and thin filaments in each half-sarcomere.

The first attempt to study this problem was that of Lebacq (1972). He briefly reported that using frog sartorius muscles the shortening heat declined too steeply when sarcomere length was increased in the range between 2 and 2.9 μm. To explain this difficulty he suggested an involvement of large changes in resting tension caused by shortening at long muscle lengths. More recently, Irving, Homsher and Lebacq (1980) studied this problem using frog semitendinosus muscles. They briefly reported that, at long sarcomere lengths over 3 μm, heat was absorbed when muscles were released without stimulation. After allowance was made for this heat absorption, the shortening heat at greater muscle lengths was progressively reduced. Linear regression with sarcomere length indicated that the corrected shortening heat would be reduced to zero at the sarcomere length of 3.9 μm. Irving and Woledge (1981) also reported that using frog sartorius muscles the shortening heat and isometric tension had the similar dependence on muscle length at sarcomere lengths between 2 and 2.6 μm. We have also studied the same problem using frog semitendinosus muscles (Yamada, Kometani and Kobayashi, 1981). It was found that the resting muscles showed a large absorption of heat when released (rubber-like elasticity) at sarcomere lengths beyond about 2.5 μm. After the shortening heat was corrected for this absorption of heat, the corrected shortening heat was almost independent of sarcomere length in the range between 2.2 and 3.1 μm.

Here we report our recent study on the same problem using frog semitendinosus muscles and also using fibre bundles dissected from them. The main reason for using the fibre bundles was that if the structure that causes the rubber-like elasticity in unstimulated muscle at long muscle lengths exists outside muscle cells, or is related to the structure outside the cells, the elasticity might be affected by the dissection. This is because connective issue might be more abundant near the surface than in the interior of the muscle.

METHODS

Ventral heads of the semitendinosus muscles of frogs (*Rana japonica*) were used. Bundles of fibres were dissected so as to reduce the length of the tendon to which muscle fibres were attached. The fibre bundles were about 1 to 1.2 mm in diameter and consisted of about 50-100 fibres. The maximum tension produced by tetanic stimulation per unit cross sectional area ranged from 563 to 1783 gwt/cm^2. This was 1360-1810 gwt/cm^2 in whole muscles. Muscles were stimulated directly via platinum electrodes with square pulses of 2 ms duration (Digitimer Type 2533 isolated stimulator) at about 10 Hz at 0°C. The muscle were connected via length and tension transducers to a Ling 200 series vibration generator. Muscle length was made to follow a control signal by a feedback network driving the vibration generator.

Heat measurements were made using an electroplated thermopile (Ricchiuti and Mommaerts, 1965) from a 9 mm region. The thermopile output was amplified by an Ancom 15C-3a chopper amplifier, low pass filtered (cut-off frequency, 200 Hz), and recorded on a Nicolet 2090 digital oscilloscope. The thermopile was thin, estimated equivilant half-thickness being 14 μm (Hill, 1965), so for the purpose of these experiments no correction was necessary for lag in heat conduction from muscle to thermopile. Heat loss from the muscle was exponential with rate constant in the range 0.07-0.11 s^{-1} for bundles or bundle pairs and 0.02-0.06 s^{-1} for whole muscles, and was corrected for by the method of Hill (1965).

The sarcomere lengths of muscles were measured each time muscle lengths were changed, by diffraction of a 1.2 mm diameter 25 mW He-Ne laser beam in unstimulated muscles on the thermopile. The length of muscle fibers which corresponded to a sarcomere length of 2.2 μm, estimated from the relation between changes in muscle length and changes in sarcomere length, was taken as the standard length (l_0) of muscle fibres. The value of l_0 was used to estimate the cross sectional area of muscle.

RESULTS

The Relation Between Tension and Sarcomere Length

The muscles were tetanically stimulated for 3 s at 10 Hz at 0°C. The isometrically contracting muscles were released at 1.2 s after the stimulation was started. Fig. 1 shows the isometric tension developed at the time of release in six bundles or bundle pairs at different sarcomere lengths. The relation between tension and sarcomere length was similar to the relation obtained during the tension plateau in single muscle fibres as reported by ter Keurs, Iwazumi and Pollack (1979). However, the

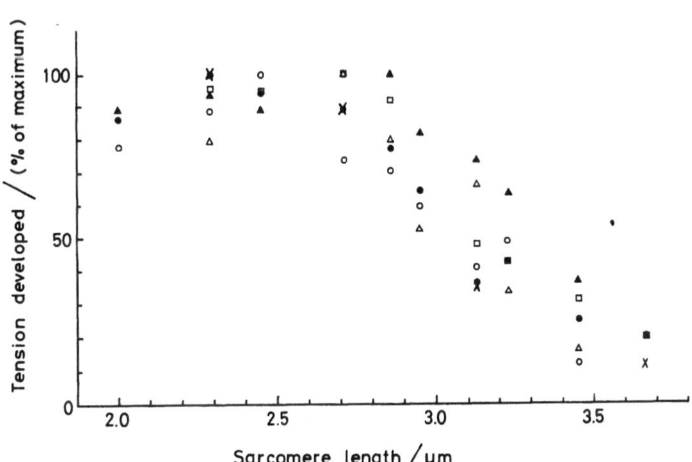

Figure 1: The relation between tension and sarcomere length in six fibre bundles or bundle pairs dissected from semitendinosus muscles. The tension was measured at 1.2 s after tetanic stimulation was started. Temperature, 0°C.

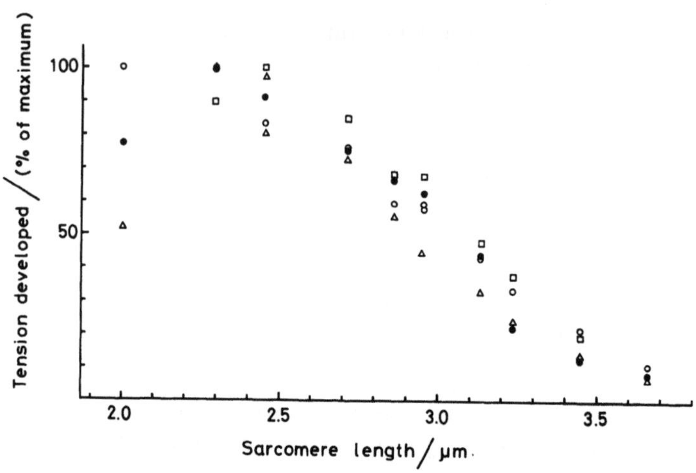

Figure 2: The tension-length relation in whole semitendinosus muscles. Conditions similar to those in Fig. 1.

length-tension relation in the intact whole muscles was different from this (Fig. 2). In whole muscles the tension declined linearly as sarcomere length was increased in the range from about 2.3 μm to 3.66 μm.

Absorption of Heat Caused by Releasing Muscles at Long Sarcomere Lengths

The muscles showed a large heat absorption when released at long muscle lengths without stimulation. This absorption of heat on releasing muscles was measured separately just before or after the measurements of the shortening heat. Then, this was used to correct the shortening heat, assuming that the same thermoelasticity holds for active as well as for resting muscles. In the case of fibre bundles the correction was necessary mostly at long sarcomere lengths over 3 μm. This is different from experiments on whole muscles as will be described below. Fig. 3-a shows the thermoelastic heat/tension ratio (Hill, 1953), $\Delta Q/\Delta P l_0$, at various sarcomere lengths in fibre bundles, where ΔQ is the amount of heat absorbed, ΔP is the change of tension on releasing muscles and l_0 the standard length of fibres. The heat absorption on release mostly appeared at sarcomere lengths over 3 μm and declined again at very long sarcomere lengths.

In the case of whole muscles on the other hand, the thermoelastic heat absorption on release started to appear at a sarcomere length of just over 2.5 μm. Moreover, the values of the thermoelastic heat/tension ratio were greater than those for fibre bundles (Fig. 3-b). Fig. 4 shows that more heat was absorbed in whole muscles than in fibre bundles for a given fall of tension and shows more clearly the differences in the thermoelastic property at long muscle lengths between whole muscles and fibre bundles.

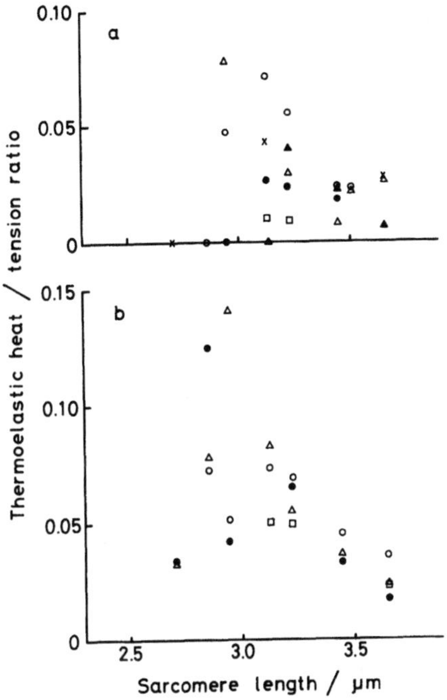

Figure 3: The thermoelastic heat/tension relation, $\Delta Q/\Delta Pl_0$, of unstimulated muscles at different sarcomere lengths in fibre bundles (a) and in whole semitendinosus muscles (b).

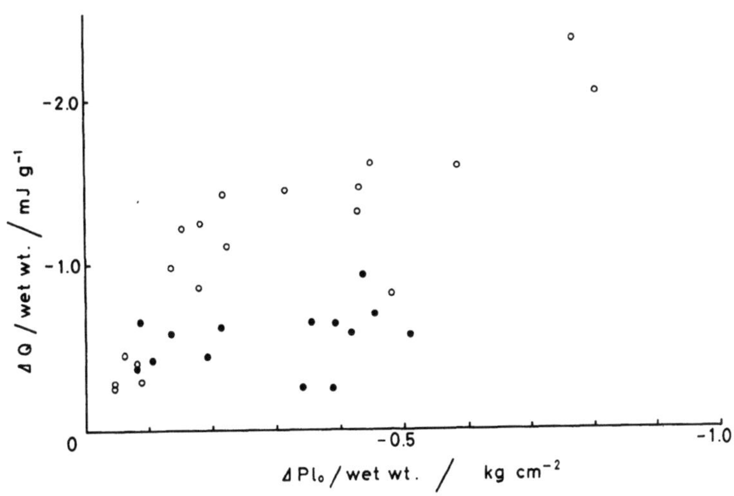

Figure 4: The thermoelastic heat absorption per unit weight of unstimulated muscles plotted against tension changes times standard length of muscle fibres l_0. Filled circles show measurements on fibre bundles and open circles those on whole semitendinosus muscles.

The Relation Between Shortening Heat and Sarcomere Length

Measurement of the shortening heat. Shortening heat was measured in a single tetanus of 3 s duration as was described by Irving, Woledge and Yamada (1979). After 1.2 s of isometric contraction the muscle was allowed to shorten by 0.22 to 0.3 μm per sarcomere in 0.114 to 0.12 s at constant velocities, which were about the maximum for each muscle. Thus the shortening heat was the heat produced in 0.4 s after the shortening was started, in excess of the heat of isometric control calculated as a weighted average of the isometric heats at the long and short lengths. To compare the results obtained from different experiments on different muscles, the shortening heat was converted to that produced by 0.25 μm shortening per sarcomere, assuming that the shortening heat is linearly related to the extent of shortening in the range between 0.2 and 0.3 μm per sarcomere (Irving and Woledge, 1981). The shortening heat produced per gram of muscle by 0.25 μm shortening per sarcomere was normalized by dividing it with the maximal tension produced per unit cross sectional area of muscle. The dimensionless quantity thus obtained was compared in different muscles and at different sarcomere lengths.

Experiments on fibre bundles. Fig. 5 shows the relation between shortening heat and sarcomere length in fibre bundles or bundle pairs. The shortening heat at each sarcomere length represents mean with ± standard error of mean for an average of measurements on five bundles or bundle pairs. The shortening heat is plotted at the centre of the 0.25

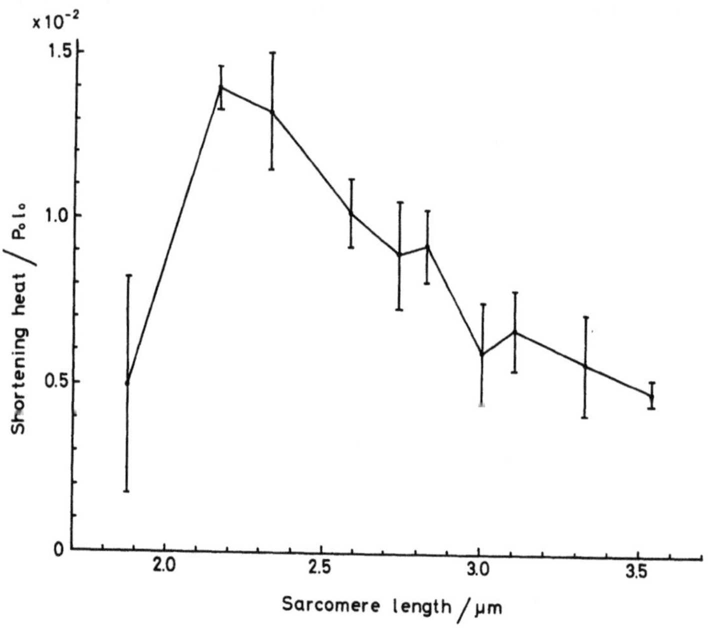

Figure 5: The relation between shortening heat and sarcomere length in fibre bundles. Each point represents mean ± S.E. of mean for an average of measurements on fibre bundles or bundle pairs. Temperature, 0°C.

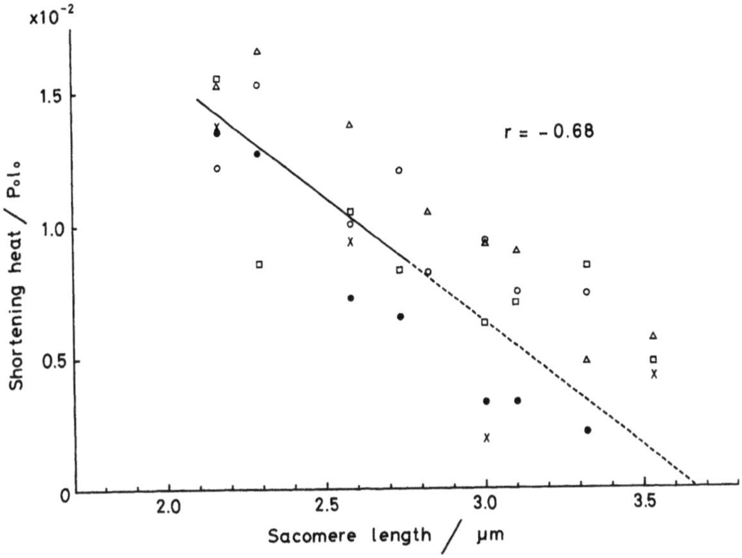

Figure 6: The relation between shortening heat and sarcomere length in fibre bundles. Each point represents shortening heat at different sarcomere lengths in different muscles. The regression line was calculated by assuming a linear relation in the sarcomere length range between 2.17 and 2.74 μm (Fig. 5). The correction for the thermoelastic heat absorption was not necessary for all the shortening heat measurements in this length range. The extrapolation of the regression line to the longer length range (dotted line) shows that points are deviated upward from the regression line in the range over 2.8 μm.

μm shortening per sarcomere. Fig. 6 summarizes all the shortening heat measurements mentioned above. The regression line was drawn assuming a linear relation between shortening heat and sarcomere length in the range between 2.17 and 2.74 μm (see Fig. 5). At this range of sarcomere length the fibre bundles did not show the large thermoelastic heat exchange. The correction for the thermoelastic heat absorption was only necessary at sarcomere lengths of mostly over 3 μm. Extrapolation of the regression line shows that shortening heat would be zero at the sarcomere length of 3.67 μm.

Experiments on whole muscles. Fig. 7 summarizes the results of shortening heat measurements on three pairs of intact whole muscles. The shortening heat at each sarcomere length represents mean with ± standard error of mean for an average of measurements on three pairs of whole semitendinosus muscles. The regression line was obtained from the mean values by assuming a linear relation between shortening heat and sarcomere length in the range over 2 μm. The regression line shows that shortening heat would be zero at the sarcomere length of 3.78 μm. However, it was necessary to correct the shortening heat for the heat absorption at sarcomere lengths longer than 2.5 μm. Therefore the dependence of shortening heat on sarcomere length was determined mostly on measurements which were corrected for the thermoelastic heat absorption

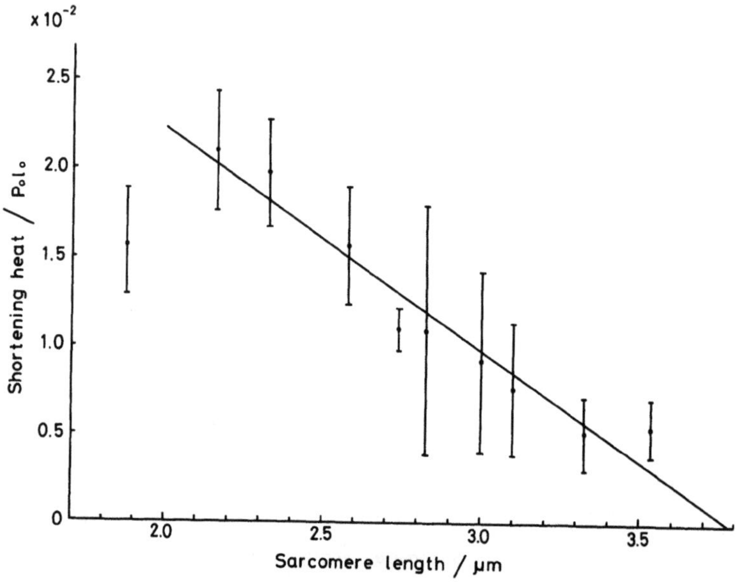

Figure 7: The relation between shortening heat and sarcomere length in whole semitendinosus muscles. The shortening heat at each sarcomere length represents mean ± S.E. of mean for an average of measurements on three pairs of muscles. The regression line was obtained from the mean values by assuming a linear relation in the range over 2 μm.

caused by releasing muscles without stimulation, which was obtained from the separate measurements.

DISCUSSION

Using bundles of fibres dissected from muscle, we were able to show without involvement of any correction for the large thermoelastic heat absorption, that shortening heat at the maximum velocity of shortening is linearly related to sarcomere length in the range between 2 and 3 μm. This result is consistent with the idea that the shortening heat represents the higher rate of cross-bridge turnover during shortening, and also consistent with the independent force-generator theory. However, there seems to be some points that deserve further discussion.

Origin of the thermoelasticity at long muscle lengths. Because the large absorption of heat was observed only at long muscle lengths, the thermoelasticity should be caused by the parallel elasticity of muscle. It is known that elasticity of fibrous proteins is rubber-like (Meyer and Haselbach, 1949). It was found that the thermoelastic heat/tension ratio was smaller in fibre bundles dissected from muscles than in intact whole muscles (Figs. 3 and 4). Moreover, the sarcomere length at which the thermoelastic heat absorption started to appear was shifted to longer lengths by at least 0.3 μm per sarcomere in fibre bundles than in whole muscles (Fig. 3). These findings indicate that the structure that causes the rubber-like elasticity observed in unstimulated muscles at long lengths exists outside muscle cells.

Validity of the correction for the thermoelastic absorption of heat. The above discussion would justify the assumption that the thermoelastic heat absorption observed in unstimulated muscles at long muscle lengths should be unaffected in active muscles at the same length. The fact that the shortening heat, after the absorption of heat was subtracted, linearly depended on sarcomere length in whole muscles, as was observed in this study and also by Irving et al. (1980), supports this. However, very large corrections may introduce errors. Fig. 6 shows that points representing the shortening heat at sarcomere lengths longer than 3 μm are somewhat deviated upwards from the regression line obtained for points in the range between 2.17 and 2.74 μm. This is probably caused by errors due to corrections. The different results reported by us previously (Yamada et al., 1981) might be caused by the same mechanism, especially when shortening heat was small by some reason.

The use of fibre bundles. The fibre bundles used consisted of comparatively large number of fibres. Therefore, the bundles contained some damaged fibres, or debris of them, on the surface. These damaged fibres did not seem to affect the present results in any essential way. However, the normalized shortening heat in fibre bundles was smaller than that in whole muscles by about 30%. The reason for this reduction is unknown. The difference in the length-tension relation between the fibre bundles and whole muscles also remains to be explained.

ACKNOWLEDGEMENTS

We thank Drs. R.C. Woledge, Nancy A. Curtin and J.A. Rall for valuable advice in constructing the thermopile. We also thank Dr. T. Kobayashi for his help in constructing the thermopile.

This work was supported in part by Grant-in-Aid for Scientific Research (548100, 56370006, 57222019) from the Ministry of Education, Science and Culture of Japan.

REFERENCES

Curtin, N.A. and Woledge, R.C. (1978). Energy changes and muscular contraction. Physiol. Rev. 58: 690-761.

Gordon, A.M., Huxley, A.F. and Julian, F.J. (1966). The variation in isometric tension with sarcomere length in vertebrate muscle fibres. J. Physiol. 184: 170-192.

Hill, A.V. (1938). The heat of shortening and the dynamic constants of muscle. Proc. Roy. Soc. B. 126: 136-195.

Hill, A.V. (1953). The 'instantaneous' elasticity of active muscle. Proc. Roy. Soc. B. 141: 161-178.

Hill, A.V. (1965). *Trails and Trials in Physiology.* London: Arnold.

Homsher, E. and Kean, C.J. (1978). Skeletal muscle energetics and metabolism. Ann. Rev. Physiol. 40: 93-131.

Huxley, A.F. (1957). Muscle structure and theories of contraction. Prog. Biophys. biophys. Chem. 7: 255-318.

Irving, M., Woledge, R.C. and Yamada, K. (1979). The heat produced by frog muscle in a series

of contractions with shortening. J. Physiol. 293: 103-118.
Irving, M., Homsher, E. and Lebacq, J. (1980). Dependence of the shortening heat on sarcomere length in frog skeletal muscle. Fedn. Proc. 39: 1730.
Irving, M. and Woledge, R.C. (1981). The dependence on extent of shortening of the extra energy liberated by rapidly shortening frog skeletal muscle. J. Physiol. 321: 411-422.
Kodama, T. and Yamada, K. (1979). An explanation of the shortening heat based on the enthalpy profile of the myosin ATPase reaction. In: *Cross-bridge Mechanism in Muscle Contraction*, eds. Sugi, H. and Pollack, G.H., pp. 481-488. Tokyo: Univ. of Tokyo Press.
Lebacq, J. (1972) La chaleur de raccourcissement musculaire à différentes longeurs du sarcomère. J. Physiol. Paris 65: 440A
Meyer, K.H. and Haselbach, C. (1949). Rubber-like properties of hair keratin. Nature 164: 33-34.
Ricchiuti, N.V. and Mommaerts, W.F.H.M. (1965). Technique for myothermic measurements. Physiologist Washington 8: 259.
ter Keurs, H.E.D.J., Iwazumi, T. and Pollack, G.H. (1979). The length-tension relation in skeletal muscle: revisited. In: *Cross-bridge Mechanism in Muscle Contraction*, eds. Sugi, H. and Pollack, G.H., pp. 277-295. Tokyo: Univ. of Tokyo Press.
Yamada, K., Kometani, K. and Kobayashi, T. (1981). The relation between shortening heat and sarcomere length in frog skeletal muscle. J. Physiol. Soc. Japan 43: 373.

DISCUSSION

POLLACK: It seems to me that the absorption of heat occurs at sarcomere lengths at which you begin to recruit resting tension. Do you think that this absorption is due, somehow, to the presence of resting tension, either within or outside the cell?

YAMADA: I didn't plot the resting tension. The thermoelastic heat absorption begins to develop at somewhat shorter sarcomere lengths than the resting tension does.

HUXLEY: Could you give a quantitative comparison of these? I mean supposing you were producing this shortening heat from splitting ATP, and you were producing one splitting of ATP for each, say, hundred angstroms of sliding past an overlapped cross-bridge. Can you make an estimate of the number of cross-bridges which would be engaged to produce the observed amount of heat?

YAMADA: I think yes. But we are not very clear how many turnovers of cross-bridges might occur during 0.25 μm of shortening per sarcomere.

HUXLEY: But I mean you have a measure for the total heat produced. So you should be able to say how many molecules of ATP this would correspond to, and that would give you a total number for the 0.25 microns of shortening that you observe.

HOMSHER: Dr. Huxley's question will be answered in the next presentation. It turns out to be about 6 per second.

POLLACK: Were these muscles released to shorten from the onset of contraction, or was there a tetanic plateau that was developed and then the muscles were released to zero load for the shortening?

YAMADA: The second was the case.

TIROSH: Continuing the question of Pollack, I'm interested in the end; did you wait for regeneration of the tetanic force, or did you end with zero force? You begin the contraction in isometric conditions.

YAMADA: Yes.

TIROSH: What about the end? Do you wait for redevelopment of isometric force?. My question relates to another aspect of heat measurement, as was done by A.V. Hill and Woledge, which is a tension-related entropic component of heat. Namely, there are two components, one is the thermo-elastic heat, which is well accounted for by doing work on elastic elements. The other is a reversible component which is related to the tension developed. And I wonder if you took these into account?

YAMADA: Yes, the methods that we use to estimate shortening heat account for the thermoelastic heat changes during the period including shortening.

TIROSH: But what about the other component? There is another component which is not explained, which is considered to be an entropic component which, in a closed cycle, should integrate to zero. Not the unexplained heat component in your terms, but the unexplained heat component in A.V. Hill's terms. Namely, the reversible contribution of heat production and absorption during a closed cycle of contraction that should not have a chemical counterpart.

HOMSHER: The thing is that Kazuhiro has measured this heat production at V_{max}, or very close to V_{max}. So he hasn't measured the tension component. He could do that if he released the muscle at slower velocity, but he hasn't done that.

TIROSH: I think that in heat measurements it is very important to take into account thermodynamic arguments about reversible contributions. This issue has been brought out in the history of heat measurements. For example, heat of shortening, on one hand, was A.V. Hill's long-standing contribution. Then came the experiments of Carlson et al. (J. Gen. Physiol., 46: 851-882, 1963), where twitch measurements rather than isometric or isotonic tetani were made. And the difference of results, indeed, had to do with a closed cycle: One kind of experiment is strictly *part* of the cycle, namely the isotonic tetanus, whereas the twitch measurements are over a closed cycle. I realize in all the debate around this, that those who like to work with tetani deny the twitch. And vice versa.

WILKIE: That's absolutely not true!

TIROSH: I'm happy to hear it. I'd like to hear more about reservations of making either kind of experiments.

HOMSHER: Well, look, the problem about shortening heat is simply this. The way Hill defines it, shortening heat is the amount of heat produced above a steady state heat production. It's a definition ... it isn't a thermodynamic entity. It's a simple, functional definition. In the twitch, one does not have the same baseline against which to define a shortening heat. You assert that when the muscle shortens in a tetanus, that after relaxation, the shortening heat may have been reabsorbed and disappears. And that simply isn't true, as John Lebacq and Xavier Aubert showed about five or six years ago. They measured shortening heat in tetani defined according to Hill's definition, allowed muscle to shorten, the muscle produced extra heat, the muscle was allowed to redevelop tension and then relax, and at the end of the contraction after the muscle had relaxed, the shortening heat was still there. So it isn't a problem that the twitches are a superior form of a complete cycle. It's simply that the way Hill defined the shortening heat cannot be applied simply to twitch.

TIROSH: Indeed, A.V. Hill (Proc. Roy. Soc. B., 159: 596-605, 1964) responded to Carlson's experiments and he went on, according to his definition, to find the shortening heat in twitches. He did it very carefully, and surprisingly he came to an unexpected result -- that in twitches, in contrast to tetanic stimulation, the net heat production decreased in *opposite* relation to the *increase* of the total heat and work obtained at various velocities or load. These results indicated that the excess heat of shortening as measured in a fraction of a cycle could be either a part of a reversible contribution due to tension as well as length changes, or due to release of prestored energy, either elastic or entropic. Thus, the excess heat of shortening as measured under tetanic contractions may have no chemical account, at least within the measured period, as was concluded by Carlson et al. from their mechanical, heat and chemical measurements on closed cycles.

HOMSHER: It's a red herring. The facts of the matter are as follows: in a twitch Hill did not have a steady state baseline against which to reference his measurements. Second, and most important, although he claims that the twitch heat production can be deduced by the equation $H = A + ax$, where A is a constant and x the distance shortened, you will find that on recalculating the data in his figures, this equation does *not* hold.

WILKIE: Those experiments you cited were in fact by Carlson, Hardy and Wilkie, so I feel responsible for their conclusions. The fact is that Hill *did* do what was described, which was to introduce the previously unheard of quantity h to make up the difference and thereby, I think, he created a great deal of confusion in people's minds.

THE EFFECT OF SHORTENING ON ENERGY LIBERATION AND HIGH ENERGY PHOSPHATE HYDROLYSIS IN FROG SKELETAL MUSCLE

Earl Homsher, M. Irving and T. Yamada

Department of Physiology, School of Medicine, University of California at Los Angeles, Los Angeles, CA 90024

ABSTRACT

It is generally assumed that the increased rate of energy liberation (as heat and work, h+w) accompanying shortening stems from an increased rate of crossbridge cycling and ATP hydrolysis. Experiments were performed to test two premises of this assumption: first, is the increased rate of heat production accompanying shortening derived from crossbridge activity? This question was answered by measuring the amount of shortening heat produced by a fixed displacement of 0.3 μm/sarcomere in the sarcomere length range of 2.25-3.75 μm. Shortening heat declines linearly with decreasing amounts of thick and thin filament overlap and becomes zero at a sarcomere spacing of ca. 3.70 μm. Secondly, the extent to which the measured consumption of high energy phosphate accounts for the measured tetanic (h+w) production during and after shortening for 300 ms at a velocity of V_{max} or ½V_{max} was examined. The results of these experiments showed that within 700 ms of the end of shortening at both velocities, all the (h+w) could be explained by the hydrolysis of ATP. At V_{max} all the (h+w) produced by the end of shortening could be explained by the measured ATP hydrolysis. However, at V_{max} less than half of the (h+w) produced by the end of shortening could be explained by the measured ATP splitting and there was a high rate of ATP splitting *after* the end of shortening. These results suggest that while shortening at velocities $\leq V_{max}$ the energy liberation is indeed derived from an increased rate of ATP hydrolysis by crossbridges, at V_{max} the crossbridge ATPase cycle differs somewhat from that at lower shortening velocities.

INTRODUCTION

It was almost 45 years ago that A.V. Hill (1938) first showed that when an isometrically contracting muscle is allowed to shorten, not only does the rate of work production increase, but the rate of heat liberation increases as well. The rate of total energy liberation (h+w) in frog skeletal muscle at 0° C increases from ca. 15 mW/g in the isometric case to about 50 mW/g during shortening at velocities of $\frac{1}{2}V_{max}$. The rate of heat production itself continues to rise to ca. 40 mW/g at velocities near V_{max} despite the fact that the power output drops to zero. Based on Hill's work, the rate of *heat* production in excess of that seen in the isometric contraction came to be called the "shortening heat." With the advent of the sliding filament model, advances in biochemical studies of isolated muscle proteins (Taylor, 1979), and the ideas stemming from Huxley's 1957 model of muscle contraction it was logical to assume that the increased rate of energy liberation with shortening described by Hill was a result of an increased rate of ATP hydrolysis subsequent to an increased rate of crossbridge attachment and detachment during shortening. This explanation contains two basic assumptions: first, that the increased energy production is itself a result of the interaction of the thick and thin filaments. The second is that the observed enthalpy production (h+w) is derived solely from high energy phosphate splitting. We tested the first assumption by measuring the extent to which shortening heat production depends on the overlap between the thick and thin filaments. We then examined the second assumption by performing energy balance experiments on muscles shortening at different shortening velocities, $\frac{1}{2}V_{max}$ and V_{max}.

Length Dependence of Shortening Heat Production

To measure shortening heat production, one often begins by tetanizing a muscle at a length near the plateau of the length-tension curve. After tension has come to a steady value (at time t_1), the muscle is allowed to shorten to a shorter length. With the cessation of shortening, stimulation continues (to a time t_2) while the muscle isometrically redevelops a steady force. After the muscle has relaxed, a second isometric tetanus (t_2 seconds in duration), at the length to which the muscle had shortened, is given. In both tetani heat production is monitored. To determine the shortening heat production, the time course of heat evolution between t_1 and t_2 for both contractions is plotted. Because the shortening heat is defined as that heat produced by a shortening muscle in excess of that produced by a non-shortening (isometric) control, the heat produced by the isometrically contracting muscle in the interval ($t_1 - t_2$) is subtracted from that produced over the same interval by the shortening muscle, yielding the shortening heat. For experiments in which the dependence of the shortening heat production on sarcomere length was examined, dorsal heads of frog semitendinosus muscles were dissected and the relationship between sarcomere spacing and resting muscle length was measured for each muscle using laser diffraction techniques (Homsher, Irving and Wallner, 1981). This muscle was chosen for these experiments because it can be reversibly stretched to lengths greater than 3.65 μm (Smith, 1972; Homsher, Mommaerts, Ricchiuti and Wallner, 1972). The muscles were then mounted on thermopiles for myothermal recording (as previously described [Homsher et al., 1981]) and experiments conducted according to the experimental protocol shown in Fig. 1. A muscle pair was brought to resting sarcomere lengths

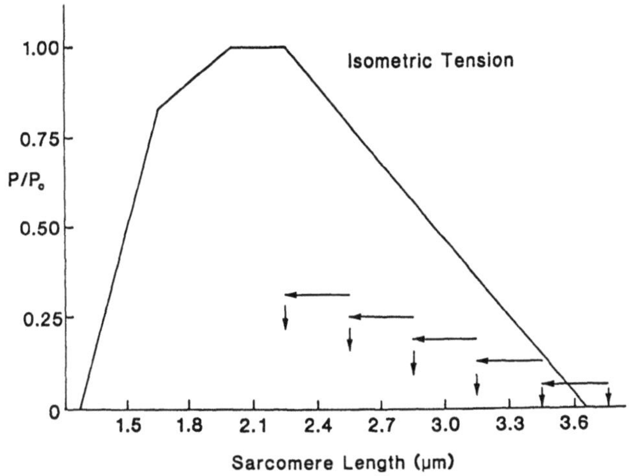

Figure 1: Experimental design of studies of the length dependence of shortening heat production. Vertical arrows indicate lengths at which muscles are stimulated and the horizontal arrows indicate the lengths over which shortening takes place.

indicated by the vertical arrows (3.75, 3.45, 3.15 μm, etc.) and tetanically stimulated for 2.0 s to obtain isometric controls (myothermal records F and G in Fig. 2). Muscles at these various lengths were also tetanically stimulated and after 0.75 s were allowed to shorten (as indicated by the horizontal arrows of Fig. 1) 0.3 μm/sarcomere/s at a velocity near V_{max} (1.9 μm/sarcomere/s), producing force, displacement, and myothermal records A,B, and D respectively in Fig. 2. When a resting muscle shortens from an initial length >2.7 μm, the decline in passive (parallel elastic element) force is accompanied by an absorption of heat. The same effect is assumed to occur in an actively shortening muscle. To allow for this effect, thermal changes accompanying the shortening of resting muscles (records G in Fig. 2) were measured and subtracted from the myothermal records produced by actively shortening muscles. After correction of myothermal records for heat loss and conduction lag (Homsher et al., 1981), the heat production of the isometric control (at the length to which the muscle had shortened) was subtracted from that of the corresponding shortening muscle to yield the time course of the shortening heat (seen in H of Fig. 2). Shortening heat (Joules) obtained for shortening over the length ranges of 3.75-3.45, 3.45-3.15, 3.15-2.85, 2.85-2.55, to 2.55-2.25 μm, was normalized for the distance shortened (meters) and tetanic force exerted at 2.25 μm (P_o, newtons), yielding the dimensionless quantity, α/P_o, the shortening heat coefficient. The shortening heat coefficient was then plotted as function of the mean sarcomere length during shortening with the result seen in Fig. 3. The result is rather simple; i.e., the shortening heat coefficient, and hence the amount of shortening heat produced per unit displacement, declines linearly with a decline in the amount of thick and thin filament overlap and reaches zero at a sarcomere length 3.75 μm which not different from the length at

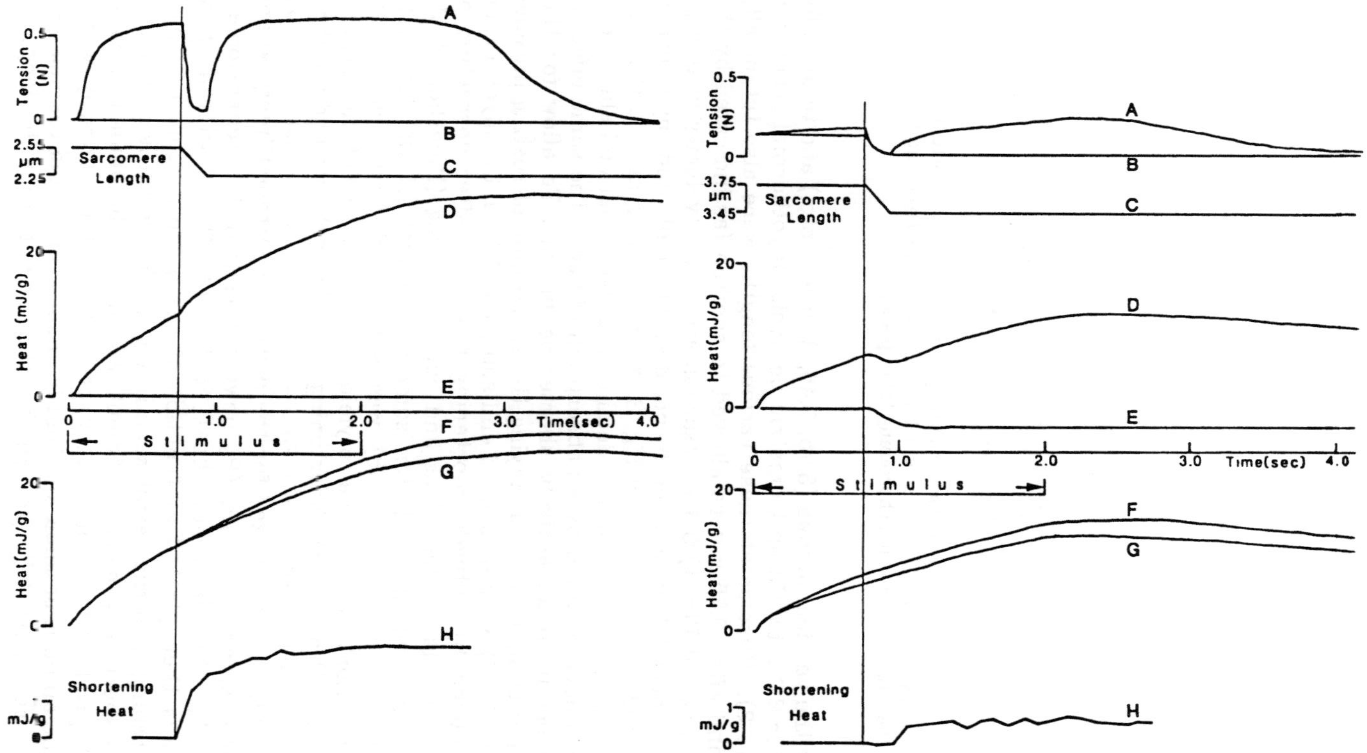

Figure 2: Traces of original records of active (A) and passive (B) tension, length (C), heat production during active (D) and passive (E) shortening, heat production in an isometric tetanus at the initial (F) and final (G) sarcomere lengths, and the shortening heat production (H) for experiments in the sarcomere range of 2.55-2.25 μm (left panel) and 3.75-3.45 μm (right panel). Muscle pair blotted weight, 77.8 mg.

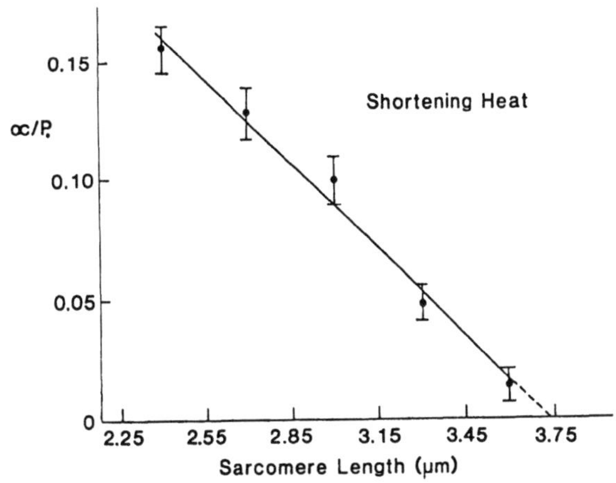

Figure 3: The effect of sarcomere length on the shortening heat coefficient. φ/P_0 located at mean sarcomere length during the shortening.

which filament overlap is zero (3.65 μm). The results in Fig. 3 show that shortening heat production, like isometric force (7), is dependent on the amount of filament overlap and is thus probably produced by the action of crossbridges.

Energy Balance During Shortening

Earlier work (Kushmerick and Davies, 1969; Curtin, Gilbert, Kretzschmar and Wilkie, 1974; Rall, Homsher, Wallner and Mommaerts, 1976) had indicated that only about 50% of the energy produced during shortening could be explained by high energy phosphate hydrolysis. However, these reports involved shortening near the beginning of a tetanus at a point in time when a large amount of unexplained enthalpy is produced by processes related to intracellular calcium circulation and binding (Curtin and Woledge, 1978; Curtin and Woledge, 1979; Homsher, Kean, Wallner, and Sarian-Garibian, 1979; Curtin and Woledge, 1981; Homsher and Kean, 1982). However, it has been shown, using R. temporaria sartorius muscles, that the isometric unexplained enthalpy is produced early (within the first 3 s) in the tetanus (Curtin and Woledge, 1979; Homsher et al., 1979) and thereafter ceases to be evolved; i.e., after about 3 seconds of tetanic stimulation, succeeding energy production in an isometric contraction can be fully explained by the measured high energy phosphate utilization. In sartorius muscles of R. pipiens (a species of from which produces less isometric unexplained enthalpy than that seen in R. temporaria [Homsher, Rall, Wallner and Ricchiuti, 1975]), there is no significant amount of unexplained enthalpy produced by muscles in the one second interval between the 2nd and 3rd second of a maintained isometric tetanus (4). These observations suggest a technique by which one may definitively examine the energy balance during shortening and the

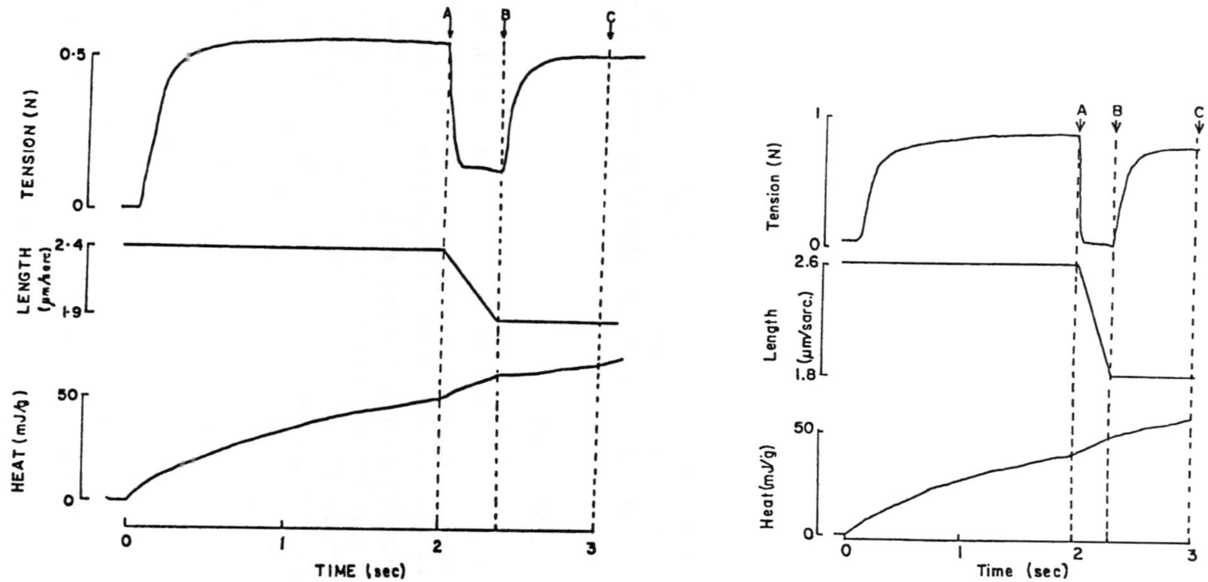

Figure 4: Traces of original recording of force, length change, and heat production during shortening at ½V_{max} (left panel) and V_{max} (right panel) in a 3 second tetanus. After 2 sec of isometric contraction (A) the muscle was allowed to shorten until the point indicated by (B) and thereafter contracted isometrically to point (C). The blotted weight of the muscle pair shortening at ½V_{max} was 141.2 mg and that at V_{max} was 173.5.

experimental protocol is shown in Fig. 4. In these experiments sartorius muscles at an initial sarcomere length of 2.4 μm (determined by laser diffraction) were tetanized isometrically for 2 s prior to shortening, to deplete the isometric unexplained enthalpy. In the time interval between 2-2.35 s (A→B in Fig. 4), the muscles shortened at a velocity of 1.33 μm/sarcomere/s ($\frac{1}{2}V_{max}$) to a sarcomere length of 1.9 μm. Stimulation was continued for an additional 0.65 s while isometric tension redeveloped to steady state levels (B→C in Fig. 4). The enthalpy (h+w) produced and the amount of high energy phosphates hydrolyzed in the time intervals A→B and B→C were measured as previously described (Homsher et al., 1981). Knowing the amount of chemical change occurring in the muscle, one can calculate how much enthalpy the muscle should produce (the explained enthalpy) which is then compared to the amount of (h+w) actually produced (the observed enthalpy). If the two values are significantly different from each other, then other energetically significant reactions must be occurring, and the simply hypothesis for the increased rate of energy liberation during shortening described above is incorrect. As shown in Fig. 4, the energy balance during and after shortening was studied with shortening at two different shortening velocities, V_{max} and $\frac{1}{2}V_{max}$. In these experiments, there were no significant changes in the muscle content of ATP, ADP, or AMP. The equality of the changes in inorganic phosphate and free creatine indicated that the only significant net measured reaction occurring was the splitting of phosphocreatine (PCr). The results of the energy balance experiments are shown in Fig. 5.

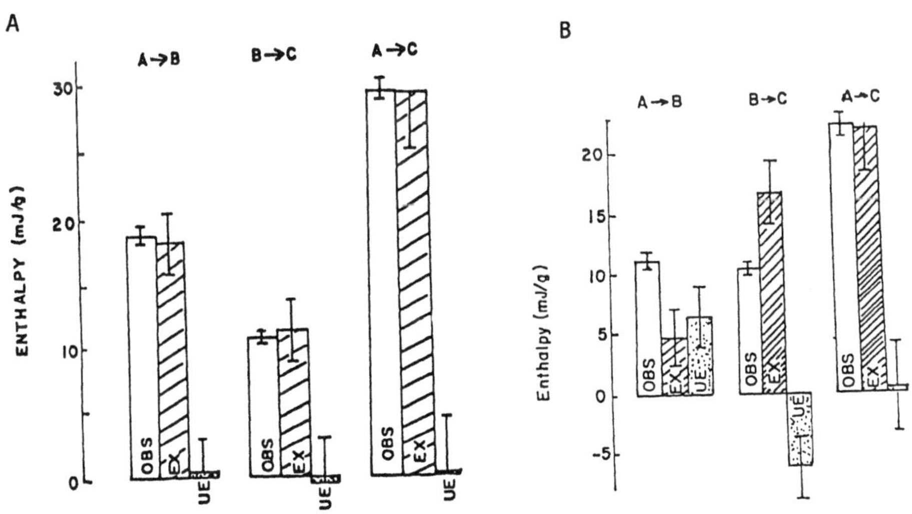

Figure 5: Results of Energy balance measurements. Enthalpy liberation given in mJ/g±S.E.M. from between 17 to 29 pairs of muscles. Unshaded bars, observed enthalpy; cross-hatched bars, explained enthalpy; stipled bars, unexplained enthalpy. Data given for period during shortening (A-B) and the period immediately after shortening (B-C), and the entire 1 sec. shortening and post-shortening period (A-C).

Considering first the results at $\frac{1}{2}V_{max}$ (Fig. 5A), it is clear that the large amounts of (h+w) produced during (A→B) and after (B→C) shortening (18.8±0.5 mJ/g and 10.8±0.5 mJ/g respectively) are accounted for by the measured changes in PCr; i.e., there is no significant unexplained enthalpy produced during either period. The mean rate of ATP splitting during shortening was 1.58±0.23 μmol/g.s. The mean rate of ATP hydrolysis in the post shortening period at a sarcomere length 1.9 μm was 0.43±0.09 μmol/g.s which is significantly less than that during shortening ($P<0.01$) and is not significantly different from that in an isometrically tetanized muscle (0.32±0.11 μmol/g.s) which had not undergone any shortening.

The data obtained at V_{max} (shown in Fig. 5B) reveal a very different behavior. During shortening (A→B) a substantial amount of (h+w) is produced (11.3±0.7 mJ/g) and more than half of it, 6.5±2.6 mJ/g, is not accounted for by the measured chemical changes. In fact the rate of ATP splitting during shortening, 0.48±0.24 μmol/g.s, is not significantly different from that seen in an isometric contraction. In the 0.7 s isometric period following shortening (B→C) the situation is the complete reverse; i.e., the amount of energy explained by observed ATP splitting significantly ($P<0.05$) exceeds, by 6.1 mJ/g, that produced by the muscle (10.6± 1.5 mJ/g), and the rate of ATP hydrolysis, 0.71±0.10 μmol/g.s, is significantly greater ($P<0.02$) than that in a comparable isometric period. Thus over the entire cycle (period A→C), the energy imbalances during and after shortening at V_{max} cancel; i.e., 21.9±0.9 mJ/g of heat+work are produced, 99% of which is accounted for by the measured chemical changes. Thus, the major conclusion resulting from these experiments is that the net energy liberated as a result of muscular shortening is accounted for by ATP hydrolysis.

DISCUSSION

The results of our studies on the length dependence of shortening heat production are in agreement with earlier work (Irving and Woledge, 1981) on frog sartorius muscles. In that work the amount of shortening heat produced, measured over the sarcomere length range of 2.05-2.60 μm, was found to decline linearly with the amount of thick and thin filament overlap. These results therefore indicate that the shortening heat production depends on the number of crossbridges available for attachment to the thin filaments. A weakness in the current experiments is that the reference sarcomere lengths were those measured in the resting muscles and are thus not exactly those which are obtained at the beginning of shortening. Further, it is known that in stretched muscle fibers, not all the sarcomeres are the same length; i.e., those at the tendonous ends are somewhat shorter than those in the central region of the fibers (Huxley and Peachey, 1961). To determine the extent to which these factors affected our results, the sarcomere spacing was monitored (using laser diffraction measurements) in semitendinosus muscles which contracted and shortened in experimental protocols similar to those employed in the myothermal experiments. These studies showed that

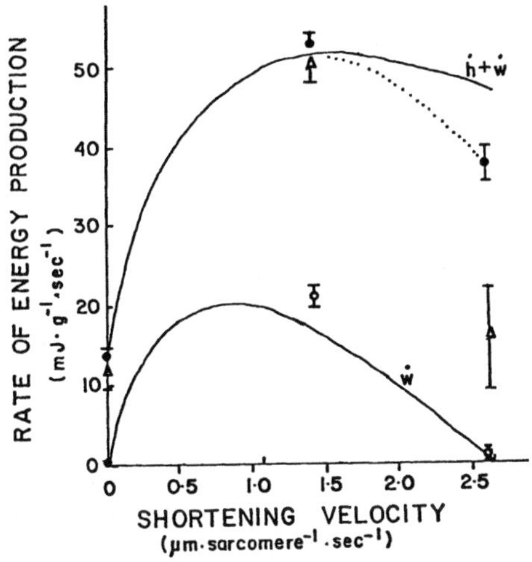

Figure 6: A plot of the rate of energy liberation as a function of relative shortening velocity. The solid line labeled w is the rate of external work production, and the line labeled h+w is the total rate of enthalpy production. The open circles are the measured rate of external work production, the filled circles are the observed rates of enthalpy production, and the open triangles the rate of explained enthalpy production.

when the initial sarcomere length was ≧3.15 μm, the isometric period prior to shortening was associated with a ca. 0.1 μm/ sarcomere lengthening of the central region of the muscles (at shorter sarcomere lengths there was a ca. 0.1 μm/sarcomere shortening during the isometric period). During isovelocity shortening the muscle sarcomere length in the central region (that from which heat is measured) decreased by the expected 0.3 μm/sarcomere, regardless of the initial sarcomere length. These experiments suggest that the decline in shortening heat with initial sarcomere length is real; the only significant consequence of the use of the resting sarcomere length as a reference is that the sarcomere length at which shortening heat becomes zero may be slightly under-estimated.

Figure 6 shows a plot of the rate of energy (and power) liberation as a function of shortening velocity; the solid lines are the results predicted by A.V. Hill's work (1938; 1964; Hill and Woledge, 1962), the circles (O), the results from our myothermal and mechanical measurements, and the triangles (Δ), the results based on our measurement of chemical change during shortening. The myothermal and mechanical results agree well with Hill's data under all conditions examined. The results from measurement of biochemical change under isometric and ½V_{max} conditions also agree with Hill's work. However at V_{max} there is a serious and significant discrepancy between the amount of observed and explained enthalpy; i.e., more enthalpy is produced than can be explained by the measured chemical changes occurring in the muscle. It is unlikely that the difference is caused by an insufficiency in the arrest of muscle metabolism or a difference in the experimental execution because the experiments at V_{max}

and $\frac{1}{2}V_{max}$ were similar with respect to a) sarcomere length range over which shortening occurred (2.6-1.8 μm and 2.4-1.9 μm respectively), b) the duration of shortening (300 and 350 ms respectively), c) the amount of shortening heat produced (5.7±0.9 and 6.3±0.5 mJ/g, respectively). In addition it is unlikely that as has been suggested by Kodama and Yamada (1978), the rate of ATP cleavage on myosin is rate limiting at V_{max}. Similarly the failure of the muscle at V_{max}. This is because ATP is hydrolyysed by the muscle shortening at $\frac{1}{2}V_{max}$ three times faster than one shortening at V_{max}. Similarly the failure of the muscle shortening near V_{max} to sustain an elevated rae of ATP hydrolysis can not be attributed to a rate limiting step in the phosphorylation of ADP by PCr and creatine phosphokinase. Indeed, the lack of measurable changes in ADP (<0.02 μmol/g) or ATP during shortening at $\frac{1}{2}V_{max}$ indicates that ADP rephosphorylation proceeds at a rate in excess of 1.6 μmol/g.s. It seems, therefore, that increasing the shortening velocity *per se* above $\frac{1}{2}V_{max}$ somehow both reduces the mean ATP hydrolysis rate and defers some ATP hydrolysis until after the cessation of shortening. This point can be better appreciated by consideration of the rate of ATP turnover by myosin during shortening at V_{max}, $\frac{1}{2}V_{max}$, and zero velocity; these rates are 1.7±0.9 s^{-1}, 5.5± 0.8 s^{-1}, and 1.1±0.4 s^{-1} respectively (assuming that the muscle contains 0.28 μmol of myosin S-1 heads per gram of muscle). The results tend to suggest that rapid shortening may create a type of crossbridge whose ability to cycle is limited by the rapid movement of the thick and thin filaments past one another. It is very likely that the explanation of the energy balance data at V_{max} involves the presence of an incomplete thermodynamic cycle. The essence of this idea is that crossbridges may exist in at least two different states (state I and II in Fig. 7 [Woledge, 1971]). Each state would have a different enthalpy content, so that in the transition from I to II heat would be evolved. The transition from II to I *per se* driven by the hydrolysis of ATP, would thus involve an absorption of enthalpy. Over a complete cycle in the steady state, one would observe only enthalpy production corresponding to ATP hydrolysis. However, if, in the transition from an isometric contraction to steady state shortening, there were a redistribution of the steady state concentration of crossbridges to state II (an incomplete thermodynamic cycle), there would be an evolution of heat *without* a corresponding ATP hydrolysis. On the return to an isometric contraction, there would be an extra ATP

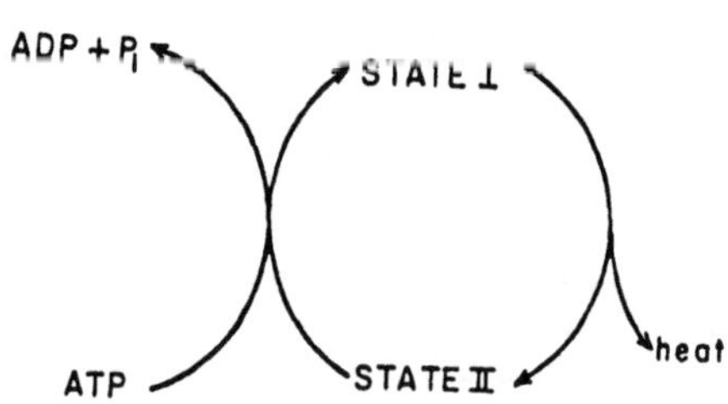

Figure 7: A mechanism for dissociation of crossbridge enthalpy production from ATP hydrolysis.

hydrolysis without a corresponding enthalpy production. Since the extra heat production at V_{max} is 6 mJ/g and because its reversal requires about 0.2 μmol/g, we estimate that the molar enthalpy change from state I to II is roughly -30 kJ/mole S-1. These results thus place restrictions on the types of models that one might use for explanations for muscle contraction.

ACKNOWLEDGEMENTS

This work was supported by grant HL 11351 from the USPHS, and C791127 of the Muscular Dystrophy Association of America.

REFERENCES

Curtin, N.A., Gilbert, C., Kretzschmar, K.M., and Wilkie, D.R. (1974). The effects of the performance of work on total energy output and metabolism during muscular contraction. J. Physiol. 238: 455-472.

Curtin, N.A. and Woledge, R.C. (1978). Energy changes and muscle contraction. Physiol. Rev. 58: 690-761.

Curtin, N.A. and Woledge, R.C. (1979). Chemical change and energy production during contraction of frog muscle: how are their time courses related. J. Physiol. 228: 353-366.

Curtin, N.A. and Woledge, R.C. (1981). Effect of muscle length on energy balance in frog skeletal muscle. J. Physiol. 316: 453-568.

Gordon, A.M., Huxley, A.F., and Julian, F.J. (1966). Tension development in highly stretched vertebrate muscle fibers. J. Physiol. 184: 143-169.

Hill, A.V. (1938). The heat of shortening and the dynamic constants of muscle. Proc. Roy. Soc. B. 126: 136-195.

Hill, A.V. and Woledge, R.C. (1962). An examination of absolute values in myothermic measurements. J. Physiol. 162: 311-333.

Hill, A.V. (1964). The effect of load on the heat of shortening of muscle. Proc. Roy. Soc. B. 150: 297-318.

Homsher, E., Mommaerts, W.F.H.M., Ricchiuti, N.V., and Wallner, A. (1972). Activation heat, activation metabolism, and tension-related heat in frog semitendinosus muscles. J. Physiol. 220: 600-625.

Homsher, E., Rall, J.A., Wallner, A., and Ricchiuti, N.V. (1975). Energy liberation and chemical change in frog skeletal muscle during single isometric tetanic contractions. J. Gen. Physiol. 65: 1-21.

Homsher, E. and Kean, C.J.C. (1982). Unexplained enthalpy production in contracting skeletal muscles. Fed. Prod. 41: 149-154.

Homsher, E., Kean, C.J., Wallner, A., Sarian-Garibian, V. (1979). The time course of energy balance in an isometric tetanus. J. Gen. Physiol. 73: 553-567.

Homsher, E. Irving, M., and Wallner, A. (1981). High-energy phosphate metabolism and energy liberation associated with rapid shortening in frog skeletal muscle. J. Physiol. 321: 423-436.

Huxley, A.F. (1957). Muscle structure and theories of contraction. Prog. Biophys. Biophys. Chem. 7: 255-318.

Huxley, A.F. and Peachey, L.D. (1961). The maximum length for contraction in vertebrate striated muscle. J. Physiol. 156: 150-165.

Irvin, M. and Woledge, R.C. (1981). The energy liberation of frog skeletal muscle in tetanic contraction containing two periods of shortening. J. Physiol. 321: 401-410.

Kodama, T. and Yamada, K. (1978). An explanation of the shortening heat based on the enthalpy profile of the myosin ATPase reaction. In: *Cross-bridge Mechanism in Muscle Contraction*, ed. Sugi, H. and Pollack, G.H. pp. 481-488. Tokyo: University of Tokyo Press.

Kushmerick, M.J., and Davies, R.E. (1969). The chemical energetics of muscle contraction. II. The chemistry, efficiency, and power of maximally working sartorius muscle. Proc. Roy.

Soc. B. 174: 315-350.
Rall, J.A., Homsher, E., Wallner, A., and Mommaerts, W.F.H.M. (1976). A temporal dissociation of energy liberation and high energy phosphate splitting during shortening in frog skeletal muscle. J. Gen. Physiol. 68: 13-27.
Smith, I.C.H. (1972). Energetics of activation in frog and toad muscle. J. Physiol. 220: 583-599.
Taylor, E.W. (1979). Mechanism of actomyosin ATPase and the problem of muscle contraction. Crit. Rev. Biochem. 6: 103-164.
Woledge, R.C. (1971). Heat production and chemical change in muscle. Prog. Biophys. Mol. Biol. 22: 39-74.

DISCUSSION

CECCHI: How does the number of completed cross-bridge cycles you found compare with the number one can calculate from the A.F. Huxley model (Prog. Biophys. Biophys Chem., 7: 255-318, 1957), or from the stiffness measurements made during the shortening by Julian and Sollins (J. Gen Physiol., 66: 287-392, 1975) and by Huxley and Simmons (CSH Symp., 37: 669-680, 1973)?

HOMSHER: That's hard to say, because they specify not the *number* of cross-bridges, but the *relative* number of cross-bridges. I think our results are in qualitative agreement with theirs. They find that as the shortening velocity increases, the percentage of attached cross-bridges decreases, i.e., compared to isometric contraction.

CECCHI: It seems to me that there is a radical difference between the number you gave and number they found.

HOMSHER: Ford, Huxley, and Simmons never mentioned the number of attached cross-bridges. They can measure stiffness, but they don't know how many cross-bridges that represents. What we calculated was that 6% of all the cross-bridges were attached *and cycling* at V_{max}, while at a slower shortening velocity, 14% were attached and cycling. Thus, as the shortening velocity declines, more and more cross-bridges are attached and cycling.

CECCHI: You say 6% attachment. Can you give a number for the isometric situation?

HOMSHER: No, I know of no way to use our data to estimate the number you wish to know.

MAGID: It appears to me that there's one way to test this idea. If it's true that the myosin-ADP form has detached from the thin filament after it's lost its phosphate, then it should have lost its ability to generate isometric force. That being the case, you ought to see if, in the time after the release, when the tension is redeveloping, a large disproportion between tension and instantaneous stiffness is found. Do you agree that's a test of your hypothesis?

HOMSHER: No, I don't think so. It may be that the myosin-ADP cross-bridge can attach fairly rapidly to stationary thin filaments (as in the isometric case). If, in the isometric case, they can reattach fairly rapidly, then the dissociation of bound ADP, the binding of ATP, and the dissociation from the thin filament, should, according to the kineticists, proceed quite rapidly and one would expect that the redevelopment of force might therefore be only slightly, if at all, affected.

There are, however, several ways to test the idea. One prediction of this hypothesis is that one should see a burst of ATP splitting immediately after the cessation of shortening. We will have to make measurements closer to the end of shortening using energy balance techniques to check this point. Another prediction of this model is that the relationship between distance shortened and the amount of shortening heat produced should be curvilinear. That is, as the muscle shortens it should produce progressively less shortening heat. That's the same kind of effect that Malcolm Irving and Roger Woledge reported last year (J. Physiol. 321, 1981). Takanori Yamada, working in my laboratory, has confirmed those results and has extended them to higher velocities. He finds that the curvature is more pronounced at high velocities. On the other hand, at slower shortening velocities, he finds that the rate of shortening heat liberation is constant; i.e. the amount of shortening heat is a linear function of time, similar to our data at ½ V_{max}.

EDMAN: It's somewhat surprising that you get this linear relationship between sarcomere length and shortening heat at zero load. I found that V_{max}, or V_o, as I prefer to call it, is very constant between 1.65 and 2.70 μm (Edman, J. Physiol. 291: 143-159, 1979) but when you exceed that length V_o increases very steeply. That is attributable to the presence of passive force you get in that range. So if you release the muscle within that range, you actually get a negative load on the fiber. I suppose you would have to take account of this recoil of the parallel elastic structures in the prediction of the evolution of shortening heat.

HOMSHER: I don't think our experiments are quite like yours. We allow the muscle to shorten *near* V_{max}, and if you remember the tension records shown for these experiments, at a velocity *near* V_{max} the force exerted by the muscle is still 0.02 P_o. And I think Kazuhiro (Yamada) probably would agree, we don't like to allow our muscles to go slack when we make our measurements. We prefer that there be a bit of force.

CURTIN: Earl, when you calculated those cross-bridge turnover rates, did you subtract anything for the ATP splitting by the calcium pump?

HOMSHER: No, I didn't.

CURTIN: Do you think it's realistic not to take account of ATP splitting by calcium?

HOMSHER: No I don't. I think you really should subtract it. However one is constrained to make an estimate of the pump-related ATP splitting based on isometric tetani of stretched muscles. It doesn't modify things much because a constant relatively small rate (about ¼ of the isometric rate) is subtracted from the values. It merely makes the change of the ratios of rates between isometric and rapid shortening rates more pronounced.

CURTIN: The absolute values don't change. I see.

TAYLOR: Could you discuss briefly your reasons for completely discarding a change in calcium release as the source of the energy imbalance at V_{max}?

HOMSHER: Well, I decided to eliminate that for several reasons. First, the result you showed the day before yesterday indicated that a sudden change in muscle length produced only a small change in sarcoplasmic calcium. We're talking about a relatively large effect here compared to what you saw. Secondly, the amount of energy associated with the cycling of calcium during an isometric contraction at stretched length corresponding to a sarcomere spacing of 3.6 μm, is only 10% of the energy liberation rate we saw at our high velocities. So this would have to change by a mammoth amount in order to have much of an effect on our results. On the other hand, it would be very useful to have more quantitative measures of the change in calcium release and sequestration during shortening.

SUGI: Many years ago you seriously considered the appropriate baseline for estimating shortening heat. As far as I remember, you used a small twitch, the peak of which was equal to the isotonic load. What is your opinion about this approach now?

HOMSHER: I am not confident that myothermal studies of twitches is a viable form of study of thick and thin filament interaction because I've looked at some of the laser diffraction patterns obtained in twitches and they are very messy. All sorts of inhomogeneities occur during relaxation. Our results were not constrained only to twitches, however, but they applied to tetani as well. We simply measured the amount of heat produced during a bout of shortening and found, as had others before, that it was greater than muscle which did not shorten. How much greater? To specify this increase, Hill took as a baseline the amount of heat produced by a muscle contracting isometrically at or near the muscle length over which shortening occurred. The difference between the two was the so called shortening heat. Hill's definition of the shortening heat led to an inconsistency in that shortening heat *(as defined by Hill)* could be shown in the tetanus but not in the twitch. Those outside the energetics field mistakenly took the internal consistency of Hill's *interpretation* of his results as an invalidation of the myothermal technique. We showed that if one used, as a baseline the heat produced by an isometric muscle *producing the same force* as the shortening muscle, the inconsistency vanished. The important point here is that Hill's interpretation of his work on shortening heat is simply incorrect. In fact those who have made measurements seem to be in reasonable agreement about the amount and time course of heat production during shortening. The problem comes when one attempts (as I now see with the aid of hindsight) to partition the energy liberation into simple mechanical components.

DREIZEN: Have you looked at the effect of temperature?

HOMSHER: Yes. Most energeticists do their studies at 0°C to slow things down. That's an advantage for us because we want to follow the time course of heat production. If you increase the temperature, say to 20°C, the rate of energy liberation increases 10 to 20 fold, and everything happens so much more rapidly that it's very difficult to obtain meaningful time-resolved experiments. So I think if anything I'd like to go to a lower temperature.

DREIZEN: The question I was leading to was whether if you went to some intermediate temperature, say 10°, and looked at less than V_{max} you might then pick up your imbalance. This maneuver would permit you to look at different cross-bridge cycling rates, while keeping the velocity constant. This could test your hypothesis about the relationship of cross-bridge cycling to heat liberation.

SHIMIZU: Based on your assumption, it is very difficult for me to understand why the difference of the shortening velocity gives a very large difference in the energy balance. When you have a very slow shortening velocity you have also a problem about the superposition of many cross-bridges on the actin filament.

HOMSHER: Yes. The idea, though, is that cross-bridges pass through an individual cycle slow relative to the rate at which myosin-ADP is detached from the thin filament. This problem shouldn't come up instantaneously; it should come up gradually. That is, what one would also want to do is look at velocities between ½ V_{max} and V_{max}. One should see this process develop gradually. In principle I agree with you, but at ½ V_{max} apparently the rate at which the filaments slide by one another is not fast enough to materially depress the rate of ADP and phosphate release from the actomyosin linkage.

HOLMES: I would like to clarify this problem a bit. The probability of finding the correct binding site is the same, independent of velocity, because there are more sites passing you and they go by faster. Your chances of finding one are the same. So you've got to say then that there's some other time-limiting effect.

HOMSHER: I see what you're saying. I suppose the explanation I am using is very similar to the one Andrew Huxley advanced for the decline in the rate of energy liberation at high shortening velocity (Proc. Roy. Soc., 1976). He postulated that there is a two step attachment process: a rapid equilibrium first step attachment to the actin, and then a second slower first order transition to a tight binding state. This kind of reaction mechanism will make the binding of a cross-bridge dependent on the speed of shortening in the fashion we have suggested.

TAWADA: To account for the "unexplained" heat production, you assume the degradation of the myosin-ADP-Pi complex.

HOMSHER: Yes.

TAWADA: Why is the rate of heat production constant during rapid shortening when muscle shortens with V_{max}? Heat production doesn't seem to decline with time when muscle shortens with V_{max} (Fig. 4). Heat production remains constant although you assume the degradation of the myosin-ADP-Pi complex is responsible for the heat production.

HOMSHER: I think you'd be hard pressed to judge from that figure how the rate of heat production is changing. This is because first it is a tracing, second it is not properly amplified to answer your question. If, however, you look at the original records you will find that the rate of production bends over.

TAWADA: Really?

HOMSHER: Takenori Yamada has a poster outside and he shows that.

TAWADA: No, I'm not asking about the rate of *shortening* heat production. To get that you must subtract some baseline. I'm asking you about the rate of *total* heat production while the muscle shortens with V_{max}.

HOMSHER: It bends over.

TAWADA: I don't think so.

EBASHI: I am very much impressed by your work on the maximum velocity of shortening, but I still cannot understand the explanation. How do you explain this rather puzzling fact that you get unexplained heat at V_{max} but not at intermediate velocities?

HOMSHER: In our experiments, we find relatively little ATP splitting during shortening, and a relatively high amount of ATP splitting after the end of shortening. We don't know the time course for the splitting after shortening; all we know is we measure for change in a high energy phosphate splitting over the 700 millisecond interval. The hydrolysis may occur very shortly after the end of shortening, but the basis of our explanation is that some cross-bridges detach before their products have dissociated. If the products dissociate and then an ATP binds to myosin, for each cycle we'd be obliged to split an ATP.

To explain this energy imbalance at high velocities we have to have the cross-bridge come off with some product on it, so that ATP can't go on right away and be split. So we have it come off in the form of myosin-ADP. We have the phosphate coming off earlier because that's what the kineticists tell us happens. Now, the next problem then is to get the myosin-ADP cross-bridge to attach to the thin filament, but to keep it from attaching too rapidly while the filaments are sliding rapidly past one another. To accomplish this you must have a two-step attachment, so that the myosin-ADP can't reattach very rapidly. I think this is why you might get the energy imbalance when the fiber shortens at V_{max}.

POLLACK: If I understand you correctly, you're implying that at V_{max} the cross-bridge cycle is incomplete, and the bridges are detaching before they have a chance to complete their cycle.

HOMSHER: That's right.

POLLACK: Then what's driving the process? What's causing the filaments to translate at V_{max}?

HOMSHER: Other cross-bridges attach along the length of the filament.

POLLACK: So you assume some of them are able to complete their cycle, while others are not?

HOMSHER: That's right. Because we do see ATP being split while muscle shortens, just not enough.

POLLACK: What puzzles me is why some should go through a full cycle and others through a part cycle. The filaments are translating at some constant velocity.

MAGID: They're independent Jerry. The cross-bridges are undergoing their own cycles.

HOMSHER: Yes, I suppose that's the only way I can answer you. The way that they would detach -- how many would be detaching without going through the cycle -- would be dependent on how fast the filaments are sliding by and what the rate limiting step is, and how fast they normally would dump their products. And so the fraction going off into the refractory pool (to coin a phrase) would be dependent on the relative rate constants.

RITZ-GOLD: I'd just like to add that perhaps the heads in the refractory pool might be thought of as those that have been forcibily detached by the rapid actin movement while still in the product-activated 90° conformational state. Once sliding has stopped, these unstable detached heads might then be "catching up" energetically by undergoing the slower exothermic conformational change. However, they would now be discharging the activation energy in the form of heat instead of in the form of work.

T. YAMADA: I want to make a brief comment. The amount of energy imbalance we are discussing is about 5-6 mJ/g of wet muscle weight. If we relate this energy imbalance to the incomplete cycling of cross-bridges during isometric to isovelocity shortening and vice versa, we can estimate the energy change per cross-bridge associated with this incomplete reaction step(s) to be about 18-22 kJ/mol of myosin head, assuming that the concentration of myosin head is 0.28 μmol/g of wet muscle weight (Ebashi et al., Quart. Rev. Biophys. 2: 351-384, 1969). Actually comparable enthalpy changes are associated with various reactions occurring during cross-bridge cycling -- 16 kJ/mol for the binding of myosin head to actin, -110 kJ/mol for the relase of bound phosphate from M·ADP·Pi and 90 kJ/mol for the dissociation of ADP from the myosin head (Curtin & Woledge, Rev. Physiol. 58: 690-761, 1978). Therefore it is possible to assign the observed energy imbalance to incomplete cycling of cross-bridges.

SHIMIZU: Let me give some comment to Dr. Homsher's explanation. We (Nishiyama & Shimizu, Mathem. Biosci. 54: 115-135, 1981; Biochem. Biophys. Acta 587: 540-555, 1979) published a kind of dynamical sarcomere model in which we give a steric structure of the sarcomeres, giving particular pitches for myosin and actin filaments. I think that the original idea is from Dr. Huxley: The different pitches give some kind of vernier effect, and if you count how many times myosin can interact with actin, based on such kinds of steric structure, you can understand what Dr. Homsher said.

THE DEPENDENCE ON THE DISTANCE OF SHORTENING OF THE ENERGY OUTPUT FROM FROG SKELETAL MUSCLE SHORTENING AT VELOCITIES OF V_{max}, $\frac{1}{2}V_{max}$ AND $\frac{1}{4}V_{max}$

Takenori Yamada and Earl Homsher

Department of Physiology, School of Medicine, University of California at Los Angeles, Los Angeles, CA 90024

ABSTRACT

Shortening heat and work were measured, during and after the shortening, in contractions in which muscles shortened various distances (all ending at 2.05 micron of sarcomere length) at various velocities. Shortening heat and work were produced as a non-linear function of the distance shortened at velocities > $\frac{1}{2}V_{max}$, which was probably caused by a redistribution of the crossbridge population to different state(s) during the approach to the steady state.

A.V. Hill (1938) reported that shortening muscle produces heat in a fashion which is linearly dependent on the distance shortened. Later he showed that the shortening heat produced per unit distance is also linearly dependent on the load (Hill, 1964). Recently Irving and Woledge (1981) found that the shortening heat is not linearly dependent on the distance shortened; shortening heat and work production per unit distance of shortening decreases as the distance of shortening is increased. In this study we investigated the effect of shortening velocity on the relationship between shortening heat and work production and the distance of shortening.

Shortening heat and work production were measured in 2.1 sec tetani in *R. pipiens* sartorius muscle. At 1.0 sec after the beginning of tetanic stimulation, the muscle shortened between 0.016 to 0.5 μm per sarcomere (all ending at a sarcomere length of 2.05 μm), at a velocity of 2.63 (V_{max}), 1.43 ($\frac{1}{2}V_{max}$) or 0.71 ($\frac{1}{4}V_{max}$) μm/sarcomere/sec. The amount of shortening heat and work produced were measured at 2.1 sec (when tension had returned to an isometric value) of the tetanus. Shortening

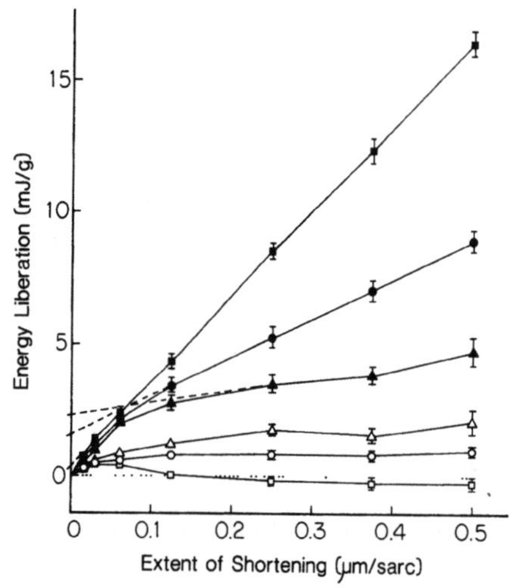

Figure 1: Shortening heat and excess energy production in the 1.0 to 2.1 sec interval of 2.1 sec tetani at various distances of shortening after correcting the load dependent shortening heat production and series elastic work done during redevelopment of tension. Shortening of various distances (all ending at 2.05 micron of sarcomere length) at a velocity of V_{max}, $\frac{1}{2}V_{max}$ or $\frac{1}{4}V_{max}$ started at 1.0 sec of 2.1 sec tetanic stimulation. Shortening heat and work production were measured at 2.1 sec. The open symbols represent the shortening heat after corrected for load dependent shortening heat production; non-steady state shortening heat. The solid symbols represent the non-steady state shortening heat plus work corrected for series elastic work done during redevelopment of tension. The velocity of shortening was V_{max} (Δ, ▲), $\frac{1}{2}V_{max}$ (○, ●), or $\frac{1}{4}V_{max}$ (□, ■).

heat and work production were plotted against the distance shortened, and the plots showed that the amount of energy released was composed of linear and non-linear components. We found that in the linear phase the rate of shortening heat and work production was 21.1 and 7.8 mJ/g/sec at V_{max}, 20.3 and 20.9 mJ/g/sec at $\frac{1}{2}V_{max}$, and 13.3 and 22.9 mJ/g/sec at $\frac{1}{4}V_{max}$, respectively. The non-linear energy liberation was obtained from the y-intercept of a straight line fitted to the linear portion of the plot. The non-linear shortening heat and work production was 2.9 and 2.8 mJ/g for V_{max}, 1.5 and 2.2 mJ/g for $\frac{1}{2}V_{max}$, and 0.9 and 1.6 mJ/g for $\frac{1}{4}V_{max}$, respectively. Those plots were then corrected for the known load dependence of shortening heat production and for the series elastic work done during tension redevelopment. While these corrections did reduce the amount of non-linear energy liberation there remained significant amounts of non-steady state shortening heat production (1.3, 0.8 and 0.2 mJ/g for V_{max}, $\frac{1}{2}V_{max}$, and $\frac{1}{4}V_{max}$, respectively) and non-steady state work production (1.0, 0.7 and 0.0 mJ/g for V_{max}, $\frac{1}{2}V_{max}$ and $\frac{1}{4}V_{max}$, respectively) as shown in Fig. 1.

These results suggest that a non-steady state population of crossbridges is created during transition to shortening. Further the shift of the crossbridge population is larger as the shortening velocity is increased.

REFERENCES

Hill, A.V. (1938). The heat of shortening and the dynamic constants of muscle. Proc. Roy. Soc, B126: 136-195.

Hill, A.V. (1964). The effect of load on the heat of shortening muscle. Proc. Roy. Soc. B159: 297-318.

Irving, M. and Woledge, R.C. (1981). The dependence on extent of shortening of the extra energy liberated by rapidly shortening frog skeletal muscle. J. Physiol. 321: 411-422.

SIMULTANEOUS HEAT AND TENSION MEASUREMENTS FROM SINGLE MUSCLE CELLS

Nancy A. Curtin, J.V. Howarth*, Jack A. Rall+, M.G.A. Wilson++ and R.C. Woledge++

Department of Physiology, Charing Cross Hospital Medical School, London, W.6, England
**Marine Biological Association, Plymouth, England*
+Department of Physiology, Ohio State University, Columbus, Ohio 43210 (to whom correspondence should be sent)
++Department of Physiology, University College London, London, W.C. 1, England

ABSTRACT

Simultaneous force and heat measurements were made in single cells from skeletal muscle of the frog during isometric twitches and tetani at 10 and 0°C. A Hill-Downing type thermopile of low heat capacity was used. In twitches, peak force development was found to be well correlated with heat production at both temperatures, during posttetanic twitch potentiation (at 10°C) and during posttetanic twitch depression (at 0°C). In a twitch at 0°C, heat production started less than 14 msec after the stimulus had begun, before force development. As in whole muscle, the heat during a tetanus could be separated into two components: an early component produced at an exponentially decreasing rate, labile heat, and a steady rate, stable maintenance heat rate. Increasing temperature from 0 to 10°C doubled the stable maintenance heat rate. At the higher temperature the time constant of labile heat production was halved and the quantity of labile heat decreased. When two tetani were given at 10°C, a 5 min rest interval was required before the second tetanus produced the same force and heat as the first. At 0°C this interval was at least 10 min. With shorter intervals, both heat and force were depressed. At 10°C both were depressed equally but at 0°C the effect on heat was greater than on force. At both temperatures labile heat was depressed to a greater extent than the stable maintenance heat rate. Results are interpreted in terms of possible calcium-parvalbumin interaction during a tetanus.

INTRODUCTION

Recently Curtin, Howarth, and Woledge (1981) have demonstrated the feasibility of measuring the time course and amount of heat production in single cells from skeletal muscle. We have extended these observations by measuring simultaneously force and heat production in isometric contractions.

There are a number of ways that studies of the energetics of contraction in single muscle cells may contribute to an overall understanding of muscle function. They will: a) allow a more detailed comparison of energetic with mechanical properties of single cells, b) allow a characterization of the energetic properties of the different cell types known to exist in Amphibian muscle (Lännergren and Smith, 1966) and c) simplify the interpretation of the effects of pharmacological agents on the energetic properties of cells.

METHODS

Single cells with intact tendons were isolated from tibialis anterior muscle, *R. temporaria*. T-shaped clips of either aluminum or platinum were folded over each tendon and used as connections. The cell was mounted on the thermopile which was fixed horizontally onto the floor of a chamber made of anodized aluminum. The temperature of the chamber was controlled by fluid circulating through channels in its walls. There was a hole in each end of the chamber; through which connections were made from one end of the cell to the force transducer (Cambridge Electronics Model 400) and from the other end of the cell to a micrometer. Cell length was adjusted with the micrometer while observing the laser diffraction pattern; striation spacing was set at 2.25 μm at the start of each experiment. Cells were stimulated through the clips. Stimulus parameters determined when the solution was drained were typically: voltage, 1-6 V; duration of square wave, 0.2-3 msec; frequency, 20-30 Hz (10°C), 10 Hz (0°C). Stimulus parameters were selected to minimize stimulus heat. With longer durations (3 msec) of lower voltage (1 V), the energy used to activate the cell was less.

Heat production was measured as the temperature change produced during contraction. The temperature changes were measured by a Hill-Downing type thermopile consisting of 40 constantan-chromel thermocouples (Curtin et al., 1981). The active region of the thermopile was 5.7 mm long and its overall length was 12 mm. Thermopile sensitivity was 2.41 μV/m°C temperature change. The thermopile was mounted on an anodized aluminum frame in such a way that it was slightly bowed so that the thermo-junctions over which the cell was placed were elevated with respect to the reference thermo-junctions. With this arrangement the Ringer solution drained away from the cell more effectively than with a flat thermopile. Thermopile output was amplified by a low noise chopper modulated amplifier (Ancom 15C-3A). In one experiment a Kipp (A80) galvanometer system was used (see Fig. 3).

During a contraction in a single cell, there is substantial flow of heat away from the preparation into the thermopile frame. The time course of this heat loss was determined by warming the cell through the thermo-junctions using the Peltier method (Kretzschmar and Wilkie, 1972, 1975). Cool-off curves were recorded and subsequently analyzed. The curves could be described adequately as a sum of two exponentials. Typical time constants were about 0.3 sec and 1.4 sec. Results from this analysis were used to correct the thermal records for heat loss.

The temperature change caused by the stimulus itself was measured in various ways. 1) When the cell responded to one polarity of stimulus but not the other, polarity could be reversed and stimulus heat recorded without activating the cell. 2) Sometimes stimulus heat was measured by altering stimulus duration (D) and frequency (F) or voltage (V) in such a way to keep $F \cdot D \cdot V^2$ constant but not activate the cell. In a twitch, stimulus heat was negligible. In 5 sec tetanic contractions at 10°C, stimulus heat averaged 6.5% (n = 4) of the peak heat produced. In the one measurement made at 0°C, stimulus was 6.5% of the peak heat in a 5 sec tetanus. Stimulus heat was not subtracted from the records.

After a cell was mounted on the thermopile and sarcomere length set at 2.25 μm, it was allowed to equilibrate to the desired temperature for 30 min to 1 hr. Twitch and tetanus force in solution were determined during this equilibration period. Then the solution was drained, threshold parameters determined and force and heat production measured. Peltier calibration curves were periodically obtained. The cell often remained in the drained humidified chamber for hours (8 hrs in one case) producing stable mechanical and thermal responses. On other occasions the solution was slowly returned to the chamber without injuring the cell. At the end of each experiment or before the Ringer was returned to the chamber, the alignment of the cell over the thermo-junctions was noted. The cell was finally cut free of its tendons and clips and the dry weight of the cell was measured on a Cahn 26 automatic electrobalance.

RESULTS AND DISCUSSION

Force and Heat Production in Isometric Twitches at 10 and 0°C

Fig. 1 shows examples from the same cell of isometric twitches at 10 (A) and 0°C (B). The observed temperature change at both temperatures is approximately 1 m°C. But the temperature changes corrected for heat loss (bottom of Fig. 1) are 1.2 m°C at 10°C and 2 m°C at 0°C. Even after correction for heat loss, the temperature changes are less than they would be if the cell alone were heated because the heat produced by the cell must also warm up the adhering fluid and thermopile. This effect can be evaluated from the relationship of the heat capacity of the cell to that of the cell, adhering fluid and thermopile. This ratio in whole muscle experiments is approximately 0.9 whereas the ratio in single cell experiments, according to preliminary results, is about 0.4. Thus the temperature changes observed in Fig. 1 would increase to 3 m°C at 10°C and 5 m°C at 0°C if this effect were taken into account. The traces at the bottom of Fig. 1 have not been corrected for the distortion (lag) in the time course of the heat records due to the finite heat capacity of the thermopile and adhering fluid. This distortion in time course is only likely to be significant in the early part of the heat traces. Nonetheless it is obvious from the record at 0°C that heat production reaches its maximum rate early in the contraction and that this rate declines to zero by the time the cell completely relaxes.

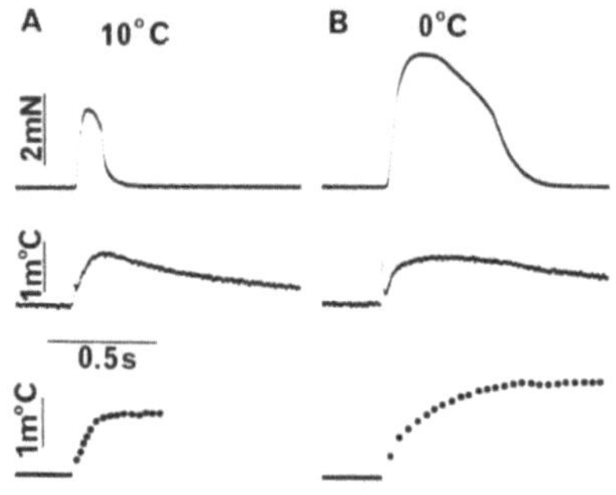

Figure 1: Force and heat production during isometric twitches at 10 (A) and 0°C (B). Above: force, Middle: temperature; Bottom: temperature change corrected for heat loss. All results from same cell. Twitch to tetanus, 0.45 at 10°C and 0.82 at 0°C. Cell length, 7.4 mm; major and minor diameters, 142 μm and 120 μm; dry wt., 29.9 μg.

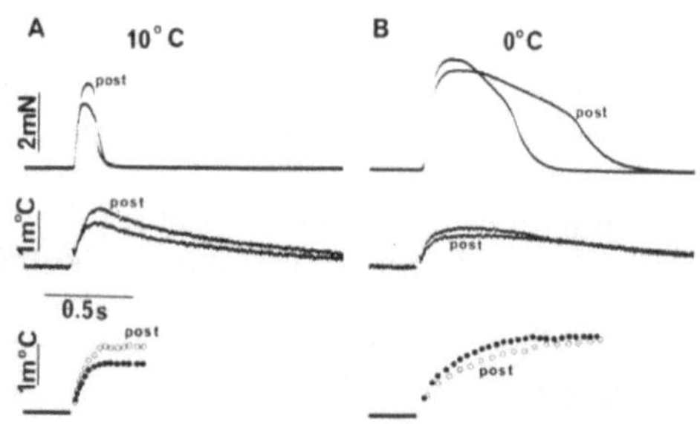

Figure 2: Force and heat production during pre- and posttetanic isometric twitches at 10 (A) and 0°C (B). Above: force; Middle: temperature; Bottom: temperature change corrected for heat loss. The pretetanic twitches are the same as those of Fig. 1. A: posttetanic twitch produced within 30 sec after a 2 sec tetanus (stimulation rate: 30 Hz). B: posttetanic twitch produced within 1 min after a 10 sec tetanus (stimulation rate: 10 Hz).

The ratio of the peak force to the peak temperature change after heat loss correction is one measure of the economy of contraction. In this cell, it changes by less than 6% between 10 and 0°C despite a large change in peak force and the time course of force development.

Force and heat production have been examined in posttetanic twitches at 10 and 0°C. Fig. 2A shows a twitch produced within 30 sec after a 2 sec tetanus at 10°C compared to a pretetanus twitch. Peak force of the posttetanic twitch increased by 31%, the temperature change, after heat loss correction, increased by 36%, and the time to peak force development increased by 47%. Again, the economy of force production has changed little during postetanic potentiation. In experiments under similar conditions, Blinks, Rüdel, and Taylor (1978) have shown that the calcium transient, as monitored by the bioluminescent protein aequorin, is depressed in amplitude and prolonged in duration during posttetanic twitch potentiation. Their interpretation is that less calcium is released during a twitch after a tetanus but that the calcium leads to greater force development because it remains effective for a longer period of time. Connolly, Gough and Winegrad (1971) have suggested a similar interpretation of posttetanic potentiation based on mechanical experiments. The energetic studies show that this increased force is accompanied by increased energy utilization by the cell.

A different kind of result is observed when a twitch is produced within 1 min after a 10 sec tetanus at 0°C (Fig. 2B). Peak twitch force, compared to a pretetanus twitch, decreased by 10%, the temperature change, corrected for heat loss, decreased by 4% whereas the time course of relaxation in the twitch was dramatically prolonged. The small change in peak force is not surprising because the twitch to tetanus ratio was 0.82 and Ramsey and Street (1941) have shown that posttetanic twitch potentiation only occurs when the twitch to tetanus ratio is less than 0.64. It is surprising that the prolongation of twitch force, predominantly due to the slowing of the linear phase of relaxation, does not lead to increased heat production. If the linear phase of relaxation was linked directly with the uptake of calcium by the cell, one might expect that the depressed rate of relaxation would lead to continued cross-bridge cycling and thus increased energy liberation. It seems that under these conditions the change in the shape of the twitch is not directly linked to the amount of energy used by the cell for contraction. All the twitch results taken together suggest that energy utilized during the contraction is well correlated with peak force rather than with the force-time integral.

The onset of heat and force production in an isometric twitch were examined in a cell at 0°C. Fig. 3 shows the time course of force and heat production at the beginning of the response. Heat records from eight single twitches produced by pulses of alternating polarity have been amplified by a galvanometer system and averaged to eliminate the stimulus artifact. The records have not been corrected for heat loss (negligible at these times) or thermopile-recording system lag (considerable at these times). The latency of the heat trace is thus less than 14

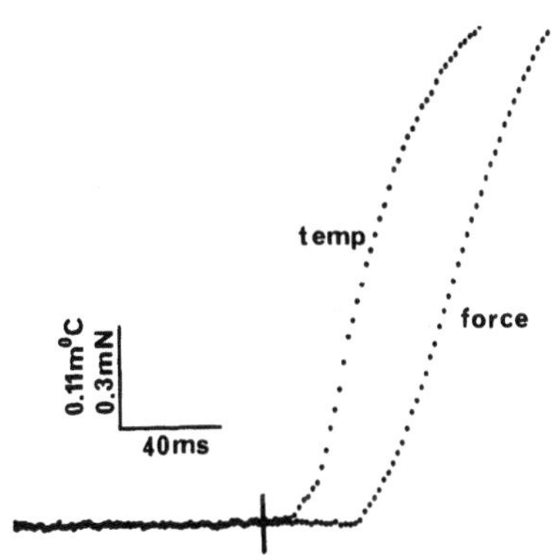

Figure 3: Onset of force and heat production in an isometric twitch at 0°C. Average of eight twitches. The moment of stimulation is indicated by the vertical mark. Twitch to tetanus ratio, 0.75. Cell length, 7.3 mm; major and minor diameters, 109 μm and 104 μm; dry wt. 27.8 μg.

msec whereas the latency of the force trace is 40 msec. Hill (1958) observed a latency of heat production of 10 msec (this latency became 5 msec after correction for instrumental delay) and of force of 24 msec at 0°C in whole frog muscles. The longer latency in the force record in our experiments may be due, at least in part, to stimulation occurring through the ends of the cell rather than uniformly along the cell length. This early heat production may reflect the thermal accompaniments of calcium movements during muscle contraction.

Force and Heat Production in Isometric Tetani at 10 and 0°C

Examples from two cells of the time course of force and heat production in tetanic contractions at 10 (A) and 0°C (B) are shown in Fig. 4. Fig. 4A shows a 5 sec tetanus during which force decreased by 6% of the peak value. The temperature change nearly reaches a plateau indicating that the rate of heat production was approaching the rate of heat loss. Observed peak temperature change was 6.3 m°C. The temperature change after correcting for heat loss was 47 m°C. In a different cell at 0°C (Fig. 4B), the force during a 10 sec tetanus decreased by 3% of the peak value and the temperature change corrected for heat loss was 31 m°C.

The following information can be derived from an examination of the shape of the temperature changes corrected for heat loss in Fig. 4. In both traces heat production has ceased by the end of mechanical relaxation. They both show a steady rate of heat production during force maintenance which is reminiscent of the stable maintenance heat rate in whole muscles. (Aubert, 1956). The stable maintenance heat rate in whole muscles is thought to reflect the steady rate of ATP splitting due to cross-bridge and calcium cycling (Curtin and Woledge, 1978 and Homsher and Kean, 1978). In both heat traces the initial rate is greater than the steady rate; this is more obvious at 0°C than at 10° C. This feature of a higher initial than steady rate is similar to the labile heat observed in whole muscle (Aubert, 1956). Labile heat is produced at an exponentially

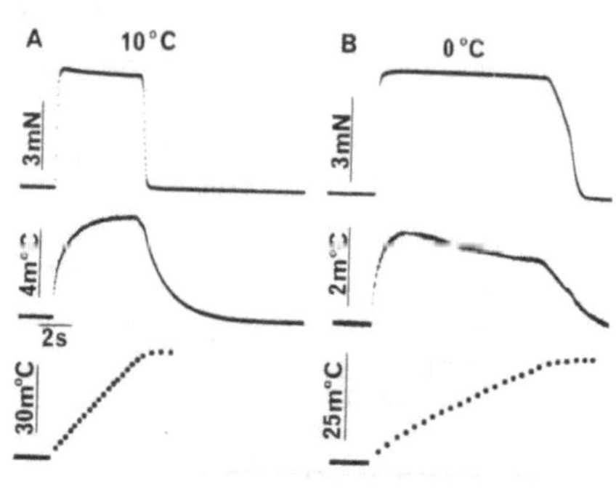

Figure 4: Force and heat production during isometric tetani at 10 (A) and 0° C (b). Above: force; Middle: temperature; Bottom: temperature change corrected for heat loss. A: 5 sec tetanus (stimulation rate: 20 Hz). Cell length, 8 mm. Stimulus heat less than 3% of heat signal at 5 sec. B: 10 sec tetanus (stimulation rate: 10 Hz). Stimulus heat 6.5% of heat signal at 5 sec. Same cell as in Fig. 1. Note differences in calibration markers on thermal traces.

decreasing rate during a tetanus. It has been suggested (Curtin and Woledge, 1978 and Homsher and Kean, 1978) that the source of the labile heat may be calcium binding to the parvalbumin, a reaction known to produce heat. Thus one would predict that muscles with a large parvalbumin content, like frog, would have a large labile heat whereas muscles containing little or no parvalbumin, as in chicks and mammals, would have little or no labile heat. Though not extensively studied, existing evidence supports this idea (Woledge, 1982).

The time course of heat production during a tetanus can be described by:

$$H_t = H_A\,(1 - e^{-t/\gamma_A}) + \dot{h}_b t$$

where H_t is the temperature change corrected for heat loss, m°C; H_A is the amount of labile heat, m °C; γ_A is the time constant of labile heat production, sec; $\dot{h}_b$ is the stable maintenance heat rate, m°C/sec and t is time from the onset of stimulation. An index of the curvature of the trace can be derived from the ratio: $H_A/\dot{h}_b\gamma_A$. The higher the ratio, the more curved the tetanus heat trace becomes.

In two cells studied at 10 and 0°C, the following effects of increasing temperature on the heat traces were observed: a) the stable maintenance heat rate doubled ($\dot{h}_b$), b) $1/\gamma_A$ doubled, and c) the amount of labile heat (H_A) decreased and d) the ratio of H_A to $\dot{h}_b\gamma_A$ decreased, i.e., the traces were less curved at 10 than at 0°C. The increase in stable maintenance heat rate with increasing temperature can be attributed to increased cross-bridge and calcium cycling rates leading to an increased rate of ATP splitting and perhaps to increased tetanus force production. The cause of a decreased labile heat at the higher temperature is unknown. This decrease in labile heat may be related to the relative capabilities of the sarcoplasmic reticulum and parvalbumin to act as calcium buffers during contraction and relaxation. Parvalbumin may be less effective as a calcium buffer during contraction and relaxation at 10 than 0°C.

Effects of Previous Activity on Force and Heat Production in Isometric Tetani at 10 and 0°C

These experiments were designed to determine at 10 and 0°C: 1) the duration of rest required after an initial tetanus before a subsequent tetanus would produce the same force and heat as the initial tetanus and 2) the relationship between force and heat when two tetani are separated by shorter intervals. To minimize any effects of changes in the response of the cells with experiment duration, pairs of tetani were produced with the first of the pair occurring after at least 10 min rest at 10°C and 20 min rest at 0°C. The second tetanus of the pair occurred at variable rest intervals after the initial tetanus. Fig. 5 displays some results for 2 sec tetani at 10°C. If the rest interval between tetani is 0.5 min (Fig. 5A), the second tetanus produces 7% less force and 8% less heat (heat loss corrected trace). If the rest interval between tetani is 5 min (Fig. 5B), both force and heat are back to the values observed in the initial tetanus.

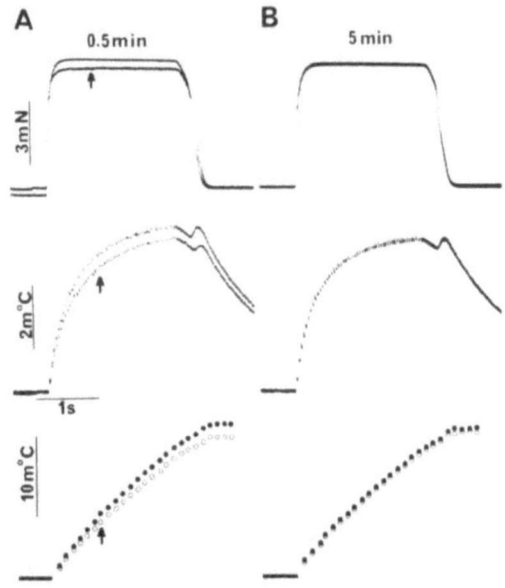

Figure 5: Effects of previous activity on force and heat production during isometric tetani at 10°C. Above: force, Middle: temperature; Bottom: temperature change corrected for heat loss. Pairs of 2 sec tetani (stimulation rate: 30Hz), the first of the pair was produced after 10 min of rest and the second was produced 0.5 min later (in A, indicated by arrows and/or open circles) or 5 min later (in B, indicated by open circles). In B, the pair of force and thermal traces are nearly superimposable. Filled circles, tetani with a 10 min rest period. Same cell as in Fig. 1. Stimulus heat at 1 sec less than 3% of the heat record at the same time.

Results from similar experiments on 5 cells can be summarized as follows: 1) if a pair of 2 sec isometric tetani at 10°C are separated by a rest interval of 5 min or more, the second tetanus will produce the same force and heat as the first, 2) if the rest interval is 2 min or less, the second tetanus will produce less force and heat than the first, 3) the shorter the rest interval the greater the reduction of force and heat in the second tetanus up to about 10% of their peak values with a 10 sec interval, and 4) force and heat are reduced approximately in parallel as the rest interval shortens. Further analysis of the experiment shown in Fig. 5 shows that the labile heat is reduced to a greater extent than the stable maintenance heat rate.

Fig. 6 shows a comparable experiment performed at 0°C on the same cell as displayed in Fig. 5. If the rest interval between 5 sec isometric tetani is 1 min (Fig. 6A), the second tetanus produces 7% less force and 22% less heat (heat loss corrected trace). If the rest interval between tetani is 9.5 min (Fig. 6B), the second tetanus produces the same force as the first and produces 6% less heat. Thus like the results at 10°C, a rest interval is required before force and heat in a second tetanus will equal that in the first tetanus. This interval appears to be greater than 9.5 min (with a 14 min interval, heat and force were the same in both tetani). Unlike the results at 10°C, heat is depressed to a greater extent than force. This result was also observed in this cell when the interval between tetani was 0.5, 2.2, and 5 min. Further analysis shows that labile heat is depressed to a greater extent than the stable maintenance heat rate as observed at 10°C. But different than the results at 10°C, stable

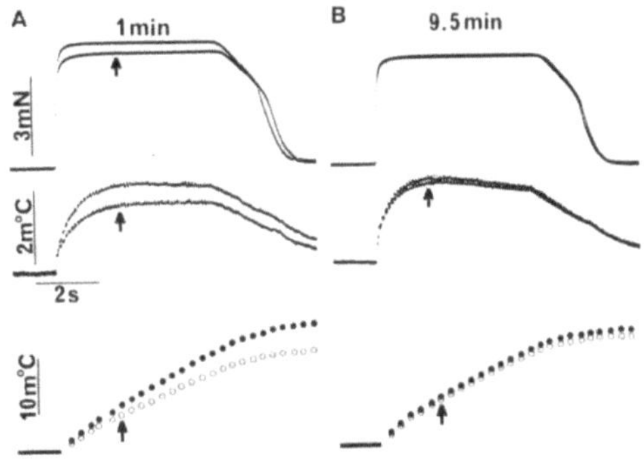

Figure 6: Effects of previous activity on force and heat production during isometric tetani at 0°C. Above: force; Middle: temperature; Bottom: temperature change corrected for heat loss. Pairs of 5 sec tetani (stimulation rate: 10Hz), the first of the pair was produced after 20 min of rest and the second was produced 1 min later (in A, indicated by arrows and/or open circles) or 9.5 min later (in B, indicated by arrows and/or open circles). Filled circles, tetani with 20 min rest period. Stimulus heat at 5 sec was less than 7% of the heat record at the same time.

maintenance heat rate is depressed more than force. More experiments will be needed at 0°C to determine the generality of this result. Nonetheless, it should be noted that in whole muscles at 0°C when the rest interval between tetani is such that a second tetanus produces less force than the first: 1) the heat produced is depressed to a greater extent than the force (Aubert, 1968), 2) the labile heat is depressed more than the stable maintenance heat rate (Aubert, 1968 and Curtin and Woledge, 1977) and 3) the stable maintenance heat rate is depressed more than the force (Curtin and Woledge, 1977).

If labile heat production is related to calcium binding to parvalbumin, the results of the repriming experiments suggest that less calcium binds to parvalbumin in a second tetanus under conditions where force is depressed after a short rest interval. This may be because less calcium is released from the terminal cisternae upon stimulation in the second tetanus and/or there are fewer available binding sites to parvalbumin possibly because some calcium released during the first tetanus still remains bound to parvalbumin.

ACKNOWLEDGEMENTS

This research was supported, in part, by United States Public Health Service grant AM-20792 from the National Institutes of Health. J.A.R. is a recipient of Research Career Development Award 1K04NS-00324.

REFERENCES

Aubert, X. (1956). Le Couplage Energétique de la Contraction Musculaire, p. 320, Brussels: Editions Arscia.

Aubert, X. (1968). In: *Symposium on Muscle*, ed. Ernst, E. and Straub, F.B., pp. 187-190, Budapest: Akadémiai Kiadó.

Blinks, J.R., Rüdel, R. and Taylor, S.R. (1978). Calcium transients in isolated amphibian skeletal muscle fibres: detection with aequorin. J. Physiol. 277: 291-323.

Connolly, R., Gough, W. and Winegrad, S. (1971). Characteristics of the isometric twitch of skeletal muscle immediately after a tetanus. A study of the influence of the distribution of calcium within the sarcoplasmic reticulum on the twitch. J. Gen. Physiol. 57: 697-709.

Curtin, N.A. and Woledge, R.C. (1977). A comparison of the energy balance in two successive isometric tetani of frog muscle. J. Physiol. 270: 455-471.

Curtin, N.A. and Woledge, R.C. (1978). Energy changes and muscular contraction. Physiol. Rev. 58: 690-761.

Curtin, N.A., Howarth, J.V. and Woledge, R.C. (1981). Measurement of heat produced by single fibres from frog skeletal muscle. J. Physiol. 313: 61-62P.

Hill, A.V. (1958). The priority of the heat production in a muscle twitch. Proc. R. Soc. B 148: 397-402.

Homsher, E. and Kean, C.J. (1978). Skeletal muscle energetics and metabolism. Ann. Rev. Physiol. 40: 90-131.

Kretzschmar, K.M. and Wilkie, D.R. (1972). A new method for absolute heat measurement, utilizing the Peltier effect. J. Physiol. 224: 18-19P.

Kretzchmar, K.M. and Wilkie, D.R. (1975). The use of the Peltier effect for simple and accurate calibration of thermoelectric devices. Proc. R. Soc. B 190: 315-321.

Lännergren, J. and Smith, R.S. (1966). Types of muscle fibres in toad skeletal muscle. Acta Physiol scand. 68: 263-274.

Ramsey, R.W. and Street, S.F. (1941). Muscle function as studied in single muscle fibers. In: *Muscle, Biological Symposia*, vol. 3, ed. Fenn, W.O., pp. 9-34, Lancaster, Pa: Cattell.

Woledge, R.C. (1982). Is labile heat characteristic of muscles with a high parvalbumin content? Observations on the retractor capitis muscle of the terrapin *Pseudemys elegans scripta*. J. Physiol. 324: 21P.

DISCUSSION

TAYLOR: In discussing the possible role of calcium in post-tetanic potentiation, you focused your remarks on the record at 0° but it seemed to me that the 10° record showed exactly the opposite effect, i.e., the rate of relaxation was increased.

RALL: At 10° the time to peak tension is somewhat prolonged. But the shoulder seems to occur at about the same time. The rate of relaxation is not prolonged, and it might well be slightly faster. The conclusion that we made is the following. It seems as if the rate of relaxation in an isometric twitch is *not* the main determinant of the amount of energy used. So at 0° we have a prolongation of the rate of relaxation, but the amount of energy doesn't change dramatically. At 10°, if anything, your observation is correct, the rate of relaxation might be slightly faster, but the amount of energy increase still seems to be best correlated with the amount of force produced.

GODT: I am interested again in the post-tetanic potentiation results. When you said that relaxation is slower, you focused on the pumping rate of calcium by the sarcoplasmic reticulum. Couldn't it be the off-rate of calcium from troponin that controls the rate of relaxation? I am just speculating. The on-rate is not going to change, as it's going to be diffusion limited as Dr. Gillis said the other day. Possibly the off-rate of calcium-troponin is affecting relaxation.

RALL: The reason that I mentioned a possible change in the pumping rate of calcium is that it's one possibility. Now this possibility does not seem to fit well with our intuitive feeling of what muscle cells do. So in a sense I wanted to mention this possibility to suggest that it doesn't fit very well. Your idea is another possibility. A third possibility is that there's a change in the rate of cross-bridge detachment. The curious thing to me, and maybe some other people might want to comment, is that I'm not sure what the linear phase of relaxation in an isometric contraction is due to. Certainly it is dramatically changed by previous activity. I'm inclined to think that it might not be associated with calcium pumping.

GODT: Since you see no differences in ATP utilization when relaxation is slowed by previous stimulation, one wuld then think perhaps it has something to do with the off-rate of calcium-troponin. And as Al Gordon pointed out the other day, cross-bridge detachment and changes of the calcium-troponin interaction are not necessarily different. They may be part of the same process.

COOKE: I think that cross-bridge cycle rates might really be playing a role here, at least at 0°, because work reported by Rüegg at this meeting and work done in our own lab (Cooke et al., FEBS Lett. 144: 33-37, 1982) indicates that when myosin is phosphorylated it's going to cycle more slowly. And this is going to change both the ATPase and the characteristics of the twitch. If the myosin holds on longer, that might prolong the twitch. It certainly might lead to a higher tension development.

RALL: I agree, that could be possible.

COOKE: And myosin is going to get phosphorylated in the tetanus. It explains everything at 0° beautifully. I don't know what's going on at 10°. But maybe the various rates of phosphorylation or dephosphorylation are such that myosin is phosphorylated more easily at 0° than at 10°. It would be very interesting to see what the level of myosin phosphorylation was under these conditions.

BRESSLER: Jack, do you think that the interval necessary for complete repriming of heat production determined in your last slide (Fig. 6) can be related to the duration of the tetanus? Have you checked?

RALL: We have not examined that. We've only examined the one tetanus duration that I've discussed. I should also say that we haven't studied the repriming phenomenon extensively at 0°. We've done more experiments at 10°. The obvious question is whether or not the repriming of heat and force production is dependent on tetanus duration.

GILLIS: Regarding questions about myosin phosphorylation, I think that there is no clear temporal relationship between dephosphorylation of the light chain of myosin and relaxation. In frog muscle, for example, we (Barany et al., J. Biol. Chem. 254: 3617-3623, 1979) found at room temperature that relaxation is finished quickly, before dephosphorylation starts. But at low temperature the two things go together.

POLLACK: I was interested to see that the temperature begins rising about 15 or 20 milliseconds before the tension develops, and also to see that there was no apparent break point in the rise of temperature at the time when the tension begins to rise. In other words, the early temperature rise is fairly steady. What do you think the initial rise in due to? Is it the same process as that which goes on during the development of tension or is it something different?

RALL: I mentioned in the talk that the records haven't been corrected for any distortion associated with the recording, which would be greatest at the earliest time. So therefore the best one can do, in part, is to relate our observations to those of A.V. Hill. We saw a 14 millisecond latency in heat production. Hill saw a 10 millisecond latency. His correction suggests that the actual latency might be half that. And that's getting close to the latency that might be expected, for example, for calcium binding to troponin, which is known to be a heat producing reaction. My personal interpretation would be that it is likely that the latency of heat production that we see might be associated with calcium binding to troponin. But more experiments need to be done to pin down the exact time course of the early heat production.

POLLACK: But then wouldn't you expect to see a change in the slope of heat production as soon as tension begins to rise? Because presumably there's a contribution to the temperature rise due to actomyosin interaction.

RALL: We really haven't examined that point. The experiment was designed only to look at two particular questions, that is, first of all, could the latency of heat production be measured in a single muscle cell? And secondly, what sort of latency do we get?

HOMSHER: Have you taken any of your muscles and tried to stretch them to non-overlap, first question? Second question, you hear these rumors from time to time that the first contraction after a long rest, a number of hours, produces more energy than in a steady state such as your 20-minute contractions. Have you looked at that? Third question, could you comment on the possibility of doing experiments like those of Paul Edman on a thermopile -- that is, looking at the motion of the muscle while recording heat production?

RALL: To answer question number one, we have not changed the length. The attitude that we have taken is that you should start out doing the simplest sort of experiments first, but we haven't changed length. The answer to question number two is also no. We haven't examined contractions after a very long rest period. Regarding the question associated with segment length measurement and clamping, one of the problems, of course, is that with this kind of system, we have to keep the temperature constant because we want to be able to measure temperature changes that are less than one thousandth of a degree. Also we have this sort of opaque thermopile. And it's not clear to me that it would be easy to also incorporate the machinery needed for segment length clamping into this system.

NOBLE: Dr. Rall, the tension and heat are lower when the second tetanus follows a short interval, and this, based on Dr. Gillis' presentation, is related to the presence of parvalbumin. I was under the impression that the contractile proteins in frog muscle were always completely saturated with calcium, but if by having some calcium remain on parvalbumin, the tension is lower, that would suggest that this is not the case, because if the contractile proteins are normally completely saturated with calcium, the tension should be constant. Is that a fair conclusion?

RALL: I didn't say why I thought the force might well have gone down. Normally one supposes that during the first tetanus, that troponin *is* saturated. I think in part we suppose that based on the aequorin measurements which show that, if anything, the light signal increases with time, suggesting it's possibly saturated. Maybe I should shift the question to Stuart Taylor and ask him what happens to the light signal when a second tetanus comes after the first. It may be more pertinent to your point.

TAYLOR: Well, as I described the other day, the aequorin signal will be diminished, but it's not necessarily something that I think is directly translatable to what the degree of saturation is. I think that the results of the potentiating agents that I described indicate the same sort of thing, that on the *descending* limb, under normal conditions, the contractile proteins are saturated, but not along the *ascending* limb.

SUGI: I just want to make a comment about the fatigue-producing mechanism. Last summer one Japanese investigator published that in the field of ciliary motion, one of the products of heart metabolism very effectively stops dyenin-tubulin interaction. He then tested this with regard to actin-myosin interaction, so I think it is very probable that one of the heart metabolism products can cause fatigue.

CONCLUDING REMARKS

CONCLUDING REMARKS

Hugh E. Huxley

According to the program I see that I am fortunately supposed only to make some "concluding remarks" rather than provide a summary of the entire meeting. So I don't have such a difficult job. We have had about four and a half days of very concentrated talks and discussions. For my own part - and I think many people would agree with me - it has really been one of the best meetings I have ever attended in many respects, but in one respect in particular. Namely, that of having had such a substantial number of important experimental measurements described so concisely and then having had these experiments and their implications fully discussed by such a high proportion of the leading workers in the particular fields in question. So it has been possible to reach a consensus on many topics, or at least to have them argued out in a rather complete way. I think we are all extremely grateful to Dr. Pollack and Professor Sugi for having arranged the scientific program so well, for having encouraged us to be concise in our talks, and for having provided plenty of time for discussion.

The meeting has covered many different topics which are closely relevant to the mechanism of muscular contraction and, as I have said, I am not going to try to summarize them all at the present moment. Nevertheless, I thought it might be useful to take a few minutes to consider some of the general lines of thought and even consensus that have been around at this meeting. For this purpose it might be useful to look back at the previous meeting of this series, which took place a little less than four years ago in Tokyo (Sugi & Pollack, Cross-Bridge Mechanism in Muscle Contraction, University of Tokyo Press/University Park Press, 1979) and to see whether there have been any significant changes or progress since then.

I think there has been a very great deal of both. The previous meeting, like this one, brought together primarily two kinds of workers on muscle -- physiologists studying the overall behavior of intact muscles or skinned fibers, and people concerned with the molecular mechanism of contraction from rather physical and structural points of view. At both meetings biochemists and people concerned with enzyme kinetics and so on have been rather in the minority.

I think the object of both the previous meeting and this one has been to see to what extent the physiological behavior of muscle could be accounted for by the straight-forward sliding filament model in which force and movement during contraction is produced by configurational changes taking place within actomyosin cross linkages (or "crossbridges"), and in which these crossbridge changes act upon actin and myosin filaments which themselves remained approximately constant in length during contraction. I think that the Tokyo meeting, perfectly reasonably, concentrated on various aspects of the behavior of muscles which did not seem to be very satisfactorily explained by the sliding filament mechanism. And not surprisingly, at that meeting, it was possible to identify a fair number of examples of such "unaccountable properties". Indeed, the number may have been sufficient for some people to wonder whether there might be something fundamentally wrong with the sliding filament model, and to enquire if perhaps one might be better off to try and find a totally new one!

And so in the final discussion at the Tokyo meeting Professor Sugi identified four specific questions in his introductory remarks; four specific questions which he indicated needed to be answered satisfactorily if we were to be able to make up our minds about the sliding filament/crossbridge concept. The first question he raised was, what is the meaning of stiffness? Does it all reside in the crossbridges, or are other myofibrillar elements involved? I think this is more a technical question rather than a fundamental one about mechanisms and, as far as I can see, even now the relative contributions of the different elements in the sarcomere structure to the measured stiffness have not been fully sorted out under all conditions. But it also seems to me that most people who are making measurements of stiffness do not now seriously doubt that a large part of the stiffness they're measuring, whether in active or in rigor muscle, is produced by the crossbridges, although other elements may be contributing too.

Secondly, Professor Sugi asked about the sites of force generation, a much more fundamental question. That is, whether the site of force generation is indeed in the crossbridges. He asked this particular question since, in the case of *Limulus* muscle, experiments had been described which suggested that shortening was produced by A-band and A-filament shortening. Therefore, by implication, if this was happening in a *Limulus* muscle, it might be thought the same process could be occurring in other muscles too and may previously have been overlooked or neglected in some way. I think at this present meeting Professor Sugi has clearly answered that question himself with his very direct optical demonstration that physiological shortening of *Limulus* muscle takes place without any significant shortening of the A-filaments. It still remains possible that there are changes in filament length taking place on a much longer time scale but, as I understand his experiments, they do not appear to be involved in the mechanism which produces active force.

The third question Professor Sugi posed was whether the force generators worked independently or cooperatively. This was with particular reference to Dr. Pollack's observations of stepwise shortening in various circumstances. In this case, and again I'm only speaking for myself, I don't feel that the problem of stepwise shortening has been settled yet, either experimentally or conceptually. However, let us suppose that we have another meeting in four years time and in the meantime everyone goes away and does a lot of experiments and comes back and says "yes", that they do indeed find stepwise synchronized shortening in the sarcomeres of vertebrate striated muscles. I don't think that would be an altogether amazing and unprecendented observation which would force us totally to change our views on how contraction was brought about. Professor Wilkie has already mentioned some other examples of synchrony taking place in biological systems, and one need not even go so far away as he did to find a very relevant one. After all, a considerable proportion of all the flying insects in the world depend on flight muscles which operate by that precise mechanism, i.e., they operate by undergoing stepwise shortening which is synchronized throughout the whole of the muscle. In the fibrillar flight muscles the synchrony is achieved, not by any kind of synchronous nerve stimulation (because the muscle is active the whole time) but as a consequence of the characteristic dynamic properties of the fully activated contraction mechanism. I think it is fair to say that the mechanism which produces this stepwise shortening and lengthening in insect muscle is not yet fully understood. But I'm also pretty sure that the people who work on these muscles are quite confident that this oscillatory shortening and lengthening is produced by the synchronized action of the crossbridges, and that it will be possible in the fullness of time to work out the characterstics of the delayed tension rise and fall in terms of varying rate constants in a crossbridge model.

Professor Sugi's fourth and final question, perhaps the most imporant of all when it comes to assessing the experimental evidence about intact muscle, concerned the famous length-tension relationship. Originally, the apparent linear fall of isometric tension with decreasing overlap seemed to be one of the most satisfactory demonstrations that force was produced in striated muscle by the action of independent tension generators adding up in parallel in the region where actin filaments overlaped myosin crossbridges. So it was a rather fundamental experiment and was certainly a very beautiful and persuasive demonstration that the model seemed to be along the right lines. However, by the end of the Tokyo meeting it appeared to many people that there really was something seriously wrong with the original linear length-tension diagram, some real difficulty, and that force might therefore be being developed in a muscle in some quite different way.

At the present meeting, on the other hand, (while I don't claim to understand all the intricacies of the discussions about creep and the contributions of different amounts of parallel structures in different types of frog muscles) it does seem to me that that difficulty has been removed. We have seen very beautiful, convincing, and approximately linear length-tension diagrams, produced now by Both Dr. Edman and Dr. ter Keurs. Personally, I would not now seriously doubt that to a close approximation, the crossbridges in the overlap zone are all involved equivalently in tension production, and that their individual contributions add up in parallel. Indeed the system behaves, perhaps almost to an unexpectedly accurate extent, in the way one might expect! Futhermore, the very nice experiments on the shortening heat that we heard about this morning seem to fall very convincingly into exactly the same sort of picture. Therefore, I feel much more free to concentrate in my remaining remarks on the question of exactly how the moving crossbridges produce force rather than in dealing with the question of whether some other different kind of process is involved.

In my closing remarks at the Tokyo meeting I suggested that there were three particular directions of work which might be important or interesting to pursue. First, one wanted to have new or additional probes, i.e. ones additional to the use of X-ray diffraction evidence, that would give independent and fairly direct evidence as to whether or not there were present myosin crossbridges attached to actin in a contracting muscle. Such probes would, if possible, also detect movement or at least different configurations of that attached state. We've heard at this meeting, and for some years previously, of a number of types of probe which now begin to give us information about what's happening to the crossbridge. I think there has been a general consensus (which was in part reached by informal discussion one evening during the week between people interested in that particular topic) that most of our observations would be well satisfied by the fairly straightforward though rather crude model of the crossbridge mechanism illustrated in Fig. 1.

The type of observations which have to be accomodated by such a model are the following: There are various probes, either spin labelled probes or optical probes (e.g. fluorescent labels), which can be attached to the myosin head. In an active muscle there is good evidence in a number of cases that the region of myosin containing those probes becomes attached to actin and immobilized. In several instances it appears that somewhere around 20 or 30% of the myosin heads are attached in this way and that they are all attached at the same angle. This angle seems closely similar to the angle at which the heads (or rather, the probes) are attached in rigor (when virtually all the myosin heads seem to be attached).

On the other hand, the X-ray evidence suggests that, when what you're looking at includes the whole mass of the myosin head, then one sees attachment over a considerable range of configurations relative to the actin filaments. The behavior of some other optical probes gives a similar picture. And so the possible way, which Ken Holmes mentioned earlier, of reconciling these observations is to suppose that the myosin head, the myosin S-1 subunit, consists of several domains. One of these domains may be rigidly attached to actin and always in the "rigor" configuration, while the other domains can adopt during the working stroke various configurations relative to the fixed part. The behavior detected by the probe would depend on which domain the the probe was attached to.

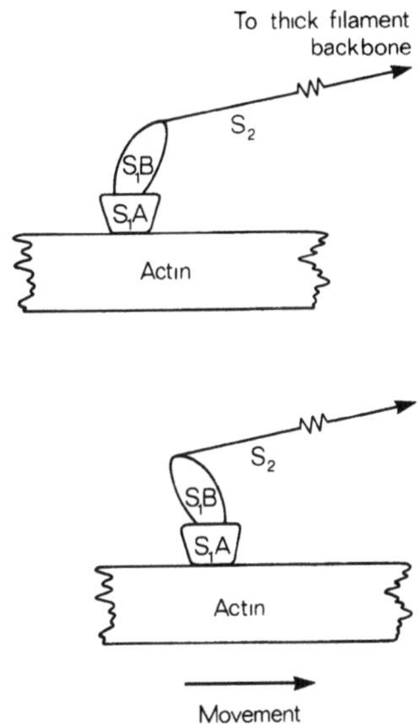

Figure 1: Diagrammatic representation of a myosin head (S_1) containing two (or more) domains (A and B) which might change their relative arrangement during the working stroke of the cross-bridge, producing relative sliding movement between the myosin and actin filaments. One of domains (A) would maintain a fixed relationship to actin, which might be the same as in the rigor state.

The thing about which I'm slightly less clear is to what extent the experiments with probes exclude the possibility that, in addition to those heads where there is a domain attached in the rigor configuration, there may be also other attached heads in which the same domain is attached at random angles. In my mind at least, that still remains to be sorted out. However, I don't think there is at present any sort of enormous internal conflict between these different types of probe measurement. But obviously it would be very advantageous to find more and more places on the S-1 head to which the various types of probe might be attached.

Another aspect of crossbridge behavior about which I think we have heard much less than we ought to have heard during the meeting is Dr. Harrington's extremely interesting and stimulating suggestion about the possibility of active shortening taking place in part of the S-2 portion of the myosin molecule. Dr. Tirosh was commenting earlier on the difficulty, if you saw a railway train and didn't know how it was working, of figuring out exactly where the force was being developed, i.e. were the wheels driving the pistons or vice-versa? The same thing could be said to apply to a myosin crossbridge model in which configuration changes were taking place in both the S-1 and S-2 part of the structure. If one could actually *see* it working, one would see the myosin head undergoing some repetitive tilting process, and one might also see the S-2 portion stretching under some conditions and shortening back under other. It wouldn't be at all self-evident where the motile force was actually being generated.

So I think it's really quite difficult at the moment in much of the evidence that we have, using various probes of the structural changes, to distinguish between models in which the head rotation is active and models in which it is passive and is being produced by an active process taking place elsewhere in the crossbridge. However, in this connection, the experiments which Professor Shimizu described, showing that S-1 on its own appears to be able to drive his "actomyosin motor" round without having the S-2 connected to it, do seem to show that S-1 is capable of producing motion on its own. One could also argue perfectly well that the experiments don't exclude the possibility that, in addition, some force generation is taking place in Professor Harrington's part of the crossbridge structure.

There were two other topics which I suggested at the last meeting might repay further study. The first of these was a comparison of the biochemical and physiological rate constants for different steps in the crossbridge cycle. We've not heard very much about that at the meeting -- it hasn't been that sort of meeting -- but there was one paper which I thought rather relevant to this topic which was concerned with the effect of ionic strength on the elasticity of relaxed muscle. It seems to be pretty clear that at low ionic strength, say 50 millimolar, even in the relaxed state, many crossbridges are attached to actin (but not cycling), whereas at higher ionic strength, say 150 millimolar, there seems to be rather good evidence that very many fewer are so attached. So there is a very large effect indeed of ionic strength on this particularly important aspect of crossbridge behavior. Most of the enzyme kinetic experiments so far have, of necessity, not only been done in solution (rather than in an organised filament structure) but have also nearly always been done at very low ionic strength in order to increase the interaction between S-1 and actin. I think that is somewhat worrying. It is clear now that a lot of them need to be repeated again under more physiological conditions to see if it makes a significant difference to the conclusions.

The second subject I thought at the time was a good one for future work was to try to obtain crystalline preparations of muscle proteins, in particular of actin and myosin, in order to be able to carry out high resolution X-ray crystallographic analysis of them. Having been at a protein crystallography workshop just before this meeting, and having seen the number of different proteins which people have crystallized and whose structures have been solved at 2 Å resolution, I feel extremely jealous that, unhappily, we're still not in that position regarding muscle proteins. However, actin has now been crystalized and its structure is being studied by a number of groups. There have been a number of unexpectedly difficult technical problems and so far no high-resolution structure is available, but I would be very much surprised if the structure was still unsolved when we have our next meeting, say in four years' time. In the case of myosin, to the best of my knowledge, no one has crystallized the myosin head or fragments of it so far. It is very, very obvious that is something all of us should be trying much harder to do.

Finally, I would like to thank Professor Sugi and Dr. Pollack once again for organizing this stimulating meeting, for taking such care with its organization, and for obtaining the necessary financial support. I would like to express all our thanks to them for the enormous amount of time and effort they expended to give us such a fine meeting and for keeping us so well entertained when we weren't in the lecture room. Thank you.

PARTICIPANTS

Altringham, John D., University of St. Andrews, Dept. of Physiology, Bute Medical Buildings, St. Andrews, Fife, KY16 9TS, Scotland

Bressler, Bernard M., University of British Columbia, Dept. of Anatomy, Vancouver, B.C. Canada. V6T 1W5

Brozovich, Frank V., University of Washington, Division of Bioengineering & Dept. of Anesthesiology WD-12, Seattle, WA. 98195

Cantino, Marie, University of Washington, Division of Bioengineering & Dept. of Anesthesiology WD-12, Seattle, WA. 98195

Cecchi, Giovanni, Universitá Degli Studi di Firenze, Interfacoltá di Fisiologia, Viale G.B. Morgagni 63, 50134 Florence, Italy

Colomo, Francesco, Universitá Degli Studi di Firenze, Interfacoltá di Fisiologia, Viale G.B. Morgagni 63, 50134 Florence, Italy

Cooke, Roger, University of California, Cardiovascular Research Institute, 841 HSW, San Francisco, CA. 94143

Curtin, Nancy A., Charing Cross Hospital Medical School, Dept. of Physiology, Fulham Palace Road, London W6 8RF, England

Denney, Michael, University of Washington, Division of Bioengineering & Dept. of Anesthesiology WD-12, Seattle, WA. 98195

Dewey, Maynard M., State University of New York, Dept. of Anatomical Sciences, Stony Brook, NY. 11794

Dreizen, Paul, State University of New York, Downstate Medical Center, 450 Clarkson Ave., Brooklyn, NY. 11203

Ebashi, Setsuro, University of Tokyo, Dept. of Pharmacology, Faculty of Medicine, 7-3-1 Hongo, Bunkyo-ku, Tokyo 113, Japan

Edman, K.A. Paul, University of Lund, Dept. of Pharmacology, S2232 62, Lund, Sweden

Fay, Fred S., University of Massachusetts Medical Center, Dept. of Physiology, 55 Lake Ave. North, Worcester, MA. 01605

Fristoe, Frank Jr., University of Washington, Division of Bioengineering & Dept. of Anesthesiology WD-12, Seattle, WA. 98195

Garcia, Maria, University of Washington, Division of Bioengineering & Dept. of Anesthesiology WD-12, Seattle, WA. 98195

Gillis, Jean M., University of Louvain, Lab. de Physiologie Generale, UCL 5540 Avenue Hippocrate 55, 1200 Brussels, Belgium

Godt, Robert E., Medical College of Georgia, Dept. of Physiology, Augusta, GA. 30912

Gordon, Albert M., University of Washington, Dept. of Physiology and

Biophysics SJ-40, Seattle, WA. 98195
Gulati, Jagdish, Albert Einstein College of Medicine, Division of Cardiology, Dept. of Medicine, 1300 Morris Park Avenue, Bronx, NY. 10461
Harrington, William F., Johns Hopkins University, Dept. of Biology, Baltimore, MD. 21218
Hashizume, Hiroo, Tokyo Institute of Technology, 4259 Nagatsuta, Midori, Yokohama 227, Japan
Hatta, Ichiro, Nagoya University, Dept. of Applied Physics, Faculty of Engineering, Chikusa-ku, Nagoya 464, Japan
Holmes, Kenneth, Max-Planck Institute, Dept. of Biophysics, Jahnstrasse 29, Heidelberg, West Germany, D-6900
Homsher, Earl, University of California, Dept. of Physiology, Los Angeles, CA. 90024
Housmans, Philippe L., The Mayo Foundation, Dept. of Pharmacology, Rochester, MN. 55901
Huntsman, Lee L., University of Washington, Division of Bioengineering WD-12, Seattle, WA. 98195
Huxley, Hugh E., Cambridge University, MRC Laboratory of Molecular Biology, Cambridge CB2 2QH, England
Ingels, Neil B. Jr., Palo Alto Medical Research Foundation, 860 Bryant St., Palo Alto, CA. 94301
Iwazumi, Tatsuo, University of Texas, Medical Branch, Dept. of Physiology, 3 Adler Circle, Galveston, TX. 77550
Janiak, Martin, Rigaku Denki USA, 3 Electronics Ave., Danvers, MA. 01923
Johnson, Dale E., University of Washington, Division of Bioengineering WD-12, Seattle, WA. 98195
Kawai, Masataka, Columbia University, Dept. of Physiology, 630 West 168th St., New York, NY. 10032
Krueger, John W., Albert Einstein College of Medicine, Division of Cardiology, 1300 Morris Park Ave., Bronx, NY. 10461
Kurihara, Satoshi, Jikei University, Dept. of Physiology, School of Medicine, Minato-ku, Tokyo 105, Japan
Lacktis, Joan W., University of Washington, Division of Bioengineering & Dept. of Anesthesiology WD-12, Seattle, WA. 98195
Levine, Rhea C., Medical College of Pennsylvania, Dept. of Anatomy, 3300 Henry Avenue, Philadelphia, PA. 19129
Luft, John H., University of Washington, Dept. of Biological Structure SM-20, Seattle, WA. 98195
Maéda, Yuichiro, Max-Planck Institute, Dept. of Biophysics, Jahnstrasse 29, Heidelberg, West Germany D-6900
Magid, Alan, Duke University, Medical Center, Dept. of Anatomy, Durham, NC. 27710
Martyn, Donald A., University of Washington, Division of Bioengineering WD-12, Seattle, WA. 98195
Mashima, Hidenobu, Juntendo University, School of Medicine, Dept. of Physiology, 2-1-1 Hongo, Bunkyo-ku, Tokyo 113, Japan
Matsubara, Ichiro, Tohoku University, Dept. of Pharmacology, Faculty of Medicine, 2-1 Seriyo-machi, Sendai, Miyagi-ken 980, Japan
Maughan, David W., University of Vermont, Dept. of Physiology/Biophysics,

Burlington, VT. 05401
Nichols, T. Richard, University of Washington, Dept. of Kinesiology DX-10, Seattle, WA. 98195
Nishiyama, Ken-ichi, Teikyo University, Dept. of Economics, 359 Otsuka, Hachioji-shi, Tokyo 192-03, Japan
Noble, Mark, Midhurst Medical Research Institute, Midhurst, Sussex GU 29 OBL, England
Okinaga, Shoichi, Teikyo University, 11-1 Kaga, 2-Chome, Itabashi-ku, Tokyo 173, Japan
Podolsky, Richard J., National Institutes of Health, Building 6 - Room 110, NIAMD, Bethesda, MD. 20805
Pollack, Gerald H., University of Washington, Division of Bioengineering and Dept. of Anesthesiology WD-12, Seattle, WA. 98195
Rall, Jack A., Ohio State University, Dept. of Physiology, 1645 Neil Ave., Columbus, OH. 43210
Reedy, Michael K., Duke University Medical Center, Dept. of Anatomy, Box 3011, Durham, NC. 27710
Ridgway, Ellis B., Medical College of Virginia, Dept. of Physiology, Box 608, Richmond, VA. 23298
Ritz-Gold, Carolyn J., 38451 Timpanogas, Fremont, CA. 94536
Robinson, Thomas F., Albert Einstein College of Medicine, Division of Cardiology, Dept. of Medicine, 1300 Morris Park Ave., Bronx, NY 10461
Rowe, Arthur, University of Leicester, School of Biological Sciences, Dept. of Biochemistry, Leicester LEI 7RH, England
Rüegg, J. Caspar, University of Heidelberg, Dept. of Physiology II, Im Neuenheimer Feld 326, Heidelberg, West Germany D-6900
Rushmer, Robert F., University of Washington, Division of Bioengineering FL-20, Seattle, WA. 98195
Saeki, Yasutake, Tsurumi University, Dept. of Physiology, Faculty of Dentistry, 20103 Tsurumi, Tsurumi-ku, Yokohama 230, Japan
Sakai, Toshio, Jikei University, School of Medicine, Dept. of Physiology, 3-25-8 Nishi-shinbashi, Minato-ku, Tokyo 105, Japan
Schoenberg, Mark, National Institutes of Health, Building 6, Room 101, Bethesda, MD. 20805
Shimizu, Hiroshi, University of Tokyo, Faculty of Pharmaceutical Sciences, 7-3-1 Hongo, Bunkyo-ku, Tokyo 113, Japan
Shimura, Hikaru, Rigaku Denki USA, 3 Electronics Ave., Danvers, MA. 01923
Squire, John J., Imperial College, Dept. of Metallurgy & Material Science, Biopolymer Group, London SW7 2BP, England
Sugi, Haruo, Teikyo University, School of Medicine, Dept. of Physiology, 2-11-1 Kaga, Itabashi-ku, Tokyo 173, Japan
Tameyasu, Tsukasa, Teikyo University, School of Medicine, Dept. of Physiology, 2-11-1 Kaga, Itabashi-ku, Tokyo 173, Japan
Tanaka, Hidehiro, Teikyo University, School of Medicine, Dept. of Physiology, 2-11-1 Kaga, Itabashi-ku, Tokyo 173, Japan
Tanaka, Hiroaki, University of Tokyo, Faculty of Pharmaceutical Sciences, 7-3-1 Hongo, Bunkyo-ku, Tokyo 113, Japan
Tawada, Katsuhisa, Kyushu University, Faculty of Science, Dept. of

Biology, 6-10-1 Hakosaki, Higashi-ku, Fukuoka-ken 812, Japan
Taylor, Stuart R., The Mayo Foundation, Dept. of Pharmacology, Rochester, MN. 55901
ter Keurs, Henk E.D.J., Afd. Exp. Cardiologie, c/o Lab. voor Fysiologie der Rijksuniversiteit Leiden, Wassenaarseweg 62, Leiden, The Netherlands
Thomas, David D., University of Minnesota, Dept. of Biochemistry, Minneapolis, MN. 55455
Tirosh, Reuven, c/o Dr. G. Berke, Weizmann Institute, POB 29, Rehovoth, Israel
Tregear, Richard, ARC Institute of Animal Physiology, Babraham Hall, Babraham, Cambridge CB2 4AT, England
Tsuchiya, Teizo, Teikyo University, School of Medicine, Dept. of Physiology, 2-11-1 Kaga, Itabashi-ku, Tokyo 173, Japan
Verdugo, Pedro J., University of Washington, Division of Bioengineering WD-12, Seattle, WA. 98195
Wakabayashi, Katsuzi, Osaka University, Faculty of Engineering Science, Dept. of Biological Engineering, 1-1 Machikaneyama, Toyonaka, Osaka 560, Japan
Wakabayashi, Takeyuki, University of Tokyo, Faculty of Science, Dept. of Physics, 7-3-1 Hongo, Bunkyo-ku, Tokyo 113, Japan
Wang, Kuan, University of Texas, Dept. of Chemistry, Austin, TX. 78712
Warshaw, David M., University of Massachusetts, Dept. of Physiology, 55 Lake Ave. North, Worcester, MA. 01605
Wilkie, Douglas, University College, Dept. of Physiology, London WC1E 6BT, England
Winegrad, Saul, University of Pennsylvania, Dept. of Physiology, Philadelphia, PA. 19174
Wray, John S., Max-Planck Institute fur Medizinische Forschung, Abteilung Biophysik, Jahnstrasse 29, Heidelberg, West Germany D-6900
Yamada, Kazuhiro, Oita Medical College, Dept. of Physiology, 1506 Toin, Hazama-cho, Oita-gun, Oita-ken 879-56, Japan
Yamada, Takenori, University of California, Dept. of Physiology, Los Angeles, CA. 90024
Yanagida, Toshio, Osaka University, Faculty of Engineering Science, Dept. of Biological Engineering, 1-1 Machikaneyama, Toyonaka, Osaka 560, Japan

INDEX

A-band
    lengthening, 107-113
    mass, 30, 42, 47
    shortening, 32-33, 37, 43, 66-69, 85, 87, 98, 114 130-134, 144-5, 151, 154
Acetylcholine, 490
Actin,
    labelled, 163, 168, 191
    polymerization, 429
Activation,
    length dependent, 453, 456, 458, 470, 504, 522, 820-822, 830-831
Actomyosin motor, 429-433, 908
Aequorin, 455-466, 469-470, 544, 553-558, 562, 565, 818, 820, 890, 899
Afterloaded contraction, 651, 836
A-I junction, 81, 29, 293-299, 305
Antibody labelling, 285-305, 328, 519, 579
Arthrin, 39-40, 45
ATP,
    and A-band shortening, 142-146, 152-154
    activation by, 69, 75, 79-81, 108, 135-137, 145, 151, 153
    analogs, 185, 274, 569-572, 587, 595, 598, 605-606, 642, 671
        ATP(γs), 251-255, 258, 261, 605, 610-611
        ε-ATP, 397-401, 405, 408-409, 423
ATP (continued)
    concentration, 75, 138-139, 141, 152-154, 336-341, 398, 401-402, 405, 418, 596-597, 605-610, 615, 723
    depletion, 127, 138, 147, 152, 283, 309, 723
    hydrolysis, 87,116, 340, 347, 397, 431, 435, 532, 575, 585-586, 594, 606-607, 611, 615, 698, 849-850, 865-866, 872, 874, 877, 880, 892
    and phosphorylation, 614
    regeneration of, 116, 152, 349, 555, 597, 606, 615
    unhydrolized, 177
AMP-PNP, 58, 177-184, 185-192, 273-275, 278, 284, 381, 413, 417, 420-423, 597-598
ATPase activity, 68, 74-75, 81, 87, 183, 256, 267, 269, 283, 393, 398, 409, 431-433, 533-534, 560, 573, 605-606, 614-615, 658, 660, 665, 694, 706, 709, 897
Autocorrelation, 82, 89
Auxotonic contraction, 467-468, 471, 766, 770-775, 821-826, 830, 834
Barnacle muscle, 463-464, 551-557, 562
Bessel function, 97, 102, 105, 239, 241

Bound water, 60
Bragg, 627, 766-768, 777-778
Brownian motion, 162, 164, 166, 168-169, 416, 431

Caffeine, 550, 565-568, 667, 725, 728, 834
Calcium,
free, 456, 458, 554, 561, 578
potentiation of release, 456-457, 463, 469-470, 818, 899
regenerative release, 566-7
release, 455-462, 549-550, 562, 565-566, 762, 807, 820, 830, 878
sensitivity, 455-456, 462-463, 471, 551, 553-554, 563, 565, 569-571, 694
sensitivity, hysteresis of, 553-557, 560
transient, 455-464
uptake, 455-457, 462-464, 562
Catastrophe theory, 738, 793, 795
Charge diameter, 56
Cilia, 434, 776, 784, 899
Cinemicrography, 737, 768, 776
Collagen, 52-53, 544, 831
Compliance,
apparatus, 386, 458, 513, 587, 606, 642
cross bridge, 117, 385, 389, 392-394, 509, 649-653
"end"/tendon, 385-389, 479, 604, 641, 644-653, 664, 686, 764
H zone, 117
muscle fiber (undefined), 117, 300, 394, 449-450, 455, 653, 687-689, 757-758, 762
segment, 508
Connectin, 320
Connecting filaments, 213, 268, 307, 312-314
Contraction bands, 66, 109, 116-118, 119-134, 148, 154
Contracture,
barium, 28, 119, 123, 127,
Contracture (continued)
barium (continued)
132, 163-169, 207, 215-218, 230-231, 236, 243, 379-380, 427, 806, 847-850
potassium, 68, 76, 93, 98-100, 843
Cooperativity, 184, 262, 433, 560, 571
Creep, 473-476, 483-503, 524, 541-542, 545, 905
Cross bridge,
attached, 29, 162-166, 174, 184, 185-192, 194, 200, 202, 220, 267, 409, 414, 420-421, 425, 462, 464, 586, 593, 597, 621, 641-642, 647, 649, 662, 671, 687-690, 706, 876, 906-907
attached, in resting muscle at low ionic strength, 269-270, 273-281, 282
attachment site, 163-164, 167
cycling rate, 611, 698, 705-706, 879, 893
detachment, 173, 463-464, 560, 593, 608-609, 897
disordered, 93, 102-104, 163-164, 172, 188, 244, 250, 371, 379, 400, 405, 413, 417, 420-421, 424-427
force, 716
hindered, 128, 697-698, 702-707
lateral elasticity, 711, 717
powerstroke, 173, 185, 189, 373, 380-381, 665
repeat, 135, 137, 145, 147, 163
rotation, 55, 170-173, 181, 189
state, 164-165, 189, 372, 426, 597-598, 621, 641, 646, 738, 793-795
stroke, 166, 668
turnover, 727-728, 851, 853, 860, 877

Cytoskeletal matrix, 4, 16, 47, 56, 132, 285, 299, 301, 440, 449, 786

Deactivation,
  by shortening, 481-484, 561, 830
Dextran, 33, 181, 184, 208-210, 694, 697-710, 721, 733
Dielectric constant, 365-367
Donnan potential, 73, 81, 332, 353-357
Double overlap of filaments, 125, 128, 468
Dual diffractometry, 766

Elastic component or element, 397, 441, 524, 542, 620, 632, 682, 689, 737, 745, 843, 863, 867
  axial elasticity, 694, 697-698, 702, 706
  parallel (PEC), 133, 287, 307, 389, 392, 524, 542, 737, 860, 867, 877 radial elasticity, 268
  series (SEC), 810-811
Elastic modulus, 387, 395, 694, 698, 702, 706, 709
Electron probe resonance (EPR), 247, 371, 392, 405, 413-427, 687
Electrostatic field, 209, 212, 316, 321, 332, 690, 716
Energy balance, 350-351, 575, 848-851, 866, 869, 871, 874, 877, 879
Enthalpy change, 351, 847, 851, 875
Ethylene glycol, 177-183

Fatigue, 338, 340, 347, 440, 899
Fibrillin, 267, 287, 320
Field theory, 527
Filament extension, 442
Filament overlap, 30, 38, 391, 395, 456, 458, 460, 469-470, 476, 519-523, 709, 739, 848, 865, 867, 869, 872
Fixed charge, 66, 81, 83, 332, 355
Force enhancement by stretch, 468, 483-484, 501, 739, 745, 747, 749-751, 763
Force velocity, 481-482, 492, 507, 536, 665-668, 728 731, 737, 757, 763-764, 778, 813-814, 831-832, 848
Frank-Starling hypothesis, 822
Fourier maps, 23, 68-69, 73, 87

Gap band, 312-313
Gap filament, 268, 286, 296, 320
Ghost fibers, 11, 267, 406, 410, 532
Glycolysis, 333, 335, 340-344

Heat
  latency of, 891-892
  labile, 887, 892-895
  maintenance, 813-814, 887, 892-895
  production, 575, 748, 849, 853, 878, 887-898
  shortening, 848, 853-864, 865-867, 872-3, 877-878, 883-884
  stimulus, 889, 892-894
  thermoelastic, 853-864
  unexplained, 577, 849, 863, 869-872, 879-880
Hill equation (cooperative binding), 557, 660, 664
Hill equation (P-V), 667, 728, 759-760, 813
Horseshoe crab, 65-66, 107-108, 111-112, 117, 221, 232, 238, 241-242, 245-247, 251, 253, 255, 258-259, 722-723
Hypertonicity, 681-685, 725-729
Hypotonicity, 693, 745, 751

I-band,
  striations within, 288-298
Immunofluorescence, 290
Interference microscopy, 38, 70, 73, 123, 132
Interfilament distance, 145, 186, 693-694, 709, 712, 715,

Interfilament distance (cont'd)
717, 720, 751
Intermediate filaments, 287, 299, 301, 303, 320, 449-450
Internal load, 133, 667-668, 698, 705, 708, 728-729, 808, 810, 821-822, 830-832
Iodoacetate rigor, 186, 191
Ionic strength, 9-12, 18, 44, 60, 207-210, 237, 245, 248, 269-273, 278-284, 327, 404, 670, 693, 714-715, 718, 725-728, 801, 908
Isotonic contraction, 420, 667, 770, 837

Latency relaxation, 777, 790
Lateral force, 711-712, 716-717, 719
Lateral spacing, 711, 719
Lattice compression, 34, 47, 122, 209, 694, 697-698, 706, 709, 721
Lattice shrinkage, 184, 249, 694, 711-712, 716
Lattice spacing, 29-30, 34, 38-39, 44, 54, 208-211, 219, 249, 355, 683, 694, 697-700, 709-719, 721-722
Length-tension relation
ascending limb, 118, 130, 456, 466-472, 544, 806
descending limb, 453-454, 455-456, 460, 469, 484, 486, 492, 495, 508, 512, 519-522, 525, 542, 544, 693, 737, 820
sarcomere length-force, 73, 118-119, 128, 130, 153, 319, 453-456, 467-471, 473-476, 481-488, 491-493, 495-498, 501-508, 524-525, 527, 532, 534, 541-542, 545, 693, 737, 742, 748, 792, 806, 839, 905
segment length-force, 822, 836
Lethocerus, 19, 29-35, 38-40, 178,

Lethocerus (continued)
184-187, 192, 605-607
Light (laser, optical) diffraction, 11, 119, 136-137, 150, 314, 473-477, 480, 507, 511-512, 524, 624, 634, 742-743, 766-770, 779, 806, 837, 872, 878
layer lines, 68-69, 74, 95, 97, 100-102
microstructure, of laser diffraction pattern, 474, 478, 779
Light scattering, 82, 87-92, 261
Limulus, 31, 40, 45, 65-67, 73-85, 89-92, 93-103, 115, 118, 136, 147, 186-192, 238, 245-246, 321, 332, 353-355, 442, 904
Local contraction, 108-111, 142
Longitudinal filaments, 20, 285-286, 290-293, 299, 301, 786,

Maintenance heat, see Heat
Mass transfer, 174-175, 216, 715
Maximum shortening velocity, 133, 280, 454, 495-496, 504-509, 657, 665, 669, 772, 758-761, 831-832, 853, 860, 863, 865-867, 871-880, 883-884
M-line, 5, 7, 11-13, 16, 20, 61, 85, 114, 138, 172, 235, 287, 290, 295, 297, 299-300, 444, 518
Myofibrillar skew, 768, 770
Myofibrillar stroma, 267-268, 321, 448
Myosin,
aggregation, 4, 93, 103
backbone, 4, 47, 50-51, 55-56, 59, 164, 169, 174, 186, 216, 219, 235-242, 247, 314, 380, 395
core, 19, 174, 215-216, 219, 286, 314, 318, 320-321, 327-328, 439-440
crown, 30, 39, 103, 105

domain, 21-29, 183, 373-384, 392-393, 405, 421-422, 425-427, 728, 906-907
light chain, 550, 605-606, 613-614, 665, 897
nose-cone of S1, 170, 373, 380-381, 425
polymerization, 6
rod, 387-391, 395, 787-790
strandedness, 40, 45, 80, 93, 97, 102-105, 230-231, 241, 251, 318, 410
subfragment-1 (myosin head), 3, 5-6, 18, 21-28, 29, 55-56, 66, 89-90, 95, 97-98, 111, 119, 128, 150, 163-164, 167-168, 170, 178-181, 258, 172, 186-189, 193, 207-208, 215-218, 234, 237-238, 246-249, 252-253, 267, 269, 273-278, 307, 313, 321, 356, 365, 371, 373-377, 380-384, 395, 397-398, 400-405, 408-410, 413-421, 426-427, 429-436, 439-440, 446, 650, 664-665, 707, 718, 875, 906-908
subfragment-2, 18, 145, 150, 164-173, 185-186, 219-220, 393-395, 717-720, 728, 787-791, 907-908

Nebulin, 40, 268, 285-295, 299-305, 449
N_2-line, 66, 135, 142, 144-145, 147, 150, 288, 290, 293-298, 305, 321
Non-overlap, 118, 186, 191-192, 254, 268, 307, 312-327, 367, 416, 444
Nuclear magnetic resonance (NMR), 331-333, 335, 339, 346-347, 376, 427, 577
Nyquist plot, 667, 671

Osmotic force, 694

Papain, 82, 87-91, 353-356
Papillary muscle, 651, 782, 795, 805-806, 821-823, 830-837, 841, 843
Paramyosin, 36, 45, 75-80, 98, 103, 105
Parvalbumin, 362, 457, 462, 550, 573-579, 850, 887, 893, 895, 898-899
Passive force, 116, 133, 268, 271, 273, 280-281, 287, 301, 307-328, 389, 411, 434, 441, 476, 484, 509, 511-512, 517, 521, 525, 529-530, 541-543, 638, 684, 739, 745-746, 750, 760, 821, 823, 826-831, 842, 854, 862, 867, 877
Passive shortening, 131-133
Patterson function, 221, 224, 226, 228, 230, 234
Pauses in sarcomere shortening, 738, 765-768, 771, 773, 776-780, 789
Phase-frequency, 657-670
Phase locked loop, 477, 769-772
Phase microscopy, 87, 119
pH, intracellular, 331-333, 339, 344, 349, 351
pH sensitive electrodes, 349
Phosphatase, alkaline, 75, 106, 353-356
Phosphorylation,
of myosin light chains, 76-77, 550, 605-616, 665, 670, 897
paramyosin, 75-79
thick filament proteins, 75-79, 897
titin and nebulin, 305
troponin I, 614
Post-tetanic potentiation, 463, 887-891
PVP, 33, 719, 731-733

Quick release, 165-166, 169, 171-174, 379, 410, 427, 459-461, 557, 586, 589-591, 601-603, 607-611, 614,

Quick release (continued)
617, 623-624, 629-635, 637-638, 643, 648-649, 652-655, 702-703, 740, 791, 835
Quick stretch, 165, 168, 171, 173, 272, 281, 319, 379, 391, 605-606, 749

Radial compression, 307, 694, 697
Radial force, 718-719
Rapid cooling contracture (RCC), 550, 565-568
Refractive index, 31, 38, 42-44, 68, 119-120, 132
Restoring force, 133, 318, 792, 833
Rigor, 14, 29, 33, 43, 58-59, 79, 117, 134, 136, 145-147, 153-154, 163, 169, 175, 178, 180, 184, 185-193, 203-206, 207-208, 211-213, 221, 231, 234, 237-250, 251-266, 269-281, 324-325, 355, 357, 368, 371, 373, 377-380
rigor complex, 17, 19, 22, 28

Sarcomere length
dispersion, 468, 471, 473-474, 477-478, 482-483, 507, 515, 521, 839
distribution, 152, 477, 543
inhomogeneity, 256, 484, 489-490, 508, 529, 545, 646, 806
instability, 481, 484, 486, 488-492, 496, 501-502, 508
nonuniformity, 471, 516, 545, 583, 624, 648, 689
Sarcomere length clamp, 513-514, 520-521, 524-525 (*see also* Segment length clamp)
Sarcoplasmic reticulum, 52, 301, 343, 456-457, 462, 543, 549-550, 555, 558, 562, 573, 798, 850, 893, 896
SDS gel electrophoresis, 7, 30, 39-
SDS gel electrophoresis (cont'd)
40, 75-77, 90, 287-289, 293, 304, 315, 387, 391, 433-435, 440
Segment length clamp, 475, 495, 498, 501-505, 509, 762, 898
Shrinkage, during EM preparation, 34, 47, 53, 55, 61
Side-struts, 268, 307, 318, 321-322
Sinusoidal length change, 311-312, 583, 592, 595, 598, 642, 649, 652-654, 657-660, 664, 668-669, 674
Smooth muscle, 10, 299, 322, 464, 534, 536, 550, 612, 617-622
Sounds generated by muscle, 737-738, 773, 779
Spin label, 405, 413-427, 738, 793, 906
Starling mechanism, *see* Frank-Starling
STEM, 4, 7, 29-40, 45, 79, 105
Stepwise shortening, 434, 634, 737-738, 765-773, 776, 778, 780-781, 783-785, 791, 793, 904-905
Steric blocking model, 251, 259, 267, 269, 278, 550
Steric hindrance, 41, 47, 128, 425, 470, 721-722, 728
Stiffness,
instantaneous, 273, 442, 601-603, 606-609, 624, 642, 757, 876
and myofilament overlap, 272-273, 309, 385, 389, 391, 394, 635, 642
radial, 318, 719-720
resting, 269-272, 277, 313, 319-320, 327, 389, 418, 441, 490, 686, 719
rigor, 385-394, 418, 615, 709, 719
and tension, 178-179, 249, 312-314, 415, 544, 601, 604, 607, 615, 620, 641-654, 673-674, 678, 681-

Stiffness (continued)
and tension (continued)
689, 733, 737, 762-764, 876
time dependent, 270
Superlattice, 222, 230-231
Synchrony of contraction, 765, 779-781, 784-785, 790-791, 837, 905
Synchrotron, 161, 178, 185, 207, 215

T_1, 588-592, 603, 643, 653, 688, 757-761
T_2, 589
Tannic acid, 35, 48-51
Thermopile, 854-855, 866, 887-891, 898
Thick filament,
isolated, 56, 59, 66, 75-86, 89, 303, 321
length, 68, 73, 84-86, 111, 105, 136, 142, 256, 320, 511-512, 518, 521
misalignment, 66, 69, 72-74, 107, 112-117, 326
mass, 7, 29, 38, 43, 79, 219
regulation, 551
shortening, 66-69, 73-83, 87, 102-105, 107-108, 118, 136, 153, 321, 353, 791-792
Thin filament
length, 121, 512, 518, 521
regulation, 550
Third filament, 286, 293-301, 304, 320, 325, 450, 484, 785
Titin, 40, 268, 285-295, 299-305, 323-324, 449
Transverse tubules, 120, 507, 550, 567
Tropomyosin, 22-29, 251-263, 267, 269, 283, 373, 376-377, 406, 410, 429
Troponin, 27, 251, 261-263, 267, 269, 376, 410, 429, 455, 462, 464, 518, 521, 550-551, 560, 563, 569-571, 573-578, 658, 664, 801, 808-809, 850, 896-899

Vanadate, 89, 251, 253, 256, 259
Viscoelasticity, 623, 631-632, 635, 638, 677, 686, 753-754, 783, 843
Viscosity, 277, 431, 598, 648-649, 652, 686, 728, 732-733
Voltage clamp, 553

X-ray diffraction,
equatorial reflections, 32, 41, 60, 147, 159, 161-162, 165, 171, 175, 179, 193-194, 197-205, 207, 210-213, 216, 221-224, 227, 230-232, 245, 249-250, 279, 307, 379, 646, 682, 688, 711-718, 721, 728
forbidden reflections, 22-23, 26, 163, 325
layer line, 151, 159, 162-163, 169, 174, 179, 183, 238-250, 646, 792
meridional reflections, 23, 97, 102, 135, 145-148, 159, 162-172, 175, 187, 237-250, 255, 320-321, 378, 444, 518, 521, 792
off meridional, 162-163, 191, 238, 240-241, 244, 249, 728,
radial density, 170
X-ray spectroscopy, 332, 359

Z-line, 10-11, 31, 85, 123, 135, 142, 144, 150, 172, 211-212, 268, 286-287, 290-295, 297, 299, 307-308, 312-313, 316, 318, 320, 326, 440-442, 450, 472, 484, 741, 786

MIX
Papier aus verantwortungsvollen Quellen
Paper from responsible sources
FSC® C105338

If you have any concerns about our products,
you can contact us on
ProductSafety@springernature.com

In case Publisher is established outside the EU,
the EU authorized representative is:
Springer Nature Customer Service Center GmbH
Europaplatz 3, 69115 Heidelberg, Germany

Printed by Libri Plureos GmbH
in Hamburg, Germany